STUDY SMARTER

CHAPTER
Test Prep
VIDEOS

Step-by-step solutions on video for all chapter test exercises from the text

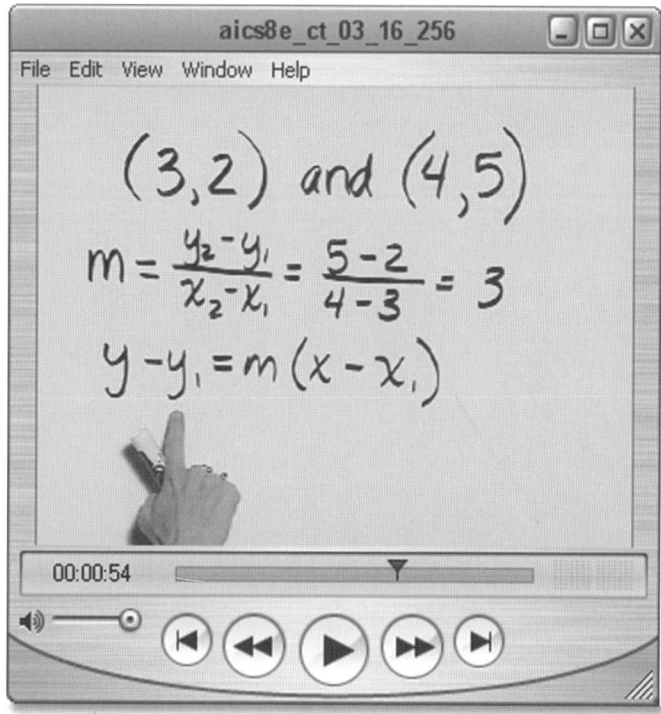

CHAPTER TEST PREP VIDEOS ARE ACCESSIBLE THROUGH THE FOLLOWING:

INTERMEDIATE ALGEBRA
FOR COLLEGE STUDENTS

EIGHTH EDITION

ALLEN R. ANGEL
MONROE COMMUNITY COLLEGE

DENNIS C. RUNDE
STATE COLLEGE OF FLORIDA

Prentice Hall

Boston   Columbus   Indianapolis   New York   San Francisco   Upper Saddle River
Amsterdam   Cape Town   Dubai   London   Madrid   Milan   Munich   Paris   Montréal   Toronto
Delhi   Mexico City   São Paulo   Sydney   Hong Kong   Seoul   Singapore   Taipei   Tokyo

Editorial Director, Mathematics: *Christine Hong*
Editor in Chief: *Paul Murphy*
Sponsoring Editor: *Mary Beckwith*
Executive Project Manager: *Kari Heen*
Associate Editor: *Joanna Doxey*
Editorial Assistant: *Kristin Rude*
Editor in Chief, Development: *Carol Trueheart*
Production Management: *Elm Street Publishing Services*
Senior Managing Editor: *Karen Wernholm*
Production Supervisor: *Patty Bergin*
Cover Designer: *Barbara T. Atkinson*
Text Design: *Kl Creative*
Digital Assets Manager: *Marianne Groth*
Media Producers: *Audra J. Walsh* and *Vicki Dreyfus*
Executive Manager, Course Production: *Peter Silvia*
Software Development: *Eileen Moore* and *Marty Wright*
Executive Marketing Manager: *Michelle Renda*
Marketing Manager: *Adam Goldstein*
Marketing Assistant: *Margaret Wheeler*
Senior Prepress Supervisor: *Caroline Fell*
Senior Manufacturing Manger: *Evelyn Beaton*
Senior Media Buyer: *Ginny Michaud*
Composition: *Prepare Inc.*
Art Studios: *Scientific Illustrators* and *Laserwords*
Cover images: *Wind turbine farm over sunset: ©TedNad/Shutterstock; Gold wheat/clouds: ©Triff/Shutterstock*

Library of Congress Cataloging-in-Publication Data
Angel, Allen R., 1942-
    Intermediate algebra for college students. —8th ed./by Allen R. Angel, Dennis Runde.
        p. cm.
Includes index.
ISBN-13: 978-0-321-62091-0 (student edition: alk. paper)
ISBN-10: 0-321-62091-7 (student edition: alk. paper)
I. Algebra. I. Runde. Dennis C. II. Title.
QA154.3.A53 2011
512.9—dc22   2009052295

1 2 3 4 5 6 7 8 9 10—QWD—14 13 12 11 10

Prentice Hall
is an imprint of

www.pearsonhighered.com

ISBN-10: 0-321-62091-7
ISBN-13: 978-0-321-62091-0

To my wife, Kathy,
and our sons, Robert and Steven
Allen R. Angel

To my wife, Kristin,
and our sons, Alex, Nick, and Max
Dennis C. Runde

Brief Contents

Contents

8   Quadratic Functions   493

9   Exponential and Logarithmic Functions   567

10   Conic Sections   633

Preface

This book was written for college students who have successfully completed a first course in elementary algebra. Our primary goal was to write a book that students can read, understand, and enjoy. To achieve this goal we have used short sentences, clear explanations, and many detailed, worked-out examples. We have tried to make the book relevant to college students by using practical applications of algebra throughout the text.

The many factors that contributed to the success of the previous editions have been retained. In preparing this revision, we considered the suggestions of instructors and students throughout the country. The *Principles and Standard for School Mathematics*, prepared by the National Council of Teachers of Mathematics (NCTM), and *Beyond Crossroads: Implementing Mathematics Standards in the First Two Years of College*, by the American Mathematical Association of Two-Year Colleges (AMATYC) together with advances in technology, influenced the writing of this text.

New to This Edition

One of the most important features of the text is its readability. The book is very readable for students of all reading skill levels. The Eighth Edition continues this emphasis and has been revised with a focus on improving accessibility and addressing the learning needs and styles of today's students. To this end, the following changes have been made:

Content Changes

- Discussions throughout the text have been thoroughly revised for brevity and accessibility. Whenever possible, a visual example or diagram is used to explain concepts and procedures.
- **Understanding Algebra** is a new feature appearing in the margin throughout the text. Placed at key points, **Understanding Algebra** draws students' attention quickly to the important concepts and facts that they need to master.
- The pedagogical use of color has been enhanced and now includes a color-coded system for variables and notation to support a more visual approach.
- Exercise sets now begin with new **Warm-up Exercises**—and include an emphasis on vocabulary. These exercises are great as a warm-up to the homework or as a 5-minute quiz. The Concept/Writing exercises (formerly found at the start of the exercise sets) are now located after the Problem Solving section in the exercise sets.
- Exercises and applications have been updated throughout.

Enhancements to Resources

- The Chapter Test Prep Video and Lecture Series Videos are now captioned in both English and Spanish. The videos are available in MyMathLab. The Chapter Test Prep videos are also available on YouTube.
- MyMathLab and MathXL have been significantly updated including:
  1. A substantial increase in exercises coverage
  2. Suggested Assignments in homework builder
  3. New study skills and math-reading connections coverage

Features of the Text

Full-Color Format

Color is used pedagogically in the following ways:

- Important definitions and procedures are color screened.
- Color screening or color type is used to make other important items stand out.
- Artwork is enhanced and clarified with use of multiple colors.
- The full-color format allows for easy identification of important features by students.
- The full-color format makes the text more appealing and interesting to students.

Accuracy

Accuracy in a mathematics text is essential. To ensure accuracy in this book, math teachers from around the country have read the pages carefully for typographical errors and have checked all the answers.

Connections

Many of our students do not thoroughly grasp new concepts the first time they are presented. In this text we encourage students to make connections. That is, we introduce a concept, then later in the text briefly reintroduce it and build upon it. Often an important concept is used in many sections of the text. Important concepts are also reinforced throughout the text in the Cumulative Review Exercises and Cumulative Review Tests.

Chapter Opening Application

Each chapter begins with a real-life application related to the material covered in the chapter. By the time students complete the chapter, they should have the knowledge to work the problem.

Goals of This Chapter

This feature on the chapter opener page gives students a preview of the chapter and also indicates where this material will be used again in other chapters of the book. This material helps students see the connections among various topics in the book and the connection to real-world situations.

The Use of Icons

At the beginning of each exercise set the icons for MathXL®, *MathXL*, and for MyMathLab, *MyMathLab*, are illustrated to remind students of these homework resources.

Keyed Section Objectives

Each section opens with a list of skills that the student should learn in that section. The objectives are then keyed to the appropriate portions of the sections with blue numbers such as **1**.

Problem Solving

Pólya's five-step problem-solving procedure is discussed in Section 2.2. Throughout the book, problem solving and Pólya's problem-solving procedure are emphasized.

Practical Applications

Practical applications of algebra are stressed throughout the text. Students need to learn how to translate application problems into algebraic symbols. The problem-solving approach used throughout this text gives students ample practice in setting up and solving application problems. The use of practical applications motivates students.

Detailed, Worked-Out Examples

A wealth of examples have been worked out in a step-by-step, detailed manner. Important steps are highlighted in color, and no steps are omitted until after the student has seen a sufficient number of similar examples.

Now Try Exercises

In each section, after each example, students are asked to work an exercise that parallels the example given in the text. These Now Try Exercises make the students *active*, rather than passive, learners and they reinforce the concepts as students work the exercises. Through these exercises, students have the opportunity to immediately apply what they have learned. After each example, Now Try Exercises are indicated in green type such as Now Try Exercise 27. They are also indicated in green type in the exercise sets, such as 27.

Study Skills Section

Students taking this course may benefit from a review of essential study skills. Such study skills are essential for success in mathematics. Section 1.1, the first section of the text, discusses such study skills. This section should be very beneficial for your students and should help them to achieve success in mathematics.

Understanding Algebra

The new **Understanding Algebra** boxes appear in the margin throughout the text. Placed at key points, **Understanding Algebra** helps students focus on the important concepts and facts that they need to master.

Helpful Hints

The Helpful Hint boxes offer useful suggestions for problem solving and other varied topics. They are set off in a special manner so that students will be sure to read them.

Avoiding Common Errors

Common student errors are illustrated. Explanations of why the shown procedures are incorrect are given. Explanations of how students may avoid such errors are also presented.

Exercise Sets

The exercise sets are broken into three main categories: Warm-Up Exercises, Practice the Skills, and Problem Solving. Many exercise sets also contain Concept/Writing, Challenge Problems, and/or Group Activities. Each exercise set is graded in difficulty. The early problems help develop the student's confidence, and then students are eased gradually into the more difficult problems. A sufficient number and variety of examples are given in each section for the student to successfully complete even the more difficult exercises. The number of exercises in each section is more than ample for student assignments and practice.

Warm-Up Exercises

Exercise sets now begin with new Warm-Up Exercises. These fill-in-the-blank exercises include an emphasis on vocabulary. They serve as a great warm-up to the homework exercises or as 5-minute quizzes.

Problem-Solving Exercises

These exercises help students become better thinkers and problem solvers. Many of these exercises involve real-life applications of algebra. It is important for students to be able to apply what they learn to real-life situations. Many problem-solving exercises help with this.

Concept/Writing Exercises

Most exercise sets include exercises that require students to write out the answers in words. These exercises improve students' understanding and comprehension of the material. Many of these exercises involve problem solving and conceptualization and help develop better reasoning and critical thinking skills. These exercises are located following the Problem-Solving exercises within the end-of-section exercise sets.

Challenge Problems

These exercises, which are part of many exercise sets, provide a variety of problems. Many were written to stimulate student thinking. Others provide additional applications of algebra or present material from future sections of the book so that students can see and learn the material on their own before it is covered in class. Others are more challenging than those in the regular exercise set.

Video Lecture Exercises

The exercises that are worked out in detail on the Lecture Videos are marked with the video icon, ▦. This will prove helpful for your students.

Cumulative Review Exercises

All exercise sets (after the first two) contain questions from previous sections in the chapter and from previous chapters. These Cumulative Review Exercises will reinforce topics that were previously covered and help students retain the earlier material, while they are learning the new material. For the students' benefit, Cumulative Review Exercises are keyed to the section where the material is covered, using brackets, such as [3.4].

Group Activities

Many exercise sets have group activity exercises that lead to interesting group discussions. Many students learn well in a cooperative learning atmosphere, and these exercises will get students talking mathematics to one another.

Mid-Chapter Tests

In the middle of each chapter is a Mid-Chapter Test. Students should take each Mid-Chapter Test to make sure they understand the material presented in the chapter up to that point. In the student answers, brackets such as [2.3] are used to indicate the section where the material was first presented.

Chapter Summary

At the end of each chapter is a comprehensive chapter summary that includes important chapter facts and examples illustrating these important facts.

Chapter Review Exercises

At the end of each chapter are review exercises that cover all types of exercises presented in the chapter. The review exercises are keyed using color numbers and brackets, such as [1.5], to the sections where the material was first introduced.

Chapter Practice Tests

The comprehensive end-of-chapter practice test will enable the students to see how well they are prepared for the actual class test. The section where the material was first introduced is indicated in brackets in the student answers.

Cumulative Review Tests

These tests, which appear at the end of each chapter after the first, test the students' knowledge of material from the beginning of the book to the end of that chapter. Students can use these tests for review, as well as for preparation for the final exam. These exams, like the Cumulative Review Exercises, will serve to reinforce topics taught earlier. In the answer section, after each answer, the section where that material was covered is given using brackets.

Answers

The *odd answers* are provided for the exercise sets. *All answers* are provided for the Cumulative Review Exercises, Mid-Chapter Test, Chapter Review Exercises, Chapter Practice Tests, and Cumulative Review Tests. Answers are not provided for the Group Activity exercises since we want students to reach agreement by themselves on the answers to these exercises.

Prerequisite

The prerequisite for this course is a working knowledge of elementary algebra. Although some elementary algebra topics are briefly reviewed, students should have a basic understanding of elementary algebra before taking this course.

Modes of Instruction

The format and readability of this book lend it to many different modes of instruction. The constant reinforcement of concepts will result in greater understanding and retention of the material by your students.

The features of the text and the large variety of supplements available make this text suitable for many types of instructional modes, including

- lecture
- hybrid or blended courses
- distance learning
- self-paced instruction
- modified lecture
- cooperative or group study
- learning laboratory

Student and Instructor Resources

STUDENT RESOURCES

Student Solutions Manual Provides complete worked-out solutions to • the odd-numbered section exercises • all exercises in Mid-Chapter Tests, Chapter Reviews, Chapter Practice Tests, and Cumulative Review Tests	**Worksheets for Classroom or Lab Practice** • Extra practice exercises for every section of the text with ample space for students to show their work
Lecture Videos • For each section of the text, there are about 20 minutes of lecture. Exercises in the text that are worked on the videos are identified in the text by the ⬛ icon. • Captioned in English and Spanish • Available in MyMathLab®	**Chapter Test Prep Videos** • Step-by-step solutions to every exercise in each Chapter Practice Test • Available in MyMathLab® • Available in YouTube. (search "Angel Intermediate Algebra" and click on "Channels")

INSTRUCTOR RESOURCES

Annotated Instructor's Edition Contains all the content found in the student edition, plus the following: • Answers to exercises on the same text page with graphing answers in the Graphing Answer section at the back of the text • Instructor Example provided in the margin paired with each student example	**Instructor's Resource Manual with Tests and Mini-Lectures** • Mini-lectures for each text section • Several forms of test per chapter (free response and multiple choice) • Answers to all items • Available for download from the IRC and in MyMathLab®
	TestGen® • Available for download from the IRC
Instructor's Solutions Manual • Available for download from the IRC and in MyMathLab®	**Online Resources** • MyMathLab® (access code required) • MathXL® (access code required)

Acknowledgments

We thank our spouses, Kathy Angel and Kris Runde, for their support and encouragement throughout the project. We are grateful for their wonderful support and understanding while we worked on the book.

We also thank our children: Robert and Steven Angel and Alex, Nick, and Max Runde. They also gave us support and encouragement and were very understanding when we could not spend as much time with them as we wished because of book deadlines. Special thanks to daughter-in-law, Kathy; mother-in-law, Patricia, and father-in-law, Scott. Without the support and understanding of our families, this book would not be a reality.

We want to thank Rafiq Ladhani and his team at Edumedia for accuracy reviewing the pages and checking all answers.

We thank Larry Gilligan, University of Cincinnati, and Donna Petrie, Monroe Community College, for their contributions to the series.

Many people at Pearson deserve thanks, including all those listed on the copyright page. In particular, we thank Paul Murphy, Editor in Chief; Mary Beckwith, Sponsoring Editor; Joanna Doxey, Associate Editor; Marketing Managers, Michelle Renda and Adam Goldstein; Debbie Meyer, Project Editor; Patty Bergin, Production Supervisor; Karen Wernholm, Senior Managing Editor; and Barbara Atkinson, Senior Designer.

We would like to thank the following reviewers and focus group participants for their thoughtful comments and suggestions:

Laura Adkins, *Missouri Southern State College, MO*
Arthur Altshiller, *Los Angeles Valley College, CA*
Jacob Amidon, *Cayuga Community College, NY*
Bhagirathi Anand, *Long Beach City College, CA*
Sheila Anderson, *Housatonic Community College, CT*
Peter Arvanites, *State University of New York–Rockland Community College, NY*
Jannette Avery, *Monroe Community College, NY*
Mary Lou Baker, *Columbia State Community College, TN*
Linda Barton, *Ball State University, IN*
Jon Becker, *Indiana University, IN*
Paul Boisvert, *Oakton Community College, IL*
Beverly Broomell, *Suffolk County Community College, NY*
Lavon Burton, *Abilene Christian University, TX*
Marc Campbell, *Daytona Beach Community College, FL*
Mitzi Chaffer, *Central Michigan University, MI*
Terry Cheng, *Irvine Valley College, CA*
Ted Corley, *Arizona State University and Glendale Community College, AZ*
Charles Curtis, *Missouri Southern State College, MO*
Joseph de Guzman, *Riverside City College (Norco), CA*
Marla Dresch Butler, *Gavilan Community College, CA*
Gary Egan, *Monroe Community College, NY*
Mark W. Ernsthausen, *Monroe Community College, NY*
Elizabeth Farber, *Bucks County Community College, PA*
Warrene Ferry, *Jones County Junior College, MS*
Christine Fogal, *Monroe Community College, NY*

Gary Glaze, *Spokane Falls Community College, WA*
James Griffiths, *San Jacinto College, TX*
Kathy Gross, *Cayuga Community College, NY*
Abdollah Hajikandi, *State University of New York–Buffalo, NY*
Cynthia Harrison, *Baton Rouge Community College, LA*
Mary Beth Headlee, *State College of Florida, FL*
Kelly Jahns, *Spokane Community College, WA*
Cheryl Kane, *University of Nebraska–Lincoln, NE*
Judy Kasabian, *El Camino College, CA*
Maryanne Kirkpatrick, *Laramie County Community College, WY*
Marcia Kleinz, *Atlantic Cape Community College, NJ*
Shannon Lavey, *Cayuga Community College, NY*
Kimberley A. Martello, *Monroe Community College, NY*
Shywanda Moore, *Meridian Community College, MS*
Catherine Moushon, *Elgin Community College, IL*
Kathy Nickell, *College of DuPage, IL*
Jean Olsen, *Pikes Peak Community College, CO*
Shelle Patterson, *Moberly Area Community College, MO*
Patricia Pifko, *Housatonic Community College, CT*
David Price, *Tarrant County College, TX*
Elise Price, *Tarrant County College, TX*
Dennis Reissig, *Suffolk County Community College, NY*
Linda Retterath, *Mission College, CA*
Dale Rohm, *University of Wisconsin–Stevens Point, WI*
Troy Rux, *Spokane Falls Community College, WA*
Hassan Saffari, *Prestonburg Community College, KY*
Rick Silvey, *St. Mary College, KS*
Julia Simms, *Southern Illinois University–Edwardsville, IL*
Linda Smoke, *Central Michigan University, MI*
Jed Soifer, *Atlantic Cape Community College, NJ*
Richard C. Stewart, *Monroe Community College, NY*
Elizabeth Suco, *Miami–Dade College, FL*
Harold Tanner, *Orangeburg–Calhoun Technological College, SC*
Dale Thielker, *Ranken Technological College, MO*
Ken Wagman, *Gavilan Community College, CA*
Patrick Ward, *Illinois Central College, IL*
Robert E. White, *Allan Hancock College, CA*
Cindy Wilson, *Henderson State University, AZ*

Focus Group Participants

Linda Barton, *Ball State University, IN*
Karen Egedy, *Baton Rogue Community College, LA*
Daniel Fahringer, *Harrisburg Area Community Collge, PA*
Sharon Hansa, *Longview Community Collge, MO*
Cynthia Harrison, *Baton Rogue Community College, LA*
Judy Kasabian, *El Camino College, CA*
Mark Molino, *Erie Community College, NY*
Kris Mudunuri, *Long Beach City College, CA*
Fred Peskoff, *Borough of Manhattan Community College, NY*
David Price, *Tarrant County College, TX*
Elise Price, *Tarrant County College, TX*
Adrian Ranic, *Erie Community College, NY*
Dale Siegel, *Kingsborough Community College, NY*
Christopher Yarish, *Harrisburg Area Community College, PA*

To the Student

Algebra is a course that requires active participation. You must read the text and pay attention in class, and, most importantly, you must work the exercises. The more exercises you work, the better.

The text was written with you in mind. Short, clear sentences are used, and many examples are given to illustrate specific points. The text stresses useful applications of algebra. Hopefully, as you progress through the course, you will come to realize that algebra is not just another math course that you are required to take, but a course that offers a wealth of useful information and applications.

This text makes full use of color. The different colors are used to highlight important information. Important procedures, definitions, and formulas are placed within colored boxes.

The boxes marked **Understanding Algebra** should be studied carefully. They emphasize concepts and facts that you need to master to succeed. **Helpful Hints** should be studied carefully, for they stress important information. Be sure to study **Avoiding Common Errors** boxes. These boxes point out common errors and provide the correct procedures for doing these problems.

After each example you will see a Now Try Exercise reference, such as Now Try Exercise 27. The exercise indicated is very similar to the example given in the book. You may wish to try the indicated exercise after you read the example to make sure you truly understand the example. In the exercise set, the Now Try Exercises are written in green, such as 27.

In the exercise sets, the exercises with a video icon, ◄█, indicate that these exercises are worked out on the Lecture Videos.

Some questions you should ask your professor early in the course include: What supplements are available for use? Where can help be obtained when the professor is not available? Supplements that may be available include: the Student Solutions Manual; the Lecture Series Videos; the Chapter Test Prep Videos; *Math*XL; and MyMathLab. All these items are discussed under the heading of Supplements in Section 1.1 and listed in the Preface.

You may wish to form a study group with other students in your class. Many students find that working in small groups provides an excellent way to learn the material. By discussing and explaining the concepts and exercises to one another, you reinforce your own understanding. Once guidelines and procedures are determined by your group, make sure to follow them.

One of the first things you should do is to read Section 1.1, Study Skills for Success in Mathematics. Read this section slowly and carefully, and pay particular attention to the advice and information given. Occasionally, refer back to this section. This could be the most important section of the book. Pay special attention to the material on doing your homework and on attending class.

At the end of all exercise sets (after the first two) are **Cumulative Review Exercises**. You should work these problems on a regular basis, even if they are not assigned. These problems are from earlier sections and chapters of the text, and they will refresh your memory and reinforce those topics. If you have a problem when working these exercises, read the appropriate section of the text or study your notes that correspond to that material. The section of the text when the Cumulative Review Exercise was introduced is indicated by brackets, [ ], to the left of the exercise. After reviewing the material, if you still have a problem, make an appointment to see your professor. Working the Cumulative Review Exercises throughout the semester will also help prepare you to take your final exam.

Near the middle of each chapter is a **Mid-Chapter Test**. You should take each Mid-Chapter Test to make sure you understand the material up to that point. The section where the material was first introduced is given in brackets after the answer in the answer section of the book.

At the end of each chapter are a **Chapter Summary, Chapter Review Exercises**, a **Chapter Practice Test**, and a **Cumulative Review Test**. Before each examination you should review this material carefully and take the Chapter Practice Test (you may want to review the *Chapter Test Prep Video* also). If you do well on the Chapter Practice Test, you should do well on the class test. The questions in the Review Exercises are marked to indicate the section in which that material was first introduced. If you have a problem with a Review Exercise question, reread the section indicated. You may also wish to take the Cumulative Review Test that appears at the end of every chapter (starting with Chapter 2).

In the back of the text there is an **answer section** that contains the answers to the *odd-numbered* exercises, including the Challenge Problems. Answers to *all* Cumulative Review Exercises, Mid-Chapter Tests, Chapter Review Exercises, Chapter Practice Tests, and Cumulative Review Tests are provided. Answers to the Group Activity exercises are not provided, for we wish students to reach agreement by themselves on answers to these exercises. The answers should be used only to check your work. For the Mid-Chapter Tests, Chapter Practice Tests, and Cumulative Review Tests, after each answer the section number where that type of exercise was covered is provided.

We have tried to make this text as clear and error free as possible. No text is perfect, however. If you find an error in the text, or an example or section that you believe can be improved, we would greatly appreciate hearing from you. If you enjoy the text, we would also appreciate hearing from you. You can submit comments to math@pearson.com, subject for Allen Angel and Dennis Runde.

Allen R. Angel
Dennis C. Runde

Basic Concepts

Goals of This Chapter

In this chapter, we review algebra concepts that are central to your success in this course. Throughout this chapter, and in the entire book, we use real-life examples to show how mathematics is relevant in your daily life. In Section 1.1, we present some advice to help you establish effective study skills and habits. Other topics discussed in this chapter are sets, real numbers, and exponents.

© Walter Arce/Dreamstime.com

Have you ever asked yourself, "When am I going to use algebra?" In this chapter and throughout this book, we use algebra to study real-life applications. These applications range from the NASCAR Cup series in Exercise 91 to natural disasters in Exercise 92, both on page 14. On page 55, we use scientific notation to determine the revenue of four football teams in the NFL. We will find that mathematics can be used in virtually every aspect of our lives.

1.1 Study Skills for Success in Mathematics, and Using a Calculator

1 Have a positive attitude.

2 Prepare for and attend class.

3 Prepare for and take examinations.

4 Find help.

5 Learn to use a calculator.

You need to acquire certain study skills that will help you to complete this course successfully. These study skills will also help you succeed in any other mathematics courses you may take.

It is important for you to realize that this course is the foundation for more advanced mathematics courses. If you have a thorough understanding of algebra, you will find it easier to be successful in later mathematics courses.

1 Have a Positive Attitude

You may be thinking to yourself, "I hate math" or "I wish I did not have to take this class." You may have heard the term *math anxiety* and feel that you fall into this category. The first thing you need to do to be successful in this course is to change your attitude to a more positive one. You must be willing to give this course and yourself a fair chance.

Based on past experiences in mathematics, you may feel this will be difficult. However, mathematics is something you need to work at. Many of you taking this course are more mature now than when you took previous mathematics courses. Your maturity and your desire to learn are extremely important and can make a tremendous difference in your ability to succeed in mathematics. I believe you can be successful in this course, but you also need to believe it.

2 Prepare for and Attend Class

Preview the Material Before class, you should spend a few minutes previewing any new material in the textbook. You do not have to understand everything you read yet. Just get a feeling for the definitions and concepts that will be discussed. This quick preview will help you to understand what your instructor is explaining during class. After the material is explained in class, read the corresponding sections of the text slowly and carefully, word by word.

Read the Text A mathematics text is not a novel. Mathematics textbooks should be read slowly and carefully. If you do not understand what you are reading, reread the material. When you come across a new concept or definition, you may wish to underline or highlight it so that it stands out. This way, when you look for it later, it will be easier to find. When you come across a worked-out example, read and follow the example carefully. Do not just skim it. Try working out the example yourself on another sheet of paper. Also, work the **Now Try Exercises** that appear after each example. The Now Try Exercises are designed so that you have the opportunity to immediately apply new ideas. Make notes of anything that you do not understand to ask your instructor.

Do the Homework *Two very important commitments that you must make to be successful in this course are to attend class and do your homework regularly.* Your assignments must be worked conscientiously and completely. Mathematics cannot be learned by observation. You need to practice what you have heard in class. By doing homework you truly learn the material.

Don't forget to check the answers to your homework assignments. Answers to the odd-numbered exercises are in the back of this book. In addition, the answers to all the Cumulative Review Exercises, Mid-Chapter Tests, Chapter Review Exercises, Chapter Practice Tests, and Cumulative Review Tests are provided. For the Mid-Chapter Tests, Chapter Practice Tests, and Cumulative Review Tests, the section where the material was first introduced is provided in brackets after each answer. Answers to the

Group Activity Exercises are not provided because we want you to arrive at the answers as a group.

If you have difficulty with some of the exercises, mark them and do not hesitate to ask questions about them in class. You should not feel comfortable until you understand all the concepts needed to work every assigned problem.

When you do your homework, make sure that you write it neatly and carefully. Pay particular attention to copying signs and exponents correctly. Do your homework in a step-by-step manner. This way you can refer back to it later and still understand what was written.

Attend and Participate in Class You should attend every class. Generally, the more absences you have, the lower your grade will be. Every time you miss a class, you miss important information. If you must miss a class, contact your instructor ahead of time and get the reading assignment and homework.

While in class, pay attention to what your instructor is saying. If you do not understand something, ask your instructor to repeat or explain the material. If you do not ask questions, your instructor will not know that you have a problem understanding the material.

In class, take careful notes. Write numbers and letters clearly so that you can read them later. It is not necessary to write down every word your instructor says. Copy down the major points and the examples that do not appear in the text. You should not be taking notes so frantically that you lose track of what your instructor is saying.

Study Study in the proper atmosphere. Study in an area where you are not constantly disturbed so that your attention can be devoted to what you are reading. The area where you study should be well ventilated and well lit. You should have sufficient desk space to spread out all your materials. Your chair should be comfortable. You should try to minimize distractions while you are studying. You should not study for hours on end. Short study breaks are a good idea.

When studying, you should not only understand how to work a problem, you should also know why you follow the specific steps you do to work the problem. If you do not have an understanding of why you follow the specific process, you will not be able to solve similar problems.

Time Management It is recommended that students spend at least 2 hours studying and doing homework for every hour of class time. Some students require more time than others. Finding the necessary time to study is not always easy. The following are some suggestions that you may find helpful.

1. Plan ahead. Determine when you will have time to study and do your homework. Do not schedule other activities for these time periods. Try to space these periods evenly over the week.

2. Be organized so that you will not have to waste time looking for your books, pen, calculator, or notes.

3. Use a calculator to perform tedious calculations.

4. When you stop studying, clearly mark where you stopped in the text.

5. Try not to take on added responsibilities. You must set your priorities. If your education is a top priority, as it should be, you may have to cut the time spent on other activities.

6. If time is a problem, do not overburden yourself with too many courses. Consider taking fewer credits. If you do not have sufficient time to study, your understanding and your grades in all of your courses may suffer.

3 Prepare for and Take Examinations

Study for an Exam If you do some studying each day, you should not need to cram the night before an exam. If you wait until the last minute, you will not have time to seek the help you may need. To review for an exam,

1. Read your class notes.

2. Review your homework assignments.

3. Study the formulas, definitions, and procedures you will need for the exam.

4. Read the Avoiding Common Errors boxes and Helpful Hint boxes carefully.

5. Read the summary at the end of each chapter.

6. Work the review exercises at the end of each chapter. If you have difficulties, restudy those sections. If you still have trouble, seek help.

7. Work the Mid-Chapter Tests and the Chapter Practice Tests.

8. Rework quizzes previously given if the material covered in the quizzes will be included on the test.

9. Work the Cumulative Review Test if material from earlier chapters will be included on the test.

Take an Exam Make sure that you get a good night's sleep the day before the test. If you studied properly, you should not have to stay up late the night before to prepare for the test. Arrive at the exam site early so that you have a few minutes to relax before the exam. If you need to rush to get to the exam, you will start out nervous and anxious. After you receive the exam, do the following:

1. Carefully write down any formulas or ideas that you want to remember.

2. Look over the entire exam quickly to get an idea of its length and to make sure that no pages are missing. You will need to pace yourself to make sure that you complete the entire exam. Be prepared to spend more time on problems worth more points.

3. Read the test directions carefully.

4. Read each problem carefully. Answer each question completely and make sure that you have answered the specific question asked.

5. Starting with number 1, work each question in order. If you come across a question that you are not sure of, do not spend too much time on it. Continue working the questions that you understand. After completing all other questions, go back and finish those questions you were not sure of. Do not spend too much time on any one question.

6. Attempt each problem. You may be able to earn at least partial credit.

7. Work carefully and write clearly so that your instructor can read your work. Also, it is easy to make mistakes when your writing is unclear.

8. Check your work and your answers if you have time.

9. Do not be concerned if others finish the test before you. Do not be disturbed if you are the last to finish. Use all your extra time to check your work.

4 Find Help

Use the Supplements This text comes with many supplements. Find out from your instructor early in the semester which of these supplements are available and which supplements might be beneficial for you to use. Reading supplements should not replace reading the text. Instead supplements should enhance your understanding of the material. If you miss a class, you may want to review the video on the topic you missed before attending the next class.

There are many supplements available. The supplements that may be available to you are: the Student Solutions Manual which works out the odd section exercises as well as all the end-of-chapter exercises; the CD Lecture Series Videos which show about 20 minutes of lecture per section and include the worked out solutions to the exercises marked with this icon ◼ ; the Chapter Test Prep Video CD, which works out every problem in every Chapter Practice Test; *Math XL* MathXL®, a powerful online tutorial and homework system, which is also available on CD; *MyMathLab* MyMathlab, the online course which houses MathXL plus a variety of other supplements; and the Prentice Hall Mathematics Tutor Center, which provides live tutorial support via phone, fax, or e-mail.

Seek Help One thing I stress with my own students is to *get help as soon as you need it!* Do not wait! In mathematics, one day's material is usually based on the previous day's material. So if you don't understand the material today, you may not be able to understand the material tomorrow.

Where should you seek help? There are often a number of places to obtain help on campus. You should try to make a friend in the class with whom you can study. Often you can help one another. You may wish to form a study group with other students in your class. Discussing the concepts and homework with your peers will reinforce your own understanding of the material.

You should not hesitate to visit your instructor when you are having problems with the material. Be sure you read the assigned material and attempt the homework before meeting with your instructor. Come prepared with specific questions to ask.

Often other sources of help are available. Many colleges have a mathematics laboratory or a mathematics learning center where tutors are available to help students. Ask your instructor early in the semester if any tutors are available, and find out where the tutors are located. Then use these tutors as needed.

5 Learn to Use a Calculator

Many instructors require their students to purchase and to use a calculator in class. You should find out as soon as possible which calculator, if any, your instructor expects you to use. If you plan on taking additional mathematics courses, you should determine which calculator will be required in those courses and consider purchasing that calculator for use in this course if its use is permitted by your instructor. Many instructors require a scientific calculator and many others require a graphing calculator.

In this book we provide information about both types of calculators. Always read and save the user's manual for whatever calculator you purchase.

EXERCISE SET 1.1

Do you know all of the following information? If not, ask your instructor as soon as possible.

1. What is your instructor's name?

2. What are your instructor's office hours?

3. Where is your instructor's office located?

4. How can you best reach your instructor?

5. Where can you obtain help if your instructor is not available?

6. What supplements are available to assist you in learning?

7. Does your instructor recommend or require a specific calculator? If so, which one?

8. When can you use a calculator? Can it be used in class, on homework, on tests?

9. What is your instructor's attendance policy?

10. Why is it important that you attend every class possible?

11. Do you know the name and phone number of a friend in class?

12. For each hour of class time, how many hours outside class are recommended for homework and studying?

13. List what you should do to be properly prepared for each class.

14. Explain how a mathematics textbook should be read.

15. Write a summary of the steps you should follow when taking an exam.

16. Having a positive attitude is very important for success in this course. Are you beginning this course with a positive attitude? It is important that you do!

17. You need to make a commitment to spend the time necessary to learn the material, to do the homework, and to attend class regularly. Explain why you believe this commitment is necessary to be successful in this course.

18. What are your reasons for taking this course?

19. What are your goals for this course?

20. Have you given any thought to studying with a friend or a group of friends? Can you see any advantages in doing so? Can you see any disadvantages in doing so?

---

indicates an exercise worked out on the Lecture Series Videos.

1.2 Sets and Other Basic Concepts

1 Identify sets.

2 Identify and use inequalities.

3 Use set builder notation.

4 Find the union and intersection of sets.

5 Identify important sets of numbers.

Understanding Algebra

Since the *time* that a car travels can *vary* or *change*, it is represented by the variable *t*.

> ### Variable
>
> When a letter is used to represent various numbers it is called a **variable**.

For instance, if t = the time, in hours, that a car is traveling, then t is a variable since the time is constantly changing as the car is traveling. We often use the letters x, y, z, and t to represent variables. However, other letters may be used.

If a letter represents one particular value it is called a **constant**. For example, if s = the number of seconds in a minute, then s represents a constant because there are always 60 seconds in a minute. The number of seconds in a minute does not vary. In this book, letters representing both variables and constants are italicized.

The term **algebraic expression**, or simply **expression**, will be used often in the text. An expression is any combination of numbers, variables, exponents, mathematical symbols (other than equals signs), and mathematical operations.

1 Identify Sets

A **set** is a collection of objects. The objects in a set are called **elements** of the set. Sets are indicated by means of braces, { }, and are often named with capital letters. When the elements of a set are listed within the braces, as illustrated below, the set is said to be in **roster form**.

Set	Number of Elements
$A = \{a, b, c\}$	3
$B = \{\text{yellow, green, blue, red}\}$	4
$C = \{1, 2, 3, 4, 5\}$	5

The symbol $\in$ is used to indicate that an item is an element of a set. Since 2 is an element of set C we may write

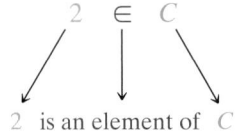

$$2 \quad \in \quad C$$

2 is an element of C

This is read "2 is an element of the set C."

A set may be finite or infinite. Sets A, B, and C each have a finite number of elements and are therefore *finite sets*. In some sets it is impossible to list all the elements. These are *infinite sets*. The following set, called the set of **natural numbers** or **counting numbers**, is an example of an infinite set.

Understanding Algebra

The symbol ... is called an *ellipsis* to indicate that the pattern continues forever in the same manner.

$$N = \{1, 2, 3, 4, 5, \ldots\}$$

The three dots after the last comma are called an *ellipsis*. They indicate that the set continues on and on in the same manner.

Another important infinite set is the integers. The set of **integers** follows.

$$I = \{\ldots, -4, -3, -2, -1, 0, 1, 2, 3, 4, \ldots\}$$

Notice that the set of integers includes both positive and negative integers and the number 0.

If we write

Understanding Algebra

The *positive integers* are

1, 2, 3, 4, 5, 6, ...

The *negative integers* are

−1, −2, −3, −4, −5, −6, ...

$$D = \{1, 2, 3, 4, 5, \ldots, 163\}$$

we mean that the set continues in the same manner until the number 163. Set D is the set of the first 163 natural numbers. D is therefore a finite set.

A special set that contains no elements is called the **null set**, or **empty set**, written
{ } or Ø. For example, the set of students in your class under 8 years of age is the null or
empty set.

2 Identify and Use Inequalities

Inequality Symbols

> is read "is greater than."

≥ is read "is greater than or equal to."

< is read "is less than."

≤ is read "is less than or equal to."

≠ is read "is not equal to."

Inequalities can be explained using the real number line (**Fig. 1.1**).

FIGURE 1.1

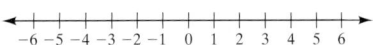

The number a is greater than the number b, $a > b$, when a is to the right of b on
the number line (**Fig. 1.2**). We can also state that the number b is less than a, $b < a$,
when b is to the left of a on the number line. The inequality $a \neq b$ means either $a < b$
or $a > b$.

Lesser   Greater

                         b a

FIGURE 1.2 $a > b$ or $b < a$

EXAMPLE 1   Insert either > or < in the shaded area between the numbers to
make each statement true.

a) 6 ☐ 2   **b)** −7 ☐ 1   **c)** −4 ☐ −5

Solution   Draw a number line and indicate the location of the numbers in parts
a), **b)**, and **c)** as illustrated in **Figure 1.3**.

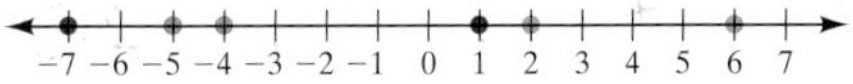

FIGURE 1.3

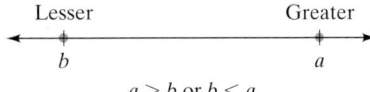

a) $6 > 2$ Note that 6 is to the right of 2 on the number line.

b) $-7 < 1$ Note that −7 is to the left of 1 on the number line.

c) $-4 > -5$ Note that −4 is to the right of −5 on the number line.

Now Try Exercise 19

Helpful Hint

Remember that the symbol used in an inequality, if it is true, always points to the smaller of
the two numbers.

Notation	Is read as
$x > 2$	x is any real number greater than 2.
$x \leq -3$	x is any real number less than or equal to −3.
$-4 \leq x < 3$	x is any real number greater than or equal to −4 and less than 3.

In the inequalities $x > 2$ and $x \leq -3$, the 2 and the -3 are called **endpoints**. In the inequality $-4 \leq x < 3$, the -4 and 3 are the endpoints. The solutions to inequalities that use either $<$ or $>$ do not include the endpoints, but the solutions to inequalities that use either $\leq$ or $\geq$ do include the endpoints. This is shown as follows:

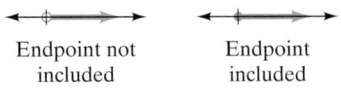

Endpoint not   Endpoint
included   included

Below are three illustrations.

Inequality	Inequality Indicated on the Number Line

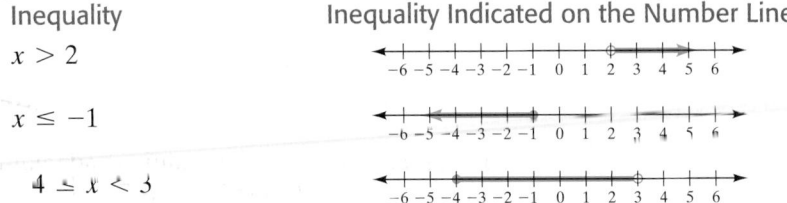

The word *between* indicates that the endpoints are not included in the answer. For example, the set of natural numbers between 2 and 6 is $\{3, 4, 5\}$. If we wish to include the endpoints, we can use the word *inclusive*. For example, the set of natural numbers between 2 and 6 inclusive is $\{2, 3, 4, 5, 6\}$.

3 Use Set Builder Notation

A second method of describing a set is called **set builder notation**. An example of set builder notation is

$$E = \{x | x \text{ is a natural number greater than } 7\}$$

This is read "Set E is the set of all elements x, such that x is a natural number greater than 7." In roster form, this set is written

$$E = \{8, 9, 10, 11, 12, \ldots\}$$

The general form of set builder notation is

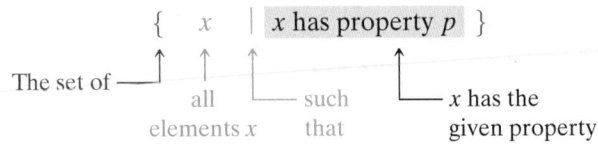

The set of ——
all   such x has the
elements x that   given property

{ x | x has property p }

We often will use the variable x when using set builder notation, although any variable can be used.

Two condensed ways of writing set $E = \{x | x \text{ is a natural number greater than } 7\}$ in set builder notation follow.

$$E = \{x | x > 7 \text{ and } x \in N\} \quad \text{or} \quad E = \{x | x \geq 8 \text{ and } x \in N\}$$

The set $A = \{x | -3 < x \leq 4 \text{ and } x \in I\}$ is the set of integers greater than -3 and less than or equal to 4. The set written in roster form is $\{-2, -1, 0, 1, 2, 3, 4\}$. Notice that the endpoint -3 is not included in the set but the endpoint 4 is included.

How do the sets $B = \{x | x > 2 \text{ and } x \in N\}$ and $C = \{x | x > 2\}$ differ? Set B contains only the natural numbers greater than 2, that is, $\{3, 4, 5, 6, \ldots\}$. Set C contains not only the natural numbers greater than 2 but also fractions and decimal numbers greater than 2. Since there is no smallest number greater than 2, this set cannot be written in roster form. On the top of the next page we illustrate these two sets on the number line. We have also illustrated two other sets.

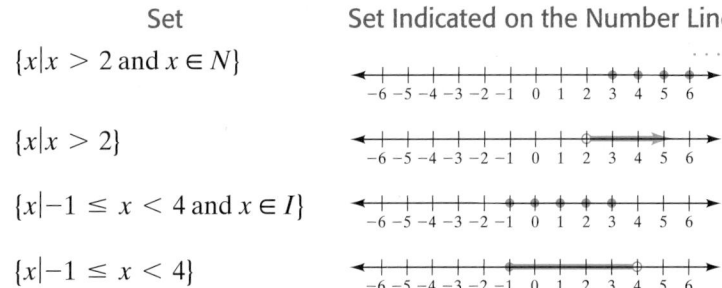

Set	Set Indicated on the Number Line
$\{x\|x > 2 \text{ and } x \in N\}$	
$\{x\|x > 2\}$	
$\{x\|-1 \le x < 4 \text{ and } x \in I\}$	
$\{x\|-1 \le x < 4\}$	

Another method of indicating inequalities, called *interval notation*, will be discussed in Section 2.5.

4 Find the Union and Intersection of Sets

Just as *operations* such as addition and multiplication are performed on numbers, operations can be performed on sets. Two set operations are *union* and *intersection*.

Union of Two Sets

The **union** of set A and set B, written $A \cup B$, is the set of elements that belong to either set A *or* set B.

Because the word *or*, as used in this context, means belonging to set A or set B or both sets, the union is formed by combining, or joining together, the elements in set A with those in set B. If an item is an element in either set A, or set B, or in both sets, then it is an element in the union of the sets. If an element appears in both sets, we only list it once when we write the union of two sets.

Examples of Union of Sets

$A = \{1, 2, 3, 4, 5\},$ $B = \{3, 4, 5, 6, 7\},$ $A \cup B = \{1, 2, 3, 4, 5, 6, 7\}$
$A = \{a, b, c, d, e\},$ $B = \{x, y, z\},$ $A \cup B = \{a, b, c, d, e, x, y, z\}$

In set builder notation we can express $A \cup B$ as

Union

$$A \cup B = \{x\|x \in A \text{ or } x \in B\}$$

Intersection of Two Sets

The **intersection** of set A and set B, written $A \cap B$, is the set of all elements that are common to both set A *and* set B.

Because the word *and*, as used in this context, means belonging to *both* set A and set B, the intersection is formed by using only those elements that are in both set A and set B. If an item is an element in only one of the two sets, then it is not an element in the intersection of the sets.

Examples of Intersection of Sets

$A = \{1, 2, 3, 4, 5\},$ $B = \{3, 4, 5, 6, 7\},$ $A \cap B = \{3, 4, 5\}$
$A = \{a, b, c, d, e\},$ $B = \{x, y, z\},$ $A \cap B = \{\ \}$

Note that in the last example, sets A and B have no elements in common. Therefore, their intersection is the empty set. In set builder notation we can express $A \cap B$ as

Intersection

$$A \cap B = \{x \mid x \in A \text{ and } x \in B\}$$

5 Identify Important Sets of Numbers

In the box below, we describe different sets of numbers and provide letters that are often used to represent these sets of numbers.

Important Sets of Numbers

Real numbers	$\mathbb{R} = \{x \mid x \text{ is a point on the number line}\}$
Natural or counting numbers	$N = \{1, 2, 3, 4, 5, \ldots\}$
Whole numbers	$W = \{0, 1, 2, 3, 4, 5, \ldots\}$
Integers	$I = \{\ldots, -3, -2, -1, 0, 1, 2, 3, \ldots\}$
Rational numbers	$Q = \left\{ \dfrac{p}{q} \,\middle|\, p \text{ and } q \text{ are integers}, q \neq 0 \right\}$
Irrational numbers	$H = \{x \mid x \text{ is a real number that is not rational}\}$

Understanding Algebra

A *rational number* can be expressed as the *ratio* of two integers:

$$\frac{\text{integer}}{\text{integer}}$$

A **rational number** is any number that can be represented as a quotient of two integers, with the denominator not 0.

Examples of Rational Numbers

$$\frac{3}{5}, \quad -\frac{2}{3}, \quad 0, \quad 1.63, \quad 7, \quad -17, \quad \sqrt{4}$$

Notice that 0, or any other integer, is also a rational number since it can be written as a fraction with a denominator of 1. For example, $0 = \dfrac{0}{1}$ and $7 = \dfrac{7}{1}$.

The number 1.63 can be written $\dfrac{163}{100}$ and is thus a quotient of two integers. Since $\sqrt{4} = 2$ and 2 is an integer, $\sqrt{4}$ is a rational number. *Every rational number when written as a decimal number will be either a repeating or a terminating decimal number.*

Understanding Algebra

A rational number whose decimal representation ends is a *terminating decimal number*. A rational number whose decimal representation repeats is a *repeating decimal number*.

Examples of Repeating Decimals	Examples of Terminating Decimals
$\dfrac{2}{3} = 0.6666\ldots$ 6 repeats.	$\dfrac{1}{2} = 0.5$
$\dfrac{1}{7} = 0.142857142857\ldots$ 142857 repeats.	$\dfrac{9}{4} = 2.25$

To show that a digit or group of digits repeats, we can place a bar above the digit or group of digits that repeat. For example, we may write

$$\frac{2}{3} = 0.\overline{6} \quad \text{and} \quad \frac{1}{7} = 0.\overline{142857}$$

An **irrational number** is a real number that is not a rational number. Some irrational numbers are $\sqrt{2}, \sqrt{3}, \sqrt{5},$ and $\sqrt{6}$. Another irrational number is pi, π. When we give a decimal value for an irrational number, we are giving only an *approximation* of the value of the irrational number. The symbol $\approx$ means "is approximately equal to."

$$\pi \approx 3.14 \qquad \sqrt{2} \approx 1.41 \qquad \sqrt{3} \approx 1.73 \qquad \sqrt{10} \approx 3.16$$

Helpful Hint

Remember in writing an *approximation,* use the symbol $\approx$.

The **real numbers** are formed by taking the *union* of the rational numbers and the irrational numbers. Therefore, any real number must be either a rational number or an irrational number. The symbol $\mathbb{R}$ is often used to represent the set of real numbers. **Figure 1.4** illustrates various real numbers on the number line.

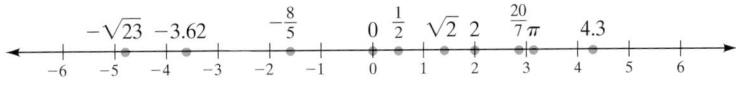

FIGURE 1.4   Real Numbers

Subset

The set A is a **subset** of the set B when every element of A is also an element of B and we write $A \subseteq B$.

For example, the set of natural numbers, $\{1, 2, 3, 4, \ldots\}$, is a subset of the set of whole numbers, $\{0, 1, 2, 3, 4, \ldots\}$, because every element in the set of natural numbers is also an element in the set of whole numbers. **Figure 1.5** illustrates the relationships between the various subsets of the real numbers. In **Figure 1.5a**, you see that the set of natural numbers is a subset of the set of whole numbers, of the set of integers, and of the set of rational numbers. Therefore, every natural number must also be a whole number, an integer, a rational number, and a real number.

Every natural number is also

- a whole number,
- an integer,
- a rational number, and
- a real number.

Using the same reasoning, we can see that the set of whole numbers is a subset of the integers, the set of rational numbers, and the set of real numbers, and that the set of integers is a subset of the set of rational numbers and the set of real numbers.

Looking at **Figure 1.5b**, we see that the positive integers, 0, and the negative integers form the integers, that the integers and noninteger rational numbers form the rational numbers, and so on.

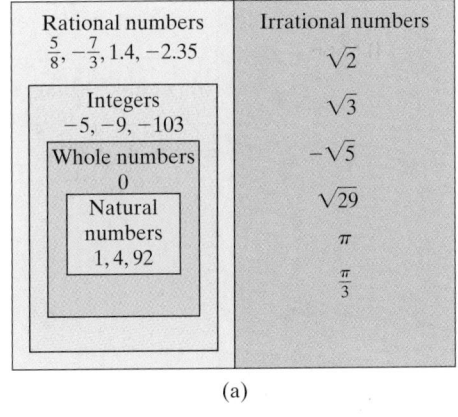

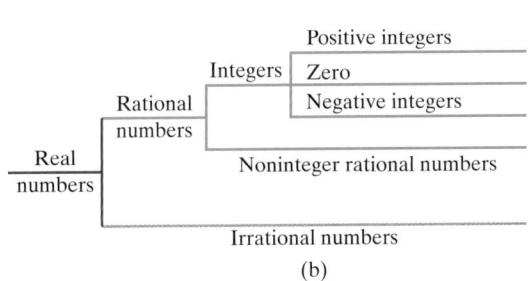

FIGURE 1.5   (a)   (b)

EXAMPLE 2 Consider the following set:

$$\left\{-8, 0, \frac{5}{9}, 12.25, \sqrt{7}, -\sqrt{11}, \frac{22}{7}, 5, 7.1, -54, \pi\right\}$$

List the elements of the set that are

a) natural numbers.   **b)** whole numbers.   **c)** integers.

d) rational numbers.   **e)** irrational numbers.   **f)** real numbers.

Solution

a) Natural numbers: 5   **b)** Whole numbers: 0, 5   **c)** Integers: −8, 0, 5, −54

d) Rational numbers can be written in the form p/q, $q \neq 0$. Each of the following can be written in this form and is a rational number.

$$-8, 0, \frac{5}{9}, 12.25, \frac{22}{7}, 5, 7.1, -54$$

e) Irrational numbers are real numbers that are not rational. The following numbers are irrational.

$$\sqrt{7}, -\sqrt{11}, \pi$$

f) All of the numbers in the set are real numbers. The union of the rational numbers and the irrational numbers forms the real numbers.

$$-8, 0, \frac{5}{9}, 12.25, \sqrt{7}, -\sqrt{11}, \frac{22}{7}, 5, 7.1, -54, \pi$$

Now Try Exercise 39

Not all numbers are real numbers. Some numbers that we discuss later in the text that are not real numbers are complex numbers and imaginary numbers.

EXERCISE SET 1.2   Math XL   MyMathLab
MathXL®   MyMathLab

Warm-Up Exercises

Fill in the blanks with the appropriate word, phrase, or symbol(s) from the following list.

empty	constant	rational	set	variable	union
approximation	elements	subset	intersection	irrational	algebraic expression

1. A letter used to represent various numbers is a _____.

2. A letter that represents one particular value is a _____.

3. Any combination of numbers, variables, exponents, math symbols, and operations is called an _____.

4. A collection of objects is a _____.

5. The objects in a set are called _____.

6. The set that contains no elements is the _____ set.

7. If every element of set A is an element of set B, then set A is a _____ of set B.

8. $A \cup B$ represents the _____ of the two sets A and B.

9. $A \cap B$ represents the _____ of the two sets A and B.

10. A number that can be represented as a quotient of two integers is a _____ number.

11. A real number that is not a rational number is an _____ number.

12. The symbol $\approx$ is called the _____ symbol.

Practice the Skills

Insert $<$, $>$, or $=$ in the shaded area to make each statement true.

13. 5 ▢ 3

14. −1 ▢ 8

15. $\dfrac{-4}{2}$ ▢ −2

16. $\dfrac{9}{-3}$ ▢ −3

17. −1 ▢ −1.01

18. 2 ▢ −3

19. −5 ▢ −3

20. −8 ▢ −1

21. −14.98 ▢ −14.99

22. −3.4 ▢ −3.2

23. 1.7 ▢ 1.9

24. −1.1 ▢ −21

25. −π ▢ −4

26. −723 ▢ −655

27. $-\dfrac{7}{8}$ ▢ $-\dfrac{10}{11}$

28. $-\dfrac{4}{7}$ ▢ $-\dfrac{5}{9}$

In Exercises 29–38, list each set in roster form.

29. $A = \{x | -1 < x < 1 \text{ and } x \in I\}$

30. $B = \{y | y \text{ is an odd natural number less than 6}\}$

31. $C = \{z | z \text{ is an even integer greater than 16 and less than or equal to 20}\}$

32. $D = \{x | x \geq -3 \text{ and } x \in I\}$

33. $E = \{x | x < 3 \text{ and } x \in W\}$

34. $F = \left\{x \middle| -\dfrac{6}{5} \leq x < \dfrac{15}{4} \text{ and } x \in N\right\}$

A green numbered exercise, such as **19.**, *indicates a Now Try Exercise.*

35. $H = \{x | x$ is a whole number multiple of 7$\}$

36. $L = \{x | x$ is an integer greater than $-5\}$

37. $J = \{x | x > 0$ and $x \in I\}$

38. $K = \{x | x$ is a whole number between 9 and 10$\}$

39. Consider the set $\left\{-2, 4, \dfrac{1}{2}, \dfrac{5}{9}, 0, \sqrt{2}, \sqrt{8}, -1.23, \dfrac{78}{79}\right\}$.
List the elements that are
a) natural numbers.
b) whole numbers.
c) integers.
d) rational numbers.
e) irrational numbers.
f) real numbers.

40. Consider the set $\left\{2, 4, -5.33, \dfrac{11}{2}, \sqrt{5}, \sqrt{2}, -100, -7, 4.7\right\}$.
List the elements that are:
a) whole numbers.
b) natural numbers.
c) rational numbers.
d) integers.
e) irrational numbers.
f) real numbers.

Find $A \cup B$ and $A \cap B$ for each set A and B.

41. $A = \{1, 2, 3\}, B = \{4, 5, 6\}$

42. $A = \{1, 2, 3, 4, 5\}, B = \{2, 4, 6, 7\}$

43. $A = \{-3, -1, 1, 3\}, B = \{-4, -3, -2, -1, 0\}$

44. $A = \{-3, -2, -1, 0\}, B = \{-1, 0, 1, 2\}$

45. $A = \{ \ \}, B = \{2, 4, 6, 8, 10\}$

46. $A = \{2, 4, 6\}, B = \{2, 4, 6, 8, \ldots\}$

47. $A = \{0, 10, 20, 30\}, B = \{5, 15, 25\}$

48. $A = \{1, 3, 5\}, B = \{1, 3, 5, 7, \ldots\}$

49. $A = \{-1, 0, 1, e, i, \pi\}, B = \{-1, 0, 1\}$

50. $A = \left\{1, \dfrac{1}{2}, \dfrac{1}{4}, \dfrac{1}{6}, \ldots\right\}, B = \left\{\dfrac{1}{4}, \dfrac{1}{6}, \dfrac{1}{8}, \dfrac{1}{10}\right\}$

Describe each set.

51. $A = \{1, 2, 3, 4, \ldots\}$

52. $B = \{2, 4, 6, 8, \ldots\}$

53. $C = \{0, 3, 6, 9, \ldots\}$

54. $A = \{a, b, c, d, \ldots, z\}$

55. $B = \{\ldots, -5, -3, -1, 1, 3, 5, \ldots\}$

56. $C = \{$Alabama, Alaska, $\ldots$, Wyoming$\}$

In Exercises 57 and 58, **a)** *write out how you would read each set;* **b)** *write the set in roster form.*

57. $A = \{x | x < 7$ and $x \in N\}$

58. $B = \{x | x$ is one of the last five capital letters in the English alphabet$\}$

Illustrate each set on a number line.

59. $\{x | x \geq 0\}$

60. $\{w | w > -5\}$

61. $\{z | z \leq 2\}$

62. $\{y | y < 4\}$

63. $\{p | -6 \leq p < 3\}$

64. $\{x | -1.67 \leq x < 5.02\}$

65. $\{q | q > -3$ and $q \in N\}$

66. $\{x | -1.93 \leq x \leq 2$ and $x \in I\}$

67. $\{r | r \leq \pi$ and $r \in W\}$

68. $\left\{x \left| \dfrac{5}{12} < x \leq \dfrac{7}{12} \text{ and } x \in N \right.\right\}$

Express in set builder notation each set of numbers that are indicated on the number line.

69.

70.

71.

72.

73.

74.

75.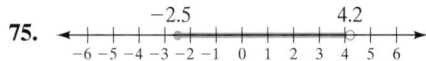

76.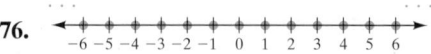

77. ![number line with -6 to 6, dots marked at -4, -2, -1, 0, 1, 2, 3]

78. number line showing $-\frac{12}{5}$ and $\frac{4}{11}$ on a line from -6 to 6

Refer to the box on page 10 for the meanings of ℝ, *N, W, I, Q, and H. Then determine whether the first set is a subset of the second set for each pair of sets.*

79. *N, W*　　　　**80.** *W, Q*　　　　**81.** *Q, I*　　　　**82.** *I, Q*

83. *Q,* ℝ　　　　**84.** *Q, H*　　　　**85.** *W, N*　　　　**86.** *H,* ℝ

Problem Solving

87. Construct a set that contains five rational numbers between 1 and 2.

88. Construct a set that contains five rational numbers between 0 and 1.

89. Determine two sets *A* and *B* such that $A \cup B = \{2, 4, 5, 6, 8, 9\}$ and $A \cap B = \{4, 5, 9\}$.

90. Determine two sets *A* and *B* such that $A \cup B = \{3, 5, 7, 8, 9\}$ and $A \cap B = \{5, 7\}$.

91. **NASCAR Sprint Cup** The 2008 NASCAR Sprint Cup series consisted of 36 races held between February and November. Two such races were the LifeLock 400 and the Dodge Challenger 500. The tables below show the top six finishers in both races.

LifeLock 400

Position	Driver
1	Dale Earnhardt Jr
2	Kasey Kahn
3	Matt Kenseth
4	Brian Vickers
5	Jimmy Johnson
6	Carl Edwards

Dodge Challenger 500

Position	Driver
1	Kyle Busch
2	Carl Edwards
3	Dale Earnhardt Jr
4	David Regan
5	Matt Kenseth
6	Denny Hamlin

Source: www.NASCAR.com

a) Find the set of drivers who had a top 6 finish in the LifeLock 400 *or* the Dodge Challenger 500.

b) Does part **a)** represent the union or intersection of the drivers?

c) Find the set of drivers who had a top 6 finish in the LifeLock 400 *and* the Dodge challenger 500.

d) Does part **c)** represent the union or intersection of the drivers?

© Jonathan Ferrey/Allsport Concepts/Getty Images

92. **Disasters** The tables below give estimates of the six deadliest earthquakes and the six deadliest natural disasters.

Six Deadliest Earthquakes

Deaths	Magnitude	Location	Year
255,000	7.8–8.2	Tangshan, China	1976
200,000	8.3	Xining, China	1927
200,000	8.6	Gansu, China	1920
175,000	9.0	Asia/Africa	2004
143,000	8.3	Kwanto, Japan	1923
110,000	7.3	Turkmenistan	1948

Six Deadliest Natural Disasters

Deaths	Event	Location	Year
3.7 million	Flood	Huang He River, China	1931
300,000	Cyclone	Bangladesh	1970
255,000	Earthquake	Tangshan, China	1976
200,000	Earthquake	Xining, China	1927
200,000	Earthquake	Gansu, China	1920
175,000	Earthquake/ Tsunami	Asia/Africa	2004

Source: www.msnbc.com/modules/tables/worstquakesofcentury, Associated Press, Reuters, U.S. Geological Survey, *The World Almanac, The Washington Post*

a) Find the set of the location of the six deadliest earthquakes *or* the location of the six deadliest natural disasters.

b) Does part **a)** represent the union or intersection of the categories?

c) Find the set of the location of the six deadliest earthquakes *and* the location of the set of the six deadliest natural disasters.

d) Does part **c)** represent the union or intersection of the categories?

93. Algebra Tests The table below shows the students who had a grade of A on the first two tests in an intermediate algebra class.

First Test	Second Test
Albert	Linda
Carmen	Jason
Frank	David
Linda	Frank
Barbara	Earl
	Kate
	Ingrid

a) Find the set of students who had a grade of A on the first *or* second tests.

b) Does part **a)** represent the union or intersection of the students?

c) Find the set of students who had a grade of A on the first *and* second tests.

d) Does part **c)** represent the union or intersection of the students?

94. Running Races The table below shows the runners who participated in a 3-kilometer (km) race and a 5-kilometer race.

3 Kilometers	5 Kilometers
Adam	Luan
Kim	Betty
Luan	Darnell
Ngo	Ngo
Carmen	Frances
Earl	George
Martha	Adam

a) Find the set of runners who participated in a 3-km *or* a 5-km race.

b) Does part **a)** represent the union or intersection of the runners?

c) Find the set of runners who participated in a 3-km *and* a 5-km race.

d) Does part **c)** represent the union or intersection of runners?

95. Populous Countries The table below shows the five most populous countries in 1950 and in 2010 and the five countries expected to be most populous in 2050. This information was taken from the U.S. Census Bureau Web site.

1950	2010	2050
China	China	India
India	India	China
United States	United States	United States
Russia	Indonesia	Indonesia
Japan	Brazil	Nigeria

a) Find the set of the five most populous countries in 2010 *or* 2050.

b) Find the set of the five most populous countries in 1950 *or* 2050.

c) Find the set of the five most populous countries in 1950 *and* 2010.

d) Find the set of the five most populous countries in 2010 *and* 2050.

e) Find the set of the five most populous countries in 1950 *and* 2010 *and* 2050.

96. Writing Contest The table below shows the students from an English class who participated in three writing contests in a local high school.

First Contest	Second Contest	Third Contest
Jill	Tom	Pat
Sam	Shirley	Richard
Tom	Bob	Arnold
Pat	Donna	Donna
Shirley	Sam	Kate
Richard	Jill	
	Kate	

a) Find the set of students who participated in the first contest *or* the second contest.

b) Find the set of students who participated in the second contest *or* the third contest.

c) Find the set of students who participated in the first contest *and* the second contest.

d) Find the set of students who participated in the first contest *and* the third contest.

e) Find the set of students who participated in the first contest *and* the second contest *and* the third contest.

97. Cub Scouts The Cub Scouts in Pack 108 must complete four achievements to earn their Wolf Badge. Doug Wedding, their den leader, has the following table in his record book. A *Yes* indicates that the Cub Scout has completed that achievement.

Scout	Achievement			
	1	2	3	4
Alex	Yes	Yes	Yes	Yes
James	Yes	Yes	No	No
George	No	Yes	No	Yes
Connor	No	Yes	No	Yes
Stephen	No	No	Yes	No

Let A = the set of scouts who have completed Achievement 1: *Feats of Skill*.

Let B = the set of scouts who have completed Achievement 2: *Your Flag*.

Let C = the set of scouts who have completed Achievement 3: *Cooking and Eating*.

Let D = the set of scouts who have completed Achievement 4: *Making Choices*.

a) Give each of the sets A, B, C, and D using the roster method.

b) Determine the set $A \cap B \cap C \cap D$, that is, find the set of elements that are common to all four sets.

c) Which scouts have met all the requirements to receive their Wolf Badge?

98. Goods and Services The following graph shows the percentage weight given to different goods and services in the consumer price index for December 2006.

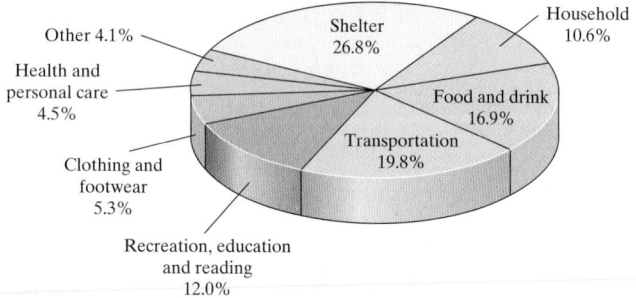

Source: U.S. Bureau of Labor Statistics

a) List the set of goods and services that have a weight of 21% or greater.

b) List the set of goods and services that have a weight of less than 6%.

99. The following diagram is called a *Venn diagram*. From the diagram determine the following sets:

a) A

b) B

c) $A \cup B$

d) $A \cap B$

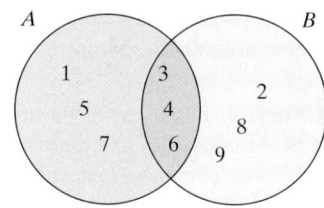

100. Use the following Venn diagram to determine the following sets:

a) A

b) B

c) $A \cup B$

d) $A \cap B$

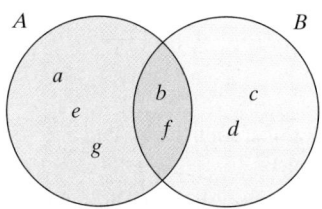

101. a) Explain the difference between the following sets of numbers: $\{x | x > 1 \text{ and } x \in N\}$ and $\{x | x > 1\}$.

b) Write the first set given in roster form.

c) Can you write the second set in roster form? Explain your answer.

102. Repeat Exercise 101 for the sets $\{x | 2 < x < 6 \text{ and } x \in N\}$ and $\{x | 2 < x < 6\}$.

103. NASCAR Cup Draw a Venn diagram for the data given in Exercise 91 on page 14.

Concept/Writing Exercises

104. Is the set of natural or counting numbers a finite or infinite set? Explain.

105. List the five inequality symbols and write down how each is read.

106. List the set of integers *between* 3 and 7.

107. List the set of integers *between* −1 and 3 *inclusive*.

108. Explain why every integer is also a rational number.

109. Describe the counting numbers, whole numbers, integers, rational numbers, irrational numbers, and real numbers. Explain the relationships among the sets of numbers.

In Exercises 110–119, indicate whether each statement is true or false.

110. Every natural number is a whole number.

111. Every whole number is a natural number.

112. Some rational numbers are integers.

113. Every integer is a rational number.

114. Every rational number is an integer.

115. The union of the set of rational numbers with the set of irrational numbers forms the set of real numbers.

116. The intersection of the set of rational numbers and the set of irrational numbers is the empty set.

117. The set of natural numbers is a finite set.

118. The set of integers between π and 4 is the null set.

119. The set of rational numbers between 3 and π is an infinite set.

Challenge Problems

120. a) Write the decimal numbers equivalent to $\frac{1}{9}, \frac{2}{9}$, and $\frac{3}{9}$.

 b) Write the fractions equivalent to $0.\overline{4}, 0.\overline{5}$, and $0.\overline{6}$.

 c) What is $0.\overline{9}$ equal to? Explain how you determined your answer.

Group Activity

121. Newspaper Preferences The Venn diagram that follows shows the results of a survey given to 45 people. The diagram shows the number of people in the survey who read the *New York Post*, the *New York Daily News*, and *The Wall Street Journal*.

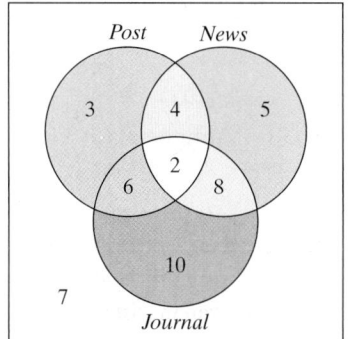

a) Group member 1: Determine the number surveyed who read *both* the *Post* and the *News* that is, *Post* ∩ *News*.

b) Group member 2: Determine the number who read both the *Post* and the *News*, that is, *Post* ∩ *News*.

c) Group member 3: Determine the number who read both the *News* and the *Journal*, that is, *News* ∩ *Journal*.

d) Share your answer with the other members of the group and see if the group agrees with your answer.

e) As a group, determine the number of people who read all three papers.

f) As a group, determine the number of people who do not read any of the three papers.

1.3 Properties of and Operations with Real Numbers

1 Evaluate absolute values.

2 Add real numbers.

3 Subtract real numbers.

4 Multiply real numbers.

5 Divide real numbers.

6 Use the properties of real numbers.

Two numbers that are the same distance from 0 on the number line but in opposite directions are called **additive inverses**, or **opposites**, of each other. For example, 3 is the additive inverse of -3, and -3 is the additive inverse of 3. The number 0 is its own additive inverse. The sum of a number and its additive inverse is 0. What are the additive inverses of -56.3 and $\frac{76}{5}$? Their additive inverses are 56.3 and $-\frac{76}{5}$, respectively.

Additive Inverses

-56.3 and 56.3

$\frac{76}{5}$ and $-\frac{76}{5}$

0 and 0

> **Helpful Hint**
>
> Notice that the additive inverse of a positive number is a negative number and the additive inverse of a negative number is a positive number.

Additive Inverse

For any real number a, its additive inverse is $-a$.

Consider the number -5. Its additive inverse is $-(-5)$. Since we know this number must be positive, this implies that $-(-5) = 5$. This is an example of the double negative property.

Double Negative Property

For any real number a, $-(-a) = a$.

By the double negative property, $-(-7.4) = 7.4$ and $-\left(-\dfrac{12}{5}\right) = \dfrac{12}{5}$.

Understanding Algebra

With additive inverses, one number is *positive* and the other is *negative*.

1 Evaluate Absolute Values

Absolute Value

The **absolute value** of a number is its distance from the number 0 on a number line. The symbol $|\ \ |$ is used to indicate absolute value.

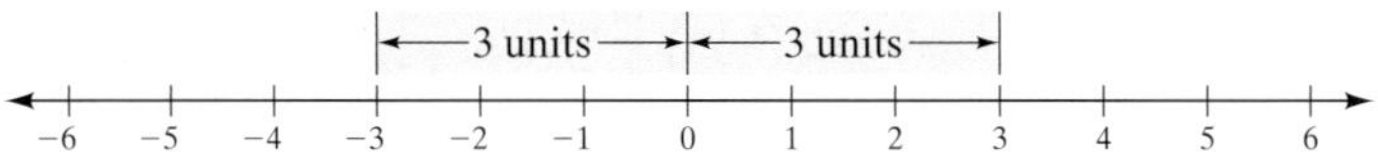

FIGURE 1.6

Consider the numbers 3 and -3 (**Fig. 1.6**). Both numbers are 3 units from 0 on the number line. Thus

$$|3| = 3 \quad \text{and} \quad |-3| = 3$$

EXAMPLE 1 Evaluate.   **a)** $|9|$ **b)** $|-8.2|$ **c)** $|0|$

Solution

a) $|9| = 9$, since 9 is 9 units from 0 on the number line.

b) $|-8.2| = 8.2$, since -8.2 is 8.2 units from 0 on the number line.

c) $|0| = 0$.

The absolute value of any nonzero number will always be a positive number, and the absolute value of 0 is 0.

To find the absolute value of a real number without using a number line, use the following definition.

Now Try Exercise 13

Absolute Value

If a represents any real number, then

$$|a| = \begin{cases} a & \text{if } a \geq 0 \\ -a & \text{if } a < 0 \end{cases}$$

The definition of absolute value indicates that the absolute value of any nonnegative number is the number itself, and the absolute value of any negative number is the additive inverse (or opposite) of the number. The absolute value of a number can be found by using the definition, as illustrated below.

$|6.3| = 6.3$	Since 6.3 is greater than or equal to 0, its absolute value is 6.3.
$|0| = 0$	Since 0 is greater than or equal to 0, its absolute value is 0.
$|-12| = -(-12) = 12$	Since -12 is less than 0, its absolute value is $-(-12)$ or 12.

EXAMPLE 2   Evaluate using the definition of absolute value.

a) $-|5|$ **b)** $-|-6.43|$

Solution

a) We are finding the opposite of the absolute value of 5. Since the absolute value of 5 is positive, its opposite must be negative.

$$-|5| = -(5) = -5$$

b) We are finding the opposite of the absolute value of -6.43. Since the absolute value of -6.43 is positive, its opposite must be negative.

$$-|-6.43| = -(6.43) = -6.43$$

Now Try Exercise 21

EXAMPLE 3   Insert $<, >$, or $=$ in the shaded area between the two values to make each statement true.

a) $|8|$ ▨ $|-8|$ **b)** $|-1|$ ▨ $-|-3|$

Solution

a) Since both $|8|$ and $|-8|$ equal 8, we have $|8| = |-8|$.

b) Since $|-1| = 1$ and $-|-3| = -3$, we have $|-1| > -|-3|$.

Now Try Exercise 29

2 Add Real Numbers

To Add Two Numbers with the Same Sign (Both Positive or Both Negative)

Add their absolute values and place the common sign before the sum.

The sum of two positive numbers will always be a positive number, and the sum of two negative numbers will always be a negative number.

EXAMPLE 4   Evaluate $-4 + (-7)$.

Solution   Since both numbers being added are negative, the sum will be negative. We need to add the absolute values of these numbers and then place a negative sign before the value. First, find the absolute value of each number.

$$|-4| = 4 \qquad |-7| = 7$$

Then add the absolute values.

$$|-4| + |-7| = 4 + 7 = 11$$

Finally, since both numbers are negative, the sum must be negative. Thus,

$$-4 + (-7) = -11$$

Now Try Exercise 45

To Add Two Numbers with Different Signs (One Positive and the Other Negative)

Subtract the smaller absolute value from the larger absolute value. The answer has the sign of the number with the larger absolute value.

The sum of a positive number and a negative number may be either positive, negative, or zero. The sign of the answer will be the same as the sign of the number with the larger absolute value.

EXAMPLE 5 Evaluate $5 + (-9)$.

Solution   Since the numbers being added are of opposite signs, we subtract the smaller absolute value from the larger absolute value. First we take each absolute value.

$$|5| = 5 \qquad |-9| = 9$$

Now we find the difference, $9 - 5 = 4$. The number -9 has a larger absolute value than the number 5, so their sum is negative.

$$5 + (-9) = -4$$

Now Try Exercise 43

EXAMPLE 6 Evaluate.   **a)** $1.3 + (-2.7)$ **b)** $-\dfrac{7}{8} + \dfrac{5}{6}$

Solution

a) $1.3 + (-2.7) = -1.4$

b) Begin by writing both fractions with the least common denominator, 24.

$$-\frac{7}{8} + \frac{5}{6} = -\frac{21}{24} + \frac{20}{24} = \frac{(-21) + 20}{24} = \frac{-1}{24} = -\frac{1}{24}$$

Now Try Exercise 49

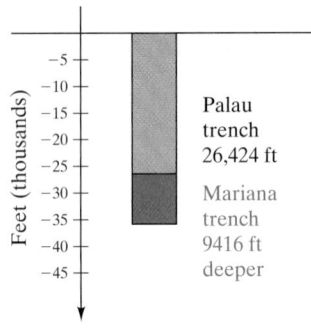

Depth Below Sea Level

Feet (thousands)

Palau trench 26,424 ft

Mariana trench 9416 ft deeper

FIGURE 1.7

EXAMPLE 7 Depth of Ocean Trenches The Palau Trench in the Pacific Ocean lies 26,424 feet below sea level. The deepest ocean trench, the Mariana Trench, is 9416 feet deeper than the Palau Trench (see **Fig. 1.7**). Find the depth of the Mariana Trench.

Solution   Consider distance below sea level to be negative. Therefore, the total depth is

$$-26,424 + (-9416) = -35,840 \text{ feet}$$

or 35,840 feet below sea level.

Now Try Exercise 127

3 Subtract Real Numbers

Every subtraction problem can be expressed as an addition problem using the following rule.

> **Subtraction of Real Numbers**
>
> $$a - b = a + (-b)$$

To subtract b from a, add the opposite (or additive inverse) of b to a.
For example, $5 - 7$ means $5 - (+7)$. To subtract $5 - 7$, add the opposite of $+7$, which is -7, to 5.

$$5 - 7 = 5 + (-7)$$

subtract positive add negative
7   7

Since $5 + (-7) = -2$, then $5 - 7 = -2$.

EXAMPLE 8 Evaluate.   **a)** $3 - 8$ **b)** $-6 - 4$

Solution   **a)** $3 - 8 = 3 + (-8) = -5$ **b)** $-6 - 4 = -6 + (-4) = -10$

Now Try Exercise 79

EXAMPLE 9 Evaluate $8 - (-15)$.

Solution   In this problem, we are subtracting a negative number. The procedure to subtract remains the same.

$$8 - (-15) = 8 + 15 = 23$$

subtract   negative   add   positive
15   15

Thus, $8 - (-15) = 23$.

Now Try Exercise 81

From Example 9 and other examples, we have the following principle.

> **Subtracting a Negative Number**
> $$a - (-b) = a + b$$

We can use this principle to evaluate problems such as $8 - (-15)$ and other problems where we *subtract a negative quantity*.

EXAMPLE 10 Evaluate $-4 - (-11)$.

Solution $-4 - (-11) = -4 + 11 = 7$

Now Try Exercise 47

EXAMPLE 11   **a)** Subtract 35 from -42.   **b)** Subtract $-\dfrac{3}{5}$ from $-\dfrac{5}{9}$.

Solution

a) $-42 - 35 = -77$

b) $-\dfrac{5}{9} - \left(-\dfrac{3}{5}\right) = -\dfrac{5}{9} + \dfrac{3}{5} = -\dfrac{25}{45} + \dfrac{27}{45} = \dfrac{2}{45}$

Now Try Exercise 99

EXAMPLE 12 **Extreme Temperatures** The hottest temperature ever recorded in the United States was 134°F, which occurred at Greenland Ranch, California, in Death Valley on July 10, 1913. The coldest temperature ever recorded in the United States was -79.8°F, which occurred at Prospect Creek Camp, Alaska, in the Endicott Mountains on January 23, 1971 (see **Fig. 1.8**). Determine the difference between these two temperatures.

Solution   To find the difference, we subtract.

$$134° - (-79.8°) = 134° + 79.8° = 213.8°$$

Now Try Exercise 125

Addition and subtraction are often combined in the same problem, as in the following examples. Unless parentheses are present, if the expression involves only addition and subtraction, we add and subtract from left to right. When parentheses are used, we add and subtract within the parentheses first. Then we add and subtract from left to right.

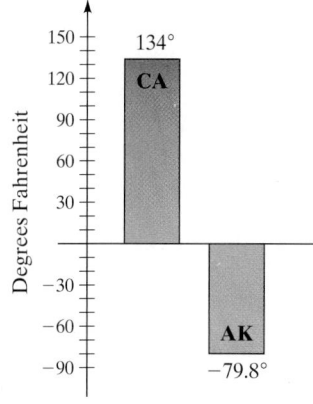

FIGURE 1.8

EXAMPLE 13 Evaluate $-15 + (-37) - (5 - 9)$.

Solution
$$\begin{aligned} -15 + (-37) - (5 - 9) &= -15 + (-37) - (-4) \\ &= -15 - 37 + 4 \\ &= -52 + 4 = -48 \end{aligned}$$

Now Try Exercise 85

EXAMPLE 14 Evaluate $2 - |-3| + 4 - (6 - |-8|)$.

Solution   Begin by replacing the numbers in absolute value signs with their numerical equivalents; then evaluate.

$$\begin{aligned} 2 - |-3| + 4 - (6 - |-8|) &= 2 - 3 + 4 - (6 - 8) \\ &= 2 - 3 + 4 - (-2) \\ &= 2 - 3 + 4 + 2 \\ &= -1 + 4 + 2 \\ &= 3 + 2 = 5 \end{aligned}$$

Now Try Exercise 59

4 Multiply Real Numbers

Understanding Algebra

Observe that:

$(-)(-) = +, (+)(+) = +$

$(-)(+) = -, (+)(-) = -$

Multiply Two Real Numbers

1. To multiply two numbers with *like signs*, either both positive or both negative, multiply their absolute values. The answer is *positive*.

2. To multiply two numbers with *unlike signs*, one positive and the other negative, multiply their absolute values. The answer is *negative*.

EXAMPLE 15 Evaluate.   **a)** $(4.2)(-1.6)$ **b)** $(-18)\left(-\dfrac{1}{2}\right)$.

Solution

a) $(4.2)(-1.6) = -6.72$ The numbers have unlike signs. The answer is negative.

b) $(-18)\left(-\dfrac{1}{2}\right) = 9$ The numbers have like signs, both negative. The answer is positive.

Now Try Exercise 65

EXAMPLE 16 Evaluate $4(-2)(-3)(1)$.

Solution $4(-2)(-3)(1) = (-8)(-3)(1) = 24(1) = 24$

Now Try Exercise 67

Understanding Algebra

When multiplying more than two negative numbers, the product will be

- *negative,* if there are an *odd* number of negative numbers.
- *positive,* if there are an *even* number of negative numbers.

When multiplying more than two numbers, the product will be *negative* when there is an *odd* number of negative numbers. The product will be *positive* when there is an *even* number of negative numbers.

Multiplicative Property of Zero

For any number a,

$$a \cdot 0 = 0 \cdot a = 0$$

By the multiplicative property of zero, $5(0) = 0$ and $(-7.3)(0) = 0$.

EXAMPLE 17 Evaluate $9(5)(-2.63)(0)(4)$.

Solution   If one or more of the factors is 0, the product is 0. Thus, $9(5)(-2.63)(0)(4) = 0$.

Now Try Exercise 101

5 Divide Real Numbers

The rules for the division of real numbers are similar to those for multiplication of real numbers.

Understanding Algebra

Observe that

$$\frac{(-)}{(-)} = +, \quad \frac{(+)}{(+)} = +$$

$$\frac{(+)}{(-)} = -, \quad \frac{(-)}{(+)} = -$$

> **Divide Two Real Numbers**
>
> 1. To divide two numbers with *like signs*, either both positive or both negative, divide their absolute values. The answer is *positive*.
> 2. To divide two numbers with *unlike signs*, one positive and the other negative, divide their absolute values. The answer is *negative*.

EXAMPLE 18 Evaluate.   **a)** $-24 \div 4$ **b)** $-6.45 \div (-0.4)$

Solution

a) $\dfrac{-24}{4} = -6$ The numbers have unlike signs. The answer is negative.

b) $\dfrac{-6.45}{-0.4} = 16.125$ The numbers have like signs. The answer is positive.

Now Try Exercise 71

EXAMPLE 19 Evaluate $\dfrac{-3}{8} \div \left| \dfrac{-2}{5} \right|$.

Solution   Since $\left| \dfrac{-2}{5} \right|$ is equal to $\dfrac{2}{5}$, we write

$$\frac{-3}{8} \div \left| \frac{-2}{5} \right| = \frac{-3}{8} \div \frac{2}{5}$$

Now invert the divisor and proceed as in multiplication.

$$\frac{-3}{8} \div \frac{2}{5} = \frac{-3}{8} \cdot \frac{5}{2} = \frac{-3 \cdot 5}{8 \cdot 2} = \frac{-15}{16} \text{ or } -\frac{15}{16}$$

Now Try Exercise 75

When the denominator of a fraction is a negative number, we usually rewrite the fraction with a positive denominator. To do this, we use the following fact.

Understanding Algebra

A negative fraction can have the minus sign in the denominator, or in the numerator, or in front of the fraction. Thus,

$$\frac{3}{-4} = \frac{-3}{4} = -\frac{3}{4}.$$

> **Sign of a Fraction**
>
> For any number a and any nonzero number b,
>
> $$\frac{a}{-b} = \frac{-a}{b} = -\frac{a}{b}$$

Thus, when we have a quotient of $\dfrac{1}{-2}$, we rewrite it as either $\dfrac{-1}{2}$ or $-\dfrac{1}{2}$.

6 Use the Properties of Real Numbers

We have already discussed the double negative property and the multiplicative property of zero. **Table 1.1** lists other basic properties for the operations of addition and multiplication on the real numbers.

TABLE 1.1

For real numbers a, b, and c	Addition	Multiplication
Commutative property	$a + b = b + a$	$ab = ba$
Associative property	$(a + b) + c = a + (b + c)$	$(ab)c = a(bc)$
Identity property	$a + 0 = 0 + a = a$ $\left(\begin{array}{c} \text{0 is called the \textbf{additive}} \\ \text{\textbf{identity element}.} \end{array} \right)$	$a \cdot 1 = 1 \cdot a = a$ $\left(\begin{array}{c} \text{1 is called the \textbf{multiplicative}} \\ \text{\textbf{identity element}.} \end{array} \right)$
Inverse property	$a + (-a) = (-a) + a = 0$ $\left(\begin{array}{c} -a \text{ is called the \textbf{additive}} \\ \textbf{inverse} \text{ or \textbf{opposite} of } a. \end{array} \right)$	$a \cdot \dfrac{1}{a} = \dfrac{1}{a} \cdot a = 1$ $\left(\begin{array}{c} \dfrac{1}{a} \text{ is called the \textbf{multiplicative}} \\ \textbf{inverse} \text{ or \textbf{reciprocal} of } a, a \neq 0. \end{array} \right)$
Distributive property (of multiplication over addition)	$a(b + c) = ab + ac$	

Note that the commutative property involves a change in *order*, and the associative property involves a change in *grouping*.

The distributive property also applies when there are more than two numbers within the parentheses.

$$a(b + c + d + \cdots + n) = ab + ac + ad + \cdots + an$$

EXAMPLE 20 Name each property illustrated.

a) $7 \cdot m = m \cdot 7$ **b)** $(a + 6) + 2b = a + (6 + 2b)$

c) $4s + 5t = 5t + 4s$ **d)** $2v(w + 3) = 2v \cdot w + 2v \cdot 3$

Solution

a) Change of order, commutative property of multiplication.

b) Change of grouping, associative property of addition.

c) Change of order, commutative property of addition.

d) $2v$ is distributed, distributive property.

Now Try Exercise 113

In Example 20 **d)** the expression $2v \cdot w + 2v \cdot 3$ can be simplified to $2vw + 6v$ using the properties of the real numbers.

EXAMPLE 21 Name each property illustrated.

a) $9 \cdot 1 = 9$ **b)** $x + 0 = x$

c) $4 + (-4) = 0$ **d)** $1(xy) = xy$

Solution

a) Identity property of multiplication

b) Identity property of addition

c) Inverse property of addition

d) Identity property of multiplication

Now Try Exercise 115

EXAMPLE 22 Write the additive inverse (or opposite) and multiplicative inverse (or reciprocal) of each of the following.

a) -3 **b)** $\dfrac{2}{3}$

Solution

a) The additive inverse is 3. The multiplicative inverse is $\dfrac{1}{-3} = -\dfrac{1}{3}$.

b) The additive inverse is $-\dfrac{2}{3}$. The multiplicative inverse is $\dfrac{1}{\frac{2}{3}} = \dfrac{3}{2}$.

Now Try Exercise 121

EXERCISE SET 1.3

Math XL
MathXL®

MyMathLab
MyMathLab

Warm-Up Exercises

Fill in the blanks with the appropriate word, phrase, or symbol(s) from the following list.

subtract	d	negative	distributive	absolute value
reflexive	positive	additive inverse	add	associative
commutative	c	any	0	$-c$

1. The sum of two positive numbers is a _____ number.

2. The sum of two negative numbers is a _____ number.

3. For any real number a, its _____ is $-a$.

4. For any real number c, $-(-c) =$ _____.

5. To add two numbers with the same sign, _____ their absolute values and keep the common sign with the sum.

6. To add two numbers with different signs, _____ the smaller absolute value from the larger absolute value and keep the sign of the number with the larger absolute value.

7. The _____ of a number is its distance from 0 on the number line.

8. The absolute value of _____ number is always nonnegative.

9. The property $a(b + c) = ab + ac$ is the _____ property.

10. The property $d + e = e + d$ is the _____ property of addition.

Practice the Skills

Evaluate each absolute value expression.

11. $|5|$ **12.** $|1.9|$ **13.** $|-7|$ **14.** $|-8|$

15. $\left|-\dfrac{7}{8}\right|$ **16.** $|-8.61|$ **17.** $|0|$ **18.** $-|1|$

19. $-|-7|$ **20.** $-|-\pi|$ **21.** $-\left|\dfrac{5}{9}\right|$ **22.** $-\left|-\dfrac{7}{19}\right|$

Insert $<, >,$ or $=$ in the shaded area to make each statement true.

23. $|-9|$ ⬜ $|9|$ **24.** $|-4|$ ⬜ $|6|$ **25.** $|-8|$ ⬜ -8 **26.** $|-10|$ ⬜ -5

27. $|-\pi|$ ⬜ -3 **28.** $-|-1|$ ⬜ -1 **29.** $|-7|$ ⬜ $-|2|$ **30.** $-|9|$ ⬜ $-|13|$

31. $-(-3)$ ⬜ $-|-3|$ **32.** $|-(-4)|$ ⬜ -4 **33.** $|19|$ ⬜ $|-25|$ **34.** $-|-1|$ ⬜ $|-5|$

List the values from smallest to largest.

35. $-1, -2, |-3|, 4, -|5|$

36. $-8, -12, -|9|, -|20|, -|-17|$

37. $-32, |-7|, 15, -|4|, 4$

38. $\pi, -\pi, |-3|, -|-3|, -2, |-2|$

39. $-6.1, |-6.3|, -|-6.5|, 6.8, |6.4|$

40. $-2.1, -2, -2.4, -|2.8|, -|2.9|$

41. $\dfrac{1}{3}, \left|-\dfrac{1}{2}\right|, -2, \left|\dfrac{3}{5}\right|, \left|-\dfrac{3}{4}\right|$

42. $\left|-\dfrac{5}{2}\right|, \dfrac{3}{5}, |-3|, \left|-\dfrac{5}{3}\right|, \left|-\dfrac{2}{3}\right|$

Evaluate each addition and subtraction problem.

43. $7 + (-4)$ **44.** $-2 + 9$ **45.** $-12 + (-10)$ **46.** $-2.18 - 3.14$

47. $-9 - (-5)$ **48.** $-12 - (-4)$ **49.** $\dfrac{4}{5} - \dfrac{6}{7}$ **50.** $-\dfrac{5}{12} - \left(-\dfrac{7}{8}\right)$

51. $-14.21 - (-13.22)$

52. $79.33 - (-16.05)$

53. $10 - (-2.31) + (-4.39)$

54. $-|7.31| - (-3.28) + 5.76$

55. $9.9 - |8.5| - |17.6|$

56. $|11 - 4| - 10$

57. $|17 - 12| - |3|$

58. $|12 - 5| - |5 - 12|$

59. $-|-3| - |7| + (6 + |-2|)$

60. $|-4| - |-4| - |-4 - 4|$

61. $\left(\dfrac{3}{5} + \dfrac{3}{4}\right) - \dfrac{1}{2}$

62. $\dfrac{4}{5} - \left(\dfrac{3}{4} - \dfrac{2}{3}\right)$

Evaluate each multiplication and division problem.

63. $-5 \cdot 8$

64. $(-9)(-3)$

65. $-4\left(-\dfrac{5}{16}\right)$

66. $-4\left(-\dfrac{3}{4}\right)\left(-\dfrac{1}{2}\right)$

67. $(-1)(-2)(-1)(2)(-3)$

68. $(-2.1)(-7.8)(-9.1)$

69. $(-1.1)(3.4)(8.3)(-7.6)$

70. $-16 \div 8$

71. $-66 \div (-6)$

72. $-4 \div \left(-\dfrac{1}{4}\right)$

73. $-\dfrac{7}{9} \div \dfrac{-7}{9}$

74. $\left|-\dfrac{1}{2}\right| \cdot \left|\dfrac{-3}{4}\right|$

75. $\left(-\dfrac{3}{4}\right) \div |-16|$

76. $\left|\dfrac{3}{8}\right| \div (-4)$

77. $\left|\dfrac{-7}{6}\right| \div \left|\dfrac{-1}{2}\right|$

78. $\dfrac{-4}{9} \div |-4|$

Evaluate.

79. $10 - 14$

80. $-12 - 15$

81. $7 - (-11)$

82. $-\dfrac{1}{8} + \left(-\dfrac{1}{16}\right)$

83. $3\left(-\dfrac{2}{3}\right)\left(-\dfrac{7}{2}\right)$

84. $(-3.2)(4.9)(-2.73)$

85. $-14.4 - (-9.6) - 15.8$

86. $(1.32 - 2.76) - (-3.85 + 4.28)$

87. $9 - (8 - 7) - (-2 - 1)$

88. $(4.2)(-1)(-9.6)(3.8)$

89. $-|12| \cdot \left|\dfrac{-1}{2}\right|$

90. $-\left|\dfrac{-24}{5}\right| \cdot \left|\dfrac{3}{8}\right|$

91. $\left|\dfrac{-9}{4}\right| \div \left|\dfrac{-4}{9}\right|$

92. $(-|3| + |5|) - (1 - |-9|)$

93. $5 - |-7| + 3 - |-2|$

94. $\left(\dfrac{3}{8} - \dfrac{4}{7}\right) - \left(-\dfrac{1}{2}\right)$

95. $\left(-\dfrac{3}{5} - \dfrac{4}{9}\right) - \left(-\dfrac{2}{3}\right)$

96. $(|-9| - 8) - (3 \cdot |-5|)$

97. $(25 - |32|)(-7 - 4)$

98. $\left[(-2)\left|-\dfrac{1}{2}\right|\right] \div \left|-\dfrac{1}{4}\right|$

99. Subtract 29 from -10.

100. Subtract $-\dfrac{1}{2}$ from $-\dfrac{2}{3}$.

101. $7(3)(0)(-193)$

102. $16(-5)(-10)(0)$

Name each property illustrated.

103. $r + s = s + r$

104. $7(v + w) = 7v + 7w$

105. $b \cdot 0 = 0$

106. $c \cdot d = d \cdot c$

107. $(x + 3) + 6 = x + (3 + 6)$

108. $x + 0 = x$

109. $x = 1 \cdot x$

110. $x(y + z) = xy + xz$

111. $2(xy) = (2x)y$

112. $(2x \cdot 3y) \cdot 6y = 2x \cdot (3y \cdot 6y)$

113. $4(x + y + 2) = 4x + 4y + 8$

114. $-(-2) = 2$

115. $5 + 0 = 5$

116. $4 \cdot \dfrac{1}{4} = 1$

117. $3 + (-3) = 0$

118. $(x + y) = 1(x + y)$

119. $-(-x) = x$

120. $x + (-x) = 0$

List both the additive inverse and the multiplicative inverse for each problem.

121. 6

122. -13

123. $-\dfrac{22}{9}$

124. $-\dfrac{3}{5}$

Problem Solving

125. Temperature Change The most unusual temperature change according to the *Guinness Book of World Records* occurred from 7:30 A.M. to 7:32 A.M. on January 22, 1943, in Spearfish, South Dakota. During these 2 minutes the temperature changed from $-4°F$ to $45°F$. Determine the increase in temperature in these 2 minutes.

126. The Film *Gold* During the production of the documentary film *Gold*, the film crew experienced severe changes in temperature. In a South African gold mine 3 miles below the surface of the earth, the temperature was $140°F$. On a mountain near Cuzco, Peru, the temperature was $40°F$. Determine the difference in temperature between these two filming sites.

Source: History Channel Web site

127. Submarine Dive A submarine dives 358.9 feet. A short time later the submarine comes up 210.7 feet. Find the submarine's final depth from its starting point. (Consider distance in a downward direction as negative.)

© Shutterstock

128. Checking Account Sharon Koch had a balance of −$32.64 in her checking account when she deposited a check for $99.38. What is her new balance?

129. Extreme Temperatures The lowest temperature ever recorded in the United States was −79.8°F on January 23, 1971, in Prospect Creek, Alaska. The lowest temperature in the contiguous states (all states except Alaska and Hawaii) was −69.7°F on January 20, 1954, in Rogers Pass, Montana. Find the difference in these temperatures.

130. Estimated Taxes In 2010, Joanne Butler made four quarterly estimated income tax payments of $3000 each. When she completed her year 2010 income tax forms, she found her total tax was $10,125.
   a) Will Joanne be entitled to a refund or will she owe more taxes? Explain.
   b) How much of a refund will she receive or how much more in taxes will she owe?

131. Stock Prices Ron Blackwood purchased 100 shares of Home Depot stock at $30.30 per share. Six months later Ron sold all 100 shares for a price of $42.37 per share. What was Ron's total gain or loss for this transaction?

132. Book Contract Samuel Pritchard signed a contract with a publishing company that called for an advance payment of $60,000 on the sale of his book *Moon Spray*. When the book is published and sales begin, the publishers will automatically deduct this advance from the author's royalties.

   a) Six months after the release of the book, the author's royalties totaled $47,600 before the advance was deducted. Determine how much money he will receive from or owe to the publisher.
   b) After 1 year, the author's royalties are $87,500. Determine how much money he will receive from or owe to the publishing company.

133. Write your own realistic word problem that involves subtracting a positive number from a negative number. Indicate the answer to your word problem.

134. Write your own realistic word problem that involves subtracting a negative number from a negative number. Indicate the answer to your problem.

135. Small Businesses The average first-year expenditures and the average first-year incomes of small start-up businesses is shown on the bar graph below. Estimate the average first-year profit by subtracting the average first-year expenditures from the average first-year income.

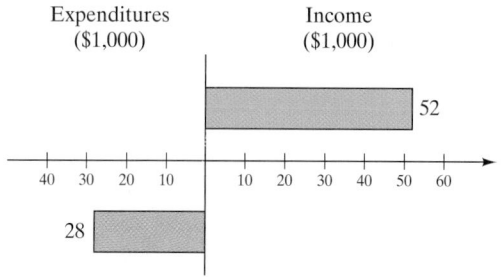

Concept/Writing Exercises

136. Give the definition of absolute value.

In Exercises 137–142, find the unknown number(s). Explain how you determined your answer.

137. All numbers a such that $|a| = |-a|$
138. All numbers a such that $|a| = a$
139. All numbers a such that $|a| = 6$
140. All numbers a such that $|a| = -a$
141. All numbers a such that $|a| = -9$
142. All numbers x such that $|x - 3| = |3 - x|$
143. Explain how to add two numbers with the same sign.
144. Explain how to add two numbers with different signs.
145. Explain how to subtract real numbers.
146. Explain how the rules for multiplication and division of real numbers are similar.

147. List two other ways that the fraction $\dfrac{a}{-b}$ may be written.
148. a) Write the associative property of multiplication.
   b) Explain the property.
149. a) Write the commutative property of addition.
   b) Explain the property.
150. a) Write the distributive property of multiplication over addition.
   b) Explain the property.
151. Using an example, explain why addition is not distributive over multiplication. That is, explain why $a + (b \cdot c) \neq (a + b) \cdot (a + c)$.

Challenge Problems

152. Evaluate $1 - 2 + 3 - 4 + \cdots + 99 - 100$. (*Hint:* Group in pairs of two numbers.)
153. Evaluate $1 + 2 - 3 + 4 + 5 - 6 + 7 + 8 - 9 + 10 + 11 - 12 + \cdots + 22 + 23 - 24$.
   (*Hint:* Examine in groups of three numbers.)
154. Evaluate $\dfrac{(1) \cdot |-2| \cdot (-3) \cdot |4| \cdot (-5)}{|-1| \cdot (-2) \cdot |-3| \cdot (4) \cdot |-5|}$.

155. Evaluate $\dfrac{(1)(-2)(3)(-4)(5) \cdots (97)(-98)}{(-1)(2)(-3)(4)(-5) \cdots (-97)(98)}$.

Cumulative Review Exercises

[1.2] **156.** Answer true or false: Every irrational number is a real number.

157. List the set of natural numbers.

158. Consider the set $\left\{ 3, 4, -2, \dfrac{5}{6}, \sqrt{11}, 0 \right\}$. List the elements that are

   a) integers,

   b) rational numbers,

   c) irrational numbers,

   d) real numbers.

159. $A = \{4, 7, 9, 12\}$; $B = \{1, 4, 7, 19\}$. Find

   a) $A \cup B$

   b) $A \cap B$

160. Illustrate $\{x \mid -4 < x \le 5\}$ on a number line.

1.4 Order of Operations

1 Evaluate exponential expressions.

2 Evaluate square and higher roots.

3 Evaluate expressions using the order of operations.

4 Evaluate expressions containing variables.

Before we discuss the order of operations we need to speak briefly about exponents and roots.

1 Evaluate Exponential Expressions

In multiplication, the numbers or expressions that are multiplied are called **factors**. If $a \cdot b = c$, then a and b are factors of c. For example, since $2 \cdot 3 = 6$, both 2 and 3 are factors of 6. The number 1 is a factor of every number and expression.

The quantity 3^2 is called an **exponential expression**. In the expression, the 3 is called the **base** and the 2 is called the **exponent**. The expression 3^2 is read "three squared" or "three to the second power."

Note that

$$3^2 \quad \substack{\longleftarrow \text{ exponent} \\ \longleftarrow \text{ base}}$$

$$3^2 = \underbrace{3 \cdot 3}_{\text{2 factors of 3}}$$

The expression 5^3 is read "five cubed" or "five to the third power." Note that

$$5^3 = \underbrace{5 \cdot 5 \cdot 5}_{\text{3 factors of 5}}$$

In general, the base b to the nth power is written b^n. For any natural number n,

$$b^n = \underbrace{b \cdot b \cdot b \cdot b \cdot \,\cdots\, \cdot b}_{n \text{ factors of } b}$$

Note that 0^0 is *undefined*.

EXAMPLE 1 Evaluate.   **a)** $(0.5)^3$ **b)** $(-3)^5$ **c)** 1^{23} **d)** $\left(-\dfrac{4}{7} \right)^3$

Solution

a) $(0.5)^3 = (0.5)(0.5)(0.5) = 0.125$

b) $(-3)^5 = (-3)(-3)(-3)(-3)(-3) = -243$

c) $1^{23} = 1$; 1 raised to any power will equal 1.

d) $\left(-\dfrac{4}{7} \right)^3 = \left(-\dfrac{4}{7} \right)\left(-\dfrac{4}{7} \right)\left(-\dfrac{4}{7} \right) = -\dfrac{64}{343}$

Now Try Exercise 19

Helpful Hint

Study Tip

Be very careful when writing or copying exponents. Since exponents are small it is very easy to write or copy an exponent and then later not recognize what you have written.

Understanding Algebra

Do not confuse -7^2 with $(-7)^2$.

-7^2 means $-(7 \cdot 7) = -49$, a *negative* number.

$(-7)^2$ means $(-7)(-7) = 49$, a *positive* number.

It is not necessary to write exponents of 1. Whenever we encounter a numerical value or a variable without an exponent, we assume that it has an exponent of 1. Thus, 3 means 3^1, x means x^1, x^3y means x^3y^1, and $-xy$ means $-x^1y^1$.

Avoiding Common Errors

Students often evaluate expressions containing $-x^2$ incorrectly. The expression $-x^2$ means $-1(x^2)$ or $-(x^2)$, not $(-x)^2$. Note that -5^2 means $-1(5)^2$ or $-(5^2) = -(5 \cdot 5) = -25$ while $(-5)^2$ means $(-5)(-5) = 25$. *In general, $-x^m$ means $-1(x^m)$ or $-(x^m)$, not $(-x)^m$.*

EXAMPLE 2 Evaluate.   **a)** -6^2 **b)** $(-6)^2$

Solution

a) $-6^2 = -(6 \cdot 6) = -36$

b) $(-6)^2 = (-6)(-6) = 36$

Now Try Exercise 41

EXAMPLE 3 Evaluate $-5^2 + (-5)^2 - 4^3 + (-4)^3$.

Solution   First, we evaluate each exponential expression. Then we add or subtract, working from left to right.

$$-5^2 + (-5)^2 - 4^3 + (-4)^3 = -(5^2) + (-5)^2 - (4^3) + (-4)^3$$
$$= -25 + 25 - 64 + (-64)$$
$$= -25 + 25 - 64 - 64$$
$$= -128$$

Now Try Exercise 59

Using Your Calculator

Evaluating Exponential Expressions on a Scientific and a Graphing Calculator

On most scientific and graphing calculators the $\boxed{x^2}$ key can be used to square a number. Below we show the sequence of keys to press to evaluate 5^2.

Scientific Calculator   5 $\boxed{x^2}$ 25 — answer displayed

Graphing Calculator   5 $\boxed{x^2}$ $\boxed{\text{ENTER}}$ 25 — answer displayed

To evaluate exponential expressions with other exponents, you can use the $\boxed{y^x}$ or $\boxed{\wedge}$ key. Most scientific calculators have a $\boxed{y^x}$ key,* whereas graphing calculators use the $\boxed{\wedge}$ key. To evaluate exponential expressions using these keys, first enter the base, then press either the $\boxed{y^x}$ or $\boxed{\wedge}$ key, and then enter the exponent. For example, to evaluate 6^4 we do the following:

Scientific Calculator   6 $\boxed{y^x}$ 4 $\boxed{=}$ 1296 — answer displayed

Graphing Calculator   6 $\boxed{\wedge}$ 4 $\boxed{\text{ENTER}}$ 1296 — answer displayed

*Some calculators have $\boxed{x^y}$ or $\boxed{a^b}$ keys instead of a $\boxed{y^x}$ key.

2 Evaluate Square and Higher Roots

The symbol used to indicate a root, $\sqrt{}$, is called a **radical sign**. The number or expression inside the radical sign is called the **radicand**. In $\sqrt{25}$, the radicand is 25. The **principal or positive square root** of a positive number a, written $\sqrt{a}$, is the positive number that when multiplied by itself gives a. For example, the principal square root of 4 is 2, written $\sqrt{4} = 2$, because $2 \cdot 2 = 4$. In general, $\sqrt{a} = b$ if $b \cdot b = a$. Whenever we use the words *square root*, we are referring to the "principal square root."

EXAMPLE 4 Evaluate.   **a)** $\sqrt{25}$ **b)** $\sqrt{\dfrac{81}{4}}$ **c)** $\sqrt{0.64}$ **d)** $-\sqrt{49}$

Solution

a) $\sqrt{25} = 5$, since $5 \cdot 5 = 25$.

b) $\sqrt{\dfrac{81}{4}} = \dfrac{9}{2}$, since $\dfrac{9}{2} \cdot \dfrac{9}{2} = \dfrac{81}{4}$.

c) $\sqrt{0.64} = 0.8$, since $(0.8)(0.8) = 0.64$.

d) $-\sqrt{49}$ means $-(\sqrt{49})$. We determine that $\sqrt{49} = 7$, since $7 \cdot 7 = 49$. Therefore, $-\sqrt{49} = -7$.

Now Try Exercise 21

The square roots of numbers such as $\sqrt{2}$, $\sqrt{3}$, and $\sqrt{5}$ are irrational numbers. The decimal values of such numbers can never be given exactly since irrational numbers are nonterminating, nonrepeating decimal numbers. The *approximate* value of $\sqrt{2}$ and other irrational numbers can be found with a calculator.

$$\sqrt{2} \approx 1.414213562 \quad \text{From a calculator}$$

In this section we introduce square roots; cube roots, symbolized by $\sqrt[3]{}$; and higher roots. The number used to indicate the root is called the **index**.

$$\text{index} \searrow \quad \swarrow \text{radical sign}$$
$$\sqrt[n]{a} \leftarrow \text{radicand}$$

The index of a square root is 2. However, we generally do not show the index 2. Therefore, $\sqrt{a} = \sqrt[2]{a}$.

The concept used to explain square roots can be expanded to explain cube roots and higher roots. The cube root of a number a is written $\sqrt[3]{a}$.

$$\sqrt[3]{a} = b \quad \text{if} \quad \underbrace{b \cdot b \cdot b}_{3 \text{ factors of } b} = a$$

For example, $\sqrt[3]{8} = 2$, because $2 \cdot 2 \cdot 2 = 8$. The expression $\sqrt[n]{a}$ is read "the nth root of a."

$$\sqrt[n]{a} = b \quad \text{if} \quad \underbrace{b \cdot b \cdot b \cdot \cdots \cdot b}_{n \text{ factors of } b} = a$$

EXAMPLE 5 Evaluate.   **a)** $\sqrt[3]{125}$ **b)** $\sqrt[4]{81}$ **c)** $\sqrt[5]{32}$

Solution

a) $\sqrt[3]{125} = 5$, since $5 \cdot 5 \cdot 5 = 125$

b) $\sqrt[4]{81} = 3$, since $3 \cdot 3 \cdot 3 \cdot 3 = 81$

c) $\sqrt[5]{32} = 2$, since $2 \cdot 2 \cdot 2 \cdot 2 \cdot 2 = 32$

Now Try Exercise 25

EXAMPLE 6 Evaluate.   **a)** $\sqrt[4]{256}$ **b)** $\sqrt[3]{\dfrac{1}{27}}$ **c)** $\sqrt[3]{-8}$ **d)** $-\sqrt[3]{8}$

Solution

a) $\sqrt[4]{256} = 4$, since $4 \cdot 4 \cdot 4 \cdot 4 = 256$.   **b)** $\sqrt[3]{\dfrac{1}{27}} = \dfrac{1}{3}$, since $\left(\dfrac{1}{3}\right)\left(\dfrac{1}{3}\right)\left(\dfrac{1}{3}\right) = \dfrac{1}{27}$.

c) $\sqrt[3]{-8} = -2$, since $(-2)(-2)(-2) = -8$.

d) $-\sqrt[3]{8}$ means $-(\sqrt[3]{8})$. We determine that $\sqrt[3]{8} = 2$, since $2 \cdot 2 \cdot 2 = 8$. Therefore, $-\sqrt[3]{8} = -2$.

Now Try Exercise 27

Using Your Calculator ▪▪▪▪▫

Evaluating Roots on a Scientific Calculator

The square roots of numbers can be found on most calculators with a square-root key, $\boxed{\sqrt{x}}$. To evaluate $\sqrt{25}$ on most calculators that have this key, press

$$25 \;\; \boxed{\sqrt{x}} \;\; 5 \quad\xleftarrow{\text{answer displayed}}$$

Higher roots can be found on calculators that contain either the $\boxed{\sqrt[x]{y}}$ key or the $\boxed{y^x}$ key.* To evaluate $\sqrt[4]{625}$ on a calculator with a $\boxed{\sqrt[x]{y}}$ key, do the following:

$$625 \;\; \boxed{\sqrt[x]{y}} \;\; 4 \;\; \boxed{=} \;\; 5 \quad\xleftarrow{\text{answer displayed}}$$

Note that the number within the radical sign (the radicand), 625, is entered, then the $\boxed{\sqrt[x]{y}}$ key is pressed, and then the root (or index) 4 is entered. When the $\boxed{=}$ key is pressed, the answer 5 is displayed.

To evaluate $\sqrt[4]{625}$ on a calculator with a $\boxed{y^x}$ key, use the inverse key as follows:

$$625 \;\; \boxed{\text{INV}} \;\; \boxed{y^x} \;\; 4 \;\; \boxed{=} \;\; 5 \quad\xleftarrow{\text{answer displayed}}$$

* Calculator keys vary. Some calculators have $\boxed{x^y}$ or $\boxed{a^b}$ keys instead of the $\boxed{y^x}$ key, and some calculators have a $\boxed{2^{\text{nd}}}$ or $\boxed{\text{shift}}$ key instead of the $\boxed{\text{INV}}$ key.

Using Your Graphing Calculator ▪▪▪▫

Evaluating Roots on a Graphing Calculator

To find the square root on a graphing calculator, use $\sqrt{}$. The $\sqrt{}$ appears above the $\boxed{x^2}$ key, so you will need to press the $\boxed{2^{\text{nd}}}$ key to evaluate square roots. For example, to evaluate $\sqrt{25}$ press

$$\boxed{2^{\text{nd}}} \; \boxed{x^2} \; 25 \; \boxed{\text{ENTER}} \;\; 5 \;\longleftarrow \text{answer displayed}$$

When you press $\boxed{2^{\text{nd}}}$ $\boxed{x^2}$, the Texas Instruments TI-84 Plus generates $\sqrt{}($. Then you insert the radicand, then the right parentheses, and press $\boxed{\text{ENTER}}$. To learn how to find cube and higher roots, refer to your graphing calculator manual. With the TI-84 Plus, you can use the $\boxed{\text{MATH}}$ key. When you press this key you get a number of options including 4 and 5, which are shown below.

$$4 : \sqrt[3]{}(\qquad 5 : \sqrt[x]{y}$$

Option 4 can be used to find cube roots and option 5 can be used to find higher roots, as shown in the following examples.

EXAMPLE Evaluate $\sqrt[3]{120}$.

Solution

$$\boxed{\text{MATH}} \; 4 \; \underset{\substack{\text{enter}\\\text{radicand}}}{120} \; \boxed{)} \; \boxed{\text{ENTER}} \; 4.932424149 \quad\xleftarrow{\text{answer displayed}}$$

select option 4

To find the root with an index greater than 3, first enter the index, then press the $\boxed{\text{MATH}}$ key, and then press option 5.

(continued on the next page)

EXAMPLE Evaluate $\sqrt[4]{625}$.

Solution

answer displayed

4 | MATH | 5 625 | ENTER | 5

index ↗　　select ↗　enter
　　　　　option　radicand

We will show another way to find roots on a graphing calculator in Section 7.2 when we discuss rational exponents.

3　Evaluate Expressions Using the Order of Operations

You will often have to evaluate expressions containing multiple operations. To do so, follow the **order of operations** indicated below.

Order of Operations

To evaluate mathematical expressions, use the following order:

1. First, evaluate the expressions within grouping symbols, including parentheses, ( ), brackets, [ ], braces, { }, and absolute value, | |. If the expression contains nested grouping symbols (one pair of grouping symbols within another pair), evaluate the expression in the innermost grouping symbols first.
2. Next, evaluate all terms containing exponents and radicals.
3. Next, evaluate all multiplications or divisions in the order in which they occur, working from left to right.
4. Finally, evaluate all additions or subtractions in the order in which they occur, working from left to right.

Helpful Hint

It should be noted that a fraction bar acts as a grouping symbol. Thus, when evaluating expressions containing a fraction bar, we work separately above and below the fraction bar.

Brackets are often used in place of parentheses to help avoid confusion. For example, the expression $7((5 \cdot 3) + 6)$ is easier to follow when written $7[(5 \cdot 3) + 6]$. Remember to evaluate the innermost group first.

EXAMPLE 7 Evaluate $6 + 3 \cdot 5^2 - 17$.

Solution We will use shading to indicate the order in which the operations are to be evaluated. Since there are no parentheses, we first evaluate 5^2.

$$6 + 3 \cdot 5^2 - 17 = 6 + 3 \cdot 25 - 17$$

Next, we perform multiplications or divisions from left to right.

$$= 6 + 75 - 17$$

Finally, we perform additions or subtractions from left to right.

$$= 81 - 17$$
$$= 64$$

Now Try Exercise 67

EXAMPLE 8 Evaluate $10 + \{6 - [4(5 - 2)]\}^2$.

Solution   First, evaluate the expression within the innermost parentheses. Then continue according to the order of operations.

$$10 + \{6 - [4(\,5 - 2\,)]\}^2 = 10 + \{6 - [4(3)]\}^2$$
$$= 10 + [\,6 - (12)\,]^2$$
$$= 10 + (-6)^2$$
$$= 10 + 36$$
$$= 46$$

Now Try Exercise 77

EXAMPLE 9 Evaluate $\dfrac{6 \div \frac{1}{2} + 5|7 - 3|}{1 + (3 - 5) \div 2}$.

Solution   Remember that the fraction bar acts as a grouping symbol. Work separately above the fraction bar and below the fraction bar.

$$\frac{6 \div \frac{1}{2} + 5|7 - 3|}{1 + (3 - 5) \div 2} = \frac{6 \div \frac{1}{2} + 5|4|}{1 + (-2) \div 2}$$
$$= \frac{12 + 20}{1 + (-1)}$$
$$= \frac{32}{0}$$

Since division by 0 is not possible, the original expression is **undefined**.

Now Try Exercise 83

Understanding Algebra

To *evaluate* mathematical expressions, be sure the *variable* is *replaced* by a specific *number*.

4 Evaluate Expressions Containing Variables

To evaluate mathematical expressions, we use the order of operations just given. Example 10 is an application problem in which we use the order of operations.

EXAMPLE 10 **Nutritional Supplements** The approximate sales of supplements between 1997 and 2009, in billions of U.S. dollars, can be estimated by the equation

$$\text{sales} = -0.063x^2 + 1.62x + 9.5$$

where x represents years since 1997. In the expression on the right side of the equals sign, substitute 1 for x to estimate the sales of supplements in 1998, 2 for x to estimate the sales of supplements in 1999, and so on.

Estimate the sales of supplements during the years **a)** 1998 and **b)** 2009.

Solution
a) We will substitute 1 for x to estimate the sales of supplements in 1998.

$$\text{sales} = -0.063x^2 + 1.62x + 9.5$$
$$= -0.063(1)^2 + 1.62(1) + 9.5$$
$$= -0.063 + 1.62 + 9.5$$
$$= 11.057$$

Therefore, in 1998 approximately $11.057 billion worth of supplements was sold in the United States.

b) The year 2009 corresponds to the number 12. We can obtain the 12 by subtracting 1997 from 2009. Therefore, to estimate the sales of supplements in 2009, we substitute 12 for x in the equation.

$$
\begin{aligned}
\text{sales} &= -0.063x^2 + 1.62x + 9.5 \\
&= -0.063(12)^2 + 1.62(12) + 9.5 \\
&= -0.063(144) + 19.44 + 9.5 \\
&= 19.868
\end{aligned}
$$

The answer is reasonable: From the information given we expected to see an increase. In 2009 approximately $19.868 billion worth of supplements was sold in the United States.

Now Try Exercise 121

EXAMPLE 11 Evaluate $-x^3 - xy - y^2$ when $x = -2$ and $y = 5$.

Solution   Substitute -2 for each x and 5 for each y in the expression. Then evaluate.

$$
\begin{aligned}
-x^3 - xy - y^2 &= -(-2)^3 - (-2)(5) - (5)^2 \\
&= -(-8) - (-10) - 25 \\
&= 8 + 10 - 25 \\
&= -7
\end{aligned}
$$

Now Try Exercise 101

EXAMPLE 12

a) Find an algebraic expression for the following statement. To the variable y, add 3. Multiply this sum by 8. Then subtract 20 from this product. Finally, divide this difference by 5.

b) Evaluate the algebraic expression in part **a)** when $y = 2$.

Solution

a) Translate the statements into algebraic expressions as shown below.

Statement	Algebraic Expression
To the variable y, add 3.	$y + 3$
Multiply this sum by 8.	$8(y + 3)$
Then subtract 20 from this product.	$8(y + 3) - 20$
Finally divide this difference by 5.	$\dfrac{8(y + 3) - 20}{5}$

b) To evaluate this algebraic expression when $y = 2$, substitute 2 for y.

$$\frac{8(y + 3) - 20}{5} \qquad \text{Algebraic expression}$$

$$= \frac{8(2 + 3) - 20}{5} \qquad \text{Substituted 2 for } y.$$

$$= \frac{8(5) - 20}{5} \qquad \text{Added } 2 + 3 = 5.$$

$$= \frac{40 - 20}{5} \qquad \text{Multiplied } 8 \cdot 5 = 40.$$

$$= \frac{20}{5} \qquad \text{Subtracted } 40 - 20 = 20.$$

$$= 4 \qquad \text{Divided } \frac{20}{5} = 4.$$

Now Try Exercise 109

EXERCISE SET 1.4

Math XL **MyMathLab**
MathXL®   MyMathLab

Warm-Up Exercises

Fill in the blanks with the appropriate word, phrase, or symbol(s) from the following list.

base	index	positive	negative	factors	radicand
exponent	exponential	radical sign	3	6	real
16	64	9	18		

1. Numbers or expressions that are multiplied together are called _____.

2. The quantity 7^2 is called an _____ expression.

3. In the expression 4^3, the 3 is called the _____.

4. In the expression 4^3, the 4 is called the _____.

5. The value of 4^3 is _____.

6. In the expression $\sqrt{81}$, the $\sqrt{\ }$ is called the _____.

7. In the expression $\sqrt{81}$, the 81 is called the _____.

8. The principal or positive square root of 81 is _____.

9. In the expression $\sqrt[4]{81}$, the 4 is called the _____.

10. The value of $\sqrt[4]{81}$ is _____.

11. The value of $\sqrt[4]{-81}$ is not a _____ number.

12. The cube root of a negative number is a _____ number.

Practice the Skills

Evaluate each expression without using a calculator.

13. 3^2

14. $(-4)^3$

15. -3^2

16. -4^3

17. $(-5)^2$

18. $\left(\dfrac{1}{2}\right)^3$

19. $-\left(\dfrac{3}{5}\right)^4$

20. $(0.3)^2$

21. $\sqrt{49}$

22. $\sqrt{169}$

23. $-\sqrt{36}$

24. $-\sqrt{0.64}$

25. $\sqrt[3]{-27}$

26. $\sqrt[3]{\dfrac{-216}{343}}$

27. $\sqrt[3]{0.001}$

28. $\sqrt[4]{\dfrac{1}{16}}$

Use a calculator to evaluate each expression. Round answers to the nearest thousandth.

29. $(0.35)^4$

30. $-(1.7)^{3.9}$

31. $\left(-\dfrac{13}{12}\right)^8$

32. $\left(\dfrac{5}{7}\right)^7$

33. $(6.721)^{5.9}$

34. $(5.382)^{6.9}$

35. $\sqrt[3]{26}$

36. $-\sqrt[4]{72.8}$

37. $\sqrt[5]{362.65}$

38. $-\sqrt{\dfrac{8}{9}}$

39. $-\sqrt[3]{\dfrac{20}{53}}$

40. $\sqrt[3]{-\dfrac{15}{19}}$

Evaluate a) x^2 and b) $-x^2$ for each given value of x.

41. 3

42. 7

43. -10

44. -2

45. -1

46. -6

47. $\dfrac{1}{3}$

48. $-\dfrac{4}{5}$

Evaluate a) x^3 and b) $-x^3$ for each given value of x.

49. 3

50. -3

51. -5

52. 5

53. -2

54. 4

55. $\dfrac{2}{5}$

56. $-\dfrac{3}{4}$

Evaluate each expression.

57. $4^2 + 2^3 - 2^2 - 3^3$

58. $(-1)^2 + (-1)^3 - 1^4 + 1^5$

59. $-2^2 - 2^3 + 1^{10} + (-2)^3$

60. $(-3)^3 - 2^2 - (-2)^2 + (9 - 9)^2$

61. $(1.5)^2 - (3.9)^2 + (-2.1)^3$

62. $(3.7)^2 - (0.8)^2 + (2.4)^3$

63. $\left(-\dfrac{1}{2}\right)^4 - \left(\dfrac{1}{2}\right)^2 + \left(-\dfrac{1}{2}\right)^3$

64. $\left(\dfrac{3}{4}\right)^2 - \dfrac{1}{4} - \left(-\dfrac{3}{8}\right)^2 + \left(\dfrac{1}{4}\right)^3$

Evaluate each expression.

65. $3 + 5 \cdot 8$

66. $(2 - 6) \div 4 + 3$

67. $18 - 7 \div 7 + 8$

68. $4 \cdot 3 \div 6 - 2^3$

69. $\dfrac{3}{4} \div \dfrac{1}{2} - 2 + 5 \div 10$

70. $3 \cdot 6 \div 18 + \dfrac{3}{5}$

71. $\dfrac{1}{2} \cdot \dfrac{2}{3} \div \dfrac{3}{4} - \dfrac{1}{6} \cdot \left(-\dfrac{1}{3}\right)$

72. $3[4 + (-2)(8)] + 3^3$

73. $10 \div [(3 + 2^2) - (2^4 - 8)]$

74. $[3 - (4 - 2^3)^2]^2$

75. $5\left(\sqrt[3]{27} + \sqrt[5]{32}\right) \div \dfrac{\sqrt{100}}{4}$

76. $\{5 + [4^2 - 3(2 - 7)] - 5\}^2$

77. $\{[(12 - 15) - 3] - 2\}^2$

78. $4\{6 - [(25 \div 5) - 2]\}^3$

79. $3[5(16 - 6) \div (25 \div 5)^2]^2$

80. $\dfrac{15 \div 3 + 7 \cdot 2}{\sqrt{25} \div 5 + 8 \div 2}$

81. $\dfrac{4 - (2 + 3)^2 - 6}{4(3 - 2) - 3^2}$

82. $-2\left|-3 - \dfrac{1}{3}\right| + 5$

83. $\dfrac{8 + 4 \div 2 \cdot 3 + 9}{5^2 - 3^2 \cdot 2 - 7}$

84. $\dfrac{6(-3) + 4 \cdot 7 - 4^2}{-6 + \sqrt{4}(2^2 - 1)}$

85. $\dfrac{8 - [4 - (3 - 1)^2]}{5 - (-3)^2 + 4 \div 2}$

86. $12 - 15 \div |5| - (|4| - 2)^2$

87. $-2|-3| - \sqrt{36} \div |2| + 3^2$

88. $\dfrac{5 - |-15| \div |3|}{2(4 - |5|) + 9}$

89. $\dfrac{6 - |-4| - 4|8 - 5|}{5 - 6 \cdot 2 \div |-6|}$

90. $-\dfrac{1}{2}[8 - |-6| \div 3 - 4]^2$

91. $\dfrac{2}{5}\left[\sqrt[3]{27} - |-9| + 4 - 3^2\right]^2$

92. $\dfrac{3(12 - 9)^2}{-3^2} - \dfrac{2(3^2 - 4^2)}{4 - (-2)}$

93. $\dfrac{24 - 5 - 4^2}{|-8| + 4 - 2(3)} + \dfrac{4 - (-3)^2 + |4|}{3^2 - 4 \cdot 3 + |-7|}$

94. $\dfrac{-2 - 8 \div 4^2 \cdot |8|}{|8| - \sqrt{64}} + \dfrac{[(8 - 3)^2 - 7]^2}{2^2 + 16}$

Evaluate each expression for the given value or values.

95. $5x^2 + 7x$ when $x = 2$

96. $5x^2 - 2x + 17$ when $x = 3$

97. $-9x^2 + 3x - 29$ when $x = -1$

98. $3(x - 2)^2$ when $x = \dfrac{1}{4}$

99. $16(x + 5)^3 - 25(x + 5)$ when $x = -4$

100. $7x + 3y^3$ when $x = 2, y = 4$

101. $6x^2 + 3y^3 - 25$ when $x = 1, y = -3$

102. $4x^2 - 3y - 10$ when $x = 4, y = -2$

103. $3(a + b)^2 + 4(a + b) - 6$ when $a = 4, b = -1$

104. $-9 - \{2x - [5x - (2x + 1)]\}$ when $x = 3$

105. $-9 - \{x - [2x - (x - 3)]\}$ when $x = 4$

106. $\dfrac{(x - 3)^2}{9} + \dfrac{(y + 5)^2}{16}$ when $x = 4, y = 3$

107. $\dfrac{-b + \sqrt{b^2 - 4ac}}{2a}$ when $a = 6, b = -11, c = 3$

108. $\dfrac{-b - \sqrt{b^2 - 4ac}}{2a}$ when $a = 2, b = 1, c = -10$

Problem Solving

In Exercises 109–114, write an algebraic expression for each problem. Then evaluate the expression for the given value of the variable or variables.

109. Multiply the variable y by 7. From this product subtract 14. Now divide this difference by 2. Find the value of this expression when $y = 6$.

110. Subtract 4 from z. Multiply this difference by 5. Now square this product. Find the value of this expression when $z = 10$.

111. Six is added to the product of 3 and x. This expression is then multiplied by 6. Nine is then subtracted from this product. Find the value of the expression when $x = 3$.

112. The sum of x and y is multiplied by 2. Then 5 is subtracted from this product. This expression is then squared. Find the value of the expression when $x = 2$ and $y = -3$.

113. Three is added to x. This sum is divided by twice y. This quotient is then squared. Finally, 3 is subtracted from this expression. Find the value of the expression when $x = 5$ and $y = 2$.

114. Four is subtracted from x. This sum is divided by $10y$. The quotient is then cubed. Finally, 19 is added to this expression. Find the value of the expression when $x = 64$ and $y = 3$.

Use a calculator to answer Exercises 115–128.

115. Riding a Bike Frank Kelso can ride his bike at a rate of 8.2 miles per hour on the *C & O* Tow Path in Maryland. The distance, in miles, traveled after riding his bike x hours is determined by

$$\text{distance} = 8.2x$$

How far has Frank traveled in

a) 3 hours?

b) 7 hours?

© Allen R. Angel

116. Salary On January 2, 2010, Mary Ferguson started a new job with an annual salary of \$32,550. Her boss agreed to give her a raise of \$1,200 per year for the next 20 years. Her salary, in dollars, is determined by

$$\text{salary} = 32{,}550 + 1{,}200x$$

where x is the number of years since 2010. Substitute 1 for x to determine her salary in 2011, 2 for x to determine her salary in 2012, and so on. Find Mary's salary in

a) 2014.

b) 2024.

117. Throwing a Ball Cuong Chapman threw a baseball upward from a dormitory window. The height of the ball above the ground, in feet, is determined by

$$\text{height} = -16x^2 + 72x + 22$$

where x is the number of seconds after the baseball is thrown from the window. Determine the height of the ball

a) 2 seconds

b) 4 seconds

after it is thrown from the window.

118. Velocity See Exercise 117. After the ball is thrown from the window, its velocity (or speed), in feet per second, is determined by

$$velocity = -32x + 72$$

Find the velocity of the ball

a) 2 seconds

b) 4 seconds

after it is thrown from the window.

119. Spending Money The amount, in dollars, spent on holiday gifts by an individual can be estimated by

$$spending = 26.865x + 488.725$$

where x is the number of years since 2002. Substitute 1 for x to find the amount spent in 2003, 2 for x to find the amount spent in 2004, and so on. Assuming this trend continues, determine the amount each consumer will spend on holiday gifts in

a) 2015.

b) 2020.

120. Centenarians People who live to be 100 years old or older are known as centenarians. The approximate number of centenarians living in the United States between the years of 1995 and 2050, in thousands, can be estimated by

$$number\ of\ centenarians = 0.30x^2 - 3.69x + 92.04$$

where x represents years since 1995. Substitute 1 for x to find the number of centenarians in 1996, 2 for x to find the number of centenarians in 1997, and so on.

Source: U.S. Census Bureau

a) Estimate the number of centenarians that were living in the United States in 2005.

b) Estimate the number of centenarians that will be living in the United States in 2050.

121. Public Transportation Between 1992 and 2008, the approximate number of public transportation trips per year in the United States, in billions, can be estimated using

$$number\ of\ trips = 0.065x^2 - 0.39x + 8.47$$

where x represents years since 1992. Substitute 1 for x to estimate the number of trips made in 1993, 2 for x to estimate the number of trips made in 1994, and so on.

Source: American Public Transportation Association

a) Estimate the number of trips made using public transportation in 2000.

b) Assume this trend continues. Estimate the number of trips that will be made in 2010.

The trolley is one form of public transportation in San Francisco.

122. Inflation Inflation was on the decline for the years from 2000 to 2002. In 2003, it was on the rise. The inflation rate, as a percent, for the even-numbered years since 2000, can be estimated by

$$inflation = 0.35x^2 - 1.09x + 3.07$$

where x is the number of 2-year periods since 2000. Substitute 1 for x to find the inflation rate in 2002, 2 for x to find the inflation rate in 2004, and so on. Assuming this trend continues, find the inflation rate in

Source: Treasury Department

a) 2006.

b) 2010.

123. Auctions Sales, in billions of dollars, from auctions can be estimated by

$$sales = 13.5x + 189.83$$

where x is the number of years since 2002. Substitute 1 for x to find the sales from auctions in 2003, 2 for x to find the sales from auctions in 2004, and so on. Assuming this trend continues, find the sales from auctions in

Source: National Auctioneers Association

a) 2010.

b) 2018.

124. Carbon Dioxide. The total production of carbon dioxide (CO_2) from all countries except the United States, Canada, and western Europe (measured in millions of metric tons) can be approximated by

$$CO_2 = 0.073x^2 - 0.39x + 0.55$$

where x represents the number of 10-year periods since 1905. Substitute 1 for x to calculate the CO_2 production in 1915, 2 for x to calculate the CO_2 production in 1925, 3 for x in 1935, and so on.

a) Find the approximate amount of CO_2 produced by all countries except the United States, Canada, and western Europe in 1945.

b) Assuming this trend continues, find the approximate amount of CO_2 produced by all countries except the United States, Canada, and western Europe in 2015.

125. Latchkey Kids The number of *latchkey kids*, children who care for themselves while their parents are working, increases with age. The percent of children of different ages, from 5 to 14 years old, who are latchkey kids can be approximated by

$$percent\ of\ children = 0.23x^2 - 1.98x + 4.42$$

The x value represents the age of the children. For example, substitute 5 for x to get the percent of all 5-year-olds who are latchkey kids, substitute 6 for x to get the percent of all 6-year-olds who are latchkey kids, and so on.

a) Find the percent of all 10-year-olds who are latchkey kids.

b) Find the percent of all 14-year-olds who are latchkey kids.

126. Newspaper Readers The number of Americans who read a daily newspaper is steadily going down. The percent reading a daily newspaper can be approximated by

$$\text{percent} = -6.2x + 82.2$$

where x represents the number of 10-year periods since 1960. Substitute 1 for x to get the percent for 1970, 2 for x to get the percent for 1980, 3 for x to get the percent for 1990, and so forth.

a) Find the percent of U.S. adults who read a daily newspaper in 1970.

b) Assuming that this trend continues, find the percent of U.S. adults who will read a daily newspaper in 2010.

127. Organically Grown Food Increasing fears of pesticides and genetically altered crops have led many people to purchase organically grown food. From 1990 to 2010, sales, in billions of U.S. dollars, of organically grown food can be estimated by

$$\text{sales} = 0.062x^2 + 0.020x + 1.18$$

where x represents years since 1990. Substitute 1 for x to estimate the sales of organically grown food in 1991, 2 for x to estimate sales in 1992, and so on.

a) Estimate the sales of organically grown food in 1991.

b) Estimate the sales of organically grown food in 2010.

© Allen R. Angel

128. Cellular Phones The number of cellular subscribers, in millions, can be approximated by

$$\text{number of subscribers} = 0.42x^2 - 3.44x + 5.80$$

where x represents years since 1982. Substitute 1 for x to get the number of subscribers in 1983, 2 for x to get the number of subscribers in 1984, and so on.

a) Find the number of people who used cell phones in 1989.

b) Find the number of people who will use cell phones in 2009.

© Shutterstock

Concept/Writing Exercises

129. What is the meaning of a^n?

130. What does it mean if $\sqrt[n]{a} = b$?

131. What is the principal square root of a positive number?

132. Explain why $\sqrt{-4}$ cannot be a real number.

133. Explain why an odd root of a negative number will be negative.

134. Explain why an odd root of a positive number will be positive.

135. Explain the order of operations to follow when evaluating a mathematical expression. See page 32.

136. a) Explain step-by-step how you would evaluate

$$\frac{5 - 18 \div 3^2}{4 - 3 \cdot 2}$$

b) Evaluate the expression.

137. a) Explain step-by-step how you would evaluate $16 \div 2^2 + 6 \cdot 4 - 24 \div 6$.

b) Evaluate the expression.

138. a) Explain step-by-step how you would evaluate $\{5 - [4 - (3 - 8)]\}^2$.

b) Evaluate the expression.

Cumulative Review Exercises

[1.2] **139.** $A = \{a, b, c, d, f\}$, $B = \{b, c, f, g, h\}$. Find

    a) $A \cap B$,

    b) $A \cup B$.

[1.3] *In Exercises 140–142, the letter a represents a real number. For what values of a will each statement be true?*

    140. $|a| = |-a|$

141. $|a| = a$

142. $|a| = 8$

143. List from smallest to largest: $-|6|, -4, |-5|, -|-2|, 0$.

144. Name the following property:

$$(7 + 3) + 9 = 7 + (3 + 9).$$

Mid-Chapter Test: 1.1–1.4

To find out how well you understand the chapter material covered to this point, take this brief test. The answers, and the section where the material was initially discussed, are given in the back of the book. Review any questions that you answered incorrectly.

1. Where is your instructor's office? What are your instructor's office hours?

2. Given $A = \{-3, -2, -1, 0, 1, 2\}$ and $B = \{-1, 1, 3, 5\}$, find $A \cup B$ and $A \cap B$.

3. Describe the set $D = \{0, 5, 10, 15, \dots\}$.

4. Illustrate the set $\{x \mid x \geq 3\}$ on a number line.

5. Insert $<$ or $>$ in the shaded area of $\dfrac{3}{5}$ ▨ $\dfrac{4}{9}$ to make a true statement.

6. Express ←|⎯|⎯|⎯|⎯|⎯|⎯|⎯|⎯|⎯|→ in set builder notation.
   $\quad\quad\quad -6\ -5\ -4\ -3\ -2\ -1\ \ 0\ \ 1\ \ 2\ \ 3$

7. Is W a subset of N? Explain.

8. List the values from smallest to largest: $-15, |-17|, |-6|, 7$.

Evaluate each expression.

9. $7 - 2.3 - (-4.5)$

10. $\left(\dfrac{2}{5} + \dfrac{1}{3}\right) - \dfrac{1}{2}$

11. $(5)(-2)(3.2)(-8)$

12. $\left|-\dfrac{8}{13}\right| \div (-2)$

13. Evaluate $(7 - |-2|) - (-8 + |16|)$.

14. Name the property illustrated by $5(x + y) = 5x + 5y$.

15. Simplify $\sqrt{0.81}$.

16. Evaluate
   a) -11^2
   b) (-11^2)

17. a) List the order of operations.
   b) Evaluate $4 - 2 \cdot 3^2$ and explain how you determined your answer.

Evaluate each expression.

18. $5 \cdot 4 \div 10 + 2^5 - 11$.

19. $\dfrac{1}{4}\{[(12 \div 4)^2 - 7]^3 \div 2\}^2$

20. $\dfrac{\sqrt{16} + \left(\sqrt{49} - 6\right)^4}{\sqrt[3]{-27} - (4 - 3^2)}$

1.5 Exponents

1 Use the product rule for exponents.

2 Use the quotient rule for exponents.

3 Use the negative exponent rule.

4 Use the zero exponent rule.

5 Use the rule for raising a power to a power.

6 Use the rule for raising a product to a power.

7 Use the rule for raising a quotient to a power.

In this section we discuss the rules of exponents.

1 Use the Product Rule for Exponents

Consider the multiplication $x^3 \cdot x^5$. We can simplify this expression as follows:

$$x^3 \cdot x^5 = (x \cdot x \cdot x) \cdot (x \cdot x \cdot x \cdot x \cdot x) = x^8$$

This problem could also be simplified using the **product rule for exponents**.*

Product Rule for Exponents

If m and n are natural numbers and a is any real number, then
$$a^m \cdot a^n = a^{m+n}$$

To multiply exponential expressions, maintain the common base and add the exponents.

$$x^3 \cdot x^5 = x^{3+5} = x^8$$

Understanding Algebra

When *multiplying* expressions with the same base, keep the base and *add the exponents*:
$$2^3 \cdot 2^5 = 2^8 = 256$$

EXAMPLE 1   Simplify.   **a)** $2^3 \cdot 2^4$ **b)** $d^2 \cdot d^5$ **c)** $h \cdot h^9$

Solution

a) $2^3 \cdot 2^4 = 2^{3+4} = 2^7 = 128$ b) $d^2 \cdot d^5 = d^{2+5} = d^7$

c) $h \cdot h^9 = h^1 \cdot h^9 = h^{1+9} = h^{10}$

Now Try Exercise 13

*The rules given in this section also apply for rational or fractional exponents.

2 Use the Quotient Rule for Exponents

Consider the division $x^7 \div x^4$. We can simplify this expression as follows:

$$\frac{x^7}{x^4} = \frac{\overset{1}{\cancel{x}}\cdot\overset{1}{\cancel{x}}\cdot\overset{1}{\cancel{x}}\cdot\overset{1}{\cancel{x}}\cdot x\cdot x\cdot x}{\underset{1}{\cancel{x}}\cdot\underset{1}{\cancel{x}}\cdot\underset{1}{\cancel{x}}\cdot\underset{1}{\cancel{x}}} = x\cdot x\cdot x = x^3$$

This problem could also be simplified using the **quotient rule for exponents**.

Understanding Algebra

When *dividing* expressions with the same base, keep the base and *subtract the exponents*:

$$\frac{5^6}{5^4} = 5^{6-4} = 5^2 \text{ (or 25)}$$

Quotient Rule for Exponents

If a is any nonzero real number and m and n are nonzero integers, then

$$\frac{a^m}{a^n} = a^{m-n}$$

To divide expressions in exponential form, maintain the common base and subtract the exponents.

$$\frac{x^7}{x^4} = x^{7-1} = x^3$$

EXAMPLE 2 Simplify.   **a)** $\frac{6^4}{6^2}$ **b)** $\frac{x^7}{x^3}$ **c)** $\frac{y^2}{y^5}$

Solution   **a)** $\frac{6^4}{6^2} = 6^{4-2} = 6^2 = 36$ **b)** $\frac{x^7}{x^3} = x^{7-3} = x^4$ **c)** $\frac{y^2}{y^5} = y^{2-5} = y^{-3}$

Now Try Exercise 15

3 Use the Negative Exponent Rule

Notice in Example 2 **c)** that the answer contains a negative exponent. Let's do part **c)** again by dividing out common factors.

$$\frac{y^2}{y^5} = \frac{\overset{1}{\cancel{y}}\cdot\overset{1}{\cancel{y}}}{\underset{1}{\cancel{y}}\cdot\underset{1}{\cancel{y}}\cdot y\cdot y\cdot y} = \frac{1}{y^3}$$

By dividing out common factors and using the result from Example 2 **c)**, we can reason that $y^{-3} = \frac{1}{y^3}$. This is an example of the negative exponent rule.

Understanding Algebra

When simplifying expressions with a base raised to a *negative exponent*, the answer is a fraction with a *numerator* of 1 and a *denominator* of the base raised to a *positive exponent*:

$$4^{-3} = \frac{1}{4^3} = \frac{1}{64}$$

Negative Exponent Rule

For any nonzero real number a and any whole number m,

$$a^{-m} = \frac{1}{a^m}$$

An expression raised to a negative exponent is equal to 1 divided by the expression with the sign of the exponent changed.

EXAMPLE 3 Write each expression without negative exponents.

a) 7^{-2} **b)** $8a^{-6}$ **c)** $\frac{1}{c^{-5}}$

Solution

a) $7^{-2} = \frac{1}{7^2} = \frac{1}{49}$ **b)** $8a^{-6} = 8\cdot\frac{1}{a^6} = \frac{8}{a^6}$

c) $\frac{1}{c^{-5}} = 1 \div c^{-5} = 1 \div \frac{1}{c^5} = \frac{1}{1}\cdot\frac{c^5}{1} = c^5$

Now Try Exercise 37

> **Helpful Hint**
>
> In Example 3 **c)** we showed that $\dfrac{1}{c^{-5}} = c^5$. In general, for any nonzero real number a and any whole number m, $\dfrac{1}{a^{-m}} = a^m$. When a factor of the numerator or the denominator is raised to any power, the factor can be moved to the other side of the fraction bar provided the sign of the exponent is changed. Thus, for example
>
> $$\frac{2a^{-3}}{b^2} = \frac{2}{a^3 b^2} \qquad \frac{a^{-2}b^4}{c^{-3}} = \frac{b^4 c^3}{a^2}$$
>
> NOTE: When using this procedure, the sign of the base does not change, only the sign of the exponent. For example,
>
> $$-c^{-3} = -(c^{-3}) = -\frac{1}{c^3}$$

Generally, we do not leave exponential expressions with negative exponents. *When we indicate that an exponential expression is to be simplified, we mean that the answer should be written without negative exponents.*

EXAMPLE 4 Simplify.   **a)** $\dfrac{5xz^2}{y^{-4}}$ **b)** $4^{-2}x^{-1}y^2$ **c)** $-3^3 x^2 y^{-6}$

Solution

a) $\dfrac{5xz^2}{y^{-4}} = 5xy^4 z^2$ *5xy⁴z²*   **b)** $4^{-2}x^{-1}y^2 = \dfrac{1}{4^2} \cdot \dfrac{1}{x^1} \cdot y^2 = \dfrac{y^2}{16x}$

c) $-3^3 x^2 y^{-6} = -(3^3)x^2 \cdot \dfrac{1}{y^6} = -\dfrac{27x^2}{y^6}$

Now Try Exercise 41

Notice that the expressions in Example 4 do not involve addition or subtraction. The presence of a plus sign or a minus sign makes for a very different problem as we will see in our next example.

EXAMPLE 5 Simplify.   **a)** $4^{-1} + 6^{-1}$ **b)** $2 \cdot 3^{-2} + 7 \cdot 6^{-2}$

Solution

a) $4^{-1} + 6^{-1} = \dfrac{1}{4} + \dfrac{1}{6}$ Negative exponent rule

$\qquad\qquad\quad = \dfrac{3}{12} + \dfrac{2}{12}$ Rewrite with the LCD, 12.

$\qquad\qquad\quad = \dfrac{3 + 2}{12} = \dfrac{5}{12}$

b) $2 \cdot 3^{-2} + 7 \cdot 6^{-2} = 2 \cdot \dfrac{1}{3^2} + 7 \cdot \dfrac{1}{6^2}$ Negative exponent rule

$\qquad\qquad\qquad\quad = \dfrac{2}{1} \cdot \dfrac{1}{9} + \dfrac{7}{1} \cdot \dfrac{1}{36}$

$\qquad\qquad\qquad\quad = \dfrac{2}{9} + \dfrac{7}{36}$

$\qquad\qquad\qquad\quad = \dfrac{8}{36} + \dfrac{7}{36}$ Rewrite with the LCD, 36.

$\qquad\qquad\qquad\quad = \dfrac{8 + 7}{36} = \dfrac{15}{36} = \dfrac{5}{12}$

Now Try Exercise 75

4 Use the Zero Exponent Rule

The next rule we will study is the **zero exponent rule**. Any nonzero number divided by itself is 1. Therefore,

$$\frac{x^5}{x^5} = 1.$$

By the quotient rule for exponents,

$$\frac{x^5}{x^5} = x^{5-5} = x^0.$$

Since $x^0 = \dfrac{x^5}{x^5}$ and $\dfrac{x^5}{x^5} = 1$, then

$$x^0 = 1.$$

> **Zero Exponent Rule**
>
> If a is any nonzero real number, then
>
> $$a^0 = 1$$

The zero exponent rule illustrates that *any nonzero real number with an exponent of 0 equals* 1. We must specify that $a \neq 0$ because 0^0 is undefined.

EXAMPLE 6 Simplify (assume that the base is not 0).

a) 162^0 **b)** $7p^0$ **c)** $-y^0$ **d)** $-(8x + 9y)^0$

Solution

a) $162^0 = 1$ **b)** $7p^0 = 7 \cdot p^0 = 7 \cdot 1 = 7$

c) $-y^0 = -1 \cdot y^0 = -1 \cdot 1 = -1$

d) $-(8x + 9y)^0 = -1 \cdot (8x + 9y)^0 = -1 \cdot 1 = -1$

Now Try Exercise 33

5 Use the Rule for Raising a Power to a Power

Consider the expression $\left(x^3\right)^2$. We can simplify this expression as follows:

$$\left(x^3\right)^2 = x^3 \cdot x^3 = x^{3+3} = x^6$$

This problem could also be simplified using the rule for **raising a power to a power** (also called the **power rule**).

> **Understanding Algebra**
>
> When raising a *power* to a *power*, keep the base and *multiply* the exponents:
>
> $$(a^4)^3 = a^{4 \cdot 3} = a^{12}$$

> **Raising a Power to a Power (the Power Rule)**
>
> If a is a real number and m and n are integers, then
>
> $$\left(a^m\right)^n = a^{m \cdot n}$$

To raise an exponential expression to a power, maintain the base and multiply the exponents.

$$\left(x^3\right)^2 = x^{3 \cdot 2} = x^6$$

EXAMPLE 7 Simplify (assume that the base is not 0).

a) $(2^2)^3$ **b)** $\left(z^{-5}\right)^4$ **c)** $\left(2^{-3}\right)^2$

Solution

a) $(2^2)^3 = 2^{2 \cdot 3} = 2^6 = 64$ **b)** $\left(z^{-5}\right)^4 = z^{-5 \cdot 4} = z^{-20} = \dfrac{1}{z^{20}}$

c) $\left(2^{-3}\right)^2 = 2^{-3 \cdot 2} = 2^{-6} = \dfrac{1}{2^6} = \dfrac{1}{64}$

Now Try Exercise 81

Students often confuse the *product rule*

$$a^m \cdot a^n = a^{m+n}$$

with the *power rule*

$$(a^m)^n = a^{m \cdot n}$$

For example, $(x^3)^2 = x^6$, not x^5, and $(y^2)^5 = y^{10}$, not y^7.

6 Use the Rule for Raising a Product to a Power

Consider the expression $(xy)^2$. We can simplify this expression as follows:

$$(xy)^2 = (xy)(xy) = x \cdot x \cdot y \cdot y = x^2 y^2$$

This expression could also be simplified using the rule for **raising a product to a power**.

Understanding Algebra

When raising a product to a power, raise each factor of the product to the power:

$$(3 \cdot 5)^2 = 3^2 \cdot 5^2 = 9 \cdot 25$$
$$= 225$$

Raising a Product to a Power

If a and b are real numbers and m is an integer, then

$$(ab)^m = a^m b^m$$

To raise a product to a power, raise all factors within the parentheses to the power outside the parentheses.

EXAMPLE 8   Simplify.   **a)** $(-9x^3)^2$ **b)** $(3x^{-5}y^4)^{-3}$

Solution

a) $(-9x^3)^2 = (-9)^2(x^3)^2 = 81x^6$

b) $(3x^{-5}y^4)^{-3} = 3^{-3}(x^{-5})^{-3}(y^4)^{-3}$ Raise a product to a power.

$$= \frac{1}{3^3} \cdot x^{15} \cdot y^{-12}$$ Negative exponent rule, power rule

$$= \frac{1}{27} \cdot x^{15} \cdot \frac{1}{y^{12}}$$ Negative exponent rule

$$= \frac{x^{15}}{27y^{12}}$$

Now Try Exercise 93

7 Use the Rule for Raising a Quotient to a Power

Consider the expression $\left(\dfrac{x}{y}\right)^2$. We can simplify this expression as follows:

$$\left(\frac{x}{y}\right)^2 = \frac{x}{y} \cdot \frac{x}{y} = \frac{x \cdot x}{y \cdot y} = \frac{x^2}{y^2}$$

This expression could also be simplified using the rule for **raising a quotient to a power**.

Understanding Algebra

When raising a quotient to an exponent, we write the numerator raised to the exponent divided by the denominator raised to the exponent.

$$\left(\frac{2}{5}\right)^3 = \frac{2^3}{5^3} = \frac{8}{125}$$

Raising a Quotient to a Power

If a and b are real numbers and m is an integer, then

$$\left(\frac{a}{b}\right)^m = \frac{a^m}{b^m}, \qquad b \neq 0$$

To raise a quotient to a power, raise all factors in the parentheses to the exponent outside the parentheses.

EXAMPLE 9 Simplify.   **a)** $\left(\dfrac{5}{x^2}\right)^3$ **b)** $\left(\dfrac{2x^{-2}}{y^3}\right)^{-4}$

Solution

a) $\left(\dfrac{5}{x^2}\right)^3 = \dfrac{5^3}{(x^2)^3} = \dfrac{125}{x^6}$

b) $\left(\dfrac{2x^{-2}}{y^3}\right)^{-4} = \dfrac{2^{-4}(x^{-2})^{-4}}{(y^3)^{-4}}$ Raise a quotient to a power.

$\qquad = \dfrac{2^{-4}x^8}{y^{-12}}$ Power rule

$\qquad = \dfrac{x^8 y^{12}}{2^4}$ Negative exponent rule

$\qquad = \dfrac{x^8 y^{12}}{16}$

Now Try Exercise 99

Consider $\left(\dfrac{a}{b}\right)^{-n}$. Using the rule for raising a quotient to a power, we get

$$\left(\dfrac{a}{b}\right)^{-n} = \dfrac{a^{-n}}{b^{-n}} = \dfrac{b^n}{a^n} = \left(\dfrac{b}{a}\right)^n$$

Using this result, we see that when we have a rational number raised to a negative exponent, we can take the reciprocal of the base and change the sign of the exponent, as follows

$$\left(\dfrac{8}{9}\right)^{-3} = \left(\dfrac{9}{8}\right)^3 \qquad \left(\dfrac{x^2}{y^3}\right)^{-4} = \left(\dfrac{y^3}{x^2}\right)^4$$

Now we will work some examples that combine a number of properties. Whenever the same variable appears above and below the fraction bar, we generally move the variable with the *lesser exponent* to the opposite side of the fraction bar. This will result in the exponent on the variable being positive when the product rule is applied. Examples 10 and 11 illustrate this procedure.

> **Understanding Algebra**
>
> $\left(\dfrac{x}{y}\right)^{-n} = \left(\dfrac{y}{x}\right)^n = \dfrac{y^n}{x^n}$

EXAMPLE 10 Simplify.   **a)** $\left(\dfrac{15x^2y^4}{5x^2y}\right)^2$ **b)** $\left(\dfrac{6x^4y^{-2}}{12xy^3z^{-1}}\right)^{-3}$

Solution Exponential expressions can often be simplified in more than one way. In general, it will be easier to first simplify the expression within the parentheses.

a) $\left(\dfrac{15x^2y^4}{5x^2y}\right)^2 = (3y^3)^2 = 9y^6$

b) $\left(\dfrac{6x^4y^{-2}}{12xy^3z^{-1}}\right)^{-3} = \left(\dfrac{x^4 \cdot x^{-1}z}{2y^3 \cdot y^2}\right)^{-3}$ Move x, y^{-2}, and z^{-1} to the other side of the fraction bar and change the signs of their exponents.

$\qquad = \left(\dfrac{x^3 z}{2y^5}\right)^{-3}$ Product rule

$\qquad = \left(\dfrac{2y^5}{x^3 z}\right)^3$ Take the reciprocal of the expression inside the parentheses and change the sign of the exponent outside the parentheses.

$\qquad = \dfrac{2^3 y^{5\cdot3}}{x^{3\cdot3} z^3}$ Raise a quotient to a power.

$\qquad = \dfrac{8y^{15}}{x^9 z^3}$

Now Try Exercise 109

EXAMPLE 11 Simplify $\dfrac{(2p^{-3}q^5)^{-2}}{(p^{-5}q^4)^{-3}}$.

Solution   First, use the power rule. Then simplify further.

$$\frac{(2p^{-3}q^5)^{-2}}{(p^{-5}q^4)^{-3}} = \frac{2^{-2}p^6 q^{-10}}{p^{15}q^{-12}} \qquad \text{Power rule}$$

$$= \frac{q^{-10} \cdot q^{12}}{2^2 p^{15} \cdot p^{-6}} \qquad \begin{array}{l}\text{Move } 2^{-2}, p^6, \text{ and } q^{-12} \text{ to the other side of the fraction}\\ \text{bar and change the signs of their exponents.}\end{array}$$

$$= \frac{q^{-10+12}}{4p^{15-6}} \qquad \text{Product rule}$$

$$= \frac{q^2}{4p^9}$$

Now Try Exercise 115

Summary of Rules of Exponents

For all real numbers a and b and all integers m and n:

Product rule	$a^m \cdot a^n = a^{m+n}$	
Quotient rule	$\dfrac{a^m}{a^n} = a^{m-n},$	$a \neq 0$
Negative exponent rule	$a^{-m} = \dfrac{1}{a^m},$	$a \neq 0$
Zero exponent rule	$a^0 = 1,$	$a \neq 0$
Raising a power to a power	$(a^m)^n = a^{m \cdot n}$	
Raising a product to a power	$(ab)^m = a^m b^m$	
Raising a quotient to a power	$\left(\dfrac{a}{b}\right)^m = \dfrac{a^m}{b^m},$	$b \neq 0$

EXERCISE SET 1.5   Math XL   MyMathLab
MathXL®   MyMathLab

Warm-Up Exercises

Fill in the blanks with the appropriate word, phrase, or symbol(s) from the following list.

additive	zero exponent	raising a power	undefined	raising a product
inverse	$\dfrac{1}{9}$	-9	quotient	$\dfrac{1}{8}$
reciprocal	product	raising a quotient	negative exponent	8

1. The rule $a^m \cdot a^n = a^{m+n}$ is called the _____ rule for exponents.

2. For $a \neq 0$, the rule $\dfrac{a^m}{a^n} = a^{m-n}$ is called the _____ rule for exponents.

3. For $a \neq 0$, the rule $a^{-m} = \dfrac{1}{a^m}$ is called the _____ rule.

4. For $a \neq 0$, the rule $a^0 = 1$ is called the _____ rule.

5. The value of 0^0 is _____.

6. The rule $(a^m)^n = a^{m \cdot n}$ is called the _____ to a power rule.

7. The rule $(ab)^m = a^m b^m$ is called the _____ to a power rule.

8. The rule $\left(\dfrac{a}{b}\right)^m = \dfrac{a^m}{b^m}$ is called the _____ to a power rule.

9. If $x \neq 0$, then $\dfrac{1}{x}$ is the _____ of x.

10. If y is any real number, then $-y$ is the additive _____ of y.

11. The simplified form of 3^{-2} is _____.

12. The simplified form of $\left(\dfrac{1}{2}\right)^{-3}$ is _____.

Practice the Skills

Evaluate each expression.

13. $2^3 \cdot 2^2$

14. $3^2 \cdot 3^3$

15. $\dfrac{3^7}{3^5}$

16. $\dfrac{8^7}{8^6}$

17. 9^{-2}

18. 7^{-2}

19. $\dfrac{1}{5^{-3}}$

20. $\dfrac{1}{3^{-2}}$

21. 15^0

22. 24^0

23. $(2^3)^2$

24. $(3^2)^2$

25. $(2 \cdot 4)^2$

26. $(6 \cdot 5)^2$

27. $\left(\dfrac{4}{7}\right)^2$

28. $\left(\dfrac{3}{5}\right)^4$

Evaluate each expression.

29. a) 3^{-2} **b)** $(-3)^{-2}$ **c)** -3^{-2} **d)** $-(-3)^{-2}$

30. a) 4^{-3} **b)** $(-4)^{-3}$ **c)** -4^{-3} **d)** $-(-4)^{-3}$

31. a) $\left(\dfrac{1}{2}\right)^{-1}$ **b)** $\left(-\dfrac{1}{2}\right)^{-1}$ **c)** $-\left(\dfrac{1}{2}\right)^{-1}$ **d)** $-\left(-\dfrac{1}{2}\right)^{-1}$

32. a) $\left(\dfrac{3}{5}\right)^{-2}$ **b)** $\left(-\dfrac{3}{5}\right)^{-2}$ **c)** $-\left(\dfrac{3}{5}\right)^{-2}$ **d)** $-\left(-\dfrac{3}{5}\right)^{-2}$

Simplify each expression and write the answer without negative exponents. Assume that all bases represented by variables are nonzero.

33. a) $5x^0$ **b)** $-5x^0$ **c)** $(-5x)^0$ **d)** $-(-5x)^0$

34. a) $7y^0$ **b)** $(7y)^0$ **c)** $-7y^0$ **d)** $(-7y)^0$

35. a) $3xyz^0$ **b)** $(3xyz)^0$ **c)** $3x(yz)^0$ **d)** $3(xyz)^0$

36. a) $x^0 + y^0$ **b)** $(x + y)^0$ **c)** $x + y^0$ **d)** $x^0 + y$

Simplify each expression and write the answer without negative exponents.

37. $7y^{-3}$

38. $\dfrac{1}{y^{-1}}$

39. $\dfrac{9}{x^{-4}}$

40. $\dfrac{8}{5x^{-2}}$

41. $\dfrac{3a}{b^{-3}}$

42. $\dfrac{10x^4}{y^{-1}}$

43. $\dfrac{17m^{-2}n^{-3}}{2}$

44. $\dfrac{13x^{-3}}{z^4}$

45. $\dfrac{5x^{-2}y^{-3}}{z^{-4}}$

46. $\dfrac{15ab^5}{3c^{-3}}$

47. $\dfrac{9^{-1}x^{-1}}{y}$

48. $\dfrac{8^{-1}z}{x^{-1}y^{-1}}$

Simplify each expression and write the answer without negative exponents.

49. $2^5 \cdot 2^{-7}$

50. $a^3 \cdot a^5$

51. $x^6 \cdot x^{-4}$

52. $x^{-4} \cdot x^3$

53. $\dfrac{8^7}{8^5}$

54. $\dfrac{4^3}{4^{-1}}$

55. $\dfrac{7^{-5}}{7^{-3}}$

56. $\dfrac{x^{-7}}{x^4}$

57. $\dfrac{m^{-6}}{m^5}$

58. $\dfrac{p^0}{p^{-3}}$

59. $\dfrac{5w^{-2}}{w^{-7}}$

60. $\dfrac{x^{-7}}{x^{-9}}$

61. $3a^{-2} \cdot 4a^{-6}$

62. $(-8v^4)(-3v^{-5})$

63. $(-3p^{-2})(-p^3)$

64. $(2x^{-3}y^{-4})(6x^{-4}y^7)$

65. $(5r^2s^{-2})(-2r^5s^2)$

66. $(-6p^{-4}q^6)(2p^3q)$

67. $(2x^4y^7)(4x^3y^{-5})$

68. $\dfrac{27x^3y^2}{9xy}$

69. $\dfrac{33x^5y^{-4}}{11x^3y^2}$

70. $\dfrac{16x^{-2}y^3z^{-2}}{-2x^4y}$

71. $\dfrac{9xy^{-4}z^3}{-3x^{-2}yz}$

72. $\dfrac{(x^{-2})(4x^2)}{x^3}$

Evaluate each expression.

73. a) $4(a + b)^0$ **b)** $4a^0 + 4b^0$ **c)** $(4a + 4b)^0$ **d)** $-4a^0 + 4b^0$

74. a) $-3^0 + (-3)^0$ **b)** $-3^0 - (-3)^0$ **c)** $-3^0 + 3^0$ **d)** $-3^0 - 3^0$

75. a) $4^{-1} - 3^{-1}$ **b)** $4^{-1} + 3^{-1}$ **c)** $2 \cdot 4^{-1} + 3 \cdot 5^{-1}$ **d)** $(2 \cdot 4)^{-1} + (3 \cdot 5)^{-1}$

76. a) $5^{-2} + 4^{-1}$ **b)** $5^{-2} - 4^{-1}$ **c)** $3 \cdot 5^{-2} + 2 \cdot 4^{-1}$ **d)** $(3 \cdot 5)^{-2} - (2 \cdot 4)^{-1}$

Simplify each expression and write the answer without negative exponents.

77. $(3^2)^2$

78. $(5^2)^{-1}$

79. $(3^2)^{-2}$

80. $(2^2)^{-3}$

81. $(b^{-3})^{-2}$

82. $(-c)^4$

83. $(-c)^3$

84. $(-x)^{-4}$

85. $(-5x^{-3})^2$

86. $-11(x^{-3})^2$

87. $4^{-2} + 8^{-1}$

88. $5^{-1} + 2^{-1}$

89. $3 \cdot 4^{-2} + 9 \cdot 8^{-1}$

90. $5 \cdot 2^{-3} + 7 \cdot 4^{-2}$

91. $\left(\dfrac{4b}{3}\right)^{-2}$

92. $\left(\dfrac{2c}{5}\right)^{-3}$

93. $(4x^2y^{-2})^2$

94. $(4x^2y^3)^{-3}$

95. $(5p^2q^{-4})^{-3}$

96. $(8s^{-3}t^{-4})^2$

97. $(-3g^{-4}h^3)^{-3}$

98. $8(x^2y^{-1})^{-4}$

99. $\left(\dfrac{5j}{4k^2}\right)^2$

100. $\left(\dfrac{3x^2y^4}{z}\right)^3$

101. $\left(\dfrac{2r^4s^5}{r^2}\right)^3$

102. $\left(\dfrac{5m^5n^6}{10m^4n^7}\right)^3$

103. $\left(\dfrac{4xy}{y^3}\right)^{-3}$

104. $\left(\dfrac{9x^{-2}}{xy}\right)^{-2}$

105. $\left(\dfrac{5x^{-2}y}{x^{-5}}\right)^3$

106. $\left(\dfrac{4x^2y}{x^{-5}}\right)^{-3}$

107. $\left(\dfrac{14x^2y}{7xz}\right)^{-3}$

108. $\left(\dfrac{3xy}{z^{-2}}\right)^3$

109. $\left(\dfrac{x^8y^{-2}}{x^{-2}y^3}\right)^2$

110. $\left(\dfrac{x^2y^{-3}z^6}{x^{-1}y^2z^4}\right)^{-1}$

111. $\left(\dfrac{4x^{-1}y^{-2}z^3}{2xy^2z^{-3}}\right)^{-2}$

112. $\left(\dfrac{9x^4y^{-6}z^4}{3xy^{-6}z^{-2}}\right)^{-2}$

113. $\left(\dfrac{-a^3b^{-1}c^{-3}}{4ab^3c^{-4}}\right)^{-3}$

114. $\dfrac{(2x^{-1}y^{-2})^{-3}}{(5x^{-1}y^3)^2}$

115. $\dfrac{(3x^{-4}y^2)^3}{(2x^3y^5)^3}$

116. $\dfrac{(2xy^2z^{-3})^2}{(9x^{-1}yz^2)^{-1}}$

Problem Solving

Simplify each expression. Assume that all variables represent nonzero integers.

117. $x^{2a} \cdot x^{5a+3}$

118. $y^{2m+3} \cdot y^{5m-7}$

119. $w^{2a-5} \cdot w^{3a-2}$

120. $d^{-4x+7} \cdot d^{5x-6}$

121. $\dfrac{x^{2w+3}}{x^{w-4}}$

122. $\dfrac{y^{5m-1}}{y^{7m-1}}$

123. $(x^{3p+5})(x^{2p-3})$

124. $(s^{2t-3})(s^{-t+5})$

125. $x^{-m}(x^{3m+2})$

126. $y^{3b+2} \cdot y^{2b+4}$

127. $\dfrac{30m^{a+b}n^{b-a}}{6m^{a-b}n^{a+b}}$

128. $\dfrac{24x^{c+3}y^{d+4}}{8x^{c-4}y^{d+6}}$

129. a) For what values of x is $x^4 > x^3$?

  b) For what values of x is $x^4 < x^3$?

  c) For what values of x is $x^4 = x^3$?

  d) Why can you not say that $x^4 > x^3$?

130. Is 3^{-8} greater than or less than 2^{-8}? Explain.

131. a) Explain why $(-1)^n = 1$ for any even number n.

  b) Explain why $(-1)^n = -1$ for any odd number n.

132. a) Explain why $(-12)^{-8}$ is positive.

  b) Explain why $(-12)^{-7}$ is negative.

133. a) Is $\left(-\dfrac{2}{3}\right)^{-2}$ equal to $\left(\dfrac{2}{3}\right)^{-2}$?

  b) Will $(x)^{-2}$ equal $(-x)^{-2}$ for all real numbers x except 0? Explain your answer.

134. a) Is $\left(-\dfrac{2}{3}\right)^{-3}$ equal to $\left(\dfrac{2}{3}\right)^{-3}$?

  b) Will $(x)^{-3}$ equal $(-x)^{-3}$ for any nonzero real number x? Explain.

  c) What is the relationship between $(-x)^{-3}$ and $(x)^{-3}$ for any nonzero real number x?

Determine what exponents must be placed in the shaded area to make each expression true. Each shaded area may represent a different exponent. Explain how you determined your answer.

135. $\left(\dfrac{x^2y^{-2}}{x^{-3}y}\right)^2 = x^{10}y^2$

136. $\left(\dfrac{x^{-2}y^3z}{x^4y^{\blacksquare}z^{-3}}\right)^3 = \dfrac{z^{12}}{x^{18}y^6}$

137. $\left(\dfrac{x^{\blacksquare}y^5z^{-2}}{x^4y^{\blacksquare}z}\right)^{-1} = \dfrac{x^5z^3}{y^2}$

Challenge Problems

We will learn in Section 7.2 that the rules of exponents given in this section also apply when the exponents are rational numbers. Using this information and the rules of exponents, evaluate each expression.

138. $\left(\dfrac{x^{1/2}}{x^{-1}}\right)^{3/2}$

139. $\left(\dfrac{x^{5/8}}{x^{1/4}}\right)^3$

140. $\left(\dfrac{x^4}{x^{-1/2}}\right)^{-1}$

141. $\dfrac{x^{1/2}y^{-3/2}}{x^5y^{5/3}}$

142. $\left(\dfrac{x^{1/2}y^4}{x^{-3}y^{5/2}}\right)^2$

Group Activity

Discuss and answer Exercise 143 as a group.

143. Doubling a Penny On day 1 you are given a penny. On each following day, you are given double the amount you were given on the previous day.

a) Write down the amounts you would be given on each of the first 6 days.

b) Express each of these numbers as an exponential expression with a base of 2.

c) By looking at the pattern, determine an exponential expression for the number of cents you will receive on day 10.

d) Write a general exponential expression for the number of cents you will receive on day n.

e) Write an exponential expression for the number of cents you will receive on day 30.

f) Calculate the value of the expression in part e). Use a calculator if one is available.

g) Determine the amount found in part f) in dollars.

h) Write a general exponential expression for the number of dollars you will receive on day n.

Cumulative Review Exercises

[1.2] **144.** If $A = \{3, 4, 6\}$ and $B = \{1, 2, 5, 9\}$, find

a) $A \cup B$ and

b) $A \cap B$.

145. Illustrate the following set on the number line: $\{x \mid -3 \le x < 2\}$.

[1.4] **146.** Evaluate $8 + |12| \div |-3| - 4 \cdot 2^2$.

147. Evaluate $\sqrt[3]{-125}$.

1.6 Scientific Notation

1 Write numbers in scientific notation.

2 Change numbers in scientific notation to decimal form.

3 Use scientific notation in problem solving.

1 Write Numbers in Scientific Notation

Scientists and engineers often deal with very large and very small numbers. For example, the frequency of an FM radio signal may be 14,200,000,000 hertz (or cycles per second) and the diameter of a hydrogen atom is about 0.0000000001 meter. Because it is difficult to work with many zeros, scientists often express such numbers with exponents. For example, the number 14,200,000,000 might be written as 1.42×10^{10}, and 0.0000000001 as 1×10^{-10}. Numbers such as 1.42×10^{10} and 1×10^{-10} are in a form called **scientific notation**. In scientific notation, numbers are expressed as $a \times 10^n$, where $1 \le a < 10$ and n is an integer. When a power of 10 has no numerical coefficient showing, as in 10^5, we assume that the numerical coefficient is 1. Thus, 10^5 means 1×10^5 and 10^{-4} means 1×10^{-4}.

The diameter of this galaxy is about 1×10^{21} meters.

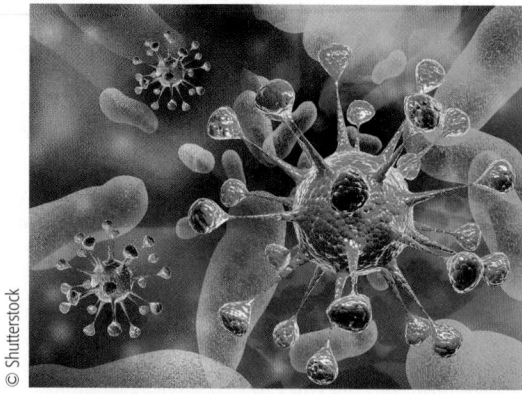

The diameter of these viruses (shown in green) is about 1×10^{-7} meters.

Examples of Numbers in Scientific Notation

$$3.2 \times 10^6 \qquad 4.176 \times 10^3 \qquad 2.64 \times 10^{-2}$$

The following shows the number 32,400 changed to scientific notation.

$$32,400 = 3.24 \times 10,000$$
$$= 3.24 \times 10^4 \qquad (10,000 = 10^4)$$

There are four zeros in 10,000, the same number as the exponent in 10^4. The procedure for writing a number in scientific notation follows.

To Write a Number in Scientific Notation

1. Move the decimal point in the number to the right of the first nonzero digit. This gives a number greater than or equal to 1 and less than 10.
2. Count the number of places you moved the decimal point in step 1. If the original number is 10 or greater, the count is to be considered positive. If the original number is less than 1, the count is to be considered negative.
3. Multiply the number obtained in step 1 by 10 raised to the count (power) found in step 2.

EXAMPLE 1   Write the following numbers in scientific notation.

a) 68,900   **b)** 0.000572   **c)** 0.0074

Solution

a) The decimal point in 68,900 is to the right of the last zero.

$$68,900. = 6.89 \times 10^4$$

The decimal point is moved four places. Since the original number is greater than 10, the exponent is positive.

b) $0.000572 = 5.72 \times 10^{-4}$

The decimal point is moved four places. Since the original number is less than 1, the exponent is negative.

c) $0.0074 = 7.4 \times 10^{-3}$

Now Try Exercise 11

2   Change Numbers in Scientific Notation to Decimal Form

Occasionally, you may need to convert a number written in scientific notation to its decimal form. The procedure to do so follows.

To Convert a Number in Scientific Notation to Decimal Form

1. Observe the exponent on the base 10.
2. **a)** If the exponent is positive, move the decimal point in the number to the right the same number of places as the exponent. It may be necessary to add zeros to the number. This will result in a number greater than or equal to 10.
   b) If the exponent is 0, the decimal point in the number does not move from its present position. Drop the factor 10^0. This will result in a number greater than or equal to 1 but less than 10.
   c) If the exponent is negative, move the decimal point in the number to the left the same number of places as the exponent. It may be necessary to add zeros. This will result in a number less than 1.

EXAMPLE 2 Write the following numbers without exponents.

a) 2.1×10^4 **b)** 8.73×10^{-3} **c)** 1.45×10^8

Solution

a) Move the decimal point four places to the right.

$$2.1 \times \boxed{10^4} = 2.1 \times \boxed{10,000} = 21,000$$

b) Move the decimal point three places to the left.

$$8.73 \times 10^{-3} = 0.00873$$

c) Move the decimal point eight places to the right.

$$1.45 \times 10^8 = 145,000,000$$

Now Try Exercise 25

3 Use Scientific Notation in Problem Solving

We can use the rules of exponents when working with numbers written in scientific notation, as illustrated in the following applications.

EXAMPLE 3 **Public Debt per Person** The public debt is the total amount owed by the U.S. federal government to lenders in the form of government bonds. On July 20, 2008, the U.S. public debt was approximately $9,525,000,000,000 (9 trillion, 525 billion dollars). The U.S. population on that date was approximately 305,000,000.

a) Find the approximate U.S. debt per person in the United States (the per capita debt).

b) On July 1, 1982, the U.S. debt was approximately $1,142,000,000,000. How much larger was the debt in 2008 than in 1982?

c) How many times greater was the debt in 2008 than in 1982?

Solution

a) To find the per capita debt, we divide the public debt by the population.

$$\frac{9,525,000,000,000}{305,000,000} = \frac{9.525 \times 10^{12}}{3.05 \times 10^8} \approx 3.12 \times 10^{12-8} \approx 3.12 \times 10^4 \approx 31,200$$

Thus, the per capita debt was about $31,200. This means that if the citizens of the United States wished to "chip in" and pay off the federal debt, it would take about $31,200 for every man, woman, and child in the United States.

b) We need to find the difference in the debt between 2008 and 1982.

$$9,525,000,000,000 - 1,142,000,000,000 = 9.525 \times 10^{12} - 1.142 \times 10^{12}$$
$$= (9.525 - 1.142) \times 10^{12}$$
$$= 8.383 \times 10^{12}$$
$$= 8,383,000,000,000$$

The U.S. public debt was $8,383,000,000,000 greater in 2008 than in 1982.

c) To find out how many times greater the 2008 public debt was, we divide the 2008 debt by the 1982 debt as follows:

$$\frac{9,525,000,000,000}{1,142,000,000,000} = \frac{9.525 \times 10^{12}}{1.142 \times 10^{12}} \approx 8.34$$

Thus, the 2008 public debt was about 8.34 times greater than the 1982 public debt.

Now Try Exercise 87

EXAMPLE 4 **Tax Collections** The data for the graph in **Figure 1.9** is taken from
the U.S. Census Bureau Web site. The graph shows the cumulative state government
tax collections in 2007. We have given the amounts collected in scientific notation.

State Government Tax Collections, by Type: 2007

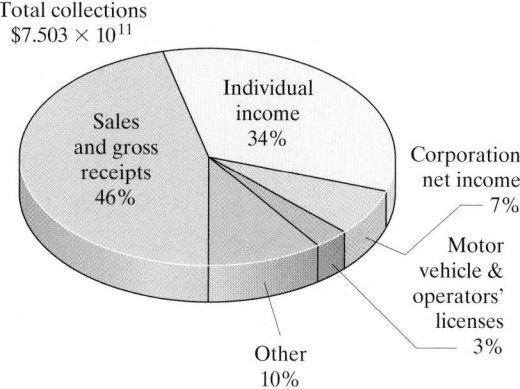

Total collections
7.503×10^{11}

Individual
income
34%

Sales
and gross
receipts
46%

Corporation
net income
7%

Motor
vehicle &
operators'
licenses
3%

Other
10%

Source: U.S. Bureau of the Census

FIGURE 1.9

a) Determine, using scientific notation, how much money was collected from individual income taxes in 2007.

b) Determine, using scientific notation, how much more money was collected from
sales and gross receipts than from corporation net income taxes.

Solution

a) In 2007, 34% of the 7.503×10^{11} was collected from individual income taxes. In
decimal form, 34% is 0.34, and in scientific notation, 34% is 3.4×10^{-1}. To determine 34% of 7.503×10^{11}, we multiply using scientific notation.

$$\text{individual income tax collected} = (3.4 \times 10^{-1})(7.503 \times 10^{11})$$
$$= (3.4 \times 7.503)(10^{-1} \times 10^{11})$$
$$= 25.5102 \times 10^{-1+11}$$
$$= 25.5102 \times 10^{10}$$
$$= 2.55102 \times 10^{11}$$

Thus, about 2.55102×10^{11} or $255,102,000,000 was collected from individual
income taxes in 2007.

b) In 2007, 46% was collected from sales and gross receipts and 7% was collected
from corporation net income taxes. To determine how much more money was
collected from sales and gross receipts than from corporation net income taxes,
we first determine the difference in the two percents.

$$\text{difference} = 46\% - 7\% = 39\%$$

To find 39% of 7.503×10^{11}, we change 39% to scientific notation and then we
multiply.

$$39\% = 0.39 = 3.9 \times 10^{-1}$$
$$\text{difference in tax collections} = (3.9 \times 10^{-1})(7.503 \times 10^{11})$$
$$= (3.9 \times 7.503)(10^{-1} \times 10^{11})$$
$$= 29.2617 \times 10^{10}$$
$$= 2.92617 \times 10^{11}$$

Thus, about 2.92617×10^{11} or $292,617,000,000 more money was collected
from sales and gross receipts than from corporation net income taxes.

Now Try Exercise 95

Using Your Calculator

On a scientific or graphing calculator the product of (8,000,000)(400,000) might be displayed as 3.2 ^{12} or 3.2E12. Both of these represent 3.2×10^{12}, which is 3,200,000,000,000.

To enter numbers in scientific notation on either a scientific or graphing calculator, you generally use the $\boxed{\text{EE}}$ or $\boxed{\text{EXP}}$ keys. To enter 4.6×10^8, you would press either 4.6 $\boxed{\text{EE}}$ 8 or 4.6 $\boxed{\text{EXP}}$ 8. Your calculator screen might then show 4.6 ^{08} or 4.6E8.

On the TI-84 Plus the EE appears above the $\boxed{,}$ key. So to enter (8,000,000)(400,000) in scientific notation you would press

answer displayed

$8 \boxed{2^{nd}} \boxed{,} , 6 \boxed{\times} 4 \boxed{2^{nd}} \boxed{,} , 5 \boxed{\text{ENTER}}$ 3.2E12

$\underbrace{\qquad}_{\text{to get EE}}$ $\underbrace{\qquad}_{\text{to get EE}}$

EXERCISE SET 1.6

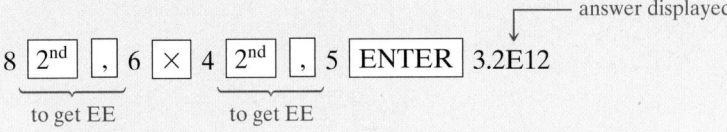
MathXL®   MyMathLab

Warm-Up Exercises

Fill in the blanks with the appropriate word, phrase, or symbol(s) from the following list.

Positive   negative   1   5   scientific notation   2   4   3   0

1. A number written as $a \times 10^n$ where $1 \le a < 10$ and n is an integer is written in _____.

2. When a number greater than 10 is written in scientific notation, the value for n in 10^n is a _____ integer.

3. To write 0.00329 in scientific notation, move the decimal point _____ places to the right.

4. To write 75,618 in scientific notation, move the decimal point _____ places to the left.

Practice the Skills

Express each number in scientific notation.

5. 3700
6. 860
7. 0.043
8. 0.000000918
9. 760,000
10. 9,260,000,000
11. 0.00000186
12. 0.00000914
13. 5,780,000
14. 0.0000723
15. 0.000106
16. 952,000,000

Express each number without exponents.

17. 3.1×10^4
18. 5×10^8
19. 2.13×10^{-5}
20. 6.78×10^{-5}
21. 9.17×10^{-1}
22. 5.4×10^1
23. 3.0×10^6
24. 7.6×10^4
25. 2.03×10^5
26. 9.25×10^{-6}
27. 1×10^6
28. 1×10^{-8}

Express each value without exponents.

29. $(4 \times 10^5)(6 \times 10^2)$
30. $(7.6 \times 10^{-3})(1.2 \times 10^{-1})$
31. $\dfrac{8.4 \times 10^{-6}}{4 \times 10^{-4}}$
32. $\dfrac{8.5 \times 10^3}{1.7 \times 10^{-2}}$
33. $\dfrac{9.45 \times 10^{-3}}{3.5 \times 10^2}$
34. $(5.2 \times 10^{-3})(4.1 \times 10^5)$
35. $(8.2 \times 10^5)(1.4 \times 10^{-2})$
36. $(6.3 \times 10^4)(3.7 \times 10^{-8})$
37. $\dfrac{1.68 \times 10^4}{5.6 \times 10^7}$
38. $\dfrac{7.2 \times 10^{-2}}{3.6 \times 10^{-6}}$
39. $(9.1 \times 10^{-4})(7.4 \times 10^{-4})$
40. $\dfrac{8.6 \times 10^{-8}}{4.3 \times 10^{-6}}$

Express each value in scientific notation.

41. $(0.03)(0.0005)$
42. $(2500)(7000)$
43. $\dfrac{35,000,000}{7000}$
44. $\dfrac{560,000}{0.0008}$
45. $\dfrac{0.00069}{23,000}$
46. $\dfrac{0.000018}{0.000009}$
47. $(47,000)(35,000,000)$
48. $\dfrac{0.0000286}{0.00143}$
49. $\dfrac{2016}{0.0021}$
50. $\dfrac{0.018}{160}$
51. $\dfrac{0.00153}{0.00051}$
52. $(0.0015)(0.00038)$

📱 *Express each value in scientific notation. Round decimal numbers to the nearest thousandth.*

53. $(4.78 \times 10^9)(1.96 \times 10^5)$

54. $\dfrac{5.55 \times 10^3}{1.11 \times 10^1}$

55. $(7.23 \times 10^{-3})(1.46 \times 10^5)$

56. $(5.71 \times 10^5)(4.7 \times 10^{-3})$

57. $\dfrac{4.36 \times 10^{-4}}{8.17 \times 10^{-7}}$

58. $\dfrac{9.675 \times 10^{25}}{3.225 \times 10^{15}}$

59. $(4.89 \times 10^{15})(6.37 \times 10^{-41})$

60. $(4.36 \times 10^{-6})(1.07 \times 10^{-6})$

61. $(4.16 \times 10^3)(9.14 \times 10^{-31})$

62. $\dfrac{3.71 \times 10^{11}}{4.72 \times 10^{-9}}$

63. $\dfrac{1.5 \times 10^{35}}{4.5 \times 10^{-26}}$

64. $(4.9 \times 10^5)(1.347 \times 10^{31})$

Scientific Notation *In Exercises 65–78, write each italicized number in scientific notation.*

65. It cost NASA more than *$850 million* to send the rovers *Spirit* and *Opportunity* to Mars.

© NASA/Jet Propulsion Laboratory

66. The distance between the sun and Earth is about *93 million* miles.

67. The average cost for a 30-second ad in Super Bowl XXLI was *$2.7 million*.

68. According to the U.S. Census Bureau, the world population in 2050 will be about *9.2 billion* people.

69. According to the 2008 *World Almanac and Fact Book*, the richest man in the world is Warren Buffett of Berkshire Hathaway, who is worth about *$ 62 billion*.

70. The 2006 U.S. federal budget was about *$2.56 trillion*.

71. In 2008, the U.S. debt was about *$9.5 trillion*.

72. The speed of light is about *186,000* miles per second.

73. One centimeter = *0.0001* hectometer.

74. One millimeter = *0.000001* kilometer.

75. One inch ≈ *0.0000158* mile.

76. One ounce ≈ *0.00003125* ton.

77. One milligram = *0.000000001* metric ton.

78. A certain computer can compute one computation in *0.0000001* second.

Problem Solving

79. Explain how you can quickly divide a number given in scientific notation by

a) 10,

b) 100,

c) 1 million.

d) Divide 6.58×10^{-4} by 1 million. Leave your answer in scientific notation.

80. Explain how you can quickly multiply a number given in scientific notation by

a) 10,

b) 100,

c) 1 million.

d) Multiply 7.59×10^7 by 1 million. Leave your answer in scientific notation.

81. Science Experiment During a science experiment you find that the correct answer is 5.25×10^4.

a) If you mistakenly write the answer as 4.25×10^4, by how much is your answer off?

b) If you mistakenly write the answer as 5.25×10^5, by how much is your answer off?

c) Which of the two errors is the more serious? Explain.

82. Earth Orbit

a) Earth completes its 5.85×10^8-mile orbit around the sun in 365 days. Find the distance traveled per day.

b) Earth's speed is about eight times faster than a bullet. Estimate the speed of a bullet in miles per hour.

© Trinacria Photo\Shutterstock

83. Distance to the Sun The distance to the sun is 93,000,000 miles. If a spacecraft travels at a speed of 3100 miles per hour, how long will it take for it to reach the sun?

84. Universe We have proof that there are at least 1 sextillion, 10^{21}, stars in the universe.

a) Write that number without exponents.

b) How many million stars is this? Explain how you determined your answer.

85. U.S. and World Population The population of the United States on July 20, 2008, was estimated to be 3.046×10^8. On this day, the population of the world was about 6.711×10^9.

Source: U.S. Census Bureau

a) How many people lived outside of the United States in 2008?

b) What percentage of the world's population lived in the United States in 2008?

86. New River Gorge Bridge The New River Gorge Bridge, shown below, is 3030.5 feet long. It was completed in 1977 near Fayetteville, West Virginia, and is the world's longest single arch steel span. Its total weight is 8.80×10^7 pounds and its heaviest single piece weighs 1.84×10^5 pounds.

a) How many times greater is the total weight of the bridge than the weight of the heaviest single piece?

b) What is the difference in weight between the total weight of the bridge and the weight of the heaviest single piece?

© Allen R. Angel

87. Gross Domestic Product Gross Domestic Product (GDP) is a measure of economic activity. GDP is the total amount of goods and services produced in a country in a year. In 2007, the GDP for the United States was about $11.750 trillion and the population of the United States was about 302.2 million.

Source: U.S. Treasury Web site

a) Write each of these two numbers in scientific notation.

b) Determine the GDP *per capita* by dividing the GDP by the population of the United States.

88. Gross Domestic Product In 2007, the GDP (see Exercise 87) of the world was about $55.500 trillion and the population of the world was about 6.6 billion people.

Source: U.S. Treasury Web site and www.en.wikipedia.org/wiki

a) Write each of these numbers in scientific notation.

b) Determine the GDP *per capita* by dividing the GDP by the population of the world.

89. Population Density The population density (people per square kilometer) is determined by dividing the population of a country by its land area. Find the population density of China if its population in 2008 was 1.32×10^9 people and its land area was 9.8×10^6 square kilometers. (Round your answer to the nearest unit.)

90. Population Density Find the population density (see Exercise 89) of India if its population in 2008 was 1.13×10^9 people and its land area is 3.2×10^6 square kilometers. (Round your answer to the nearest unit.)

91. Recycling Plastic In the United States only about 5% of the 4.2×10^9 pounds of used plastic is recycled annually.

a) How many pounds are recycled annually?

b) How many pounds are not recycled annually?

92. Distance to Proxima Centauri The distance from Earth to the sun is approximately 150 million kilometers. The next closest star to Earth is Proxima Centauri. It is about 268,000 times farther away from Earth than is the sun. Approximate the distance of Proxima Centauri from Earth. Write your answer in scientific notation.

Source: NASA Web site

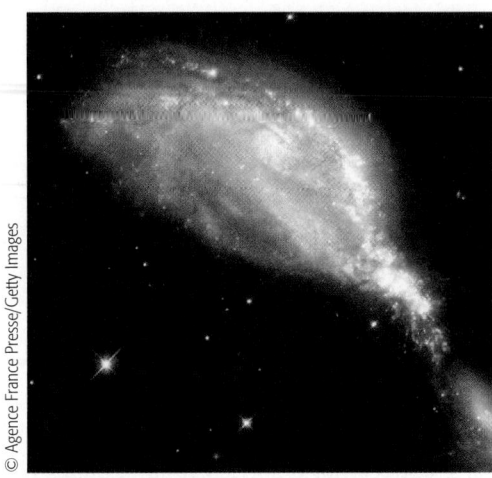

© Agence France Presse/Getty Images

Proxima Centauri

93. Most Populous Countries In 2007, the six most populous countries accounted for 3,347,000,000 people out of the world's total population of 6,600,000,000. The six most populous countries in 2007 are shown in the following graph, along with each country's population.

Six Most Populous Countries (population in millions)

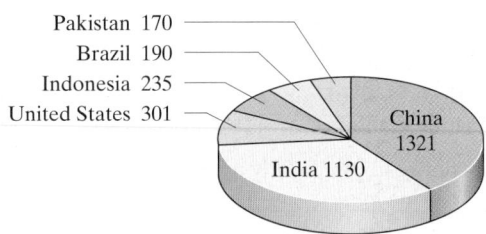

Source: U.S. Census Bureau

Note: China includes mainland China and Taiwan.

a) How many more people lived in China than in the United States?

b) What percent of the world's population lived in China?

c) If the area of China is 3.70×10^6 square miles, determine China's population density (people per square mile).

d) If the area of the United States is 3.62×10^6 square miles, determine the population density in the United States.*

*As of June 2008, the region with the greatest population density was Macau (China), with a population density of 48,459 people per square mile. The country with the greatest population density was Monaco, with a population density of 42,689 people per square mile.

94. World Population The entire span of human history was required for the world's population to reach 6.711×10^9 people in the year 2008. At current rates, the world's population will double in about 62 years.

a) Estimate the world's population in 2070.

b) Assuming 365 days in a year, estimate the average number of additional people added to the world's population each day between 2008 and 2070.

95. Federal Outlays The following graph appeared on page 86 of the 2007 Internal Revenue Service Form 1040 tax booklet. The graph shows the distribution of outlays (expenditures) of the federal government in Fiscal Year (FY) 2006. The total for the outlay of the federal government in FY 2006 was 2.655×10^{12}.

Outlays

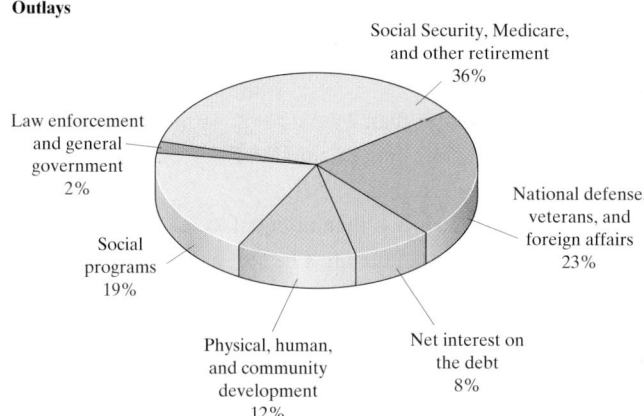

Use this circle graph to answer the following questions. Write all answers in scientific notation.

a) What was the outlay in FY 2006 for law enforcement and general government?

b) What was the outlay in FY 2006 for Social Security, Medicare, and other retirement programs?

c) What was the total outlay in FY 2006 for all programs other than the net interest on the national debt?

96. Football Revenue in the NFL In 2007, the 32 professional football teams in the NFL generated more than $5 billion in revenue. The four teams generating the most revenue were the Washington Redskins, Dallas Cowboys, New England Patriots, and Houston Texans. The total revenue from these four teams was 1.034×10^9. The graph shows the percent distribution of the 1.034×10^9 among these four teams.

Use this graph to answer the following questions.

Four NFL Teams Generating the Most Revenue, (total of 1.034×10^9)

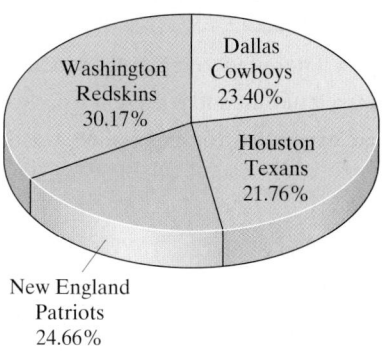

Source: NFL

a) Determine the revenue for the Dallas Cowboys and for the Houston Texans. Express answers in scientific notation.

b) What is the difference in revenue between the Dallas Cowboys and the Houston Texans?

c) If the total revenue from the 32 teams was $5 billion in 2007, what percent of the total revenue did these 4 teams have? Express your answer to the nearest percent.

97. Land Area The land area, in square kilometers, for the five largest countries on our planet is given in the graph below.

Land Area (in millions of square kilometers) of the Five Largest Countries

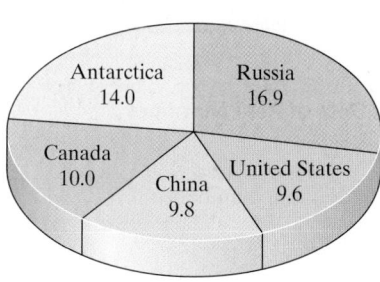

Source: www.world-gazetteer.com

a) What is the total land area of the five largest countries? Write the answer in scientific notation.

b) How much more land area is there in Antarctica than in the United States? Write the answer in scientific notation.

Challenge Problem

98. Light-Year A *light-year* is the distance that light travels in 1 year.

a) Find the number of miles in a light-year if light travels at 1.86×10^5 miles *per second*.

b) If Earth is 93,000,000 miles from the sun, how long does it take for light from the sun to reach Earth?

c) Our galaxy, the Milky Way, is about 6.25×10^{16} miles across. If a spacecraft could travel at half the speed of light, how long would it take for the craft to travel from one end of the galaxy to the other?

Chapter 1 Summary

IMPORTANT FACTS AND CONCEPTS	EXAMPLES

Section 1.2

A **variable** is a letter used to represent various numbers.	x and y are commonly used for variables.
A **constant** is a letter used to represent a particular value.	If h is the number of hours in a day, then $h = 24$, a constant.
An **algebraic expression** (or **expression**) is any combination of numbers, variables, exponents, mathematical symbols, and operations.	$3x^2(x - 2) + 2x$ is an algebraic expression.

A **set** is a collection of objects. The objects are called **elements**.	If $A = \{$blue, green, red$\}$, then blue, green, and red are elements of A.
Roster form is a set having elements listed inside a pair of braces.	
A first set is a **subset** of a second set when every element of the first set is also an element of the second set.	$\{1, 3, 5\}$ is a subset of $\{1, 2, 3, 4, 5\}$
Null set, or **empty set**, symbolized $\{\ \}$ or $\varnothing$, has no elements.	The set of living people over 200 years of age is an empty set.

Inequality Symbols

$>$ is read "is greater than."	$6 > 2$ is read 6 is greater than 2
$\geq$ is read "is greater than or equal to."	$5 \geq 5$ is read 5 is greater than or equal to 5
$<$ is read "is less than."	$-4 < 3$ is read -4 is less than 3
$\leq$ is read "is less than or equal to."	$-10 \leq -1$ is read -10 is less than or equal to -1
$\neq$ is read "is not equal to."	$-5 \neq 17$ is read -5 is not equal to 17

Inequalities can be graphed on a real number line.

$x > 3$

Set builder notation has the form

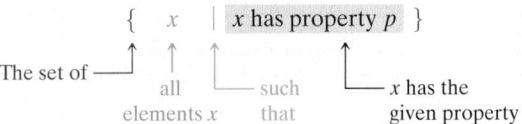

$\{x | -1 \leq x < 2\}$

Important Sets of Real Numbers

Real numbers	$\mathbb{R} = \{x | x$ is a point on a number line$\}$
Natural or counting numbers	$N = \{1, 2, 3, 4, 5, \ldots\}$
Whole numbers	$W = \{0, 1, 2, 3, 4, 5, \ldots\}$
Integers	$I = \{\ldots, -3, -2, -1, 0, 1, 2, 3, \ldots\}$
Rational numbers	$Q = \left\{\dfrac{p}{q} \middle| p \text{ and } q \text{ are integers}, q \neq 0\right\}$
Irrational numbers	$H = \{x | x$ is a real number that is not rational$\}$

The **union** of set A and set B, written $A \cup B$, is the set of elements that belong to either set A *or* set B.	Given $A = \{1, 2, 3, 5, 7\}$ and $B = \{3, 4, 5, 6, 7\}$, then
	$A \cup B = \{1, 2, 3, 4, 5, 6, 7\}$
The **intersection** of set A and set B, written $A \cap B$, is the set of all elements that are common to both set A *and* set B.	$A \cap B = \{3, 5, 7\}$.

Section 1.3

Additive Inverse

For any real number a, its additive inverse is $-a$.	-8 is the additive inverse of 8.

Double Negative Property

For any real number a, $-(-a) = a$.	$-(-5) = 5$

Absolute Value

If a represents any real number, then

$$|a| = \begin{cases} a & \text{if } a \geq 0 \\ -a & \text{if } a < 0 \end{cases}$$

$|9| = 9, \quad |-9| = 9$

IMPORTANT FACTS AND CONCEPTS	EXAMPLES
Section 1.3 (cont.)	

Adding Real Numbers

To add two numbers with the *same sign* (both positive or both negative), add their absolute values and place the common sign before the sum.

Add $-6 + (-8)$.
$$|-6| = 6 \ \text{ and } \ |-8| = 8$$
$$|-6| + |-8| = 6 + 8 = 14$$
Thus, $-6 + (-8) = -14$.

To add two numbers with *different signs* (one positive and the other negative) subtract the smaller absolute value from the larger absolute value. The answer has the sign of the number with the larger absolute value.

Add $8 + (-2)$.
$$8 + (-2) = |8| - |-2|$$
$$= 8 - 2$$
$$= 6$$
Thus, $8 + (-2) = 6$.

Subtracting Real Numbers

$$a - b = a + (-b)$$

$$-14 - 10 = -14 + (-10) = -24$$

Multiplying Real Numbers

To multiply two numbers with *like signs*, either both positive or both negative, multiply their absolute values. The answer is *positive*.

$$(-1.6)(-8.9) = 14.24$$

To multiply two numbers with *unlike signs*, one positive and the other negative, multiply their absolute values. The answer is *negative*.

$$21\left(-\frac{1}{7}\right) = -3$$

Multiplicative Property of Zero

For any number a,
$$a \cdot 0 = 0 \cdot a = 0$$

$$0 \cdot 5 = 0$$

Dividing Two Real Numbers

1. To divide two numbers with *like signs*, either both positive or both negative, divide their absolute values. The answer is *positive*.

$$\frac{-8}{-2} = 4$$

2. To divide two numbers with *unlike signs*, one positive and the other negative, divide their absolute values. The answer is *negative*.

$$\frac{-21}{7} = -3$$

Division By Zero

For any real number $a \neq 0$, then $\frac{a}{0}$ is not defined.

$\frac{7}{0}$ is not defined.

Properties of Real Numbers

For real numbers a, b, and c,

Commutative Property

$$a + b = b + a$$
$$a \cdot b = b \cdot a$$

$$6 + 7 = 7 + 6$$
$$3 \cdot 16 = 16 \cdot 3$$

Associative Property

$$(a + b) + c = a + (b + c)$$
$$(ab)c = a(bc)$$

$$(5 + 4) + 11 = 5 + (4 + 11)$$
$$(8 \cdot 2) \cdot 15 = 8 \cdot (2 \cdot 15)$$

Identity Property

$$a + 0 = 0 + a = a$$
$$a \cdot 1 = 1 \cdot a = a$$

$$31 + 0 = 0 + 31 = 31$$
$$6 \cdot 1 = 1 \cdot 6 = 6$$

Inverse Property

$$a + (-a) = (-a) + a = 0$$
$$a \cdot \frac{1}{a} = \frac{1}{a} \cdot a = 1$$

$$18 + (-18) = -18 + 18 = 0$$
$$14 \cdot \frac{1}{14} = \frac{1}{14} \cdot 14 = 1$$

Distributive Property

$$a(b + c) = ab + ac$$

$$9(x + 10) = 9 \cdot x + 9 \cdot 10$$

IMPORTANT FACTS AND CONCEPTS	EXAMPLES

Section 1.4

Factors are numbers or expressions that are multiplied.	In $3 \cdot 5 = 15$, the 3 and 5 are factors of 15.
For any natural number n, b^n is an exponential expression such that $$b^n = \underbrace{b \cdot b \cdot b \cdot \,\cdots\, \cdot b}_{n \text{ factors}}$$	$(-2)^5 = (-2)(-2)(-2)(-2)(-2) = -32$

Square root of a number $$\sqrt{a} = b \text{ if } b^2 = a$$	$\sqrt{36} = 6$ since $6^2 = 36$
Cube root of a number $$\sqrt[3]{a} = b \quad \text{if} \quad b^3 = a$$	$\sqrt[3]{64} = 4 \quad \text{since} \quad 4^3 = 64$
nth root of a number $$\sqrt[n]{a} = b \quad \text{if} \quad b^n = a$$	$\sqrt[4]{625} = 5 \quad \text{since} \quad 5^4 = 625$

Order of Operations

To evaluate mathematical expressions, use the following order:

1. First, evaluate the expressions within grouping symbols, including parentheses, ( ), brackets, [ ], braces, { }, and absolute value, | |. If the expression contains nested grouping symbols (one pair of grouping symbols within another pair), evaluate the expression in the innermost grouping symbols first.

2. Next, evaluate all terms containing exponents and radicals.

3. Next, evaluate all multiplications or divisions in the order in which they occur, working from left to right.

4. Finally, evaluate all additions or subtractions in the order in which they occur, working from left to right.

Evaluate $4 + 3 \cdot 9^2 - \sqrt{121}$.

$$\begin{aligned} 4 + 3 \cdot 9^2 - \sqrt{121} &= 4 + 3 \cdot 81 - 11 \\ &= 4 + 243 - 11 \\ &= 247 - 11 \\ &= 236 \end{aligned}$$

Section 1.5

Product Rule for Exponents

If m and n are natural numbers and a is any real number, then

$$a^m \cdot a^n = a^{m+n}$$

$x^8 \cdot x^{15} = x^{8+15} = x^{23}$

Quotient Rule for Exponents

If a is any nonzero real number and m and n are nonzero integers, then

$$\frac{a^m}{a^n} = a^{m-n}$$

$\dfrac{z^{21}}{z^{14}} = z^{21-14} = z^7$

Negative Exponent Rule

For any nonzero real number a and any whole number m,

$$a^{-m} = \frac{1}{a^m}$$

$y^{-13} = \dfrac{1}{y^{13}}$

Zero Exponent Rule

If a is any nonzero real number, then

$$a^0 = 1$$

$7x^0 = 7 \cdot 1 = 7$

Raising a Power to a Power (the Power Rule)

If a is a real number and m and n are integers, then

$$(a^m)^n = a^{m \cdot n}$$

$(c^{-8})^{-5} = c^{(-8)(-5)} = c^{40}$

IMPORTANT FACTS AND CONCEPTS	EXAMPLES
Section 1.5 (cont.)	

Raising a Product to a Power

If a and b are real numbers and m is an integer, then

$$(ab)^m = a^m b^m$$

$(8x^6)^2 = 8^2(x^6)^2 = 64x^{12}$

Raising a Quotient to a Power

If a and b are real numbers and m is an integer, then

$$\left(\frac{a}{b}\right)^m = \frac{a^m}{b^m}, \qquad b \neq 0$$

and

$$\left(\frac{a}{b}\right)^{-m} = \left(\frac{b}{a}\right)^m, \quad a \neq 0, b \neq 0$$

$\left(\dfrac{2}{r}\right)^3 = \dfrac{2^3}{r^3} = \dfrac{8}{r^3}$

$\left(\dfrac{6}{x^3}\right)^{-5} = \left(\dfrac{x^3}{6}\right)^5 = \dfrac{(x^3)^5}{6^5} = \dfrac{x^{15}}{7776}$

| **Section 1.6** | |

A number written in **scientific notation** has the form $a \times 10^n$ where $1 \leq a < 10$ and n is an integer.

$5.2 \times 10^7, \quad 1.036 \times 10^{-8}$

To Write a Number in Scientific Notation

1. Move the decimal point in the number to the right of the first nonzero digit.

2. Count the number of places you moved the decimal point in step 1. If the original number is 10 or greater, the count is positive. If the original number is less than 1, the count is negative.

3. Multiply the number obtained in step 1 by 10 raised to the count (power) found in step 2.

$12,900 = 1.29 \times 10^4$

$0.035 = 3.5 \times 10^{-2}$

To Convert a Number in Scientific Notation to Decimal Form

1. Observe the exponent on the base 10.

2. a) If the exponent is positive, move the decimal point in the number to the right the same number of places as the exponent.

   b) If the exponent is negative, move the decimal point in the number to the left the same number of places as the exponent.

$3.08 \times 10^3 = 3080$

$8.76 \times 10^{-4} = 0.000876$

Chapter 1 Review Exercises

[1.2] *List each set in roster form.*

1. $A = \{x \mid x$ is a natural number between 3 and 10$\}$

2. $B = \{x \mid x$ is a whole number multiple of 3$\}$

Let N = set of natural numbers, W = set of whole numbers, I = set of integers, Q = set of rational numbers, H = set of irrational numbers, and $\mathbb{R}$ = set of real numbers. Determine whether the first set is a subset of the second set for each pair of sets.

3. $Q, \mathbb{R}$ **4.** N, W **5.** Q, H **6.** $H, \mathbb{R}$

Consider the set of numbers $\left\{-2, 4, 6, \dfrac{1}{2}, \sqrt{7}, \sqrt{3}, 0, \dfrac{15}{27}, -\dfrac{1}{5}, 1.47\right\}$. *List the elements of the set that are*

7. natural numbers.   **8.** whole numbers.   **9.** integers.

10. rational numbers.   **11.** irrational numbers.   **12.** real numbers.

Indicate whether each statement is true or false.

13. $\frac{0}{1}$ is not a real number.

14. A real number cannot be divided by 0.

15. $0, \frac{3}{5}, -2,$ and 4 are all rational numbers.

16. Every rational number and every irrational number is a real number.

Find $A \cup B$ and $A \cap B$ for each set A and B.

17. $A = \{1, 2, 3, 4, 5, 6\}, B = \{2, 4, 6, 8, 10\}$

18. $A = \{3, 5, 7, 9\}, B = \{2, 4, 6, 8\}$

19. $A = \{1, 3, 5, 7, \ldots\}, B = \{2, 4, 6, 8, \ldots\}$

20. $A = \{4, 6, 9, 10, 11\}, B = \{3, 5, 9, 10, 12\}$

Illustrate each set on the number line.

21. $\{x | x > 5\}$

22. $\{x | x \leq -2\}$

23. $\{x | -1.3 < x \leq 2.4\}$

24. $\left\{ x \left| \frac{2}{3} \leq x < 4 \text{ and } x \in N \right. \right\}$

[1.3] *Insert either $<, >,$ or $=$ in the shaded area between the two numbers to make each statement true.*

25. $-3 \quad 0$

26. $-4 \quad -3.9$

27. $1.06 \quad 1.6$

28. $|-8| \quad 8$

29. $|-4| \quad |-10|$

30. $13 \quad |-9|$

31. $\left| -\frac{2}{3} \right| \quad \frac{3}{5}$

32. $-|-2| \quad -6$

Write the numbers in each list from smallest to largest.

33. $\pi, -\pi, -3, 3$

34. $0, \frac{3}{5}, 2.7, |-3|$

35. $|-10|, |-5|, 3, -2$

36. $|-3|, -7, |-7|, -3$

37. $-4, 6, -|-3|, 5$

38. $|1.6|, |-2.3|, -2, 0$

Name each property illustrated.

39. $-7(x + 5) = -7x - 35$

40. $rs = sr$

41. $(x + 4) + 2 = x + (4 + 2)$

42. $p + 0 = 0$

43. $8(rs) = (8r)s$

44. $-(-6) = 6$

45. $11(0) = 0$

46. $b + (-b) = 0$

47. $x \cdot \frac{1}{x} = 1$

48. $k + l = 1 \cdot (k + l)$

[1.3, 1.4] *Evaluate.*

49. $5 + 3^2 - \sqrt{36} \div 2$

50. $-4 \div (-2) + 16 - \sqrt{81}$

51. $(7 - 9) - (-3 + 5) + 16$

52. $2|-7| - 4|-6| + 7$

53. $(6 - 9) \div (9 - 6) + 3$

54. $|6 - 3| \div 3 + 4 \cdot 8 - 12$

55. $\sqrt{9} + \sqrt[3]{64} + \sqrt[5]{32}$

56. $3^2 - 6 \cdot 9 + 4 \div 2^2 - 15$

57. $4 - (2 - 9)^0 + 3^2 \div 1 + 3$

58. $5^2 + (-2 + 2^2)^3 + 9$

59. $-3^2 + 14 \div 2 \cdot 3 - 8$

60. $\{[(12 \div 4)^2 - 1]^2 \div 16\}^3$

61. $\dfrac{9 + 7 \div (3^2 - 2) + 6 \cdot 8}{\sqrt{81} + \sqrt{1} - 10}$

62. $\dfrac{-(5 - 7)^2 - 3(-2) + |-6|}{18 - 9 \div 3 \cdot 5}$

Evaluate.

63. Evaluate $2x^2 + 3x + 8$ when $x = 2$

64. Evaluate $5a^2 - 7b^2$ when $a = -3$ and $b = -4$.

65. Political Campaigning The cost of political campaigning has changed dramatically since 1952. The amount spent, in millions of dollars, on all U.S. elections—including local, state, and national offices; political parties; political action committees; and ballot issues—is approximated by

$$\text{dollars spent} = 50.86x^2 - 316.75x + 541.48$$

where x represents the number of 4-year periods since 1948. Substitute 1 for x to get the amount spent in 1952, 2 for x to get the amount spent in 1956, 3 for x to get the amount spent in 1960, and so on.

a) Find the amount spent for elections in 1976.

b) Find the amount projected to be spent for elections in 2008.

66. Railroad Traffic Railroad traffic has been increasing steadily since 1965. Most of this is due to the increase in trains used to transport goods by container. We can approximate the amount of freight carried in ton-miles (1 ton-mile equals 1 ton of freight hauled 1 mile) by

$$\text{freight hauled} = 14.04x^2 + 1.96x + 712.05$$

where x represents the number of 5-year periods since 1960. Substitute 1 for x to get the amount hauled in 1965, 2 for x to get the amount hauled in 1970, 3 for x for 1975, and so forth.

a) Find the amount of freight hauled by trains in 1980.

b) Find the amount of freight projected to be hauled by trains in 2010.

[1.5] *Simplify each expression and write the answer without negative exponents.*

67. $2^3 \cdot 2^2$

68. $x^2 \cdot x^3$

69. $\dfrac{a^{12}}{a^4}$

70. $\dfrac{y^{12}}{y^5}$

71. $\dfrac{b^7}{b^{-2}}$

72. $c^3 \cdot c^{-6}$

73. $5^{-2} \cdot 5^{-1}$

74. $8x^0$

75. $(-9m^3)^2$

76. $\left(\dfrac{5}{7}\right)^{-1}$

77. $\left(\dfrac{2}{3}\right)^{-3}$

78. $\left(\dfrac{x}{y^2}\right)^{-1}$

79. $(5xy^3)(-3x^2y)$

80. $(2v^3w^{-4})(7v^{-6}w)$

81. $\dfrac{6x^{-3}y^5}{2x^2y^{-2}}$

82. $\dfrac{12x^{-3}y^{-4}}{4x^{-2}y^5}$

83. $\dfrac{g^3h^{-6}j^{-9}}{g^{-2}h^{-1}j^5}$

84. $\dfrac{21m^{-3}n^{-2}}{7m^{-4}n^2}$

85. $\left(\dfrac{4a^2b}{a}\right)^3$

86. $\left(\dfrac{x^5y}{-3y^2}\right)^2$

87. $\left(\dfrac{p^3q^{-1}}{p^{-4}q^5}\right)^2$

88. $\left(\dfrac{-2ab^{-3}}{c^2}\right)^3$

89. $\left(\dfrac{5xy^3}{z^2}\right)^{-2}$

90. $\left(\dfrac{9m^{-2}n}{3mn}\right)^{-3}$

91. $(-2m^2n^{-3})^{-2}$

92. $\left(\dfrac{15x^5y^{-3}z^{-2}}{-3x^4y^{-4}z^3}\right)^4$

93. $\left(\dfrac{2x^{-1}y^5z^4}{3x^4y^{-2}z^{-2}}\right)^{-2}$

94. $\left(\dfrac{10x^{-2}y^{-2}z}{-x^4y^{-4}z^3}\right)^{-1}$

[1.6] *Express each number in scientific notation.*

95. 0.0000742

96. 460,000

97. 183,000

98. 0.000002

Simplify each expression and express the answer without exponents.

99. $(25 \times 10^{-3})(1.2 \times 10^6)$

100. $\dfrac{21 \times 10^3}{7 \times 10^5}$

101. $\dfrac{6,000,000}{0.02}$

102. $(0.004)(500,000)$

103. Online Adverstising The three companies with the greatest number of people who read their ads in 2007 are listed below.

Company	Number of People
Google	1.107×10^9
Double Click	1.079×10^9
Yahoo	3.62×10^8

a) How many more people read ads on Google than on Double Click?

b) How many more people read ads on Google than on Yahoo?

c) What is the total number of people who read ads on all three companies?

104. Voyager On February 17, 1998, the *Voyager 1* spacecraft became the most distant explorer in the solar system, breaking the record of the *Pioneer 10*. The 28-year-old *Voyager 1* had traveled more than 1.4×10^{10} kilometers from Earth (about 150 times the distance from Earth to the sun).

a) Represent 1.4×10^{10} as an integer.

b) How many billion kilometers had *Voyager 1* traveled?

c) Assuming that *Voyager 1* traveled about the same number of kilometers in each of the 28 years, how many kilometers did it average in a year?

d) If 1 kilometer $\approx$ 0.6 miles, how far, in miles, did *Voyager 1* travel?

Chapter 1 Practice Test

Chapter Test Prep Videos provide fully worked-out solutions to any of the exercises you want to review. Chapter Test Prep Videos are available via **MyMathLab**, *or on* **You Tube** *(search "Angel Intermediate Algebra" and click on "Channels").*

1. List $A = \{x|x$ is a natural number greater than or equal to 6$\}$ in roster form.

Indicate whether each statement is true or false.

2. Every real number is a rational number.

3. The union of the set of rational numbers and the set of irrational numbers is the set of real numbers.

Consider the set of numbers
$$\left\{-\frac{3}{5}, 2, -4, 0, \frac{19}{12}, 2.57, \sqrt{8}, \sqrt{2}, \;1.92\right\}, \text{ List the elements of}$$
the set that are

4. rational numbers.

5. real numbers.

Find $A \cup B$ and $A \cap B$ for sets A and B.

6. $A = \{8, 10, 11, 14\}, B = \{5, 7, 8, 9, 10\}$

7. $A = \{1, 3, 5, 7, \ldots\}, B = \{3, 5, 7, 9, 11\}$

In Exercises 8 and 9, illustrate each set on the number line.

8. $\{x| -2.3 \le x < 5.2\}$

9. $\left\{x\left| -\frac{5}{2} < x < \frac{6}{5} \text{ and } x \in I\right.\right\}$

10. List from smallest to largest: $|3|, -|4|, -2, 9$.

Name each property illustrated.

11. $(x + y) + 8 = x + (y + 8)$

12. $3x + 4y = 4y + 3x$

Evaluate each expression.

13. $\{6 - [7 - 3^2 \div (3^2 - 2 \cdot 3)]\}$

14. $2^4 + 4^2 \div 2^3 \cdot \sqrt{25} + 7$

15. $\dfrac{-3|4 - 8| \div 2 + 6}{-\sqrt{36} + 18 \div 3^2 + 4}$

16. $\dfrac{-6^2 + 3(4 - |6|) \div 6}{4 - (-3) + 12 \div 4 \cdot 5}$

17. Evaluate $-x^2 + 2xy + y^2$ when $x = 2$ and $y = 3$.

18. **Cannonball** To celebrate July 4, a cannon is fired upwards from a fort overlooking the ocean below. The height, h, in feet, of the cannonball above sea level at any time t, in seconds, can be determined by the formula $h = -16t^2 + 120t + 200$. Find the height of the cannonball above sea level **a)** 1 second after the cannon is fired, **b)** 5 seconds after the cannon is fired.

Simplify each expression and write the answer without negative exponents

19. 3^{-2}

20. $\left(\dfrac{4m^{-3}}{n^2}\right)^2$

21. $\dfrac{24a^2b^{-3}c^0}{30a^3b^2c^{-2}}$

22. $\left(\dfrac{-3x^3y^{-2}}{x^{-1}y^5}\right)^{-3}$

23. Convert 389,000,000 to scientific notation.

24. Simplify $\dfrac{3.12 \times 10^6}{1.2 \times 10^{-2}}$ and write the number without exponents.

25. **World Population**

a) In 2050, the world population is expected to be about 9.2 billion people. Write this number in scientific notation.

b) The graph below shows the expected distribution of the 2050 world population for the three age groups 0–14, 15–64, and 65 and older. Use scientific notation to determine the number of people in each of these age groups in 2050.

Expected Distribution of World Population by Age

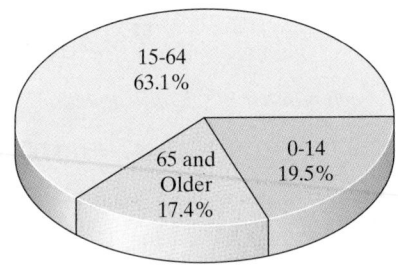

Equations and Inequalities

Goals for This Chapter

In this chapter, we focus on solving linear equations and inequalities, and on using linear equations, formulas, and inequalities to solve real-life problems. We will also introduce a powerful problem-solving technique that we will use throughout this textbook. We will witness the power of algebra as a problem-solving tool in a multitude of areas, including real estate, chemistry, business, banking, physics, and personal finance.

© killerb10/iStockphoto

A big part of the American dream is to own your own home. Many factors are involved in the cost of obtaining a loan to purchase a house. These factors will vary between different lenders. In Example 8 on page 89, we will compare the loan costs offered by two different lenders.

2.1 Solving Linear Equations

1 Identify the Reflexive, Symmetric, and Transitive Properties

We begin by reviewing how to solve linear equations. First, we introduce three properties of equality.

Properties of Equality

For all real numbers a, b, and c:

1. $a = a$.	Reflexive property
2. If $a = b$, then $b = a$.	Symmetric property
3. If $a = b$ and $b = c$, then $a = c$.	Transitive property

Examples of the Reflexive Property

$$7 = 7$$

$$x + 5 = x + 5$$

Examples of the Symmetric Property

If $x = 3$, then $3 = x$.

If $y = x + 9$, then $x + 9 = y$.

Examples of the Transitive Property

If $x = a$ and $a = 4y$, then $x = 4y$.

If $a + b = c$ and $c = 4d$, then $a + b = 4d$.

We will often use these properties without referring to them by name.

2 Combine Like Terms

When an algebraic expression consists of several parts, the parts that are added are called the **terms** of the expression.

The expression $3x^2 - 6x - 2$

which can be written as: $\underbrace{3x^2}_{\text{term}} + \underbrace{(-6x)}_{\text{term}} + \underbrace{(-2)}_{\text{term}}$

has three terms.

Expression	Terms
$\frac{1}{2}x^2 - 3x - 7$	$\frac{1}{2}x^2$, $\;-3x$, $\;-7$
$-5x^3 + 3x^2y - 2$	$-5x^3$, $\;3x^2y$, $\;-2$
$4(x + 3) + 2x + \frac{1}{5}(x - 2) + 1$	$4(x + 3)$, $\;2x$, $\;\frac{1}{5}(x - 2)$, $\;1$

The numerical part of a term that precedes the variable is called its **numerical coefficient** or simply its **coefficient**. In the term $6x^2$, the 6 is the numerical coefficient.

Term	Numerical Coefficient
$x = 1 \cdot x$	1
$-a^2 = -1 \cdot a^2$	-1
$\dfrac{5k}{9} = \dfrac{5}{9}k$	$\dfrac{5}{9}$
$\dfrac{-6xyz}{7} = -\dfrac{6}{7} \cdot xyz$	$-\dfrac{6}{7}$
$8 = 8x^0$	8

When a term consists of only a number, that number is called a **constant**. For example, in the expression $x^2 - 4$, the -4 is a constant.

The **degree of a term** with whole number exponents is the sum of the exponents on the variables in the term. For example, $3x^2$ is a second-degree term, and $-4x$ is a first-degree term

Term	Degree
x^2	2
$3x = 3x^1$	1
$6 = 6x^0$	0
$4xy^5 = 4x^1y^5$	$1 + 5 = 6$
$6x^3y^5$	$3 + 5 = 8$

Like terms are terms that have the same variables with the same exponents. For example, $3x$ and $5x$ are like terms, $2x^2$ and $-3x^2$ are like terms, and $3x^2y$ and $-2x^2y$ are like terms. Terms that are not like terms are said to be **unlike terms**. All constants are considered like terms.

To **simplify an expression** means to combine all like terms in the expression. To combine like terms, we can use the distributive property.

Examples of Combining Like Terms

$8x - 2x = (8 - 2)x = 6x$

$3x^2 - 5x^2 = (3 - 5)x^2 = -2x^2$

$-7x^2y + 3x^2y = (-7 + 3)x^2y = -4x^2y$

When simplifying expressions, we rearrange the terms by using the commutative and associative properties.

EXAMPLE 1 Simplify by combining like terms.

a) $-2x + 5 + 3x - 7$ **b)** $7x^2 - 2x^2 + 3x + 4$ **c)** $2x - 3y + 4 + 5x - 6y - 3$

Solution

a) $-2x + 5 + 3x - 7 = \underbrace{-2x + 3x}_{x} + \underbrace{5 - 7}_{-2}$ Place like terms together.

This expression simplifies to $x - 2$.

b) $7x^2 - 2x^2 + 3x + 4 = 5x^2 + 3x + 4$

c) $2x - 3y + 4 + 5x - 6y - 3 = 2x + 5x - 3y - 6y + 4 - 3$ Place like terms together.

$= 7x - 9y + 1$

Now Try Exercise 39

EXAMPLE 2 Simplify $-2(a + 7) - [-3(a - 1) + 8]$.

Solution

$$
\begin{aligned}
-2(a + 7) - [-3(a - 1) + 8] &= -2(a + 7) - 1[-3(a - 1) + 8] \\
&= -2a - 14 - 1[-3a + 3 + 8] \quad \text{Distributive property} \\
&= -2a - 14 - 1[-3a + 11] \quad \text{Combine like terms.} \\
&= -2a - 14 + 3a - 11 \quad \text{Distributive property} \\
&= a - 25 \quad \text{Combine like terms.}
\end{aligned}
$$

Now Try Exercise 55

3 Solve Linear Equations

Understanding Algebra

Every equation must contain an equals sign.

Equation

An **equation** is a mathematical statement of equality. *An equation must contain an equals sign and a mathematical expression on each side of the equals sign.*

Examples of Equations

$$x + 8 = -7$$
$$2x^2 - 4 = -3x + 13$$

The numbers that make an equation a true statement are called the **solutions** of the equation. The **solution set** of an equation is the set of real numbers that make the equation true.

Equation	Solution	Solution Set
$2x + 3 = 9$	3	$\{3\}$

Two or more equations with the same solution set are called **equivalent equations**. Equations are generally solved by starting with the given equation and producing a series of simpler equivalent equations.

Example of Equivalent Equations

Equations	Solution Set
$2x + 3 = 9$	$\{3\}$
$2x = 6$	$\{3\}$
$x = 3$	$\{3\}$

In this section, we will discuss how to solve **linear equations in one variable**. A linear equation is an equation that can be written in the form $ax + b = c, a \neq 0$.

Understanding Algebra

To solve linear equations, we use the addition and multiplication properties of equality to *isolate the variable* on one side of the equation.

Addition Property of Equality

If $a = b$, then $a + c = b + c$ for any $a, b,$ and c.

The addition property of equality states that the same number can be added to or subtracted from both sides of an equation without changing the solution to the original equation.

Multiplication Property of Equality

If $a = b$, then $a \cdot c = b \cdot c$ for any $a, b,$ and c.

The multiplication property of equality states that both sides of an equation can be multiplied by or divided by the same nonzero number without changing the solution.

Our goal when solving equations is to get the variable on one side of the equation by itself, or to *isolate the variable*.

To Solve Linear Equations

1. **Clear fractions.** If the equation contains fractions, eliminate the fractions by multiplying both sides of the equation by the least common denominator.
2. **Simplify each side separately.** Simplify each side of the equation as much as possible. Use the distributive property to clear parentheses and combine like terms as needed.
3. **Isolate the variable term on one side.** Use the addition property to get all terms with the variable on one side of the equation and all constant terms on the other side. It may be necessary to use the addition property a number of times to accomplish this.
4. **Solve for the variable.** Use the multiplication property to get the variable (with a coefficient of 1) on one side.
5. **Check.** Check by substituting the value obtained in step 4 back into the original equation.

EXAMPLE 3 Solve the equation $2x + 9 = 14$.

Solution

$$2x + 9 = 14$$
$$2x + 9 \;-9 = 14 \;-9 \qquad \text{Subtract 9 from both sides.}$$
$$2x = 5$$
$$\frac{\overset{1}{\cancel{2}}x}{\underset{1}{\cancel{2}}} = \frac{5}{2} \qquad \text{Divide both sides by 2.}$$
$$x = \frac{5}{2}$$

Check

$$2x + 9 = 14$$
$$\cancel{2}\left(\frac{5}{\cancel{2}}\right) + 9 \overset{?}{=} 14$$
$$5 + 9 \overset{?}{=} 14$$
$$14 = 14 \qquad \text{True}$$

Since the value checks, the solution is $\frac{5}{2}$.

Now Try Exercise 61

EXAMPLE 4 Solve the equation $-2b + 8 = 3b - 7$.

Solution

$$-2b + 8 = 3b - 7$$
$$-2b + 2b + 8 = 3b + 2b - 7 \qquad \text{Add } 2b \text{ to both sides.}$$
$$8 = 5b - 7$$
$$8 + 7 = 5b - 7 + 7 \qquad \text{Add 7 to both sides.}$$
$$15 = 5b$$
$$\frac{15}{5} = \frac{5b}{5} \qquad \text{Divide both sides by 5.}$$
$$3 = b$$

Now Try Exercise 63

Example 5 contains decimal numbers. We work this problem following the procedure given earlier. Also, whenever an equation contains like terms on the same side of the equation, combine the like terms before using the addition or multiplication properties.

EXAMPLE 5 Solve the equation $4(x - 3.1) = 2.1(x - 4) + 3.5x$.

Solution

$$4(x - 3.1) = 2.1(x - 4) + 3.5x$$

$$4(x) - 4(3.1) = 2.1(x) - 2.1(4) + 3.5x \qquad \text{Distributive property}$$

$$4x - 12.4 = 2.1x - 8.4 + 3.5x$$

$$4x - 12.4 = 5.6x - 8.4 \qquad \text{Combine like terms.}$$

$$4x - 12.4 + 8.4 = 5.6x - 8.4 + 8.4 \qquad \text{Add 8.4 to both sides.}$$

$$4x - 4.0 = 5.6x$$

$$4x - 4x - 4.0 = 5.6x - 4x \qquad \text{Subtract } 4x \text{ from both sides.}$$

$$-4.0 = 1.6x$$

$$\frac{-4.0}{1.6} = \frac{1.6x}{1.6} \qquad \text{Divide both sides by 1.6.}$$

$$-2.5 = x$$

The solution is -2.5.

Now Try Exercise 111

To save space, we will not always show the check of our answers. You should, however, check all your answers. When the equation contains decimal numbers, using a calculator to solve and check the equation may save you some time.

Using Your Calculator

Checking Solutions by Substitution

To check solutions using a calculator, substitute the solution into both sides of the equation to see whether you get the same value. The graphing calculator screen in **Figure 2.1** shows that both sides of the equation given in Example 5 equal -22.4 when -2.5 is substituted for x. Thus the solution -2.5 checks.

```
4(-2.5-3.1)
            -22.4
2.1(-2.5-4)+3.5*
-2.5
            -22.4
```

← Value of the left side of the equation

← Value of the right side of the equation

$$4(x - 3.1) = 2.1(x - 4) + 3.5x$$

$$4(-2.5 - 3.1) = 2.1(-2.5 - 4) + 3.5(-2.5)$$

FIGURE 2.1

Now we will work an example that contains nested grouping symbols.

EXAMPLE 6 Solve the equation $7c - 15 = -2[6(c - 3) - 4(2 - c)]$.

Solution

$$7c - 15 = -2[6(c - 3) - 4(2 - c)]$$

$$7c - 15 = -2[6c - 18 - 8 + 4c] \qquad \text{Distributive property}$$

$$7c - 15 = -2[10c - 26] \qquad \text{Combine like terms.}$$

$$7c - 15 = -20c + 52 \qquad \text{Distributive property}$$

$$7c + 20c - 15 = -20c + 20c + 52 \qquad \text{Add } 20c \text{ to both sides.}$$
$$27c - 15 = 52$$
$$27c - 15 + 15 = 52 + 15 \qquad \text{Add } 15 \text{ to both sides.}$$
$$27c = 67$$
$$\frac{27c}{27} = \frac{67}{27} \qquad \text{Divide both sides by } 27.$$
$$c = \frac{67}{27}$$

Now Try Exercise 91

In solving equations, some of the intermediate steps can often be omitted. Now we will illustrate how this may be done.

Solution

a) $x + 4 = 6$
$x + 4 - 4 = 6 - 4$ ⟵ Do this step mentally.
$x = 2$

b) $3x = 6$
$\dfrac{3x}{3} = \dfrac{6}{3}$ ⟵ Do this step mentally.
$x = 2$

Shortened Solution

a) $x + 4 = 6$
$x = 2$

b) $3x = 6$
$x = 2$

4 Solve Equations Containing Fractions

When an equation contains fractions, we begin by multiplying *both* sides of the equation by the least common denominator.

> ### Least Common Denominator
>
> The **least common denominator** (LCD) of a set of denominators is the smallest number that each of the denominators divides into without remainder.

For example, if the denominators of two fractions are 4 and 6, then 12 is the least common denominator since 12 is the smallest number that both 4 and 6 divide into without remainder.

EXAMPLE 7 Solve the equation $5 - \dfrac{2a}{3} = -9$.

Solution   The least common denominator is 3. Multiply both sides of the equation by 3 and then use the distributive property on the left side of the equation. *This process will eliminate all fractions from the equation.*

> ### Understanding Algebra
>
> After multiplying both sides of an equation by the LCD, the equation should not contain any fractions.

$$5 - \frac{2a}{3} = -9$$
$$3\left(5 - \frac{2a}{3}\right) = 3(-9) \qquad \text{Multiply both sides by } 3.$$
$$3(5) - \overset{1}{\cancel{3}}\left(\frac{2a}{\underset{1}{\cancel{3}}}\right) = -27 \qquad \text{Distributive property}$$
$$15 - 2a = -27$$
$$15 - 15 - 2a = -27 - 15 \qquad \text{Subtract 15 from both sides.}$$
$$-2a = -42$$
$$\frac{-2a}{-2} = \frac{-42}{-2} \qquad \text{Divide both sides by } -2.$$
$$a = 21$$

Now Try Exercise 97

EXAMPLE 8 Solve the equation $\frac{1}{2}(x + 4) = \frac{1}{3}x$.

Solution Begin by multiplying both sides of the equation by 6, the LCD of 2 and 3.

$$6\left[\frac{1}{2}(x + 4)\right] = 6\left(\frac{1}{3}x\right) \qquad \text{Multiply both sides by 6.}$$

$$3(x + 4) = 2x \qquad \text{Simplify.}$$

$$3x + 12 = 2x \qquad \text{Distributive property}$$

$$3x - 2x + 12 = 2x - 2x \qquad \text{Subtract } 2x \text{ from both sides.}$$

$$x + 12 = 0$$

$$x + 12 - 12 = 0 - 12 \qquad \text{Subtract 12 from both sides.}$$

$$x = -12$$

Now Try Exercise 99

We will be discussing equations containing fractions further in Section 6.4.

5 Identify Conditional Equations, Contradictions, and Identities

All equations discussed so far have been true for only specific values of the variable. Such equations are called **conditional equations**. Some equations are never true and have no solution; these are called **contradictions**. Other equations, called **identities** are always true and have an infinite number of solutions. **Table 2.1** summarizes these types of linear equations and their corresponding number of solutions.

TABLE 2.1

Type of linear equation	Number of solutions
Conditional equation	One
Contradiction	None (solution set: $\varnothing$)
Identity	Infinite number (solution set: $\mathbb{R}$)

The solution set of a conditional equation contains the solution given in set braces. For example, the solution set to Example 8 is $\{-12\}$. The solution set of a contradiction is the empty set or null set, $\{\ \}$ or $\varnothing$. The solution set of an identity is the set of real numbers, $\mathbb{R}$.

EXAMPLE 9 Determine whether the equation $5(a - 3) - 3(a - 6) = 2(a + 1) + 1$ is a conditional equation, a contradiction, or an identity. Give the solution set for the equation.

Solution

$$5(a - 3) - 3(a - 6) = 2(a + 1) + 1$$

$$5a - 15 - 3a + 18 = 2a + 2 + 1 \qquad \text{Distributive property}$$

$$2a + 3 = 2a + 3 \qquad \text{Combine like terms.}$$

Since we obtain the same expression on both sides of the equation, it is an identity. This equation is true for all real numbers. The solution set is $\mathbb{R}$.

Now Try Exercise 125

In Example 9, had we continued solving the equation and subtracted $2a$ from both sides, we would obtain the equation $3 = 3$. This equation also is an identity and also indicates that the solution set is $\mathbb{R}$.

EXAMPLE 10 Determine whether $2(3m + 1) = 6m + 3$ is a conditional equation, a contradiction, or an identity. Give the solution set for the equation.

Solution

$$2(3m + 1) = 6m + 3$$
$$6m + 2 = 6m + 3 \qquad \text{Distributive property}$$
$$6m - 6m + 2 = 6m - 6m + 3 \qquad \text{Subtract } 6m \text{ from both sides.}$$
$$2 = 3$$

Since $2 = 3$ is never a true statement, this equation is a contradiction. Its solution set is $\varnothing$.

Now Try Exercise 119

> **Understanding Algebra**
>
> If while solving an equation you get an equation with no variables, then the original equation is either an identity or a contradiction. For example $3 = 3$ is an identity and means the solution set is $\mathbb{R}$; $2 = 3$ is a contradiction and means the solution set is $\varnothing$.

6   Understand the Concepts to Solve Equations

The numbers or variables that appear in equations do not affect the procedures used to solve the equations. In the following example we will solve the equation using the concepts and procedures that have been presented.

EXAMPLE 11 Assume in the following equation that $\odot$ represents the variable that we are solving for and that all the other symbols represent nonzero real numbers. Solve the equation for $\odot$.

$$\square \odot + \triangle = \#$$

Solution   To solve for $\odot$ we need to isolate the $\odot$. We use the addition and multiplication properties to solve for $\odot$.

$$\square \odot + \triangle = \#$$
$$\square \odot + \triangle - \triangle = \# - \triangle \qquad \text{Subtract } \triangle \text{ from both sides.}$$
$$\square \odot = \# - \triangle$$
$$\frac{\square \odot}{\square} = \frac{\# - \triangle}{\square} \qquad \text{Divide both sides by } \square.$$
$$\odot = \frac{\# - \triangle}{\square}$$

Thus the solution is $\odot = \dfrac{\# - \triangle}{\square}$.

Now Try Exercise 133

Consider the equation $5x + 7 = 12$. If we let $5 = \square$, $x = \odot$, $7 = \triangle$, and $12 = \#$, the equation has the same form as the equation in Example 11. Therefore, the solution will be of the same form.

Equation	Solution
$\square \odot + \triangle = \#$	$\odot = \dfrac{\# - \triangle}{\square}$
$5x + 7 = 12$	$x = \dfrac{12 - 7}{5} = \dfrac{5}{5} = 1$

If you solve the equation $5x + 7 = 12$, you will see that its solution is 1.

EXERCISE SET 2.1

Math XL
MathXL®

MyMathLab
MyMathLab

Fill in the blanks with the appropriate word, phrase, or symbol(s) from the following list.

conditional   Ø   like terms   identity   least common denominator $\mathbb{R}$

terms   contradiction   degree   unlike terms   isolate

1. The parts that are added in an algebraic expression are called the _____ of the expression.

2. Terms that have identical variable parts are called _____.

3. The goal in solving equations is to _____ the variable on one side of the equation.

4. We can eliminate fractions from an equation by multiplying both sides of the equation by the _____.

5. An equation that is always true is known as a(n) _____.

6. An equation that is true for only specific values of the variable is known as a(n) _____ equation.

7. An equation that is never true is known as a(n) _____.

8. The _____ of a term is the sum of the exponents on the variables in the term.

9. The symbol _____ is used to indicate the solution set of a contradiction.

10. The symbol _____ is used to indicate the solution set of an identity.

Practice the Skills

Name each indicated property.

11. If $x = 13$, then $13 = x$.

12. If $m + 2 = 3$, then $3 = m + 2$.

13. If $b = c$ and $c = 9$, then $b = 9$.

14. If $x + 1 = a$ and $a = 2y$, then $x + 1 = 2y$.

15. $a + c = a + c$

16. If $r = 4$, then $r + 3 = 4 + 3$.

17. If $x = 8$, then $x - 8 = 8 - 8$.

18. If $2x = 4$, then $3(2x) = 3(4)$.

19. If $5x = 4$, then $\frac{1}{5}(5x) = \frac{1}{5}(4)$.

20. If $a + 2 = 4$, then $a + 2 - 2 = 4 - 2$.

21. If $\frac{t}{4} + \frac{1}{3} = \frac{5}{6}$, then $12\left(\frac{t}{4} + \frac{1}{3}\right) = 12\left(\frac{5}{6}\right)$.

22. If $x - 3 = x + y$ and $x + y = z$, then $x - 3 = z$.

23. If $x + 3 = 7$, then $x = 4$.

24. If $5x = 35$ then $x = 7$.

Give the degree of each term.

25. $5y$

26. $-2z$

27. $5c^3$

28. $-6y^2$

29. $3ab$

30. $\frac{1}{2}x^4y$

31. 6

32. -3

33. $-5r$

34. $18p^2q^3$

35. $5a^2b^4c$

36. m^4n^6

37. $3x^5y^6z$

38. $-2x^4y^7z^8$

Simplify each expression. If an expression cannot be simplified, so state.

39. $7r + 3b - 11x + 12y$

40. $3x^2 + 4x + 5$

41. $-2x^2 - 5x + 7x - 3$

42. $2a^2 - 4ab + 5ab - 10b^2$

43. $10.6c^2 - 2.3c + 5.9c - 1.9c^2$

44. $7y + 3x - 7 + 5x - 2y$

45. $w^3 + w^2 - w + 1$

46. $b + b^2 - 4b + b^2 + 3b$

47. $8pq - 9pq + p + q$

48. $7x^3y^2 + 11y^3x^2$

49. $12\left(\frac{1}{6} + \frac{d}{4}\right) + 5d$

50. $4.3 - 3.2x - 2(x - 2)$

51. $3\left(x + \frac{1}{2}\right) - \frac{1}{3}x + 5$

52. $6n + 0.6(n - 3) - 5(n + 0.7)$

53. $4 - [6(3x + 2) - x] + 4$

54. $3(a + c) - 4(a + c) - 3$

55. $9x - [3x - (5x - 4y)] - 2y$

56. $-2[3x - (2y - 1) - 5x] + y$

57. $5b - \{7[2(3b - 2) - (4b + 9)] - 2\}$

58. $2\{[3a - (2b - 5a)] - 3(2a - b)\}$

59. $-\{[2rs - 3(r + 2s)] - 2(2r^2 - s)\}$

60. $p^2q + 4pq - [-(pq + 4p^2q) + pq]$

Solve each equation.

61. $5a - 1 = 14$

62. $7x - 6 - 5x = -8$

63. $4x - 5 = 2(x + 5)$

64. $5s - 3 = 2s + 6$

65. $4x - 8 = -4(2x - 3) + 4$

66. $8w + 7 = -3w - 15$

67. $-6(z - 1) = -5(z + 2)$

68. $7(x - 1) = 3(x + 2)$

69. $-3(t - 5) = 2(t - 5)$

70. $4(2x - 4) = -2(x + 3)$

71. $3x + 4(2 - x) = 4x + 5$

72. $6(3 - q) = -4(q + 1)$

73. $2 - (x + 5) = 4x - 8$

74. $4x - 2(3x - 7) = 2x - 6$

75. $p - (p + 4) = 4(p - 1) + 2p$

76. $8x + 2(x - 4) = 8x + 12$

77. $-3(y - 1) + 2y = 4(y - 3)$

78. $5r - 13 - 6r = 3(r + 5) - 16$

79. $6 - (n + 3) = 3n + 5 - 2n$

80. $8 - 3(2a - 4) = 5 + 3a - 4a$

81. $4(2x - 2) - 3(x + 7) = -4$

82. $-2(3w + 6) - (4w - 3) = 21$

83. $-4(3 - 4x) - 2(x - 1) = 12x$

84. $-4(2z - 6) = -3(z - 4) + z$

85. $5(a + 3) - a = -(4a - 6) + 1$

86. $3(2x - 4) + 3(x + 1) = 9$

87. $5(x - 2) - 14x = x - 5$

88. $3[6 - (h + 2)] - 6 = 4(-h + 7)$

89. $2[3x - (4x - 6)] = 5(x - 6)$

90. $-z - 6z + 3 = 4 - [6 - z - (3 - 2z)]$

91. $4\{2 - [3(c + 1) - 2(c + 1)]\} = -2c$

92. $3\{[(x - 2) + 4x] - (x - 3)\} = 4 - (x - 12)$

93. $-\{4(d + 3) - 5[3d - 2(2d + 7)] - 8\} = -10d - 6$

94. $-3(6 - 4x) = 4 - \{5x - [6x - (4x - (3x + 2))]\}$

Solve each equation. Leave your answer as a fraction if it is not an integer value.

95. $\dfrac{d}{5} = -7$

96. $\dfrac{7m + 9}{6} = 5$

97. $\dfrac{4x - 2}{3} = -6$

98. $\dfrac{1}{2}(6r - 10) = 7$

99. $\dfrac{3}{4}t + \dfrac{7}{8}t = 39$

100. $\dfrac{1}{4}(x - 2) = \dfrac{1}{3}(2x + 6)$

101. $\dfrac{1}{2}(x - 2) = \dfrac{1}{3}(x + 2)$

102. $\dfrac{1}{2}x + 2 = \dfrac{3}{8}x - 1$

103. $4 - \dfrac{3}{4}a = 7$

104. $x - 2 = \dfrac{3}{4}(x + 4)$

105. $\dfrac{1}{2} = \dfrac{4}{5}x - \dfrac{1}{4}$

106. $\dfrac{1}{3}x + \dfrac{5}{6} = 2x$

107. $\dfrac{1}{4}(x + 3) = \dfrac{1}{3}(x - 2) + 1$

108. $\dfrac{5}{6}m - \dfrac{5}{12} = \dfrac{7}{8}m + \dfrac{2}{3}$

Solve each equation. Round answers to the nearest hundredth.

109. $0.4n + 4.7 = 5.1n$

110. $0.2(x - 30) = 1.6x$

111. $4.7x - 3.6(x - 1) = 4.9$

112. $6.1p - 4.5(3 - 2p) = 15.7$

113. $5(z + 3.41) = -7.89(2z - 4) - 5.67$

114. $0.05(2000 + 2x) = 0.04(2500 - 6x)$

115. $0.6(500 - 2.4x) = 3.6(2x - 4000)$

116. $0.42x - x = 5.1(x + 3)$

117. $1000(7.34q + 14.78) = 100(3.91 - 4.21q)$

118. $0.6(14x - 8000) = -0.4(20x + 12{,}000) + 20.6x$

Find the solution set for each exercise. Then indicate whether the equation is conditional, an identity, or a contradiction.

119. $3(y + 3) - 4(2y - 7) = -5y + 2$

120. $7x + 5 - 5(x - 3) = 5(x + 4) - 3x$

121. $7 + 3(x - 2) + 8x = 6(x + 1) + 2x - 9$

122. $-5(c + 3) + 4(c - 2) = 2(c + 2)$

123. $4 - \left(\dfrac{2}{3}x + 2\right) = 2\left(-\dfrac{1}{3}x + 1\right)$

124. $7 - \left(\dfrac{1}{2}x + 4\right) = 3\left(-\dfrac{1}{6}x + 2\right)$

125. $6(x - 1) = -3(2 - x) + 3x$

126. $0.6(z + 5) - 0.5(z + 2) = 0.1(z - 23)$

127. $0.8z - 0.3(z + 10) = 0.5(z + 1)$

128. $4(2 - 3x) = -[6x - (8 - 6x)]$

Problem Solving

129. Population Density The population density of the United States has been steadily increasing since 2000. The population density of the United States can be estimated using the equation

$$P = 0.82t + 78.5$$

where P is the population density, measured in people per square mile, and t is the number of years since 2000. Use $t = 1$ for 2001, $t = 2$ for 2002, and so on. If the population density continues to increase at its current rate,

a) determine the population density of the United States in 2008.

b) during what year will the population density of the United States reach 100 people per square mile?

130. Sleeping Babies Dr. Richard Ferber, a pediatric sleep expert, has developed a method[*] to help children, 6 months of age or older, sleep through the night. Often called "Ferberizing," it calls for parents to wait for increasing lengths of time before entering the child's room at night to comfort the crying child. The suggested waiting time depends on

---

[*]Before trying this method, parents should first consult with their pediatrician.

how many nights the parents have been using the method and may be found using the equation

$$W = 5n + 5$$

where W is the waiting time in minutes and n is the number of the night. For example, on the first night, $n = 1$, on the second night, $n = 2$, and so on.

a) How long should parents wait on the first night?

b) How long should parents wait on the fourth night?

c) On what night should parents wait 30 minutes?

d) On what night should parents wait 40 minutes?

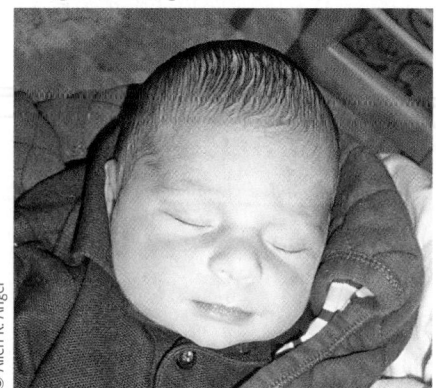

© Allen R. Angel

131. Rising Health Care Costs Health care spending in the United States is projected to grow approximately according to the equation $C = 0.2x + 2.8$, where C represents the total amount spent on health care in trillions of dollars and x represents the years since 2008. Use $x = 1$ for 2009, $x = 2$ for 2010, and so on.

Source: National Coalition on Health Care

a) How much was spent on health care in the United States in 2009?

b) If this trend continues, during what year will health care spending reach $4 trillion?

132. Aging Population The percentage of the American population that is older than age 65 is projected to grow according to the equation $P = 1.5x + 38.7$. In this equation P represents the percentage of the American population older than 65 and x represents the years since 2008. Use $x = 1$ for 2009, $x = 2$ for 2010, and so on.

Source: National Academy of Sciences

© Christian Wheatley/Shutterstock

a) What is the percentage of Americans older than age 65 in 2009?

b) During what year is the percentage of Americans older than age 65 projected to reach 50%?

Solve each equation for the given symbol. Assume that the symbol you are solving for represents the variable and that all other symbols represent nonzero real numbers. See Example 11.

133. Solve $\ast\triangle - \square = \odot$ for $\triangle$.

134. Solve $\triangle(\odot + \square) = \otimes$ for $\triangle$.

135. Solve $\odot\square + \triangle = \otimes$ for $\odot$.

136. Solve $\triangle(\odot + \square) = \otimes$ for $\square$.

Concept/Writing Exercises

137. Consider the equation $2x = 5$. Give three equivalent equations. Explain why the equations are equivalent.

138. Consider the equation $x = 4$. Give three equivalent equations. Explain why the equations are equivalent.

139. Make up an equation that is a contradiction. Explain how you created the contradiction.

140. Make up an equation that is an identity. Explain how you created the equation.

141. Create an equation with two terms to the left of the equals sign and three terms to the right of the equals sign that is equivalent to the equation $\frac{1}{2}p + 3 = 6$.

142. Create an equation with three terms to the left of the equals sign and two terms to the right of the equals sign that is equivalent to the equation $3m + 1 = m + 5$.

143. Consider the equation $2(a + 5) + n = 4a - 8$. What real number must n be for the solution of the equation to be -2? Explain how you determined your answer.

144. Consider the equation $-3(x + 2) + 5x + 12 = n$. What real number must n be for the solution of the equation to be 6? Explain how you determined your answer.

Cumulative Review Exercises

[1.3] **145. a)** Explain how to find the absolute value of a number.

    b) Write the definition of absolute value.

[1.4] *Evaluate.*

146. a) -3^2

    b) $(-3)^2$

147. $\sqrt[3]{-125}$

148. $\left(-\dfrac{2}{7}\right)^2$

2.2 Problem Solving and Using Formulas

1 Use the problem-solving procedure.

2 Solve for a variable in an equation or formula.

1 Use the Problem-Solving Procedure

One of the main reasons for studying mathematics is that we can use it to solve everyday problems. To solve most real-life application problems mathematically, we need to be able to express the problem in mathematical symbols using expressions or equations, and when we do this, we are creating a **mathematical model** of the situation.

In this section, we present a problem-solving procedure and discuss formulas. A **formula** is an equation that is a mathematical model of a real-life situation. Throughout the book we will be problem solving. When we do so, we will determine an equation or formula that represents or models the real-life situation.

You can approach any problem using the general five-step problem-solving procedure developed by George Pólya and presented in his book *How to Solve It.*

Guidelines for Problem Solving

1. **Understand the problem.**
   - Read the problem **carefully** at least twice. In the first reading, get a general overview of the problem. In the second reading, determine (a) exactly what you are being asked to find and (b) what information the problem provides.
   - If possible, make a sketch to illustrate the problem. Label the information given.
   - List the information in a table if it will help in solving the problem.

2. **Translate the problem into mathematical language.**
   - This will generally involve expressing the problem algebraically.
   - Sometimes this involves selecting a particular formula to use, whereas other times it is a matter of generating your own equation. It may be necessary to check other sources for the appropriate formula to use.

3. **Carry out the mathematical calculations necessary to solve the problem.**

4. **Check the answer found in step 3.**
   - Ask yourself: "Does the answer make sense?" "Is the answer reasonable?" If the answer is not reasonable, recheck your method for solving the problem and your calculations.
   - Check the solution in the original problem if possible.

5. **Answer the question.** Make sure you have answered the question asked. State the answer clearly in a sentence.

The following examples show how to apply the guidelines for problem solving. When necessary, we will provide the steps in the examples to illustrate the five-step process.

As was stated in step 2 of the problem-solving process—*translate the problem into mathematical language*—we will sometimes need to find and use a *formula*. We will show how to do that in this section.

EXAMPLE 1 **A Personal Loan** Diane Basile makes a $5000, 4% simple interest personal loan to her brother, Bob Basile, for a period of 5 years.

a) At the end of 5 years, what interest will Bob pay Diane?

b) When Bob settles his loan at the end of 5 years, how much money, in total, must he pay Diane?

Solution   **a) Understand**   When a person borrows money using a simple interest loan, the person must repay the simple interest and principal (the original amount borrowed) at the maturity date of the loan. Here the simple interest rate is 4% and the loan is for 5 years.

Translate   The **simple interest formula** is:

$$\text{interest} = \text{principal} \cdot \text{rate} \cdot \text{time or } i = prt$$

where

$i = $ the simple interest

$p = $ principal

$r = $ interest rate written in decimal form

$t = $ time

Note that the rate and the time must both use the same time measure. We will usually use years. In this problem, $p = \$5000$, $r = 0.04$, and $t = 5$. We obtain the simple interest, i, by substituting these values in the simple interest formula.

Carry Out
$$
\begin{aligned}
i &= prt \\
&= 5000(0.04)(5) \\
&= 1000
\end{aligned}
$$

Check   The answer appears reasonable in that Bob will pay $1000 for the use of $5000 for 5 years.

Answer   The simple interest owed is $1000.

b) Bob must pay the principal he borrowed, $5000, plus the interest determined in part **a)**, $1000. Thus, when Bob settles his loan, he must pay Diane $6000.

Now Try Exercise 67

EXAMPLE 2   Certificate of Deposit   Pola Sommers received a holiday bonus of $1350 and invests the money in a certificate of deposit (CD) at a 3.6% annual interest rate compounded monthly for 18 months.

a) How much money will the CD be worth in 18 months?

b) How much interest will she earn during the 18 months?

Solution   **a) Understand**   **Compound interest** means that you get interest on your investment for one period of time. Then in the next period you get interest paid on your investment, plus you get interest paid on the interest that was paid in the first period. This process continues for each period.

Translate   The **compound interest formula** is:

$$A = p\left(1 + \frac{r}{n}\right)^{nt}$$

where $A = $ the accumulated amount, or balance, in the account

$p = $ the principal, or the initial investment

$r = $ the interest rate written in decimal form

$n = $ the number of times per year the interest is compounded

$t = $ time measured in years

In this problem, we have $p = \$1350$, $r = 3.6\%$, $n = 12$ (since there are 12 months in a year), and $t = 1.5 \left(18 \text{ months} = \dfrac{18}{12} = 1.5 \text{ years}\right)$. Substitute these values into the formula and evaluate.

$$A = p\left(1 + \frac{r}{n}\right)^{nt}$$

Carry Out

$$= 1350\left(1 + \frac{.036}{12}\right)^{12(1.5)}$$

$$= 1350(1 + 0.003)^{18}$$

$$= 1350(1.003)^{18}$$

$$\approx 1350(1.05539928) \qquad \text{From a calculator}$$

$$\approx 1424.79 \qquad\qquad \text{Rounded to the nearest cent}$$

Check The answer $1424.79 is reasonable, since it is more than Pola originally invested.

Answer Pola's CD will be worth $1424.79 at the end of 18 months.

b) Understand The interest will be the difference between the original amount invested and the value of the certificate of deposit at the end of 18 months.

Translate $\text{interest} = \left(\begin{array}{c}\text{value of the certificate}\\ \text{of deposit after 18 months}\end{array}\right) - \left(\begin{array}{c}\text{amount originally}\\ \text{invested}\end{array}\right)$

Carry Out $= 1424.79 - 1350 = 74.79$

Check The amount of interest is reasonable and the arithmetic is easily checked.

Answer The interest gained in the 18-month period will be $74.79.

Now Try Exercise 77

Our next example involves a formula containing subscripts. **Subscripts** are numbers (or other variables) placed below and to the right of variables. For example, if a formula contains an original velocity and a final velocity, these velocities are symbolized as V_0 and V_f, respectively. Subscripts are read using the word "sub." For example, V_f is read "V sub f" and x_2 is read "x sub 2."

EXAMPLE 3 **Comparing Investments** Sharon Griggs is in the 25% federal income tax bracket. She is trying to decide whether to invest in tax-free municipal bonds with a rate of 2.24% or in a taxable certificate of deposit with a rate of 3.70%.

a) Determine the taxable rate equivalent to a 2.24% tax-free rate for Sharon.

b) If both investments were for the same period of time, which investment would provide Sharon with the greater return on her investment?

Solution **a)** Understand Some interest we receive, such as from municipal bonds, is tax free. Other interest we receive, such as from savings accounts or certificates of deposit, is taxable on our federal income tax returns. Paying taxes on the interest has the effect of reducing the amount of money we actually get to keep from the interest. We need to find the taxable interest rate that is equivalent to a 2.24% tax-free rate for Sharon, who is in a 25% federal income tax bracket.

Translate A formula used to compare taxable and tax-free interest rates is

$$T_f = T_a(1 - F)$$

where T_f is the tax-free rate, T_a is the taxable rate, and F is the federal income tax bracket. To determine the equivalent taxable rate, T_a, we substitute the appropriate values in the formula and solve for T_a.

$$T_f = T_a(1 - F)$$

$$0.0224 = T_a(1 - 0.25)$$

Carry Out

$$0.0224 = T_a(0.75)$$

$$\frac{0.0224}{0.75} = T_a$$

$$0.0299 \approx T_a \qquad \text{Rounded to four places}$$

Check   The answer, 0.0299 or 2.99%, appears reasonable because it is larger than 2.24%, which is what we expected.

Answer   A taxable investment yielding about 2.99% would give Sharon about the same interest as a 2.24% tax-free investment.

b) We are asked to determine which investment will provide Sharon with the greater return on her investment.

As we saw in part **a)**, the equivalent taxable rate of the municipal bonds is 2.99%. The taxable rate of the certificate of deposit is 3.70%. Therefore the certificate of deposit paying 3.70% will give Sharon a greater return on her investment than the tax-free municipal bond paying 2.24%.

Now Try Exercise 83

2 Solve for a Variable in an Equation or Formula

There are many occasions when you might be given an equation or formula that is solved for one variable but you want to solve it for a different variable. Since formulas are equations, the same procedure we use to solve for a variable in an equation will be used to solve for a variable in a formula.

When you are given an equation (or formula) that is solved for one variable and you want to solve it for a different variable, treat each variable in the equation, except the one you are solving for, as if it were a constant. Then *isolate the variable* you are solving for using the procedures similar to those previously used to solve equations.

EXAMPLE 4   Solve the equation $5x - 8y = 32$ for y.

Solution   We will solve for the variable y by isolating the term containing the y on the left side of the equation.

$$5x - 8y = 32$$
$$5x - 5x - 8y = -5x + 32 \qquad \text{Subtract } 5x \text{ from both sides.}$$
$$-8y = -5x + 32$$
$$\frac{-8y}{-8} = \frac{-5x + 32}{-8} \qquad \text{Divide both sides by } -8.$$
$$y = \frac{-5x + 32}{-8}$$
$$y = \frac{-1(-5x + 32)}{-1(-8)} \qquad \text{Multiply the numerator and the denominator by } -1.$$
$$y = \frac{5x - 32}{8} \quad \text{or} \quad y = \frac{5}{8}x - 4$$

Now Try Exercise 29

EXAMPLE 5   Solve the equation $2y - 3 = \frac{1}{2}(x + 3y)$ for y.

Solution   We begin by multiplying both sides of the equation by the least common denominator, 2.

$$2y - 3 = \frac{1}{2}(x + 3y)$$
$$2(2y - 3) = 2\left[\frac{1}{2}(x + 3y)\right] \qquad \text{Multiply both sides by the LCD, 2.}$$
$$4y - 6 = x + 3y \qquad \text{Distributive property}$$
$$4y - 3y - 6 = x + 3y - 3y \qquad \text{Subtract } 3y \text{ from both sides.}$$
$$y - 6 = x$$
$$y - 6 + 6 = x + 6 \qquad \text{Add 6 to both sides.}$$
$$y = x + 6$$

Now Try Exercise 35

Now let's solve for a variable in a formula. Remember: Our goal is to isolate the variable for which we are solving.

EXAMPLE 6   The formula for the perimeter of a rectangle is $P = 2l + 2w$, where l is the length and w is the width of the rectangle (see **Fig. 2.2**). Solve this formula for the width, w.

Solution   Since we are solving for w, we must isolate the w on one side of the equation.

Rectangle

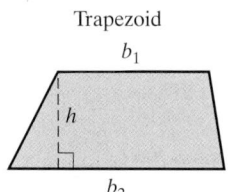

l

FIGURE 2.2

$$P = 2l + 2w$$
$$P - 2l = 2l - 2l + 2w \qquad \text{Subtract } 2l \text{ from both sides.}$$
$$P - 2l = 2w$$
$$\frac{P - 2l}{2} = \frac{2w}{2} \qquad \text{Divide both sides by 2.}$$
$$\frac{P - 2l}{2} = w$$

Thus, $w = \dfrac{P - 2l}{2}$ or $w = \dfrac{P}{2} - \dfrac{2l}{2} = \dfrac{P}{2} - l$.

Now Try Exercise 49

EXAMPLE 7   A formula used to find the area of a trapezoid is $A = \dfrac{1}{2}h(b_1 + b_2)$, where h is the height and b_1 and b_2 are the lengths of the bases of the trapezoid (see **Fig. 2.3**). Solve this formula for b_2.

Solution   We begin by multiplying both sides of the equation by the LCD, 2, to clear fractions.

Trapezoid

b_1

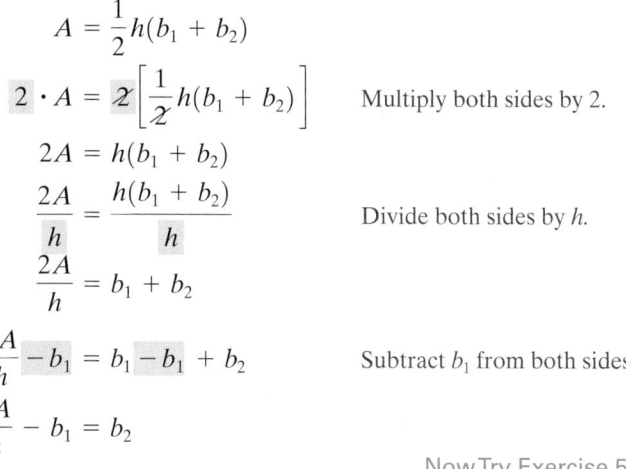

b_2

FIGURE 2.3

$$A = \frac{1}{2}h(b_1 + b_2)$$
$$2 \cdot A = 2\left[\frac{1}{2}h(b_1 + b_2)\right] \qquad \text{Multiply both sides by 2.}$$
$$2A = h(b_1 + b_2)$$
$$\frac{2A}{h} = \frac{h(b_1 + b_2)}{h} \qquad \text{Divide both sides by } h.$$
$$\frac{2A}{h} = b_1 + b_2$$
$$\frac{2A}{h} - b_1 = b_1 - b_1 + b_2 \qquad \text{Subtract } b_1 \text{ from both sides.}$$
$$\frac{2A}{h} - b_1 = b_2$$

Now Try Exercise 57

EXAMPLE 8   In Example 3 on page 77 we introduced the formula $T_f = T_a(1 - F)$.

a) Solve this formula for T_a.

b) John and Dorothy Cutter are in the 33% income tax bracket. What is the equivalent taxable yield of a 2.6% tax-free yield?

Solution

a) We wish to solve this formula for T_a. Therefore we treat all other variables in the equation as if they were constants. Since T_a is multiplied by $(1 - F)$, to isolate T_a we divide both sides of the equation by $1 - F$.

$$T_f = T_a(1 - F)$$
$$\frac{T_f}{1 - F} = \frac{T_a(1 - F)}{1 - F} \qquad \text{Divide both sides by } 1 - F.$$
$$\frac{T_f}{1 - F} = T_a \quad \text{or} \quad T_a = \frac{T_f}{1 - F}$$

b) Substitute the appropriate values into the formula found in part **a)**.

$$T_a = \frac{T_f}{1 - F}$$

$$T_a = \frac{0.026}{1 - 0.33} = \frac{0.026}{0.67} \approx 0.039$$

Thus, the equivalent taxable yield would be about 3.9%.

Now Try Exercise 63

EXERCISE SET 2.2

Math XL MathXL®   **MyMathLab** MyMathLab

Fill in the blanks with the appropriate word, phrase, or symbol(s) from the following list.

formula	mathematical model	understand	translate
check	subscript	superscript	

1. To express a problem using mathematical symbols is to create a _____.

2. A number or variable placed below and to the right of a variable is a _____.

3. To express a problem algebraically is to _____ the problem into mathematical language.

4. The first step in our problem-solving procedure is to _____ the problem.

5. To _____ an answer we first ask, "Does the answer make sense?"

6. A _____ is an equation that is a mathematical model of a real-life situation.

Practice the Skills

Evaluate the following formulas for the values given. Use the $\boxed{\pi}$ key on your calculator for π when needed. Round answers to the nearest hundredth.

7. $W = Fd$ when $F = 20, d = 15$ (a physics formula used to measure work)

8. $A = lw$ when $l = 7, w = 6$ (a formula for finding the area of a rectangle)

9. $R = R_1 + R_2$ when $R_1 = 100, R_2 = 200$ (a formula used when studying electricity)

10. $A = \frac{1}{2}bh$ when $b = 7, h = 6$ (a formula for finding the area of a triangle)

11. $A = \pi r^2$ when $r = 8$ (a formula for finding the area of a circle)

12. $P_1 = \frac{T_1 P_2}{T_2}$ when $T_1 = 150, T_2 = 300, P_2 = 200$ (a chemistry formula that relates temperature and pressure of gases)

13. $\bar{x} = \frac{x_1 + x_2 + x_3}{3}$ when $x_1 = 40, x_2 = 90, x_3 = 80$ (a formula for finding the average of three numbers)

14. $A = \frac{1}{2}h(b_1 + b_2)$ when $h = 15, b_1 = 20, b_2 = 28$ (a formula for finding the area of a trapezoid)

15. $A = P + Prt$ when $P = 160, r = 0.05, t = 2$ (a banking formula that yields the total amount in an account after the interest is added)

16. $E = a_1 p_1 + a_2 p_2$ when $a_1 = 10, p_1 = 0.2, a_2 = 100, p_2 = 0.3$ (a statistics formula for finding the expected value of an event)

17. $m = \frac{y_2 - y_1}{x_2 - x_1}$ when $y_2 = 4, y_1 = -3, x_2 = -2, x_1 = -6$ (a formula for finding the slope of a straight line; we will discuss this formula in Chapter 3)

18. $F = G\frac{m_1 m_2}{r^2}$ when $G = 0.5, m_1 = 100, m_2 = 200, r = 4$ (a physics formula that gives the force of attraction between two masses that are separated by a distance, r)

19. $R_T = \frac{R_1 R_2}{R_1 + R_2}$ when $R_1 = 100, R_2 = 200$ (a formula in electronics for finding the total resistance in a parallel circuit containing two resistors)

20. $d = \sqrt{(x_2 - x_1)^2 + (y_2 - y_1)^2}$ when $x_2 = 5, x_1 = -3, y_2 = -6, y_1 = 3$ (a formula for finding the distance between two points on a straight line; we will discuss this formula in Chapter 10)

21. $x = \frac{-b + \sqrt{b^2 - 4ac}}{2a}$ when $a = 2, b = -5, c = -12$ (from the quadratic formula; we will discuss the quadratic formula in Chapter 8)

22. $x = \frac{-b - \sqrt{b^2 - 4ac}}{2a}$ when $a = 2, b = -5, c = -12$ (from the quadratic formula)

23. $A = p\left(1 + \frac{r}{n}\right)^{nt}$ when $p = 100, r = 0.06, n = 1, t = 3$ (the compound interest formula; see Example 2)

24. $z = \frac{\bar{x} - \mu}{\frac{\sigma}{\sqrt{n}}}$ when $\bar{x} = 78, \mu = 66, \sigma = 15, n = 25$ (a statistics formula for finding the standard, or z score, of a sample mean, $\bar{x}$)

Solve each equation for y (see Examples 4 and 5).

25. $3x + y = 5$

26. $3x + 4y = 8$

27. $3x + 2y = 6$

28. $-6x + 5y = 25$

29. $6x - 2y = 16$

30. $9x = 7y + 23$

31. $\frac{3}{4}x - y = 5$

32. $\frac{x}{4} - \frac{y}{6} = 2$

33. $3(x - 2) + 3y = 6x$

34. $y - 4 = \frac{2}{3}(x + 6)$

35. $y + 1 = -\frac{4}{3}(x - 9)$

36. $\frac{1}{5}(x + 3y) = \frac{4}{7}(2x - 1)$

Solve each equation for the indicated variable (see Examples 6–8).

37. $E = IR$, for I

38. $C = 2\pi r$, for r

39. $C = \pi d$, for d

40. $A = lw$, for l

41. $P = 2l + 2w$, for l

42. $P = 2l + 2w$, for w

43. $V = lwh$, for h

44. $V = \pi r^2 h$, for h

45. $A = P + Prt$, for r

46. $Ax + By = C$, for y

47. $V = \frac{1}{3}lwh$, for l

48. $A = \frac{1}{2}bh$, for b

49. $y = mx + b$, for m

50. $IR + Ir = E$, for R

51. $y - y_1 = m(x - x_1)$, for m

52. $z = \frac{x - \mu}{\sigma}$, for σ

53. $z = \frac{x - \mu}{\sigma}$, for μ

54. $y = \frac{kx}{z}$, for z

55. $P_1 = \frac{T_1 P_2}{T_2}$, for T_2

56. $F = \frac{mv^2}{r}$, for m

57. $A = \frac{1}{2}h(b_1 + b_2)$, for h

58. $D = \frac{x_1 + x_2 + x_3}{n}$, for n

59. $S = \frac{n}{2}(f + l)$ for n

60. $S = \frac{n}{2}(f + l)$, for l

61. $C = \frac{5}{9}(F - 32)$, for F

62. $F = \frac{9}{5}C + 32$, for C

63. $F = \frac{km_1m_2}{d^2}$, for m_1

64. $F = \frac{km_1m_2}{d^2}$ for m_2

Problem Solving

In Exercises 65–88, round your answer to two decimal places, when appropriate.

65. Currency Exchange

  a) According to the Universal Converter Web site, on August 16, 2008, $1 of U.S. currency could be exchanged for 0.68 euros. Write a formula using d for U.S. dollars and e for euros that can be used to convert from dollars to euros.

  b) Write a formula that can be used to convert from euros to dollars.

66. Currency Exchange

  a) According to the Universal Converter Web site, on August 16, 2008, $1 of U.S. currency could be exchanged for 110.54 Japanese yen. Write a formula using d for U.S. dollars and y for yen that can be used to convert from dollars to yen.

  b) Write a formula that can be used to convert from yen to dollars.

In Exercises 67–70, use the simple interest formula i = prt. See Example 1.

67. A Personal Loan Edison Tan loaned his colleague, Ken Pothoven, $1100 at a simple interest rate of 7% per year for 4 years. Determine the simple interest Ken must pay Edison when he repays the loan at the end of 4 years.

68. Determine the Rate Steve Marino borrowed $500 from his credit union for 2 years. The simple interest that he paid was $52.90. What simple interest rate was Steve charged?

69. Determine the Length of the Loan Mary Haran loaned her daughter, Dawn, $20,000 at a simple interest rate of 3.75% per year. At the end of the loan period, Dawn repaid Mary the original $20,000 plus $4875 interest. Determine the length of the loan.

70. A Certificate of Deposit Erin Grabish received $2000 for speaking at a financial planning seminar. Erin invested the money in a certificate of deposit for 2 years. When she redeemed the certificate of deposit, she received $2166. What simple interest rate did Erin receive on this certificate of deposit?

In Exercises 71–76, if you are not sure of the formula to use, refer to Appendix A.

71. Dartboard Area Marc Mazzoni, dart-throwing champion in the state of Michigan, practices on a dartboard with concentric circles as shown in the figure.

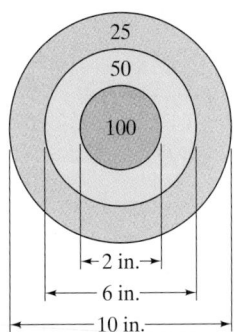

a) Find the area of the circle marked 100.

b) Find the area of the entire dartboard.

72. Planning a Sandbox Betsy Nixon is planning to build a rectangular sandbox for her daughter. She has 38 feet of wood to use for the sides. If the length is to be 11 feet, what is the width to be?

73. Concrete Driveway Volume Anthony Palmiotto is laying concrete for a driveway. The driveway is to be 15 feet long by 10 feet wide by 6 inches deep.

a) Find the volume of concrete needed in cubic feet.

b) If 1 cubic yard = 27 cubic feet, how many cubic yards of concrete are needed?

c) If the concrete costs $35 per cubic yard, what is the cost of the concrete? Concrete must be purchased in whole cubic yards.

74. Helipad Area A helipad in Raleigh, North Carolina, has two concentric circles as shown in the figure above and to the right.

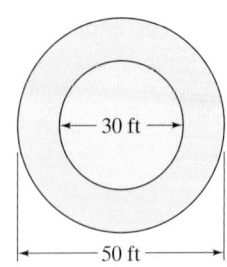

Find the area of the blue region in the figure.

75. Ice Cream Containers Gil and Lori's Delicious Ice Cream Company sells ice cream in two containers, a cylindrical tub and a rectangular box as shown in the figure. Which container holds more ice cream, and what is the difference in their volumes?

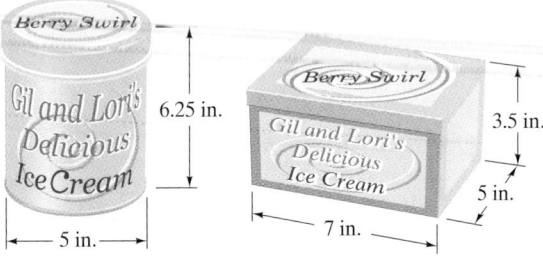

76. Bucket Capacity Sandra Hakanson has a bucket in which she wishes to mix some detergent. The dimensions of the bucket are shown in the figure.

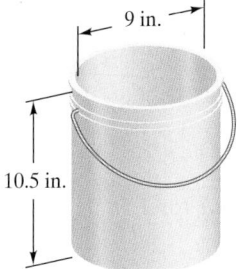

a) Find the capacity of the bucket in cubic inches.

b) If 231 cubic inches = 1 gallon, what is the capacity of the bucket in gallons?

c) If the instructions on the bottle of detergent say to add 1 ounce per gallon of water, how much detergent should Sandra add to a full bucket of water?

For Exercises 77–80, refer to Example 2.

77. A Savings Account Beth Rechsteiner invested $10,000 in a savings account paying 6% interest compounded quarterly. How much money will she have in her account at the end of 2 years?

78. Monthly Compounding Vigay Patel invested $8500 in a savings account paying 3.2% interest compounded monthly. How much money will he have in his account at the end of 4 years?

79. A Certificate of Deposit Heather Kazakoff invests $4390 in a certificate of deposit paying 4.1% interest compounded semiannually. How much will the certificate of deposit be worth after 36 months?

80. Comparing Accounts James Misenti has $1500 to invest for 1 year. He has a choice of a credit union account that pays 4.5% annual simple interest and a bank account that pays 4% interest compounded quarterly. Determine which account would pay more interest and by how much.

For Exercises 81–84, refer to Example 3.

81. Equivalent Taxable Rate Kimberly Morse-Austin is a student and is in the 15% federal income tax bracket. She is considering investing $825 in a tax-free bond mutual fund paying 3.5% simple interest. Determine the taxable rate equivalent to a 3.5% tax-free rate.

82. Comparing Investments Dave Ostrow is in the 35% federal income tax bracket and is considering two investments: a tax-free municipal bond paying 3% simple interest or a taxable certificate of deposit paying 4.5% simple interest. Which investment yields the greatest return?

83. Father and Son Investing Anthony Rodriquez is in the 35% federal income tax bracket and his son Angelo is in the 28% federal income tax bracket. They are each considering a tax-free mutual fund currently yielding 4.6% simple interest.
   a) Determine the taxable rate equivalent to a 4.6% tax-free rate for Anthony.
   b) Determine the taxable rate equivalent to a 4.6% tax-free rate for Angelo.

84. Investment Comparison Marissa Felberty is considering investing $9200 in a 6.75% taxable account or in a 5.5 % tax-free account. If she is in the 25% tax bracket, which investment will yield the greater return?

Exercises 85–88 contain a variety of situations. Solve each exercise.

85. Weight Loss A nutritionist explains to Robin Thomas that a person loses weight by burning more calories than he or she eats. If Robin burns more than 2400 calories daily, her weight loss can be approximated by the mathematical model $w = 0.02c$, where w is the *weekly* weight loss and c is the number of calories per *day* burned *above* 2400 calories.
   a) Find Robin's weekly weight loss if she exercises and burns off 2600 calories per day.
   b) How many calories would Robin have to burn off in a day to lose 2 pounds in a week?

© Dreamstime LLC -Royalty Free

86. Stress Test During a stress test, the maximum allowable heart rate, m, in beats per minute, can be approximated by the equation $m = -0.875x + 190$, where x represents the patient's age from 1 through 99. Using this mathematical model, find
   a) the maximum heart rate for a 50-year-old.
   b) the age of a person whose maximum heart rate is 160 beats per minute.

87. Balancing a Portfolio Some financial planners recommend the following rule of thumb for investors. The percent of stocks in your total portfolio should be equal to 100 minus your age. The remainder is to be put into bonds or cash.
   a) Construct a mathematical model for the percent to be kept in stocks (use S for percent in stock and a for a person's age).
   b) Using this rule of thumb, find the percent to be kept in stocks for a 60-year-old.

88. Body Mass Index The body mass index is a standard way of evaluating a person's body weight in relation to their height. To determine your body mass index (BMI) using metric measurements, divide your weight, in kilograms, by your height, in meters squared. To calculate BMI using pounds and inches, multiply your weight in pounds by 705, then divide by the square of your height, in inches.
   a) Create a formula for finding a person's BMI using kilograms and meters.
   b) Create a formula for finding a person's BMI when the weight is given in pounds and the height is given in inches.
   c) Determine your BMI.

Challenge Problem

89. Solve the formula $r = \dfrac{s/t}{t/u}$ for **a)** s, **b)** u.

Cumulative Review Exercises

[1.4]   **90.** Evaluate $-\sqrt{3^2 + 4^2} + |3 - 4| - 6^2$.

   91. Evaluate $\dfrac{7 + 9 \div (2^3 + 4 \div 4)}{|3 - 7| + \sqrt{5^2 - 3^2}}$.

92. Evaluate $a^3 - 3a^2b + 3ab^2 - b^3$ when $a = -2$, $b = 3$.

[2.1]   **93.** Solve the equation $\dfrac{1}{4}t + \dfrac{1}{2} = 1 - \dfrac{1}{8}t$.

2.3 Applications of Algebra

1 Translate a verbal statement into an algebraic expression or equation.

2 Use the problem-solving procedure.

1 Translate a Verbal Statement into an Algebraic Expression or Equation

Translating an application problem into an equation is perhaps the most difficult part of solving word problems. We begin this section by giving some examples of phrases that translate into algebraic expressions.

Phrase	Algebraic Expression
a number increased by 8	$x + 8$
twice a number	$2x$
7 less than a number	$x - 7$
one-ninth of a number	$\dfrac{1}{9}x$ or $\dfrac{x}{9}$
2 more than 3 times a number	$3x + 2$
4 less than 6 times a number	$6x - 4$
12 times the sum of a number and 5	$12(x + 5)$

The variable x was used in these algebraic expressions, but any variable could have been used to represent the unknown quantity.

EXAMPLE 1 Express each phrase as an algebraic expression.

a) the radius, r, decreased by 9 centimeters

b) 5 less than twice the distance, d

c) 7 times a number, n, increased by 8

Solution

a) $r - 9$ **b)** $2d - 5$ **c)** $7n + 8$

Now Try Exercise 9

Helpful Hint

Study Tip

Success solving word problems requires hard work! Be sure to:

- Read the book and examples carefully.
- Attend class every day.
- Work all of the exercises assigned to you.

   As you read through the examples in the rest of the chapter, think about how they can be expanded to other, similar problems. For example, in Example 1 **a)** we stated that the radius, r, decreased by 9 centimeters, can be represented by $r - 9$. You can generalize this to other, similar problems. For example, a weight, w, decreased by 15 pounds, can be represented as $w - 15$.

EXAMPLE 2 Write each phrase as an algebraic expression.

a) the cost of purchasing x shirts at $4 each

b) the distance traveled in t hours at 65 miles per hour

c) the number of cents in n nickels

d) an 8% commission on sales of x dollars

Solution

a) We can reason like this: one shirt would cost 1(4) dollars; two shirts, 2(4) dollars; three shirts, 3(4) dollars; four shirts, 4(4) dollars, and so on. Continuing this reasoning process, we can see that x shirts would cost $x(4)$ or $4x$ dollars. We can use the same reasoning process to complete each of the other parts.

b) $65t$

c) $5n$

d) $0.08x$ (8% is written as 0.08 in decimal form.)

Now Try Exercise 7

> **Helpful Hint**
>
> When we are asked to find a percent, we are always finding the percent of some quantity. Therefore, when a percent is listed, it is *always* multiplied by a number or a variable. In the following examples we use the variable c, but any letter could be used to represent the variable.
>
> | Phrase | How Written |
> |---|---|
> | 6% of a number | $0.06c$ |
> | the cost of an item increased by a 7% tax | $c + 0.07c$ |
> | the cost of an item reduced by 35% | $c - 0.35c$ |

Sometimes in a problem two numbers are related to each other. We often represent one of the numbers as a variable and the other as an expression containing that variable. We generally let the less complicated description be represented by the variable and write the second (more complex expression) in terms of the variable. In the following examples, we use x for the variable.

Phrase	One Number	Second Number
Dawn's age now and Dawn's age in 3 years	x	$x + 3$
one number is 9 times the other	x	$9x$
a second number is 4 less than a number	x	$x - 4$
a number and the number increased by 16%	x	$x + 0.16x$
a number and the number decreased by 10%	x	$x - 0.10x$
the sum of two numbers is 10	x	$10 - x$
a 6-foot board cut into two lengths	x	$6 - x$
\$10,000 shared by two people	x	$10,000 - x$

The last three examples may not be obvious. Consider "The sum of two numbers is 10." When we add x and $10 - x$, we get $x + (10 - x) = 10$. When a 6-foot board is cut into two lengths, the two lengths will be x and $6 - x$. For example, if one length is 2 feet, the other must be $6 - 2$ or 4 feet.

> **Helpful Hint**
>
> Suppose you read the following sentence in an application problem: "A 12-foot rope is cut into two pieces." You probably know you should let x represent the length of the first piece. What you may not be sure of is whether you should let $x - 12$ or $12 - x$ represent the length of the second piece. To help you decide, it can be helpful to use specific numbers to establish a pattern. In this example, you might use a pattern similar to the one below to help you.
>
> | If the First Piece Is ... | Then the Second Piece Is ... |
> |---|---|
> | 2 feet | 10 feet = 12 feet − 2 feet |
> | 5 feet | 7 feet = 12 feet − 5 feet |
>
> From this pattern you can see that if the first piece is x feet, then the second piece is $12 - x$ feet.

EXAMPLE 3 For each relation, select a variable to represent one quantity and express the second quantity in terms of the first.

a) The speed of the second train is 1.8 times the speed of the first.

b) $90 is shared by David and his brother.

c) It takes Tom 3 hours longer than Roberta to complete the task.

d) Hilda has $5 more than twice the amount of money Hector has.

e) The length of a rectangle is 7 units less than 3 times its width.

Solution

a) speed of first train, s; speed of second train, $1.8s$

b) amount David has, x; amount brother has, $90 - x$

c) Roberta, t; Tom, $t + 3$

d) Hector, x; Hilda, $2x + 5$

e) width, w; length, $3w - 7$

Now Try Exercise 11

The word *is* often translates to an equals sign.

Verbal Statement	Algebraic Equation
4 less than 6 times a number *is* 17	$6x - 4 = 17$
a number decreased by 4 *is* 5 more than twice the number	$x - 4 = 2x + 5$
the product of two consecutive integers *is* 72	$x(x + 1) = 72$
a number increased by 15% *is* 90	$x + 0.15x = 90$
a number decreased by 12% *is* 52	$x - 0.12x = 52$
the sum of a number and the number increased by 4% *is* 324	$x + (x + 0.04x) = 324$
the cost of renting a car for x days at $24 per day *is* $120	$24x = 120$

Understanding Algebra

When translating word problems into algebraic symbols, the word *is* often translates to *is equal to* and is represented with an equals sign, =.

2 Use the Problem-Solving Procedure

There are many types of word problems, and the general problem-solving procedure given in Section 2.2 can be used to solve all types. We now present the five-step problem-solving procedure again so you can easily refer to it. We have included some additional information under step 2, since in this section we are going to emphasize translating word problems into equations.

Problem-Solving Procedure for Solving Application Problems

1. **Understand the problem.** Identify the quantity or quantities you are being asked to find.

2. **Translate the problem into mathematical language** (express the problem as an equation).

   a) Choose a variable to represent one quantity, and *write down exactly what it represents*. Represent any other quantity to be found in terms of this variable.

   b) Using the information from step a), write an equation that represents the word problem.

3. **Carry out the mathematical calculations** (solve the equation).

4. **Check the answer** (using the original wording of the problem).

5. **Answer the question asked.**

Sometimes we will combine steps or not show some steps in the problem-solving procedure due to space limitations. Even if we do not show a check to a problem, you should always check the answer to make sure that your answer is reasonable and makes sense.

EXAMPLE 4 **Long-Distance Plans** AT&T's One Rate Nationwide Plan requires customers to pay a $5.00 monthly fee and 5¢ per minute for any long-distance call made. AT&T's Unlimited Nationwide Calling Plan has a $22 monthly fee for unlimited calling—in other words, there is no per-minute fee. How many minutes of long-distance calls would a customer need to use for the two plans to cost the same amount?

Solution   Understand   We are asked to find the *number of minutes* of long-distance calls that would result in both plans having the same total cost. To solve the problem we will write algebraic expressions for each plan and then set these expressions equal to each other.

Translate   Let n = number of minutes of long-distance calls.

Then $0.05n$ = cost for n minutes at 5¢ per minute.

Cost of one Rate Nationwide Plan = Cost of Unlimited Nationwide Calling Plan

monthly fee + cost for n minutes = monthly fee

$$5 \quad + \quad 0.05n \quad = \quad 22$$

Carry Out

$$5 + 0.05n = 22$$
$$0.05n = 17$$
$$\frac{0.05n}{0.05} = \frac{17}{0.05}$$
$$n = 340$$

Check   The number of minutes is reasonable and the arithmetic is easily checked.

Answer   If 340 minutes were used per month, both plans would have the same total cost.

Now Try Exercise 33

EXAMPLE 5 **Corvette Cost** The price of a 2008 Chevrolet Corvette Z06 was $72,125. This was an 8.5% increase from the price of a 2007 Chevrolet Corvette Z06. Determine the price of a 2007 Chevrolet Corvette Z06.

Solution   Understand   We need to determine the price of a 2007 Corvette. We will use the facts that the price increased by 8.5% from 2007 to 2008 and that the 2008 price was $72,125 to solve the problem.

Translate   Let x = the price of a 2007 Corvette.

Then $0.085x$ = the increase in the Corvette price from 2007 to 2008.

$$\begin{pmatrix} 2007 \text{ Corvette Z06} \\ \text{price} \end{pmatrix} + \begin{pmatrix} \text{increase in price} \\ \text{from 2007 to 2008} \end{pmatrix} = \begin{pmatrix} 2008 \text{ Corvette Z06} \\ \text{price} \end{pmatrix}$$

$$x \quad + \quad 0.085x \quad = \quad 72,125$$

Carry Out

$$x + 0.085x = 72,125$$
$$1.085x = 72,125$$
$$x \approx 66,474.65$$

Check and Answer   The number obtained is less than the price of a 2008 Corvette Z06, which is what we expected. The price of the 2007 Corvette Z06 was about $66,474.65.

Now Try Exercise 41

EXAMPLE 6 **Land Area** The total land area of Gibraltar, Nauru, Bermuda, and Norfolk Island is 116 km². The land area of Gibraltar is $\frac{1}{3}$ the land area of Nauru. The land area of Norfolk Island is $\frac{5}{3}$ the land area of Nauru. The land area of Bermuda is 10 km² less than 3 times the land area of Nauru. Determine the land area of each of these four countries.

Solution   Understand   We need to determine the land area (in km²) of Gibraltar, Nauru, Bermuda, and Norfolk Island. Observe that the land area of Gibraltar, Norfolk Island, and Bermuda is described in terms of the land area of Nauru. Therefore, we will let the unknown variable be the land area of Nauru. Also, notice that the total land area of the four countries is 116 km².

Translate   Let a = the land area of Nauru

and $\frac{1}{3}a$ = the land area of Gibraltar

and $\frac{5}{3}a$ = the land area of Norfolk Island

and $3a - 10$ = the land area of Bermuda.

$$\begin{pmatrix}\text{land area of}\\\text{Nauru}\end{pmatrix} + \begin{pmatrix}\text{land area of}\\\text{Gibraltar}\end{pmatrix} + \begin{pmatrix}\text{land area of}\\\text{Norfolk Island}\end{pmatrix} + \begin{pmatrix}\text{land area of}\\\text{Bermuda}\end{pmatrix} = \begin{pmatrix}\text{total}\\\text{land area}\end{pmatrix}$$

$$a \quad + \quad \frac{1}{3}a \quad + \quad \frac{5}{3}a \quad + \quad (3a - 10) \quad = \quad 116$$

Carry Out $a + \frac{1}{3}a + \frac{5}{3}a + (3a - 10) = 116$

$$a + 2a + 3a - 10 = 116$$
$$6a - 10 = 116$$
$$6a = 126$$
$$a = 21$$

Check and Answer   The land area of Nauru is 21 km². The land area of Gibraltar is $\frac{1}{3}(21) = 7$ km². The land area of Norfolk Island is $\frac{5}{3}(21) = 35$ km². The land area of Bermuda is $(3 \cdot 21) - 10 = 63 - 10 = 53$ km². The total land area is $(21 + 7 + 35 + 53) = 116$ km², so the answer checks.
Source: www.gazetteer.com

Now Try Exercise 53

EXAMPLE 7 **Daytona Beach** Erin Grabish took her family to visit Daytona Beach, Florida. They stayed for one night at a Holiday Inn. When they made their hotel reservation, they were quoted a rate of $95 per night before tax. When they checked out, their total bill was $110.85, which included the room tax and a $3.50 charge for a candy bar from the in-room bar. Determine the tax rate for the room.

Solution   Understand   Their total bill consists of their room rate, the room tax, and the $3.50 cost for the candy bar. The room tax is determined by multiplying the cost of the room by the tax rate for the room. We are asked to find the tax rate for the room.

Translate   Let t = tax rate for the room.

Then $0.01t$ = room tax rate as a decimal.

room cost + room tax + candy bar = total
$$95 \quad + \quad 95(0.01t) + \quad 3.50 \quad = 110.85$$

Carry Out $95 + 0.95t + 3.50 = 110.85$
$$0.95t + 98.50 = 110.85$$
$$0.95t = 12.35$$
$$t = 13$$

Check and Answer   If you substitute 13 for t in the equation, you will see that the answer checks. The room tax rate is 13%.

Now Try Exercise 47

The Daytona 500 race

EXAMPLE 8 **Home Mortgage** Mary Shapiro is buying her first home and she is considering two banks for a $60,000 mortgage. Citibank is charging a 6.50% interest rate with no points for a 30-year loan. (A point is a one-time charge of 1% of the amount of the mortgage.) The monthly mortgage payments for the Citibank mortgage would be $379.24. Citibank is also charging a $200 application fee. Bank of America is charging a 6.00% interest rate with 2 points for a 30-year loan. The monthly mortgage payments for Bank of America would be $359.73 and the cost of the points that Mary would need to pay at the time of closing is $0.02($60,000) = 1200. Bank of America has waived its application fee.

a) How long would it take for the total payments of the Citibank mortgage to equal the total payments of the Bank of America mortgage?

b) If Mary plans to keep her house for 20 years, which mortgage would result in the lower total cost?

Solution   **a)** Understand   Citibank is charging a higher interest rate and a small application fee but no points. Bank of America is charging a lower rate and no application fee but 2 points. We need to determine the number of months when the total payments of the two loans would be equal.

Translate   Let x = number of months.

Then $379.24x$ = cost of mortgage payments for x months
with the Citibank mortgage

and $359.73x$ = cost of mortgage payments for x months
with the Bank of America.

total cost with Citibank = total cost with Bank of America

mortgage payments	+	application fee	=	mortgage payments	+	points
$379.24x$	+	200	=	$359.73x$	+	1200

Carry Out

$$379.24x + 200 = 359.73x + 1200$$
$$379.24x = 359.73x + 1000$$
$$19.51x = 1000$$
$$x \approx 51.26$$

Answer   The cost would be the same in about 51.26 months or about 4.3 years.

b) The total cost would be the same in about 4.3 years. Prior to the 4.3 years, the cost of the loan with Bank of America would be more because of the initial $1200 charge for points. However, after the 4.3 years the cost with Bank of America would be less because of the lower monthly payment. If we evaluate the total cost with Citibank over 20 years (240 monthly payments), we obtain $91,217.60. If we evaluate the total cost with Bank of America over 20 years, we obtain $87,535.20. Therefore, Mary will save $3682.40 over the 20-year period with Bank of America.

Now Try Exercise 49

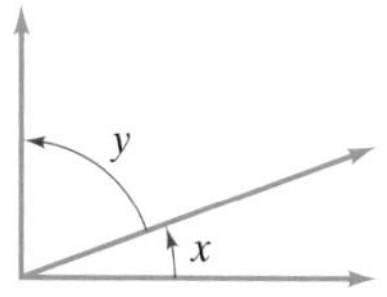

Let's now look at two examples that involve angles. In Example 9 we use complementary angles. **Complementary angles** are two angles whose sum measures 90° (see **Fig. 2.4**).
In **Figure 2.4**, angle x (symbolized $\sphericalangle x$) and angle y ($\sphericalangle y$) are complementary angles since their sum measures 90°.

FIGURE 2.4

EXAMPLE 9 **Complementary Angles** If angles A and B are complementary angles and angle B is 42° greater than angle A, determine the measures of angle A and angle B.

Solution   Understand   The sum of the measures of the two angles must be 90° because they are complementary angles. We will use this fact to set up an equation. Since angle B is described in terms of angle A, we will let x represent the measure of angle A.

Translate   Let x = the measure of angle A.

Then $x + 42$ = the measure of angle B.

the measure of angle A + the measure of angle B = 90°

$x \quad + \quad x + 42 \quad = 90$

Carry Out $2x + 42 = 90$

$2x = 48$

$x = 24$

Check and Answer   Since $x = 24$, the measure of angle A is 24°. The measure of angle $B = x + 42 = 24 + 42 = 66$. Thus, angle B has a measure of 66°. Note that angle B is 42° greater than angle A, and the sum of the measures of both angles is $24° + 66° = 90°$.

Now Try Exercise 21

In Example 10 we use supplementary angles. **Supplementary angles** are two angles whose sum measures 180° (see **Fig. 2.5**).

In **Figure 2.5**, angle x and angle y are supplementary angles since their sum measures 180°.

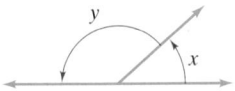

FIGURE 2.5

EXAMPLE 10 **Supplementary Angles** If angles C and D are supplementary angles and the measure of angle C is 6° greater than twice the measure of angle D, determine the measures of angle C and angle D.

Solution   Understand   The sum of the measures of the two angles must be 180° because they are supplementary angles. Since angle C is described in terms of angle D, we will let x represent the measure of angle D.

Translate   Let x = the measure of angle D.

Then $2x + 6$ = the measure of angle C.

the measure of angle C + the measure of angle D = 180°

$2x + 6 \quad + \quad x \quad = 180$

Carry Out $3x + 6 = 180$

$3x = 174$

$x = 58$

Check and Answer   Since $x = 58$, the measure of angle D is 58°. The measure of angle $C = 2x + 6 = 2(58) + 6 = 122$. Thus the measure of angle $C = 122°$. Note that the measure of angle C is 6° more than twice the measure of angle D, and the sum of the measures of the angles is $122° + 58° = 180°$.

Now Try Exercise 23

Helpful Hint

Study Tip

Here are some suggestions if you are having some difficulty with application problems.

1. Instructor—Make an appointment to see your instructor. Make sure you have read the material in the book and attempted all the homework problems. *Go with specific questions for your instructor.*

2. Tutoring—If your college learning center offers free tutoring, you may wish to take advantage of tutoring.

3. Study Group—Form a study group with classmates. Exchange phone numbers and email addresses. You may be able to help one another.

4. Student's Solutions Manual—If you get stuck on an exercise, you may want to use the Student's Solutions Manual to help you understand a problem. Do not use the Solutions Manual in place of working the exercises. In general, the Solutions Manual should be used only to check your work.

(continued on next page)

5. MyMathLab—MyMathLab provides exercises correlated to the text. In addition, online tools such as video lectures, animations, and a multimedia textbook are available to help you understand the material.

6. MathXL®—MathXL is a powerful online homework, tutorial, and assessment system correlated specifically to this text. You can take chapter tests in MathXL and receive a personalized study plan based on your test results. The study plan links directly to tutorial exercises for the objectives you need to study or retest.

7. Pearson Tutor Center—Once the program has been initiated by your instructor, you can get individual tutoring by phone, fax, or email.

It is important that you keep trying! Remember, the more you practice, the better you will become at solving application problems.

EXERCISE SET 2.3

MathXL MathXL® MyMathLab MyMathLab

Warm-Up Exercises

Fill in the blanks with the appropriate word, phrase, or symbol(s) from the following list.

less than   greater than   is $x + 3$ $x - 3$ $x - 7$ $7 - x$

1. The phrase "a number increased by 3" can be represented by the algebraic expression _____.

2. The phrase "a number decreased by 3" can be represented by the algebraic expression _____.

3. A seven-foot rope is cut into two pieces. If we let x = the length of the first piece, then _____ = the length of the second piece.

4. The word "_____" in a word problem often means "is equal to."

5. The phrase "6 _____ a number" can be represented by the algebraic expression $x - 6$.

6. The phrase "5 _____ a number" can be represented by the algebraic expression $x + 5$.

Practice the Skills

In Exercises 7–10, express each phrase as an algebraic expression.

7. the cost of buying y books at $19.95 each

8. 17 more than 4 times a number, m

9. 11 times a number n, decreased by 7.5

10. 7 times a number, p, increased by 8

In Exercises 11–20, select one variable to represent one quantity and express the second quantity in terms of the first.

11. A 12-foot piece of wood is cut into 2 pieces.

12. One angle of a triangle is $7°$ more than another angle.

13. The length of a rectangle is 29 meters longer than the width.

14. A 17-hour task is shared between Robin and Tom.

15. $165 is shared between Max and Lora.

16. George can paint a house twice as fast as Jason.

17. Nora can jog 1.3 miles per hour faster than Betty.

18. The speed limit on an expressway is 30 miles per hour faster than the speed limit on a local road.

19. The cost of electricity has increased by 22%.

20. The price of a refrigerator has increased by 6%.

© Dreamstime LLC -Royalty Free

See Exercise 16.

Problem Solving

In Exercises 21–72, write an equation that can be used to solve the problem. Find the solution to the problem.

21. **Complementary Angles** Angles A and B are complementary angles. Determine the measures of angles A and B if angle A is four times the size of angle B. See Example 9.

22. **Complementary Angles** Angles C and D are complementary angles. Determine the measures of angles C and D if angle D is $15°$ less than twice angle C.

23. **Supplementary Angles** Angles A and B are supplementary angles. Find the measures of angles A and B if angle B is 4 times the size of angle A. See Example 10.

24. **Supplementary Angles** Angle A and angle B are supplementary angles. Find the measure of each angle if angle A is $30°$ greater than angle B.

25. Angles in a Triangle The sum of the measures of the angles of a triangle is 180°. Find the three angles of a triangle if one angle is 20° greater than the smallest angle and the third angle is twice the smallest angle.

26. Angles in a Triangle Find the measures of the three angles of a triangle if one angle is twice the smallest angle and the third angle is 60° greater than the smallest angle.

27. History Honor Society One benefit of membership in a national honor society is a 25% discount on all history magazine subscriptions. Thomas used this discount to order an annual subscription to *American Heritage* magazine and paid $24. What was the cost of a regular subscription?

28. A New Suit Matthew Stringer is shopping for a new suit. At K & G Menswear, he finds that the sale price of a suit that was reduced by 25% is $187.50. Find the regular price of the suit.

29. Bus Pass Kate Spence buys a monthly bus pass, which entitles her to unlimited bus travel, for $45 per month. Without the pass each bus ride costs $1.80. How many rides per month would Kate have to take so that the cost of the rides without the bus pass is equal to the total cost of the rides with the bus pass?

30. Laundry Costs It costs Bill Winschief $12.50 a week to wash and dry his clothes at the corner laundry. If a washer and dryer cost a total of $940, how many weeks will it take for the laundry cost to equal the cost of a washer and dryer? (Disregard energy cost.)

31. Truck Rental The cost of renting a truck is $35 a day plus 20¢ per mile. How far can Tanya Richardson drive in 1 day if she has only $80?

32. Waitress Pay Candice Colton is a banquet waitress. She is paid $3.25 per hour plus 15% of the total cost of the food and beverages she serves during the banquet. If, during a 5-hour shift, Candice earns $331.25, what was the total cost of the food and beverages she served?

33. Playing Golf Albert Sanchez has two options for membership in a golf club. A social membership costs $1775 in annual dues. In addition, he would pay a $50 green fee and a $25 golf cart fee every time he played. A golf membership costs $2425 in annual dues. With this membership, Albert would only pay a $25 golf cart fee when he played. How many times per year would Albert need to golf for the two options to cost the same?

34. George Washington Bridge Tolls Travelers going into New York (during off-peak hours) using the George Washington Bridge must pay a toll by paying either $8 cash or $6 using the EZ Pass system. The EZ Pass system is a prepaid plan that also requires a one-time $10 activation fee. How many

trips to New York would a person need to make so that the total amount spent using the EZ Pass system would be equal to the amount spent on tolls if paying with cash?

Source: Port Authority of NY and NJ Web site

35. Toll Bridge Mr. and Mrs. Morgan live in a resort island community attached to the mainland by a toll bridge. The toll is $2.50 per car going to the island, but there is no toll coming from the island. Island residents can purchase a monthly pass for $20, which permits them to cross the toll bridge from the mainland for only 50¢ each time. How many times a month would the Morgans have to go to the island from the mainland for the cost of the monthly pass to equal the regular toll cost?

36. Sales Tax The sales tax rate in North Carolina is 4.25%. What is the maximum price that Don and Betty Lichtenberg can spend on a new computer desk if the total cost of the desk, including sales tax, is $650?

37. Renting an Apartment The DuVall family is renting an apartment in Southern California. For 2010, the rent is $1720 per month. The monthly rent in 2010 is 7.5% higher than the monthly rent in 2009. Determine the monthly rent in 2009.

38. Retirement Funds Eva Chang makes regular contributions of $5000 annually to a retirement plan. She has some of her contribution going to the growth stock fund and some going to the global equities fund. Her contribution to the growth stock fund is $250 less than twice what she contributes to the global equities fund. How much does she contribute to each fund?

39. Girl Scouts To make money for the organization, the Girl Scouts have their annual cookie drive. This year, the total sales from two districts, the southeast district and the northwest district, totaled $4.6 million. If the sales from the southeast district were $0.31 million more than the sales from the northwest district, find the sales from each district.

40. NFL Franchise Values As of July 13, 2008, the Dallas Cowboys and the Washington Redskins had the highest franchise values among all professional sports teams. The total value of the two franchises was $2.967 billion. The value of the Cowboys was about 2.25% higher than the value of the Redskins. Determine the value of the two teams.

Source: ESPN Magazine

41. Fiber Comparison One serving of canned blackberries contains 4.4 grams of fiber. This is 10% more fiber than in a medium-sized apple. How many grams of fiber are in a medium-sized apple?

Source: Continuum Health Partners

42. Lycopene Comparison One pink grapefruit contains 3 milligrams (mg) of lycopene. This is 50% more lycopene than in a tablespoon of ketchup. How many milligrams of lycopene are in a tablespoon of ketchup?

Source: Advance Physician Formulas

43. Minimum Wage Increase From 2008 to 2009, the federal minimum hourly wage increased about 10.69% to $7.25. What was the minimum hourly wage in 2008?

Source: U.S. Department of Labor

44. Bones and Steel According to *Health* magazine, the stress a bone can withstand in pounds per square inch is 6000 pounds more than 3 times the amount that steel can withstand. If the difference between the amount of stress a bone and steel can withstand is 18,000 pounds per square inch, find the stress that both steel and a bone can withstand.

45. Pollen There are 57 major sources of pollen in the United States. These pollen sources are categorized as grasses, weeds, and trees. If the number of weeds is 5 less than twice the number of grasses, and the number of trees is 2 more than twice the number of grasses, find the number of grasses, weeds, and trees that are major pollen sources.

46. Car Antitheft System By purchasing and installing a LoJack antitheft system, Janet Samuels can save 15% of the price of her auto insurance. The LoJack system costs $743.65 to purchase and install. If Janet's annual insurance before installing the LoJack system is $849.44, in how many years would the LoJack system pay for itself?

47. Ordering Lunch After Valerie Fandl is seated in a restaurant, she realizes that she only has $20.00. If she must pay 7% sales tax and wishes to leave a 15% tip on the total bill (meal plus tax), what is the maximum price of the lunch she can order?

48. Hotel Tax Rate While on vacation in Milwaukee, the Ahmeds were quoted a hotel room price of $85 per night plus tax. They stayed one night and watched a movie that cost $9.25. Their total bill came to $106.66. What was the tax rate?

49. Comparing Mortgages The Chos are purchasing a new home and are considering a 30-year $70,000 mortgage with two different banks. Madison Savings is charging 9.0% with 0 points and First National is charging 8.5% with 2 points. First National is also charging a $200 application fee, whereas Madison is charging none. The monthly mortgage payments with Madison would be $563.50 and the monthly mortgage payments with First National would be $538.30.

   a) After how many months would the total payments for the two banks be the same?

   b) If the Chos plan to keep their house for 30 years, which mortgage would have the lower total cost? (See Example 8.)

50. Payment Plan The Midtown Tennis Club offers two payment plans for its members. Plan 1 is a monthly fee of $25 plus $10 per hour of court rental time. Plan 2 has no monthly fee, but court time costs $18.50 per hour. How many hours would Mrs. Levin have to play per month so that plan 1 becomes advantageous?

51. Refinancing a Mortgage Dung Nguyen is considering refinancing his house at a lower interest rate. He has an 11.875% mortgage, is presently making monthly principal and interest payments of $510, and has 20 years left on his mortgage. Because interest rates have dropped, Countrywide Mortgage Corporation is offering him a rate of 9.5%, which would result in principal and interest payments of $420.50 for 20 years. However, to get this mortgage, his closing cost would be $2500.

　a) In how many months after refinancing will he have spent the same amount on his new mortgage plus closing cost as he would have spent on the original mortgage?

　b) If he plans to spend the next 20 years in the house, would he save money by refinancing?

52. Dinner Seminars Heather Jockson, a financial planner, is sponsoring dinner seminars. She must pay for the dinners of those attending out of her own pocket. She chooses a restaurant that seats 40 people and charges her $9.50 per person. If she earns 12% commission of sales made, determine how much in sales she must make from these 40 people

　a) to break even.

　b) to make a profit of $500.

53. 2008 Olympic Swimmers The top four medal winners in swimming for the United States during the 2008 Summer Olympics were Michael Phelps, Natalie Coughlin, Ryan Lochte, and Matt Grevers. The total number of medals won by these swimmers was 21. Lochte won one more medal than Grevers. Coughlin won twice as many medals as Grevers. Phelps won two more than twice as many medals as Grevers. How many medals did each swimmer win?

Source: United States Olympic Committee

54. Test Grades On a recent test in an intermediate algebra class, 34 students received grades of A, B, C, or D. There were twice as many C's as D's. There were 2 more B's than D's. The number of A's was two more than twice the number of D's. Determine the number of A's, B's, C's, and D's from this test.

55. Plants and Animals Approximately 1,500,000 species worldwide have been categorized as either plants, animals, or insects. Insects are often subdivided into beetles and insects that are not beetles. There are about 100,000 more plant than animal species. There are 290,000 more nonbeetle insects than animals. The number of beetles is 140,000 less than twice the number of animals. Find the number of animal, plant, nonbeetle insect, and beetle species.

56. Sides of a Triangle The sum of the lengths of the sides of a triangle is 30 inches. The length of the first side is 3 inches more than twice the shortest side. The length of the second side is 2 inches more than twice the shortest side. Find the lengths of the three sides of the triangle.

57. Perimeter of a Triangle John is developing a game that contains a triangular game board. The perimeter of the triangular board is 36 inches. Find the length of the three sides if one side is 3 inches greater than the smallest side, and the third side is 3 inches less than twice the length of the smallest side.

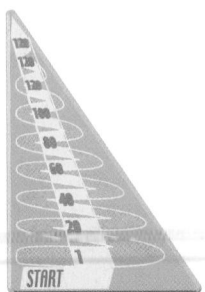

58. Angles of a Triangle A rectangular piece of paper is cut from opposite corners to form a triangle. One angle of the triangle measures 12° greater than the smallest angle. The third angle measures 27° less than three times the smallest angle. If the sum of the interior angles of a triangle measure 180°, determine the measures of the three angles.

59. Triangular Garden The perimeter of a triangular garden is 60 feet. Find the length of the three sides if one side is 4 feet greater than twice the length of the smallest side, and the third side is 4 feet less than 3 times the length of the smallest side.

60. Stairway Railing A stairway railing has a design that contains triangles. In one of the triangles one angle measures 20° less than twice the smallest angle. The third angle measures 25° greater than twice the smallest angle. Determine the measures of the three angles.

61. Fence Dimensions Greg Middleton, a landscape architect, wishes to fence in two equal areas as illustrated in the figure. If both areas are squares and the total length of fencing used is 91 meters, find the dimensions of each square.

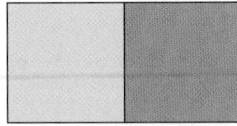

62. Sandbox Construction Edie Hall is planning to build a rectangular sandbox for her children. She wants its length to be 3 feet more than its width. Find the length and width of the sandbox if only 22 feet of lumber are available to form the frame. Use $P = 2l + 2w$.

63. Bookcase Dimensions Eric Krassow wishes to build a bookcase with four shelves (including the top) as shown in the figure. The width of the bookcase is to be 3 feet more than the height. If only 30 feet of wood are available to build the bookcase, what will be the dimensions of the bookcase?

64. Fence Dimensions Collette Siever wishes to fence in three rectangular areas along a river bank as illustrated in the figure. Each rectangle is to have the same dimensions, and the length of each rectangle is to be 1 meter greater than its width (along the river). Find the length and width of each rectangle if the total amount of fencing used is 114 meters.

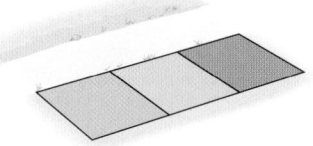

65. Price Reductions During the first week of a going-out-of-business sale, Sam's General Store reduces all prices by 10%. The second week of the sale, Sam's reduces all items by an additional $5. If Jim Condor bought a calculator for $49 during the second week of the sale, find the original price of the calculator.

66. Farm Divisions Deborah Schmidt's farm is divided into three regions. The area of one region is twice as large as the area of the smallest region, and the area of the third region is 4 acres less than three times the area of the smallest region. If the total acreage of the farm is 512 acres, find the area of each of the three regions.

67. Selling Paintings J. P. Richardson sells each of his paintings for $500. The gallery where he displays his work charges him $1350 a month plus a 10% commission on sales. How many paintings must J. P. sell in a month to break even?

68. Comparing Toy Sales Kristen Hodge is shopping for a certain bicycle for her niece and knows that Toys "R" Us and Wal-Mart sell the bike for the same price. On December 26, Toys "R" Us has the bike on sale for 37% off the original price and Wal-Mart has the bike on sale for $50 off the original price. After visiting both stores, Kristen discovers that the sale prices for the bike at the two stores are also the same.

a) Determine the original price of the bike.

b) Determine the sale price of the bike.

69. Incandescent Bulbs The cost of purchasing incandescent bulbs for use over a 9750-hour period is $9.75. The energy cost of incandescent bulbs over this period is $73. The cost of one equivalent fluorescent bulb that lasts about 9750 hours is $20. By using a fluorescent bulb instead of incandescent bulbs for 9750 hours, the total savings of purchase price plus energy cost is $46.75. What is the energy cost of using the fluorescent bulb over this period?

70. Dinner Bill The five members of the Newton family are going out to dinner with the three members of the Lee family. Before dinner, they decide that the Newtons will pay $\frac{5}{8}$ of the bill (before tip) and the Lees will pay $\frac{3}{8}$ plus the entire 15% tip. If the total bill including the 15% tip comes to $184.60, how much will be paid by each family?

71. Earning an A To find the average of a set of test scores, we divide the sum of the test scores by the number of test scores. On her first four algebra tests, Paula West's scores were 88, 92, 97, and 96.

a) Write an equation that can be used to determine the grade Paula needs to obtain on her fifth test to have a 90 average.

b) Explain how you determined your equation.

c) Solve the equation and determine the score.

72. Physics Exam Average Francis Timoney's grades on five physics exams were 70, 83, 97, 84, and 74.

a) If the final exam will count twice as much as each exam, what grade does Francis need on the final exam to have an 80 average?

b) If the highest possible grade on the final exam is 100 points, is it possible for Francis to obtain a 90 average? Explain.

73. a) Make up your own realistic word problem involving percents. Represent this word problem as an equation.

b) Solve the equation and answer the word problem.

74. a) Make up your own realistic word problem involving money. Represent this word problem as an equation.

b) Solve the equation and answer the word problem.

Challenge Problems

75. Truck Rental The Elmers Truck Rental Agency charges $28 per day plus 15¢ a mile. If Martina Estaban rented a small truck for 3 days and the total bill was $121.68, including a 4% sales tax, how many miles did she drive?

76. Money Market On Monday Sophia Murkovic purchased shares in a money market fund. On Tuesday the value of the shares went up 5%, and on Wednesday the value fell 5%. How much did Sophia pay for the shares on Monday if she sold them on Thursday for $59.85?

Group Activity

Discuss and answer Exercise 77 as a group.

77. a) Have each member of the group pick a number. Then multiply the number by 2, add 33, subtract 13, divide by 2, and subtract the number each started with. Record each answer.

b) Now compare answers. If you did not all get the same answer, check each other's work.

c) As a group, explain why this procedure will result in an answer of 10 for any real number n selected.

Cumulative Review Exercises

[1.3]   *Evaluate.*

78. $7 - \left| -\dfrac{3}{5} \right|$

79. $-6.4 - (-3.7)$

80. $\left| -\dfrac{5}{8} \right| \div |-4|$

81. $5 - |-3| - |12|$

[1.5]   **82.** Simplify $(2x^4 y^{-6})^{-3}$.

Mid-Chapter Test: 2.1–2.3

To find out how well you understand the chapter material to this point, take this brief test. The answers, and the section where the material was initially discussed, are given in the back of the book. Review any questions that you answered incorrectly.

1. Give the degree of $6x^5 y^7$.

Simplify each expression.

2. $3x^2 + 7x - 9x + 2x^2 - 11$

3. $2(a - 1.3) + 4(1.1a - 6) + 17$

Solve each equation.

4. $7x - 9 = 5x - 21$

5. $\dfrac{3}{4} y + \dfrac{1}{2} = \dfrac{7}{8} y - \dfrac{5}{4}$

6. $3p - 2(p + 6) = 4(p + 1) - 5$

7. $0.6(a - 3) - 3(0.4a + 2) = -0.2(5a + 9) - 4$

Find the solution set for each equation. Then indicate whether the equation is conditional, an identity, or a contradiction.

8. $4x + 15 - 9x = -7(x - 2) + 2x + 1$

9. $-3(3x + 1) = -[4x + (6x - 5)] + x + 7$

In Exercises 10 and 11, perform the indicated calculations.

10. Evaluate $A = \dfrac{1}{2} hb$, where $h = 10$ and $b = 16$.

11. Evaluate $R_T = \dfrac{R_1 R_2}{R_1 + R_2}$, where $R_1 = 100$ and $R_2 = 50$.

12. Solve $y = 7x + 13$ for x.

13. Solve $A = \dfrac{2x_1 + x_2 + x_3}{n}$ for x_3.

Solve each exercise.

14. Robert invested $700 into a certificate of deposit earning 6% interest compounded quarterly. How much is the certificate of deposit worth 5 years later?

15. Angles A and B are complementary angles. Determine the measures of angles A and B if angle A is 6° more than twice angle B.

16. The cost of renting a ladder is $15 plus $1.75 per day. How many days did Tom Lang rent the ladder if the total cost was $32.50?

17. The perimeter of a triangle is 100 feet. The longest side is two times the length of the shortest side and the other side is 20 feet longer than the shortest side. Find the lengths of the three sides of the triangle.

18. Tien bought a pair of shoes for $36.00. With tax, the cost was $37.62. Find the tax rate.

19. The population of a small town is increasing by 52 people per month. If the current population is 5693 people, how many months ago was the population 3613 people?

20. When asked to solve the equation $\dfrac{1}{2} x + \dfrac{1}{3} = \dfrac{1}{4} x - \dfrac{1}{2}$, Mary Dunwell claimed that to eliminate fractions, the left side should be multiplied by 6 and the right should be multiplied by 8. This is incorrect. Why is this incorrect? Explain your answer. What *single* number should be used to eliminate fractions from the *entire* equation? Solve the equation correctly.

2.4 Additional Application Problems

 Solve motion problems.

 Solve mixture problems.

In this section we discuss two additional types of application problems, motion and mixture problems.

1 Solve Motion Problems

A formula with many useful applications is

Motion Formula

$$\text{amount} = \text{rate} \cdot \text{time}$$

The "amount" in this formula can be a measure of many different quantities, depending on the rate. If the rate is *miles* per hour, the amount will be *distance*. If the rate is *gallons* of water per minute, the amount will be *volume*, and so on.

The motion formula can be used in many applications. A nurse may use the formula when giving a patient medicine intravenously. A company drilling for oil may use the formula to determine the time it takes to reach oil. When applying the formula the units must be consistent. For example, if a photocopier's rate is 45 copies per minute, time needs to be measured in minutes.

When the motion formula is used to calculate distance, the word *amount* is replaced with the word *distance* and the formula is called the **distance formula**.

Understanding Algebra

Problems that use the formula amount = rate · time are called *motion problems* because they involve movement, at a constant rate, for a certain period of time.

Distance Formula

The distance formula is

$$\text{distance} = \text{rate} \cdot \text{time}$$
$$\text{or} \quad d = rt$$

When a motion problem has two different rates, it is often helpful to put the information into a table to help analyze the problem.

EXAMPLE 1   **Ships at Sea**   The aircraft carrier USS *John F. Kennedy* and the nuclear-powered submarine USS *Memphis* leave from the Puget Sound Naval Yard at the same time heading for the same destination in the Indian Ocean. The aircraft carrier travels at its maximum speed of 34.5 miles per hour and the submarine travels submerged at its maximum speed of 20.2 miles per hour. The aircraft carrier and submarine are to travel at these speeds until they are 100 miles apart. How long will it take for the aircraft carrier and submarine to be 100 miles apart? (See **Fig. 2.6**.)

Solution   Understand   We wish to find out how long it will take for the difference between their distances to be 100 miles. To solve this problem, we will use the distance formula, $d = rt$. To help understand the problem, we will put the information into a table.

Let $t = $ time.

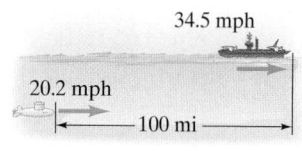

34.5 mph

20.2 mph

100 mi

FIGURE 2.6

	Rate	Time	Distance
Aircraft carrier	34.5	t	$34.5t$
Submarine	20.2	t	$20.2t$

Translate   The difference between their distances is 100 miles. Thus,

$$\text{aircraft carrier's distance} - \text{submarine's distance} = 100$$
$$34.5t \qquad - \qquad 20.2t \qquad = 100$$

Carry Out
$$14.3t = 100$$
$$t \approx 6.99$$

Answer   The aircraft carrier and submarine will be 100 miles apart in about 7 hours.

Now Try Exercise 3

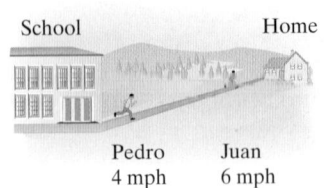

School   Home

Pedro   Juan
4 mph   6 mph

Juan arrives at home
$\frac{1}{2}$ hour before Pedro

FIGURE 2.7

EXAMPLE 2 Running Home To get in shape for the upcoming track season, Juan and Pedro Santiago begin running home after school. Juan runs at a rate of 6 mph and Pedro runs at a rate of 4 mph. When they leave the same school at the same time, Juan arrives home $\frac{1}{2}$ hour before Pedro (see **Fig. 2.7**).

a) How long does it take for Pedro to reach home?

b) How far do Juan and Pedro live from school?

Solution   **a)** Understand   Both boys will run the same distance. However, since Juan runs at a faster rate than Pedro does, Juan's time will be less than Pedro's time by $\frac{1}{2}$ hour.

Let $t =$ Pedro's time to run home.

Then $t - \frac{1}{2} =$ Juan's time to run home.

Runner	Rate	Time	Distance
Pedro	4	t	$4t$
Juan	6	$t - \frac{1}{2}$	$6\left(t - \frac{1}{2}\right)$

Translate   When the boys are home they will both have run the same distance from school. So

Pedro's distance $=$ Juan's distance

$$4t = 6\left(t - \frac{1}{2}\right)$$

Carry Out

$$4t = 6t - 3$$
$$-2t = -3$$
$$t = \frac{3}{2}$$

Answer   Pedro will take $1\frac{1}{2}$ hours to reach home.

b) The distance can be determined using either Juan's or Pedro's rate and time. We will multiply Pedro's rate by Pedro's time to find the distance.

$$d = rt = 4\left(\frac{3}{2}\right) = \frac{12}{2} = 6 \text{ miles}$$

Therefore, Juan and Pedro live 6 miles from their school.

Now Try Exercise 9

Helpful Hint

In Example 2, we could have let t be Juan's time and then $t + \frac{1}{2}$ would represent Pedro's time. Although this would lead to a different table and a different equation, the final answer would still be the same. Rework the problem now to confirm this.

EXAMPLE 3 Soda Pop Production A soda pop bottling machine fills and caps bottles. The machine can be run at two different rates. At the faster rate the machine fills and caps 600 more bottles per hour than it does at the slower rate. The machine is turned on for 4.8 hours on the slower rate, then it is changed to the faster rate for another 3.2 hours. During these 8 hours a total of 25,920 bottles were filled and capped. Find the slower rate and the faster rate.

Solution   Understand   This problem uses a number of bottles, an amount, in place of a distance. However, the problem is worked in a similar manner. We will use the formula, amount = rate · time. We are given that there are two different rates, and we are asked to find these rates. We will use the fact that the amount filled at the slower rate plus the amount filled at the faster rate equals the total amount filled.

Let r = the slower rate.

Then $r + 600$ = the faster rate.

	Rate	Time	Amount
Slower rate	r	4.8	$4.8r$
Faster rate	$r + 600$	3.2	$3.2(r + 600)$

Translate   amount filled at slower rate + amount filled at faster rate = 25,920

$$4.8r \qquad + \qquad 3.2(r + 600) \qquad = 25{,}920$$

Carry Out
$$4.8r + 3.2r + 1920 = 25{,}920$$
$$8r + 1920 = 25{,}920$$
$$8r = 24{,}000$$
$$r = 3000$$

Answer   The slower rate is 3000 bottles per hour. The faster rate is $r + 600$ or $3000 + 600 = 3600$ bottles per hour.

Now Try Exercise 11

2   Solve Mixture Problems

As we did with motion problems, we will use tables to help organize the information in mixture problems. Examples 4 and 5 are mixture problems that involve money.

EXAMPLE 4   **Two Investments** Bettie Truitt sold her boat for $15,000. Bettie loaned some of this money to her friend Kathy Testone. The loan was for 1 year with a simple interest rate of 4.5%. Bettie put the balance into a money market account at her credit union that yielded 3.75% simple interest. After 1 year, Bettie had earned a total of $637.50 from the two investments. Determine the amount Bettie loaned to Kathy.

Solution   Understand and Translate   To work this problem, we will use the simple interest formula, interest = principal · rate · time. We know that part of the investment is made at 4.5% and the rest is made at 3.75% simple interest. We are asked to determine the amount that Bettie loaned to Kathy.

Let p = amount loaned to Kathy at 4.5%.

Then $15{,}000 - p$ = amount invested at 3.75%.

Notice that the sum of the two amounts equals the total amount invested, $15,000. We will find the amount loaned to Kathy with the aid of a table.

Investment	Principal	Rate	Time	Interest
Loan to Kathy	p	0.045	1	$0.045p$
Money market	$15{,}000 - p$	0.0375	1	$0.0375(15{,}000 - p)$

Since the total interest collected is $637.50, we write

interest from 4.5% loan +   interest from 3.75% account = total interest

$$0.045p \qquad + \qquad 0.0375(15{,}000 - p) \qquad = \qquad 637.50$$

> ## Understanding Algebra
>
> Any problem where two or more quantities are combined to produce a different quantity or where a single quantity is separated into two or more different quantities may be considered a *mixture problem*.

Carry Out

$$0.045p + 0.0375(15,000 - p) = 637.50$$
$$0.045p + 562.50 - 0.0375p = 637.50$$
$$0.0075p + 562.50 = 637.50$$
$$0.0075p = 75$$
$$p = 10,000$$

Answer   Therefore, the loan to Kathy was for $10,000, and $15,000 - p$ or $15,000 - \$10,000 = \$5000$ was invested in the money market account.

Now Try Exercise 15

EXAMPLE 5 Hot Dog Stand Sales Matt's Hot Dog Stand in Chicago sells hot dogs for $2.00 each and beef tacos for $2.25 each. If the sales for the day total $585.50 and 278 items were sold, how many of each item were sold?

Solution   Understand and Translate   We are asked to find the number of hot dogs and beef tacos sold.

Let x = number of hot dogs sold.

Then $278 - x$ = number of beef tacos sold.

Item	Cost of Item	Number of Items	Total Sales
Hot dogs	2.00	x	$2.00x$
Beef tacos	2.25	$278 - x$	$2.25(278 - x)$

total sales of hot dogs + total sales of beef tacos = total sales

$$2.00x \qquad + \qquad 2.25(278 - x) \qquad = \qquad 585.50$$

Carry Out

$$2.00x + 625.50 - 2.25x = 585.50$$
$$-0.25x + 625.50 = 585.50$$
$$-0.25x = -40$$
$$x = \frac{-40}{-0.25} = 160$$

Answer   Therefore, 160 hot dogs and $278 - 160 = 118$ beef tacos were sold.

Now Try Exercise 17

In Example 5 we could have multiplied both sides of the equation by 100 to eliminate the decimal numbers, and then solved the equation.

Example 6 is a mixture problem that involves the mixing of two solutions.

EXAMPLE 6 Medicine Mixture Chemistry professor Tony Gambino has both 6% and 15% lithium citrate solutions. He wishes to make 0.5 liter of an 8% lithium citrate solution. How much of each solution must he mix?

Solution   Understand and Translate   We are asked to find the amount of each solution to be mixed.

Let x = number of liters of 6% solution.

Then $0.5 - x$ = number of liters of 15% solution.

The amount of lithium citrate in a solution is found by multiplying the percent strength of lithium citrate in the solution by the volume of the solution. We will draw a sketch of the problem (see **Fig. 2.8** on page 101) and then construct a table.

Understanding Algebra

The concentration of a mixed solution must always be between the concentrations of the two solutions used in making it. For example, if you mix a 5% acid solution with a 10% acid solution, the mixture must have a concentration between 5% and 10%.

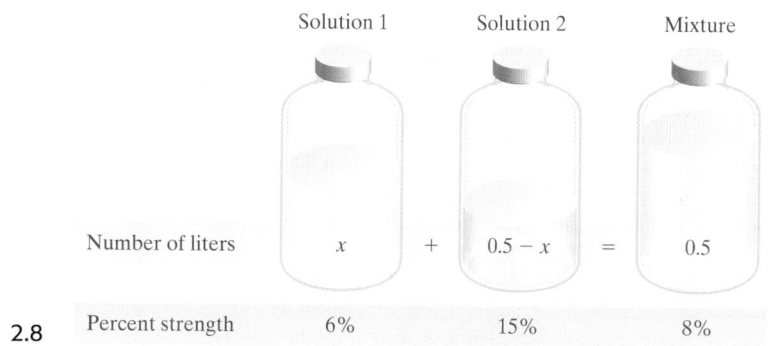

	Solution 1	Solution 2	Mixture
Number of liters	x	$+$ $0.5 - x$	$=$ 0.5
Percent strength	6%	15%	8%

FIGURE 2.8

Solution	Strength of Solution	Number of Liters	Amount of Lithium Citrate
1	0.06	x	$0.06x$
2	0.15	$0.5 - x$	$0.15(0.5 - x)$
Mixture	0.08	0.5	$0.08(0.5)$

$$\left(\begin{array}{c}\text{amount of}\\\text{lithium citrate}\\\text{in 6\% solution}\end{array}\right) + \left(\begin{array}{c}\text{amount of}\\\text{lithium citrate}\\\text{in 15\% solution}\end{array}\right) = \left(\begin{array}{c}\text{amount of lithium citrate}\\\text{in mixture}\end{array}\right)$$

$$0.06x \qquad + \qquad 0.15(0.5 - x) \qquad = \qquad 0.08(0.5)$$

Carry Out

$$0.06x + 0.15(0.5 - x) = 0.08(0.5)$$
$$0.06x + 0.075 - 0.15x = 0.04$$
$$0.075 - 0.09x = 0.04$$
$$-0.09x = -0.035$$
$$x = \frac{-0.035}{-0.09} \approx 0.39 \quad \left(\begin{array}{c}\text{to nearest}\\\text{hundredth}\end{array}\right)$$

Tony must mix 0.39 liter of the 6% solution and $0.5 - x$ or $0.5 - 0.39 = 0.11$ liter of the 15% solution to make 0.5 liter of an 8% solution.

Now Try Exercise 21

EXERCISE SET 2.4    *Math XL* MathXL®   **MyMathLab** MyMathLab

Practice the Skills and Problem Solving

In Exercises 1–14, write an equation that can be used to solve the motion problem. Solve the equation and answer the question asked.

1. A Hike in the Rockies Two friends, Don O'Neal and Judy McElroy, go hiking in the Rocky Mountains. While hiking they come across Bear Lake. They wonder what the distance around the lake is and decide to find out. Don knows he walks at 5 mph and Judy knows she walks at 4.5 mph. If they start walking at the same time in opposite directions around the lake, and meet in 1.2 hours, what is the distance around the lake?

2. Photocopier Speed Maria Hannaseck has two copiers on which to produce fliers. One copier has a rate of 45 copies per minute and the other has a rate of 15 copies per minute. If Darrell starts both copiers at the same time and lets them print fliers for 10 minutes, how many fliers will the copiers produce?

© iStockphoto

3. **Balloon Flights** Each year Albuquerque, New Mexico, has a hot air balloon festival during which people can obtain rides in hot air balloons. Suppose part of the Diaz family goes in one balloon and other members of the family go in another balloon. Because they fly at different altitudes and carry different weights, one balloon travels at 14 miles per hour while the other travels in the same direction at 11 miles per hour. In how many hours will they be 12 miles apart?

4. **Bikes** Paul and Frank are on the same bike path 39.15 miles apart. They are going to ride their bikes toward each other until they meet. Frank starts pedaling $1\frac{1}{2}$ hours before Paul. Paul rides 1.8 miles per hour faster than Frank. If they meet 3 hours after Paul starts riding, find the speed of each biker.

5. **Cornfield** Rodney and Dennis are gleaning (collecting) corn from a cornfield that is 1.5 miles long. Rodney starts on one end and is gleaning corn at a rate of 0.15 miles per hour. Dennis starts on the other end and is gleaning corn at a rate of 0.10 miles per hour. If they start at the same time and continue to work at the same rate, how long will it be before Rodney and Dennis meet?

6. **Photocopying** To make a large number of copies, Eileen Jones uses two photocopiers. One copier can produce copies at a rate of 42 copies per minute. The other copier can produce copies at a rate of 52 copies per minute. If Eileen starts both machines at the same time, how long will it take the two machines to produce a total of 1316 copies?

7. **Charity Race** The Alpha Delta Pi Sorority raises money for the Ronald McDonald House by holding an annual "Roll for Ronald" race. Mary Lou Baker rides a bicycle and travels at twice the speed of Wayne Siegert, who is on rollerblades. Mary and Wayne begin the race at the same time and after 3 hours, Mary is 18 miles ahead of Wayne.

   a) What is Wayne's speed?

   b) What is Mary's speed?

8. **Canyon Hiking** Jennifer Moyers hikes down to the bottom of Bryce Canyon, camps overnight, and returns the next day. Her hiking speed down averages 3.5 miles per hour and her return trip averages 2.1 miles per hour. If she spent a total of 16 hours hiking, find

   a) how long it took her to reach the bottom of the canyon.

   b) the total distance traveled.

See Exercise 8.

9. **Catching Up** Luis Nunez begins walking at a rate of 4 mph. Forty-five minutes after he leaves, his wife, Kristin, realizes that Luis has forgotten his wallet. Kristin gets on her bicycle and begins riding at a rate of 24 mph along the same path that Luis took.

   a) How long will it take for Kristin to catch Luis?

   b) How far from their house will Kristin catch Luis?

10. **Walking on the Beach** Max leaves his beach condo and begins walking down the beach at a rate of 3 miles per hour. Thirty minutes later, Rhiannon leaves the same condo and begins walking down the same beach at a rate of 4 miles per hour.

   a) How long will it take for Rhiannon to catch Max?

   b) How far from the condo will Rhiannon catch Max?

11. **Packing Spaghetti** Two machines are packing spaghetti into boxes. The smaller machine can package 400 boxes per hour and the larger machine can package 600 boxes per hour. If the larger machine is on for 2 hours before the smaller machine is turned on, how long after the smaller machine is turned on will a total of 15,000 boxes of spaghetti be boxed?

12. **Snail Races** As part of their preschool science project, Mrs. Joy Pribble's class is holding snail races. The first snail, Zippy, is known to move at a rate of 5 inches per hour. The second snail, Lightning, is known to move at 4.5 inches per hour. If the snails race along a straight path, and if Zippy finishes the race 0.25 hour before Lightning,

   a) determine the time it takes Lightning to run the race.

   b) determine the time it takes Zippy to run the race.

   c) what is the distance of the race?

13. **Meeting for Lunch** Ena and Jana live 385 miles apart and would like to meet somewhere in between their houses and then go to a restaurant for lunch. If Ena drives at 60 miles per hour and Jana drives at 50 miles per hour, how long will it take for them to meet?

14. **Walkie-Talkie Range** A set of Maxon RS446 walkie-talkies has a range of about two miles. Alice Burstein and Mary Kalscheur begin walking along a nature trail heading in opposite directions carrying their walkie-talkies. If Alice walks at a rate of 3.8 mph and Mary walks at a rate of 4.2 mph, how long will it take for them to be out of the range of the walkie-talkies?

In Exercises 15–28, write an equation that can be used to solve the mixture problem. Solve each equation and answer the question asked.

15. Two Investments Bill Palow invested $30,000 for one year in two separate accounts paying 3% and 4.1% annual simple interest. If Bill earned a total of $1091.73 from the two investments, how much was invested in each account?

16. Two Investments Terry Edwards invested $3000 for two years, part at 3.5% simple interest and the rest at 2.5% simple interest. After two years, she earned a total interest of $190. How much was invested at each rate?

17. Mixing Coffee Joan Smith is the owner of a Starbucks Coffee Shop. She sells Kona coffee for $6.20 per pound and amaretto coffee that sells for $5.80 per pound. She finds that by mixing these two blends she creates a coffee that sells well. If she uses 18 pounds of amaretto coffee in the blend and wishes to sell the mixture at $6.10 per pound, how many pounds of the Kona coffee should she mix with the amaretto coffee?

18. Mixing Nuts J. B. Davis owns a nut shop. He sells almonds for $6 per pound and walnuts for $5.20 per pound. He receives a special request from a customer who wants to purchase 30 pounds of a mixture of almonds and walnuts for $165. Determine how many pounds of almonds and walnuts should be mixed.

19. Investing in Webkinz Nicholas is a toy collector and is investing in Webkinz stuffed animals. He purchased some basset hounds for $50 each and twice as many black cats for $13 each. If Nicholas spent a total of $304, how many basset hounds and how many black cats did Nicholas purchase?

20. Sulfuric Acid Solutions Read Wickham, a chemistry teacher, needs a 5% sulfuric acid solution for an upcoming chemistry laboratory. When checking the storeroom, he realizes he has only 8 ounces of a 25% sulfuric acid solution. There is not enough time for him to order more, so he decides to make a 5% sulfuric acid solution by very carefully adding water to the 25% sulfuric acid solution. Determine how much water Read must add to the 25% solution to reduce it to a 5% solution.

21. Vinegar Solutions Distilled white vinegar purchased in supermarkets generally has a 5% acidity level. To make her sauerbraten, Chef Judy Ackermary marinates veal overnight in a special 8% distilled vinegar that she creates herself. To create the 8% solution she mixes a regular 5% vinegar solution with a 12% vinegar solution that she purchases by mail. How many ounces of the 12% vinegar solution should she add to 40 ounces of the 5% vinegar solution to get the 8% vinegar solution?

22. Hydrogen Peroxide Solution David Robertson works as a chemical engineer for US Peroxide Corporation. David has 2500 gallons of a commercial-grade hydrogen peroxide solution that is 60% pure hydrogen peroxide. How much distilled water (which is 0% hydrogen peroxide) will David need to add to this solution to create a new solution that is 25% pure hydrogen peroxide?

23. Horseradish Sauces Sally Finkelstein has a recipe that calls for horseradish sauce that is 45% pure horseradish. At the grocery store she finds one horseradish sauce that is 30% pure horseradish and another that is 80% pure horseradish. How many teaspoons of each of these horseradish sauces should Sally mix together to get 4 teaspoons of horseradish sauce that is 45% pure horseradish?

24. Grass Seed Mixture The Pearlman Nursery sells two types of grass seeds in bulk. The lower quality seeds have a germination rate of 76%, but the germination rate of the higher quality seeds is not known. Seven pounds of the higher quality seeds are mixed with 14 pounds of the lower quality seeds. If a later analysis of the mixture reveals that the mixture's germination rate was 80%, what is the germination rate of the higher quality seed?

25. Acid Solutions Two acid solutions are available to a chemist. One is a 20% sulfuric acid solution, but the label that indicates the strength of the other sulfuric acid solution is missing. Two hundred milliliters of the 20% acid solution and 100 milliliters of the solution with the unknown strength are mixed together. Upon analysis, the mixture was found to have a 25% sulfuric acid concentration. Find the strength of the solution with the missing label.

26. Diluting Vinegar Alex wishes to use vinegar that is 13% acetic acid as a natural weed killer. He has some table vinegar that is 5% acetic acid and two cups of pickling vinegar that is 15% acetic acid. How many cups of table vinegar should Alex mix with the two cups of pickling vinegar to obtain vinegar that is 13% acetic acid?

27. Candy Mixture A supermarket is selling two types of candies, orange slices and strawberry leaves. The orange slices cost $1.29 per pound and the strawberry leaves cost $1.79 per pound. How many pounds of each should be mixed to get a 12-pound mixture that sells for $17.48?

28. Octane Ratings The octane rating of gasoline indicates the percent of pure octane in the gasoline. For example, most regular gasoline has an 87-octane rating, which means that this gasoline is 87% octane (and 13% of some other non-oc- tane fuel such as pentane). Blake De Young owns a gas sta- tion and has 850 gallons of 87-octane gasoline. How many gallons of 93-octane gasoline must he mix with this 87-octane gasoline to obtain 89-octane gasoline?

In Exercises 29–46, write an equation that can be used to solve the motion or mixture problem. Solve each equation and answer the questions asked.

29. Route 66 The famous highway U.S. Route 66 connects Chicago and Los Angeles and extends 2448 miles. Julie Tur- ley begins in Chicago and drives at an average rate of 45 mph on Route 66 toward Los Angeles. At the same time Kamilia Nemri begins in Los Angeles and drives on Route 66 at an average rate of 50 mph toward Chicago. If Julie and Kamilia maintain these average speeds, how long after they start will they meet?

30. Meeting at a Restaurant Mike Mears and Scott Greenhalgh live 110 miles from each other. They frequently meet for lunch at a restaurant that is between Mike's house and Scott's house. Leaving at the same time from their respective houses, Mike takes 1 hour and 30 minutes and Scott takes 1 hour and 15 minutes to get to the restaurant. If they each drive at the same speed,

a) determine their speed.

b) how far from Scott's house is the restaurant?

31. Sump Pump Rates Gary Egan needs to drain his 15,000-gal- lon inground swimming pool to have it resurfaced. He uses two pumps to drain the pool. One drains 10 gallons of water a minute while the other drains 20 gallons of water a minute. If the pumps are turned on at the same time and remain on until the pool is drained, how long will it take for the pool to be drained?

32. Two Investments Chuy Carreon invested $8000 for 1 year, part at 3% and part at 5% simple interest. How much was in- vested in each account if the same amount of interest was re- ceived from each account?

33. Antifreeze Solution How many quarts of pure antifreeze should Doreen Kelly add to 10 quarts of a 20% antifreeze so- lution to make a 50% antifreeze solution?

34. Trip to Hawaii A jetliner flew from Chicago to Los Angeles at an average speed of 500 miles per hour. Then it continued on over the Pacific Ocean to Hawaii at an average speed of 550 miles per hour. If the entire trip covered 5200 miles and the part over the ocean took twice as long as the part over land, how long did the entire trip take?

35. Refueling a Jet An Air Force jet is going on a long-distance flight and will need to be refueled in midair over the Pacific Ocean. A refueling plane that carries fuel can travel much farther than the jet but flies at a slower speed. The refueling plane and jet will leave from the same base, but the refueling plane will leave 2 hours before the jet. The jet will fly at 800 mph and the refueling plane will fly at 520 miles per hour.

a) How long after the jet takes off will the two planes meet?

b) How far from the base will the refueling take place?

36. Working Two Jobs Hal Turziz works two part-time jobs. One job pays $7.50 an hour and the other pays $8.25 an hour. Last week Hal earned a total of $190.50 and worked for a total of 24 hours. How many hours did Hal work at each job?

37. Artwork Sales Joseph DeGuizman, an artist, sells both large paintings and small paintings. He sells his small paintings for $60 and his large paintings for $180. At the end of the week he determined that the total amount he made by selling 12 paintings was $1200. Determine the number of small and the number of large paintings that he sold.

38. Trip to Work Vince Jansen lives 35 miles from work. Due to construction, he must drive the first 15 minutes at a speed 10 mph slower than he does on the rest of the trip. If the entire trip takes him 45 minutes, determine Vince's speed on each part of the trip to work.

39. Alcohol Solution Herb Garrett has an 80% methyl alcohol solution. He wishes to make a gallon of windshield washer solution by mixing his methyl alcohol solution with water. If 128 ounces, or a gallon, of windshield washer fluid should contain 6% methyl alcohol, how much of the 80% methyl al- cohol solution and how much water must be mixed?

40. Mowing the Lawn Richard Stewart mows part of his lawn in second gear and part in third gear. It took him 2 hours to mow the entire lawn and the odometer on his tractor shows that he covered 13.8 miles while cutting the grass. If he aver- ages 4.2 miles per hour in second gear and 7.8 miles per hour in third gear, how long did he cut in each gear?

41. Concert Tickets Full-price tickets for a Jonas Brothers concert cost $56.50. Students can purchase a discounted ticket for $49.50. If a total of 3250 tickets are sold and if the total amount of ticket sales was $162,611,

a) how many full-price tickets were sold?

b) how many student tickets were sold?

42. Milk Mixture Sundance Dairy has 200 quarts of whole milk containing 6% butterfat. How many quarts of low-fat milk containing 1.5% butterfat should be added to produce milk containing 2.4% butterfat?

43. Comparing Transportation George Young can ride his bike to work in $\frac{3}{4}$ hour. If he takes his car to work, the trip takes $\frac{1}{6}$ hour. If George drives his car an average of 14 miles per hour faster than he rides his bike, determine the distance he travels to work.

44. Milk Carton Machine An old machine that folds and seals milk cartons can produce 50 milk cartons per minute. A new machine can produce 70 milk cartons per minute. The old machine has made 1000 milk cartons when the new machine is turned on. If both machines then continue working, how long after the new machine is turned on will the new machine have produced the same total number of milk cartons as the old machine?

45. Ocean Salinity The salinity (salt content) of the Atlantic Ocean averages 37 parts per thousand. If 64 ounces of saltwater is collected and placed in the sun, how many ounces of pure water would need to evaporate to raise the salinity to 45 parts per thousand? (Only the pure water is evaporated; the salt is left behind.)

46. Two Rockets Two rockets are launched from the Kennedy Space Center. The first rocket, launched at noon, will travel at 8000 miles per hour. The second rocket will be launched some time later and travel at 9500 miles per hour. When should the second rocket be launched if the rockets are to meet at a distance of 38,000 miles from earth?

a) Explain how to find the solution to this problem.

b) Find the solution to the problem.

47. a) Make up your own realistic motion problem that can be represented as an equation.

b) Write the equation that represents your problem.

c) Solve the equation, and then find the answer to your problem.

48. a) Make up your own realistic mixture problem that can be represented as an equation.

b) Write the equation that represents your problem.

c) Solve the equation, and then find the answer to your problem.

Challenge Problems

49. Distance to Calais The Chunnel (the underwater tunnel from Folkestone, England, to Calais, France) is 31 miles long. A person can board France's TGV bullet train in Paris and travel nonstop through the Chunnel and arrive in London 3 hours later. The TGV averages about 130 miles per hour from Paris to Calais. It then reduces its speed to an average of 90 miles per hour through the 31-mile Chunnel. When leaving the Chunnel in Folkestone it travels only at an average of about 45 miles per hour for the 68-mile trip from Folkestone to London because of outdated tracks. Using this information, determine the distance from Paris to Calais, France.

50. Race Cars Two cars labeled A and B are in a 500-lap race. Each lap is 1 mile. The lead car, A, is averaging 125 miles per hour when it reaches the halfway point. Car B is exactly 6.2 laps behind.

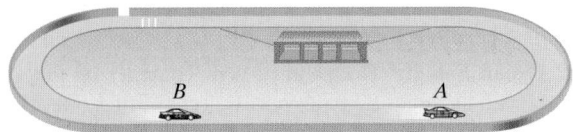

a) Find the average speed of car B.

b) When car A reaches the halfway point, how far behind, in seconds, is car B from car A?

51. Antifreeze Solution The radiator of an automobile has a capacity of 16 quarts. It is presently filled with a 20% antifreeze solution. How many quarts must be drained and replaced with pure antifreeze to make the radiator contain a 50% antifreeze solution?

Cumulative Review Exercises

[1.6] **52.** Express the quotient in scientific notation.
$$\frac{2.16 \times 10^5}{3.6 \times 10^8}$$

Solve.

[2.1] **53.** $0.6x + 0.22 = 0.4(x - 2.3)$

54. $\dfrac{2}{3}x + 8 = x + \dfrac{25}{4}$

[2.2] **55.** Solve the equation $\dfrac{3}{5}(x - 2) = \dfrac{2}{7}(2x + 3y)$ for y.

[2.3] **56. Truck Rental** Hertz/Penske Truck Rental Agency charges \$35 per day plus 75¢ a mile. Budget Truck Rental Agency charges \$20 per day plus 80¢ a mile for the same truck. What distance would you have to drive in 1 day to make the cost of renting from Hertz/Penske equal to the cost of renting from Budget?

2.5 Solving Linear Inequalities

1 Solve inequalities.

2 Graph solutions on a number line, interval notation, and solution sets.

3 Solve compound inequalities involving *and*.

4 Solve compound inequalities involving *or*.

1 Solve Inequalities

Inequalities and set builder notation were introduced in Section 1.2. You may wish to review that section now. The inequality symbols follow.*

Inequality Symbols

$>$	is greater than
$\geq$	is greater than or equal to
$<$	is less than
$\leq$	is less than or equal to

A mathematical expression containing one or more of these symbols is called an **inequality**.

Examples of Inequalities in One Variable

$$2x + 3 \leq 5 \qquad 4x > 3x - 5 \qquad 1.5 \leq -2.3x + 4.5 \qquad \frac{1}{2}x + 3 \geq 0$$

To solve an inequality, we must isolate the variable on one side of the inequality symbol. To isolate the variable, we use the same basic techniques used in solving equations.

---

*$\neq$, is not equal to, is also an inequality. $\neq$ means $<$ or $>$. Thus, $2 \neq 3$ means $2 < 3$ or $2 > 3$.

Properties Used to Solve Linear Inequalities

1. If $a > b$, then $a + c > b + c$.
2. If $a > b$, then $a - c > b - c$.
3. If $a > b$, and $c > 0$, then $ac > bc$.
4. If $a > b$, and $c > 0$, then $\dfrac{a}{c} > \dfrac{b}{c}$.
5. If $a > b$, and $c < 0$, then $ac < bc$.
6. If $a > b$, and $c < 0$, then $\dfrac{a}{c} < \dfrac{b}{c}$.

Helpful Hint

The properties used to solve linear inequalities can be summarized in three sentences:

1. The same number can be added to or subtracted from both sides of an inequality.
2. Both sides of an inequality can be multiplied or divided by any positive number.
3. *When both sides of an inequality are multiplied or divided by a negative number, the direction of the inequality symbol reverses.*

Example of Multiplication by a Negative Number	Example of Division by a Negative Number

Multiply both sides of the inequality by -1 and reverse the direction of the inequality symbol.

$$4 > -2$$
$$-1\,(4) < -1\,(-2)$$
$$-4 < 2$$

$$10 \geq -4$$
$$\frac{10}{-2} \leq \frac{-4}{-2}$$
$$-5 \leq 2$$

Divide both sides of the inequality by -2 and reverse the direction of the inequality symbol.

Helpful Hint

Do not forget to reverse the direction of the inequality symbol when multiplying or dividing both sides of the inequality by a negative number.

Inequality	Direction of Inequality Symbol
$-3x < 6$	$\dfrac{-3x}{-3} > \dfrac{6}{-3}$
$-\dfrac{x}{2} > 5$	$(-2)\left(-\dfrac{x}{2}\right) < (-2)(5)$

EXAMPLE 1 Solve the inequalities. **a)** $5x - 7 \geq -17$ **b)** $-6x + 4 < -14$

Solution

a)
$$5x - 7 \geq -17$$
$$5x - 7 + 7 \geq -17 + 7 \qquad \text{Add 7 to both sides.}$$
$$5x \geq -10$$
$$\frac{5x}{5} \geq \frac{-10}{5} \qquad \text{Divide both sides by 5.}$$
$$x \geq -2$$

The solution set is $\{x \mid x \geq -2\}$. Any real number greater than or equal to -2 will satisfy the inequality.

b)
$$-6x + 4 < -14$$
$$-6x + 4 - 4 < -14 - 4 \qquad \text{Subtract 4 from both sides.}$$
$$-6x < -18$$
$$\frac{-6x}{-6} > \frac{-18}{-6} \qquad \begin{array}{l}\text{Divide both sides by } -6 \text{ and reverse} \\ \text{the direction of the inequality.}\end{array}$$
$$x > 3$$

The solution set is $\{x \mid x > 3\}$. Any number greater than 3 will satisfy the inequality.

Now Try Exercise 17

2 Graph Solutions on a Number Line, Interval Notation, and Solution Sets

The solution to an inequality can be shown on a number line or written as a solution set. The solution can also be written in interval notation, as illustrated next.

The following symbols will be used when representing solutions to inequalities.

Inequality Symbol	Number Line Endpoints	Interval Notation Symbols	Is endpoint included in solution?
$<$ or $>$	○	(or)	No
$\leq$ or $\geq$	●	[or]	Yes

The symbol ∞ is read "infinity"; it indicates that the solution set continues indefinitely.

<table>
<tr><td>Understanding Algebra

Whenever ∞ is used in interval notation, a parenthesis must be used on the corresponding side of the interval notation.</td><td colspan="3">

Solution of Inequality	Solution Set Indicated on Number Line	Solution Set Represented in Interval Notation
$x \geq 5$		$[5, \infty)$
$x < 3$		$(-\infty, 3)$
$2 < x \leq 6$		$(2, 6]$
$-6 \leq x \leq -1$		$[-6, -1]$
$x > a$		(a, ∞)
$x \geq a$		$[a, \infty)$
$x < a$		$(-\infty, a)$
$x \leq a$		$(-\infty, a]$
$a < x < b$		(a, b)
$a \leq x \leq b$		$[a, b]$
$a < x \leq b$		$(a, b]$
$a \leq x < b$		$[a, b)$

</td></tr>
</table>

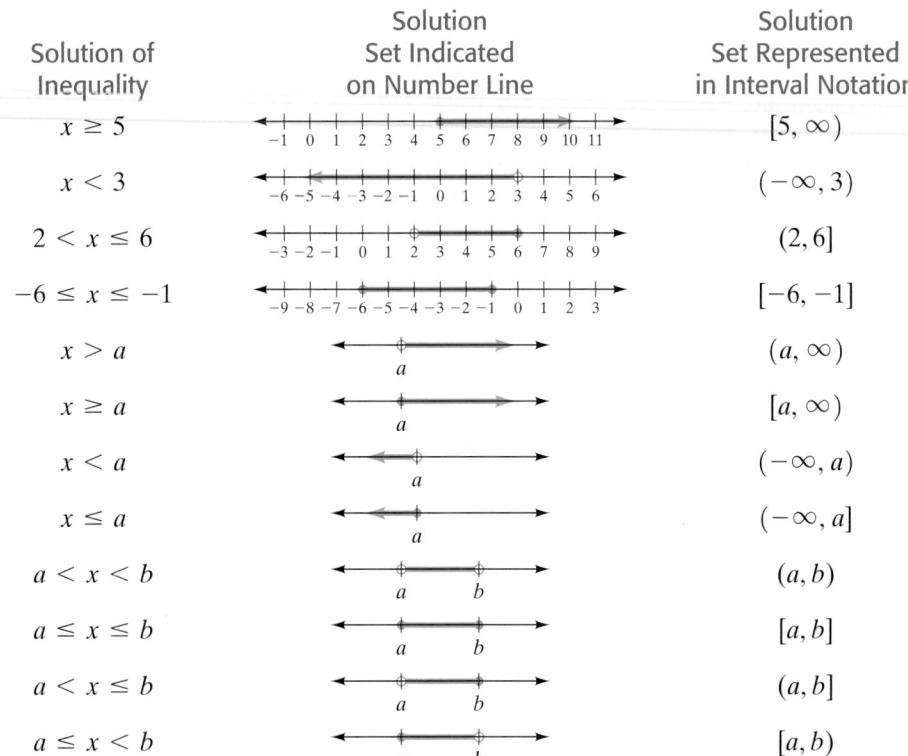

EXAMPLE 2 Solve the following inequality and give the solution both on a number line and in interval notation.

$$\frac{1}{4}z - \frac{1}{2} < \frac{2z}{3} + 2$$

Solution   We can eliminate fractions from an inequality by multiplying both sides of the inequality by the least common denominator, LCD, of the fractions.

$$\frac{1}{4}z - \frac{1}{2} < \frac{2z}{3} + 2$$

$$12\left(\frac{1}{4}z - \frac{1}{2}\right) < 12\left(\frac{2z}{3} + 2\right) \qquad \text{Multiply both sides by the LCD, 12.}$$

$$3z - 6 < 8z + 24 \qquad \text{Distributive property}$$

$$3z - 8z - 6 < 8z - 8z + 24 \qquad \text{Subtract } 8z \text{ from both sides.}$$

$$-5z - 6 < 24$$

$$-5z - 6 + 6 < 24 + 6 \qquad \text{Add 6 to both sides.}$$

$$-5z < 30$$

$$\frac{-5z}{-5} > \frac{30}{-5} \qquad \text{Divide both sides by } -5 \text{ and reverse the direction of inequality symbol.}$$

$$z > -6$$

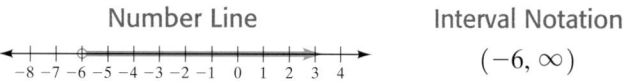

Number Line

Interval Notation

$(-6, \infty)$

The solution set is $\{z | z > -6\}$.

Now Try Exercise 31

In Example 2 we illustrated the solution on a number line, in interval notation, and as a solution set.

EXAMPLE 3 Solve the inequality $2(3p - 5) + 9 \leq 8(p + 1) - 2(p - 3)$.

Solution
$$2(3p - 5) + 9 \leq 8(p + 1) - 2(p - 3)$$
$$6p - 10 + 9 \leq 8p + 8 - 2p + 6$$
$$6p - 1 \leq 6p + 14$$
$$6p - 6p - 1 \leq 6p - 6p + 14$$
$$-1 \leq 14$$

Since -1 is always less than or equal to 14, the inequality is true for all real numbers. When an inequality is true for all real numbers, the solution set is *the set of all real numbers*, $\mathbb{R}$. The solution set to this example can also be indicated on a number line or given in interval notation.

or $(-\infty, \infty)$

Now Try Exercise 23

If Example 3 had resulted in the expression $-1 \geq 14$, the inequality would never have been true, since -1 is never greater than or equal to 14. When an inequality is never true, it has no solution. The solution set of an inequality that has no solution is the *empty* or *null set*, { } or $\emptyset$. We will represent the empty set on the number line as follows, ←——|——→ .
                                                        0

Understanding Algebra

If, while solving an inequality, you get an inequality with no variables that is always

- true, such as $-1 \leq 14$, then the solution set is the set of *all real numbers*, $\mathbb{R}$.

- false, such as $5 < -3$, then the solution set is the empty set, $\emptyset$.

> **Helpful Hint**
>
> Generally, when writing a solution to an inequality, we write the variable on the left. For example, when solving an inequality, if we obtain $5 \geq y$ we would write the solution as $y \leq 5$. For example,
>
> $-6 < x$ means $x > -6$ (inequality symbol points to -6 in both cases)
> $4 > x$ means $x < 4$ (inequality symbol points to x in both cases)
> $a < x$ means $x > a$ (inequality symbol points to a in both cases)
> $a > x$ means $x < a$ (inequality symbol points to x in both cases)

EXAMPLE 4 **Packages on a Boat** A small boat has a maximum weight load of 750 pounds. Millie Harrison has to transport packages weighing 42.5 pounds each.

a) Write an inequality that can be used to determine the maximum number of packages that Millie can safely place on the boat if she weighs 128 pounds.

b) Find the maximum number of packages that Millie can transport.

Solution

a) Understand and Translate   Let n = number of packages.

Millie's weight + weight of n packages ≤ 750
$$128 \quad + \quad 42.5n \quad\quad \leq 750$$

b) Carry Out $128 + 42.5n \leq 750$
$$42.5n \leq 622$$
$$n \leq 14.6$$

Answer   Therefore, Millie can transport up to 14 packages on the boat.

Now Try Exercise 65

© Tereshchenko Dmitry/Shutterstock

EXAMPLE 5 Bowling Alley Rates At the Corbin Bowl bowling alley in Tarzana, California, it costs $2.50 to rent bowling shoes and it costs $4.00 per game bowled.

a) Write an inequality that can be used to determine the maximum number of games that Ricky Olson can bowl if he has only $20.

b) Find the maximum number of games that Ricky can bowl.

Solution   **a)**   Understand and Translate

Let g = number of games bowled.

Then $4.00g$ = cost of bowling g games.

cost of shoe rental + cost of bowling g games $\leq$ money Ricky has

$$2.50 \qquad + \qquad 4.00g \qquad\qquad \leq \qquad 20$$

b) Carry Out

$$2.50 + 4.00g \leq 20$$
$$4.00g \leq 17.50$$
$$\frac{4.00g}{4.00} \leq \frac{17.50}{4.00}$$
$$g \leq 4.375$$

Answer and Check   Since he can't play a portion of a game, the maximum number of games that he can afford to bowl is 4. If Ricky were to bowl 5 games, he would owe $2.50 + 5($4.00$) = 22.50, which is more money than the $20 that he has.

Now Try Exercise 67

EXAMPLE 6 Profit For a business to realize a profit, its revenue (or income), R, must be greater than its cost, C. That is, a profit will be obtained when $R > C$ (the company breaks even when $R = C$). A company that produces playing cards has a weekly cost equation of $C = 1525 + 1.7x$ and a weekly revenue equation of $R = 4.2x$, where x is the number of decks of playing cards produced and sold in a week. How many decks of cards must be produced and sold in a week for the company to make a profit?

Solution   Understand and Translate   The company will make a profit when $R > C$, or

$$4.2x > 1525 + 1.7x$$

Carry Out
$$2.5x > 1525$$
$$x > \frac{1525}{2.5}$$
$$x > 610$$

Answer   The company will make a profit when more than 610 decks are produced and sold in a week.

Now Try Exercise 69

EXAMPLE 7 Tax Tables The 2008 tax rate schedule for married couples who file a joint tax return is shown below.

Schedule Y-1 Use if your filing status is **Married filing jointly or Qualifying widow(er)**

If the taxable income is over–	But not over–	The tax is:	Of the amount over–
$0	$16,050	10%	$0
$16,050	$65,100	$1,605.00 + 15%	$16,050
$65,100	$131,450	$8,962.50 + 25%	$65,100
$131,450	$200,300	$25,550.00 + 28%	$131,450
$200,300	$357,700	$44,828.00 + 33%	$200,300
$357,700	∞	$96,770.00 + 35%	$357,700

a) Write, in interval notation, the amounts of taxable income that make up each of the six listed tax brackets, that is, the 10%, 15%, 25%, 28%, 33%, and 35% tax brackets.

b) Determine the tax for a married couple filing jointly if their taxable income is $13,500.

c) Determine the tax for a married couple filing jointly if their taxable income is $136,000.

Solution

a) The words *But Not Over* mean "less than or equal to." The taxable incomes that make up the six tax brackets are

$(0, 16{,}050]$ for the 10% tax bracket

$(16{,}050, 65{,}100]$ for the 15% tax bracket

$(65{,}100, 131{,}450]$ for the 25% tax bracket

$(131{,}450, 200{,}300]$ for the 28% tax bracket

$(200{,}300, 357{,}700]$ for the 33% tax bracket

$(357{,}700, \infty)$ for the 35% tax bracket

b) The tax for a married couple filing jointly with taxable income of $13,500 is 10% of $13,500. Therefore,

$$\text{tax} = 0.10(13{,}500) = \$1{,}350$$

The tax is $1350.

c) A taxable income of $136,000 places a married couple filing jointly in the 28% tax bracket. The tax is $25,500 + 28% of the taxable income over $131,450. The taxable income over $131,450 is $136,000 − $131,450 = $4550. Therefore,

$$\text{tax} = 25{,}550.00 + 0.28(4550) = 25{,}550 + 1274 = 26{,}824$$

The tax is $26,824.

Now Try Exercise 79

3   Solve Compound Inequalities Involving *And*

A **compound inequality** is formed by joining two inequalities with the word *and* or *or*. Sometimes the word *and* is implied without being written.

Examples of Compound Inequalities

$$3 < x \quad \text{and} \quad x < 5$$
$$x + 4 > 2 \quad \text{or} \quad 2x - 3 < 6$$
$$4x - 6 \geq -3 \quad \text{and} \quad x - 6 < 17$$

The solution of a compound inequality using the word *and* is all the numbers that make *both* parts of the inequality true. Consider

$$3 < x \quad \text{and} \quad x < 5$$

The numbers that satisfy both inequalities may be easier to see if we graph the solution to each inequality on a number line (see **Fig. 2.9**). Now we can see that the numbers that satisfy both inequalities are the numbers between 3 and 5. The solution set is $\{x|3 < x < 5\}$.

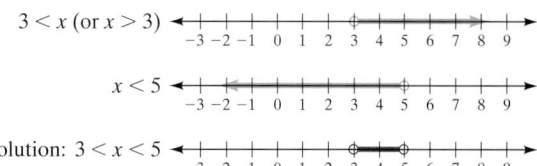

FIGURE 2.9

The intersection of two sets is the set of elements common to both sets. *To find the solution set of an inequality containing the word **and**, take the **intersection** of the solution sets of the two inequalities.*

EXAMPLE 8 Solve $x + 5 \leq 8$ and $2x - 9 > -7$

Solution   Begin by solving each inequality separately.

$$x + 5 \leq 8 \quad \text{and} \quad 2x - 9 > -7$$
$$x \leq 3 \qquad\qquad\qquad 2x > 2$$
$$x > 1$$

Now take the intersection of the sets $\{x | x \leq 3\}$ and $\{x | x > 1\}$. When we find $\{x | x \leq 3\} \cap \{x | x > 1\}$, we are finding the values of x common to both sets. **Figure 2.10** illustrates that the solution set is $\{x | 1 < x \leq 3\}$. In interval notation, the solution is $(1, 3]$.

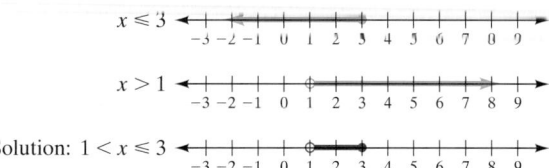

FIGURE 2.10

Now Try Exercise 57

Sometimes a compound inequality using the word *and* can be written in a shorter form.

$$3 < x \text{ and } x < 5 \quad \text{is written} \quad 3 < x < 5,$$
$$-1 < x + 3 \text{ and } x + 3 \leq 5 \quad \text{is written} \quad -1 < x + 3 \leq 5$$

EXAMPLE 9 Solve $-1 < x + 3 \leq 5$

Solution $-1 < x + 3 \leq 5$ means $-1 < x + 3$ and $x + 3 \leq 5$. Solve each inequality separately.

$$-1 < x + 3 \quad \text{and} \quad x + 3 \leq 5$$
$$-4 < x \qquad\qquad\qquad x \leq 2$$

Remember that $-4 < x$ means $x > -4$. **Figure 2.11** illustrates that the solution set is $\{x | -4 < x \leq 2\}$. In interval notation, the solution is $(-4, 2]$.

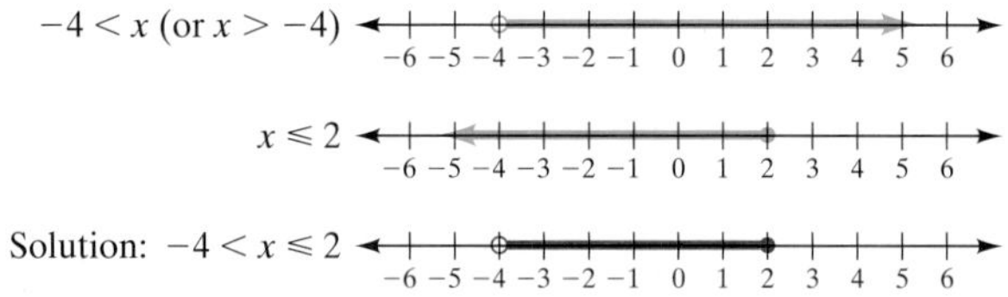

FIGURE 2.11

Now Try Exercise 35

The inequality in Example 9, $-1 < x + 3 \leq 5$, can be solved in another way. We could have subtracted 3 from all three parts to isolate the variable in the middle and solve the inequality.

$$-1 < x + 3 \leq 5$$
$$-1 - 3 < x + -3 \leq 5 - 3$$
$$-4 < x \leq 2$$

Note that this is the same solution as obtained in Example 9.

Understanding Algebra

When solving an inequality with more than two *parts,* whatever we do to one part, we must do to *all* parts.

EXAMPLE 10 Solve the inequality $-3 \le 2t - 7 < 8$. —————————

Solution   We wish to isolate the variable t. We begin by adding 7 to all three parts of the inequality.

$$-3 \le 2t - 7 < 8$$
$$-3 + 7 \le 2t - 7 + 7 < 8 + 7$$
$$4 \le 2t < 15$$

Now divide all three parts of the inequality by 2.

$$\frac{4}{2} \le \frac{2t}{2} < \frac{15}{2}$$
$$2 \le t < \frac{15}{2}$$

The solution may also be illustrated on a number line, written in interval notation, or written as a solution set. Below we show each form.

The answer in interval notation is $\left[2, \dfrac{15}{2}\right)$. The solution set is $\left\{t \,\middle|\, 2 \le t < \dfrac{15}{2}\right\}$.

Now Try Exercise 41

EXAMPLE 11 Solve the inequality $-2 < \dfrac{4 - 3x}{5} < 8$. —————————

Solution   Multiply all three parts by 5 to eliminate the denominator.

$$-2 < \frac{4 - 3x}{5} < 8$$
$$-2(5) < 5\left(\frac{4 - 3x}{5}\right) < 8(5)$$
$$-10 < 4 - 3x < 40$$
$$-10 - 4 < 4 - 4 - 3x < 40 - 4$$
$$-14 < -3x < 36$$

Now divide all three parts of the inequality by -3. Remember that when we multiply or divide an inequality by a negative number, the direction of the inequality symbols reverse.

$$\frac{-14}{-3} > \frac{-3x}{-3} > \frac{36}{-3}$$
$$\frac{14}{3} > x > -12$$

Although $\dfrac{14}{3} > x > -12$ is correct, we generally write compound inequalities with the smaller value on the left. We will, therefore, rewrite the solution as

$$-12 < x < \frac{14}{3}$$

The solution may also be illustrated on a number line, written in interval notation, or written as a solution set.

The solution in interval notation is $\left(-12, \dfrac{14}{3}\right)$. The solution set is $\left\{x \,\middle|\, -12 < x < \dfrac{14}{3}\right\}$.

Now Try Exercise 43

Avoiding Common Errors

You must be very careful when writing the solution to a compound inequality. In Example 11 we can change the solution from

$$\frac{14}{3} > x > -12 \quad \text{to} \quad -12 < x < \frac{14}{3}$$

This is correct since both say that x is greater than -12 and less than $\frac{14}{3}$. Notice that the inequality symbol in both cases is pointing to the smaller number.

In Example 11, had we written the answer $\frac{14}{3} < x < -12$, we would have given the incorrect solution. Remember that the inequality $\frac{14}{3} < x < -12$ means that $\frac{14}{3} < x$ and $x < -12$. There is no number that is both greater than $\frac{14}{3}$ and less than -12. Also, by examining the inequality $\frac{14}{3} < x < -12$, it appears as if we are saying that -12 is a greater number than $\frac{14}{3}$, which is obviously incorrect.

It would also be incorrect to write the answer as

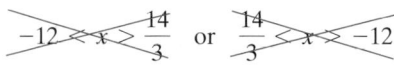

EXAMPLE 12 **Calculating Grades** In an anatomy and physiology course, an average score greater than or equal to 80 and less than 90 will result in a final grade of B. Steve Reinquist received scores of 85, 90, 68, and 70 on his first four exams. For Steve to receive a final grade of B in the course, between which two scores must his fifth (and last) exam fall?

Solution   Let x = Steve's last exam score.

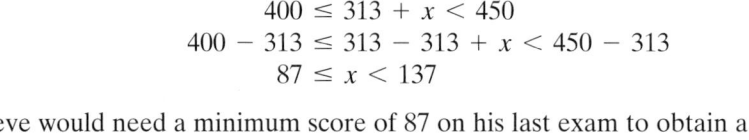

$$80 \leq \text{average of five exams} < 90$$
$$80 \leq \frac{85 + 90 + 68 + 70 + x}{5} < 90$$
$$80 \leq \frac{313 + x}{5} < 90$$
$$400 \leq 313 + x < 450$$
$$400 - 313 \leq 313 - 313 + x < 450 - 313$$
$$87 \leq x < 137$$

Steve would need a minimum score of 87 on his last exam to obtain a final grade of B. If the highest score he could receive on the test is 100, is it possible for him to obtain a final grade of A (90 average or higher)? Explain.

Now Try Exercise 75

4 **Solve Compound Inequalities Involving** *Or*

The solution to a compound inequality using the word *or* is all the numbers that make *either* of the inequalities a true statement. Consider the compound inequality

$$x > 3 \quad \text{or} \quad x < 5$$

What are the numbers that satisfy the compound inequality? Let's graph the solution to each inequality on the number line (see **Fig. 2.12**). Note that every real number satisfies at least one of the two inequalities. Therefore, the solution set to the compound inequality is the set of all real numbers, $\mathbb{R}$.

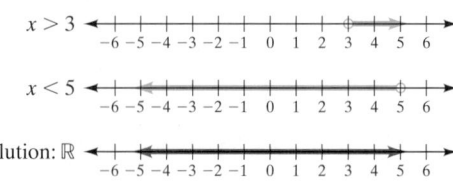

FIGURE 2.12

The *union* of two sets is the set of elements that belong to *either* of the sets. *To find the solution set of an inequality containing the word* **or**, *take the* **union** *of the solution sets of the two inequalities.*

EXAMPLE 13 Solve $r - 2 \leq -6$ or $-4r + 3 < -5$.

Solution   Solve each inequality separately.

$$r - 2 \leq -6 \quad \text{or} \quad -4r + 3 < -5$$
$$r \leq -4 \qquad\qquad -4r < -8$$
$$\qquad\qquad\qquad r > 2$$

Now graph each solution on number lines and then find the union (see **Fig. 2.13**). The union is $r \leq -4$ or $r > 2$.

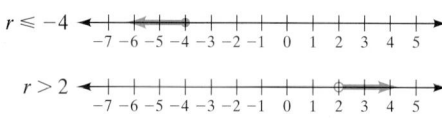

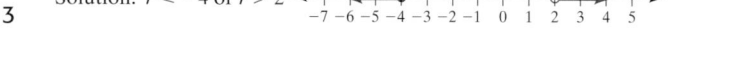

FIGURE 2.13   Solution: $r \leq -4$ or $r > 2$

The solution set is $\{r \mid r \leq -4\} \cup \{r \mid r > 2\}$, which can be written as $\{r \mid r \leq -4$ or $r > 2\}$. In interval notation, the solution is $(-\infty, -4] \cup (2, \infty)$.

Now Try Exercise 59

Helpful Hint

There are various ways to write the solution to an inequality problem. Be sure to indicate the solution to an inequality problem in the form requested by your professor. Examples of various forms follow.

Inequality	Number Line	Interval Notation	Solution Set
$x < \dfrac{5}{3}$	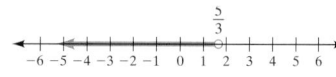	$\left(-\infty, \dfrac{5}{3}\right)$	$\left\{ x \,\middle|\, x < \dfrac{5}{3} \right\}$
$-4 < t \leq \dfrac{5}{3}$		$\left(-4, \dfrac{5}{3}\right]$	$\left\{ t \,\middle|\, -4 < t \leq \dfrac{5}{3} \right\}$

EXERCISE SET 2.5

Math XL   MyMathLab
MathXL®   MyMathLab

Warm-up Exercises

Fill in the blanks with the appropriate word, phrase, or symbol(s) from the following list.

direction   open   closed   intersection   union   compound   simple

1. A _____ inequality is formed by joining two inequalities with the word *and* or *or*.

2. A(n) _____ circle on the number line indicates that the endpoint is not part of the solution.

3. A(n) _____ circle on the number line indicates that the endpoint is part of the solution.

4. To find the solution set of an inequality containing the word *and*, take the _____ of the solution sets of the two inequalities.

5. To find the solution set of an inequality containing the word *or*, take the _____ of the solution sets of the two inequalities.

6. Whenever you multiply or divide both sides of an inequality by a negative number, you must change the _____ of the inequality symbol.

Practice the Skills

Express each inequality **a)** *using a number line,* **b)** *in interval notation, and* **c)** *as a solution set (use set builder notation)*

7. $x > -3$

8. $t > \dfrac{3}{4}$

9. $w \le \pi$

10. $-4 < x < 3$

11. $-3 < q \le \dfrac{4}{5}$

12. $x \ge -\dfrac{6}{5}$

13. $-7 < x \le -4$

14. $-2\dfrac{7}{8} \le k < -1\dfrac{2}{3}$

Solve each inequality and graph the solution on the number line.

15. $x + 8 > 10$

16. $2x + 3 > 4$

17. $3 - x < -4$

18. $12b - 5 \le 8b + 7$

19. $4.7x - 5.48 \ge 11.44$

20. $1.4x + 2.2 < 2.6x - 0.2$

21. $4(x + 2) \le 4x + 8$

22. $15.3 > 3(a - 1.4)$

23. $5b - 6 \ge 3(b + 3) + 2b$

24. $-6(d + 2) < -9d + 3(d - 1)$

25. $2y - 6y + 8 \le 2(-2y + 9)$

26. $\dfrac{y}{2} + \dfrac{4}{5} \le 3$

Solve each inequality and give the solution in interval notation.

27. $4 + \dfrac{4x}{3} < 6$

28. $4 - 3x < 5 + 2x + 17$

29. $\dfrac{v - 5}{3} - v \ge -3(v - 1)$

30. $\dfrac{h}{2} - \dfrac{5}{6} < \dfrac{7}{8} + h$

31. $\dfrac{t}{3} - t + 7 \le -\dfrac{4t}{3} + 8$

32. $\dfrac{6(x - 2)}{5} > \dfrac{10(2 - x)}{3}$

33. $-3x + 1 < 3[(x + 2) - 2x] - 1$

34. $4[x - (3x - 2)] > 3(x + 5) - 15$

Solve each inequality and give the solution in interval notation.

35. $-2 \le t + 3 < 4$

36. $-7 < p - 6 \le -5$

37. $-15 \le -3z \le 12$

38. $-16 < 5 - 3n \le 13$

39. $4 \le 2x - 4 < 7$

40. $-12 < 3x - 5 \le -1$

41. $14 \le 2 - 3g < 15$

42. $\dfrac{1}{2} < 3x + 4 < 13$

Solve each inequality and give the solution set.

43. $5 \le \dfrac{3x + 1}{2} < 11$

44. $\dfrac{3}{5} < \dfrac{-x-5}{3} < 2$

45. $-6 \le -3(2x - 4) < 12$

46. $-6 < \dfrac{4 - 3x}{2} < \dfrac{2}{3}$

47. $0 \le \dfrac{3(u - 4)}{7} \le 1$

48. $-15 < \dfrac{3(x - 2)}{5} \le 0$

Solve each inequality and indicate the solution set.

49. $c \le 1$ and $c > -3$

50. $d > 0$ or $d \le 8$

51. $x < 2$ and $x > 4$

52. $w \le -1$ or $w > 6$

53. $x + 1 < 3$ and $x + 1 > -4$

54. $5x - 3 \le 7$ or $-2x + 5 < -3$

Solve each inequality and give the solution in interval notation.

55. $2s + 3 < 7$ or $-3s + 4 \le -17$

56. $4a + 7 \ge 9$ and $-3a + 4 \le -17$

57. $4x + 5 \ge 5$ and $3x - 7 \le -1$

58. $5 - 3x < -3$ and $5x - 3 > 10$

59. $4 - r < -2$ or $3r - 1 < -1$

60. $-x + 3 < 0$ or $2x - 5 \ge 3$

61. $2k + 5 > -1$ and $7 - 3k \le 7$

62. $2q - 11 \le -7$ or $2 - 3q < 11$

Problem Solving

63. UPS Packages To avoid a surcharge, the length plus the girth of a package to be shipped by United Parcel Service (UPS) can be no larger than 130 inches.

a) Write an inequality that expresses this information, using l for the length and g for the girth.

b) UPS has defined girth as twice the width plus twice the depth. Write an inequality using length, l, width, w, and

depth, d, to indicate the maximum allowable dimensions of a package that may be shipped without paying the surcharge.

c) If the length of a package is 40 inches and the width of a package is 20.5 inches, find the maximum allowable depth of the package.

64. Carry-On Luggage Many airlines have limited the size of luggage passengers may carry onboard on domestic flights. The length, l, plus the width, w, plus the depth, d, of the carry-on luggage must not exceed 45 inches.

    a) Write an inequality that describes this restriction, using l, w, and d as described above.

    b) If Ryan McHenry's luggage is 23 inches long and 12 inches wide, what is the maximum depth it can have and still be carried on the plane?

© Allen R. Angel

In Exercises 65–78, set up an inequality that can be used to solve the problem. Solve the problem and find the desired value.

65. Weight Limit Cal Worth, a janitor, must move a large shipment of books from the first floor to the fifth floor. The sign on the elevator reads "maximum weight 800 pounds." If each box of books weighs 70 pounds, find the maximum number of boxes that Cal should place in the elevator, if he does not ride.

66. Elevator Limit If the janitor in Exercise 65, weighing 195 pounds, must ride up with the boxes of books, find the maximum number of boxes that can be placed into the elevator.

67. Text Messages The Verizon Wireless Message Pack 200 plan includes 200 text messages per month for $5.00. Additional text messages are $0.15 each. If Berma Williams uses this plan, how many text messages can she purchase for $20?

© iStockphoto

68. Parking Garage A downtown parking garage in Austin, Texas, charges $1.25 for the first hour and $0.75 for each additional hour or part thereof. What is the maximum length of time you can park in the garage if you wish to pay no more than $3.75?

69. Book Profit April Lemons is considering writing and publishing her own book. She estimates her revenue equation as $R = 6.42x$, and her cost equation as $C = 10{,}025 + 1.09x$, where x is the number of books she sells. Find the minimum number of books she must sell to make a profit. See Example 6.

70. Dry Cleaning Profit Peter Collinge is opening a dry cleaning store. He estimates his cost equation as $C = 8000 + 0.08x$ and his revenue equation as $R = 1.85x$, where x is the number of garments dry cleaned in a year. Find the minimum number of garments that must be dry cleaned in a year for Peter to make a profit.

© Allen R. Angel

71. Mailing Large Envelopes The cost to mail a large envelope is $0.83 for the first ounce and $0.17 for each additional ounce. What is the maximum weight of a large envelope that Joni Burnette can mail for $2.70?

72. Presorted First-Class Mail Companies can send pieces of mail weighing up to 1 ounce by using *presorted first-class mail*. The company must first purchase a bulk permit for $180 per year, and then pay $0.394 per piece sent. Without the permit, each piece would cost $0.42. Determine the minimum number of pieces of mail that would have to be mailed for it to be financially worthwhile for a company to use presorted first-class mail.

73. Comparing Payment Plans Melissa Pfistner recently accepted a sales position in Ohio. She can select between two payment plans. Plan 1 is a salary of $300 per week plus a 10% commission on sales. Plan 2 is a salary of $400 per week plus an 8% commission on sales. For what amount of weekly sales would Melissa earn more by plan 1?

74. College Employment To be eligible to continue her financial assistance for college, Katie Hanenberg can earn no more than $2000 during her 8-week summer employment. She already earns $90 per week as a day-care assistant. She is considering adding an evening job at a fast-food restaurant, where she will earn $6.25 per hour. What is the maximum number of hours she can work at the restaurant without jeopardizing her financial assistance?

75. A Passing Grade To pass a course, Corrina Schultz needs an average score of 60 or more. If Corrina's scores are 66, 72, 90, 49, and 59, find the minimum score that she can get on her sixth and last exam and pass the course.

76. Minimum Grade To receive an A in a course, Stephen Heasley must obtain an average score of 90 or higher on five exams. If Stephen's first four exam scores are 92, 87, 96, and 77, what is the minimum score that Stephen can receive on the fifth exam to get an A in the course?

77. Averaging Grades Calisha Mahoney's grades on her first four exams are 85, 92, 72, and 75. An average greater than or equal to 80 and less than 90 will result in a final grade of B. What range of grades on Calisha's fifth and last exam will result in a final grade of B? Assume a maximum grade of 100.

78. Clean Air For air to be considered "clean," the average of three pollutants must be less than 3.2 parts per million. If the first two pollutants are 2.7 and 3.42 ppm, what values of the third pollutant will result in clean air?

79. Income Taxes Refer to Example 7 on page 110. Su-hua and Ting-Fang Zheng file a joint tax return. Determine the 2008 income tax Su-hua and Ting-Fang will owe if their taxable income is

a) $78,221.   **b)** $301,233.

80. Income Taxes Refer to Example 7 on page 110. Jose and Mildred Battiste file a joint tax return. Determine the 2008 income tax Jose and Mildred will owe if their taxable income is

a) $128,479.   **b)** $275,248.

Velocity *In physics, an object that is traveling upwards has a positive velocity ($v > 0$) and an object that is traveling downwards has a negative velocity ($v < 0$). In Exercises 81–86, the velocity, v, is given for an object t seconds after it is projected upward. Using interval notation, determine the intervals of time when the object is traveling* **a)** *upward or* **b)** *downward.*

81. $v = -32t + 96$, $0 \le t \le 10$

82. $v = -32t + 172.8$, $0 \le t \le 12$

83. $v = -9.8t + 49$, $0 \le t \le 13$

84. $v = -9.8t + 31.36$, $0 \le t \le 6$

85. $v = -32t + 320$, $0 \le t \le 8$

86. $v = -9.8t + 68.6$, $0 \le t \le 5$

87. Water Acidity Thomas Hayward is checking the water acidity in a swimming pool. The water acidity is considered normal when the average pH reading of three daily measurements is greater than 7.2 and less than 7.8. If the first two pH readings are 7.48 and 7.15, find the range of pH values for the third reading that will result in the acidity level being normal.

88. If $a > b$, will a^2 always be greater than b^2? Explain and give an example to support your answer.

89. Insurance Policy A Blue Cross/Blue Shield insurance policy has a $100 deductible, after which it pays 80% of the total medical cost, c. The customer pays 20% until the customer has paid a total of $500, after which the policy pays 100% of the medical cost. We can describe this policy as follows:

Blue Cross Pays

$$\begin{cases} 0, & \text{if } c \le \$100 \\ 0.80(c - 100), & \text{if } \$100 < c \le \$2100 \\ c - 500, & \text{if } c > \$2100 \end{cases}$$

Explain why this set of inequalities describes Blue Cross/Blue Shield's payment plan.

90. Explain why the inequality $a < bx + c < d$ cannot be solved for x unless additional information is given.

Growth Charts *Exercises 91 and 92 show growth charts for children from birth to age 36 months. In general, the nth percentile represents that value that is above n% and below (100 − n)% of the items being measured. For example, a 24-month-old boy at the 60th percentile for weight weighs more than 60% and less than 40% of 24-month-old boys.*

91. The following chart shows the weight-for-age percentiles for boys from birth to age 36 months. The red curve is the 50th percentile. The orange region is between the 10th percentile (blue curve) and the 90th percentile (green curve). That is, 80% of the weights are between the values represented by the blue curve and the green curve. Use this graph to determine, in interval notation, where 80% of the weights occur for boys of age

a) 9 months.   **b)** 21 months.

c) 36 months.

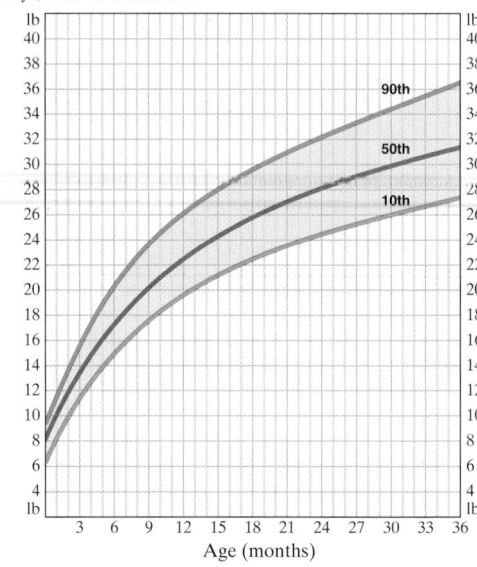

Weight-for-age percentiles:
Boys, birth to 36 months

Source: National Center for Health Statistics

92. The following chart shows the weight-for-age percentiles for girls from birth to age 36 months. The orange region is between the 10th percentile (blue curve) and the 90th percentile (green curve) and 80% of the weights are in this region. Use this graph to determine, in interval notation, where 80% of the weights occur for girls of age

a) 9 months.   **b)** 21 months.

c) 36 months.

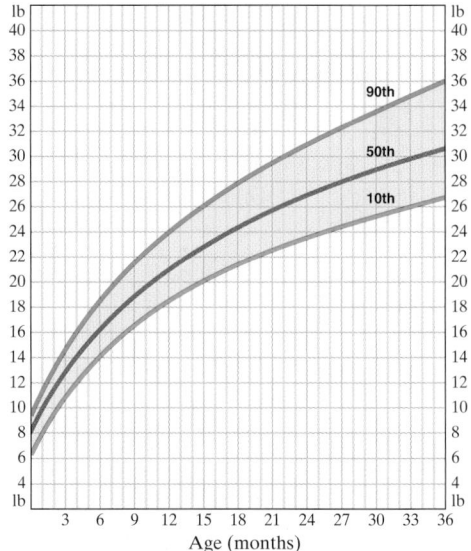

Weight-for-age percentiles:
Girls, birth to 36 months

Source: National Center for Health Statistics

Challenge Problems

93. Calculating Grades Stephen Heasley's first five scores in European history were 82, 90, 74, 76, and 68. The final exam for the course is to count one-third in computing the final average. A final average greater than or equal to 80 and less than 90 will result in a final grade of B. What range of final exam scores will result in Stephen's receiving a final grade of B in the course? Assume that a maximum score of 100 is possible.

In Exercises 94–96, **a)** *explain how to solve the inequality, and* **b)** *solve the inequality and give the solution in interval notation.*

94. $x < 3x - 10 < 2x$

95. $x < 2x + 3 < 2x + 5$

96. $x + 5 < x + 3 < 2x + 2$

Cumulative Review Exercises

[1.2]   **97.** For $A = \{1, 2, 6, 8, 9\}$ and $B = \{1, 3, 4, 5, 8\}$, find

      a) $A \cup B$.

      b) $A \cap B$.

      98. For $A = \left\{-3, 4, \dfrac{5}{2}, \sqrt{7}, 0, -\dfrac{13}{29}\right\}$, list the elements that are

    a) counting numbers

    b) whole numbers

    c) rational numbers

    d) real numbers

[1.3]   *Name each illustrated property.*

      99. $(3x + 8) + 4y = 3x + (8 + 4y)$

      100. $5x + y = y + 5x$

[2.2]   **101.** Solve the formula $R = L + (V - D)r$ for V.

2.6 Solving Equations and Inequalities Containing Absolute Values

1 Understand absolute value on the number line.

2 Solve equations of the form $|x| = a, a > 0$.

3 Solve inequalities of the form $|x| < a, a > 0$.

4 Solve inequalities of the form $|x| > a, a > 0$.

5 Solve inequalities of the form $|x| < a$ or $|x| > a$, $a < 0$.

6 Solve inequalities of the form $|x| < 0, |x| \le 0$, $|x| > 0$, or $|x| \ge 0$.

7 Solve equations of the form $|x| = |y|$.

1 Understand Absolute Value on the Number Line

Absolute value

The **absolute value** of a number x, symbolized $|x|$, is the distance x is from 0 on the number line.

$|3| = 3$ because the number 3 is 3 units from 0 on the number line.

$|-3| = 3$ because the number -3 is 3 units from 0 on the number line.

Now consider the equation $|x| = 3$. We are looking for values of x that *are exactly* 3 units from 0 on the number line. Thus, the solutions to $|x| = 3$ are $x = 3$ and $x = -3$ (see **Fig. 2.14a**).

Next, consider the inequality $|x| < 3$. We are looking for values of x that *are less than* 3 units from 0 on the number line. Thus, the solutions to $|x| < 3$ are the numbers that are between -3 and 3 on the number line (see **Fig. 2.14b**).

Finally, consider the inequality $|x| > 3$. We are looking for values of x that *are greater than* 3 units from 0 on the number line. Thus, the solutions to $|x| > 3$ are the numbers that are less than -3 or greater than 3 on the number line (see **Fig. 2.14c**).

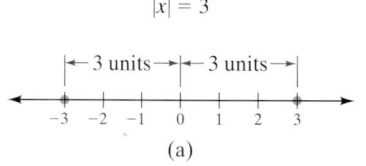

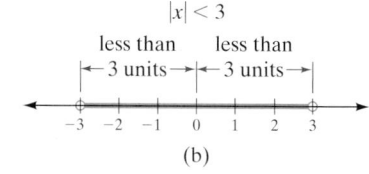

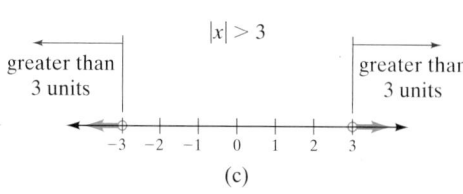

FIGURE 2.14

We will use these examples and their illustrations on the number line to develop methods for solving equations and inequalities involving absolute value.

2 Solve Equations of the Form $|x| = a, a > 0$

When solving an equation of the form $|x| = a, a > 0$, we are finding the values that are exactly a units from 0 on the number line.

> **To Solve Equations of the Form $|x| = a$**
>
> If $|x| = a$ and $a > 0$, then $x = a$ or $x = -a$.

EXAMPLE 1 Solve each equation.

a) $|x| = 2$ **b)** $|x| = 0$ **c)** $|x| = -2$

Solution

a) Using the procedure, we get $x = 2$ or $x = -2$. The solution set is $\{-2, 2\}$.

b) The only real number whose absolute value equals 0 is 0. Thus, the solution set for $|x| = 0$ is $\{0\}$.

c) The absolute value of a number is never negative, so there are no solutions to this equation. The solution set is $\varnothing$.

Now Try Exercise 13

Understanding Algebra

The absolute value of a number is never negative.

EXAMPLE 2 Solve the equation $|2w - 1| = 5$.

Solution   We are looking for the values of w such that $2w - 1$ is exactly 5 units from 0 on a number line. Thus, the quantity $2w - 1$ must be equal to 5 or -5.

$$
\begin{array}{ccc}
2w - 1 = 5 & \text{or} & 2w - 1 = -5 \\
2w = 6 & & 2w = -4 \\
w = 3 & & w = -2
\end{array}
$$

Check

$$
\begin{array}{cc}
w = 3 & w = -2 \\
|2w - 1| = 5 & |2w - 1| = 5 \\
|2(3) - 1| \overset{?}{=} 5 & |2(-2) - 1| \overset{?}{=} 5 \\
|6 - 1| \overset{?}{=} 5 & |-4 - 1| \overset{?}{=} 5 \\
|5| \overset{?}{=} 5 & |-5| \overset{?}{=} 5 \\
5 = 5 \quad \text{True} & 5 = 5 \quad \text{True}
\end{array}
$$

The solutions 3 and -2 each result in $2w - 1$ being 5 units from 0 on the number line. The solution set is $\{-2, 3\}$.

Now Try Exercise 19

Consider the equation $|2w - 1| - 3 = 2$. The first step in solving this equation is to isolate the absolute value term. We do this by adding 3 to both sides of the equation. This results in the equation $|2w - 1| = 5$, which we solved in Example 2.

3 Solve Inequalities of the Form $|x| < a, a > 0$

Recall our discussion earlier regarding $|x| < 3$. The solutions to $|x| < 3$ are the numbers that are between -3 and 3 on the number line (See **Fig. 2.14b** on page 119). Similarly, the solutions to $|x| < a$ are the numbers that are between $-a$ and a on the number line.

To solve inequalities of the form $|x| < a$, we can use the following procedure.

> **To Solve Inequalities of the Form $|x| < a$**
>
> If $|x| < a$ and $a > 0$, then $-a < x < a$.

EXAMPLE 3 Solve the inequality $|2x - 3| < 5$.

Solution   The solution to this inequality will be the set of values such that the distance between $2x - 3$ and 0 on a number line will be less than 5 units (see **Fig. 2.15**). Using **Figure 2.15**, we can see that $-5 < 2x - 3 < 5$.

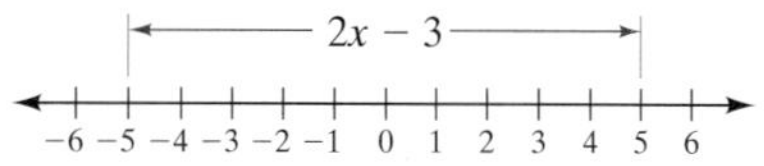

FIGURE 2.15

Solving, we get

$$-5 < 2x - 3 < 5$$
$$-2 < 2x < 8$$
$$-1 < x < 4$$

The solution set is $\{x | -1 < x < 4\}$.

Now Try Exercise 33

EXAMPLE 4 Solve the inequality $|2x + 1| \leq 9$ and graph the solution on a number line.

Solution   Since this inequality is of the form $|x| \leq a$, we write

$$-9 \leq 2x + 1 \leq 9$$
$$-10 \leq 2x \leq 8$$
$$-5 \leq x \leq 4$$

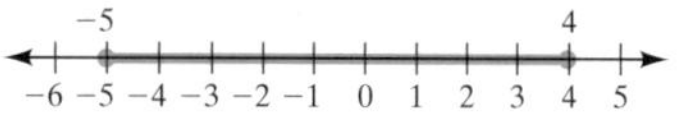

Now Try Exercise 75

EXAMPLE 5 Solve the inequality $|7.8 - 4x| - 5.3 < 14.1$ and graph the solution on a number line.

Solution   First isolate the absolute value by adding 5.3 to both sides of the inequality. Then solve as in the previous examples.

$$|7.8 - 4x| - 5.3 < 14.1$$
$$|7.8 - 4x| < 19.4$$
$$-19.4 < 7.8 - 4x < 19.4$$
$$-27.2 < -4x < 11.6$$
$$\frac{-27.2}{-4} > \frac{-4x}{-4} > \frac{11.6}{-4}$$
$$6.8 > x > -2.9 \quad \text{or} \quad -2.9 < x < 6.8$$

The solution set is $\{x | -2.9 < x < 6.8\}$.

Now Try Exercise 43

Understanding Algebra

In Example 5, the solution

$$-2.9 < x < 6.8$$

written in interval notation is

$$(-2.9, 6.8).$$

4 Solve Inequalities of the Form $|x| > a, a > 0$

Recall our discussion earlier regarding $|x| > 3$. The solutions to $|x| > 3$ are the numbers that are less than -3 or greater than 3 on the number line (See **Fig. 2.14c** on page 119). Similarly, the solutions to $|x| > a$ are the numbers that are less than $-a$ or greater than a on the number line.

To solve inequalities of the form $|x| > a$, we can use the following procedure.

To Solve Inequalities of the Form $|x| > a$

If $|x| > a$ and $a > 0$, then $x < -a$ or $x > a$.

EXAMPLE 6 Solve the inequality $|2x - 3| > 5$ and graph the solution on a number line.

Solution   The solution to $|2x - 3| > 5$ is the set of values such that the distance between $2x - 3$ and 0 on a number line will be greater than 5. The quantity $2x - 3$ must be either less than -5 or greater than 5 (see **Fig. 2.16**).

FIGURE 2.16

Since $2x - 3$ must be either less than -5 or greater than 5, we set up and solve the following compound inequality:

$$2x - 3 < -5 \quad \text{or} \quad 2x - 3 > 5$$
$$2x < -2 \qquad\qquad 2x > 8$$
$$x < -1 \qquad\qquad x > 4$$

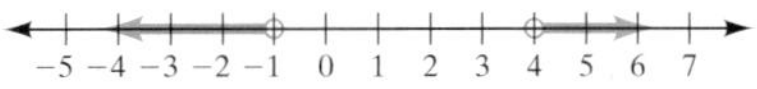

The solution set to $|2x - 3| > 5$ is $\{x | x < -1 \text{ or } x > 4\}$.

Now Try Exercise 51

EXAMPLE 7 Solve the inequality $|2x - 1| \geq 7$ and graph the solution on a number line.

Solution   Since this inequality is of the form $|x| \geq a$, we use the procedure given above.

$$2x - 1 \leq -7 \quad \text{or} \quad 2x - 1 \geq 7$$
$$2x \leq -6 \qquad\qquad 2x \geq 8$$
$$x \leq -3 \qquad\qquad x \geq 4$$

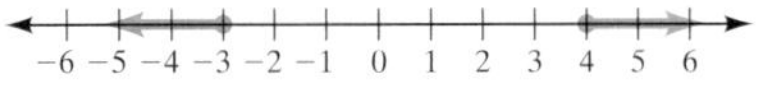

Now Try Exercise 53

Understanding Algebra

In Example 7, any value of x less than or equal to -3, or greater than or equal to 4, would result in $2x - 1$ representing a number that is greater than or equal to 7 units from 0 on a number line. The solution set is $\{x | x \leq -3 \text{ or } x \geq 4\}$. In interval notation, the solution is $(-\infty, -3] \cup [4, \infty)$.

EXAMPLE 8 Solve the inequality $\left| \dfrac{3x - 4}{2} \right| \geq 9$ and graph the solution on a number line.

Solution   Since this inequality is of the form $|x| \geq a$, we write

$$\frac{3x - 4}{2} \leq -9 \quad \text{or} \quad \frac{3x - 4}{2} \geq 9$$

Now multiply both sides of each inequality by the least common denominator, 2. Then solve each inequality.

$$2\left(\frac{3x - 4}{2}\right) \leq -9 \cdot 2 \quad \text{or} \quad 2\left(\frac{3x - 4}{2}\right) \geq 9 \cdot 2$$
$$3x - 4 \leq -18 \qquad\qquad 3x - 4 \geq 18$$
$$3x \leq -14 \qquad\qquad 3x \geq 22$$
$$x \leq -\frac{14}{3} \qquad\qquad x \geq \frac{22}{3}$$

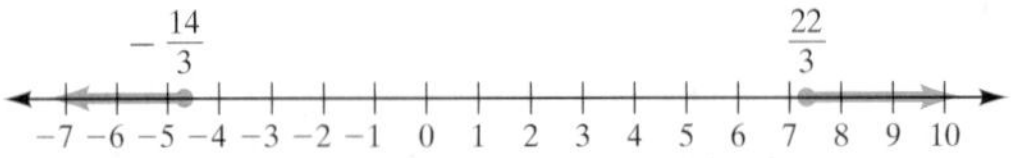

Now Try Exercise 57

> **Helpful Hint**
>
> Some general information about equations and inequalities containing absolute value follows. For real numbers a, b, and c, where $a \neq 0$ and $c > 0$:
>
> | Form of Equation or Inequality | The Solution Will Be: | Solution on a Number Line: |
> |---|---|---|
> | $\|ax + b\| = c$ | Two distinct numbers, p and q | |
> | $\|ax + b\| < c$ | The set of numbers between two numbers, $p < x < q$ | |
> | $\|ax + b\| > c$ | The set of numbers less than one number or greater than a second number, $x < p$ or $x > q$ | |

> **Understanding Algebra**
>
> Any inequality of the form $\|x\| < a$, where a is a negative number, will have solution set $\varnothing$.

5 Solve Inequalities of the Form $\|x\| < a$ or $\|x\| > a$, $a < 0$

Consider the inequality $\|x\| < -3$. Since $\|x\|$ will always have a value greater than or equal to 0 for any real number x, this inequality can never be true, and the solution is the empty set, $\varnothing$.

EXAMPLE 9 Solve the inequality $\|6x - 8\| + 5 < 3$.

Solution Begin by subtracting 5 from both sides of the inequality.

$$\|6x - 8\| + 5 < 3$$
$$\|6x - 8\| < -2$$

Since $\|6x - 8\|$ will always be greater than or equal to 0 for any real number x, this inequality can never be true. Therefore, the solution is the empty set, $\varnothing$.

Now Try Exercise 41

Now consider the inequality $\|x\| > -3$. Since $\|x\|$ will always have a value greater than or equal to 0 for any real number x, this inequality will always be true and the solution is the set of all real numbers, $\mathbb{R}$.

EXAMPLE 10 Solve the inequality $\|5x + 3\| + 4 \geq -9$.

Solution Begin by subtracting 4 from both sides of the inequality.

$$\|5x + 3\| + 4 \geq -9$$
$$\|5x + 3\| \geq -13$$

> **Understanding Algebra**
>
> Any inequality of the form $\|x\| > a$, where a is a negative number, will have solution set $\mathbb{R}$.

Since $\|5x + 3\|$ will always be greater than or equal to 0 for any real number x, this inequality is true for all real numbers. Thus, the solution is the set of all real numbers, $\mathbb{R}$.

Now Try Exercise 59

6 Solve Inequalities of the Form $\|x\| < 0$, $\|x\| \leq 0$, $\|x\| > 0$, or $\|x\| \geq 0$

Each of the examples on the top of page 124 requires you to recall that the absolute value of a number can never be negative.

Inequality	Solution Set	Explanation
$\|x - 5\| < 0$	$\emptyset$	The absolute value of a number can never be < 0.
$\|x - 5\| \le 0$	$\{5\}$	The absolute value of a number can never be < 0 but it can be $= 0$. When $x = 5$, we get

$$\|x - 5\| \le 0$$
$$\|5 - 5\| \le 0$$
$$0 \le 0 \quad \text{True}$$

| $\|x - 5\| > 0$ | $\{x \| x \ne 5\}$ | Substituting any real number for x except for 5 makes $\|x - 5\|$ positive. When $x = 5$, we get |

$$\|x - 5\| > 0$$
$$\|5 - 5\| > 0$$
$$0 > 0 \quad \text{False}$$

| $\|x - 5\| \ge 0$ | $\mathbb{R}$ | The absolute value of any number is always ≥ 0. |

EXAMPLE 11 Solve each inequality. **a)** $\|x + 2\| > 0$ **b)** $\|3x - 8\| < 0$

Solution

a) The inequality will be true for every value of x except -2. The solution set is $\{x \| x < -2 \text{ or } x > -2\}$.

b) Determine the number that makes the absolute value equal to 0 by setting the expression within the absolute value sign equal to 0 and solving for x.

$$3x - 8 = 0$$
$$3x = 8$$
$$x = \frac{8}{3}$$

The inequality will be true only when $x = \frac{8}{3}$. The solution set is $\left\{\frac{8}{3}\right\}$.

Now Try Exercise 61

7 Solve Equations of the Form $\|x\| = \|y\|$

Now we will discuss absolute value equations where an absolute value appears on both sides of the equation.

When solving an absolute value equation with an absolute value expression on each side of the equals sign, the two expressions must have the same absolute value. Therefore, *the expressions must be equal to each other or be opposites of each other.*

> **To Solve Equations of the Form $\|x\| = \|y\|$**
>
> If $\|x\| = \|y\|$, then $x = y$ or $x = -y$.

EXAMPLE 12 Solve the equation $\|z + 3\| = \|2z - 7\|$.

Solution If we let $z + 3$ be x and $2z - 7$ be y, this equation is of the form $\|x\| = \|y\|$. Using the procedure given above, we obtain the two equations

$$z + 3 = 2z - 7 \quad \text{or} \quad z + 3 = -(2z - 7)$$

Now solve each equation.

$$z + 3 = 2z - 7 \quad \text{or} \quad z + 3 = -(2z - 7)$$
$$3 = z - 7 \qquad\qquad z + 3 = -2z + 7$$
$$10 = z \qquad\qquad 3z + 3 = 7$$
$$3z = 4$$
$$z = \frac{4}{3}$$

Understanding Algebra

If $\|x\| = \|y\|$, then $x = y$ or $x = -y$.

Check $z = 10$ $|z + 3| = |2z - 7|$ $z = \dfrac{4}{3}$ $|z + 3| = |2z - 7|$

$$|10 + 3| \stackrel{?}{=} |2(10) - 7|$$ $$\left|\dfrac{4}{3} + 3\right| \stackrel{?}{=} \left|2\left(\dfrac{4}{3}\right) - 7\right|$$

$$|13| \stackrel{?}{=} |20 - 7|$$ $$\left|\dfrac{13}{3}\right| \stackrel{?}{=} \left|\dfrac{8}{3} - \dfrac{21}{3}\right|$$

$$|13| \stackrel{?}{=} |13|$$ $$\left|\dfrac{13}{3}\right| \stackrel{?}{=} \left|-\dfrac{13}{3}\right|$$

$$13 = 13 \quad \text{True}$$ $$\dfrac{13}{3} = \dfrac{13}{3} \quad \text{True}$$

The solution set is $\left\{10, \dfrac{4}{3}\right\}$.

Now Try Exercise 63

EXAMPLE 13 Solve the equation $|4x - 7| = |6 - 4x|$.

Solution $4x - 7 = 6 - 4x$ or $4x - 7 = -(6 - 4x)$

$8x - 7 = 6$ $4x - 7 = -6 + 4x$

$8x = 13$ $-7 = -6$ False

$x = \dfrac{13}{8}$

Since the equation $4x - 7 = -(6 - 4x)$ results in a false statement, the absolute value equation has only one solution. A check will show that the solution set is $\left\{\dfrac{13}{8}\right\}$.

Now Try Exercise 69

Summary of Procedures for Solving Equations and Inequalities Containing Absolute Value

For $a > 0$,

If $|x| = a$, then $x = a$ or $x = -a$.

If $|x| < a$, then $-a < x < a$.

If $|x| > a$, then $x < -a$ or $x > a$.

If $|x| = |y|$, then $x = y$ or $x = -y$.

EXERCISE SET 2.6   *Math XL*   *MyMathLab*
MathXL®   MyMathLab

Warm-Up Exercises

Fill in the blanks with the appropriate word, phrase, or symbol(s) from the following list.

$\|x\| = 4$	$\|x\| < 4$	$\|x\| \le 4$	$\|x\| > 4$	$\|x\| \ge 4$	$\|x\| = 5$
$\|x\| < 5$	$\|x\| \le 5$	$\|x\| > 5$	$\|x\| \ge 5$	$\|x\| < -6$	$\|x\| > -6$

1. The graph of the solution to _____ on the number line is .

3. The graph of the solution to _____ on the number line is .

2. The graph of the solution to _____ on the number line is .

4. The graph of the solution to _____ on the number line is .

5. The graph of the solution to _____ on the number line is ←+—+—◆—+—+—+—+—+—+—+—◆—+—+→ .
  −6 −5 −4 −3 −2 −1 0 1 2 3 4 5 6

6. The solution set of _____ is $\{x|\ 5 \leq x \leq 5|$.

7. The solution set of _____ is $\{x| x \leq -5 \text{ or } x \geq 5\}$.

8. The solution set of _____ is $\{-5, 5\}$.

9. The solution set of _____ is $\{x|-5 < x < 5\}$.

10. The solution set of _____ is $\{x| x < -5 \text{ or } x > 5\}$.

11. The solution set of _____ is $\mathbb{R}$.

12. The solution set of _____ is $\varnothing$.

Practice the Skills

Find the solution set for each equation.

13. $|a| = 7$

14. $|b| = 17$

15. $|c| = \dfrac{1}{2}$

16. $|x| = 0$

17. $|d| = -\dfrac{5}{6}$

18. $|l + 4| = 6$

19. $|x + 5| = 8$

20. $|3 + y| = \dfrac{3}{5}$

21. $|4.5q + 31.5| = 0$

22. $|4.7 - 1.6z| = 14.3$

23. $|5 - 3x| = \dfrac{1}{2}$

24. $|6(y + 4)| = 24$

25. $\left|\dfrac{x - 3}{4}\right| = 5$

26. $\left|\dfrac{3z + 5}{6}\right| - 2 = 7$

27. $\left|\dfrac{x - 3}{4}\right| + 8 = 8$

28. $\left|\dfrac{5x - 3}{2}\right| + 5 = 9$

29. $|x - 5| + 4 = 3$

30. $|2x + 3| - 5 = -8$

Find the solution set for each inequality.

31. $|w| < 1$

32. $|p| \leq 9$

33. $|q + 5| \leq 8$

34. $|7 - x| < 6$

35. $|5b - 15| < 10$

36. $|x - 3| - 7 < -2$

37. $|2x + 3| - 5 \leq 10$

38. $|4 - 3x| - 4 < 11$

39. $|3x - 7| + 8 < 14$

40. $\left|\dfrac{2x - 1}{9}\right| \leq \dfrac{5}{9}$

41. $|2x - 6| + 5 \leq 1$

42. $|2x - 3| < -10$

43. $\left|\dfrac{1}{2}j + 4\right| < 7$

44. $\left|\dfrac{k}{4} - \dfrac{3}{8}\right| < \dfrac{7}{16}$

45. $\left|\dfrac{x - 3}{2}\right| - 4 \leq -2$

46. $\left|7x - \dfrac{1}{2}\right| < 0$

Find the solution set for each inequality.

47. $|y| > 8$

48. $|a| \geq 13$

49. $|x + 4| > 5$

50. $|2b - 7| > 3$

51. $|7 - 3b| > 5$

52. $\left|\dfrac{6 + 2z}{3}\right| > 2$

53. $|2h - 5| > 3$

54. $|2x - 1| \geq 12$

55. $|0.1x - 0.4| + 0.4 > 0.6$

56. $|3.7d + 6.9| - 2.1 > -5.4$

57. $\left|\dfrac{x}{2} + 4\right| \geq 5$

58. $\left|4 - \dfrac{3x}{5}\right| \geq 9$

59. $|7w + 3| - 12 \geq -12$

60. $|2.6 - x| \geq 0$

61. $|4 - 2x| > 0$

62. $|2c - 8| > 0$

Find the solution set for each equation.

63. $|3p - 5| = |2p + 10|$

64. $|6n + 3| = |4n - 13|$

65. $|6x| = |3x - 9|$

66. $|5t - 10| = |10 - 5t|$

67. $\left|\dfrac{2r}{3} + \dfrac{5}{6}\right| = \left|\dfrac{r}{2} - 3\right|$

68. $|3x - 8| = |3x + 8|$

69. $\left|-\dfrac{3}{4}m + 8\right| = \left|7 - \dfrac{3}{4}m\right|$

70. $\left|\dfrac{3}{2}r + 2\right| = \left|8 - \dfrac{3}{2}r\right|$

Find the solution set for each equation or inequality.

71. $|h| = 9$

72. $|y| \leq 8$

73. $|q + 6| > 2$

74. $|9d + 7| \leq -9$

75. $|2w - 7| \leq 9$

76. $|2z - 7| + 5 > 8$

77. $|5a - 1| = 9$

78. $|2x - 4| + 5 = 13$

79. $|5 + 2x| > 0$

80. $|7 - 3b| = |5b + 15|$

81. $|4 + 3x| \leq 9$

82. $|2.4x + 4| + 4.9 > 3.9$

83. $|3n + 8| - 4 = -10$

84. $|4 - 2x| - 3 = 7$

85. $\left|\dfrac{w + 4}{3}\right| + 5 < 9$

86. $\left| \dfrac{5t - 10}{6} \right| > \dfrac{5}{3}$

87. $\left| \dfrac{3x - 2}{4} \right| - \dfrac{1}{3} \geq -\dfrac{1}{3}$

88. $\left| \dfrac{2x - 4}{5} \right| = 14$

89. $|2x - 8| = \left| \dfrac{1}{2}x + 3 \right|$

90. $\left| \dfrac{1}{3}y + 3 \right| = \left| \dfrac{2}{3}y - 1 \right|$

91. $|2 - 3x| = \left| 4 - \dfrac{5}{3}x \right|$

92. $\left| \dfrac{-2u + 3}{7} \right| \leq 5$

Problem Solving

93. Glass Thickness Certain types of glass manufactured by PPG Industries ideally will have a thickness of 0.089 inches. However, due to limitations in the manufacturing process, the thickness is allowed to vary from the ideal thickness by up to 0.004 inch. If t represents the actual thickness of the glass, then the allowable range of thicknesses can be represented using the inequality $|t - 0.089| \leq 0.004$.

Source: www.ppg.com

a) Solve this inequality for t (use interval notation).

b) What is the smallest thickness the glass is allowed to be?

c) What is the largest thickness the glass is allowed to be?

94. Plywood Guarantee Certain plywood manufactured by Lafor International is guaranteed to be $\dfrac{5}{8}$ inch thick with a tolerance of plus or minus $\dfrac{1}{56}$ of an inch. If t represents the actual thickness of the plywood, then the allowable range of thicknesses can be represented using the inequality $\left| t - \dfrac{5}{8} \right| \leq \dfrac{1}{56}$.

Source: www.sticktrade.com

a) Solve this inequality for t (use interval notation).

b) What is the smallest thickness the plywood is allowed to be?

c) What is the largest thickness the plywood is allowed to be?

95. Submarine Depth A submarine is 160 feet below sea level. It has rock formations above and below it, and should not change its depth by more than 28 feet. Its distance below sea level, d, can be described by the inequality $|d - 160| \leq 28$.

a) Solve this inequality for d. Write your answer in interval notation.

b) Between what vertical distances, measured from sea level, may the submarine move?

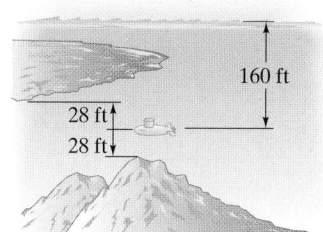

96. A Bouncing Spring A spring hanging from a ceiling is bouncing up and down so that its distance, d, above the ground satisfies the inequality $|d - 4| \leq \dfrac{1}{2}$ foot (see the figure).

a) Solve this inequality for d. Write your answer in interval notation.

b) Between what distances, measured from the ground, will the spring oscillate?

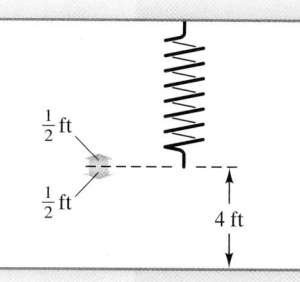

97. How many solutions are there to the following equations or inequalities if $a \neq 0$ and $k > 0$?

a) $|ax + b| = k$

b) $|ax + b| < k$

c) $|ax + b| > k$

98. Suppose $|x| < |y|$ and $x < 0$ and $y < 0$.

a) Which of the following must be true: $x < y$, $x > y$, or $x = y$?

b) Give an example to support your answer to part **a)**.

99. How many solutions will $|ax + b| = k$, $a \neq 0$ have if

a) $k < 0$,

b) $k = 0$,

c) $k > 0$?

100. Suppose m and n ($m < n$) are two distinct solutions to the equation $|ax + b| = c$. Indicate the solutions, using both inequality symbols and the number line, to each inequality. (See the Helpful Hint on page 123.)

a) $|ax + b| < c$

b) $|ax + b| > c$

Concept/Writing Exercises

101. For what value of x will the inequality $|ax + b| \leq 0$ be true? Explain.

102. For what value of x will the inequality $|ax + b| > 0$ *not* be true? Explain.

103. a) Explain how to find the solution to the equation $|ax + b| = c$. (Assume that $c > 0$ and $a \neq 0$.)

b) Solve this equation for x.

104. a) Explain how to find the solution to the inequality $|ax + b| < c$. (Assume that $a > 0$ and $c > 0$.)

   b) Solve this inequality for x.

105. a) Explain how to find the solution to the inequality $|ax + b| > c$. (Assume that $a > 0$ and $c > 0$.)

   b) Solve this inequality for x.

106. a) What is the first step in solving the inequality $-4|3x - 5| \le -12$?

   b) Solve this inequality and give the solution in interval notation.

Determine what values of x will make each equation true. Explain your answer.

107. $|x - 4| = |4 - x|$ **108.** $|x - 4| = -|x - 4|$ **109.** $|x| = x$ **110.** $|x + 2| = x + 2$

Solve. Explain how you determined your answer.

111. $|x + 1| = 2x - 1$ **112.** $|3x + 1| = x - 3$ **113.** $|x - 4| = -(x - 4)$

Challenge Problems

Solve by considering the possible signs for x.

114. $|x| + x = 8$ **115.** $x + |-x| = 8$ **116.** $|x| - x = 8$ **117.** $x - |x| = 8$

Group Activity

Discuss and answer Exercise 118 as a group.

118. Consider the equation $|x + y| = |y + x|$.

   a) Have each group member select an x value and a y value and determine whether the equation holds. Repeat for two other pairs of x- and y-values.

   b) As a group, determine for what values of x and y the equation is true. Explain your answer.

   c) Now consider $|x - y| = -|y - x|$. Under what conditions will this equation be true?

Cumulative Review Exercises

Evaluate.

[1.4] **119.** $\dfrac{1}{3} + \dfrac{1}{4} \div \dfrac{2}{5}\left(\dfrac{1}{3}\right)^2$

   120. $4(x + 3y) - 5xy$ when $x = 1, y = 3$

[2.4] **121. Swimming** Terry Chong swims across a lake averaging 2 miles an hour. Then he turns around and swims back across the lake, averaging 1.6 miles per hour. If his total swimming time is 1.5 hours, what is the width of the lake?

[2.5] **122.** Find the solution set to the inequality
$$-7(x - 3) + 5(x + 1) \ge 20$$

Chapter 2 Summary

IMPORTANT FACTS AND CONCEPTS	EXAMPLES
Section 2.1	

Properties of Equality

For all real numbers $a, b,$ and c:

1. $a = a$.	Reflexive property	$9 = 9$
2. If $a = b$, then $b = a$.	Symmetric property	If $x = 10$, then $10 = x$.
3. If $a = b$ and $b = c$, then $a = c$.	Transitive property	If $y = a + b$ and $a + b = 4t$, then $y = 4t$.

Terms are the parts being added in an algebraic expression.	In the expression $9x^2 - 2x + \dfrac{1}{5}$, the terms are $9x^2$, $-2x$, and $\dfrac{1}{5}$.

The **coefficient** is the numerical part of a term that precedes the variable.	Term Coefficient $15x^4y$ 15

The **degree of a term** with whole number exponents is the sum of the exponents on the variables.	Term Degree $17xy^5$ $1 + 5 = 6$

Like terms are terms that have the same variables with the same exponents. **Unlike terms** are terms that are not like terms.	Like Terms Unlike Terms $2x, 7x$ $3x, 4y$ $9x^2, -5x^2$ $10x^2, 2x^{10}$

To **simplify an expression** means to combine all the like terms.	$3x^2 + 12x - 5 + 7x^2 - 12x + 1 = 10x^2 - 4$

IMPORTANT FACTS AND CONCEPTS	EXAMPLES

Section 2.1 (cont.)

An **equation** is a mathematical statement of equality.	$x + 15 = 36$
The **solution** of an equation is the number(s) that make the equation a true statement.	The solution to $\frac{1}{2}x + 1 = 7$ is 12.

A **linear equation in one variable** is an equation that has the form $$ax + b = c, a \neq 0$$	$8x - 3 = 17$

Addition Property of Equality

If $a = b$, then $a + c = b + c$ for any $a, b,$ and c.	If $5x - 7 = 19$, then $5x - 7 + 7 = 19 + 7$.

Multiplication Property of Equality

If $a = b$, then $a \cdot c = b \cdot c$ for any $a, b,$ and c.	If $\frac{1}{3}x = 2$, then $3 \cdot \frac{1}{3}x = 3 \cdot 2$.

To Solve Linear Equations

1. Clear fractions.
2. Simplify each side separately.
3. Isolate the variable term on one side.
4. Solve for the variable.
5. Check.

See page 67 for more detail.

Solve the equation $\frac{1}{2}x + 7 = \frac{4}{3}x - 3$.

$$\frac{1}{2}x + 7 = \frac{4}{3}x - 3$$
$$6\left(\frac{1}{2}x + 7\right) = 6\left(\frac{4}{3}x - 3\right)$$
$$3x + 42 = 8x - 18$$
$$42 = 5x - 18$$
$$60 = 5x$$
$$12 = x$$

A check shows that 12 is the solution.

A **conditional equation** is an equation that is true only for specific values of the variable.	$2x + 4 = 5$ Solution is $x = \frac{1}{2}$
A **contradiction** is an equation that has no solution (solution set is $\varnothing$).	$2x + 6 = 2x + 8$ Solution set is $\varnothing$
An **identity** is an equation that has an infinite number of solutions (solution set is $\mathbb{R}$).	$3x + 6 = 3(x + 2)$ Solution set is $\mathbb{R}$

Section 2.2

A **mathematical model** is a real-life application expressed mathematically.	The speed of a car, s, increased by 20 mph is 60 mph. Model: $s + 20 = 60$.
A **formula** is an equation that is a mathematical model for a real-life situation.	$A = l \cdot w$

Guidelines for Problem Solving

1. Understand the problem.
2. Translate the problem into mathematical language.
3. Carry out the mathematical calculations necessary to solve the problem.
4. Check the answer found in step 3.
5. Answer the question.

See page 75 for more detail.

Max Johnson made a $2000, 3% simple interest personal loan to Jill Johnson for 6 years. At the end of 6 years, what interest will Jill pay to Max?

Understand	This is a simple interest problem.
Translate	$i = prt$
Carry Out	$= 2000(0.03)(6)$
	$= 360$
Check	The answer appears reasonable.
Answer	The simple interest owed is $360.

Simple interest formula is $i = prt$.	Find the simple interest on a 2-year, $1000, 6% simple interest loan. $$i = (1000)(0.06)(2) = 120$$ The simple interest is $120.

Compound interest formula is $A = p\left(1 + \dfrac{r}{n}\right)^{nt}$.	Find the amount in a savings account for a deposit of $6500 paying 4.8% interest compounded semiannually for 10 years. $$A = 6500\left(1 + \frac{0.048}{2}\right)^{2 \cdot 10}$$ $$\approx 10{,}445.10$$ The amount in the savings account is $10,445.10.

IMPORTANT FACTS AND CONCEPTS	EXAMPLES

Section 2.2 (cont.)

To solve an equation (or formula) for a variable means to isolate that variable.	Solve the equation $3x + 7y = 2$ for y. $$7y = -3x + 2$$ $$y = -\frac{3}{7}x + \frac{2}{7}$$

Section 2.3

Phrases can be translated into algebraic expressions.	Phrase Algebraic Expression 4 more than 7 times a number $7x + 4$
Complementary angles are two angles whose sum measures 90°.	If angle $A = 62°$ and angle $B = 28°$, then angles A and B are complementary angles.
Supplementary angles are two angles whose sum measures 180°.	If angle $A = 103°$ and angle $B = 77°$, then angles A and B are supplementary angles.

Section 2.4

A **general motion problem formula** is amount = rate · time.	Find the amount of gas pumped when gas is pumped for 3 minutes at 6 gallons per minute. $$A = 6 \cdot 3 = 18 \text{ gallons}$$
The **distance formula** is distance = rate · time.	Find the distance traveled when a car travels at 60 miles per hour for 5 hours. $$D = 60 \cdot 5 = 300 \text{ miles}$$
A **mixture problem** is any problem where two or more quantities are combined to produce a different quantity or where a single quantity is separated into two or more different quantities.	If 4 liters of a 10% solution is mixed with 8 liters of a 16% solution, find the strength of the mixture. $$4(0.10) + 8(0.16) = 12(x)$$ $$x = 0.14 \quad \text{or} \quad x = 14\%$$

Section 2.5

Properties Used to Solve Linear Inequalities **1.** If $a > b$, then $a + c > b + c$. **2.** If $a > b$, then $a - c > b - c$. **3.** If $a > b$, and $c > 0$, then $ac > bc$. **4.** If $a > b$, and $c > 0$, then $\dfrac{a}{c} > \dfrac{b}{c}$. **5.** If $a > b$, and $c < 0$, then $ac < bc$. **6.** If $a > b$, and $c < 0$, then $\dfrac{a}{c} < \dfrac{b}{c}$.	**1.** If $6 > 5$, then $6 + 3 > 5 + 3$. **2.** If $6 > 5$, then $6 - 3 > 5 - 3$. **3.** If $7 > 3$, then $7 \cdot 4 > 3 \cdot 4$. **4.** If $7 > 3$, then $\dfrac{7}{4} > \dfrac{3}{4}$. **5.** If $9 > 2$, then $9(-3) < 2(-3)$. **6.** If $9 > 2$, then $\dfrac{9}{-3} < \dfrac{2}{-3}$.
A **compound inequality** is formed by joining two inequalities with the word *and* or *or*.	$$x \le 7 \quad \text{and} \quad x > 5$$ $$x < -1 \quad \text{or} \quad x \ge 4$$
To find the solution set of an inequality containing the word *and* take the **intersection** of the solution sets of the two inequalities.	Solve $x \le 7$ and $x > 5$. The intersection of $\{x \mid x \le 7\}$ and $\{x \mid x > 5\}$ is $\{x \mid 5 < x \le 7\}$ or $(5, 7]$.
To find the solution set of an inequality containing the word *or*, take the **union** of the solution sets of the two inequalities.	Solve $x < -1$ or $x \ge 4$. The union of $\{x \mid x < -1\}$ or $\{x \mid x \ge 4\}$ is $\{x \mid x < -1 \text{ or } x \ge 4\}$ or $(-\infty, -1) \cup [4, \infty)$.

Section 2.6

To Solve Equations of the Form $\lvert x \rvert = a$ If $\lvert x \rvert = a$ and $a > 0$, then $x = a$ or $x = -a$.	Solve $\lvert x \rvert = 6$. $\lvert x \rvert = 6$ gives $x = 6$ or $x = -6$.
To Solve Inequalities of the Form $\lvert x \rvert < a$ If $\lvert x \rvert < a$ and $a > 0$, then $-a < x < a$.	Solve $\lvert 3x + 1 \rvert < 13$. $$-13 < 3x + 1 < 13$$ $$-\frac{14}{3} < x < 4$$ $$\left\{ x \,\middle|\, -\frac{14}{3} < x < 4 \right\} \quad \text{or} \quad \left(-\frac{14}{3}, 4 \right)$$

IMPORTANT FACTS AND CONCEPTS	EXAMPLES

Section 2.6 (cont.)

To Solve Inequalities of the Form $|x| > a$

If $|x| > a$ and $a > 0$, then $x < -a$ or $x > a$.

Solve $|2x - 3| \geq 5$.

$$2x - 3 \leq -5 \qquad \text{or} \qquad 2x - 3 \geq 5$$
$$2x \leq -2 \qquad\qquad\qquad 2x \geq 8$$
$$x \leq -1 \qquad\qquad\qquad x \geq 4$$
$$\{x | x \leq -1 \text{ or } x \geq 4\} \qquad \text{or} \qquad (-\infty, -1] \cup [4, \infty)$$

If $|x| > a$ and $a < 0$, the solution set is $\mathbb{R}$.
If $|x| < a$ and $a < 0$, the solution set is $\varnothing$.

$|x| > -7$, the solution set is $\mathbb{R}$.
$|x| < -7$, the solution set is $\varnothing$.

To Solve Equations of the Form $|x| = |y|$.

If $|x| = |y|$, then $x = y$ or $x = -y$.

Solve $|x| = |3|$.

$$x = 3 \text{ or } x = -3.$$

Chapter 2 Review Exercises

[2.1]   *State the degree of each term.*

1. $9a^2b^6$

2. $2y$

3. $-21xyz^5$

Simplify each expression. If an expression cannot be simplified, so state.

4. $7(z + 3) - 2(z + 4)$

5. $x^2 + 2xy + 6x^2 - 13$

6. $b^2 + b - 9$

7. $2[-(x - y) + 3x] - 5y + 10$

Solve each equation. If an equation has no solution, so state.

8. $4(a + 3) - 6 = 2(a + 1)$

9. $3(x + 1) - 3 = 4(x - 5)$

10. $3 + \dfrac{x}{2} = \dfrac{5}{6}$

11. $\dfrac{1}{2}(3t + 4) = \dfrac{1}{3}(4t + 1)$

12. $2\left(\dfrac{x}{2} - 4\right) = 3\left(x + \dfrac{1}{3}\right)$

13. $3x - 7 = 9x + 8 - 6x$

14. $2(x - 6) = 5 - \{2x - [4(x - 2) - 9]\}$

[2.2]   *Evaluate each formula for the given values.*

15. $r = \sqrt{x^2 + y^2}$ when $x = 3$ and $y = -4$

16. $x = \dfrac{-b + \sqrt{b^2 - 4ac}}{2a}$ when $a = 8, b = 10, c = -3$

17. $h = \dfrac{1}{2}at^2 + v_0t + h_0$ when $a = -32, v_0 = 0, h_0 = 85, t = 1$

18. $z = \dfrac{\bar{x} - \mu}{\dfrac{\sigma}{\sqrt{n}}}$ when $\bar{x} = 50, \mu = 54, \sigma = 5, n = 25$

Solve each equation for the indicated variable.

19. $D = r \cdot t$, for t

20. $P = 2l + 2w$, for w

21. $A = \pi r^2 h$, for h

22. $A = \dfrac{1}{2}bh$, for h

23. $y = mx + b$, for m

24. $2x - 3y = 5$, for y

25. $R_T = R_1 + R_2 + R_3$, for R_2

26. $S = \dfrac{3a + b}{2}$, for a

27. $K = 2(d + l)$, for l

[2.3]   *In Exercises 28–32, write an equation that can be used to solve each problem. Solve the problem and check your answer.*

28. GPS Device Anna Conn purchased a GPS device for $630, which was 10% off of the original price. Determine the original price.

29. Population Increase A small town's population is increasing by 350 people per year. If the present population is 4750, how long will it take for the population to reach 7200?

30. Commission Salary Celeste Nossiter's salary is $300 per week plus 6% commission of sales. How much in sales must Celeste make to earn $708 in a week?

31. Car Rental Comparison At the Kansas City airport, the cost to rent a Ford Focus from Hertz is $24.99 per day with unlimited mileage. The cost to rent the same car from Avis is $19.99 per day plus $0.10 per mile that the car is driven. If Cathy Panik needs to rent a car for 3 days, determine the number of miles she would need to drive in order for the cost of the car rental to be the same from both companies.

32. Sale At a going-out-of-business sale, furniture is selling at 40% off the regular price. In addition, green-tagged items are reduced by an additional $20. If Alice Barr purchased a green-tagged item and paid $136, find the item's regular price.

[2.4]   *In Exercises 33–37, solve the following motion and mixture problems.*

33. Investing a Bonus After Ty Olden received a $5000 bonus at work, he invested some of the money in a money market account yielding 3.5% simple interest and the rest in a certificate of deposit yielding 4.0% simple interest. If the total amount of interest that Mr. Olden earned for the year was $187.15, determine the amount invested in each investment.

34. Fertilizer Solutions Dale Klitzke has liquid fertilizer solutions that are 20% and 60% nitrogen. How many gallons of each of these solutions should Dale mix to obtain 250 gallons of a solution that is 30% nitrogen?

35. Two Trains Two trains leave Portland, Oregon, at the same time traveling in opposite directions. One train travels at 60 miles per hour and the other at 80 miles per hour. In how many hours will they be 910 miles apart?

36. Space Shuttles Space Shuttle 2 takes off 0.5 hour after Shuttle 1 takes off. If Shuttle 2 travels 300 miles per hour faster than Shuttle 1 and overtakes Shuttle 1 exactly 5 hours after Shuttle 2 takes off, find

    a) the speed of Shuttle 1 and

    b) the distance from the launch pad when Shuttle 2 overtakes Shuttle 1.

37. Mixing Coffee Tom Tomlins, the owner of a gourmet coffee shop, has two coffees, one selling for $6.00 per pound and the other for $6.80 per pound. How many pounds of each type of coffee should he mix to make 40 pounds of coffee to sell for $6.50 per pound?

[2.3, 2.4]   *Solve.*

38. Electronics Sale At Best Buy, the price of a cordless telephone has been reduced by 20%. If the sale price is $28.80, determine the original price.

39. Jogging Nicolle Ryba jogged for a distance and then turned around and walked back to her starting point. While jogging she averaged 7.2 miles per hour, and while walking she averaged 2.4 miles per hour. If the total time spent jogging and walking was 4 hours, find

    a) how long she jogged, and

    b) the total distance she traveled.

40. Angle Measures Find the measures of three angles of a triangle if one angle measures 25° greater than the smallest angle and the other angle measures 5° less than twice the smallest angle.

41. Swimming Pool Two hoses are filling a swimming pool. The hose with the larger diameter supplies 1.5 times as much water as the hose with the smaller diameter. The larger hose is on for 2 hours before the smaller hose is turned on. If 5 hours after the larger hose is turned on there are 3150 gallons of water in the pool, find the rate of flow from each hose.

42. Complementary Angles One complementary angle has a measure that is 30° less than twice the measure of the other angle. Determine the measures of the two angles.

43. Blue Dye A clothier has two blue dye solutions, both made from the same concentrate. One solution is 6% blue dye and the other is 20% blue dye. How many ounces of the 20% solution must be mixed with 10 ounces of the 6% solution to result in the mixture being a 12% blue dye solution?

44. Two Investments David Alevy invests $12,000 in two savings accounts. One account is paying 10% simple interest and the other account is paying 6% simple interest. If in one year, the same interest is earned on each account, how much was invested at each rate?

45. Fitness Center The West Ridge Fitness Center has two membership plans. The first plan is a flat $40-per-month fee plus $1.00 per visit. The second plan is $25 per month plus a $4.00 per visit charge. How many visits would Jeff Feazell have to make per month to make it advantageous for him to select the first plan?

46. Trains in Alaska Two trains leave Anchorage at the same time along parallel tracks heading in opposite directions. The faster train travels 10 miles per hour faster than the slower train. Find the speed of each train if the trains are 270 miles apart after 3 hours.

[2.5]   *Solve the inequality. Graph the solution on a real number line.*

47. $3z + 9 \le 15$

48. $8 - 2w > -4$

49. $2x + 1 > 6$

50. $26 \le 4x + 5$

51. $\dfrac{4x + 3}{3} > -5$

52. $2(x - 1) > 3x + 8$

53. $-4(x - 2) \ge 6x + 8 - 10x$

54. $\dfrac{x}{2} + \dfrac{3}{4} > x - \dfrac{x}{2} + 1$

Write an inequality that can be used to solve each problem. Solve the inequality and answer the question.

55. Weight Limit A canoe can safely carry a total weight of 560 pounds. If Bob and Kathy together weigh a total of 300 pounds, what is the maximum number of 40-pound boxes they can carry in their canoe?

56. Bicycling Pat and Janie Wetter wish to rent a bicycle built for two while visiting a state park. The charge is \$14 for the first hour and \$7 for every hour after the first hour. How many hours can the Wetters bike if they have \$63 to spend?

57. Fitness Center A fitness center guarantees that customers will lose a minimum of 5 pounds the first week and $1\frac{1}{2}$ pounds each additional week. Find the maximum amount of time needed to lose 27 pounds.

58. Exam Scores Patrice Lee's first four exam scores are 94, 73, 72, and 80. If a final average greater than or equal to 80 and less than 90 is needed to receive a final grade of B in the course, what range of scores on the fifth and last exam will result in Patrice's receiving a B in the course? Assume a maximum score of 100.

Solve each inequality. Write the solution in interval notation.

59. $-2 < z - 5 < 3$

60. $8 < p + 11 \le 16$

61. $3 < 2x - 4 < 12$

62. $-12 < 6 - 3x < -2$

63. $-1 < \frac{5}{9}x + \frac{2}{3} \le \frac{11}{9}$

64. $-8 < \frac{4 - 2x}{3} < 0$

Find the solution set to each compound inequality.

65. $2x + 1 \le 7$ and $7x - 3 > 11$

66. $2x - 1 > 5$ or $3x - 2 \le 10$

67. $4x - 5 < 11$ and $-3x - 4 \ge 8$

68. $\frac{7 - 2g}{3} \le -5$ or $\frac{3 - g}{9} > 1$

[2.5, 2.6] *Find the solution set to each equation or inequality.*

69. $|h| = 4$

70. $|x| < 8$

71. $|x| \ge 9$

72. $|l + 5| = 13$

73. $|x - 2| \ge 5$

74. $|4 - 2x| = 5$

75. $|-2q + 9| < 7$

76. $\left|\frac{2x - 3}{5}\right| = 1$

77. $\left|\frac{x - 4}{3}\right| < 6$

78. $|4d - 1| = |6d + 9|$

79. $|2x - 3| + 4 \ge -17$

Solve each inequality. Give the solution in interval notation.

80. $|2x - 3| \ge 5$

81. $3 < 2x - 5 \le 11$

82. $-6 \le \frac{3 - 2x}{4} < 5$

83. $2p - 5 < 7$ and $9 - 3p \le 15$

84. $x - 3 \le 4$ or $2x - 5 > 7$

85. $-10 < 3(x - 4) \le 18$

Chapter 2 Practice Test

Chapter Test Prep Videos provide fully worked-out solutions to any of the exercises you want to review. Chapter Test Prep Videos are available via MyMathLab, or on YouTube (search "Angel Intermediate Algebra" and click on "Channels")

1. State the degree of the term $-3a^2bc^4$.

Simplify.

2. $2p - 3q + 2pq - 6p(q - 3) - 4p$

3. $7q - \{2[3 - 4(q + 7)] + 5q\} - 8$

In Exercises 4–8, solve the equation.

4. $7(d + 2) = 3(2d - 4)$

5. $\frac{r}{12} + \frac{1}{3} = \frac{4}{9}$

6. $-2(x + 3) = 4\{3[x - (3x + 7)] + 2\}$

7. $7x - 6(2x - 4) = 3 - (5x - 6)$

8. $-\frac{1}{2}(4x - 6) = \frac{1}{3}(3 - 6x) + 2$

9. Find the value of S_n for the given values.
$$S_n = \frac{a_1(1 - r^n)}{1 - r}, a_1 = 3, r = \frac{1}{3}, n = 3$$

10. Solve $c = \frac{a - 5b}{2}$ for b.

11. Solve $A = \frac{1}{2}h(b_1 + b_2)$ for b_2.

In Exercises 12–16, write an equation that can be used to solve each problem. Solve the equation and answer the question asked.

12. Golf Club Discount Find the cost of a set of golf clubs, before tax, if the cost of the clubs plus 7% tax is \$668.75.

13. Health Club Costs The cost of joining a health club is $240 per year, plus $2 per visit (for towel cleaning and toiletry expenses). If Bill Rush wishes to spend a total of $400 per year for the health club, how many visits can he make?

14. Bicycle Travel Jeffrey Chang and Roberto Fernandez start at the same point and bicycle in opposite directions. Jeffrey's speed is 15 miles per hour and Roberto's speed is 20 miles per hour. In how many hours will the two men be 147 miles apart?

15. Salt Solution How many liters of a 12% salt solution must be added to 10 liters of a 25% salt solution to get a 20% salt solution?

16. Two Investments June White has $12,000 to invest. She places part of her money in a savings account paying 8% simple interest and the balance in a savings account paying 7% simple interest. If the total interest from the two accounts at the end of 1 year is $910, find the amount placed in each account.

Solve each inequality and graph the solution on a number line.

17. $3(2q + 4) < 5(q - 1) + 7$

18. $\dfrac{6 - 2x}{5} \geq -12$

Solve each inequality and write the solution in interval notation.

19. $x - 3 \leq 4$ and $2x + 1 > 10$

20. $7 \leq \dfrac{2u - 5}{3} < 9$

Find the solution set to the following equations.

21. $|2b + 5| = 9$

22. $|2x - 3| = \left|\dfrac{1}{2}x - 10\right|$

Find the solution set to the following inequalities.

23. $|4z + 12| \leq 0$

24. $|2x - 3| + 6 > 11$

25. $\left|\dfrac{2x - 3}{8}\right| \leq \dfrac{1}{4}$

Cumulative Review Test

Take the following test and check your answers with those given in the back of the book. Review any questions that you answered incorrectly. The section where the material was covered is indicated after the answer.

1. If $A = \{1, 3, 5, 7, 9, 11, 13, 15\}$ and $B = \{2, 3, 5, 7, 11, 13\}$, find

a) $A \cup B$

b) $A \cap B$

2. Name each indicated property.

a) $9x + y = y + 9x$

b) $(2x)y = 2(xy)$

c) $4(x + 3) = 4x + 12$

Evaluate.

3. $-4^3 + (-6)^2 \div (2^3 - 2)^2$

4. $a^2b^3 + ab^2 - 3b$ when $a = -1$ and $b = -2$

5. $\dfrac{8 - \sqrt[3]{27} \cdot 3 \div 9}{|-5| - [5 - (12 \div 4)]^2}$

In Exercises 6 and 7, simplify.

6. $(5x^4y^3)^{-2}$

7. $\left(\dfrac{4m^2n^{-4}}{m^{-3}n^2}\right)^2$

8. Comparing State Sizes Rhode Island has a land area of about 1.045×10^3 square miles. Alaska has a land area of about 5.704×10^5 square miles. How many times larger is the land area of Alaska than that of Rhode Island?

In Exercises 9–11, solve the equation.

9. $-3(y + 7) = 2(-2y - 8)$

10. $1.2(x - 3) = 2.4x - 4.98$

11. $\dfrac{2m}{3} - \dfrac{1}{6} = \dfrac{4}{9}m$

12. Explain the difference between a conditional linear equation, an identity, and a contradiction. Give an example of each.

13. Evaluate the formula $x = \dfrac{-b + \sqrt{b^2 - 4ac}}{2a}$ for $a = 3$, $b = -8$, and $c = -3$.

14. Solve the formula $y - y_1 = m(x - x_1)$ for x.

15. Solve the inequality $-4 < \dfrac{5x - 2}{3} < 2$ and give the answer

a) on a number line,

b) as a solution set, and

c) in interval notation.

In Exercises 16 and 17, find the solution set.

16. $|3h - 1| = 8$

17. $|2x - 4| - 6 \geq 18$

18. Baseball Sale One week after the World Series, Target marks the price of all baseball equipment down by 40%. If Maxwell Allen purchases a Louisville Slugger baseball bat for $21 on sale, what was the original price of the bat?

19. Two Cars Two cars leave Newark, New Jersey, at the same time traveling in opposite directions. The car traveling north is moving 20 miles per hour faster than the car traveling south. If the two cars are 300 miles apart after 3 hours, find the speed of each car.

20. Mixed Nuts Molly Fitzgerald, owner of Molly's Nut House, has cashews that cost $6.50 per pound and peanuts that cost $2.50 per pound. If she wishes to make 40 pounds of a mixture of cashews and peanuts that sells for $4.00 per pound, how many pounds of cashews and how many pounds of peanuts should Molly mix together?

3

Graphs and Functions

Goals of This Chapter

The two primary goals of this chapter are to provide you with a good understanding of graphing and functions. Graphing is a key element to this and many mathematics courses. Functions are closely related to graphing and are a unifying concept throughout mathematics. We will use both graphing and functions throughout the rest of this book.

© jdana\iStockphoto

We see graphs daily in newspapers, magazines, and on the Internet. You will see many such graphs in this chapter. For example, in Exercise 74 on page 161, a graph is used to show inflation-adjusted gasoline prices.

3.1 Graphs

1 Plot points in the Cartesian coordinate system.

2 Draw graphs by plotting points.

3 Graph nonlinear equations.

4 Interpret graphs.

René Descartes

1 Plot Points in the Cartesian Coordinate System

Many algebraic relationships are easier to understand if we can see a visual picture of them. A **graph** is a picture that shows the relationship between two or more variables in an equation.

The **Cartesian** (or **rectangular**) **coordinate system**, named after the French mathematician and philosopher René Descartes (1596–1650), consists of two axes (or number lines) in a plane drawn perpendicular to each other (**Fig. 3.1**). Note how the two axes yield four **quadrants**, labeled with capital Roman numerals I, II, III, and IV.

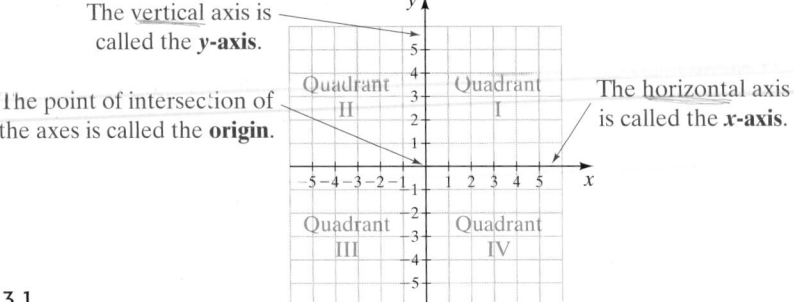

The vertical axis is called the **y-axis**.

The point of intersection of the axes is called the **origin**.

The horizontal axis is called the **x-axis**.

FIGURE 3.1

To locate a point in the Cartesian coordinate system, we will use an **ordered pair** of the form (x, y). The first number, x, is called the **x-coordinate** and the second number, y, is called the **y-coordinate**.

To plot the point $(3, 5)$, start at the origin,

$$(3, 5)$$

x-coordinate is 3 means go *right* 3 units

y-coordinate is 5 means go *up* 5 units

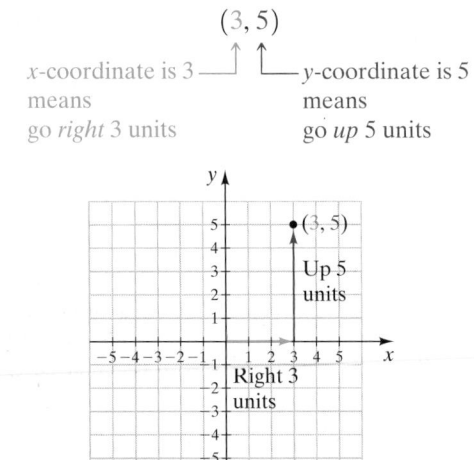

FIGURE 3.2

The ordered pairs A at $(-2, 3)$, B at $(0, 2)$, C at $(4, -1)$, and D at $(-4, 0)$ are plotted in **Figure 3.3**.

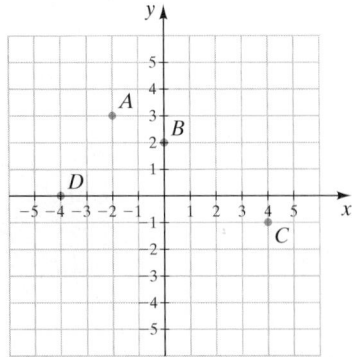

FIGURE 3.3

EXAMPLE 1   Plot the following points on the same set of axes.

a) $A(1, 4)$ **b)** $B(4, 1)$ **c)** $C(0, 2)$

d) $D(-3, 0)$ **e)** $E(-3, -1)$ **f)** $F(2, -4)$

Solution

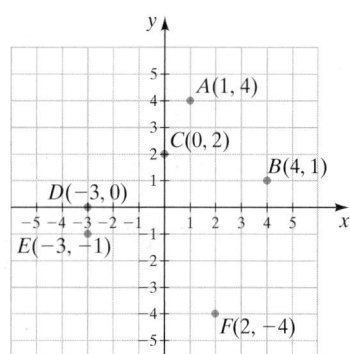

FIGURE 3.4

Now Try Exercise 7

Understanding Algebra

From **Figure 3.4** observe the following:

- $(1, 4)$ is a different point than $(4, 1)$
- When the *x*-coordinate is 0 (point C), the point is on the *y*-axis.
- When the *y*-coordinate is 0 (point D) the point is on the *x*-axis.

2 Draw Graphs by Plotting Points

In Chapter 2, we solved equations that contained one variable. Now we will discuss equations that contain two variables. If an equation contains two variables, its solutions are ordered pairs of numbers.

EXAMPLE 2   Determine whether the following ordered pairs are solutions to the equation $y = 2x - 3$.

a) $(1, -1)$ **b)** $\left(\dfrac{1}{2}, -2\right)$

c) $(4, 6)$ **d)** $(-1, -5)$

Solution   We substitute the first number in the ordered pair for *x* and the second number for *y*. If the result is a true statement, the ordered pair is a solution to the equation.

a) $y = 2x - 3$ **b)** $y = 2x - 3$

$-1 \stackrel{?}{=} 2(1) - 3$ $-2 \stackrel{?}{=} 2\left(\dfrac{1}{2}\right) - 3$

$-1 \stackrel{?}{=} 2 - 3$ $-2 \stackrel{?}{=} 1 - 3$

$-1 = -1$ True $-2 = -2$ True

c) $y = 2x - 3$ **d)** $y = 2x - 3$

$6 \stackrel{?}{=} 2(4) - 3$ $-5 \stackrel{?}{=} 2(-1) - 3$

$6 \stackrel{?}{=} 8 - 3$ $-5 \stackrel{?}{=} -2 - 3$

$6 = 5$ False $-5 \stackrel{?}{=} -5$ True

Thus the ordered pairs $(1, -1)$, $\left(\dfrac{1}{2}, -2\right)$, and $(-1, -5)$ are solutions to the equation $y = 2x - 3$. The ordered pair $(4, 6)$ is not a solution.

Now Try Exercise 17

There are an infinite number of solutions to the equation in Example 2. One method to find solutions to $y = 2x - 3$ is to substitute values for *x* and find the corresponding values of y. For example, to find the solution when $x = 0$, substitute 0 for x and solve for y.

$$y = 2x - 3$$
$$y = 2(0) - 3$$
$$y = 0 - 3$$
$$y = -3$$

Thus, another solution to the equation is $(0, -3)$.

Graph of an Equation

A **graph of an equation** is an illustration of the set of points whose ordered pairs are solutions to the equation.

The table below shows the four ordered pair solutions we have found to $y = 2x - 3$.

x	y	(x, y)
-1	-5	$(-1, -5)$
0	-3	$(0, -3)$
$\dfrac{1}{2}$	-2	$\left(\dfrac{1}{2}, -2\right)$
1	-1	$(1, -1)$

When plotting these points we see they all lie on the same line; thus, the points are said to be **collinear** (see **Fig. 3.5a**). The graph of the equation is a line through these points (see **Fig. 3.5b**). The line continues indefinitely in both directions as indicated by the arrows.

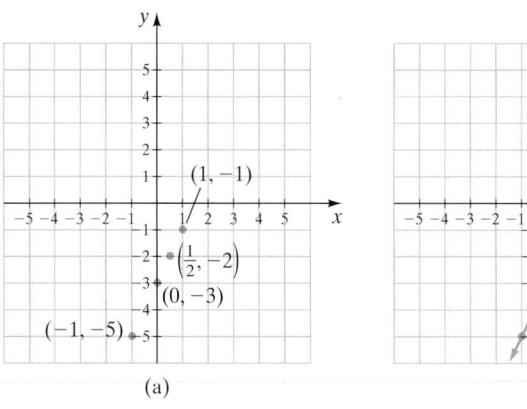
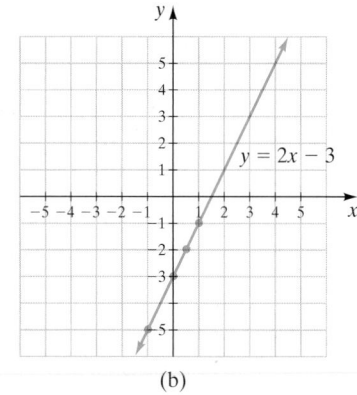

(a)   (b)

FIGURE 3.5

Understanding Algebra

The equation $y = 2x - 3$ has an infinite number of solutions. Each solution is represented by a point on the line in **Fig. 3.5b**. Also, each point on the line represents a solution to the equation $y = 2x - 3$.

Because the graph is a line, the graph is said to be **linear** and the equation is called a **linear equation**. The equation is also called a **first-degree equation** since the greatest exponent on any variable is 1.

Helpful Hint

Study Tip

In this chapter, and in several upcoming chapters, you will be plotting points and drawing graphs using the Cartesian coordinate system. The following suggestions can improve the quality of the graphs you produce.

1. For your homework, using graph paper will help you maintain a consistent scale throughout your graph.

2. Your axes and lines will look much better and will be much more accurate if drawn with a ruler or straightedge.

(continued on next page)

3. If you are not using graph paper, use a ruler to make a consistent scale on your axes. It is impossible to get an accurate graph when axes are marked with an uneven scale.

4. Use a pencil instead of a pen. A mistake can often be fixed quickly with a pencil and eraser, and you will not have to start over from the beginning.

5. Work all of the homework problems that you are assigned.

EXAMPLE 3   Graph $y = x$.

Solution   We first find some ordered pairs that are solutions by selecting values of x and finding the corresponding values of y. We will select 0, some positive values, and some negative values for x. In general, we will choose numbers close to 0, so that the ordered pairs will fit on the axes. The graph is illustrated in **Figure 3.6**.

x	y	(x, y)
−2	−2	$(−2, −2)$
−1	−1	$(−1, −1)$
0	0	$(0, 0)$
1	1	$(1, 1)$
2	2	$(2, 2)$

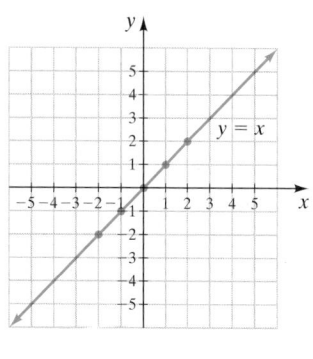

FIGURE 3.6

1. Select values for x.
2. Compute y.
3. Ordered pairs.
4. Plot the points and draw the graph.

Now Try Exercise 27

EXAMPLE 4   Graph $y = -\dfrac{1}{3}x + 1$.

Solution   When we select values for x, we will select some positive values, some negative values, and 0. We will select multiples of 3 so the values of y are integer values. The graph is illustrated in **Figure 3.7**.

Understanding Algebra

When graphing linear equations in which the coefficient on x is a fraction, choose values of x that are multiples of the denominator. This may result in integer values for y.

x	y
−6	3
−3	2
0	1
3	0
6	−1

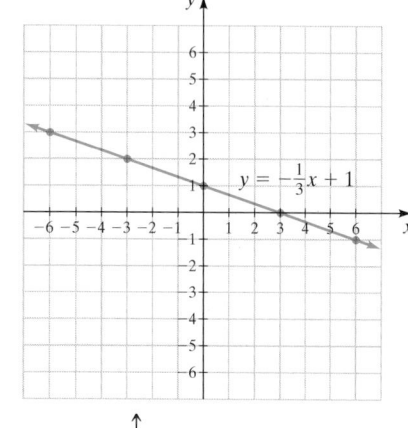

FIGURE 3.7

1. Select values for x.
2. Compute y.
3. Plot the points and draw the graph.

Now Try Exercise 35

If we are asked to graph an equation not solved for y, such as $x + 3y = 3$, our first step will be to solve the equation for y. For example, if we solve $x + 3y = 3$ for y we obtain

$$x + 3y = 3$$

$$3y = -x + 3 \qquad \text{Subtract } x \text{ from both sides.}$$

$$y = \frac{-x + 3}{3} \qquad \text{Divide both sides by 3.}$$

$$y = \frac{-x}{3} + \frac{3}{3} = -\frac{1}{3}x + 1$$

The resulting equation, $y = -\frac{1}{3}x + 1$, is the same equation we graphed in Example 4. Therefore, the graph of $x + 3y = 3$ is also illustrated in **Figure 3.7** on page 139.

3 Graph Nonlinear Equations

Equations whose graphs are not straight lines are called **nonlinear equations**. To graph nonlinear equations by plotting points, we follow the same procedure used to graph linear equations. However, since the graphs are not straight lines, we may need to plot more points to draw the graphs.

EXAMPLE 5 Graph $y = x^2 - 4$.

Solution   We select some values for x and find the corresponding values of y. Then we plot the points and connect them with a smooth curve. The graph is shown in **Figure 3.8**.

x	y
-3	5
-2	0
-1	-3
0	-4
1	-3
2	0
3	5

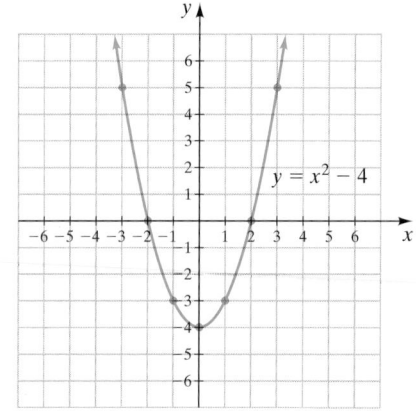

FIGURE 3.8

> ### Understanding Algebra
>
> When substituting values for x, we must follow the order of operations discussed in Section 1.4.

Now Try Exercise 41

EXAMPLE 6 Graph $y = \frac{1}{x}$.

Solution   When selecting values for x, notice that $x = 0$ gives us $y = \frac{1}{0}$, which is undefined. Therefore, there will be no part of the graph at $x = 0$. Also notice that when $x = \frac{1}{2}$, we get $\dfrac{1}{\frac{1}{2}} = 1 \div \frac{1}{2} = 1 \cdot \frac{2}{1} = 2$. This graph has two branches, one to the left of the y-axis and one to the right of the y-axis, as shown in **Figure 3.9** on page 141.

x	y
-3	$-\dfrac{1}{3}$
-2	$-\dfrac{1}{2}$
-1	-1
$-\dfrac{1}{2}$	-2
$\dfrac{1}{2}$	2
1	1
2	$\dfrac{1}{2}$
3	$\dfrac{1}{3}$

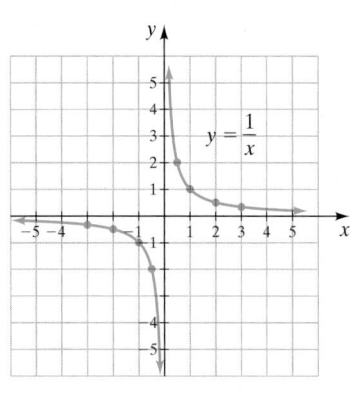

FIGURE 3.9

Now Try Exercise 51

In the graph for Example 6, notice that for values of x far to the right of 0, or far to the left of 0, the graph approaches the x-axis but does not touch it. For example when $x = 1000$, $y = 0.001$ and when $x = -1000$, $y = -0.001$.

EXAMPLE 7   Graph $y = |x|$.

Solution   To graph this absolute value equation, we select some values for x and find the corresponding values of y. Then we plot the points and draw the graph.
Notice that this graph is V-shaped, as shown in **Figure 3.10**.

x	y
-4	4
-3	3
-2	2
-1	1
0	0
1	1
2	2
3	3
4	4

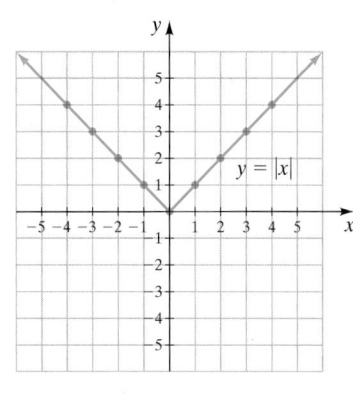

FIGURE 3.10

Now Try Exercise 45

Understanding Algebra

The absolute value of a number x, written $|x|$, is the distance that x is from 0 on the number line.

Avoiding Common Errors

When graphing nonlinear equations, be sure to plot enough points to get a true picture of the graph. For example, when graphing $y = \dfrac{1}{x}$ many students consider only integer values of x. Following is a table of values for the equation and two graphs that contain the points indicated in the table.

x	-3	-2	-1	1	2	3
y	$-\dfrac{1}{3}$	$-\dfrac{1}{2}$	-1	1	$\dfrac{1}{2}$	$\dfrac{1}{3}$

(continued next page)

CORRECT

INCORRECT

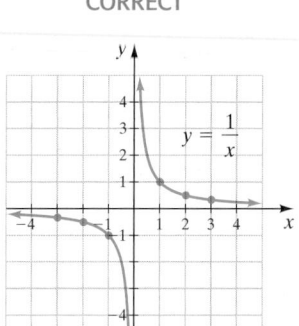

FIGURE 3.11

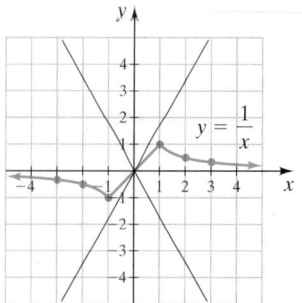

FIGURE 3.12

If you select and plot fractional values of x near 0, as was done in Example 6, you get the graph in **Figure 3.11**. The graph in **Figure 3.12** cannot be correct because the equation is not defined when x is 0 and therefore the graph cannot cross the y-axis. Whenever you plot a graph that contains a variable in the denominator, select values for the variable that are very close to the value that makes the denominator 0 and observe what happens. For example, when graphing $y = \dfrac{1}{x - 3}$ you should use values of x close to 3, such as 2.9 and 3.1 or 2.99 and 3.01, and see what values you obtain for y.

Also, when graphing nonlinear equations, it is a good idea to consider both positive and negative values. For example, if you used only positive values of x when graphing $y = |x|$, the graph would appear to be a straight line going through the origin, instead of the V-shaped graph shown in **Figure 3.10** on page 141.

Using Your Graphing Calculator

Throughout this book, we will introduce some of the basic uses of the graphing calculator. More detailed information is available in your calculator manual. A **window** is the screen in which a graph is displayed. In this book all windows and keystrokes will be from a TI-84 Plus graphing calculator. **Figure 3.13** shows the standard window and **Figure 3.14** shows the standard window settings of the **TI-84** Plus.

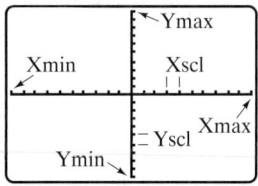

FIGURE 3.13

FIGURE 3.14

To graph the equation $y = -2x + 3$, press

$$\boxed{Y=}\ \boxed{(-)}\ \boxed{2}\ \boxed{X,T,\theta,n}\ \boxed{+}\ \boxed{3}\ \boxed{GRAPH}$$

The graph of the equation then appears in the window as shown in **Figure 3.15**.

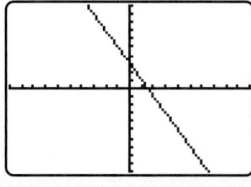

FIGURE 3.15

4 Interpret Graphs

We see many different types of graphs daily in newspapers, in magazines, on the Internet and so on. Throughout this book, we present a variety of graphs. Since being able to draw and interpret graphs is very important, we will study this further in Section 3.2. In Example 8 you must understand and interpret graphs to answer the question.

EXAMPLE 8   When Jim Herring went to see his mother in Cincinnati, he boarded a Southwest Airlines plane. The plane sat on the runway for 20 minutes and then took off. The plane flew at about 600 miles per hour for about 2 hours. It then reduced its speed to about 300 miles per hour and circled the Cincinnati Airport for about 15 minutes before it came in for a landing. After landing, the plane taxied to the gate and stopped. Which graph, **Figure 3.16, 3.17, 3.18,** or **3.19,** best illustrates this situation?

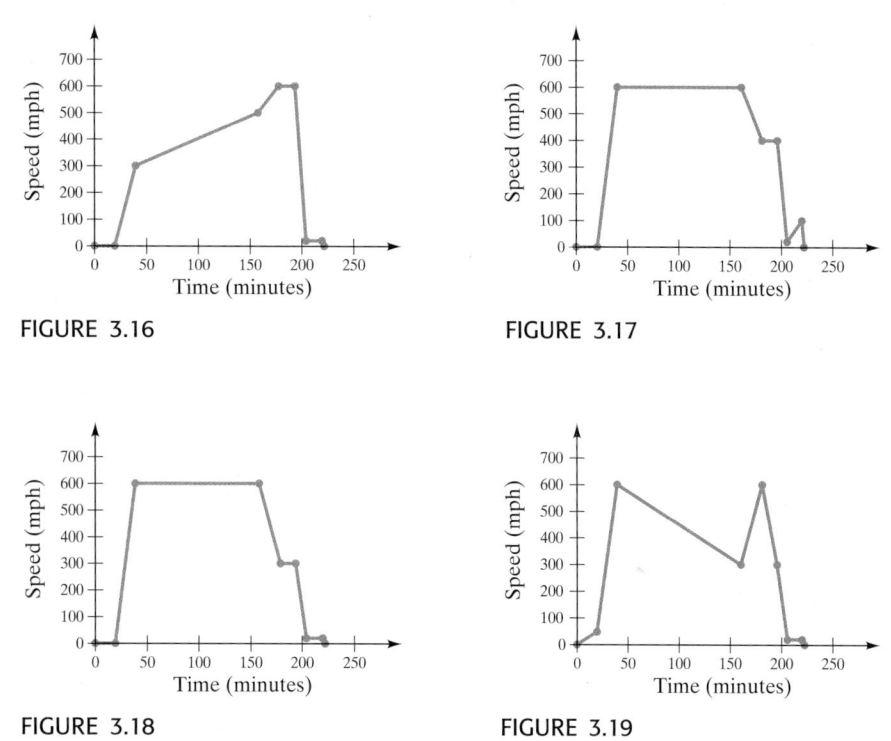

FIGURE 3.16

FIGURE 3.17

FIGURE 3.18

FIGURE 3.19

Solution   The graph that depicts the situation described is **Figure 3.18**, reproduced with annotations in **Figure 3.20**. The graph shows speed versus time, with time on the horizontal axis.

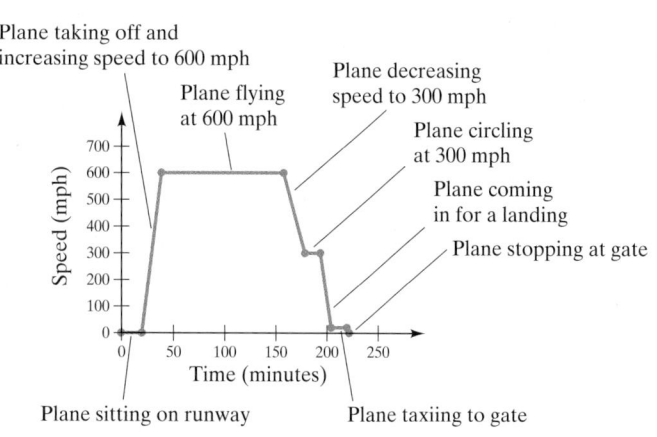

FIGURE 3.20

Now Try Exercise 75

EXERCISE SET 3.1

Math XL
MathXL®

MyMathLab
MyMathLab

Warm-Up Exercises

Fill in the blanks with the appropriate word, phrase, or symbol(s) from the following list.

x-coordinate   collinear   graph   y-coordinate   solution   ordered pair

1. A solution to an equation with two variables is an _____.

2. A _____ of an equation is an illustration of the set of points whose ordered pairs are solutions to the equation.

3. Three or more points that lie on the same line are said to be _____.

4. The first coordinate in an ordered pair is the _____ and the second coordinate is the y-coordinate.

Practice the Skills

List the ordered pairs corresponding to the indicated points.

5.

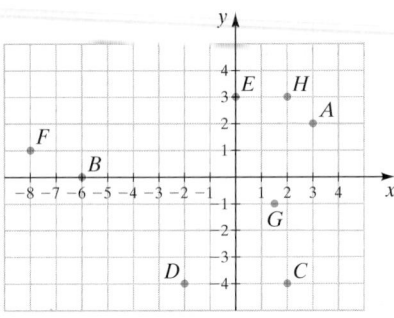

6.

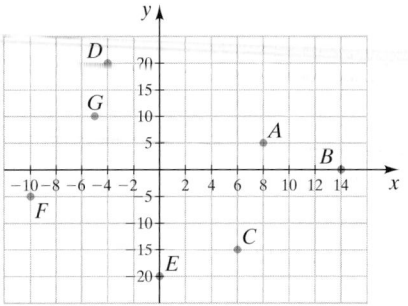

7. Graph the following points on the same axes.
$A(4, 2)$ $B(-6, 2)$ $C(0, -1)$ $D(-2, 0)$

8. Graph the following points on the same axes.
$A(-4, -2)$ $B(3, 2)$ $C(2, -3)$ $D(-3, 3)$

Determine the quadrant in which each point is located.

9. $(1, 3)$ **10.** $(-9, 1)$ **11.** $(4, -3)$ **12.** $(36, 43)$

13. $(-12, 18)$ **14.** $(-31, -8)$ **15.** $(-11, -19)$ **16.** $(8, -52)$

Determine whether the given ordered pair is a solution to the given equation.

17. $(2, -1)$; $y = 2x - 5$ **18.** $(1, 3)$; $2x + 3y = 6$ **19.** $(-4, -2)$; $y = |x| + 3$ **20.** $(1, -5)$; $y = x^2 + x - 7$

21. $(-2, 5)$; $s = 2r^2 - r - 5$ **22.** $\left(\frac{1}{4}, \frac{11}{4}\right)$; $y = |x - 3|$ **23.** $(2, 1)$; $-a^2 + 2b^2 = -2$

24. $(-10, -2)$; $|p| - 3|q| = 4$ **25.** $\left(\frac{1}{2}, \frac{5}{2}\right)$; $2x^2 + 6x - y = 0$ **26.** $\left(-3, \frac{7}{2}\right)$; $2m^2 + 3n = 2$

Graph each equation.

27. $y = x + 1$ **28.** $y = 3x$ **29.** $y = -3x - 5$ **30.** $y = -2x + 2$

31. $y = 2x + 4$ **32.** $y = x + 2$ **33.** $y = \frac{1}{2}x$ **34.** $y = -\frac{1}{3}x$

35. $y = \frac{1}{2}x - 1$ **36.** $y = -\frac{1}{2}x - 3$ **37.** $y = -\frac{1}{3}x + 2$ **38.** $y = -\frac{1}{3}x + 4$

39. $y = x^2$ **40.** $y = x^2 - 2$ **41.** $y = -x^2$ **42.** $y = -x^2 + 4$

43. $y = |x| + 1$ **44.** $y = |x| + 2$ **45.** $y = -|x|$ **46.** $y = -|x| - 3$

47. $y = x^3$ **48.** $y = -x^3$ **49.** $y = x^3 + 1$ **50.** $y = \frac{1}{x}$

51. $y = -\frac{1}{x}$ **52.** $x^2 = 1 + y$ **53.** $x = |y|$ **54.** $x = y^2$

In Exercises 55–62, use a calculator to obtain at least eight points that are solutions to the equation. Then graph the equation by plotting the points.

55. $y = x^3 - x^2 - x + 1$ **56.** $y = -x^3 + x^2 + x - 1$ **57.** $y = \frac{1}{x + 1}$ **58.** $y = \frac{1}{x} + 1$

59. $y = \sqrt{x}$ **60.** $y = \sqrt{x + 4}$ **61.** $y = \frac{1}{x^2}$ **62.** $y = \frac{|x^2|}{2}$

63. Is the point represented by the ordered pair $\left(\frac{1}{3}, \frac{1}{12}\right)$ on the graph of the equation $y = \frac{x^2}{x + 1}$? Explain.

64. Is the point represented by the ordered pair $\left(-\frac{1}{2}, -\frac{3}{5}\right)$ on the graph of the equation $y = \frac{x^2 + 1}{x^2 - 1}$? Explain.

65. a) Plot the points $A(2,7)$, $B(2,3)$, and $C(6,3)$, and then draw $\overline{AB}$, $\overline{AC}$, and $\overline{BC}$. ($\overline{AB}$ represents the line segment from A to B.)

b) Find the area of the figure.

66. a) Plot the points $A(-4,5)$, $B(2,5)$, $C(2,-3)$, and $D(-4,-3)$, and then draw $\overline{AB}$, $\overline{BC}$, $\overline{CD}$, and $\overline{DA}$.

b) Find the area of the figure.

67. UK to US Travel The following graph shows the number of passengers that traveled by ship from the United Kingdom to the United States for each 5-year period from 1890 to 1959.

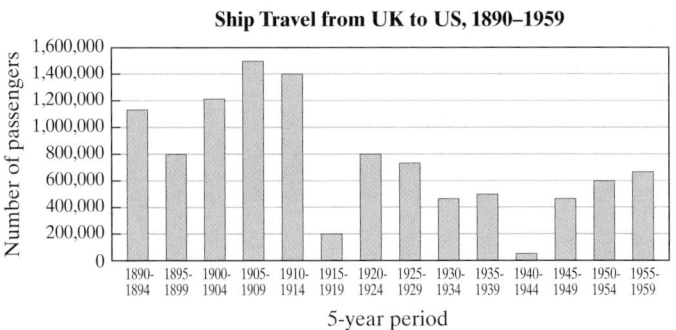

Ship Travel from UK to US, 1890–1959

Source: BT27 Passenger Lists on findmypast.com

a) Estimate the number of passengers during 1895–1899.

b) Estimate the number of passengers during 1955–1959.

c) During which 5-year periods was the number of passengers greater than 1,000,000 passengers?

d) Does this graph appear to be linear?

68. Scotland Temperature The following graph shows the average temperature in degrees Celsius for each month in Glasgow, Scotland.

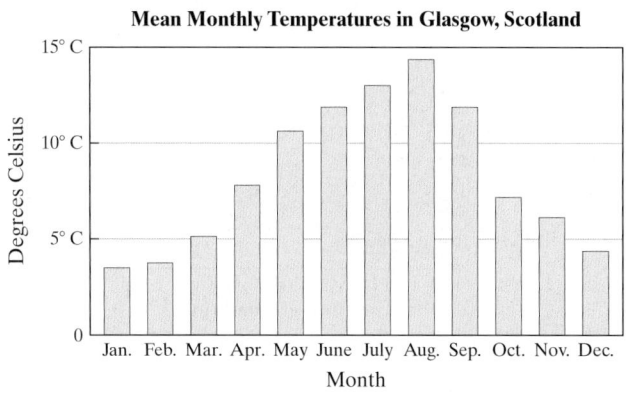

Mean Monthly Temperatures in Glasgow, Scotland

Source: www.wunderground.com

a) Estimate the average temperature in February.

b) Estimate the average temperature in July.

c) During which months was the average temperature greater than 10°C?

d) Does this graph appear to be linear?

Match Exercises 69–72 with the corresponding graph of elevation above sea level versus time, labeled a–d, below.

69. Nancy Johnson started out by walking up a steep hill for 5 minutes. For the next 5 minutes she walked down a steep hill to an elevation lower than her starting point. For the next 10 minutes she walked on level ground. For the next 10 minutes she walked up a slight hill, at which time she reached her starting elevation.

70. James Condor started out by walking up a hill for 5 minutes. For the next 10 minutes he walked down a hill to an elevation equal to his starting elevation. For the next 10 minutes he walked on level ground. For the next 5 minutes he walked downhill.

71. Mary Beth Headlee walked for 5 minutes on level ground. Then for 5 minutes she climbed a slight hill. Then she walked on level ground for 5 minutes. Then for the next 5 minutes she climbed a steep hill. During the next 10 minutes she descended uniformly until she reached the height at which she had started.

72. Don Ransford walked on level ground for 5 minutes. Then he walked down a steep hill for 10 minutes. For the next 5 minutes he walked on level ground. For the next 5 minutes he walked back up to his starting height. For the next 5 minutes he walked on level ground.

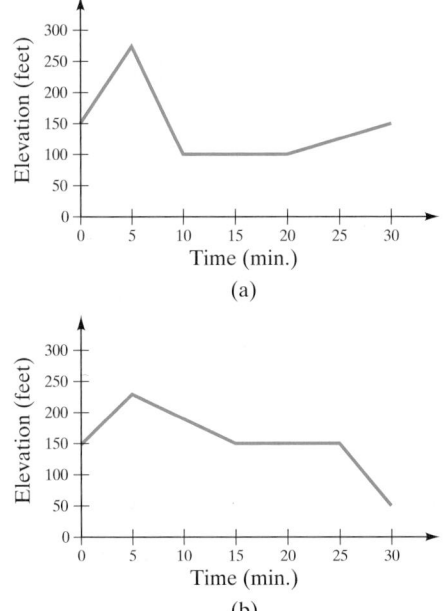

(a)

(b)

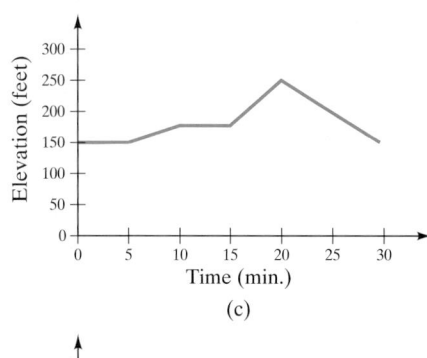

(c)

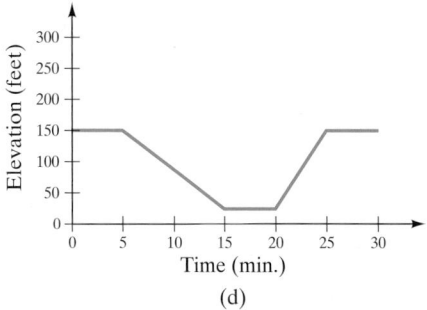

(d)

Match Exercises 73–76 with the corresponding graph of speed versus time, labeled a–d, below.

73. To go to work, Donna Clark drove on a country road for 10 minutes, then drove on a highway for 12 minutes, then drove in stop-and-go traffic for 8 minutes.

74. To go to work, Bob Plough drove in stop-and-go traffic for 5 minutes, then drove on the expressway for 20 minutes, then drove in stop-and-go traffic for 5 minutes.

75. To go to work, Ron Breitfelder walked for 3 minutes, waited for the train for 5 minutes, rode the train for 15 minutes, then walked for 7 minutes.

76. To go to work, Kim Ghiselin rode her bike uphill for 10 minutes, then rode downhill for 15 minutes, then rode on a level street for 5 minutes.

(a)

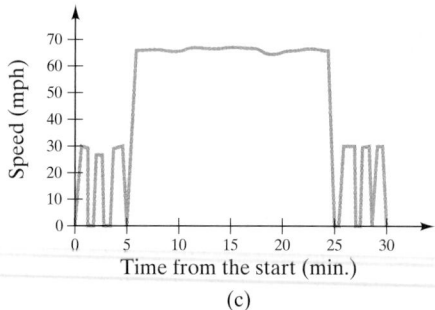

(c)

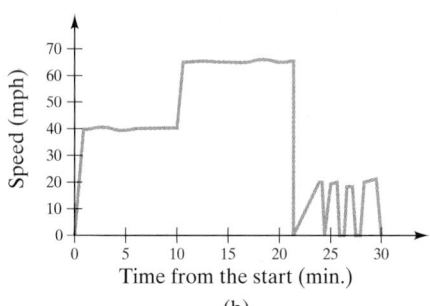

(b)

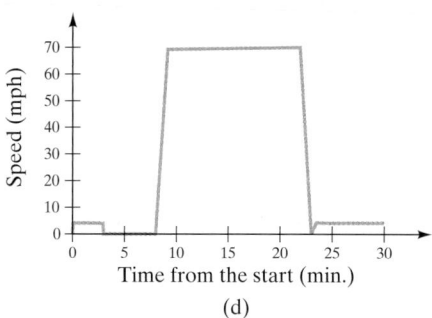

(d)

Match Exercises 77–80 with the corresponding graph of speed versus time, labeled a–d below.

77. Christina Dwyer walked for 5 minutes to warm up, jogged for 20 minutes, and then walked for 5 minutes to cool down.

78. Annie Droullard went for a leisurely bike ride at a constant speed for 30 minutes.

79. Michael Odu took a 30-minute walk through his neighborhood. He stopped very briefly on 7 occasions to pick up trash.

80. Richard Dai walked through his neighborhood and stopped 3 times to chat with his neighbors. He was gone from his house a total of 30 minutes.

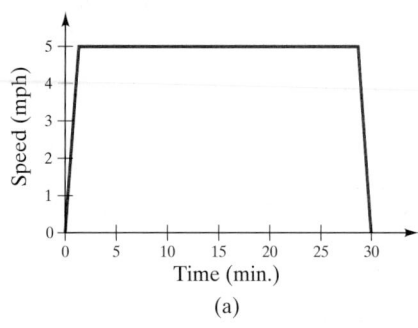

(a)

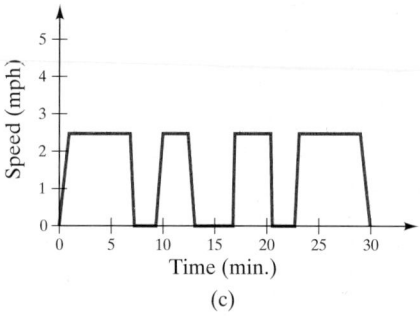

(c)

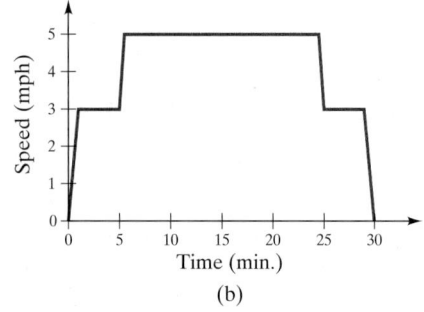

(b)

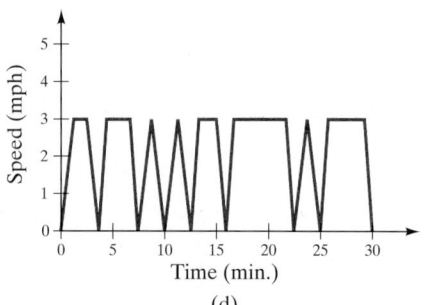

(d)

Match Exercises 81–84 with the corresponding graph of distance traveled versus time, labeled a–d. Recall from Chapter 2 that distance = rate × time. Selected distances are indicated on the graphs.

81. Train A traveled at a speed of 40 mph for 1 hour, then 80 mph for 2 hours, and then 60 mph for 3 hours.

82. Train C traveled at a speed of 80 mph for 2 hours, then stayed in a station for 1 hour, and then traveled 40 mph for 3 hours.

83. Train B traveled at a speed of 20 mph for 2 hours, then 60 mph for 3 hours, and then 80 mph for 1 hour.

84. Train D traveled at 30 mph for 1 hour, then 65 mph for 2 hours, and then 30 mph for 3 hours.

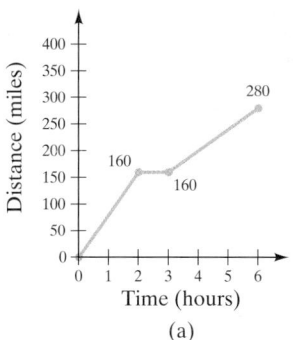

(a)

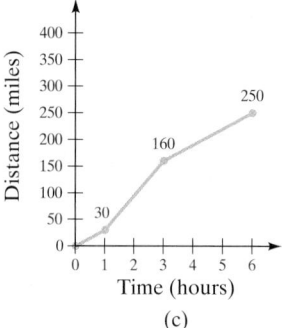

(c)

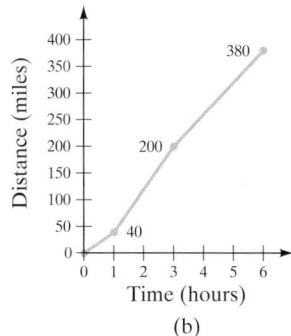

(b)

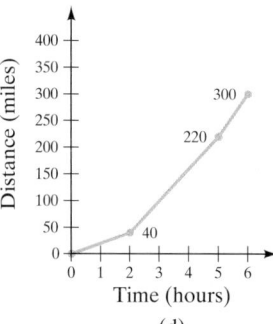

(d)

Use a graphing calculator to graph each function. Make sure you select values for the window that will show the curvature of the graph. Then, if your calculator can display tables, display a table of values in which the x-values extend by units, from 0 to 6.

85. $y = 2x - 3$

86. $y = \frac{1}{3}x + 2$

87. $y = x^2 - 2x - 8$

88. $y = -x^2 + 16$

89. $y = x^3 - 2x + 4$

90. $y = 2x^3 - 6x^2 - 1$

Challenge Problems

We will discuss many of the concepts introduced in Exercises 91–98 in Section 3.4.

91. Graph $y = x + 1$, $y = x + 3$, and $y = x - 1$ on the same axes.

   a) What do you notice about the graphs of the equations and the values where the graphs intersect the *y*-axis?

   b) Do all the graphs seem to have the same slant (or slope)?

92. Graph $y = \frac{1}{2}x$, $y = \frac{1}{2}x + 3$, and $y = \frac{1}{2}x - 4$ on the same axes.

   a) What do you notice about the graphs of equations and the values where the graphs intersect the *y*-axis?

   b) Do all of these graphs seem to have the same slant (or slope)?

93. Graph $y = 2x$. Determine the *rate of change* of *y* with respect to *x*. That is, by how many units does *y* change compared to each unit change in *x*?

94. Graph $y = 4x$. Determine the rate of change of *y* with respect to *x*.

95. Graph $y = 3x + 2$. Determine the rate of change of *y* with respect to *x*.

96. Graph $y = \frac{1}{2}x$. Determine the rate of change of *y* with respect to *x*.

97. The ordered pair $(3, -7)$ represents one point on the graph of a linear equation. If *y* increases 4 units for each unit increase in *x* on the graph, find two other solutions to the equation.

98. The ordered pair $(1, -4)$ represents one point on the graph of a linear equation. If *y* increases 3 units for each unit increase in *x* on the graph, find two other solutions to the equation.

Graph each equation.

99. $y = |x - 2|$

100. $x = y^2 + 2$

Group Activity

Discuss and work Exercises 101–102 as a group.

101. a) Group member 1: Plot the points $(-2, 4)$ and $(6, 8)$. Determine the *midpoint* of the line segment connecting these points.

   Group member 2: Follow the above instructions for the points $(-3, -2)$ and $(5, 6)$.

   Group member 3: Follow the above instructions for the points $(4, 1)$ and $(-2, 4)$.

   b) As a group, determine a formula for the midpoint of the line segment connecting the points (x_1, y_1) and (x_2, y_2). (*Note*: We will discuss the midpoint formula further in Chapter 10.)

102. Three points on a parallelogram are $A(3, 5)$, $B(8, 5)$, and $C(-1, -3)$.

   a) Individually determine a fourth point D that completes the parallelogram.

   b) Individually compute the area of your parallelogram.

   c) Compare your answers. Did you all get the same answers? If not, why not?

   d) Is there more than one point that can be used to complete the parallelogram? If so, give the points and find the corresponding areas of each parallelogram.

Cumulative Review Exercises

[2.2] **103.** Evaluate $\dfrac{-b + \sqrt{b^2 - 4ac}}{2a}$ for $a = 2, b = 7,$ and $c = -15$.

[2.3] **104. Truck Rental** Hertz Truck Rental charges a daily fee of \$60 plus 10¢ a mile. National Automobile Rental Agency charges a daily fee of \$50 plus 24¢ a mile for the same truck. What distance would you have to drive in 1 day to make the cost of renting from Hertz equal to the cost of renting from National?

[2.5] **105.** Solve the inequality $-1 \le \dfrac{4 - 3x}{2} < 5$. Write the solution in set builder notation.

[2.6] **106.** Find the solution set for the inequality $|3x + 2| > 7$.

3.2 Functions

1 Understand relations.

2 Recognize functions.

3 Use the vertical line test.

4 Understand function notation.

5 Applications of functions in daily life.

1 Understand Relations

We often find that one quantity is related to a second quantity. For example, the amount you spend for oranges is related to the number of oranges you purchase.

Suppose oranges cost 30 cents apiece. Then one orange costs 30 cents, two oranges cost 60 cents, three oranges cost 90 cents, and so on. The ordered pairs that represent this situation are (1, 30), (2, 60), (3, 90), and so on. An equation that represents this situation is

$$C = 30n$$

The cost of the oranges *depends* on the number of oranges purchased. Thus, cost is the *dependent* variable.

The number of oranges is the independent variable.

Now consider the equation $y = 2x + 3$. Some ordered pairs that satisfy this equation are $(-2, -1), (-1, 1), (0, 3), (1, 5), (2, 7),$ and so on.

$$y = 2x + 3$$

The value of y *depends* on the value of x. Thus, y is the *dependent* variable.

x is the independent variable.

Dependent and Independent Variables

For an equation in variables x and y, if the value of y depends on the value of x, then y is the **dependent variable** and x is the **independent variable**.

Since related quantities can be represented as ordered pairs, the concept of a *relation* can be defined as follows.

Relation, Domain, Range

A **relation** is any set of ordered pairs of the form (x, y). The set of x-coordinates is called the **domain** of the relation. The set of y-coordinates is called the **range** of the relation.

Since the equation $y = 2x + 3$ can be represented as a set of ordered pairs, it is a relation.

2 Recognize Functions

We now discuss functions—one of the most important concepts in mathematics.

Function

A **function** is a relation in which each element in the domain corresponds to exactly one element in the range.

Consider the oranges that cost 30 cents apiece that we just discussed. We can illustrate the number of oranges and the cost of the oranges using **Figure 3.21**.

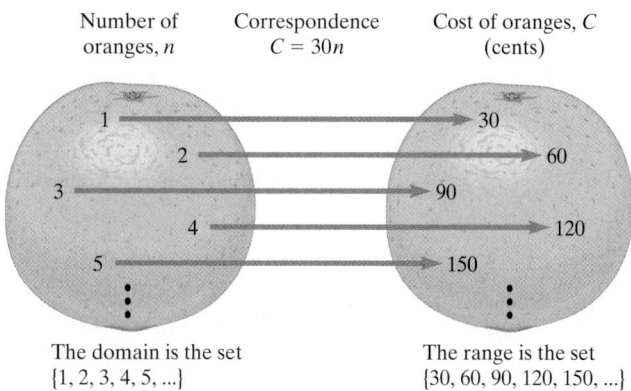

| Number of oranges, n | Correspondence $C = 30n$ | Cost of oranges, C (cents) |

The domain is the set
$\{1, 2, 3, 4, 5, ...\}$

The range is the set
$\{30, 60, 90, 120, 150, ...\}$

FIGURE 3.21

Notice that each number in the set of numbers of oranges, n, corresponds to (or is paired with) exactly one number in the set of cost of oranges, C. Therefore, this correspondence is a function.

EXAMPLE 1 Determine whether each correspondence is a function.

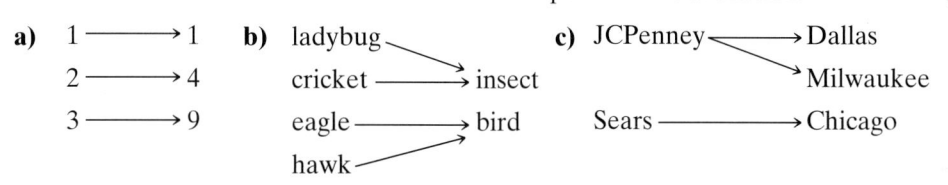

a) 1 ⟶ 1
   2 ⟶ 4
   3 ⟶ 9

b) ladybug
   cricket ⟶ insect
   eagle ⟶ bird
   hawk

c) JCPenney ⟶ Dallas
   ⟶ Milwaukee
   Sears ⟶ Chicago

Understanding Algebra

In a relation (or function), the set of values for the independent variable is called the *domain*. The set of values for the dependent variable is called the *range*.

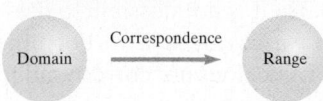

Domain — Correspondence → Range

Solution

a) For a correspondence to be a function, each element in the domain must correspond with exactly one element in the range. Here the domain is {1, 2, 3} and the range is {1, 4, 9}. Since each element in the domain corresponds to exactly one element in the range, this correspondence is a function.

b) Here the domain is {ladybug, cricket, eagle, hawk} and the range is {insect, bird}. Even though the domain has four elements and the range has two elements, each element in the domain corresponds with exactly one element in the range. Thus, this correspondence is a function.

c) Here the domain is {JCPenney, Sears} and the range is {Dallas, Milwaukee, Chicago}. Notice that JCPenney corresponds to both Dallas and Milwaukee. Therefore each element in the domain *does not* correspond to exactly one element in the range. Thus, this correspondence is a relation but *not* a function.

Now Try Exercise 17

EXAMPLE 2 Give the domain and range, then determine whether the relation is a function.

a) $\{(1, 4), (2, 3), (3, 5), (-1, 3), (0, 6)\}$

b) $\{(-1, 3), (4, 2), (3, 1), (2, 6), (3, 5)\}$

Solution

a) The domain is $\{1, 2, 3, -1, 0\}$ and the range is $\{4, 3, 5, 6\}$. Notice that when listing the range, we only include the number 3 once, even though it appears in both $(2, 3)$ and $(-1, 3)$. Each number in the domain corresponds with exactly one number in the range. For example, the 1 in the domain corresponds with only the 4 in the range, and so on. Since no *x*-value corresponds to more than one *y*-value, this relation *is a function*.

b) The domain is $\{-1, 4, 3, 2\}$ and the range is $\{3, 2, 1, 6, 5\}$. Since the ordered pairs $(3, 1)$ and $(3, 5)$ have *the same first coordinate* and a different second coordinate, each value in the domain does not correspond to exactly one value in the range. Therefore, this relation is *not a function*.

Now Try Exercise 23

Example 2 leads to an alternate definition of function.

Function

A **function** is a set of ordered pairs in which no *first* coordinate is repeated.

Recall the equation $y = 2x + 3$, introduced on page 148. Some ordered pairs that satisfy this equation are $(-2, -1), (-1, 1), (0, 3), (1, 5)$, and $(2, 7)$. Notice that each value of *x* gives a unique value of *y*. Therefore, the equation $y = 2x + 3$ is not only a relation, it is also a function.

3 Use the Vertical Line Test

Graph of Function or Relation

The **graph of a function or relation** is the graph of its set of ordered pairs.

The two sets of ordered pairs in Example 2 **a)** and **b)** are graphed in **Figures 3.22a** and **3.22b**, respectively. Notice that in the function in **Figure 3.22a** it is not possible to draw a vertical line that intersects two points. In **Figure 3.22b** we *can* draw a vertical line through the points $(3, 1)$ and $(3, 5)$. This shows that each *x*-value does not correspond to exactly one *y*-value, and the graph does not represent a function.

Function

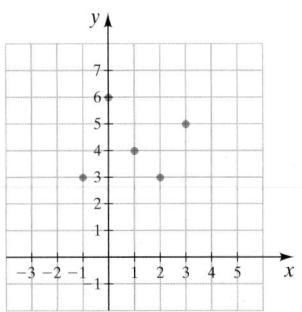

(a) First set of ordered pairs

Not a function

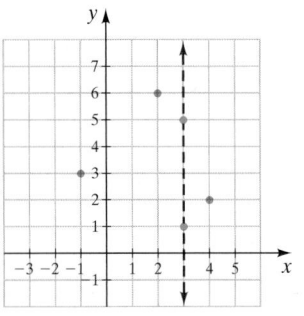

(b) Second set of ordered pairs

FIGURE 3.22

This method of determining whether a graph represents a function is called the **vertical line test**.

> **Vertical Line Test**
>
> If a vertical line can be drawn so that it intersects a graph at more than one point, then the graph does not represent a function. If a vertical line cannot be drawn so that it intersects a graph at more than one point, then the graph represents a function.

The vertical line test shows that **Figure 3.23b** represents a function and **Figures 3.23a** and **3.23c** do not represent functions.

Not a function

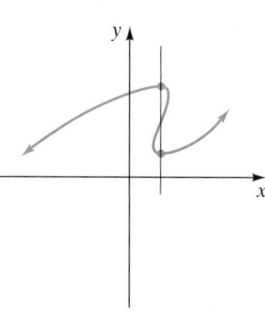

(a)

Function

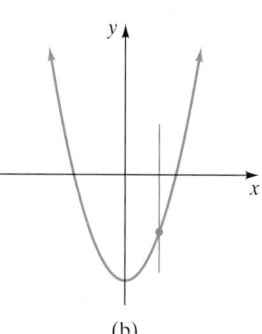

(b)

Not a function

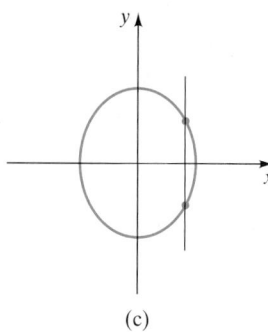

(c)

FIGURE 3.23

EXAMPLE 3 Use the vertical line test to determine whether the following graphs represent functions. Also determine the domain and range of each function or relation.

a)

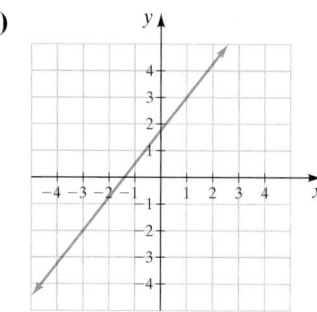

FIGURE 3.24

b)

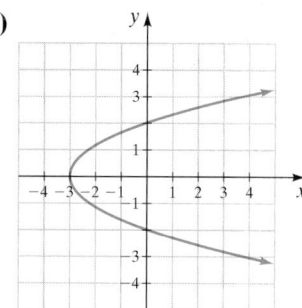

FIGURE 3.25

Solution

a) A vertical line cannot be drawn to intersect the graph in **Figure 3.24** at more than one point. Thus this is the graph of a function. Since the line extends indefinitely in both directions, every value of x will be included in the domain. The domain is the set of real numbers.

$$\text{Domain:} \quad \mathbb{R} \quad \text{or} \quad (-\infty, \infty)$$

Since all values of y are included on the graph, the range is also the set of real numbers.

$$\text{Range:} \quad \mathbb{R} \quad \text{or} \quad (-\infty, \infty)$$

b) Since a vertical line can be drawn to intersect the graph in **Figure 3.25** at more than one point, this is *not* the graph of a function. The domain of this relation is the set of values greater than or equal to -3.

$$\text{Domain:} \quad \{x | x \geq -3\} \quad \text{or} \quad [-3, \infty)$$

The range is the set of y-values, which can be any real number.

$$\text{Range:} \quad \mathbb{R} \quad \text{or} \quad (-\infty, \infty)$$

Now Try Exercise 33

EXAMPLE 4 Consider the function graphed in **Figure 3.26**.

a) What element of the range is paired with 4 in the domain?

b) What elements of the domain are paired with −2 in the range?

c) What is the domain of the function?

d) What is the range of the function?

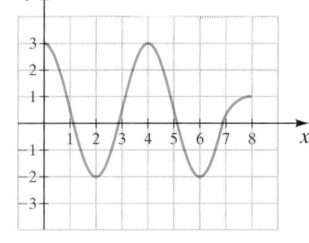

FIGURE 3.26

Solution

a) The range is the set of *y*-values. The *y*-value paired with the *x*-value 4 is 3.

b) The domain is the set of *x*-values. The *x*-values paired with the *y*-value −2 are 2 and 6.

c) The domain is the set of *x*-values, 0 through 8. Thus the domain is

$$\{x \mid 0 \leq x \leq 8\} \quad \text{or} \quad [0, 8]$$

d) The range is the set of *y*-values, −2 through 3. Thus, the range is

$$\{y \mid -2 \leq y \leq 3\} \quad \text{or} \quad [-2, 3]$$

Now Try Exercise 39

EXAMPLE 5 **Figure 3.27** illustrates a graph of speed versus time of a man out for a walk and run. Write a story about the man's outing that corresponds to this function.

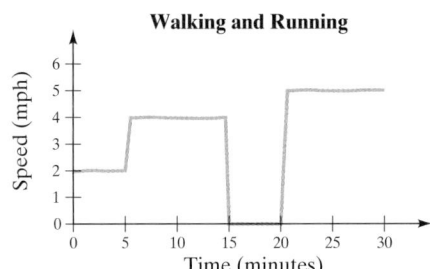

Walking and Running

FIGURE 3.27

Solution **Understand** The horizontal axis is time and the vertical axis is speed. When the graph is horizontal it means the person is traveling at the constant speed indicated on the vertical axis. The steep portions of the graph that increase with time indicate an increase in speed, whereas the steep portions of the graph that decrease with time indicate a decrease in speed.

Answer Here is one possible interpretation of the graph. The man walks for about 5 minutes at a speed of about 2 miles per hour. Then the man speeds up to about 4 miles per hour and walks fast or runs at about this speed for about 10 minutes. Then the man slows down and stops, and then rests for about 5 minutes. Finally, the man speeds up to about 5 miles per hour and runs at this speed for about 10 minutes.

Now Try Exercise 67

4 Understand Function Notation

In Section 3.1 we graphed the equations shown in **Table 3.1** on page 153. Notice that each of these graphs passes the vertical line test. Therefore, the *graphs* represent functions and we say the *equations* define functions. We will use specific notation to name functions that are defined by equations.

TABLE 3.1   Example

Section 3.1 example	Equation graphed	Graph	Does the graph represent a function?	Domain	Range
3	$y = x$		Yes	$(-\infty, \infty)$	$(-\infty, \infty)$
4	$y = -\dfrac{1}{3}x + 1$		Yes	$(-\infty, \infty)$	$(-\infty, \infty)$
5	$y = x^2 - 4$		Yes	$(-\infty, \infty)$	$[-4, \infty)$
6	$y = \dfrac{1}{x}$		Yes	$(-\infty, 0) \cup (0, \infty)$	$(-\infty, 0) \cup (0, \infty)$
7	$y = |x|$		Yes	$(-\infty, \infty)$	$[0, \infty)$

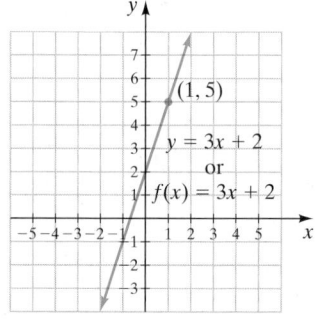

FIGURE 3.28

Function Notation

If an equation involving x as the independent variable and y as the dependent variable defines a function, we say **y is a function of x** and we write **$y = f(x)$** (pronounced "f of x.")

Consider the equation $y = 3x + 2$ as seen in **Figure 3.28**. Since this graph passes the vertical line test, this equation defines a function and we can write the equation using function notation as $f(x) = 3x + 2$.

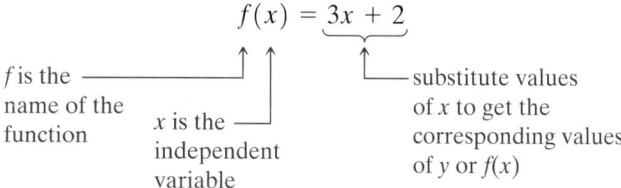

For example, if $f(x) = 3x + 2$, then $f(1)$, read "f of one" is found as follows:

$$f(x) = 3x + 2$$
$$f(1) = 3(1) + 2 = 5$$

Therefore, when x is 1, y is 5. The ordered pair $(1, 5)$ appears on the graph of $y = 3x + 2$ in **Figure 3.28**.

Understanding Algebra

Although we most frequently use f as the name of a function, we can use other letters as well. For example, $g(x)$ and $h(x)$ will also be used to represent functions of x.

Helpful Hint

Linear equations that are not solved for y can be written using function notation by solving the equation for y, then replacing y with $f(x)$. For example, the equation $-9x + 3y - 6$ becomes $y = 3x + 2$ and we can write $f(x) = 3x + 2$.

EXAMPLE 6 If $f(x) = -4x^2 + 3x - 2$, find

a) $f(2)$ **b)** $f(-1)$ **c)** $f(a)$

Solution

a) $f(x) = -4x^2 + 3x - 2$

$f(2) = -4(2)^2 + 3(2) - 2 = -4(4) + 6 - 2 = -16 + 6 - 2 = -12$

b) $f(-1) = -4(-1)^2 + 3(-1) - 2 = -4(1) - 3 - 2 = -4 - 3 - 2 = -9$

c) To evaluate the function at a, we replace each x in the function with an a.

$$f(x) = -4x^2 + 3x - 2$$
$$f(a) = -4a^2 + 3a - 2$$

Now Try Exercise 45

EXAMPLE 7 Determine each indicated function value.

a) $g(-2)$ for $g(t) = \dfrac{1}{t + 8}$

b) $h(5)$ for $h(s) = 2|s - 6|$

c) $j(-3)$ for $j(r) = \sqrt{22 - r}$

Solution   In each part, substitute the indicated value into the function and evaluate.

a) $g(-2) = \dfrac{1}{-2 + 8} = \dfrac{1}{6}$

b) $h(5) = 2|5 - 6| = 2|-1| = 2(1) = 2$

c) $j(-3) = \sqrt{22 - (-3)} = \sqrt{22 + 3} = \sqrt{25} = 5$

Now Try Exercise 49

5 Applications of Functions in Daily Life

Now we examine additional applications of functions.

EXAMPLE 8 Business Jets   The graph in **Figure 3.29** shows the number of business jets manufactured for the years from 1994 through 2008, projected through 2013.

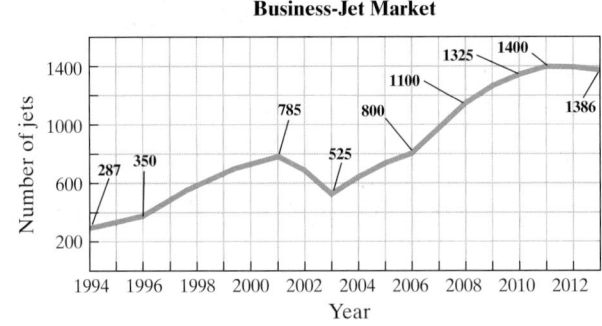

Business-Jet Market

FIGURE 3.29   *Source:* Forecast International

a) Explain why the graph in **Figure 3.29** represents a function.

b) Determine the number of business jets projected to be manufactured in 2010.

c) Determine the projected percent increase in the number of business jets to be manufactured from 2003 to 2011.

d) Determine the percent decrease in the number of business jets manufactured from 2001 to 2003.

Solution

a) The graph represents a function since each year corresponds to a specific number of business jets manufactured. Notice that the graph passes the vertical line test.

b) In 2010, the graph shows that 1325 business jets should be manufactured. If we let the function be represented by J, then $J(2010) = 1325$.

c) We will follow the problem-solving procedure to solve this problem.

Understand and Translate We need to determine the percent increase in the number of business jets to be manufactured from 2003 to 2011. To do this, use the formula

$$\text{percent change (increase or decrease)} = \frac{\left(\begin{array}{c}\text{value in}\\\text{latest period}\end{array}\right) - \left(\begin{array}{c}\text{value in}\\\text{previous period}\end{array}\right)}{\text{value in previous period}}$$

The latest period is 2011 and the previous period is 2003. Substituting the values, we get

$$\text{percent change} = \frac{1400 - 525}{525}$$

Carry Out

$$= \frac{875}{525} \approx 1.667 = 166.7\%$$

Check and Answer Our calculations appear correct. There is projected to be about a 166.7% increase in the number of business jets manufactured from 2003 to 2011.

d) To find the percent decrease from 2001 to 2003, we follow the same procedure as in part **c)**. The latest period is 2003 and the previous period is 2001.

$$\text{percent change (increase or decrease)} = \frac{\left(\begin{array}{c}\text{value in}\\\text{latest period}\end{array}\right) - \left(\begin{array}{c}\text{value in}\\\text{previous period}\end{array}\right)}{\text{value in previous period}}$$

$$= \frac{525 - 785}{785} = \frac{-260}{785} \approx -0.331 = -33.1\%$$

The negative sign preceding the 33.1% indicates a percent decrease. Thus, there was about a 33.1% decrease in the manufacture of business jets from 2001 to 2003.

Now Try Exercise 73

EXAMPLE 9 30-Year Mortgage Rates The graph in **Figure 3.30** shows the 30-year mortgage rate from 1971 to 2007.

a) Using the graph in **Figure 3.30**, explain why this set of points represents a function.

b) Using the graph in **Figure 3.31**, estimate the 30-year mortgage rate in 2006.

Source: www.data360.org

FIGURE 3.30

Source: www.data360.org

FIGURE 3.31

Solution

a) Since each year corresponds with exactly one 30-year mortgage rate, the set of points represents a function. Notice that this graph passes the vertical line test.

b) If we connect the points with line segments as in **Figure 3.31**, we can estimate from the graph that the 30-year mortgage rate in 2006 was about 6%. If we call this function f, then $f(2006) \approx 6\%$.

Now Try Exercise 75

Often formulas are written using function notation as shown next.

EXAMPLE 10   The Celsius temperature, C, is a function of the Fahrenheit temperature, F.

$$C(F) = \frac{5}{9}(F - 32)$$

Determine the Celsius temperature that corresponds to 50°F.

Solution   We need to find $C(50)$. We do so by substitution.

$$C(F) = \frac{5}{9}(F - 32)$$

$$C(50) = \frac{5}{9}(50 - 32)$$

$$= \frac{5}{9}(18) = 10$$

Therefore, 50°F = 10°C.

Now Try Exercise 55

EXERCISE SET 3.2    *Math* **XL**   **MyMathLab**
                         MathXL®   MyMathLab

Warm-Up Exercises

Fill in the blanks with the appropriate word, phrase, or symbol(s) from the following list.

| relation | function | domain | graph | range | x-coordinate | y-coordinates |
| $f(x)$ | vertical line | vertical line test | dependent | independent | ordered pairs | |

1. The set of x-coordinates of a relation is called the _____ of the relation.

2. The set of _____ of a relation is called the range of the relation.

3. A relation is any set of _____.

4. A function is a _____ in which each element of the domain corresponds to exactly one element of the range.

5. If a _____ intersects a graph in more than one point, then this graph is not the graph of a function.

6. In a function, every _____ corresponds to exactly one y-coordinate.

7. In a function, every element in the domain must correspond to exactly one element in the _____.

8. The notation _____ is pronounced "f of x."

9. The notation $y = f(x)$, means that y is a _____ of x.

10. In the notation $y = f(x)$, y is the _____ variable.

11. In the notation $y = f(x)$, x is the _____ variable.

12. The _____ of a function or relation is the graph of its set of ordered pairs.

Practice the Skills

In Exercises 13–20, **a)** *determine if the relation illustrated is a function.* **b)** *Give the domain and range of each function or relation.*

13.   minimum driving age

Idaho ⟶ 15

Texas ⟶ 16

Georgia ⟶ 18

14.   minimum driving age

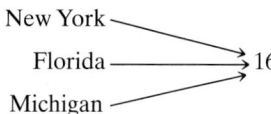

15.   twice a number

3 ⟶ 6

5 ⟶ 10

11 ⟶ 22

16. nicknames

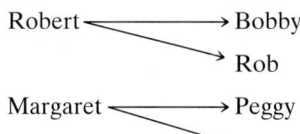

17. number of siblings

Cameron ⟶ 3

Tyrone ⟶ 6

Vishnu

18.   a number squared

4 ⟶ 16

5 ⟶ 25

7 ⟶ 49

19.   cost of a stamp

1990 ⟶ 20

2001 ⟶ 34

2002 ⟶ 37

20.   absolute value

$|-8|$ ⟶ 8

$|8|$

$|0|$ ⟶ 0

In Exercises 21–28, **a)** *determine which of the following relations are also functions.* **b)** *Give the domain and range of each relation or function.*

21. $\{(1,4),(2,2),(3,5),(4,3),(5,1)\}$

22. $\{(1,0),(4,2),(9,3),(1,-1),(4,-2),(9,-3)\}$

23. $\{(3,-1),(5,0),(1,2),(4,4),(2,2),(7,9)\}$

24. $\{(-1,1),(0,-3),(3,4),(4,5),(-2,-2)\}$

25. $\{(1,4),(2,5),(3,6),(2,2),(1,1)\}$

26. $\{(6,3),(-3,4),(0,3),(5,2),(3,5),(2,8)\}$

27. $\{(0,3),(1,3),(2,2),(1,-1),(2,-7)\}$

28. $\{(3,5),(2,5),(1,5),(0,5),(-1,5)\}$

In Exercises 29–40, **a)** *determine whether the graph illustrated represents a function.* **b)** *Give the domain and range of each function or relation.* **c)** *Approximate the value or values of x where y = 2.*

29.

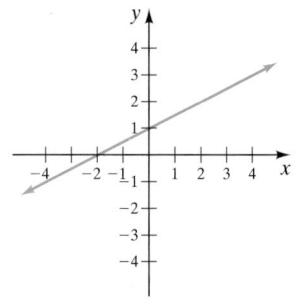

30.

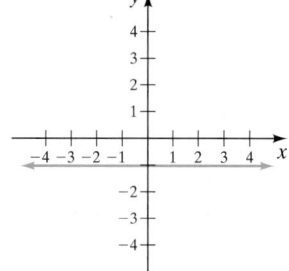

31.

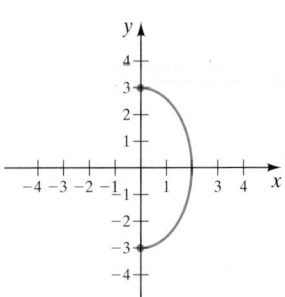

32.

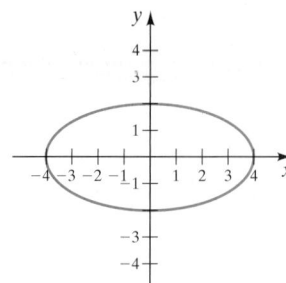

33.

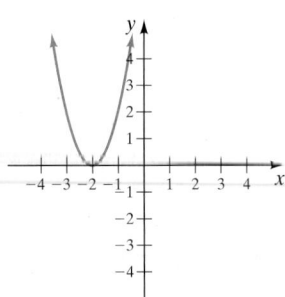

34.

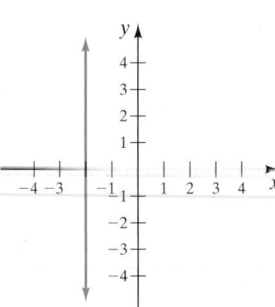

35.

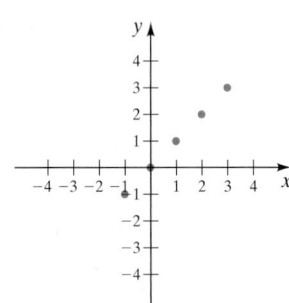

36.

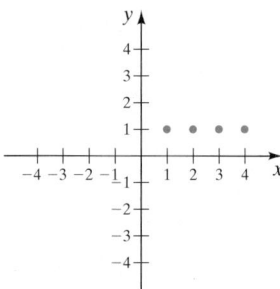

37.

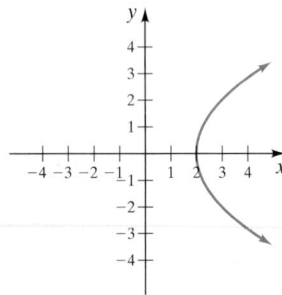

38.

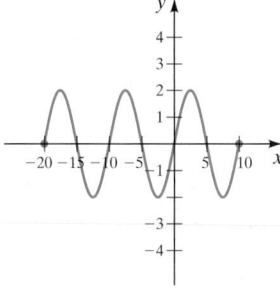

39.

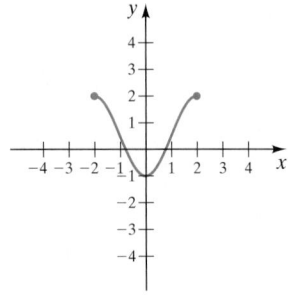

40.

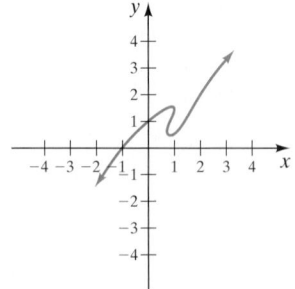

Evaluate each function at the indicated values.

41. $f(x) = 4x - 3$; find
   a) $f(2)$.   **b)** $f(-2)$.

42. $f(a) = \frac{1}{3}a + 4$; find
   a) $f(0)$.   **b)** $f(-12)$.

43. $h(x) = x^2 - x - 6$; find
   a) $h(0)$.   **b)** $h(-1)$.

44. $g(x) = -2x^2 + 7x - 11$; find

   a) $g(2)$.   **b)** $g\left(\dfrac{1}{2}\right)$.

45. $r(t) = -t^3 - 2t^2 + t + 4$; find

   a) $r(1)$.   **b)** $r(-2)$.

46. $g(t) = 4 - 3t + 16t^2 - 2t^3$; find

   a) $g(0)$.   **b)** $g(3)$.

47. $h(z) = |5 - 2z|$; find

   a) $h(6)$.   **b)** $h\left(\dfrac{5}{2}\right)$.

48. $q(x) = -2|x + 8| + 13$; find

   a) $q(0)$.   **b)** $q(-4)$.

49. $s(t) = \sqrt{t + 3}$; find

   a) $s(-3)$.   **b)** $s(6)$.

50. $f(t) = \sqrt{5 - 2t}$; find

   a) $f(-2)$.   **b)** $f(2)$.

51. $g(x) = \dfrac{x^3 - 2}{x - 2}$; find

   a) $g(0)$.   **b)** $g(2)$.

52. $h(x) = \dfrac{x^2 + 4x}{x + 6}$; find

   a) $h(-3)$.   **b)** $h\left(\dfrac{2}{5}\right)$.

Problem Solving

53. Area of a Rectangle The formula for the area of a rectangle is $A = lw$. If the length of a rectangle is 4 inches, then the area is a function of its width, $A(w) = 4w$. Find the area when the width is

   a) 4 feet.

   b) 6.5 feet.

54. Simple Interest The formula for the simple interest earned for a period of 1 year is $i = pr$, where p is the principal invested and r is the simple interest rate. If \$1000 is invested, the simple interest earned in 1 year is a function of the simple interest rate, $i(r) = 1000r$. Determine the simple interest earned in 1 year if the interest rate is

   a) 2.5%.

   b) 4.25%.

55. Area of a Circle The formula for the area of a circle is $A = \pi r^2$. The area is a function of the radius.

   a) Write this function using function notation.

   b) Determine the area when the radius is 12 yards.

56. Perimeter of a Square The formula for the perimeter of a square is $P = 4s$ where s represents the length of any one of the sides of the square.

   a) Write this function using function notation.

   b) Determine the perimeter of a square with sides of length 7 meters.

57. Temperature The formula for changing Fahrenheit temperature into Celsius temperature is $C = \dfrac{5}{9}(F - 32)$. The Celsius temperature is a function of Fahrenheit temperature.

   a) Write this function using function notation.

   b) Find the Celsius temperature that corresponds to $-31°F$

58. Volume of a Cylinder The formula for the volume of a right circular cylinder is $V = \pi r^2 h$. If the height, h, is 3 feet, then the volume is a function of the radius, r.

   a) Write this formula in function notation, where the height is 3 feet.

   b) Find the volume if the radius is 2 feet.

59. Sauna Temperature The temperature, T, in degrees Celsius, in a sauna n minutes after being turned on is given by the function $T(n) = -0.03n^2 + 1.5n + 14$. Find the sauna's temperature after

   a) 3 minutes.   **b)** 12 minutes.

60. Stopping Distance The stopping distance, d, in meters for a car traveling v kilometers per hour is given by the function $d(v) = 0.18v + 0.01v^2$. Find the stopping distance for the following speeds:

   a) 60 km/hr   **b)** 25 km/hr

61. Air Conditioning When an air conditioner is turned on maximum in a bedroom at 80°, the temperature, T, in the room after A minutes can be approximated by the function $T(A) = -0.02A^2 - 0.34A + 80, 0 \le A \le 15$.

   a) Estimate the room temperature 4 minutes after the air conditioner is turned on.

   b) Estimate the room temperature 12 minutes after the air conditioner is turned on.

62. Accidents The number of accidents, n, in 1 month involving drivers x years of age can be approximated by the function $n(x) = 2x^2 - 150x + 4000$. Find the approximate number of accidents in 1 month that involved

   a) 18-year-olds.   **b)** 25-year-olds.

63. Oranges The total number of oranges, T, in a square pyramid whose base is n by n oranges is given by the function

$$T(n) = \frac{1}{3}n^3 + \frac{1}{2}n^2 + \frac{1}{6}n$$

Find the number of oranges if the base is

   a) 6 by 6 oranges.   **b)** 8 by 8 oranges.

64. Rock Concert If the cost of a ticket to a rock concert is increased by x dollars, the estimated increase in revenue, R, in thousands of dollars is given by the function $R(x) = 24 + 5x - x^2$, $x < 8$. Find the increase in revenue if the cost of the ticket is increased by

a) $1.

b) $4.

65. Supply and Demand The price of a bushel of soybeans can be estimated by the function

$$f(Q) = -0.00004Q + 4.25, \quad 10{,}000 \le Q \le 60{,}000$$

where $f(Q)$ is the price of a bushel of soybeans and Q is the annual number of bushels of soybeans produced.

a) Construct a graph showing the relationship between the number of bushels of soybeans produced and the price of a bushel of soybeans.

b) Estimate the cost of a bushel of soybeans if 40,000 bushels of soybeans are produced in a given year.

66. Household Expenditures The average annual household expenditure is a function of the average annual household income. The average expenditure can be estimated by the function

$$f(i) = 0.6i + 5000, \quad \$3500 \le i \le \$50{,}000$$

where $f(i)$ is the average household expenditure and i is the average household income.

a) Draw a graph showing the relationship between average household income and the average household expenditure.

b) Estimate the average household expenditure for a family whose average household income is $30,000.

Concept/Writing Exercises

Review Example 5 before working Exercises 67–72.

67. Heart Rate The following graph shows a person's heart rate while doing exercise. Write a story that this graph may represent.

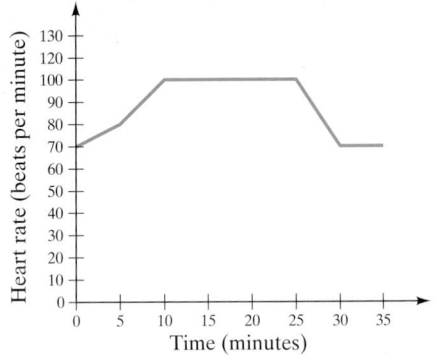

68. Water Level The following graph shows the water level at a certain point during a flood. Write a story that this graph may represent.

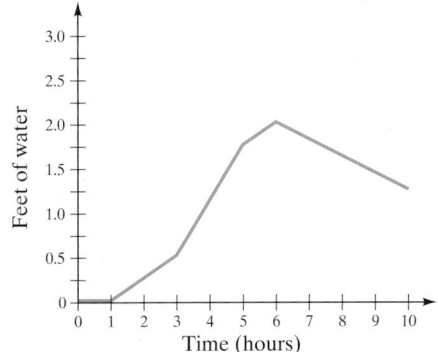

69. Height above Sea Level The following graph shows height above sea level versus time when a man leaves his house and goes for a walk. Write a story that this graph may represent.

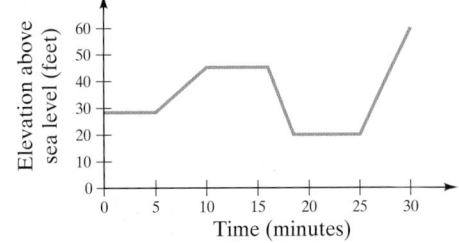

70. Water Level in a Bathtub The following graph shows the level of water in a bathtub versus time. Write a story that this graph may represent.

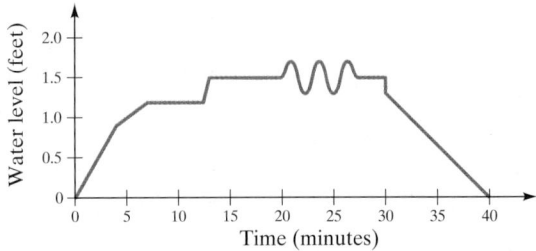

71. Speed of a Car The following graph shows the speed of a car versus time. Write a story that this graph may represent.

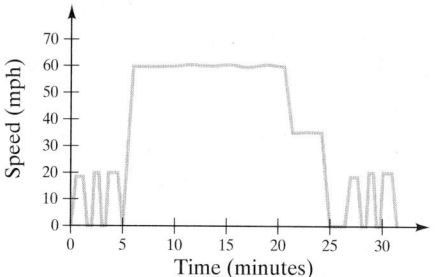

73. College Costs The following graph compares the average cost (in 2006 dollars) of attending college for one year at private colleges and public colleges.

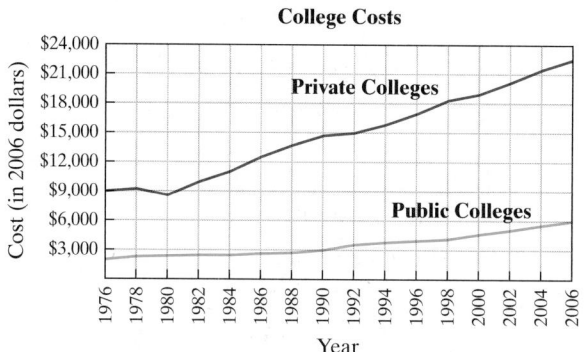

College Costs

Source: www.data360.org

a) Does each line shown represent a function? Explain.

b) In these graphs, which variable is the independent variable?

c) If *f* represents the average cost of attending private college for one year, determine $f(2006)$.

d) If *g* represents the average cost of attending public college for one year, determine $g(2006)$.

e) Determine the percent increase from 1990 to 2006 in the average cost of one year of public college.

75. Super Bowl Commercials The average price of the cost of a 30-second commercial during the Super Bowl has been increasing over the years. The following chart gives the approximate cost of a 30-second commercial for selected years from 1981 through 2009.

Year	Cost ($1000s)
1981	280
1985	500
1989	740
1993	970
1997	1200
2001	2000
2005	2400
2009	3000

a) Draw a line graph that displays this information.

b) Does the graph appear to be approximately linear? Explain.

c) From the graph, estimate the cost of a 30-second commercial in 2004.

72. Distance Traveled The following graph shows the distance traveled by a person in a car versus time. Write a story that this graph may represent.

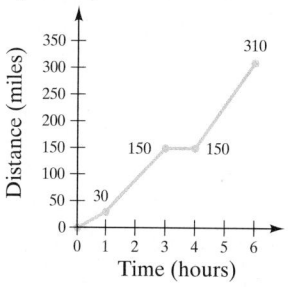

74. Price of Gasoline The following graph shows the inflation-adjusted average price of a gallon (in 2008 dollars) of gasoline for the years from 1918 to 2008.

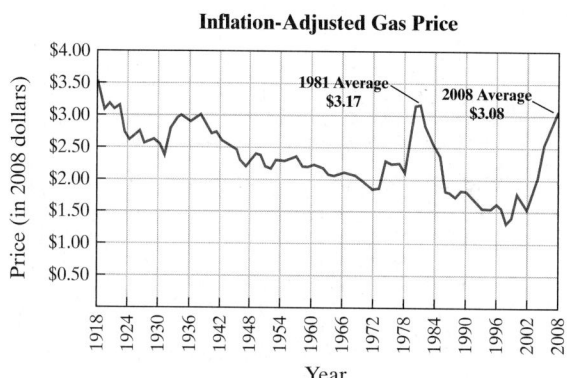

Inflation-Adjusted Gas Price

Source: US Energy Information Administration

a) Does this graph represent a function?

b) In this graph, what is the dependent variable?

c) If *p* represents this function, determine $p(2006)$.

d) Determine the percent decrease from 1981 to 2008 (round your answer to the nearest tenth of a percent)

76. Shipments of LCD Monitors The following graph shows the shipments of LCD monitors, in millions of units, for the years from 2002 to 2008.

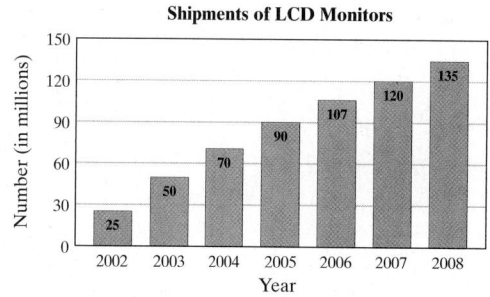

Shipments of LCD Monitors

Source: DisplaySearch, Market Intelligence Center, *Wall Street Journal*

a) Draw a line graph that displays this information.

b) Does the graph you drew in part **a)** appear to be approximately linear? Explain.

c) Assuming this trend continues, from the line graph you drew, estimate the number of LCD monitors to be shipped in 2009.

d) Does the bar graph represent a function?

e) Does the line graph you drew in part **a)** represent a function?

Group Activity

In many real-life situations, more than one function may be needed to represent a problem. This often occurs where two or more different rates are involved. For example, when discussing federal income taxes, there are different tax rates. When two or more functions are used to represent a problem, the function is called a **piecewise function***. Following are two examples of piecewise functions and their graphs.*

$$f(x) = \begin{cases} -x + 2, & 0 \le x < 4 \\ 2x - 10, & 4 \le x < 8 \end{cases}$$

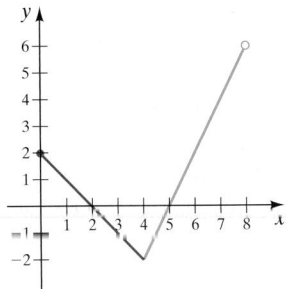

$$f(x) = \begin{cases} 2x - 1, & -2 \le x < 2 \\ x - 2, & 2 \le x < 4 \end{cases}$$

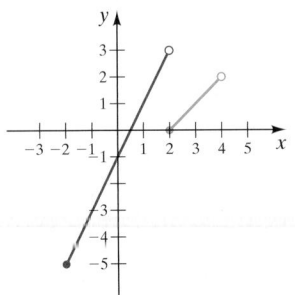

As a group, graph the following piecewise functions.

77. $f(x) = \begin{cases} x + 3, & -1 \le x < 2 \\ 7 - x, & 2 \le x < 4 \end{cases}$

78. $g(x) = \begin{cases} 2x + 3, & -3 < x < 0 \\ -3x + 1, & 0 \le x < 2 \end{cases}$

Cumulative Review Exercises

[2.1] **79.** Solve $3x - 2 = \dfrac{1}{3}(3x - 3)$.

[2.2] **80.** Solve the following formula for p_2.

$$E = a_1 p_1 + a_2 p_2 + a_3 p_3$$

[2.5] **81.** Solve the inequality $\dfrac{3}{5}(x - 3) > \dfrac{1}{4}(3 - x)$ and indicate the solution

 a) on the number line;

 b) in interval notation; and

 c) in set builder notation.

[2.6] **82.** Solve $\left| \dfrac{x - 4}{3} \right| + 9 = 11$.

3.3 Linear Functions: Graphs and Applications

1 Graph linear functions.

2 Graph linear functions using intercepts.

3 Graph equations of the form $x = a$ and $y = b$.

4 Study applications of functions.

1 Graph Linear Functions

In Section 3.1 we graphed linear equations. To graph the linear equation $y = 2x + 4$, we can make a table of values, plot the points, and draw the graph, as shown in **Figure 3.32**. Since this graph passes the vertical line test, the equation defines a function and we may write $f(x) = 2x + 4$.

x	y
−2	0
0	4
1	6

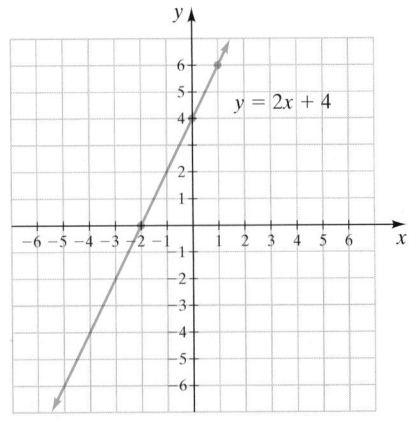

FIGURE 3.32

This is an example of a linear function.

Linear Function

A linear function is a function of the form $f(x) = ax + b$.

- The graph of any linear equation is a straight line.
- The domain of any linear function is all real numbers, $\mathbb{R}$.
- If $a \neq 0$, then the range of any linear function is all real numbers, $\mathbb{R}$.

Helpful Hint

To graph linear functions, we treat $f(x)$ as y and follow the same procedure used to graph linear equations discussed in Section 3.1.

EXAMPLE 1 Graph $f(x) = \dfrac{1}{2}x - 1$.

Solution   We construct a table of values by substituting values for x and finding corresponding values of $f(x)$ or y. Then we plot the points and draw the graph, as illustrated in **Figure 3.33**.

x	$f(x)$
-2	-2
0	-1
2	0

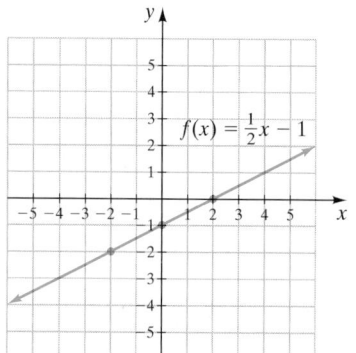

FIGURE 3.33

Now Try Exercise 31

Note that the vertical axis in **Figure 3.33** may also be labeled as $f(x)$ instead of y. In this book we will continue to label it y.

2 Graph Linear Functions Using Intercepts

Linear equations are not always given in the form $y = ax + b$. The equation $2x + 3y = 6$ is an example of a linear equation given in *standard form*.

Standard Form of a Linear Equation

The **standard form of a linear equation** is

$$ax + by = c$$

where $a, b,$ and c are real numbers, and a and b are not both 0.

Examples of Linear Equations in Standard Form

$$2x + 3y = 4 \qquad -x + 5y = -2$$

Examine the graph in Figure 3.32 on page 162. The graph crosses the x-axis at the point $(-2, 0)$ and crosses the y-axis at $(0, 4)$. These important points are called the **intercepts** of the graph. When a linear equation is given in standard form, it may be easier to graph the equation by finding the x-intercept and the y-intercept.

x- and y-intercepts

The **x-intercept** is the point at which a graph crosses the *x*-axis.

- The *x*-intercept will always be of the form $(x, 0)$.

The **y-intercept** is the point at which a graph crosses the *y*-axis.

- The *y*-intercept will always be of the form $(0, y)$.

The following procedure will help us graph linear equations using the intercepts.

To Graph Linear Equations Using the x- and y-intercepts

1. *Find the y-intercept.* Set *x* equal to 0 and find the corresponding value for *y*.
2. *Find the x-intercept.* Set *y* equal to 0 and find the corresponding value for *x*.
3. *Plot the intercepts.*
4. *Draw the line.* Using a straightedge, draw a line through the points. Draw an arrowhead at both ends of the line.

EXAMPLE 2 Graph $5x = 10y - 20$ using the *x*- and *y*-intercepts.

Solution   To find the *y*-intercept, set $x = 0$ and solve for *y*.

$$5x = 10y - 20$$
$$5(0) = 10y - 20$$
$$0 = 10y - 20$$
$$20 = 10y$$
$$2 = y$$

The *y*-intercept is $(0, 2)$.

To find the *x*-intercept, set $y = 0$ and solve for *x*.

$$5x = 10y - 20$$
$$5x = 10(0) - 20$$
$$5x = -20$$
$$x = -4$$

The *x*-intercept is $(-4, 0)$. Now plot the intercepts and draw the graph (**Fig. 3.34**).

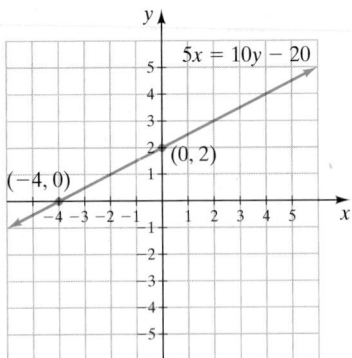

FIGURE 3.34

Now Try Exercise 23

Helpful Hint

When graphing a linear equation by using the *x*-intercept and *y*-intercept, it is helpful to plot a checkpoint. For example, in Example 2, if we substitute $x = 2$ into the equation and solve for *y*, we get $y = 3$. Thus, a third point on the graph should be $(2, 3)$. Looking at **Figure 3.34**, we see that this point is on the graph and we are reassured that our graph is correct.

EXAMPLE 3   Graph $f(x) = -\frac{1}{3}x - 1$ using the x- and y-intercepts.

Solution   Treat $f(x)$ the same as y. To find the y-intercept, set $x = 0$ and solve for $f(x)$.

$$f(x) = -\frac{1}{3}x - 1$$

$$f(x) = -\frac{1}{3}(0) - 1 = -1$$

The y-intercept is $(0, -1)$.

To find the x-intercept, set $f(x) = 0$ and solve for x.

$$f(x) = -\frac{1}{3}x - 1$$

$$0 = -\frac{1}{3}x - 1$$

$$3(0) = 3\left(-\frac{1}{3}x - 1\right) \qquad \text{Multiply both sides by 3.}$$

$$0 = -x - 3 \qquad \text{Distributive property}$$

$$x = -3 \qquad \text{Add } x \text{ to both sides.}$$

The x-intercept is $(-3, 0)$. The graph is shown in **Figure 3.35**.

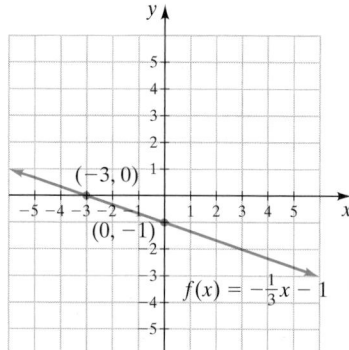

FIGURE 3.35

Now Try Exercise 15

EXAMPLE 4   Graph $-6x + 4y = 0$.

Solution   If we substitute $x = 0$ we find that $y = 0$. Thus the graph goes through the origin. We will select $x = -2$ and $x = 2$ and substitute these values into the equation to find two other points on the graph.

Let $x = -2$.	Let $x = 2$.
$-6x + 4y = 0$	$-6x + 4y = 0$
$-6(-2) + 4y = 0$	$-6(2) + 4y = 0$
$12 + 4y = 0$	$-12 + 4y = 0$
$4y = -12$	$4y = 12$
$y = -3$	$y = 3$
ordered pairs: $(-2, -3)$	$(2, 3)$

Two other points on the graph are at $(-2, -3)$ and $(2, 3)$. The graph of $-6x + 4y = 0$ is shown in **Figure 3.36**.

Understanding Algebra

The graph of every linear equation in standard form with a constant of 0 (equations of the form, $ax + by = 0$) will pass through the origin. Both the x-intercept and y-intercept will be $(0, 0)$.

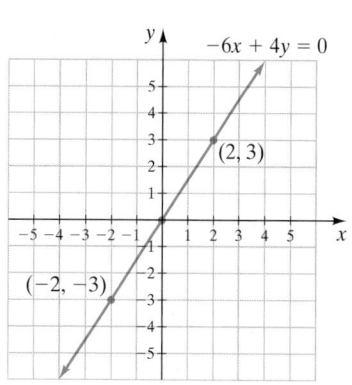

FIGURE 3.36

Now Try Exercise 35

Using Your Graphing Calculator

We will find the intercepts of the graph of $y = 4x - 5$. First, graph the function by pressing the keys shown below. The graph is shown in **Figure 3.37a**.

$$\boxed{Y=}\ \boxed{4}\ \boxed{X,T,Q,n}\ \boxed{-}\ \boxed{5}\ \boxed{GRAPH}$$

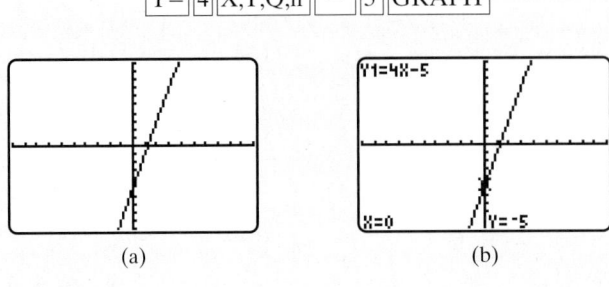

(a)   (b)

FIGURE 3.37

To get the y-intercept, press $\boxed{TRACE}$. The y-intercept is $(0, -5)$ as shown in **Figure 3.37b**. To get the x-intercept, we use the "zero" feature by first pressing the keys

$$\boxed{2nd}\ \boxed{TRACE}\ \boxed{2}$$

You are asked for a left-bound. Move the flashing cursor left or right by using the keys $\boxed{<}$ or $\boxed{>}$. Move the cursor to the left of the x-intercept and press $\boxed{ENTER}$.

Next, you are asked for a right-bound. Move the cursor to the right of the x-intercept and press $\boxed{ENTER}$. Finally, you are asked for a guess. Move the cursor close to the x-intercept and press $\boxed{ENTER}$. The screen is shown in **Figure 3.38**. The x-intercept is $(1.25, 0)$.

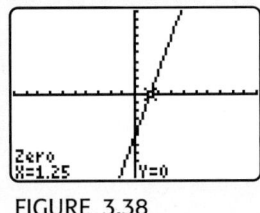

FIGURE 3.38

3 Graph Equations of the Form $x = a$ and $y = b$

Examples 5 and 6 illustrate how equations of the form $x = a$ and $y = b$, where a and b are constants, are graphed.

EXAMPLE 5 Graph the equation $y = -3$.

Solution   This equation can be written as $y = -3 + 0x$. Thus, for any value of x selected, y is -3. The graph of $y = -3$ is illustrated in **Figure 3.39**.

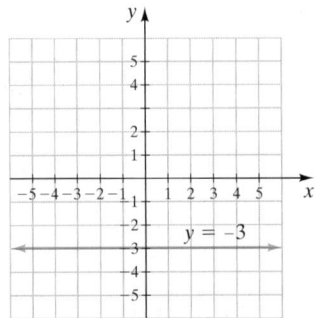

FIGURE 3.39

Now Try Exercise 43

Understanding Algebra

Any equation of the form $y = b$ or $f(x) = b$, where b represents a constant, is a constant function whose graph is a horizontal line.

Equation of a Horizontal Line

The graph of any equation of the form $y = b$ will always be a horizontal line for any real number b.

Notice that the graph of $y = -3$ is a function since it passes the vertical line test. For each value of x selected, the value of y, or the value of the function, is -3. This is an example of a **constant function**. We may write

$$f(x) = -3$$

EXAMPLE 6   Graph the equation $x = 2$.

Solution   This equation can be written as $x = 2 + 0y$. Thus, for every value of y selected, x will have a value of 2 (**Fig. 3.40**).

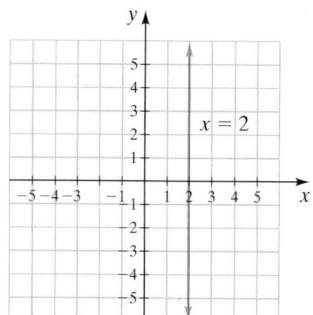

FIGURE 3.40

Now Try Exercise 41

Understanding Algebra

An equation of the form $x = a$ will always be a vertical line, and will never define a function.

Equation of a Vertical Line

The graph of any equation of the form $x = a$ will always be a vertical line for any real number a.

Notice that the graph of $x = 2$ does not represent a function since it does not pass the vertical line test.

4   Study Applications of Functions

Graphs are often used to show the relationship between variables.

EXAMPLE 7   Tire Store Profit   The yearly profit, p, of a tire store can be estimated by the function $p(n) = 20n - 30{,}000$, where n is the number of tires sold per year.

a) Draw a graph of profit versus tires sold for up to and including 6000 tires.

b) Estimate the number of tires that must be sold for the company to break even.

c) Estimate the number of tires sold if the company has a $70,000 profit.

Solution   **a)** Understand

Understanding Algebra

A company's break-even point is the point where the profit is equal to 0. In other words, the company neither makes money nor loses money.

p is a function of n

p is the profit.
p is the dependent variable.
The vertical axis will be labeled p.

n is the number of tires sold.
n is the independent variable.
The horizontal axis will be labeled n.

Since the minimum number of tires that can be sold is 0, the horizontal axis will go from 0 to 6000 tires. We will graph this equation by determining and plotting the intercepts.

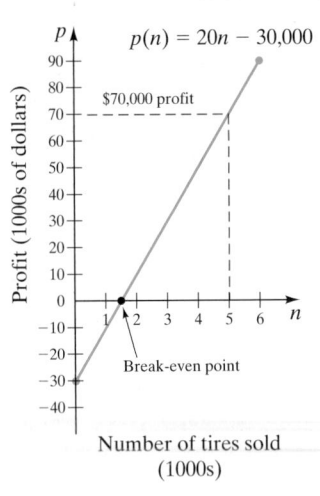

$p(n) = 20n - 30,000$

$70,000 profit

Break-even point

Profit (1000s of dollars)

Number of tires sold
(1000s)

FIGURE 3.41

Translate and Carry Out To find the p-intercept, we set $n = 0$ and solve for $p(n)$.

$$p(n) = 20n - 30,000$$
$$p(n) = 20(0) - 30,000 = -30,000$$

Thus, the p-intercept is $(0, -30,000)$.
    To find the n-intercept, we set $p(n) = 0$ and solve for n.

$$p(n) = 20n - 30,000$$
$$0 = 20n - 30,000$$
$$30,000 = 20n$$
$$1500 = n$$

Thus the n-intercept is $(1500, 0)$.

Answer   Now we use the p- and n-intercepts to draw the graph (see **Fig. 3.41**).

b) The break-even point is where the graph intersects the n-axis, for this is where the profit, p, is 0. To break even, approximately 1500 tires must be sold.

c) To make $70,000, approximately 5000 tires must be sold (shown by the dashed red line in **Fig. 3.41**).

Now Try Exercise 51

> **Helpful Hint**
>
> Sometimes it is difficult to read an exact answer from a graph. To determine the exact number of tires needed to break even in Example 7, substitute 0 for $p(n)$ in the function $p(n) = 20n - 30,000$ and solve for n. To determine the exact number of tires needed to obtain a $70,000 profit, substitute 70,000 for $p(n)$ and solve the equation for n.

EXAMPLE 8 Toy Store Sales Rob Kimball is the owner of a toy store. His monthly salary consists of $200 plus 10% of the store's sales for that month.

a) Write a function expressing his monthly salary, m, in terms of the store's sales, s.

b) Draw a graph of his monthly salary for sales up to and including $20,000.

c) If the store's sales for the month of April are $15,000, what will Rob's salary be for April?

Solution

a) Since Rob's monthly salary depends on the store's sales, m depends on s, and m is a function of s.

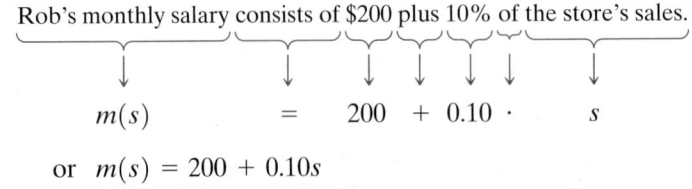

Rob's monthly salary consists of $200 plus 10% of the store's sales.

$$m(s) \qquad = \qquad 200 \ + \ 0.10 \ \cdot \qquad s$$

or $m(s) = 200 + 0.10s$

s	m
0	200
10,000	1200
20,000	2200

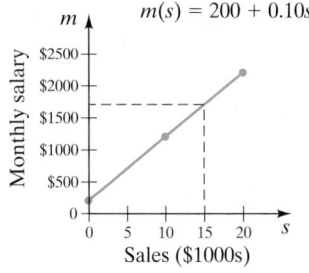

FIGURE 3.42

b) Since m is a function of s, s will be on the horizontal axis and m will be on the vertical axis. We will graph this function by plotting points. We select values for s and find the corresponding values for m. The table in the margin shows three ordered pairs which we will plot and then draw the graph shown in **Figure 3.42**.

c) By reading our graph carefully, we can estimate that when the store's sales are $15,000, Rob's monthly salary is about $1700.

Now Try Exercise 53

EXERCISE SET 3.3

Math XL   MathXL®   MyMathLab   MyMathLab

Warm-Up Exercises

Fill in the blanks with the appropriate word, phrase, or symbol(s) from the following list.

linear	domain	range
vertical	constant	standard form

x-intercept	y-intercept	horizontal
fail	pass	real numbers

1. The domain, or set of x-coordinates, of every linear function is all _____, symbolized $\mathbb{R}$.

2. The _____ or set of y-coordinates, of every linear function $f(x) = ax + b, a \neq 0$, is all real numbers, symbolized $\mathbb{R}$.

3. The _____ is the point where the graph crosses the x-axis.

4. The _____ is the point where the graph crosses the y-axis.

5. The graph of an equation of the form $y = b$, where b is a real number, is a _____ line.

6. The graph of an equation of the form $x = a$, where a is a real number, is a _____ line.

7. An equation of the form $x = a$ will always _____ the vertical line test and therefore will never define a function.

8. A function of the form $f(x) = b$, where b is a real number, is known as a _____ function.

9. A function of the form $f(x) = ax + b$, where a and b are real numbers, is known as a _____ function.

10. The _____ of a linear equation is $ax + by = c$.

Practice the Skills

Write each equation in standard form.

11. $y = -4x + 3$

12. $7x = 3y - 6$

13. $3(x - 2) = 4(y - 5)$

14. $\frac{1}{2}y = 2(x - 3) + 4$

Graph each equation using the x- and y-intercepts.

15. $f(x) = 2x + 3$

16. $y = x - 5$

17. $y = -2x + 1$

18. $f(x) = -6x + 5$

19. $2y = 4x + 6$

20. $2x - 3y = 12$

21. $\frac{4}{3}x = y - 3$

22. $\frac{1}{4}x + y = 2$

23. $15x + 30y = 60$

24. $6x + 12y = 24$

25. $0.25x + 0.50y = 1.00$

26. $-1.6y = 0.4x + 9.6$

27. $120x - 360y = 720$

28. $250 = 50x - 50y$

29. $\frac{1}{3}x + \frac{1}{4}y = 12$

30. $\frac{1}{2}x + \frac{3}{2}y = -3$

Graph each equation.

31. $f(x) = \frac{1}{3}x$

32. $y = \frac{1}{2}x$

33. $y = -2x$

34. $g(x) = 4x$

35. $2x + 4y = 0$

36. $-10x + 5y = 0$

37. $6x - 9y = 0$

38. $18x + 6y = 0$

Graph each equation.

39. $y = 4$

40. $y = -4$

41. $x = -4$

42. $x = 4$

43. $y = -1.5$

44. $f(x) = -3$

45. $x = 0$

46. $g(x) = 0$

47. $x = \frac{5}{2}$

48. $x = -3.25$

Problem Solving

49. Distance Using the distance formula

$$\text{distance} = \text{rate} \cdot \text{time, or } d = rt$$

draw a graph of distance versus time for a constant rate of 30 miles per hour.

50. Simple Interest Using the simple interest formula

$$\text{interest} = \text{principal} \cdot \text{rate} \cdot \text{time, or } i = prt$$

draw a graph of interest versus time for a principal of $1000 and a rate of 3%.

51. Bicycle Profit The profit of a bicycle manufacturer can be approximated by the function $p(x) = 60x - 80{,}000$, where x is the number of bicycles produced and sold.

a) Draw a graph of profit versus the number of bicycles sold (for up to and including 5000 bicycles).

b) Estimate the number of bicycles that must be sold for the company to break even.

c) Estimate the number of bicycles that must be sold for the company to make $150,000 profit.

52. Taxi Operating Costs Raul Lopez's weekly cost of operating a taxi is $75 plus 15¢ per mile.

a) Write a function expressing Raul's weekly cost, c, in terms of the number of miles, m.

b) Draw a graph illustrating weekly cost versus the number of miles, for up to and including 200, driven per week.

c) If during 1 week, Raul drove the taxi 150 miles, what would be the cost?

d) How many miles would Raul have to drive for the weekly cost to be $135?

53. Salary Plus Commission Jayne Haydack's weekly salary at Charter Network is $500 plus 15% commission on her weekly sales.

a) Write a function expressing Jayne's weekly salary, s, in terms of her weekly sales, x.

b) Draw a graph of Jayne's weekly salary versus her weekly sales, for up to and including $5000 in sales.

c) What is Jayne's weekly salary if her sales were $3000?

d) If Jayne's weekly salary for the week was $1100, what were her weekly sales?

54. Salary Plus Commission Christina Miller, a real estate agent, makes $100 per week plus a 3% sales commission on each property she sells.

a) Write a function expressing her weekly salary, s, in terms of sales, x.

b) Draw a graph of her salary versus her weekly sales, for sales up to $100,000.

c) If she sells one house per week for $75,000, what will her weekly salary be?

© Morgan Lane Photography/Shutterstock

Concept/Writing Exercises

55. Weight of Girls The following graph shows weight, in kilograms, for girls (up to 36 months of age) versus length (or height), in centimeters. The red line is the average weight for all girls of the given length, and the green lines represent the upper and lower limits of the normal range.

Girls: Birth to 36 Months Physical Growth

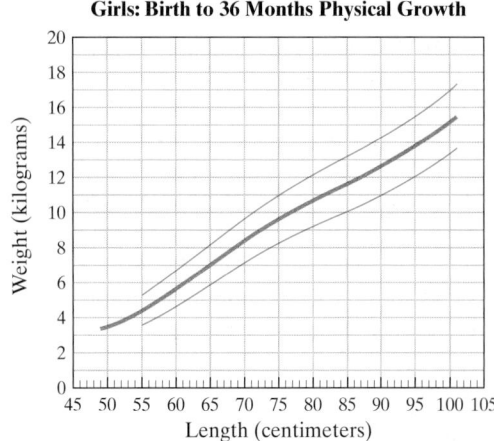

Length (centimeters)

Source: National Center for Health Statistics

a) Explain why the red line represents a function.

b) What is the independent variable? What is the dependent variable?

c) Is the graph of weight versus length approximately linear?

d) What is the weight in kilograms of the average girl who is 85 centimeters long?

e) What is the average length in centimeters of the average girl with a weight of 7 kilograms?

f) What weights are considered normal for a girl 95 centimeters long?

g) What is happening to the normal range as the lengths increase? Is this what you would expect to happen? Explain.

56. Compound Interest The following graph shows the effect of compound interest.

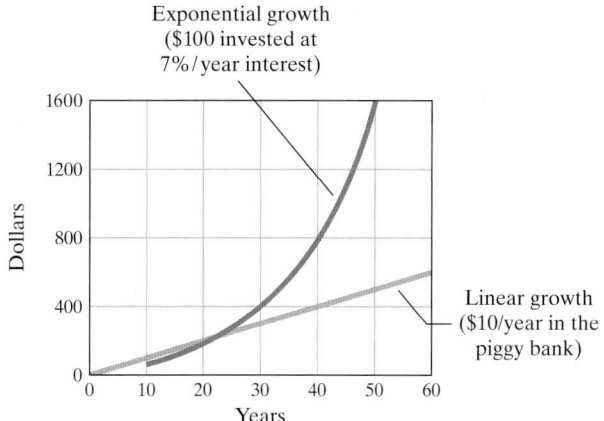

Exponential growth
($100 invested at
7%/year interest)

Linear growth
($10/year in the
piggy bank)

If a child puts $10 each year in a piggy bank, the savings will grow linearly, as shown by the lower curve. If, at age 10, the child invests $100 at 7% interest compounded annually, that $100 will grow exponentially.

a) Explain why both graphs represent functions.

b) What is the independent variable? What is the dependent variable?

c) Using the linear growth curve, determine how long it would take to save $600.

d) Using the exponential growth curve, which begins at year 10, how long after the account is opened would the amount reach $600?

e) Starting at year 20, how long would it take for the money growing at a linear rate to double?

f) Starting at year 20, how long would it take for the money growing exponentially to double? (Exponential growth will be discussed at length in Chapter 9.)

57. When, if ever, will the x- and y-intercepts of a graph be the same? Explain.

58. Write two linear functions whose x- and y-intercepts are both $(0, 0)$. Explain.

59. Write a function whose graph will have no x-intercept but will have a y-intercept at $(0, 4)$. Explain.

60. Write an equation whose graph will have no y-intercept but will have an x-intercept at -5. Explain.

Challenge Problems

61. If the x- and y-intercepts of a linear function are at 1 and -3, respectively, what will be the new x- and y-intercepts if the graph is moved (or translated) up 3 units?

62. If the x- and y-intercepts of a linear function are -1 and 3, respectively, what will be the new x- and y-intercepts if the graph is moved (or translated) down 4 units?

 Find the x- and y-intercepts of the graph of each equation using your graphing calculator.

63. $y = 2(x + 3.2)$

64. $5x - 2y = 7$

65. $-4x - 3.2y = 8$

66. $y = \dfrac{3}{5}x - \dfrac{1}{2}$

Group Activities

In Exercises 67 and 68, we give two ordered pairs, which are on a graph. **a)** *Plot the points and draw the line through the points.* **b)** *Find the change in y, or the vertical change, between the points.* **c)** *Find the change in x, or the horizontal change, between the points.* **d)** *Find the ratio of the vertical change to the horizontal change between these two points. Do you know what this ratio represents? (We will discuss this further in Section 3.4.)*

67. $(0, 2)$ and $(-4, 0)$

68. $(3, 5)$ and $(-1, -1)$

Cumulative Review Exercises

[1.4] **69.** Evaluate $4\{2 - 3[(1 - 4) - 5]\} - 8$.

[2.1] **70.** Solve $\dfrac{1}{3}y - 3y = 6(y + 2)$.

[2.6] *In Exercises 71–73,* **a)** *explain the procedure to solve the equation or inequality for x (assume that b > 0) and* **b)** *solve the equation or inequality.*

71. $|x - a| = b$

72. $|x - a| < b$

73. $|x - a| > b$

74. Solve the equation $|x - 4| = |2x - 2|$.

3.4 The Slope-Intercept Form of a Linear Equation

1   Understand translations of graphs.

2   Find the slope of a line.

3   Recognize slope as a rate of change.

4   Write linear equations in slope-intercept form.

5   Graph linear equations using the slope and the y-intercept.

6   Use the slope-intercept form to construct models from graphs.

1   Understand Translations of Graphs

Consider the three equations

$$y = 2x + 3$$
$$y = 2x$$
$$y = 2x - 3$$

Observe the graphs shown in **Figure 3.43a** and **3.43b**. The graph of $y = 2x$ is shown in blue in both figures and has a y-intercept of $(0, 0)$. Notice that the graph of $y = 2x + 3$ has y-intercept $(0, 3)$ and that the graph of $y = 2x - 3$ has y-intercept $(0, -3)$. Notice also that the lines are **parallel**—that is, the lines do not intersect.

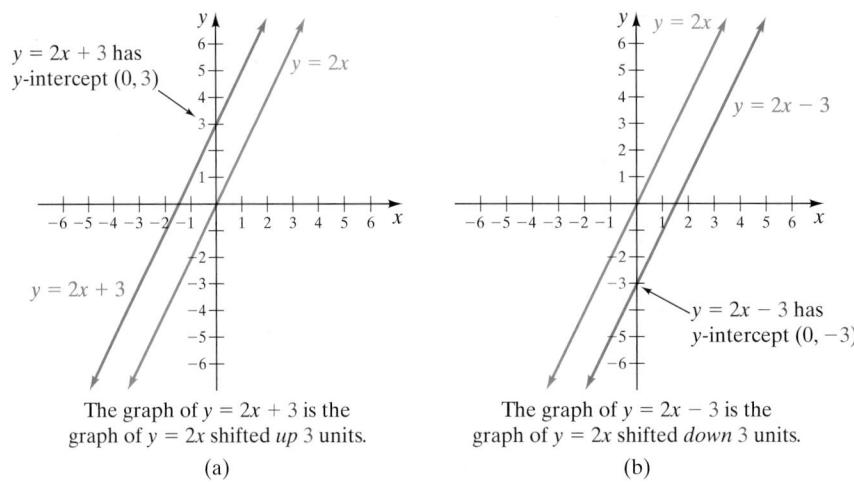

The graph of $y = 2x + 3$ is the graph of $y = 2x$ shifted *up* 3 units.
(a)

The graph of $y = 2x - 3$ is the graph of $y = 2x$ shifted *down* 3 units.
(b)

FIGURE 3.43

Understanding Algebra

In general, the graph of $y = mx + b$ will be parallel to the graph of $y = mx$ and will have y-intercept $(0, b)$.

The graphs of $y = 2x + 3$ and $y = 2x - 3$ are identical to $y = 2x$ except for the y-intercept. We say that the graphs of $y = 2x + 3$ and $y = 2x - 3$ are **vertical transla-tions** of the graph of $y = 2x$.

After viewing the graphs shown in **Figure 3.43**, can you predict what the graph of $y = 2x + 4$ would look like? The graph of $y = 2x + 4$ is parallel to $y = 2x$ and has y-intercept $(0, 4)$. We say that the graph of $y = 2x + 4$ is the graph of $y = 2x$ shifted or **translated** up 4 units (see **Figure 3.44**)

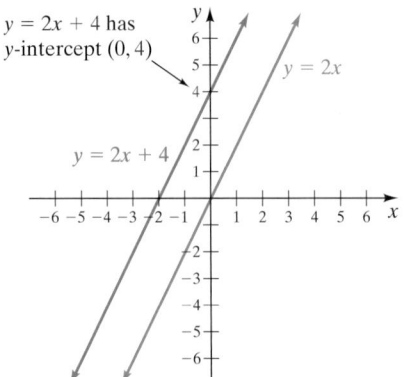

FIGURE 3.44

Notice in the equations $y = 2x, y = 2x + 3, y = 2x - 3$, and $y = 2x + 4$ that the coefficient on x is 2 and that all four lines are parallel. When lines are parallel they have the same *steepness* or *slope*.

Understanding Algebra

Slope is often described with the phrase *rise over run*.

2 Find the Slope of a Line

The *slope of a line* is a measure of the *steepness* of the line. The slope of a line is an important concept in many areas of mathematics.

> ### Slope of a Line
>
> - The **slope of a line**, *m*, is the ratio of the vertical change, or *rise*, to the horizontal change, or *run*, between any two selected points on the line.
> - $m^* = \text{slope} = \dfrac{\text{vertical change}}{\text{horizontal change}} = \dfrac{\text{rise}}{\text{run}}$

As an example, consider the graph of $y = 2x$ from our previous discussion. This line goes through two points, $(1, 2)$ and $(3, 6)$ (see **Fig. 3.45a**). From **Figure 3.45b**, we can see that the vertical change (rise) is $6 - 2$, or 4 units. The horizontal change (run), is $3 - 1$, or 2 units.

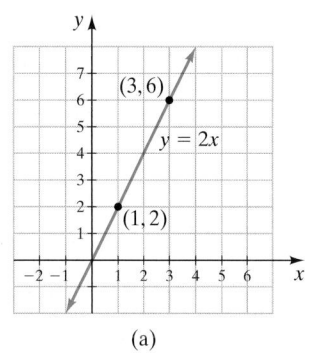

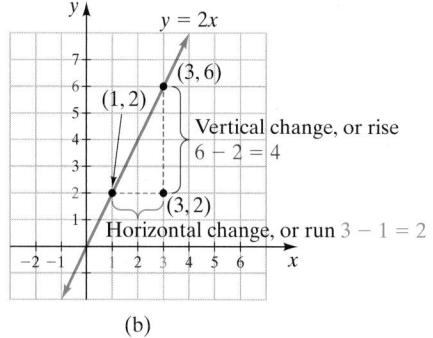

FIGURE 3.45   (a)   (b)

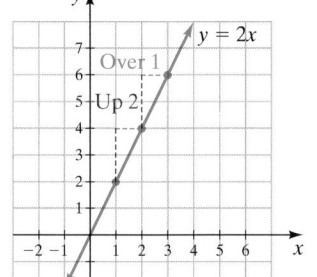

FIGURE 3.46

$$m = \text{slope} = \frac{\text{vertical change}}{\text{horizontal change}} = \frac{\text{rise}}{\text{run}} = \frac{4}{2} = 2$$

Thus, the slope of the line through these two points is 2. By examining the line connecting these two points, we can see that as the graph moves up 2 units it moves to the right 1 unit (**Fig. 3.46**).

We have determined that the slope of the graph of $y = 2x$ is 2. If you were to compute the slope of the other two lines in **Figure 3.43** on page 172, you would find that the graphs of $y = 2x + 3$ and $y = 2x - 3$ also have a slope of 2.

Can you guess what the slope of the graphs of the equations $y = -3x + 2$, $y = -3x$, and $y = -3x - 2$ is? The slope of all three lines is -3.

Now we present the formula to find the slope of a line given any two points (x_1, y_1) and (x_2, y_2) on the line. Look at **Figure 3.47**. The *rise* is the difference between y_2 and y_1 and the *run* is the difference between x_2 and x_1.

Understanding Algebra

In general, the slope of an equation of the form $y = mx + b$ is *m*.

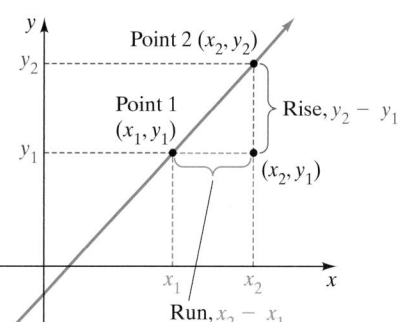

FIGURE 3.47

---

*The letter *m* is traditionally used for slope. It is believed *m* comes from the French word *monter*, which means "to climb."

Understanding Algebra

The Greek letter delta, Δ, is often used to represent the phrase "the change in." So, Δy represents "the change in y" and Δx represents "the change in x" and the formula for slope can be given as

$$m = \frac{\Delta y}{\Delta x} = \frac{y_2 - y_1}{x_2 - x_1}$$

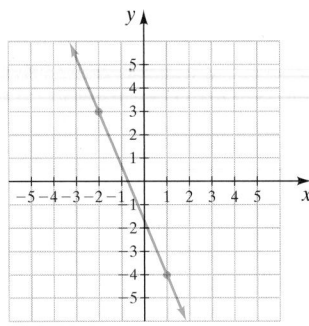

FIGURE 3.48

Slope of a Line Through the Points (x_1, y_1) and (x_2, y_2)

$$m = \text{slope} = \frac{\text{vertical change}}{\text{horizontal change}} = \frac{\text{rise}}{\text{run}} = \frac{y_2 - y_1}{x_2 - x_1}$$

> **Helpful Hint**
>
> It makes no difference which two points on a line are selected when finding the slope of the line. It also makes no difference which point you choose as (x_1, y_1) or (x_2, y_2).

EXAMPLE 1 Find the slope of the line in **Figure 3.48**.

Solution Two points on the line are $(-2, 3)$ and $(1, -4)$. Let $(x_2, y_2) = (-2, 3)$ and $(x_1, y_1) = (1, -4)$. Then

$$m = \frac{y_2 - y_1}{x_2 - x_1} = \frac{3 - (-4)}{-2 - 1} = \frac{3 + 4}{-3} = -\frac{7}{3}$$

The slope of the line is $-\frac{7}{3}$. Note that if we had let $(x_1, y_1) = (-2, 3)$ and $(x_2, y_2) = (1, -4)$, the slope would still be $-\frac{7}{3}$. Try it and see.

Now Try Exercise 35

A line that rises going from left to right (**Fig. 3.49a**) has a **positive slope**. A line that neither rises nor falls going from left to right (**Fig. 3.49b**) has **zero slope**. A line that falls going from left to right (**Fig. 3.49c**) has a **negative slope**.

Positive slope   Zero slope   Negative slope

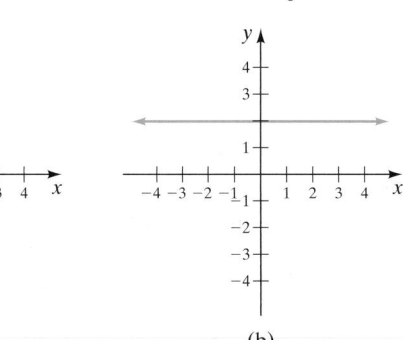
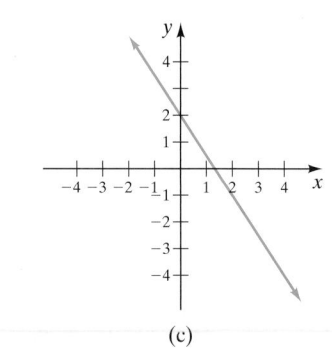

FIGURE 3.49   (a)   (b)   (c)

Understanding Algebra

- The slope of any horizontal line is 0.
- The slope of any vertical line is *undefined*.

Consider the graph of $x = 3$ (**Fig. 3.50**). What is its slope? The graph is a vertical line and goes through the points $(3, 2)$ and $(3, 5)$. Then the slope of the line is

$$m = \frac{y_2 - y_1}{x_2 - x_1} = \frac{5 - 2}{3 - 3} = \frac{3}{0}$$

Since we cannot divide by 0, we say that the slope of this line is undefined. *The slope of any vertical line is* **undefined**.

Slope is undefined.

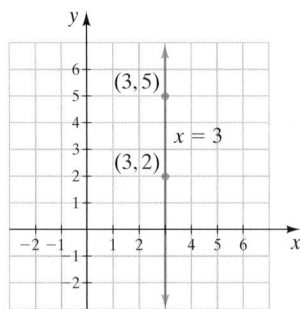

FIGURE 3.50

3 Recognize Slope as a Rate of Change

Sometimes it is helpful to describe slope as a *rate of change*. Consider a slope of $\frac{3}{4}$. This means that the y-value increases 3 units for each 4-unit increase in x. Equivalently, we can say that the y-value increases $\frac{3}{4}$ units, or 0.75 units, for each 1-unit increase in x.

When we give the change in y per unit change in x we are giving the slope as a **rate of change**. When discussing real-life situations or when creating mathematical models, it is often useful to discuss slope as a rate of change. In these models, the independent variable is often *time*.

EXAMPLE 2 **Public Debt** The following table and corresponding graph illustrate the U.S. public debt in trillions of dollars from 1980 through 2008.

Year	U.S. Public Debt (trillions of dollars)
1980	0.91
1984	1.57
1988	2.60
1992	4.06
1996	5.22
2000	5.67
2004	7.38
2008	9.67

Source: U.S. Department of the Treasury

FIGURE 3.51

a) Determine the slope of the line segments between 1996 and 2000 and between 2004 and 2008.

b) Compare the two slopes found in part a) and explain what this means in terms of the U.S. public debt.

Solution   a) Understand   To find the slope between any two years, find the ratio of the change in debt to the change in time.

Slope from 1996 to 2000

$$m = \frac{5.67 - 5.22}{2000 - 1996} = \frac{0.45}{4} = 0.1125$$

The U.S. public debt from 1996 to 2000 increased at a rate of $0.1125 trillion (or $112.5 billion) per year.

Slope from 2004 to 2008

$$m = \frac{9.67 - 7.38}{2008 - 2000} = \frac{2.29}{4} = 0.5725$$

The U.S. public debt from 2004 to 2008 increased at a rate of $0.5725 trillion (or $572.5 billion) per year.

b) Slope measures a rate of change. There was a much greater (over 5 times greater) increase in the average rate of change in the public debt from 2004 to 2008 than from 1996 to 2000.

Now Try Exercise 69

4 Write Linear Equations in Slope-Intercept Form

A linear equation written in the form $y = mx + b$ is said to be in **slope-intercept form**.

> ### Slope-Intercept Form
>
> The **slope-intercept form of a linear equation** is
> $$y = mx + b$$
> where **m is the slope** of the line and **$(0, b)$** is the **y-intercept** of the line.

Understanding Algebra

To write a linear equation in slope-intercept form, solve the equation for y.

Examples of Equations in Slope-Intercept Form

$$y = 3x - 6 \qquad y = \frac{1}{2}x + \frac{3}{2}$$

Slope ⎯⎯⎯⎯   ⎯⎯⎯ y-intercept is $(0, b)$

$$y = mx + b$$

Equation	Slope	y-Intercept
$y = 3x - 6$	3	$(0, -6)$
$y = \frac{1}{2}x + \frac{3}{2}$	$\frac{1}{2}$	$\left(0, \frac{3}{2}\right)$

EXAMPLE 3 Determine the slope and y-intercept of the graph of the equation $-5x + 2y = 8$.

Solution   Write the equation in slope-intercept form by solving the equation for y.

$$-5x + 2y = 8$$
$$2y = 5x + 8$$
$$y = \frac{5x + 8}{2}$$
$$y = \frac{5x}{2} + \frac{8}{2}$$
$$y = \frac{5}{2}x + 4$$

The slope is $\frac{5}{2}$; the y-intercept is $(0, 4)$.

Now Try Exercise 43

5 Graph Linear Equations Using the Slope and the y-Intercept

The slope-intercept form of a line is useful in drawing the graph of a linear equation, as illustrated in Example 4.

EXAMPLE 4 Graph $2y + 4x = 6$ using the y-intercept and slope.

Solution   Begin by solving for y to get the equation in slope-intercept form.

$$2y + 4x = 6$$
$$2y = -4x + 6$$
$$y = -2x + 3$$

The slope is -2 and the y-intercept is $(0, 3)$. Place a point at 3 on the y-axis (**Fig. 3.52**). Since the slope is $-2 = \frac{-2}{1}$, the ratio of the vertical change to the horizontal change must be -2 to 1. Thus, if you start at $(0, 3)$ and move down 2 units and to the right 1 unit, you will obtain a second point on the graph.

Continue this process of moving 2 units down and 1 unit to the right to get a third point. Now draw a line through the three points to get the graph.

Now Try Exercise 45

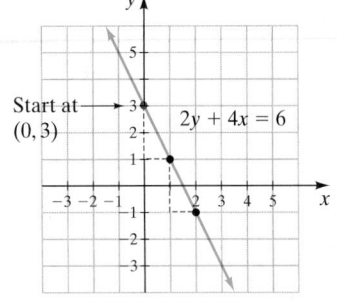

Start at $(0, 3)$ $2y + 4x = 6$

FIGURE 3.52

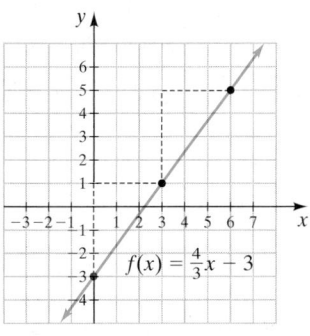

FIGURE 3.53

In Example 4, we chose to move down and to the right to get the second and third points. Since -2 also equals $\dfrac{2}{-1}$, we could have also chosen to move up and to the left to get the second and third points.

EXAMPLE 5   Graph $f(x) = \dfrac{4}{3}x - 3$ using the y-intercept and slope.

Solution   Since $f(x)$ is the same as y, this function is in slope-intercept form. The y-intercept is $(0, -3)$ and the slope is $\dfrac{4}{3}$. Place a point at -3 on the y-axis. Then, since the slope is positive, obtain the second and third points by moving up 4 units and to the right 3 units. The graph is shown in **Figure 3.53**.

Now Try Exercise 51

6   Use the Slope-Intercept Form to Construct Models from Graphs

We can often use the slope-intercept form of a linear equation to determine a function that models a real-life situation.

EXAMPLE 6   **Newspapers**   Consider the green graph in **Figure 3.54**, which shows the declining number of adults who read a daily newspaper. Notice that the graph is somewhat linear. The dashed red line is a linear function which was drawn to approximate the green graph.

a) Write a linear function to represent the dashed red line.

b) Assuming this trend continues, use the function determined in part **a)** to estimate the percent of adults who will read a newspaper in 2015.

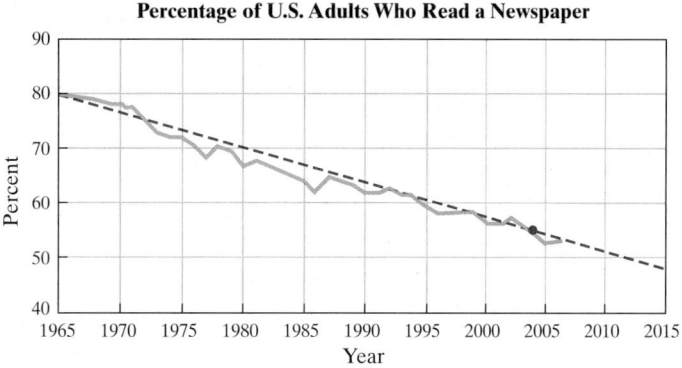

FIGURE 3.54   *Source:* NAA Market & Business Analysis

Solution

a) We will let $x =$ the number of years since 1965. Then on the x-axis we can replace 1965 with 0, 1966 with 1, 1967 with 2, and so on. Then 2004 would be 39 and 2005 would be 40 (see **Fig. 3.55**). We will let $y =$ percent.

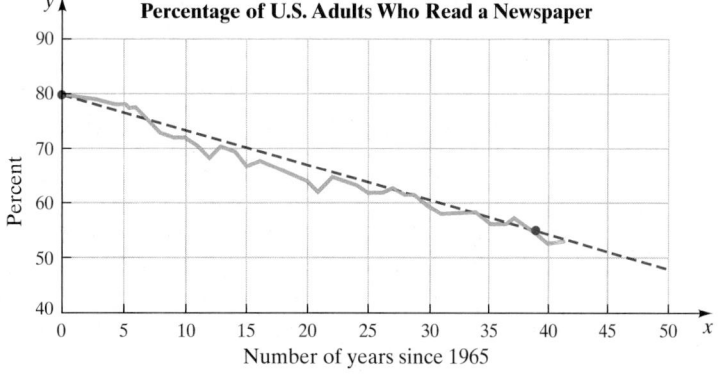

FIGURE 3.55   *Source:* NAA Market & Business Analysis

We will select two points on the graph that will allow us to find the slope of the graph. The y intercept is at 80. Thus, one point on the graph is $(0, 80)$. In 2004, or year 39 in **Figure 3.55**, it appears that about 55% of the adult population read a daily newspaper. Let's select $(39, 55)$ as a second point.

$$\text{slope} = \frac{y_2 - y_1}{x_2 - x_1} = \frac{55 - 80}{39 - 0} = \frac{-25}{39} \approx -0.641$$

Since the slope is approximately -0.641 and the y-intercept is $(0, 80)$, the equation of the straight line is $y = -0.641x + 80$. This equation in function notation is $f(x) = -0.641x + 80$. To use this function remember that $x = 0$ represents 1965, $x = 1$ represents 1966, and so on. Note that $f(x)$, the percent, is a function of x, the number of years since 1965.

b) To determine the approximate percent of readers in 2015, we substitute $2015 - 1965$, or 50, for x in the function.

$$f(x) = -0.641x + 80$$
$$f(50) = -0.641(50) + 80$$
$$= -32.05 + 80$$
$$= 47.95$$

Thus, if the current trend continues, about 47.95% of adults will read a daily newspaper in 2015.

Now Try Exercise 73

EXERCISE SET 3.4   Math XL   MyMathLab
MathXL®   MyMathLab

Warm-Up Exercises

Fill in the blanks with the appropriate word, phrase, or symbol(s) from the following list.

| translation | parallel | slope | vertical | horizontal | rise | run |
| positive | negative | solve | function | standard form | slope-intercept form | rate of change |

1. The _____ _____ is the measure of the steepness of a line.

2. The graph of $y = 2x + 3$ is a _____ of the graph of $y = 2x$.

3. A linear equation written as $ax + by = c$ is in _____.

4. A linear equation written as $y = mx + b$ is in _____.

5. Slope is often described as _____ over _____.

6. A line that rises going left to right has a _____ slope.

7. A line that falls going left to right has a _____ slope.

8. A _____ line has zero slope.

9. A _____ line has undefined slope.

10. Two lines that have the same slope are _____ lines.

11. When we give the change in y per unit change in x we are giving the slope as a _____.

12. To write an equation in slope-intercept form, _____ the equation for y.

Practice the Skills

Find the slope of the line through the given points. If the slope of the line is undefined, so state.

13. $(3, 5)$ and $(1, 9)$

14. $(3, 4)$ and $(6, 5)$

15. $(5, 2)$ and $(1, 4)$

16. $(-3, 7)$ and $(7, -3)$

17. $(-3, 5)$ and $(1, 1)$

18. $(2, 6)$ and $(2, -3)$

19. $(4, 2)$ and $(4, -6)$

20. $(8, -4)$ and $(-1, -2)$

21. $(-3, 4)$ and $(-1, 4)$

22. $(2, 8)$ and $(-5, 8)$

23. $(0, 3)$ and $(9, -3)$

24. $(0, -6)$ and $(-5, -3)$

Solve for the given variable if the line through the two given points is to have the given slope.

25. $(3, 2)$ and $(4, j)$, $m = 1$

26. $(-4, 3)$ and $(-2, r)$, $m = -3$

27. $(5, 0)$ and $(1, k)$, $m = \frac{1}{2}$

28. $(5, d)$ and $(9, 2)$, $m = -\frac{3}{4}$

29. $(x, 2)$ and $(3, -4)$, $m = 2$

30. $(-2, -3)$ and $(x, 5)$, $m = \frac{1}{2}$

31. $(12, -4)$ and $(r, 2)$, $m = -\frac{1}{2}$

32. $(-4, -4)$ and $(x, -1)$, $m = -\frac{3}{5}$

63. The graph of $3x - 2y = 6$ is translated down 4 units. Find the equation of the translated graph.

64. The graph of $-3x - 5y = 15$ is translated up 3 units. Find the equation of the translated graph.

65. If a line passes through the points $(6, 4)$ and $(-4, 2)$, find the change of y with respect to a 1-unit change in x.

66. If a line passes through the points $(-3, -4)$ and $(5, 2)$, find the change of y with respect to a 1-unit change in x.

TV Sales *For Exercises 67 and 68, use the graphs below. The graph on the left shows digital TV sales (in millions) and the graph on the right shows analog TV sales (in millions) for the years from 2004 to 2008.*

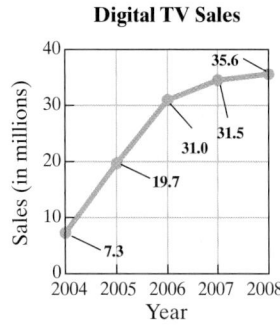

Digital TV Sales

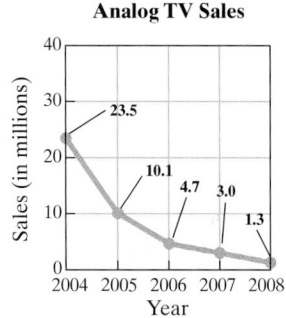

Analog TV Sales

Source: Consumer Electronics Association

67. a) For the graph of digital TV sales, determine the slope of the line segment from 2005 to 2006.

b) Is the slope of the line segment positive or negative?

c) Find the average rate of change from 2004 to 2008.

68. a) For the graph of analog TV sales, determine the slope of the line segment from 2005 to 2006.

b) Is the slope of the line segment positive or negative?

c) Find the average rate of change from 2004 to 2008.

69. Amtrak Expenses The following table gives the expenses, in millions of dollars, of Amtrak for selected years.

Year	Amtrak Expenses (in millions of dollars)
1995	$ 2257
2000	$ 2876
2004	$ 3133
2008	$ 3260

Source: Amtrak

a) Plot these points on a graph.

b) Connect these points using line segments.

c) Determine the slopes of each of the three line segments.

d) During which period was there the greatest average rate of change? Explain.

70. Demand for Steel The table above and to the right gives the world demand for steel, in millions of metric tons, for the years from 2004 to 2007.

a) Plot these points on a graph.

b) Determine the slope of each line segment.

c) Is this graph an example of a linear function? Explain.

Year	World Demand for Steel (in millions of metric tons)
2004	950
2005	1029
2006	1121
2007	1179

Source: International Iron and Steel Institute

d) During which period was rate of change highest? Explain.

71. Heart Rate The following bar graph shows the maximum recommended heart rate, in beats per minute, under stress for men of different ages. The bars are connected by a straight line.

a) Use the straight line to determine a function that can be used to estimate the maximum recommended heart rate, h, for $0 \le x \le 50$, where x is the number of years after age 20.

b) Using the function from part **a)**, determine the maximum recommended heart rate for a 34-year-old man.

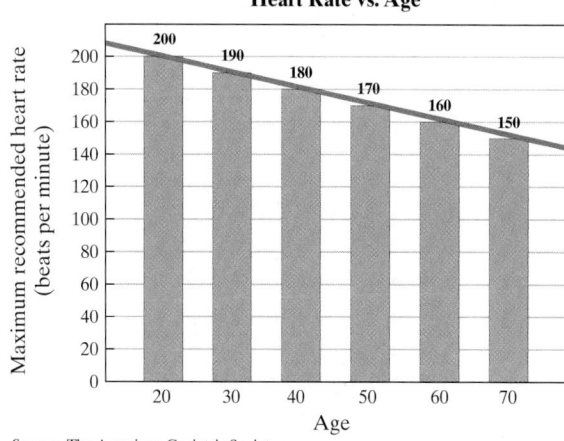

Heart Rate vs. Age

Source: The American Geriatric Society

72. Poverty Threshold The poverty threshold is an estimate of the annual family income necessary to have a minimally acceptable standard of living. The following bar graph shows the poverty threshold for a family of four for the years 2003 through 2007.

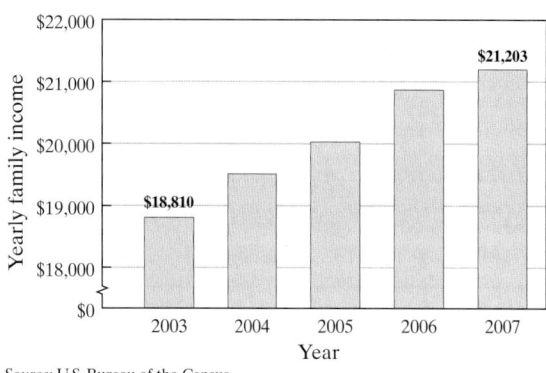

U.S. Poverty Threshold for a Family of Four

Source: U.S. Bureau of the Census

a) Determine a linear function that can be used to estimate the poverty threshold for a family of four, P, from 2003 through 2007. Let t represent the number of years since 2003.

b) Using the function from part **a)**, determine the poverty threshold in 2004. Compare your answer with the graph to see whether the graph supports your answer.

c) Assuming this trend continues, determine the poverty threshold for a family of four in the year 2015.

d) Assuming this trend continues, in which year will the poverty threshold for a family of four reach $22,997.75?

73. Teacher Salaries The following graph shows teacher salaries for the 2008–2009 school year in the Manatee County, Florida, school system for teachers whose highest degree is a bachelor's degree. Teachers with 0 years of experience earn $37,550 per year and teachers with 5 years of experience earn $38,600. Let S represent the annual teacher salary and let t represent years of experience.

Teacher Salary Schedule

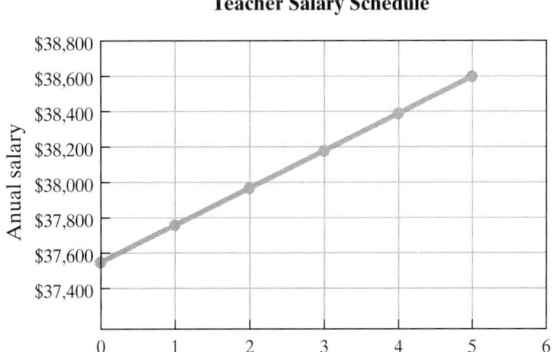

Source: Manatee County School Board

a) Determine a linear function $S(t)$ that fits this data.

b) Using the function from part **a)**, estimate the annual salary for a teacher with 3 years of experience. Compare your answer with the graph to see whether the graph supports your answer.

c) Assuming this trend continues, what will be the annual salary for a teacher with 10 years of experience?

d) Assuming this trend continues, how many years of experience must a teacher have to earn $40,070 per year?

74. Firefighter Salaries In Livonia, Michigan, firefighters with 0 years of experience earn an annual salary of $33,259 and firefighters with 5 years of experience earn an annual salary of $47,091. Let S represent the annual firefighters salary and let t represent years of experience.

Source: www.firehouse.com

a) Determine a linear function $S(t)$ that fits this data.

b) Use the function from part **a)** to estimate the annual salary for a firefighter with 3 years of experience.

c) Assuming this trend continues, what will be the annual salary for a firefighter with 10 years of experience?

d) Assuming this trend continues, how many years of experience must a firefighter have to earn an annual salary of $52,623.80?

75. Park Ranger Salaries In Maryland, state park rangers with 0 years of experience earn an annual salary of $37,855 and park rangers with 5 years of experience earn an annual salary of $47,123. Let S represent the annual park ranger salary and let t represent years of experience.

Source: www.dbm.maryland.gov

a) Determine a linear function $S(t)$ that fits this data.

b) Use the function from part **a)** to estimate the annual salary for a park ranger with 3 years of experience.

c) Assuming this trend continues, what will be the annual salary for a park ranger with 10 years of experience?

d) Assuming this trend continues, how many years of experience must a park ranger have to earn an annual salary of $52,683.80?

76. Social Security The number of workers per Social Security beneficiary has been declining approximately linearly since 1970. In 1970 there were 3.7 workers per beneficiary. In 2050 it is projected there will be 2.0 workers per beneficiary. Let W be the workers per Social Security beneficiary and t be the number of years since 1970.

a) Find a function $W(t)$ that fits the data.

b) Estimate the number of workers per beneficiary in 2020.

Suppose you are attempting to graph the equations shown and you get the screens shown. Explain how you know that you have made a mistake in entering each equation. The standard window setting is used on each graph.

77. $y = 3x + 6$

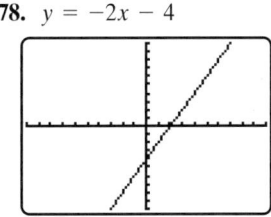

78. $y = -2x - 4$

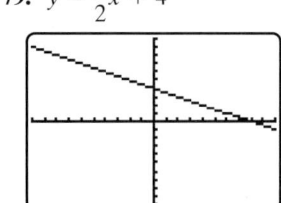

79. $y = \frac{1}{2}x + 4$

80. $y = -4x - 1$

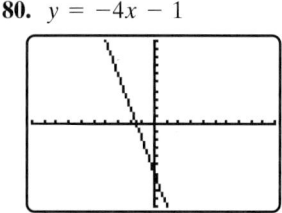

Challenge Problems

81. Castle The photo below is the Castle at Chichén Itzá, Mexico. Each side of the castle has a stairway consisting of 91 steps. The steps of the castle are quite narrow and steep, which makes them hard to climb. The total vertical distance of the 91 steps is 1292.2 inches. If a straight line were to be drawn connecting the tips of the steps, the absolute value of the slope of this line would be 2.21875. Find the average height and width of a step.

82. A tangent line is a straight line that touches a curve at a single point (the tangent line may cross the curve at a different point if extended). **Figure 3.56** shows three tangent lines to the curve at points *a*, *b*, and *c*. Note that the tangent line at point *a* has a positive slope, the tangent line at point *b* has a slope of 0, and the tangent line at point *c* has a negative slope. Now consider the curve in **Figure 3.57**. Imagine that tangent lines are drawn at all points on the curve except at endpoints *a* and *e*. Where on the curve in **Figure 3.57** would the tangent lines have a positive slope, a slope of 0, and a negative slope?

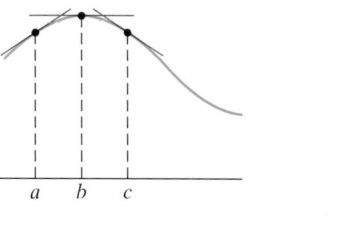

FIGURE 3.56   FIGURE 3.57

Group Activity

83. The following graph from *Consumer Reports* shows the depreciation on a typical car. The initial purchase price is represented as 100%.

a) Group member 1: Determine the 1-year period in which a car depreciates most. Estimate from the graph the percent a car depreciates during this period.

b) Group member 2: Determine between which years the depreciation appears linear or nearly linear.

c) Group member 3: Determine between which 2 years the depreciation is the lowest.

d) As a group, estimate the slope of the line segment from year 0 to year 1. Explain what this means in terms of rate of change.

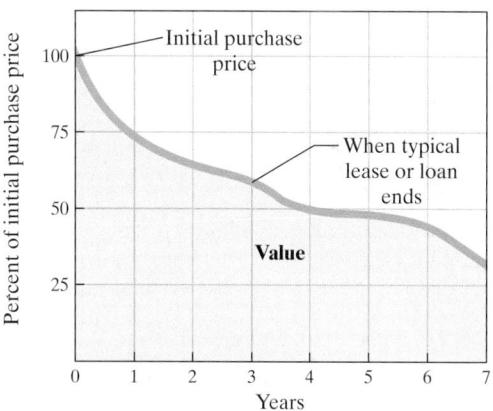

Cumulative Review Exercises

[1.4] **84.** Evaluate $\dfrac{-6^2 - 32 \div 2 \div |-8|}{5 - 3 \cdot 2 - 4 \div 2^2}$.

Solve each equation.

[2.1] **85.** $\dfrac{1}{4}(x + 3) + \dfrac{1}{5}x = \dfrac{2}{3}(x - 2) + 1$

   86. $2.6x - (-1.4x + 3.4) = 6.2$

[2.4] **87. Trains** Two trains leave Chicago, Illinois, traveling in the same direction along parallel tracks. The first train leaves 3 hours before the second, and its speed is 15 miles per hour faster than the second. Find the speed of each train if they are 270 miles apart 3 hours after the second train leaves Chicago.

[2.6] **88.** Solve
   a) $|2x + 1| > 5$.   **b)** $|2x + 1| < 5$.

Mid-Chapter Test: 3.1–3.4

To find out how well you understand the chapter material to this point, take this brief test. The answers, and the section where the material was initially discussed, are given in the back of the book. Review any questions you answered incorrectly.

1. In which quadrant is the point $(-3.5, -4.2)$ located?

Graph each equation in Exercises 2–5.

2. $y = 3x + 2$

3. $y = -x^2 + 3$

4. $y = |x| - 4$

5. $y = \sqrt{x - 4}$

6. a) What is a relation?
   b) What is a function?
   c) Is every relation a function? Explain.
   d) Is every function a relation? Explain.

In Exercises 7–9, determine which of the following relations are also functions. Give the domain and range of each relation or function.

7. $\{(1, 5), (2, -3), (7, -1), (-5, 6)\}$

8.

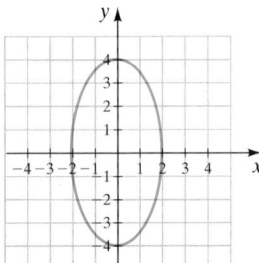

9.

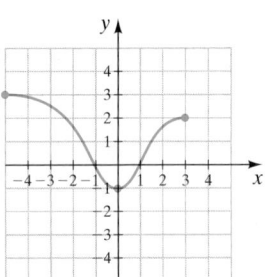

10. If $g(x) = 2x^2 + 8x - 13$, find $g(-2)$.

11. The height, h, in feet of an apple thrown from the top of a building is
$$h(t) = -6t^2 + 3t + 150$$
where t is time in seconds. Find the height of the apple 3 seconds after it is thrown.

12. Write the equation $7(x + 3) + 2y = 3(y - 1) + 18$ in standard form.

Graph each equation in Exercises 13–15.

13. $x + 3y = -3$

14. $x = -4$

15. $y = 5$

16. **Profit** The daily profit, in dollars, for a shoe company is $p(x) = 30x - 660$, where x is the number of pairs of shoes manufactured and sold.

   a) Draw a graph of profit versus the number of pairs of shoes sold (for up to 40 pairs).
   b) Determine the number of pairs of shoes that must be sold for the company to break even.
   c) Determine the number of pairs of shoes that must be sold for the company to make a daily profit of $360.

17. Find the slope of the line passing through $(9, -2)$ and $(-7, 8)$.

18. Write the equation of the line given in the graph shown.

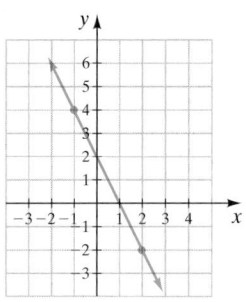

19. Write the equation $-3x + 2y = 18$ in slope-intercept form. Determine the slope and y-intercept.

20. If the graph of $y = 5x - 3$ is translated up 4 units, determine
   a) the slope of the translated graph.
   b) the y-intercept of the translated graph.
   c) the equation of the translated graph.

3.5 The Point-Slope Form of a Linear Equation

1 Understand the point-slope form of a linear equation.

2 Use the point-slope form to construct models from graphs.

3 Recognize parallel and perpendicular lines.

1 Understand the Point-Slope Form of a Linear Equation

The **point-slope form** of a line is used to determine the equation of a line when a *point* on the line and the *slope* of the line are known. The point-slope form can be developed from the equation for the slope between any two points (x, y) and (x_1, y_1) on a line, as shown in **Figure 3.58**.

$$m = \frac{y - y_1}{x - x_1}$$

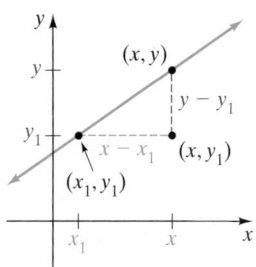

FIGURE 3.58

Multiplying both sides of the equation by $x - x_1$, we obtain

$$y - y_1 = m(x - x_1)$$

Point-Slope Form

The **point-slope form of a linear equation** is

$$y - y_1 = m(x - x_1)$$

where m is the *slope* of the line and (x_1, y_1) is a *specific point* on the line.

EXAMPLE 1   Write, in slope-intercept form, the equation of the line that passes through the point $(1, 4)$ and has slope -3.

Solution   Since we are given a point on the line and the slope of the line, we can write the equation in point-slope form. We can then solve the equation for y to write the equation in slope-intercept form. The slope is -3 and the point on the line is $(1, 4)$, so we have $m = -3$, $x_1 = 1$, and $y_1 = 4$.

$$y - y_1 = m(x - x_1)$$
$$\downarrow \quad \downarrow \quad \downarrow$$

$y - 4 = -3(x - 1)$ Point-slope form

$y - 4 = -3x + 3$ Distributive property

$y = -3x + 7$ Slope-intercept form

The graph of $y = -3x + 7$ has a slope of -3 and passes through the point $(1, 4)$.

Now Try Exercise 5

The point-slope form can also be used to find the equation of a line when we are given two points on the line. We show this in Example 2.

EXAMPLE 2   Write, in slope-intercept form, the equation of the line that passes through the points $(2, 3)$ and $(1, 4)$.

Solution   Although we are not given the slope of the line, we can use the two given points to determine the slope. We can then proceed as we did in Example 1. We will let $(2, 3)$ be (x_1, y_1) and $(1, 4)$ be (x_2, y_2).

$$m = \frac{y_2 - y_1}{x_2 - x_1} = \frac{4 - 3}{1 - 2} = \frac{1}{-1} = -1$$

The slope, m, is -1. Now we must choose one of the two given points to use as (x_1, y_1) in the point-slope form of the equation of a line. We will choose $(2, 3)$. We have $m = -1$, $x_1 = 2$, and $y_1 = 3$.

$$y - y_1 = m(x - x_1)$$
$$\downarrow \quad \downarrow \quad \downarrow$$

$y - 3 = -1(x - 2)$ Point-slope form

$y - 3 = -x + 2$ Distributive property

$y = -x + 5$ Slope-intercept form

The graph of $y = -x + 5$ is shown in **Figure 3.59**. Notice that the y-intercept of this line is at 5, the slope is -1, and the line goes through the points $(2, 3)$ and $(1, 4)$.

Now Try Exercise 11

Understanding Algebra

Forms of Linear Equations:

1. Standard Form

   $ax + by = c$

2. Slope-Intercept Form

   $y = mx + b$

3. Point-Slope Form

   $y - y_1 = m(x - x_1)$

Understanding Algebra

When using the point-slope formula and you have a choice of points to use, you can use any one of the given points. You should try to use the point that makes your calculations the easiest.

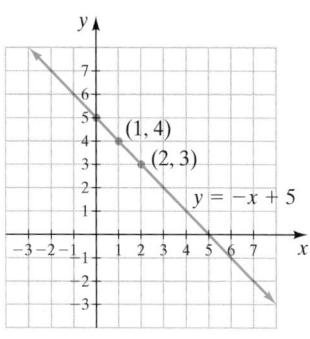

FIGURE 3.59

2 Use the Point-Slope Form to Construct Models from Graphs

Now let's look at an application where we use the point-slope form to determine a function that models a given situation.

EXAMPLE 3 Burning Calories The number of calories burned in 1 hour of bicycle riding is a linear function of the speed of the bicycle. A person riding at 12 mph will burn about 564 calories in 1 hour and while riding at 18 mph will burn about 846 calories in 1 hour. This information is shown in **Figure 3.60**.

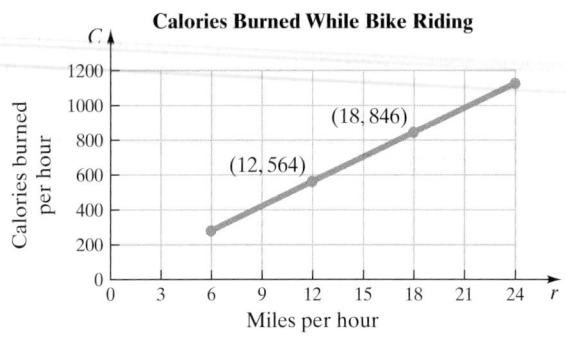

FIGURE 3.60   *Source:* American Heart Association

a) Determine a linear function that can be used to estimate the number of calories, C, burned in 1 hour when a bicycle is ridden at r mph, for $6 \le r \le 24$.

b) Use the function determined in part **a)** to estimate the number of calories burned in 1 hour when a bicycle is ridden at 20 mph.

c) Use the function determined in part **a)** to estimate the speed at which a bicycle should be ridden to burn 800 calories in 1 hour.

Solution

a) Understand and Translate   We will use the variables r (for rate) and C (for calories) instead of x and y, respectively. To find the necessary function, we will use the points $(12, 564)$ and $(18, 846)$ and proceed as we did in Example 2. We will first calculate the slope and then use the point-slope form to determine the equation of the line.

Carry Out
$$m = \frac{C_2 - C_1}{r_2 - r_1}$$
$$= \frac{846 - 564}{18 - 12} = \frac{282}{6} = 47$$

Now we write the equation using the point-slope form. We will choose the point $(12, 564)$ for (r_1, C_1).

$$C - C_1 = m(r - r_1)$$
$$C - 564 = 47(r - 12) \quad \text{Point-slope form}$$
$$C - 564 = 47r - 564$$
$$C = 47r \quad \text{Slope-intercept form}$$

Answer   Since the number of calories burned, C, is a function of the rate, r, the function we are seeking is

$$C(r) = 47r$$

b) To estimate the number of calories burned in 1 hour while riding at 20 mph, we substitute 20 for r in the function.

$$C(r) = 47r$$
$$C(20) = 47(20) = 940$$

Therefore, 940 calories are burned while riding at 20 mph for 1 hour.

- Parallel lines never intersect and have the same slope.
- Perpendicular lines intersect at right angles and have slopes that are negative reciprocals.

c) To estimate the speed at which a bicycle should be ridden to burn 800 calories in 1 hour, we substitute 800 for $C(r)$ in the function.

$$C(r) = 47r$$
$$800 = 47r$$
$$\frac{800}{47} = r$$
$$r \approx 17.02$$

Thus the person would need to ride the bicycle at about 17.02 mph to burn 800 calories in 1 hour.

Now Try Exercise 53

In Example 3, the function was $C(r) = 47r$. The graph of this function has a slope of 47 and a y-intercept at $(0, 0)$. If the graph in **Figure 3.60** on page 186 was extended to the left, it would intersect the origin. This makes sense since a rate of 0 miles per hour would result in 0 calories being burned.

3 Recognize Parallel and Perpendicular Lines

Figure 3.61 illustrates two *parallel* lines.

Parallel lines

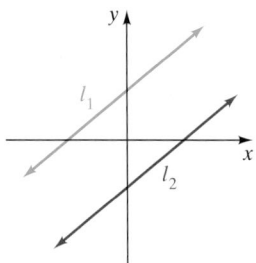

FIGURE 3.61

> **Parallel Lines**
>
> Two lines are **parallel** when they have the same slope. Any two vertical lines are parallel to each other.

All vertical lines are parallel even though their slope is undefined.

Figure 3.62 illustrates perpendicular lines. Two lines are *perpendicular* when they intersect at right (or 90°) angles.

> **Perpendicular Lines**
>
> Two lines are **perpendicular** when their slopes are *negative reciprocals*. Any vertical line is perpendicular to any horizontal line.

Perpendicular lines

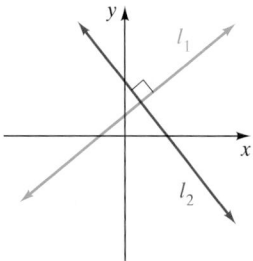

FIGURE 3.62

For any nonzero number a, its **negative reciprocal** is $\frac{-1}{a}$ or $-\frac{1}{a}$. For example, the negative reciprocal of 2 is $\frac{-1}{2}$ or $-\frac{1}{2}$. The product of any nonzero number and its negative reciprocal is -1.

EXAMPLE 4 Two points on line l_1 are $(0, 2)$ and $(3, 4)$. Two points on line l_2 are $(-2, 4)$ and $(2, -2)$. By comparing their slopes, determine whether l_1 and l_2 are parallel lines, perpendicular lines, or neither parallel nor perpendicular lines.

Solution   First, determine the slope m_1 of l_1.

$$m_1 = \frac{4 - 2}{3 - 0} = \frac{2}{3}$$

Thus, the slope of the first line is $\frac{2}{3}$. Next, determine the slope m_2 of l_2.

$$m_2 = \frac{-2 - 4}{2 - (-2)} = \frac{-6}{4} = -\frac{3}{2}$$

The product of a nonzero number, a, and its negative reciprocal, $-\frac{1}{a}$, is always -1.

$$a\left(-\frac{1}{a}\right) = -1$$

Thus, the slope of the second line is $-\frac{3}{2}$. Since the slopes are not equal, the lines are not parallel. The slopes are negative reciprocals of each other. Note that

$$m_1 \cdot m_2 = \frac{2}{3} \cdot \left(-\frac{3}{2}\right) = -1$$

Since the slopes are negative reciprocals of each other, the lines are perpendicular.

Now Try Exercise 15

EXAMPLE 5 Consider the equation $2x + 4y = 8$. Determine the equation of the line that has a y-intercept of 5 and is **a)** parallel to the given line and **b)** perpendicular to the given line.

Solution

a) If we know the slope, m, of a line and its y-intercept, $(0, b)$, we can use the slope-intercept form, $y = mx + b$, to write the equation. We begin by solving the given equation for y.

$$2x + 4y = 8$$
$$4y = -2x + 8$$
$$y = \frac{-2x + 8}{4}$$
$$y = -\frac{1}{2}x + 2$$

Two lines are parallel when they have the same slope. Therefore, the slope of the line parallel to the given line must be $-\frac{1}{2}$. Since its slope is $-\frac{1}{2}$ and its y-intercept is 5, its equation must be

$$y = -\frac{1}{2}x + 5.$$

The graphs of $2x + 4y = 8$ (in blue) and $y = -\frac{1}{2}x + 5$ (in green) are shown in **Figure 3.63**.

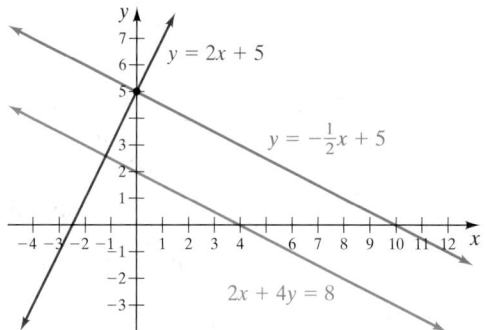

FIGURE 3.63

b) Two lines are perpendicular when their slopes are negative reciprocals. We know that the slope of the given line is $-\frac{1}{2}$. Therefore, the slope of the perpendicular line must be $\frac{-1}{-\frac{1}{2}}$ or 2. The line perpendicular to the given line has a y-intercept of 5. Thus the equation is

$$y = 2x + 5.$$

Figure 3.63 also shows the graph of $y = 2x + 5$ (in red).

Now Try Exercise 35

EXAMPLE 6 Consider the equation $5y = -10x + 7$.

a) Determine the equation of a line that passes through $\left(4, \frac{1}{3}\right)$ that is perpendicular to the graph of the given equation. Write the equation in standard form.

b) Write the equation determined in part **a)** using function notation.

Solution

a) Determine the slope of the given line by solving the equation for y.

$$5y = -10x + 7$$
$$y = \frac{-10x + 7}{5}$$
$$y = -2x + \frac{7}{5}$$

Since the slope of the given line is -2, the slope of a line perpendicular to it must be the negative reciprocal of -2, which is $\frac{1}{2}$. Since the line whose equation we are seeking contains point $\left(4, \frac{1}{3}\right)$ and has slope $\frac{1}{2}$, we will use the point-slope from to obtain

$$y - y_1 = m(x - x_1)$$
$$y - \frac{1}{3} = \frac{1}{2}(x - 4) \qquad \text{Point-slope form}$$

Now multiply both sides of the equation by the least common denominator, 6, to eliminate fractions.

$$6\left(y - \frac{1}{3}\right) = 6\left[\frac{1}{2}(x - 4)\right]$$
$$6y - 2 = 3(x - 4)$$
$$6y - 2 = 3x - 12$$

Now write the equation in standard form.

$$-3x + 6y - 2 = -12$$
$$-3x + 6y = -10 \qquad \text{Standard form}$$

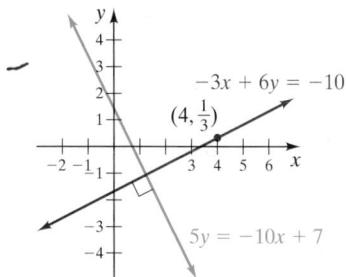

Note that $3x - 6y = 10$ is also an acceptable answer. **Figure 3.64** shows the graph of the given equation, $5y = -10x + 7$ (in blue), and the graph of the equation of the perpendicular line, $-3x + 6y = -10$ (in red).

FIGURE 3.64

b) To write the equation using function notation, we solve the equation determined in part **a)** for y, and then replace y with $f(x)$.

We will leave it to you to show that the function is $f(x) = \frac{1}{2}x - \frac{5}{3}$.

Now Try Exercise 39

The following chart summarizes the three forms of a linear equation we have studied and mentions when each may be useful.

Forms of Linear Equations

Standard form: $ax + by = c$	Used to graph a linear equation by finding the intercepts of a graph
Slope-intercept form: $y = mx + b$	Used to find the slope and y-intercept of a line
	Used to find the equation of a line given its slope and y-intercept
	Used to determine if two lines are parallel or perpendicular
	Used to graph a linear equation by using the slope and y-intercept
Point-slope form: $y - y_1 = m(x - x_1)$	Used to find the equation of a line when given the slope of a line and a point (x_1, y_1) on the line
	Used to find the equation of a line when given two points on a line

EXERCISE SET 3.5

Math XL
MathXL®

MyMathLab
MyMathLab

Warm-Up Exercises

Fill in the blanks with the appropriate word, phrase, or symbol(s) from the following list.

standard form   point-slope form   slope-intercept form   parallel   perpendicular   negative reciprocals   the same

1. Two lines that have the same slope are _____ lines.

2. Perpendicular lines have slopes that are _____ of each other.

3. The _____ of the equation of a line is $y - y_1 = m(x - x_1)$.

4. The _____ form of the equation of a line is $y = mx + b$.

Practice the Skills

Use the point-slope form to find the equation of a line with the properties given. Then write the equation in slope-intercept form.

5. Slope = 3, through $(2, 1)$

6. Slope = -3, through $(1, -2)$

7. Slope = $-\dfrac{1}{2}$, through $(4, -1)$

8. Slope = $-\dfrac{7}{8}$, through $(-8, -2)$

9. Slope = $\dfrac{1}{2}$, through $(-1, -5)$

10. Slope = $-\dfrac{3}{2}$, through $(7, -4)$

11. Through $(2, -3)$ and $(-6, 9)$

12. Through $(4, -2)$ and $(1, 9)$

13. Through $(4, -3)$ and $(6, -2)$

14. Through $(1, 0)$ and $(-4, -1)$

Two points on l_1 and two points on l_2 are given. Determine whether l_1 is parallel to l_2, l_1 is perpendicular to l_2, or neither.

15. l_1: $(-2, 0)$ and $(0, 2)$; l_2: $(-3, 0)$ and $(0, 3)$

16. l_1: $(7, 6)$ and $(3, 9)$; l_2: $(5, -1)$ and $(9, -4)$

17. l_1: $(4, 6)$ and $(5, 7)$; l_2: $(-1, -1)$ and $(1, 4)$

18. l_1: $(-3, 4)$ and $(4, -3)$; l_2: $(-5, -6)$ and $(6, -5)$

19. l_1: $(3, 2)$ and $(-1, -2)$; l_2: $(2, 0)$ and $(3, -1)$

20. l_1: $(3, 5)$ and $(9, 1)$; l_2: $(4, 0)$ and $(6, 3)$

Determine whether the two equations represent lines that are parallel, perpendicular, or neither.

21. $y = x + 9$
    $y = -x + 2$

22. $2x + 3y = 11$
    $y = -\dfrac{2}{3}x + 4$

23. $4x + 2y = 8$
    $8x = 4 - 4y$

24. $2x - y = 4$
    $3x + 6y = 18$

25. $2x - y = 4$
    $-x + 4y = 4$

26. $6x + 2y = 8$
    $4x - 5 = -y$

27. $y = \dfrac{1}{2}x - 6$
    $-4y = 8x + 15$

28. $2y - 8 = -5x$
    $y = -\dfrac{5}{2}x - 2$

29. $y = \dfrac{1}{2}x + 6$
    $-2x + 4y = 8$

30. $-4x + 6y = 11$
    $2x - 3y = 5$

31. $x - 2y = -9$
    $y = x + 6$

32. $\dfrac{1}{2}x - \dfrac{3}{4}y = 1$
    $\dfrac{3}{5}x + \dfrac{2}{5}y = -1$

Find the equation of a line with the properties given. Write the equation in the form indicated.

33. Through $(2, 5)$ and parallel to the graph of $y = 2x + 4$ (slope-intercept form)

34. Through $(-1, 6)$ and parallel to the graph of $4x - 2y = 6$ (slope-intercept form)

35. Through $(-3, -5)$ and parallel to the graph of $2x - 5y = 7$ (standard form)

36. Through $(-1, 4)$ and perpendicular to the graph of $y = -2x - 1$ (standard form)

37. With x-intercept $(3, 0)$ and y-intercept $(0, 5)$ (slope-intercept form)

38. Through $(-2, -1)$ and perpendicular to the graph of $f(x) = -\dfrac{1}{5}x + 1$ (function notation)

39. Through $(1, 2)$ and perpendicular to the graph of $y = -\dfrac{1}{4}x + 5$ (function notation)

40. Through $(-3, 5)$ and perpendicular to the line with x-intercept $(2, 0)$ and y-intercept $(0, 2)$ (standard form)

41. Through $(6, 2)$ and perpendicular to the line with x-intercept $(2, 0)$ and y-intercept $(0, -3)$ (slope-intercept form)

42. Through the point $(1, 2)$ and parallel to the line through the points $(3, 5)$ and $(-2, 3)$ (function notation)

Problem Solving

43. Treadmill The number of calories burned in 1 hour on a treadmill is a function of the speed of the treadmill. A person walking on a treadmill at a speed of 2.5 miles per hour will burn about 210 calories. At 6 miles per hour the person will burn about 370 calories. Let C be the calories burned in 1 hour and s be the speed of the treadmill.

a) Determine a linear function $C(s)$ that fits the data.

b) Estimate the calories burned by the person on a treadmill in 1 hour at a speed of 5 miles per hour.

44. Inclined Treadmill The number of calories burned for 1 hour on a treadmill going at a constant speed is a function of the incline of the treadmill. At 4 miles per hour a person on a 5° incline will burn 525 calories. At 4 mph on a 15° incline the person will burn 880 calories. Let C be the calories burned and d be the degrees of incline of the treadmill.

a) Determine a linear function $C(d)$ that fits the data.

b) Determine the number of calories burned by the person in 1 hour on a treadmill going 4 miles per hour and at a 9° incline.

45. Demand for iPods The *demand* for a product is the number of items the public is willing to buy at a given price. Suppose the demand, d, for iPods sold in 1 month is a linear function of the price, p, for $150 \leq p \leq 400$. If the price is $200, then 50 iPods will be sold each month. If the price is $300, only 30 iPods will be sold.

a) Using ordered pairs of the form (p, d), write an equation for the demand, d, as a function of price, p.

b) Using the function from part **a)**, determine the demand when the price of the iPods is $260.

c) Using the function from part **a)**, determine the price charged if the demand for iPods is 45.

© Jaimie Duplass/Shutterstock

46. Demand for New Sandwiches The marketing manager of Arby's restaurants determines that the demand, d, for a new sandwich is a linear function of the price, p, for $0.80 \leq p \leq 4.00$. If the price is $1.00, then 530 sandwiches will be sold each month. If the price is $2.00, only 400 sandwiches will be sold each month.

a) Using ordered pairs of the form (p, d), write an equation for the demand, d, as a function of price, p.

b) Using the function from part **a)**, determine the demand when the price of the sandwich is $2.60.

c) Using the function from part **a)**, determine the price charged if the demand for sandwiches is 244 sandwiches.

47. Supply of Kites The *supply* of a product is the number of items a seller is willing to sell at a given price. The maker of a new kite for children determines that the number of kites she is willing to supply, s, is a linear function of the selling price p for $2.00 \leq p \leq 4.00$. If a kite sells for $2.00, then 130 per month will be supplied. If a kite sells for $4.00, then 320 per month will be supplied.

a) Using ordered pairs of the form (p, s), write an equation for the supply, s, as a function of price, p.

b) Using the function from part **a)**, determine the supply when the price of a kite is $2.80.

c) Using the function from part **a)**, determine the price paid if the supply is 225 kites.

48. Supply of Baby Strollers The manufacturer of baby strollers determines that the supply, s, is a linear function of the selling price, p, for $200 \leq p \leq 300$. If a stroller sells for $210.00, then 20 strollers will be supplied per month. If a stroller sells for $230.00, then 30 strollers will be supplied per month.

© Nick Thompson/iStockphoto

a) Using ordered pairs of the form (p, s), write an equation for the supply, s, as a function of price, p.

b) Using the function from part **a)**, determine the supply when the price of a stroller is $220.00.

c) Using the function from part **a)**, determine the selling price if the supply is 35 strollers.

49. High School Play The income, i, from a high school play is a linear function of the number of tickets sold, t. When 80 tickets are sold, the income is $1000. When 200 tickets are sold, the income is $2500.

a) Use these data to write the income, i, as a function of the number of tickets sold, t.

b) Using the function from part **a)**, determine the income if 120 tickets are sold.

c) If the income is $2200, how many tickets were sold?

50. Gas Mileage of a Car The gas mileage, m, of a specific car is a linear function of the speed, s, at which the car is driven, for $30 \leq s \leq 60$. If the car is driven at a rate of 30 mph, the car's gas mileage is 35 miles per gallon. If the car is driven at 60 mph, the car's gas mileage is 20 miles per gallon.

a) Use this data to write the gas mileage, m, as a function of speed, s.

b) Using the function from part **a)**, determine the gas mileage if the car is driven at a speed of 48 mph.

c) Using the function from part **a)**, determine the speed at which the car must be driven to get gas mileage of 40 miles per gallon.

51. Auto Registration The registration fee, r, for a vehicle in a certain region is a linear function of the weight of the vehicle, w, for $1000 \leq w \leq 6000$ pounds. When the weight is 2000 pounds, the registration fee is $30. When the weight is 4000 pounds, the registration fee is $50.

a) Use these data to write the registration fee, r, as a function of the weight of the vehicle, w.

b) Using the function from part **a)**, determine the registration fee for a 2006 Ford Mustang if the weight of the vehicle is 3613 pounds.

c) If the cost of registering a vehicle is $60, determine the weight of the vehicle.

52. Lecturer Salary Suppose the annual salary of a lecturer at Chaumont University is a linear function of the number of years of teaching experience. A lecturer with 9 years of teaching experience is paid $41,350. A lecturer with 15 years of teaching experience is paid $46,687.

a) Use this data to write the annual salary of a lecturer, s, as a function of the number of years of teaching experience, n.

b) Using the function from part **a)**, determine the annual salary of a lecturer with 10 years of teaching experience.

c) Using the function from part **a)**, estimate the number of years of teaching experience a lecturer must have to obtain an annual salary of $44,908.

53. Life Expectancy As seen in the following graph, the expected number of remaining years of life of a person, y, *approximates* a linear function. The expected number of remaining years is a function of the person's current age, a, for $30 \leq a \leq 80$.

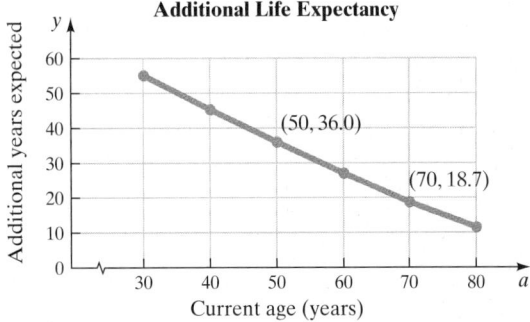

Source: TIAA/CREF

a) Using the two points on the graph, determine the function $y(a)$ that can be used to approximate the graph.

b) Using the function from part **a)**, estimate the additional life expectancy of a person who is currently 37 years old.

c) Using the function from part **a)**, estimate the current age of a person who has an additional life expectancy of 25 years.

54. Guarneri del Gesù Violin The graph below shows that the projected value, v, of a Guarneri del Gesù violin is a linear function of the age, a, in years, of the violin, for $261 \leq a \leq 290$.

Source: Machold Rare Violins, LTD

a) Determine the function $v(a)$ represented by this line.

b) Using the function from part **a)**, determine the projected value of a 265-year-old Guarneri del Gesù violin.

c) Using the function from part **a)**, determine the age of a Guarneri del Gesù violin with a projected value of $15 million.

Guarneri del Gesù, "Sainton," 1741

55. Boys' Weights The diagram on page 193 shows percentiles for boys' lengths (heights) and weights from birth to age 36 months. Certain portions of the graphs can be approximated with a linear function. For example, the graph representing the 95th percentile of boys' weights (the top red line) from age 18 months to age 36 months is approximately linear.

Boys: Birth to 36 months
Length-for-Age and Weight-for-Age Percentiles

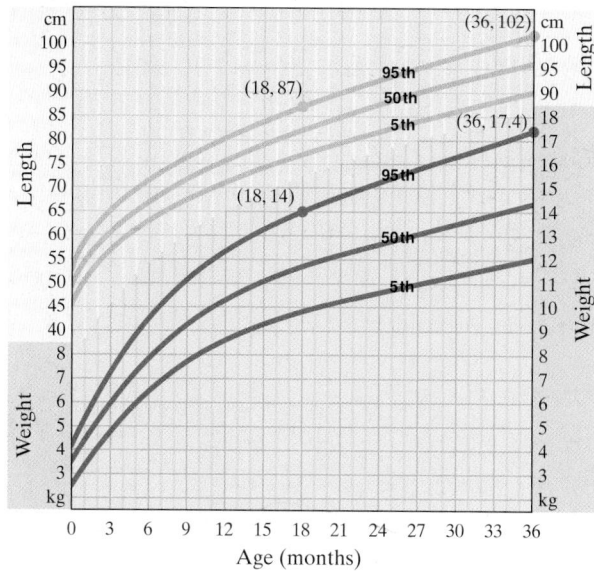

Source: National Center for Health Statistics

a) Use the points shown on the graph of the 95th percentile to write weight, *w*, as a linear function of age, *a*, for boys between 18 and 36 months old.

b) Using the function from part **a)**, estimate the weight of a 22-month-old boy who is in the 95th percentile for weight. Compare your answer with the graph to see whether the graph supports your answer.

56. Boys' Lengths The diagram in Exercise 55 shows that the graph representing the 95th percentile of boys' lengths (the top yellow line) from age 18 months to age 36 months is approximately linear.

a) Use the points shown on the graph of the 95th percentile to write length, *l*, as a linear function of age, *a*, for boys between ages 18 and 36 months.

b) Using the function from part **a)**, estimate the length of a 21-month-old boy who is in the 95th percentile. Compare your answer with the graph to see whether the graph supports your answer.

Group Activity

57. The graph on the right shows the growth of the circumference of a girl's head. The orange line is the average head circumference of all girls for the given age while the green lines represent the upper and lower limits of the normal range. Discuss and answer the following questions as a group.

a) Explain why the graph of the average head circumference represents a function.

b) What is the independent variable? What is the dependent variable?

c) What is the domain of the graph of the average head circumference? What is the range of the average head circumference graph?

d) What interval is considered normal for girls of age 18?

e) For this graph, is head circumference a function of age or is age a function of head circumference? Explain your answer.

f) Estimate the average girl's head circumference at age 10 and at age 14.

g) This graph appears to be nearly linear. Determine an equation or function that can be used to estimate the orange line between (2, 48) and (18, 55).

Head Circumference

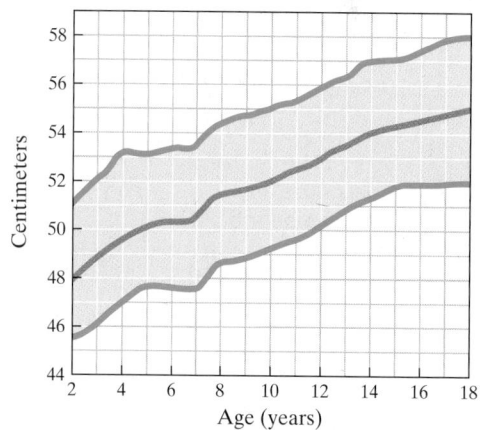

Source: National Center for Health Statistics

Cumulative Review Exercises

[2.5] **58.** Solve the inequality $6 - \frac{1}{2}x > 2x + 5$ and indicate the solution in interval notation.

59. What must you do when multiplying or dividing both sides of an inequality by a negative number?

[3.2] **60. a)** What is a relation?

b) What is a function?

c) Draw a graph that is a relation but not a function.

61. Find the domain and range of the function $\{(4, 7), (5, -4), (3, 2), (6, -1)\}$.

3.6 The Algebra of Functions

1 Find the sum, difference, product, and quotient of functions.

2 Graph the sum of functions.

1 Find the Sum, Difference, Product, and Quotient of Functions

In this section we introduce some ways that functions can be combined. Consider the functions $f(x) = x - 3$ and $g(x) = x^2 + 2x$. We find $f(5)$ and $g(5)$ as follows.

$$f(x) = x - 3 \qquad g(x) = x^2 + 2x$$
$$f(5) = 5 - 3 = 2 \qquad g(5) = 5^2 + 2(5) = 35$$

Next, if we look at the sum of the two functions, we get

$$f(x) \quad + \quad g(x)$$
$$\downarrow \qquad \downarrow \qquad \downarrow$$
$$(x - 3) + (x^2 + 2x) = x^2 + 3x - 3$$

This new function is designated as $(f + g)(x)$ and we write

$$(f + g)(x) = x^2 + 3x - 3$$

We find $(f + g)(5)$ as follows.

$$(f + g)(\ ) = 5^2 + 3(5) - 3$$
$$= 25 + 15 - 3 = 37$$

Notice that

$$f(5) + g(5) = (f + g)(5)$$
$$2 + 35 = 37 \qquad \text{True}$$

In fact, for any real number substituted for x you will find that

$$f(x) + g(x) = (f + g)(x)$$

Similar notation exists for subtraction, multiplication, and division of functions.

Understanding Algebra

The domain of a function is the set of values that can be used for the independent variable. For example, the domain of

- $f(x) = 2x^2 - 6x + 5$ is all real numbers since any real number can be substituted for x.

- $g(x) = \dfrac{1}{x - 8}$ is all real numbers except 8, since $x = 8$ leads to $\dfrac{1}{0}$, which is undefined.

Operations on Functions

If $f(x)$ represents one function, $g(x)$ represents a second function, and x is in the domain of both functions, then the following operations on functions may be performed:

Sum of functions: $(f + g)(x) = f(x) + g(x)$

Difference of functions: $(f - g)(x) = f(x) - g(x)$

Product of functions: $(f \cdot g)(x) = f(x) \cdot g(x)$

Quotient of functions: $(f/g)(x) = \dfrac{f(x)}{g(x)}$, provided that $g(x) \neq 0$

EXAMPLE 1   If $f(x) = x^2 + x - 6$ and $g(x) = x - 3$, find

a) $(f + g)(x)$

b) $(f - g)(x)$

c) $(g - f)(x)$

d) Does $(f - g)(x) = (g - f)(x)$?

Solution   To answer parts **a)**–**c)**, we perform the indicated operation.

a) $(f + g)(x) = f(x) + g(x)$
$$= (x^2 + x - 6) + (x - 3)$$
$$= x^2 + x - 6 + x - 3$$
$$= x^2 + 2x - 9$$

b) $(f - g)(x) = f(x) - g(x)$
$$= (x^2 + x - 6) - (x - 3)$$
$$= x^2 + x - 6 - x + 3$$
$$= x^2 - 3$$

c) $(g - f)(x) = g(x) - f(x)$
$$= (x - 3) - (x^2 + x - 6)$$
$$= x - 3 - x^2 - x + 6$$
$$= -x^2 + 3$$

d) By comparing the answers to parts **b)** and **c)**, we see that
$$(f - g)(x) \neq (g - f)(x)$$

Now Try Exercise 11

EXAMPLE 2 If $f(x) = x^2 - 4$ and $g(x) = x - 2$, find

a) $(f - g)(6)$ **b)** $(f \cdot g)(5)$ **c)** $(f/g)(8)$

Solution

a) $(f - g)(x) = f(x) - g(x)$
$$= (x^2 - 4) - (x - 2)$$
$$= x^2 - x - 2$$
$$(f - g)(6) = 6^2 - 6 - 2$$
$$= 36 - 6 - 2$$
$$= 28$$

We could have also found the solution as follows:

$$f(x) = x^2 - 4 \qquad\qquad g(x) = x - 2$$
$$f(6) = 6^2 - 4 = 32 \qquad g(6) = 6 - 2 = 4$$
$$(f - g)(6) = f(6) - g(6)$$
$$= 32 - 4 = 28$$

b) We will find $(f \cdot g)(5)$ using the fact that

$$(f \cdot g)(5) = f(5) \cdot g(5)$$
$$f(x) = x^2 - 4 \qquad\qquad g(x) = x - 2$$
$$f(5) = 5^2 - 4 = 21 \qquad g(5) = 5 - 2 = 3$$

Thus $f(5) \cdot g(5) = 21 \cdot 3 = 63$. Therefore, $(f \cdot g)(5) = 63$.

c) We will find $(f/g)(8)$ by using the fact that

$$(f/g)(8) = f(8)/g(8)$$
$$f(x) = x^2 - 4 \qquad\qquad g(x) = x - 2$$
$$f(8) = 8^2 - 4 = 60 \qquad g(8) = 8 - 2 = 6$$

Therefore, $(f/g)(8) = f(8)/g(8) = 60/6 = 10$. Therefore, $(f/g)(8) = 10$.

Now Try Exercise 31

2 Graph the Sum of Functions

Now we will explain how we can graph the sum, difference, product, or quotient of two functions. **Figure 3.65** on page 196 shows two functions, $f(x)$, illustrated in blue, and $g(x)$, illustrated in red.

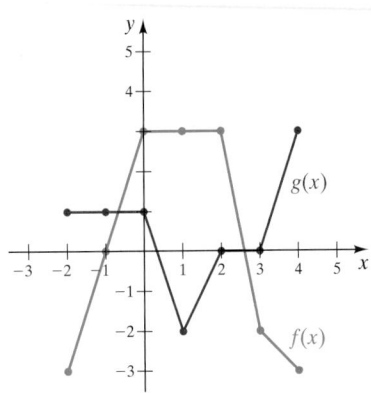

FIGURE 3.65

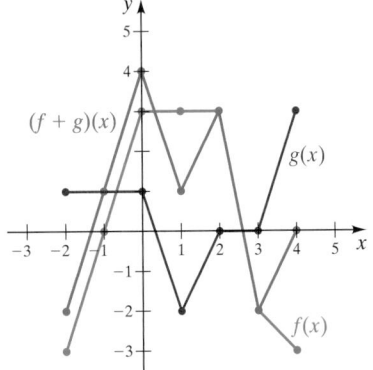

FIGURE 3.66

The table below gives the integer values of x from -2 to 4, the values of $f(-2)$ through $f(4)$, and the values of $g(-2)$ through $g(4)$. These values are taken directly from **Figure 3.65**. The values of $(f + g)(-2)$ through $(f + g)(4)$ are determined by adding the values of $f(x)$ and $g(x)$. The graph of $(f + g)(x) = f(x) + g(x)$ is illustrated in green in **Figure 3.66**.

x	$f(x)$	$g(x)$	$(f + g)(x)$
-2	-3	1	$-3 + 1 = -2$
-1	0	1	$0 + 1 = 1$
0	3	1	$3 + 1 = 4$
1	3	-2	$3 + (-2) = 1$
2	3	0	$3 + 0 = 3$
3	-2	0	$-2 + 0 = -2$
4	-3	3	$-3 + 3 = 0$

We could graph the difference, product, or quotient of the two functions using a similar technique. For example, to graph the product function $(f \cdot g)(x)$, we would evaluate $(f \cdot g)(-2)$ as follows:

$$(f \cdot g)(-2) = f(-2) \cdot g(-2)$$
$$= (-3)(1) = -3$$

Thus, the graph of $(f \cdot g)(x)$ would have an ordered pair at $(-2, -3)$. Other ordered pairs would be determined by the same procedure.

In newspapers, magazines, and on the Internet we often find graphs that show the sum of two or more functions. Graphs that show the sum of functions are generally indicated in one of three ways: line graphs, bar graphs, or stacked (or cumulative) line graphs. Examples 3 through 5 show the three general methods. Each of these examples will use the same data pertaining to cholesterol.

EXAMPLE 3 Line Graph Ray Hundley has kept a record of his bad cholesterol (low-density lipoprotein, or LDL) and his good cholesterol (high-density lipoprotein or HDL) from 2006 through 2010. **Table 3.2** shows his LDL and his HDL for these years.

TABLE 3.2   Cholesterol

	2006	2007	2008	2009	2010
LDL	220	240	140	235	130
HDL	30	40	70	35	40

a) Explain why the data consisting of the years and the LDL values are a function, and the data consisting of the years and the HDL values are also a function.

b) Draw a line graph that shows the LDL, the HDL, and the total cholesterol from 2006 through 2010. The total cholesterol is the sum of the LDL and the HDL.

c) If L represents the amount of LDL and H represents the amount of HDL, show that $(L + H)(2010) = 170$.

d) By looking at the graph drawn in part **b)**, determine the years in which the LDL was less than 180.

Solution

a) The data consisting of the years and the LDL values are a function because for each year there is exactly one LDL value. Note that the year is the independent variable, and the LDL value is the dependent variable. The data consisting of the years and the HDL values are a function for the same reason.

b) For any given year, the total cholesterol is the sum of the LDL and HDL for that year. For example, for 2009, to find the cholesterol, we add $235 + 35 = 270$. The graph in **Figure 3.67** shows LDL, HDL, and total cholesterol for the years 2006 through 2010.

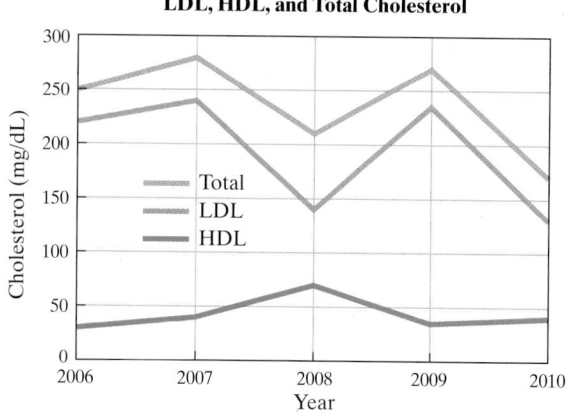

FIGURE 3.67

c) To find LDL + HDL, or the total cholesterol, we add the two values for 2010.

$$(L + H)(2010) = L(2010) + H(2010)$$
$$= 130 + 40 = 170$$

d) By looking at the graph drawn in part **b)**, we see that the years in which the LDL was less than 180 are 2008 and 2010.

Now Try Exercise 63a

EXAMPLE 4 Bar Graph

a) Using the data given in **Table 3.2** on page 196, draw a bar graph that shows the LDL, HDL, and total cholesterol for the years 2006 through 2010.

b) If L represents the amount of LDL and H represents the amount of HDL, use the graph drawn in part **a)** to determine $(L + H)(2007)$.

c) By observing the graph drawn in part **a)**, determine in which years the total cholesterol was less than 220.

d) By observing the graph drawn in part **a)**, estimate the HDL in 2008.

Solution

a) To obtain a bar graph showing the total cholesterol, we add the HDL to the LDL for each given year. For example, for 2006, we start by drawing a bar up to 220 to represent the LDL. Directly on top of that bar we add a second bar of 30 units to represent the HDL. This brings the total bar to $220 + 30$ or 250 units. We use the same procedure for each year from 2006 to 2010. The bar graph is shown in **Figure 3.68**.

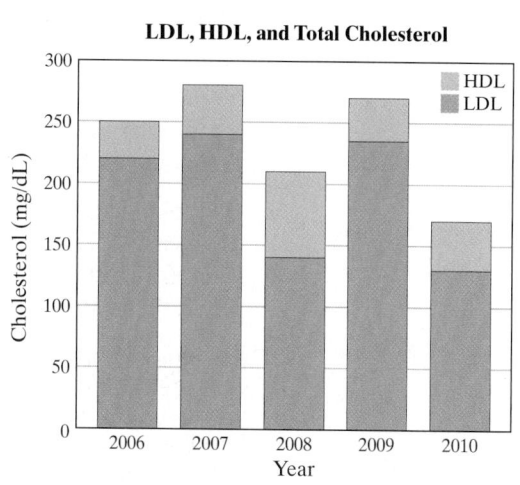

FIGURE 3.68

b) By observing the graph in **Figure 3.68**, we see that the $(L + H)(2007)$, or the total cholesterol for 2007, is about 280.

c) By observing the graph, we see that the total cholesterol was less than 220 in 2008 and 2010.

d) For 2008, the HDL bar begins at about 140 and ends at about 210. The difference in these amounts, $210 - 140 = 70$, represents the amount of HDL in 2008. Therefore, the HDL in 2008 was about 70.

Now Try Exercise 63b

EXAMPLE 5 Stacked Line Graph

a) Using the data from **Table 3.2** on page 196, draw a stacked (or cumulative) line graph that shows the LDL, HDL, and total cholesterol for the years 2006–2010.

b) Using the graph drawn in part **a)**, determine which years the total cholesterol was greater than or equal to 200.

c) Using the graph drawn in part **a)**, estimate the amount of HDL in 2010.

d) Using the graph from part **a)**, determine the years in which the LDL was greater than or equal to 180 and the total cholesterol was less than or equal to 250.

Solution

a) To obtain a stacked line graph, begin by drawing a line graph to represent the LDL. This is the same line graph used in **Figure 3.67** on page 197. The area under this line graph (the blue area in **Figure 3.69**) represents the LDL. Next, going year by year, we add the HDL to the LDL to create a second line graph. For example, in year 2006, the second line graph has a value at $220 + 30$ or 250. The area between the two line graphs (the green area in **Figure 3.69**) represents the HDL and the total area under the top line graph represents the total cholesterol.

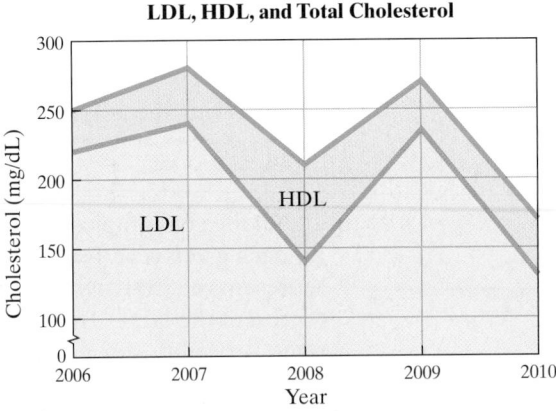

FIGURE 3.69

b) By looking at the graph, we see that the total cholesterol, indicated by the dark green line, was greater than or equal to 200 in 2006, 2007, 2008, and 2009.

c) By looking at the HDL area of the graph, we can see that in 2010 the HDL starts at around 130 and ends at about 170. If we subtract, we obtain $170 - 130 = 40$. Therefore, the HDL in 2010 is about 40. If we let H represent the amount of HDL, then $H(2010) \approx 40$.

d) By looking at the graph, we can determine that the only year in which the LDL was greater than or equal to 180 and the total cholesterol was less than or equal to 250 was 2006.

Now Try Exercise 63c

EXERCISE SET 3.6

Math XL
MathXL®

MyMathLab
MyMathLab

Warm-Up Exercises

Fill in the blanks with the appropriate word, phrase, or symbol(s) from the following list.

domain	range	$\dfrac{x^2}{2x-5}$	$x^2 + 2x - 5$	$x^2 - 2x + 5$	$x^2 - 2x - 5$

$2x^3 - 5x^2$	all real numbers	3	5	7	all real numbers except 3

For Exercises 1–6, let $f(x) = x^2$ and $g(x) = 2x - 5$.

1. $(f - g)(x) = $ _____.

2. $(f - g)(2) = $ _____.

3. $(f + g)(x) = $ _____.

4. $(f + g)(2) = $ _____.

5. $(f \cdot g)(x) = $ _____.

6. $(f/g)(x) = $ _____.

7. The domain of the function $f(x) = x - 3$ is _____.

8. The domain of the function $f(x) = \dfrac{1}{x - 3}$ is _____.

Practice the Skills

*For each pair of functions, find **a)** $(f + g)(x)$, **b)** $(f + g)(a)$, and **c)** $(f + g)(2)$.*

9. $f(x) = x + 5, g(x) = 3x - 2$

10. $f(x) = x^2 - x - 8, g(x) = x^2 + 1$

11. $f(x) = -3x^2 + x - 4, g(x) = x^3 + 3x^2$

12. $f(x) = 4x^3 + 2x^2 - x - 1, g(x) = x^3 - x^2 + 2x + 6$

13. $f(x) = 4x^3 - 3x^2 - x, g(x) = 3x^2 + 4$

14. $f(x) = 3x^2 - x + 2, g(x) = 6 - 4x^2$

Let $f(x) = x^2 - 4$ and $g(x) = -5x + 3$. Find the following.

15. $f(3) + g(3)$

16. $f(5) + g(5)$

17. $f(4) - g(4)$

18. $f\left(\dfrac{1}{4}\right) - g\left(\dfrac{1}{4}\right)$

19. $f(3) \cdot g(3)$

20. $f(-1) \cdot g(-1)$

21. $\dfrac{f\left(\dfrac{3}{5}\right)}{g\left(\dfrac{3}{5}\right)}$

22. $f(-1)/g(-1)$

23. $g(-3) - f(-3)$

24. $g(6) \cdot f(6)$

25. $g(0)/f(0)$

26. $f(2)/g(2)$

Let $f(x) = 2x^2 - x$ and $g(x) = x - 6$. Find the following.

27. $(f + g)(x)$

28. $(f + g)(a)$

29. $(f + g)(1)$

30. $(f + g)(-3)$

31. $(f - g)(-2)$

32. $(f - g)(1)$

33. $(f \cdot g)(0)$

34. $(f \cdot g)(3)$

35. $(f/g)(-1)$

36. $(f/g)(6)$

37. $(g/f)(5)$

38. $(g - f)(4)$

39. $(g - f)(x)$

40. $(g - f)(r)$

Problem Solving

Using the graph on the right, find the value of the following.

41. $(f + g)(0)$

42. $(f - g)(0)$

43. $(f \cdot g)(2)$

44. $(f/g)(1)$

45. $(g - f)(-1)$

46. $(g + f)(-3)$

47. $(g/f)(4)$

48. $(g \cdot f)(-1)$

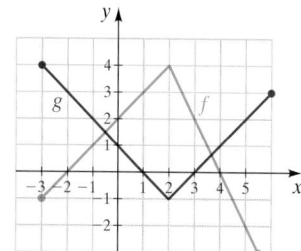

Using the graph on the right, find the value of the following.

49. $(f + g)(-2)$ **50.** $(f - g)(-1)$

51. $(f \cdot g)(1)$ **52.** $(g - f)(3)$

53. $(f/g)(4)$ **54.** $(g/f)(5)$

55. $(g/f)(2)$ **56.** $(g \cdot f)(0)$

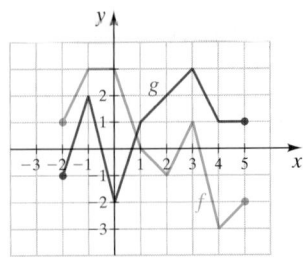

57. Retirement Account The following graph shows the amount of money Sharon and Frank Dangman have contributed to a joint retirement account for the years 2006 to 2010.

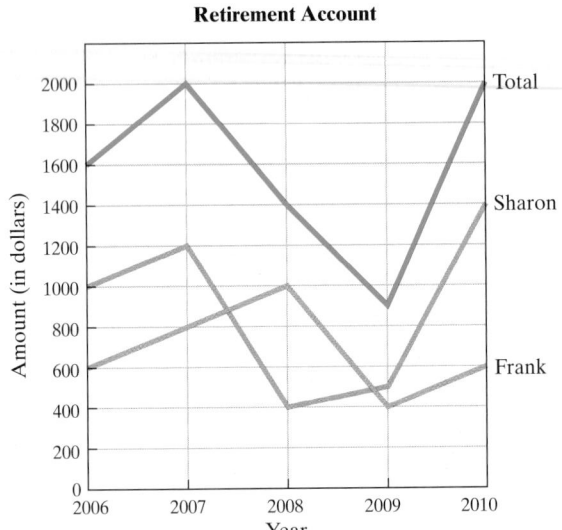

Retirement Account

a) In which year did Frank contribute $1000?

b) In 2010, estimate how much more Sharon contributed to the retirement account than Frank contributed.

c) For this five-year period, estimate the total amount Sharon and Frank contributed to the joint retirement account.

d) If Frank's income is a function $F(x)$ and Sharon's income is a function $S(x)$, estimate $(F + S)(2009)$.

58. College Employment Kelly Housman is a college student who works as a waitress on weekends and as a tutor during the week. The following graph shows her income for the months of August, September, October, November, and December.

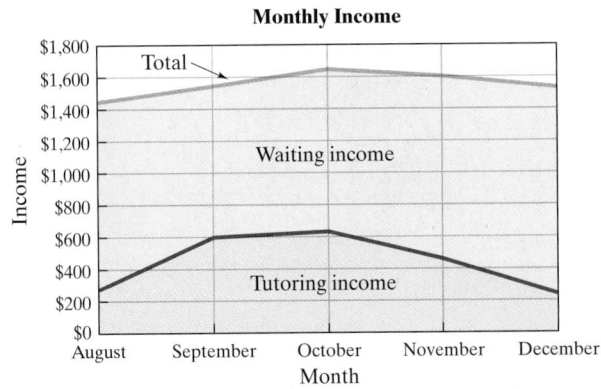

Monthly Income

a) In which month did Kelly earn a total of over $1600?

b) In September, estimate Kelly's income from waiting.

c) For this 5-month period, estimate Kelly's total income.

d) If the tutoring income is a function $T(x)$ and waiting income is a function $W(x)$, estimate $(W + T)(October)$.

59. Gender Makeup of Classes The following bar graph shows the total number of students in Dr. James Condor's Intermediate Algebra class for each semester shown. The blue bars on the bottom of the graph represent the number of female students, F, and the red bars on the top of the graph represent the number of male students, M.

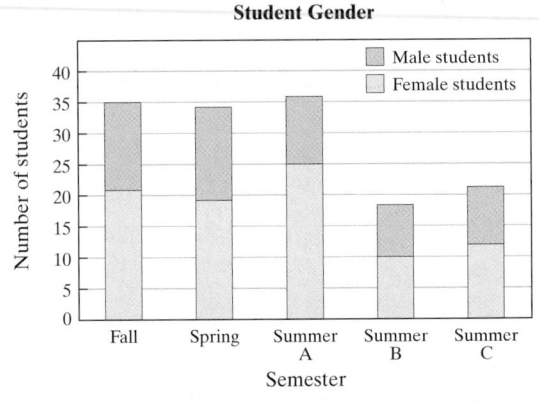

Student Gender

a) In what semester did Dr. Condor have the greatest number of Intermediate Algebra students? How many students were in that class?

b) In what semester did Dr. Condor have the fewest number of female students? How many female students did he have in class that semester?

c) Estimate $M(Summer\,A)$

d) Estimate $F(Spring)$

60. Global Population The following graph shows the projected total global population and the projected population of children 0–14 years of age from 2002 to 2050.

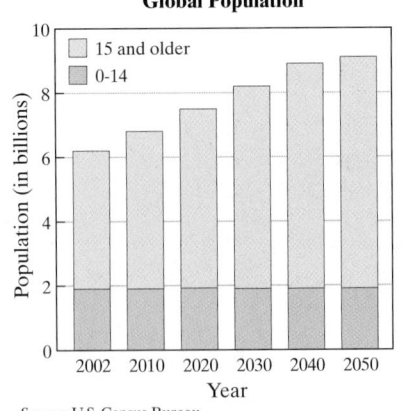

Global Population

Source: U.S. Census Bureau

a) Estimate the projected global population in 2050.

b) Estimate the projected number of children 0–14 years of age in 2050.

c) Estimate the projected number of people 15 years of age and older in 2050.

d) Estimate the projected difference in the total global population between 2002 and 2050.

61. House Sales In many regions of the country, houses sell better in the summer than at other times of the year. The graph below shows the total sales of houses in the town of Mineral Point from 2006 to 2010. The graph also shows the sale of houses in the summer, *S*, and in other times of the year, *Y*.

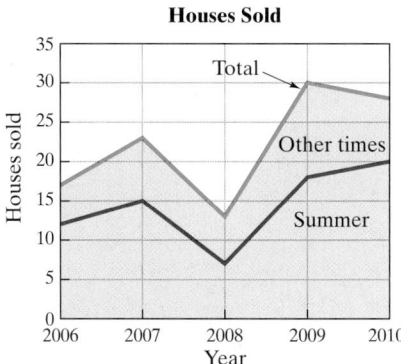

Houses Sold

Source: U.S. Census Bureau

a) Estimate the number of houses sold in the summer of 2010.

b) Estimate the number of houses sold at other times in 2010.

c) Estimate *Y* (2009).

d) Estimate (*S* + *Y*) (2007).

62. Income Rod Sac deCrasse owns a business where he does landscaping in the summer and snow removal in the winter. The graph below shows the total income, *T*, for the years 2006–2010 broken down into his landscaping income, *L*, and his snow removal income, *S*.

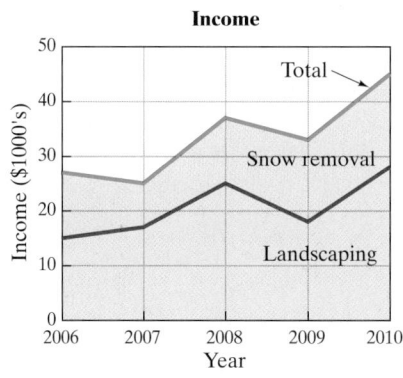

Income

a) Estimate the total income for 2010.

b) Estimate *L*(2006).

c) Estimate *S*(2009).

d) Estimate (*L* + *S*)(2007).

63. Income The chart below shows Mr. and Mrs. Abrams's income for the years from 2006 to 2010.

	2006	2007	2008	2009	2010
Mr. Abrams	$15,500	$17,000	$8,000	$25,000	$20,000
Mrs. Abrams	$4,500	$18,000	$28,000	$7,000	$22,500

a) Draw a line graph illustrating Mr. Abram's income, Mrs. Abram's income, and their total income for the years 2006–2010. See Example 3.

b) Draw a bar graph illustrating the given information. See Example 4.

c) Draw a stacked line graph illustrating the given information. See Example 5.

64. Telephone Bills The chart below shows Kelly Lopez's home telephone bills and cellular telephone bills (rounded to the nearest $10) for the years from 2006 to 2010.

	2006	2007	2008	2009	2010
Home	$40	$50	$60	$50	$0
Cellular	$80	$50	$20	$50	$60

a) Draw a line graph illustrating the home telephone bills, the cellular phone bills, and the total phone bills for the years 2006–2010.

b) Draw a bar graph illustrating the given information.

c) Draw a stacked line graph illustrating the given information.

65. Taxes Maria Cisneros pays both federal and state income taxes. The chart shows the amount of income taxes she paid to the federal government and to her state government for the years from 2006 to 2010.

	2006	2007	2008	2009	2010
Federal	$4000	$5000	$3000	$6000	$6500
State	$1600	$2000	$0	$1700	$1200

a) Draw a line graph illustrating the amount spent on federal taxes, the amount spent on state taxes, and the total amount spent on these two taxes for the years 2006–2010.

b) Draw a bar graph illustrating the given information.

c) Draw a stacked line graph illustrating the given information.

66. College Tuition The Olmert family has twin children, Justin and Kelly, who are attending different colleges. The tuition for Justin's and Kelly's colleges are given in the chart below for the years from 2006 to 2009.

	2006	2007	2008	2009
Justin	$12,000	$6000	$8000	$9000
Kelly	$2000	$8000	$8000	$5000

a) Draw a line graph illustrating the given information, including the total tuition spent on college for both Justin and Kelly for the years 2006–2009.

b) Draw a bar graph illustrating the given information.

c) Draw a stacked line graph illustrating the given information.

Concept/Writing Exercises

For Exercises 67–76, let f and g represent two functions that are graphed on the same axes.

67. What restriction is placed on the property $f(x)/g(x) = (f/g)(x)$? Explain.

68. Does $f(x) - g(x) = (f - g)(x)$ for all values of x? Explain.

69. Does $(f - g)(x) = (g - f)(x)$ for all values of x? Explain and give an example to support your answer.

70. Does $(f + g)(x) = (g + f)(x)$ for all values of x? Explain and give an example to support your answer.

71. If, at a, $(f + g)(a) = 0$, what must be true about $f(a)$ and $g(a)$?

72. If, at a, $(f \cdot g)(a) = 0$, what must be true about $f(a)$ and $g(a)$?

73. If, at a, $(f - g)(a) = 0$, what must be true about $f(a)$ and $g(a)$?

74. If, at a, $(f - g)(a) < 0$, what must be true about $f(a)$ and $g(a)$?

75. If, at a, $(f/g)(a) < 0$, what must be true about $f(a)$ and $g(a)$?

76. If, at a, $(f \cdot g)(a) < 0$, what must be true about $f(a)$ and $g(a)$?

Group Activity

77. SAT Scores The graph shows the average math and reading/verbal scores of entering college classes on the SAT college entrance exam for the years 2000 through 2006. Let f represent the math scores, and g represent the reading/verbal scores, and let t represent the year. As a group, draw a graph that represents $(f + g)(t)$.

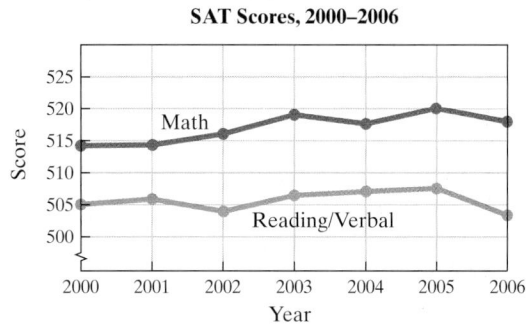

Cumulative Review Exercises

[1.5] **78.** Evaluate $(-4)^{-3}$.

[1.6] **79.** Express 2,960,000 in scientific notation.

[2.2] **80.** Solve the formula $A = \dfrac{1}{2}bh$ for h.

[2.3] **81. Washing Machine** The cost of a washing machine, including a 6% sales tax, is $477. Determine the pre-tax cost of the washing machine.

[3.1] **82.** Graph $y = |x| - 2$.

[3.3] **83.** Graph $3x - 4y = 12$.

3.7 Graphing Linear Inequalities

1 Graph linear inequalities in two variables.

1 Graph Linear Inequalities in Two Variables

Understanding Algebra

The following symbols are used in linear inequalities:

- $<$ is less than
- $>$ is greater than
- $\leq$ is less than or equal to
- $\geq$ is greater than or equal to

Linear Inequality in Two Variables

A **linear inequality in two variables** can be written in one of the following forms:

$$ax + by < c, \quad ax + by > c, \quad ax + by \leq c, \quad ax + by \geq c$$

where a, b, and c are real numbers and a and b are not both 0.

Examples of Linear Inequalities in Two Variables

$$2x + 3y > 2 \qquad\qquad 3y < 4x - 9$$

$$-x - 2y \leq 3 \qquad\qquad 5x \geq 2y - 7$$

Consider the graph of the equation $x + y = 3$ shown in **Figure 3.70**. The line acts as a **boundary** between two *half-planes* and divides the plane into three distinct sets of points: the points on the line itself, the points in the half-plane on one side of the line, and the points in the half-plane on the other side of the line.

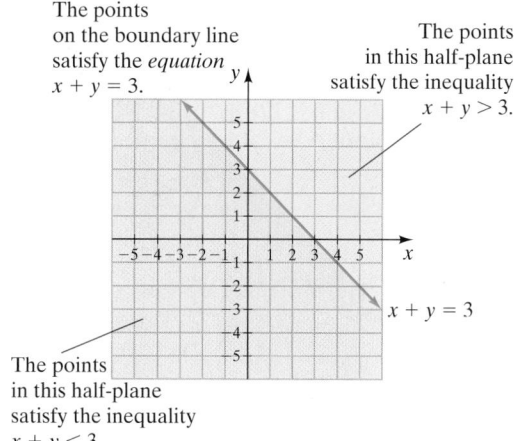

The points on the boundary line satisfy the *equation* $x + y = 3$.

The points in this half-plane satisfy the inequality $x + y > 3$.

The points in this half-plane satisfy the inequality $x + y < 3$.

FIGURE 3.70

Understanding Algebra

If a linear inequality has

- $<$ or $>$, draw a *dashed* boundary line.

- $\leq$ or $\geq$, draw a *solid* boundary line.

When we graph linear inequalities, we generally will shade only one of the two half-planes established by the boundary line. If the inequality is written using $<$ or $>$, we draw a dashed line to indicate that the boundary line is not part of the solution set. If the inequality is written using $\leq$ or $\geq$, we will draw a solid line to indicate that the boundary line is part of the solution set.

> **To Graph a Linear Inequality in Two Variables**
>
> 1. To get the equation of the boundary line, replace the inequality symbol with an equals sign.
> 2. Draw the graph of the equation in step 1. If the original inequality contains a $\geq$ or $\leq$ symbol, draw the boundary line using a solid line. If the original inequality contains a $>$ or $<$ symbol, draw the boundary line using a dashed line.
> 3. Select any point not on the boundary line and determine if this point is a solution to the original inequality. If the point selected is a solution, shade the half-plane on the side of the line containing this point. If the selected point does not satisfy the inequality, shade the half-plane on the side of the line not containing this point.

   In step 3 we are deciding which half-plane contains points that satisfy the given inequality.

EXAMPLE 1 Graph the inequality $y < \dfrac{2}{3}x - 3$.

Solution   First graph the equation $y = \dfrac{2}{3}x - 3$. Since the original inequality contains a less than sign, $<$, use a dashed line when drawing the graph (**Fig. 3.71**). The dashed line indicates that the points on this line are not solutions to the inequality $y < \dfrac{2}{3}x - 3$. Select a point not on the line and determine if this point satisfies the inequality. Often the easiest point to use is the origin, $(0, 0)$.

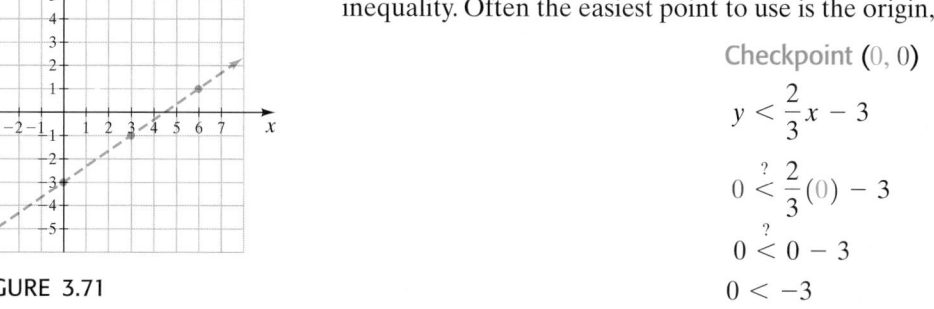

FIGURE 3.71

$$\text{Checkpoint } (0, 0)$$

$$y < \frac{2}{3}x - 3$$

$$0 \overset{?}{<} \frac{2}{3}(0) - 3$$

$$0 \overset{?}{<} 0 - 3$$

$$0 < -3 \qquad \text{False}$$

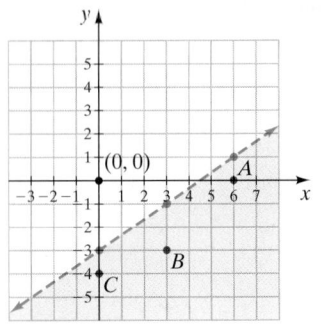

FIGURE 3.72

Since 0 is not less than -3, the point $(0, 0)$ does not satisfy the inequality. The solution will be all points in the half-plane that does *not* contain point $(0, 0)$. Shade in this half-plane (**Fig. 3.72**). Every point in the shaded half-plane satisfies the given inequality. Let's check a few selected points A, B, and C.

Point A	Point B	Point C
$(6, 0)$	$(3, -3)$	$(0, -4)$
$y < \dfrac{2}{3}x - 3$	$y < \dfrac{2}{3}x - 3$	$y < \dfrac{2}{3}x - 3$
$0 \overset{?}{<} \dfrac{2}{3}(6) - 3$	$-3 \overset{?}{<} \dfrac{2}{3}(3) - 3$	$-4 \overset{?}{<} \dfrac{2}{3}(0) - 3$
$0 \overset{?}{<} 4 - 3$	$-3 \overset{?}{<} 2 - 3$	$-4 \overset{?}{<} 0 - 3$
$0 < 1$ True	$-3 < -1$ True	$-4 < -3$ True

Now Try Exercise 9

EXAMPLE 2 Graph the inequality $y \geq -\dfrac{1}{2}x$.

Solution   First, we graph the equation $y = -\dfrac{1}{2}x$. Since the inequality is $\geq$, we use a solid boundary line to indicate that the points on the line are solutions to the inequality (**Fig. 3.73**). Since the point $(0, 0)$ is on the line, we cannot select that point as our checkpoint. Let's select the point $(3, 1)$.

$$\text{Checkpoint } (3, 1)$$

$$y \geq -\frac{1}{2}x$$

$$1 \overset{?}{\geq} -\frac{1}{2}(3)$$

$$1 \geq -\frac{3}{2} \qquad \text{True}$$

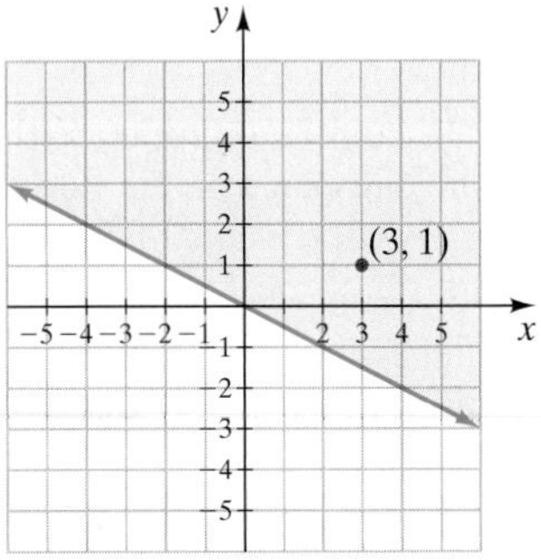

Since the point $(3, 1)$ satisfies the inequality, every point in the same half-plane as $(3, 1)$ will also satisfy the inequality $y \geq -\dfrac{1}{2}x$. Shade this half-plane as indicated.

Every point in the shaded half-plane, as well as every point on the boundary line, satisfies the inequality.

Now Try Exercise 19

FIGURE 3.73

EXAMPLE 3 Graph the inequality $3x - 2y < -6$.

Solution   First, we graph the equation $3x - 2y = -6$. Since the inequality is $<$, we use a dashed boundary line (**Fig. 3.74**). Substituting the checkpoint $(0, 0)$ into the inequality results in a false statement.

$$\text{Checkpoint } (0, 0)$$

$$3x - 2y < -6$$

$$3(0) - 2(0) \overset{?}{<} -6$$

$$0 < -6 \qquad \text{False}$$

The solution is, therefore, the half-plane that does not contain the origin.

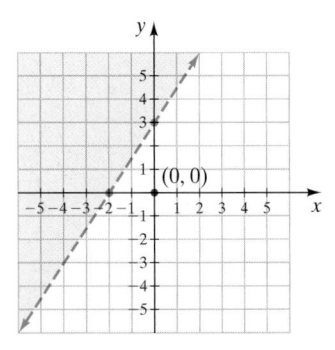

FIGURE 3.74

Now Try Exercise 17

📟 *Using Your Graphing Calculator* ⬛⬛⬛

We will graph the inequality $3x - 2y < -6$ that we graphed in Example 3. First, solve the inequality for y to get $y > \frac{3}{2}x + 3$.

We begin by entering the equation of the boundary line $y = \frac{3}{2}x + 3$. **Figure 3.75a** shows the TI-84 Plus screen.

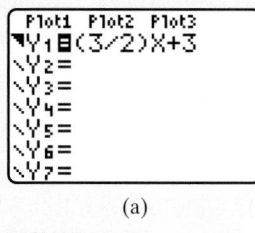   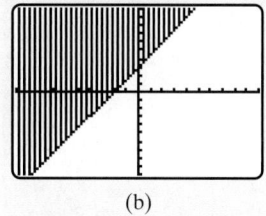

(a)   (b)

FIGURE 3.75

Notice the symbol to the left of Y_1. The symbol indicates the shading will go above the boundary line since the inequality used is $>$. To get this symbol, use the left arrow key until the cursor is in this position on the screen. Then press ENTER until the symbol appears. After pressing GRAPH the window shows the graph in **Figure 3.75b**. Compare **Figure 3.75b** to **Figure 3.74**. A word of caution: notice that the window does not show a dashed boundary line.

⬛⬛⬛

EXERCISE SET 3.7   MathXL®   MyMathLab

Warm-Up Exercises

Fill in the blanks with the appropriate word, phrase, or symbol(s) from the following list.

solid   checkpoint   boundary line   dashed   half-plane

1. When graphing a linear inequality, the _____ divides the plane into two half-planes.

2. If the linear inequality contains $<$ or $>$ draw a _____ boundary line.

3. If the linear inequality contains $\leq$ or $\geq$ draw a _____ boundary line.

4. To determine which half-plane to shade, pick a _____ that is not on the boundary line.

Practice the Skills

Graph each inequality.

5. $y < 2x + 1$

6. $y \geq 3x - 1$

7. $y > 2x - 1$

8. $y \leq -x + 4$

9. $y \geq \frac{1}{2}x - 3$

10. $y < 3x + 2$

11. $2x + 3y > 6$

12. $2x - 3y \geq 12$

**13.** $y \leq -3x + 5$

14. $y \leq \frac{2}{3}x + 3$

**15.** $2x + y < 4$

16. $3x - 4y \leq 12$

17. $10 \geq 5x - 2y$

18. $-x - 2y > 4$

19. $y \geq -\frac{1}{2}x$

20. $y < \frac{1}{2}x$

21. $y < \frac{2}{3}x$

22. $y \geq -\frac{3}{2}x$

23. $y < -2$

24. $y < x$

25. $x > 1$

26. $x \geq 4$

27. Stock-Car Races Patrick Cunningham is taking some friends and their families to the stock-car races. Tickets cost $8 for children and $15 for adults and Patrick only has $175 to spend. Let x represent the number of children's tickets purchased and let y represent the number of adults' tickets purchased.

a) Write a linear inequality in which the total cost for the tickets is less than or equal to $175.

b) Does Patrick have enough money to purchase tickets for 8 children and 6 adults?

c) Does Patrick have enough money to purchase tickets for 10 children and 8 adults?

28. Canoe Weight John and Robyn Pearse are using a canoe to carry jugs of water and boxes of food to flood victims. Their canoe has a maximum weight load of 800 pounds. John weighs 175 pounds and Robyn weighs 145 pounds. Each jug of water weighs 8.5 pounds and each box of food weighs 12 pounds. Let x represent the number of jugs of water and let y represent the number of boxes of food.

a) Write a linear inequality in which the total weight in the canoe, including John and Robyn, is less than or equal to 800 pounds

b) With John and Robyn both in the canoe, can they carry 20 jugs of water and 20 boxes of food without exceeding the canoe's weight limit?

c) With John and Robyn both in the canoe, can they carry 25 jugs of water and 25 boxes of food without exceeding the canoe's weight limit?

29. a) Graph $f(x) = 2x - 4$.

b) On the graph, shade the region bounded by $f(x)$, $x = 2$, $x = 4$, and the x-axis.

30. a) Graph $g(x) = -x + 4$.

b) On the graph, shade the region bounded by $g(x)$, $x = 1$, and the x- and y-axes.

Concept/Writing Exercises

31. When graphing an inequality containing $>$ or $<$, why are points on the line not solutions to the inequality?

32. When graphing an inequality containing $\geq$ or $\leq$, why are points on the line solutions to the inequality?

33. When graphing a linear inequality, when can $(0, 0)$ not be used as a checkpoint?

34. When graphing a linear inequality of the form $y > ax + b$ where a and b are real numbers, will the solution always be above the line? Explain.

Challenge Problems

Graph each inequality.

35. $y < |x|$

36. $y \geq x^2$

37. $y < x^2 - 4$

Cumulative Review Exercises

[2.1] **38.** Solve the equation $9 - \dfrac{5x}{3} = -6$.

[2.2] **39.** If $C = \bar{x} + Z\dfrac{\sigma}{\sqrt{n}}$, find C when $\bar{x} = 80$, $Z = 1.96$, $\sigma = 3$, and $n = 25$.

[2.3] **40. Store Sale** Olie's Records and Stuff is going out of business. The first week all items are being reduced by 10%. The second week all items are being reduced by an additional $2. If during the second week Bob Frieble purchases a CD for $12.15, find the original cost of the CD.

[3.2] **41.** $f(x) = -x^2 + 5$; find $f(-3)$.

[3.3] **42.** Write an equation of the line that passes through the point $(8, -2)$ and is perpendicular to the line whose equation is $2x - y = 4$.

[3.4] **43.** Determine the slope of the line through $(-2, 7)$ and $(2, -1)$.

Chapter 3 Summary

IMPORTANT FACTS AND CONCEPTS	EXAMPLES

Section 3.1

The **Cartesian** (or **rectangular**) **coordinate** system consists of two axes drawn perpendicular to each other. The **x-axis** is the horizontal axis. The **y-axis** is the vertical axis. The **origin** is the point of intersection of the two axes. The two axes yield four **quadrants** (I, II, III, IV). An **ordered pair (x, y)** is used to give the two coordinates of a point.

Plot the following points on the same set of axes.

$$A(2,3), B(-2,3), C(-4,-1), D(3,-3), E(5,0)$$

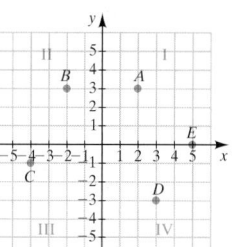

A **graph of an equation** is an illustration of the set of points whose coordinates satisfy the equation. Points on the same straight line are said to be **collinear**.

A **linear equation** is an equation whose graph is a straight line. A linear equation is also called a **first-degree equation**.

$y = 2x + 1$ is a linear equation whose graph is illustrated below.

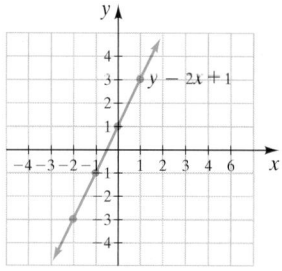

The points $(1,3), (0,1), (-1,-1),$ and $(-2,-3)$ are collinear.

A **nonlinear equation** is an equation whose graph is not a straight line.

$y = x^2 + 2$ is a nonlinear equation whose graph is illustrated below.

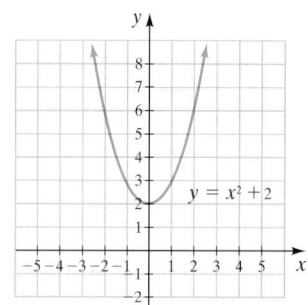

Section 3.2

For an equation in variables x and y, if the value of y depends on the value of x, then y is the **dependent variable** and x is the **independent variable**.

In the equation $y = 2x^2 + 3x - 4$, x is the independent variable and y is the dependent variable.

A **relation** is any set of ordered pairs of the form (x, y).

The set of x-coordinates of a relation is the **domain**. The set of y-coordinates of a relation is the **range**.

A **function** is a relation in which each element in the domain corresponds to exactly one element in the range.

Alternate Definition:

A **function** is a set of ordered pairs in which no first coordinate is repeated.

$\{(1,2),(2,3),(1,4)\}$ is a relation, but not a function.

$\{(1,6),(2,7),(3,10)\}$ is a relation. It is also a function since each element in the domain corresponds to exactly one element in the range.

$$\text{domain: } \{1,2,3\}, \text{ range } = \{6,7,10\}$$

IMPORTANT FACTS AND CONCEPTS	EXAMPLES

Section 3.2 (cont.)

The **vertical line test** can be used to determine if a graph represents a function.

If a vertical line can be drawn through any part of the graph and the line intersects another part of the graph, the graph does not represent a function. If a vertical line cannot be drawn to intersect the graph at more than one point, the graph represents a function.

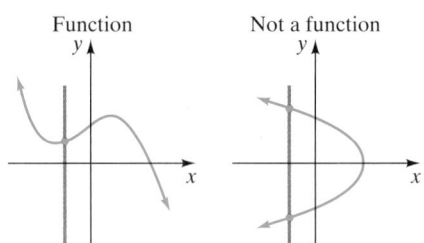

Function notation can be used to write an equation when y is a function of x. For function notation, replace y with $f(x)$, $g(x)$, $h(x)$, and so on.

$y = 7x - 9$ can be written as $f(x) = 7x - 9$

Given $y = f(x)$, to find $f(a)$, replace each x with a.

Let
$$f(x) = x^2 + 2x - 8.$$
Then
$$f(1) = 1^2 + 2(1) - 8 = -5$$
$$f(a) = a^2 + 2a - 8.$$

Section 3.3

A **linear function** is a function of the form $f(x) = ax + b$. The graph of a linear function is a straight line.

Graph $f(x) = \dfrac{1}{3}x - 2$.

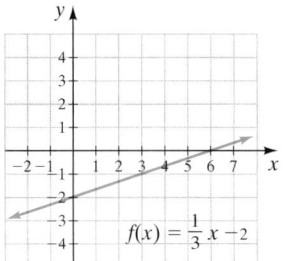

The **standard form of a linear equation** is $ax + by = c$, where a, b, and c are real numbers, and a and b are not both zero.

$$3x + 5y = 7, \qquad -2x + \frac{1}{3}y = \frac{1}{8}$$

The **x-intercept** is the point where the graph crosses the x-axis.

To find the x-intercept, set $y = 0$ and solve for x.

The **y-intercept** is the point where the graph crosses the y-axis.

To find the y-intercept, set $x = 0$ and solve for y.

Graph $2x - 3y = 12$ using the x- and y- intercepts.

For x-intercept, let $y = 0$.

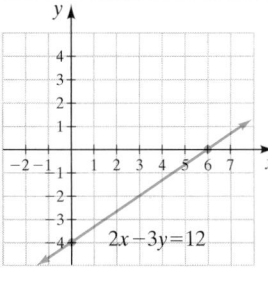

$$2x - 3y = 12$$
$$2x - 3(0) = 12$$
$$2x = 12$$
$$x = 6$$

Therefore, the x-intercept is $(6, 0)$.

For y-intercept, let $x = 0$.

$$2x - 3y = 12$$
$$2(0) - 3y = 12$$
$$-3y = 12$$
$$y = -4$$

Therefore, the y-intercept is $(0, -4)$.

IMPORTANT FACTS AND CONCEPTS	EXAMPLES

Section 3.3 (cont.)

The graph of any equation of the form $y = b$ (or function of the form $f(x) = b$) will always be a horizontal line for any real number b. The function $f(x) = b$ is called the **constant function**.	Graph $y = 5$ (or $f(x) = 5$). 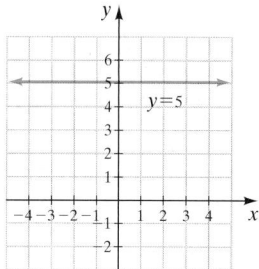
The graph of any equation of the form $x = a$ will always be a vertical line for any real number a.	Graph $x = -3.5$. 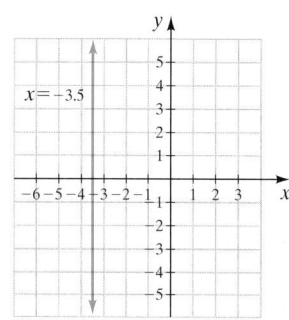

Section 3.4

The **slope of a line** is the ratio of the vertical change (or rise) to the horizontal change (or run) between any two points. The slope of the line through the distinct points (x_1, y_1) and (x_2, y_2) is $$m = \text{slope} = \frac{\text{rise}}{\text{run}} = \frac{\text{change in } y \text{ (vertical change)}}{\text{change in } x \text{ (horizontal change)}} = \frac{y_2 - y_1}{x_2 - x_1}$$ provided that $x_1 \neq x_2$.	The slope of the line through $(-1, 3)$ and $(7, 5)$ is $$m = \frac{5 - 3}{7 - (-1)} = \frac{2}{8} = \frac{1}{4}.$$
A line that rises from left to right has a **positive slope**. A line that falls from left to right has a **negative slope**. A horizontal line has **zero slope**. The slope of a vertical line is **undefined**.	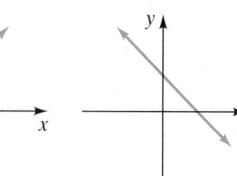 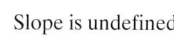 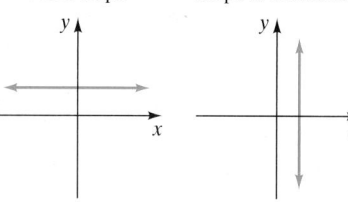
The **slope-intercept form of a linear equation** is $$y = mx + b$$ where m is the slope of the line and $(0, b)$ is the y-intercept of the line.	$y = 7x - 1$, $y = -3x + 10$

IMPORTANT FACTS AND CONCEPTS	EXAMPLES

Section 3.5

The **point-slope form of a linear equation** is

$$y - y_1 = m(x - x_1)$$

where m is the slope of the line and (x_1, y_1) is a point on the line.

If $m = 9$ and (x_1, y_1) is $(5, 2)$, then

$$y - 2 = 9(x - 5)$$

Two lines are **parallel** if they have the same slope.

Two lines are **perpendicular** if their slopes are negative reciprocals. For any real number $a \neq 0$, its negative reciprocal is $-\dfrac{1}{a}$.

The graphs of $y = 2x + 4$ and $y = 2x + 7$ are parallel since both graphs have the same slope, 2, but different y-intercepts.

The graphs of $y = 3x - 5$ and $y = -\dfrac{1}{3}x + 8$ are perpendicular since one graph has a slope of 3 and the other graph has a slope of $-\dfrac{1}{3}$. The number $-\dfrac{1}{3}$ is the negative reciprocal of 3.

Section 3.6

Operations on Functions

Sum of functions: $(f + g)(x) = f(x) + g(x)$

Difference of functions: $(f - g)(x) = f(x) - g(x)$

Product of functions: $(f \cdot g)(x) = f(x) \cdot g(x)$

Quotient of functions: $(f/g)(x) = \dfrac{f(x)}{g(x)}, g(x) \neq 0$

If $f(x) = x^2 + 2x - 5$ and $g(x) = x - 3$, then

$$\begin{aligned}(f + g)(x) = f(x) + g(x) &= (x^2 + 2x - 5) + (x - 3) \\ &= x^2 + 3x - 8\end{aligned}$$

$$\begin{aligned}(f - g)(x) = f(x) - g(x) &= (x^2 + 2x - 5) - (x - 3) \\ &= x^2 + x - 2\end{aligned}$$

$$\begin{aligned}(f \cdot g)(x) &= f(x) \cdot g(x) \\ &= (x^2 + 2x - 5)(x - 3) \\ &= x^3 - x^2 - 11x + 15\end{aligned}$$

$$(f/g)(x) = \dfrac{f(x)}{g(x)} = \dfrac{x^2 + 2x - 5}{x - 3}, \quad x \neq 3$$

Section 3.7

A **linear inequality in two variables** can be written in one of the following forms:

$$ax + by < c, ax + by > c, ax + by \leq c, ax + by \geq c$$

where a, b, and c are real numbers and a and b are not both 0.

$$3x - 4y > 1, \quad 2x + 5y \leq -4$$

To Graph a Linear Inequality in Two Variables

1. To get the equation of the boundary line, replace the inequality symbol with an equals sign.

2. Draw the graph of the equation in step 1. If the original inequality is $\geq$ or $\leq$ draw a solid boundary line. If the original inequality is a $>$ or $<$ draw a dashed boundary line.

3. Select any point not on the boundary line. If the point selected is a solution, shade the half-plane on the side of the line containing this point. If the selected point does not satisfy the inequality, shade the half-plane on the side of the line not containing this point.

Graph $y > -x + 1$.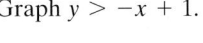

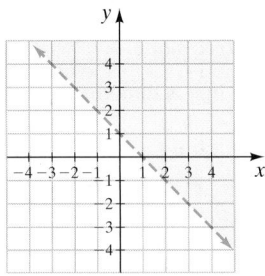

Chapter 3 Review Exercises

[3.1]   **1.** Plot the ordered pairs on the same axes.

    a) $A(5, 3)$ **b)** $B(0, -3)$ **c)** $C\left(5, \dfrac{1}{2}\right)$ **d)** $D(-4, 2)$ **e)** $E(-6, -1)$ **f)** $F(-2, 0)$

Graph each equation.

2. $y = \dfrac{1}{2}x$ **3.** $y = -2x - 1$ **4.** $y = \dfrac{1}{2}x + 3$ **5.** $y = -\dfrac{3}{2}x + 1$ **6.** $y = x^2$

7. $y = x^2 - 1$ **8.** $y = |x|$ **9.** $y = |x| - 1$ **10.** $y = x^3$ **11.** $y = x^3 + 4$

[3.2]   **12.** Define function.

13. Is every relation a function? Is every function a relation? Explain.

Determine whether the following relations are functions.

14.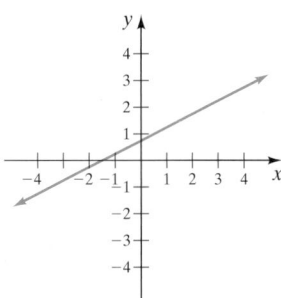

15. $\{(1, 1), (1, -1), (4, 2), (-4, -2)\}$

For Exercises 16–19, **a)** *determine whether the following graphs represent functions;* **b)** *determine the domain and range of each relation or function.*

16.

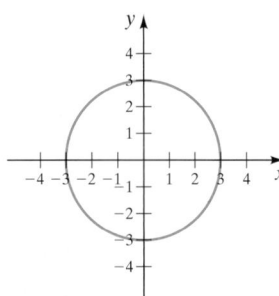

17.

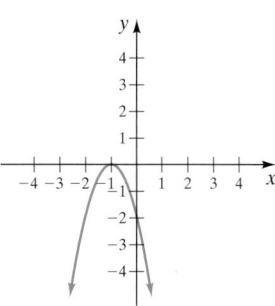

18.

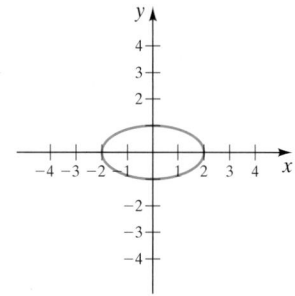

19.

20. If $f(x) = 3x - 7$, find

   a) $f(2)$ and

   b) $f(h)$.

21. If $g(t) = 2t^3 - 3t^2 + 6$, find

   a) $g(-1)$ and

   b) $g(a)$.

22. Speed of Car Deb Exum goes for a ride in a car. The following graph shows the car's speed as a function of time. Make up a story that corresponds to this graph.

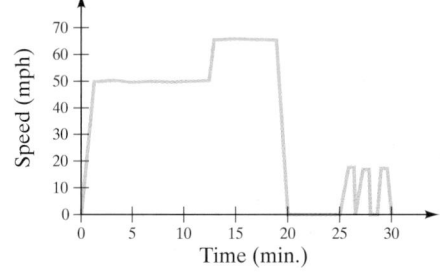

23. Apple Orchard The number of baskets of apples, N, that are produced by x trees in a small orchard ($x \leq 100$) is given by the function $N(x) = 40x - 0.2x^2$. How many baskets of apples are produced by

   a) 30 trees?

   b) 50 trees?

24. Falling Ball If a ball is dropped from the top of a 196-foot building, its height above the ground, h, at any time, t, can be found by the function $h(t) = -16t^2 + 196, 0 \leq t \leq 3.5$. Find the height of the ball at

   a) 1 second.

   b) 3 seconds.

[3.3]   *Graph each equation using intercepts.*

25. $3x - 4y = 6$

26. $\dfrac{1}{3}x = \dfrac{1}{8}y + 10$

Graph each equation or function.

27. $f(x) = 4$

28. $x = -2$

29. Bagel Company The yearly profit, p, of a bagel company can be estimated by the function $p(x) = 0.1x - 5000$, where x is the number of bagels sold per year.

   a) Draw a graph of profits versus bagels sold for up to and including 250,000 bagels.

   b) Estimate the number of bagels that must be sold for the company to break even.

   c) Estimate the number of bagels sold if the company has $22,000 profit.

30. Interest Draw a graph illustrating the interest on a $12,000 loan for a 1-year period for various interest rates up to and including 20%. Use interest = principal · rate · time.

[3.4]   *Determine the slope and y-intercept of the graph represented by the given equation.*

31. $y = \dfrac{1}{2}x + 6$

32. $f(x) = -2x + 3$

33. $3x + 5y = 13$

34. $3x + 4y = 10$

35. $x = -7$

36. $f(x) = 8$

Determine the slope of the line through the two given points.

37. $(-1, 3), (2, 9)$

38. $(-2, 3)(4, 1)$

Find the slope of each line. If the slope is undefined, so state. Then write the equation of the line.

39.

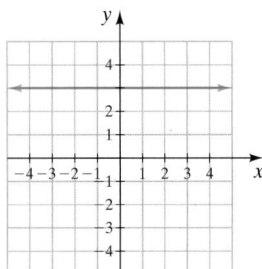

40.

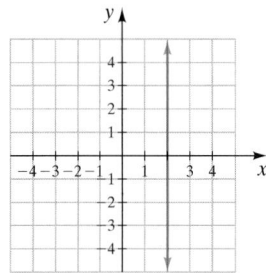

41.

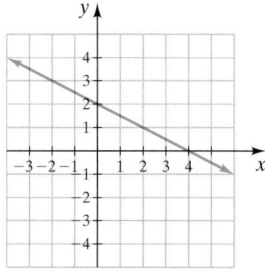

42. If the graph of $y = -2x + 5$ is translated down 4 units, determine

   a) the slope of the translated graph.

   b) the y-intercept of the translated graph.

   c) the equation of the translated graph.

43. If one point on a graph is $(-6, -4)$ and the slope is $\dfrac{2}{3}$, find the y-intercept of the graph.

44. Typhoid Fever The following chart shows the number of reported cases of typhoid fever in the United States for select years from 1970 through 2000.

   a) Plot each point and draw line segments from point to point.

   b) Compute the slope of the line segments.

   c) During which 10-year period did the number of reported cases of typhoid fever increase the most?

Year	Number of reported typhoid fever cases
1970	346
1980	510
1990	552
2000	317

Source: U.S. Dept. of Health and Human Services

45. Social Security The following graph shows the number of Social Security beneficiaries from 1980 projected through 2070. Use the slope-intercept form to find the function $n(t)$ (represented by the red dashed line) that can be used to approximate this data.

Social Security Beneficiaries

[3.5]   *Determine whether the two given lines are parallel, perpendicular, or neither.*

46. $-3x + 4y = 7$
$\qquad y = \dfrac{3}{4}x + 7$

47. $2x - 3y = 7$
$\qquad -3x - 2y = 8$

48. $4x - 2y = 13$
$\qquad -2x + 4y = -9$

Find the equation of the line with the properties given. Write each answer in slope–intercept form.

49. Slope $= \frac{1}{2}$, through $(-2, 1)$

50. Through $(-3, 1)$ and $(4, -6)$

51. Through $(0, 6)$ and parallel to the graph of $y = -\frac{2}{3}x + 1$

52. Through $(2, 8)$ and parallel to the graph whose equation is $5x - 2y = 7$

53. Through $(-3, 1)$ and perpendicular to the graph whose equation is $y = \frac{3}{5}x + 5$

54. Through $(4, 5)$ and perpendicular to the graph whose equation is $4x - 2y = 8$

Two points on l_1 and two points on l_2 are given. Determine whether l_1 is parallel to l_2, l_1 is perpendicular to l_2, or neither.

55. l_1: $(5, 3)$ and $(0, -3)$; l_2: $(1, -1)$ and $(2, -2)$

56. l_1: $(3, 2)$ and $(2, 3)$; l_2: $(4, 1)$ and $(1, 4)$

57. l_1: $(7, 3)$ and $(4, 6)$; l_2: $(5, 2)$ and $(6, 3)$

58. l_1: $(-3, 5)$ and $(2, 3)$; l_2: $(-4, -2)$ and $(-1, 2)$

59. Insurance Rates The monthly rates for $100,000 of life insurance from the General Financial Group for men increases approximately linearly from age 35 through age 50. The rate for a 35-year-old man is $10.76 per month and the rate for a 50-year-old man is $19.91 per month. Let r be the rate and let a be the age of a man between 35 and 50 years of age.

    a) Determine a linear function $r(a)$ that fits these data.

    b) Using the function in part **a)**, estimate the monthly rate for a 40-year-old man.

60. Burning Calories The number of calories burned in 1 hour of swimming, when swimming between 20 and 50 yards per minute, is a linear function of the speed of the swimmer. A person swimming at 30 yards per minute will burn about 489 calories in 1 hour. While swimming at 50 yards per minute a person will burn about 525 calories in 1 hour. This information is shown in the following graph.

Calories Burned while Swimming

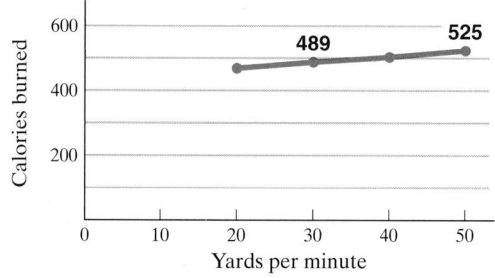

Source: Health Magazine Web Site, www.health.com

    a) Determine a linear function that can be used to estimate the number of calories, C, burned in 1 hour when a person swims at r yards per minute.

    b) Use the function determined in part **a)** to determine the number of calories burned in 1 hour when a person swims at 40 yards per minute.

    c) Use the function determined in part **a)** to estimate the speed at which a person needs to swim to burn 600 calories in 1 hour.

[3.6] *Given $f(x) = x^2 - 3x + 4$ and $g(x) = 2x - 5$, find the following.*

61. $(f + g)(x)$ **62.** $(f + g)(4)$

63. $(g - f)(x)$ **64.** $(g - f)(-1)$

65. $(f \cdot g)(-1)$ **66.** $(f \cdot g)(3)$

67. $(f/g)(1)$ **68.** $(f/g)(2)$

69. Female Population According to the U.S. Census, the female population is expected to grow worldwide. The following graph shows the female population worldwide for selected years from 2002 to 2050.

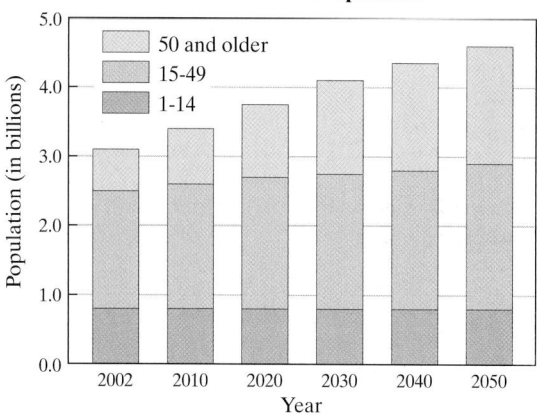

Source: U.S. Census Bureau

    a) Estimate the projected female population worldwide in 2050.

    b) Estimate the projected number of women 15–49 years of age in 2050.

    c) Estimate the number of women who are projected to be in the 50 years and older age group in 2010.

    d) Estimate the projected percent increase in the number of women 50 years and older from 2002 to 2010.

70. Retirement Income Joni Burnette recently retired from her full-time job. The following graph shows her retirement income for the years 2007–2010.

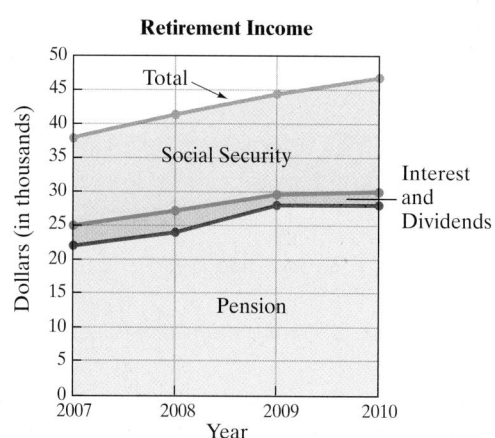

    a) Estimate Joni's total retirement income in 2010.

    b) Estimate Joni's pension income in 2009.

    c) Estimate Joni's interest and dividend income in 2007.

Graph each inequality.

71. $y \geq -5$ **72.** $x < 4$ **73.** $y \leq 4x - 3$ **74.** $y < \dfrac{1}{3}x - 2$

Chapter 3 Practice Test

CHAPTER
Test Prep
VIDEOS

Chapter Test Prep Videos provide fully worked-out solutions to any of the exercises you want to review. *Chapter Test Prep Videos are available via* **MyMathLab** *, or on* **YouTube** *(search "Angel Intermediate Algebra" and click on "Channels")*

1. Graph $y = -2x + 1$.

2. Graph $y = \sqrt{x}$.

3. Graph $y = x^2 - 4$.

4. Graph $y = |x|$.

5. Define *function*.

6. Is the following set of ordered pairs a function? Explain your answer.

$$\{(3,1), (-2,6), (4,6), (5,2), (7,3)\}$$

In Exercises 7 and 8, determine whether the graphs represent functions. Give the domain and range of the relation or function.

7.

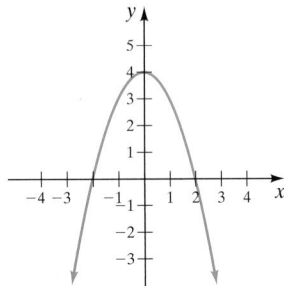

8.

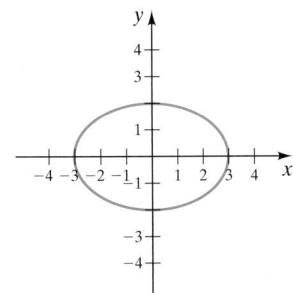

9. If $f(x) = 3x^2 - 6x + 5$, find $f(-2)$.

In Exercises 10 and 11, graph the equation using the x- and y-intercepts.

10. $-20x + 10y = 40$

11. $\dfrac{x}{5} - \dfrac{y}{4} = 1$

12. Graph $f(x) = -3$

13. Graph $x = 4$.

14. **Profit Graph** The yearly profit, p, for Zico Publishing Company on the sales of a particular book can be estimated by the function $p(x) = 10.2x - 50,000$, where x is the number of books produced and sold.

   a) Draw a graph of profit versus books sold for up to and including 30,000 books.

   b) Use the function $p(x)$ to estimate the number of books that must be sold for the company to break even.

   c) Use the function $p(x)$ to estimate the number of books the company must sell to make a $100,000 profit.

15. Determine the slope and y-intercept of the graph of the equation $4x - 3y = 15$

16. Write the equation, in slope-intercept form, of the line that goes through the points $(3, 2)$ and $(4, 5)$.

17. Determine the equation, in slope-intercept form, of the line that goes through the point $(6, -5)$ and is perpendicular to the graph of $y = \dfrac{1}{2}x + 1$.

18. **U.S. Population** Determine the function represented by the red line on the graph that can be used to estimate the projected U.S. population, p, from 2000 to 2050. Let 2000 be the reference year so that 2000 corresponds to $t = 0$.

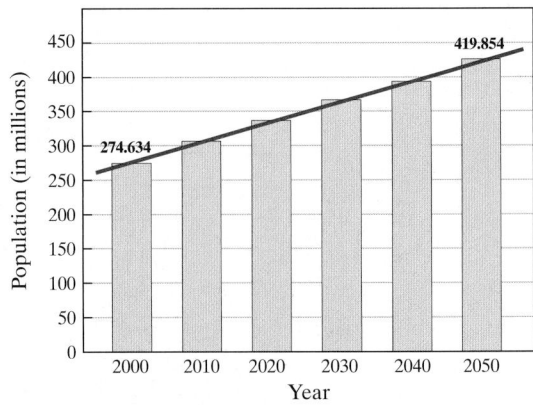

U.S. Population Projections 2000–2050

419.854

274.634

Source: U.S. Bureau of the Census

19. Determine whether the graphs of the two equations are parallel, perpendicular, or neither. Explain your answer.

$$2x - 3y = 12$$
$$4x + 10 = 6y$$

20. Heart Disease Deaths due to heart disease has been declining approximately linearly since the year 2000. The bar graph below shows the number of deaths, per 100,000 deaths, due to heart disease in selected years from 2000 to 2010.

Heart Disease Death Rate

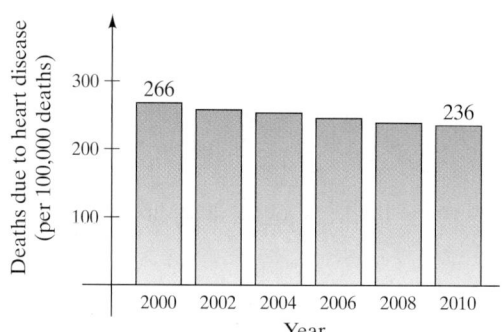

Source: U.S. Dept. of Health and Human Services

a) Let r be the number of deaths due to heart disease per 100,000 deaths, and let t represent the years since 2000. Write the linear function $r(t)$ that can be used to approximate the data.

b) Use the function from part **a)** to estimate the death rate due to heart disease in 2006.

c) Assuming this trend continues until the year 2020, estimate the death rate due to heart disease in 2020.

In Exercises 21–23, if $f(x) = 2x^2 - x$ and $g(x) = x - 6$, find

21. $(f + g)(3)$

22. $(f/g)(-1)$

23. $f(a)$

24. Paper Use The following graph shows paper use from 1995 and projected through 2015.

Paper Use

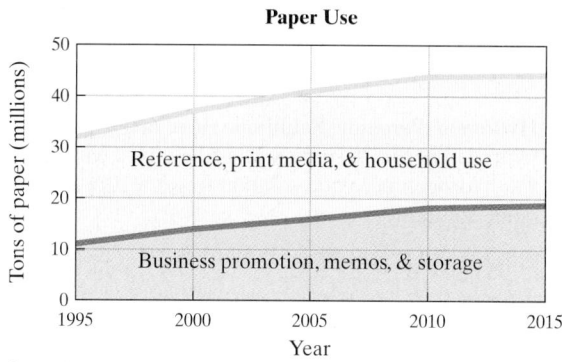

Source: CAP Ventures

a) Estimate the total number of tons of paper to be used in 2010.

b) Estimate the number of tons of paper to be used by businesses in 2010.

c) Estimate the number of tons of paper to be used for reference, print media, and household use in 2010.

25. Graph $y < 3x - 2$.

Cumulative Review Test

Take the following test and check your answers with those given in the back of the book. Review any questions that you answered incorrectly. The section where the material was covered is indicated after the answer.

1. For $A = \{1, 3, 5, 7, 9\}$ and $B = \{2, 3, 5, 7, 11, 14\}$, determine

a) $A \cap B$.

b) $A \cup B$.

2. Consider the set $\{-6, -4, \frac{1}{3}, 0, \sqrt{3}, 4.67, \frac{37}{2}, -\sqrt{5}\}$. List the elements of the set that are

a) natural numbers.

b) real numbers.

3. Evaluate $10 - \{3[6 - 4(6^2 \div 4)]\}$.

Simplify.

4. $\left(\dfrac{5x^2}{y^{-3}}\right)^2$

5. $\left(\dfrac{3x^4 y^{-2}}{6xy^3}\right)^3$

6. Natural Gas Consumption The total consumption of natural gas in 2008 was 21.8 trillion cubic feet (2.18×10^{13}). The following pie chart shows the breakdown of consumption by sector.

Natural Gas Consumption by Sector
(2.18×10^{13} cubic feet)

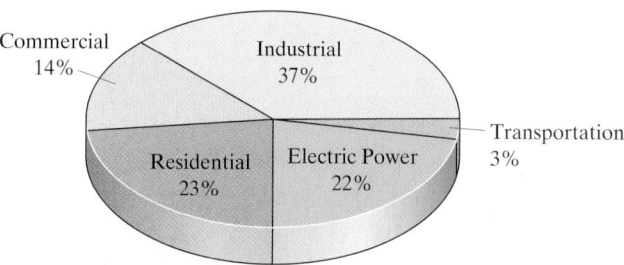

(Total is 99% due to rounding.)

Source: Energy Information Administration

Answer the following questions using scientific notation.

a) What was the amount of natural gas consumption by the commercial sector in 2008?

b) How much more natural gas was consumed by the industrial sector than by the transportation sector in 2008?

c) If the consumption of natural gas is expected to increase by a total of 10% from 2008 to 2011, what will be the consumption of natural gas in 2011?

In Exercises 7 and 8, solve the equations.

7. $2(x + 4) - 5 = -3[x - (2x + 1)]$

8. $\dfrac{4}{5} - \dfrac{x}{3} = 10$

9. Simplify $7x - \{4 - [2(x - 4)] - 5\}$.

10. Solve $A = \dfrac{1}{2}h(b_1 + b_2)$ for b_1.

11. Hydrogen Peroxide Solutions How many gallons of 15% hydrogen peroxide solution must be mixed with 10 gallons of 4% hydrogen peroxide solution to get a 10% hydrogen peroxide solution?

12. Solve the inequality $4(x - 4) < 8(2x + 3)$.

13. Solve the inequality $-1 < 3x - 7 < 11$

14. Determine the solution set of $|3x + 5| = |2x - 10|$.

15. Determine the solution set of $|2x - 1| \le 3$.

16. Graph $y = -\dfrac{3}{2}x - 4$.

17. a) Determine whether the following graph represents a function.

   b) Find the domain and range of the graph.

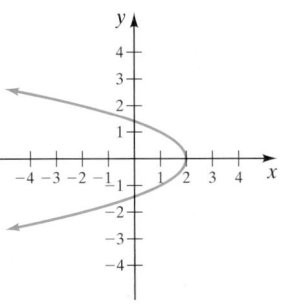

18. Determine the slope of the line through the points $(-5, 3)$ and $(4, -1)$.

19. Determine whether the graphs of the two given equations are parallel, perpendicular, or neither.

$$2x - 5y = 8$$
$$5x - 2y = 12$$

20. If $f(x) = x^2 + 3x - 2$ and $g(x) = 4x - 9$, find $(f + g)(x)$.

4

Systems of Equations and Inequalities

Goals of This Chapter

In this chapter we solve systems of linear equations using the following methods: by graphing, by substitution, by the addition method, by using matrices, and by using determinants and Cramer's rule. We also solve systems of linear *inequalities*. Throughout the chapter there are many real-life applications. The chapter covers essential topics used by businesses to consider the relationships among variables involved in the day-to-day operations of a business.

© Allen R. Angel

Systems of equations are frequently used to solve real-life problems. For example, in Example 6 on page 240 we use a system of equations to study business loans taken out by a toy store.

4.1 Solving Systems of Linear Equations in Two Variables

1 Solve systems of linear equations graphically.

2 Solve systems of linear equations by substitution.

3 Solve systems of linear equations using the addition method.

System of Linear Equations

When two or more linear equations are considered simultaneously, the equations are called a **system of linear equations**.

For example,

$$\left.\begin{array}{l} (1)\ y = x + 5 \\ (2)\ y = 2x + 4 \end{array}\right\} \quad \text{System of linear equations}$$

Solution to a System of Linear Equations in Two Variables

A **solution to a system of linear equations in two variables** is an ordered pair that satisfies each equation in the system.

The only solution to the system above is $(1, 6)$.

Check in Equation (1)	Check in Equation (2)
$(1, 6)$	$(1, 6)$
$y = x + 5$	$y = 2x + 4$
$6 \overset{?}{=} 1 + 5$	$6 \overset{?}{=} 2(1) + 4$
$6 = 6$ True	$6 = 6$ True

The ordered pair $(1, 6)$ satisfies *both* equations and is a solution to the system of equations.

1 Solve Systems of Linear Equations Graphically

To solve a system of linear equations in two variables graphically, graph both equations in the system on the same axes. If the system has a unique solution, then the solution will be the ordered pair common to both lines, or the point of intersection of both lines in the system.

When two lines are graphed, three situations are possible, as illustrated in **Figure 4.1** below.

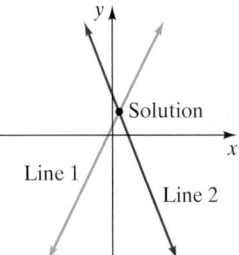

Line 1 *Intersects* Line 2

- Exactly one solution
- The solution is at the point of intersection.
- **Consistent system**
- Lines have different slopes

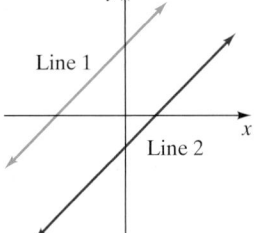

Line 1 is *Parallel* to Line 2

- No solution
- Since parallel lines do not intersect, there is no solution.
- **Inconsistent system**
- Lines have same slopes and different y-intercepts

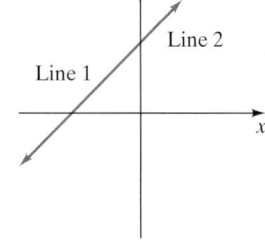

Line 1 is the *Same Line* as Line 2

- Infinite number of solutions
- Every point on the common line is a solution.
- **Dependent system**
- Lines have same slope and same y-intercept

FIGURE 4.1

We can determine if a system of linear equations is consistent, inconsistent, or dependent by writing each equation in slope-intercept form and comparing the slopes and y-intercepts (see **Fig. 4.1**).

EXAMPLE 1 Without graphing the equations, determine whether the following system of equations is consistent, inconsistent, or dependent.

$$3x - 4y = 8$$
$$-9x + 12y = -24$$

Solution   Write each equation in slope-intercept form.

$$3x - 4y = 8 \qquad\qquad -9x + 12y = -24$$
$$-4y = -3x + 8 \qquad\qquad 12y = 9x - 24$$
$$y = \frac{3}{4}x - 2 \qquad\qquad y = \frac{3}{4}x - 2$$

Since the equations have the same slope, $\frac{3}{4}$, and the same y-intercept, $(0, -2)$, the equations represent the same line. Therefore, the system is dependent, and there are an infinite number of solutions.

Now Try Exercise 19

EXAMPLE 2 Solve the following system of equations graphically.

$$y = x + 2$$
$$y = -x + 4$$

Solution   Graph both equations on the same axes (**Fig. 4.2**). The solution is the point of intersection of the two lines, $(1, 3)$.

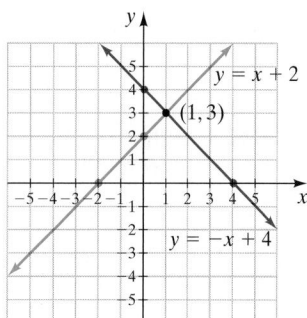

FIGURE 4.2

Now Try Exercise 25

Helpful Hint

When solving equations by graphing, it is always a good idea to check the solution by substituting the values for x and y into *each* of the original equations. Check the solution to Example 2 now.

✎ *Using Your Graphing Calculator* ▦▦▦▦

In Chapter 3 we learned how to use a graphing calculator to graph the equations of lines. Here we will use a graphing calculator to solve a system of linear equations graphically.

EXAMPLE   Use your graphing calculator to solve the systems of equations.

$$y = 2x - 3$$
$$y = -\frac{3}{2}x + 4$$

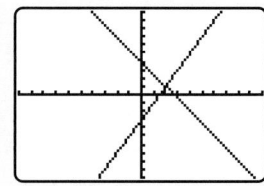

FIGURE 4.3

Solution   Press $\boxed{Y=}$ and enter these equations as Y_1 and Y_2, respectively, and press $\boxed{GRAPH}$. The graphs of the equations are shown in **Figure 4.3**.

The solution is at the point of intersection of the two lines. To determine the point, press $\boxed{2nd}$ $\boxed{TRACE}$ $\boxed{5}$ to activate the "intersect" feature. The calculator will ask you to choose the curves. Do so by pressing $\boxed{ENTER}$ $\boxed{ENTER}$. Now you are asked to make a guess. Press $\boxed{\blacktriangleleft}$ or $\boxed{\blacktriangleright}$ until you see the flashing cursor near the intersection point and then press $\boxed{ENTER}$. **Figure 4.4** shows that the intersection of the two lines is at the point $(2, 1)$. Substitute $x = 2$ and $y = 1$ into each of the original equations to confirm that $(2, 1)$ is a solution to the system of equations.

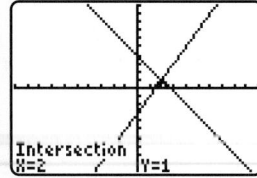

FIGURE 4.4

2   Solve Systems of Linear Equations by Substitution

Although solving systems of equations graphically helps us visualize a system of equations and its solution, an exact solution is sometimes difficult to determine from the graph. For this reason we introduce two algebraic methods to find the solution to a system of linear equations in two variables. The first of these is **substitution**.

To Solve a Linear System of Equations by Substitution

1. Solve for a variable in either equation. (If possible, solve for a variable with a numerical coefficient of 1 to avoid working with fractions.)
2. Substitute the expression found for the variable in step 1 *into the other equation*. This will result in an equation containing only one variable.
3. Solve the equation obtained in step 2.
4. Substitute the value found in step 3 into the equation from step 1. Solve the equation to find the remaining variable.
5. Check your solution in *all* equations in the system.

EXAMPLE 3 Solve the following system of equations by substitution.

$$y = 3x - 5$$
$$y = -4x + 9$$

Solution   Since both equations are already solved for y, we can substitute $3x - 5$ for y in the second equation and then solve for the remaining variable, x.

$$3x - 5 = -4x + 9$$
$$7x - 5 = 9$$
$$7x = 14$$
$$x = 2$$

> **Understanding Algebra**
>
> In Example 3, once we found that $x = 2$, we could have substituted $x = 2$ into *either* of the original equations.

Now find y by substituting $x = 2$ into the first equation.

$$y = 3x - 5$$
$$y = 3(2) - 5$$
$$y = 6 - 5 = 1$$

Thus, we have $x = 2$ and $y = 1$, or the ordered pair $(2, 1)$. A check will show that the solution to the system of equations is $(2, 1)$.

Now Try Exercise 39

EXAMPLE 4 Solve the following system of equations by substitution.

$$2x + y = 11$$
$$x + 3y = 18$$

Solution Begin by solving for one of the variables in either of the equations. If possible, you should solve for a variable with a numerical coefficient of 1, so you may avoid working with fractions.

Let's solve for y in $2x + y = 11$.

$$2x + y = 11$$
$$y = -2x + 11$$

Understanding Algebra

In Example 4, we could have also begun by solving for x in the second equation.

Next, substitute $(-2x + 11)$ for y in the *other equation*, $x + 3y = 18$, and solve for the remaining variable, x.

$$x + 3y \qquad = 18$$
$$x + 3(\overbrace{-2x + 11}) = 18 \qquad \text{Substitute } (-2x + 11) \text{ for } y.$$
$$x - 6x + 33 = 18$$
$$-5x + 33 = 18$$
$$-5x = -15$$
$$x = 3$$

Finally, substitute $x = 3$ into the equation $y = -2x + 11$ and solve for y.

$$y = -2x + 11$$
$$y = -2(3) + 11 = 5$$

The solution is the ordered pair $(3, 5)$. Check this solution.

Now Try Exercise 41

Helpful Hint

Students sometimes successfully solve for one of the variables and forget to solve for the other. Remember that a solution must contain a numerical value for each variable in the system.

---

3 Solve Systems of Linear Equations Using the Addition Method

A third method of solving a system of equations is the **addition** (or elimination) **method**. The goal of this process is to obtain two equations whose sum will be an equation containing only one variable.

EXAMPLE 5 Solve the following system of equations using the addition method.

$$2x + 5y = 3$$
$$3x - 5y = 17$$

Solution Note that one equation contains $+5y$ and the other contains $-5y$. By adding the equations, we can eliminate the variable y and obtain one equation containing only one variable, x.

$$\begin{array}{r} 2x + 5y = 3 \\ 3x - 5y = 17 \\ \hline 5x = 20 \end{array}$$

Now solve for the remaining variable, x.

$$\frac{5x}{5} = \frac{20}{5}$$
$$x = 4$$

Finally, solve for y by substituting 4 for x in either of the original equations.

$$2x + 5y = 3$$
$$2(4) + 5y = 3$$
$$8 + 5y = 3$$
$$5y = -5$$
$$y = -1$$

A check will show that the solution is $(4, -1)$.

Now Try Exercise 53

To Solve a Linear System of Equations Using the Addition (or Elimination) Method

1. If necessary, rewrite each equation in standard form, $ax + by = c$.
2. If necessary, multiply one or both equations by a constant(s) so that when the equations are added, the sum will contain only one variable.
3. Add the respective sides of the equations. This will result in a single equation containing only one variable.
4. Solve the equation obtained in step 3.
5. Substitute the value found in step 4 into either of the original equations. Solve that equation to find the value of the remaining variable.
6. Check your solution in all equations in the system.

In step 2 of the procedure, we indicate that it may be necessary to multiply both sides of an equation by a constant. To help avoid confusion, we will number our equations using parentheses, such as (*eq. 1*) or (*eq. 2*).

In Example 6 we will solve the same system we solved in Example 4, but this time we will use the addition method.

EXAMPLE 6 Solve the following system of equations using the addition method.

$$2x + y = 11 \quad (eq. 1)$$
$$x + 3y = 18 \quad (eq. 2)$$

Solution   Our goal is to obtain two equations whose sum will be an equation containing only one variable. To eliminate the variable x, we multiply (*eq. 2*) by -2 and then add the two equations.

$$2x + y = \quad 11 \quad (eq. 1)$$
$$-2x - 6y = -36 \quad (eq. 2) \quad \text{Multiplied by } -2$$

Now add.

$$2x + y = \quad 11$$
$$\underline{-2x - 6y = -36}$$
$$-5y = -25$$
$$y = \quad 5$$

Now solve for x by substituting 5 for y in either of the original equations.

$$2x + y = 11$$
$$2x + 5 = 11 \quad \text{Substitute 5 for } y.$$
$$2x = 6$$
$$x = 3$$

The solution is $(3, 5)$.

Now Try Exercise 61

In Example 6, we could have first eliminated the variable y by multiplying (*eq.* 1) by -3 and then adding.

Sometimes both equations must be multiplied by different numbers in order for one of the variables to be eliminated. This procedure is illustrated in Example 7.

EXAMPLE 7   Solve the following system of equations using the addition method.

$$4x + 3y = 7 \quad (eq.\,1)$$
$$3x - 7y = -3 \quad (eq.\,2)$$

Solution   We can eliminate the variable x by multiplying (*eq.* 1) by -3 and (*eq.* 2) by 4.

$$
\begin{array}{rll}
-12x - 9y = -21 & (eq.\,1)\ \text{Multiplied by } -3 \\
\underline{12x - 28y = -12} & (eq.\,2)\ \text{Multiplied by } 4 \\
-37y = -33 & \text{Sum of equations} \\
y = \dfrac{33}{37} &
\end{array}
$$

We could find x by substituting $\dfrac{33}{37}$ for y in one of the original equations and solving for x. An easier method to solve for x is to go back to the original equations and eliminate the variable y by multiplying (*eq.* 1) by 7 and (*eq.* 2) by 3.

$$
\begin{array}{rll}
28x + 21y = 49 & (eq.\,1)\ \text{Multiplied by } 7 \\
\underline{9x - 21y = -9} & (eq.\,2)\ \text{Multiplied by } 3 \\
37x = 40 & \text{Sum of equations} \\
x = \dfrac{40}{37} &
\end{array}
$$

The solution is $\left(\dfrac{40}{37}, \dfrac{33}{37} \right)$.

Now Try Exercise 67

In Example 7, the same solution could be obtained by multiplying (*eq.* 1) by 3 and (*eq.* 2) by -4 and then adding. Try it now and see.

EXAMPLE 8   Solve the following system of equations using the addition method.

$$2x + y = 11 \quad (eq.\,1)$$
$$\frac{1}{18}x + \frac{1}{6}y = 1 \quad (eq.\,2)$$

Solution   When a system of equations contains an equation with fractions, it is generally best to *clear*, or remove, fractions from that equation. In (*eq.* 2), if we multiply both sides of the equation by the least common denominator, 18, we obtain

$$18\left(\frac{1}{18}x + \frac{1}{6}y \right) = 18\,(1)$$

$$18\left(\frac{1}{18}x \right) + 18\left(\frac{1}{6}y \right) = 18\,(1)$$

$$x + 3y = 18 \quad (eq.\,3)$$

The system is now

$$2x + y = 11 \quad (eq.\,1)$$
$$x + 3y = 18 \quad (eq.\,3)$$

This is the same system of equations we solved in Example 6. Thus the solution to this system is $(3, 5)$, the same as obtained in Example 6.

Now Try Exercise 51

Understanding Algebra

- If an equation contains fractions, clear the fractions by multiplying both sides of the equation by the LCD of the fractions.

- If an equation contains decimal numbers, clear the decimal numbers by multiplying both sides of the equation by an appropriate power of 10.

EXAMPLE 9 Solve the following system of equations using the addition method.

$$0.2x + 0.1y = 1.1 \quad (eq.\,1)$$
$$x + 3y = 18 \quad (eq.\,2)$$

Solution   When a system of equations contains an equation with decimal numbers, it is generally best to *clear*, or remove, the decimal numbers from that equation. In (*eq. 1*), if we multiply both sides of the equation by 10 we obtain

$$10\,(0.2x) + 10\,(0.1y) = 10\,(1.1)$$
$$2x + y = 11 \quad (eq.\,3)$$

The system is now

$$2x + y = 11 \quad (eq.\,3)$$
$$x + 3y = 18 \quad (eq.\,2)$$

This is the same system of equations we solved in Example 6. Thus the solution to this system is $(3, 5)$, the same as obtained in Example 6.

Now Try Exercise 69

EXAMPLE 10 Solve the following system of equations using the addition method.

$$x - 3y = 4 \quad (eq.\,1)$$
$$-2x + 6y = 1 \quad (eq.\,2)$$

Solution   We begin by multiplying (*eq. 1*) by 2.

$$
\begin{array}{rl}
2x - 6y = 8 & (eq.\,1) \quad \text{Multiplied by 2}\\
-2x + 6y = 1 & (eq.\,2)\\
\hline
0 = 9 & \text{False}
\end{array}
$$

Since $0 = 9$ *is a false statement, this system has no solution. The system is inconsistent and the graphs of these equations are parallel lines.*

Now Try Exercise 59

EXAMPLE 11 Solve the following system of equations using the addition method.

$$x - \frac{1}{2}y = 2$$
$$y = 2x - 4$$

Solution   First, rewrite each equation in standard form.

$$x - \frac{1}{2}y = 2 \quad (eq.\,1)$$
$$-2x + y = -4 \quad (eq.\,2)$$

Now proceed as in previous examples.

$$
\begin{array}{rl}
2x - y = 4 & (eq.\,1) \quad \text{Multiplied by 2}\\
-2x + y = -4 & (eq.\,2)\\
\hline
0 = 0 & \text{True}
\end{array}
$$

Since $0 = 0$ *is a true statement, the system is dependent and has an infinite number of solutions. Both equations represent the same line.* Notice that if you multiply both sides of (*eq. 1*) by -2 you obtain (*eq. 2*).

Now Try Exercise 63

We have illustrated three methods that can be used to solve a system of linear equations: graphing, substitution, and the addition method. When you need an exact solution, graphing may not be the best to use. Of the two algebraic methods, the addition method may be the easiest to use if there are no numerical coefficients of 1 in the system.

EXERCISE SET 4.1   Math XL   MyMathLab
MathXL®   MyMathLab

Warm-Up Exercises

Fill in the blanks with the appropriate word, phrase, or symbol(s) from the following list.

| unique solution | dependent | consistent | inconsistent | intersecting lines | ordered triple |
| one variable | two variables | parallel lines | the same line | ordered pair | |

1. The goal of the addition method is to obtain an equation that only contains _____.

2. For a system with three equations and three variables, the solution is an _____.

3. A system of equations that has an infinite number of solutions is a(n) _____ system.

4. A system of equations that has no solution is a(n) _____ system.

5. A system of equations that has at least one solution is a(n) _____ system.

6. The _____ to a system of two linear equations is located at the intersection point of the graphs of the equations.

7. The lines that represent the equations in an inconsistent system of linear equations are _____.

8. The lines that represent the equations in a dependent system of linear equations are _____.

9. For a consistent linear system with two equations and two variables, the solution is an _____.

10. The lines that represent the equations in a consistent system of linear equations are _____.

Practice the Skills

Determine which, if any, of the given ordered pairs or ordered triples satisfy the system of linear equations.

11. $y = 2x + 4$
$y = 2x - 1$
a) $(0, 4)$
b) $(3, 10)$

12. $4x + 3y = 30$
$y = \dfrac{3}{4}x - 3$
a) $(6, 2)$ **b)** $(4, 0)$

13. $x + y = 25$
$0.25x + 0.45y = 7.50$
a) $(5, 20)$
b) $(18.75, 6.25)$

14. $y = \dfrac{x}{3} - \dfrac{7}{3}$
$5x - 35 = 15y$
a) $(1, -2)$
b) $(7, 0)$

15. $x + 2y - z = -5$
$2x - y + 2z = 8$
$3x + 3y + 4z = 5$
a) $(3, 1, -2)$
b) $(1, -2, 2)$

16. $4x + y - 3z = 1$
$2x - 2y + 6z = 11$
$-6x + 3y + 12z = -4$
a) $(2, -1, -2)$
b) $\left(\dfrac{1}{2}, 2, 1\right)$

Write each equation in slope-intercept form. Without graphing the equations, state whether the system of equations is consistent, inconsistent, or dependent. Also indicate whether the system has exactly one solution, no solution, or an infinite number of solutions.

17. $x + y = -2$
$3y + 12 = -6x$

18. $x - \dfrac{1}{2}y = 4$
$2x - y = 7$

19. $\dfrac{x}{3} + \dfrac{y}{4} = 1$
$4x + 3y = 12$

20. $\dfrac{x}{3} + \dfrac{y}{4} = 1$
$2x - 3y = 12$

21. $3x - 3y = 9$
$2x - 2y = -4$

22. $2x = 3y + 4$
$6x - 9y = 12$

23. $y = \dfrac{3}{2}x + \dfrac{1}{2}$
$3x - 2y = -\dfrac{5}{2}$

24. $x - y = 3$
$\dfrac{1}{4}x - 2y = -6$

Determine the solution to each system of equations graphically. If the system is inconsistent or dependent, so state.

25. $y = -x + 3$
$y = x + 5$

26. $y = 2x + 8$
$y = -3x - 12$

27. $y = 4x - 1$
$3y = 12x + 9$

28. $x + y = 1$
$3x - y = -5$

29. $2x + 3y = 6$
$4x = -6y + 12$

30. $y = -2x - 1$
$x + 2y = 4$

31. $5x + 3y = 13$
$x = 2$

32. $2x - 5y = 10$
$y = \dfrac{2}{5}x - 2$

33. $y = -5x + 5$
$y = 2x - 2$

34. $4x - y = 9$
$x - 3y = 16$

35. $x - \dfrac{1}{2}y = -2$
$2y = 4x - 6$

36. $y = -\dfrac{1}{3}x - 1$
$3y = 4x - 18$

Find the solution to each system of equations by substitution.

37. $x - 3y = 1$
$y = -2x + 2$

38. $3x - 2y = -7$
$y = 2x + 6$

39. $x = 2y + 3$
$y = x$

40. $y = 3x - 16$
$x = y$

41. $a + 3b = 5$
$2a - b = 3$

42. $m + 2n = 4$
$m + \dfrac{1}{2}n = 4$

43. $5x + 6y = 6.7$
$3x - 2y = 0.1$

44. $x = 0.5y + 1.7$
$10x - y = 1$

45. $a - \dfrac{1}{2}b = 2$
$b = 2a - 4$

46. $x + 3y = -2$
$y = -\dfrac{1}{3}x - \dfrac{2}{3}$

47. $5x - 2y = -7$
$y = \dfrac{5}{2}x + 1$

48. $y = \dfrac{2}{3}x - 1$
$2x - 3y = 5$

49. $5x - 4y = -7$
$x - \dfrac{3}{5}y = -2$

50. $6s + 3t = 4$
$s = \dfrac{1}{2}t$

51. $\dfrac{1}{2}x - \dfrac{1}{3}y = 2$
$\dfrac{1}{4}x + \dfrac{2}{3}y = 6$

52. $\dfrac{1}{2}x + \dfrac{1}{3}y = 3$
$\dfrac{1}{5}x + \dfrac{1}{8}y = 1$

Solve each system of equations using the addition method.

53. $x + y = 7$
$x - y = 3$

54. $-x + y = 4$
$x - 2y = 6$

55. $4x - 3y = 1$
$5x + 3y = -10$

56. $2x - 5y = 6$
$-4x + 10y = -1$

57. $10m - 2n = 6$
$-5m + n = -3$

58. $4r - 3s = 2$
$2r + s = 6$

59. $2c - 5d = 1$
$-4c + 10d = 6$

60. $2v - 3w = 8$
$3v - 6w = 1$

61. $7p - 3q = 4$
$2p + 5q = 7$

62. $5s - 3t = 7$
$t = s + 1$

63. $5a - 10b = 15$
$a = 2b + 3$

64. $2x - 7y = 3$
$-5x + 3y = 7$

65. $2x - y = 8$
$3x + y = 6$

66. $5x + 4y = 6$
$2x = -5y - 1$

67. $3x - 4y = 5$
$2x = 5y - 3$

68. $4x + 5y = 3$
$2x - 3y = 4$

69. $0.2x - 0.5y = -0.4$
$-0.3x + 0.4y = -0.1$

70. $0.15x - 0.40y = 0.65$
$0.60x + 0.25y = -1.1$

71. $2.1m - 0.6n = 8.4$
$-1.5m - 0.3n = -6.0$

72. $-0.25x + 0.10y = 1.05$
$-0.40x - 0.625y = -0.675$

73. $\dfrac{1}{2}x - \dfrac{1}{3}y = 1$
$\dfrac{1}{4}x - \dfrac{1}{9}y = \dfrac{2}{3}$

74. $\dfrac{1}{5}x + \dfrac{1}{2}y = 4$
$\dfrac{2}{3}x - y = \dfrac{8}{3}$

75. $\dfrac{1}{3}x = 4 - \dfrac{1}{4}y$
$3x = 4y$

76. $\dfrac{2}{3}x - 4 = \dfrac{1}{2}y$
$x - 3y = \dfrac{1}{3}$

Problem Solving

77. Salaries In January 2010, Mary Bennett started a new job with an annual salary of \$38,000. Her boss agreed to increase her salary by \$1000 each January in the years to come. Her salary is determined by the equation $y = 38,000 + 1000t$, where t is the number of years since 2010. (See red line in graph.) Also, in January 2010, Wynn Nguyen started a new job with an annual salary of \$45,500. Her boss agreed to increase her salary by \$500 each January in the years to come. Her salary is determined by the equation $y = 45,500 + 500t$, where t is the number of years since 2010. (See blue line in graph.) Solve the system of equations to determine the year both salaries will be the same. What will be the salary in that year?

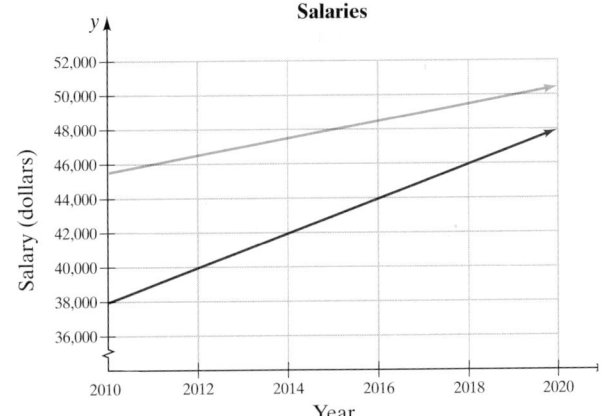

78. Truck Rental Hope Duncan is planning to move and needs to rent a 24-foot truck for one day. Budget charges $50 plus 79 cents per mile. Ryder charges $40 plus 99 cents per mile. The costs for each company can be represented by the following system of equations where x represents the number of miles driven and y represents the total cost for renting the truck for the day.

$$y = 50 + 0.79x$$
$$y = 40 + 0.99x$$

Solve the system to determine the number of miles for the costs of both companies to be equal. What will be that cost?

In Exercises 79 and 80, **a)** *create a system of linear equations that has the solution indicated and* **b)** *explain how you determined your solution.*

79. $(2, 5)$

80. $(-3, 4)$

81. The solution to the following system of equations is $(2, -3)$. Find A and B.

$$Ax + 4y = -8$$
$$3x - By = 21$$

82. The solution to the following system of equations is $(-5, 3)$. Find A and B.

$$3x + Ay = -3$$
$$Bx - 2y = -16$$

83. If $(2, 6)$ and $(-1, -6)$ are two solutions of $f(x) = mx + b$, find m and b.

84. If $(3, -5)$ and $(-2, 10)$ are two solutions of $f(x) = mx + b$, find m and b.

Concept/Writing Exercises

85. Explain how you can determine, without graphing or solving, whether a system of two linear equations is consistent, inconsistent or dependent.

86. When solving a system of linear equations using the addition (or elimination) method, what is the object of the process?

87. When solving a linear system by addition, how can you tell if the system is dependent?

88. When solving a linear system by addition, how can you tell if the system is inconsistent?

89. Explain how you can tell by observation that the following system is dependent.

$$2x + 3y = 1$$
$$4x + 6y = 2$$

90. Explain how you can tell by observation that the following system is inconsistent.

$$-x + 3y = 5$$
$$2x - 6y = -13$$

91. **a)** Write a system of equations that would be most easily solved by substitution.

b) Explain why substitution would be the easiest method to use.

c) Solve the system by substitution.

92. **a)** Write a system of equations that would be most easily solved using the addition method.

b) Explain why the addition method would be the easiest method to use.

c) Solve the system using the addition method.

93. The solutions of a system of linear equations include $(-4, 3)$ and $(-6, 11)$.

a) How many other solutions does the system have? Explain.

b) Determine the slope of the line containing $(-4, 3)$ and $(-6, 11)$. Determine the equation of the line containing these points. Then determine the y-intercept.

c) Does this line represent a function?

94. The solutions of a system of linear equations include $(-5, 1)$ and $(-5, -4)$.

a) How many other solutions does the system have? Explain.

b) Determine the slope of the line containing $(-5, 1)$ and $(-5, -4)$. Determine the equation of the line containing these points. Does this graph have a y-intercept? Explain.

c) Does this line represent a function?

95. Construct a system of equations that is dependent. Explain how you created your system.

96. Construct a system of equations that is inconsistent. Explain how you created your system.

97. Suppose you graph a system of two linear equations on your graphing calculator, but only one line shows in the window. What are two possible explanations for this?

98. Suppose you graph a system of linear equations on your graphing calculator and get the following.

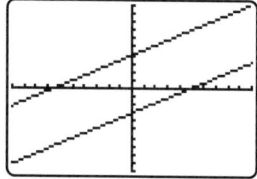

a) By observing the window, can you be sure that this system is inconsistent? Explain.

b) What can you do on your graphing calculator to determine whether the system is inconsistent?

Challenge Problems

Solve each system of equations.

99. $\dfrac{x+2}{2} - \dfrac{y+4}{3} = 4$

$\dfrac{x+y}{2} = \dfrac{1}{2} + \dfrac{x-y}{3}$

100. $\dfrac{5x}{2} + 3y = \dfrac{9}{2} + y$

$\dfrac{1}{4}x - \dfrac{1}{2}y = 6x + 12$

Solve each system of equations. (Hint: Let $u = \dfrac{1}{a}$, *then* $\dfrac{3}{a} = 3 \cdot \dfrac{1}{a} = 3u$).

101. $\dfrac{3}{a} + \dfrac{4}{b} = -1$

$\dfrac{1}{a} + \dfrac{6}{b} = 2$

102. $\dfrac{6}{x} + \dfrac{1}{y} = -1$

$\dfrac{3}{x} - \dfrac{2}{y} = -3$

By solving for x and y, determine the solution to each system of equations. In all equations, a ≠ 0 and b ≠ 0. The solution will contain either a, b, or both letters.

103. $4ax + 3y = 19$

$-ax + y = 4$

104. $ax = 2 - by$

$-ax + 2by - 1 = 0$

Group Activity

Discuss and answer Exercise 105 as a group.

105. Trends In the graph below, the red line indicates the long-term trend of firearm deaths, and the purple line indicates the long-term trend in motor vehicle deaths. The black lines indicate the short-term trends in deaths from firearms and motor vehicles.

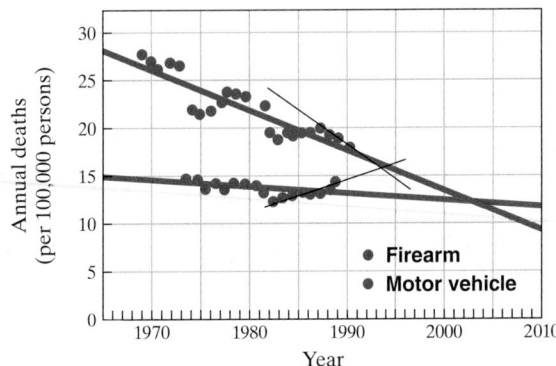

Mortality Trends

Source: Scientific American

a) Discuss the long-term trend in motor vehicle deaths.

b) Discuss the long-term trend in firearm deaths.

c) Discuss the short-term trend in motor vehicle deaths compared with the long-term trend in motor vehicle deaths.

d) Discuss the short-term trend in firearm deaths compared with the long-term trend in firearm deaths.

e) Using the long-term trends, estimate when the number of deaths from firearms will equal the number of deaths from motor vehicles.

f) Repeat part **e)** using the short-term trends.

g) Determine a function, $M(t)$, that can be used to estimate the number of deaths per 100,000 people (long-term) from motor vehicles from 1965 through 2010.

h) Determine a function, $F(t)$, that can be used to estimate the number of deaths per 100,000 people (long-term) from firearms from 1965 through 2010.

i) Solve the system of equations formed by parts **g)** and **h)**. Does the solution agree with the solution in part **e)**? If not, explain why.

Cumulative Review Exercises

[1.2] **106.** Explain the difference between a rational number and an irrational number.

[1.2] **107. a)** Are all rational numbers real numbers?
      b) Are all irrational numbers real numbers?

[2.1] **108.** Solve the equation $\dfrac{1}{2}(x-7) = \dfrac{3}{4}(2x+1)$.

[2.2] **109.** Find all numbers such that $|x-6| = |6-x|$.

[2.2] **110.** Evaluate $A = p\left(1 + \dfrac{r}{n}\right)^{nt}$, when $p = 500$, $r = 0.04$, $n = 2$, and $t = 1$.

[3.5] **111.** Is the following relation a function? Explain your answer. $\{(-3, 4), (7, 2), (-4, 5), (5, 0), (-3, -1)\}$

[3.6] **112.** Let $f(x) = x + 3$ and $g(x) = x^2 - 9$. Find $(f/g)(3)$.

4.2 Solving Systems of Linear Equations in Three Variables

1 Solve systems of linear equations in three variables.

2 Learn the geometric interpretation of a system of equations in three variables.

3 Recognize inconsistent and dependent systems.

1 Solve Systems of Linear Equations in Three Variables

The equation $2x - 3y + 4z = 8$ is an example of a **linear equation in three variables**. The solution to a linear equation in three variables is an *ordered triple* of the form (x, y, z). One solution to the equation given is $(1, 2, 3)$. Check now to verify that $(1, 2, 3)$ is a solution to the equation.

To solve systems of linear equations with three variables, we can use either substitution or the addition method, both of which were discussed in Section 4.1.

EXAMPLE 1 Solve the following system by substitution.

$$x = -3$$
$$3x + 4y = 7$$
$$-2x - 3y + 5z = 19$$

Solution   Since we know that $x = -3$, we substitute -3 for x in the equation $3x + 4y = 7$ and solve for y.

$$3x + 4y = 7$$
$$3(-3) + 4y = 7$$
$$-9 + 4y = 7$$
$$4y = 16$$
$$y = 4$$

Now we substitute $x = -3$ and $y = 4$ into the last equation and solve for z.

$$-2x - 3y + 5z = 19$$
$$-2(-3) - 3(4) + 5z = 19$$
$$6 - 12 + 5z = 19$$
$$-6 + 5z = 19$$
$$5z = 25$$
$$z = 5$$

Check $x = -3, y = 4, z = 5$. The solution must be checked in *all three* original equations.

$$x = -3 \qquad\qquad 3x + 4y = 7 \qquad\qquad -2x - 3y + 5z = 19$$
$$-3 = -3 \;\; \text{True} \quad 3(-3) + 4(4) \stackrel{?}{=} 7 \qquad -2(-3) - 3(4) + 5(5) \stackrel{?}{=} 19$$
$$7 = 7 \;\; \text{True} \qquad\qquad 19 = 19 \;\; \text{True}$$

The solution is the ordered triple $(-3, 4, 5)$. Remember that the ordered triple lists the x-value first, the y-value second, and the z-value third.

Now Try Exercise 5

Not every system of linear equations in three variables can be solved easily by substitution. We can also find the solution by the addition method, as illustrated in Example 2.

EXAMPLE 2 Solve the following system of equations using the addition method.

$$3x + 2y + z = 4 \quad (eq.\,1)$$
$$2x - 3y + 2z = -7 \quad (eq.\,2)$$
$$x + 4y - z = 10 \quad (eq.\,3)$$

Solution   Our first step is to use the addition method to write two new equations where each has the same two variables. We do this by eliminating one of the three variables x, y, or z. Here we will eliminate the variable z. Use (*eq.* 1) and (*eq.* 3) to eliminate z and call this new equation (*eq.* 4).

$$\begin{array}{rl} 3x + 2y + z = & 4 \quad (eq.\,1) \\ \underline{x + 4y - z = 10} \quad (eq.\,3) \\ 4x + 6y \quad\;\; = 14 \quad \text{Sum of equations, } (eq.\,4) \end{array}$$

Now, use (*eq.* 1) and (*eq.* 2) to eliminate z and call this new equation (*eq.* 5). Notice that we multiplied (*eq.* 1) by -2 so that after adding (*eq.* 1) and (*eq.* 2), z would be eliminated from (*eq.* 5).

$$\begin{array}{rl} -6x - 4y - 2z = & -8 \quad (eq.\,1) \text{ Multiplied by } -2 \\ \underline{2x - 3y + 2z = -7} \quad (eq.\,2) \\ -4x - 7y \quad\;\; = -15 \quad \text{Sum of equations, } (eq.\,5) \end{array}$$

We now have a system of two equations with two variables, (*eq.* 4), and (*eq.* 5). We can solve this system using the addition method.

$$\begin{array}{rl} 4x + 6y = & 14 \quad (eq.\,4) \\ \underline{-4x - 7y = -15} \quad (eq.\,5) \\ -y = -1 \quad \text{Sum of equations} \\ y = \;\;\; 1 \end{array}$$

Next we substitute $y = 1$ into either one of the two equations containing only two variables [(*eq.* 4) or (*eq.* 5)] and solve for x.

$$\begin{array}{rl} 4x + 6y = 14 & (eq.\,4) \\ 4x + 6(1) = 14 & \text{Substitute 1 for } y \text{ in } (eq.\,4). \\ 4x + 6 = 14 \\ 4x = 8 \\ x = 2 \end{array}$$

Finally, we substitute $x = 2$ and $y = 1$ into any of the original equations and solve for z.

$$\begin{array}{rl} 3x + 2y + z = & 4 \quad (eq.\,1) \\ 3(2) + 2(1) + z = & 4 \quad \text{Substitute 2 for } x \text{ and 1 for } y \text{ in } (eq.\,1). \\ 6 + 2 + z = & 4 \\ 8 + z = & 4 \\ z = & -4 \end{array}$$

The solution is the ordered triple $(2, 1, -4)$. Check this solution in *all three* original equations.

Now Try Exercise 17

EXAMPLE 3 Solve the following system of equations.

$$\begin{array}{rl} 2x - 3y + 2z = & -1 \quad (eq.\,1) \\ x + 2y \quad\quad\;\; = & 14 \quad (eq.\,2) \\ x \quad\;\; - 5z = & -11 \quad (eq.\,3) \end{array}$$

> **Helpful Hint**
>
> If an equation in a system contains fractions, eliminate the fractions by multiplying each term in the equation by the least common denominator. Then continue to solve the system. If, for example, one equation in the system is $\frac{3}{4}x - \frac{5}{8}y + z = \frac{1}{2}$, multiply both sides of the equation by 8 to obtain the equivalent equation $6x - 5y + 8z = 4$.

EXERCISE SET 4.2

Math XL
MathXL®

MyMathLab
MyMathLab

Warm-Up Exercises

Fill in the blanks with the appropriate word, phrase, or symbol(s) from the following list.

ordered pair   ordered triple   plane   line   dependent   inconsistent

1. The solution to a system of linear equations in three variables is an _____.

2. A linear equation in three variables is represented by a _____ in a 3-dimensional coordinate system.

3. If, while solving a system of linear equations in three variables, you obtain a statement that is always false, the system is _____ and has no solution.

4. If, while solving a system of linear equations in three variables, you obtain a statement that is always true, the system is _____ and has an infinite number of solutions.

Practice the Skills

Solve by substitution.

5. $x = 2$
$2x - y = 4$
$3x + 2y - 2z = 4$

6. $-x + 3y - 5z = -7$
$2y - z = -1$
$z = 3$

7. $5x - 6z = -17$
$3x - 4y + 5z = -1$
$2z = -6$

8. $2x - 5y = 12$
$-3y = -9$
$2x - 3y + 4z = 8$

9. $x + 2y = 6$
$3y = 9$
$x + 2z = 12$

10. $x - y + 5z = -4$
$3x - 2z = 6$
$4z = 2$

Solve using the addition method.

11. $x - 2y = -1$
$3x + 2y = 5$
$2x - 4y + z = -1$

12. $x - y + 2z = 1$
$y - 4z = 2$
$-2x + 2y - 5z = 2$

13. $2y + 4z = 2$
$x + y + 2z = -2$
$2x + y + z = 2$

14. $2x + y - 8 = 0$
$3x - 4z = -3$
$2x - 3z = 1$

15. $3p + 2q = 11$
$4q - r = 6$
$6p + 7r = 4$

16. $3s + 5t = -12$
$2t - 2u = 2$
$-s + 6u = -2$

17. $p + q + r = 4$
$p - 2q - r = 1$
$2p - q - 2r = -1$

18. $x - 2y + 3z = -7$
$2x - y - z = 7$
$-4x + 3y + 2z = -14$

19. $2x - 2y + 3z = 5$
$2x + y - 2z = -1$
$4x - y - 3z = 0$

20. $2x - y - 2z = 3$
$x - 3y - 4z = 2$
$x + y + 2z = -1$

21. $r - 2s + t = 2$
$2r + 3s - t = -3$
$2r - s - 2t = 1$

22. $3a - 3b + 4c = -1$
$a - 2b + 2c = 2$
$2a - 2b - c = 3$

23. $2a + 2b - c = 2$
$3a + 4b + c = -4$
$5a - 2b - 3c = 5$

24. $x - 2y + 2z = 3$
$2x - 3y + 2z = 5$
$x + y + 6z = -2$

25. $-x + 3y + z = 0$
$-2x + 4y - z = 0$
$3x - y + 2z = 0$

26. $x + y + z = 0$
$-x - y + z = 0$
$-x + y + z = 0$

27. $-\dfrac{1}{4}x + \dfrac{1}{2}y - \dfrac{1}{2}z = -2$
$\dfrac{1}{2}x + \dfrac{1}{3}y - \dfrac{1}{4}z = 2$
$\dfrac{1}{2}x - \dfrac{1}{2}y + \dfrac{1}{4}z = 1$

28. $\dfrac{2}{3}x + y - \dfrac{1}{3}z = \dfrac{1}{3}$
$\dfrac{1}{2}x + y + z = \dfrac{5}{2}$
$\dfrac{1}{4}x - \dfrac{1}{4}y + \dfrac{1}{4}z = \dfrac{3}{2}$

29. $x - \dfrac{2}{3}y - \dfrac{2}{3}z = -2$
$\dfrac{2}{3}x + y - \dfrac{2}{3}z = \dfrac{1}{3}$
$-\dfrac{1}{4}x + y - \dfrac{1}{4}z = \dfrac{3}{4}$

30. $\dfrac{1}{8}x + \dfrac{1}{4}y + z = 2$
$\dfrac{1}{3}x + \dfrac{1}{4}y + z = \dfrac{17}{6}$
$-\dfrac{1}{4}x + \dfrac{1}{3}y - \dfrac{1}{2}z = -\dfrac{5}{6}$

31. $0.2x + 0.3y + 0.3z = 1.1$
$0.4x - 0.2y + 0.1z = 0.4$
$-0.1x - 0.1y + 0.3z = 0.4$

32. $0.6x - 0.4y + 0.2z = 2.2$
$-0.1x - 0.2y + 0.3z = 0.9$
$-0.2x - 0.1y - 0.3z = -1.2$

Determine whether the following systems are inconsistent, dependent, or neither.

33. $2x + y + 2z = 1$
$x - 2y - z = 0$
$3x - y + z = 2$

34. $2p - 4q + 6r = 5$
$-p + 2q - 3r = -3$
$3p + 4q + 5r = 8$

35. $x - 4y - 3z = -1$
$-3x + 12y + 9z = 3$
$2x - 10y - 7z = 5$

36. $5a - 4b + 2c = 5$
$-10a + 8b - 4c = -10$
$-7a - 4b + c = 7$

37. $x + 3y + 2z = 6$
$x - 2y - z = 8$
$-3x - 9y - 6z = -7$

38. $2x - 2y + 4z = 2$
$-3x + y = -9$
$2x - y + z = 5$

Problem Solving

39. Three solutions to the equation $Ax + By + Cz = 1$ are $(-1, 2, -1)$, $(-1, 1, 2)$, and $(1, -2, 2)$. Determine the values of A, B, and C and write the equation using the numerical values found.

40. Three solutions to the equation $Ax + By + Cz = 14$ are $(3, -1, 2)$, $(2, -2, 1)$, and $(-5, 3, -24)$. Find the values of A, B, and C and write the equation using the numerical values found.

In Exercises 41 and 42, write a system of linear equations in three variables that has the given solution.

41. $(3, 1, 6)$

42. $(-2, 5, 3)$

43. a) Find the values of a, b, and c such that the points $(1, -1)$, $(-1, -5)$, and $(3, 11)$ lie on the graph of $y = ax^2 + bx + c$.

  b) Find the quadratic equation whose graph passes through the three points indicated. Explain how you determined your answer.

44. a) Find the values of a, b, and c such that the points $(1, 7)$, $(-2, -5)$, and $(3, 5)$ lie on the graph of $y = ax^2 + bx + c$.

  b) Find the quadratic equation whose graph passes through the three points indicated. Explain how you determined your answer.

Concept/Writing Exercises

An equation in three variables represents a plane. Consider a system of equations consisting of three equations in three variables. Answer the following questions.

45. If the three planes are parallel to one another as illustrated in the figure, how many points will be common to all three planes? Is the system consistent or inconsistent? Explain your answer.

46. If two of the planes are parallel to each other and the third plane intersects each of the other two planes, how many points will be common to all three planes? Is the system consistent or inconsistent? Explain your answer.

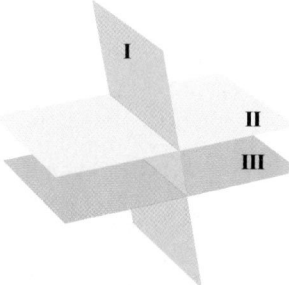

47. If the three planes are as illustrated in the figure, how many points will be common to all three planes? Is the system consistent or inconsistent? Explain your answer.

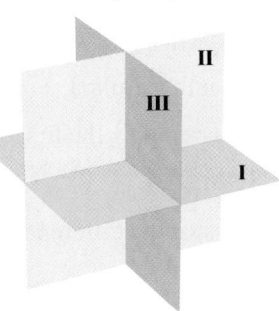

48. If the three planes are as illustrated in the figure, how many points will be common to all three planes? Is the system dependent? Explain your answer.

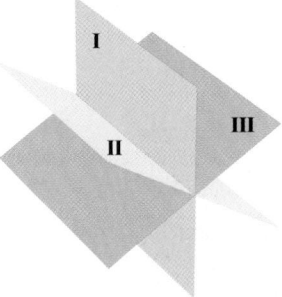

49. Is it possible for a system of linear equations in three variables to have exactly

  a) no solution,

  b) one solution,

  c) two solutions? Explain your answer.

50. In a system of linear equations in three variables, if the graphs of two equations are parallel planes, is it possible for the system to be

  a) consistent,   **b)** dependent,

  c) inconsistent? Explain your answer.

Challenge Problems

Find the solution to the following systems of equations.

51. $3p + 4q = 11$
$2p + r + s = 9$
$q - s = -2$
$p + 2q - r = 2$

52. $3a + 2b - c = 0$
$2a + 2c + d = 5$
$a + 2b - d = -2$
$2a - b + c + d = 2$

Cumulative Review Exercises

[2.2]   **53. Cross-Country Skiing** Margie Steiner begins skiing along a trail at 3 miles per hour. Ten minutes $\left(\frac{1}{6}\text{ hour}\right)$ later, her husband, David, begins skiing along the same trail at 5 miles per hour.

  a) How long after David leaves will he catch up to Margie?

  b) How far from the starting point will they be when they meet?

[2.6]   *Determine each solution set.*

54. $\left| 4 - \dfrac{2x}{3} \right| > 5$

55. $\left| \dfrac{3x - 4}{2} \right| + 1 < 7$

56. $\left| 3x + \dfrac{1}{5} \right| = -5$

© Photos.com/Jupiter Unlimited

4.3 Systems of Linear Equations: Applications and Problem Solving

1 Use systems of equations to solve applications.

2 Use linear systems in three variables to solve applications.

1 Use Systems of Equations to Solve Applications

Many of the applications solved in earlier chapters using only one variable can now be solved using two variables. Whenever we use two variables to solve an application problem, we must write a system of two equations.

EXAMPLE 1 **Land Area of the Carolinas** The combined land area of North Carolina and South Carolina is 78,820 square miles. The difference between the two state's land areas is 18,602 square miles. If North Carolina has the larger land area, find the land area of each state.

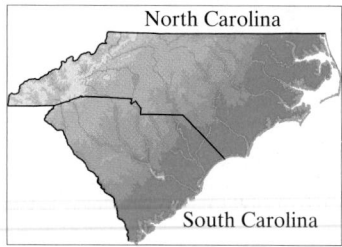

North Carolina

South Carolina

Solution   Understand   We need to determine the land area of North Carolina and the land area of South Carolina. We will use two variables and therefore will need to determine two equations.

Translate   Let N = the land area of North Carolina.

                Let S = the land area of South Carolina.

Since the total area of the two states is 78,820 square miles, the first equation is

$$N + S = 78,820$$

Since North Carolina has a larger land area and since the difference in the land area is 18,602 square miles, the second equation is

$$N - S = 18,602$$

The system of two equations is

$$N + S = 78,820 \quad (eq.\,1)$$
$$N - S = 18,602 \quad (eq.\,2)$$

Carry Out   We will use the addition method to solve this system of equations.

$$
\begin{array}{lll}
N + S = 78,820 & & \\
\underline{N - S = 18,602} & & \\
2N \qquad\; = 97,422 & & \text{Sum of the two equations} \\
N = 48,711 & & \text{Divided both sides by 2.}
\end{array}
$$

Thus, $N = 48,711$. To determine the value for S, substitute 48,711 into (*eq.* 1).

$$
\begin{array}{lll}
N + S = 78,820 & & \\
48,711 + S = 78,820 & & \\
S = 30,109 & & \text{Subtracted 48,711 from both sides.}
\end{array}
$$

Answer   The land area of North Carolina is 48,711 square miles and the land area of South Carolina is 30,109 square miles.

Now Try Exercise 1

EXAMPLE 2 Canoe Speed The Burnhams are canoeing on the Suwannee River. They travel at an average speed of 4.75 miles per hour when paddling with the current and 2.25 miles per hour when paddling against the current. Determine the speed of the canoe in still water and the speed of the current.

Solution   Understand   When they are traveling with the current, the canoe's speed is the canoe's speed in still water *plus* the current's speed. When traveling against the current, the canoe's speed is the canoe's speed in still water *minus* the current's speed.

Translate   Let s = speed of the canoe in still water.

                Let c = speed of the current.

The system of equations is:

speed of the canoe traveling with the current: $s + c = 4.75$

speed of the canoe traveling against the current: $s - c = 2.25$

Carry Out   We will use the addition method, as we discussed in Section 4.1, to solve this system of equations.

$$
\begin{array}{l}
s + c = 4.75 \\
\underline{s - c = 2.25} \\
2s \qquad\; = 7.00 \\
s = 3.5
\end{array}
$$

The speed of the canoe in still water is 3.5 miles per hour. We now determine the speed of the current.

$$s + c = 4.75$$
$$3.5 + c = 4.75$$
$$c = 1.25$$

Answer   The speed of the current is 1.25 miles per hour, and the speed of the canoe in still water is 3.5 miles per hour.

Now Try Exercise 13

EXAMPLE 3 **Salary** Yamil Bermudez, a salesman at Hancock Appliances, receives a weekly salary plus a commission, which is a percentage of his sales. One week, with sales of $3000, his total take-home pay was $850. The next week, with sales of $4000, his total take-home pay was $1000. Find his weekly salary and his commission rate.

Solution   Understand   Yamil's take-home pay consists of his weekly salary plus commission. We are given information about two specific weeks that we can use to find his weekly salary and his commission rate.

Translate   Let s = his weekly salary.

Let r = his commission rate.

In week 1, his commission on $3000 is $3000r$, and in week 2 his commission on $4000 is $4000r$. Thus the system of equations is

$$\text{salary} + \text{commission} = \text{take-home salary}$$

1st week $s + 3000r = 850$ ⎱
2nd week $s + 4000r = 1000$ ⎰   System of equations

Carry Out
$$-s - 3000r = -850 \quad \text{1st week multiplied by } -1$$
$$\underline{s + 4000r = 1000} \quad \text{2nd week}$$
$$1000r = 150 \quad \text{Sum of equations}$$
$$r = \frac{150}{1000}$$
$$r = 0.15$$

Yamil's commission rate is 15%. Now we find his weekly salary by substituting 0.15 for r in either equation.

$$s + 3000r = 850$$
$$s + 3000(0.15) = 850 \quad \text{Substitute 0.15 for } r \text{ in the 1st-week equation.}$$
$$s + 450 = 850$$
$$s = 400$$

Answer   Yamil's weekly salary is $400 and his commission rate is 15%.

Now Try Exercise 15

EXAMPLE 4 **Riding Horses** Ben Campbell leaves his ranch riding on his horse at 5 miles per hour. One-half hour later, Joe Campbell leaves the same ranch and heads along the same route on his horse at 8 miles per hour.

a) How long after Joe leaves the ranch will he catch up to Ben?

b) When Joe catches up to Ben, how far from the ranch will they be?

Solution   **a)** Understand   When Joe catches up to Ben, they both will have traveled the same distance. Joe will have traveled the distance in $\frac{1}{2}$ hour less time since he left $\frac{1}{2}$ hour after Ben. We will use the formula distance = rate · time to solve this problem.

Translate

Let b = time traveled by Ben.

Let j = time traveled by Joe.

We will set up a table to organize the given information.

	Rate	Time	Distance
Ben	5	b	$5b$
Joe	8	j	$8j$

Since both Ben and Joe cover the same distance, we write

$$\text{Ben's distance} = \text{Joe's distance}$$
$$5b = 8j$$

Our second equation comes from the fact that Joe is traveling for $\frac{1}{2}$ hour less time than Ben. Therefore, $j = b - \frac{1}{2}$. Thus our system of equations is:

$$5b = 8j$$
$$j = b - \frac{1}{2}$$

Carry Out   We will solve this system of equations using substitution. Since $j = b - \frac{1}{2}$, substitute $b - \frac{1}{2}$ for j in the first equation and solve for b.

$$5b = 8j$$
$$5b = 8\left(b - \frac{1}{2}\right)$$
$$5b = 8b - 4$$
$$-3b = -4$$
$$b = \frac{-4}{-3} = 1\frac{1}{3}$$

Therefore, the time Ben has been traveling is $1\frac{1}{3}$ hours. To get the time Joe has been traveling, we will subtract $\frac{1}{2}$ hour from Ben's time.

$$j = b - \frac{1}{2}$$
$$j = 1\frac{1}{3} - \frac{1}{2}$$
$$j = \frac{4}{3} - \frac{1}{2} = \frac{8}{6} - \frac{3}{6} = \frac{5}{6}$$

Answer   Joe will catch up to Ben $\frac{5}{6}$ of an hour (or 50 minutes) after Joe leaves the ranch.

b) We can use either Ben's or Joe's distance to determine the distance traveled from the ranch. We will use Joe's distance.

$$d = 8j = 8\left(\frac{5}{6}\right) = \frac{\overset{4}{\cancel{8}}}{1} \cdot \frac{5}{\underset{3}{\cancel{6}}} = \frac{20}{3} = 6\frac{2}{3}$$

Answer   Thus, Joe will catch up to Ben when they are $6\frac{2}{3}$ miles from the ranch.

Now Try Exercise 33

EXAMPLE 5 **Mixing Solutions** Chung Song, a chemist with Johnson and Johnson, wishes to create a new household cleaner containing 30% trisodium phosphate (TSP). Chung needs to mix a 16% TSP solution with a 72% TSP solution to get 6 liters of a 30% TSP solution. How many liters of the 16% solution and of the 72% solution will he need to mix?

Solution   Understand   To solve this problem we use the fact that the amount of TSP in a solution is found by multiplying the percent strength of the solution by the number of liters (the volume) of the solution. Chung needs to mix a 16% solution and a 72% solution to obtain 6 liters of a solution whose strength, 30%, is between the strengths of the two solutions being mixed.

Translate   Let x = number of liters of the 16% solution.
                       Let y = number of liters of the 72% solution.

We will draw a sketch (**Fig. 4.6**) and then use a table to help analyze the problem.

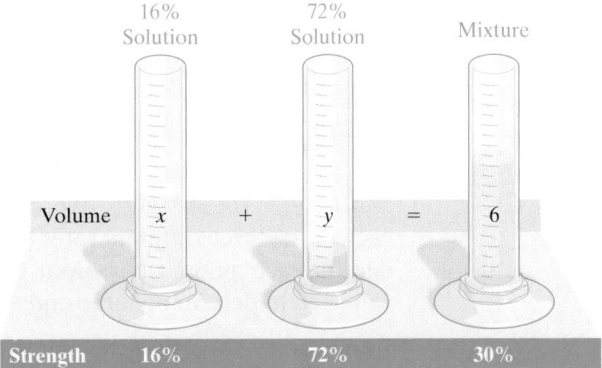

FIGURE 4.6

Solution	Strength of Solution	Number of Liters	Amount of TSP
16% solution	0.16	x	$0.16x$
72% solution	0.72	y	$0.72y$
Mixture	0.30	6	$0.30(6)$

Since the sum of the volumes of the 16% solution and the 72% solution is 6 liters, our first equation is

$$x + y = 6$$

The second equation comes from the fact that the solutions are mixed.

$$\left(\begin{array}{c}\text{amount of TSP}\\\text{in 16\% solution}\end{array}\right) + \left(\begin{array}{c}\text{amount of TSP}\\\text{in 72\% solution}\end{array}\right) = \left(\begin{array}{c}\text{amount of TSP}\\\text{in mixture}\end{array}\right)$$
$$0.16x \qquad + \qquad 0.72y \qquad = \qquad 0.30(6)$$

Therefore, the system of equations is

$$x + y = 6$$
$$0.16x + 0.72y = 0.30(6)$$

Carry Out   Solving $x + y = 6$ for y, we get $y = -x + 6$. Substituting $-x + 6$ for y in the second equation gives us

$$0.16x + 0.72y = 0.30(6)$$
$$0.16x + 0.72(-x + 6) = 0.30(6)$$
$$0.16x - 0.72x + 4.32 = 1.8$$
$$-0.56x + 4.32 = 1.8$$
$$-0.56x = -2.52$$
$$x = \frac{-2.52}{-0.56} = 4.5$$

Therefore, Chung must use 4.5 liters of the 16% solution. Since the two solutions must total 6 liters, he must use 6 − 4.5 or 1.5 liters of the 72% solution.

Now Try Exercise 17

> **Helpful Hint**
>
> In Example 5, the equation $0.16x + 0.72y = 0.30(6)$ could have been simplified by multiplying both sides of the equation by 100. This would give the equation $16x + 72y = 30(6)$ or $16x + 72y = 180$. Then the system of equations would be $x + y = 6$ and $16x + 72y = 180$. If you solve this system, you should obtain the same solution. Try it and see.

2 Use Linear Systems in Three Variables to Solve Applications

Now let us look at some applications that involve three equations and three variables.

EXAMPLE 6 **Bank Loans** Tiny Tots Toys must borrow $25,000 to pay for an expansion. It is not able to obtain a loan for the total amount from a single bank, so it takes out loans from three different banks. It borrows some of the money at a bank that charges it 8% interest. At the second bank, it borrows $2000 more than one-half the amount borrowed from the first bank. The interest rate at the second bank is 10%. The balance of the $25,000 is borrowed from a third bank, where Tiny Tots pays 9% interest. The total annual interest Tiny Tots Toys pays for the three loans is $2220. How much does it borrow at each rate?

Solution   Understand   We are asked to determine how much is borrowed at each of the three different rates. Therefore, this problem will contain three variables, one for each amount borrowed. Since the problem will contain three variables, we will need to determine three equations to use in our system of equations.

Translate   Let x = amount borrowed at first bank.

Let y = amount borrowed at second bank.

Let z = amount borrowed at third bank.

Since the total amount borrowed is $25,000 we know that

$$x + y + z = 25,000 \qquad \text{Total amount borrowed is \$25,000.}$$

At the second bank, Tiny Tots Toys borrows $2000 more than one-half the amount borrowed from the first bank. Therefore, our second equation is

$$y = \frac{1}{2}x + 2000 \qquad \text{Second, } y, \text{ is \$2000 more than } \frac{1}{2} \text{ of first, } x.$$

Our last equation comes from the fact that the total annual interest charged by the three banks is $2220. The interest at each bank is found by multiplying the interest rate by the amount borrowed.

$$0.08x + 0.10y + 0.09z = 2220 \qquad \text{Total interest is \$2220.}$$

Thus, our system of equations is

$$x + y + z = 25,000 \qquad (1)$$

$$y = \frac{1}{2}x + 2000 \qquad (2)$$

$$0.08x + 0.10y + 0.09z = 2220 \qquad (3)$$

Both sides of equation (2) can be multiplied by 2 to remove fractions.

$$2(y) = 2\left(\frac{1}{2}x + 2000\right)$$

$$2y = x + 4000 \qquad \qquad \text{Distributive property}$$

$$-x + 2y = 4000 \qquad \qquad \text{Subtract } x \text{ from both sides.}$$

The decimals in equation (3) can be removed by multiplying both sides of the equation by 100. This gives

$$8x + 10y + 9z = 222,000$$

> **Understanding Algebra**
>
> Whenever we use *three* variables to solve an application problem, we must determine *three* equations to solve the problem.

Our simplified system of equations is therefore

$$x + y + z = 25{,}000 \quad (eq.\,1)$$
$$-x + 2y \qquad = 4000 \quad (eq.\,2)$$
$$8x + 10y + 9z = 222{,}000 \quad (eq.\,3)$$

Carry Out There are various ways of solving this system. Let's use (*eq.* 1) and (*eq.* 3) to eliminate the variable z.

$$
\begin{array}{ll}
-9x - 9y - 9z = -225{,}000 & (eq.\,1) \quad \text{Multiplied by } -9 \\
\underline{8x + 10y + 9z = \;\;222{,}000} & (eq.\,3) \\
-x + \;\;y \qquad\quad = \;\;-3{,}000 & \text{Sum of equations, } (eq.\,4)
\end{array}
$$

Now we use (*eq.* 2) and (*eq.* 4) to eliminate the variable x and solve for y.

$$
\begin{array}{ll}
x - 2y = -4000 & (eq.\,2) \quad \text{Multiplied by } -1 \\
\underline{-x + \;\;y = -3000} & (eq.\,4) \\
-y = -7000 & \text{Sum of equations} \\
y = \;\;7000 &
\end{array}
$$

Now that we know the value of y we can solve for x.

$$
\begin{array}{ll}
-x + 2y = 4000 & (eq.\,2) \\
-x + 2(7000) = 4000 & \text{Substitute 7000 for } y \text{ in } (eq.\,2). \\
-x + 14{,}000 = 4000 & \\
-x = -10{,}000 & \\
x = \;\;10{,}000 &
\end{array}
$$

Finally, we solve for z.

$$
\begin{array}{l}
x + y + z = 25{,}000 \quad (eq.\,1) \\
10{,}000 + 7000 + z = 25{,}000 \\
17{,}000 + z = 25{,}000 \\
z = \;\;8000
\end{array}
$$

Answer Tiny Tots Toys borrows $10,000 at 8%, $7000 at 10%, and $8000 at 9% interest.

Now Try Exercise 55

EXAMPLE 7 **Inflatable Boats** Hobson, Inc., has a small manufacturing plant that makes three types of inflatable boats: one-person, two-person, and four-person models. Each boat requires the service of three departments: cutting, assembly, and packaging. The cutting, assembly, and packaging departments are allowed to use a total of 380, 330, and 120 person-hours per week, respectively. The time requirements for each boat and department are specified in the following table. Determine how many of each type of boat Hobson must produce each week for its plant to operate at full capacity.

| | Time (person-hr) per Boat | | |
Department	One-Person Boat	Two-Person Boat	Four-Person Boat
Cutting	0.6	1.0	1.5
Assembly	0.6	0.9	1.2
Packaging	0.2	0.3	0.5

Solution **Understand** We are told that three different types of boats are produced and we are asked to determine the number of each type produced. Since this problem involves three amounts to be found, the system will contain three equations in three variables.

Translate   We will use the information given in the table.

Let x = number of one-person boats.

Let y = number of two-person boats.

Let z = number of four-person boats.

The total number of cutting hours for the three types of boats must equal 380 person-hours.

$$0.6x + 1.0y + 1.5z = 380$$

The total number of assembly hours must equal 330 person-hours.

$$0.6x + 0.9y + 1.2z = 330$$

The total number of packaging hours must equal 120 person-hours.

$$0.2x + 0.3y + 0.5z = 120$$

Therefore, the system of equations is

$$0.6x + 1.0y + 1.5z = 380$$
$$0.6x + 0.9y + 1.2z = 330$$
$$0.2x + 0.3y + 0.5z = 120$$

Multiplying each equation in the system by 10 will eliminate the decimal numbers and give a simplified system of equations.

$$6x + 10y + 15z = 3800 \quad (eq.\,1)$$
$$6x + 9y + 12z = 3300 \quad (eq.\,2)$$
$$2x + 3y + 5z = 1200 \quad (eq.\,3)$$

Carry Out   Let's first eliminate the variable x using ($eq.\,1$) and ($eq.\,2$), and then ($eq.\,1$) and ($eq.\,3$).

$$6x + 10y + 15z = 3800 \quad (eq.\,1)$$
$$\underline{-6x - 9y - 12z = -3300} \quad (eq.\,2) \text{ Multiplied by } -1$$
$$y + 3z = 500 \quad \text{Sum of equations, } (eq.\,4)$$

$$6x + 10y + 15z = 3800 \quad (eq.\,1)$$
$$\underline{-6x - 9y - 15z = -3600} \quad (eq.\,3) \text{ Multiplied by } -3$$
$$y = 200 \quad \text{Sum of equations, } (eq.\,5)$$

Note that when we added the last two equations, both variables x and z were eliminated at the same time. Now we know the value of y and can solve for z.

$$y + 3z = 500 \quad (eq.4)$$
$$200 + 3z = 500 \quad \text{Substitute 200 for } y.$$
$$3z = 300$$
$$z = 100$$

Finally, we find x.

$$6x + 10y + 15z = 3800 \quad (eq.\,1)$$
$$6x + 10(200) + 15(100) = 3800$$
$$6x + 2000 + 1500 = 3800$$
$$6x + 3500 = 3800$$
$$6x = 300$$
$$x = 50$$

Answer   Hobson should produce 50 one-person boats, 200 two-person boats, and 100 four-person boats per week.

Now Try Exercise 59

EXERCISE SET 4.3

MathXL®   MyMathLab

Practice the Skills/Problem Solving

1. **Land Area** The combined land area of the countries of Georgia and Ireland is 139,973 square kilometers. The difference between the two countries land area is 573 square kilometers. If Ireland has the larger land area, determine the land area of each country.

Cliffs of Moher, Ireland

2. **Daytona 500 Wins** As of this writing, Richard Petty has won the Daytona 500 race the greatest number of times and Dale Yarborough has won the second greatest number of Daytona 500 races. Petty's number of wins is one less than twice Yarborough's number of wins. The total number of wins by the two drivers is 11. Determine the number of wins by Petty and by Yarborough.

Richard Petty

3. **Fat Content** A nutritionist finds that a large order of fries at McDonald's has more fat than a McDonald's quarter-pound hamburger. The fries have 4 grams more than three times the amount of fat that the hamburger has. The difference in the fat content between the fries and the hamburger is 46 grams. Find the fat content of the hamburger and of the fries.

4. **Theme Parks** The two most visited theme parks in the United States in 2007 were Walt Disney's Magic Kingdom in Florida and Disneyland in California. The total number of visitors to these parks was 31.8 million people. The number of people who visited the Magic Kingdom was 2.2 million more than the number of people who visited Disneyland. How many people visited each of these parks in 2004? *Source:* www.coastergrotto.com

5. **Hot Dog Stand** At Big Al's hot dog stand, 2 hot dogs and 3 sodas cost $7. The cost of 4 hot dogs and 2 sodas is $10. Determine the cost of a hot dog and the cost of a soda.

6. **Water and Pretzel** At a professional football game, the cost of 2 bottles of water and 3 pretzels is $16.50. The cost of 4 bottles of water and 1 pretzel is $15.50. Determine the cost of a bottle of water and the cost of a pretzel.

7. **Memory Cards** For her digital camera, Judy Ackerman purchased a 512-MB memory card and a 4-GB memory card. Together the two memory cards can store a total of 996 photos. The 4-GB memory card can store 15 more than eight times the number of photos the 512-MB memory card can store. How many photos can each memory card store?

8. **Washer and Dryer** Beth Rechsteiner purchased a new washer and dryer for a total price of $1650. The dryer cost $150 less than the washer. Determine the price of the washer and the price of the dryer.

9. **Complementary Angles** Two angles are **complementary angles** if the sum of their measures is 90.° (See Section 2.3.) If the measure of the larger of two complementary angles is 15° more than two times the measure of the smaller angle, find the measures of the two angles.

10. **Complementary Angles** The difference between the measures of two complementary angles is 46°. Determine the measures of the two angles.

11. **Supplementary Angles** Two angles are **supplementary angles** if the sum of their measures is 180°. (See Section 2.3.) Find the measures of two supplementary angles if the measure of one angle is 28° less than three times the measure of the other.

12. **Supplementary Angles** Determine the measures of two supplementary angles if the measure of one angle is three and one half times larger than the measure of the other angle.

13. **Rowing Speed** The Heart O'Texas Rowing Team, while practicing in Austin, Texas, rowed an average of 15.6 miles per hour with the current and 8.8 miles per hour against the current. Determine the team's rowing speed in still water and the speed of the current.

14. **Flying Speed** Jung Lee, in his Piper Cub airplane, flew an average of 121 miles per hour with the wind and 87 miles per hour against the wind. Determine the speed of the airplane in still air and the speed of the wind.

15. **Salary Plus Commission** Don Lavigne, an office equipment sales representative, earns a weekly salary plus a commission on his sales. One week his total compensation on sales of $4000 was $660. The next week his total compensation on sales of $6000 was $740. Find Don's weekly salary and his commission rate.

16. **Truck Rental** A truck rental agency charges a daily fee plus a mileage fee. Hugo was charged $85 for 2 days and 100 miles and Christina was charged $165 for 3 days and 400 miles. What is the agency's daily fee, and what is the mileage fee?

17. **Lavender Oil** Pola Sommers, a massage therapist, needs 3 ounces of a 20% lavender oil solution. She has only 5% and 30% lavender oil solutions available. How many ounces of each should Pola mix to obtain the desired solution?

18. **Fertilizer Solutions** Frank Ditlman needs to apply a 10% liquid nitrogen solution to his rose garden, but he only has a 4% liquid nitrogen solution and a 20% liquid nitrogen solution available. How much of the 4% solution and how much of the 20% solution should Frank mix together to get 10 gallons of the 10% solution?

19. **Weed Killer** Round-Up Concentrate Grass and Weed Killer consists of an 18% active ingredient glyphosate (and 82% inactive ingredients). The concentrate is to be mixed with water and the mixture applied to weeds. If the final mixture is to contain 0.9% active ingredient, how much concentrate and how much water should be mixed to make 200 gallons of the final mixture?

20. **Lawn Fertilizer** Scott's Winterizer Lawn Fertilizer is 22% nitrogen. Schultz's Lime with Lawn Fertilizer is 4% nitrogen. William Weaver, owner of Weaver's Nursery, wishes to mix these two fertilizers to make 400 pounds of a special 10% nitrogen mixture for midseason lawn feeding. How much of each fertilizer should he mix?

21. **Birdseed** Birdseed costs $0.59 a pound and sunflower seeds cost $0.89 a pound. Angela Leinenbachs' pet store wishes to make a 40-pound mixture of birdseed and sunflower seeds that sells for $0.76 per pound. How many pounds of each type of seed should she use?

22. **Coffee** Franco Manue runs a grocery store. He wishes to mix 30 pounds of coffee to sell for a total cost of $170. To obtain the mixture, he will mix coffee that sells for $5.20 per pound with coffee that sells for $6.30 per pound. How many pounds of each coffee should he use?

23. **Amtrak** Ann Marie Whittle has been pricing Amtrak fares for a group to visit New York. Three adults and four children would cost a total of $159. Two adults and three children would cost a total of $112. Determine the price of an adult's ticket and a child's ticket.

24. **Buffalo Wings** The Wing House sells both regular size and jumbo size orders of Buffalo chicken wings. Three regular orders and five jumbo orders of wings cost $67. Four regular and four jumbo orders of wings cost $64. Determine the cost of a regular order of wings and a jumbo order of wings.

25. **Savings Accounts** Mr. and Mrs. Gamton invest a total of $10,000 in two savings accounts. One account pays 5% interest and the other 6%. Find the amount placed in each account if the accounts receive a total of $540 in interest after 1 year. Use interest = principal · rate · time.

26. **Investments** Louis Okonkwo invested $30,000, part at 9% and part at 5%. If he had invested the entire amount at 6.5%, his total annual interest would be the same as the sum of the annual interest received from the two other accounts. How much was invested at each interest rate?

27. **Milk** Becky Slaats is a plant engineer at Velda Farms Dairy Cooperative. She wishes to mix whole milk, which is 3.25% fat, and skim milk, which has no fat, to obtain 260 gallons of a mixture of milk that contains 2% fat. How many gallons of whole milk and how many gallons of skim milk should Becky mix to obtain the desired mixture?

28. **Quiche Lorraine** Lambert Marx's recipe for quiche lorraine calls for 2 cups (16 ounces) of light cream that is 20% butterfat. It is often difficult to find light cream with 20% butterfat at the supermarket. What is commonly found is heavy cream, which is 36% butterfat, and half-and-half, which is 10.5% butterfat. How much of the heavy cream and how much of the half-and-half should Lambert mix to obtain the mixture necessary for the recipe?

29. **Birdseed** By ordering directly through www.birdseed.com, the Carters can purchase Season's Choice birdseed for $1.79 per pound and Garden Mix birdseed for $1.19 per pound. If they wish to purchase 20 pounds and spend $28 on birdseed, how many pounds of each type should they buy?

30. **Juice** The Healthy Favorites Juice Company sells apple juice for 8.3¢ an ounce and raspberry juice for 9.3¢ an ounce. The company wishes to market and sell 8-ounce cans of apple-raspberry juice for 8.7¢ an ounce. How many ounces of each should be mixed?

31. **Car Travel** Two cars start at the same point in Alexandria, Virginia, and travel in opposite directions. One car travels 5 miles per hour faster than the other car. After 4 hours, the two cars are 420 miles apart. Find the speed of each car.

32. **Road Construction** Kip Ortiz drives from Atlanta to Louisville, a distance of 430 miles. Due to road construction and heavy traffic, during the first part of his trip, Kip drives at an average rate of 50 miles per hour. During the rest of his trip he drives at an average rate of 70 miles per hour. If his total trip takes 7 hours, how many hours does he drive at each speed?

33. **Avon Conference** Cabrina Wilson and Dabney Jefferson are Avon representatives who are attending a conference in Seattle. After the conference, Cabrina drives home to Boise at an average speed of 65 miles per hour and Dabney drives home to Portland at an average speed of 50 miles per hour. If the sum of their driving times is 11.4 hours and if the sum of the distances driven is 690 miles, determine the time each representative spent driving home.

34. **Exercise** For her exercise routine, Cynthia Harrison rides a bicycle for half an hour and then rollerblades for half an hour. Cynthia rides the bicycle at a speed that is twice the speed at which she rollerblades. If the total distance covered is 12 miles, determine the speed at which she bikes and rollerblades.

35. **Dog Diet** Bill Lutes's dog is on a strict diet. The dog is to receive, among other nutrients, 20 grams of protein and 6 grams of carbohydrates. Bill has only two food mixes available of the following compositions. How many grams of each mix should be used to obtain the right diet for his dog?

Mix	Protein (%)	Carbohydrate (%)
Mix A	10	6
Mix B	20	2

© Micicmakin/Shutterstock

36. **Chair Manufacturing** A company makes two models of chairs. Information about the construction of the chairs is given in the table shown. On a particular day the company allocated 46.4 person-hours for assembling and 8.8 person-hours for painting. How many of each chair can be made?

Model	Time to Assemble	Time to Paint
Model A	1 hr	0.5 hr
Model B	3.2 hr	0.4 hr

37. **Brass Alloy** By weight, one alloy of brass is 70% copper and 30% zinc. Another alloy of brass is 40% copper and 60% zinc. How many grams of each of these alloys need to be melted and combined to obtain 300 grams of a brass alloy that is 60% copper and 40% zinc?

38. **Silver Alloy** Sterling silver is 92.5% pure silver. How many grams of pure (100%) silver and how many grams of sterling silver must be mixed to obtain 250 g of a 94% silver alloy?

39. **Internal Revenue Service** The following graph shows the number of paper tax returns and the number of online tax returns filed with the IRS in the years from 2002 to 2010. If t represents the number of years since 2002, the number of paper tax returns, in millions, filed with the IRS can be estimated by the function $P(t) = -2.73t + 58.37$ and the number of online tax returns, in millions, filed with the IRS can be estimated by the function $o(t) = 1.95t + 10.58$. Assuming this trend continues, solve this system of equations to determine the year that the number of paper tax returns will be the same as the number of online tax returns.

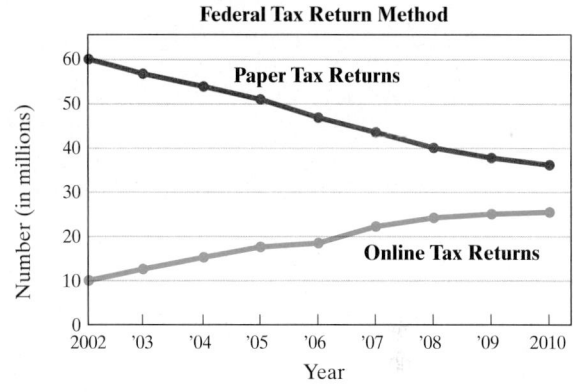

Federal Tax Return Method

Source: www.irs.gov/pubs

40. **Walking and Jogging** Cuong Tham tries to exercise every day. He walks at 3 miles per hour and then jogs at 5 miles per hour. If it takes him 0.9 hours to travel a total of 3.5 miles, how long does he jog?

41. **Texas Driving** Tom Johnson and Melissa Acino started driving at the same time in different cars from Oklahoma City. They both traveled south on Route 35. When Melissa reached the Dallas/Ft. Worth area, a distance of 150 miles, Tom had only reached Denton, Texas, a distance of 120 miles. If Melissa averaged 15 miles per hour faster than Tom, find the average speed of each car.

42. **Photocopy Costs** At a local copy center two plans are available.
    Plan 1: 10¢ per copy
    Plan 2: an annual charge of $120 plus 4¢ per copy
    a) Represent this information as a system of equations.
    b) Graph the system of equations for up to 4000 copies made.
    c) From the graph, estimate the number of copies a person would have to make in a year for the two plans to have the same total cost.
    d) Solve the system algebraically. If your answer does not agree with your answer in part **c)**, explain why.

In Exercises 43–62, solve each problem using a system of three equations in three unknowns.

43. **Mail Volume** The average American household receives 24 pieces of mail each week. The number of bills and statements is two less than twice the number of pieces of personal mail. The number of advertisements is two more than five times the number of pieces of personal mail. How many pieces of personal mail, bills and statements, and advertisements does the average family get each week?
    Source: Arthur D. Little, Inc.

44. Submarine Personnel A 141-person crew is standard on a Los Angeles class submarine. The number of chief petty officers (enlisted) is four more than the number of commissioned officers. The number of other enlisted men is three less than eight times the number of commissioned officers. Determine the number of commissioned officers, chief petty officers, and other enlisted people on the submarine.

USS *Greenville*

45. College Football Bowl Games Through 2007, the Universities of Alabama, Tennessee, and Texas have had the most appearances in college football bowl games. These three schools have had a total of 147 bowl appearances. Alabama has had 8 more appearances than Texas. Together, the number of appearances by Tennessee and Texas is 39 more than the number of appearances by Alabama. Determine the number of bowl appearances for each school.

46. Summer Olympics In the 2008 Summer Olympics in Beijing, China, the countries earning the most medals were the United States, China, and Russia. Together, these three countries earned a total of 282 medals. The United States earned 10 more medals than China. Together the number of medals earned by the United States and Russia is 18 less than twice the number of medals earned by China. Determine the number of medals each country earned.

Source: en.beijing2008.cn

47. NPF Softball The three teams with the most wins in the 2008 National Pro Fastpitch (NPF) softball league regular season were the Chicago Bandits, the Philadelphia Force, and the Washington Glory. These three teams had a total of 93 wins. The Bandits had two more wins than the Glory and the Force had one less win than the Bandits. Determine the number of wins each team had.

Source: www.profastpitch.com

48. Wettest Cities The three cities within the continental United States that receive the highest average annual rainfall are Mobile, Alabama, Pensacola, Florida, and New Orleans, Louisiana. The total average annual rainfall for all three cities is 196 inches. Mobile has an average rainfall that is 3 inches more than that of New Orleans. New Orleans has an average rainfall that is 1 inch less than that of Pensacola. Determine the average annual rainfall for each of the three cities.

Source: www.livescience.com

49. Top-Selling U.S. Albums As of 2008, the best-selling music albums based on sales in the United States are The Eagles—*Their Greatest Hits 1971–1975,* Michael Jackson—*Thriller,* and Pink Floyd—*The Wall.* The total number of the three albums sold in the United States is 79.5 million. The number of *Thriller* albums sold was 2 million less than the number *of Their Greatest Hits 1971–1975* albums sold. The number of *The Wall* albums sold was 3.5 million less than the number of *Thriller* albums sold. Determine the number of each album sold.

Source: Recording Industry Association of America

The Eagles—*Their Greatest Hits 1971–1975*

50. Best-Selling Movies As of 2008, The best-selling movies based on box-offices sales of all time are *Titanic, The Lord of the Rings: The Return of the King,* and *Pirates of the Caribbean: Dead Man's Chest.* The total sales of all three movies are $4024 million. The combined sales of *The Return of the King* and *Dead Man's Chest* are $354 million more than sales of *Titanic.* Sales of *Return of the King* are $69 million more than sales of *Dead Man's Chest.* Determine the sales for each of the movies.

Source: All-Time Worldwide Box Office

51. Super Bowls As of 2010, the states of Florida, California, and Louisiana have hosted the most Super Bowls. These three states hosted a total of 35 Super Bowls. Florida hosted 6 more Super Bowls than Louisiana. Together, Florida and Louisiana hosted two more than twice the number California hosted. Determine the number of Super Bowls hosted by each of these three states.

Source: www.nfl.com

52. Concert Tickets Three kinds of tickets for a Soggy Bottom Boys concert are available: up-front, main-floor, and balcony. The most expensive tickets, the up-front tickets, are twice as expensive as balcony tickets. Balcony tickets are $10 less than main-floor tickets and $30 less than up-front tickets. Determine the price of each kind of ticket.

53. Triangle The sum of the measures of the angles of a triangle is 180°. The smallest angle of the triangle has a measure $\frac{2}{3}$ the measure of the second smallest angle. The largest angle has a measure that is 30° less than three times the measure of the second smallest angle. Determine the measure of each angle.

54. Triangle The largest angle of a triangle has a measure that is 10° less than three times the measure of the second smallest angle. The measure of the smallest angle is equal to the difference between the measure of the largest angle and twice the measure of the second smallest angle. Determine the measures of the three angles of the triangle.

55. Investments Tam Phan received a check for $10,000. She decided to divide the money (not equally) into three different investments. She placed part of her money in a savings account paying 3% interest. The second amount, which was twice the first amount, was placed in a certificate of deposit paying 5% interest. She placed the balance in a money market fund paying 6% interest. If Tam's total interest over the period of 1 year was $525.00, how much was placed in each account?

56. Bonus Nick Pfaff, an attorney, divided his $15,000 holiday bonus check among three different investments. With some of the money, he purchased a municipal bond paying 5.5% simple interest. He invested twice the amount of money that he paid for the municipal bond in a certificate of deposit paying 4.5% simple interest. Nick placed the balance of the money in a money market account paying 3.75% simple interest. If Nick's total interest for 1 year was $692.50, how much was placed in each account?

57. Hydrogen Peroxide A 10% solution, a 12% solution, and a 20% solution of hydrogen peroxide are to be mixed to get 8 liters of a 13% solution. How many liters of each must be mixed if the volume of the 20% solution must be 2 liters less than the volume of the 10% solution?

58. Sulfuric Acid An 8% solution, a 10% solution, and a 20% solution of sulfuric acid are to be mixed to get 100 milliliters of a 12% solution. If the *volume of acid* from the 8% solution is to equal half the *volume of acid* from the other two solutions, how much of each solution is needed?

59. Furniture Manufacturing Donaldson Furniture Company produces three types of rocking chairs: the children's model, the standard model, and the executive model. Each chair is made in three stages: cutting, construction, and finishing. The time needed for each stage of each chair is given in the following chart. During a specific week the company has available a maximum of 154 hours for cutting, 94 hours for construction, and 76 hours for finishing. Determine how many of each type of chair the company should make to be operating at full capacity.

Stage	Children's	Standard	Executive
Cutting	5 hr	4 hr	7 hr
Construction	3 hr	2 hr	5 hr
Finishing	2 hr	2 hr	4 hr

60. Bicycle Manufacturing The Jamis Bicycle Company produces three models of bicycles: Dakar, Komodo, and Aragon. Each bicycle is made in three stages: welding, painting, and assembling. The time needed for each stage of each bicycle is given in the chart below. During a specific week, the company has available a maximum of 133 hours for welding, 78 hours for painting, and 96 hours for assembling. Determine how many of each type of bicycle the company should make to be operating at full capacity.

Stage	Dakar	Komodo	Aragon
Welding	2	3	4
Painting	1	2	2.5
Assembling	1.5	2	3

61. Current Flow In electronics it is necessary to analyze current flow through paths of a circuit. In three paths (A, B, and C) of a circuit, the relationships are the following:

$$
\begin{aligned}
I_A + I_B + I_C &= 0 \\
-8I_B + 10I_C &= 0 \\
4I_A - 8I_B &= 6
\end{aligned}
$$

where I_A, I_B, and I_C represent the current in paths A, B, and C, respectively. Determine the current in each path of the circuit.

62. Forces on a Beam In physics we often study the forces acting on an object. For three forces, F_1, F_2, and F_3, acting on a beam, the following equations were obtained.

$$
\begin{aligned}
3F_1 + F_2 - F_3 &= 2 \\
F_1 - 2F_2 + F_3 &= 0 \\
4F_1 - F_2 + F_3 &= 3
\end{aligned}
$$

Find the three forces.

Group Activity

Discuss and answer Exercise 63 as a group.

63. Two Cars A *nonlinear system of equations* is a system of equations containing at least one equation that is not linear. (Nonlinear systems of equations will be discussed in Chapter 10.) The graph shows a nonlinear system of equations. The curves represent speed versus time for two cars.

a) Are the two curves functions? Explain.

b) Discuss the meaning of this graph.

c) At time $t = 0.5$ hr, which car is traveling at a greater speed? Explain your answer.

d) Assume the two cars start at the same position and are traveling in the same direction. Which car, A or B, traveled farther in 1 hour? Explain your answer.

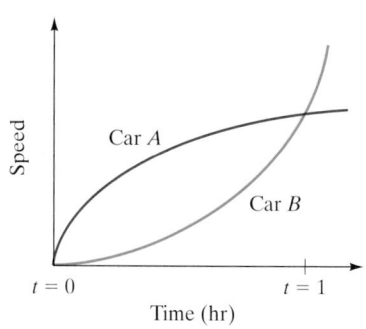

Cumulative Review Exercises

[1.4]   **64.** Evaluate $\frac{1}{2}x + \frac{2}{5}xy + \frac{1}{8}y$ when $x = -2$, $y = 5$.

[2.1]   **65.** Solve $4 - 2[(x - 5) + 2x] = -(x + 6)$.

[3.2]   **66.** Explain how to determine whether a graph represents a function.

[3.5]   **67.** Write an equation of the line that passes through points $(6, -4)$ and $(2, -8)$.

Mid-Chapter Test: 4.1–4.3

To find out how well you understand the chapter material to this point, take this brief test. The answers, and the section where the material was initially discussed, are given in the back of the book. Review any questions that you answered incorrectly.

1. For the following system of equations:

a) Write each equation in slope-intercept form.

b) Without graphing the equations, state whether the system is consistent, inconsistent, or dependent.

c) Indicate whether the system has exactly one solution, no solution, or an infinite number of solutions.

$$7x - y = 13$$
$$2x + 3y = 9$$

Solve each system of equations using the graphing method.

2. $y = 2x$
   $y = -x + 3$

3. $x + y = -4$
   $3x - 2y = 3$

Solve each system of equations using the substitution method.

4. $2x + 5y = -3$
   $x - 2y = -6$

5. $4x - 3y = 8$
   $2x + y = -1$

Solve each system of equations using the addition method.

6. $x = 4y - 19$
   $7x + 5y = -1$

7. $3x + 4y = 3$
   $9x + 5y = \dfrac{11}{2}$

Solve each system of equations by any method. If the system is inconsistent or dependent, so state.

8. $\dfrac{1}{3}a - \dfrac{1}{4}b = -1$

   $\dfrac{1}{2}a + \dfrac{1}{6}b = 5$

9. $3m - 2n = 1$

   $n = \dfrac{3}{2}m - 7$

10. $8x - 16y = 24$
   $x = 2y + 3$

Solve each system of equations.

11. $x + y + z = 2$
   $2x - y + 2z = -2$
   $3x + 2y + 6z = 1$

12. $2x - y - z = 1$
   $3x + 5y + 2z = 12$
   $-6x - 4y + 5z = 3$

13. When asked to solve the system of equations

$$\begin{aligned} x + \ y + \ z &= \ \ 4 \\ -x + 2y + 2z &= \ \ 5 \\ 7x + 5y - \ z &= -2 \end{aligned}$$

Frank Dumont claimed that the solution was only $x = 1$. This is incorrect. Why is it incorrect? Explain your answer. Then solve the system completely.

14. Cashews and Pecans A local nut shop sells cashews for $12 per pound and pecans for $6 per pound. How many pounds of each type should William Pritchard buy to have a 15-pound mixture that sells for $10 per pound?

15. Sum of Numbers The sum of three numbers is 32. The largest number is four times the smallest number. The sum of the two smaller numbers is 8 less than the largest number. Find the three numbers.

4.4 Solving Systems of Equations Using Matrices

1 Write an augmented matrix.

2 Solve systems of linear equations.

3 Solve systems of linear equations in three variables.

4 Recognize inconsistent and dependent systems.

1 Write an Augmented Matrix

Matrix

- A **matrix** is a rectangular array of numbers written within brackets. The plural of *matrix* is **matrices**.
- The numbers in a matrix are called the **elements** of the matrix.
- The **dimensions** of a matrix are the number of rows by the number of columns.

An example of a matrix is

3 columns

$$\downarrow \quad \downarrow \quad \downarrow$$

2 rows $\longrightarrow$ $\begin{bmatrix} 5 & 7 & 2 \\ -1 & 3 & 4 \end{bmatrix}$

The dimensions of this matrix are 2 rows by 3 columns. We write 2×3 and we say "2 by 3."

A **square matrix** is a matrix with the same number of rows and columns. An example of a 2×2 square matrix is

2 columns

$$\downarrow \quad \downarrow$$

2 rows $\longrightarrow$ $\begin{bmatrix} 4 & 6 \\ 9 & -2 \end{bmatrix}$

We will solve systems of linear equations using an *augmented matrix*. An **augmented matrix** is a matrix made up of two smaller matrices separated by a vertical line. An augmented matrix that represents a system of linear equations will have the numerical coefficients to the left of the vertical line and the constants to the right of the vertical line as illustrated below. The system of equations

$$2x - 3y = 10$$
$$4x - 5y = 9$$

is represented with the augmented matrix

coefficients
on x \qquad constants

$$\downarrow \qquad \downarrow$$

$$\begin{bmatrix} 2 & -3 & 10 \\ 4 & -5 & 9 \end{bmatrix}$$

$$\uparrow$$

coefficients
on y

The 2×2 matrix $\begin{bmatrix} 2 & -3 \\ 4 & -5 \end{bmatrix}$ is the **coefficient matrix**. The 2×1 matrix $\begin{bmatrix} 10 \\ 9 \end{bmatrix}$ is the **constant matrix**. So, the matrix that represents the system of equations $\begin{bmatrix} 2 & -3 & 10 \\ 4 & -5 & 9 \end{bmatrix}$ is the coefficient matrix *augmented* by the constant matrix.

When we represent a system of linear equations with an augmented matrix, each equation needs to be in the form $ax + by = c$. For example, the system of equations

$$y = -2x + 5$$
$$x = 7y - 4$$
would be rewritten as
$$2x + y = 5$$
$$x - 7y = -4$$

This system can be represented with the augmented matrix

$$\begin{bmatrix} 2 & 1 & 5 \\ 1 & -7 & -4 \end{bmatrix}$$

When using an augmented matrix to solve a system of linear equations, we will use a method very similar to the addition method discussed in Section 4.1

2 Solve Systems of Linear Equations

To solve a system of two linear equations using matrices, we perform certain procedures discussed below so that we can rewrite the augmented matrix in **row echelon** (or **triangular**) **form**:

$$\left[\begin{array}{cc|c} 1 & a & p \\ 0 & 1 & q \end{array}\right]$$

where the a, p, and q represent numbers. From this type of augmented matrix we can write an equivalent system of equations. This matrix represents the linear system

$$
\begin{array}{cc}
\begin{array}{l} 1x + ay = p \\ 0x + 1y = q \end{array}
& \text{or} \quad
\begin{array}{l} x + ay = p \\ y = q \end{array}
\end{array}
$$

For example,

$$\left[\begin{array}{cc|c} 1 & 2 & 4 \\ 0 & 1 & 5 \end{array}\right] \quad \text{represents} \quad \begin{array}{l} x + 2y = 4 \\ y = 5 \end{array}$$

Note that the system above on the right can be easily solved by substitution. Its solution is $(-6, 5)$.

We use **row transformations** to rewrite the augmented matrix in row echelon form. We will use three row transformation procedures. Row transformations on a matrix are equivalent to operations you would perform on the equations in the system.

Understanding Algebra

When working with an augmented matrix

- to obtain a 1, we usually use the first row transformation procedure.

- to obtain a 0, we usually use the second row transformation procedure.

Procedures for Row Transformations

1. All the numbers in a row may be multiplied (or divided) by any nonzero real number. (This is the same as multiplying both sides of an equation by a nonzero real number.)

2. All the numbers in a row may be multiplied by any nonzero real number. These products may then be added to the corresponding numbers in any other row. (This is equivalent to eliminating a variable from a system of equations using the addition method.)

3. The order of the rows may be switched. (This is equivalent to switching the order of the equations in a system of equations.)

Generally, when obtaining a 1 in the augmented matrix we use row transformation procedure 1, and when obtaining a 0 we use row transformation procedure 2. *Work by columns starting from the left.* Start with the first column, first row.

EXAMPLE 1 Solve the following system of equations using matrices.

$$
\begin{array}{l}
2x + 4y = -2 \\
3x - 2y = 5
\end{array}
$$

Solution   First we write the augmented matrix.

$$\left[\begin{array}{cc|c} 2 & 4 & -2 \\ 3 & -2 & 5 \end{array}\right]$$

Our goal is to obtain a matrix of the form $\left[\begin{array}{cc|c} 1 & a & p \\ 0 & 1 & q \end{array}\right]$. We begin by using row transformation procedure 1 to replace the 2 in the first column, first row, with 1. To do so, we multiply the first row of numbers by $\dfrac{1}{2}$. We abbreviate this multiplication as $\dfrac{1}{2}R_1$ and place it to the right of the matrix in the same row where the operation was performed.

$$\left[\begin{array}{cc|c} \left(\frac{1}{2}\right)2 & \left(\frac{1}{2}\right)4 & \left(\frac{1}{2}\right)(-2) \\ 3 & -2 & 5 \end{array}\right] \quad \frac{1}{2}R_1$$

or

$$\left[\begin{array}{cc|c} 1 & 2 & -1 \\ 3 & -2 & 5 \end{array}\right]$$

We next use the row transformations to produce a 0 in the first column, second row where currently a 3 is in this position. We will multiply the elements in row 1 by -3 and add the products to row 2. We abbreviate this row transformation as $-3R_1 + R_2$.

The elements in the first row multiplied by -3 are:

$$-3(1) \qquad -3(2) \qquad -3(-1)$$

or

$$-3 \qquad -6 \qquad 3$$

Now add these products to their respective elements in the second row to get

$$\begin{bmatrix} 1 & 2 & -1 \\ -3+3 & -6+(-2) & 3+5 \end{bmatrix} \quad -3R_1 + R_2$$

or

$$\begin{bmatrix} 1 & 2 & -1 \\ 0 & -8 & 8 \end{bmatrix}$$

We next use row transformations to produce a 1 in the second row, second column where currently a -8 is in this position. We do this by multiplying the second row by $-\dfrac{1}{8}$.

$$\begin{bmatrix} 1 & 2 & -1 \\ \left(-\dfrac{1}{8}\right)0 & \left(-\dfrac{1}{8}\right)(-8) & \left(-\dfrac{1}{8}\right)8 \end{bmatrix} \quad -\dfrac{1}{8}R_2$$

or

$$\begin{bmatrix} 1 & 2 & -1 \\ 0 & 1 & -1 \end{bmatrix}$$

The matrix is now in row echelon form and the equivalent system of equations is

$$x + 2y = -1$$
$$y = -1$$

Now we can solve for x using substitution.

$$x + 2y = -1$$
$$x + 2(-1) = -1$$
$$x - 2 = -1$$
$$x = 1$$

A check will shown that $(1, -1)$ is the solution to the original system.

Now Try Exercise 19

3 Solve Systems of Linear Equations in Three Variables

Now we will use matrices to solve a system of three linear equations in three variables. We use the same row transformation procedures used when solving a system of two linear equations. Our goal is to produce an augmented matrix in the row echelon form

$$\begin{bmatrix} 1 & a & b & p \\ 0 & 1 & c & q \\ 0 & 0 & 1 & r \end{bmatrix}$$

where a, b, c, p, q and r represent numbers. This matrix represents the following system of equations.

$$\begin{array}{ccc} 1x + ay + bz = p & & x + ay + bz = p \\ 0x + 1y + cz = q & \text{or} & y + cz = q \\ 0x + 0y + 1z = r & & z = r \end{array}$$

Study Tip

When using matrices, be careful to keep all the numbers lined up neatly in rows and columns. One slight mistake in copying numbers from one matrix to another will lead to an incorrect attempt at solving a system of equations.

$$x - 3y + z = 3$$

For example, the system of equations, $4x + 2y - 5z = 20$, when correctly represented

$$-5x - y - 4z = 13$$

with the augmented matrix, $\begin{bmatrix} 1 & -3 & 1 & | & 3 \\ 4 & 2 & -5 & | & 20 \\ -5 & -1 & -4 & | & 13 \end{bmatrix}$, leads to the solution $(1, -2, -4)$.

However, a matrix that looks quite similar, $\begin{bmatrix} 1 & -3 & 1 & | & 3 \\ 4 & -1 & -5 & | & 20 \\ -5 & 2 & -4 & | & 13 \end{bmatrix}$, leads to the incorrect ordered triple of $\left(-\dfrac{25}{53}, -\dfrac{130}{53}, -\dfrac{206}{53} \right)$.

EXAMPLE 2 Solve the following system of equations using matrices.

$$x - 2y + 3z = -7$$
$$2x - y - z = 7$$
$$-x + 3y + 2z = -8$$

Solution   First write the augmented matrix.

$$\begin{bmatrix} 1 & -2 & 3 & | & -7 \\ 2 & -1 & -1 & | & 7 \\ -1 & 3 & 2 & | & -8 \end{bmatrix}$$

Next we use row transformations to produce a first column of $\begin{smallmatrix}1\\0\\0\end{smallmatrix}$. Since the number in the first column, first row, is already a 1, we will work with the 2 in the first column, second row. Multiplying the numbers in the first row by -2 and adding those products to the respective numbers in the second row will replace 2 with 0. The matrix is now

$$\begin{bmatrix} 1 & -2 & 3 & | & -7 \\ 0 & 3 & -7 & | & 21 \\ -1 & 3 & 2 & | & -8 \end{bmatrix} \quad -2R_1 + R_2$$

Continuing down the first column, we wish to replace the -1 with 0. By multiplying the numbers in the first row by 1, and then adding the products to the third row, we get

$$\begin{bmatrix} 1 & -2 & 3 & | & -7 \\ 0 & 3 & -7 & | & 21 \\ 0 & 1 & 5 & | & -15 \end{bmatrix} \quad 1R_1 + R_3$$

Now we work with the second column. We wish to produce a second column of the form $\begin{smallmatrix}a\\1\\0\end{smallmatrix}$ where a represents a number. Since there is presently a 1 in the third row, second column, and we want a 1 in the second row, second column, we switch the second and third rows of the matrix. This gives

$$\begin{bmatrix} 1 & -2 & 3 & | & -7 \\ 0 & 1 & 5 & | & -15 \\ 0 & 3 & -7 & | & 21 \end{bmatrix} \quad \text{Switch } R_2 \text{ and } R_3.$$

Continuing down the second column, we wish to replace the 3 in the third row with 0 by multiplying the numbers in the second row by -3 and adding those products to the third row. This gives

$$\begin{bmatrix} 1 & -2 & 3 & | & -7 \\ 0 & 1 & 5 & | & -15 \\ 0 & 0 & -22 & | & 66 \end{bmatrix} \quad -3R_2 + R_3$$

Now we work with the third column. We wish to produce a third column of the form $\begin{matrix} b \\ c \\ 1 \end{matrix}$ where b and c represent numbers. We can multiply the numbers in the third row by $-\dfrac{1}{22}$ to replace -22 with 1.

$$\begin{bmatrix} 1 & -2 & 3 & | & -7 \\ 0 & 1 & 5 & | & -15 \\ 0 & 0 & 1 & | & -3 \end{bmatrix} \quad -\tfrac{1}{22}R_3$$

This matrix is now in row echelon form. From this matrix we obtain the system of equations

$$x - 2y + 3z = -7$$
$$y + 5z = -15$$
$$z = -3$$

The third equation gives us the value of z in the solution. Now we can solve for y by substituting -3 for z in the second equation.

$$y + 5z = -15$$
$$y + 5(-3) = -15$$
$$y - 15 = -15$$
$$y = 0$$

Now we solve for x by substituting 0 for y and -3 for z in the first equation.

$$x - 2y + 3z = -7$$
$$x - 2(0) + 3(-3) = -7$$
$$x - 0 - 9 = -7$$
$$x - 9 = -7$$
$$x = 2$$

The solution is $(2, 0, -3)$. Check this now by substituting the appropriate values into each of the original equations.

Now Try Exercise 33

Understanding Algebra

When working with augmented matrices to solve a system of linear equations, if you produce

- A row with 0's to the left and a non-zero number to the right of the vertical bar, the system is inconsistent and has no solution.
- A row with all 0's, the system is dependent and has an infinite number of solutions.

4 Recognize Inconsistent and Dependent Systems

When solving a system of two equations, if you obtain an augmented matrix in which one row of numbers on the left side of the vertical line is all zeros but a zero does not appear in the same row on the right side of the vertical line, the system is inconsistent and has no solution. For example, a system of equations that yields the following augmented matrix is an inconsistent system.

$$\begin{bmatrix} 1 & 2 & | & 5 \\ 0 & 0 & | & 3 \end{bmatrix} \quad \longleftarrow \text{Inconsistent system}$$

The second row of the matrix represents the equation

$$0x + 0y = 3$$

which is never true.

If you obtain a matrix in which a 0 appears across an entire row, the system of equations is dependent. For example, a system of equations that yields the following augmented matrix is a dependent system.

$$\left[\begin{array}{cc|c} 1 & -3 & -4 \\ 0 & 0 & 0 \end{array}\right]$$ $\longleftarrow$ Dependent system

The second row of the matrix represents the equation

$$0x + 0y = 0$$

which is always true.

Similar rules hold for systems with three equations.

$$\left[\begin{array}{ccc|c} 1 & 3 & 7 & 5 \\ 0 & 0 & 0 & -1 \\ 0 & 1 & -2 & 3 \end{array}\right]$$ $\longleftarrow$ Inconsistent system

$$\left[\begin{array}{ccc|c} 1 & 3 & -1 & 2 \\ 0 & 0 & 0 & 0 \\ 0 & 5 & 6 & -4 \end{array}\right]$$ $\longleftarrow$ Dependent system

EXERCISE SET 4.4   *Math XL*   *MyMathLab*
MathXL®   MyMathLab

Warm-Up Exercises

Fill in the blanks with the appropriate word, phrase, or symbol(s) from the following list.

row transformations   augmented matrix   square matrix   dimensions   inconsistent   elements
row echelon   functions   rectangular form   column transformations   dependent

1. To solve a system of linear equations using matrices, we use _____ to rewrite the augmented matrix in row echelon form.

2. The numbers inside the matrix are called the _____.

3. The number of rows by the number of columns refer to the _____ of the matrix.

4. When a matrix consists of two smaller matrices separated by a vertical line it is called a(n) _____.

5. If, while solving a system of equations using an augmented matrix, you produce the matrix $\left[\begin{array}{cc|c} -2 & 7 & 1 \\ 0 & 0 & 6 \end{array}\right]$, you can conclude (assuming that all of your calculations are correct) that the system is _____ and has no solution.

6. If, while solving a system of equations using an augmented matrix, you produce the matrix $\left[\begin{array}{cc|c} -2 & 7 & 1 \\ 0 & 0 & 0 \end{array}\right]$, you can conclude (assuming all of your calculations are correct) that the system is _____ and has an infinite number of solutions.

7. The matrix $\left[\begin{array}{ccc} 2 & 4 & 0 \\ -1 & 1 & 5 \\ 3 & 6 & -2 \end{array}\right]$ is an example of a _____.

8. The augmented matrix $\left[\begin{array}{ccc|c} 1 & 4 & 6 & 1 \\ 0 & 1 & 3 & -2 \\ 0 & 0 & 1 & 7 \end{array}\right]$ is in _____ form.

Practice the Skills

Perform each row transformation indicated and write the new matrix.

9. $\left[\begin{array}{cc|c} 7 & -14 & -35 \\ 3 & -7 & -4 \end{array}\right]$ Multiply numbers in the first row by $\frac{1}{7}$.

10. $\left[\begin{array}{cc|c} 1 & 8 & 3 \\ 0 & 4 & -3 \end{array}\right]$ Multiply numbers in the second row by $\frac{1}{4}$.

11. $\left[\begin{array}{ccc|c} 4 & 7 & 2 & -1 \\ 3 & 2 & 1 & -5 \\ 1 & 1 & 3 & -8 \end{array}\right]$ Switch row 1 and row 3.

12. $\begin{bmatrix} 1 & 5 & 7 & | & 2 \\ 0 & 8 & -1 & | & -6 \\ 0 & 1 & 3 & | & -4 \end{bmatrix}$ Switch row 2 and row 3.

13. $\begin{bmatrix} 1 & 3 & | & 12 \\ -4 & 11 & | & -6 \end{bmatrix}$ Multiply numbers in the first row by 4 and add the products to the second row.

14. $\begin{bmatrix} 1 & 5 & | & 6 \\ \frac{1}{2} & 10 & | & -4 \end{bmatrix}$ Multiply numbers in the first row by $-\frac{1}{2}$ and add the products to the second row.

15. $\begin{bmatrix} 1 & 0 & 8 & | & \frac{1}{4} \\ 5 & 2 & 2 & | & -2 \\ 6 & -3 & 1 & | & 0 \end{bmatrix}$ Multiply numbers in the first row by -5 and add the products to the second row.

16. $\begin{bmatrix} 1 & 2 & -1 & | & 6 \\ 0 & 1 & 5 & | & 0 \\ 0 & 0 & 3 & | & 12 \end{bmatrix}$ Multiply numbers in the third row by $\frac{1}{3}$.

Solve each system using matrices.

17. $x + 3y = 3$
$-x + y = -3$

18. $x - 2y = 0$
$2x + 5y = 9$

19. $x + 3y = -2$
$-2x - 7y = 3$

20. $2x + 4y = -8$
$3x - 5y = -1$

21. $5a - 10b = -10$
$2a + b = 1$

22. $3s - 2t = 1$
$-2s + 4t = -6$

23. $2x - 5y = -6$
$-4x + 10y = 12$

24. $-2m - 4n = 7$
$3m + 6n = -8$

25. $12x + 2y = 2$
$6x - 3y = -11$

26. $4r + 2s = -10$
$-2r + s = -7$

27. $-3x + 6y = 5$
$2x - 4y = 7$

28. $8x = 4y + 12$
$-2x + y = -3$

29. $12x - 8y = 6$
$-3x + 4y = -1$

30. $2x - 3y = 3$
$-5x + 9y = -7$

31. $10m = 8n + 15$
$16n = -15m - 2$

32. $8x = 9y + 4$
$16x - 27y = 11$

Solve each system using matrices.

33. $x + 2y - 4z = 5$
$3x - y + z = 1$
$2x + 2y - 3z = 6$

34. $a - 3b + 4c = 7$
$4a + b + c = -2$
$-2a - 3b + 5c = 12$

35. $x + 2y = 5$
$y - z = -1$
$2x - 3z = 0$

36. $3a - 5c = 3$
$a + 2b = -6$
$7b - 4c = 5$

37. $x - 2y + 4z = 5$
$-3x + 4y - 2z = -8$
$4x + 5y - 4z = -3$

38. $3x + 5y + 2z = 3$
$-x - y - z = -2$
$2x - 2y + 5z = 11$

39. $2x - 5y + z = 1$
$3x - 5y + z = 3$
$-4x + 10y - 2z = -2$

40. $x + 2y + 3z = 1$
$4x + 5y + 6z = -3$
$7x + 8y + 9z = 0$

41. $4p - q + r = 4$
$-6p + 3q - 2r = -5$
$2p + 5q - r = 7$

42. $-4r + 3s - 6t = 14$
$4r + 2s - 2t = -3$
$2r - 5s - 8t = -23$

43. $2x - 4y + 3z = -12$
$3x - y + 2z = -3$
$-4x + 8y - 6z = 10$

44. $3x - 2y + 4z = -1$
$5x + 2y - 4z = 9$
$-6x + 4y - 8z = 2$

45. $5x - 3y + 4z = 22$
$-x - 15y + 10z = -15$
$-3x + 9y - 12z = -6$

46. $9x - 4y + 5z = -2$
$-9x + 5y - 10z = -1$
$9x + 3y + 10z = 1$

Problem Solving

Solve using matrices.

47. Angles of a Roof In a triangular cross section of a roof, the largest angle is 55° greater than the smallest angle. The largest angle is 20° greater than the remaining angle. Find the measure of each angle.

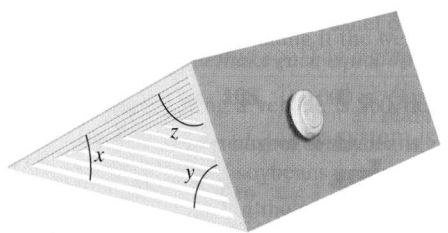

48. Right Angle A right angle is divided into three smaller angles. The largest of the three angles is twice the smallest. The remaining angle is 10° greater than the smallest angle. Find the measure of each angle.

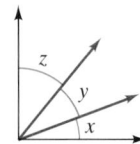

49. Most Valuable Sports Franchises As of 2008, the three most valuable National Football League franchises are in Washington, D.C., Dallas, and Houston, respectively. The total value of these three franchises is $2928 million. The Washington franchise is worth $177 million more than the Dallas franchise.

The Houston franchise is worth $18 million less than the Dallas franchise. Determine the value of each franchise.

Source: www.espn.com

50. Baseball Player Autograph Collection Alex Runde has a large collection of baseball players' autographs from players from the Tampa Bay Rays, the Milwaukee Brewers, and the Colorado Rockies. The total number of autographs from players from all three teams is 42. The number of autographs from the Rays is 5 more than twice the number of autographs from the Rockies. The number of autographs from the Brewers is 8 less than the number of autographs from the Rays. Determine the number of autographs Alex has from players on each team.

© Dennis C. Runde

Concept/Writing Exercises

51. When solving a system of linear equations by matrices, if two rows are identical, will the system be consistent, dependent, or inconsistent?

52. When solving a system of equations using matrices, how will you know if the system is

a) dependent,

b) inconsistent?

53. When solving a system of linear equations using matrices, if two rows of matrices are switched, will the solution to the system change? Explain.

54. You can tell whether a system of two equations in two variables is consistent, dependent, or inconsistent by comparing the slopes and y-intercepts of the graphs of the equations. Can you tell, without solving, if a system of three equations in three variables is consistent, dependent, or inconsistent? Explain.

Cumulative Review Exercises

[1.2]   **55.** $A = \{1, 2, 4, 6, 9\}$; $B = \{3, 4, 5, 6, 10\}$. Find
a) $A \cup B$;
b) $A \cap B$.

[2.5]   **56.** Indicate the inequality $-1 < x \le 4$
a) on a number line,

b) as a solution set, and

c) in interval notation.

[3.2]   **57.** What does a graph represent?

[3.4]   **58.** If $f(x) = -2x^2 + 3x - 6$, find $f(-5)$.

4.5 Solving Systems of Equations Using Determinants and Cramer's Rule

1 Evaluate the determinant of a 2 × 2 matrix.

2 Use Cramer's rule.

3 Evaluate the determinant of a 3 × 3 matrix.

4 Use Cramer's rule with systems in three variables.

Understanding Algebra

To evaluate a 2 × 2 determinant, subtract the product of the two diagonals as shown below:

$$\begin{vmatrix} a_1 & b_1 \\ a_2 & b_2 \end{vmatrix} = \begin{vmatrix} a_1 & b_1 \\ a_2 & b_2 \end{vmatrix}$$

$$= a_1b_2 - a_2b_1$$

1 Evaluate the Determinant of a 2 × 2 Matrix

Associated with every square matrix is a number called its **determinant**. For a 2 × 2 matrix, its determinant is defined as follows.

Determinant

The **determinant** of a 2 × 2 matrix $\begin{bmatrix} a_1 & b_1 \\ a_2 & b_2 \end{bmatrix}$ is denoted $\begin{vmatrix} a_1 & b_1 \\ a_2 & b_2 \end{vmatrix}$ and is evaluated as

$$\begin{vmatrix} a_1 & b_1 \\ a_2 & b_2 \end{vmatrix} = a_1b_2 - a_2b_1$$

A system of linear equations may also be solved using determinants.

EXAMPLE 1 Evaluate each determinant.

a) $\begin{vmatrix} 2 & -1 \\ 3 & -5 \end{vmatrix}$ **b)** $\begin{vmatrix} 2 & 3 \\ -1 & 4 \end{vmatrix}$

Solution

a) $a_1 = 2, a_2 = 3, b_1 = -1, b_2 = -5$

$$\begin{vmatrix} 2 & -1 \\ 3 & -5 \end{vmatrix} = 2(-5) - (3)(-1) = -10 + 3 = -7$$

b) $\begin{vmatrix} 2 & 3 \\ -1 & 4 \end{vmatrix} = (2)(4) - (-1)(3) = 8 + 3 = 11$

Now Try Exercise 7

2 Use Cramer's Rule

If we begin with the equations

$$a_1x + b_1y = c_1$$
$$a_2x + b_2y = c_2$$

we can use the addition method to show that

$$x = \frac{c_1b_2 - c_2b_1}{a_1b_2 - a_2b_1} \quad \text{and} \quad y = \frac{a_1c_2 - a_2c_1}{a_1b_2 - a_2b_1}$$

(see Challenge Problem 67 on page 264). Notice that the *denominators* of x and y are both $a_1b_2 - a_2b_1$. Following is the determinant that yields this denominator. We have labeled this denominator D.

$$D = \begin{vmatrix} a_1 & b_1 \\ a_2 & b_2 \end{vmatrix} = a_1b_2 - a_2b_1$$

The *numerators* of x and y are different. Following are two determinants, labeled D_x and D_y, that yield the numerators of x and y.

$$D_x = \begin{vmatrix} c_1 & b_1 \\ c_2 & b_2 \end{vmatrix} = c_1b_2 - c_2b_1 \qquad D_y = \begin{vmatrix} a_1 & c_1 \\ a_2 & c_2 \end{vmatrix} = a_1c_2 - a_2c_1$$

We use determinants $D, D_x,$ and D_y in *Cramer's rule*. **Cramer's rule** can be used to solve systems of linear equations.

Cramer's Rule for System of Linear Equations

For a system of linear equations of the form

$$a_1x + b_1y = c_1$$
$$a_2x + b_2y = c_2$$

with

$$D = \begin{vmatrix} a_1 & b_1 \\ a_2 & b_2 \end{vmatrix}, \quad D_x = \begin{vmatrix} c_1 & b_1 \\ c_2 & b_2 \end{vmatrix}, \text{ and } D_y = \begin{vmatrix} a_1 & c_1 \\ a_2 & c_2 \end{vmatrix}$$

then

$$x = \frac{D_x}{D} \quad \text{and} \quad y = \frac{D_y}{D}, \quad D \neq 0.$$

Helpful Hint

The elements in determinant D are the coefficients of the x and y terms.

$$a_1x + b_1y = c_1$$
$$a_2x + b_2y = c_2$$
$$\downarrow \quad \downarrow$$

$$D = \begin{vmatrix} a_1 & b_1 \\ a_2 & b_2 \end{vmatrix}$$

To obtain D_x replace the values in the first column with the constants.

$$a_1x + b_1y = c_1$$
$$a_2x + b_2y = c_2$$

$$D_x = \begin{vmatrix} c_1 & b_1 \\ c_2 & b_2 \end{vmatrix}$$

To obtain D_y replace the values in the second column with the constants.

$$a_1x + b_1y = c_1$$
$$a_2x + b_2y = c_2$$

$$D_y = \begin{vmatrix} a_1 & c_1 \\ a_2 & c_2 \end{vmatrix}$$

EXAMPLE 2 Solve the following system of equations using Cramer's rule.

$$3x + 5y = 7$$
$$4x - y = -6$$

Solution   Both equations are given in the desired form, $ax + by = c$. When labeling a, b, and c, we will refer to $3x + 5y = 7$ as equation 1 and $4x - y = -6$ as equation 2 for the subscripts.

$$\begin{array}{ccc} a_1 & b_1 & c_1 \\ \downarrow & \downarrow & \downarrow \\ 3x + & 5y = & 7 \\ 4x - & 1y = & -6 \\ \uparrow & \uparrow & \uparrow \\ a_2 & b_2 & c_2 \end{array}$$

We now determine D, D_x, D_y.

$$D = \begin{vmatrix} a_1 & b_1 \\ a_2 & b_2 \end{vmatrix} = \begin{vmatrix} 3 & 5 \\ 4 & -1 \end{vmatrix} = 3(-1) - 4(5) = -3 - 20 = -23$$

$$D_x = \begin{vmatrix} c_1 & b_1 \\ c_2 & b_2 \end{vmatrix} = \begin{vmatrix} 7 & 5 \\ -6 & -1 \end{vmatrix} = 7(-1) - (-6)(5) = -7 + 30 = 23$$

$$D_y = \begin{vmatrix} a_1 & c_1 \\ a_2 & c_2 \end{vmatrix} = \begin{vmatrix} 3 & 7 \\ 4 & -6 \end{vmatrix} = 3(-6) - 4(7) = -18 - 28 = -46$$

Now we find the values of x and y.

$$x = \frac{D_x}{D} = \frac{23}{-23} = -1$$

$$y = \frac{D_y}{D} = \frac{-46}{-23} = 2$$

Thus the solution is $x = -1$, $y = 2$ or the ordered pair $(-1, 2)$. A check will show that this ordered pair satisfies both equations.

Now Try Exercise 15

When the determinant $D = 0$, Cramer's rule cannot be used since division by 0 is undefined. You must then use a different method to solve the system. However, you may evaluate D_x and D_y to determine whether the system is dependent or inconsistent.

When $D = 0$

For a system of two linear equations with two variables,

    if $D = 0$, $D_x = 0$, and $D_y = 0$, then the system is dependent.

    if $D = 0$ and either $D_x \neq 0$ or $D_y \neq 0$, then the system is inconsistent.

---

3 Evaluate the Determinant of a 3 × 3 Matrix

For the determinant

$$\begin{vmatrix} a_1 & b_1 & c_1 \\ a_2 & b_2 & c_2 \\ a_3 & b_3 & c_3 \end{vmatrix}$$

the **minor determinant** of a_1 is found by crossing out the elements in the same row and column in which the element a_1 appears. The remaining elements form the minor determinant of a_1. The minor determinants of other elements are found similarly.

$$\begin{vmatrix} \cancel{a_1} & \cancel{b_1} & \cancel{c_1} \\ a_2 & b_2 & c_2 \\ a_3 & b_3 & c_3 \end{vmatrix} \qquad \begin{vmatrix} b_2 & c_2 \\ b_3 & c_3 \end{vmatrix} \qquad \text{Minor determinant of } a_1$$

$$\begin{vmatrix} a_1 & b_1 & c_1 \\ \cancel{a_2} & \cancel{b_2} & \cancel{c_2} \\ a_3 & b_3 & c_3 \end{vmatrix} \qquad \begin{vmatrix} b_1 & c_1 \\ b_3 & c_3 \end{vmatrix} \qquad \text{Minor determinant of } a_2$$

$$\begin{vmatrix} a_1 & b_1 & c_1 \\ a_2 & b_2 & c_2 \\ \cancel{a_3} & \cancel{b_3} & \cancel{c_3} \end{vmatrix} \qquad \begin{vmatrix} b_1 & c_1 \\ b_2 & c_2 \end{vmatrix} \qquad \text{Minor determinant of } a_3$$

To evaluate determinants of a 3×3 matrix, we use minor determinants. The following box shows how such a determinant may be evaluated by **expansion by the minors of the first column**.

Expansion of the Determinant by the Minors of the First Column

$$\begin{vmatrix} a_1 & b_1 & c_1 \\ a_2 & b_2 & c_2 \\ a_3 & b_3 & c_3 \end{vmatrix} = a_1 \overbrace{\begin{vmatrix} b_2 & c_2 \\ b_3 & c_3 \end{vmatrix}}^{\substack{\text{Minor} \\ \text{determinant} \\ \text{of } a_1}} - a_2 \overbrace{\begin{vmatrix} b_1 & c_1 \\ b_3 & c_3 \end{vmatrix}}^{\substack{\text{Minor} \\ \text{determinant} \\ \text{of } a_2}} + a_3 \overbrace{\begin{vmatrix} b_1 & c_1 \\ b_2 & c_2 \end{vmatrix}}^{\substack{\text{Minor} \\ \text{determinant} \\ \text{of } a_3}}$$

EXAMPLE 3 Evaluate $\begin{vmatrix} 4 & -2 & 6 \\ 3 & 5 & 0 \\ 1 & -3 & -1 \end{vmatrix}$ using expansion by the minors of the first column.

Solution   We will follow the procedure given in the box.

$$\begin{vmatrix} 4 & -2 & 6 \\ 3 & 5 & 0 \\ 1 & -3 & -1 \end{vmatrix} = 4 \begin{vmatrix} 5 & 0 \\ -3 & -1 \end{vmatrix} - 3 \begin{vmatrix} -2 & 6 \\ -3 & -1 \end{vmatrix} + 1 \begin{vmatrix} -2 & 6 \\ 5 & 0 \end{vmatrix}$$

$$= 4[5(-1) - (-3)0] - 3[(-2)(-1) - (-3)6] + 1[(-2)0 - 5(6)]$$

$$= 4(-5 + 0) - 3(2 + 18) + 1(0 - 30)$$

$$= 4(-5) - 3(20) + 1(-30)$$

$$= -20 - 60 - 30$$

$$= -110$$

The determinant has a value of -110.

Now Try Exercise 13

4 Use Cramer's Rule with Systems in Three Variables

Cramer's rule can also be used to solve systems of equations in three variables as follows.

Cramer's Rule for a System of Equations in Three Variables

To solve the system

$$a_1x + b_1y + c_1z = d_1$$
$$a_2x + b_2y + c_2z = d_2$$
$$a_3x + b_3y + c_3z = d_3$$

with

$$D = \begin{vmatrix} a_1 & b_1 & c_1 \\ a_2 & b_2 & c_2 \\ a_3 & b_3 & c_3 \end{vmatrix} \qquad D_x = \begin{vmatrix} d_1 & b_1 & c_1 \\ d_2 & b_2 & c_2 \\ d_3 & b_3 & c_3 \end{vmatrix}$$

$$D_y = \begin{vmatrix} a_1 & d_1 & c_1 \\ a_2 & d_2 & c_2 \\ a_3 & d_3 & c_3 \end{vmatrix} \qquad D_z = \begin{vmatrix} a_1 & b_1 & d_1 \\ a_2 & b_2 & d_2 \\ a_3 & b_3 & d_3 \end{vmatrix}$$

then

$$x = \frac{D_x}{D} \qquad y = \frac{D_y}{D} \qquad z = \frac{D_z}{D}, \qquad D \neq 0$$

Note that the denominators of the expressions for x, y, and z are all the same determinant, D. Note that the d's replace the a's, the numerical coefficients of the x-terms, in D_x. The d's replace the b's, the numerical coefficients of the y-terms, in D_y. And the d's replace the c's, the numerical coefficients of the z-terms, in D_z.

EXAMPLE 4 Solve the following system of equations using determinants.

$$3x - 2y - z = -6$$
$$2x + 3y - 2z = 1$$
$$x - 4y + z = -3$$

Solution

$$a_1 = 3 \quad b_1 = -2 \quad c_1 = -1 \quad d_1 = -6$$
$$a_2 = 2 \quad b_2 = 3 \quad c_2 = -2 \quad d_2 = 1$$
$$a_3 = 1 \quad b_3 = -4 \quad c_3 = 1 \quad d_3 = -3$$

We will use expansion by the minor determinants of the first column to evaluate D, D_x, D_y, and D_z.

$$D = \begin{vmatrix} 3 & -2 & -1 \\ 2 & 3 & -2 \\ 1 & -4 & 1 \end{vmatrix} = 3\begin{vmatrix} 3 & -2 \\ -4 & 1 \end{vmatrix} - 2\begin{vmatrix} -2 & -1 \\ -4 & 1 \end{vmatrix} + 1\begin{vmatrix} -2 & -1 \\ 3 & -2 \end{vmatrix}$$

$$= 3(-5) - 2(-6) + 1(7)$$
$$= -15 + 12 + 7 = 4$$

$$D_x = \begin{vmatrix} -6 & -2 & -1 \\ 1 & 3 & -2 \\ -3 & -4 & 1 \end{vmatrix} = -6\begin{vmatrix} 3 & -2 \\ -4 & 1 \end{vmatrix} - 1\begin{vmatrix} -2 & -1 \\ -4 & 1 \end{vmatrix} + (-3)\begin{vmatrix} -2 & -1 \\ 3 & -2 \end{vmatrix}$$

$$= -6(-5) - 1(-6) - 3(7)$$
$$= 30 + 6 - 21 = 15$$

$$D_y = \begin{vmatrix} 3 & -6 & -1 \\ 2 & 1 & -2 \\ 1 & -3 & 1 \end{vmatrix} = 3\begin{vmatrix} 1 & -2 \\ -3 & 1 \end{vmatrix} - 2\begin{vmatrix} -6 & -1 \\ -3 & 1 \end{vmatrix} + 1\begin{vmatrix} -6 & -1 \\ 1 & -2 \end{vmatrix}$$

$$= 3(-5) - 2(-9) + 1(13)$$
$$= -15 + 18 + 13 = 16$$

$$D_z = \begin{vmatrix} 3 & -2 & -6 \\ 2 & 3 & 1 \\ 1 & -4 & -3 \end{vmatrix} = 3\begin{vmatrix} 3 & 1 \\ -4 & -3 \end{vmatrix} - 2\begin{vmatrix} -2 & -6 \\ -4 & -3 \end{vmatrix} + 1\begin{vmatrix} -2 & -6 \\ 3 & 1 \end{vmatrix}$$

$$= 3(-5) - 2(-18) + 1(16)$$
$$= -15 + 36 + 16 = 37$$

We found that $D = 4$, $D_x = 15$, $D_y = 16$, and $D_z = 37$. Therefore,

$$x = \frac{D_x}{D} = \frac{15}{4} \qquad y = \frac{D_y}{D} = \frac{16}{4} = 4 \qquad z = \frac{D_z}{D} = \frac{37}{4}$$

The solution to the system is $\left(\frac{15}{4}, 4, \frac{37}{4}\right)$. Note the ordered triple lists x, y, and z in this order.

Understanding Algebra

When we have a system of equations in three variables in which one or more equations are missing a variable, we represent the missing variable with a coefficient of 0. Thus,

$$2x - 3y + 2z = -1$$
$$x + 2y \quad\quad = 14$$
$$x \quad\quad - 3z = -5$$

is written

$$2x - 3y + 2z = -1$$
$$x + 2y + 0z = 14$$
$$x + 0y - 3z = -5$$

Now Try Exercise 33

> **Helpful Hint**
>
> When evaluating determinants, if any two rows (or columns) are identical, or identical except for opposite signs, the determinant has a value of 0. For example,
>
> $$\begin{vmatrix} 5 & -2 \\ 5 & -2 \end{vmatrix} = 0 \quad \text{and} \quad \begin{vmatrix} 5 & -2 \\ -5 & 2 \end{vmatrix} = 0$$
>
> $$\begin{vmatrix} 5 & -3 & 4 \\ 2 & 6 & 5 \\ 5 & -3 & 4 \end{vmatrix} = 0 \quad \text{and} \quad \begin{vmatrix} 5 & -3 & 4 \\ -5 & 3 & -4 \\ 6 & 8 & 2 \end{vmatrix} = 0$$

As with determinants of a 2×2 matrix, when the determinant $D = 0$, Cramer's rule cannot be used since division by 0 is undefined. You must then use a different method to solve the system. However, you may evaluate D_x, D_y, and D_z to determine whether the system is dependent or inconsistent.

> **When $D = 0$**
>
> For a system of three linear equations with three variables,
>
> if $D = 0$, $D_x = 0$, $D_y = 0$, and $D_z = 0$, then the system is dependent.
>
> if $D = 0$ and $D_x \neq 0$, $D_y \neq 0$, or $D_z \neq 0$, then the system is inconsistent.

EXERCISE SET 4.5

MathXL MathXL®   MyMathLab MyMathLab

Warm-Up Exercises

Fill in the blanks with the appropriate word, phrase, or symbol(s) from the following list.

D D_x D_y D_z determinant dependent inconsistent unique

1. Associated with every square matrix is a number called its _____.

For Exercises 2–4, refer to the system of equations
$$2x - 3y = 2$$
$$x - 7y = 6$$

2. The determinant _____ $= \begin{vmatrix} 2 & -3 \\ 1 & -7 \end{vmatrix}$.

3. The determinant _____ $= \begin{vmatrix} 2 & 2 \\ 1 & 6 \end{vmatrix}$.

4. The determinant _____ $= \begin{vmatrix} 2 & -3 \\ 6 & -7 \end{vmatrix}$.

5. In a system of two linear equations with two variables, if $D = 0$, $D_x = 0$, and $D_y = 0$, then the system is _____ and the system has an infinite number of solutions.

6. In a system of two linear equations with two variables, if $D = 0$, and either $D_x \neq 0$ or $D_y \neq 0$, then the system is _____ and the system has no solutions.

Practice the Skills

Evaluate each determinant.

7. $\begin{vmatrix} 3 & 2 \\ 5 & 7 \end{vmatrix}$

8. $\begin{vmatrix} 3 & 5 \\ -1 & -2 \end{vmatrix}$

 **9.** $\begin{vmatrix} \frac{1}{2} & 3 \\ 2 & -4 \end{vmatrix}$

10. $\begin{vmatrix} 13 & -\frac{2}{3} \\ -1 & 0 \end{vmatrix}$

11. $\begin{vmatrix} 3 & 2 & 0 \\ 0 & 5 & 3 \\ -1 & 4 & 2 \end{vmatrix}$

12. $\begin{vmatrix} 4 & 1 & 1 \\ 0 & 0 & 3 \\ 2 & 2 & 9 \end{vmatrix}$

13. $\begin{vmatrix} 2 & 3 & 1 \\ 1 & -3 & -6 \\ -4 & 5 & 9 \end{vmatrix}$

14. $\begin{vmatrix} 5 & -8 & 6 \\ 3 & 0 & 4 \\ -5 & -2 & 1 \end{vmatrix}$

Solve each system of equations using determinants.

15. $3x + 4y = 10$
$x + 3y = 5$

16. $2x + 4y = -2$
$-5x - 2y = 13$

17. $-x - 2y = 2$
$x + 3y = -6$

18. $2r + 3s = -9$
$3r + 5s = -16$

19. $6x = 4y + 7$
$8x - 1 = -3y$

20. $6x + 3y = -4$
$9x + 5y = -6$

21. $5p - 7q = -21$
$-4p + 3q = 22$

22. $4x = -5y - 2$
$-2x = y + 4$

23. $x + 5y = 3$
$2x - 6 = -10y$

24. $9x + 6y = -3$
$6x + 4y = -2$

25. $3r = -4s - 6$
$3s = -5r + 1$

26. $x = y - 1$
$3y = 2x + 9$

27. $5x - 5y = 3$
$-x + y = -4$

28. $2x - 5y = -3$
$-4x + 10y = 7$

29. $6.3x - 4.5y = -9.9$
$-9.1x + 3.2y = -2.2$

30. $-1.1x + 8.3y = 36.5$
$3.5x + 1.6y = -4.1$

Solve each system using determinants.

31. $x + y - z = 2$
$2x + 3y - 2z = 6$
$5x - 2y + 3z = 4$

32. $2x + 3y = 4$
$3x + 7y - 4z = -3$
$x - y + 2z = 9$

33. $3x - 5y - 4z = -4$
$4x + 2y = 1$
$6y - 4z = -11$

34. $2x + 5y + 3z = 2$
$6x - 9y = 5$
$3y + 2z = 1$

35. $x + 4y - 3z = -6$
$2x - 8y + 5z = 12$
$3x + 4y - 2z = -3$

36. $2x + y - 2z = 4$
$2x + 2y - 4z = 1$
$-6x + 8y - 4z = 1$

37. $a - b + 2c = 3$
$a - b + c = 1$
$2a + b + 2c = 2$

38. $-2x + y + 8 = -2$
$3x + 2y + z = 3$
$x - 3y - 5z = 5$

39. $a + 2b + c = 1$
$a - b + 3c = 2$
$2a + b + 4c = 3$

40. $4x - 2y + 6z = 2$
$-6x + 3y - 9z = -3$
$2x - 7y + 11z = -5$

41. $1.1x + 2.3y - 4.0z = -9.2$
$-2.3x + 4.6z = 6.9$
$-8.2y - 7.5z = -6.8$

42. $4.6y - 2.1z = 24.3$
$-5.6x + 1.8y = -5.8$
$2.8x - 4.7y - 3.1z = 7.0$

43. $-6x + 3y - 12z = -13$
$5x + 2y - 3z = 1$
$2x - y + 4z = -5$

44. $x - 2y + z = 2$
$4x - 6y + 2z = 3$
$2x - 3y + z = 0$

45. $2x + \dfrac{1}{2}y - 3z = 5$
$-3x + 2y + 2z = 1$
$4x - \dfrac{1}{4}y - 7z = 4$

46. $\dfrac{1}{4}x - \dfrac{1}{2}y + 3z = -3$
$2x - 3y + 2z = -1$
$\dfrac{1}{6}x + \dfrac{1}{3}y - \dfrac{1}{3}z = 1$

47. $0.3x - 0.1y - 0.3z = -0.2$
$0.2x - 0.1y + 0.1z = -0.9$
$0.1x + 0.2y - 0.4z = 1.7$

48. $0.6u - 0.4v + 0.5w = 3.1$
$0.5u + 0.2v + 0.2w = 1.3$
$0.1u + 0.1v + 0.1w = 0.2$

Problem Solving

Solve for the given letter.

49. $\begin{vmatrix} 4 & 6 \\ -2 & y \end{vmatrix} = 32$

50. $\begin{vmatrix} b - 3 & -4 \\ b + 2 & -6 \end{vmatrix} = 14$

51. $\begin{vmatrix} 4 & 7 & y \\ 3 & -1 & 2 \\ 4 & 1 & 5 \end{vmatrix} = -35$

52. $\begin{vmatrix} 3 & 2 & -2 \\ 0 & 5 & -6 \\ -1 & x & -7 \end{vmatrix} = -31$

Solve using determinants.

53. Ticket Prices Dayton Sinkia purchased 5 adult tickets and 8 student tickets to see the play *A Servant of two Masters* at State College of Florida for $90. Danielle Zoller purchased 3 adult tickets and 7 student tickets for $65. Determine the price of an adult ticket and the price of a student ticket.

54. Little League Concession Stand Prices At Braden River Little League, Beth Van Vranken purchased 3 hot dogs, 4 bottles of water, and 2 bags of sunflower seeds for $11. Jeff Parrill purchased 5 hot dogs, 3 bottles of water, and 4 bags of sunflower seeds for $14.25. Dave Hauck purchased 1 hot dog, 2 bottles of water, and 5 bags of sunflower seeds for $7.75. Determine the price of a hot dog, a bottle of water, and a bag of sunflower seeds.

Concept/Writing Exercises

55. Given a determinant of the form $\begin{vmatrix} a_1 & b_1 \\ a_2 & b_2 \end{vmatrix}$, how will the value of the determinant change if the *a*'s are switched with each other and the *b*'s are switched with each other, $\begin{vmatrix} a_2 & b_2 \\ a_1 & b_1 \end{vmatrix}$? Explain your answer.

56. Given a determinant of the form $\begin{vmatrix} a_1 & b_1 \\ a_2 & b_2 \end{vmatrix}$, how will the value of the determinant change if the *a*'s are switched with the *b*'s, $\begin{vmatrix} b_1 & a_1 \\ b_2 & a_2 \end{vmatrix}$? Explain your answer.

57. In a 2×2 determinant, if the rows are the same, what is the value of the determinant?

58. If all the numbers in one row or one column of a 2×2 determinant are 0, what is the value of the determinant?

59. If all the numbers in one row or one column of a 3×3 determinant are 0, what is the value of the determinant?

60. Given a 3×3 determinant, if all the numbers in one row are multiplied by -1, will the value of the new determinant change? Explain.

61. Given a 3×3 determinant, if the first and second rows are switched, will the value of the new determinant change? Explain.

62. In a 3×3 determinant, if any two rows are the same, can you make a generalization about the value of the determinant?

63. In a 3×3 determinant, if the numbers in the first row are multiplied by -1 and the numbers in the second row are multiplied by -1, will the value of the new determinant change? Explain.

64. In a 3×3 determinant, if the numbers in the second row are multiplied by -1 and the numbers in the third row are multiplied by -1, will the value of the new determinant change? Explain.

65. In a 3×3 determinant, if the numbers in the second row are multiplied by 2, will the value of the new determinant change? Explain.

66. In a 3×3 determinant, if the numbers in the first row are multiplied by 3 and the numbers in the third row are multiplied by 4, will the value of the new determinant change? Explain.

Challenge Problems

67. Use the addition method to solve the following system for **a)** x and **b)** y.

$$a_1 x + b_1 y = c_1$$

$$a_2 x + b_2 y = c_2$$

Cumulative Review Exercises

[2.5]   **68.** Solve the inequality $3(x - 2) < \dfrac{4}{5}(x - 4)$ and indicate the solution in interval notation.

Graph $3x + 4y = 8$ *using the indicated method.*

[3.2]   **69.** By plotting points

　　　　70. Using the x- and y-intercepts

[3.3]   **71.** Using the slope and y-intercept

4.6 Solving Systems of Linear Inequalities

1 Solve systems of linear inequalities.

2 Solve linear programming problems.

3 Solve systems of linear inequalities containing absolute value.

1 Solve Systems of Linear Inequalities

In Section 3.7 we showed how to graph linear inequalities in two variables. In Section 4.1 we learned how to solve systems of equations graphically. In this section we show how to solve **systems of linear inequalities** graphically.

> **To Solve a System of Linear Inequalities**
>
> Graph each inequality on the same axes. The solution is the set of points whose coordinates satisfy all the inequalities in the system.

EXAMPLE 1   Determine the solution to the following system of inequalities.

$$y < -\frac{1}{2}x + 2$$

$$x - y \le 4$$

Solution   First graph the inequality $y < -\dfrac{1}{2}x + 2$ (**Fig. 4.7** on page 265). Now on the same axes graph the inequality $x - y \le 4$ (**Fig. 4.8** on page 265). The solution is the set of points common to the graphs of both inequalities. It is the part of the graph that contains both shadings. The dashed line is not part of the solution, but the part of the solid line that satisfies both inequalities is part of the solution.

Understanding Algebra

Recall from Section 3.7, that a dashed line is used when the inequality is $<$ or $>$ and a solid line is used when the inequality is $\leq$ or $\geq$. Also, the shading is placed on the side of the boundary line that contains the solutions to the inequality.

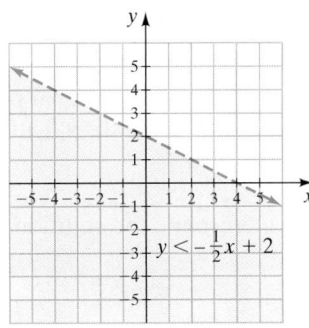

FIGURE 4.7

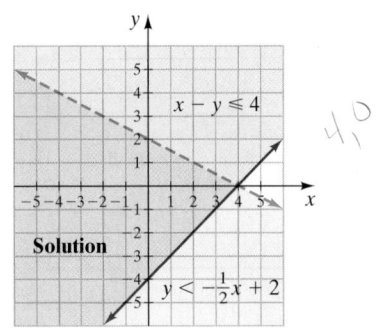

FIGURE 4.8

Now Try Exercise 5

EXAMPLE 2 Determine the solution to the following system of inequalities.

$$3x - y < 6$$
$$2x + 2y \geq 5$$

Solution   Graph $3x - y < 6$ (see **Fig. 4.9**). Graph $2x + 2y \geq 5$ on the same axes (**Fig. 4.10**). The solution is the part of the graph that contains both shadings and the part of the solid line that satisfies both inequalities.

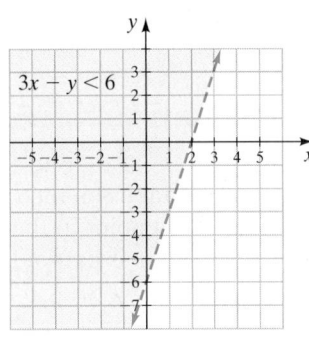

FIGURE 4.9

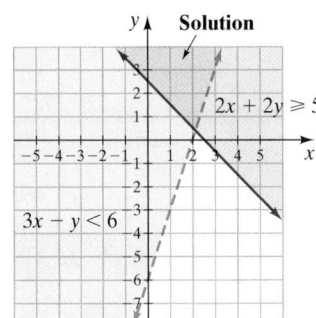

FIGURE 4.10

Now Try Exercise 7

EXAMPLE 3 Determine the solution to the following system of inequalities.

$$y > -1$$
$$x \leq 4$$

Solution   The solution is illustrated in **Figure 4.11**.

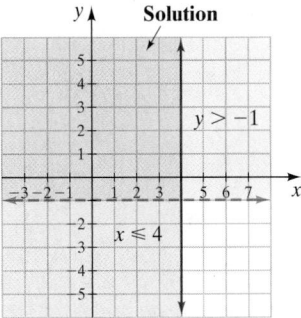

FIGURE 4.11

Now Try Exercise 15

2 Solve Linear Programming Problems

Linear programming is a mathematical process that includes graphing more than two linear inequalities on the same axes. The inequalities involved in linear programming are called the **constraints**.

EXAMPLE 4 Determine the solution to the following system of inequalities.

$$x \geq 0$$
$$y \geq 0$$
$$2x + 3y \leq 12$$
$$2x + \ y \leq 8$$

Solution   The first two inequalities, $x \geq 0$ and $y \geq 0$, indicate that the solution must be in the first quadrant because that is the only quadrant where both x and y are positive. **Figure 4.12** illustrates the graphs of the four inequalities.

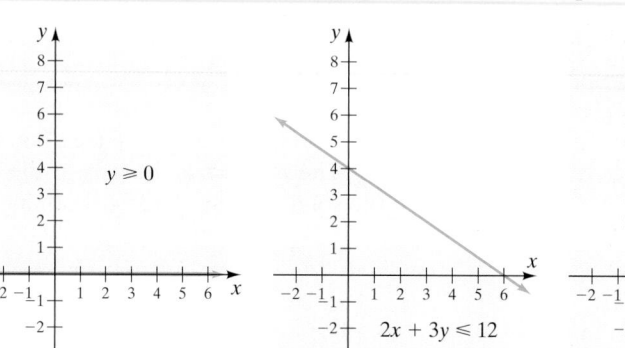

FIGURE 4.12

Figure 4.13 illustrates the graphs on the same axes and the solution to the system of inequalities. Note that every point in the shaded area and every point on the lines that form the polygonal region is part of the answer.

Now Try Exercise 23

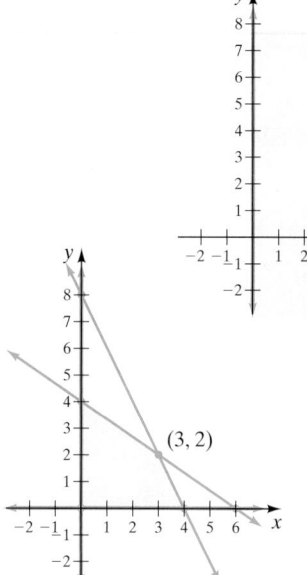

FIGURE 4.13

EXAMPLE 5 Determine the solution to the following system of inequalities.

$$x \geq 0$$
$$y \geq 0$$
$$x \leq 15$$
$$8x + 8y \leq 160$$
$$4x + 12y \leq 180$$

Solution   The first two inequalities indicate that the solution must be in the first quadrant. The third inequality indicates that x must be a value less than or equal to 15. **Figure 4.14a** indicates the graphs of corresponding equations and shows the region that satisfies all the inequalities in the system. **Figure 4.14b** indicates the solution to the system of inequalities.

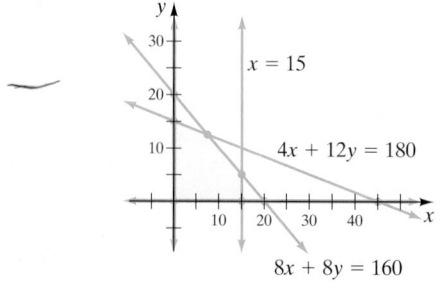

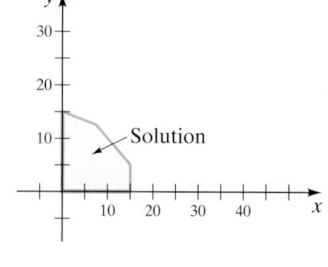

FIGURE 4.14   (a)   (b)

Now Try Exercise 29

3 Solve Systems of Linear Inequalities Containing Absolute Value

Now we will graph systems of linear inequalities containing absolute value in the Cartesian coordinate system. Before we do some examples, let us recall the rules for absolute value inequalities that we learned in Section 2.6.

> **Solving Absolute Value Inequalities**
>
> If $|x| < a$ and $a > 0$, then $-a < x < a$.
> If $|x| > a$ and $a > 0$, then $x < -a$ or $x > a$.

EXAMPLE 6   Graph $|x| < 3$ in the Cartesian coordinate system.

Solution   From the rules given, we know that $|x| < 3$ means $-3 < x < 3$. We draw dashed vertical lines through -3 and 3 and shade the area between the two (**Fig. 4.15**).

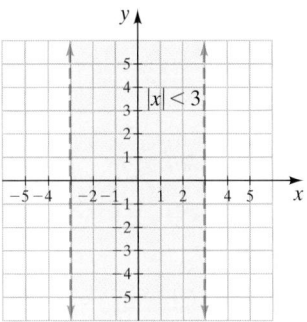

FIGURE 4.15

Now Try Exercise 33

EXAMPLE 7   Graph $|y + 1| > 3$ in the Cartesian coordinate system.

Solution   From the rules given, we know that $|y + 1| > 3$ means $y + 1 < -3$ or $y + 1 > 3$. First we solve each inequality.

$$y + 1 < -3 \quad \text{or} \quad y + 1 > 3$$
$$y < -4 \qquad\qquad y > 2$$

Now we graph both inequalities and take the *union* of the two graphs. The solution is the shaded area in **Fig. 4.16**.

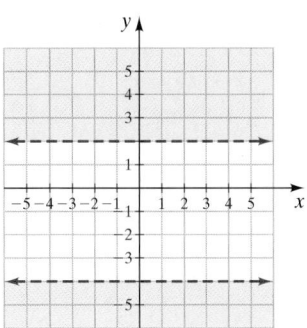

FIGURE 4.16

Now Try Exercise 35

EXAMPLE 8 Graph the following system of inequalities.

$$|x| < 3$$
$$|y + 1| > 3$$

Solution   We draw both inequalities on the same axes. Therefore, we combine the graph drawn in Example 6 with the graph drawn in Example 7 (see **Fig. 4.17**). The points common to both inequalities form the solution to the system.

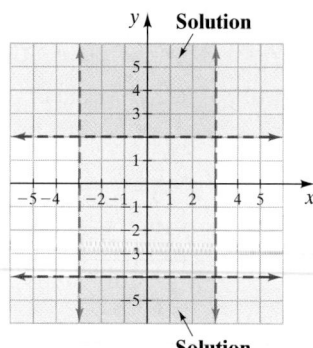

FIGURE 4.17

Now Try Exercise 41

EXERCISE SET 4.6      *MathXL* *MyMathLab*

Warm-Up Exercises

Fill in the blanks with the appropriate word, phrase, or symbol(s) from the following list.

solid   constraints   constants   dashed   linear programming   satisfy   geometry

1. The solution to a system of linear inequalities is the set of points whose coordinates _____ all the inequalities in the system.

2. A mathematical process for which you often must graph more than two linear inequalities on the same axes is called _____.

3. The inequalities in a linear programming problem are called the _____.

4. When graphing a linear inequality, if the inequality is < or >, use a _____ line. If the inequality is ≤ or ≥, use a _____ line.

Practice the Skills

Determine the solution to each system of inequalities.

5. $2x - y < 4$
   $y \geq -x + 2$

6. $y \leq -2x + 1$
   $y > -3x$

7. $y < 3x - 2$
   $y \leq -2x + 3$

8. $y \geq 2x - 5$
   $y > -3x + 5$

9. $y < x$
   $y \geq 3x + 2$

10. $-3x + 2y \geq -5$
   $y \leq -4x + 7$

11. $-2x + 3y < -5$
   $3x - 8y > 4$

12. $-4x + 3y \geq -4$
   $y > -3x + 3$

13. $-4x + 5y < 20$
   $x \geq -3$

14. $y \geq -\dfrac{2}{3}x + 1$
   $y > -4$

15. $x \leq 4$
   $y \geq -2$

16. $x \geq 0$
   $x - 3y < 6$

17. $5x + 2y > 10$
   $3x - y > 3$

18. $3x + 2y > 8$
   $x - 5y < 5$

19. $-2x > y + 4$
   $-x < \dfrac{1}{2}y - 1$

20. $y \leq 3x - 2$
   $\dfrac{1}{3}y < x + 1$

21. $y < 3x - 4$
   $6x \geq 2y + 8$

22. $\dfrac{1}{2}x + \dfrac{1}{2}y \geq 2$
   $2x - 3y \leq -6$

Determine the solution to each system of inequalities. Use the method discussed in Examples 4 and 5.

23. $x \geq 0$
   $y \geq 0$
   $2x + 3y \leq 6$
   $4x + y \leq 4$

24. $x \geq 0$
   $y \geq 0$
   $x + y \leq 6$
   $7x + 4y \leq 28$

25. $x \geq 0$
   $y \geq 0$
   $2x + 3y \leq 8$
   $4x + 2y \leq 8$

26. $x \geq 0$
   $y \geq 0$
   $3x + 2y \leq 18$
   $2x + 4y \leq 20$

27. $x \geq 0$
$y \geq 0$
$3x + y \leq 9$
$2x + 5y \leq 10$

28. $x \geq 0$
$y \geq 0$
$5x + 4y \leq 16$
$x + 6y \leq 18$

29. $x \geq 0$
$y \geq 0$
$x \leq 4$
$x + y \leq 6$
$x + 2y \leq 8$

30. $x \geq 0$
$y \geq 0$
$x \leq 4$
$2x + 3y \leq 18$
$4x + 2y \leq 20$

31. $x \geq 0$
$y \geq 0$
$x \leq 15$
$30x + 25y \leq 750$
$10x + 40y \leq 800$

32. $x \geq 0$
$y \geq 0$
$x \leq 15$
$40x + 25y \leq 1000$
$5x + 30y \leq 900$

Determine the solution to each inequality.

33. $|x| < 2$

34. $|x| > 1$

35. $|y - 2| \leq 4$

36. $|y| \geq 2$

Determine the solution to each system of inequalities.

37. $|y| > 2$
$y \leq x + 3$

38. $|x| > 1$
$y \leq 3x + 2$

39. $|y| < 4$
$y \geq -2x + 2$

40. $|x - 2| \leq 3$
$x - y > 2$

41. $|x + 2| < 3$
$|y| > 4$

42. $|x - 2| > 1$
$y > -2$

43. $|x - 3| \leq 4$
$|y + 2| \leq 1$

44. $|x + 1| \leq 2$
$|y - 3| \leq 1$

Concept/Writing Exercises

45. If in a system of two inequalities, one inequality contains $<$ and the other inequality contains $\geq$, is the point of intersection of the two boundary lines of the inequalities in the solution set? Explain.

46. If in a system of two inequalities, one inequality contains $\leq$ and the other inequality contains $\geq$, is the point of intersection of the two boundary lines of the inequalities in the solution set? Explain.

47. If in a system of two inequalities, one inequality contains $<$ and the other inequality contains $>$, is the point of intersec-

tion of the two boundary lines of the inequalities in the solution set? Explain.

48. a) Is it possible for a system of linear inequalities to have no solution? Explain. Make up an example to support your answer.

b) Is it possible for a system of two linear inequalities to have exactly one solution? Explain. If you answer yes, make up an example to support your answer.

Without graphing, determine the number of solutions in each indicated system of inequalities. Explain your answers.

49. $3x - y \leq 4$
$3x - y > 4$

50. $2x + y < 6$
$2x + y > 6$

51. $5x - 2y \leq 3$
$5x - 2y \geq 3$

52. $5x - 3y > 5$
$5x - 3y > -1$

53. $2x - y < 7$
$3x - y < -2$

54. $x + y \leq 0$
$x - y \geq 0$

Challenge Problems

Determine the solution to each system of inequalities.

55. $y \geq x^2$
$y \leq 4$

56. $y < 4 - x^2$
$y > -5$

57. $y < |x|$
$y < 4$

58. $y \geq |x - 2|$
$y \leq -|x - 2|$

Cumulative Review Exercises

[2.2] **59.** A formula for levers in physics is $f_1 d_1 + f_2 d_2 = f_3 d_3$. Solve this formula for f_2.

[3.2] *State the domain and range of each function.*

60. $\{(4, 3), (5, -2), (-1, 2), (0, -5)\}$

61. $f(x) = \dfrac{2}{3}x - 4$

62.

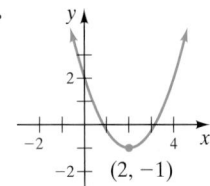

Chapter 4 Summary

IMPORTANT FACTS AND CONCEPTS	EXAMPLES

Section 4.1

A **system of linear equations** is a system having two or more linear equations.

A **solution** to a system of linear equations is the ordered pair or pairs that satisfy all equations in the system.

System of equations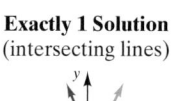

The solution to the above system of equations is $(2, 7)$.

A **consistent system of equations** is a system of equations that has one solution.

An **inconsistent system of equations** is a system of equations that has no solution.

A **dependent system of equations** is a system of equations that has an infinite number of solutions.

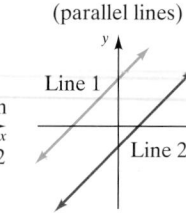
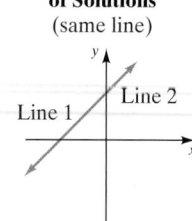

To Solve a System of Linear Equations by Graphing

1. Graph both lines on the same set of axes.
2. Determine the point(s) of intersection, if it exists.
3. Check your solution in all equations of the system.

Solve the system of equations graphically.

$$y = x - 4$$
$$y = -x + 6$$

Graph both lines on the same set of axes.

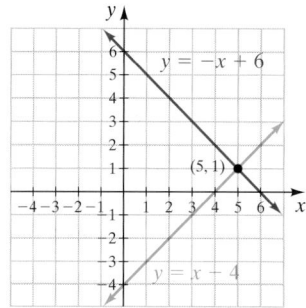

A check shows that $(5, 1)$ is a solution to the system of equations.

To Solve a Linear System of Equations by Substitution

1. Solve for a variable in either equation.
2. Substitute the expression found for the variable in step 1 into the other equation.
3. Solve the equation obtained in step 2.
4. Substitute the value found in step 3 into the equation from step 1. Solve the equation to find the remaining variable.
5. Check your solution in all equations in the system.

Solve the system of equations by substitution.

$$y = -2x - 1$$
$$5x + 6y = 8$$

Substitute $y = -2x - 1$ in the second equation:

$$5x + 6y = 8$$
$$5x + 6(-2x - 1) = 8$$
$$5x - 12x - 6 = 8$$
$$-7x - 6 = 8$$
$$-7x = 14$$
$$x = -2.$$

Substitute $x = -2$ into $y = -2x - 1$ to obtain

$$y = -2x - 1$$
$$y = -2(-2) - 1 = 4 - 1 = 3$$

A check shows that $(-2, 3)$ is a solution to the system of equations.

IMPORTANT FACTS AND CONCEPTS	EXAMPLES

Section 4.1 (cont.)

To Solve a Linear System of Equations Using the Addition (or Elimination) Method

1. If necessary, rewrite each equation in standard form.
2. If necessary, multiply one or both equations by a constant(s) so that when the equations are added, the sum will contain only one variable.
3. Add the respective sides of the equations.
4. Solve the equation obtained in step 3.
5. Substitute the value found in step 4 into either of the original equations. Solve that equation to find the value of the remaining variable.
6. Check your solution in all equations in the system.

Solve the system of equations using the addition method.

$$2x + y = 4 \quad (eq.\,1)$$
$$x - 2y = 2 \quad (eq.\,2)$$

$$
\begin{aligned}
4x + 2y &= 8 \quad (eq.\,1)\text{ Multiplied by 2}\\
x - 2y &= 2\\
\hline
5x &= 10 \quad \text{Sum of equations}\\
x &= 2
\end{aligned}
$$

Now solve for y using $(eq.\,1)$.

$$2(2) + y = 4$$
$$y = 0$$

The solution is $(2, 0)$.

Section 4.2

To solve a system of three linear equations, use the substitution method or addition method.

Solve the system of equations.

$$
\begin{aligned}
x - y + 3z &= -1 \quad (eq.\,1)\\
4y - 7z &= 2 \quad (eq.\,2)\\
z &= 2 \quad (eq.\,3)
\end{aligned}
$$

Substitute 2 for z in $(eq.\,2)$ to obtain the value for y.

$$
\begin{aligned}
4y - 7z &= 2\\
4y - 7(2) &= 2\\
4y &= 16\\
y &= 4.
\end{aligned}
$$

Substitute 4 for y and 2 for z in $(eq.\,1)$ to obtain the value for x.

$$
\begin{aligned}
x - y + 3z &= -1\\
x - 4 + 3(2) &= -1\\
x &= -3.
\end{aligned}
$$

A check shows that $(-3, 4, 2)$ is a solution to the system of equations.

Section 4.3

Applications:

Systems of two linear equations in two unknowns.

The sum of the areas of two circles is 180 square meters. The difference of their areas is 20 square meters. Determine the area of each circle.

Solution

Let x be the area of the larger circle and y be the area of the smaller circle.

The two equations for this systems are

$$
\begin{aligned}
x + y &= 180 \quad \text{Sum of areas}\\
x - y &= 20 \quad \text{Difference of areas}\\
\hline
2x &= 200\\
x &= 100
\end{aligned}
$$

Substitute 100 for x in the first equation to get

$$
\begin{aligned}
x + y &= 180\\
100 + y &= 180\\
y &= 80
\end{aligned}
$$

The area of the larger circle is 100 square meters and the area of the smaller circle is 80 square meters.

IMPORTANT FACTS AND CONCEPTS	EXAMPLES

Section 4.4

A **matrix** is a rectangular array of numbers within brackets. The numbers inside the brackets are called **elements**.

A **square matrix** has the same number of rows and columns.

$$\begin{bmatrix} 8 & 1 & 4 \\ -3 & 0 & 2 \end{bmatrix}, \quad \begin{bmatrix} 5 & 0 \\ -2 & 8 \\ 6 & -11 \end{bmatrix}$$

$$\begin{bmatrix} 5 & -1 \\ 8 & 2 \end{bmatrix}, \quad \begin{bmatrix} 3 & 0 & -6 \\ -1 & 5 & 2 \\ 9 & 10 & -7 \end{bmatrix}$$

An **augmented matrix** is a matrix separated by a vertical line. For a system of equations, the coefficients of the variables are placed on the left side of the vertical line and the constants are placed on the right side in the augmented matrix.

The **row echelon** (or **triangular**) **form** of an augmented matrix is

$$\begin{bmatrix} 1 & a & | & p \\ 0 & 1 & | & q \end{bmatrix}$$

System　　　　Augmented Matrix

$$2x - 3y = 8$$
$$5x + 7y = -4$$

$$\begin{bmatrix} 2 & -3 & | & 8 \\ 5 & 7 & | & -4 \end{bmatrix}$$

row echelon form

$$\begin{bmatrix} 1 & -6 & | & 2 \\ 0 & 1 & | & 9 \end{bmatrix}$$

Row transformations can be used to rewrite a matrix into row echelon form.

Procedures for Row Transformations

1. All the numbers in a row may be multiplied (or divided) by any nonzero real number.
2. All the numbers in a row may be multiplied by any nonzero real number. These products may then be added to the corresponding numbers in any other row.
3. The order of the rows may be switched.

Solve the system of equations

$$x + 4y = -7$$
$$6x - 5y = 16$$

The augmented matrix is

$$\begin{bmatrix} 1 & 4 & | & -7 \\ 6 & -5 & | & 16 \end{bmatrix} = \begin{bmatrix} 1 & 4 & | & -7 \\ 0 & -29 & | & 58 \end{bmatrix} -6R_1 + R_2$$

$$= \begin{bmatrix} 1 & 4 & | & -7 \\ 0 & 1 & | & -2 \end{bmatrix} -\frac{1}{29}R_2$$

The equivalent system of equations is

$$x + 4y = -7$$
$$y = -2$$

Substitute -2 for y into the first equation.

$$x + 4(-2) = -7$$
$$x - 8 = -7$$
$$x = 1.$$

The solution is $(1, -2)$.

A system of equations is **inconsistent** and has **no solution** if you obtain an augmented matrix in which one row of numbers has zeros on the left side of the vertical line and a nonzero number on the right side of the vertical line.

A system of equations is **dependent** and has an **infinite number of solutions** if you obtain an augmented matrix in which a 0 appears across an entire row.

$$\begin{bmatrix} 1 & 2 & -3 & | & 23 \\ 0 & 0 & 0 & | & 8 \\ -1 & 7 & 6 & | & 9 \end{bmatrix}$$

The second row shows this system is inconsistent and has no solution.

$$\begin{bmatrix} 1 & 6 & -1 & | & 15 \\ 0 & 0 & 0 & | & 0 \\ 3 & 5 & 8 & | & -12 \end{bmatrix}$$

The second row shows this system is dependent and has an infinite number of solutions.

IMPORTANT FACTS AND CONCEPTS	EXAMPLES

Section 4.5

The **determinant** of a 2×2 matrix $\begin{bmatrix} a_1 & b_1 \\ a_2 & b_2 \end{bmatrix}$ is denoted $\begin{vmatrix} a_1 & b_1 \\ a_2 & b_2 \end{vmatrix}$ and is evaluated as

$$\begin{vmatrix} a_1 & b_1 \\ a_2 & b_2 \end{vmatrix} = a_1 b_2 - a_2 b_1$$

$$\begin{vmatrix} 3 & -2 \\ 5 & 1 \end{vmatrix} = (3)(1) - (5)(-2) = 3 + 10 = 13$$

Cramer's Rule for Systems of Linear Equations

For a system of linear equations of the form

$$a_1 x + b_1 y = c_1$$
$$a_2 x + b_2 y = c_2$$

with

$$D = \begin{vmatrix} a_1 & b_1 \\ a_2 & b_2 \end{vmatrix}, D_x = \begin{vmatrix} c_1 & b_1 \\ c_2 & b_2 \end{vmatrix}, \text{ and } D_y = \begin{vmatrix} a_1 & c_1 \\ a_2 & c_2 \end{vmatrix}$$

then $\quad x = \dfrac{D_x}{D} \quad$ and $\quad y = \dfrac{D_y}{D}, D \neq 0$

Solve the system of equations.
$$2x + y = 6$$
$$4x - 3y = -13$$

$$D = \begin{vmatrix} 2 & 1 \\ 4 & -3 \end{vmatrix} = -10$$

$$D_x = \begin{vmatrix} 6 & 1 \\ -13 & -3 \end{vmatrix} = -5 \qquad D_y = \begin{vmatrix} 2 & 6 \\ 4 & -13 \end{vmatrix} = -50$$

Then

$$x = \frac{D_x}{D} = \frac{-5}{-10} = \frac{1}{2}, \quad y = \frac{D_y}{D} = \frac{-50}{-10} = 5$$

The solution is $\left(\dfrac{1}{2}, 5 \right)$.

For the determinant
$$\begin{vmatrix} a_1 & b_1 & c_1 \\ a_2 & b_2 & c_2 \\ a_3 & b_3 & c_3 \end{vmatrix}$$

The **minor determinant of a_1** is found by crossing out the elements in the same row and column containing the element a_1.

Expansion of the Determinant by the Minors of the First Column

$$\begin{array}{ccc} \text{Minor} & \text{Minor} & \text{Minor} \\ \text{determinant} & \text{determinant} & \text{determinant} \\ \text{of } a_1 & \text{of } a_2 & \text{of } a_3 \\ \downarrow & \downarrow & \downarrow \end{array}$$

$$\begin{vmatrix} a_1 & b_1 & c_1 \\ a_2 & b_2 & c_2 \\ a_3 & b_3 & c_3 \end{vmatrix} = a_1 \begin{vmatrix} b_2 & c_2 \\ b_3 & c_3 \end{vmatrix} - a_2 \begin{vmatrix} b_1 & c_1 \\ b_3 & c_3 \end{vmatrix} + a_3 \begin{vmatrix} b_1 & c_1 \\ b_2 & c_2 \end{vmatrix}$$

For $\begin{vmatrix} 6 & 2 & -1 \\ 0 & 3 & 5 \\ 7 & 1 & 9 \end{vmatrix}$, the minor determinant of a_1 is $\begin{vmatrix} 3 & 5 \\ 1 & 9 \end{vmatrix}$.

Evaluate $\begin{vmatrix} 2 & 0 & 3 \\ -1 & -5 & 2 \\ 3 & 6 & -4 \end{vmatrix}$ using expansion by minors of the first column.

$$\begin{vmatrix} 2 & 0 & 3 \\ -1 & -5 & 2 \\ 3 & 6 & -4 \end{vmatrix} = 2 \begin{vmatrix} -5 & 2 \\ 6 & -4 \end{vmatrix} - (-1) \begin{vmatrix} 0 & 3 \\ 6 & -4 \end{vmatrix} + 3 \begin{vmatrix} 0 & 3 \\ -5 & 2 \end{vmatrix}$$
$$= 2(8) + 1(-18) + 3(15)$$
$$= 16 - 18 + 45$$
$$= 43$$

Cramer's Rule for a System of Equations in Three Variables

To solve the system

$$a_1 x + b_1 y + c_1 z = d_1$$
$$a_2 x + b_2 y + c_2 z = d_2$$
$$a_3 x + b_3 y + c_3 z = d_3$$

with

$$D = \begin{vmatrix} a_1 & b_1 & c_1 \\ a_2 & b_2 & c_2 \\ a_3 & b_3 & c_3 \end{vmatrix} \qquad D_x = \begin{vmatrix} d_1 & b_1 & c_1 \\ d_2 & b_2 & c_2 \\ d_3 & b_3 & c_3 \end{vmatrix}$$

$$D_y = \begin{vmatrix} a_1 & d_1 & c_1 \\ a_2 & d_2 & c_2 \\ a_3 & d_3 & c_3 \end{vmatrix} \qquad D_z = \begin{vmatrix} a_1 & b_1 & d_1 \\ a_2 & b_2 & d_2 \\ a_3 & b_3 & d_3 \end{vmatrix}$$

then

$$x = \frac{D_x}{D} \quad y = \frac{D_y}{D} \quad z = \frac{D_z}{D}, \quad D \neq 0$$

Solve the system of equations.
$$2x + y + z = 0$$
$$4x - y + 3z = -9$$
$$6x + 2y + 5z = -8$$

$$D = \begin{vmatrix} 2 & 1 & 1 \\ 4 & -1 & 3 \\ 6 & 2 & 5 \end{vmatrix} = -10 \qquad D_x = \begin{vmatrix} 0 & 1 & 1 \\ -9 & -1 & 3 \\ -8 & 2 & 5 \end{vmatrix} = -5$$

$$D_y = \begin{vmatrix} 2 & 0 & 1 \\ 4 & -9 & 3 \\ 6 & -8 & 5 \end{vmatrix} = -20 \qquad D_z = \begin{vmatrix} 2 & 1 & 0 \\ 4 & -1 & -9 \\ 6 & 2 & -8 \end{vmatrix} = 30$$

Then

$$x = \frac{D_x}{D} = \frac{-5}{-10} = \frac{1}{2}, \quad y = \frac{D_y}{D} = \frac{-20}{-10} = 2 \quad z = \frac{D_z}{D} = \frac{30}{-10} = -3$$

The solution is $\left(\dfrac{1}{2}, 2, -3 \right)$.

IMPORTANT FACTS AND CONCEPTS	EXAMPLES

Section 4.6

To solve a system of linear inequalities, graph each inequality on the same axes. The solution is the set of points whose coordinates satisfy all the inequalities in the system.	Determine the solution to the system of inequalities. $$y < -\frac{1}{3}x + 1$$ $$x - y \leq 2$$ 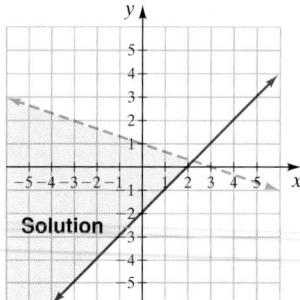
Linear programming is a process where more than two linear inequalities are graphed on the same axes.	Determine the solution to the system of inequalities. $$x \geq 0$$ $$y \geq 0$$ $$x \leq 8$$ $$x + y \leq 10$$ $$x + 2y \leq 16$$ 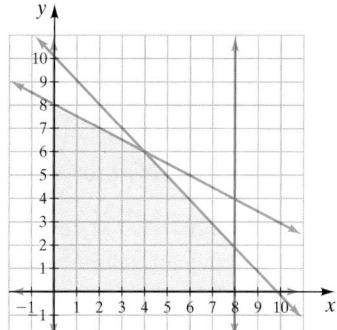
For systems of linear inequalities with absolute values: If $\lvert x \rvert < a$ and $a > 0$, then $-a < x < a$. If $\lvert x \rvert > a$ and $a > 0$, then $x < -a$ or $x > a$.	Determine the solution to the system of inequalities. $$\lvert x \rvert < 2$$ $$\lvert y - 1 \rvert > 3$$ 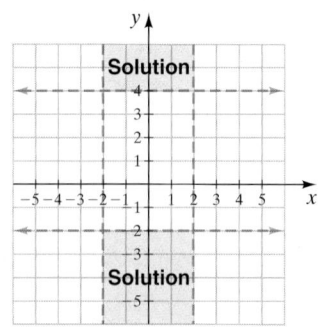

Chapter 4 Review Exercises

[4.1] *Write each equation in slope-intercept form. Without graphing or solving the system of equations, state whether the system of linear equations is consistent, inconsistent, or dependent. Also indicate whether the system has exactly one solution, no solution, or an infinite number of solutions.*

1. $x - 4y = -2$
$2x - 8y = 1$

2. $4x - 5y = 8$
$3x + 4y = 9$

3. $y = \dfrac{1}{3}x + 4$
$x + 2y = 8$

4. $6x = 5y - 8$
$4x = 6y + 10$

Determine the solution to each system of equations graphically. If the system is inconsistent or dependent, so state.

5. $y = -2x - 3$
$y = 3x + 7$

6. $x = -5$
$y = 3$

7. $3x + 3y = 12$
$2x - y = -4$

8. $3y - 3x = -9$
$\dfrac{1}{2}x - \dfrac{1}{2}y = \dfrac{3}{2}$

Find the solution to each system of equations by subsitution.

9. $4x + 7y = -3$
$x = 5y + 6$

10. $4x - 3y = -1$
$y = 2x + 1$

11. $a = 2b - 8$
$2b - 5a = 0$

12. $2x + y = 12$
$\dfrac{1}{2}x - \dfrac{3}{4}y = 1$

Find the solution to each system of equations using the addition method.

13. $-3x + 4y = -2$
$x - 5y = -3$

14. $-2x - y = 5$
$2x + 2y = 6$

15. $2a + 3b = 7$
$a - 2b = -7$

16. $0.4x - 0.3y = 1.8$
$-0.7x + 0.5y = -3.1$

17. $4r - 3s = 8$
$2r + 5s = 8$

18. $-2m + 3n = 15$
$3m + 3n = 10$

19. $x + \dfrac{3}{5}y = \dfrac{11}{5}$
$x - \dfrac{3}{2}y = -2$

20. $4x + 4y = 16$
$y = 4x - 3$

21. $y = -\dfrac{3}{4}x + \dfrac{5}{2}$
$x + \dfrac{5}{4}y = \dfrac{7}{2}$

22. $2x - 5y = 12$
$x - \dfrac{4}{3}y = -2$

23. $2x + y = 4$
$3x + \dfrac{3}{2}y = 6$

24. $2x = 4y + 5$
$2y = x - 7$

[4.2] *Determine the solution to each system of equations using substitution or the addition method.*

25. $5x - 9y + 2z = 12$
$4y - 3z = -3$
$5z = 5$

26. $2a + b - 2c = 5$
$3b + 4c = 1$
$3c = -6$

27. $x + 2y + 3z = 3$
$-2x - 3y - z = 5$
$3x + 3y + 7z = 2$

28. $-x - 4y + 2z = 1$
$2x + 2y + z = 0$
$-3x - 2y - 5z = 5$

29. $3y - 2z = -4$
$3x - 5z = -7$
$2x + y = 6$

30. $a + 2b - 5c = 19$
$2a - 3b + 3c = -15$
$5a - 4b - 2c = -2$

31. $x - y + 3z = 1$
$-x + 2y - 2z = 1$
$x - 3y + z = 2$

32. $-2x + 2y - 3z = 6$
$4x - y + 2z = -2$
$2x + y - z = 4$

[4.3] *Express each problem as a system of linear equations and use the method of your choice to find the solution to the problem.*

33. Ages Luan Baker is 10 years older than his niece, Jennifer Miesen. If the sum of their ages is 66, find Luan's age and Jennifer's age.

34. Air Speed An airplane can travel 560 miles per hour with the wind and 480 miles per hour against the wind. Determine the speed of the plane in still air and the speed of the wind.

35. Mixing Solutions Sally Dove has two acid solutions, as illustrated. How much of each must she mix to get 6 liters of a 40% acid solution?

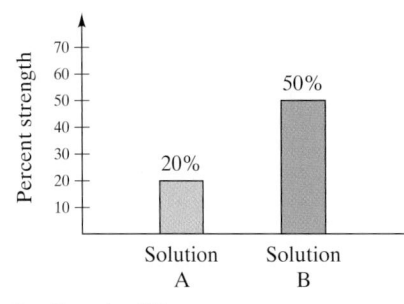

See Exercise 35.

36. Ice Hockey The admission at an ice hockey game is $15 for adults and $11 for children. A total of 650 tickets were sold. Determine how many children's tickets and how many adults' tickets were sold if a total of $8790 was collected.

37. Return to Space John Glenn was the first American astronaut to go into orbit around Earth. Many years later he returned to space. The second time he returned to space he was 5 years younger than twice his age when he went into space for the first time. The sum of his ages for both times he was in space is 118. Find his age each time he was in space.

38. Savings Accounts Jorge Minez has a total of $40,000 invested in three different savings accounts. He has some money invested in one account that gives 7% interest. The second account has $5000 less than the first account and gives 5% interest. The third account gives 3% interest. If the total annual interest that Jorge receives in a year is $2300, find the amount in each account.

© NASA

See Exercise 37.

[4.4]   *Solve each system of equations using matrices.*

39. $x + 4y = -5$
$5x - 3y = 21$

40. $2x - 5y = 1$
$2x + 4y = 10$

41. $3y = 6x - 12$
$4x = 2y + 8$

42. $2x - y - z = 5$
$x + 2y + 3z = -2$
$3x - 2y + z = 2$

43. $3a - b + c = 2$
$2a - 3b + 4c = 4$
$a + 2b - 3c = -6$

44. $x + y + z = 3$
$3x + 4y = -1$
$y - 3z = -10$

[4.5]   *Solve each system of equations using determinants.*

45. $7x - 2y = -3$
$-8x + 3y = 7$

46. $x + 4y = 5$
$5x + 3y = -9$

47. $9m + 4n = -1$
$7m - 2n = -11$

48. $p + q + r = 5$
$2p + q - r = -5$
$3p + 2q - 3r = -12$

49. $-2a + 3b - 4c = -7$
$2a + b + c = 5$
$-2a - 3b + 4c = 3$

50. $y + 3z = 4$
$-x - y + 2z = 0$
$x + 2y + z = 1$

[4.6]   *Graph the solution to each system of inequalities.*

51. $-x + 3y > 6$
$2x - y \leq 2$

52. $5x - 2y \leq 10$
$3x + 2y > 6$

53. $y > 2x + 3$
$y < -x + 4$

54. $x > -2y + 4$
$y < -\dfrac{1}{2}x - \dfrac{3}{2}$

Determine the solution to the system of inequalities.

55. $x \geq 0$
$y \geq 0$
$x + y \leq 6$
$4x + y \leq 8$

56. $x \geq 0$
$y \geq 0$
$2x + y \leq 6$
$4x + 5y \leq 20$

57. $|x| \leq 3$
$|y| > 2$

58. $|x| > 4$
$|y - 2| \leq 3$

Chapter 4 Practice Test

Chapter Test Prep Videos provide fully worked-out solutions to any of the exercises you want to review. Chapter Test Prep Videos are available via **MyMathLab**, or on **You Tube** (search "Angel Intermediate Algebra" and click on "Channels").

1. Define **a)** a consistent system of equations, **b)** a dependent system of equations, and **c)** an inconsistent system of equations.

Determine, without solving the system, whether the system of equations is consistent, inconsistent, or dependent. State whether the system has exactly one solution, no solution, or an infinite number of solutions.

2. $5x + 2y = 4$
$6x = 3y - 7$

3. $5x + 3y = 9$
$2y = -\dfrac{10}{3}x + 6$

4. $5x - 4y = 6$
$-10x + 8y = -10$

Solve each system of equations by the method indicated.

5. $y = 3x - 2$
$y = -2x + 8$
graphically

6. $y = -x + 6$
$y = 2x + 3$
graphically

7. $y = 4x - 3$
$y = 5x - 4$
substitution

8. $4a + 7b = 2$
$5a + b = -13$
substitution

9. $8x + 3y = 8$
$6x + y = 1$
addition

10. $0.3x = 0.2y + 0.4$
$-1.2x + 0.8y = -1.6$
addition

11. $\dfrac{3}{2}a + b = 6$

$a - \dfrac{5}{2}b = -4$

addition

12. $x + y + z = 2$
$-2x - y + z = 1$
$x - 2y - z = 1$
addition

13. Write the augmented matrix for the following system of equations.

$-2x + 3y + 7z = 5$
$3x - 2y + z = -2$
$x - 6y + 9z = -13$

14. Consider the following augmented matrix.

$$\begin{bmatrix} 6 & -2 & 4 & 4 \\ 4 & 3 & 5 & 6 \\ 2 & -1 & 4 & -3 \end{bmatrix}$$

Show the results obtained by multiplying the elements in the third row by -2 and adding the products to their corresponding elements in the second row.

Solve each system of equations using matrices.

15. $2x + 7y = 1$
$3x + 5y = 7$

16. $x - 2y + z = 7$
$-2x - y - z = -7$
$4x + 5y - 2z = 3$

Evaluate each determinant.

17. $\begin{vmatrix} 3 & -1 \\ 5 & -2 \end{vmatrix}$

18. $\begin{vmatrix} 8 & 2 & -1 \\ 3 & 0 & 5 \\ 6 & -3 & 4 \end{vmatrix}$

Solve each system of equations using determinants and Cramer's rule.

19. $4x + 3y = -6$
$-2x + 5y = 16$

20. $2r - 4s + 3t = -1$
$-3r + 5s - 4t = 0$
$-2r + s - 3t = -2$

Use the method of your choice to find the solution to each problem.

21. Bird Seed Mixture Agway Gardens has sunflower seeds, in a barrel, that sell for \$0.49 per pound and gourmet bird seed mix that sells for \$0.89 per pound. How much of each must be mixed to get a 20-pound mixture that sells for \$0.73 per pound?

© Allen R. Angel

22. Mixing Solutions Tyesha Blackwell, a chemist, has 6% and 15% solutions of sulfuric acid. How much of each solution should she mix to get 10 liters of a 9% solution?

23. Sum of Numbers The sum of three numbers is 29. The greatest number is four times the smallest number. The remaining number is 1 more than twice the smallest number. Find the three numbers.

Determine the solution to each system of inequalities.

24. $3x + 2y < 9$
$-2x + 5y \le 10$

25. $|x| > 3$
$|y| \le 1$

Cumulative Review Test

Take the following test and check your answers with those given in the back of the book. Review any questions that you answered incorrectly. The section where the material was covered is indicated after the answer.

1. Evaluate $48 \div \left\{ 4 \left[3 + \left(\dfrac{5 + 10}{5} \right)^2 \right] - 32 \right\}$.

2. Consider the following set of numbers.

$$\left\{ \frac{1}{2}, -4, 9, 0, \sqrt{3}, -4.63, 1 \right\}$$

List the elements of the set that are

a) natural numbers;

b) rational numbers;

c) real numbers.

3. Write the following numbers in order from smallest to largest.

$$-1, |-4|, \frac{3}{4}, \frac{5}{8}, -|-8|, |-12|$$

Solve.

4. $-[3 - 2(x - 4)] = 3(x - 6)$

5. $\dfrac{2}{3}x - \dfrac{5}{6} = 2$

6. $|2x - 3| - 5 = 4$

7. Solve the formula $M = \dfrac{1}{2}(a + x)$ for x.

8. Find the solution set of the inequality.

$$0 < \frac{3x - 2}{4} \leq 8$$

9. Simplify $\left(\dfrac{3x^2 y^{-2}}{y^3} \right)^{-2}$.

10. Graph $2y = 3x - 8$.

11. Write in slope-intercept form the equation of the line that is parallel to the graph of $2x - 3y = 8$ and passes through the point $(2, 3)$

12. Graph the inequality $6x - 3y < 12$.

13. Determine which of the following graphs represent functions. Explain.

a)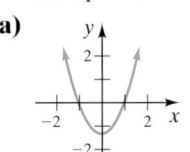

b)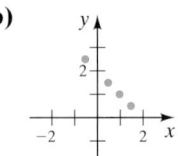

c)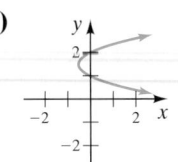

14. If $f(x) = \dfrac{x + 3}{x^2 - 9}$, find

a) $f(-4)$ **b)** $f(h)$ **c)** $f(3)$.

Solve each system of equations.

15. $3x + y = 6$
$y = 4x - 1$

16. $2p + 3q = 11$
$-3p - 5q = -16$

17. $x - 2y = 0$
$2x + z = 7$
$y - 2z = -5$

18. **Angles of a Triangle** If the largest angle of a triangle is nine times the measure of the smallest angle, and the middle-sized angle is 70° greater than the measures of the smallest angle, find the measures of the three angles.

19. **Walking and Jogging** Mark Simmons walks at 4 miles per hour and Judy Bolin jogs at 6 miles per hour. Mark begins walking $\dfrac{1}{2}$ hour before Judy starts jogging. If Judy jogs on the same path that Mark walks, how long after Judy begins jogging will she catch up to Mark?

20. **Rock Concert** There are two different prices of seats at a rock concert. The higher-priced seats sell for $20 and the less expensive seats sell for $16. If a total of 1000 tickets are sold and the total ticket sales are $18,400, how many of each type of ticket are sold?

5 Polynomials and Polynomial Functions

Goals of This Chapter

In the first part of this chapter, we discuss polynomials and polynomial functions. We then turn our attention to factoring. *You must have a thorough understanding of factoring to work the problems in many of the remaining chapters.* Pay particular attention to how to use factoring to find the *x*-intercepts of a quadratic function. We will refer back to this topic later in the course.

© Jupiter Unlimited

Quite often students participate in competitions such as essay writing contests, math contests, or spelling bees. In Exercise 82 on page 286, we use a polynomial function to determine the number of different ways students can finish first, second, or third in a spelling bee.

5.1 Addition and Subtraction of Polynomials

1 Find the degree of a polynomial.

2 Evaluate polynomial functions.

3 Understand graphs of polynomial functions.

4 Add and subtract polynomials.

1 Find the Degree of a Polynomial

Recall from Chapter 2 that the parts that are added or subtracted in a mathematical expression are called **terms**. The **degree of a term** with whole number exponents is the sum of the exponents of the variables, if there are variables. Nonzero constants have degree 0, and the term 0 has no degree.

> **Polynomial**
>
> A **polynomial** is a finite sum of terms in which all variables have whole number exponents and no variable appears in a denominator.

Understanding Algebra

+ or − signs separate terms in a polynomial.

- $3x^2 + 2x + 6$ is a *polynomial in one variable x.*
- $x^2y - 7x + 3$ is a *polynomial in two variables x and y.*
- $x^{1/2}$ is *not* a polynomial because the variable does not have a whole number exponent.
- $\dfrac{1}{x}$ (or x^{-1}) is *not* a polynomial because the variable does not have a whole number exponent.
- $\dfrac{1}{x+1}$ is *not* a polynomial because the variable is in the denominator.

The **leading term** of a polynomial is the term of highest degree. The **leading coefficient** is the coefficient of the leading term.

EXAMPLE 1 For each polynomial give the number of terms, the degree of the polynomial, the leading term, and the leading coefficient.

a) $2x^5 - 3x^2 + 6x - 9$ **b)** $8x^2y^4 - 6xy^3 + 3xy^2z^4$

Solution   We will organize the answers in tabular form.

Polynomial	Number of terms	Degree of polynomial	Leading term	Leading coefficient
a) $2x^5 - 3x^2 + 6x - 9$	4	5 (from $2x^5$)	$2x^5$	2
b) $8x^2y^4 - 6xy^3 + 3xy^2z^4$	3	7 (from $3xy^2z^4$)	$3xy^2z^4$	3

Now Try Exercise 23

Polynomials are classified according to the number of terms they have, as indicated in the following chart.

Understanding Algebra

The prefix *mono* means *one.*
The prefix *bi* means *two.*
The prefix *poly* means *many.*

Polynomial type	Description	Examples
Monomial	A polynomial of one term	$4x^2, 6x^2y, 3, -2xyz^5, 7$
Binomial	A polynomial of two terms	$x^2 + 1, 2x^2 - y, 6x^3 - 5y^2$
Trinomial	A polynomial of three terms	$x^3 + 6x - 8, x^2y - 9x + y^2$

Polynomials containing more than three terms are not given specific names. *Poly* is a prefix meaning *many.* A polynomial is called **linear** if it is degree 0 or 1. A polynomial in one variable is called **quadratic** if it is degree 2, and **cubic** if it is degree 3.

Type of Polynomial	Examples
Linear	$2x - 4,\quad 5$
Quadratic	$3x^2 + x - 6,\quad 4x^2 - 8$
Cubic	$-4x^3 + 3x^2 + 5,\quad 2x^3 + 7x$

Understanding Algebra

In *descending order,* the exponents *descend* or get *lower.*

The polynomials $2x^3 + 4x^2 - 6x + 3$ and $4x^2 - 3xy + 5y^2$ are examples of polynomials in **descending order** of the variable x because the exponents on the variable x descend or get lower as the terms go from left to right. Polynomials are usually written in descending order of a given variable.

EXAMPLE 2 Write each polynomial in descending order of the variable x.

a) $5x + 4x^2 - 6$ 　　　　　　　　　**b)** $xy - 6x^2 + 8y^2$

Solution

a) $5x + 4x^2 - 6 = 4x^2 + 5x - 6$

b) $xy - 6x^2 + 8y^2 = -6x^2 + xy + 8y^2$

Now Try Exercise 19

2 Evaluate Polynomial Functions

The expression $2x^3 + 6x^2 + 3$ is a polynomial. If we write $P(x) = 2x^3 + 6x^2 + 3$, then we have a polynomial function. In a **polynomial function**, the expression used to describe the function is a polynomial.

EXAMPLE 3 For the polynomial function $P(x) = 4x^3 - 6x^2 - 2x + 9$, find

a) $P(0)$ 　　　　　**b)** $P(3)$ 　　　　**c)** $P(-2)$

Solution

a) $P(x) = 4x^3 - 6x^2 - 2x + 9$

$\quad P(0) = 4(0)^3 - 6(0)^2 - 2(0) + 9$

$\qquad\quad = 0 - 0 - 0 + 9 = 9$

b) $P(3) = 4(3)^3 - 6(3)^2 - 2(3) + 9$

$\qquad\quad = 4(27) - 6(9) - 6 + 9 = 57$

c) $P(-2) = 4(-2)^3 - 6(-2)^2 - 2(-2) + 9$

$\qquad\qquad = 4(-8) - 6(4) + 4 + 9 = -43$

Now Try Exercise 29

Businesses, governments, and other organizations often need to track and make projections about things such as sales, profits, changes in the population, effectiveness of new drugs, and so on. To do so, they often use graphs and functions.

EXAMPLE 4 **Online Prescriptions** More and more physicians are ordering prescriptions online (e-prescriptions) for their patients. **Figure 5.1** shows the number of e-prescriptions for the years from 2004 to 2008. The polynomial function that can be used to approximate the number of e-prescriptions, in millions, is

$$P(t) = 10.5t^2 - 18.3t + 3.4$$

where t is the number of years since 2004 and $0 \le t \le 4$.

a) Use the function to estimate the number of e-prescriptions in 2008.

b) Compare your answer from part **a)** to the graph. Does the graph support your answer?

c) If this trend continues beyond 2008 estimate the number of e-prescriptions in 2010.

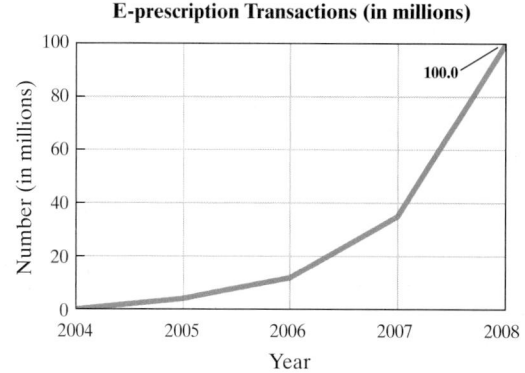

E-prescription Transactions (in millions)

Source: Pharmacy Health Information Exchange

FIGURE 5.1

Solution

a) **Understand**   We need to determine the value of t to substitute into this function. Since t is the number of years since 2004, the year 2008 corresponds to $t = 4$. Thus, to estimate the number of e-prescriptions, we compute $P(4)$.

Translate and Carry Out $P(t) = 10.5t^2 - 18.3t + 3.4$

$$P(4) = 10.5(4)^2 - 18.3(4) + 3.4$$
$$= 168 - 73.2 + 3.4$$
$$= 98.2$$

Check and Answer   The number of e-prescriptions in 2008 was about 98.2 million, or 98,200,000.

b) From part **a)** we see there were about 98.2 million e-prescriptions in 2008. The line graph in **Figure 5.1** shows that there were 100 million e-prescriptions in 2008. Since both values are very close, we conclude that the graph supports the outcome from part **a)**.

c) **Understand**   To estimate the number of e-prescriptions in 2010, observe that 2010 is 6 years after 2004. Thus, $t = 6$, and we substitute 6 for t in the polynomial function.

Translate and Carry Out $P(t) = 10.5t^2 - 18.3t + 3.4$

$$P(6) = 10.5(6)^2 - 18.3(6) + 3.4$$
$$= 271.6$$

Check and Answer   If this trend continues, in 2010 there will be about 271.6 million, or 271,600,000, e-prescriptions.

Now Try Exercise 93

Understanding Algebra

A function *increases* if your pencil goes up as you trace the graph from left to right. A function *decreases* if your pencil goes down as you trace the graph from left to right.

3 Understand Graphs of Polynomial Functions

The graphs of all polynomial functions are smooth, continuous curves. **Figure 5.2** shows a graph of a quadratic polynomial function. **Figure 5.3** and **Figure 5.4** show the graphs of cubic polynomial functions. Each of the functions in these graphs has a *positive leading coefficient*. Notice that in each graph, the function continues to *increase* (the green part of the graph) to the right of some value of x.

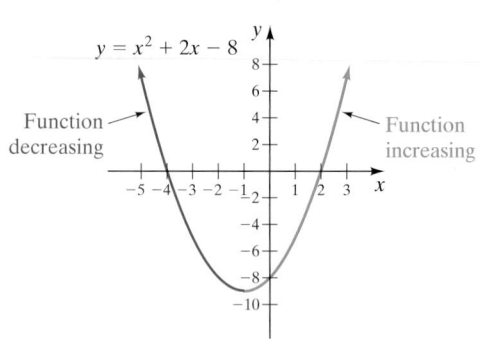

FIGURE 5.2

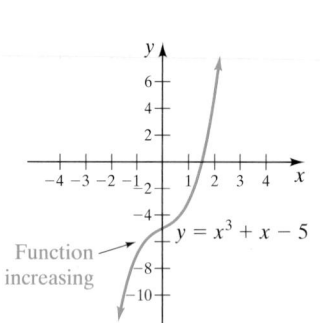

FIGURE 5.3

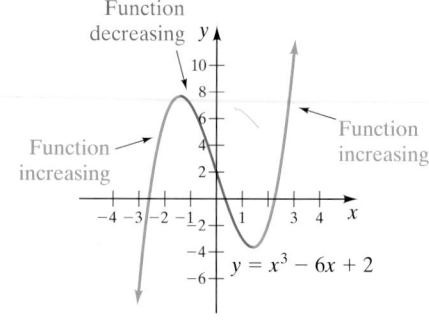

FIGURE 5.4

Figure 5.5 on page 283 shows a graph of a quadratic polynomial function. **Figure 5.6** and **Figure 5.7** show the graphs of cubic polynomial functions. Each of the functions in these graphs has a *negative leading coefficient*. Notice that in each graph, the function continues to *decrease* (the red part of the graph) to the right of some value of x.

Why does the leading coefficient determine whether a function will increase or decrease to the right of some value of x? The leading coefficient is the coefficient of the

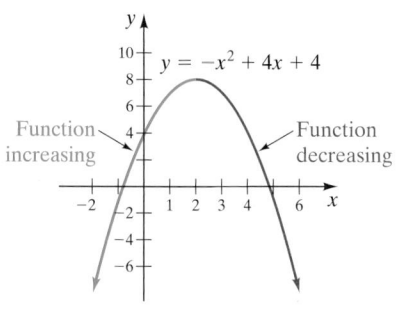

FIGURE 5.5

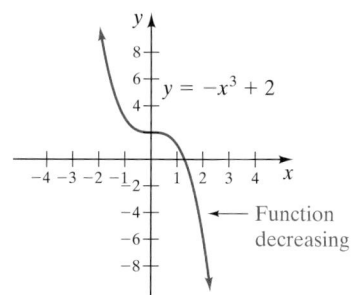

FIGURE 5.6

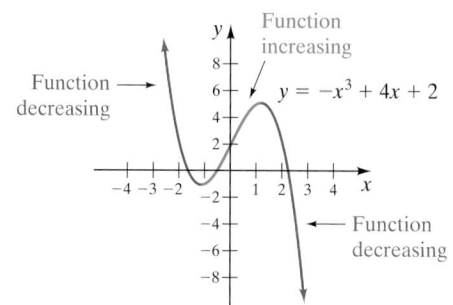

FIGURE 5.7

Understanding Algebra

Polynomial functions with a *positive leading coefficient* will *increase* to the right of some value of x. Polynomial functions with a *negative leading coefficient* will *decrease* to the right of some value x.

term with the greatest exponent on the variable. As x increases, this term will eventually dominate all the other terms in the function. So if the coefficient of this term is positive, the function will *eventually* increase as x increases. If the leading coefficient is negative, the function will *eventually* decrease as x increases. This information, along with checking the y-intercept of the graph, can be useful in determining whether a graph is correct or complete.

📟 *Using Your Graphing Calculator* ▮▮▮

Whenever you graph a polynomial function on your calculator, make sure your screen shows every change in direction of your graph. For example, suppose you graph $y = 0.1x^3 - 2x^2 + 5x - 8$ on your calculator. Using the standard window, you get the graph shown in **Figure 5.8**.

However, from our preceding discussion you should realize that since the leading coefficient, 0.1, is positive, the graph must increase to the right of some value of x. The graph in **Figure 5.8** does not show this. If you change your window as shown in **Figure 5.9,** you will get the graph shown. Now you can see how the graph increases to the right of about $x = 12$. When graphing, the y-intercept is often helpful in determining the values to use for the range. Recall that to find the y-intercept, we set $x = 0$ and solve for y. For example, if graphing $y = 4x^3 + 6x^2 + x - 180$, the y-intercept will be at -180.

$y = 0.1x^3 - 2x^2 + 5x - 8$

FIGURE 5.8

$y = 0.1x^3 - 2x^2 + 5x - 8$

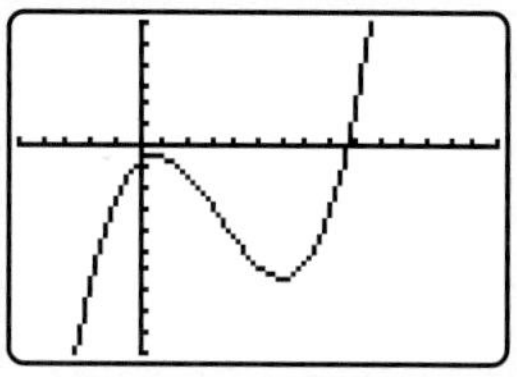

$[-10, 30, 2, -100, 60, 10]$

FIGURE 5.9

4 Add and Subtract Polynomials

To *add or subtract polynomials,* remove parentheses if any are present. Then combine like terms.

EXAMPLE 5 Simplify $(4x^2 - 6x + 8) + (2x^2 + 5x - 1)$.

Solution

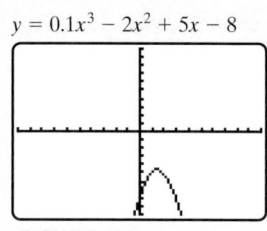

$$(4x^2 - 6x + 8) + (2x^2 + 5x - 1)$$
$$= 4x^2 - 6x + 8 + 2x^2 + 5x - 1 \qquad \text{Remove parentheses.}$$
$$= \underline{4x^2 + 2x^2} \; \underline{-6x + 5x} \; \underline{+8 - 1} \qquad \text{Rearrange terms.}$$
$$= \qquad 6x^2 \qquad -x \qquad +7 \qquad \text{Combine like terms.}$$

Now Try Exercise 39

EXAMPLE 6 Simplify $(3x^2y - 4xy + y) + (x^2y + 2xy + 8y - 5)$.

Solution

$$(3x^2y - 4xy + y) + (x^2y + 2xy + 8y - 5)$$

$$= 3x^2y - 4xy + y + x^2y + 2xy + 8y - 5 \qquad \text{Remove parentheses.}$$

$$= \underline{3x^2y + x^2y} \ \underline{- 4xy + 2xy} \ \underline{+ y + 8y} - 5 \qquad \text{Rearrange terms.}$$

$$= \qquad 4x^2y \qquad -2xy \qquad +9y \ - 5 \qquad \text{Combine like terms.}$$

Now Try Exercise 45

Helpful Hint

Recall that $-x$ means $-1 \cdot x$. Thus $-(2x^2 - 4x + 6)$ means $-1(2x^2 - 4x + 6)$, and the distributive property applies. When you subtract one polynomial from another, the *signs of every term* of the polynomial being subtracted must change. For example,

$$x^2 - 6x + 3 - (2x^2 - 4x + 6) = x^2 - 6x + 3 - 1(2x^2 - 4x + 6)$$

$$= x^2 - 6x + 3 - 2x^2 + 4x - 6$$

$$= -x^2 - 2x - 3$$

EXAMPLE 7 Subtract $(-x^2 - 2x + 11)$ from $(x^3 + 4x + 6)$.

Solution

$$(x^3 + 4x + 6) - (-x^2 - 2x + 11)$$

$$= (x^3 + 4x + 6) - 1(-x^2 - 2x + 11) \qquad \text{Insert 1.}$$

$$= x^3 + 4x + 6 + x^2 + 2x - 11 \qquad \text{Distributive property}$$

$$= x^3 + x^2 + 4x + 2x + 6 - 11 \qquad \text{Rearrange terms.}$$

$$= x^3 + x^2 + 6x - 5 \qquad \text{Combine like terms.}$$

Now Try Exercise 61

EXAMPLE 8 Simplify $x^2y - 4xy^2 + 5 - (2x^2y - 3y^2 + 11)$.

Solution

$$x^2y - 4xy^2 + 5 - 1(2x^2y - 3y^2 + 11) \qquad \text{Insert 1.}$$

$$= x^2y - 4xy^2 + 5 - 2x^2y + 3y^2 - 11 \qquad \text{Distributive property}$$

$$= x^2y - 2x^2y - 4xy^2 + 3y^2 + 5 - 11 \qquad \text{Rearrange terms.}$$

$$= -x^2y - 4xy^2 + 3y^2 - 6 \qquad \text{Combine like terms.}$$

Note that $-x^2y$ and $-4xy^2$ are not like terms since the variables have different exponents. Also, $-4xy^2$ and $3y^2$ are not like terms since $3y^2$ does not contain the variable x.

Now Try Exercise 49

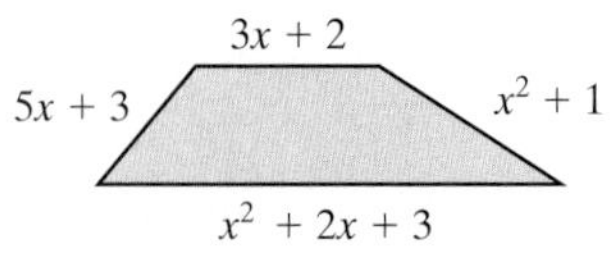

FIGURE 5.10

EXAMPLE 9 Perimeter Find an expression for the perimeter of the quadrilateral in **Figure 5.10**.

Solution   The perimeter is the sum of the lengths of the sides of the figure.

$$\text{Perimeter} = (x^2 + 2x + 3) + (x^2 + 1) + (5x + 3) + (3x + 2) \qquad \text{Sum of the sides}$$

$$= x^2 + 2x + 3 + x^2 + 1 + 5x + 3 + 3x + 2 \qquad \text{Remove parentheses.}$$

$$= x^2 + x^2 + 2x + 5x + 3x + 3 + 1 + 3 + 2 \qquad \text{Rearrange terms.}$$

$$= 2x^2 + 10x + 9 \qquad \text{Combine like terms.}$$

The perimeter of the quadrilateral is $2x^2 + 10x + 9$.

Now Try Exercise 73

EXERCISE SET 5.1

Warm-Up Exercises

Fill in the blanks with the appropriate word, phrase, or symbol(s) from the following list.

leading   polynomial   degree   trinomial   terms   5   monomial   factors   binomial   7   1   0

1. The parts that are added or subtracted in a polynomial are called _____.

2. The sum of the exponent(s) of the variables of a term is called the _____ of the term.

3. The degree of $-7x^3y^2$ is _____.

4. The degree of $6x$ is _____.

5. In a polynomial, the term of highest degree is called the _____ term.

6. A polynomial with one term is a _____.

7. A polynomial with two terms is a _____.

8. A polynomial with three terms is a _____.

Practice the Skills

Determine whether each expression is a polynomial. If the polynomial has a specific name, for example, "monomial" or "binomial," give the name. If the expression is not a polynomial, explain why it is not.

9. -4

10. $5z^{-3}$

11. $7z$

12. $5x^2 - 6x + 9$

13. $4x^{-1}$

14. $8x^2 - 4x + 9y^2$

15. $3x^{1/2} + 2xy$

16. $10xy + 5y^2$

Write each polynomial in descending order of the variable x. If the polynomial is already in descending order, so state. Give the degree of each polynomial.

17. $-5 + 2x - x^2$

18. $-3x - 9 + 8x^2$

19. $9y^2 + 3xy + 10x^2$

20. $-2 + x - 7x^2 + 4x^3$

21. $-2x^4 + 5x^2 - 4$

22. $15xy^2 + 3x^2y - 9 - 2x^3$

*Give **a)** the degree of each polynomial and **b)** its leading coefficient.*

23. $x^4 + 3x^6 - 2x - 13$

24. $17x^4 + 13x^5 - x^7 + 4x^3$

25. $4x^2y^3 + 6xy^4 + 9xy^5$

26. $-a^4b^3c^2 + 9a^8b^9c^4 - 8a^7c^{20}$

27. $-\dfrac{1}{3}m^4n^5p^8 + \dfrac{3}{5}m^3p^6 - \dfrac{5}{9}n^4p^6q$

28. $-0.6x^2y^3z^2 + 2.9xyz^9 - 1.7x^8y^4$

Evaluate each polynomial function at the given value.

29. Find $P(2)$ if $P(x) = x^2 - 6x + 3$

30. Find $P(-1)$ if $P(x) = 4x^2 + 6x - 21$.

31. Find $P\left(\dfrac{1}{2}\right)$ if $P(x) = 2x^2 - 3x - 6$.

32. Find $P\left(\dfrac{1}{3}\right)$ if $P(x) = \dfrac{1}{2}x^3 - x^2 + 6$.

33. Find $P(0.4)$ if $P(x) = 0.2x^3 + 1.6x^2 - 2.3$.

34. Find $P(-1.2)$ if $P(x) = -1.6x^3 - 4.6x^2 - 0.1x$.

In Exercises 35–56, simplify.

35. $(x^2 + 3x - 1) + (6x - 5)$

36. $(5b^2 - 8b + 7) - (2b^2 - 3b - 15)$

37. $(x^2 - 8x + 11) - (5x + 9)$

38. $(2x - 13) - (3x^2 - 4x + 26)$

39. $(4y^2 + 9y - 1) - (2y^2 + 10)$

40. $(5n^2 - 7) + (9n^2 + 8n + 12)$

41. $\left(-\dfrac{5}{9}a + 6\right) + \left(-\dfrac{2}{3}a^2 - \dfrac{1}{4}a - 1\right)$

42. $(6y^2 - 9y + 14) - (-2y^2 - y - 8)$

43. $(1.4x^2 + 1.6x - 8.3) - (4.9x^2 + 3.7x + 11.3)$

44. $(-12.4x^2y - 6.2xy + 9.3y^2) - (-5.3x^2y + 1.6xy - 10.4y^2)$

45. $\left(-\dfrac{1}{3}x^3 + \dfrac{1}{4}x^2y + 8xy^2\right) + \left(-x^3 - \dfrac{1}{2}x^2y + xy^2\right)$

46. $\left(-\dfrac{3}{5}xy^2 + \dfrac{5}{8}\right) - \left(-\dfrac{1}{2}xy^2 + \dfrac{3}{5}\right)$

47. $(3a - 6b + 5c) - (-2a + 4b - 8c)$

48. $(9r + 7s - t) + (-2r - 6s - 3t)$

49. $(3a^2b - 6ab + 5b^2) - (4ab - 6b^2 + 5a^2b)$

50. $(3x^2 - 5y^2 - 2xy) - (4x^2 + 8y^2 - 9xy)$

51. $(8r^2 - 5t^2 + 2rt) + (-6rt + 2t^2 - r^2)$

52. $(a^2 - b^2 + 5ab) + (-3b^2 - 2ab + a^2)$

53. $6x^2 - 5x - [3x - (4x^2 - 9)]$

54. $3xy^2 - 2x - [-(4xy^2 + 3x) - 5xy]$

55. $5w - 6w^2 - [(3w - 2w^2) - (4w + w^2)]$

56. $-[-(5r^2 - 3r) - (2r - 3r^2) - 2r^2]$

57. Subtract $(4x - 11)$ from $(7x + 8)$.

58. Subtract $(-x^2 + 3x + 5)$ from $(4x^2 - 6x + 2)$.

59. Add $-2x^2 - 4x - 12$ and $-x^2 - 2x$.

60. Subtract $(5x^2 - 6)$ from $(2x^2 - 9x + 8)$.

61. Subtract $0.2a^2 - 3.9a + 26.4$ from $-5.2a^2 - 9.6a$.

62. Add $6x^2 + 12xy$ and $-2x^2 + 4xy + 3y$.

63. Subtract $\left(5x^2y + \dfrac{5}{9}\right)$ from $\left(-\dfrac{1}{2}x^2y + xy^2 + \dfrac{3}{5}\right)$

64. Subtract $(6x^2y + 7xy)$ from $(2x^2y + 12xy)$.

Simplify. Assume that all exponents represent natural numbers.

65. $(3x^{2r} - 7x^r + 1) + (2x^{2r} - 3x^r + 2)$

66. $(8x^{2r} - 5x^r + 4) + (6x^{2r} + x^r + 3)$

67. $(x^{2s} - 8x^s + 6) - (2x^{2s} - 4x^s - 13)$

68. $(5a^{2m} - 6a^m + 4) - (2a^{2m} + 7)$

69. $(7b^{4n} - 5b^{2n} + 1) - (3b^{3n} - b^{2n})$

70. $(-3r^{3a} + r^a - 6) - (-2r^{3a} - 8r^{2a} + 6)$

Problem Solving

Perimeter   *In Exercises 71–76, find an expression for the perimeter of each figure. See Example 9.*

71.
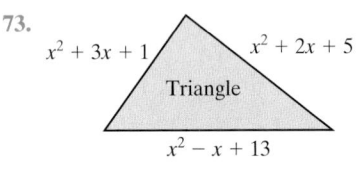
Square
$x^2 + 2x + 6$

72.
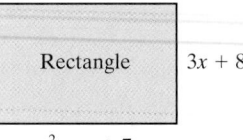
Rectangle $3x + 8$
$x^2 - x + 7$

73.
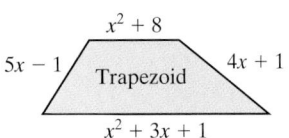
$x^2 + 3x + 1$ $x^2 + 2x + 5$
Triangle
$x^2 - x + 13$

74.
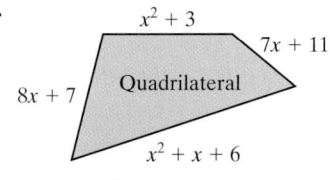
$x^2 + 3$
$7x + 11$
$8x + 7$ Quadrilateral
$x^2 + x + 6$

75.
$x^2 + 8$
$5x - 1$ Trapezoid $4x + 1$
$x^2 + 3x + 1$

76.
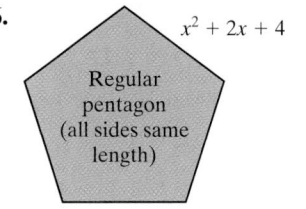
$x^2 + 2x + 4$
Regular pentagon (all sides same length)

In Exercises 77–86, if necessary, round answers to the nearest hundredth

77. **Area** The area of a square is a function of its side, s, where $A(s) = s^2$. Find the area of a square if its side is 12 meters.

78. **Volume** The volume of a cube is a function of its side, s, where $V(s) = s^3$. Find the volume of a cube if its side is 7 centimeters.

79. **Area** The area of a circle is a function of its radius, where $A(r) = \pi r^2$. Find the area of a circle if its radius is 6 inches. Use the $\boxed{\pi}$ key on your calculator.

80. **Volume** The volume of a sphere is a function of its radius where $V(r) = \dfrac{4}{3}\pi r^3$. Find the volume of a sphere when its radius is 4 inches.

© Kirill R/Shutterstock

81. **Height** When an object is dropped from the Empire State Building (height 1250 feet), the object's height, h, in feet, from the ground at time, t, in seconds, after being dropped can be determined by

$$h = P(t) = -16t^2 + 1250$$

Find the height an object is from the ground 6 seconds after being dropped.

82. **Spelling Bee** The number of ways that the first-, second-, and third-place winners in a spelling bee can be selected from n participants is given by $P(n) = n^3 - 3n^2 + 2n$. If there are seven participants, how many ways can the winner and first and second runners-up be selected?

83. **Committees** The number of different committees of 2 students where the 2 students are selected from a class of n students is given by $c(n) = \dfrac{1}{2}(n^2 - n)$. If a biology class has 15 students, how many different committees having 2 students can be selected?

84. **Committees** The number of different committees of 3 students where the 3 students are selected from a class of n students is given by $c(n) = \dfrac{1}{6}n^3 - \dfrac{1}{2}n^2 + \dfrac{1}{3}n$. If an art class has 10 students, how many different committees having 3 students can be selected?

85. Savings Account On January 2, 2010, Jorge Sanchez deposited $650 into an account that pays simple interest at a rate of $24 each year. The amount in the account is a function of time given by $A(t) = 650 + 24t$, where t is the number of years after 2010. Find the amount in the account in **a)** 2011. **b)** 2025.

86. Financing Frank Gunther just bought a new Ford Edge. After making the down payment, the amount to be financed is $28,250. Using a 0% (or interest-free) loan, the monthly payment is $387.50. The amount still owed on the car is a function of time give by $A(t) = \$28,250 - \$387.50t$, where t is the number of months after Frank bought the car. How much is still owed **a)** 2 months, **b)** 15 months after Frank bought the car?

See Exercise 86.

Profit   *The profit of a company is found by subtracting its cost from its revenue. In Exercises 87 and 88 R(x) represents the company's revenue when selling x items and C(x) represents the company's cost when producing x items.* **a)** *Determine a profit function P(x).* **b)** *Evaluate P(x) when x* = 100.

87. $R(x) = 2x^2 - 60x,$
  $C(x) = 8050 - 420x$

88. $R(x) = 5.5x^2 - 80.3x$
  $C(x) = 1.2x^2 + 16.3x + 12,040.6$

*In Exercises 89–92, determine which of the graphs—***a)**, **b)**, *or* **c)**—*is the graph of the given equation. Explain how you determined your answer.*

89. $y = x^2 + 3x - 4$

a)

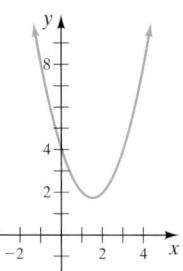

b)

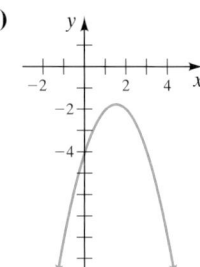

c)

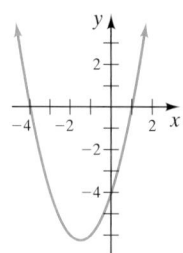

90. $y = x^3 + 2x^2 - 4$

a)

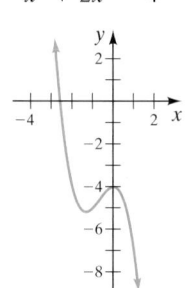

b)

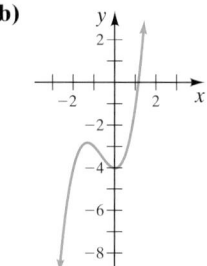

c)

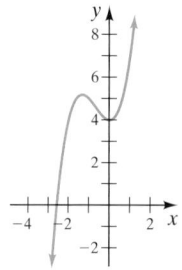

91. $y = -x^3 + 2x - 6$

a)

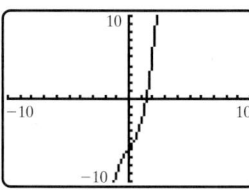

b)

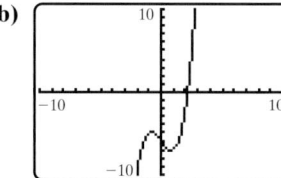

c)

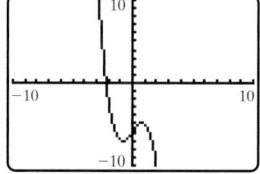

92. $y = x^3 + 4x^2 - 5$

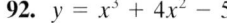

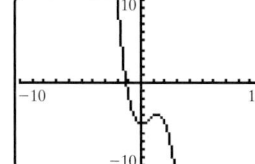

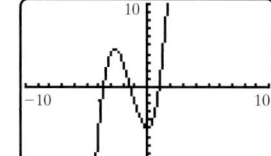

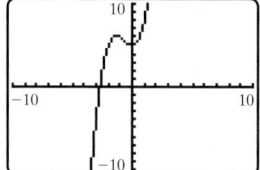

93. The graph shows the percent increase in condo fees for the Annandale Gardens Association for the years from 2007 to 2009. The percent increase, $P(t)$, can be approximated by the function

$$P(t) = 1.25t^2 - 5.75t + 10$$

where t is the number of years since 2007.

a) Use this function to estimate the percent increase in condo fees in 2009.

b) Compare your answer in part a) to the graph. Does the graph support your answer?

c) If this trend continues estimate the percent increase in condo fees in 2012.

Percent Increase in Condo Fees

Source: Annandale Gardens Association budgets.

94. Inclined Plane A ball rolls down an inclined plane. The distance, $d(t)$, in feet, the ball has traveled is given by the function

$$d(t) = 2.36t^2$$

where t is time in seconds $0 < t < 5$.

   Find the distance the ball has traveled down the inclined plane in

a) 1 second.

b) 3 seconds.

c) 5 seconds.

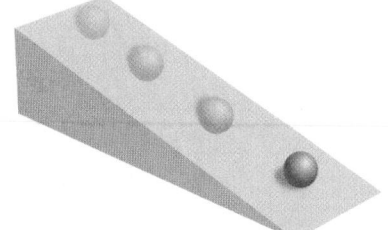

95. Inflation The function $C(t) = 0.31t^2 + 0.59t + 9.61$, where t is years since 1997, approximates the cost, in thousands of dollars, for purchasing in the future what $10,000 would purchase in 1997. This function is based on a 6% annual inflation rate and $0 \leq t \leq 25$. Estimate the cost in 2012 for goods that cost $10,000 in 1997.

96. Population. The population of a town is determined by the function $P(t) = 6t^2 + 7t + 6500$ where t is the number of years since 2001. Find the population of the town in 2011.

*Answer Exercises 97 and 98 using a graphing calculator if you have one. If you do not have a graphing calculator, draw the graphs in part **a)** by plotting points. Then answer parts **b)** through **e)**.*

97. a) Graph

$$y_1 = x^3$$
$$y_2 = x^3 - 3x^2 - 3$$

b) In both graphs, for values of $x > 3$, do the functions increase or decrease as x increases?

c) When the leading term of a polynomial function is x^3, the polynomial must increase for $x > a$, where a is some real number greater than 0. Explain why this must be so.

d) In both graphs, for values of $x < -3$, do the functions increase or decrease as x decreases?

e) When the leading term of a polynomial function is x^3, the polynomial must decrease for $x < a$, where a is some real number less than 0. Explain why this must be so.

98. a) Graph

$$y_1 = x^4$$
$$y_2 = x^4 - 6x^2$$

b) In each graph, for values of $x > 3$, are the functions increasing or decreasing as x increases?

c) When the leading term of a polynomial function is x^4, the polynomial must increase for $x > a$, where a is some real number greater than 0. Explain why this must be so.

d) In each graph, for values of $x < -3$, are the functions increasing or decreasing as x decreases?

e) When the leading term of a polynomial function is x^4, the polynomial must increase for $x < a$, where a is some real number less than 0. Explain why this must be so.

Concept/Writing Exercises

99. What is the degree of a nonzero constant?

100. What is the leading term of a polynomial?

101. What is the leading coefficient of a polynomial?

102. a) How do you determine the degree of a term?

   b) What is the degree of $6x^4y^3z$?

103. a) How do you determine the degree of a polynomial?

   b) What is the degree of $-4x^4 + 6x^3y^4 + z^5$?

104. What does it mean when a polynomial is in descending order of the variable x?

105. a) When is a polynomial linear?

   b) Give an example of a linear polynomial.

106. a) When is a polynomial quadratic?

   b) Give an example of a quadratic polynomial.

107. a) When is a polynomial cubic?

   b) Give an example of a cubic polynomial.

108. When one polynomial is being subtracted from another, what happens to the signs of all the terms of the polynomial being subtracted?

109. Write a fifth-degree trinomial in x in descending order that lacks fourth-, third-, and second-degree terms.

110. Write a seventh-degree polynomial in y in descending order that lacks fifth-, third-, and second-degree terms.

111. Is the sum of two trinomials always a trinomial? Explain, and give an example to support your answer.

112. Is the sum of two binomials always a binomial? Explain, and give an example to support your answer.

113. Is the sum of two quadratic polynomials always a quadratic polynomial? Explain, and give an example to support your answer.

114. Is the difference of two cubic polynomials always a cubic polynomial? Explain, and give an example to support your answer.

Challenge Problems

*Determine which of the graphs—**a)**, **b)**, or **c)**—is the graph of the given equation. Explain how you determined your answer.*

115. $y = -x^4 + 3x^3 - 5$

a)
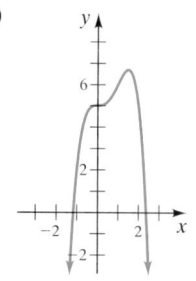

b)
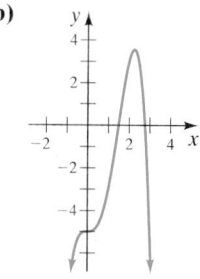

c)
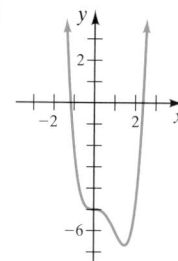

116. $y = 8x^2 + x - 5$

a)
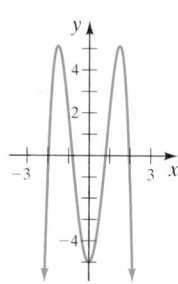

b)
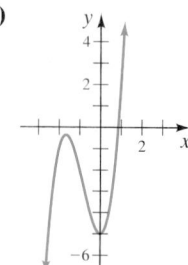

c)
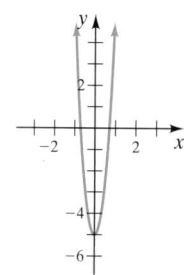

Group Activity

Discuss and answer Exercises 117 and 118 as a group.

117. If the leading term of a polynomial function is $3x^3$, which of the following could possibly be the graph of the polynomial? Explain. Consider what happens for large positive values of x and for large negative values of x.

a)
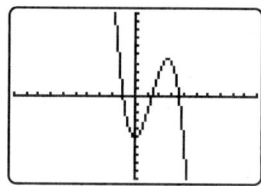

b)
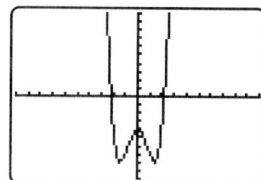

c)
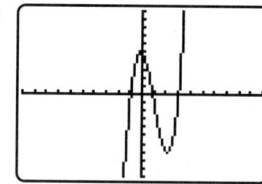

118. If the leading term of a polynomial is $-2x^4$, which of the following could possibly be the graph of the polynomial? Explain.

a)
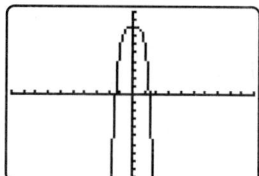

b)
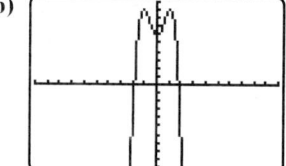

c)
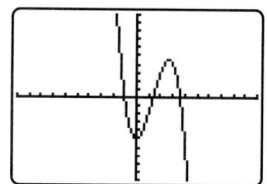

Cumulative Review Exercises

[1.4] **119.** Evaluate $\sqrt[4]{81}$.

[2.1] **120.** Solve $1 = \dfrac{8}{5}x - \dfrac{1}{2}$.

[2.4] **121. Molding Machines** An older molding machine can produce 40 plastic buckets in 1 hour. A newer machine can produce 50 buckets in 1 hour. How long will it take the two machines to produce a total of 540 buckets?

[3.4] **122.** Find the slope of the line through the points $(10, -4)$ and $(-1, -2)$.

[4.2] **123.** Solve the system of equations.

$$-4s + 3t = 16$$
$$4t - 2u = 2$$
$$-s + 6u = -2$$

5.2 Multiplication of Polynomials

1 Multiply a monomial by a polynomial.

2 Multiply a binomial by a binomial.

3 Multiply a polynomial by a polynomial.

4 Find the square of a binomial.

5 Find the product of the sum and difference of the same two terms.

6 Find the product of polynomial functions.

Understanding Algebra

When multiplying expressions with the same base, *add the exponents.*

1 Multiply a Monomial by a Polynomial

To multiply polynomials, you must remember that *each term of one polynomial must multiply each term of the other polynomial.* This results in monomials multiplying monomials. To multiply monomials, we use the product rule for exponents.

Product Rule for Exponents

$$a^m \cdot a^n = a^{m+n}$$

In Example 1, we use the word *factors.* Recall that any expressions that are *multiplied* are called *factors.*

Multiply a Monomial by a Monomial

EXAMPLE 1 Multiply. ───────────

a) $(4x^2)(5x^3)$ **b)** $(3x^2y)(4x^5y^3)$ **c)** $(-2a^4b^7)(-3a^8b^3c^5)$

Solution   We use the Product Rule for Exponents to multiply the factors.

a) $(4x^2)(5x^3) = 4 \cdot 5 \cdot x^2 \cdot x^3$ Remove parentheses and rearrange terms.

$\qquad\qquad\quad = 20x^{2+3}$ Product Rule, $x^2 \cdot x^3 = x^{2+3}$

$\qquad\qquad\quad = 20x^5$

b) $(3x^2y)(4x^5y^3) = 3 \cdot 4 \cdot x^2 \cdot x^5 \cdot y \cdot y^3$ Remove parentheses, rearrange terms.

$\qquad\qquad\qquad = 12x^{2+5}y^{1+3}$ Product Rule

$\qquad\qquad\qquad = 12x^7y^4$

c) $(-2a^4b^7)(-3a^8b^3c^5) = (-2)(-3)a^4 \cdot a^8 \cdot b^7 \cdot b^3 \cdot c^5$ Remove parentheses, rearrange terms.

$\qquad\qquad\qquad\qquad = 6a^{4+8}b^{7+3}c^5$ Product rule

$\qquad\qquad\qquad\qquad = 6a^{12}b^{10}c^5$

Now Try Exercise 9

In Example 1a), both $4x^2$ and $5x^3$ are *factors* of the product $20x^5$. In Example 1b), both $3x^2y$ and $4x^5y^3$ are *factors* of the product $12x^7y^4$. In Example 1c), both $-2a^4b^7$ and $-3a^8b^3c^5$ are *factors* of the product $6a^{12}b^{10}c^5$.

Multiply a Monomial by a Polynomial   When multiplying a monomial by a binomial, we can use the distributive property. When multiplying a monomial by a polynomial that contains more than two terms we can use the **expanded form of the distributive property**.

Distributive Property, Expanded Form

$$a(b + c + d + \cdots + n) = ab + ac + ad + \cdots + an$$

In Example 2a) we multiply a monomial by a binomial and in Examples 2b) and 2c) we multiply a monomial by a trinomial.

EXAMPLE 2 Multiply. ───────────

a) $3x^2\left(\dfrac{1}{6}x^3 - 5x^2\right)$ **b)** $2xy(3x^2y + 6xy^2 + 9)$ **c)** $0.4x(0.3x^3 + 0.7xy^2 - 0.2y^4)$

Solution

a) $3x^2\left(\dfrac{1}{6}x^3 - 5x^2\right) = 3x^2\left(\dfrac{1}{6}x^3\right) - 3x^2(5x^2) = \dfrac{1}{2}x^5 - 15x^4$

b) $2xy(3x^2y + 6xy^2 + 9) = (2xy)(3x^2y) + (2xy)(6xy^2) + (2xy)(9)$
$$= 6x^3y^2 + 12x^2y^3 + 18xy$$

c) $0.4x(0.3x^3 + 0.7xy^2 - 0.2y^4)$
$$= (0.4x)(0.3x^3) + (0.4x)(0.7xy^2) - (0.4x)(0.2y^4)$$
$$= 0.12x^4 + 0.28x^2y^2 - 0.08xy^4$$

Now Try Exercise 13

2 Multiply a Binomial by a Binomial

Consider multiplying $(a + b)(c + d)$. Treat $(a + b)$ as a single term and use the distributive property, to get

$$(a + b)(c + d) = (a + b)c + (a + b)d$$
$$= ac + bc + ad + bd$$

When multiplying a binomial by a binomial, each term of the first binomial must be multiplied by each term of the second binomial, and then all like terms are combined.

Binomials can be multiplied vertically as well as horizontally.

EXAMPLE 3   Multiply $(3x + 2)(x - 5)$.

Solution   We will multiply vertically. List the binomials in descending order of the variable one beneath the other. It makes no difference which one is placed on top.

Multiply each term of the top binomial by each term of the bottom binomial, as shown. Remember to align like terms so that they can be added.

$$
\begin{array}{r}
3x + 2 \\
x - 5 \\
\hline
\end{array}
$$

$-5(3x + 2) \longrightarrow -15x - 10$ Multiply the top binomial by -5.
$x\,(3x + 2) \longrightarrow 3x^2 + 2x$ Multiply the top binomial by x.
$\overline{\qquad\quad 3x^2 - 13x - 10}$ Add like terms in columns.

In Example 3, the binomials $3x + 2$ and $x - 5$ are *factors* of the trinomial $3x^2 - 13x - 10$.

Now Try Exercise 21

Understanding Algebra

The acronym *FOIL* helps you remember how to multiply *two binomials*. In this process there are *four* products:

F irst
O uter
I nner
L ast

The FOIL Method   A convenient way to multiply two binomials is called the **FOIL method**. To multiply two binomials using the FOIL method, list the binomials side by side. The word FOIL indicates that you multiply the F irst terms, O uter terms, I nner terms, and L ast terms of the two binomials, in that order. This procedure is illustrated in Example 4, where we multiply the same two binomials we multiplied in Example 3.

EXAMPLE 4   Multiply $(3x + 2)(x - 5)$ using the FOIL method.

Solution

$$(3x + 2)(x - 5)$$

$$\overset{F}{(3x)(x)} + \overset{O}{(3x)(-5)} + \overset{I}{(2)(x)} + \overset{L}{(2)(-5)}$$

$$= 3x^2 - 15x + 2x - 10 = 3x^2 - 13x - 10$$

Now Try Exercise 25

3 Multiply a Polynomial by a Polynomial

When multiplying a trinomial by a binomial or a trinomial by a trinomial, every term of the first polynomial must be multiplied by every term of the second polynomial.

EXAMPLE 5 Multiply $x^2 + 1 - 4x$ by $2x^2 - 3$.

Solution   Since the trinomial is not in descending order, rewrite it as $x^2 - 4x + 1$.
   Place the longer polynomial on top, then multiply. Make sure you align like terms as you multiply so that the terms can be added more easily.

$$
\begin{array}{rl}
& x^2 - 4x \;+\; 1 \quad \text{Trinomial written in descending order} \\
& \underline{\qquad 2x^2 \;-\; 3} \\
-3(x^2 - 4x + 1)\longrightarrow \quad & -3x^2 + 12x \;-\; 3 \quad \text{Multiply top expression by } -3. \\
2x^2(x^2 - 4x + 1)\longrightarrow \quad 2x^4 - 8x^3 + 2x^2 & \qquad\qquad\qquad\quad \text{Multiply top expression by } 2x^2. \\
\hline
2x^4 - 8x^3 - \;\;& x^2 + 12x \;-\; 3 \quad \text{Add like terms in columns.}
\end{array}
$$

Now Try Exercise 35

EXAMPLE 6 Multiply $3x^2 + 6xy - 5y^2$ by $x + 4y$.

Solution

$$
\begin{array}{rl}
& 3x^2 + 6xy - 5y^2 \\
& \underline{\qquad\qquad x + 4y} \\
4y(3x^2 + 6xy - 5y^2)\longrightarrow \quad & 12x^2y + 24xy^2 - 20y^3 \quad \text{Multiply top expression by } 4y. \\
x(3x^2 + 6xy - 5y^2)\longrightarrow \quad 3x^3 + \;\;6x^2y - \;\;5xy^2 & \qquad\qquad\qquad\qquad\quad \text{Multiply top expression by } x. \\
\hline
3x^3 + 18x^2y + 19xy^2 - 20y^3 & \quad \text{Add like terms in columns.}
\end{array}
$$

Now Try Exercise 31

Now we will study some special formulas.

4 Find the Square of a Binomial

We must often *square a binomial,* so we have special formulas for doing so.

Square of a Binomial

$$(a + b)^2 = a^2 + 2ab + b^2$$
$$(a - b)^2 = a^2 - 2ab + b^2$$

If you forget the formulas, you can easily derive them by multiplying $(a + b)(a + b)$ and $(a - b)(a - b)$.
   Examples 7 and 8 illustrate the use of the square of binomial formulas.

EXAMPLE 7 Expand.

a) $(3x + 7)^2$ **b)** $(4x^2 - 5y)^2$

Solution

a) $(3x + 7)^2 = (3x)^2 + 2(3x)(7) + (7)^2$
$$= 9x^2 + 42x + 49$$

b) $(4x^2 - 5y)^2 = (4x^2)^2 - 2(4x^2)(5y) + (5y)^2$
$$= 16x^4 - 40x^2y + 25y^2$$

Now Try Exercise 45

> **Helpful Hint**
>
> Squaring binomials, as in Example 7, can also be done using the FOIL method.

> **Avoiding Common Errors**
>
> Remember the middle term when squaring a binomial.
>
> | CORRECT | INCORRECT |
> |---------|-----------|
> | $(x + 2)^2 = (x + 2)(x + 2)$ | ~~$(x + 2)^2 = x^2 + 4$~~ |
> | $\quad\quad\;\; = x^2 + 4x + 4$ | |
> | $(x - 3)^2 = (x - 3)(x - 3)$ | ~~$(x - 3)^2 = x^2 + 9$~~ |
> | $\quad\quad\;\; = x^2 - 6x + 9$ | |

EXAMPLE 8 Expand $[x + (y - 1)]^2$.

Solution   This example is worked the same way as any other square of a binomial. Treat x as the first term and $(y - 1)$ as the second term.

$$[x + (y - 1)]^2 = (x)^2 + 2(x)(y - 1) + (y - 1)^2$$
$$= x^2 + (2x)(y - 1) + y^2 - 2y + 1$$
$$= x^2 + 2xy - 2x + y^2 - 2y + 1$$

None of the six terms are like terms, so no terms can be combined. Note that $(y - 1)^2$ is also the square of a binomial and was expanded as such.

Now Try Exercise 51

5 Find the Product of the Sum and Difference of the Same Two Terms

Below we multiply $(x + 6)(x - 6)$ using the FOIL method.

$$(x + 6)(x - 6) = x^2 - 6x + 6x - (6)(6) = x^2 - 6^2$$

Note that the outer and inner products add to 0. By examining this example, we see that the product of the sum and difference of the same two terms is the difference of the squares of the two terms.

> **Product of the Sum and Difference of the Same Two Terms**
>
> $$(a + b)(a - b) = a^2 - b^2$$

In other words, to multiply two binomials that differ only in the sign between their two terms, take the difference between the square of the first term and the square of the second term. Note that $a^2 - b^2$ represents a **difference of two squares**.

EXAMPLE 9 Multiply.

a) $\left(3x + \dfrac{4}{5}\right)\left(3x - \dfrac{4}{5}\right)$ **b)** $(0.2x + 0.3z^2)(0.2x - 0.3z^2)$

Solution   Each is a product of the sum and difference of the same two terms. Therefore,

a) $\left(3x + \dfrac{4}{5}\right)\left(3x - \dfrac{4}{5}\right) = (3x)^2 - \left(\dfrac{4}{5}\right)^2 = 9x^2 - \dfrac{16}{25}$

b) $(0.2x + 0.3z^2)(0.2x - 0.3z^2) = (0.2x)^2 - (0.3z^2)^2$
$$= 0.04x^2 - 0.09z^4$$

Now Try Exercise 27

EXAMPLE 10 Multiply $(5x + y^4)(5x - y^4)$.

Solution $(5x + y^4)(5x - y^4) = (5x)^2 - (y^4)^2 = 25x^2 - y^8$

Now Try Exercise 49

EXAMPLE 11 Multiply $[4x + (3y + 2)][4x - (3y + 2)]$.

Solution   We treat $4x$ as the first term and $3y + 2$ as the second term. Then we have the sum and difference of the same two terms.

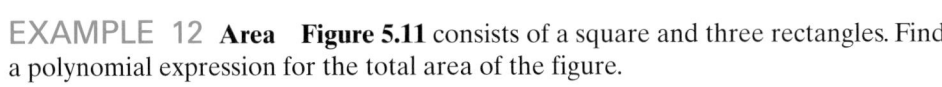

$$[4x + (3y + 2)][4x - (3y + 2)] = (4x)^2 - (3y + 2)^2$$
$$= 16x^2 - (9y^2 + 12y + 4)$$
$$= 16x^2 - 9y^2 - 12y - 4$$

Now Try Exercise 55

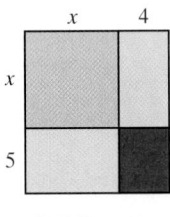

FIGURE 5.11

EXAMPLE 12 **Area** **Figure 5.11** consists of a square and three rectangles. Find a polynomial expression for the total area of the figure.

Solution   To find the total area, we find the areas of the four regions and then add.

$$\text{Area of the orange square } = x \cdot x = x^2$$
$$\text{Area of the blue rectangle } = x \cdot 4 = 4x$$
$$\text{Area of the green rectangle } = x \cdot 5 = 5x$$
$$\text{Area of the red rectangle } = 4 \cdot 5 = 20$$

The total area is the sum of these four quantities.

$$\text{Total area } = x^2 + 4x + 5x + 20 = x^2 + 9x + 20.$$

Now Try Exercise 85

Note that **Figure 5.11** shows a geometric representation that $(x + 4)(x + 5) = x^2 + 9x + 20$.

6 Find the Product of Polynomial Functions

Earlier we learned that for functions $f(x)$ and $g(x)$, $(f \cdot g)(x) = f(x) \cdot g(x)$. Let's work one example involving multiplication of polynomial functions now.

EXAMPLE 13 Let $f(x) = x + 4$ and $g(x) = x - 2$. Find

a) $f(3) \cdot g(3)$ **b)** $(f \cdot g)(x)$ **c)** $(f \cdot g)(3)$

Solution

a) Both $f(x)$ and $g(x)$ are polynomial functions since the expressions to the right of the equals signs are polynomials.

$$f(x) = x + 4 \qquad g(x) = x - 2$$
$$f(3) = 3 + 4 = 7 \qquad g(3) = 3 - 2 = 1$$
$$f(3) \cdot g(3) = 7 \cdot 1 = 7$$

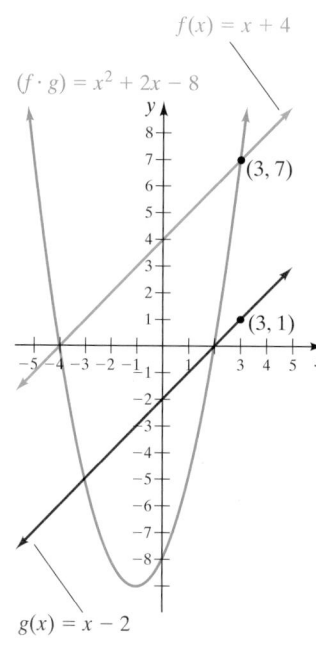

$f(x) = x + 4$

$(f \cdot g) = x^2 + 2x - 8$

(3, 7)

(3, 1)

$g(x) = x - 2$

FIGURE 5.12

b) We know that

$$(f \cdot g)(x) = f(x) \cdot g(x)$$
$$= (x + 4)(x - 2)$$
$$= x^2 - 2x + 4x - 8$$
$$= x^2 + 2x - 8$$

c) To evaluate $(f \cdot g)(3)$, substitute 3 for each x in $(f \cdot g)(x)$.

$$(f \cdot g)(x) = x^2 + 2x - 8$$
$$(f \cdot g)(3) = 3^2 + 2(3) - 8$$
$$= 9 + 6 - 8 = 7$$

Now Try Exercise 79

In Example 13, we found that if $f(x) = x + 4$ and $g(x) = x - 2$, then $(f \cdot g)(x) = x^2 + 2x - 8$. The graphs of $y = f(x) = x + 4$, $y = g(x) = x - 2$, and $y = (f \cdot g)(x) = x^2 + 2x - 8$ are shown in **Figure 5.12**.

We see from the graphs that $f(3) = 7, g(3) = 1$, and $(f \cdot g)(3) = 7$, which is what we expected from Example 13. Notice in **Figure 5.12** that the product of the two linear functions $f(x) = x + 4$ and $g(x) = x - 2$ is the quadratic function $(f \cdot g)(x) = x^2 + 2x - 8$.

EXERCISE SET 5.2

 Math XL MathXL®   **MyMathLab** MyMathLab

Warm-Up Exercises

Fill in the blanks with the appropriate word, phrase, or symbol(s) from the following list.

inside	difference	associative	product	distributive	third	inner
final	square	sum	factors	second	commutative	first

1. To multiply two polynomials, every term of the first polynomial must be multiplied by every term of the _____ polynomial.

2. The rule $a^m \cdot a^n = a^{m+n}$ is called the _____ rule for exponents.

3. Expressions that are multiplied are called _____.

4. The rule $a(b + c + d + \cdots + n) = ab + ac + ad + \cdots + an$ is called the expanded form of the _____ property.

5. In the FOIL method, F stands for _____.

6. In the FOIL method, I stands for _____.

7. The rule $(a + b)^2 = a^2 + 2ab + b^2$ is called the _____ of a binomial.

8. The expression $a^2 - b^2$ represents the _____ of two squares.

Practice the Skills

Multiply.

9. $(7xy)(6xy^4)$

10. $(-5xy^4)(9x^4y^6)$

11. $\left(\frac{5}{9}x^2y^5\right)\left(\frac{1}{5}x^5y^3z^2\right)$

12. $2y^3(3y^2 + 2y - 10)$

13. $-3x^2y(-2x^4y^2 + 5xy^3 + 6)$

14. $3x^4(2xy^2 + 5x^7 - 9y)$

15. $\frac{2}{3}yz(3x + 4y - 12y^2)$

16. $\frac{1}{2}x^2y(4x^5y^2 + 3x - 7y^2)$

17. $0.3(2x^2 - 5x + 11y)$

18. $0.8(0.2a + 0.9b - 1.3c)$

19. $0.3a^5b^4(9.5a^6b - 4.6a^4b^3 + 1.2ab^5)$

20. $4.6m^2n(1.3m^4n^2 - 2.6m^3n^3 + 5.9n^4)$

Multiply the following binomials.

21. $(4x - 6)(3x - 5)$

22. $(2x - 1)(7x + 5)$

23. $(4 - x)(3 + 2x^2)$

24. $(5x + y)(6x - y)$

25. $\left(\frac{1}{2}x + 2y\right)\left(2x - \frac{1}{3}y\right)$

26. $\left(\frac{1}{3}a + \frac{1}{4}b\right)\left(\frac{1}{2}a - b\right)$

27. $(0.3a + 0.5b)(0.3a - 0.5b)$

28. $(4.6r - 5.8s)(0.2r - 2.3s)$

Multiply the following polynomials.

29. $(x^2 + 3x + 1)(x - 4)$

30. $(x + 3)(2x^2 - x - 8)$

31. $(a - 3b)(2a^2 - ab + 2b^2)$

32. $(7p - 3)(-2p^2 - 4p + 1)$

33. $(x^3 - x^2 + 3x + 7)(x + 1)$

34. $(2x - 1)(x^3 + 3x^2 - 5x + 6)$

35. $(5x^3 + 4x^2 - 6x + 2)(x + 5)$

36. $(a^3 - 2a^2 + 5a - 6)(2a^2 - 5a - 4)$

37. $(3m^2 - 2m + 4)(m^2 - 3m - 5)$

38. $(2a^2 - 6a + 3)(3a^2 - 5a - 2)$

39. $(2x - 1)^3$

40. $(3x + y)^3$

41. $(5r^2 - rs + 2s^2)(2r^2 - s^2)$

42. $(4x^2 - 5xy + y^2)(x^2 - 2y^2)$

Multiply using either the formula for the square of a binomial or for the product of the sum and difference of the same two terms.

43. $(x + 2)(x + 2)$

44. $(y - 6)(y - 6)$

45. $(2x - 9)(2x - 9)$

46. $(3z + 4)(3z + 4)$

47. $(4x - 3y)^2$

48. $(2a + 5b)^2$

49. $(5m^2 + 2n)(5m^2 - 2n)$

50. $(5p^2 + 6q^2)(5p^2 - 6q^2)$

51. $[y + (4 - 2x)]^2$

52. $[(a + b) + 6]^2$

53. $[5x + (2y + 1)]^2$

54. $[4 - (p - 3q)]^2$

55. $[a + (b + 4)][a - (b + 4)]$

56. $[2x + (y + 5)][2x - (y + 5)]$

Multiply.

57. $2xy(x^2 + xy + 12y^2)$

58. $4a^2b^2\left(\dfrac{1}{4}ab - \dfrac{1}{9}b^6\right)$

59. $\dfrac{1}{2}xy^2(4x^2 + 3xy - 7y^4)$

60. $-\dfrac{3}{5}x^2y\left(-\dfrac{2}{3}xy^4 + \dfrac{1}{9}xy^2 + 8\right)$

61. $-\dfrac{3}{5}xy^3z^2\left(-xy^2z^5 - 5xy + \dfrac{1}{9}xz^7\right)$

62. $\dfrac{2}{3}x^2y^4\left(\dfrac{3}{5}xy^3 - \dfrac{1}{8}x^4y + 4xy^3z^5\right)$

63. $(3a + 4)(7a - 6)$

64. $(5p - 9q)(4p - 11q)$

65. $\left(8x + \dfrac{1}{5}\right)\left(8x - \dfrac{1}{5}\right)$

66. $\left(6a - \dfrac{1}{7}\right)\left(6a + \dfrac{1}{7}\right)$

67. $\left(x - \dfrac{1}{2}y\right)^3$

68. $\left(\dfrac{1}{2}m - n\right)^3$

69. $(x + 3)(2x^2 + 4x - 3)$

70. $(5a + 4)(a^2 - a + 2)$

71. $(2p - 3q)(3p^2 + 4pq - 2q^2)$

72. $(2m + n)(3m^2 - mn + 2n^2)$

73. $[(3x + 2) + y][(3x + 2) - y]$

74. $[a + (3b + 5)][a - (3b + 5)]$

75. $(a + b)(a - b)(a^2 - b^2)$

76. $(2a + 3)(2a - 3)(4a^2 + 9)$

77. $(x - 4)(6 + x)(2x - 8)$

78. $(3x - 5)(5 - 2x)(3x + 8)$

For the functions given, find **a)** $(f \cdot g)(x)$ *and* **b)** $(f \cdot g)(4)$.

79. $f(x) = x - 5, g(x) = x + 6$

80. $f(x) = 2x - 3, g(x) = x - 1$

81. $f(x) = 2x^2 + 6x - 4, g(x) = 5x + 3$

82. $f(x) = 4x^2 + 7, g(x) = 2 - x$

83. $f(x) = -x^2 + 3x, g(x) = x^2 + 2$

84. $f(x) = -x^2 + 2x + 7, g(x) = x^2 - 1$

Problem Solving

Area   *In Exercises 85–88, find a polynomial expression for the total area of each figure.*

85.

86.

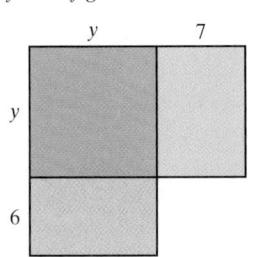

87.

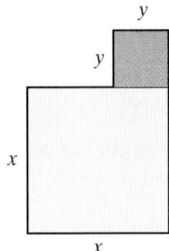

88.

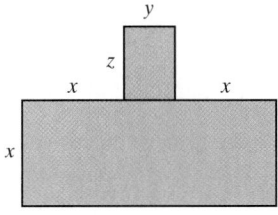

Area   *In Exercises 89 and 90,* **a)** *find the area of the rectangle by finding the area of the four sections and adding them, and* **b)** *multiply the two sides and compare the product with your answer to part* **a)**.

89.

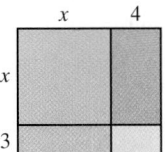

90.

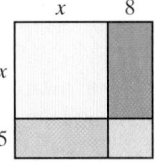

Area   *Write a polynomial expression for the area of each figure. All angles in the figures are right angles.*

91.

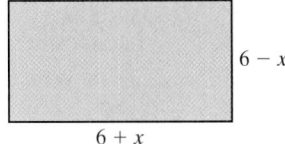

92.

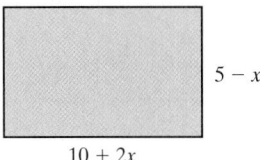

Area   *In Exercises 93 and 94,* **a)** *write a polynomial expression for the area of the shaded portion of the figure.* **b)** *The area of the shaded portion is indicated above each figure. Find the area of the larger and smaller rectangles.*

93.

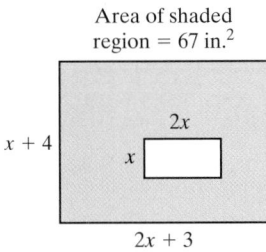

94.

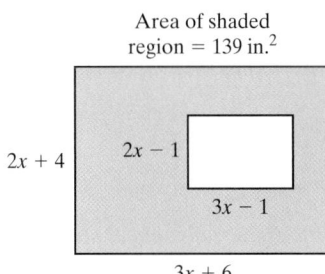

95. Write two binomials whose product is $x^2 - 49$. Explain how you determined your answer.

96. Write two binomials whose product is $4x^2 - 9$. Explain how you determined your answer.

97. Write two binomials whose product is $x^2 + 12x + 36$. Explain how you determined your answer.

98. Write two binomials whose product is $16y^2 - 8y + 1$. Explain how you determined your answer.

99. Consider the expression $a(x - n)^3$. Write this expression as a product of factors.

100. Consider the expression $P(1 - r)^4$. Write this expression as a product of factors.

101. Area The expression $(a + b)^2$ can be represented by the following figure.

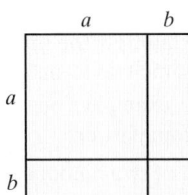

a) Explain why this figure represents $(a + b)^2$.

b) Find $(a + b)^2$ using the figure by finding the area of each of the four parts of the figure, then adding the areas together.

c) Simplify $(a + b)^2$ by multiplying $(a + b)(a + b)$.

d) How do the answers in parts **b)** and **c)** compare? If they are not the same, explain why.

102. Volume The expression $(a + b)^3$ can be represented by the following figure.

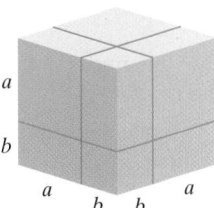

a) Explain why this figure represents $(a + b)^3$.

b) Find $(a + b)^3$ by adding the volume of the eight parts of the figure.

c) Simplify $(a + b)^3$ by multiplying.

d) How do the answers in parts **b)** and **c)** compare? If they are not the same, explain why.

103. Compound Interest The compound interest formula is

$$A = P\left(1 + \frac{r}{n}\right)^{nt}$$

where A is the amount, P is the principal invested, r is the annual rate of interest, n is the number of times the interest is compounded annually, and t is the time in years.

a) Simplify this formula for $n = 1$.

b) Find the value of A if $P = \$1000$, $n = 1$, $r = 6\%$, and $t = 2$ years.

104. Compound Interest Use the formula given in Exercise 103 to find A if $P = \$4000$, $n = 2$, $r = 8\%$, and $t = 2$ years.

105. Line-Up The number of ways a teacher can award different prizes to 2 students in a class having n students is given by the formula $P(n) = n(n - 1)$.

a) Use this formula to determine the number of ways a teacher can award different prizes to 2 students in a class having 11 students.

b) Rewrite the formula by multiplying the factors.

c) Use the result from part **b)** to determine the number of ways a teacher can award different prizes to 2 students in a class having 11 students.

d) Are the results from parts **a)** and **c)** the same? Explain.

106. Horse Racing The number of ways horses can finish first, second, and third, in a race having n horses, is given by the formula

$$P(n) = n(n - 1)(n - 2)$$

a) Use this formula to determine the number of ways horses can finish first, second, and third in a race having 7 horses.

b) Rewrite the formula by multiplying the factors.

c) Use the result from part **b)** to determine the number of ways horses can finish first, second, and third in a race having 7 horses.

d) Are the results from parts **a)** and **c)** the same? Explain.

© Cindy Haggerty\Shutterstock

107. If $f(x) = x^2 - 3x + 5$, find $f(a + b)$ by substituting $(a + b)$ for each x in the function.

108. If $f(x) = 2x^2 - x + 3$, find $f(a + b)$.

In Exercises 109–114, simplify. Assume that all variables represent natural numbers.

109. $3x^t(5x^{2t-1} + 6x^{3t})$

110. $5k^{r+2}(4k^{r+2} - 3k^r - k)$

111. $(6x^m - 5)(2x^{2m} - 3)$

112. $(x^{3n} - y^{2n})(x^{2n} + 2y^{4n})$

113. $(y^{a-b})^{a+b}$

114. $(a^{m+n})^{m+n}$

In Exercises 115 and 116, perform the polynomial multiplication.

115. $(x - 3y)^4$

116. $(2a - 4b)^4$

117. a) Explain how a multiplication in one variable, such as $(x^2 + 2x + 3)(x + 2) = x^3 + 4x^2 + 7x + 6$, may be checked using a graphing calculator.

b) Check the multiplication given in part **a)** using your graphing calculator.

118. a) Show that the multiplication $(x^2 - 4x - 5)(x - 1) \neq x^3 + 6x^2 - 5x + 5$ using your graphing calculator.

b) Multiply $(x^2 - 4x - 5)(x - 1)$.

c) Check your answer in part **b)** on your graphing calculator.

Concept/Writing Exercises

119. a) Explain how to multiply two binomials using the FOIL method.

b) Make up two binomials and multiply them using the FOIL method.

c) Multiply the same two binomials using LIOF (last, inner, outer, first).

d) Compare the results of parts **b)** and **c)**. If they are not the same, explain why.

120. a) Explain how to multiply a monomial by a polynomial.

b) Multiply $3x(4x^2 - 6x - 7)$ using your procedure from part **a)**.

121. a) Explain how to multiply a polynomial by a polynomial.

b) Multiply $4 + x$ by $x^2 - 6x + 3$ using your procedure from part **a)**.

122. a) Explain how to expand $(2x - 3)^2$ using the formula for the square of a binomial.

b) Expand $(2x - 3)^2$ using your procedure from part **a)**.

123. a) What is meant by the product of the sum and difference of the same two terms?

b) Give an example of a problem that is the product of the sum and difference of the same two terms.

c) How do you multiply the product of the sum and difference of the same two terms?

d) Multiply the example you gave in part **b)** using your procedure from part **c)**.

124. Will the product of two binomials always be a

a) binomial?

b) trinomial? Explain.

125. Will the product of two first-degree polynomials always be a second-degree polynomial? Explain.

126. a) Given $f(x)$ and $g(x)$, explain how you would find $(f \cdot g)(x)$.

  b) If $f(x) = x - 8$ and $g(x) = x + 8$, find $(f \cdot g)(x)$.

Challenge Problems

Multiply.

127. $[(y + 1) - (x + 2)]^2$

128. $[(a - 2) - (a + 3)]^2$

Cumulative Review Exercises

[1.3] **129.** Evaluate $\dfrac{4}{5} - \left(\dfrac{3}{4} - \dfrac{2}{3}\right)$.

[1.5] **130.** Simplify $\left(\dfrac{2r^4 s^5}{r^2}\right)^3$.

[2.5] **131.** Solve the inequality $-12 < 3x - 5 \le -1$ and give the solution in interval notation.

[3.2] **132.** If $g(x) = -x^2 + 2x + 3$, find $g\left(\dfrac{1}{2}\right)$.

5.3 Division of Polynomials and Synthetic Division

1 Divide a polynomial by a monomial.

2 Divide a polynomial by a binomial.

3 Divide polynomials using synthetic division.

4 Use the Remainder Theorem.

1 Divide a Polynomial by a Monomial

To divide a polynomial by a monomial, we use the fact that

$$\frac{A + B}{C} = \frac{A}{C} + \frac{B}{C}$$

If the polynomial has more than two terms, we expand this procedure.

Because division by zero is undefined, when we are given a division problem containing a variable in the denominator *we always assume that the denominator is nonzero.*

> **To Divide a Polynomial by a Monomial**
>
> Divide each term of the polynomial by the monomial.

To divide a polynomial by a monomial, we will need to use the quotient rule for exponents and the zero exponent rule.

Quotient rule for exponents: $\dfrac{a^m}{a^n} = a^{m-n}, \quad a \ne 0$

Zero exponent rule: $a^0 = 1, \quad a \ne 0$

EXAMPLE 1 Divide. ────

a) $\dfrac{x^{10}}{x^4}$ **b)** $\dfrac{5x^3 y^5}{4xy^2}$

Solution   We will use the quotient rule to divide.

a) $\dfrac{x^{10}}{x^4} = x^{10-4}$ Quotient rule

$= x^6$

b) $\dfrac{5x^3 y^5}{4xy^2} = \dfrac{5}{4} \cdot \dfrac{x^3}{x} \cdot \dfrac{y^5}{y^2}$

$= \dfrac{5}{4} x^{3-1} y^{5-2}$ Quotient rule

$= \dfrac{5x^2 y^3}{4}$

Now Try Exercise 11

EXAMPLE 2 Divide.

a) $\dfrac{p^4}{p^4}$ **b)** $\dfrac{8r^6 s^7}{3rs^7}$

Solution   We will use both the quotient rule and the zero exponent rule to divide.

a) $\dfrac{p^4}{p^4} = p^{4-4}$ Quotient rule

$\phantom{\dfrac{p^4}{p^4}} = p^0$

$\phantom{\dfrac{p^4}{p^4}} = 1$ Zero exponent rule

b) $\dfrac{8r^6 s^7}{3rs^7} = \dfrac{8}{3} \cdot \dfrac{r^6}{r} \cdot \dfrac{s^7}{s^7}$

$\phantom{\dfrac{8r^6 s^7}{3rs^7}} = \dfrac{8}{3} r^{6-1} s^{7-7}$ Quotient rule

$\phantom{\dfrac{8r^6 s^7}{3rs^7}} = \dfrac{8}{3} r^5 s^0$

$\phantom{\dfrac{8r^6 s^7}{3rs^7}} = \dfrac{8}{3} r^5 (1)$ Zero exponent rule

$\phantom{\dfrac{8r^6 s^7}{3rs^7}} = \dfrac{8}{3} r^5$ or $\dfrac{8r^5}{3}$

> Now Try Exercise 17

In Example 2, both $\dfrac{8}{3} r^5$ and $\dfrac{8r^5}{3}$ are acceptable answers. Now we will divide a polynomial by a monomial.

EXAMPLE 3 Divide $\dfrac{4x^2 - 8x - 7}{2x}$.

Solution $\dfrac{4x^2 - 8x - 7}{2x} = \dfrac{4x^2}{2x} - \dfrac{8x}{2x} - \dfrac{7}{2x}$

$\phantom{\dfrac{4x^2 - 8x - 7}{2x}} = 2x - 4 - \dfrac{7}{2x}$

> Now Try Exercise 25

EXAMPLE 4 Divide $\dfrac{4y - 6x^4 y^3 - 3x^5 y^2 + 5x}{2xy^2}$.

Solution $\dfrac{4y - 6x^4 y^3 - 3x^5 y^2 + 5x}{2xy^2} = \dfrac{4y}{2xy^2} - \dfrac{6x^4 y^3}{2xy^2} - \dfrac{3x^5 y^2}{2xy^2} + \dfrac{5x}{2xy^2}$

$ = \dfrac{2}{xy} - 3x^3 y - \dfrac{3x^4}{2} + \dfrac{5}{2y^2}$

> Now Try Exercise 31

2 Divide a Polynomial by a Binomial

We divide a polynomial by a binomial in much the same way as we perform long division. In a division problem, the expression we are dividing is called the *dividend* and the expression we are dividing by is called the *divisor*.

EXAMPLE 5 Divide $\dfrac{x^2 + 7x + 10}{x + 2}$. $\leftarrow$ Dividend
                                 $\leftarrow$ Divisor

Solution   Rewrite the division problem as

$$x + 2 \overline{) x^2 + 7x + 10} \quad \leftarrow \text{Dividend}$$

Divisor $\rightarrow$

Divide x^2 (the first term in the dividend $x^2 + 7x + 10$) by x (the first term in the divisor $x + 2$).

$$\frac{x^2}{x} = x$$

Place the quotient, x, above the term containing x in the dividend.

$$x + 2 \overline{)x^2 + 7x + 10}$$

with x above.

Next, multiply the x by $x + 2$ as you would do in long division and place the product under the dividend, aligning like terms.

Times

$$x + 2\overline{)x^2 + 7x + 10}$$

Equals $\rightarrow$ $x^2 + 2x$ $\longleftarrow$ $x(x + 2)$

Now subtract $x^2 + 2x$ from $x^2 + 7x$.

$$
\begin{array}{r}
x \\
x + 2 \overline{)x^2 + 7x + 10} \\
-(x^2 + 2x) \\
\hline
5x
\end{array}
$$

Now bring down the next term, $+10$.

$$
\begin{array}{r}
x \\
x + 2 \overline{)x^2 + 7x + 10} \\
x^2 + 2x \\
\hline
5x + 10
\end{array}
$$

Divide $5x$ by x.

$$\frac{5x}{x} = +5$$

Place $+5$ above the constant in the dividend and multiply 5 by $x + 2$. Finish the problem by subtracting.

Times

$$
\begin{array}{r}
x + 5 \\
x + 2 \overline{)x^2 + 7x + 10} \\
x^2 + 2x \\
\hline
5x + 10 \\
\end{array}
$$

Equals $\rightarrow$ $5x + 10$ $\longleftarrow$ $5(x + 2)$

0 $\longleftarrow$ Remainder

Thus, $\dfrac{x^2 + 7x + 10}{x + 2} = x + 5$.

Now Try Exercise 35

In Example 5, the remainder is zero. Therefore, $x^2 + 7x + 10 = (x + 2)(x + 5)$. Note that $x + 2$ and $x + 5$ are both *factors* of $x^2 + 7x + 10$. In a division problem, if the remainder is zero, then both the divisor and quotient are factors of the dividend.

When writing an answer in a division problem when the remainder is not zero, write the remainder over the divisor and add this expression to the quotient. For example, suppose that the remainder in Example 5 was 4. Then the answer would be written $x + 5 + \dfrac{4}{x + 2}$. If the remainder in Example 5 was -7, the answer would be written $x + 5 + \dfrac{-7}{x + 2}$, which we would rewrite as $x + 5 - \dfrac{7}{x + 2}$.

EXAMPLE 6 Divide $\dfrac{6x^2 - 7x + 3}{2x + 1}$.

Solution   In this example we will subtract mentally and not show the change of sign in the subtractions.

$$
\begin{array}{r}
3x - 5 \quad\longleftarrow \text{ Quotient} \\
\text{Divisor} \longrightarrow 2x + 1 \overline{\smash{\big)}\, 6x^2 - 7x + 3} \quad\longleftarrow \text{ Dividend} \\
\underline{6x^2 + 3x} \qquad\longleftarrow 3x(2x + 1) \\
-10x + 3 \\
\underline{-10x - 5} \quad\longleftarrow -5(2x + 1) \\
8 \quad\longleftarrow \text{ Remainder}
\end{array}
$$

Thus, $\dfrac{6x^2 - 7x + 3}{2x + 1} = 3x - 5 + \dfrac{8}{2x + 1}$.

Now Try Exercise 45

When dividing a polynomial by a binomial, the answer may be *checked* by multiplying the divisor by the quotient and then adding the remainder. You should obtain the dividend. To check Example 6, we do the following:

$$(2x + 1)(3x - 5) + 8 = 6x^2 - 10x + 3x - 5 + 8 = 6x^2 - 7x + 3$$

$\uparrow$ Divisor $\uparrow$ Quotient $\uparrow$ Remainder $\uparrow$ Dividend

Since we got the dividend, our division is correct.

> **Helpful Hint**
>
> *When you are dividing a polynomial by a binomial, you should list both the polynomial and binomial in descending order. If a term of any degree is missing, it is often helpful to include that term with a numerical coefficient of 0. For example, when dividing $(6x^2 + x^3 - 4) \div (x - 2)$, we rewrite the problem as $(x^3 + 6x^2 + 0x - 4) \div (x - 2)$ before beginning the division.*

EXAMPLE 7 Divide $(4x^2 - 12x + 3x^5 - 17)$ by $(-2 + x^2)$.

Solution   Write both the dividend and divisor in descending powers of the variable x. This gives $(3x^5 + 4x^2 - 12x - 17) \div (x^2 - 2)$. Where a power of x is missing, add that power of x with a coefficient of 0, then divide.

$$
\begin{array}{r}
3x^3 \qquad\qquad + 6x + 4 \\
x^2 + 0x - 2 \overline{\smash{\big)}\, 3x^5 + 0x^4 + 0x^3 + 4x^2 - 12x - 17} \\
\underline{3x^5 + 0x^4 - 6x^3} \qquad\qquad\qquad\quad\longleftarrow 3x^3(x^2 + 0x - 2) \\
6x^3 + 4x^2 - 12x \\
\underline{6x^3 + 0x^2 - 12x} \qquad\quad\longleftarrow 6x(x^2 + 0x - 2) \\
4x^2 + 0x - 17 \\
\underline{4x^2 + 0x - 8} \quad\longleftarrow 4(x^2 + 0x - 2) \\
-9 \longleftarrow \text{ Remainder}
\end{array}
$$

In obtaining the answer, we performed the divisions

$$\frac{3x^5}{x^2} = 3x^3 \qquad \frac{6x^3}{x^2} = 6x \qquad \frac{4x^2}{x^2} = 4$$

The quotients $3x^3$, $6x$, and 4 were placed above their like terms in the dividend. The answer is $3x^3 + 6x + 4 - \dfrac{9}{x^2 - 2}$. You can check this answer by multiplying the divisor by the quotient and adding the remainder.

Now Try Exercise 55

3   Divide Polynomials Using Synthetic Division

When a polynomial is divided by a binomial of the form $x - a$, the division process can be greatly shortened by a process called **synthetic division**. Consider the following examples. In the example on the right, we use only the numerical coefficients.

$$
\begin{array}{r}
2x^2 + 5x - 4 \\
x - 3\overline{)2x^3 - x^2 - 19x + 18} \\
\underline{2x^3 - 6x^2} \\
5x^2 - 19x \\
\underline{5x^2 - 15x} \\
-4x + 18 \\
\underline{-4x + 12} \\
6
\end{array}
\qquad
\begin{array}{r}
2 + 5 - 4 \\
1 - 3\overline{)2 - 1 - 19 + 18} \\
\underline{2 - 6} \\
5 - 19 \\
\underline{5 - 15} \\
-4 + 18 \\
\underline{-4 + 12} \\
6
\end{array}
$$

Note that the variables do not play a role in determining the numerical coefficients of the quotient. This division problem can be done more quickly and easily using synthetic division.

Following is an explanation of how we use synthetic division. Consider again the division

$$
\frac{2x^3 - x^2 - 19x + 18}{x - 3}
$$

1. Write the dividend in descending powers of x. Then list the numerical coefficients of each term in the dividend. If a term of any degree is missing, place 0 in the appropriate position to serve as a placeholder. In the problem above, the numerical coefficients of the dividend are

$$
2 \quad -1 \quad -19 \quad 18
$$

2. When dividing by a binomial of the form $x - a$, place a to the left of the line of numbers from step 1. In this problem, we are dividing by $x - 3$; thus, $a = 3$. We write

$$
\underline{3}\,\big|\; 2 \quad -1 \quad -19 \quad 18
$$

3. Leave some space under the row of coefficients, then draw a horizontal line. Bring down the first coefficient on the left as follows:

$$
\begin{array}{r|rrrr}
3 & 2 & -1 & -19 & 18 \\
& & & & \\
\hline
& 2 & & &
\end{array}
$$

4. Multiply the 3 by the number brought down, the 2, to get 6. Place the 6 under the next coefficient, the -1. Then add $-1 + 6$ to get 5.

$$
\begin{array}{r|rrrr}
3 & 2 & -1 & -19 & 18 \\
& & 6 & & \\
\hline
& 2 & 5 & &
\end{array}
$$

5. Multiply the 3 by the sum 5 to get 15. Place 15 under -19. Then add to get -4. Repeat this procedure as illustrated.

$$
\begin{array}{r|rrrr}
3 & 2 & -1 & -19 & 18 \\
& & 6 & 15 & -12 \\
\hline
& 2 & 5 & -4 & 6
\end{array}
$$
$\longleftarrow$ Dividend coefficients

$\longleftarrow$ Remainder is 6

Quotient coefficients

In the last row, the first three numbers are the numerical coefficients of the quotient, as shown in the long division. The last number, 6, is the remainder obtained by long division. The degree of the quotient is always one less than the degree of the dividend. Thus, the quotient is $2x^2 + 5x - 4$ and the remainder is 6.

We have $\dfrac{2x^3 - x^2 - 19x + 18}{x - 3} = 2x^2 + 5x - 4 + \dfrac{6}{x - 3}$.

Understanding Algebra

When dividing polynomials using synthetic division, the divisor must be of the form $x - a$.

When the divisor is	$a =$
$x - 1$	1
$x - 2$	2
$x - 3$	3
$x + 1$	-1
$x + 2$	-2
$x + 3$	-3

EXAMPLE 8 Use synthetic division to divide.

$$(5 - x^2 + x^3) \div (x + 2)$$

Solution First, list the terms of the dividend in descending order of x.

$$(x^3 - x^2 + 5) \div (x + 2)$$

Since there is no first-degree term, insert 0 as a placeholder when listing the numerical coefficients. Since $x + 2 = x - (-2)$, $a = -2$.

$$
\begin{array}{r|rrrr}
-2 & 1 & -1 & 0 & 5 \\
   &   & -2 & 6 & -12 \\
\hline
   & 1 & -3 & 6 & -7 \quad \longleftarrow \text{Remainder is } -7 \\
\end{array}
$$
$$\underbrace{}_{\text{Quotient coefficients}}$$

Since the dividend is a third-degree polynomial, the quotient must be a second-degree polynomial. The answer is $x^2 - 3x + 6 - \dfrac{7}{x + 2}$.

Now Try Exercise 61

EXAMPLE 9 Use synthetic division to divide.

$$(3x^4 + 11x^3 - 20x^2 + 7x + 35) \div (x + 5)$$

Solution
$$
\begin{array}{r|rrrrr}
-5 & 3 & 11 & -20 & 7 & 35 \\
   &   & -15 & 20 & 0 & -35 \\
\hline
   & 3 & -4 & 0 & 7 & 0 \\
\end{array}
$$

Since the dividend is of the fourth degree, the quotient must be of the third degree. The quotient is $3x^3 - 4x^2 + 0x + 7$ with a remainder of 0. The quotient can be simplified to $3x^3 - 4x^2 + 7$.

Now Try Exercise 71

In Example 9, since the remainder is 0, both $x + 5$ and $3x^3 - 4x^2 + 7$ are *factors* of $3x^4 + 11x^3 - 20x^2 + 7x + 35$. Furthermore, since both are factors,

$$(x + 5)(3x^3 - 4x^2 + 7) = 3x^4 + 11x^3 - 20x^2 + 7x + 35$$

4 Use the Remainder Theorem

In Example 8, when we divided $x^3 - x^2 + 5$ by $x + 2$, we found that the remainder was -7. If we write $x + 2$ as $x - (-2)$ and evaluate the polynomial function $P(x) = x^3 - x^2 + 5$ at -2, we obtain -7.

$$P(x) = x^3 - x^2 + 5$$

$$P(-2) = (-2)^3 - (-2)^2 + 5 = -8 - 4 + 5 = -7$$

It can be shown that for any polynomial function $P(x)$, the value of the function at a, $P(a)$, has the same value as the remainder when $P(x)$ is divided by $x - a$.

To obtain the remainder when a polynomial $P(x)$ is divided by a binomial of the form $x - a$, we can use the **Remainder Theorem**.

Remainder Theorem

If the polynomial $P(x)$ is divided by $x - a$, the remainder is equal to $P(a)$.

Understanding Algebra

Remember the remainder in synthetic division is equal to $P(a)$.

EXAMPLE 10 Use the Remainder Theorem to find the remainder when $3x^4 + 6x^3 - 2x + 4$ is divided by $x + 4$.

Solution First we write the divisor $x + 4$ in the form $x - a$. Since $x + 4 = x - (-4)$, we evaluate $P(-4)$.

$$P(x) = 3x^4 + 6x^3 - 2x + 4$$
$$P(-4) = 3(-4)^4 + 6(-4)^3 - 2(-4) + 4$$
$$= 3(256) + 6(-64) + 8 + 4$$
$$= 768 - 384 + 8 + 4 = 396$$

Thus, when $3x^4 + 6x^3 - 2x + 4$ is divided by $x + 4$, the remainder is 396.

Now Try Exercise 87

Using synthetic division, we will show that the remainder in Example 10 is indeed 396.

$$
\begin{array}{r|rrrrr}
-4 & 3 & 6 & 0 & -2 & 4 \\
   &   & -12 & 24 & -96 & 392 \\
\hline
   & 3 & -6 & 24 & -98 & 396
\end{array}
$$
$\longleftarrow$ Remainder

If we were to graph the polynomial $P(x) = 3x^4 + 6x^3 - 2x + 4$, the value of $P(x)$, or y, at $x = -4$ would be 396.

EXAMPLE 11 Use the Remainder Theorem to determine whether $x - 5$ is a factor of $6x^2 - 25x - 25$.

Solution Let $P(x) = 6x^2 - 25x - 25$. If $P(5) = 0$, then the remainder of $(6x^2 - 25x - 25)/(x - 5)$ is 0 and $x - 5$ is a factor of the polynomial. If $P(5) \neq 0$, then there is a remainder and $x - 5$ is not a factor.

$$P(x) = 6x^2 - 25x - 25$$
$$P(5) = 6(5)^2 - 25(5) - 25$$
$$= 6(25) - 25(5) - 25$$
$$= 150 - 125 - 25 = 0$$

Since $P(5) = 0$, $x - 5$ is a factor of $6x^2 - 25x - 25$. Note that $6x^2 - 25x - 25 = (x - 5)(6x + 5)$.

Now Try Exercise 89

EXERCISE SET 5.3

Math XL
MathXL®

MyMathLab
MyMathLab

Warm-Up Exercises

Fill in the blanks with the appropriate word, phrase, or symbol(s) from the following list.

| first | factors | artificial | quotient rule | remainder | term | $P(3)$ |
| quotient | $P(7)$ | synthetic | dividend | zero exponent | divisor | |

1. In a division problem, the expression we are dividing is called the _____.

2. In a division problem, the expression we are dividing by is called the _____.

3. The rule $\dfrac{a^m}{a^n} = a^{m-n}$ is called the _____ for exponents.

4. The rule $a^0 = 1$ is called the _____ rule.

5. To divide a polynomial by a monomial, divide each _____ of the polynomial by the monomial.

6. To check the answer to a division problem, multiply the divisor by the quotient and add the _____.

7. When dividing a polynomial by $x - a$, instead of using long division, you can use a shorter process called _____ division.

8. In a division problem, if the remainder is 0, then the quotient and divisor are _____ of the dividend.

9. If the polynomial $P(x)$ is divided by $x - 7$, then the remainder is equal to _____ .

10. In the division problem $\dfrac{x^2 - 9x + 18}{x - 3} = x - 6$, the $x - 6$ is called the _____ .

Practice the Skills

Divide.

11. $\dfrac{x^9}{x^7}$

12. $\dfrac{m^{13}}{m^3}$

13. $\dfrac{a^{11}}{a^7}$

14. $\dfrac{b^{13}}{b^8}$

15. $\dfrac{z^{17}}{z^9}$

16. $\dfrac{q^6}{q^6}$

17. $\dfrac{12r^7 s^{10}}{3rs^8}$

18. $\dfrac{7y^{14} z^7}{4y^{11} z}$

19. $\dfrac{25x^{18} y^{19}}{5x^{10} y^8}$

20. $\dfrac{21a^8 b^{17}}{9a^7 b^{10}}$

Divide.

21. $\dfrac{4x + 18}{2}$

22. $\dfrac{9x + 5}{3}$

23. $\dfrac{4x^2 + 2x}{2x}$

24. $\dfrac{12x^2 - 8x - 28}{4}$

25. $\dfrac{5y^3 + 6y^2 - 12y}{3y}$

26. $\dfrac{14y^5 + 21y^2}{7y^4}$

27. $\dfrac{4x^5 - 6x^4 + 12x^3 - 8x^2}{4x^2}$

28. $\dfrac{15x^3 y - 25xy^3}{5xy}$

29. $\dfrac{8x^2 y^2 - 10xy^3 - 5y}{2y^2}$

30. $\dfrac{4x^{13} + 12x^9 - 11x^7}{4x^6}$

31. $\dfrac{9x^2 y - 12x^3 y^2 + 25y^3}{2xy^2}$

32. $\dfrac{a^2 b^2 c - 6abc^2 + 5a^3 b^5}{2abc^2}$

33. $\dfrac{3xyz + 6xyz^2 - 9x^3 y^5 z^7}{6xy}$

34. $\dfrac{6abc^3 - 5a^2 b^3 c^4 + 8ab^5 c}{3ab^2 c^3}$

Divide using long division.

35. $\dfrac{x^2 + 3x + 2}{x + 1}$

36. $\dfrac{x^2 + x - 20}{x + 5}$

37. $\dfrac{6x^2 + 16x + 8}{3x + 2}$

38. $\dfrac{2x^2 + 3x - 20}{x + 4}$

39. $\dfrac{6x^2 + x - 2}{2x - 1}$

40. $\dfrac{12x^2 - 17x - 7}{3x + 1}$

41. $\dfrac{x^2 + 6x + 3}{x + 1}$

42. $\dfrac{a^2 - a - 27}{a + 3}$

43. $\dfrac{2b^2 + b - 8}{b - 2}$

44. $\dfrac{2c^2 + c + 3}{2c + 5}$

45. $\dfrac{8x^2 + 6x - 25}{2x - 3}$

46. $\dfrac{8z^2 - 18z - 11}{4z + 1}$

47. $\dfrac{4x^2 - 36}{2x - 6}$

48. $\dfrac{16p^2 - 25}{4p + 5}$

49. $\dfrac{x^3 + 3x^2 + 5x + 9}{x + 1}$

50. $\dfrac{-a^3 - 6a^2 + 2a - 4}{a - 1}$

51. $\dfrac{4y^3 + 12y^2 + 7y - 9}{2y + 3}$

52. $\dfrac{9b^3 - 3b^2 - 3b + 6}{3b + 2}$

53. $(4a^3 - 5a) \div (2a - 1)$

54. $(2x^3 + 6x + 33) \div (x + 4)$

55. $\dfrac{3x^5 + 2x^2 - 12x - 4}{x^2 - 2}$

56. $\dfrac{4b^5 - 18b^3 + 14b^2 + 18b - 21}{2b^2 - 3}$

57. $\dfrac{3x^4 + 4x^3 - 32x^2 - 5x - 20}{3x^3 - 8x^2 - 5}$

58. $\dfrac{3a^4 - 9a^3 + 13a^2 - 11a + 4}{a^2 - 2a + 1}$

59. $\dfrac{2c^4 - 8c^3 + 19c^2 - 33c + 15}{c^2 - c + 5}$

60. $\dfrac{2y^5 + 2y^4 - 3y^3 - 17y^2 + 21}{2y^2 - 3}$

Use synthetic division to divide.

61. $(x^2 + 7x + 6) \div (x + 1)$

62. $(x^2 - 7x + 6) \div (x - 6)$

63. $(x^2 + 5x + 6) \div (x + 2)$

64. $(x^2 - 5x + 6) \div (x - 3)$

65. $(x^2 - 11x + 28) \div (x - 4)$

66. $(x^2 + 17x + 72) \div (x + 8)$

67. $(x^2 + 5x - 14) \div (x - 3)$

68. $(x^2 - 2x - 39) \div (x + 5)$

69. $(3x^2 - 7x - 10) \div (x - 4)$

70. $(2b^2 - 9b + 2) \div (b - 6)$

71. $(4x^3 - 3x^2 + 2x) \div (x - 1)$

72. $(z^3 - 7z^2 - 13z + 30) \div (z - 2)$

73. $(3c^3 + 7c^2 - 4c + 16) \div (c + 3)$

74. $(3y^4 - 25y^2 - 29) \div (y - 3)$

75. $(y^4 - 1) \div (y - 1)$

76. $(a^4 - 16) \div (a - 2)$

77. $\dfrac{x^4 + 16}{x + 4}$

78. $\dfrac{z^4 + 81}{z + 3}$

79. $\dfrac{x^5 + x^4 - 9}{x + 1}$

80. $\dfrac{a^7 - 2a^6 + 15}{a - 2}$

81. $\dfrac{b^5 + 4b^4 - 14}{b + 1}$

82. $\dfrac{z^5 - 3z^3 - 7z}{z - 2}$

83. $(3x^3 + 2x^2 - 4x + 1) \div \left(x - \dfrac{1}{3} \right)$

84. $(8x^3 - 6x^2 - 5x + 12) \div \left(x + \dfrac{3}{4} \right)$

85. $(2x^4 - x^3 + 2x^2 - 3x + 7) \div \left(x - \dfrac{1}{2} \right)$

86. $(9y^3 + 9y^2 - y + 2) \div \left(y + \dfrac{2}{3} \right)$

Determine the remainder for the following divisions using the Remainder Theorem. If the divisor is a factor of the dividend, so state.

87. $(4x^2 - 5x + 6) \div (x - 2)$

88. $(-2x^2 + 3x - 2) \div (x + 3)$

89. $(x^3 - 2x^2 + 4x - 8) \div (x - 2)$

90. $(x^4 + 3x^3 + x^2 + 22x + 8) \div (x + 4)$

91. $(-2x^3 - 6x^2 + 2x - 4) \div \left(x - \dfrac{1}{2} \right)$

92. $(-5x^3 - 6) \div \left(x - \dfrac{1}{5} \right)$

Problem Solving

93. Area The area of a rectangle is $6x^2 - 8x - 8$. If the length is $2x - 4$, find the width.

94. Area The area of a rectangle is $15x^2 - 29x - 14$. If the width is $5x + 2$, find the length.

Area and Volume *In Exercises 95 and 96, how many times greater is the area or volume of the figure on the right than the figure on the left? Explain how you determined your answer.*

95.

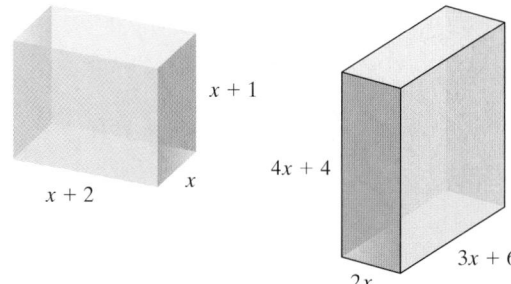

96.

97. If $\dfrac{P(x)}{x - 4} = x + 2$, find $P(x)$.

98. If $\dfrac{P(x)}{2x + 4} = 2x - 3$, find $P(x)$.

99. If $\dfrac{P(x)}{x + 4} = x + 5 + \dfrac{6}{x + 4}$, find $P(x)$.

100. If $\dfrac{P(x)}{2x - 3} = 2x - 1 - \dfrac{8}{2x - 3}$, find $P(x)$.

In Exercises 101 and 102, divide.

101. $\dfrac{2x^3 - x^2 y - 7xy^2 + 2y^3}{x - 2y}$

102. $\dfrac{x^3 + y^3}{x + y}$

In Exercises 103 and 104, divide. The answers contain fractions.

103. $\dfrac{2x^2 + 2x - 2}{2x - 3}$

104. $\dfrac{3x^3 - 7}{3x - 2}$

105. Volume The volume of the box that follows is $2r^3 + 4r^2 + 2r$. Find w in terms of r.

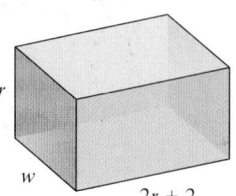

106. Volume The volume of the box that follows is $6a^3 + a^2 - 2a$. Find b in terms of a.

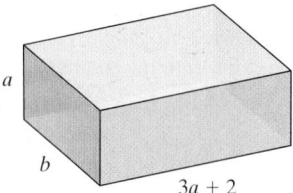

107. When a polynomial is divided by $x - 3$, the quotient is $x^2 - 3x + 4 + \dfrac{5}{x - 3}$. What is the polynomial? Explain how you determined your answer.

108. When a polynomial is divided by $2x - 3$, the quotient is $2x^2 + 6x - 5 + \dfrac{5}{2x - 3}$. What is the polynomial? Explain how you determined your answer.

In Exercises 109 and 110, divide. Assume that all variables in the exponents are natural numbers.

109. $\dfrac{4x^{n+1} + 2x^n - 3x^{n-1} - x^{n-2}}{2x^n}$

110. $\dfrac{3x^n + 6x^{n-1} - 2x^{n-2}}{2x^{n-1}}$

111. Is $x - 1$ a factor of $x^{100} + x^{99} + \cdots + x^1 + 1$? Explain.

112. Is $x + 1$ a factor of $x^{100} + x^{99} + \cdots + x^1 + 1$? Explain.

113. Is $x + 1$ a factor of $x^{99} + x^{98} + \cdots + x^1 + 1$? Explain.

114. Divide $0.2x^3 - 4x^2 + 0.32x - 0.64$ by $x - 0.4$.

115. a) Explain how to divide a polynomial by a monomial.

    b) Divide $\dfrac{5x^4 - 6x^3 - 4x^2 - 12x + 1}{3x}$ using the procedure you gave in part **a)**.

116. a) Explain how to divide a trinomial in x by a binomial in x.

    b) Divide $2x^2 - 12 + 5x$ by $x + 4$ using the procedure you gave in part **a)**.

117. A trinomial divided by a binomial has a remainder of 0. Is the quotient a factor of the trinomial? Explain.

118. a) Explain how the answer may be checked when dividing a polynomial by a binomial.

    b) Use your explanation in part **a)** to check whether the following division is correct.
$$\frac{8x^2 + 2x - 15}{4x - 5} = 2x + 3$$

    c) Check to see whether the following division is correct.
$$\frac{6x^2 - 23x + 13}{3x - 4} = 2x - 5 - \frac{8}{3x - 4}$$

119. When dividing a polynomial by a polynomial, before you begin the division, what should you do to the polynomials?

120. Explain why $\dfrac{x - 1}{x}$ is not a polynomial.

121. a) Describe how to divide a polynomial by $(x - a)$ using synthetic division.

    b) Divide $x^2 + 3x - 4$ by $x - 5$ using the procedure you gave in part **a)**.

122. a) State the Remainder Theorem.

    b) Find the remainder when $x^2 - 6x - 4$ is divided by $x - 1$, using the procedure you stated in part **a)**.

123. In the division problem $\dfrac{x^2 + 11x + 21}{x + 2} = x + 9 + \dfrac{3}{x + 2}$, is $x + 9$ a factor of $x^2 + 11x + 21$? Explain.

124. In the division problem $\dfrac{x^2 - 3x - 28}{x + 4} = x - 7$, is $x - 7$ a factor of $x^2 - 3x - 28$? Explain.

125. Is it possible to divide a binomial by a monomial and obtain a monomial as a quotient? Explain.

126. a) Is the sum, difference, and product of two polynomials always a polynomial?

    b) Is the quotient of two polynomials always a polynomial? Explain.

127. Explain how you can determine using synthetic division if an expression of the form $x - a$ is a factor of a polynomial in x.

128. Given $P(x) = ax^2 + bx + c$ and a value d such that $P(d) = 0$, explain why d is a solution to the equation $ax^2 + bx + c = 0$.

Cumulative Review Exercises

[1.6] **129.** Divide $\dfrac{8.45 \times 10^{23}}{4.225 \times 10^{13}}$ and express the answer in scientific notation.

[2.3] **130.** **Triangle** Find the three angles of a triangle if one angle is twice the smallest angle and the third angle is $60°$ greater than the smallest angle.

[2.6] **131.** Find the solution set for $\left| \dfrac{5x - 3}{2} \right| + 4 = 8$.

[3.6] **132.** Let $f(x) = x^2 - 4$ and $g(x) = -5x + 3$. Find $f(6) \cdot g(6)$.

[5.1] **133.** Add $(6r + 5s - t) + (-3r - 2s - 7t)$.

5.4 Factoring a Monomial from a Polynomial and Factoring by Grouping

1 Find the greatest common factor.

2 Factor a monomial from a polynomial.

3 Factor a common binomial factor.

4 Factor by grouping.

Factoring is the opposite of multiplying. To factor an expression means to write it as a product of other expressions. For example, in Section 5.2 we learned to perform the following multiplications:

$$3x^2(6x + 3y + 4x^3) = 18x^3 + 9x^2y + 12x^5$$
Multiplying

and

$$(6x + 3y)(2x - 5y) = 12x^2 - 24xy - 15y^2$$
Multiplying

In this section, we learn how to determine the *factors* of a given expression. For example, we will learn how to perform each factoring illustrated below.

$$18x^3 + 9x^2y + 12x^5 = 3x^2(6x + 3y + 4x^3)$$

Factoring

and

$$12x^2 - 24xy - 15y^2 = (6x + 3y)(2x - 5y)$$

Factoring

1 Find the Greatest Common Factor

To factor a monomial from a polynomial, we factor out the *greatest common factor* from each term in the polynomial. The **greatest common factor (GCF)** is the product of the factors common to all terms in the polynomial.

For example, the GCF for $6x + 21$ is 3 since 3 is the largest number that is a factor of both $6x$ and 21. To factor, we use the distributive property.

$$6x + 21 = 3(2x + 7)$$

The 3 and the $2x + 7$ are *factors* of the polynomial $6x + 21$.

Consider the terms x^3, x^4, x^5, and x^6. The GCF of these terms is x^3, since x^3 is the highest power of x that divides all four terms.

The first step in any factoring problem is to determine whether all the terms have a common factor.

EXAMPLE 1 Find the GCF of the following terms.

a) y^{12}, y^4, y^9, y^7 **b)** x^3y^2, xy^4, x^5y^6 **c)** $6x^2y^3z, 9x^3y^4, 24x^2z^5$

Solution

a) Note that y^4 is the highest power of y common to all four terms. The GCF is, therefore, y^4.

b) The highest power of x that is common to all three terms is x (or x^1). The highest power of y that is common to all three terms is y^2. Thus, the GCF of the three terms is xy^2.

c) The GCF is $3x^2$. Since y does not appear in $24x^2z^5$, it is not part of the GCF. Since z does not appear in $9x^3y^4$, it is not part of the GCF.

Now Try Exercise 93

EXAMPLE 2 Find the GCF of the following terms.
$$6(x - 2)^2, 5(x - 2), 18(x - 2)^7$$

Solution   The three numbers 6, 5, and 18 have no common factor other than 1. The highest power of $(x - 2)$ common to all three terms is $(x - 2)$. Thus, the GCF of the three terms is $(x - 2)$.

Now Try Exercise 95

2 Factor a Monomial from a Polynomial

When we factor a monomial from a polynomial, we are factoring out the greatest common factor. *The first step in any factoring problem is to factor out the GCF.*

Understanding Algebra

In any factoring problem, the first step is to factor out the GCF.

To Factor a Monomial from a Polynomial

1. Determine the greatest common factor of all terms in the polynomial.

2. Write each term as the product of the GCF and another factor.

3. Use the distributive property to *factor out* the GCF.

EXAMPLE 3 Factor $15x^4 - 5x^3 + 25x^2$.

Solution The GCF is $5x^2$. Write each term as the product of the GCF and another product. Then factor out the GCF.

$$15x^4 - 5x^3 + 25x^2 = 5x^2 \cdot 3x^2 - 5x^2 \cdot x + 5x^2 \cdot 5$$
$$= 5x^2(3x^2 - x + 5)$$

Now Try Exercise 15

Helpful Hint

To check the factoring process, multiply the factors using the distributive property. The product should be the polynomial with which you began. For instance, in Example 3,

Check
$$5x^2(3x^2 - x + 5) = 5x^2(3x^2) + 5x^2(-x) + 5x^2(5)$$
$$= 15x^4 - 5x^3 + 25x^2$$

EXAMPLE 4 Factor $20x^3y^3 + 6x^2y^4 - 12xy^7$.

Solution The GCF is $2xy^3$. Write each term as the product of the GCF and another product. Then factor out the GCF.

$$20x^3y^3 + 6x^2y^4 - 12xy^7 = 2xy^3 \cdot 10x^2 + 2xy^3 \cdot 3xy - 2xy^3 \cdot 6y^4$$
$$= 2xy^3(10x^2 + 3xy - 6y^4)$$

Check $\quad 2xy^3(10x^2 + 3xy - 6y^4) = 20x^3y^3 + 6x^2y^4 - 12xy^7$

Now Try Exercise 19

When the leading coefficient of a polynomial is negative, we generally factor out a common factor with a negative coefficient. This results in the leading coefficient of the remaining polynomial being positive.

EXAMPLE 5 Factor.

a) $-12a - 18$ **b)** $-2b^3 + 6b^2 - 16b$

Solution Since the leading coefficients in parts **a)** and **b)** are negative, we factor out common factors with a negative coefficient.

a) $-12a - 18 = -6(2a + 3)$ Factor out -6.

b) $-2b^3 + 6b^2 - 16b = -2b(b^2 - 3b + 8)$ Factor out $-2b$.

Now Try Exercise 27

EXAMPLE 6 Throwing a Ball When a ball is thrown upward with a velocity of 32 feet per second from the top of a 160-foot-tall building, its distance, d, from the ground at any time, t, can be determined by the function $d(t) = -16t^2 + 32t + 160$.

a) Determine the ball's distance from the ground after 3 seconds—that is, find $d(3)$.

b) Factor out the GCF from the right side of the function.

c) Evaluate $d(3)$ in factored form.

d) Compare your answers to parts **a)** and **c)**.

Solution

a) $d(t) = -16t^2 + 32t + 160$

$d(3) = -16(3)^2 + 32(3) + 160$ Substitute 3 for t.
$= -16(9) + 96 + 160$
$= 112$

The distance is 112 feet.

b) Factor -16 from the three terms on the right side of the equals sign.

$$d(t) = -16(t^2 - 2t - 10)$$

c) $d(t) = -16(t^2 - 2t - 10)$

$d(3) = -16[3^2 - 2(3) - 10]$ Substitute 3 for t.

$= -16(9 - 6 - 10)$

$= -16(-7)$

$= 112$

d) The answers are the same. You may find the calculations in part **c)** easier than the calculations in part **a)**.

Now Try Exercise 65

3 Factor a Common Binomial Factor

Sometimes factoring involves factoring a binomial as the greatest common factor, as illustrated in Examples 7 through 10.

EXAMPLE 7 Factor $3x(5x - 6) + 4(5x - 6)$.

Solution   The GCF is $(5x - 6)$. Factoring out the GCF gives

$$3x(5x - 6) + 4(5x - 6) = (5x - 6)(3x + 4)$$

Now Try Exercise 37

> **Understanding Algebra**
>
> In Example 7, the greatest common factor $(5x - 6)$ is a binomial factor.

In Example 7, we could have also placed the common factor on the right to obtain

$$3x(5x - 6) + 4(5x - 6) = (3x + 4)(5x - 6)$$

The factored forms $(5x - 6)(3x + 4)$ and $(3x + 4)(5x - 6)$ are equivalent because of the commutative property of multiplication and both are correct. Generally, when we list the answer to an example or exercise, we will place the common term that has been factored out on the left.

EXAMPLE 8 Factor $9(2x - 5) + 6(2x - 5)^2$.

Solution   The GCF is $3(2x - 5)$. Rewrite each term as the product of the GCF and another factor.

$9(2x - 5) + 6(2x - 5)^2 = 3(2x - 5) \cdot 3 + 3(2x - 5) \cdot 2(2x - 5)$

$= 3(2x - 5)[3 + 2(2x - 5)]$ Factor out $3(2x - 5)$.

$= 3(2x - 5)[3 + 4x - 10]$ Distributive property

$= 3(2x - 5)(4x - 7)$ Simplify.

Now Try Exercise 39

EXAMPLE 9 Factor $(3x - 4)(a + b) - (x - 1)(a + b)$.

Solution   The binomial $(a + b)$ is the GCF. We therefore factor it out.

$(3x - 4)(a + b) - (x - 1)(a + b) = (a + b)[(3x - 4) - (x - 1)]$ Factor out $(a + b)$.

$= (a + b)(3x - 4 - x + 1)$ Simplify.

$= (a + b)(2x - 3)$ Factors

Now Try Exercise 43

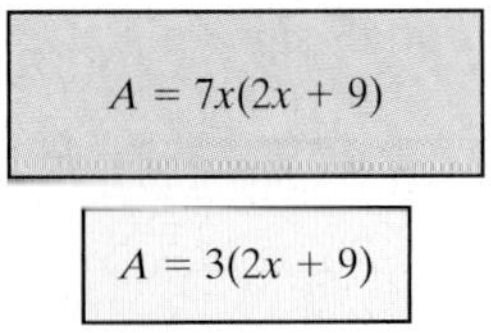

$A = 7x(2x + 9)$

$A = 3(2x + 9)$

FIGURE 5.13

EXAMPLE 10 Area In **Figure 5.13,** the area of the large rectangle is $7x(2x + 9)$ and the area of the small rectangle is $3(2x + 9)$. Find an expression, in factored form, for the difference of the areas for these two rectangles.

Solution   To find the difference of the areas, subtract the area of the small rectangle from the area of the large rectangle.

$$7x(2x + 9) - 3(2x + 9) \qquad \text{Subtract areas.}$$
$$= (2x + 9)(7x - 3) \qquad \text{Factor out } (2x + 9).$$

The difference of the areas for the two rectangles is $(2x + 9)(7x - 3)$.

Now Try Exercise 59

4 Factor by Grouping

When a polynomial contains *four terms,* it may be possible to factor the polynomial by grouping. To **factor by grouping,** remove common factors from groups of terms. This procedure is illustrated in the following example.

EXAMPLE 11 Factor $ax + ay + bx + by$.

Solution   There is no factor (other than 1) common to all four terms. However, a is common to the first two terms and b is common to the last two terms. Factor a from the first two terms and b from the last two terms.

$$ax + ay + bx + by = a(x + y) + b(x + y)$$

Now $(x + y)$ is common to both terms. Factor out $(x + y)$.

$$a(x + y) + b(x + y) = (x + y)(a + b)$$

Thus, $ax + ay + bx + by = (x + y)(a + b)$ or $(a + b)(x + y)$.

Now Try Exercise 49

To Factor Four Terms by Grouping

1. Determine if all four terms have a common factor. If so, factor out the greatest common factor from each term.
2. Arrange the four terms into two groups of two terms each. Each group of two terms must have a GCF.
3. Factor the GCF from each group of two terms.
4. If the two terms formed in step 3 have a GCF, factor it out.

EXAMPLE 12 Factor $x^3 - 5x^2 + 2x - 10$ by grouping.

Solution   There are no factors common to all four terms. However, x^2 is common to the first two terms and 2 is common to the last two terms. Factor x^2 from the first two terms and factor 2 from the last two terms.

$$x^3 - 5x^2 + 2x - 10 = x^2(x - 5) + 2(x - 5)$$
$$= (x - 5)(x^2 + 2)$$

Now Try Exercise 55

In Example 12, $(x^2 + 2)(x - 5)$ is also an acceptable answer. Would the answer to Example 12 change if we switch the order of $2x$ and $-5x^2$? Let's try it in Example 13.

EXAMPLE 13 Factor $x^3 + 2x - 5x^2 - 10$.

Solution   There is no factor common to all four terms. Factor x from the first two terms and -5 from the last two terms.

$$x^3 + 2x - 5x^2 - 10 = x(x^2 + 2) - 5(x^2 + 2)$$
$$= (x^2 + 2)(x - 5)$$

Notice that we got equivalent results in Examples 12 and 13.

Now Try Exercise 51

> **Helpful Hint**
>
> When factoring four terms by grouping, if the *first* and *third* terms are positive, you must factor a positive expression from both the first two terms and the last two terms to obtain a factor common to the remaining two terms (see Example 12). If the *first* term is positive and the *third* term is negative, you must factor a positive expression from the first two terms and a negative expression from the last two terms to obtain a factor common to the remaining two terms (see Example 13).

Remember, the first step in any factoring problem is to determine whether all the terms have a common factor. If so, begin by factoring out the GCF. For instance, to factor $x^4 - 5x^3 + 2x^2 - 10x$, we first factor out x from each term. Then we factor the remaining four terms by grouping, as was done in Example 12.

$$x^4 - 5x^3 + 2x^2 - 10x = x(x^3 - 5x^2 + 2x - 10) \qquad \text{Factor out the GCF, } x, \text{ from all four terms.}$$

$$= x(x - 5)(x^2 + 2) \qquad \text{Factors from Example 12}$$

EXERCISE SET 5.4   Math XL   MyMathLab

Warm-Up Exercises

Fill in the blanks with the appropriate word, phrase, or symbol(s) from the following list.

| associative | common | $6x^3y^4$ | commutative | -2 | $24(x-2)^4$ | dividing |
| distributive | grouping | $4(x-2)^3$ | -1 | greatest common factor | multiplying | $2x^2y$ |

1. Factoring is the opposite of _____.

2. In a factoring problem, the first step is to factor out the _____.

3. When a polynomial contains four terms, it may be possible to factor the polynomial by _____.

4. The first step in factoring $-x^2 + 2x - 13$ is to factor out _____.

5. To factor by grouping, remove _____ factors from groups of terms.

6. The GCF of $2x^2y^4$ and $6x^3y$ is _____.

7. The GCF of $12(x-2)^3$ and $8(x-2)^4$ is _____.

8. To check the answer to a factoring problem, multiply the factors using the _____ property.

Practice the Skills

Factor out the greatest common factor.

9. $7n + 14$

10. $15p + 25$

11. $2x^2 - 4x + 16$

12. $6x^2 - 12x + 27$

13. $12y^2 - 16y + 28$

14. $12x^3 - 8x^2 - 10x$

15. $9x^4 - 3x^3 + 11x^2$

16. $45y^{12} + 60y^{10}$

17. $-24a^7 + 9a^6 - 3a^2$

18. $-16c^5 - 12c^4 + 6c^3$

19. $3x^2y + 6x^2y^2 + 3xy$

20. $24a^2b^2 + 16ab^4 + 72ab^3$

21. $80a^5b^4c - 16a^4b^2c^2 + 24a^2c$

22. $36xy^2z^3 + 36x^3y^2z + 9x^2yz$

23. $9p^4q^5r - 3p^2q^2r^2 + 12pq^5r^3$

24. $24m^6 + 8m^4 - 4m^3n$

25. $-22p^2q^2 - 16pq^3 + 26r$

26. $-15y^3z^5 - 28y^3z^6 + 9xy^2z^2$

Factor out a factor with a negative coefficient.

27. $-8x + 4$

28. $-20a - 50$

29. $-x^2 - 4x + 23$

30. $-y^5 - 6y^2 - 7$

31. $-3r^2 - 6r + 9$

32. $-12t^2 + 48t - 72$

33. $-6r^4s^3 + 4r^2s^4 + 2rs^5$

34. $-5p^6q^3 - 10p^4q^4 + 25pq^7$

35. $-a^4b^2c + 5a^3bc^2 + 3a^2b$

36. $-20x^5y^3z - 4x^4yz^2 - 8x^2y^5$

Factor.

37. $x(a + 9) + 1(a + 9)$

38. $y(b - 2) - 6(b - 2)$

39. $7x(x - 4) + 2(x - 4)^2$

40. $4y(y + 1) - 7(y + 1)^2$

41. $(x - 2)(3x + 5) - (x - 2)(5x - 4)$

42. $(z + 4)(z + 3) + (z - 1)(z + 3)$

43. $(2a + 4)(a - 3) - (2a + 4)(2a - 1)$

44. $(6b - 1)(b + 4) + (6b - 1)(2b + 5)$

45. $x^2 + 4x - 5x - 20$

46. $a^2 + 3a - 6a - 18$

47. $8y^2 - 4y - 20y + 10$

48. $18m^2 + 30m + 9m + 15$

49. $am + an + bm + bn$

50. $cx - cy - dx + dy$

51. $x^3 - 3x^2 + 4x - 12$

52. $2z^3 + 4z^2 - 5z - 10$

53. $10m^2 - 12mn - 25mn + 30n^2$

54. $12x^2 + 9xy - 4xy - 3y^2$

55. $5a^3 + 15a^2 - 10a - 30$

56. $2r^4 - 2r^3 - 9r^2 + 9r$

57. $c^5 - c^4 + c^3 - c^2$

58. $b^4 - b^3 - b + b^2$

Problem Solving

Area *In Exercises 59–62, A represents an expression for the area of the figure. Find an expression, in factored form, for the difference of the areas of the geometric figures. See Example 10.*

59.

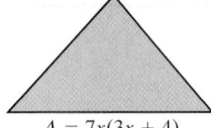

$A = 6x(2x + 1)$ $A = 5(2x + 1)$

60.

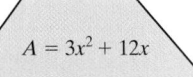

$A = 7x(3x + 4)$ $A = 2(3x + 4)$

61.

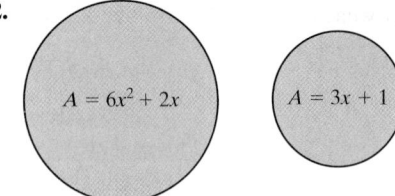

$A = 3x^2 + 12x$ $A = 2x + 8$

62.

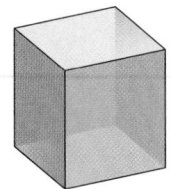 

$A = 6x^2 + 2x$ $A = 3x + 1$

Volume *In Exercises 63 and 64, V represents an expression for the volume of the figure. Find an expression, in factored form, for the difference of the volumes of the geometric solids.*

63.

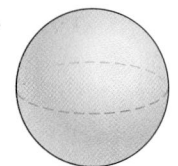

$V = 9x(3x + 2)$ $V = 5(3x + 2)$

64.

$V = 18x^2 + 24x$ $V = 3x + 4$

65. Flare When a flare is shot upward with a velocity of 80 feet per second, its height, h, in feet, above the ground at t seconds can be found by the function $h(t) = -16t^2 + 80t$.

a) Find the height of the flare 3 seconds after it was shot.

b) Express the function with the right side in factored form.

c) Evaluate $h(3)$ using the factored form from part **b).**

66. Jump Shot When a basketball player shoots a jump shot, the height, h, in feet, of the ball above the ground at any time t, under certain circumstances, may be found by the function $h(t) = -16t^2 + 20t + 8$.

a) Find the height of the ball at 1 second.

b) Express the function with the right side in factored form.

c) Evaluate $h(1)$ using the factored form in part **b).**

© Jupiter Unlimited

67. Skating Rink The area of the skating rink with semicircular ends shown is $A = \pi r^2 + 2rl$.

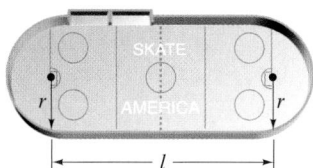

a) Find A when $r = 20$ feet and $l = 40$ feet.

b) Write the area A in factored form.

c) Find A when $r = 20$ feet and $l = 40$ feet using the factored form in part **b).**

68. Area The formula for finding the area of a trapezoid may be given as $A = \dfrac{1}{2}hb_1 + \dfrac{1}{2}hb_2$. Express this formula in factored form.

69. Purchase a Stereo Fred Yang just bought a stereo for $975. With a 0% (interest-free) loan, Fred must pay $75 each month until the stereo is paid off. The amount still owed on the stereo is a function of time, where

$$A(t) = 975 - 75t$$

and t is the number of months after Fred bought the stereo.

a) Determine the amount still owed 6 months after Fred bought the stereo.

b) Write the function in factored form.

c) Use the result from part **b)** to determine the amount still owed 6 months after Fred bought the stereo.

© Allen R. Angel

70. Salary On January 2, 2010, Jill Ferguson started a new job with an annual salary of $33,000. Her salary will increase by $1500 each year. Thus, her salary is a function of the number of years she works, where

$$S(n) = 33{,}000 + 1500n$$

and n is the number of years after 2010.
a) Determine Jill's salary in 2014
b) Write the function in factored form.
c) Use the result from part **b)** to determine Jill's salary in 2014.

71. Price of Cars When the 2010 cars came out, the list price of one model increased by 6% over the list price of the 2009 model. Then in a special sale, the prices of all 2010 cars were reduced by 6%. The sale price can be represented by $(x + 0.06x) - 0.06(x + 0.06x)$, where x is the list price of the 2009 model.
a) Factor out $(x + 0.06x)$ from each term.
b) Is the sale price more or less than the price of the 2009 model?

Read Exercise 71 before working Exercises 72–74.

72. Price of Dress A dress is reduced by 10%, and then the sale price is reduced by another 10%.
a) Write an expression for the final price of the dress.
b) How does the final price compare with the regular price of the dress? Use factoring in obtaining your answer.

73. Price of Mower The price of a Toro lawn mower is increased by 15%. Then at a Fourth of July sale the price is reduced by 20%.
a) Write an expression for the final price of the lawn mower.
b) How does the sale price compare with the regular price? Use factoring in obtaining your answer.

74. Find Price In which of the following, **a)** or **b)**, will the final price be lower, and by how much?
a) Decreasing the price of an item by 6%, then increasing that price by 8%
b) Increasing the price of an item by 6%, then decreasing that price by 8%

Factor.

75. $5a(3x + 2)^5 + 4(3x + 2)^4$

76. $4p(2r - 3)^7 - 3(2r - 3)^6$

77. $4x^2(x - 3)^3 - 6x(x - 3)^2 + 4(x - 3)$

78. $12(p + 2q)^4 - 40(p + 2q)^3 + 12(p + 2q)^2$

79. $ax^2 + 2ax - 4a + bx^2 + 2bx - 4b$

80. $6a^2 - a^2c + 18a - 3ac + 6ab - abc$

Factor. Assume that all variables in the exponents represent natural numbers.

81. $x^{6m} - 5x^{4m}$

82. $x^{2mn} + x^{6mn}$

83. $3x^{4m} - 2x^{3m} + x^{2m}$

84. $r^{y+4} + r^{y+3} + r^{y+2}$

85. $a^r b^r + c^r b^r - a^r d^r - c^r d^r$

86. $6a^k b^k - 2a^k c^k - 9b^k + 3c^k$

87. a) Does $6x^3 - 3x^2 + 9x = 3x(2x^2 - x + 3)$?
b) If the above factoring is correct, what should be the value of $6x^3 - 3x^2 + 9x - [3x(2x^2 - x + 3)]$ for any value of x? Explain.
c) Select a value for x and evaluate the expression in part **b)**. Did you get what you expected? If not, explain why.

88. a) Determine whether the following factoring is correct.

$$3(x - 2)^2 - 6(x - 2) = 3(x - 2)[(x - 2) - 2]$$
$$= 3(x - 2)(x - 4)$$

b) If the above factoring is correct, what should be the value of $3(x - 2)^2 - 6(x - 2) - [3(x - 2)(x - 4)]$ for any value of x? Explain.
c) Select a value for x and evaluate the expression in part **b)**. Did you get what you expected? If not, explain why.

89. Consider the factoring $8x^3 - 16x^2 - 4x = 4x(2x^2 - 4x - 1)$.
a) If we let

$$y_1 = 8x^3 - 16x^2 - 4x$$
$$y_2 = 4x(2x^2 - 4x - 1)$$

and graph each function, what should happen? Explain.
b) On your graphing calculator, graph y_1 and y_2 as given in part **a)**.
c) Did you get the results you expected?
d) When checking a factoring process by this technique, what does it mean if the graphs do not overlap? Explain.

90. Consider the factoring $2x^4 - 6x^3 - 8x^2 = 2x^2(x^2 - 3x - 4)$.
a) Enter

$$y_1 = 2x^4 - 6x^3 - 8x^2$$
$$y_2 = 2x^2(x^2 - 3x - 4)$$

into your calculator.
b) If you use the TABLE feature of your calculator, how would you expect the table of values for y_1 to compare with the table of values for y_2? Explain.
c) Use the TABLE feature to show the values for y_1 and y_2 for values of x from 0 to 6.
d) Did you get the results you expected?
e) When checking a factoring process using the TABLE feature, what does it mean if the values of y_1 and y_2 are different?

Concept/Writing Exercises

91. What is the first step in *any* factoring problem?

92. What is the greatest common factor of the terms of an expression?

93. a) Explain how to find the greatest common factor of the terms of a polynomial.

   b) Using your procedure from part **a)**, find the greatest common factor of the polynomial

$$6x^2y^5 - 2x^3y + 12x^9y^3$$

   c) Factor the polynomial in part **b)**.

94. Determine the GCF of the following terms:

$$x^4y^6, x^3y^5, xy^6, x^2y^4$$

   Explain how you determined your answer.

95. Determine the GCF of the following terms:

$$12(x - 4)^3, 6(x - 4)^6, 3(x - 4)^9$$

   Explain how you determined your answer.

96. When a term of a polynomial is itself the GCF, what is written in place of that term when the GCF is factored out? Explain.

97. a) Explain how to factor a polynomial of four terms by grouping.

   b) Factor $6x^3 - 2xy^3 + 3x^2y^2 - y^5$ by your procedure from part **a)**.

98. What is the first step when factoring $-x^2 + 8x - 15$? Explain your answer.

Cumulative Review Exercises

[1.4] **99.** Evaluate $\dfrac{\left(\left|\frac{1}{2}\right| - \left|-\frac{1}{3}\right|\right)^2}{-\left|\frac{1}{3}\right| \cdot \left|-\frac{2}{5}\right|}$.

[2.1] **100.** Solve $3(2x - 4) + 3(x + 1) = 9$.

[3.1] **101.** Graph $y = x^2 - 1$.

[4.3] **102. Exercise** Jason Richter tries to exercise every day. He walks at 3 mph and then jogs at 5 mph. If it takes him 0.9 hr to travel a total of 3.5 mi, how long does he jog?

[5.2] **103.** Multiply $(7a - 3)(-2a^2 - 4a + 1)$.

Mid-Chapter Test: 5.1–5.4

To find out how well you understand the chapter material to this point, take this brief test. The answers, and the section where the material was initially discussed, are given in the back of the book. Review any questions that you answered incorrectly.

1. Write the polynomial $-7 + 2x + 5x^4 - 1.5x^3$ in descending order. Give the degree of this polynomial

2. Evaluate $P\left(\dfrac{1}{2}\right)$ given $P(x) = 8x^2 - 7x + 3$.

3. Simplify $(2n^2 - n - 12) + (-3n^2 - 6n + 8)$.

4. Subtract $(7x^2y - 10xy)$ from $(-9x^2y + 4xy)$.

5. Find a polynomial expression, in simplified form, for the perimeter of the triangle.

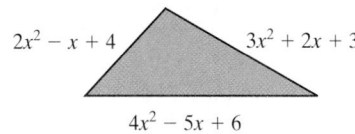

Multiply.

6. $2x^5(3xy^4 + 5x^2 - 7x^3y)$

7. $(7x - 6y)(3x + 2y)$

8. $(3x + 1)(2x^3 - x^2 + 5x + 9)$

9. $\left(8p - \dfrac{1}{5}\right)\left(8p + \dfrac{1}{5}\right)$

10. $(4m - 3n)(3m^2 + 2mn - 6n^2)$

11. Write $x^2 - 14x + 49$ as the square of a binomial. Explain how you determined your answer.

Divide.

12. $\dfrac{4x^4y^3 + 6x^2y^2 - 11x}{2x^2y^2}$

13. $\dfrac{12x^2 + 23x + 7}{4x + 1}$

14. $\dfrac{2y^3 - y^2 + 7y - 10}{2y - 3}$

Use synthetic division to divide.

15. $\dfrac{x^2 - x - 72}{x + 8}$

16. $\dfrac{3a^4 - 2a^3 - 14a^2 + 11a + 2}{a - 2}$

17. Factor out the greatest common factor in $32b^3c^3 + 16b^2c + 24b^5c^4$.

Factor completely.

18. $7b(2x + 9) - 3c(2x + 9)$

19. $2b^4 - b^3c + 4b^3c - 2b^2c^2$

20. $5a(3x - 2)^5 - 4(3x - 2)^6$

5.5 Factoring Trinomials

1 Factor trinomials of the form $x^2 + bx + c$.

2 Factor out a common factor.

3 Factor trinomials of the form $ax^2 + bx + c, a \neq 1$, using trial and error.

4 Factor trinomials of the form $ax^2 + bx + c, a \neq 1$, using grouping.

5 Factor trinomials using substitution.

1 Factor Trinomials of the Form $x^2 + bx + c$

In this section we learn how to **factor trinomials** of the form $ax^2 + bx + c, a \neq 0$. Notice that a represents the coefficient of the x-squared term, b represents the coefficient of the x-term, and c represents the constant term.

Trinomials	Coefficients
$3x^2 + 2x - 5$	$a = 3, \quad b = 2, \quad c = -5$
$-\dfrac{1}{2}x^2 - 4x + 3$	$a = -\dfrac{1}{2}, \quad b = -4, \quad c = 3$

To Factor Trinomials of the Form $x^2 + bx + c$ (note: $a = 1$)

1. Find two numbers (or factors) whose product is c and whose sum is b.
2. The factors of the trinomial will be of the form

$$(x + \boxed{})(x + \boxed{})$$

One factor determined in step 1 / Other factor determined in step 1

If the numbers determined in step 1 are, for example, 3 and -5, the factors would be written $(x + 3)(x - 5)$. This procedure is illustrated in the following examples.

EXAMPLE 1 Factor $x^2 - x - 12$.

Solution $a = 1, b = -1, c = -12$. We must find two numbers whose product is c, which is -12, and whose sum is b, which is -1. We begin by listing the factors of -12, trying to find a pair whose sum is -1.

Factors of -12	Sum of Factors
$(1)(-12)$	$1 + (-12) = -11$
$(2)(-6)$	$2 + (-6) = -4$
$(3)(-4)$	$3 + (-4) = -1$
$(4)(-3)$	$4 + (-3) = 1$
$(6)(-2)$	$6 + (-2) = 4$
$(12)(-1)$	$12 + (-1) = 11$

The numbers we are seeking are 3 and -4 because their product is -12 and their sum is -1. Now we factor the trinomial using the 3 and -4.

$$x^2 - x - 12 = (x + 3)(x - 4)$$

One factor of -12 / Other factor of -12

Now Try Exercise 13

Notice in Example 1 that we listed all the factors of -12. However, after the two factors whose product is c and whose sum is b are found, there is no need to go further in listing the factors. The factors were listed here to show, for example, that $(2)(-6)$ is a different set of factors than $(-2)(6)$. Note that as the positive factor increases the sum of the factors increases.

> **Helpful Hint**
>
> Consider the factors $(2)(-6)$ and $(-2)(6)$ and the sums of these factors.
>
> | Factors | Sum of Factors |
> |---------|----------------|
> | $2(-6)$ | $2 + (-6) = -4$ |
> | $-2(6)$ | $-2 + 6 = 4$ |
>
> Notice that if the signs of each number in the product are changed, the sign of the sum of factors is changed. We can use this fact to more quickly find the factors we are seeking. If, when seeking a specific sum, you get the opposite of that sum, change the sign of each factor to get the sum you are seeking.

EXAMPLE 2 Factor $p^2 - 7p + 6$.

Solution We must find two numbers whose product is 6 and whose sum is -7. Since the sum of two negative numbers is a negative number, and the product of two negative numbers is a positive number, both numbers must be negative numbers. The negative factors of 6 are $(-1)(-6)$ and $(-2)(-3)$. As shown below, the numbers we are looking for are -1 and -6.

Factors of 6	Sum of Factors
$(-1)(-6)$	$-1 + (-6) = -7$
$(-2)(-3)$	$-2 + (-3) = -5$

Therefore,

$$p^2 - 7p + 6 = (p - 1)(p - 6)$$

Since the factors may be placed in any order, $(p - 6)(p - 1)$ is also an acceptable answer.

Now Try Exercise 23

> **Helpful Hint**
>
> **Checking Factoring**
>
> Factoring problems can be checked by multiplying the factors obtained. If the factoring is correct, you will obtain the polynomial you started with. To check Example 2, we will multiply the factors using the FOIL method.
>
> $$(p - 1)(p - 6) = p^2 - 6p - p + 6 = p^2 - 7p + 6$$
>
> Since the product of the factors is the trinomial we began with, our factoring is correct. You should always check your factoring.

The procedure used to factor trinomials of the form $x^2 + bx + c$ can be used on other trinomials, as in the following example.

EXAMPLE 3 Factor $x^2 + 2xy - 15y^2$.

Solution We must find two numbers whose product is -15 and whose sum is 2. The two numbers are 5 and -3.

Factors of -15	Sum of Factors
$5(-3)$	$5 + (-3) = 2$

Since the last term of the trinomial contains y^2, the second term of each factor must contain y.

Check

$$x^2 + 2xy - 15y^2 = (x + 5y)(x - 3y)$$

$$(x + 5y)(x - 3y) = x^2 - 3xy + 5xy - 15y^2$$

$$= x^2 + 2xy - 15y^2$$

Now Try Exercise 75

2 Factor Out a Common Factor

The first step when factoring any trinomial is to determine whether all three terms have a common factor. If so, factor out the GCF. Then factor the remaining polynomial.

EXAMPLE 4   Factor $3x^4 - 6x^3 - 72x^2$.

Solution   The factor $3x^2$ is the GCF of all three terms of the trinomial. Factor it out first.

$$3x^4 - 6x^3 - 72x^2 = 3x^2(x^2 - 2x - 24) \quad \text{Factor out } 3x^2.$$

Now continue to factor $x^2 - 2x - 24$. Find two numbers whose product is -24 and whose sum is -2. The numbers are -6 and 4.

$$3x^2(x^2 - 2x - 24) = 3x^2(x - 6)(x + 4)$$

Therefore, $3x^4 - 6x^3 - 72x^2 = 3x^2(x - 6)(x + 4)$.

Now Try Exercise 33

3 Factor Trinomials of the Form $ax^2 + bx + c, a \neq 1$, Using Trial and Error

The first method we will use to factor trinomials of the form

$$ax^2 + bx + c, \quad a \neq 1$$

is the Trial and Error method. This method is sometimes called the FOIL (or Reverse FOIL) method. We will begin by multiplying $(2x + 3)(x + 1)$ using the FOIL method.

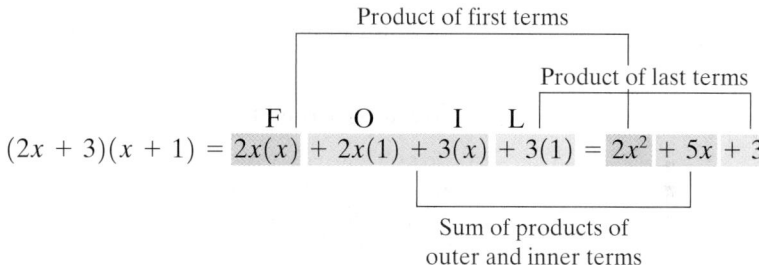

Therefore, if you are factoring the trinomial $2x^2 + 5x + 3$, you should realize that the product of the first terms of the factors must be $2x^2$, the product of the last terms must be 3, and the sum of the products of the outer and inner terms must be $5x$.

To factor $2x^2 + 5x + 3$, we begin as shown here.

$$2x^2 + 5x + 3 = (2x \qquad)(x \qquad) \quad \text{The product of the first terms is } 2x^2.$$

Now we fill in the second terms using positive integers whose product is 3. Only positive integers will be considered since the product of the last terms is positive, and the sum of the products of the outer and inner terms is also positive. The two possibilities are

$$\left.\begin{array}{l} (2x + 1)(x + 3) \\ (2x + 3)(x + 1) \end{array}\right\} \quad \begin{array}{l} \text{The product of} \\ \text{the last terms is } 3. \end{array}$$

To determine which factoring is correct, we find the sum of the products of the outer and inner terms. If either has a sum of $5x$, the middle term of the trinomial, that factoring is correct.

$$(2x + 1)(x + 3) = 2x^2 + 6x + x + 3 = 2x^2 + 7x + 3 \qquad \text{Wrong middle term}$$

$$(2x + 3)(x + 1) = 2x^2 + 2x + 3x + 3 = 2x^2 + 5x + 3 \qquad \text{Correct middle term}$$

Therefore, the factors of $2x^2 + 5x + 3$ are $2x + 3$ and $x + 1$. Thus,

$$2x^2 + 5x + 3 = (2x + 3)(x + 1).$$

Following are guidelines for the **trial and error** method of factoring a trinomial where $a \neq 1$ and the three terms have no common factors.

To Factor Trinomials of the Form $ax^2 + bx + c$, $a \neq 1$, Using Trial and Error

1. Write all pairs of factors of the coefficient of the squared term, a.
2. Write all pairs of factors of the constant, c.
3. Try various combinations of these factors until the correct middle term, bx, is found.

EXAMPLE 5 Factor $3t^2 - 13t + 10$.

Solution First we determine that the three terms have no common factor. Next we determine that a is 3 and the only factors of 3 are 1 and 3. Therefore, we write

$$3t^2 - 13t + 10 = (3t \qquad)(t \qquad)$$

The number 10 has both positive and negative factors. However, since the product of the last terms must be positive $(+10)$, and the sum of the products of the outer and inner terms must be negative (-13), the two factors of 10 must both be negative. (Why?) The negative factors of 10 are $(-1)(-10)$ and $(-2)(-5)$. Below is a list of the possible factors. We look for the factors that give us the correct middle term, $-13t$.

Possible Factors	Sum of Products of Outer and Inner Terms	
$(3t - 1)(t - 10)$	$-31t$	
$(3t - 10)(t - 1)$	$-13t$	⟵ Correct middle term
$(3t - 2)(t - 5)$	$-17t$	
$(3t - 5)(t - 2)$	$-11t$	

Thus, $3t^2 - 13t + 10 = (3t - 10)(t - 1)$.

Now Try Exercise 35

The following Helpful Hint is very important. Study it carefully.

Understanding Algebra

When using the *trial and error* method to factor, you may have to try several possibilities before you get the *correct factors*.

Understanding Algebra

When choosing factors, we generally start with *medium-sized* factors; that is the factors whose terms have coefficients that are closest to each other.

Helpful Hint

Factoring by Trial and Error

When factoring a trinomial of the form $ax^2 + bx + c$, the sign of the constant term, c, is very helpful in finding the solution. If $a > 0$, then:

1. When the constant term, c, is positive, and the numerical coefficient of the x term, b, is positive, both numerical factors will be positive.

   Example

   $$x^2 + 7x + 12 = (x + 3)(x + 4)$$
   $$\uparrow \quad \uparrow \qquad \uparrow \quad \uparrow$$

   Positive Positive   Positive Positive

2. When c is positive and b is negative, both numerical factors will be negative.

   Example

   $$x^2 - 5x + 6 = (x - 2)(x - 3)$$
   $$\uparrow \quad \uparrow \qquad \uparrow \quad \uparrow$$

   Negative Positive   Negative Negative

   Whenever the constant, c, is positive (as in the two examples above) the sign in both factors will be the same as the sign in the x-term of the trinomial.

3. When c is negative, one of the numerical factors will be positive and the other will be negative.

   Example

   $$x^2 + x - 6 = (x + 3)(x - 2)$$
   $$\uparrow \qquad \uparrow \quad \uparrow$$

   Negative   Positive Negative

EXAMPLE 6 Factor $8x^2 + 8x - 30$.

Solution First we notice that the GCF, 2, can be factored out.

$$8x^2 + 8x - 30 = 2(4x^2 + 4x - 15)$$

The factors of 4, the leading coefficient, are $4 \cdot 1$ and $2 \cdot 2$. Therefore, the factoring will be of the form $(4x \quad)(x \quad)$ or $(2x \quad)(2x \quad)$.

Let's start with $(2x \quad)(2x \quad)$. The factors of -15 are $(1)(-15)$, $(3)(-5)$, $(5)(-3)$, and $(15)(-1)$. We want our middle term to be $4x$.

Possible Factors	Sum of Products of Outer and Inner Terms
$(2x + 1)(2x - 15)$	$-28x$
$(2x + 3)(2x - 5)$	$-4x$
$(2x + 5)(2x - 3)$	$4x$

Because we found the set of factors that gives the correct x-term, we can stop. Thus,

$$8x^2 + 8x - 30 = 2(2x + 5)(2x - 3)$$

Now Try Exercise 37

In Example 6, if we compare the second and third set of factors we see they are the same except for the signs of the second terms. Notice that when the signs of the second term in each factor are switched the sum of the products of the outer and inner terms also changes sign.

Using Your Graphing Calculator ▣▣▣

A graphing calculator can be used to check factoring problems. To check the factoring in Example 6,

$$8x^2 + 8x - 30 = 2(2x + 5)(2x - 3)$$

we let $Y_1 = 8x^2 + 8x - 30$ and $Y_2 = 2(2x + 5)(2x - 3)$. Then we use the TABLE feature to compare results, as in **Figure 5.14**.

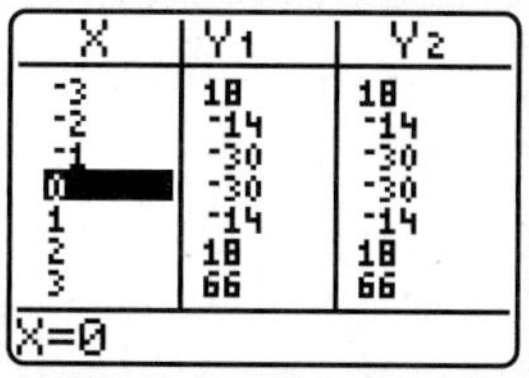

X	Y₁	Y₂
-3	18	18
-2	-14	-14
-1	-30	-30
0	-30	-30
1	-14	-14
2	18	18
3	66	66

X=0

FIGURE 5.14

Since Y_1 and Y_2 have the same values for each value of X, a mistake has not been made. This procedure can only tell you if a mistake has been made; it cannot tell you if you have factored completely. For example, $8x^2 + 8x - 30$ and $(4x + 10)(2x - 3)$ will give the same set of values.

EXAMPLE 7 Factor $6x^2 - 11xy - 10y^2$.

Solution The factors of 6 are either $6 \cdot 1$ or $2 \cdot 3$. Therefore, the factors of the trinomial may be of the form $(6x \quad)(x \quad)$ or $(2x \quad)(3x \quad)$. We will start as follows.

$$6x^2 - 11xy - 10y^2 = (2x \quad)(3x \quad)$$

The factors of -10 are $(-1)(10)$, $(1)(-10)$, $(-2)(5)$, and $(2)(-5)$. Since there are eight factors of -10, there will be eight pairs of possible factors to try.

The correct factorization is

$$6x^2 - 11xy - 10y^2 = (2x - 5y)(3x + 2y)$$

Now Try Exercise 51

In Example 7, we were fortunate to find the correct factors by using the form $(2x \quad)(3x \quad)$. If we had not found the correct factors using these, we would have tried $(6x \quad)(x \quad)$.

When factoring a trinomial whose leading coefficient is negative, we start by factoring out a negative number. For example,

$$-24x^3 - 60x^2 + 36x = -12x(2x^2 + 5x - 3) \qquad \text{Factor out } -12x.$$
$$= -12x(2x - 1)(x + 3)$$

and

$$-3x^2 + 8x + 16 = -1(3x^2 - 8x - 16) \qquad \text{Factor out } -1.$$
$$= -(3x + 4)(x - 4)$$

EXAMPLE 8 Area of Shaded Region In **Figure 5.15,** find an expression, in factored form, for the area of the shaded region.

Solution   To find the area of the shaded region, we need to subtract the area of the small rectangle from the area of the large rectangle. Recall that the area of a rectangle is length · width.

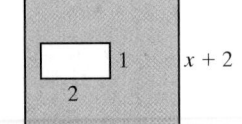

$$\text{Area of large rectangle} = (x + 3)(x + 2)$$
$$= x^2 + 2x + 3x + 6$$
$$= x^2 + 5x + 6$$

FIGURE 5.15

$$\text{Area of small rectangle} = (2)(1) = 2$$
$$\text{Area of shaded region} = \text{large area} - \text{small area}$$
$$= x^2 + 5x + 6 - 2$$
$$= x^2 + 5x + 4 \qquad \text{Simplify.}$$
$$= (x + 4)(x + 1) \qquad \text{Factor.}$$

The area of the shaded region is $(x + 4)(x + 1)$.

Now Try Exercise 89

4 Factor Trinomials of the Form $ax^2 + bx + c, a \neq 1,$ Using Grouping

Now we will discuss the **grouping** method of factoring trinomials of the form $ax^2 + bx + c, a \neq 1$.

> **To Factor Trinomials of the Form $ax^2 + bx + c, a \neq 1$, Using Grouping**
>
> **1.** Find two numbers whose product is $a \cdot c$ and whose sum is b.
> **2.** Rewrite the middle term, bx, using the numbers found in step 1.
> **3.** Factor by grouping.

EXAMPLE 9 Factor $2x^2 - 5x - 12$.

Solution   We see that $a = 2, b = -5$, and $c = -12$. We must find two numbers whose product is $a \cdot c$ or $2(-12) = -24$, and whose sum is b, -5. The two numbers are -8 and 3 because $(-8)(3) = -24$ and $-8 + 3 = -5$. Now rewrite the middle term, $-5x$, using $-8x$ and $3x$.

$$2x^2 - 5x - 12 = 2x^2 \overbrace{- 8x + 3x}^{-5x} - 12$$

Now factor by grouping as explained in Section 5.4.

$$2x^2 - 5x - 12 = 2x^2 - 8x + 3x - 12$$
$$= 2x(x - 4) + 3(x - 4)$$
$$= (x - 4)(2x + 3) \qquad \text{Factored out } (x - 4).$$

Thus $2x^2 - 5x - 12 = (x - 4)(2x + 3)$.

Now Try Exercise 61

> **Helpful Hint**
>
> In Example 9 we wrote $-5x$ as $-8x + 3x$. As we show below, the same factors would be obtained if we wrote $-5x$ as $3x - 8x$. Therefore, it makes no difference which factor is listed first when factoring by grouping.
>
> $$2x^2 - 5x - 12 = 2x^2 \overbrace{+ 3x - 8x}^{-5x} - 12$$
> $$= x(2x + 3) - 4(2x + 3)$$
> $$= (2x + 3)(x - 4) \qquad \text{Factored out } (2x + 3).$$

EXAMPLE 10   Factor $12a^2 - 19ab + 5b^2$.

Solution   We must find two numbers whose product is $(12)(5) = 60$ and whose sum is -19. Since the product of the numbers is positive and their sum is negative, the two numbers must both be negative.

The two numbers are -15 and -4 because $(-15)(-4) = 60$ and $-15 + (-4) = -19$. Now rewrite the middle term, $-19ab$, using $-15ab$ and $-4ab$. Then factor by grouping.

$$12a^2 - 19ab + 5b^2 = 12a^2 \overbrace{- 15ab - 4ab}^{-19ab} + 5b^2$$
$$= 3a(4a - 5b) - b(4a - 5b)$$
$$= (4a - 5b)(3a - b)$$

Now Try Exercise 45

> **Understanding Algebra**
>
> If a polynomial is not factorable, it is said to be *prime*.

Try Example 10 again, this time writing $-19ab$ as $-4ab - 15ab$. If you do it correctly, you should get the same factors.

It is important for you to realize that not every trinomial can be factored by the methods presented in this section. A polynomial that cannot be factored (over a specific set of numbers) is called a **prime polynomial**.

EXAMPLE 11   Factor $2x^2 + 6x + 5$.

Solution   When you try to factor this polynomial, you will see that it cannot be factored using either trial and error or grouping. This polynomial is prime over the set of integers.

Now Try Exercise 47

5   Factor Trinomials Using Substitution

Sometimes a more complicated trinomial can be factored by substituting one variable for another. The next three examples illustrate **factoring using substitution**.

EXAMPLE 12   Factor $y^4 - y^2 - 6$.

Solution   If we can rewrite this expression in the form $ax^2 + bx + c$, it will be easier to factor. Since $(y^2)^2 = y^4$, if we substitute x for y^2, the trinomial becomes

$$y^4 - y^2 - 6 = (y^2)^2 - y^2 - 6$$
$$= x^2 - x - 6 \qquad \text{Substitute } x \text{ for } y^2.$$

Now factor $x^2 - x - 6$.

$$= (x + 2)(x - 3)$$

Finally, substitute y^2 in place of x to obtain

$$= (y^2 + 2)(y^2 - 3) \qquad \text{Substitute } y^2 \text{ for } x.$$

Thus, $y^4 - y^2 - 6 = (y^2 + 2)(y^2 - 3)$. Note that x was substituted for y^2, and then y^2 was substituted back for x.

Now Try Exercise 65

EXAMPLE 13 Factor $3z^4 - 17z^2 - 28$.

Solution   Let $x = z^2$. Then the trinomial can be written

$$3z^4 - 17z^2 - 28 = 3(z^2)^2 - 17z^2 - 28$$
$$= 3x^2 - 17x - 28 \qquad \text{Substitute } x \text{ for } z^2.$$
$$= (3x + 4)(x - 7) \qquad \text{Factor.}$$

Now substitute z^2 for x.

$$= (3z^2 + 4)(z^2 - 7) \qquad \text{Substitute } z^2 \text{ for } x.$$

Thus, $3z^4 - 17z^2 - 28 = (3z^2 + 4)(z^2 - 7)$.

Now Try Exercise 69

EXAMPLE 14 Factor $2(x + 5)^2 - 5(x + 5) - 12$.

Solution   Again use a substitution, as in Examples 12 and 13. Substitute $a = x + 5$ in the equation, to obtain

$$2(x + 5)^2 - 5(x + 5) - 12$$
$$= 2a^2 - 5a - 12 \qquad \text{Substitute } a \text{ for } (x + 5).$$

Now factor $2a^2 - 5a - 12$.

$$= (2a + 3)(a - 4)$$

Finally, replace a with $x + 5$ to obtain

$$= [2(x + 5) + 3][(x + 5) - 4] \quad \text{Substitute } (x + 5) \text{ for } a.$$
$$= [2x + 10 + 3][x + 1]$$
$$= (2x + 13)(x + 1)$$

Thus, $2(x + 5)^2 - 5(x + 5) - 12 = (2x + 13)(x + 1)$. Note that a was substituted for $x + 5$, and then $x + 5$ was substituted back for a.

Now Try Exercise 73

In Examples 12 and 13 we used x in our substitution, whereas in Example 14 we used a. The letter selected does not affect the final answer.

EXERCISE SET 5.5

MathXL MathXL®   *MyMathLab* MyMathLab

Warm-Up Exercises

Fill in the blanks with the appropriate word, phrase, or symbol(s) from the following list.

$-1, x - 10$	$a = x - 6$	prime	composite	using substitution	$3x + 7$
medium-sized	trial and error	GCF	-2	$x + 10$	-3

1. A polynomial that cannot be factored is called a _____ polynomial.

2. In any factoring problem, the first step is to factor out the _____.

3. When factoring using the _____ method, you may have to try several possibilities to get the correct factors.

4. In choosing factors for a factoring problem, generally try the _____ factors first.

5. A method of factoring a trinomial that can be factored by substituting one variable for another is called factoring _____.

6. To factor $2(x - 6)^2 - 5(x - 6) - 12$, use the substitution _____.

7. The first step in factoring $-3x^2 + 15x - 12$ is to factor out the GCF _____.

8. If one factor of $x^2 + 3x - 70$ is $x - 7$, the other factor is _____.

Practice the Skills

When factoring a trinomial of the form $ax^2 + bx + c$, determine if the signs will be both $+$, both $-$, or one $+$, one $-$, if:

9. $a > 0, b > 0,$ and $c > 0$

10. $a > 0, b > 0,$ and $c < 0$

11. $a > 0, b < 0,$ and $c < 0$

12. $a > 0, b < 0,$ and $c > 0$

Factor each trinomial completely. If the polynomial is prime, so state.

13. $x^2 + 7x + 12$

14. $a^2 + 2a - 15$

15. $b^2 + 8b - 9$

16. $y^2 - y - 20$

17. $z^2 + 4z + 4$

18. $c^2 - 14c + 49$

19. $r^2 + 24r + 144$

20. $y^2 - 18y + 81$

21. $x^2 + 30x - 64$

22. $x^2 - 11x - 210$

23. $x^2 - 13x - 30$

24. $p^2 - 6p - 24$

25. $-a^2 + 18a - 45$

26. $-x^2 - 15x - 56$

27. $x^2 + xy + 7y^2$

28. $a^2 + 10ab + 24b^2$

29. $-2m^2 - 14m - 20$

30. $-4x^2 - 16x - 12$

31. $4r^2 + 12r - 16$

32. $b^2 - 12bc - 45c^2$

33. $x^3 + 3x^2 - 18x$

34. $x^4 + 14x^3 + 33x^2$

35. $5a^2 - 8a + 3$

36. $5w^2 + 11w + 2$

37. $3x^2 - 3x - 6$

38. $-3b^2 - 14b + 5$

39. $6c^2 - 13c - 63$

40. $30z^2 - 71z + 35$

41. $8b^2 - 2b - 3$

42. $4a^2 + 47a + 33$

43. $6c^2 + 11c - 10$

44. $5z^2 - 16z + 12$

45. $16p^2 - 16pq - 12q^2$

46. $6r^4 + 5r^3 - 4r^2$

47. $4x^2 + 4xy + 9y^2$

48. $6r^2 + 7rs + 8s^2$

49. $18a^2 + 18ab - 8b^2$

50. $18y^2 - 28y - 16$

51. $8x^2 + 30xy - 27y^2$

52. $32x^2 - 22xy + 3y^2$

53. $100b^2 - 90b + 20$

54. $x^5y - 3x^4y - 18x^3y$

55. $a^3b^5 - a^2b^5 - 12ab^5$

56. $a^3b + 2a^2b - 35ab$

57. $3b^4c - 18b^3c^2 + 27b^2c^3$

58. $6p^3q^2 - 24p^2q^3 - 30pq^4$

59. $8m^8n^3 + 4m^7n^4 - 24m^6n^5$

60. $18x^2 + 27x - 35$

61. $30x^2 - x - 20$

62. $36x^2 - 23x - 8$

63. $8x^4y^5 + 24x^3y^5 - 32x^2y^5$

64. $8b^3c^2 + 28b^2c^3 + 12bc^4$

Factor each trinomial completely.

65. $x^4 + x^2 - 6$

66. $y^4 + y^2 - 12$

67. $b^4 + 9b^2 + 20$

68. $c^4 - 8c^2 + 12$

69. $6a^4 + 5a^2 - 25$

70. $(2x + 1)^2 + 2(2x + 1) - 15$

71. $4(x + 1)^2 + 8(x + 1) + 3$

72. $(2y + 3)^2 - (2y + 3) - 6$

73. $6(a + 2)^2 - 7(a + 2) - 5$

74. $6(p - 5)^2 + 11(p - 5) + 3$

75. $x^2y^2 + 9xy + 14$

76. $a^2b^2 + 8ab - 33$

77. $2x^2y^2 - 9xy - 11$

78. $3b^2c^2 - bc - 2$

79. $2y^2(2 - y) - 7y(2 - y) + 5(2 - y)$

80. $2y^2(y + 3) + 13y(y + 3) + 15(y + 3)$

81. $2p^2(p - 4) + 7p(p - 4) + 6(p - 4)$

82. $3x^2(x - 1) + 5x(x - 1) - 2(x - 1)$

83. $a^6 - 7a^3 - 30$

84. $2y^6 - 9y^3 - 5$

85. $x^2(x + 5) + 3x(x + 5) + 2(x + 5)$

86. $x^2(x + 6) - x(x + 6) - 30(x + 6)$

87. $5a^5b^2 - 8a^4b^3 + 3a^3b^4$

88. $2x^4y^6 + 3x^3y^5 - 9x^2y^4$

Problem Solving

Area *In Exercises 89–92, find an expression, in factored form, for the area of the shaded region. See Example 8.*

89.

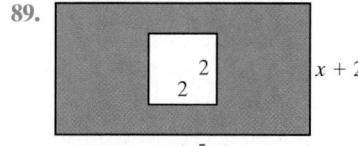

90.

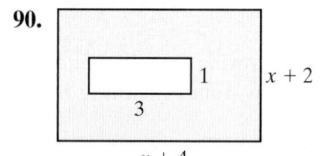

91.

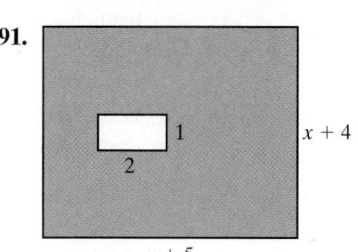

92.

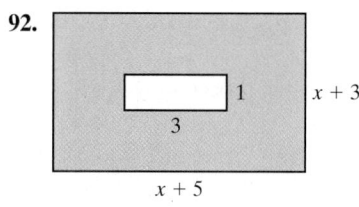

93. If the factors of a polynomial are $(2x + 3y)$ and $(x - 4y)$, find the polynomial. Explain how you determined your answer.

94. If the factors of a polynomial are $3, (4x + 5)$, and $(2x + 3)$, find the polynomial. Explain how you determined your answer.

95. If we know that one factor of the polynomial $x^2 + 4x - 21$ is $x - 3$, how can we find the other factor? Find the other factor.

96. If we know that one factor of the polynomial $x^2 - xy - 6y^2$ is $x - 3y$, how can we find the other factor? Find the other factor.

97. a) Which of the following do you think would be more difficult to factor by trial and error? Explain your answer.

$$30x^2 + 23x - 40 \quad \text{or} \quad 49x^2 - 98x + 13$$

b) Factor both trinomials.

98. a) Which of the following do you think would be more difficult to factor by trial and error? Explain your answer.

$$48x^2 + 26x - 35 \quad \text{or} \quad 35x^2 - 808x + 69$$

b) Factor both trinomials.

99. Find all integer values of b for which $2x^2 + bx - 5$ is factorable.

100. Find all integer values of b for which $3x^2 + bx - 7$ is factorable.

101. If $x^2 + bx + 5$ is factorable, what are the only two possible values of b? Explain.

102. If $x^2 + bx + c$ is factorable and c is a prime number, what are the only two possible values of b? Explain.

Challenge Problems

*Consider the trinomial $ax^2 + bx + c$. If the expression $b^2 - 4ac$, called the **discriminant**, is not a perfect square, the trinomial cannot be factored over the set of integers. **Perfect squares** are 1, 4, 9, 16, 25, 49, and so on. For Exercises 103–106, **a)** find the value of $b^2 - 4ac$. **b)** If $b^2 - 4ac$ is a perfect square, factor the polynomial; if $b^2 - 4ac$ is not a perfect square, indicate that the polynomial cannot be factored.*

103. $x^2 - 8x + 15$

104. $6y^2 - 5y - 6$

105. $x^2 - 4x + 6$

106. $3t^2 - 6t + 2$

107. Construct a trinomial of the form $x^2 + (c + 1)x + c$, where c is a real number, that is factorable.

108. Construct a trinomial of the form $x^2 - (c + 1)x + c$, where c is a real number, that is factorable.

In Exercises 109–114, factor completely. Assume that the variables in the exponents represent positive integers.

109. $4a^{2n} - 4a^n - 15$

110. $a^2(a + b) - 2ab(a + b) - 3b^2(a + b)$

111. $x^2(x + y)^2 - 7xy(x + y)^2 + 12y^2(x + y)^2$

112. $3m^2(m - 2n) - 4mn(m - 2n) - 4n^2(m - 2n)$

113. $x^{2n} + 3x^n - 10$

114. $9r^{4y} + 3r^{2y} - 2$

115. Consider $x^2 + 2x - 8 = (x + 4)(x - 2)$.

a) Explain how you can check this factoring using graphs on your graphing calculator.

b) Check the factoring as you explained in part **a)** to see whether it is correct.

116. Consider $6x^3 - 11x^2 - 10x = x(2x - 5)(3x + 2)$.

a) Explain how you can check this factoring using the TABLE feature of a graphing calculator.

b) Check the factoring as you explained in part **a)** to see whether it is correct.

Concept/Writing Exercises

117. When factoring any trinomial, what should the first step always be?

118. On a test, Tom Phu wrote the following factoring and did not receive full credit. Explain why Tom's factoring is not complete.

$$15x^2 - 21x - 18 = (5x + 3)(3x - 6)$$

119. a) Explain the step-by-step procedure to factor $6x^2 + x - 12$.

b) Factor $6x^2 + x - 12$ using the procedure you explained in part **a)**.

120. a) Explain the step-by-step procedure to factor $8x^2 - 20x - 12$.

b) Factor $8x^2 - 20x - 12$ using the procedure you explained in part **a)**.

121. Has $2x^2 + 8x + 6 = (x + 3)(2x + 2)$ been factored completely? If not, give the complete factorization. Explain.

122. Has $x^3 - 3x^2 - 10x = (x^2 + 2x)(x - 5)$ been factored completely? If not, give the complete factorization. Explain.

123. Has $3x^3 + 6x^2 - 24x = x(x + 4)(3x - 6)$ been factored completely? If not, give the complete factorization. Explain.

124. Has $x^4 + 11x^3 + 30x^2 = x^2(x + 5)(x + 6)$ been factored completely? If not, give the complete factorization. Explain.

Cumulative Review Exercises

[2.2] **125.** Solve $F = \dfrac{9}{5}C + 32$ for C.

[3.3] **126.** Graph $y = -3x + 4$.

[4.5] **127.** Evaluate the determinant $\begin{vmatrix} 3 & -2 & -1 \\ 2 & 3 & -2 \\ 1 & -4 & 1 \end{vmatrix}$.

[5.2] **128.** Multiply $[(x + y) + 6]^2$.

[5.3] **129.** Factor $2x^3 + 4x^2 - 5x - 10$.

5.6 Special Factoring Formulas

1 Factor the difference of two squares.

2 Factor perfect square trinomials.

3 Factor the sum and difference of two cubes.

Understanding Algebra

$a^2 - b^2$ is the *difference of two squares*, since the *minus sign* is between the two squares a^2 and b^2.

In this section we present some special formulas for factoring the difference of two squares, perfect square trinomials, and the sum and difference of two cubes. It is important to memorize these formulas.

1 Factor the Difference of Two Squares

The expression $x^2 - 9$ is an example of the difference of two squares.

$$x^2 - 9 = (x)^2 - (3)^2$$

To factor the difference of two squares, it is convenient to use the **difference of two squares formula**.

> **Difference of Two Squares**
>
> $$a^2 - b^2 = (a + b)(a - b)$$

EXAMPLE 1 Factor the following differences of squares.

a) $x^2 - 16$ **b)** $25x^2 - 36y^2$

Solution   Rewrite each expression as a difference of two squares. Then use the formula.

a) $x^2 - 16 = (x)^2 - (4)^2$
$$= (x + 4)(x - 4)$$

b) $25x^2 - 36y^2 = (5x)^2 - (6y)^2$
$$= (5x + 6y)(5x - 6y)$$

Now Try Exercise 13

EXAMPLE 2 Factor the following differences of squares.

a) $x^6 - y^4$ **b)** $2z^4 - 162x^6$

Solution   Rewrite each expression as a difference of two squares. Then use the formula.

a) $x^6 - y^4 = (x^3)^2 - (y^2)^2$
$$= (x^3 + y^2)(x^3 - y^2)$$

b) $2z^4 - 162x^6 = 2(z^4 - 81x^6)$
$$= 2[(z^2)^2 - (9x^3)^2]$$
$$= 2(z^2 + 9x^3)(z^2 - 9x^3)$$

Now Try Exercise 21

EXAMPLE 3 Factor $x^4 - 81y^4$.

Solution
$$x^4 - 81y^4 = (x^2)^2 - (9y^2)^2$$
$$= (x^2 + 9y^2)(x^2 - 9y^2)$$

Note that $(x^2 - 9y^2)$ is also a difference of two squares. We use the difference of two squares formula a second time to obtain

$$= (x^2 + 9y^2)[(x)^2 - (3y)^2]$$
$$= (x^2 + 9y^2)(x + 3y)(x - 3y)$$

Now Try Exercise 69

EXAMPLE 4 Factor $(x - 5)^2 - 4$ using the formula for the difference of two squares.

Solution   First we express $(x - 5)^2 - 4$ as a difference of two squares.
$$(x - 5)^2 - 4 = (x - 5)^2 - 2^2$$
$$= [(x - 5) + 2][(x - 5) - 2]$$
$$= (x - 3)(x - 7)$$

Now Try Exercise 25

Understanding Algebra

The sum of two squares $a^2 + b^2$ cannot be factored using real numbers. Therefore, the following are examples of polynomials that cannot be factored using real numbers.

$$x^2 + 9$$
$$4y^2 + 25$$
$$16z^2 + 49$$

Note: *It is not possible to factor the sum of two squares of the form $a^2 + b^2$ using real numbers.*

For example, it is not possible to factor $x^2 + 4$ since $x^2 + 4 = x^2 + 2^2$, which is a sum of two squares.

2 Factor Perfect Square Trinomials

In Section 5.2 we saw that
$$(a + b)^2 = a^2 + 2ab + b^2$$
$$(a - b)^2 = a^2 - 2ab + b^2$$

If we reverse the left and right sides of these two formulas, we obtain two **special factoring formulas**.

Perfect Square Trinomials
$a^2 + 2ab + b^2 = (a + b)^2$
$a^2 - 2ab + b^2 = (a - b)^2$

Understanding Algebra

A *perfect square trinomial* occurs when you *square* a *binomial*.

$$\underbrace{(x + 4)^2}_{\text{square of a binomial}} = \underbrace{x^2 + 8x + 16}_{\text{perfect square trinomial}}$$

These two trinomials are called **perfect square trinomials** since each is the square of a binomial. *To be a perfect square trinomial, the first and last terms must be the squares of some expression and the middle term must be twice the product of those two expressions.*

If a trinomial is a perfect square trinomial, you can factor it using the formulas given above.

Examples of Perfect Square Trinomials
$$y^2 + 6y + 9 \quad \text{or} \quad y^2 + 2(y)(3) + 3^2$$
$$9a^2b^2 - 24ab + 16 \quad \text{or} \quad (3ab)^2 - 2(3ab)(4) + 4^2$$
$$(r + s)^2 + 10(r + s) + 25 \quad \text{or} \quad (r + s)^2 + 2(r + s)(5) + 5^2$$

Now let's factor some perfect square trinomials.

EXAMPLE 5 Factor $x^2 - 8x + 16$.

Solution   Since the first term, x^2, and the last term, 16, or 4^2, are squares, this trinomial might be a perfect square trinomial. To determine whether it is, take twice the product of x and 4 to see if you obtain $8x$.

$$2(x)(4) = 8x$$

Since $8x$ is the middle term and since the sign of the middle term is negative, factor as follows:

$$x^2 - 8x + 16 = (x - 4)^2$$

Now Try Exercise 29

EXAMPLE 6   Factor $9x^4 - 12x^2 + 4$. —————————————

Solution   The first term is a square, $(3x^2)^2$, as is the last term, 2^2. Since $2(3x^2)(2) = 12x^2$, we factor as follows:

$$9x^4 - 12x^2 + 4 = (3x^2 - 2)^2$$

Now Try Exercise 37

EXAMPLE 7   Factor $(a + b)^2 + 12(a + b) + 36$. —————————

Solution   The first term, $(a + b)^2$, is a square. The last term, 36 or 6^2, is a square. The middle term is $2(a + b)(6) = 12(a + b)$. Therefore, this is a perfect square trinomial. Thus,

$$(a + b)^2 + 12(a + b) + 36 = [(a + b) + 6]^2 = (a + b + 6)^2$$

Now Try Exercise 39

EXAMPLE 8   Factor $x^2 - 6x + 9 - y^2$. —————————————

Solution   Since $x^2 - 6x + 9$ is a perfect square trinomial, which can be expressed as $(x - 3)^2$, we write

$$(x - 3)^2 - y^2$$

Now $(x - 3)^2 - y^2$ is a difference of two squares; therefore

$$(x - 3)^2 - y^2 = [(x - 3) + y][(x - 3) - y]$$
$$= (x - 3 + y)(x - 3 - y)$$

Thus, $x^2 - 6x + 9 - y^2 = (x - 3 + y)(x - 3 - y)$.

Now Try Exercise 45

The polynomial in Example 8 has four terms. In Section 5.4 we learned to factor polynomials with four terms by grouping. If you study Example 8, you will see that no matter how you arrange the four terms they cannot be arranged so that the first two terms have a common factor and the last two terms have a common factor.

> **Helpful Hint**
>
> Whenever a polynomial of four terms cannot be factored by grouping, try to rewrite three of the terms as the square of a binomial and then factor using the difference of two squares formula.

EXAMPLE 9   Factor $4a^2 + 12ab + 9b^2 - 25$. —————————

Solution   We first notice that this polynomial of four terms cannot be factored by grouping. We next see that the first three terms of the polynomial can be expressed as the square of a binomial. We complete our factoring using the difference of two squares.

$$4a^2 + 12ab + 9b^2 - 25 = (2a + 3b)^2 - 5^2$$
$$= [(2a + 3b) + 5][(2a + 3b) - 5]$$
$$= (2a + 3b + 5)(2a + 3b - 5)$$

Now Try Exercise 47

3 Factor the Sum and Difference of Two Cubes

Earlier in this section we factored the difference of two squares. Now we will factor the sum and difference of two cubes. Consider the product of $(a + b)(a^2 - ab + b^2)$.

$$
\begin{array}{r}
a^2 - ab + b^2 \\
a + b \\
\hline
a^2b - ab^2 + b^3 \\
a^3 - a^2b + ab^2 \\
\hline
a^3 \qquad\qquad + b^3
\end{array}
$$

Thus, $a^3 + b^3 = (a + b)(a^2 - ab + b^2)$. Using multiplication, we can also show that $a^3 - b^3 = (a - b)(a^2 + ab + b^2)$. Formulas for factoring **the sum and the difference of two cubes** appear in the following boxes.

> ### Understanding Algebra
>
> $a^3 + b^3$ is the *sum* of two cubes since the *plus sign* is between the two cubes a^3 and b^3.

Sum of Two Cubes

$$a^3 + b^3 = (a + b)(a^2 - ab + b^2)$$

Difference of Two Cubes

$$a^3 - b^3 = (a - b)(a^2 + ab + b^2)$$

EXAMPLE 10 Factor the sum of cubes $x^3 + 64$.

Solution   Rewrite $x^3 + 64$ as a sum of two cubes, $x^3 + 4^3$. Let x correspond to a and 4 to b. Then factor using the sum of two cubes formula.

$$
\begin{array}{ccccccccc}
a^3 & + & b^3 & = & (a & + & b)(a^2 & - & a\ b + b^2) \\
\downarrow & & \downarrow & & \downarrow & & \downarrow\ \downarrow & & \downarrow\ \downarrow \qquad \downarrow \\
x^3 & + & 4^3 & = & (x & + & 4)[x^2 & - & x(4) + 4^2] \\
\end{array}
$$
$$= (x + 4)(x^2 - 4x + 16)$$

Thus, $x^3 + 64 = (x + 4)(x^2 - 4x + 16)$.

Now Try Exercise 51

EXAMPLE 11 Factor the difference of cubes $27x^3 - 8y^6$.

Solution   We first observe that $27x^3$ and $8y^6$ have no common factors other than 1. Since we can express both $27x^3$ and $8y^6$ as cubes, we can factor using the difference of two cubes formula.

$$27x^3 - 8y^6 = (3x)^3 - (2y^2)^3$$
$$= (3x - 2y^2)[(3x)^2 + (3x)(2y^2) + (2y^2)^2]$$
$$= (3x - 2y^2)(9x^2 + 6xy^2 + 4y^4)$$

Thus, $27x^3 - 8y^6 = (3x - 2y^2)(9x^2 + 6xy^2 + 4y^4)$.

Now Try Exercise 57

EXAMPLE 12 Factor $8y^3 - 64x^6$.

Solution   First factor out 8, which is the GCF.

$$8y^3 - 64x^6 = 8(y^3 - 8x^6)$$

Next factor $y^3 - 8x^6$ by writing it as a difference of two cubes.

$$8(y^3 - 8x^6) = 8[(y)^3 - (2x^2)^3]$$
$$= 8(y - 2x^2)[y^2 + y(2x^2) + (2x^2)^2]$$
$$= 8(y - 2x^2)(y^2 + 2x^2y + 4x^4)$$

Thus, $8y^3 - 64x^6 = 8(y - 2x^2)(y^2 + 2x^2y + 4x^4)$.

Now Try Exercise 59

EXAMPLE 13 Factor $(x - 2)^3 + 125$.

Solution   Write $(x - 2)^3 + 125$ as a sum of two cubes, then use the sum of two cubes formula to factor.

$$(x - 2)^3 + (5)^3 = [(x - 2) + 5][(x - 2)^2 - (x - 2)(5) + (5)^2]$$
$$= (x - 2 + 5)(x^2 - 4x + 4 - 5x + 10 + 25)$$
$$= (x + 3)(x^2 - 9x + 39)$$

Now Try Exercise 65

Helpful Hint

The square of a binomial has a 2 as part of the middle term of the trinomial.

$$(a + b)^2 = a^2 + 2ab + b^2$$
$$(a - b)^2 = a^2 - 2ab + b^2$$

The sum or the difference of two cubes has a factor similar to the trinomial in the square of the binomial. However, the middle term does not contain a 2.

$$a^3 + b^3 = (a + b)(a^2 - ab + b^2)$$
$$a^3 - b^3 = (a - b)(a^2 + \underbrace{ab}_{\text{not } 2ab} + b^2)$$

EXAMPLE 14 **Volume** Using the cubes in **Figure 5.16,** find an expression, in factored form, for the difference of the volumes.

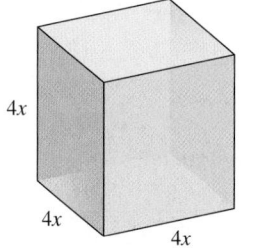
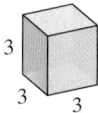

FIGURE 5.16   Large cube   Small cube

Solution   To find the difference of the volumes, subtract the volume of the small cube from the volume of the large cube.

Volume of large cube $= (4x)^3$

Volume of small cube $= 3^3$

Difference of the volumes $= (4x)^3 - 3^3$	Subtract volumes.
$= (4x - 3)[(4x)^2 + (4x)3 + 3^2]$	Factor.
$= (4x - 3)(16x^2 + 12x + 9)$	Simplify.

The difference of the volumes for the two cubes is $(4x - 3)(16x^2 + 12x + 9)$.

Now Try Exercise 87

EXERCISE SET 5.6

MathXL®   MyMathLab

Warm-up Exercises

Fill in the blanks with the appropriate word, phrase, or symbol(s) from the following list.

perfect	$(9x + 4y)$	$(x - 7)^2$	5	$(x + 7)^2$	difference	$(a - b)^2$
two	10	$(a + b)^2$	-10	two	$(9x + 4y)(9x - 4y)$	
sum	perfect square	$(a + b)(a^2 - ab + b^2)$	$(a^2 - b^2)$	$(a^2 + b^2)$	$(a - b)(a^2 + ab + b^2)$	

1. The formula $a^2 - b^2 = (a + b)(a - b)$ is the formula for factoring the difference of _____ squares.

2. $81x^2 - 16y^2$ factors into _____.

3. The trinomial $a^2 + 2ab + b^2$ is called a _____ trinomial.

4. $a^2 + 2ab + b^2$ factors into _____.

5. The perfect square trinomial $x^2 + 14x + 49$ factors into _____.

6. In order for $x^2 + bx + 25$ to be a perfect square trinomial, the value for b is _____.

7. The expression $a^3 + b^3$ is the _____ of two cubes.

8. The expression $a^3 - b^3$ is the _____ of two cubes.

9. $a^3 + b^3$ factors into _____.

10. $a^3 - b^3$ factors into _____.

Practice the Skills

Use the difference of two squares formula or the perfect square trinomial formula to factor each polynomial.

11. $x^2 - 81$

12. $x^2 - 121$

13. $a^2 - 100$

14. $1 - 16x^2$

15. $1 - 49b^2$

16. $x^2 - 81z^2$

17. $25 - 16y^4$

18. $49 - 144b^4$

19. $\dfrac{1}{100} - y^2$

20. $\dfrac{1}{36} - z^2$

21. $x^2y^2 - 121c^2$

22. $6a^2c^2 - 24x^2y^2$

23. $0.04x^2 - 0.09$

24. $0.16p^2 - 0.81q^2$

25. $36 - (x - 6)^2$

26. $144 - (a + b)^2$

27. $a^2 - (3b + 2)^2$

28. $(2c + 3)^2 - 9$

29. $x^2 + 10x + 25$

30. $b^2 - 18b + 81$

31. $49 - 14t + t^2$

32. $4 + 4a + a^2$

33. $36p^2q^2 + 12pq + 1$

34. $9x^2 - 30xy + 25y^2$

35. $0.81x^2 - 0.36x + 0.04$

36. $0.25x^2 - 0.40x + 0.16$

37. $y^4 + 4y^2 + 4$

38. $b^4 - 16b^2 + 64$

39. $(a + b)^2 + 6(a + b) + 9$

40. $(x - y)^2 + 2(x - y) + 1$

41. $(y - 3)^2 + 8(y - 3) + 16$

42. $a^4 - 2a^2b^2 + b^4$

43. $x^2 + 6x + 9 - y^2$

44. $p^2 + 2pq + q^2 - 25r^2$

45. $25 - (x^2 + 4x + 4)$

46. $49 - (c^2 - 8c + 16)$

47. $9a^2 - 12ab + 4b^2 - 9$

48. $(4a - 3b)^2 - (2a + 5b)^2$

49. $y^4 - 6y^2 + 9$

50. $z^6 + 14z^3 + 49$

Factor using the sum or difference of two cubes formula.

51. $a^3 + 125$

52. $x^3 - 8$

53. $64 - a^3$

54. $27 - b^3$

55. $p^3 - 27a^3$

56. $w^3 - 216$

57. $27y^3 - 8x^3$

58. $5x^3 + 40y^3$

59. $16a^3 - 54b^3$

60. $2b^3 - 250c^3$

61. $x^6 + y^9$

62. $16x^6 - 250y^3$

63. $(x + 1)^3 + 1$

64. $(a - 3)^3 + 8$

65. $(a - b)^3 - 27$

66. $(2x + y)^3 - 64$

67. $b^3 - (b + 3)^3$

68. $(m - n)^3 - (m + n)^3$

Factor using a special factoring formula.

69. $a^4 - 4b^4$

70. $121y^4 - 49x^2$

71. $49 - 64x^2y^2$

72. $25y^2 - 81x^2$

73. $(x + y)^2 - 16$

74. $25x^4 - 81y^6$

75. $x^3 - 64$

76. $3a^2 - 36a + 108$

77. $9x^2y^2 + 24xy + 16$

78. $a^4 + 12a^2 + 36$

79. $a^4 + 2a^2b^2 + b^4$

80. $8y^3 - 125x^6$

81. $x^2 - 2x + 1 - y^2$

82. $16x^2 - 8xy + y^2 - 4$

83. $(x + y)^3 + 1$

84. $4r^2 + 4rs + s^2 - 9$

85. $(m + n)^2 - (2m - n)^2$

86. $(r + p)^3 + (r - p)^3$

Problem Solving

Volume   *In Exercises 87–90, find an expression, in factored form, for the difference of the volumes of the two cubes. See Example 14.*

87.
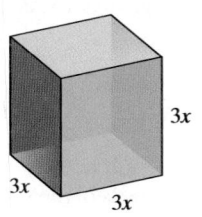 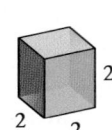

88.
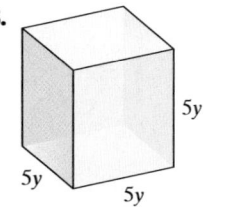 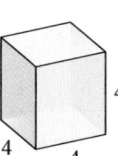

89.
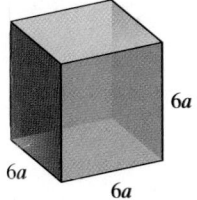 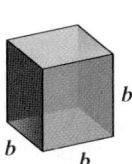

90.
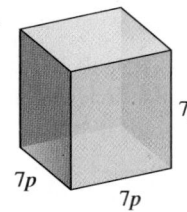 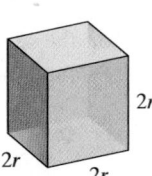

Volume   *In Exercises 91 and 92, find an expression, in factored form, for the sum of the volumes of the two cubes.*

91.
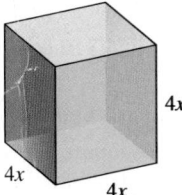 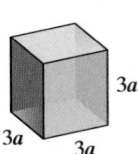

92.
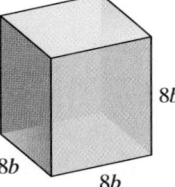 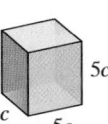

Area or Volume   *In Exercises 93–97,* **a)** *find the area or volume of the shaded figure by subtracting the smaller area or volume from the larger. The formula to find the area or volume is given under the figure.* **b)** *Write the expression obtained in part* **a)** *in factored form. Part of the GCF in Exercises 94, 96, and 97 is π.*

93.   Squares

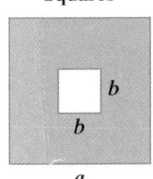

$$A = s^2$$

94.   Circles

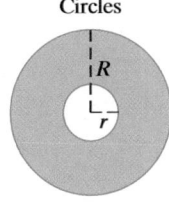

$$A = \pi r^2$$

95.   Rectangular solid

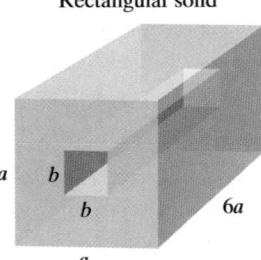

$$V = lwh$$

96.   Cylinder

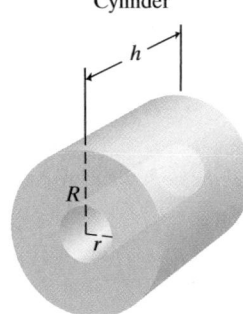

$$V = \pi r^2 h$$

97.   Sphere

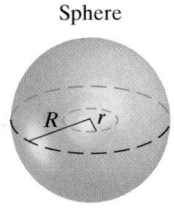

$$V = \frac{4}{3}\pi r^3$$

98. Area and Volume A circular hole is cut from a cube of wood, as shown in the figure.

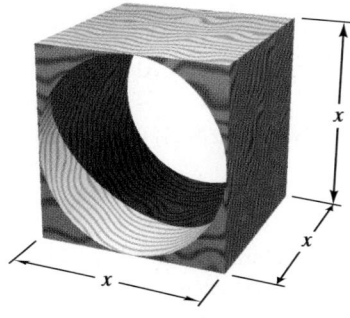

a) Write an expression in factored form in terms of x for the cross-sectional area of the remaining wood.

b) Write an expression in factored form in terms of x for the volume of the remaining wood.

99. Find two values of b that will make $4x^2 + bx + 9$ a perfect square trinomial. Explain how you determined your answer.

100. Find two values of c that will make $16x^2 + cx + 4$ a perfect square trinomial. Explain how you determined your answer.

101. Find the value of c that will make $25x^2 + 20x + c$ a perfect square trinomial. Explain how you determined your answer.

102. Find the value of d that will make $49x^2 - 42x + d$ a perfect square trinomial. Explain how you determined your answer.

103. Area A formula for the area of a square is $A = s^2$, where s is a side. Suppose the area of a square is as given below.

$$A(x) = 25x^2 - 30x + 9$$

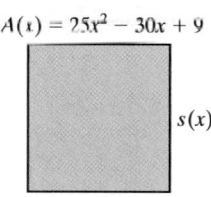

$s(x)$

a) Explain how to find the length of side x, $s(x)$.

b) Find $s(x)$.

c) Find $s(2)$.

104. Area The formula for the area of a circle is $A = \pi r^2$, where r is the radius. Suppose the area of a circle is as given below.

$$A(x) = 9\pi x^2 + 12\pi x + 4\pi$$

$r(x)$

a) Explain how to find the radius, $r(x)$.

b) Find $r(x)$.

c) Find $r(4)$.

105. Factor $x^4 + 64$ by writing the expression as $(x^4 + 16x^2 + 64) - 16x^2$, which is a difference of two squares.

Factor completely.

111. $64x^{4a} - 9y^{6a}$

113. $a^{2n} - 16a^n + 64$

115. $x^{3n} - 8$

106. Factor $x^4 + 4$ by adding and subtracting $4x^2$. (See Exercise 105.)

107. If $P(x) = x^2$, use the difference of two squares to simplify $P(a + h) - P(a)$.

108. If $P(x) = x^2$, use the difference of two squares to simplify $P(a + 1) - P(a)$.

109. Sum of Areas The figure shows how we *complete the square*. The sum of the areas of the three parts of the square that are shaded in blue and purple is

$$x^2 + 4x + 4x \quad \text{or} \quad x^2 + 8x$$

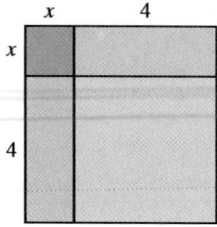

a) Find the area of the fourth part (in pink) to complete the square.

b) Find the sum of the areas of the four parts of the square.

c) This process has resulted in a perfect square trinomial in part **b)**. Write this perfect square trinomial as the square of a binomial.

110. Factor $(m - n)^3 - (9 - n)^3$.

112. $16p^{8w} - 49p^{6w}$

114. $144r^{8k} + 48r^{4k} + 4$

116. $27x^{3m} + 64x^{6m}$

In Exercises 117 and 118, indicate if the factoring is correct or not correct. Explain your answers.

117. $2x^2 - 18 \overset{?}{=} 2(x + 3)(x - 3)$

118. $8x^3 + 27 \overset{?}{=} 2x(4x^2 + 5x + 9)$

Concept/Writing Exercises

119. a) Explain how to factor the difference of two squares.

    b) Factor $x^2 - 16$ using the procedure you explained in part **a)**.

120. a) Explain how to factor a perfect square trinomial.

    b) Factor $x^2 + 12x + 36$ using the procedure you explained in part **a)**.

121. Give the formula for factoring the sum of two cubes.

122. Give the formula for factoring the difference of two cubes.

123. Explain why a sum of two squares, $a^2 + b^2$, cannot be factored over the set of real numbers.

124. Explain how to determine whether a trinomial is a perfect square trinomial.

125. Is $x^2 + 14x - 49 = (x + 7)(x - 7)$ factored correctly? Explain.

126. Is $x^2 + 14x + 49 = (x + 7)^2$ factored correctly? Explain.

127. Is $x^2 - 81 = (x - 9)^2$ factored correctly? Explain.

128. Is $x^2 - 64 = (x + 8)(x - 8)$ factored correctly? Explain.

Challenge Problem

129. The expression $x^6 - 1$ can be factored using either the difference of two squares or the difference of two cubes. At first the factors do not appear the same. But with a little algebraic manipulation they can be shown to be equal. Factor $x^6 - 1$ using **a)** the difference of two squares and **b)** the difference of two cubes. **c)** Show that these two answers are equal by factoring the answer obtained in part **a)** completely. Then multiply the two binomials together and the two trinomials together.

Group Activity

Discuss and answer Exercise 130 as a group.

130. Later in the book we will need to construct perfect square trinomials. Examine some perfect square trinomials with a leading coefficient of 1.

    a) Explain how b and c are related if the trinomial $x^2 + bx + c$ is a perfect square trinomial.

    b) Construct a perfect square trinomial if the first two terms are $x^2 + 6x$.

    c) Construct a perfect square trinomial if the first two terms are $x^2 - 10x$.

    d) Construct a perfect square trinomial if the first two terms are $x^2 - 14x$.

Cumulative Review Exercises

[2.1] **131.** Simplify $-2[3x - (2y - 1) - 5x] + 3y$.

[3.6] **132.** If $f(x) = x^2 - 3x + 6$ and $g(x) = 5x - 2$, find $(g - f)(-1)$.

[4.4] **133. Angles** A right angle is divided into three smaller angles. The largest of the three angles is twice the smallest. The remaining angle is $10°$ greater than the smallest angle. Find the measure of each angle.

[5.4] **134.** Factor out the greatest common factor of $45y^{12} + 60y^{10}$.

135. Factor $12x^2 - 9xy + 4xy - 3y^2$.

5.7 A General Review of Factoring

1 Factor polynomials using a combination of techniques

1 Factor Polynomials Using a Combination of Techniques

We now present a general procedure to factor any polynomial.

> ### To Factor a Polynomial
>
> 1. Determine whether all the terms in the polynomial have a greatest common factor other than 1. If so, factor out the GCF.
> 2. If the polynomial has two terms, determine whether it is a difference of two squares or a sum or difference of two cubes. If so, factor using the appropriate formula from Section 5.6.
> 3. If the polynomial has three terms, determine whether it is a perfect square trinomial. If so, factor accordingly. If it is not, factor the trinomial using trial and error, grouping, or substitution as explained in Section 5.5.
> 4. If the polynomial has more than three terms, try factoring by grouping. If that does not work, see if three of the terms are the square of a binomial.
> 5. As a final step, examine your factored polynomial to see if any factors listed have a common factor and can be factored further. If you find a common factor, factor it out at this point.
> 6. Check the answer by multiplying the factors.

The following examples illustrate how to use the procedure.

EXAMPLE 1 Factor $2x^4 - 50x^2$. ─────────────

Solution   First, check for a greatest common factor other than 1. Since $2x^2$ is common to both terms, factor it out.

$$2x^4 - 50x^2 = 2x^2(x^2 - 25) = 2x^2(x + 5)(x - 5)$$

Note that $x^2 - 25$ is factored as a difference of two squares.

Now Try Exercise 3

EXAMPLE 2 Factor $3x^2y^2 - 24xy^2 + 48y^2$.

Solution   Begin by factoring the GCF, $3y^2$, from each term.

$$3x^2y^2 - 24xy^2 + 48y^2 = 3y^2(x^2 - 8x + 16) = 3y^2(x - 4)^2$$

Note that $x^2 - 8x + 16$ is a perfect square trinomial. If you did not recognize this, you would still obtain the correct answer by factoring the trinomial into $(x - 4)(x - 4)$.

Now Try Exercise 27

EXAMPLE 3 Factor $24x^2 - 6xy + 40xy - 10y^2$.

Solution   As always, begin by determining if all the terms in the polynomial have a common factor. In this example, 2 is common to all terms. Factor out the 2; then factor the remaining four-term polynomial by grouping.

$$24x^2 - 6xy + 40xy - 10y^2 = 2(12x^2 - 3xy + 20xy - 5y^2)$$
$$= 2[3x(4x - y) + 5y(4x - y)]$$
$$= 2(4x - y)(3x + 5y)$$

Now Try Exercise 31

EXAMPLE 4 Factor $12a^2b - 18ab + 24b$.

Solution $$12a^2b - 18ab + 24b = 6b(2a^2 - 3a + 4)$$

Since $2a^2 - 3a + 4$ cannot be factored, we stop here.

Now Try Exercise 7

EXAMPLE 5 Factor $2x^4y + 54xy$.

Solution $$2x^4y + 54xy = 2xy(x^3 + 27)$$
$$= 2xy(x + 3)(x^2 - 3x + 9)$$

Note that $x^3 + 27$ was factored as a sum of two cubes.

Now Try Exercise 19

EXAMPLE 6 Factor $3x^2 - 18x + 27 - 3y^2$.

Solution   Factor out 3 from all four terms.

$$3x^2 - 18x + 27 - 3y^2 = 3(x^2 - 6x + 9 - y^2)$$

Now try factoring by grouping. Since the four terms within parentheses cannot be factored by grouping, check to see whether any three of the terms can be written as the square of a binomial. Since this can be done, express $x^2 - 6x + 9$ as $(x - 3)^2$ and then use the difference of two squares formula. Thus,

$$3x^2 - 18x + 27 - 3y^2 = 3[(x - 3)^2 - y^2]$$
$$= 3[(x - 3 + y)(x - 3 - y)]$$
$$= 3(x - 3 + y)(x - 3 - y)$$

Now Try Exercise 43

Helpful Hint

Study Tip

In this section, we have reviewed all the techniques for factoring expressions. If you are still having difficulty with factoring, you should study the material in Sections 5.4–5.6 again.

EXERCISE SET 5.7

Math XL   MyMathLab
MathXL®   MyMathLab

Warm-Up Exercises

Fill in the blanks with the appropriate word, phrase, or symbol(s) from the following list.

-1 -5 $21x^2y^3$ $7x^2y^3$ $7xy$

1. The first step to factor $-5x^3 + 40y^3$ is to factor out _____.

2. The greatest common factor of $14x^2y + 21xy^3$ is _____.

Practice the Skills

Factor each polynomial completely.

3. $3x^2 - 75$

4. $4x^2 - 24x + 36$

5. $10s^2 + 19s - 15$

6. $-8r^2 + 30r - 18$

7. $6x^3y^2 + 10x^2y^3 + 14x^2y^2$

8. $24m^3n - 12m^2n^2 + 16mn^3$

9. $0.8x^2 - 0.072$

10. $0.5x^2 - 0.32$

11. $6x^5 - 54x$

12. $7x^2y^2z^2 - 28x^2y^2$

13. $3x^6 - 3x^5 + 12x^5 - 12x^4$

14. $2x^2y^2 + 6xy^2 - 10xy^2 - 30y^2$

15. $5x^4y^2 + 20x^3y^2 + 15x^3y^2 + 60x^2y^2$

16. $6x^2 - 15x - 9$

17. $x^4 - x^2y^2$

18. $4x^3 + 108$

19. $x^7y^2 - x^4y^2$

20. $x^4 - 81$

21. $x^5 - 16x$

22. $20x^2y^2 + 55xy^2 - 15y^2$

23. $4x^6 + 32y^3$

24. $8x^4 - 4x^3 - 4x^3 + 2x^2$

25. $5(a + b)^2 - 20$

26. $12x^3y^2 + 4x^2y^2 - 40xy^2$

27. $6x^2 + 36xy + 54y^2$

28. $3x^2 - 30x + 75$

29. $(x + 2)^2 - 4$

30. $7y^4 - 63x^6$

31. $6x^2 + 24xy - 3xy - 12y^2$

32. $pq + 8q + pr + 8r$

33. $(y + 5)^2 + 4(y + 5) + 4$

34. $(x + 1)^2 - (x + 1) - 6$

35. $b^4 + 2b^2 + 1$

36. $45a^4 - 30a^3 + 5a^2$

37. $x^3 + \dfrac{1}{64}$

38. $27y^3 - \dfrac{1}{8}$

39. $6y^3 + 14y^2 + 4y$

40. $3x^3 + 2x^2 - 27x - 18$

41. $a^3b - 81ab^3$

42. $x^6 + y^6$

43. $49 - (x^2 + 2xy + y^2)$

44. $x^2 - 2xy + y^2 - 25$

45. $24x^2 - 34x + 12$

46. $40x^2 + 52x - 12$

47. $18x^2 + 39x - 15$

48. $7(a - b)^2 + 4(a - b) - 3$

49. $x^4 - 16$

50. $(x + 4)^2 - 12(x + 4) + 36$

51. $5bc - 10cx - 7by + 14xy$

52. $16y^4 - 9y^2$

53. $3x^4 - x^2 - 4$

54. $x^2 + 16x + 64 - 100y^2$

55. $z^2 - (x^2 - 12x + 36)$

56. $7a^3 + 56$

57. $2(y + 4)^2 + 5(y + 4) - 12$

58. $x^6 + 15x^3 + 54$

59. $a^2 + 12ab + 36b^2 - 16c^2$

60. $y^3 - y^5$

61. $10x^4y + 25x^3y - 15x^2y$

62. $4x^2y^2 + 12xy + 9$

63. $x^4 - 2x^2y^2 + y^4$

64. $12r^2s^2 + rs - 1$

Problem Solving

*Match Exercises 65–72 with the items labeled **a)** through **h)** on the right.*

65. $a^2 + b^2$

66. $a^2 - b^2$

67. $a^2 + 2ab + b^2$

68. $a^3 + b^3$

69. $a^3 - b^3$

70. $a^2 - 2ab + b^2$

71. a factor of $a^3 + b^3$

72. a factor of $a^3 - b^3$

a) $(a + b)(a^2 - ab + b^2)$ **b)** $(a - b)^2$

c) $a^2 - ab + b^2$ **d)** $(a + b)^2$

e) not factorable   **f)** $(a - b)(a^2 + ab + b^2)$

g) $(a + b)(a - b)$ **h)** $a^2 + ab + b^2$

Perimeter   *In Exercises 73 and 74, find an expression, in factored form, for the perimeter of each figure.*

73.

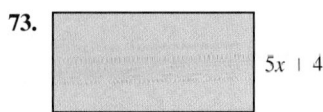

$5x + 4$

$x^2 + 2$

74.

$7x + 13$ $5x + 12$

$x^2 + 11$

Area   *In Exercises 75–78, find an expression, in factored form, for the area of the shaded region for each figure.*

75.

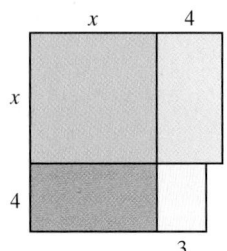

76.

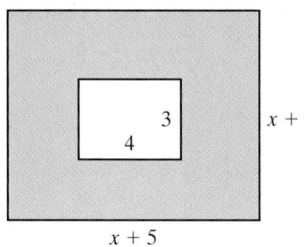

77.

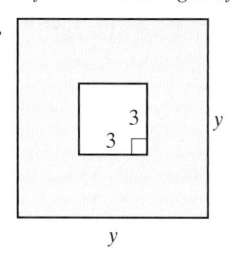

78.

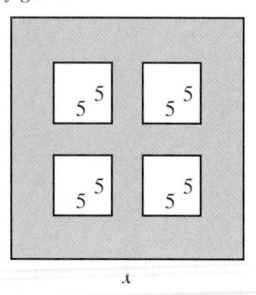

Volume   *In Exercises 79 and 80, find an expression, in factored form, for the difference in the volumes of the cubes.*

79.

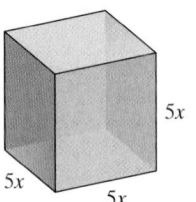

80.

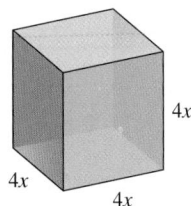

Area   *In Exercises 81–84,* **a)** *write an expression for the shaded area of the figure, and* **b)** *write the expression in factored form.*

81.

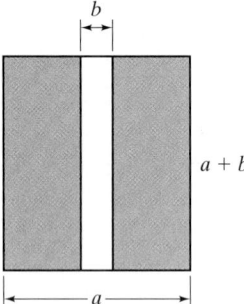

82.

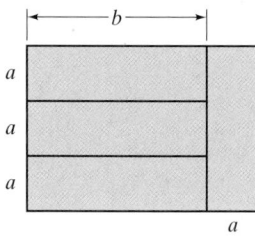

83.

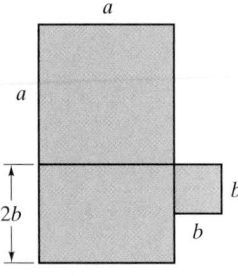

84.

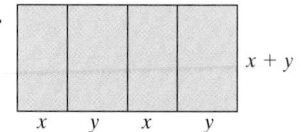

85. Surface Area

  a) Write an expression for the surface area of the four sides of the box shown (omit top and bottom).

  b) Write the expression in factored form.

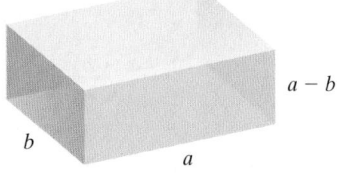

86. Explain how the formula for factoring the *difference* of two cubes can be used to factor $x^3 + 27$.

87. a) Explain how to construct a perfect square trinomial.

  b) Construct a perfect square trinomial and then show its factors.

Challenge Problems

We have worked only with positive integer exponents in this chapter. However, fractional exponents and negative integer exponents may also be factored out of an expression. The expressions below are not polynomials. **a)** *Factor out the variable with the lowest (or most negative) exponent from each expression. (Fractional exponents are discussed in Section 7.2.)* **b)** *Factor completely.*

88. $x^{-2} - 5x^{-3} + 6x^{-4}$, factor out x^{-4}

89. $x^{-3} - 2x^{-4} - 3x^{-5}$, factor out x^{-5}

90. $x^{5/2} + 3x^{3/2} - 4x^{1/2}$, factor out $x^{1/2}$

91. $5x^{1/2} + 2x^{-1/2} - 3x^{-3/2}$, factor out $x^{-3/2}$

Cumulative Review Exercises

[2.1] **92.** Solve $6(x + 4) - 4(3x + 3) = 6$.

[2.6] **93.** Find the solution set for $\left| \dfrac{6 + 2z}{3} \right| > 2$.

[4.3] **94. Mixing Coffees** Dennis Reissig runs a grocery store. He wishes to mix 30 pounds of coffee to sell for a total cost of $170. To obtain the mixture, he will mix coffee that sells for $5.20 per pound with coffee that sells for $6.30 per pound. How many pounds of each coffee should he use?

[5.2] **95.** Multiply $(5x + 4)(x^2 - x + 4)$.

[5.4] **96.** Factor $2x^3 + 6x^2 - 5x - 15$.

5.8 Polynomial Equations

1 Use the zero-factor property to solve equations.

2 Use factoring to solve equations.

3 Use factoring to solve applications.

4 Use factoring to find the x-intercepts of a quadratic function.

Whenever two polynomials are set equal to each other, we have a **polynomial equation**.

Examples of Polynomial Equations

$$x^2 + 2x = x - 5$$
$$y^3 + 3y - 2 = 0$$
$$4x^4 + 2x^2 = -3x + 2$$

The **degree of a polynomial equation** is the same as the degree of its highest term. For example, the three equations above have degree 2, 3, and 4, respectively.

> **Quadratic Equation**
>
> A second-degree equation in one variable is called a **quadratic equation**.

Examples of Quadratic Equations

$$3x^2 + 6x - 4 = 0$$
$$5x = 2x^2 - 4$$
$$(x + 4)(x - 3) = 0$$

Any quadratic equation can be written in **standard form**.

> **Standard Form of a Quadratic Equation**
>
> $$ax^2 + bx + c = 0, \quad a \neq 0$$
>
> where a, b, and c are real numbers.

Understanding Algebra

Remember that an *equation* must contain an *equals sign* (=). For a quadratic equation to be in standard form, $ax^2 + bx + c$ must be on one side of the equals sign and 0 must be on the other side.

Before going any further, make sure that you can rewrite each of the three quadratic equations given above in standard form, with $a > 0$.

1 Use the Zero-Factor Property to Solve Equations

To solve equations using factoring, we use the **zero-factor property**.

> **Zero-Factor Property**
>
> For all real numbers a and b, if $a \cdot b = 0$, then either $a = 0$ or $b = 0$, or both a and $b = 0$.

The zero-factor property states that *if the product of two factors equals 0, one (or both) of the factors must be 0.*

EXAMPLE 1 Solve the equation $(x + 5)(x - 3) = 0$.

Solution   Since the product of the factors equals 0, according to the zero-factor property, one or both factors must equal 0. Set each factor equal to 0 and solve each equation separately.

$$x + 5 = 0 \quad \text{or} \quad x - 3 = 0$$
$$x = -5 \qquad\qquad x = 3$$

Thus, if x is either -5 or 3, the equation is a true statement.

Check $x = -5$ $x = 3$

$$(x + 5)(x - 3) = 0 \qquad\qquad\qquad (x + 5)(x - 3) = 0$$
$$(-5 + 5)(-5 - 3) \overset{?}{=} 0 \qquad\qquad (3 + 5)(3 - 3) \overset{?}{=} 0$$
$$0(-8) \overset{?}{=} 0 \qquad\qquad\qquad\qquad 8(0) \overset{?}{=} 0$$
$$0 = 0 \quad \text{True} \qquad\qquad\qquad\qquad 0 = 0 \quad \text{True}$$

Now Try Exercise 11

2   Use Factoring to Solve Equations

Following is a procedure that can be used to obtain the solution to an equation by factoring.

To Solve an Equation by Factoring

1. Use the addition property to remove all terms from one side of the equation. This will result in one side of the equation being equal to 0.
2. Combine like terms in the equation and then factor.
3. Set each factor *containing a variable* equal to 0, solve the equations, and find the solutions.
4. Check the solutions in the *original* equation.

> **Helpful Hint**
> If you do not remember how to factor, review Sections 5.3–5.7.

EXAMPLE 2 Solve the equation $4x^2 = 24x$.

Solution   First, make the right side of the equation equal to 0 by subtracting $24x$ from both sides of the equation. Then factor the left side of the equation.

$$4x^2 - 24x = 0$$
$$4x(x - 6) = 0$$

Now set each factor equal to 0 and solve for x.

$$4x = 0 \quad \text{or} \quad x - 6 = 0$$
$$x = 0 \qquad\qquad x = 6$$

A check will show that the numbers 0 and 6 both satisfy the equation $4x^2 = 24x$.

Now Try Exercise 17

> **Avoiding Common Errors**
>
> The zero-factor property can be used only when one side of the equation is equal to 0.
>
> | CORRECT | INCORRECT |
> |---|---|
> | $(x - 4)(x + 3) = 0$ | $(x - 4)(x + 3) = 2$ |
> | $x - 4 = 0$ or $x + 3 = 0$ | $x - 4 = 2$ or $x + 3 = 2$ |
> | $x = 4$ $x = -3$ | $x = 6$ $x = -1$ |
>
> In the incorrect process illustrated on the right, the zero-factor property cannot be used since the right side of the equation is not equal to 0. Example 3 shows how to solve such problems correctly.

EXAMPLE 3 Solve the equation $(x - 1)(3x + 2) = 4x$.

Solution   We cannot use the zero-factor property until we get 0 on one side of the equation.

$$(x - 1)(3x + 2) = 4x$$

$$3x^2 - x - 2 = 4x \qquad \text{Multiply the factors.}$$

$$3x^2 - 5x - 2 = 0 \qquad \text{Make one side 0.}$$

$$(3x + 1)(x - 2) = 0 \qquad \text{Factor the trinomial.}$$

$$3x + 1 = 0 \quad \text{or} \quad x - 2 = 0 \qquad \text{Zero-factor property}$$

$$3x = -1 \qquad\qquad x = 2 \qquad \text{Solve the equations.}$$

$$x = -\frac{1}{3}$$

The solutions are $-\dfrac{1}{3}$ and 2. Check these values in the original equation.

Now Try Exercise 31

EXAMPLE 4 Solve the equation $3x^2 + 2x + 12 = -13x$

Solution

$$3x^2 + 2x + 12 = -13x$$

$$3x^2 + 15x + 12 = 0 \qquad \text{Make one side 0.}$$

$$3(x^2 + 5x + 4) = 0 \qquad \text{Factor out 3.}$$

$$3(x + 4)(x + 1) = 0 \qquad \text{Factor the trinomial.}$$

$$x + 4 = 0 \quad \text{or} \quad x + 1 = 0 \qquad \text{Zero-factor property}$$

$$x = -4 \qquad\qquad x = -1 \qquad \text{Solve for } x.$$

Since the factor 3 does not contain a variable, we do not have to set it equal to 0. Only the numbers -4 and -1 satisfy the equation $3x^2 + 2x + 12 = -13x$.

Now Try Exercise 35

> **Helpful Hint**
>
> When solving an equation whose leading term has a negative coefficient, we generally make the coefficient positive by multiplying both sides of the equation by -1. This makes the factoring process easier, as in the following example.
>
> $$-x^2 + 5x + 6 = 0$$
>
> $$-1(-x^2 + 5x + 6) = -1 \cdot 0$$
>
> $$x^2 - 5x - 6 = 0$$
>
> Now we can solve the equation $x^2 - 5x - 6 = 0$ by factoring.
>
> $$(x - 6)(x + 1) = 0$$
>
> $$x - 6 = 0 \quad \text{or} \quad x + 1 = 0$$
>
> $$x = 6 \qquad\qquad x = -1$$
>
> The numbers 6 and -1 both satisfy the original equation, $-x^2 + 5x + 6 = 0$.

The zero-factor property can be extended to three or more factors, as illustrated in Example 5.

EXAMPLE 5 Solve the equation $2p^3 + 5p^2 - 3p = 0$.

Solution   First factor, then set each factor containing a p equal to 0.

$$2p^3 + 5p^2 - 3p = 0$$
$$p(2p^2 + 5p - 3) = 0 \qquad \text{Factor out } p.$$
$$p(2p - 1)(p + 3) = 0 \qquad \text{Factor the trinomial.}$$
$$p = 0 \quad \text{or} \quad 2p - 1 = 0 \quad \text{or} \quad p + 3 = 0 \qquad \text{Zero-factor property}$$
$$2p = 1 \qquad\qquad p = -3 \quad \text{Solve for } p.$$
$$p = \frac{1}{2}$$

The numbers 0, $\frac{1}{2}$, and 3 arc all solutions to the equation.

Now Try Exercise 39

Note that the equation in Example 5 is not a quadratic equation because the exponent in the leading term is 3, not 2. This is a **cubic**, or **third-degree, equation**.

EXAMPLE 6 For the function $f(x) = 2x^2 - 13x - 16$, find all values of a for which $f(a) = 8$.

Solution   First we rewrite the function as $f(a) = 2a^2 - 13a - 16$. Since $f(a) = 8$, we write

$$2a^2 - 13a - 16 = 8 \qquad \text{Set } f(a) \text{ equal to 8.}$$
$$2a^2 - 13a - 24 = 0 \qquad \text{Make one side 0.}$$
$$(2a + 3)(a - 8) = 0 \qquad \text{Factor the trinomial.}$$
$$2a + 3 = 0 \quad \text{or} \quad a - 8 = 0 \quad \text{Zero-factor property}$$
$$2a = -3 \qquad\qquad a = 8 \quad \text{Solve for } a.$$
$$a = -\frac{3}{2}$$

If you check these answers, you will find that $f\left(-\frac{3}{2}\right) = 8$ and $f(8) = 8$.

Now Try Exercise 59

3 Use Factoring to Solve Applications

Now let us look at some applications that use factoring in their solution.

EXAMPLE 7 **Triangle** A large canvas tent has an entrance in the shape of a triangle (see **Fig. 5.17**).
Find the base and height of the entrance if the height is to be 3 feet less than twice the base and the total area of the entrance is 27 square feet.

Solution

Understand   Let's draw a picture of the entrance and label it with the given information (**Fig. 5.18**).

FIGURE 5.17

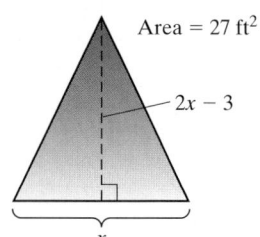

Area = 27 ft²

$2x - 3$

FIGURE 5.18 x

Translate   We use the formula for the area of a triangle to solve the problem.

$$A = \frac{1}{2}(\text{base})(\text{height})$$

$$27 = \frac{1}{2}(x)(2x - 3)$$ Substitute expressions for base, height, and area.

Carry Out $$2(27) = \cancel{2}\left[\frac{1}{\cancel{2}}(x)(2x - 3)\right]$$ Multiply both sides by 2 to remove fractions.

$$54 = x(2x - 3)$$

$$54 = 2x^2 - 3x$$ Distributive property.

$$0 = 2x^2 - 3x - 54$$ Make one side 0.

or $$2x^2 - 3x - 54 = 0$$

$$(2x + 9)(x - 6) = 0$$ Factor the trinomial.

$$2x + 9 = 0 \quad \text{or} \quad x - 6 = 0$$ Zero-factor property

$$2x = -9 \qquad\qquad x = 6$$ Solve for x.

$$x = -\frac{9}{2}$$

Answer   Since the dimensions of a geometric figure cannot be negative, we can eliminate $x = -\frac{9}{2}$ as an answer to our problem. Therefore,

$$\text{base} = x = 6 \text{ feet}$$

$$\text{height} = 2x - 3 = 2(6) - 3 = 9 \text{ feet}$$

Now Try Exercise 89

EXAMPLE 8 **Height of a Cannonball** A cannon sits on top of a 288-foot cliff overlooking a lake. A cannonball is fired upward at a speed of 112 feet per second. The height, h, in feet, of the cannonball above the lake at any time, t, is determined by the function

$$h(t) = -16t^2 + 112t + 288$$

Find the time it takes for the cannonball to hit the water after the cannon is fired.

Solution

Understand   We will draw a picture to help analyze the problem (see **Fig. 5.19**). When the cannonball hits the water, its height above the water is 0 feet.

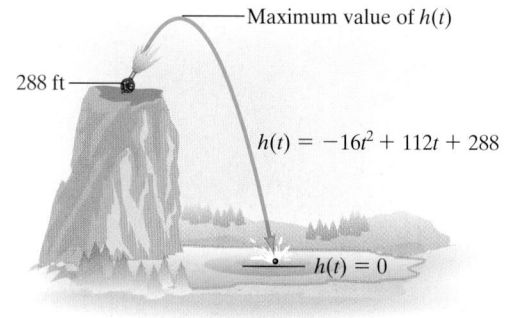

Translate   To solve the problem we need to find the time, t, when $h(t) = 0$. To do so we set the given function equal to 0 and solve for t.

FIGURE 5.19

$$-16t^2 + 112t + 288 = 0$$ Set $h(t) = 0$.

$$-16(t^2 - 7t - 18) = 0$$ Factor out -16.

$$-16(t + 2)(t - 9) = 0$$ Factor the trinomial.

$$t + 2 = 0 \quad \text{or} \quad t - 9 = 0$$ Zero-factor property

$$t = -2 \qquad\qquad t = 9$$ Solve for t.

Answer   Since t is the number of seconds, -2 is not a possible answer. The cannonball will hit the water in 9 seconds.

Now Try Exercise 95

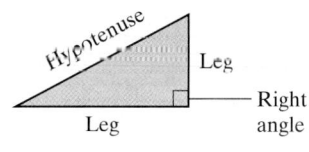

FIGURE 5.20

Pythagorean Theorem   Consider a right triangle (see **Fig. 5.20**). The two shorter sides of a right triangle are called the **legs** and the side opposite the right angle is called the **hypotenuse**. The **Pythagorean Theorem** expresses the relationship between the legs of the triangle and its hypotenuse.

Pythagorean Theorem

The square of the length of the hypotenuse of a right triangle is equal to the sum of the squares of the lengths of the two legs; that is,

$$\text{leg}^2 + \text{leg}^2 = \text{hyp}^2$$

If a and b represent the lengths of the legs and c represents the length of the hypotenuse, then

$$a^2 + b^2 = c^2$$

EXAMPLE 9   Tree Wires Jack Keating places a guy wire on a tree to help it grow straight. The location of the stake and where the wire attaches to the tree are given in **Figure 5.21**. Find the length of the wire.

Solution

Understand   Notice that the length of the wire is the hypotenuse of a right triangle formed by the tree and the ground. To solve this problem we use the Pythagorean Theorem. From the figure, we see that the legs are x and $x + 1$ and the hypotenuse is $x + 2$.

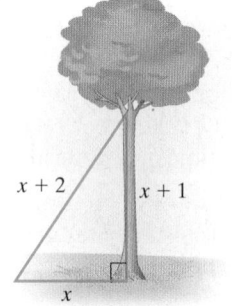

FIGURE 5.21

Translate	$\text{leg}^2 + \text{leg}^2 = \text{hyp}^2$	Pythagorean Theorem
	$x^2 + (x + 1)^2 = (x + 2)^2$	Substitute expressions for legs and hypotenuse.
Carry Out	$x^2 + x^2 + 2x + 1 = x^2 + 4x + 4$	Square terms.
	$2x^2 + 2x + 1 = x^2 + 4x + 4$	Simplify.
	$x^2 - 2x - 3 = 0$	Make one side 0.
	$(x - 3)(x + 1) = 0$	Factor.
	$x - 3 = 0 \quad \text{or} \quad x + 1 = 0$	Solve.
	$x = 3 \qquad\qquad x = -1$	

Answer   From the diagram we know that x cannot be a negative value. Therefore the only possible answer is 3. The stake is placed 3 feet from the tree. The wire attaches to the tree $x + 1$, or 4, feet from the ground. The length of the wire is $x + 2$, or 5, feet.

Now Try Exercise 99

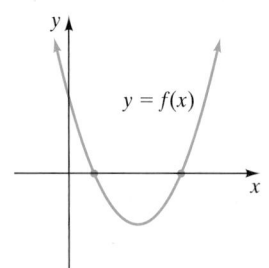

FIGURE 5.22

4 Use Factoring to Find the *x*-Intercepts of a Quadratic Function

Consider the graph in **Figure 5.22**.

At the x-intercepts, the value of the function, or y, is 0. Thus if we wish to find the **x-intercepts of a graph**, we can set the function equal to 0 and solve for x.

EXAMPLE 10 Find the x-intercepts of the graph of $y = x^2 - 2x - 8$.

Solution   At the x-intercepts y has a value of 0. Thus to find the x-intercepts we write

$$x^2 - 2x - 8 = 0$$

$$(x - 4)(x + 2) = 0$$

$$x - 4 = 0 \quad \text{or} \quad x + 2 = 0$$

$$x = 4 \qquad\qquad x = -2$$

> **Understanding Algebra**
>
> An *x-intercept* is a point where the graph crosses the *x*-axis. The *y* coordinate of an *x*-intercept is always 0.

The solutions of $x^2 - 2x - 8 = 0$ are 4 and -2. The x-intercepts of the graph of $y = x^2 - 2x - 8$ are $(4, 0)$ and $(-2, 0)$, as illustrated in **Figure 5.23**.

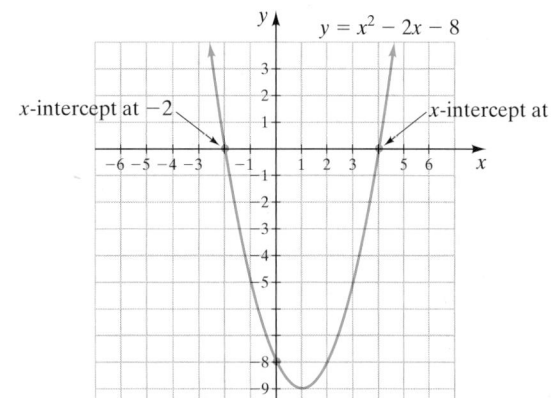

FIGURE 5.23

Now Try Exercise 65

If we know the x-intercepts of a graph, we can work backward to find the equation of the graph. Read the Using Your Graphing Calculator box that follows to learn how this is done.

Using Your Graphing Calculator

Determine the equation of the graph in **Figure 5.24**.

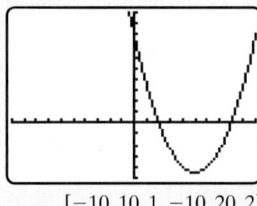

FIGURE 5.24 $[-10, 10, 1, -10, 20, 2]$

If we assume the intercepts are integer values, then the x-intercepts are at 2 and 8. Therefore,

x-Intercepts At	Factors	Possible Equation of Graph
2 and 8	$(x - 2)(x - 8)$	$y = (x - 2)(x - 8)$
		or $y = x^2 - 10x + 16$

Since the y-intercept of the graph in **Figure 5.24** is at 16, $y = x^2 - 10x + 16$ is the equation of the graph. Example 11 explains why we used the words *possible equation of the graph*.

EXAMPLE 11 Write an equation whose graph will have x-intercepts at -2 and 4.

Solution   If the x-intercepts are at -2 and 4, then one set of factors that yields these x-intercepts is $(x + 2)$ and $(x - 4)$, respectively. Therefore, one equation that will have x-intercepts at -2 and 4 is

$$y = (x + 2)(x - 4) \text{ or } y = x^2 - 2x - 8.$$

Note that other equations may have graphs with the same x-intercepts. For example, the graph of $y = 2(x^2 - 2x - 8)$ or $y = 2x^2 - 4x - 16$ also has x-intercepts at -2 and 4. In fact, the graph of $y = a(x^2 - 2x - 8)$, for any nonzero real number a, will have x-intercepts at -2 and 4.

Now Try Exercise 83

In Example 11, although the x-intercepts of the graph of $y = a(x^2 - 2x - 8)$ will always be at -2 and 4, the y-intercept of the graph will depend on the value of a. For example, if $a = 1$, the y-intercept will be at $1(8)$ or 8. If $a = 2$, the y-intercept will be at $2(8)$ or 16 and so on.

EXERCISE SET 5.8

MathXL
MathXL®

MyMathLab
MyMathLab

Warm-Up Exercises

Fill in the blanks with the appropriate word, phrase, or symbol(s) from the following list.

$a = 0$	leg	x-intercept	$b = 0$	expression	$ax^2 + bx = -c$
equation	hypotenuse	side	term	y-intercept	$ax^2 + bx + c = 0$

1. When two polynomials are set equal to each other, we have a polynomial _____.

2. The degree of a polynomial equation is the degree of the highest _____ of the polynomial.

3. The standard form of a quadratic equation is _____.

4. The zero-factor property states that if $a \cdot b = 0$, $a = 0$ or _____ or both $a = 0$ and $b = 0$.

5. A point where the graph crosses the x-axis is called an _____.

6. In a right triangle, the side opposite the right angle is the _____.

Practice the Skills

Solve.

7. $x(x + 3) = 0$

8. $x(x - 2) = 0$

9. $4x(x - 1) = 0$

10. $5x(x + 6) = 0$

11. $2(x + 1)(x - 7) = 0$

12. $3(a - 5)(a + 2) = 0$

13. $x(x - 9)(x - 4) = 0$

14. $2a(a + 3)(a + 10) = 0$

15. $(3x - 2)(7x - 1) = 0$

16. $(2x + 3)(4x + 5) = 0$

17. $4x^2 = 12x$

18. $3y^2 = -21y$

19. $x^2 + 5x = 0$

20. $4a^2 - 32a = 0$

21. $-x^2 + 6x = 0$

22. $-3x^2 - 24x = 0$

23. $3x^2 = 27x$

24. $18a^2 = -36a$

25. $a^2 + 6a + 5 = 0$

26. $x^2 - 6x + 5 = 0$

27. $x^2 + x - 12 = 0$

28. $b^2 + b - 72 = 0$

29. $x^2 + 8x + 16 = 0$

30. $c^2 - 12c = -36$

31. $(2x + 5)(x - 1) = 12x$

32. $a(a + 2) = 48$

33. $2y^2 = -y + 6$

34. $3a^2 = -a + 2$

35. $3x^2 - 6x - 72 = 0$

36. $2a^2 + 18a + 40 = 0$

37. $x^3 - 3x^2 = 18x$

38. $x^3 = -19x^2 + 42x$

39. $4c^3 + 4c^2 - 48c = 0$

40. $3b^3 - 8b^2 - 3b = 0$

41. $18z^3 = 15z^2 + 12z$

42. $12a^3 = 16a^2 + 3a$

43. $x^2 - 25 = 0$

44. $2y^2 = 72$

45. $16x^2 = 9$

46. $49c^2 = 81$

47. $4y^3 - 36y = 0$

48. $3x^4 - 48x^2 = 0$

49. $-x^2 = 2x - 99$

50. $-x^2 + 18x = 80$

51. $(x + 7)^2 - 16 = 0$

52. $(x - 6)^2 - 4 = 0$

53. $(2x + 5)^2 - 9 = 0$

54. $(x + 1)^2 - 3x = 7$

55. $6a^2 - 12 - 4a = 19a - 32$

56. $4(a^2 - 3) = 6a + 4(a + 3)$

57. $2b^3 + 16b^2 = -30b$

58. $(a - 1)(3a + 2) = 4a$

59. For $f(x) = 3x^2 + 7x + 9$, find all values of a for which $f(a) = 7$.

60. For $f(x) = 4x^2 - 11x + 2$, find all values of a for which $f(a) = -4$.

61. For $g(x) = 10x^2 - 31x + 16$ find all values of a for which $g(a) = 1$.

62. For $g(x) = 6x^2 + x - 3$, find all values of a for which $g(a) = -2$.

63. For $r(x) = x^2 - x$, find all values of a for which $r(a) = 30$.

64. For $r(x) = 10x^2 - 11x - 17$, find all values of a for which $r(a) = -11$

Use factoring to find the x-intercepts of the graphs of each equation (see Example 10).

65. $y = x^2 - 10x + 24$

66. $y = x^2 - x - 42$

67. $y = x^2 + 16x + 64$

68. $y = 15x^2 - 14x - 8$

69. $y = 12x^3 - 46x^2 + 40x$

70. $y = 12x^3 - 39x^2 + 30x$

Right Triangle   *In Exercises 71–76, use the Pythagorean Theorem to find x.*

71.

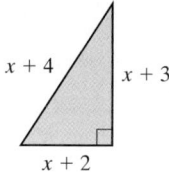

72.

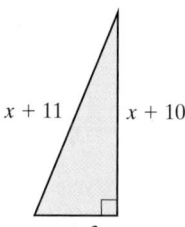

73.

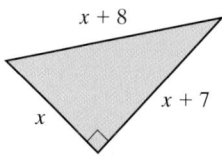

74.

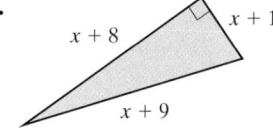

75.

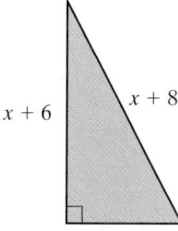

76.

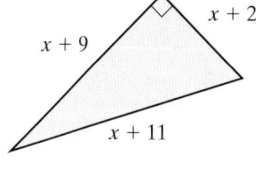

Problem Solving

*In Exercises 77–80, determine the x-intercepts of each graph; then match the equation with the appropriate graph labeled **a)–d**).*

77. $y = x^2 - 5x + 6$

78. $y = x^2 - x - 6$

79. $y = x^2 + 5x + 6$

80. $y = x^2 + x - 6$

a)

b)

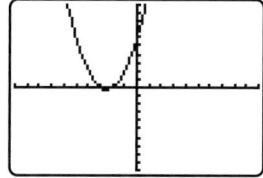

c)

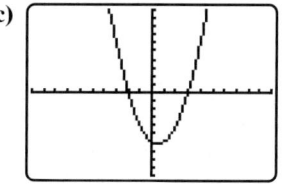

d)

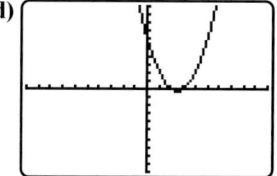

Write an equation whose graph will have x-intercepts at the given values.

81. 1 and 5

82. 3 and -7

83. 4 and -2

84. $\dfrac{3}{2}$ and 6

85. $-\dfrac{5}{6}$ and 2

86. -0.4 and 2.6

87. Rectangular Coffee Table A coffee table is rectangular. If the length of its surface area is 1 foot greater than twice its width and the surface area of the tabletop is 10 square feet, find its length and width.

88. Rectangular Shed The floor of a shed has an area of 60 square feet. Find the length and width if the length is 2 feet less than twice its width.

89. Triangular Sail A sailboat sail is triangular with a height 6 feet greater than its base. If the sail's area is 80 square feet, find its base and height.

90. Triangular Tent A triangular tent has a height that is 4 feet less than its base. If the area of a side is 70 square feet, find the base and height of the tent.

91. Rectangle Frank Bullock's garden is surrounded by a uniform-width walkway. The garden and the walkway together cover an area of 320 square feet. If the dimensions of the garden are 12 feet by 16 feet, find the width of the walkway.

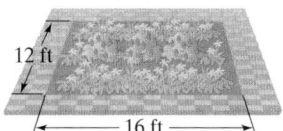

92. Picture Frame The outside dimensions of a picture frame are 28 cm and 23 cm. The area of the picture itself is 414 square centimeters. Find the width of the frame itself.

93. Vegetable Garden Sally Yang's rectangular vegetable garden is 20 feet by 30 feet. In addition to mulching her garden, she wants to put mulch around the outside of her garden in a uniform width. If she has enough mulch to cover an area of 936 square feet, how wide should the mulch border be?

94. Square Garden Ronnie Tucker has a square garden. He adds a 2-foot-wide walkway around his garden. If the total area of the walkway and garden is 196 square feet, find the dimensions of the garden.

95. Water Sculpture In a building at Navy Pier in Chicago, a water fountain jet shoots short spurts of water over a walkway. The water spurts reach a maximum height, then come down into a pond of water on the other side of the walkway. The height above the jet, h, of a spurt of water t seconds after leaving the jet can be found by the function $h(t) = -16t^2 + 32t$. Find the time it takes for the spurt of water to return to the jet's height; that is, when $h(t) = 0$.

See Exercise 95.

96. Projectile A model rocket will be launched from a hill 80 feet above sea level. The launch site is next to the ocean (sea level), and the rocket will fall into the ocean. The rocket's distance, s, above sea level at any time, t, is found by the equation $s(t) = -16t^2 + 64t + 80$. Find the time it takes for the rocket to strike the ocean.

97. Bicycle Riding Two cyclists, Bob and Tim, start at the same point. Bob rides west and Tim rides north. At some time, they are 13 miles apart. If Bob traveled 7 miles farther than Tim, determine how far each person traveled.

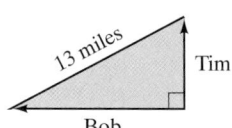

98. Picture Frame April is making a rectangular picture frame for her mother. The diagonal of the frame is 20 inches. Find the dimensions of the frame if its length is 4 inches greater than its width.

99. Tent Wires A tent has wires attached to it to help stabilize it. A wire is attached to the ground 12 feet from the tent. The length of wire used is 8 feet greater than the height from the ground to where the wire is attached to the tent. How long is the wire?

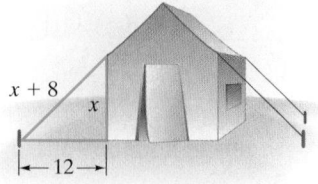

100. Car in Mud Suppose two cars, indicated by points A and B in the figure, are pulling a third car, C, out of the mud. Determine the distance from car A to car B.

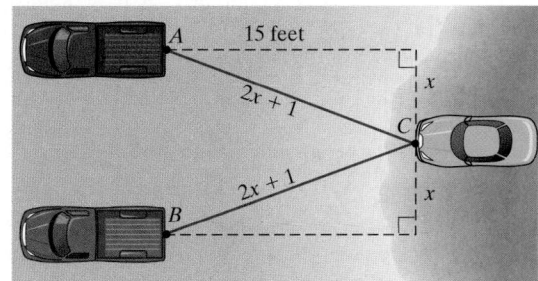

101. Bicycle Shop The Energy Conservatory Bicycle Shop has a monthly revenue equation $R(x) = 70x - x^2$ and a monthly cost equation $C(x) = 17x + 150$, where x is the number of bicycles sold and $x \geq 10$. Find the number of bicycles that must be sold for the company to break even; that is, where revenue equals costs.

102. Silk Plant Edith Hall makes silk plants and sells them to various outlets. Her company has a revenue equation $R(x) = 40x - x^2$ and a cost equation $C(x) = 14x + 25$, where x is the number of plants sold and $x \geq 5$. Find the number of plants that must be sold for the company to break even.

103. Making a Box Monique Siddiq is making a box by cutting out 2-in.-by-2-in. squares from a square piece of cardboard and folding up the edges to make a 2-inch-high box. What size piece of cardboard does Monique need to make a 2-inch-high box with a volume of 162 cubic inches?

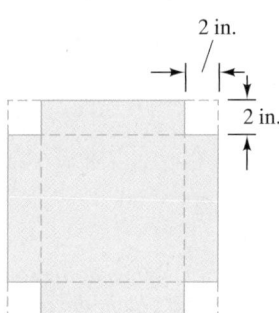

104. Making a Box A rectangular box is to be formed by cutting squares from each corner of a rectangular piece of tin and folding up the sides. The box is to be 3 inches high, the length is to be twice the width, and the volume of the box is to be 96 cubic inches. Find the length and width of the box.

105. Cube A solid cube with dimensions a^3 has a rectangular solid with dimensions ab^2 removed.

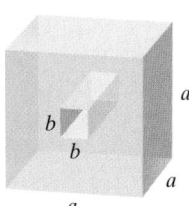

a) Write a formula for the remaining volume, V.

b) Factor the right side of the formula in part **a)**.

c) If the volume is 1620 cubic inches and a is 12 in., find b.

106. Circular Steel Blade A circular steel blade has a hole cut out of its center as shown in the figure.

a) Write a formula for the remaining area of the blade.

b) Factor the right side of the formula in part **a)**.

c) Find A if $R = 10$ cm and $r = 3$ cm.

107. Consider the following graph of a quadratic function.

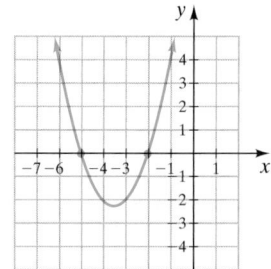

a) Write a quadratic function that has the x-intercepts indicated.

b) Write a quadratic equation in one variable that has solutions of -2 and -5.

c) How many different quadratic functions can have x-intercepts of -2 and -5? Explain.

d) How many different quadratic equations in one variable can have solutions of -2 and -5? Explain.

108. The graph of the equation $y = x^2 + 4$ is illustrated below.

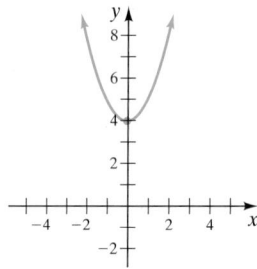

a) How many x-intercepts does the graph have?

b) How many real solutions does the equation $x^2 + 4 = 0$ have? Explain your answer.

109. Consider the quadratic function
$$P(x) = ax^2 + bx + c, a > 0.$$

a) The graph of this type of function may have no x-intercepts, one x-intercept, or two x-intercepts. Sketch each of these possibilities.

b) How many possible real solutions may the equation $ax^2 + bx + c = 0, a > 0$ have? Explain your answer to part **b)** by using the sketches in part **a)**.

110. Stopping Distance A typical car's stopping distance on dry pavement, d, in feet, can be approximated by the function $d(s) = 0.034s^2 + 0.56s - 17.11$, where s is the speed of the car before braking and $60 \leq s \leq 80$ miles per hour. How fast is the car going if it requires 190 feet to stop after the brakes are applied?

111. Stopping Distance A typical car's stopping distance on wet pavement, d, in feet, can be approximated by the function $d(s) = -0.31s^2 + 59.82s - 2180.22$, where s is the speed of the car before braking and $60 \leq s \leq 80$ miles per hour. How fast is the car going if it requires 545 feet for the car to stop after the brakes are applied?

Concept/Writing Exercises

112. How do you determine the degree of a polynomial function?

113. What is a quadratic equation?

114. What is the standard form of a quadratic equation?

115. a) Explain the zero-factor property.

    b) Solve the equation $(3x - 7)(2x + 3) = 0$ using the zero-factor property.

116. a) Explain why the equation $(x + 3)(x + 4) = 2$ *cannot* be solved by writing $x + 3 = 2$ or $x + 4 = 2$.

    b) Solve the equation $(x + 3)(x + 4) = 2$.

117. When a constant is factored out of an equation, why is it not necessary to set that constant equal to 0 when solving the equation?

118. a) Explain how to solve a polynomial equation using factoring.

    b) Solve the equation $-x - 20 = -12x^2$ using the procedure in part **a)**.

119. a) What is the first step in solving the equation $-x^2 + 2x + 35 = 0$?

    b) Solve the equation in part **a)**.

120. a) What are the two shorter sides of a right triangle called?

    b) What is the longest side of a right triangle called?

121. Give the Pythagorean Theorem and explain its meaning.

122. If the graph of $y = x^2 + 10x + 16$ has x-intercepts at -8 and -2, what is the solution to the equation $x^2 + 10x + 16 = 0$? Explain.

123. If the solutions to the equation $2x^2 - 15x + 18 = 0$ are $\frac{3}{2}$ and 6, what are the x-intercepts of the graph of $y = 2x^2 - 15x + 18$? Explain.

124. Is it possible for a quadratic function to have no x-intercepts? Explain.

125. Is it possible for a quadratic function to have only one x-intercept? Explain.

126. Is it possible for a quadratic function to have two x-intercepts? Explain.

127. Is it possible for a quadratic function to have three x-intercepts? Explain.

Challenge Problems

Solve.

128. $x^4 - 17x^2 + 16 = 0$

129. $x^4 - 13x^2 = -36$

130. $x^6 - 9x^3 + 8 = 0$

Group Activity

In more advanced mathematics courses you may need to solve an equation for y' (read "y prime"). When doing so, treat the y' as a different variable from y. Individually solve each equation for y'. Compare your answers and as a group obtain the correct answers.

131. $xy' + yy' = 1$

132. $xy - xy' = 3y' + 2$

133. $2xyy' - xy = x - 3y'$

Cumulative Review Exercises

[1.5] **134.** Simplify $(4x^{-2}y^3)^{-2}$.

[2.5] **135.** Solve the inequality and graph the solution on the number line.

$$-1 < \frac{4(3x - 2)}{3} \leq 5$$

[4.1] **136.** Solve the system of equations.

$$3x + 4y = 2$$
$$2x = -5y - 1$$

[5.2] **137.** If $f(x) = -x^2 + 3x$ and $g(x) = x^2 + 5$, find $(f \cdot g)(4)$.

[5.7] **138.** Factor $(x + 1)^2 - (x + 1) - 6$.

Chapter 5 Summary

IMPORTANT FACTS AND CONCEPTS	EXAMPLES

Section 5.1

Terms are parts that are added or subtracted in a mathematical expression.	The terms of $-3x^2 + 1.6x + 15$ are $$-3x^2, 1.6x, \text{ and } 15.$$
A **polynomial** is a finite sum of terms in which all variables have whole number exponents and no variable appears in the denominator.	$9x^7 - 3x^5 + 4x - \dfrac{1}{2}$ is a polynomial.
The **degree of a term** is the sum of the exponents of the variables.	The term $3x^2y^9$ has degree 11.
The **leading term** of a polynomial is the term of highest degree. The **leading coefficient** is the coefficient of the leading term.	In the polynomial $9x^7 - 3x^5 + 4x - \dfrac{1}{2}$, the leading term is $9x^7$ and the leading coefficient is 9.
A **monomial** is a polynomial with one term. A **binomial** is a polynomial with two terms. A **trinomial** is a polynomial with three terms.	$-13mn^2p^3$ $x^4 - 1$ $1.9x^3 - 28.3x^2 - 101.5x$
A polynomial is **linear** if it is degree 0 or 1. A polynomial in one variable is **quadratic** if it is degree 2. A polynomial in one variable is **cubic** if it is degree 3.	$19, 8y + 17$ $x^2 - 5x + 16$ $-4x^3 + 11x^2 - 9x + 6$
A **polynomial function** has the form $y = P(x)$. To evaluate $P(a)$, replace x by a.	$P(x) = 2x^2 - x + 3$ is a polynomial function. To evaluate $P(x)$ at $x = 10$, $$P(10) = 2(10)^2 - 10 + 3$$ $$= 200 - 10 + 3 = 193$$
To **add or subtract polynomials**, combine like terms.	$(5x^2 - 9x + 10) + (2x^2 + 17x - 8)$ $= 5x^2 - 9x + 10 + 2x^2 + 17x - 8 = 7x^2 + 8x + 2$

Section 5.2

To **multiply polynomials**, multiply each term of one polynomial by each term of the other polynomial.	$3a(a - 2) = 3a \cdot a - 3a \cdot 2$ $= 3a^2 - 6a$
Distributive Property, Expanded Form $a(b + c + d + \cdots + n) = ab + ac + ad + \cdots + an$	$x(2x^2 + 8x - 5) = 2x^3 + 8x^2 - 5x$
To **multiply two binomials**, use the **FOIL method:** multiply the **F**irst terms, **O**uter terms, **I**nner terms, **L**ast terms.	$(3x - 1)(4x + 9) = 12x^2 + 27x - 4x - 9$ $= 12x^2 + 23x - 9$
To multiply a polynomial by a polynomial, you can use the vertical format.	Multiply $(2x^2 - x + 8)(5x + 1)$. $$\begin{array}{r} 2x^2 - x + 8 \\ 5x + 1 \\ \hline 2x^2 - x + 8 \\ 10x^3 - 5x^2 + 40x \\ \hline 10x^3 - 3x^2 + 39x + 8 \end{array}$$
Square of a Binomial $$(a + b)^2 = a^2 + 2ab + b^2$$ $$(a - b)^2 = a^2 - 2ab + b^2$$	$(7x + 4)^2 = (7x)^2 + 2(7x)(4) + 4^2 = 49x^2 + 56x + 16$ $\left(\dfrac{1}{2}m - 3\right)^2 = \left(\dfrac{1}{2}m\right)^2 - 2\left(\dfrac{1}{2}m\right)(3) + 3^2 = \dfrac{1}{4}m^2 - 3m + 9$
Product of the Sum and Difference of the Same Two Terms $$(a + b)(a - b) = a^2 - b^2$$	$(5c + 6)(5c - 6) = (5c)^2 - 6^2 = 25c^2 - 36$

IMPORTANT FACTS AND CONCEPTS	EXAMPLES

Section 5.3

To divide a polynomial by a monomial, divide each term of the polynomial by the monomial.	$$\frac{6y + 10x^2y^5 - 17x^9y^8}{2xy^2} - \frac{6y}{2xy^2} + \frac{10x^2y^5}{2xy^2} - \frac{17x^9y^8}{2xy^2}$$ $$= \frac{3}{xy} + 5xy^3 - \frac{17x^8y^6}{2}$$
To divide two polynomials, use long division.	Divide $(8x^2 + 6x - 9) \div (2x + 1)$. $$\begin{array}{r} 4x + 1 \\ 2x + 1 \overline{)8x^2 + 6x - 9} \\ \underline{8x^2 + 4x} \\ 2x - 9 \\ \underline{2x + 1} \\ -10 \end{array}$$ Thus, $\dfrac{8x^2 + 6x - 9}{2x + 1} = 4x + 1 - \dfrac{10}{2x + 1}$.
To divide a polynomial by a binomial of the form $x - a$, use **synthetic division**.	Use synthetic division to divide $$(x^3 + 2x^2 - 11x + 5) \div (x + 4)$$ $$\begin{array}{r|rrrr} -4 & 1 & 2 & -11 & 5 \\ & & -4 & 8 & 12 \\ \hline & 1 & -2 & -3 & 17 \end{array}$$ Thus, $\dfrac{x^3 + 2x^2 - 11x + 5}{x + 4} = x^2 - 2x - 3 + \dfrac{17}{x + 4}$
Remainder Theorem If the polynomial $P(x)$ is divided by $x - a$, the remainder is $P(a)$.	Find the remainder when $$2x^3 - 6x^2 - 11x + 29 \text{ is divided by } x + 2.$$ Let $P(x) = 2x^3 - 6x^2 - 11x + 29$; then $$P(-2) = 2(-2)^3 - 6(-2)^2 - 11(-2) + 29$$ $$= -16 - 24 + 22 + 29$$ $$= 11.$$ The remainder is 11.

Section 5.4

The **greatest common factor** (GCF) is the product of the factors common to all terms in the polynomial.	The GCF of z^5, z^4, z^9, z^2 is z^2. The GCF of $9(x - 4)^3, 6(x - 4)^{10}$ is $3(x - 4)^3$.
To Factor a Monomial from a Polynomial **1.** Determine the greatest common factor of all terms in the polynomial. **2.** Write each term as the product of the GCF and another factor. **3.** Use the distributive property to *factor out* the GCF.	$$35x^6 + 15x^4 + 5x^3 = 5x^3(7x^3) + 5x^3(3x) + 5x^3(1)$$ $$= 5x^3(7x^3 + 3x + 1)$$ $$4n(7n + 10) - 13(7n + 10) = (7n + 10)(4n - 13)$$
To Factor Four Terms by Grouping **1.** Determine if all four terms have a common factor. If so, factor out the GCF from each term. **2.** Arrange the four terms into two groups of two terms each. Each group of two terms must have a GCF. **3.** Factor the GCF from each group of two terms. **4.** If the two terms formed in step 3 have a GCF, factor it out.	$$cx + cy + dx + dy = c(x + y) + d(x + y)$$ $$= (x + y)(c + d)$$ $$x^3 + 6x^2 - 5x - 30 = x^2(x + 6) - 5(x + 6)$$ $$= (x + 6)(x^2 - 5)$$

IMPORTANT FACTS AND CONCEPTS	EXAMPLES

Section 5.5

To Factor Trinomials of the Form $x^2 + bx + c$

1. Find two numbers (or factors) whose product is c and whose sum is b.

2. The factors of the trinomial will be of the form

$$(x + \boxed{})(x + \boxed{})$$

↑   ↑

One factor   Other factor
determined in   determined in
step 1   step 1

Factor $m^2 - m - 42$.

The factors of -42 whose sum is -1 are -7 and 6. Note that $(-7)(6) = -42$ and $-7 + 6 = -1$. Therefore,

$$m^2 - m - 42 = (m - 7)(m + 6)$$

To Factor Trinomials of the Form $ax^2 + bx + c, a \neq 1$, Using Trial and Error

1. Write all pairs of factors of the coefficient of the squared term, a.

2. Write all pairs of factors of the constant, c.

3. Try various combinations of these factors until the correct middle term, bx, is found.

$$4t^2 + 9t + 5 = (4t + 5)(t + 1)$$

Notice that $4t + 5t = 9t$.

$$2a^2 - 15ab + 28b^2 = (2a - 7b)(a - 4b)$$

Notice that $-8ab - 7ab = -15ab$.

To Factor Trinomials of the Form $ax^2 + bx + c, a \neq 1$, Using Grouping

1. Find two numbers whose product is $a \cdot c$ and whose sum is b.

2. Rewrite the middle term, bx, using the numbers found in step 1.

3. Factor by grouping.

Factor $2y^2 + 9y - 18$ by grouping.
Two numbers whose product is -36 and whose sum is 9 are 12 and -3. Therefore;

$$\begin{aligned} 2y^2 + 9y - 18 &= 2y^2 + 12y - 3y - 18 \\ &= 2y(y + 6) - 3(y + 6) \\ &= (y + 6)(2y - 3) \end{aligned}$$

A **prime polynomial** is a polynomial that cannot be factored.

$x^2 + 5x + 9$ is a prime polynomial.

Factoring by substitution occurs when one variable is substituted for another variable or expression.

Factor $a^6 - 2a^3 - 3$.

$$\begin{aligned} a^6 - 2a^3 - 3 &= (a^3)^2 - 2a^3 - 3 \\ &= x^2 - 2x - 3 \qquad \text{Substitute } x \text{ for } a^3. \\ &= (x - 3)(x + 1) \\ &= (a^3 - 3)(a^3 + 1) \quad \text{Substitute } a^3 \text{ for } x. \end{aligned}$$

Section 5.6

Difference of Two Squares

$$a^2 - b^2 = (a + b)(a - b)$$

$$x^2 - 49 = x^2 - 7^2 = (x + 7)(x - 7)$$

Perfect Square Trinomials

$$a^2 + 2ab + b^2 = (a + b)^2$$
$$a^2 - 2ab + b^2 = (a - b)^2$$

$$d^2 + 8d + 16 = d^2 + 2(d)(4) + 4^2 = (d + 4)^2$$
$$4m^2 - 12m + 9 = (2m)^2 - 2(2m)(3) + 3^2 = (2m - 3)^2$$

Sum of Two Cubes

$$a^3 + b^3 = (a + b)(a^2 - ab + b^2)$$

$$y^3 + 8 = y^3 + 2^3 = (y + 2)(y^2 - 2y + 4)$$

Difference of Two Cubes

$$a^3 - b^3 = (a - b)(a^2 + ab + b^2)$$

$$27z^3 - 64x^3 = (3z)^3 - (4x)^3 = (3z - 4x)(9z^2 + 12xz + 16x^2)$$

IMPORTANT FACTS AND CONCEPTS	EXAMPLES

Section 5.7

To Factor a Polynomial

1. Determine whether all the terms in the polynomial have a greatest common factor other than 1. If so, factor out the GCF.
2. If the polynomial has two terms, determine whether it is a difference of two squares or a sum or difference of two cubes. If so, factor using the appropriate formula.
3. If the polynomial has three terms, determine whether it is a perfect square trinomial. If so, factor accordingly. If it is not, factor the trinomial using trial and error, grouping, or substitution as explained in Section 5.5.
4. If the polynomial has more than three terms, try factoring by grouping. If that does not work, see if three of the terms are the square of a binomial.
5. As a final step, examine your factored polynomial to see if any factors listed have a common factor and can be factored further. If you find a common factor, factor it out at this point.
6. Check the answer by multiplying the factors together.

$$2x^7 + 16x^6 + 24x^5 = 2x^5(x^2 + 8x + 12)$$
$$= 2x^5(x + 6)(x + 2)$$

$$36a^6 - 100a^4b^2 = 4a^4(9a^2 - 25b^2)$$
$$= 4a^4[(3a)^2 - (5b)^2]$$
$$= 4a^4(3a + 5b)(3a - 5b)$$

$$125m^3 - 64 = (5m)^3 - 4^3$$
$$= (5m - 4)(25m^2 + 20m + 16)$$

Section 5.8

A **polynomial equation** is formed when two polynomials are set equal to each other.

$$x^2 - 5x = 2x + 7$$

A **quadratic equation** is a second-degree polynomial equation in one variable.

Standard Form of a Quadratic Equation
$$ax^2 + bx + c = 0, \quad a \neq 0$$
where a, b, and c are real numbers.

$$2x^2 - 6x + 11 = 0$$
$$x^2 - 4 = x + 2$$

$x^2 - 3x + 5 = 0$ is a quadratic equation in standard form.

Zero-Factor Property

For all real numbers a and b, if $a \cdot b = 0$, then either $a = 0$ or $b = 0$, or both a and $b = 0$.

Solve $(x + 6)(x - 1) = 0$.
$$x + 6 = 0 \quad \text{or} \quad x - 1 = 0$$
$$x = -6 \qquad\qquad x = 1$$
The solutions are -6 and 1.

To Solve an Equation by Factoring

1. Use the addition property to remove all terms from one side of the equation. This will result in one side of the equation being equal to 0.
2. Combine like terms in the equation and then factor.
3. Set each factor *containing a variable* equal to 0, solve the equations, and find the solutions.
4. Check the solutions in the *original* equation.

Solve $3x^2 + 13x - 4 = 2x$.
$$3x^2 + 11x - 4 = 0$$
$$(3x - 1)(x + 4) = 0$$
$$3x - 1 = 0 \quad \text{or} \quad x + 4 = 0$$
$$x = \frac{1}{3} \quad \text{or} \qquad x = -4$$

A check shows $\frac{1}{3}$ and -4 are the solutions.

Pythagorean Theorem

In a right triangle, if a and b represent the lengths of the legs and c represents the length of the hypotenuse, then
$$\text{leg}^2 + \text{leg}^2 = \text{hyp}^2$$
$$a^2 + b^2 = c^2$$

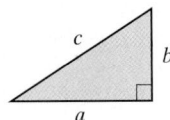

Find the length of the hypotenuse in the following right triangle.

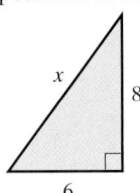

$$\text{leg}^2 + \text{leg}^2 = \text{hyp}^2$$
$$6^2 + 8^2 = x^2$$
$$36 + 64 = x^2$$
$$100 = x^2$$
$$10 = x$$

Note: -10 is not a possible answer.

Chapter 5 Review Exercises

[5.1]

Determine whether each expression is a polynomial. If the expression is a polynomial, **a)** *give the special name of the polynomial if it has one,* **b)** *write the polynomial in descending order of the variable x, and* **c)** *give the degree of the polynomial.*

1. $3x^2 + 9$

2. $5x + 4x^3 - 7$

3. $8x - x^{-1} + 6$

4. $-3 - 10x^2y + 6xy^3 + 2x^4$

Perform each indicated operation.

5. $(x^2 - 5x + 8) + (2x + 6)$

6. $(7x^2 + 2x - 5) - (2x^2 - 9x - 1)$

7. $(2a - 3b - 2) - (-a + 5b - 9)$

8. $(4x^3 - 4x^2 - 2x) + (2x^3 + 4x^2 - 7x + 13)$

9. $(3x^2y + 6xy - 5y^2) - (4y^2 + 3xy)$

10. $(-8ab + 2b^2 - 3a) + (-b^2 + 5ab + a)$

11. Add $x^2 - 3x + 12$ and $4x^2 + 10x - 9$.

12. Subtract $3a^2b - 2ab$ from $-7a^2b - ab$.

13. Find $P(2)$ if $P(x) = 2x^2 - 3x + 19$

14. Find $P(-3)$ if $P(x) = x^3 - 3x^2 + 4x - 10$

Perimeter   *In Exercises 15 and 16, find a polynomial expression for the perimeter of each figure.*

15.

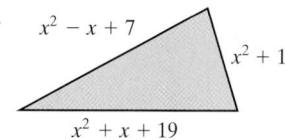

$x^2 - x + 7$

$x^2 + 1$

$x^2 + x + 19$

16.

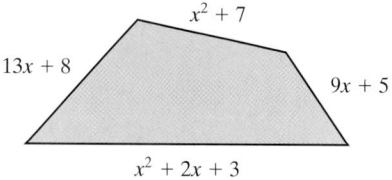

$x^2 + 7$

$13x + 8$

$9x + 5$

$x^2 + 2x + 3$

Use the following graph to work Exercises 17 and 18. The graph shows Social Security receipts and outlays from 1997 through 2025.

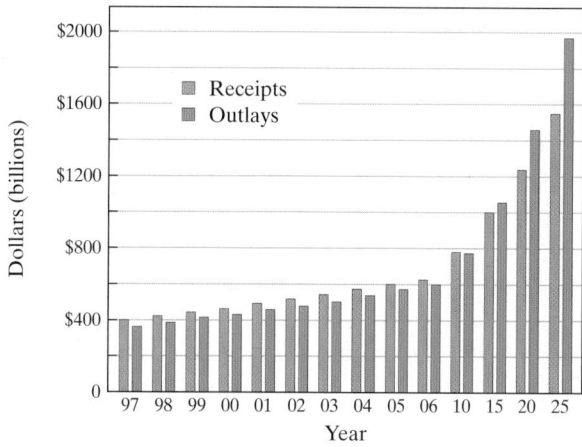

Social Security Receipts and Outlays

- Receipts
- Outlays

Dollars (billions)

Year

Source: Social Security Administration

17. Social Security Receipts The function $R(t) = 0.78t^2 + 20.28t + 385.0$, where t is years since 1997 and $0 \le t \le 28$, gives an approximation of Social Security receipts, $R(t)$, in billions of dollars.

   a) Using the function provided, estimate the receipts in 2010.

   b) Compare your answer in part **a)** with the graph. Does the graph support your answer?

18. Social Security Outlays The function $G(t) = 1.74t^2 + 7.32t + 383.91$, where t is years since 1997 and $0 \le t \le 28$, gives an approximation of Social Security outlays, $G(t)$, in billions of dollars.

   a) Using the function provided, estimate the outlays in 2010.

   b) Compare your answer in part **a)** with the graph. Does the graph support your answer?

[5.2]

Multiply.

19. $2x(3x^2 - 7x + 5)$

20. $-3xy^2(x^3 + xy^4 - 4y^5)$

21. $(3x - 5)(2x + 9)$

22. $(5a + 1)(10a - 3)$

23. $(x + 8y)^2$

24. $(a - 11b)^2$

25. $(2xy - 1)(5x + 4y)$

26. $(2pq - r)(3pq + 7r)$

27. $(2a + 9b)^2$

28. $(4x - 3y)^2$

29. $(7x + 5y)(7x - 5y)$

30. $(2a - 5b^2)(2a + 5b^2)$

31. $(4xy + 6)(4xy - 6)$

32. $(9a^2 - 2b^2)(9a^2 + 2b^2)$

33. $[(x + 3y) + 2]^2$

34. $[(2p - q) - 5]^2$

35. $(3x^2 + 4x - 6)(2x - 3)$

36. $(4x^3 + 6x - 2)(x + 3)$

Area   *In Exercises 37 and 38, find an expression for the total area of each figure.*

37.

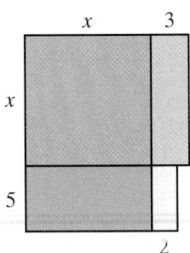

38.

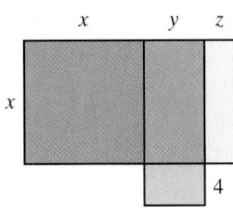

For each pair of functions, find **a)** $(f \cdot g)(x)$ *and* **b)** $(f \cdot g)(3)$.

39. $f(x) = x + 1, g(x) = x - 3$

40. $f(x) = 2x - 4, g(x) = x^2 - 3$

41. $f(x) = x^2 + x - 3, g(x) = x - 2$

42. $f(x) = x^2 - 2, g(x) = x^2 + 2$

[5.3]   *Divide.*

43. $\dfrac{4x^7y^5}{20xy^3}$

44. $\dfrac{3s^5t^8}{12s^5t^3}$

45. $\dfrac{45pq - 25q^2 - 15q}{5q}$

46. $\dfrac{7a^2 - 16a + 32}{4}$

47. $\dfrac{2x^3y^2 + 8x^2y^3 + 12xy^4}{8xy^3}$

48. $(8x^2 + 14x - 15) \div (2x + 5)$

49. $(2x^4 - 3x^3 + 4x^2 + 17x + 7) \div (2x + 1)$

50. $(4a^4 - 7a^2 - 5a + 4) \div (2a - 1)$

51. $(x^2 + x - 22) \div (x - 3)$

52. $(4x^3 + 12x^2 + x - 9) \div (2x + 3)$

Use synthetic division to obtain each quotient.

53. $(3x^3 - 2x^2 + 10) \div (x - 3)$

54. $(2y^5 - 10y^3 + y - 2) \div (y + 1)$

55. $(x^5 - 18) \div (x - 2)$

56. $(2x^3 + x^2 + 5x - 3) \div \left(x - \dfrac{1}{2}\right)$

Determine the remainder of each division problem using the Remainder Theorem. If the divisor is a factor of the dividend, so state.

57. $(x^2 - 4x + 13) \div (x - 3)$

58. $(2x^3 - 6x^2 + 3x) \div (x + 4)$

59. $(3x^3 - 6) \div \left(x - \dfrac{1}{3}\right)$

60. $(2x^4 - 6x^2 - 8) \div (x + 2)$

[5.4]   *Factor out the greatest common factor in each expression.*

61. $4x^2 + 8x + 32$

62. $15x^5 + 6x^4 - 12x^5y^3$

63. $10a^3b^3 - 14a^2b^6$

64. $24xy^4z^3 + 12x^2y^3z^2 - 30x^3y^2z^3$

Factor by grouping.

65. $5x^2 - xy + 30xy - 6y^2$

66. $12a^2 + 8ab + 15ab + 10b^2$

67. $(2x - 5)(2x + 1) - (2x - 5)(x - 8)$

68. $7x(3x - 7) + 3(3x - 7)^2$

Area   *In Exercises 69 and 70, A represents the area of the figure. Find an expression, in factored form, for the difference of the areas of the geometric figures.*

69.

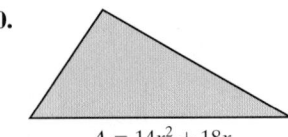

$A = 13x(5x + 2)$

$A = 7(5x + 2)$

70.

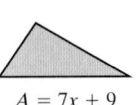

$A = 14x^2 + 18x$

$A = 7x + 9$

Volume *In Exercises 71 and 72, V represents the volume of the figure. Find an expression, in factored form, for the difference of the volumes of the geometric figures.*

71.

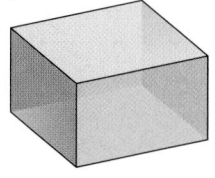

$V = 9x(17x + 3)$ $V = 7(17x + 3)$

72.

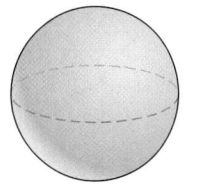

$V = 20x^2 + 25x$ $V = 8x + 10$

[5.5] *Factor each trinomial.*

73. $x^2 + 9x + 18$

74. $x^2 + 3x - 10$

75. $x^2 - 3x - 28$

76. $x^2 - 10x + 16$

77. $-x^2 + 12x + 45$

78. $-x^2 + 13x - 12$

79. $2x^3 + 13x^2 + 6x$

80. $8x^4 + 10x^3 - 25x^2$

81. $4a^5 - 9a^4 + 5a^3$

82. $12y^5 + 61y^4 + 5y^3$

83. $x^2 - 15xy - 54y^2$

84. $6p^2 - 19pq + 10q^2$

85. $x^4 + 10x^2 + 21$

86. $x^4 + 2x^2 - 63$

87. $(x + 3)^2 + 10(x + 3) + 24$

88. $(x - 4)^2 - (x - 4) - 20$

Area *In Exercises 89 and 90, find an expression, in factored form, for the area of the shaded region in each figure.*

89.

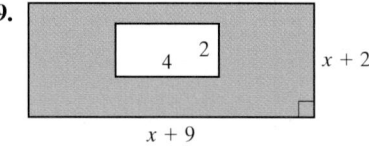

90.

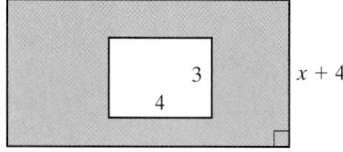

[5.6] *Use a special factoring formula to factor the following.*

91. $x^2 - 36$

92. $x^2 - 121$

93. $x^4 - 81$

94. $x^4 - 16$

95. $4a^2 + 4a + 1$

96. $16y^2 - 24y + 9$

97. $(x + 2)^2 - 16$

98. $(3y - 1)^2 - 36$

99. $p^4 + 18p^2 + 81$

100. $m^4 - 20m^2 + 100$

101. $x^2 + 8x + 16 - y^2$

102. $a^2 + 6ab + 9b^2 - 36c^2$

103. $16x^2 + 8xy + y^2$

104. $36b^2 - 60bc + 25c^2$

105. $x^3 - 27$

106. $y^3 + 64z^3$

107. $125x^3 - 1$

108. $8a^3 + 27b^3$

109. $y^3 - 64z^3$

110. $(x - 2)^3 - 27$

111. $(x + 1)^3 - 8$

112. $(a + 4)^3 + 1$

Area *In Exercises 113 and 114, find an expression, in factored form, for the area of the shaded region in each figure.*

113.

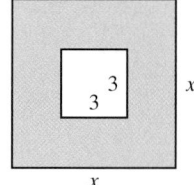

114.

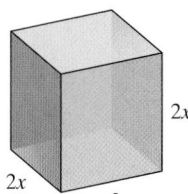

115. Volume Find an expression, in factored form, for the difference in volumes of the two cubes below.

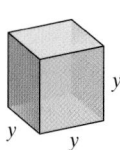

116. Volume Find an expression, in factored form, for the volume of the shaded region in the figure below.

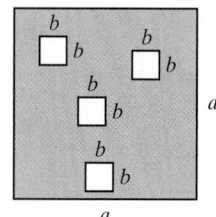

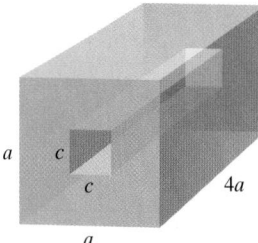

[5.4–5.7]

Factor completely.

117. $x^2y^4 - 2xy^4 - 15y^4$

118. $5x^3 - 30x^2 + 40x$

119. $3x^3y^4 + 18x^2y^4 - 6x^2y^4 - 36xy^4$

120. $3y^5 - 75y$

121. $4x^3y + 32y$

122. $5x^4y + 20x^3y + 20x^2y$

123. $6x^3 - 21x^2 - 12x$

124. $x^2 + 10x + 25 - z^2$

125. $5x^3 + 40y^3$

126. $x^2(x + 6) + 3x(x + 6) - 4(x + 6)$

127. $4(2x + 3)^2 - 12(2x + 3) + 5$

128. $4x^4 + 4x^2 - 3$

129. $(x + 1)x^2 - (x + 1)x - 2(x + 1)$

130. $9ax - 3bx + 21ay - 7by$

131. $6p^2q^2 - 5pq - 6$

132. $9x^4 - 12x^2 + 4$

133. $16y^2 - (x^2 + 4x + 4)$

134. $6(2a + 3)^2 - 7(2a + 3) - 3$

135. $6x^4y^5 + 9x^3y^5 - 27x^2y^5$

136. $x^3 - \dfrac{8}{27}y^6$

Area   *In Exercises 137–142, find an expression, in factored form, for the area of the shaded region in each figure.*

137.

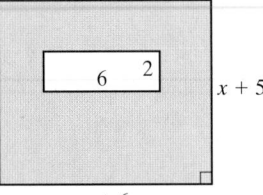

138.

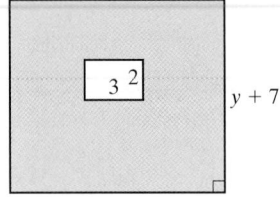

139.

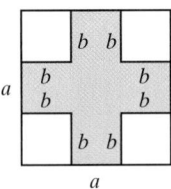

140.

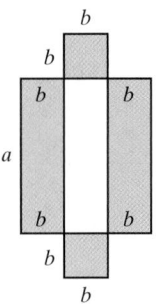

141.

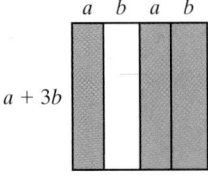

142.

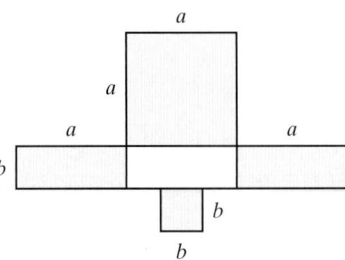

[5.8]   *Solve.*

143. $(x - 2)(4x + 1) = 0$

144. $(2x + 5)(3x + 10) = 0$

145. $4x^2 = 8x$

146. $12x^2 + 16x = 0$

147. $x^2 + 7x + 12 = 0$

148. $a^2 + a - 30 = 0$

149. $x^2 = 8x - 7$

150. $c^3 - 6c^2 + 8c = 0$

151. $5x^2 = 80$

152. $x(x + 3) = 2(x + 4) - 2$

153. $12d^2 = 13d + 4$

154. $20p^2 - 6 = 7p$

Use factoring to find the x-intercepts of the graph of each equation.

155. $y = 2x^2 - 6x - 36$

156. $y = 20x^2 - 49x + 30$

Write an equation whose graph will have x-intercepts at the given values.

157. -4 and 6

158. $-\dfrac{5}{2}$ and $-\dfrac{1}{6}$

In Exercises 159–163, answer the question.

159. Carpeting The area of Fred Bank's rectangular carpet is 108 square feet. Find the length and width of the carpet if the length is 3 feet greater than the width.

160. Triangular Sign The base of a large triangular sign is 5 feet more than twice the height. Find the base and height if the area of the triangle is 26 square feet.

161. Square One square has a side 4 inches longer than the side of a second square. If the area of the larger square is 49 square inches, find the length of a side of each square.

162. Velocity A rocket is projected upward from the top of a 144-foot-tall building with a velocity of 128 feet per second. The rocket's distance from the ground, s, at any time, t, in seconds, is given by the formula $s(t) = -16t^2 + 128t + 144$. Find the time it takes for the rocket to strike the ground.

163. Telephone Pole Two guy wires are attached to a telephone pole to help stabilize it. One wire is attached to the ground x feet from the base of the pole. The height of the pole is $x + 31$ and the length of the wire is $x + 32$. Find x.

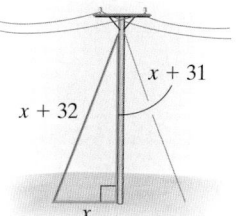

Chapter 5 Practice Test

Chapter Test Prep Videos provide fully worked-out solutions to any of the exercises you want to review. Chapter Test Prep Videos are available via **MyMathLab**, *or on* **YouTube** *(search "Angel Intermediate Algebra" and click on "Channels").*

1. a) Give the specific name of the following polynomial.

$$-4x^2 + 3x - 6x^4$$

b) Write the polynomial in descending powers of the variable x.

c) State the degree of the polynomial.

d) What is the leading coefficient of the polynomial?

Perform each operation.

2. $(7x^2y - 5y^2 + 4x) - (3x^2y + 9y^2 - 6y)$

3. $2x^3y^2(-4x^5y + 12x^3y^2 - 6x)$

4. $(2a - 3b)(5a + b)$

5. $(2x^2 + 3xy - 6y^2)(2x + y)$

6. $(12x^6 - 15x^2y + 21) \div 3x^2$

7. $(2x^2 - 7x + 9) \div (2x + 3)$

8. Use synthetic division to obtain the quotient. $(3x^4 - 12x^3 - 60x + 1) \div (x - 5)$

9. Use the Remainder Theorem to find the remainder when $2x^3 - 6x^2 - 5x + 8$ is divided by $x + 3$.

Factor completely.

10. $12x^3y + 10x^2y^4 - 14xy^3$

11. $x^3 - 2x^2 - 3x$

12. $2a^2 + 4ab + 3ab + 6b^2$

13. $2b^4 + 5b^2 - 18$

14. $4(x - 5)^2 + 20(x - 5)$

15. $(x + 4)^2 + 2(x + 4) - 3$

16. $27p^3q^6 - 8q^6$

17. If $f(x) = 3x - 4$ and $g(x) = x - 5$, find **a)** $(f \cdot g)(x)$ and **b)** $(f \cdot g)(2)$

Area *In Exercises 18 and 19, find an expression, in factored form, for the area of the shaded region.*

18.

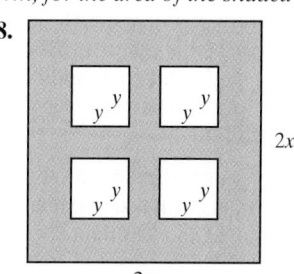

19.

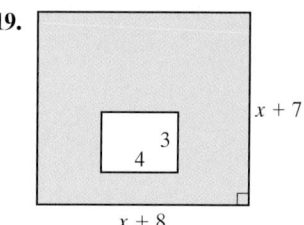

Solve.

20. $7x^2 + 25x - 12 = 0$

21. $x^3 + 3x^2 - 10x = 0$

22. Use factoring to find the x-intercepts of the graph of the equation $y = 8x^2 + 10x - 3$.

23. Find an equation whose graph has x-intercepts at 2 and 7.

24. Area The area of a triangle is 22 square meters. If the base of the triangle is 3 meters greater than 2 times the height, find the base and height of the triangle.

25. Baseball A baseball is projected upward from the top of a 448-foot-tall building with an initial velocity of 48 feet per second. The distance, s, of the baseball from the ground at any time, t, in seconds, is given by the equation $s(t) = -16t^2 + 48t + 448$. Find the time it takes for the baseball to strike the ground.

Cumulative Review Test

Take the following test and check your answers with those given in the back of the book. Review any questions that you answered incorrectly. The section where the material was covered is indicated after the answer.

1. Find $A \cup B$ for $A = \{2, 4, 6, 8\}$ and $B = \{3, 5, 6, 8\}$.

2. Illustrate $\{x | x \le -5\}$ on a real number line.

3. Divide $\left|\dfrac{3}{8}\right| \div (-4)$.

4. Evaluate $(-3)^3 - 2^2 - (-2)^2 + (9 - 8)^2$.

5. Simplify $\left(\dfrac{2r^4 s^5}{r^2}\right)^3$.

6. Solve $4(2x - 2) - 3(x + 7) = -4$.

7. Solve $k = 2(d + e)$ for e.

8. **Landscaping** Craig Campanella, a landscape architect, wishes to fence in two equal areas as illustrated in the figure. If both areas are squares and the total length of fencing used is 91 meters, find the dimensions of each square.

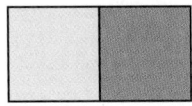

9. **Making Copies** Cecil Winthrop has a manuscript. He needs to make 6 copies before sending it to his editor in Boston. The first copy costs 15 cents per page and each additional copy costs 5 cents per page. If the total bill before tax is $248, how many pages are in the manuscript?

10. **Test Average** Todd Garner's first four test grades are 68, 72, 90, and 86. What range of grades on his fifth test will result in an average greater than or equal to 70 and less than 80?

11. Is $(4, 1)$ a solution to the equation $3x + 2y = 13$?

12. Write the equation $2 = 6x - 3y$ in standard form.

13. Find the slope of the line passing through the points $(8, -4)$ and $(-1, -2)$.

14. If $f(x) = 2x^3 - 4x^2 + x + 16$, find $f(-4)$.

15. Graph the inequality $2x - y \le 6$.

16. Solve the system of equations.

$$\frac{1}{5}x + \frac{1}{2}y = 4$$
$$\frac{2}{3}x - y = \frac{8}{3}$$

17. Solve the system of equations.

$$x - 2y = 2$$
$$2x + 3y = 11$$
$$-y + 4z = 7$$

18. Evaluate the determinant.

$$\begin{vmatrix} 8 & 5 \\ -2 & 1 \end{vmatrix}$$

19. Divide $(2x^3 - 9x + 15) \div (x - 6)$.

20. Factor $64x^3 - 27y^3$.

6

Rational Expressions and Equations

Goals of This Chapter

Rational expressions are expressions that contain fractions and *rational equations* are equations that contain rational expressions. In this chapter you will learn how to work with rational expressions, and solve rational equations. To be successful in this chapter, you must have an understanding of the factoring techniques discussed in Chapter 5.

© Jupiter Unlimited

When two or more people perform a task, it takes less time than for each person to do the task separately. For example, on page 405, we will determine how long it takes two people to mow lawns (Exercise 10), pick apples (Exercise 11), or plow a field (Exercise 15) when we know how long it takes each person to complete the task alone.

6.1 The Domains of Rational Functions and Multiplication and Division of Rational Expressions

1 Find the domains of rational functions.

2 Simplify rational expressions.

3 Multiply rational expressions.

4 Divide rational expressions.

Understanding Algebra

To understand rational expressions and rational functions, you must have a thorough understanding of the factoring techniques discussed in Chapter 5.

1 Find the Domains of Rational Functions

Rational Expression

A **rational expression** is an expression of the form $\frac{p}{q}$, where p and q are polynomials and $q \neq 0$.

Examples of Rational Expressions

$$\frac{2}{x}, \qquad \frac{t+3}{t}, \qquad \frac{p^2+4p}{p-6}, \qquad \frac{a}{a^2-4}$$

Note that the denominator of a rational expression cannot equal 0 because the fraction bar is a division symbol and division by 0 is undefined. For example,

- In the expression $\frac{2}{x}$, $x \neq 0$.

- In the expression $\frac{t+3}{t}$, $t \neq 0$.

- In the expression $\frac{p^2+4p}{p-6}$, $p - 6 \neq 0$ or $p \neq 6$.

- In the expression $\frac{a}{a^2-4}$, $a^2 - 4 \neq 0$ or $a \neq 2$ and $a \neq -2$.

Whenever we write a rational expression, we will assume that any value(s) that make the denominator 0 are excluded. For example, if we write $\frac{5}{x-3}$, we assume that $x \neq 3$, even though we do not specifically write it.

In Chapter 3 we introduced functions and in Chapter 5 we discussed polynomial functions. Next, we introduce rational functions.

Rational Function

A **rational function** is a function of the form $y = f(x) = \frac{p}{q}$ where p and q are polynomials and $q \neq 0$.

Understanding Algebra

The notation $y = f(x)$ means that the equation defines a function in which x is the independent variable and y is the dependent variable. Note that $f(x)$ does *not* mean f times x.

Examples of Rational Functions

$$f(x) = \frac{4}{x} \qquad y = \frac{x^2+2}{x+3} \qquad T(a) = \frac{a+9}{a^2-4} \qquad h(x) = \frac{7x-8}{2x+1}$$

Recall from Chapter 3 that the **domain** of a function is the set of values that can be used to replace the independent variable, usually x, in the function. Also recall from Chapter 5 that the domain of a polynomial function is the set of all real numbers. Since we cannot divide by 0, we have the following definition.

Domain of a Rational Function

The **domain of a rational function** $y = f(x) = \frac{p}{q}$ is the set of all real numbers for which the denominator, q, is *not* equal to 0.

For example,

- For the function $f(x) = \dfrac{2}{x}$, the domain is $\{x | x \neq 0\}$.

- For the function $f(t) = \dfrac{t + 3}{t}$, the domain is $\{t | t \neq 0\}$.

- For the function $g(p) = \dfrac{p^2 + 4p}{p - 6}$, the domain is $\{p | p \neq 6\}$.

- For the function $h(a) = \dfrac{a}{a^2 - 4}$, the domain is $\{a | a \neq 2 \text{ and } a \neq -2\}$.

EXAMPLE 1 Determine the domain of the following rational functions.

a) $f(x) = \dfrac{x + 1}{x - 6}$ 　　　　　　　　　**b)** $f(x) = \dfrac{x^2}{x^2 - 4}$

c) $f(x) = \dfrac{x - 3}{x^2 + 2x - 15}$ 　　　　　　**d)** $f(x) = \dfrac{x}{x^2 + 8}$

Solution

a) Since $f(x)$ is a rational function, the domain is all real numbers x for which the denominator, $x - 6$, is not equal to 0. Therefore, the domain is all real numbers except for 6. The domain is written

$$\{x | x \neq 6\}$$

b) The domain is the set of all real numbers x for which the denominator, $x^2 - 4$, is not equal to 0. First, rewrite the denominator in factored form:

$$f(x) = \dfrac{x^2}{x^2 - 4}$$

$$= \dfrac{x^2}{(x + 2)(x - 2)} \qquad \text{Factor the denominator of } f(x)$$

We see that x cannot be -2 or 2. The domain is written

$$\{x | x \neq -2 \text{ and } x \neq 2\}$$

c) The domain is the set of all real numbers x for which the denominator, $x^2 + 2x - 15$, is not equal to 0. First, rewrite the denominator in factored form:

$$f(x) = \dfrac{x - 3}{x^2 + 2x - 15}$$

$$= \dfrac{x - 3}{(x + 5)(x - 3)} \qquad \text{Factor the denominator of } f(x)$$

Although the rational function has a factor of $(x - 3)$ common to both numerator and denominator, a value of $x = 3$ would lead to an undefined function. Thus, x cannot be -5 or 3. The domain is written

$$\{x | x \neq -5 \text{ and } x \neq 3\}$$

d) The domain is all real numbers x for which the denominator, $x^2 + 8$, is not equal to 0. Since, $x^2 + 8$ is always positive, the denominator can never equal 0, and the domain is all real numbers. The domain is written

$$\{x | x \text{ is a real number}\}$$

Now Try Exercise 21

Using Your Graphing Calculator ▪▪▪▪

Experimenting with a graphing calculator will give you some idea of the wide variety of graphs of rational functions. For Example, **Figure 6.1** shows the graph of $f(x) = \dfrac{x + 1}{x - 6}$ from Example 1a) on the TI-84 Plus. The domain of $f(x)$ is $\{x \mid x \neq 6\}$. Note that there is no point on the graph that corresponds to $x = 6$.

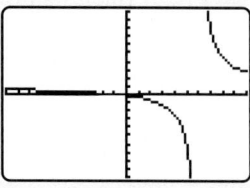

FIGURE 6.1   FIGURE 6.2

Another way to explore the domains of rational functions is by using the TABLE feature. **Figure 6.2** shows the ordered pairs of the function. Notice that when $x = 6$ the y value displayed is ERROR. This is because $f(6)$ is undefined, so $x = 6$ is excluded from the domain of $f(x)$.

2 Simplify Rational Expressions

A rational expression is **simplified** when the numerator and denominator have no common factors other than 1. The fraction $\dfrac{6}{9}$ is not simplified because the 6 and 9 both contain the common factor of 3. When the 3 is factored out, the simplified fraction is $\dfrac{2}{3}$.

$\dfrac{6}{9}$ is not a simplified fraction $\longrightarrow$ $\dfrac{6}{9} = \dfrac{\overset{1}{\cancel{3}} \cdot 2}{\underset{1}{\cancel{3}} \cdot 3} = \dfrac{2}{3}$ $\longleftarrow$ $\dfrac{2}{3}$ is a simplified fraction

The rational expression $\dfrac{ab - b^2}{2b}$ is not simplified because both the numerator and denominator have a common factor, b. To simplify this expression, factor b from each term in the numerator; then divide it out.

$$\frac{ab - b^2}{2b} = \frac{\cancel{b}(a - b)}{2\cancel{b}} = \frac{a - b}{2}$$

Thus $\dfrac{ab - b^2}{2b}$ becomes $\dfrac{a - b}{2}$ when simplified.

To Simplify Rational Expressions

1. Factor both numerator and denominator as completely as possible.
2. Divide both the numerator and the denominator by any common factors.

EXAMPLE 2 Simplify.   **a)** $\dfrac{x^2 + 5x + 4}{x + 4}$ **b)** $\dfrac{3x^3 - 3x^2}{x^3 - x}$

Solution

a) $\dfrac{x^2 + 5x + 4}{x + 4}$

$= \dfrac{(x + 4)(x + 1)}{x + 4}$ Factor the numerator.

$= \dfrac{\cancel{(x + 4)}(x + 1)}{\cancel{x + 4}}$ Divide out the common factor.

$= x + 1$

b) $\dfrac{3x^3 - 3x^2}{x^3 - x}$

$= \dfrac{3x^2(x - 1)}{x(x^2 - 1)}$ Factor out GCF from numerator and denominator.

$= \dfrac{3x^2(x - 1)}{x(x + 1)(x - 1)}$ Factor $x^2 - 1$ as the difference of two squares.

$= \dfrac{3 \overset{x}{\cancel{x^2}} \cancel{(x - 1)}}{\cancel{x}(x + 1)\cancel{(x - 1)}}$ Divide out common factors.

$= \dfrac{3x}{x + 1}$

Now Try Exercise 33

Understanding Algebra

When a factor in the numerator has terms that are *opposites* of a factor in the denominator, those two factors have a ratio of -1. For example:

$$\dfrac{3x - 2}{2 - 3x}$$

$$= \dfrac{3x - 2}{-1(-2 + 3x)}$$

$$= \dfrac{\cancel{(3x - 2)}}{-1\cancel{(3x - 2)}}$$

$$= \dfrac{1}{-1}$$

$$= -1$$

When the terms in a numerator differ only in sign from the terms in a denominator, we can factor out -1 from either the numerator or denominator. For example,

$$-2x + 3 = -1(2x - 3) = -(2x - 3)$$

$$6 - 5x = -1(-6 + 5x) = -(5x - 6)$$

$$-3x^2 + 8x - 6 = -1(3x^2 - 8x + 6) = -(3x^2 - 8x + 6)$$

EXAMPLE 3 Simplify $\dfrac{27x^3 - 8}{2 - 3x}$.

Solution

$\dfrac{27x^3 - 8}{2 - 3x} = \dfrac{(3x)^3 - (2)^3}{2 - 3x}$ Write the numerator as a difference of two cubes.

$= \dfrac{(3x - 2)(9x^2 + 6x + 4)}{2 - 3x}$ Factor; recall that $a^3 - b^3 = (a - b)(a^2 + ab + b^2)$.

$= \dfrac{\cancel{(3x - 2)}(9x^2 + 6x + 4)}{-1\cancel{(3x - 2)}}$ Factor -1 from the denominator and divide out common factors.

$= \dfrac{9x^2 + 6x + 4}{-1}$

$= -(9x^2 + 6x + 4)$ or $-9x^2 - 6x - 4$

Now Try Exercise 41

Avoiding Common Errors

| INCORRECT | INCORRECT |

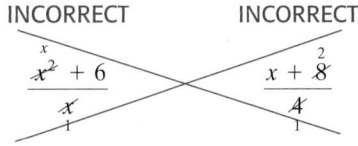

Remember that you can divide out only common *factors*. Therefore, the expressions $\dfrac{x^2 + 6}{x}$ and $\dfrac{x + 8}{4}$ cannot be simplified. Only when expressions are *multiplied* can they be factors. Neither of the expressions above can be simplified from their original form.

| CORRECT | INCORRECT |

$\dfrac{x^2 - 4}{x - 2} = \dfrac{(x + 2)\cancel{(x - 2)}}{\cancel{x - 2}}$

$= x + 2$

3 Multiply Rational Expressions

Now that we know how to simplify a rational expression, we can discuss multiplication of rational expressions.

To Multiply Rational Expressions

To multiply rational expressions, use the following rule:

$$\frac{a}{b} \cdot \frac{c}{d} = \frac{a \cdot c}{b \cdot d}, \quad b \neq 0, d \neq 0$$

To multiply rational expressions, follow these steps.

1. Factor all numerators and denominators as far as possible.
2. Divide out any common factors.
3. Multiply using the above rule.
4. Simplify the answer when possible.

If all common factors were factored out in step 2, your answer in step 4 should be in simplified form. However, if you missed a common factor in step 2, you can factor it out in step 4 to obtain an answer that is simplified.

EXAMPLE 4 Multiply. **a)** $\dfrac{x-5}{6x} \cdot \dfrac{x^2 - 2x}{x^2 - 7x + 10}$ **b)** $\dfrac{2x-3}{x-4} \cdot \dfrac{x^2 - 8x + 16}{3 - 2x}$

Solution

a) $\dfrac{x-5}{6x} \cdot \dfrac{x^2 - 2x}{x^2 - 7x + 10} = \dfrac{x-5}{6x} \cdot \dfrac{x(x-2)}{(x-2)(x-5)}$ Factor; divide out common factors.

$\qquad = \dfrac{1}{6}$

b) $\dfrac{2x-3}{x-4} \cdot \dfrac{x^2 - 8x + 16}{3 - 2x} = \dfrac{2x-3}{x-4} \cdot \dfrac{(x-4)(x-4)}{3 - 2x}$ Factor.

$\qquad = \dfrac{2x-3}{x-4} \cdot \dfrac{(x-4)(x-4)}{-1(2x-3)}$ Factor -1 from denominator; divide out common factors.

$\qquad = \dfrac{x-4}{-1}$

$\qquad = -(x-4) \quad \text{or} \quad -x + 4 \quad \text{or} \quad 4 - x$

Now Try Exercise 61

EXAMPLE 5 Multiply $\dfrac{x^2 - y^2}{x + y} \cdot \dfrac{x + 4y}{2x^2 - xy - y^2}$.

Solution

$\dfrac{x^2 - y^2}{x + y} \cdot \dfrac{x + 4y}{2x^2 - xy - y^2} = \dfrac{(x+y)(x-y)}{x+y} \cdot \dfrac{x + 4y}{(2x + y)(x-y)}$ Factor: divide out common factors.

$\qquad = \dfrac{x + 4y}{2x + y}$

Now Try Exercise 55

7. The rational expression $\dfrac{x^2 - 2x - 15}{x^2 + 3x - 40}$ simplifies to _____ .

8. The product $\dfrac{x^2}{y} \cdot \dfrac{y^2}{x^4}$ is _____ .

9. The quotient $\dfrac{x^5}{y^4} \div \dfrac{x^2}{y^2}$ is _____ .

10. The domain of the function $f(x) = \dfrac{x + 5}{x - 4}$ is _____ .

Practice the Skills

Determine the values that are excluded in the following expressions.

11. $\dfrac{7}{x}$

12. $\dfrac{x + 2}{x^2 - 64}$

13. $\dfrac{4}{2x^2 - 15x + 25}$

14. $\dfrac{2}{(x - 6)^2}$

15. $\dfrac{x - 3}{x^2 + 12}$

16. $\dfrac{-2}{49 - r^2}$

17. $\dfrac{x^2 + 81}{x^2 - 81}$

18. $\dfrac{x^2 - 36}{x^2 + 36}$

Determine the domain of each function.

19. $f(x) = \dfrac{x - 4}{x - 5}$

20. $f(z) = \dfrac{3}{-18z + 9}$

21. $y = \dfrac{5}{x^2 + x - 6}$

22. $y = \dfrac{9}{x^2 + 4x - 21}$

23. $f(a) = \dfrac{a^2 + 3a + 2}{a^2 + 4a + 3}$

24. $f(x) = \dfrac{10 - 3x}{x^3 + 8x}$

25. $g(x) = \dfrac{x^2 - x + 8}{x^2 + 4}$

26. $h(x) = \dfrac{x^3 - 64x}{x^2 + 81}$

27. $m(a) = \dfrac{a^2 + 36}{a^2 - 36}$

28. $k(b) = \dfrac{b^2 - 36}{b^2 + 36}$

Simplify each rational expression.

29. $\dfrac{x^2 + x}{x}$

30. $\dfrac{x^2 - 5x}{x}$

31. $\dfrac{5x^2 - 20xy}{15x}$

32. $\dfrac{x^2 + 7x}{x^2 - 2x}$

33. $\dfrac{x^3 - x}{x^2 - 1}$

34. $\dfrac{4x^2y + 12xy + 18x^3y^3}{10xy^2}$

35. $\dfrac{5r - 8}{8 - 5r}$

36. $\dfrac{4x^2 - 16x^4 + 6x^5y}{14x^3y^2}$

37. $\dfrac{p^2 - 2p - 24}{6 - p}$

38. $\dfrac{4x^2 - 9}{2x^2 - x - 3}$

39. $\dfrac{a^2 - 3a - 10}{a^2 + 5a + 6}$

40. $\dfrac{y^2 - 10yz + 24z^2}{y^2 - 5yz + 4z^2}$

41. $\dfrac{8x^3 - 125y^3}{2x - 5y}$

42. $\dfrac{64x^3 - 27z^3}{3z - 4x}$

43. $\dfrac{(x + 6)(x - 3) + (x + 6)(x - 2)}{2(x + 6)}$

44. $\dfrac{(2x - 1)(x + 4) + (2x - 1)(x + 1)}{3(2x - 1)}$

45. $\dfrac{a^2 + 7a - ab - 7b}{a^2 - ab + 5a - 5b}$

46. $\dfrac{xy - yw + xz - zw}{xy + yw + xz + zw}$

47. $\dfrac{x^2 - x - 12}{x^3 + 27}$

48. $\dfrac{a^3 - b^3}{a^2 - b^2}$

Multiply or divide as indicated. Simplify all answers.

49. $\dfrac{3x}{5y^2} \cdot \dfrac{y}{9}$

50. $\dfrac{32x^2}{y^4} \cdot \dfrac{5x^3}{8y^2}$

51. $\dfrac{9x^3}{4} \div \dfrac{3}{16y^2}$

52. $\dfrac{10m^4}{49x^5y^7} \div \dfrac{25m^5}{21x^{12}y^5}$

53. $\dfrac{3 - r}{r - 3} \cdot \dfrac{r - 9}{9 - r}$

54. $\dfrac{7a + 7b}{5} \div \dfrac{a^2 - b^2}{a - b}$

55. $\dfrac{x^2 + 3x - 10}{4x} \cdot \dfrac{x^2 - 3x}{x^2 - 5x + 6}$

56. $\dfrac{p^2 + 7p + 10}{p + 5} \cdot \dfrac{1}{p + 2}$

57. $\dfrac{r^2 + 10r + 21}{r + 7} \div \dfrac{(r^2 - 5r - 24)}{r^3}$

58. $(x - 3) \div \dfrac{x^2 + 3x - 18}{x^3}$

59. $\dfrac{x^2 + 12x + 35}{x^2 + 4x - 5} \div \dfrac{x^2 + 3x - 28}{7x - 7}$

60. $\dfrac{x + 1}{x^2 - 17x + 30} \div \dfrac{8x + 8}{x^2 + 7x - 18}$

61. $\dfrac{a - b}{9a + 9b} \div \dfrac{a^2 - b^2}{a^2 + 2a + 1}$

62. $\dfrac{2x^2 + 8xy + 8y^2}{x^2 + 4xy + 4y^2} \cdot \dfrac{2x^2 + 7xy + 6y^2}{4x^2 + 14xy + 12y^2}$

63. $\dfrac{3x^2 - x - 4}{4x^2 + 5x + 1} \cdot \dfrac{2x^2 - 5x - 12}{6x^2 + x - 12}$

64. $\dfrac{6x^3 - x^2 - x}{2x^2 + x - 1} \cdot \dfrac{x^2 - 1}{x^3 - 2x^2 + x}$

65. $\dfrac{x + 2}{x^3 - 8} \cdot \dfrac{(x - 2)^2}{x^2 + 4}$

66. $\dfrac{x^4 - y^8}{x^2 + y^4} \div \dfrac{x^2 - y^4}{x^2}$

67. $\dfrac{x^2 - y^2}{x^2 - 2xy + y^2} \div \dfrac{(x + y)^2}{(x - y)^2}$

68. $\dfrac{(x^2 - y^2)^2}{(x^2 - y^2)^3} \div \dfrac{x^2 + y^2}{x^4 - y^4}$

69. $\dfrac{2x^4 + 4x^2}{6x^2 + 14x + 4} \div \dfrac{x^2 + 2}{3x^2 + x}$

70. $\dfrac{8a^3 - 1}{4a^2 + 2a + 1} \div \dfrac{a^2 - 2a + 1}{(a - 1)^2}$

71. $\dfrac{(a - b)^3}{a^3 - b^3} \cdot \dfrac{a^2 - b^2}{(a - b)^2}$

72. $\dfrac{r^2 - 16}{r^3 - 64} \div \dfrac{r^2 + 8r + 16}{r^2 + 4r + 16}$

73. $\dfrac{4x + y}{5x + 2y} \cdot \dfrac{10x^2 - xy - 2y^2}{8x^2 - 2xy - y^2}$

74. $\dfrac{2x^3 - 7x^2 + 3x}{x^2 + 2x - 3} \cdot \dfrac{x^2 + 3x}{(x - 3)^2}$

75. $\dfrac{ac - ad + bc - bd}{ac + ad + bc + bd} \cdot \dfrac{pc + pd - qc - qd}{pc - pd + qc - qd}$

76. $\dfrac{2p^2 + 2pq - pq^2 - q^3}{p^3 + p^2 + pq^2 + q^2} \div \dfrac{p^3 + p + p^2q + q}{p^3 + p + p^2 + 1}$

77. $\dfrac{3r^2 + 17rs + 10s^2}{6r^2 + 13rs - 5s^2} \div \dfrac{6r^2 + rs - 2s^2}{6r^2 - 5rs + s^2}$

78. $\dfrac{x^3 - 4x^2 + x - 4}{x^5 - x^4 + x^3 - x^2} \cdot \dfrac{2x^3 + 2x^2 + x + 1}{2x^3 - 8x^2 + x - 4}$

Problem Solving

Determine the polynomial to be placed in the shaded area to give a true statement.

79. $\dfrac{\rule{1.5cm}{0.3cm}}{x^2 + 2x - 15} = \dfrac{1}{x - 3}$

80. $\dfrac{\rule{1.5cm}{0.3cm}}{3x + 2} = x - 3$

81. $\dfrac{y^2 - y - 20}{\rule{1.5cm}{0.3cm}} = \dfrac{y + 4}{y + 1}$

82. $\dfrac{\rule{1.5cm}{0.3cm}}{6p^2 + p - 15} = \dfrac{2p - 1}{2p - 3}$

Determine the polynomial to be placed in the shaded area to give a true statement.

83. $\dfrac{x^2 - x - 12}{x^2 + 2x - 3} \cdot \dfrac{\rule{1.5cm}{0.3cm}}{x^2 - 2x - 8} = 1$

84. $\dfrac{x^2 - 4}{(x + 2)^2} \cdot \dfrac{2x^2 + x - 6}{\rule{1.5cm}{0.3cm}} = \dfrac{x - 2}{2x + 5}$

85. $\dfrac{x^2 - 9}{2x^2 + 3x - 2} \div \dfrac{2x^2 - 9x + 9}{\rule{1.5cm}{0.3cm}} = \dfrac{x + 3}{2x - 1}$

86. $\dfrac{4r^2 - r - 18}{\rule{1.5cm}{0.3cm}} \div \dfrac{4r^3 - 9r^2}{6r^2 - 9r + 3} = \dfrac{3(r - 1)}{r^2}$

87. Area Consider the rectangle below. Its area is $3a^2 + 7ab + 2b^2$ and its length is $2a + 4b$. Find its width, w, in terms of a and b, by dividing its area by its length.

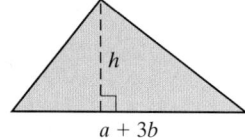

$2a + 4b$

88. Area Consider the rectangle below. Its area is $a^2 + 2ab + b^2$ and its length is $3a + 3b$. Find its width, w, in terms of a and b, by dividing its area by its length.

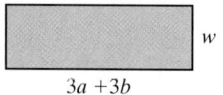

$3a + 3b$

89. Area Consider the triangle below. If its area is $a^2 + 4ab + 3b^2$ and its base is $a + 3b$, find its height h. Use area $= \dfrac{1}{2}$ (base)(height).

h

$a + 3b$

90. Area Consider the trapezoid below. If its area is $a^2 + 2ab + b^2$, find its height, h. Use area $= \dfrac{1}{2}h(a + b)$.

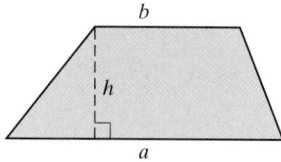

b

h

a

Challenge Problems

Perform each indicated operation.

91. $\left(\dfrac{2x^2 - 3x - 14}{2x^2 - 9x + 7} \div \dfrac{6x^2 + x - 15}{3x^2 + 2x - 5}\right) \cdot \dfrac{6x^2 - 7x - 3}{2x^2 - x - 3}$

92. $\left(\dfrac{a^2 - b^2}{2a^2 - 3ab + b^2} \cdot \dfrac{2a^2 - 7ab + 3b^2}{a^2 + ab}\right) \div \dfrac{ab - 3b^2}{a^2 + 2ab + b^2}$

93. $\dfrac{5x^2(x - 1) - 3x(x - 1) - 2(x - 1)}{10x^2(x - 1) + 9x(x - 1) + 2(x - 1)} \cdot \dfrac{2x + 1}{x + 3}$

94. $\dfrac{x^2(3x - y) - 5x(3x - y) - 24(3x - y)}{x^2(3x - y) - 9x(3x - y) + 8(3x - y)} \cdot \dfrac{x - 1}{x + 3}$

95. $\dfrac{(x - p)^n}{x^{-2}} \div \dfrac{(x - p)^{2n}}{x^{-6}}$

96. $\dfrac{x^{-3}}{(a - b)^r} \div \dfrac{x^{-5}}{(a - b)^{r+2}}$

Simplify.

97. $\dfrac{x^{5y} + 3x^{4y}}{3x^{3y} + x^{4y}}$

98. $\dfrac{m^{2x} - m^{x} - 2}{m^{2x} - 4}$

For Exercises 99–102,

    a) *Determine the domain of the function.*

    b) *Graph the function in connected mode.*

99. $f(x) = \dfrac{1}{x - 2}$ **100.** $f(x) = \dfrac{x}{x - 2}$ **101.** $f(x) = \dfrac{x^2}{x - 2}$ **102.** $f(x) = \dfrac{x - 2}{x - 2}$

103. Consider the rational function $f(x) = \dfrac{1}{x}$.

    a) Determine the domain of the function.

    b) Complete the following table for the function.

x	−10	−1	−0.5	−0.1	−0.01	0.01	0.1	0.5	1	10
y										

    c) Draw the graph of $f(x) = \dfrac{1}{x}$. Consider what happens to the function as x gets closer and closer to 0, approaching 0 from both the left and right sides.

    d) Can this fraction ever have a value of 0? Explain your answer.

Concept/Writing Exercises

104. Make up a rational expression that is undefined at $x = 4$ and $x = -5$. Explain how you determined your answer.

105. Make up a rational expression that is undefined at $x = 2$ and $x = -3$. Explain how you determined your answer.

106. Consider the rational function $g(x) = \dfrac{2}{x + 3}$. Explain why this function can never equal 0.

107. Consider the rational function $f(x) = \dfrac{1}{x}$. Explain why this function can never equal 0.

108. Consider the function $f(x) = \dfrac{x - 2}{x^2 - 81}$. For what value of x, if any, will this function **a)** equal 0? **b)** be undefined? Explain.

109. Consider the rational function $f(x) = \dfrac{x - 4}{x^2 - 36}$. For what value of x, if any, will this function **a)** equal 0? **b)** be undefined? Explain.

110. Give a function that is undefined at $x = -4$ and $x = -2$ and has a value of 0 at $x = 5$. Explain how you determined your answer.

111. Give a function that is undefined at $x = 3$ and $x = -1$ and has a value of 0 at $x = 2$. Explain how you determined your answer.

Group Activity

112. Consider the rational function $f(x) = \dfrac{x^2 - 4}{x - 2}$.

    a) As a group, determine the domain of this function.

    b) Have each member of the group individually complete the following table for the function.

x	−2	−1	0	1	1.9	1.99	2.01	2.1	3	4	5	6
y												

    c) Compare your answers to part **b)** and agree on the correct table values.

    d) As a group draw the graph of $f(x) = \dfrac{x^2 - 4}{x - 2}$. Is the function defined when $x = 2$?

    e) Can this function ever have a value of 0? If so, for what value(s) of a is $f(a) = 0$?

Cumulative Review Exercises

[2.2]   **113.** Solve $6(x - 2) + 6y = 12x$ for y.

[2.5]   **114.** Solve $4 + \dfrac{4x}{3} < 6$ and give the answer in interval notation.

[2.6]   **115.** Solve $\left| \dfrac{2x - 4}{12} \right| = 5$.

[3.2]   **116.** Let $f(x) = |6 - 3x| - 2$. Find $f(1.3)$.

[4.1]   **117.** Solve the system of equations.

$$3x + 4y = 2$$
$$2x + 5y = -1$$

[5.6]   **118.** Factor $9x^2 + 6xy + y^2 - 4$.

6.2 Addition and Subtraction of Rational Expressions

1 Add and subtract expressions with a common denominator.

2 Find the least common denominator (LCD).

3 Add and subtract expressions with different denominators.

4 Study an application of rational expressions.

1 Add and Subtract Expressions with a Common Denominator

When adding or subtracting two rational expressions with a common denominator, we add or subtract the numerators while keeping the common denominator.

To Add or Subtract Rational Expressions with a Common Denominator

To add or subtract rational expressions, use the following rules.

ADDITION	SUBTRACTION
$\dfrac{a}{c} + \dfrac{b}{c} = \dfrac{a+b}{c}, \quad c \neq 0$	$\dfrac{a}{c} - \dfrac{b}{c} = \dfrac{a-b}{c}, \quad c \neq 0$

To add or subtract rational expressions with a common denominator,

1. Add or subtract the expressions using the rules given above.
2. Simplify the expression if possible.

EXAMPLE 1 Add.

a) $\dfrac{3}{x+6} + \dfrac{x-4}{x+6}$

b) $\dfrac{x^2 + 3x - 2}{(x+5)(x-3)} + \dfrac{4x+12}{(x+5)(x-3)}$

Solution

a) Since the denominators are the same, we add the numerators and keep the common denominator.

$$\frac{3}{x+6} + \frac{x-4}{x+6} = \frac{3 + (x-4)}{x+6} \qquad \text{Add numerators.}$$

$$= \frac{x-1}{x+6}$$

b)
$$\frac{x^2+3x-2}{(x+5)(x-3)} + \frac{4x+12}{(x+5)(x-3)} = \frac{x^2+3x-2+(4x+12)}{(x+5)(x-3)} \qquad \text{Add numerators.}$$

$$= \frac{x^2+7x+10}{(x+5)(x-3)} \qquad \text{Combine like terms.}$$

$$= \frac{\cancel{(x+5)}(x+2)}{\cancel{(x+5)}(x-3)} \qquad \text{Factor; divide out common factors.}$$

$$= \frac{x+2}{x-3}$$

Now Try Exercise 11

Avoiding Common Errors

How do you simplify this problem?

$$\frac{4x}{x-2} - \frac{2x+1}{x-2}$$

CORRECT

$$\frac{4x}{x-2} - \frac{2x+1}{x-2} = \frac{4x-(2x+1)}{x-2}$$

$$= \frac{4x-2x-1}{x-2}$$

$$= \frac{2x-1}{x-2}$$

INCORRECT

$$\frac{4x}{x-2} - \frac{2x+1}{x-2} = \frac{4x-2x+1}{x-2}$$

$$= \frac{2x+1}{x-2}$$

The procedure on the right side is incorrect because the *entire numerator*, $2x+1$, must be subtracted from $4x$. Instead, only $2x$ was subtracted. Note that *the sign of each term (not just the first term) in the numerator of the fraction being subtracted must change*. Note that $-(2x+1) = -2x - 1$, by the distributive property.

EXAMPLE 2 Subtract $\dfrac{a}{a-6} - \dfrac{a^2 - 4a - 6}{a-6}$.

Solution

$$\frac{a}{a-6} - \frac{a^2-4a-6}{a-6} = \frac{a-(a^2-4a-6)}{a-6} \qquad \text{Subtract numerators.}$$

$$= \frac{a-a^2+4a+6}{a-6} \qquad \text{Distributive property.}$$

$$= \frac{-a^2+5a+6}{a-6} \qquad \text{Combine like terms.}$$

$$= \frac{-(a^2-5a-6)}{a-6} \qquad \text{Factor out } -1.$$

$$= \frac{-\cancel{(a-6)}(a+1)}{\cancel{a-6}} \qquad \begin{array}{l}\text{Factor; divide out}\\\text{common factors.}\end{array}$$

$$= -(a+1) \text{ or } -a-1$$

Now Try Exercise 13

Understanding Algebra

When subtracting rational expressions, be sure to subtract the entire numerator of the fraction being subtracted. As seen in Example 2, the distributive property is used to change the sign of each term being subtracted.

---

2 Find the Least Common Denominator (LCD)

To add or subtract two fractions with different denominators, we first obtain a least common denominator (LCD). Obtaining an LCD may involve writing numbers as a product of prime numbers. For example, the numbers 36 and 48 can be written as

$$36 = 2 \cdot 2 \cdot 3 \cdot 3 = 2^2 \cdot 3^2$$
$$48 = 2 \cdot 2 \cdot 2 \cdot 2 \cdot 3 = 2^4 \cdot 3$$

We may need to write numerical coefficients as products of prime numbers to find the LCD.

Understanding Algebra

A *prime* number is a natural number greater than 1 that has only two divisors, itself and 1. The first 10 prime numbers are 2, 3, 5, 7, 11, 13, 17, 19, 23, and 29.

To Find the Least Common Denominator (LCD) of Rational Expressions

1. Write each nonprime coefficient (other than 1) of monomials that appear in denominators as a product of prime numbers.
2. Factor each denominator completely. Any factors that occur more than once should be expressed as powers. For example, $(x+5)(x+5)$ should be expressed as $(x+5)^2$.
3. List all different factors (other than 1) that appear in any of the denominators. When the same factor appears in more than one denominator, write the factor with the highest power that appears.
4. The least common denominator is the product of all the factors found in step 3.

EXAMPLE 3 Find the LCD of each expression.

a) $\dfrac{3}{5x} - \dfrac{2}{x^2}$ **b)** $\dfrac{1}{18x^3y} + \dfrac{5}{27x^2y^3}$ **c)** $\dfrac{3}{x} - \dfrac{2y}{x+5}$ **d)** $\dfrac{7}{x^2(x+1)} + \dfrac{3z}{x(x+1)^3}$

Solution

a) The factors that appear in the denominators are 5 and x. List each factor with its highest power. The LCD is the product of these factors.

Highest power of 5 ⌐   ⌐ Highest power of x

$$\text{LCD} = 5^1 \cdot x^2 = 5x^2$$

b) The numerical coefficients written as products of prime numbers are $18 = 2 \cdot 3^2$ and $27 = 3^3$. The variable factors are x and y. Using the highest powers of the factors, we obtain the LCD.

Highest power of 3 ⌐   ⌐ Highest power of x
Highest power of 2 ⌐   ⌐ Highest power of y

$$\text{LCD} = 2^1 \cdot 3^3 \cdot x^3 \cdot y^3 = 54x^3y^3$$

c) The factors are x and $x + 5$. Note that the x in the second denominator, $x + 5$, is not a factor of that denominator since the operation is addition rather than multiplication.

$$\text{LCD} = x(x + 5)$$

d) The factors are x and $x + 1$. The highest power of x is 2 and the highest power of $x + 1$ is 3.

$$\text{LCD} = x^2(x + 1)^3$$

Now Try Exercise 31

Sometimes it is necessary to factor all denominators to obtain the LCD. This is illustrated in the next example.

EXAMPLE 4 Find the LCD of each expression.

a) $\dfrac{3}{2x^2 - 4x} + \dfrac{8x}{x^2 - 4x + 4}$ **b)** $\dfrac{4x}{x^2 - x - 12} - \dfrac{6x^2}{x^2 - 7x + 12}$

Solution

a) Factor both denominators.

$$\frac{3}{2x^2 - 4x} + \frac{8x}{x^2 - 4x + 4} = \frac{3}{2x(x - 2)} + \frac{8x}{(x - 2)^2}$$

The factors are 2, x, and $x - 2$. Multiply the factors raised to the highest power that appears for each factor.

$$\text{LCD} = 2 \cdot x \cdot (x - 2)^2 = 2x(x - 2)^2$$

b) Factor both denominators.

$$\frac{4x}{x^2 - x - 12} - \frac{6x^2}{x^2 - 7x + 12} = \frac{4x}{(x + 3)(x - 4)} - \frac{6x^2}{(x - 3)(x - 4)}$$

$$\text{LCD} = (x + 3)(x - 4)(x - 3)$$

Note that although $x - 4$ is a common factor of each denominator, the highest power of the factor that appears in either denominator is 1.

Now Try Exercise 29

3 Add and Subtract Expressions with Different Denominators

The procedure used to add or subtract rational expressions with different denominators is given below.

> **To Add or Subtract Rational Expressions with Different Denominators**
>
> 1. Determine the least common denominator (LCD).
> 2. Rewrite each fraction as an equivalent fraction with the LCD. This is done by multiplying both the numerator and denominator of each fraction by any factors needed to obtain the LCD.
> 3. Leave the denominator in factored form, but multiply out the numerator.
> 4. Add or subtract the numerators while maintaining the LCD.
> 5. When it is possible to reduce the fraction by factoring the numerator, do so.

EXAMPLE 5 Add. **a)** $\dfrac{2}{x} + \dfrac{9}{y}$ **b)** $\dfrac{5}{4a^2} + \dfrac{3}{14ab^3}$

Solution

a) First determine the LCD.

$$LCD = xy$$

Now write each fraction with the LCD. Do this by multiplying *both* numerator and denominator of each fraction by any factors needed to obtain the LCD.

In this problem, the first fraction must be multiplied by $\dfrac{y}{y}$ and the second fraction must be multiplied by $\dfrac{x}{x}$.

$$\frac{2}{x} + \frac{9}{y} = \frac{y}{y} \cdot \frac{2}{x} + \frac{9}{y} \cdot \frac{x}{x} = \frac{2y}{xy} + \frac{9x}{xy}$$

Now, add the numerators while maintaining the LCD.

$$\frac{2y}{xy} + \frac{9x}{xy} = \frac{2y + 9x}{xy} \quad \text{or} \quad \frac{9x + 2y}{xy}$$

Therefore, $\dfrac{2}{x} + \dfrac{9}{y} = \dfrac{9x + 2y}{xy}$.

b) The LCD of 4 and 14 is 28. The LCD of the two fractions is $28a^2b^3$. First, write each fraction with the denominator $28a^2b^3$. To do this, multiply the first fraction by $\dfrac{7b^3}{7b^3}$ and the second fraction by $\dfrac{2a}{2a}$.

$$\frac{5}{4a^2} + \frac{3}{14ab^3} = \frac{7b^3}{7b^3} \cdot \frac{5}{4a^2} + \frac{3}{14ab^3} \cdot \frac{2a}{2a} \quad \text{Multiply to obtain LCD.}$$

$$= \frac{35b^3}{28a^2b^3} + \frac{6a}{28a^2b^3}$$

$$= \frac{35b^3 + 6a}{28a^2b^3} \quad \text{Add numerators.}$$

Now Try Exercise 39

Understanding Algebra

When we multiply the numerator and denominator of a rational expression by the same factor, we are in effect multiplying by 1. Thus, an equivalent fraction is obtained but the *value* of the fraction does not change. In Example 5**a)**,

$$\frac{2}{x} \text{ is equivalent to } \frac{2y}{xy}$$

and

$$\frac{9}{y} \text{ is equivalent to } \frac{9x}{xy}$$

EXAMPLE 6 Subtract $\dfrac{x+2}{x-4} - \dfrac{x+5}{x+4}$.

Solution The LCD is $(x-4)(x+4)$. Write each fraction with the denominator $(x-4)(x+4)$.

$$\frac{x+2}{x-4} - \frac{x+5}{x+4} = \boxed{\frac{x+4}{x+4}} \cdot \frac{x+2}{x-4} - \frac{x+5}{x+4} \cdot \boxed{\frac{x-4}{x-4}} \qquad \text{Multiply to obtain LCD.}$$

$$= \frac{(x+4)(x+2)}{(x+4)(x-4)} - \frac{(x+5)(x-4)}{(x+4)(x-4)}$$

$$= \frac{x^2+6x+8}{(x+4)(x-4)} - \frac{x^2+x-20}{(x+4)(x-4)} \qquad \begin{array}{l}\text{Multiply binomials in}\\\text{numerators.}\end{array}$$

$$= \frac{x^2+6x+8-(x^2+x-20)}{(x+4)(x-4)} \qquad \text{Subtract numerators.}$$

$$= \frac{x^2+6x+8-x^2-x+20}{(x+4)(x-4)} \qquad \text{Distributive property}$$

$$= \frac{5x+28}{(x+4)(x-4)} \qquad \text{Combine like terms.}$$

Now Try Exercise 45

EXAMPLE 7 Add $\dfrac{2}{x-3} + \dfrac{x+5}{3-x}$.

Solution Note that each denominator is the opposite, or additive inverse, of the other. We can multiply the numerator and denominator of either one of the fractions by -1 to obtain the LCD.

$$\frac{2}{x-3} + \frac{x+5}{3-x} = \frac{2}{x-3} + \boxed{\frac{-1}{-1}} \cdot \frac{(x+5)}{(3-x)} \qquad \text{Multiply to obtain LCD.}$$

$$= \frac{2}{x-3} + \frac{-x-5}{x-3} \qquad \text{Distributive property}$$

$$= \frac{2-x-5}{x-3} \qquad \text{Add numerators.}$$

$$= \frac{-x-3}{x-3} \qquad \text{Combine like terms.}$$

Understanding Algebra

When two rational expressions have denominators that are *opposites* of each other, the LCD can be obtained by multiplying the numerator and the denominator of either rational expression by -1.

Because there are no common factors in both the numerator and denominator, $\dfrac{-x-3}{x-3}$ cannot be simplified further.

Now Try Exercise 43

EXAMPLE 8 Subtract $\dfrac{3x+4}{2x^2-5x-12} - \dfrac{2x-3}{5x^2-18x-8}$.

Solution Factor the denominator of each expression.

$$\frac{3x+4}{2x^2-5x-12} - \frac{2x-3}{5x^2-18x-8} = \frac{3x+4}{(2x+3)(x-4)} - \frac{2x-3}{(5x+2)(x-4)}$$

The LCD is $(2x + 3)(x - 4)(5x + 2)$.

$$\frac{3x + 4}{(2x + 3)(x - 4)} - \frac{2x - 3}{(5x + 2)(x - 4)}$$

$$= \frac{5x + 2}{5x + 2} \cdot \frac{3x + 4}{(2x + 3)(x - 4)} - \frac{2x - 3}{(5x + 2)(x - 4)} \cdot \frac{2x + 3}{2x + 3}$$ Multiply to obtain LCD.

$$= \frac{15x^2 + 26x + 8}{(5x + 2)(2x + 3)(x - 4)} - \frac{4x^2 - 9}{(5x + 2)(2x + 3)(x - 4)}$$ Multiply numerators.

$$= \frac{15x^2 + 26x + 8 - (4x^2 - 9)}{(5x + 2)(2x + 3)(x - 4)}$$ Subtract numerators.

$$= \frac{15x^2 + 26x + 8 - 4x^2 + 9}{(5x + 2)(2x + 3)(x - 4)}$$ Distributive property

$$= \frac{11x^2 + 26x + 17}{(5x + 2)(2x + 3)(x - 4)}$$ Combine like terms.

Now Try Exercise 49

EXAMPLE 9 Perform the indicated operations.

$$\frac{x - 1}{x - 2} - \frac{x + 1}{x + 2} + \frac{x - 6}{x^2 - 4}$$

Solution   First, factor $x^2 - 4$. The LCD of the three fractions is $(x + 2)(x - 2)$.

$$\frac{x - 1}{x - 2} - \frac{x + 1}{x + 2} + \frac{x - 6}{x^2 - 4}$$

$$= \frac{x - 1}{x - 2} - \frac{x + 1}{x + 2} + \frac{x - 6}{(x + 2)(x - 2)}$$

$$= \frac{x + 2}{x + 2} \cdot \frac{x - 1}{x - 2} - \frac{x + 1}{x + 2} \cdot \frac{x - 2}{x - 2} + \frac{x - 6}{(x + 2)(x - 2)}$$ Multiply to obtain LCD.

$$= \frac{x^2 + x - 2}{(x + 2)(x - 2)} - \frac{x^2 - x - 2}{(x + 2)(x - 2)} + \frac{x - 6}{(x + 2)(x - 2)}$$ Multiply numerators.

$$= \frac{x^2 + x - 2 - (x^2 - x - 2) + (x - 6)}{(x + 2)(x - 2)}$$ Subtract and add numerators.

$$= \frac{x^2 + x - 2 - x^2 + x + 2 + x - 6}{(x + 2)(x - 2)}$$ Distributive property

$$= \frac{3x - 6}{(x + 2)(x - 2)}$$ Combine like terms.

$$= \frac{3(x - 2)}{(x + 2)(x - 2)}$$ Factor; divide out common factors.

$$= \frac{3}{x + 2}$$

Now Try Exercise 67

Helpful Hint

Study Tip

Now that we have discussed the operations of addition, subtraction, multiplication, and division of rational expressions, let's quickly summarize the procedures.

To add or subtract rational expressions, obtain the LCD. Express each fraction with the LCD. Then add or subtract the numerators and write this result over the LCD.

To multiply rational expressions, factor each expression completely, divide out common factors, multiply numerators, and multiply denominators.

To divide rational expressions, multiply the first (or top) fraction by the reciprocal of the second (or bottom) fraction. Then, factor each expression completely, divide out common factors, multiply the numerators, and multiply the denominators.

4 Study an Application of Rational Expressions

In economics, we study revenue, cost, and profit. If $R(x)$ is a revenue function and $C(x)$ is a cost function, then the profit function, $P(x)$, is

$$P(x) = R(x) - C(x)$$

where x is the number of items manufactured and sold by a company.

EXAMPLE 10 Sailboats The Don Perrione Sailboat Company builds and sells at least six sailboats each week.

Suppose
$$R(x) = \frac{6x - 7}{x + 2} \quad \text{and} \quad C(x) = \frac{4x - 13}{x + 3}$$

where x is the number of sailboats sold. Determine the profit function.

Solution   Understand and Translate   To determine the profit function, we subtract the cost function from the revenue function.

$$P(x) = R(x) - C(x)$$

$$P(x) = \frac{6x - 7}{x + 2} - \frac{4x - 13}{x + 3}$$

The LCD is $(x + 2)(x + 3)$.

Carry Out
$$= \frac{x + 3}{x + 3} \cdot \frac{6x - 7}{x + 2} - \frac{4x - 13}{x + 3} \cdot \frac{x + 2}{x + 2} \qquad \text{Multiply to obtain LCD.}$$

$$= \frac{6x^2 + 11x - 21}{(x + 3)(x + 2)} - \frac{4x^2 - 5x - 26}{(x + 3)(x + 2)} \qquad \text{Multiply numerators.}$$

$$= \frac{(6x^2 + 11x - 21) - (4x^2 - 5x - 26)}{(x + 3)(x + 2)} \qquad \text{Subtract numerators.}$$

$$= \frac{6x^2 + 11x - 21 - 4x^2 + 5x + 26}{(x + 3)(x + 2)} \qquad \text{Distributive property}$$

$$= \frac{2x^2 + 16x + 5}{(x + 3)(x + 2)} \qquad \text{Combine like terms.}$$

Answer   The profit function is $P(x) = \dfrac{2x^2 + 16x + 5}{(x + 3)(x + 2)}$.

Now Try Exercise 77

© Elena Elisseeva\Shutterstock

EXERCISE SET 6.2

MathXL®   MyMathLab

Warm-Up Exercises

Fill in the blanks with the appropriate word, phrase, or symbol(s) from the following list.

$15(x + 1)^2$ least common denominator   opposites $5(x + 1)$ greatest common divisor   highest   lowest

1. To add or subtract two fractions with different denominators, we first obtain a _____.

2. The least common denominator of $\dfrac{3}{5(x + 1)^2}$ and $\dfrac{1}{15(x + 1)}$ is _____.

3. The LCD of rational expressions will contain the _____ power of the common factors found in the denominators.

4. When two rational expressions have denominators that are _____ of each other, the LCD can be obtained by multiplying the numerator and the denominator of either rational expression by −1.

Practice the Skills

Add or subtract.

5. $\dfrac{4x}{x+7} + \dfrac{1}{x+7}$

6. $\dfrac{3x}{x+4} + \dfrac{12}{x+4}$

7. $\dfrac{7x}{x-5} - \dfrac{2}{x-5}$

8. $\dfrac{10x}{x-6} - \dfrac{60}{x-6}$

9. $\dfrac{x}{x+3} + \dfrac{9}{x+3} - \dfrac{2}{x+3}$

10. $\dfrac{2x}{x+7} + \dfrac{17}{x+7} - \dfrac{3}{x+7}$

11. $\dfrac{5x-6}{x-8} + \dfrac{2x-5}{x-8}$

12. $\dfrac{-4x+6}{x^2+x-6} + \dfrac{5x-3}{x^2+x-6}$

13. $\dfrac{x^2-2}{x^2+6x-7} - \dfrac{-4x+19}{x^2+6x-7}$

14. $\dfrac{-x^2}{x^2+5xy-14y^2} + \dfrac{x^2+xy-2y^2}{x^2+5xy-14y^2}$

15. $\dfrac{x^3-12x^2+45x}{x(x-8)} - \dfrac{x^2+5x}{x(x-8)}$

16. $\dfrac{3r^2+15r}{r^3+2r^2-8r} + \dfrac{2r^2+5r}{r^3+2r^2-8r}$

17. $\dfrac{3x^2-x}{2x^2-x-21} + \dfrac{2x-8}{2x^2-x-21} - \dfrac{x^2-2x+27}{2x^2-x-21}$

18. $\dfrac{2x^2+9x-15}{2x^2-13x+20} - \dfrac{3x+10}{2x^2-13x+20} - \dfrac{3x-5}{2x^2-13x+20}$

Find the least common denominator.

19. $\dfrac{7}{6a^2} + \dfrac{3}{4a}$

20. $\dfrac{1}{9x^2} - \dfrac{8}{6x^5}$

21. $\dfrac{-4}{8x^2y^2} + \dfrac{7}{5x^4y^6}$

22. $\dfrac{x+12}{16x^2y} - \dfrac{x^2}{3x^3}$

23. $\dfrac{2}{3a^4b^2} + \dfrac{7}{2a^3b^5}$

24. $\dfrac{1}{x-1} - \dfrac{x}{x-3}$

25. $\dfrac{4x}{x+3} + \dfrac{6}{x+9}$

26. $\dfrac{4}{(r-7)(r+2)} - \dfrac{r+8}{r-7}$

27. $5z^2 + \dfrac{9z}{z-6}$

28. $\dfrac{b^2+3}{18b} - \dfrac{b-7}{12(b+8)}$

29. $\dfrac{x}{x^4(x-2)} - \dfrac{x+9}{x^2(x-2)^3}$

30. $\dfrac{x+2}{(x-3)^3(x+4)^2} + \dfrac{x-7}{(x+4)^4(x-9)}$

31. $\dfrac{a-2}{a^2-5a-24} + \dfrac{3}{a^2+11a+24}$

32. $\dfrac{3x-5}{6x^2+13xy+6y^2} + \dfrac{3}{3x^2+5xy+2y^2}$

33. $\dfrac{x}{2x^2-7x+3} + \dfrac{x-3}{4x^2+4x-3} - \dfrac{x^2+1}{2x^2-3x-9}$

34. $\dfrac{3}{x^2+3x-4} - \dfrac{4}{4x^2+5x-9} + \dfrac{x+2}{4x^2+25x+36}$

Add or subtract.

35. $\dfrac{3}{2x} + \dfrac{5}{x}$

36. $\dfrac{9}{x^2} + \dfrac{3}{2x}$

37. $\dfrac{5}{12x} - \dfrac{1}{4x^2}$

38. $\dfrac{5x}{4y} + \dfrac{7}{6xy}$

39. $\dfrac{3}{8x^4y} + \dfrac{1}{5x^2y^3}$

40. $\dfrac{7}{4xy^3} + \dfrac{1}{6x^2y}$

41. $\dfrac{b}{a-b} - \dfrac{a+b}{b}$

42. $\dfrac{4x}{3xy} + 11$

43. $\dfrac{a}{a-b} - \dfrac{a}{b-a}$

44. $\dfrac{9}{b-2} + \dfrac{3b}{2-b}$

45. $\dfrac{4x}{x-4} + \dfrac{x+3}{x+1}$

46. $\dfrac{x}{x^2-9} - \dfrac{4(x-3)}{x+3}$

47. $\dfrac{3}{a+2} + \dfrac{3a+1}{a^2+4a+4}$

48. $\dfrac{2m+9}{m-5} - \dfrac{4}{m^2-3m-10}$

49. $\dfrac{x}{x^2+2x-8} + \dfrac{x+1}{x^2-3x+2}$

50. $\dfrac{-x^2+5x}{(x-5)^2} + \dfrac{x+8}{x-5}$

51. $\dfrac{5x}{x^2-9x+8} - \dfrac{3(x+2)}{x^2-6x-16}$

52. $\dfrac{2}{(2p-3)(p+4)} - \dfrac{3}{(p+4)(p-4)}$

53. $4 - \dfrac{x-1}{x^2+3x-10}$

54. $\dfrac{3x}{2x-3} + \dfrac{3x+6}{2x^2+x-6}$

55. $\dfrac{3a+2}{4a+1} - \dfrac{3a+6}{4a^2+9a+2}$

56. $\dfrac{7}{3q^2+q-4} + \dfrac{9q+2}{3q^2-2q-8}$

57. $\dfrac{x-y}{x^2-4xy+4y^2} + \dfrac{x-3y}{x^2-4y^2}$

58. $\dfrac{x+2y}{x^2-xy-2y^2} - \dfrac{y}{x^2-3xy+2y^2}$

59. $\dfrac{2r}{r-4} - \dfrac{2r}{r+4} + \dfrac{64}{r^2-16}$

60. $\dfrac{4}{p+1} + \dfrac{3}{p-1} + \dfrac{p+4}{p^2-1}$

61. $\dfrac{-4}{x^2+2x-3} - \dfrac{1}{x+3} + \dfrac{1}{x-1}$

62. $\dfrac{2}{x^2-16} + \dfrac{x+1}{x^2+8x+16} + \dfrac{3}{x-4}$

63. $\dfrac{3}{3x-2} - \dfrac{1}{x-4} + 5$

64. $\dfrac{x}{3x+4} + \dfrac{3x+2}{x-5} - \dfrac{7x^2+24x+28}{3x^2-11x-20}$

65. $2 - \dfrac{1}{8r^2+2r-15} + \dfrac{r+2}{4r-5}$

66. $\dfrac{x}{x^2-10x+24} - \dfrac{3}{x-6} + 1$

67. $\dfrac{3}{5x+6} + \dfrac{x^2-x}{5x^2-4x-12} - \dfrac{4}{x-2}$

68. $\dfrac{3}{x^2-13x+36} + \dfrac{4}{2x^2-7x-4} + \dfrac{1}{2x^2-17x-9}$

69. $\dfrac{3m}{6m^2+13mn+6n^2} + \dfrac{2m}{4m^2+8mn+3n^2}$

70. $\dfrac{(x-y)^2}{x^3-y^3} + \dfrac{2}{x^2+xy+y^2}$

71. $\dfrac{5r-2s}{25r^2-4s^2} - \dfrac{2r-s}{10r^3-rs-2s^2}$

72. $\dfrac{6}{(2r-1)^2} + \dfrac{2}{2r-1} - 3$

73. $\dfrac{2}{2x+3y} - \dfrac{4x^2-6xy+9y^2}{8x^3+27y^3}$

74. $\dfrac{4}{4x-5y} - \dfrac{3x^2+2y^2}{64x^3-125y^3}$

Problem Solving

For Exercises 75–76, recall that $(f+g)(x) = f(x) + g(x)$.

75. If $f(x) = \dfrac{x+2}{x-3}$ and $g(x) = \dfrac{x}{x+4}$, find

    a) the domain of $f(x)$.

    b) the domain of $g(x)$.

    c) $(f+g)(x)$.

    d) the domain of $(f+g)(x)$.

76. If $f(x) = \dfrac{x+1}{x^2-9}$ and $g(x) = \dfrac{x}{x-3}$, find

    a) the domain of $f(x)$.

    b) the domain of $g(x)$.

    c) $(f+g)(x)$.

    d) the domain of $(f+g)(x)$.

Profit   *In Exercises 77–80, find the profit function, P(x). (See Example 10.)*

77. $R(x) = \dfrac{4x-5}{x+1}$ and $C(x) = \dfrac{2x-7}{x+2}$

78. $R(x) = \dfrac{5x-2}{x+2}$ and $C(x) = \dfrac{3x-4}{x+1}$

79. $R(x) = \dfrac{8x-3}{x+2}$ and $C(x) = \dfrac{5x-8}{x+3}$

80. $R(x) = \dfrac{7x-10}{x+3}$ and $C(x) = \dfrac{5x-8}{x+4}$

In Exercises 81–84, use $f(x) = \dfrac{x}{x^2-4}$ and $g(x) = \dfrac{2}{x^2+x-6}$. Find the following.

81. $(f+g)(x)$

82. $(f-g)(x)$

83. $(f \cdot g)(x)$

84. $(f/g)(x)$

85. Show that $\dfrac{a}{b} + \dfrac{c}{d} = \dfrac{ad+bc}{bd}$.

86. Show that $x^{-1} + y^{-1} = \dfrac{x+y}{xy}$.

Area and Perimeter   *Consider the rectangles below. Find **a)** the perimeter, and **b)** the area.*

87.

$\dfrac{a+b}{a}$

$\dfrac{a-b}{a}$

88.

$\dfrac{a+2b}{b}$

$\dfrac{-a+2b}{b}$

Determine the polynomial to be placed in the shaded area to give a true statement.

89. $\dfrac{5x^2-6}{x^2-x-1} - \dfrac{\rule{2cm}{0.3cm}}{x^2-x-1} = \dfrac{-2x^2+6x-12}{x^2-x-1}$

90. $\dfrac{r^2-6}{r^2-5r+6} - \dfrac{\rule{2cm}{0.3cm}}{r^2-5r+6} = \dfrac{1}{r-2}$

Perform the indicated operations.

91. $\left(3 + \dfrac{1}{x+3}\right)\left(\dfrac{x+3}{x-2}\right)$

92. $\left(\dfrac{3}{r+1} - \dfrac{4}{r-2}\right)\left(\dfrac{r-2}{r+10}\right)$

93. $\left(\dfrac{5}{a-5} - \dfrac{2}{a+3}\right) \div (3a + 25)$

94. $\left(\dfrac{x^2+4x-5}{2x^2+x-3} \cdot \dfrac{2x+3}{x+1}\right) - \dfrac{2}{x+2}$

95. $\left(\dfrac{x+5}{x-3} - x\right) \div \dfrac{1}{x-3}$

96. $\left(\dfrac{x+5}{x^2-25} + \dfrac{1}{x+5}\right)\left(\dfrac{2x^2-13x+15}{4x^2-6x}\right)$

97. The weighted average of two values a and b is given by $a\left(\dfrac{x}{n}\right) + b\left(\dfrac{n-x}{n}\right)$, where $\dfrac{x}{n}$ is the weight given to a and $\dfrac{n-x}{n}$ is the weight given to b.

a) Express this sum as a single fraction.

b) On exam a you received a grade of 60 and on exam b you received a grade of 92. If exam a counts $\dfrac{2}{5}$ of your final grade and exam b counts $\dfrac{3}{5}$, determine your weighted average.

98. Show that $\left(\dfrac{x}{y}\right)^{-1} + \left(\dfrac{y}{x}\right)^{-1} + (xy)^{-1} = \dfrac{x^2+y^2+1}{xy}$.

In Exercises 99 and 100, perform the indicated operation.

99. $(a-b)^{-1} + (a-b)^{-2}$

100. $\left(\dfrac{a-b}{a}\right)^{-1} - \left(\dfrac{a+b}{a}\right)^{-1}$

Concept/Writing Exercises

In Exercises 101 and 102, **a)** *explain why the subtraction is not correct, and* **b)** *show the correct subtraction.*

101. $\dfrac{x^2-4x}{(x+3)(x-2)} - \dfrac{x^2+x-2}{(x+3)(x-2)} \neq \dfrac{x^2-4x-x^2+x-2}{(x+3)(x-2)}$

102. $\dfrac{x-5}{(x+4)(x-3)} - \dfrac{x^2-6x+5}{(x+4)(x-3)} \neq \dfrac{x-5-x^2-6x+5}{(x+4)(x-3)}$

103. When two rational expressions are being added or subtracted, should the numerators of the expressions being added or subtracted be factored? Explain.

104. Are the fractions $\dfrac{x-3}{4-x}$ and $-\dfrac{x-3}{x-4}$ equivalent? Explain.

105. Are the fractions $\dfrac{8-x}{3-x}$ and $\dfrac{x-8}{x-3}$ equivalent? Explain.

106. If $f(x)$ and $g(x)$ are both rational functions, will $(f+g)(x)$ always be a rational function?

Challenge Problems

107. Express each sum as a single fraction.

a) $1 + \dfrac{1}{x}$

b) $1 + \dfrac{1}{x} + \dfrac{1}{x^2}$

c) $1 + \dfrac{1}{x} + \dfrac{1}{x^2} + \dfrac{1}{x^3} + \dfrac{1}{x^4}$

d) $1 + \dfrac{1}{x} + \dfrac{1}{x^2} + \cdots + \dfrac{1}{x^n}$

108. Let $f(x) = \dfrac{1}{x}$. Find $f(a+h) - f(a)$.

109. Let $g(x) = \dfrac{1}{x+1}$. Find $g(a+h) - g(a)$.

Cumulative Review Exercises

[2.4] **110. Filling Boxes** A cereal box machine fills cereal boxes at a rate of 80 per minute. Then the machine is slowed down and fills cereal boxes at a rate of 60 per minute. If the sum of the two time periods is 14 minutes and the number of cereal boxes filled at the higher rate is the same as the number filled at the lower rate, determine **a)** how long the machine is used at the faster rate, and **b)** the total number of cereal boxes filled over the 14-minute period.

[2.6] **111.** Solve for x and give the solution in set notation.
$|x-3| - 6 < -1$

[3.4] **112.** Find the slope of the line passing through the points $(-2, 3)$ and $(7, -3)$.

[4.5] **113.** Evaluate the determinant $\begin{vmatrix} -1 & 3 \\ 5 & -4 \end{vmatrix}$.

© Pearson Science

See Exercise 110.

[5.3] **114.** Divide $\dfrac{6x^2-5x+6}{2x+3}$.

[5.8] **115.** Solve $3p^2 = 22p - 7$.

6.3 Complex Fractions

1 Recognize complex fractions.

2 Simplify complex fractions by multiplying by the LCD.

3 Simplify complex fractions by simplifying the numerator and denominator.

1 Recognize Complex Fractions

Complex Fraction

A **complex fraction** is a fraction that has a rational expression in its numerator or its denominator or both its numerator and denominator.

Examples of Complex Fractions

$$\frac{\frac{2}{3}}{5}, \quad \frac{\frac{x+1}{x}}{4x}, \quad \frac{\frac{x}{y}}{x+1}, \quad \frac{\frac{a+b}{a}}{\frac{a-b}{b}}, \quad \frac{9+\frac{1}{x}}{\frac{1}{x^2}+\frac{8}{x}}$$

The expression above the **main fraction line** is the numerator, and the expression below the main fraction line is the denominator of the complex fraction.

$$\left.\frac{a+b}{a}\right\} \text{ numerator of complex fraction}$$
$$\longleftarrow \text{ main fraction line}$$
$$\left.\frac{a-b}{b}\right\} \text{ denominator of complex fraction}$$

To **simplify a complex fraction** means to write the expression without a fraction in its numerator and its denominator. We will explain two methods that can be used to simplify complex fractions: multiplying by the least common denominator (LCD) and simplifying the numerator and denominator.

2 Simplify Complex Fractions by Multiplying by the Least Common Denominator

To Simplify a Complex Fraction by Multiplying by the Least Common Denominator

1. Find the least common denominator of all fractions appearing within the complex fraction. This is the LCD of the complex fraction.
2. Multiply both the numerator and denominator of the complex fraction by the LCD of the complex fraction found in step 1.
3. Simplify when possible.

In step 2, you are actually multiplying the complex fraction by $\dfrac{\text{LCD}}{\text{LCD}}$, which is equivalent to multiplying the fraction by 1.

EXAMPLE 1 Simplify $\dfrac{\dfrac{4}{x^2}-\dfrac{3}{x}}{\dfrac{x^2}{5}}$.

Solution   The denominators in the complex fraction are x^2, x, and 5. Therefore, the LCD of the complex fraction is $5x^2$. Multiply the numerator and denominator by $5x^2$.

$$\frac{\dfrac{4}{x^2} - \dfrac{3}{x}}{\dfrac{x^2}{5}} = \frac{5x^2\left(\dfrac{4}{x^2} - \dfrac{3}{x}\right)}{5x^2\left(\dfrac{x^2}{5}\right)}$$

Multiply the numerator and denominator by $5x^2$.

$$= \frac{5x^2\left(\dfrac{4}{x^2}\right) - 5x^2\left(\dfrac{3}{x}\right)}{5x^2\left(\dfrac{x^2}{5}\right)}$$

Distributive property

$$= \frac{20 - 15x}{x^4}$$

Simplify.

Now Try Exercise 13

EXAMPLE 2 Simplify $\dfrac{a + \dfrac{3}{b}}{b + \dfrac{3}{a}}$.

Solution   The LCD of the complex fraction is ab.

$$\frac{a + \dfrac{3}{b}}{b + \dfrac{3}{a}} = \frac{ab\left(a + \dfrac{3}{b}\right)}{ab\left(b + \dfrac{3}{a}\right)}$$

Multiply the numerator and denominator by ab.

$$= \frac{a^2b + 3a}{ab^2 + 3b}$$

Distributive property

$$= \frac{a(ab + 3)}{b(ab + 3)} = \frac{a}{b}$$

Factor and simplify.

Now Try Exercise 17

EXAMPLE 3 Simplify $\dfrac{a^{-1} + ab^{-2}}{ab^{-2} - a^{-2}b^{-1}}$.

Solution   First rewrite each expression without negative exponents.

$$\frac{a^{-1} + ab^{-2}}{ab^{-2} - a^{-2}b^{-1}} = \frac{\dfrac{1}{a} + \dfrac{a}{b^2}}{\dfrac{a}{b^2} - \dfrac{1}{a^2b}}$$

$$= \frac{a^2b^2\left(\dfrac{1}{a} + \dfrac{a}{b^2}\right)}{a^2b^2\left(\dfrac{a}{b^2} - \dfrac{1}{a^2b}\right)}$$

Multiply the numerator and denominator by a^2b^2, the LCD of the complex fraction.

$$= \frac{a^2b^2\left(\dfrac{1}{a}\right) + a^2b^2\left(\dfrac{a}{b^2}\right)}{a^2b^2\left(\dfrac{a}{b^2}\right) - a^2b^2\left(\dfrac{1}{a^2b}\right)}$$

Distributive property

$$= \frac{ab^2 + a^3}{a^3 - b}$$

Now Try Exercise 43

In Example 3, although we could factor an a from both terms in the numerator of the answer, we could not simplify the answer further by dividing out common factors. So, we leave the answer in the form given.

3 Simplify Complex Fractions by Simplifying the Numerator and Denominator

> **To Simplify a Complex Fraction by Simplifying the Numerator and the Denominator**
>
> 1. Add or subtract as necessary to get one rational expression in the numerator.
> 2. Add or subtract as necessary to get one rational expression in the denominator.
> 3. Multiply the numerator of the complex fraction by the reciprocal of the denominator.
> 4. Simplify when possible.

Example 4 will show how Example 1 can be simplified by this second method.

EXAMPLE 4 Simplify $\dfrac{\dfrac{4}{x^2} - \dfrac{3}{x}}{\dfrac{x^2}{5}}$.

Solution   Subtract the fractions in the numerator to get one rational expression in the numerator. The common denominator of the fractions in the numerator is x^2.

$$\frac{\dfrac{4}{x^2} - \dfrac{3}{x}}{\dfrac{x^2}{5}} = \frac{\dfrac{4}{x^2} - \dfrac{3}{x} \cdot \dfrac{x}{x}}{\dfrac{x^2}{5}} \qquad \text{Obtain common denominator in numerator.}$$

$$= \frac{\dfrac{4}{x^2} - \dfrac{3x}{x^2}}{\dfrac{x^2}{5}}$$

$$= \frac{\dfrac{4 - 3x}{x^2}}{\dfrac{x^2}{5}}$$

$$= \frac{4 - 3x}{x^2} \cdot \frac{5}{x^2} \qquad \text{Multiply the numerator by the reciprocal of the denominator.}$$

$$= \frac{5(4 - 3x)}{x^4}$$

$$\text{or} \qquad \frac{20 - 15x}{x^4}$$

This is the same answer obtained in Example 1.

Now Try Exercise 13

> **Helpful Hint**
>
> Some students prefer to use the second method when the complex fraction consists of a single fraction over a single fraction, such as
>
> $$\frac{\dfrac{x + 3}{18}}{\dfrac{x - 8}{6}}$$
>
> For more complex fractions, many students prefer the first method because you do not have to add fractions.

EXERCISE SET **6.3**

Math XL
MathXL®

MyMathLab
MyMathLab

Warm-Up Exercises

Fill in the blanks with the appropriate word, phrase, or symbol(s) from the following list.

numerator   denominator   complex fraction   reciprocal   opposite

1. A _____ is a fraction that has a rational expression in its numerator or its denominator or both its numerator and denominator.

2. The expression above the main fraction line is the _____ of the complex fraction.

3. The expression below the main fraction line is the _____ of the complex fraction.

4. A complex fraction of the form $\dfrac{\dfrac{a}{b}}{\dfrac{c}{d}}$ can be simplified by multiplying the numerator by the _____ of the denominator.

Practice the Skills

Simplify.

5. $\dfrac{\dfrac{15a}{b^2}}{\dfrac{b^3}{5}}$

6. $\dfrac{\dfrac{6x}{y^2}}{\dfrac{3y}{5}}$

7. $\dfrac{\dfrac{36x^4}{5y^4z^5}}{\dfrac{9xy^2}{15z^5}}$

8. $\dfrac{\dfrac{40x^3}{7y^5z^5}}{\dfrac{8x^2y^2}{28x^4z^5}}$

9. $\dfrac{\dfrac{10x^3y^2}{9yz^4}}{\dfrac{40x^4y^7}{27y^2z^8}}$

10. $\dfrac{\dfrac{3a^4b^3}{7b^4c}}{\dfrac{15a^2b^6}{14ac^7}}$

11. $\dfrac{x - \dfrac{x}{y}}{8 + x}$

12. $\dfrac{a + \dfrac{2a}{b}}{7 + a}$

13. $\dfrac{x + \dfrac{5}{y}}{1 + \dfrac{x}{y}}$

14. $\dfrac{\dfrac{4}{x} + \dfrac{2}{x^2}}{2 + \dfrac{1}{x}}$

15. $\dfrac{\dfrac{2}{a} + \dfrac{1}{2a}}{a + \dfrac{a}{2}}$

16. $\dfrac{3 - \dfrac{1}{y}}{2 - \dfrac{1}{y}}$

17. $\dfrac{\dfrac{a^2}{b} - b}{\dfrac{b^2}{a} - a}$

18. $\dfrac{x - \dfrac{4}{y}}{y - \dfrac{4}{x}}$

19. $\dfrac{\dfrac{x}{y} - \dfrac{y}{x}}{\dfrac{x + y}{x}}$

20. $\dfrac{\dfrac{1}{m} + \dfrac{9}{m^2}}{2 + \dfrac{1}{m^2}}$

21. $\dfrac{\dfrac{a}{b} - 6}{\dfrac{-a}{b} + 6}$

22. $\dfrac{7 - \dfrac{x}{y}}{\dfrac{x}{y} - 7}$

23. $\dfrac{\dfrac{4x + 8}{3x^2}}{\dfrac{4x^3}{9}}$

24. $\dfrac{\dfrac{x^2 - y^2}{x}}{\dfrac{x + y}{x^4}}$

25. $\dfrac{\dfrac{a}{a + 1} - 1}{\dfrac{2a + 1}{a - 1}}$

26. $\dfrac{\dfrac{x}{4} - \dfrac{1}{x}}{1 + \dfrac{x + 4}{x}}$

27. $\dfrac{1 + \dfrac{x}{x + 1}}{\dfrac{2x + 1}{x - 1}}$

28. $\dfrac{\dfrac{2}{x - 1} + 2}{\dfrac{2}{x + 1} - 2}$

29. $\dfrac{\dfrac{a + 1}{a - 1} + \dfrac{a - 1}{a + 1}}{\dfrac{a + 1}{a - 1} - \dfrac{a - 1}{a + 1}}$

30. $\dfrac{\dfrac{a - 2}{a + 2} - \dfrac{a + 2}{a - 2}}{\dfrac{a - 2}{a + 2} + \dfrac{a + 2}{a - 2}}$

31. $\dfrac{\dfrac{5}{5 - x} + \dfrac{6}{x - 5}}{\dfrac{3}{x} + \dfrac{2}{x - 5}}$

32. $\dfrac{\dfrac{2}{m} + \dfrac{1}{m^2} + \dfrac{3}{m - 1}}{\dfrac{6}{m - 1}}$

33. $\dfrac{\dfrac{3}{x^2} - \dfrac{1}{x} + \dfrac{2}{x - 2}}{\dfrac{1}{x}}$

34. $\dfrac{\dfrac{2}{x^2 + x - 20} + \dfrac{3}{x^2 - 6x + 8}}{\dfrac{2}{x^2 + 3x - 10} + \dfrac{3}{x^2 + 2x - 24}}$

35. $\dfrac{\dfrac{2}{a^2 - 3a + 2} + \dfrac{2}{a^2 - a - 2}}{\dfrac{2}{a^2 - 1} + \dfrac{2}{a^2 + 4a + 3}}$

36. $\dfrac{\dfrac{1}{x^2 + 5x + 4} + \dfrac{2}{x^2 + 2x - 8}}{\dfrac{2}{x^2 - x - 2} + \dfrac{1}{x^2 - 5x + 6}}$

Simplify.

37. $\left(a^{-1} + b^{-1}\right)^{-1}$

38. $\left(a^{-2} + b\right)^{-1}$

39. $\dfrac{\dfrac{2}{xy}}{x^{-1} - y^{-1}}$

40. $\dfrac{a^{-1} + b^{-1}}{\dfrac{5}{ab}}$

41. $\dfrac{a^{-1} + 1}{b^{-1} - 1}$

42. $\dfrac{x^{-1} - y^{-1}}{x^{-1} + y^{-1}}$

43. $\dfrac{a^{-2} - ab^{-1}}{ab^{-2} + a^{-1}b^{-1}}$

44. $\dfrac{xy^{-1} + x^{-1}y^{-2}}{x^{-1} - x^{-2}y^{-1}}$

45. $\dfrac{\dfrac{9a}{b} + a^{-1}}{\dfrac{b}{a} + a^{-1}}$

46. $\dfrac{x^{-2} + \dfrac{3}{x}}{3x^{-1} + x^{-2}}$

47. $\dfrac{a^{-1} + b^{-1}}{(a + b)^{-1}}$

48. $\dfrac{4a^{-1} - b^{-1}}{(a - b)^{-1}}$

49. $5x^{-1} - (3y)^{-1}$

50. $\dfrac{\dfrac{7}{x} + \dfrac{1}{y}}{(x - y)^{-1}}$

51. $\dfrac{\dfrac{2}{xy} - \dfrac{8}{y} + \dfrac{5}{x}}{3x^{-1} - 4y^{-2}}$

52. $\dfrac{4m^{-1} + 3n^{-1} + (2mn)^{-1}}{\dfrac{5}{m} + \dfrac{7}{n}}$

Problem Solving

Area   *For Exercises 53–56, the area and width of each rectangle are given. Find the length, l, by dividing the area, A, by the width, w.*

53.

$A = \dfrac{x^2 + 12x + 35}{x + 3}$ $w = \dfrac{x^2 + 6x + 5}{x^2 + 5x + 6}$

l

54.

$A = \dfrac{x^2 + 10x + 16}{x + 4}$ $w = \dfrac{x^2 + 11x + 24}{x^2 + 3x - 4}$

$(x > 1)$

l

55.

$A = \dfrac{x^2 + 11x + 28}{x + 5}$ $w = \dfrac{x^2 + 8x + 7}{x^2 + 4x - 5}$

$(x > 1)$

l

56.

$A = \dfrac{x^2 + 17x + 72}{x + 3}$ $w = \dfrac{x^2 + 11x + 18}{x^2 + x - 6}$

$(x > 2)$

l

57. Automobile Jack   The efficiency of a jack, E, is given by the formula

$$E = \dfrac{\dfrac{1}{2}h}{h + \dfrac{1}{2}}$$

where h is determined by the pitch of the jack's thread.

Pitch

Determine the efficiency of a jack whose values of h are:

a) $\dfrac{2}{5}$

b) $\dfrac{1}{3}$

58. Resistors   If two resistors with resistances R_1 and R_2 are connected in parallel, their combined resistance, R_T, can be found from the formula

$$R_T = \dfrac{1}{\dfrac{1}{R_1} + \dfrac{1}{R_2}}$$

Simplify the right side of the formula.

59. Resistors   If three resistors with resistances R_1, R_2, and R_3 are connected in parallel, their combined resistance, R_T, can be found from the formula

$$R_T = \dfrac{1}{\dfrac{1}{R_1} + \dfrac{1}{R_2} + \dfrac{1}{R_3}}$$

Simplify the right side of this formula.

60. Optics   A formula used in the study of optics is

$$f = \left(p^{-1} + q^{-1}\right)^{-1}$$

where p is the object's distance from a lens, q is the image distance from the lens, and f is the focal length of the lens. Express the right side of the formula without any negative exponents.

61. If $f(x) = \dfrac{1}{x}$, find $f(f(a))$.

62. If $f(x) = \dfrac{2}{x + 2}$, find $f(f(a))$.

Challenge Problems

For each function, find $\dfrac{f(a+h)-f(a)}{h}$.

63. $f(x)=\dfrac{1}{x}$

64. $f(x)=\dfrac{5}{x}$

65. $f(x)=\dfrac{1}{x+1}$

66. $f(x)=\dfrac{6}{x-1}$

67. $f(x)=\dfrac{1}{x^2}$

68. $f(x)=\dfrac{3}{x^2}$

Simplify.

69. $\dfrac{1}{2+\dfrac{1}{2+\dfrac{1}{2}}}$

70. $\dfrac{1}{x+\dfrac{1}{x+\dfrac{1}{x+1}}}$

71. $\dfrac{1}{2a+\dfrac{1}{2a+\dfrac{1}{2a}}}$

Cumulative Review Exercises

[1.4]   **72.** Evaluate $\dfrac{\left|-\dfrac{3}{9}\right|-\left(-\dfrac{5}{9}\right)\cdot\left|-\dfrac{3}{8}\right|}{|-5-(-3)|}$.

[2.5]   **73.** Solve $\dfrac{3}{5}<\dfrac{-x-5}{3}<6$ and give the solution in interval notation.

[2.6]   **74.** Solve $|x-1|=|2x-4|$.

[3.5]   **75.** Determine if the two lines represented by the following equations are parallel, perpendicular, or neither.

$$6x+2y=5$$
$$4x-9=-2y$$

6.4 Solving Rational Equations

1. Solve rational equations.
2. Check solutions.
3. Solve proportions.
4. Solve problems involving rational functions.
5. Solve applications using rational expressions.
6. Solve for a variable in a formula containing rational expressions.

1 Solve Rational Equations

A **rational equation** is an equation that contains at least one rational expression.

To Solve Rational Equations

1. Determine the LCD of all rational expressions in the equation.
2. Multiply *both* sides of the equation by the LCD. This will result in every term in the equation being multiplied by the LCD.
3. Remove any parentheses and combine like terms on each side of the equation.
4. Solve the equation using the properties discussed in earlier sections.
5. Check the solution in the *original* equation.

In step 2, we multiply both sides of the equation by the LCD to eliminate the fractions from the equation.

EXAMPLE 1 Solve $\dfrac{3x}{4}+\dfrac{1}{2}=\dfrac{2x-3}{4}$.

Solution   Multiply both sides of the equation by the LCD, 4. Then use the distributive property, which results in fractions being eliminated from the equation.

$$4\left(\frac{3x}{4}+\frac{1}{2}\right)=\frac{2x-3}{4}\cdot 4 \qquad \text{Multiply both sides by 4.}$$

$$4\left(\frac{3x}{4}\right)+4\left(\frac{1}{2}\right)=2x-3 \qquad \text{Distributive property}$$

$$3x+2=2x-3$$

$$x+2=-3 \qquad \text{Subtract } 2x \text{ from both sides.}$$

$$x=-5 \qquad \text{Subtract 2 from both sides.}$$

A check will show that -5 is the solution.

Now Try Exercise 15

2 Check Solutions

When solving a rational equation with a variable denominator, it is possible to produce a value that makes a denominator 0. Since division by 0 is undefined, the value that makes a denominator 0 is not a solution and is called an **extraneous solution**. *Whenever solving a rational equation with a variable in the denominator, you must check the value(s) obtained in the original equation.*

EXAMPLE 2 Solve $4 - \dfrac{3}{x} = \dfrac{5}{2}$.

Solution   The LCD is $2x$.

$$4 - \frac{3}{x} = \frac{5}{2}$$

$$2x\left(4 - \frac{3}{x}\right) = \left(\frac{5}{2}\right)2x \qquad \text{Multiply both sides by the LCD, } 2x.$$

$$2x(4) - 2x\left(\frac{3}{x}\right) = \left(\frac{5}{2}\right)2x \qquad \text{Distributive property}$$

$$8x - 6 = 5x$$

$$3x - 6 = 0 \qquad \text{Subtract } 5x \text{ from both sides.}$$

$$3x = 6 \qquad \text{Add 6 to both sides.}$$

$$x = 2 \qquad \text{Divide both sides by 3.}$$

Check $$4 - \frac{3}{x} = \frac{5}{2}$$

$$4 - \frac{3}{2} = \frac{5}{2} \qquad \text{Substitute 2 for } x.$$

$$\frac{8}{2} - \frac{3}{2} = \frac{5}{2} \qquad \text{Rewrite 4 as } \frac{8}{2}.$$

$$\frac{5}{2} = \frac{5}{2} \qquad \text{True}$$

Since $x = 2$ checks in the original equation, it is a solution.

Now Try Exercise 21

EXAMPLE 3 Solve $x - \dfrac{6}{x} = -5$.

Solution

$$x \cdot \left(x - \frac{6}{x}\right) = -5 \cdot x \qquad \text{Multiply both sides by the LCD, } x.$$

$$x(x) - x\left(\frac{6}{x}\right) = -5x \qquad \text{Distributive property}$$

$$x^2 - 6 = -5x$$

$$x^2 + 5x - 6 = 0$$

$$(x - 1)(x + 6) = 0$$

$$x - 1 = 0 \qquad \text{or} \qquad x + 6 = 0$$

$$x = 1 \qquad\qquad\qquad x = -6$$

Checks of 1 and -6 will show that they are both solutions to the equation.

Now Try Exercise 35

EXAMPLE 4   Solve $\dfrac{3x}{x^2 - 4} + \dfrac{1}{x - 2} = \dfrac{2}{x + 2}$.

Solution   First factor the denominator $x^2 - 4$, then find the LCD.

$$\frac{3x}{(x + 2)(x - 2)} + \frac{1}{x - 2} = \frac{2}{x + 2}$$

The LCD is $(x + 2)(x - 2)$. Multiply both sides of the equation by the LCD, and then use the distributive property. This process will eliminate the fractions from the equation.

$$(x + 2)(x - 2) \cdot \left[\frac{3x}{(x + 2)(x - 2)} + \frac{1}{x - 2} \right] = \frac{2}{x + 2} \cdot (x + 2)(x - 2)$$

$$\cancel{(x + 2)}\,\cancel{(x - 2)} \cdot \frac{3x}{\cancel{(x + 2)}\,\cancel{(x - 2)}} + (x + 2)\cancel{(x - 2)} \cdot \frac{1}{\cancel{x - 2}} = \frac{2}{\cancel{x + 2}} \cdot \cancel{(x + 2)}(x - 2)$$

$$3x + (x + 2) = 2(x - 2)$$
$$4x + 2 = 2x - 4$$
$$2x + 2 = -4$$
$$2x = -6$$
$$x = -3$$

A check will show that -3 is the solution.

Now Try Exercise 39

EXAMPLE 5   Solve $\dfrac{22}{2p^2 - 9p - 5} - \dfrac{3}{2p + 1} = \dfrac{2}{p - 5}$.

Solution   Factor the denominator, then determine the LCD.

$$\frac{22}{(2p + 1)(p - 5)} - \frac{3}{2p + 1} = \frac{2}{p - 5}$$

Multiply both sides of the equation by the LCD, $(2p + 1)(p - 5)$.

$$\cancel{(2p + 1)}\,\cancel{(p - 5)} \cdot \frac{22}{\cancel{(2p + 1)}\,\cancel{(p - 5)}} - \cancel{(2p + 1)}(p - 5) \cdot \frac{3}{\cancel{2p + 1}} = \frac{2}{\cancel{p - 5}} \cdot (2p + 1)\cancel{(p - 5)}$$

$$22 - 3(p - 5) = 2(2p + 1)$$
$$22 - 3p + 15 = 4p + 2$$
$$37 - 3p = 4p + 2$$
$$35 = 7p$$
$$5 = p$$

The solution appears to be 5. However, since a variable appears in a denominator, this solution must be checked.

Check
$$\frac{22}{2p^2 - 9p - 5} - \frac{3}{2p + 1} = \frac{2}{p - 5}$$

$$\frac{22}{2(5)^2 - 9(5) - 5} - \frac{3}{2(5) + 1} \overset{?}{=} \frac{2}{5 - 5} \qquad \text{Substitute 5 for } p.$$

$$\text{Undefined} \longrightarrow \frac{22}{0} - \frac{3}{11} = \frac{2}{0} \longleftarrow \text{Undefined}$$

Since 5 makes a denominator 0 and division by 0 is undefined, 5 is an extraneous solution. Therefore, you should write "no solution" as your answer.

Now Try Exercise 43

> **Helpful Hint**
>
> Remember, whenever you solve an equation where a variable appears in any denominator, you must check any apparent solution to make sure it is not an extraneous solution. If an apparent solution makes any denominator 0, then it is an extraneous solution and not a true solution to the equation.

3 Solve Proportions

> **Proportion**
>
> A **proportion** is an equation of the form $\dfrac{a}{b} = \dfrac{c}{d}$, $b \neq 0$, $d \neq 0$.

> **Understanding Algebra**
>
> Remember that proportions are rational equations. Therefore, whenever you solve a proportion that contains one or more variable denominators, you must check to make sure that the value(s) you obtain are not extraneous solutions.

Proportions are rational equations and therefore may be solved by multiplying both sides of the equation by the LCD. Proportions may also be solved by **cross-multiplication** as follows:

$$\text{If } \frac{a}{b} = \frac{c}{d}, \text{ then } ad = bc, b \neq 0, d \neq 0$$

Proportions are used when working with similar figures.

> **Similar Figures**
>
> **Similar figures** are figures whose corresponding angles are equal and whose corresponding sides are in proportion.

Figure 6.3 illustrates two sets of similar figures.

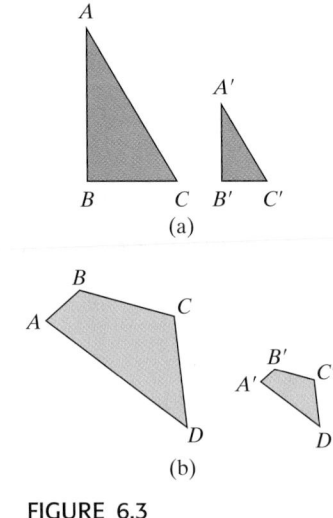

FIGURE 6.3

In **Figure 6.3a**, the ratio of the length of side AB to the length of side BC is the same as the ratio of the length of side $A'B'$ to the length of side $B'C'$. That is,

$$\frac{AB}{BC} = \frac{A'B'}{B'C'}$$

In a pair of similar figures, if the length of a side is unknown, it can often be found by using proportions.

EXAMPLE 6   Similar Triangles   Triangles ABC and $A'B'C'$ in **Figure 6.4** are similar figures. Find the length of sides AB and $B'C'$.

Solution   We can set up a proportion and then solve for x. Then we can find the lengths.

$$\frac{AB}{BC} = \frac{A'B'}{B'C'}$$

$$\frac{x-1}{5} = \frac{6}{x}$$

$$5x \cdot \frac{x-1}{5} = \frac{6}{x} \cdot 5x \qquad \text{Multiply both sides by the LCD, } 5x.$$

$$x(x-1) = 6 \cdot 5$$

$$x^2 - x = 30$$

$$x^2 - x - 30 = 0$$

$$(x-6)(x+5) = 0 \qquad \text{Factor the trinomial.}$$

$$x - 6 = 0 \qquad \text{or} \qquad x + 5 = 0$$

$$x = 6 \qquad\qquad\qquad x = -5$$

Since the length of the side of a triangle cannot be a negative number, -5 is not a possible answer. Substituting 6 for x, we see that the length of side $B'C'$ is 6 and the length of side AB is $6 - 1$ or 5.

Check
$$\frac{AB}{BC} = \frac{A'B'}{B'C'}$$

$$\frac{5}{5} \stackrel{?}{=} \frac{6}{6}$$

$$1 = 1 \qquad \text{True}$$

Now Try Exercise 49

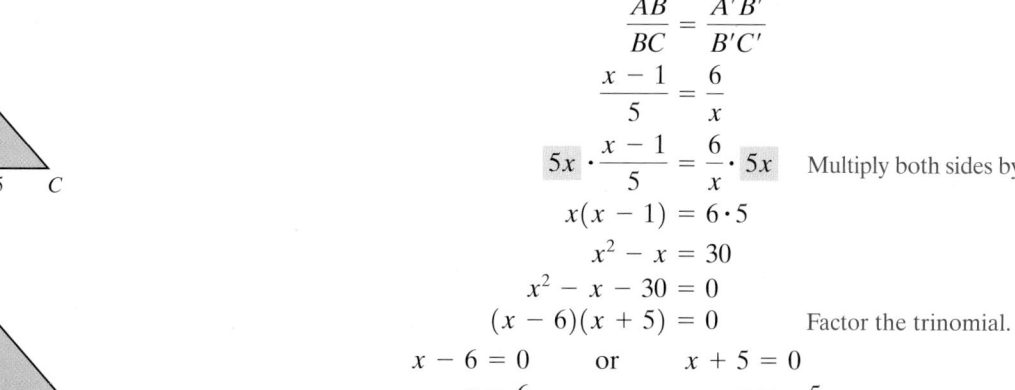

FIGURE 6.4

    The answer to Example 6 could also be obtained using cross-multiplication. Try solving Example 6 using cross-multiplication now.

EXAMPLE 7   Solve $\dfrac{x^2}{x-3} = \dfrac{9}{x-3}$.

Solution   This equation is a proportion. We will solve this equation by multiplying both sides of the equation by the LCD, $x - 3$.

$$(x-3) \cdot \frac{x^2}{x-3} = \frac{9}{x-3} \cdot (x-3)$$

$$x^2 = 9$$

$$x^2 - 9 = 0$$

$$(x+3)(x-3) = 0 \qquad\qquad \text{Factor the difference of two squares.}$$

$$x + 3 = 0 \qquad \text{or} \qquad x - 3 = 0$$

$$x = -3 \qquad\qquad\qquad x = 3$$

Check
$$x = -3 \qquad\qquad\qquad\qquad x = 3$$

$$\frac{x^2}{x-3} = \frac{9}{x-3} \qquad\qquad \frac{x^2}{x-3} = \frac{9}{x-3}$$

$$\frac{(-3)^2}{-3-3} \stackrel{?}{=} \frac{9}{-3-3} \qquad\qquad \frac{3^2}{3-3} \stackrel{?}{=} \frac{9}{3-3}$$

$$\frac{9}{-6} \stackrel{?}{=} \frac{9}{-6} \qquad\qquad\qquad \frac{9}{0} \stackrel{?}{=} \frac{9}{0} \quad \longleftarrow \text{Undefined}$$

$$-\frac{3}{2} = -\frac{3}{2} \qquad \text{True}$$

Since $x = 3$ results in a denominator of 0, 3 is *not* a solution to the equation. It is an extraneous solution. The only solution to the equation is -3.

Now Try Exercise 45

In Example 7, what would you obtain if you began by cross multiplying? Try it and see.

4 Solve Problems Involving Rational Functions

EXAMPLE 8 Consider the function $f(x) = x - \dfrac{2}{x}$. Find all a for which $f(a) = 1$.

Solution Since $f(a) = a - \dfrac{2}{a}$, we need to find all values for which $a - \dfrac{2}{a} = 1, a \neq 0$.

We begin by multiplying both sides of the equation by a, the LCD.

$$a \cdot \left(a - \frac{2}{a}\right) = a \cdot 1$$
$$a^2 - 2 = a$$
$$a^2 - a - 2 = 0$$
$$(a - 2)(a + 1) = 0$$
$$a - 2 = 0 \quad \text{or} \quad a + 1 = 0$$
$$a = 2 \qquad\qquad a = -1$$

Check

$$f(x) = x - \frac{2}{x}$$

$$f(2) = 2 - \frac{2}{2} = 2 - 1 = 1$$

$$f(-1) = -1 - \frac{2}{(-1)} = -1 + 2 = 1$$

For $a = 2$ or $a = -1, f(a) = 1$.

Now Try Exercise 53

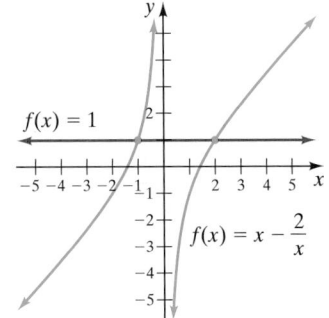

$f(x) = 1$

$f(x) = x - \dfrac{2}{x}$

FIGURE 6.5

In **Figure 6.5**, we illustrated the graph of $f(x) = x - \dfrac{2}{x}$ to show the answers obtained in Example 8. Notice that when $x = -1$ or $x = 2, f(x) = 1$.

5 Solve Applications Using Rational Expressions

Now let's look at an application of rational equations.

EXAMPLE 9 **Total Resistance** In electronics, the total resistance, R_T, of resistors connected in a parallel circuit is determined by the formula

$$\frac{1}{R_T} = \frac{1}{R_1} + \frac{1}{R_2} + \frac{1}{R_3} + \cdots + \frac{1}{R_n}$$

where $R_1, R_2, R_3, \ldots, R_n$ are the resistances of the individual resistors (measured in ohms) in the circuit. Find the total resistance if two resistors, one of 100 ohms and the other of 300 ohms, are connected in a parallel circuit. See **Figure 6.6**.

Solution Since there are only two resistances, use the formula

$$\frac{1}{R_T} = \frac{1}{R_1} + \frac{1}{R_2}$$

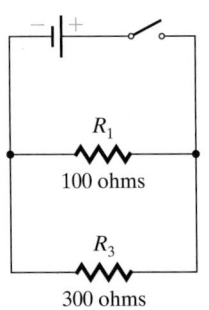

R_1

100 ohms

R_3

300 ohms

FIGURE 6.6

Let $R_1 = 100$ ohms and $R_2 = 300$ ohms; then

$$\frac{1}{R_T} = \frac{1}{100} + \frac{1}{300}$$

Multiply both sides of the equation by the LCD, $300R_T$.

$$300R_T \cdot \frac{1}{R_T} = 300R_T\left(\frac{1}{100} + \frac{1}{300}\right)$$

$$300 \, \cancel{R_T} \cdot \frac{1}{\cancel{R_T}} = \overset{3}{\cancel{300}}R_T\left(\frac{1}{\cancel{100}}\right) + \cancel{300}R_T\left(\frac{1}{\cancel{300}}\right)$$

$$300 = 3R_T + R_T$$

$$300 = 4R_T$$

$$R_T = \frac{300}{4} = 75$$

Thus, the total resistance of the parallel circuit is 75 ohms.

Now Try Exercise 93

6 Solve for a Variable in a Formula Containing Rational Expressions

If, while solving for a variable in a formula, the variable appears in more than one term, you may need to use factoring to solve for the variable. This will be demonstrated in Examples 10, 11, and 12.

EXAMPLE 10 Optics A formula used in the study of optics is

$$\frac{1}{p} + \frac{1}{q} = \frac{1}{f},$$

where

  p is the distance of an object from a lens or a mirror,

  q is the distance of the image from the lens or mirror, and

  f is the focal length of the lens or mirror.

For a person wearing glasses, q is the distance from the lens to the retina (see **Fig. 6.7**). Solve this formula for f.

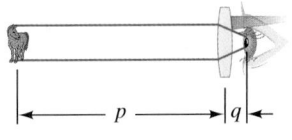

FIGURE 6.7

Solution   Our goal is to isolate the variable f. We begin by multiplying both sides of the equation by the least common denominator, pqf, to eliminate fractions.

$$\frac{1}{p} + \frac{1}{q} = \frac{1}{f}$$

$$pqf\left(\frac{1}{p} + \frac{1}{q}\right) = pqf\left(\frac{1}{f}\right) \qquad \text{Multiply both sides by the LCD, } pqf.$$

$$\cancel{p}qf\left(\frac{1}{\cancel{p}}\right) + p\cancel{q}f\left(\frac{1}{\cancel{q}}\right) = pq\cancel{f}\left(\frac{1}{\cancel{f}}\right) \qquad \text{Distributive property}$$

$$qf + pf = pq \qquad \text{Simplify.}$$

$$f(q + p) = pq \qquad \text{Factor out } f.$$

$$\frac{f\cancel{(q + p)}}{\cancel{q + p}} = \frac{pq}{q + p} \qquad \text{Divide both sides by } q + p.$$

$$f = \frac{pq}{q + p} \quad \text{or} \quad f = \frac{pq}{p + q}$$

Now Try Exercise 69

EXAMPLE 11 Banking A formula used in banking is $A = P + Prt$, where A represents the amount that must be repaid to the bank when P dollars are borrowed at simple interest rate, r, for time, t, in years. Solve this equation for P.

Solution   Since both terms containing P are by themselves on the right side of the equation, we factor P from both terms.

$$A = P + Prt \qquad \text{P is in both terms.}$$

$$A = P(1 + rt) \qquad \text{Factor out P.}$$

$$\frac{A}{1 + rt} = \frac{P\cancel{(1 + rt)}}{\cancel{1 + rt}} \qquad \text{Divide both sides by $1 + rt$ to isolate P.}$$

$$\frac{A}{1 + rt} = P$$

Thus, $P = \dfrac{A}{1 + rt}$.

Now Try Exercise 73

EXAMPLE 12 Physics A formula used for levers in physics is $d = \dfrac{fl}{f + w}$. Solve this formula for f.

Solution   We begin by multiplying both sides of the formula by $f + w$ to clear fractions. Then we will rewrite the expression with all terms containing f on one side of the equals sign and all terms not containing f on the other side of the equals sign.

$$d = \frac{fl}{f + w}$$

$$d(f + w) = \frac{fl}{\cancel{(f + w)}} \cancel{(f + w)} \qquad \text{Multiply by $f + w$ to clear fractions.}$$

$$d(f + w) = fl$$

$$df + dw = fl \qquad \text{Distributive property}$$

$$df - df + dw = fl - df \qquad \text{Isolate terms containing f on the right side of the equation.}$$

$$dw = fl - df$$

$$dw = f(l - d) \qquad \text{Factor out f.}$$

$$\frac{dw}{l - d} = \frac{f\cancel{(l - d)}}{\cancel{l - d}} \qquad \text{Isolate f by dividing both sides by $l - d$.}$$

$$\frac{dw}{l - d} = f$$

Thus, $f = \dfrac{dw}{l - d}$.

Now Try Exercise 79

Understanding Algebra

Sometimes when solving a formula for a variable, the variable appears in two or more terms that are not on the same side of the equation. When this happens, first collect all of the terms that contain the variable on the same side of the equation. Then factor out the variable for which you are solving.

Avoiding Common Errors

Remember when solving equations that contain fractions, we multiply both sides of an *equation* by the LCD to eliminate the fractions from the equation. If we are *adding or subtracting rational expressions*, we write the fractions with the LCD then add or subtract the numerators while *keeping the common denominator*.

For example, consider the addition problem

$$\frac{x}{x + 7} + \frac{3}{x + 7}$$

CORRECT

$$\frac{x}{x + 7} + \frac{3}{x + 7} = \frac{x + 3}{x + 7}$$

INCORRECT

$$\frac{x}{x + 7} + \frac{3}{x + 7} = (x + 7)\left(\frac{x}{x + 7} + \frac{3}{x + 7}\right)$$

$$= x + 3$$

EXERCISE SET 6.4

Warm-Up Exercises

Fill in the blanks with the appropriate word, phrase, or symbol(s) from the following list.

proportion   rational equation   extraneous solution   similar figures   least common denominator

cross-multiplication   no solution   factoring   an infinite number of solutions

1. To eliminate rational expressions from an equation, multiply both sides of the equation by the _____.

2. Whenever solving a _____ with a variable denominator, you must check the value(s) obtained in the original equation.

3. When solving a rational equation with a variable denominator, a value obtained that makes a denominator 0 is called a(n) _____.

4. A _____ is an equation of the form $\frac{a}{b} = \frac{c}{d}, b \neq 0, d \neq 0$.

5. Proportions may be solved by _____ as follows: If $\frac{a}{b} = \frac{c}{d}$, then $ad = bc, b \neq 0, d \neq 0$.

6. Figures whose corresponding angles are equal and whose corresponding sides are in proportion are called _____.

7. If, while solving for a variable in a formula, the variable appears in more than one term, you may need to use _____ to solve for the variable.

8. If the only value you obtain from solving a rational equation is an extraneous solution, then the equation has _____.

Practice the Skills

Solve each equation and check your solution.

9. $\frac{15}{x} = 3$

10. $\frac{12}{x} = 4$

11. $\frac{11}{b} = 2$

12. $\frac{1}{4} = \frac{z + 2}{12}$

13. $\frac{6x + 7}{5} = \frac{2x + 9}{3}$

14. $\frac{a + 2}{7} = \frac{a - 3}{2}$

15. $\frac{3x}{8} + \frac{1}{4} = \frac{2x - 3}{8}$

16. $\frac{3x}{10} + \frac{2}{5} = \frac{4x - 3}{5}$

17. $\frac{z}{3} - \frac{3z}{4} = -\frac{5z}{12}$

18. $\frac{w}{2} + \frac{2w}{3} = \frac{7w}{6}$

19. $\frac{3}{4} - x = 2x$

20. $\frac{2}{y} + \frac{1}{2} = \frac{5}{2y}$

21. $\frac{2}{r} + \frac{5}{3r} = 1$

22. $3 + \frac{2}{x} = \frac{1}{4}$

23. $\frac{x - 2}{x - 5} = \frac{3}{x - 5}$

24. $\frac{c + 3}{c + 1} = \frac{5}{2}$

25. $\frac{5y - 2}{7} = \frac{15y - 2}{28}$

26. $\frac{3}{x + 1} = \frac{2}{x - 3}$

27. $\frac{5.6}{-p - 6.2} = \frac{2}{p}$

28. $\frac{4.5}{y - 3} = \frac{6.9}{y + 3}$

29. $\frac{m + 1}{m + 10} = \frac{m - 2}{m + 4}$

30. $\frac{x - 3}{x + 1} = \frac{x - 6}{x + 5}$

31. $x - \frac{4}{3x} = -\frac{1}{3}$

32. $x + \frac{2}{x} = \frac{27}{x}$

33. $\frac{2x - 1}{3} - \frac{x}{4} = \frac{7.4}{6}$

34. $\frac{15}{x} + \frac{9x - 7}{x + 2} = 9$

35. $x + \frac{6}{x} = -7$

36. $b - \frac{8}{b} = -7$

37. $2 - \frac{5}{2b} = \frac{2b}{b + 1}$

38. $\frac{3z - 2}{z + 1} = 4 - \frac{z + 2}{z - 1}$

39. $\frac{1}{w - 3} + \frac{1}{w + 3} = \frac{-5}{w^2 - 9}$

40. $\frac{6}{x + 3} + \frac{5}{x + 4} = \frac{12x + 31}{x^2 + 7x + 12}$

41. $\frac{8}{x^2 - 9} = \frac{2}{x - 3} - \frac{4}{x + 3}$

42. $a - \frac{a}{4} + \frac{a}{5} = 19$

43. $\frac{y}{2y + 2} + \frac{2y - 16}{4y + 4} = \frac{2y - 3}{y + 1}$

44. $\frac{2}{w - 5} = \frac{22}{2w^2 - 9w - 5} - \frac{3}{2w + 1}$

45. $\frac{x^2}{x - 5} = \frac{25}{x - 5}$

46. $\frac{x^2}{x - 9} = \frac{81}{x - 9}$

47. $\frac{5}{x^2 + 4x + 3} + \frac{2}{x^2 + x - 6} = \frac{3}{x^2 - x - 2}$

48. $\frac{2}{x^2 + 2x - 8} - \frac{1}{x^2 + 9x + 20} = \frac{4}{x^2 + 3x - 10}$

Similar Figures   *For each pair of similar figures, find the length of the two unknown sides (that is, those two sides whose lengths involve the variable x).*

49.

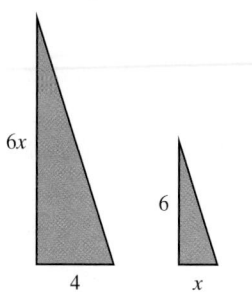

50.

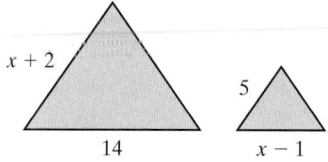

51.

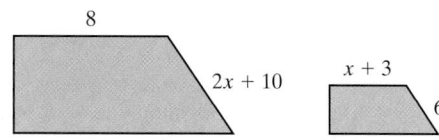

52.

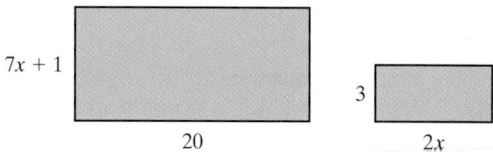

For each rational function given, find all values a for which f(a) has the indicated value.

53. $f(x) = x - \dfrac{3}{x}, f(a) = 2$

54. $f(x) = 3x - \dfrac{5}{x}, f(a) = -14$

55. $f(x) = \dfrac{x - 2}{x + 5}, f(a) = \dfrac{3}{5}$

56. $f(x) = \dfrac{x + 3}{x + 5}, f(a) = \dfrac{4}{7}$

57. $f(x) = \dfrac{6}{x} + \dfrac{6}{2x}, f(a) = 6$

58. $f(x) = \dfrac{4}{x} - \dfrac{3}{2x}, f(a) = 4$

Solve each formula for the indicated variable.

59. $\dfrac{V_1}{V_2} = \dfrac{P_2}{P_1}$, for P_1 (chemistry)

60. $T_a = \dfrac{T_f}{1 - f}$, for f (investment formula)

61. $\dfrac{V_1}{V_2} = \dfrac{P_2}{P_1}$, for V_2 (chemistry)

62. $S = \dfrac{a}{1 - r}$, for r (mathematics)

63. $m = \dfrac{y - y_1}{x - x_1}$, for y (slope)

64. $m = \dfrac{y - y_1}{x - x_1}$, for x_1 (slope)

65. $z = \dfrac{x - \bar{x}}{s}$, for x (statistics)

66. $z = \dfrac{x - \bar{x}}{s}$, for s (statistics)

67. $d = \dfrac{fl}{f + w}$, for w (physics)

68. $\dfrac{1}{p} + \dfrac{1}{q} = \dfrac{1}{f}$, for p (optics)

69. $\dfrac{1}{p} + \dfrac{1}{q} = \dfrac{1}{f}$, for q (optics)

70. $\dfrac{1}{R_T} = \dfrac{1}{R_1} + \dfrac{1}{R_2}$, for R_T (electronics)

71. $at_2 - at_1 + v_1 = v_2$, for a (physics)

72. $2P_1 - 2P_2 - P_1P_c = P_2P_c$, for P_c (economics)

73. $a_n = a_1 + nd - d$, for d (mathematics)

74. $S_n - S_nr = a_1 - a_1r^n$, for S_n (mathematics)

75. $F = \dfrac{Gm_1m_2}{d^2}$, for G (physics)

76. $\dfrac{P_1V_1}{T_1} = \dfrac{P_2V_2}{T_2}$, for T_2 (physics)

77. $\dfrac{P_1V_1}{T_1} = \dfrac{P_2V_2}{T_2}$, for T_1 (physics)

78. $A = \dfrac{1}{2}h(a + b)$, for h (mathematics)

79. $\dfrac{S - S_0}{V_0 + gt} = t$, for V_0 (physics)

80. $\dfrac{E}{e} = \dfrac{R + r}{r}$, for e (engineering)

Simplify each expression in **a)** *and solve each equation in* **b)**.

81. a) $\dfrac{1}{x+3} + \dfrac{2}{x+1}$

  b) $\dfrac{1}{x+3} + \dfrac{2}{x+1} = 0$

82. a) $\dfrac{4}{x+3} + \dfrac{5}{2x+6} + \dfrac{1}{2}$

  b) $\dfrac{4}{x+3} + \dfrac{5}{2x+6} = \dfrac{1}{2}$

83. a) $\dfrac{b+3}{b} - \dfrac{b+4}{b+5} - \dfrac{15}{b^2+5b}$

  b) $\dfrac{b+3}{b} - \dfrac{b+4}{b+5} = \dfrac{15}{b^2+5b}$

84. a) $\dfrac{4x+3}{x^2+11x+30} - \dfrac{3}{x+6} + \dfrac{2}{x+5}$

  b) $\dfrac{4x+3}{x^2+11x+30} - \dfrac{3}{x+6} = \dfrac{2}{x+5}$

Problem Solving

85. Tax-Free Investment The formula $T_a = \dfrac{T_f}{1-f}$ can be used to find the equivalent taxable yield, T_a, of a tax-free investment, T_f. In this formula, f is the individual's federal income tax bracket. Lucy Alfonso is in the 28% tax bracket.

  a) Determine the equivalent taxable yield of a 9% tax-free investment for Lucy.

  b) Solve this equation for T_f.

  c) Determine the equivalent tax-free yield for a 12% taxable investment for Lucy.

86. Tax-Free Investment See Exercise 85. Kim Ghiselin is in the 25% tax bracket.

  a) Determine the equivalent taxable yield of a 6% tax-free investment for Kim.

  b) Determine the equivalent tax-free yield for a 10% taxable investment for Kim.

87. Insurance When a homeowner purchases a homeowner's insurance policy in which the dwelling is insured for less than 80% of the replacement value, the insurance company will not reimburse the homeowner in total for their loss. The following formula is used to determine the insurance company payout, I, when the dwelling is insured for less than 80% of the replacement value.

$$I = \frac{AC}{0.80R}$$

In the formula, A is the amount of insurance carried, C is the cost of repairing the damaged area, and R is the replacement value of the home. (There are some exceptions to when this formula is used.)

  a) Suppose Jan Burdett had a fire in her kitchen that caused $10,000 worth of damage. If she carried $50,000 of insurance on a home with $100,000 replacement value, what would the insurance company pay for the repairs?

  b) Solve this formula for R, the replacement value.

88. Average Velocity Average velocity is defined as a change in distance divided by a change in time, or

$$v = \frac{d_2 - d_1}{t_2 - t_1}$$

This formula can be used when an object at distance d_1 at time t_1 travels to a distance d_2 at time t_2.

  a) Assume $t_1 = 2$ hours, $d_1 = 118$ miles, $t_2 = 9$ hours, and $d_2 = 412$ miles. Find the average velocity.

  b) Solve the formula for t_2.

89. Average Acceleration Average acceleration is defined as a change in velocity divided by a change in time, or

$$a = \frac{v_2 - v_1}{t_2 - t_1}$$

This formula can be used when an object at velocity v_1 at time t_1 accelerates (or decelerates) to velocity v_2 at time t_2.

  a) Assume $v_1 = 20$ feet per minute, $t_1 = 20$ minutes, $v_2 = 60$ feet per minute, and $t_2 = 22$ minutes. Find the average acceleration. The units will be ft/min^2.

  b) Solve the formula for t_1.

90. Economics A formula for break-even analysis in economics is

$$Q = \frac{F + D}{R - V}$$

This formula is used to determine the number of units, Q, in an apartment building that must be rented for an investor to break even. In the formula, F is the monthly fixed expenses for the entire building, D is the monthly debt payment on the building, R is the rent per unit, and V is the variable expense per unit.

  Assume an investor is considering investing in a 50-unit building. Each two-bedroom apartment can be rented for $500 per month. Variable expenses are estimated to be $200 per month per unit, fixed expenses are estimated to be $2500 per month, and the monthly debt payment is $8000. How many apartments must be rented for the investor to break even?

91. Rate of Discount The *rate of discount, P*, expressed as a fraction or decimal, can be found by the formula

$$P = 1 - \frac{R - D}{R}$$

where R is the regular price of an item and D is the discount (the amount saved off the regular price).

a) Determine the rate of discount on a purse with a regular price of $39.99 that is on sale for $30.99

b) Solve the formula above for D.

c) Solve the formula above for R.

Refer to Example 9 for Exercises 92–94.

92. Total Resistance What is the total resistance in the circuit if resistors of 300 ohms, 500 ohms, and 3000 ohms are connected in parallel?

93. Total Resistance What is the total resistance in the circuit if resistors of 200 ohms and 600 ohms are connected in parallel?

94. Total Resistance Three resistors of identical resistance are to be connected in parallel. What should be the resistance of each resistor if the circuit is to have a total resistance of 700 ohms?

Refer to Example 10 for Exercises 95 and 96.

95. Focal Length In a slide or movie projector, the film acts as the object whose image is projected on a screen. If a 100-mm-focal length (or 0.10 meter) lens is to project an image on a screen 7.5 meters away, how far from the lens should the film be?

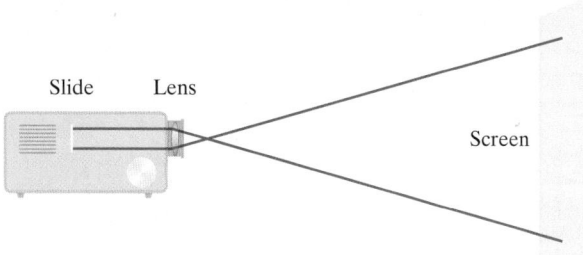

96. Curved Mirror A diamond ring is placed 20.0 cm from a concave (curved in) mirror whose focal length is 15.0 cm. Find the position of the image (or the *image distance*).

97. Investments Some investments, such as certain municipal bonds and municipal bond funds, are not only federally tax free but are also state and county or city tax free. When you wish to compare a taxable investment, T_a, with an investment that is federal, state, and county tax free, T_f, you can use the formula

$$T_a = \frac{T_f}{1 - [f + (s + c)(1 - f)]}$$

In the formula, s is your state tax bracket, c is your county or local tax bracket, and f is your federal income tax bracket. Howard Levy, who lives in Detroit, Michigan, is in a 4.6% state tax bracket, a 3% city tax bracket, and a 33% federal tax bracket. He is choosing between the Fidelity Michigan Triple Tax-Free Money Market Portfolio yielding 6.01% and the Fidelity Taxable Cash Reserve Money Market Fund yielding 7.68%.

a) Using his tax brackets, determine the taxable equivalent of the 6.01% tax-free yield.

b) Which investment should Howard make? Explain your answer.

98. Periods of Planets The synodic period of Mercury is the time required for swiftly moving Mercury to gain one lap on Earth in their orbits around the Sun. If the orbital periods (in Earth days) of the two planets are designated P_m and P_e, Mercury will be seen on the average to move $1/P_m$ of a revolution per day, while Earth moves $1/P_e$ of a revolution per day in pursuit. Mercury's daily gain on Earth is $(1/P_m) - (1/P_e)$ of a revolution, so that the time for Mercury to gain one complete revolution on Earth, the synodic period, s, may be found by the formula

$$\frac{1}{s} = \frac{1}{P_m} - \frac{1}{P_e}$$

If P_e is 365 days and P_m is 88 days, find the synodic period in Earth days.

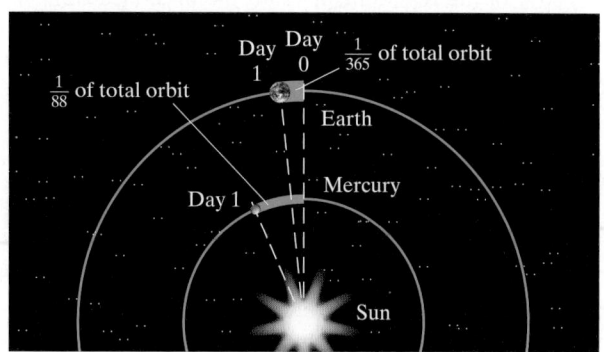

Concept/Writing Exercises

99. What restriction must be added to the statement "If $ac = bc$, then $a = b$."? Explain.

100. To the right are two graphs. One is the graph of $f(x) = \dfrac{x^2 - 9}{x - 3}$ and the other is the graph of $g(x) = x + 3$. Determine which graph is $f(x)$ and which graph is $g(x)$. Explain how you determined your answer.

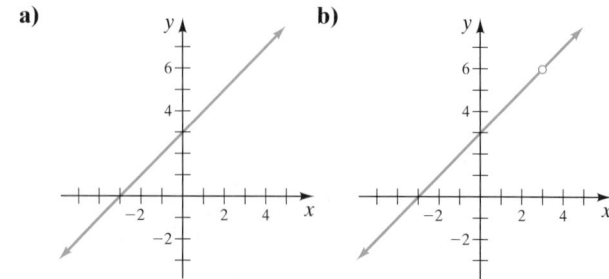

101. Make up an equation that cannot have 4 or −2 as a solution. Explain how you determined your answer.

102. Make up an equation containing the sum of two rational expressions in the variable x whose solution is the set of *real numbers*. Explain how you determined your answer.

103. Make up an equation in the variable x containing the sum of two rational expressions whose solution is the set of all real numbers except 0. Explain how you determined your answer.

Group Activity

104. Focal Length An 80-mm focal length lens is used to focus an image on the film of a camera. The maximum distance allowed between the lens and the film plane is 120 mm.

    a) Group member 1: Determine how far ahead of the film the lens should be if the object to be photographed is 10.0 meters away.

    b) Group member 2: Repeat part **a)** for a distance of 3 meters away.

    c) Group member 3: Repeat part **a)** for a distance of 1 meter away.

    d) Individually, determine the closest distance that an object could be photographed sharply.

    e) Compare your answers to see whether they appear reasonable and consistent.

Cumulative Review Exercises

[2.5] **105.** Solve the inequality $-1 \le 5 - 2x < 7$.

[3.4] **106.** Find the slope and y-intercept of the graph of the equation $3(y - 4) = -(x - 2)$.

[5.1] **107.** Simplify $3x^2y - 4xy + 2y^2 - (3xy + 6y^2 + 9x)$.

[5.8] **108. Landscaping** Jessyca Nino Aquino's garden is surrounded by a uniform-width walkway. The garden and the walkway together cover an area of 320 square feet. If the dimensions of the garden are 12 feet by 16 feet, find the width of the walkway.

Mid-Chapter Test: 6.1–6.4

To find out how well you understand the chapter material to this point, take this brief test. The answers, and the section where the material was initially discussed, are given in the back of the book. Review any questions that you answered incorrectly.

1. Determine the domain of $h(x) = \dfrac{2x + 13}{x^3 - 25x}$.

2. Simplify the rational expression $\dfrac{x^2 + 9x + 20}{2x^2 + 5x - 12}$.

Multiply or divide as indicated.

3. $\dfrac{11a + 11b}{3} \div \dfrac{a^3 + b^3}{15b}$

4. $\dfrac{x^2 + 4x - 21}{x^2 - 5x - 6} \cdot \dfrac{x^2 - 2x - 24}{x^2 + 11x + 28}$

5. $\dfrac{4a^2 + 4a + 1}{4a^2 + 6a - 2a - 3} \div \dfrac{2a^2 - 17a - 9}{(2a + 3)^2}$

6. Rectangle The area of a rectangle is $12a^2 + 13ab + 3b^2$. If the length is $18a + 6b$, find an expression for the width by dividing the area by the length.

7. Find the least common denominator for $\dfrac{x^2 - 5x - 7}{x^2 - x - 30} + \dfrac{3x^2 + 19}{x^2 - 4x - 12}$.

Add or subtract. Simplify all answers.

8. $\dfrac{5x}{x - 5} - \dfrac{25}{x - 5}$

9. $\dfrac{10}{3x^2y} + \dfrac{a}{6xy^3}$

10. $\dfrac{4}{2x^2 + 5x - 12} - \dfrac{3}{x^2 - 16}$

Simplify each complex fraction.

11. $\dfrac{9 + \dfrac{a}{b}}{\dfrac{3 - c}{b}}$

12. $\dfrac{\dfrac{5}{x} - \dfrac{8}{x^2}}{6 - \dfrac{1}{x}}$

13. $\dfrac{y^{-2} + 7y^{-1}}{7y^{-3} + y^{-4}}$

14. What is an extraneous solution? Explain under what conditions you must check for extraneous solutions.

Solve each equation and check your solutions.

15. $\dfrac{3x - 1}{7} = \dfrac{-x + 9}{2}$

16. $\dfrac{m - 7}{m - 11} = \dfrac{4}{m - 11}$

17. $x = 1 + \dfrac{12}{x}$

18. Solve $\dfrac{1}{a} - \dfrac{1}{b} = \dfrac{1}{c}$ for a.

19. Solve $x = \dfrac{4}{1 - r}$ for r.

20. Triangles The two triangles are similar triangles. Find the lengths of the two unknown sides involving the variable x.

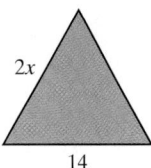

6.5 Rational Equations: Applications and Problem Solving

1 Solve work problems.

2 Solve number problems.

3 Solve motion problems.

Understanding Algebra

Problems where two or more people or machines work together to complete a certain task are referred to as *work problems*.

Understanding Algebra

In general, if a person (or machine) can complete a job in x units of time, the rate is $\frac{1}{x}$ of the task per unit of time.

In this section we will study some more applications, starting with work problems.

1 Solve Work Problems

To solve work problems we use the fact summarized in the following diagram:

$$\begin{pmatrix} \text{part of task done} \\ \text{by first person} \\ \text{or machine} \end{pmatrix} + \begin{pmatrix} \text{part of task done} \\ \text{by second person} \\ \text{or machine} \end{pmatrix} = \begin{pmatrix} 1 \\ \text{(one whole task} \\ \text{completed)} \end{pmatrix}$$

To determine the part of the task done by each person or machine, we use the formula

$$\text{rate of work} \cdot \text{time worked} = \text{part of the task completed}$$

To determine the rate of work, consider the following examples.

- If Joe can do a task by himself in 5 hours, his rate is $\frac{1}{5}$ of the task per hour.

- If Yoko can do a task by herself in 4 hours, her rate is $\frac{1}{4}$ of the task per hour.

- Similarly, if Julian can do a task by himself in x hours, his rate is $\frac{1}{x}$ of the task per hour.

EXAMPLE 1 Shoveling a Driveway After a snowfall, it takes Bud 3 hours to shovel the driveway. It takes Tina 5 hours to shovel the same driveway. If Bud and Tina work together, how long will it take them to shovel the driveway?

Solution Understand We need to find the number of hours it takes for Bud and Tina to shovel the driveway if they work together. Let $x =$ the number of hours for Bud and Tina to shovel the driveway working together.

Worker	Rate of Work	Time Worked	Part of Task Completed
Bud	$\frac{1}{3}$	x	$\frac{x}{3}$
Tina	$\frac{1}{5}$	x	$\frac{x}{5}$

Translate

$$\begin{pmatrix} \text{part of driveway shoveled} \\ \text{by Bud in } x \text{ hours} \end{pmatrix} + \begin{pmatrix} \text{part of driveway shoveled} \\ \text{by Tina in } x \text{ hours} \end{pmatrix} = 1(\text{entire driveway shoveled})$$

$$\frac{x}{3} \qquad + \qquad \frac{x}{5} \qquad = \quad 1$$

Carry Out We multiply both sides of the equation by the LCD, 15. Then we solve for x, the number of hours.

$$15\left(\frac{x}{3} + \frac{x}{5}\right) = 15 \cdot 1 \qquad \text{Multiply by the LCD, 15.}$$

$$15\left(\frac{x}{3}\right) + 15\left(\frac{x}{5}\right) = 15 \qquad \text{Distributive property}$$

$$5x + 3x = 15$$

$$8x = 15$$

$$x = \frac{15}{8}$$

Answer   Bud and Tina together can shovel the driveway in $\dfrac{15}{8}$ hours, or about 1.88 hours. This answer is reasonable because this time is less than it takes either person to shovel the driveway by himself or herself.

Now Try Exercise 15

EXAMPLE 2   **Filling a Tub** Jim McEnroy turns on the faucet and opens the drain of his tub at the same time. The faucet can fill the tub in 7.6 minutes and the drain can empty the tub in 10.3 minutes. If the faucet is on and the drain is open, how long will it take for the tub to fill?

Solution   Understand   As water from the faucet is filling the tub, water going down the drain is emptying the tub. Thus, the faucet and drain are working against each other. Let x = amount of time needed to fill the tub.

	Rate of Work	Time Worked	Part of Tub Filled or Emptied
Faucet filling tub	$\dfrac{1}{7.6}$	x	$\dfrac{x}{7.6}$
Drain emptying tub	$\dfrac{1}{10.3}$	x	$\dfrac{x}{10.3}$

Translate   Since the faucet and drain are working against each other, we will *subtract* the part of the water being emptied from the part of water being added to the tub.

$$\begin{pmatrix} \text{part of tub filled} \\ \text{in } x \text{ minutes} \end{pmatrix} - \begin{pmatrix} \text{part of tub emptied} \\ \text{in } x \text{ minutes} \end{pmatrix} = 1(\text{whole tub filled})$$

$$\frac{x}{7.6} - \frac{x}{10.3} = 1$$

Carry Out   We can clear the fractions by multiplying both sides of the equation by the LCD, $(7.6)(10.3) = 78.28$.

$$78.28\left(\frac{x}{7.6} - \frac{x}{10.3}\right) = 78.28\,(1)$$

$$\overset{10.3}{\cancel{78.28}}\left(\frac{x}{\cancel{7.6}}\right) - \overset{7.6}{\cancel{78.28}}\left(\frac{x}{\cancel{10.3}}\right) = 78.28(1)$$

$$10.3x - 7.6x = 78.28$$
$$2.7x = 78.28$$
$$x \approx 28.99$$

Answer   The tub will fill in about 29 minutes.

Now Try Exercise 25

EXAMPLE 3   **Working at a Vineyard** Chris Burditt and Mark Greenhalgh work at a vineyard. When Chris and Mark work together, they can check all the plants in a given field in 24 minutes. When Chris checks the plants by himself, it takes him 36 minutes. How long does it take Mark to check the plants by himself?

Solution   Understand   Let x = amount of time for Mark to check the plants by himself. We know that when working together they can check the plants in 24 minutes. We organize this information in the table as follows.

Worker	Rate of Work	Time Worked	Part of Plants Checked
Chris	$\dfrac{1}{36}$	24	$\dfrac{24}{36} = \dfrac{2}{3}$
Mark	$\dfrac{1}{x}$	24	$\dfrac{24}{x}$

Translate

$$\begin{pmatrix} \text{part of plants} \\ \text{checked by Chris} \end{pmatrix} + \begin{pmatrix} \text{part of plants} \\ \text{checked by Mark} \end{pmatrix} = 1(\text{whole field checked})$$

$$\frac{2}{3} \qquad + \qquad \frac{24}{x} \qquad = \quad 1$$

Carry Out

$$3x\left(\frac{2}{3} + \frac{24}{x}\right) = 3x \cdot 1 \qquad \text{Multiply both sides by the LCD, } 3x.$$

$$2x + 72 = 3x$$

$$72 = x$$

Answer   Mark can check the plants by himself in 72 minutes.

Now Try Exercise 23

Note that in Example 3, we used $\frac{2}{3}$ rather than $\frac{24}{36}$ for the part of the plants checked by Chris. Always use simplified fractions when setting up and solving equations.

2   Solve Number Problems

Now let us look at a **number problem**, where we must find a number related to one or more other numbers.

EXAMPLE 4   **Number Problem**   When the reciprocal of 3 times a number is subtracted from 7, the result is the reciprocal of twice the number. Find the number.

Solution   Understand   Let $x =$ unknown number. Then $3x$ is 3 times the number, and $\frac{1}{3x}$ is the reciprocal of 3 times the number. Twice the number is $2x$, and $\frac{1}{2x}$ is the reciprocal of twice the number.

Translate

$$7 - \frac{1}{3x} = \frac{1}{2x}$$

Carry Out

$$6x\left(7 - \frac{1}{3x}\right) = 6x \cdot \frac{1}{2x} \qquad \text{Multiply by the LCD, } 6x.$$

$$6x(7) - 6x\left(\frac{1}{3x}\right) = 6x\left(\frac{1}{2x}\right)$$

$$42x - 2 = 3$$

$$42x = 5$$

$$x = \frac{5}{42}$$

Answer   A check will verify that the number is $\frac{5}{42}$.

Now Try Exercise 33

Understanding Algebra

The distance formula is usually written as

distance = rate · time

However, it is sometimes convenient to solve the formula for *time*:

$$\frac{\text{distance}}{\text{rate}} = \frac{\text{rate} \cdot \text{time}}{\text{rate}}$$

$$\frac{\text{distance}}{\text{rate}} = \text{time}$$

or   time $= \dfrac{\text{distance}}{\text{rate}}$

3   Solve Motion Problems

The last type of problem we will look at is **motion problems**. Recall that distance = rate · time. Sometimes it is convenient to solve for time when solving motion problems.

$$\text{time} = \frac{\text{distance}}{\text{rate}}$$

EXAMPLE 5 Flying an Airplane Sally Sestani is making her preflight plan and she finds that there is a 20-mile-per-hour wind moving from east to west at the altitude she plans to fly. If she travels west (with the wind), she will be able to travel 400 miles in the same amount of time that she would be able to travel 300 miles flying east (against the wind). See **Figure 6.8.** Assuming that if it were not for the wind, the plane would fly at the same speed going east or west, find the speed of the plane in still air.

Solution   Understand   Let x = speed of the plane in still air. We will set up a table to help answer the question.

Plane	Distance	Rate	Time
Against wind	300	$x - 20$	$\dfrac{300}{x - 20}$
With wind	400	$x + 20$	$\dfrac{400}{x + 20}$

Translate   Since the times are the same, we set up and solve the following equation:

$$\frac{300}{x - 20} = \frac{400}{x + 20}$$

Carry Out
$$300(x + 20) = 400(x - 20) \quad \text{Cross-multiply.}$$
$$300x + 6000 = 400x - 8000$$
$$6000 = 100x - 8000$$
$$14{,}000 = 100x$$
$$140 = x$$

Answer   The speed of the plane in still air is 140 miles per hour.

Now Try Exercise 41

EXAMPLE 6 Paddling a Water Bike Marty and Betty McKane go out on a water bike. When paddling against the current (going out from shore), they average 2 miles per hour. Coming back (going toward shore), paddling with the current, they average 3 miles per hour. If it takes $\frac{1}{4}$ hour longer to paddle out from shore than to paddle back, how far out did they paddle?

Solution   Understand   In this problem, the times going out and coming back are not the same. It takes $\frac{1}{4}$ hour longer to paddle out than to paddle back. Therefore, to make the times equal, we can add $\frac{1}{4}$ hour to the time to paddle back (or subtract $\frac{1}{4}$ hour from the time to paddle out). Let x = the distance they paddle out from shore.

Bike	Distance	Rate	Time
Going out	x	2	$\dfrac{x}{2}$
Coming back	x	3	$\dfrac{x}{3}$

Translate
$$\text{time coming back} + \frac{1}{4}\text{hour} = \text{time going out}$$
$$\frac{x}{3} + \frac{1}{4} = \frac{x}{2}$$

Wind (20 mph)

West ← → East

Flying with the wind, 400 miles   Flying against the wind, 300 miles

FIGURE 6.8

Carry Out

$$12\left(\frac{x}{3} + \frac{1}{4}\right) = 12 \cdot \frac{x}{2} \quad \text{Multiply by the LCD, 12.}$$

$$12\left(\frac{x}{3}\right) + 12\left(\frac{1}{4}\right) = 12\left(\frac{x}{2}\right) \quad \text{Distributive property}$$

$$4x + 3 = 6x$$

$$3 = 2x$$

$$1.5 = x$$

Answer   Therefore, they paddle out 1.5 miles from shore.

Now Try Exercise 53

EXAMPLE 7 **Taking a Trip** Dawn Puppel lives in Buffalo, New York, and travels to college in South Bend, Indiana. The speed limit on some of the roads she travels is 55 miles per hour, while on others it is 65 miles per hour. The total distance traveled by Dawn is 490 miles. If Dawn follows the speed limits and the total trip takes 8 hours, how long does she drive at 55 miles per hour and how long does she drive at 65 miles per hour?

Solution   Understand and Translate   Let x = number of miles driven at 55 mph. Then $490 - x$ = number of miles driven at 65 mph.

Understanding Algebra

In Example 7, we also could have used two variables to solve the problem. We could let x = number of miles driven at 55 mph and let y = number of miles driven at 65 mph.

Then, our two equations in the system would be

$$x + y = 490$$

$$\frac{x}{55} + \frac{y}{65} = 8$$

This system can then be solved by either the substitution or the addition method discussed in Chapter 4.

Speed Limit	Distance	Rate	Time
55 mph	x	55	$\dfrac{x}{55}$
65 mph	$490 - x$	65	$\dfrac{490 - x}{65}$

Since the total time is 8 hours, we write

$$\frac{x}{55} + \frac{490 - x}{65} = 8$$

Carry Out   The LCD of 55 and 65 is 715.

$$715\left(\frac{x}{55} + \frac{490 - x}{65}\right) = 715 \cdot 8$$

$$715\left(\frac{x}{55}\right) + 715\left(\frac{490 - x}{65}\right) = 5720$$

$$13x + 11(490 - x) = 5720$$

$$13x + 5390 - 11x = 5720$$

$$2x + 5390 = 5720$$

$$2x = 330$$

$$x = 165$$

Answer   The number of miles driven at 55 mph is 165 miles. Then the time driven at 55 mph is $\dfrac{165}{55} = 3$ hours, and the time driven at 65 mph is $\dfrac{490 - 165}{65} = \dfrac{325}{65} = 5$ hours.

Now Try Exercise 59

Helpful Hint

Notice that in Example 7 the answer to the question was not the value obtained for x. The value obtained was a distance, and the question asked us to find the time. *When working word problems, you must read and work the problems very carefully and make sure you answer the questions that were asked.*

EXERCISE SET 6.5

MathXL
MathXL®

MyMathLab
MyMathLab

Warm-Up Exercises

Fill in the blanks with the appropriate word, phrase, or symbol(s) from the following list.

part of the task completed $\frac{1}{7}$ of the task $\dfrac{\text{distance}}{\text{rate}}$ one whole task $\dfrac{\text{rate}}{\text{distance}}$ 7 tasks

1. To solve work problems we use the fact that the part of the task done by the first person (or machine) plus the part of the task done by the second person (or machine) is equal to _____ completed.

2. If Paul can complete a task in 7 hours, then his rate of work is _____ per hour.

3. To determine the part of the task done by each person or machine, we use the formula rate of work · time worked = _____.

4. If we solve the distance formula for time, we get time = _____.

Problem Solving

Exercises 5–30 involve work problems. See Examples 1–3. When necessary, round answers to the nearest hundredth.

5. **Baseball Fields** It takes Richard Semmler 2 hours to prepare a Little League baseball field. It takes Larry Gilligan 6 hours to prepare the same field. How long will it take Richard and Larry working together to prepare a Little League baseball field?

© Dennis C. Runde

6. **Window Washers** Fran Thompson can wash windows in the lobby at the Days Inn in 3 hours. Jill Franks can wash the same windows in 4.5 hours. How long will it take them working together to wash the windows?

7. **Shampooing a Carpet** Jason La Rue can shampoo the carpet on the main floor of the Sheraton Hotel in 3 hours. Tom Lockheart can shampoo the same carpet in 6 hours. If they work together, how long will it take them to shampoo the carpet?

8. **Printing Checks** At the Merck Corporation it takes one computer 3 hours to print checks for its employees and a second computer 7 hours to complete the same job. How long will it take the two computers working together to complete the job?

9. **Dairy Farm** On a small dairy farm, Jin Chenge can milk 10 cows in 30 minutes. His son, Ming, can milk the same cows in 50 minutes. How long will it take them working together to milk the 10 cows?

10. **Mowing Lawns** Julio and Marcella Lopez mow lawns during the summer months. Using a self-propelled hand mower, Julio can mow a large area in 9 hours. Using a riding mower, Marcella can mow the same lawn in 4 hours. How long will it take them working together to mow the lawn?

11. **Picking Apples** Kevin Bamard can pick 25 bushels of apples in 6 hours. His young son takes twice as long to pick 25 bushels of apples. How long will it take them working together to pick 25 bushels of apples?

12. **Picking Strawberries** Amanda Heinl can pick 80 quarts of strawberries in 10 hours. Her sister, Emily, can pick 80 quarts of strawberries in 15 hours. How long will it take them working together to pick 80 quarts of strawberries?

© Allen R. Angel

13. **Cleaning Gutters** Olga Palmieri can clean the gutters on 28 houses in 4.5 days. Her co-worker, Jien-Ping, can clean the same gutters in 5.5 days. How long will it take them working together to clean the gutters on the 28 houses?

14. **Pulling Weeds** Val Short can pull weeds from a row of potatoes in 70 minutes. His friend, Jason, can pull the same weeds in 80 minutes. How long will it take them working together to pull the weeds from this row of potatoes?

15. **Plowing a Field** Wanda Garner can plow a field in 4 hours. Shawn Robinson can plow the same field in 6 hours. How long will it take them working together to plow the field?

16. **Painting** Karen can paint a living room in 6 hours. Hephner can paint the same room in 4.5 hours. How long will it take them working together to paint the living room?

17. **Filling a Pool** A small hose can fill a swimming pool in 8 hours. A large hose can fill the same pool in 5 hours. How long will it take to fill the pool when both hoses are used?

18. **Milk Tank** At a dairy plant, a milk tank can be filled in 6 hours (using the in-valve). Using the out-valve, the tank can be emptied in 8 hours. If both valves are open and milk is being pumped into the tank, how long will it take to fill the tank?

19. Water Treatment At a water treatment plant, an inlet valve can fill a large water tank in 20 hours and an outlet valve can drain the tank in 25 hours. If the tank is empty and both inlet and outlet valves are open, how long will it take to fill the tank?

20. Cabinet Makers Laura can make a set of kitchen cabinets by herself in 10 hours. If Laura and Marcia work together, they can make the same set of cabinets in 8 hours. How long will it take Marcia working by herself to make the cabinets?

21. Archeology Dr. Indiana Jones and his father, Dr. Henry Jones, are working on a dig near the Forum in Rome. Indiana and his father working together can unearth a specific plot of land in 2.6 months. Indiana can unearth the entire area by himself in 3.9 months. How long would it take Henry to unearth the entire area by himself?

© pippa west,Shutterstock

22. Digging a Trench Arthur Altshiller and Sally Choi work for General Telephone. Together it takes them 2.4 hours to dig a trench where a cable is to be laid. If Arthur can dig the trench by himself in 3.2 hours, how long would it take Sally by herself to dig the trench?

23. Jellyfish Tanks Wade Martin and Shane Wheeler work together at the Monterey Aquarium. It takes Wade 50 minutes to clean the jellyfish tanks. Since Shane is new to the job, it takes him longer to perform the same task. When working together, they can perform the task in 30 minutes. How long does it take Shane to do the task by himself?

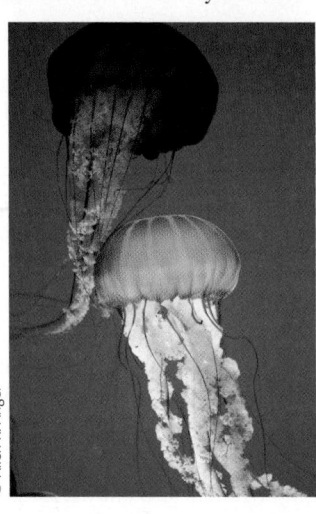

© Allen R. Angel

24. Planting Flowers Maria Vasquez and LaToya Johnson plant petunias in a botanical garden. It takes Maria twice as long as LaToya to plant the flowers. Working together, they can plant the flowers in 10 hours. How long does it take LaToya to plant the flowers by herself?

25. Filling a Washtub When only the cold water valve is opened, a washtub will fill in 8 minutes. When only the hot water valve is opened, the washtub will fill in 12 minutes. When the drain

of the washtub is open, it will drain completely in 7 minutes. If both the hot and cold water valves are open and the drain is open, how long will it take for the washtub to fill?

26. Irrigating Crops A large tank is being used on Jed Saifer's farm to irrigate the crops. The tank has two inlet pipes and one outlet pipe. The two inlet pipes can fill the tank in 8 and 12 hours, respectively. The outlet pipe can empty the tank in 15 hours. If the tank is empty, how long would it take to fill the tank when all three valves are open?

27. Pumping Water The Rushville fire department uses three pumps to remove water from flooded basements. The three pumps can remove all the water from a flooded basement in 6 hours, 5 hours, and 4 hours, respectively. If all three pumps work together how long will it take to empty the basement?

28. Installing Windows Adam, Frank, and Willy are experts at installing windows in houses. Adam can install five living room windows in a house in 10 hours. Frank can do the same job in 8 hours, and Willy can do it in 6 hours. If all three men work together, how long will it take them to install the windows?

29. Roofing a House Gary Glaze requires 15 hours to put a new roof on a house. His apprentice, Anna Gandy, can reroof the house by herself in 20 hours. After working alone on a roof for 6 hours, Gary leaves for another job. Anna takes over and completes the job. How long does it take Anna to complete the job?

30. Filling a Tank Two pipes are used to fill an oil tanker. When the larger pipe is used alone, it takes 60 hours to fill the tanker. When the smaller pipe is used alone, it takes 80 hours to fill the tanker. The large pipe begins filling the tanker. After 20 hours, the large pipe is closed and the smaller pipe is opened. How much longer will it take to finish filling the tanker using only the smaller pipe?

Exercises 31–40 involve number problems. See Example 4.

31. What number multiplied by the numerator and added to the denominator of the fraction $\frac{2}{5}$ makes the resulting fraction $\frac{3}{4}$?

32. What number added to the numerator and multiplied by the denominator of the fraction $\frac{4}{5}$ makes the resulting fraction $\frac{1}{15}$?

33. One number is twice another. The sum of their reciprocals is $\frac{3}{4}$. Find the numbers.

34. The sum of the reciprocals of two consecutive integers is $\frac{11}{30}$. Find the two integers.

35. The sum of the reciprocals of two consecutive even integers is $\frac{5}{12}$. Find the two integers.

36. When a number is added to both the numerator and denominator of the fraction $\frac{7}{9}$, the resulting fraction is $\frac{5}{6}$. Find the number added.

37. When 3 is added to twice the reciprocal of a number, the sum is $\frac{31}{10}$. Find the number.

38. The reciprocal of 3 less than a certain number is twice the reciprocal of 6 less than twice the number. Find the number(s).

39. If three times a number is added to twice the reciprocal of the number, the answer is 5. Find the number(s).

40. If three times the reciprocal of a number is subtracted from twice the reciprocal of the square of the number, the difference is −1. Find the number(s).

Exercises 41–63 involve motion problems. See Examples 5–7. When necessary, round answers to the nearest hundredth.

41. Gondola When Angelo Burnini rows his gondola in still water (with no current) in Venice, Italy, he travels at 3 mph. When he rows at the same rate in the Grand Canal, it takes him the same amount of time to travel 2.4 miles downstream (with the current) as it does to travel 2.3 miles upstream (against the current). Find the current of the canal.

42. Driving to Florida Christine Abbott lives in Bangor, Maine, and her friend Denise Brown lives in Sioux City, Iowa. They both drive to Dade City, Florida, for the winter. The trip is 1600 miles for Christine and 1500 miles for Denise. If Christine drives 5 miles per hour faster than Denise, and if they take the same amount of time to complete the trip, determine the rate for Christine and the rate for Denise.

43. Moving Sidewalk The moving sidewalk at Chicago's O'Hare International Airport moves at a speed of 2.0 feet per second. Walking on the moving sidewalk, Nancy Killian walks 120 feet in the same time that it takes her to walk 52 feet without the moving sidewalk. How fast does Nancy walk?

44. Moving Sidewalk The moving sidewalk at the Philadelphia International Airport moves at a speed of 1.8 feet per second. Nathan Trotter walks 100 feet on the moving sidewalk, then turns around on the moving sidewalk and walks at the same speed 40 feet in the opposite direction. If the time walking in each direction was the same, find Nathan's walking speed.

45. Skiing Bonnie Hellier and Clide Vincent go cross-country skiing in the Adirondack Mountains. Clide is an expert skier who averages 10 miles per hour. Bonnie averages 6 miles per hour. If it takes Bonnie $\frac{1}{2}$ hour longer to ski the trail, how long is the trail?

46. Outing in a Park Ruth and Jerry Mackin go for an outing in Memorial Park in Houston, Texas. Ruth jogs while Jerry rollerblades. Jerry rollerblades 2.9 miles per hour faster than Ruth jogs. When Jerry has rollerbladed 5.7 miles, Ruth has jogged 3.4 miles. Find Ruth's jogging speed.

47. Visiting a Resort Phil Mahler drove 60 miles to Yosemite National Park. He spent twice as much time visiting the park as it took him to drive to the park. The total time for driving and visiting the park was 5 hours. Find the average speed driving to Yosemite National Park.

48. Boating Trip Ray Packerd starts out on a boating trip at 8 A.M. Ray's boat can go 20 miles per hour in still water. How far downstream can Ray go if the current is 5 miles per hour and he wishes to go down and back in 4 hours?

49. Football Game At a football game the Carolina Panthers have the ball on their own 20-yard line. Jake Delhomme passes the ball to Steve Smith, who catches the ball and runs into the end zone for a touchdown. Assume that the ball, when passed, traveled at 14.7 yards per second and that after Steve caught the ball, he ran at 5.8 yards per second into the end zone. If the play, from the time Jake released the ball to the time Steve reached the end zone, took 10.6 seconds, how far from where Jake threw the ball did Steve catch it? Assume the entire play went down the center of the field.

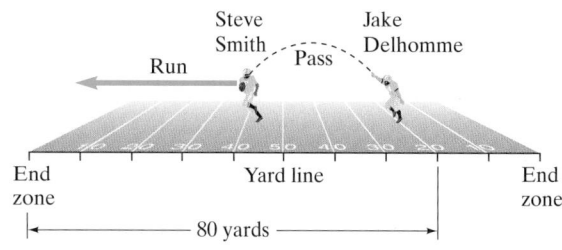

50. Skyline Drive In one day, Pauline Shannon drove from Front Royal, Virginia, to Asheville, North Carolina, a distance of 492 miles. For part of the trip she drove at a steady rate of 50 miles per hour, but in some areas she drove at a steady rate of 35 miles per hour. If the total time of the trip was 11.13 hours, how far did she travel at each speed?

51. Subway Trains The number 4 train in the New York City subway system goes from Woodlawn/Jerome Avenue in the Bronx to Flatbush Avenue in Brooklyn. The one-way distance between these two stops is 24.2 miles. On this route, two tracks run parallel to each other, one for the local train and the other for the express train. The local and express trains leave Woodlawn/Jerome Avenue at the same time. When the express reaches the end of the line at Flatbush Avenue, the local is at Wall Street, 7.8 miles from Flatbush. If the express averages 5.2 miles per hour faster than the local, find the speeds of the two trains.

52. Riding a Horse Each morning, Diane Hauck takes her horse, Beauty, for a ride on Pfeiffer Beach in Big Sur, California. She typically rides Beauty for 5.4 miles, then walks Beauty 2.3 miles. Her speed when riding is 4.2 times her speed when walking. If her total outing takes 1.5 hours, find the rate at which she walks Beauty.

53. Traveling A car and a train leave Union Station in Washington, D.C., at the same time for Rochester, New York, 390 miles away. If the speed of the car averages twice the speed of the train and the car arrives 6.5 hours before the train, find the speed of the car and the speed of the train.

54. Traveling A train and car leave from the Pasadena, California, railroad station at the same time headed for the state fair in Sacramento. The car averages 50 miles per hour and the train averages 70 miles per hour. If the train arrives at the fair 2 hours ahead of the car, find the distance from the railroad station to the state fair.

55. Traveling Two friends drive from Dallas going to El Paso, a distance of 600 miles. Mary Ann Zilke travels by highway and arrives at the same destination 2 hours ahead of Carla Canola, who took a different route. If the average speed of Mary Ann's car was 10 miles per hour faster than Carla's car, find the average speed of Mary Ann's car.

56. Sailboat Race In a 30-mile sailboat race, the winning boat, the Buccaneer, finishes 10 minutes ahead of the second-place boat, the Raven. If the Buccaneer's average speed was 2 miles per hour faster than the Raven's, find the average speed of the Buccaneer.

57. Helicopter Ride Kathy Angel took a helicopter ride to the top of the Mt. Cook Glacier in New Zealand. The flight covered a distance of 60 kilometers. Kathy stayed on top of the glacier for $\frac{1}{2}$ hour. She then flew to the town of Te Anu, 140 kilometers away. The helicopter averaged 20 kilometers per hour faster going to Te Anu than on the flight up to the top of the glacier. The total time involved in the outing was 2 hours. Find the average speed of the helicopter going to the glacier.

See Exercise 57.

58. Sailboats Two sailboats, the Serendipity and the Zerwilliker, start at the same point and at the same time on Lake Michigan, heading to the same restaurant on the lake. The Serendipity sails an average of 5.2 miles per hour and the Zerwilliker sails an average of 4.6 miles per hour. If the Serendipity arrives at the restaurant 0.4 hour ahead of the Zerwilliker, find the distance from their starting point to the restaurant.

59. Riding a Bike Robert Wiggins rides his bike from DuPont Circle in Washington, D.C., to Mount Vernon in Virginia. It takes $2\frac{1}{2}$ hours to complete the 17-mile trip from DuPont circle to Mount Vernon. On the slow part of the trip, Robert pedals his bike at a rate of 6 miles per hour. On the fast part of the trip, he pedals at 10 miles per hour. How long does Robert pedal at 6 miles per hour and how long does he pedal at 10 miles per hour?

60. Roller Blading and Jogging Sharon McGhee roller blades and jogs on a trail that is 38 miles long. On the part that is paved, she roller blades at a rate of 11 miles per hour. On the part that is not paved, she jogs at a rate of 7 miles per hour. The entire trip takes 4 hours to complete. How long does Sharon rollerblade and how long does she jog?

61. Suspension Bridge The Capilano suspension bridge in Vancouver, Canada, is 450 feet long. Phil and Heim started walking across the bridge at the same time. Heim's speed was 2 feet per minute faster than Phil's. If Heim finished walking across the bridge $2\frac{1}{2}$ minutes before Phil, find Phil's average speed in feet per minute.

62. Inclined Railroad A trip up Mount Pilatus, near Lucerne, Switzerland, involves riding an inclined railroad up to the top of the mountain, spending time at the top, then coming down the opposite side of the mountain in an aerial tram. The distance traveled up the mountain is 7.5 kilometers and the distance traveled down the mountain is 8.7 kilometers. The speed coming down the mountain is 1.2 times the speed going up. If the Lieblichs stayed at the top of the mountain for 3 hours and the total time of their outing was 9 hours, find the speed of the inclined railroad.

63. Launching Rockets Two rockets are to be launched at the same time from NASA headquarters in Houston, Texas, and are to meet at a space station many miles from Earth. The first rocket is to travel at 20,000 miles per hour and the second rocket will travel at 18,000 miles per hour. If the first rocket is scheduled to reach the space station 0.6 hour before the second rocket, how far is the space station from NASA headquarters?

64. Make up your own word problem and find the solution.

65. Make up your own motion problem and find the solution.

66. Make up your own number problem and find the solution.

Challenge Problem

67. An officer flying a California Highway Patrol aircraft determines that a car 10 miles ahead of her is speeding at 90 miles per hour.

   a) If the aircraft is traveling 150 miles per hour, how long, in minutes, will it take for the aircraft to reach the car?

   b) How far will the car have traveled in the time it takes the aircraft to reach it?

   c) If the pilot wishes to reach the car in exactly 8 minutes, how fast must the airplane fly?

Cumulative Review Exercises

[1.5] **68.** Simplify $\dfrac{(3x^4y^{-3})^{-2}}{(2x^{-1}y^6)^3}$.

[1.6] **69.** Express 9,260,000,000 in scientific notation.

[2.3] **70. Weekly Salary** Sandy Ivey receives a flat weekly salary of $240 plus 12% commission on the total dollar volume of all sales she makes. What must her dollar sales volume be in a week for her to earn $540?

[3.1] **71.** Graph $y = |x| - 2$.

[5.4] **72.** Factor $2a^4 - 2a^3 - 5a^2 + 5a$.

6.6 Variation

1 Solve direct variation problems.

2 Solve inverse variation problems.

3 Solve joint variation problems.

4 Solve combined variation problems.

Variation equations show how one quantity changes in relation to another quantity or quantities. In this section we will discuss three types of variation: *direct, inverse,* and *joint* variation.

1 Solve Direct Variation Problems

Direct variation involves two variables that increase together or decrease together. For example, consider a car traveling 80 miles per hour on an interstate highway. The car travels

- 80 miles in 1 hour,
- 160 miles in 2 hours,
- 240 miles in 3 hours, and so on.

As the *time* increases, the *distance* also increases.

The formula used to calculate distance traveled is

$$\text{distance} = \text{rate} \cdot \text{time}$$

Since the rate in the examples above is constant, the formula can be writtten

$$d = 80t$$

We say distance *varies directly* as time or that distance is *directly proportional* to time.

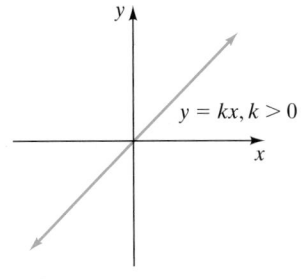

FIGURE 6.9

> **Direct Variation**
>
> If a variable y varies directly as a variable x, then
>
> $$y = kx$$
>
> where k is the **constant of proportionality** or the **variation constant.**

The graph of $y = kx$, $k > 0$, is always a straight line that goes through the origin (see **Fig. 6.9**). The slope of the line depends on the value of k.

EXAMPLE 1 Circle The circumference of a circle, C, is directly proportional to (or varies directly as) its radius, r. Write the equation for the circumference of a circle if the constant of proportionality, k, is 2π.

Solution

$$C = kr \qquad \text{\textit{C} varies directly as \textit{r}}$$
$$C = 2\pi r \qquad \text{Constant of proportionality is } 2\pi$$

Now Try Exercise 11

EXAMPLE 2 Administering a Drug The amount, a, of the drug theophylline given to patients is directly proportional to the patient's mass, m, in kilograms.

a) Write the variation equation.

b) If 150 mg is given to a boy whose mass is 30 kg, find the constant of proportionality.

c) How much of the drug should be given to a patient whose mass is 62 kg?

Solution

a) We are told the two quantities are directly proportional. We therefore set up a direct variation equation.

$$a = km$$

b) Understand and Translate   To determine the value of the constant of proportionality, we substitute the given values for the amount and mass. We then solve for k.

$$a = km$$
$$150 = k(30) \qquad \text{Substitute the given values.}$$

Carry Out $\qquad\qquad 5 = k$

Answer   Thus, $k = 5$ mg. Five milligrams of the drug should be given for each kilogram of a person's mass.

c) Understand and Translate   Now that we know the constant of proportionality we can use it to determine the amount of the drug to use for a person's mass. We set up the variation equation and substitute the values of k and m.

$$a = km$$
$$a = 5(62) \qquad \text{Substitute the given values.}$$

Carry Out $\qquad\qquad a = 310$

Answer   Thus, 310 mg of theophylline should be given to a person whose mass is 62 kg.

Now Try Exercise 57

Understanding Algebra

Direct variation involves two variables that increase together or decrease together.
   The phrases

- y varies directly as x, and
- y is directly proportional to x,

are represented by the direct variation eqation

$$y = kx.$$

EXAMPLE 3 y varies directly as the square of z. If y is 80 when z is 20, find y when z is 45.

Solution   Since y varies directly as the *square of z*, we begin with the formula $y = kz^2$. Since the constant of proportionality is not given, we must first find k using the given information.

$$y = kz^2$$
$$80 = k(20)^2 \qquad \text{Substitute the given values.}$$
$$80 = 400k \qquad \text{Solve for } k.$$
$$\frac{80}{400} = \frac{400k}{400}$$
$$0.2 = k$$

We now use $k = 0.2$ to find y when z is 45.

$$y = kz^2$$
$$y = 0.2(45)^2 \quad \text{Substitute the given values.}$$
$$y = 405$$

Thus, when z equals 45, y equals 405.

Now Try Exercise 35

2 Solve Inverse Variation Problems

Inverse variation involves two variables in which one variable increases as the other decreases and vice versa. For example, consider traveling 120 miles in a car. If the car is traveling

- 30 miles per hour, the trip takes 4 hours,
- 40 miles per hour, the trip takes 3 hours,
- 60 miles per hour, the trip takes 2 hours, and so on.

As the *rate* increases, the *time* to travel 120 miles decreases.
    The formula used to calculate time given the distance and the rate is

$$\text{time} = \frac{\text{distance}}{\text{rate}}$$

Since the distance in the examples above is constant, the formula can be rewritten

$$\text{time} = \frac{120}{\text{rate}}$$

We say time *varies inversely* as rate or that time is *inversely proportional* to rate.

Inverse Variation

If a variable y varies inversely as a variable x, then

$$y = \frac{k}{x} \quad (\text{or } xy = k)$$

where k is the constant of proportionality.

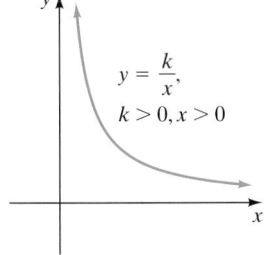

$$y = \frac{k}{x},$$
$$k > 0, x > 0$$

FIGURE 6.10

    The graph of $y = \dfrac{k}{x}$, for $k > 0$ and $x > 0$, will have the shape illustrated in **Figure 6.10**. The graph of an inverse variation equation is not defined at $x = 0$ because 0 is not in the domain of the function $y = \dfrac{k}{x}$.

EXAMPLE 4 **Melting Ice** The amount of time, t, it takes a block of ice to melt in water is inversely proportional to the water's temperature, T.

a) Write the variation equation.

b) If a block of ice takes 15 minutes to melt in 60°F water, determine the constant of proportionality.

c) Determine how long it will take a block of ice of the same size to melt in 50°F water.

Solution

a) The hotter the water temperature, the shorter the time for the block of ice to melt. The inverse variation is

$$t = \frac{k}{T}$$

b) Understand and Translate   To determine the constant of proportionality, we substitute the values for the temperature and time and solve for k.

$$t = \frac{k}{T}$$

$$15 = \frac{k}{60} \qquad \text{Substitute the given values.}$$

Carry Out $900 = k$

Answer   The constant of proportionality is 900.

c) Understand and Translate   Now that we know the constant of proportionality, we can use it to determine how long it will take for the same size block of ice to melt in 50°F water. We set up the proportion, substitute the values for k and T, and solve for t.

$$t = \frac{k}{T}$$

$$t = \frac{900}{50} \qquad \text{Substitute the given values.}$$

Carry Out $t = 18$

Answer   It will take 18 minutes for the block of ice to melt in the 50°F water.

Now Try Exercise 61

EXAMPLE 5 Lighting   The illuminance, I, of a light source varies inversely as the square of the distance, d, from the source. Assuming that the illuminance is 75 units at a distance of 4 meters, find the formula that expresses the relationship between the illuminance and the distance.

Solution   Understand and Translate   Since the illuminance varies inversely as the *square* of the distance, the general form of the equation is

$$I = \frac{k}{d^2}$$

To find k, we substitute the given values for I and d.

$$75 = \frac{k}{4^2} \qquad \text{Substitute the given values.}$$

Carry Out $$75 = \frac{k}{16} \qquad \text{Solve for } k.$$

$$(75)(16) = k$$

$$1200 = k$$

Answer   The formula is $I = \dfrac{1200}{d^2}$.

Now Try Exercise 65

3 Solve Joint Variation Problems

Joint variation involves a variable that varies directly as the product of two or more other variables.

Joint Variation

If y varies jointly as x and z, then

$$y = kxz$$

where k is the constant of proportionality.

EXAMPLE 6 Area of a Triangle The area, A, of a triangle varies jointly as its base, b, and height, h. If the area of a triangle is 48 square inches when its base is 12 inches and its height is 8 inches, find the area of a triangle whose base is 15 inches and height is 40 inches.

Solution   Understand and Translate   First write the joint variation equation and then substitute the known values and solve for k.

$$A = kbh$$

$$48 = k(12)(8) \qquad \text{Substitute the given values.}$$

Carry Out $\qquad\qquad\qquad\;\; 48 = k(96) \qquad\qquad \text{Solve for } k.$

$$\frac{48}{96} = k$$

$$k = \frac{1}{2}$$

Now solve for the unknown area using the given values.

$$A = kbh$$

$$= \frac{1}{2}(15)(40) \qquad \text{Substitute the given values.}$$

$$= 300$$

Answer   The area of the triangle is 300 square inches.

Now Try Exercise 69

Summary of Variation Equations

DIRECT	INVERSE	JOINT
$y = kx$	$y = \dfrac{k}{x}$	$y = kxz$

4 Solve Combined Variation Problems

In **combined variation**, direct and inverse variation occurs among three or more variables at the same time.

EXAMPLE 7 Pretzel Shop The owners of an Auntie Anne's Pretzel Shop find that their weekly sales of pretzels, S, vary directly as their advertising budget, A, and inversely as their pretzel price, P. When their advertising budget is $400 and the price is $1, they sell 6200 pretzels.

a) Write a variation equation expressing S in terms of A and P. Include the value of the constant.

b) Find the expected sales if the advertising budget is $600 and the price is $1.20.

Solution

a) Understand and Translate   We begin with the equation

$$S = \frac{kA}{P}$$

$$6200 = \frac{k(400)}{1} \qquad \text{Substitute the given values.}$$

Carry Out $\qquad 6200 = 400k \qquad \text{Solve for } k.$

$$15.5 = k$$

Answer   Therefore, the equation for the sales of pretzels is $S = \dfrac{15.5A}{P}$.

b) Understand and Translate   Now that we know the combined variation equation, we can use it to determine the expected sales for the given values.

$$S = \frac{15.5A}{P}$$

$$= \frac{15.5(600)}{1.20} \qquad \text{Substitute the given values.}$$

Carry Out $\qquad = 7750$

Answer   They can expect to sell 7750 pretzels.

Now Try Exercise 71

EXAMPLE 8 **Electrostatic Force** The electrostatic force, F, of repulsion between two positive electrical charges is jointly proportional to the two charges, q_1 and q_2, and inversely proportional to the square of the distance, d, between the two charges. Express F in terms of q_1, q_2, and d.

Solution

$$F = \frac{kq_1q_2}{d^2}$$

Now Try Exercise 75

EXERCISE SET 6.6

MathXL®   MyMathLab

Warm-Up Exercises

Fill in the blanks with the appropriate word, phrase, or symbol(s) from the following list.

| joint | direct | combined | inverse | constant | $y = kx$ | $y = \dfrac{k}{x}$ |

1. In _____ variation, direct and inverse variation occurs among three or more variables at the same time.

2. _____ variation involves two variables in which one variable increases as the other decreases and vice versa.

3. _____ variation involves two variables that increase together or decrease together.

4. _____ variation involves a variable that varies directly as the product of two or more other variables.

5. When y varies directly as x, we write $y = kx$, where k is the _____ of proportionality.

6. If y varies inversely as x, then the variation equation is _____.

Practice the Skills

In Exercises 7–24, determine whether the variation between the indicated quantities is direct or inverse.

7. The speed of a car and the distance traveled in one hour.

8. The number of pages Tom can read in a 2-hour period and his reading speed

9. The speed of an athlete and the time it takes him to run a 10-kilometer race

© Lowe Llaguno/Shutterstock

10. Barbara's weekly salary and the amount of money withheld for state income taxes

11. The radius of a circle and its area

12. The side of a cube and its volume

13. The radius of a balloon and its volume

14. The diameter of a circle and its circumference

15. The diameter of a hose and the volume of water coming out of the hose

16. The air temperature and the thickness of the clothing needed for a person to stay warm.

17. The time it takes an ice cube to melt in water and the temperature of the water

18. The distance between two cities on a map and the actual distance between the two cities

19. The area of a lawn and the time it takes to mow the lawn.

20. The displacement, measured in liters, of an engine and the horsepower of the engine.

© Selena/Shutterstock

21. The speed of the typist and the time taken to type a 500-word essay.

22. The number of calories eaten and the amount of exercise required to burn off those calories

23. The light illuminating an object and the distance the light is from the object

24. The number of calories in a cheeseburger and the size of the cheeseburger

For Exercises 25–32, **a)** *write the variation equation and* **b)** *find the quantity indicated.*

25. x varies directly as y. Find x when $y = 12$ and $k = 6$.

26. t varies inversely as r. Find t when $r = 20$ and $k = 60$.

27. y varies directly as R. Find y when $R = 180$ and $k = 1.7$.

28. x varies inversely as y. Find x when $y = 25$ and $k = 5$.

29. R varies inversely as W. Find R when $W = 160$ and $k = 8$.

30. L varies inversely as the square of P. Find L when $P = 4$ and $k = 100$.

31. A varies directly as B and inversely as C. Find A when $B = 12$, $C = 4$, and $k = 3$.

32. A varies jointly as R_1 and R_2 and inversely as the square of L. Find A when $R_1 = 120$, $R_2 = 8$, $L = 5$, and $k = \dfrac{3}{2}$.

For Exercises 33–42, **a)** *write the variation equation and* **b)** *find the quantity indicated.*

33. x varies directly as y. If x is 12 when y is 3, find x when y is 5.

34. Z varies directly as W. If Z is 7 when W is 28, find Z when W is 140.

35. y varies directly as the square of R. If y is 5 when R is 5, find y when R is 10.

36. P varies directly as the square of Q. If P is 32 when Q is 4, find P when Q is 7.

37. S varies inversely as G. If S is 12 when G is 0.4, find S when G is 5.

38. C varies inversely as J. If C is 7 when J is 0.7, find C when J is 12.

39. x varies inversely as the square of P. If x is 4 when P is 5, find x when P is 2.

40. R varies inversely as the square of T. If R is 3 when T is 6, find R when T is 2.

41. F varies jointly as M_1 and M_2 and inversely as d. If F is 20 when $M_1 = 5$, $M_2 = 10$, and $d = 0.2$, find F when $M_1 = 10$, $M_2 = 20$, and $d = 0.4$.

42. F varies jointly as q_1 and q_2 and inversely as the square of d. If F is 8 when $q_1 = 2$, $q_2 = 8$, and $d = 4$, find F when $q_1 = 28$, $q_2 = 12$, and $d = 2$.

Problem Solving

43. Assume a varies directly as b. If b is doubled, how will it affect a? Explain.

44. Assume a varies directly as b^2. If b is doubled, how will it affect a? Explain.

45. Assume y varies inversely as x. If x is doubled, how will it affect y? Explain.

46. Assume y varies inversely as a^2. If a is doubled, how will it affect y? Explain.

In Exercises 47–52, use the formula $F = \dfrac{km_1m_2}{d^2}$.

47. If m_1 is doubled, how will it affect F?

48. If m_1 is quadrupled and d is doubled, how will it affect F?

49. If m_1 is doubled and m_2 is halved, how will it affect F?

50. If d is halved, how will it affect F?

51. If m_1 is halved and m_2 is quadrupled, how will it affect F?

52. If m_1 is doubled, m_2 is quadrupled, and d is quadrupled, how will it affect F?

In Exercises 53 and 54, determine if the variation is of the form $y = kx$ *or* $y = \dfrac{k}{x}$, *and find* k.

53.

x	y
2	$\dfrac{5}{2}$
5	1
10	$\dfrac{1}{2}$
20	$\dfrac{1}{4}$

54.

x	y
6	2
9	3
15	5
27	9

55. Profit The profit from selling lamps is directly proportional to the number of lamps sold. When 150 lamps are sold, the profit is \$2542.50. Find the profit when 520 lamps are sold.

56. Profit The profit from selling stereos is directly proportional to the number of stereos sold. When 65 stereos are sold, the profit is \$4056. Find the profit when 80 stereos are sold.

57. Antibiotic The recommended dosage, d, of the antibiotic drug vancomycin is directly proportional to a person's weight. If Phuong Kim, who is 132 pounds, is given 2376 milligrams, find the recommended dosage for Nathan Brown, who weighs 172 pounds.

58. Dollars and Pesos American dollars vary directly to Mexican pesos. The more dollars you convert the more pesos you receive. Last week, Carlos Manuel converted \$275 into 3507 pesos. Today, he received \$400 from his aunt. If the conversion rate is unchanged when he converts the \$400 into pesos, how many pesos (round your answer to the nearest peso) will he receive?

© Natalie Speliers Ufermann/Shutterstock

59. Hooke's Law Hooke's law states that the length a spring will stretch, S, varies directly with the force (or weight), F, attached to the spring. If a spring stretches 1.4 inches when 20 pounds is attached, how far will it stretch when 15 pounds is attached?

60. Distance When a car travels at a constant speed, the distance traveled, d, is directly proportional to the time, t. If a car travels 150 miles in 2.5 hours, how far will the same car travel in 4 hours?

61. Pressure and Volume The volume of a gas, V, varies inversely as its pressure, P. If the volume, V, is 800 cubic centimeters when the pressure is 200 millimeters (mm) of mercury, find the volume when the pressure is 25 mm of mercury.

62. Building a Brick Wall The time, t, required to build a brick wall varies inversely as the number of people, n, working on it. If it takes 8 hours for five bricklayers to build a wall, how long will it take four bricklayers to build a wall?

63. Running a Race The time, t, it takes a runner to cover a specified distance is inversely proportional to the runner's speed. If Jann Avery runs at an average of 6 miles per hour, she will finish a race in 2.6 hours. How long will it take Jackie Donofrio, who runs at 5 miles per hour, to finish the same race?

64. Pitching a Ball When a ball is pitched in a professional baseball game, the time, t, it takes for the ball to reach home plate varies inversely with the speed, s, of the pitch.* A ball pitched at 90 miles per hour takes 0.459 second to reach the plate. How long will it take a ball pitched at 75 miles per hour to reach the plate?

© Richard Paul Kane/Shutterstock

---

*A ball slows down on its way to the plate due to air resistance. For a 95-mph pitch, the ball is about 8 mph faster when it leaves the pitcher's hand than when it crosses the plate.

65. **Intensity of Light** The intensity, I, of light received at a source varies inversely as the square of the distance, d, from the source. If the light intensity is 20 foot-candles at 15 feet, find the light intensity at 10 feet.

66. **Tennis Ball** When a tennis player serves the ball, the time it takes for the ball to hit the ground in the service box is inversely proportional to the speed the ball is traveling. If Andy Roddick serves at 122 miles per hour, it takes 0.21 second for the ball to hit the ground after striking his racquet. How long will it take the ball to hit the ground if he serves at 80 miles per hour?

67. **Stopping Distance** Assume that the stopping distance of a van varies directly with the square of the speed. A van traveling 40 miles per hour can stop in 60 feet. If the van is traveling 56 miles per hour, what is its stopping distance?

68. **Falling Rock** A rock is dropped from the top of a cliff. The distance it falls in feet is directly proportional to the square of the time in seconds. If the rock falls 4 feet in $\frac{1}{2}$ second, how far will it fall in 3 seconds?

69. **Volume of a Pyramid** The volume, V, of a pyramid varies jointly as the area of its base, B, and its height, h (see the figure). If the volume of the pyramid is 160 cubic meters when the area of its base is 48 square meters and its height is 10 meters, find the volume of a pyramid when the area of its base is 42 square meters and its height is 9 meters.

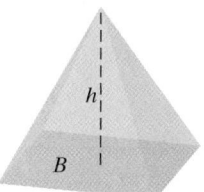

70. **Mortgage Payment** The monthly mortgage payment, P, you pay on a mortgage varies jointly as the interest rate, r, and the amount of the mortgage, m. If the monthly mortgage payment on a $50,000 mortgage at a 7% interest rate is $332.50, find the monthly payment on a $66,000 mortgage at 7%.

71. **DVD Rental** The weekly DVD rentals, R, at Hollywood Video vary directly with their advertising budget, A, and inversely with the daily rental price, P. When their advertising budget is $400 and the rental price is $2 per day, they rent 4600 DVDs per week. How many DVDs would they rent per week if they increased their advertising budget to $500 and raised their rental price to $2.50?

72. **Electrical Resistance** The electrical resistance of a wire, R, varies directly as its length, L, and inversely as its cross-sectional area, A. If the resistance of a wire is 0.2 ohm when the length is 200 feet and its cross-sectional area is 0.05 square inch, find the resistance of a wire whose length is 5000 feet with a cross-sectional area of 0.01 square inch.

73. **Weight of an Object** The weight, w, of an object in Earth's atmosphere varies inversely with the square of the distance, d, between the object and the center of Earth. A 140-pound person standing on Earth is approximately 4000 miles from Earth's center. Find the weight (or gravitational force of attraction) of this person at a distance 100 miles from Earth's surface.

74. **Wattage Rating** The wattage rating of an appliance, W, varies jointly as the square of the current, I, and the resistance, R. If the wattage is 3 watts when the current is 0.1 ampere and the resistance is 100 ohms, find the wattage when the current is 0.4 ampere and the resistance is 250 ohms.

75. **Phone Calls** The number of phone calls between two cities during a given time period, N, varies directly as the populations p_1 and p_2 of the two cities and inversely as the distance, d, between them. If 100,000 calls are made between two cities 300 miles apart and the populations of the cities are 60,000 and 200,000, how many calls are made between two cities with populations of 125,000 and 175,000 that are 450 miles apart?

© Jupiter Unlimited

76. **Water Bill** In a specific region of the country, the amount of a customer's water bill, W, is directly proportional to the average daily temperature for the month, T, the lawn area, A, and the square root of F, where F is the family size, and inversely proportional to the number of inches of rain, R.

In one month, the average daily temperature is 78°F and the number of inches of rain is 5.6. If the average family of four, who has 1000 square feet of lawn, pays $68 for water, estimate the water bill in the same month for the average family of six, who has 1500 square feet of lawn.

77. **Intensity of Illumination** An article in the magazine *Outdoor and Travel Photography* states, "If a surface is illuminated by a point-source of light (a flash), the intensity of illumination produced is inversely proportional to the square of the distance separating them."

If the subject you are photographing is 4 feet from the flash, and the illumination on this subject is $\frac{1}{16}$ of the light of the flash, what is the intensity of illumination on an object that is 7 feet from the flash?

© Photos.com

78. Force of Attraction One of Newton's laws states that the force of attraction, F, between two masses is directly proportional to the masses of the two objects, m_1 and m_2, and inversely proportional to the square of the distance, d, between the two masses.

a) Write the formula that represents Newton's law.

b) What happens to the force of attraction if one mass is doubled, the other mass is tripled, and the distance between the objects is halved?

79. Pressure on an Object The pressure, P, in pounds per square inch (psi) on an object x feet below the sea is 14.70 psi plus the product of a constant of proportionality, k, and the number of feet, x, the object is below sea level (see the figure). The 14.70 represents the weight, in pounds, of the column of air (from sea level to the top of the atmosphere) standing over a 1-inch-by-1-inch square of ocean. The kx represents the weight, in pounds, of a column of water 1 inch by 1 inch by x feet.

a) Write a formula for the pressure on an object x feet below sea level.

b) If the pressure gauge in a submarine 60 feet deep registers 40.5 psi, find the constant k.

c) A submarine is built to withstand a pressure of 160 psi. How deep can the submarine go?

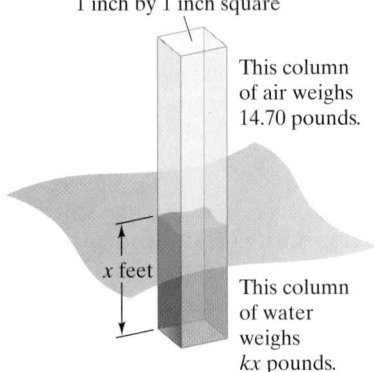

1 inch by 1 inch square

This column of air weighs 14.70 pounds.

x feet

This column of water weighs kx pounds.

Cumulative Review Exercises

[2.2] **80.** Solve the formula $V = \dfrac{4}{3}\pi r^2 h$ for h.

[3.6] **81.** Let $f(x) = x^2 - 4$ and $g(x) = -5x + 1$.
Find $f(-4) \cdot g(-2)$.

[5.2] **82.** Multiply $(7x - 3)(-2x^2 - 4x + 5)$.

[5.7] **83.** Factor $(x + 1)^2 - (x + 1) - 6$.

Chapter 6 Summary

IMPORTANT FACTS AND CONCEPTS	EXAMPLES
Section 6.1	
A **rational expression** is an expression of the form $\dfrac{p}{q}$, where p and q are polynomials and $q \neq 0$.	$\dfrac{7}{x}, \ \dfrac{x^2 - 5}{x + 1}, \ \dfrac{t^2 - t + 1}{3t^2 + 5t - 7}$
A **rational function** is a function of the form $y = f(x) = \dfrac{p}{q}$, where p and q are polynomials and $q \neq 0$.	$y = \dfrac{x - 8}{x + 9}, \quad f(x) = \dfrac{2x^2 + 4x + 1}{9x^2 - x + 3}$
The **domain of a rational function** is the set of real numbers for which the denominator is not equal to 0.	The domain of $f(x) = \dfrac{x + 9}{x - 2}$ is $\{x \mid x \neq 2\}$.
To Simplify Rational Expressions **1.** Factor both numerator and denominator completely. **2.** Divide both the numerator and the denominator by any common factors.	$\dfrac{x^3 - 8}{x^2 - 4} = \dfrac{\cancel{(x - 2)}(x^2 + 2x + 4)}{(x + 2)\cancel{(x - 2)}} = \dfrac{x^2 + 2x + 4}{x + 2}$
To Multiply Rational Expressions To multiply rational expressions, factor all numerators and denominators, and then use the following rule: $\dfrac{a}{b} \cdot \dfrac{c}{d} = \dfrac{a \cdot c}{b \cdot d}, \quad b \neq 0, d \neq 0$	$\dfrac{2x^2 + x - 1}{x^2 - 1} \cdot \dfrac{x + 1}{2x - 1} = \dfrac{\cancel{(x + 1)}\cancel{(2x - 1)}}{\cancel{(x + 1)}(x - 1)} \cdot \dfrac{x + 1}{\cancel{2x - 1}} = \dfrac{x + 1}{x - 1}$
To Divide Rational Expressions $\dfrac{a}{b} \div \dfrac{c}{d} = \dfrac{a}{b} \cdot \dfrac{d}{c} = \dfrac{a \cdot d}{b \cdot c}, \quad b \neq 0, c \neq 0, d \neq 0$	$\dfrac{x^2 + 4x + 3}{x + 1} \div \dfrac{x + 3}{x} = \dfrac{\cancel{(x + 3)}\cancel{(x + 1)}}{\cancel{x + 1}} \cdot \dfrac{x}{\cancel{x + 3}} = x$

IMPORTANT FACTS AND CONCEPTS	EXAMPLES

Section 6.2

To Add or Subtract Rational Expressions with a Common Denominator

Addition

$$\frac{a}{c} + \frac{b}{c} = \frac{a+b}{c}, \quad c \neq 0$$

Subtraction

$$\frac{a}{c} - \frac{b}{c} = \frac{a-b}{c}, \quad c \neq 0$$

$$\frac{x}{x^2 - 49} - \frac{7}{x^2 - 49} = \frac{x-7}{x^2-49} = \frac{x-7}{(x+7)(x-7)} = \frac{1}{x+7}$$

To Find the Least Common Denominator (LCD) of Rational Expressions

1. Write each nonprime coefficient (other than 1) of monomials that appear in denominators as a product of prime numbers.
2. Factor each denominator completely.
3. List all different factors that appear in any denominator. When the same factor appears in more than one denominator, write the factor with the highest power that appears.
4. The least common denominator is the product of all the factors found in step 3.

The LCD of $\dfrac{7}{9x^2y} + \dfrac{17}{3xy^3}$ is $3 \cdot 3 \cdot x^2 \cdot y^3 = 9x^2y^3$.

The LCD of $\dfrac{1}{x^2-36} - \dfrac{4x+3}{x^2+13x+42}$ is
$(x+6)(x-6)(x+7)$. Note that $x^2 - 36 = (x+6)(x-6)$
and $x^2 + 13x + 42 = (x+6)(x+7)$.

To Add or Subtract Rational Expressions with Different Denominators

1. Determine the LCD.
2. Rewrite each fraction as an equivalent fraction with the LCD.
3. Leave the denominator in factored form, but multiply out the numerator.
4. Add or subtract the numerators while maintaining the LCD.
5. When it is possible to reduce the fraction by factoring the numerator, do so.

Add $\dfrac{2a}{x^2y} + \dfrac{b}{xy^3}$.

The LCD is x^2y^3.

$$\frac{2a}{x^2y} + \frac{b}{xy^3} = \frac{y^2}{y^2} \cdot \frac{2a}{x^2y} + \frac{b}{xy^3} \cdot \frac{x}{x}$$

$$= \frac{2ay^2}{x^2y^3} + \frac{bx}{x^2y^3}$$

$$= \frac{2ay^2 + bx}{x^2y^3}$$

Section 6.3

A **complex fraction** is a fraction that has a rational expression in its numerator or its denominator or both its numerator and denominator.

$$\frac{\dfrac{2x}{x-1}}{\dfrac{x^2}{x+1}}, \qquad \frac{7 - \dfrac{6}{y}}{\dfrac{1}{y^2} + \dfrac{8}{y^3}}$$

To Simplify a Complex Fraction by Multiplying by the Least Common Denominator

1. Find the LCD of all fractions appearing within the complex fraction.
2. Multiply both the numerator and denominator of the complex fraction by the LCD of the complex fraction found in step 1.
3. Simplify when possible.

Simplify $\dfrac{1 + \dfrac{1}{x}}{x}$.

The LCD is x.

$$\frac{1 + \dfrac{1}{x}}{x} = \frac{x}{x} \cdot \frac{1 + \dfrac{1}{x}}{x} = \frac{x(1) + x\left(\dfrac{1}{x}\right)}{x(x)} = \frac{x+1}{x^2}$$

To Simplify a Complex Fraction by Simplifying the Numerator and the Denominator

1. Add or subtract as necessary to get one rational expression in the numerator.
2. Add or subtract as necessary to get one rational expression in the denominator.
3. Multiply the numerator of the complex fraction by the reciprocal of the denominator.
4. Simplify when possible.

Simplify $\dfrac{1 + \dfrac{1}{x}}{x}$.

$$\frac{1 + \dfrac{1}{x}}{x} = \frac{\dfrac{x+1}{x}}{x} = \frac{x+1}{x} \cdot \frac{1}{x} = \frac{x+1}{x^2}$$

IMPORTANT FACTS AND CONCEPTS	EXAMPLES

Section 6.4

To Solve Rational Equations

1. Determine the LCD of all rational expressions in the equation.
2. Multiply *both* sides of the equation by the LCD.
3. Remove any parentheses and combine like terms on each side of the equation.
4. Solve the equation using the properties discussed in earlier sections.
5. Check the solution in the *original* equation.

Solve $\dfrac{5}{x} + 1 = \dfrac{11}{x}$.

Multiply both sides by the LCD, x.

$$x\left(\frac{5}{x} + 1\right) = x\left(\frac{11}{x}\right)$$

$$x \cdot \frac{5}{x} + x \cdot 1 = x \cdot \frac{11}{x}$$

$$5 + x = 11$$

$$x = 6$$

The answer checks.

Proportions are equations of the form $\dfrac{a}{b} = \dfrac{c}{d}$.

$\dfrac{2}{7} = \dfrac{9}{x}$ is a proportion.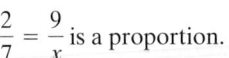

Similar figures are figures whose corresponding angles are equal and whose corresponding sides are in proportion.

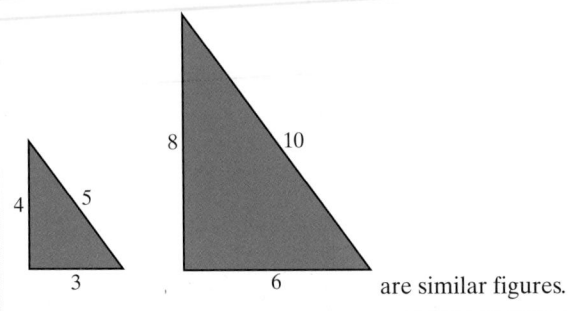

are similar figures.

Section 6.5

Applications

Work Problems:
A work problem is a problem where two or more machines or people work together to complete a task.

Carlos and Ali plant flowers in a garden. Carlos can plant a tray of flowers in 30 minutes. Ali can plant the same tray in 20 minutes. How long will it take them working together to plant the tray of flowers?

Number Problems:
A number problem is a problem where one number is related to another number.

When the reciprocal of a number is subtracted from 5, the result is the reciprocal of twice the number. Find the number.

Motion Problems:
A motion problem is a problem that involves time, rate, and distance.

Tom starts out on a canoe trip at noon. He can row at 5 miles per hour in still water. How far downstream can he go if the current is 2 miles per hour and he goes down and back in 4 hours?

Section 6.6

Direct Variation

If a variable y varies directly as a variable x, then $y = kx$, where k is the constant of proportionality.

$$y = 3x$$

Inverse Variation

If a variable y varies inversely as a variable x, then

$$y = \frac{k}{x} \quad (\text{or } xy = k)$$

where k is the constant of proportionality.

$$y = \frac{3}{x}$$

Joint Variation

If y varies jointly as x and z, then

$$y = kxz$$

where k is the constant of proportionality.

$$y = 3xz$$

Chapter 6 Review Exercises

[6.1]   *Determine the value or values of the variable that must be excluded in each rational expression.*

1. $\dfrac{2}{x-4}$

2. $\dfrac{x}{x+1}$

3. $\dfrac{-2x}{x^2+9}$

Determine the domain of each rational function.

4. $y = \dfrac{7}{(x-1)^2}$

5. $f(x) = \dfrac{x+6}{x^2}$

6. $f(x) = \dfrac{x^2-2}{x^2+4x-12}$

Simplify each expression.

7. $\dfrac{2x+6}{5x+15}$

8. $\dfrac{x^2-36}{x+6}$

9. $\dfrac{7-5x}{5x-7}$

10. $\dfrac{x^2+5x-6}{x^2+4x-12}$

11. $\dfrac{2x^2-6x+5x-15}{2x^2+7x+5}$

12. $\dfrac{a^3-8b^3}{a^2-4b^2}$

13. $\dfrac{27x^3+y^3}{9x^2-y^2}$

14. $\dfrac{2x^2+x-6}{x^3+8}$

[6.2]   *Find the least common denominator of each expression.*

15. $\dfrac{x^2}{x-4} - \dfrac{3}{x}$

16. $\dfrac{3x+1}{x+2y} + \dfrac{7x-2y}{x^2-4y^2}$

17. $\dfrac{19x-5}{x^2+2x-35} + \dfrac{3x-2}{x^2-3x-10}$

18. $\dfrac{3}{(x+2)^2} - \dfrac{6(x+3)}{x^2-4} - \dfrac{4x}{x+3}$

[6.1, 6.2]   *Perform each indicated operation.*

19. $\dfrac{10xy^2}{7z} \cdot \dfrac{14z^2}{15y}$

20. $\dfrac{x}{x-9} \cdot \dfrac{9-x}{6}$

21. $\dfrac{18x^2y^4}{xz^5} \div \dfrac{2x^2y^4}{x^4z^{10}}$

22. $\dfrac{11}{3x} + \dfrac{2}{x^2}$

23. $\dfrac{4x-4y}{x^2y} \cdot \dfrac{y^3}{16x}$

24. $\dfrac{4x^2-11x+4}{x-3} - \dfrac{x^2-4x+10}{x-3}$

25. $\dfrac{6}{xy} + \dfrac{3y}{5x^2}$

26. $\dfrac{x+2}{x-1} \cdot \dfrac{x^2+3x-4}{x^2+6x+8}$

27. $\dfrac{3x^2-7x+4}{3x^2-14x-5} - \dfrac{x^2+2x+9}{3x^2-14x-5}$

28. $5 + \dfrac{a+2}{a+1}$

29. $7 - \dfrac{b+1}{b-1}$

30. $\dfrac{a^2-b^2}{a+b} \cdot \dfrac{a^2+2ab+b^2}{a^3+a^2b}$

31. $\dfrac{1}{a^2+8a+15} \div \dfrac{3}{a+5}$

32. $\dfrac{a+c}{c} - \dfrac{a-c}{a}$

33. $\dfrac{4x^2+8x-5}{2x+5} \cdot \dfrac{x+1}{4x^2-4x+1}$

34. $(a+b) \div \dfrac{a^2-2ab-3b^2}{a-3b}$

35. $\dfrac{x^2-3xy-10y^2}{6x} \div \dfrac{x+2y}{24x^2}$

36. $\dfrac{a+1}{2a} + \dfrac{3}{4a+8}$

37. $\dfrac{x-2}{x-5} - \dfrac{3}{x+5}$

38. $\dfrac{x+4}{x^2-4} - \dfrac{3}{x-2}$

39. $\dfrac{x+1}{x-3} \cdot \dfrac{x^2+2x-15}{x^2+7x+6}$

40. $\dfrac{2}{x^2-x-6} - \dfrac{3}{x^2-4}$

41. $\dfrac{4x^2-16y^2}{9} \div \dfrac{(x+2y)^2}{12}$

42. $\dfrac{a^2+5a+6}{a^2+4a+4} \cdot \dfrac{3a+6}{a^4+3a^3}$

43. $\dfrac{x+5}{x^2-15x+50} - \dfrac{x-2}{x^2-25}$

44. $\dfrac{x+2}{x^2-x-6} + \dfrac{x-3}{x^2-8x+15}$

45. $\dfrac{1}{x+3} - \dfrac{2}{x-3} + \dfrac{6}{x^2-9}$

46. $\dfrac{a-4}{a-5} - \dfrac{3}{a+5} - \dfrac{10}{a^2-25}$

47. $\dfrac{x^3+64}{2x^2-32} \div \dfrac{x^2-4x+16}{2x+12}$

48. $\dfrac{a^2-b^4}{a^2+2ab^2+b^4} \div \dfrac{3a-3b^2}{a^2+3ab^2+2b^4}$

49. $\left(\dfrac{x^2-x-56}{x^2+14x+49} \cdot \dfrac{x^2+4x-21}{x^2-9x+8}\right) + \dfrac{3}{x^2+8x-9}$

50. $\left(\dfrac{x^2-8x+16}{2x^2-x-6} \cdot \dfrac{2x^2-7x-15}{x^2-2x-24}\right) \div \dfrac{x^2-9x+20}{x^2+2x-8}$

51. If $f(x) = \dfrac{x+1}{x+2}$ and $g(x) = \dfrac{x}{x+4}$, find

    a) the domain of $f(x)$.

    b) the domain of $g(x)$.

    c) $(f+g)(x)$.

    d) the domain of $(f+g)(x)$.

52. If $f(x) = \dfrac{x}{x^2-9}$ and $g(x) = \dfrac{x+4}{x-3}$, find

    a) the domain of $f(x)$.

    b) the domain of $g(x)$.

    c) $(f+g)(x)$.

    d) the domain of $(f+g)(x)$.

[6.3] *Simplify each complex fraction.*

53. $\dfrac{\dfrac{9a^2b}{2c}}{\dfrac{6ab^4}{4c^3}}$

54. $\dfrac{\dfrac{2}{x}+\dfrac{4}{y}}{\dfrac{x}{y}+y^2}$

55. $\dfrac{\dfrac{3}{y}-\dfrac{1}{y^2}}{7+\dfrac{1}{y^2}}$

56. $\dfrac{a^{-1}+5}{a^{-1}+\dfrac{1}{a}}$

57. $\dfrac{x^{-2}+\dfrac{3}{x}}{\dfrac{1}{x^2}-\dfrac{1}{x}}$

58. $\dfrac{\dfrac{1}{x^2-3x-18}+\dfrac{2}{x^2-2x-15}}{\dfrac{3}{x^2-11x+30}+\dfrac{1}{x^2-9x+20}}$

Area *In Exercises 59 and 60, the area and width of each rectangle are given. Find the length, l, by dividing the area, A, by the width, w.*

59.

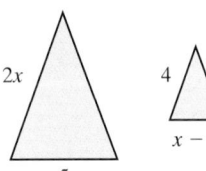

$A=\dfrac{x^2+5x+6}{x+4}$ $w=\dfrac{x^2+8x+15}{x^2+5x+4}$

l

60.

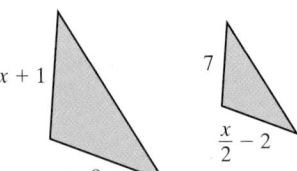

$A=\dfrac{x^2+10x+24}{x+5}$ $w=\dfrac{x^2+9x+18}{x^2+7x+10}$

l

[6.4] *In Exercises 61–70, solve each equation.*

61. $\dfrac{3}{4}=\dfrac{12}{x}$

62. $\dfrac{x}{1.5}=\dfrac{x-4}{4.5}$

63. $\dfrac{3x+4}{5}=\dfrac{2x-8}{3}$

64. $\dfrac{x}{4.8}+\dfrac{x}{2}=1.7$

65. $\dfrac{2}{y}+\dfrac{1}{5}=\dfrac{3}{y}$

66. $\dfrac{2}{x+4}-\dfrac{3}{x-4}=\dfrac{-11}{x^2-16}$

67. $\dfrac{x}{x^2-9}+\dfrac{2}{x+3}=\dfrac{4}{x-3}$

68. $\dfrac{7}{x^2-5}+\dfrac{3}{x+5}=\dfrac{4}{x-5}$

69. $\dfrac{x-3}{x-2}+\dfrac{x+1}{x+3}=\dfrac{2x^2+x+1}{x^2+x-6}$

70. $\dfrac{x+1}{x+3}+\dfrac{x+2}{x-4}=\dfrac{2x^2-18}{x^2-x-12}$

71. Solve $\dfrac{1}{a}+\dfrac{1}{b}=\dfrac{1}{c}$ for b.

72. Solve $z=\dfrac{x-\bar{x}}{s}$ for $\bar{x}$.

73. Resistors Three resistors of 100, 200, and 600 ohms are connected in parallel. Find the total resistance of the circuit. Use the formula $\dfrac{1}{R_T}=\dfrac{1}{R_1}+\dfrac{1}{R_2}+\dfrac{1}{R_3}$.

74. Focal Length What is the focal length, f, of a curved mirror if the object distance, p, is 6 centimeters and the image distance, q, is 3 centimeters? Use the formula $\dfrac{1}{p}+\dfrac{1}{q}=\dfrac{1}{f}$.

Triangles *In Exercises 75 and 76, each pair of triangles is similar. Find the lengths of the unknown sides.*

75.

$2x$ 4 $x-3$ 5

76.

$2x+1$ 7 $\dfrac{x}{2}-2$ 9

[6.5] *In Exercises 77–82, answer the question asked. When necessary, round answers to the nearest hundredth.*

77. Picking String Beans Sanford and Jerome work on a farm near Oklahoma City, Oklahoma. Sanford can pick a basket of string beans in 40 minutes while Jerome can do the same task in 30 minutes. How long will it take them to pick a basket of string beans if they work together?

78. Planting a Garden Sam and Fran plan to plant a flower garden. Together, they can plant a garden in 4.2 hours. If Sam by himself can plant the same garden in 6 hours, how long will it take Fran to plant the garden by herself?

79. Fractions What number added to the numerator and subtracted from the denominator of the fraction $\dfrac{1}{11}$ makes the result equal to $\dfrac{1}{2}$?

80. Fractions When the reciprocal of twice a number is subtracted from 1, the result is the reciprocal of three times the number. Find the number.

81. Traveling by Boat Paul Webster's motorboat can travel 15 miles per hour in still water. Traveling with the current of a river, the boat can travel 20 miles in the same time it takes to go 10 miles against the current. Find the speed of the current.

© Jupiter Unlimited

82. Flying a Plane A small plane and a car start from the same location, at the same time, heading toward the same town 450 miles away. The speed of the plane is three times the speed of the car. The plane arrives at the town 6 hours ahead of the car. Find the speeds of the car and the plane.

[6.6] *Find each indicated quantity.*

83. x is directly proportional to the square of y. If $x = 45$ when $y = 3$, find x when $y = 2$.

84. W is directly proportional to the square of L and inversely proportional to A. If $W = 4$ when $L = 2$ and $A = 10$, find W when $L = 5$ and $A = 20$.

85. z is jointly proportional to x and y and inversely proportional to the square of r. If $z = 12$ when $x = 20$, $y = 8$, and $r = 8$, find z when $x = 10$, $y = 80$, and $r = 3$.

86. Electric Surcharge The Potomac Electric Power Company places a space occupancy surcharge, s, on electric bills that is directly proportional to the amount of energy used, e. If the surcharge is \$7.20 when 3600 kilowatt-hours are used, what is the surcharge when 4200 kilowatt-hours are used?

87. Free Fall The distance, d, an object falls in free fall is directly proportional to the square of the time, t. If a person falls 16 feet in 1 second, how far will the person fall in 10 seconds? Disregard wind resistance.

88. Area The area, A, of a circle varies directly with the square of its radius, r. If the area is 78.5 when the radius is 5, find the area when the radius is 8.

89. Melting Ice Cube The time, t, for an ice cube to melt is inversely proportional to the temperature of the water it is in. If it takes an ice cube 1.7 minutes to melt in 70°F water, how long will it take an ice cube of the same size to melt in 50°F water?

© Mark Aplet\ Shutterstock

Chapter 6 Practice Test

CHAPTER **Test Prep** VIDEOS   *Chapter Test Prep Videos provide fully worked-out solutions to any of the exercises you want to review. Chapter Test Prep Videos are available via* **MyMathLab**, *or on* **YouTube** *(search "Angel Intermediate Algebra" and click on "Channels")*

1. Determine the values that are excluded in the expression
$$\frac{x + 4}{x^2 + 3x - 28}.$$

2. Determine the domain of the function
$$f(x) = \frac{x^2 + 7}{2x^2 + 7x - 4}.$$

Simplify each expression.

3. $\dfrac{10x^7 y^2 + 16x^2 y + 22x^3 y^3}{2x^2 y}$

4. $\dfrac{x^2 - 4xy - 12y^2}{x^2 + 3xy + 2y^2}$

In Exercises 5–14, perform the indicated operation.

5. $\dfrac{3xy^4}{6x^2 y^3} \cdot \dfrac{2x^2 y^4}{x^5 y^7}$

6. $\dfrac{x + 1}{x^2 - 7x - 8} \cdot \dfrac{x^2 - x - 56}{x^2 + 9x + 14}$

7. $\dfrac{7a + 14b}{a^2 - 4b^2} \div \dfrac{a^3 + a^2 b}{a^2 - 2ab}$

8. $\dfrac{x^3 + y^3}{x + y} \div \dfrac{x^2 - xy + y^2}{x^2 + y^2}$

9. $\dfrac{5}{x + 1} + \dfrac{2}{x^2}$

10. $\dfrac{x - 1}{x^2 - 9} - \dfrac{x}{x^2 - 2x - 3}$

11. $\dfrac{m}{12m^2 + 4mn - 5n^2} + \dfrac{2m}{12m^2 + 28mn + 15n^2}$

12. $\dfrac{x + 1}{4x^2 - 4x + 1} + \dfrac{3}{2x^2 + 5x - 3}$

13. $\dfrac{x^3 - 8}{x^2 + 5x - 14} \div \dfrac{x^2 + 2x + 4}{x^2 + 10x + 21}$

14. If $f(x) = \dfrac{x - 3}{x + 5}$ and $g(x) = \dfrac{x}{2x + 3}$, find

a) $(f + g)(x)$.

b) the domain of $(f + g)(x)$.

15. Area If the area of a rectangle is $\dfrac{x^2 + 11x + 30}{x + 2}$ and the length is $\dfrac{x^2 + 9x + 18}{x + 3}$, find the width.

For Exercises 16–18, simplify.

16. $\dfrac{\dfrac{1}{x} + \dfrac{2}{y}}{\dfrac{1}{x} - \dfrac{3}{y}}$

17. $\dfrac{\dfrac{a^2 - b^2}{ab}}{\dfrac{a + b}{b^2}}$

18. $\dfrac{\dfrac{7}{x} - \dfrac{6}{x^2}}{4 - \dfrac{1}{x}}$

Solve each equation.

19. $\dfrac{x}{5} - \dfrac{x}{4} = -1$

20. $\dfrac{x}{x - 8} + \dfrac{6}{x - 2} = \dfrac{x^2}{x^2 - 10x + 16}$

21. Solve $A = \dfrac{2b}{C - d}$ for C.

22. Wattage Rating The wattage rating of an appliance, W, varies jointly as the square of the current, I, and the resistance, R. If the wattage is 10 when the current is 1 ampere and the resistance is 1000 ohms, find the wattage when the current is 0.5 ampere and the resistance is 300 ohms.

23. R varies directly as P and inversely as the square of T. If $R = 30$ when $P = 40$ and $T = 2$, find R when $P = 50$ and $T = 5$.

24. Washing Windows Paul Weston can wash the windows of a house in 10 hours. His friend, Nancy Delaney, can wash the same windows in 8 hours. How long will it take them together to wash the windows of this house?

25. Rollerblading Cameron Barnette and Ashley Elliot start rollerblading at the same time at the beginning of a trail. Cameron averages 8 miles per hour, while Ashley averages 5 miles per hour. If it takes Ashley $\dfrac{1}{2}$ hour longer than Cameron to reach the end of the trail, how long is the trail?

© Allen R. Angel

Cumulative Review Test

Take the following test and check your answers with those given in the back of the book. Review any questions that you answered incorrectly. The section where the material was covered is indicated after the answer.

1. Illustrate the set $\left\{ x \middle| -\dfrac{5}{3} < x \le \dfrac{19}{4} \right\}$ on the number line.

2. Evaluate $-3x^3 - 2x^2y + \dfrac{1}{2}xy^2$ when $x = 2$ and $y = \dfrac{1}{2}$.

3. Solve the equation $2(x + 1) = \dfrac{1}{2}(x - 5)$

4. Distance Learning The Internet has made it possible for schools to offer degrees through online distance learning. The following diagram shows the degrees most often granted through online programs in 2003.

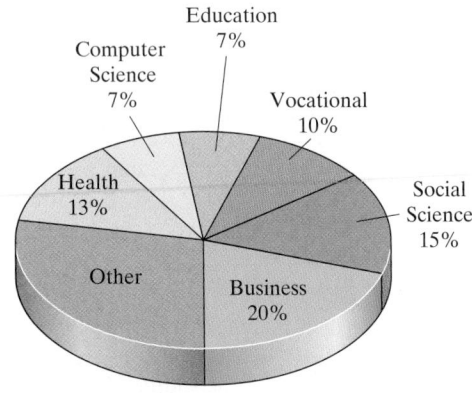

Source: The CEO Forum and Market Data Research

a) What percent makes up the "Other" category?

b) If approximately 220,000 degrees were granted through online programs, approximately how many were awarded in business?

5. Evaluate $4x^2 - 3y - 8$ when $x = 4$ and $y = -2$.

6. Simplify $\left(\dfrac{6x^5y^6}{12x^4y^7} \right)^3$.

7. Solve $F = \dfrac{mv^2}{r}$ for m.

8. Simple Interest Carmalla Banjanie invested $3000 in a certificate of deposit for 1 year. When she redeemed the certificate, she received $3180. What was the simple interest rate?

9. Meeting for a Picnic Dawn and Paula leave their homes at 8 A.M. planning to meet for a picnic at a point between them. If Dawn travels at 60 miles per hour and Paula travels at 50 miles per hour, and they live 330 miles apart, how long will it take for them to meet for the picnic?

10. Solve $\left| \dfrac{3x + 5}{3} \right| - 3 = 6$.

11. Graph $y = x^2 - 2$.

12. Let $f(x) = \sqrt{2x + 7}$. Evaluate $f(9)$.

13. Find the slope of the line passing through $(2, -4)$ and $(-5, -3)$.

14. Find an equation of a line through $\left(\dfrac{1}{2}, 1 \right)$ that is parallel to the graph of $2x + 3y - 9 = 0$. Write the equation in standard form.

15. Solve the system of equations:
$$10x - y = 2$$
$$4x + 3y = 11$$

16. Multiply $(3x^2 - 5y)(3x^2 + 5y)$.

17. Factor $3x^2 - 30x + 75$.

18. Graph $y = |x| + 2$.

19. Add $\dfrac{7}{3x^2 + x - 4} + \dfrac{9x + 2}{3x^2 - 2x - 8}$.

20. Solve $\dfrac{3y - 2}{y + 1} = 4 - \dfrac{y + 2}{y - 1}$.

7 Roots, Radicals, and Complex Numbers

Goals of This Chapter

In this chapter, we explain how to add, subtract, multiply, and divide radical expressions. We also graph radical functions, solve radical equations, and introduce imaginary numbers and complex numbers.

Make sure that you understand the three requirements for simplifying radical expressions as discussed in Section 7.5.

© Epic Stock /Shutterstock

Many scientific formulas, including many that pertain to real-life situations, contain radical expressions. In Exercise 106 on page 432, we will see how a radical is used to determine the relationship between wind speed and the height of waves in certain areas of the ocean.

7.1 Roots and Radicals

1. Find square roots.
2. Find cube roots.
3. Understand odd and even roots.
4. Evaluate radicals using absolute value.

In this chapter we expand on the concept of radicals introduced in Chapter 1. An example of a **radical expression** is $\sqrt[3]{x}$

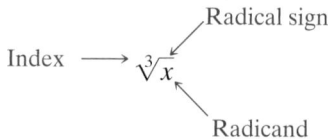

Index $\longrightarrow$ $\sqrt[3]{x}$

Radical sign

Radicand

The symbol $\sqrt{}$ is called the **radical sign** and is present in all radical expressions. The expression under the radical sign is called the **radicand**. The number directly to the left of the radical sign is called the **index** and gives us the "root" of the expression.

1 Find Square Roots

A **square root** is a radical expression that has an index of 2. The index of a square root is generally not written. Thus,

$$\sqrt{x} \text{ means } \sqrt[2]{x}$$

Every positive real number has two square roots—a principal or positive square root and a negative square root.

> ### Square Roots
>
> For any positive real number a,
>
> - The **principal** or **positive square root** of a, written $\sqrt{a}$, is the *positive* number b such that $b^2 = a$.
> - The **negative square root** of a, written $-\sqrt{a}$, is the *opposite of the principal square root of a*.

For example, the real number 25 has two square roots:

- the principal or positive square root of 25, written $\sqrt{25}$, is 5 since 5 is a positive number such that $5^2 = 25$.
- The negative square root of 25, written $-\sqrt{25}$, is -5 since it is the opposite of the principal square root, 5. Also note that $(-5)^2 = (-5)(-5) = 25$.

In this book, whenever we use the words "square root," we are referring to the principal or positive square root. If you are asked to find the *square root* of 36, you are being asked to find the positive square root of 36, $\sqrt{36}$, and your answer will be 6.

When we evaluate square roots using our calculator, if we get a decimal number we can usually see if the decimal number terminates or repeats. In Chapter 1, we learned that a rational number can be written as either a terminating or a repeating decimal number. If a decimal number does not terminate and does not repeat, then it is an irrational number. The table below gives some examples.

Radical	Calculator Result	Terminating, Repeating, or Neither?	Rational or Irrational?
$\sqrt{49}$	7	Terminating	Rational
$\sqrt{0.01}$	.1	Terminating	Rational
$\sqrt{\dfrac{25}{4}}$	2.5	Terminating	Rational
$\sqrt{\dfrac{4}{9}}$	.6666666666	Repeating	Rational
$\sqrt{2}$	1.414213562	Neither	Irrational
$\sqrt{2.5}$	1.58113883	Neither	Irrational
$\sqrt{\dfrac{1}{2}}$	.7071067812	Neither	Irrational

Understanding Algebra

When we say "square root" we refer to a radical expression with an index of 2. Furthermore, we usually do not write the index of 2. Thus, *"square root of x"* means

$\sqrt[2]{x}$

but we write

$\sqrt{x}$

Now let's consider $\sqrt{-25}$. Since the square of any real number will always be greater than or equal to 0, there is no real number that when squared equals −25. For this reason, $\sqrt{-25}$ is *not a real number*. Since the square of any real number cannot be negative, *the square root of a negative number is not a real number*. If you evaluate $\sqrt{-25}$ on most calculators, you will get an error message. We will discuss numbers like $\sqrt{-25}$ later in this chapter.

Helpful Hint

Do not confuse $-\sqrt{36}$ with $\sqrt{-36}$. Because $\sqrt{36} = 6$, $-\sqrt{36} = -6$. However, $\sqrt{-36}$ is not a real number. The square root of a negative number is not a real number.

$$\sqrt{36} = 6$$
$$-\sqrt{36} = -6$$
$$\sqrt{-36} \text{ is not a real number.}$$

The Square Root Function When graphing square root functions, functions of the form $f(x) = \sqrt{x}$, we must always remember that the radicand, x, cannot be negative. Thus, the domain of $f(x) = \sqrt{x}$ is $\{x | x \geq 0\}$, or $[0, \infty)$ in interval notation. To graph $f(x) = \sqrt{x}$, we can select some convenient values of x and find the corresponding values of $f(x)$ or y and then plot the ordered pairs, as shown in **Figure 7.1**.

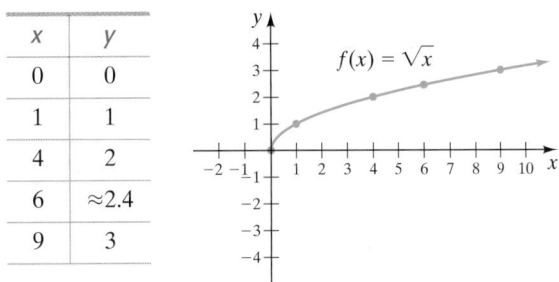

x	y
0	0
1	1
4	2
6	≈2.4
9	3

FIGURE 7.1

Since the value of $f(x)$ can never be negative, the range of $f(x) = \sqrt{x}$ is $\{y | y \geq 0\}$, or $[0, \infty)$ in interval notation.

EXAMPLE 1 For each function, find the indicated value(s).

a) $f(x) = \sqrt{11x - 2}$, $f(6)$ **b)** $g(r) = -\sqrt{-3r + 1}$, $g(-5)$ and $g(7)$

Solution

a) $f(6) = \sqrt{11(6) - 2}$ Substitute 6 for x.

$\quad\quad = \sqrt{64}$

$\quad\quad = 8$

b) $g(-5) = -\sqrt{-3(-5) + 1}$ Substitute −5 for r.

$\quad\quad\quad = -\sqrt{16}$

$\quad\quad\quad = -4$

$\quad g(7) = -\sqrt{-3(7) + 1}$ Substitute 7 for r.

$\quad\quad\quad = -\sqrt{-20}$ Not a real number.

Thus, $g(7)$ is not a real number.

Now Try Exercise 77

2 Find Cube Roots

The index of a cube root is 3. We introduced cube roots in Section 1.4. You may wish to review that material now.

> **Cube Root**
>
> The **cube root** of a number a, written $\sqrt[3]{a}$, is the number b such that $b^3 = a$.

Examples

$$\sqrt[3]{8} = 2 \qquad\qquad \text{since } 2^3 = 8$$
$$\sqrt[3]{-27} = -3 \qquad\qquad \text{since } (-3)^3 = -27$$

For each real number, there is only one cube root. The cube root of a positive number is positive and the cube root of a negative number is negative.

EXAMPLE 2 For each function, find the indicated value(s). ⎯⎯⎯⎯⎯

a) $f(x) = \sqrt[3]{10x + 34}, f(3)$ **b)** $g(r) - \sqrt[3]{12r \quad 20}, g(-4) \text{ and } g(1)$

Solution

a) $f(3) = \sqrt[3]{10(3) + 34}$ Substitute 3 for x.

$\quad\quad = \sqrt[3]{64} = 4$

b) $g(-4) = \sqrt[3]{12(-4) - 20}$ Substitute -4 for r.

$\quad\quad = \sqrt[3]{-68}$

$\quad\quad \approx -4.081655102$ From a calculator

$g(1) = \sqrt[3]{12(1) - 20}$ Substitute 1 for r.

$\quad\quad = \sqrt[3]{-8}$

$\quad\quad = -2$

Now Try Exercise 83

The Cube Root Function   **Figure 7.2** shows the graph of $y = \sqrt[3]{x}$. To obtain this graph, we substituted values for x and found the corresponding values of $f(x)$ or y.

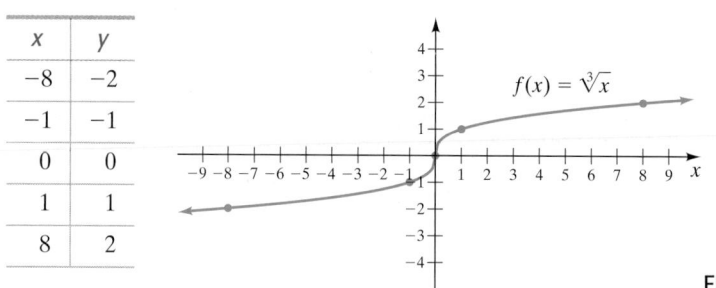

x	y
-8	-2
-1	-1
0	0
1	1
8	2

FIGURE 7.2

Notice that both the domain and range are all real numbers, $\mathbb{R}$.

3 Understand Odd and Even Roots

Up to this point we have discussed square and cube roots. Other radical expressions have different indices. For example, in the expression $\sqrt[5]{xy}$, (read "the fifth root of xy") the index is 5 and the radicand is xy.

Radical expressions that have an even-number index such as $2, 4, 6, \ldots$, are **even roots**. Square roots are even roots since their index is 2. Radical expressions that have an odd-number index such as $1, 3, 5, \ldots$, are **odd roots**.

Even Root

The nth root of a, $\sqrt[n]{a}$, where n is an *even index* and a is a nonnegative real number, is called an **even root** and is the nonnegative real number b such that $b^n = a$.

Examples of Even Roots

$$\sqrt{9} = 3 \qquad \text{since } 3^2 = 3 \cdot 3 = 9$$
$$\sqrt[4]{16} = 2 \qquad \text{since } 2^4 = 2 \cdot 2 \cdot 2 \cdot 2 = 16$$
$$\sqrt[6]{729} = 3 \qquad \text{since } 3^6 = 3 \cdot 3 \cdot 3 \cdot 3 \cdot 3 \cdot 3 = 729$$
$$\sqrt[4]{\frac{1}{256}} = \frac{1}{4} \qquad \text{since } \left(\frac{1}{4}\right)^4 = \left(\frac{1}{4}\right)\left(\frac{1}{4}\right)\left(\frac{1}{4}\right)\left(\frac{1}{4}\right) = \frac{1}{256}$$

Any real number when raised to an even power results in a positive real number. Thus, *the even root of a negative number is not a real number.*

> **Helpful Hint**
>
> There is an important difference between $-\sqrt[4]{16}$ and $\sqrt[4]{-16}$. The number $-\sqrt[4]{16}$ is the opposite of $\sqrt[4]{16}$. Because $\sqrt[4]{16} = 2$, $-\sqrt[4]{16} = -2$. However, $\sqrt[4]{-16}$ is not a real number since no real number when raised to the fourth power equals -16.
>
> $$-\sqrt[4]{16} = -(\sqrt[4]{16}) = -2$$
> $$\sqrt[4]{-16} \text{ is not a real number.}$$

Understanding Algebra

- A radical with an even index must have a nonnegative radicand if it is to be a real number.
- A radical with an odd index will be a real number with any real number as its radicand.
- Note that $\sqrt[n]{0} = 0$ regardless of whether n is an odd or even index.

Odd Root

The nth root of a, $\sqrt[n]{a}$, where n is an *odd index* and a is *any real number*, is called an **odd root** and is the real number b such that $b^n = a$.

Examples of Odd Roots

$$\sqrt[3]{8} = 2 \qquad \text{since } 2^3 = 2 \cdot 2 \cdot 2 = 8$$
$$\sqrt[3]{-8} = -2 \qquad \text{since } (-2)^3 = (-2)(-2)(-2) = -8$$
$$\sqrt[5]{243} = 3 \qquad \text{since } 3^5 = 3 \cdot 3 \cdot 3 \cdot 3 \cdot 3 = 243$$
$$\sqrt[5]{-243} = -3 \qquad \text{since } (-3)^5 = (-3)(-3)(-3)(-3)(-3) = -243$$

An odd root of a positive number is a positive number, and an odd root of a negative number is a negative number.

EXAMPLE 3 Indicate whether or not each radical expression is a real number. If the expression is a real number, find its value.

a) $\sqrt[4]{-81}$ **b)** $-\sqrt[4]{81}$ **c)** $\sqrt[5]{-32}$ **d)** $-\sqrt[5]{-32}$

Solution

a) Not a real number. Even roots of negative numbers are not real numbers.

b) Real number, $-\sqrt[4]{81} = -(\sqrt[4]{81}) = -(3) = -3$

c) Real number, $\sqrt[5]{-32} = -2$ since $(-2)^5 = -32$

d) Real number, $-\sqrt[5]{-32} = -(-2) = 2$

Now Try Exercise 21

Table 7.1 summarizes the information about even and odd roots.

TABLE 7.1

	n Is Even	n Is Odd
$a > 0$	$\sqrt[n]{a}$ is a positive real number.	$\sqrt[n]{a}$ is a positive real number.
$a < 0$	$\sqrt[n]{a}$ is not a real number.	$\sqrt[n]{a}$ is a negative real number.
$a = 0$	$\sqrt[n]{0} = 0$	$\sqrt[n]{0} = 0$

4 Evaluate Radicals Using Absolute Value

You may think that $\sqrt{a^2} = a$, but this is not necessarily true. Below we evaluate $\sqrt{a^2}$ for $a = 2$ and $a = -2$. You will see that when $a = -2$, $\sqrt{a^2} \neq a$.

$$a = 2: \qquad \sqrt{a^2} = \sqrt{2^2} = \sqrt{4} = 2 \qquad \text{Note that } \sqrt{2^2} = 2.$$
$$a = -2: \qquad \sqrt{a^2} = \sqrt{(-2)^2} = \sqrt{4} = 2 \qquad \text{Note that } \sqrt{(-2)^2} \neq -2.$$

By examining these examples, we can reason that $\sqrt{a^2}$ will *always be a positive real number* for any nonzero real number a. Recall from Section 1.3 that the *absolute value* of any real number a, or $|a|$, is also a positive number for any nonzero number. We use these facts to reason that

---

Radicals and Absolute Value

For any real number a,

$$\sqrt{a^2} = |a|$$

This indicates that the principal square root of a^2 is the absolute value of a.

EXAMPLE 4 Use absolute value to evaluate. ———————

a) $\sqrt{9^2}$ **b)** $\sqrt{0^2}$ **c)** $\sqrt{(-15.7)^2}$

Solution

a) $\sqrt{9^2} = |9| = 9$ **b)** $\sqrt{0^2} = |0| = 0$ **c)** $\sqrt{(-15.7)^2} = |-15.7| = 15.7$

Now Try Exercise 41

When simplifying a square root, if the radicand contains a variable and we are not sure that the radicand is positive, we need to use absolute value signs when simplifying.

EXAMPLE 5 Simplify. ———————

a) $\sqrt{(x + 8)^2}$ **b)** $\sqrt{16x^2}$ **c)** $\sqrt{25y^6}$ **d)** $\sqrt{a^2 - 6a + 9}$

Solution   Each square root has a radicand that contains a variable. Since we do not know the value of the variable, we do not know whether it is positive or negative. Therefore, we must use absolute value signs when simplifying.

a) $\sqrt{(x + 8)^2} = |x + 8|$

b) Write $16x^2$ as $(4x)^2$, then simplify.

$$\sqrt{16x^2} = \sqrt{(4x)^2} = |4x|$$

c) Write $25y^6$ as $(5y^3)^2$, then simplify.

$$\sqrt{25y^6} = \sqrt{(5y^3)^2} = |5y^3|$$

d) Notice that $a^2 - 6a + 9$ is a perfect square trinomial. Write the trinomial as the square of a binomial, then simplify.

$$\sqrt{a^2 - 6a + 9} = \sqrt{(a - 3)^2} = |a - 3|$$

Now Try Exercise 63

If you have a square root whose radicand contains a variable and are given instructions like "Assume all variables represent positive values and the radicand is nonnegative," then it is not necessary to use the absolute value sign when simplifying.

EXAMPLE 6 Simplify. Assume all variables represent positive values and the radicand is nonnegative.

a) $\sqrt{64x^2}$ **b)** $\sqrt{81p^4}$ **c)** $\sqrt{49x^6}$ **d)** $\sqrt{4x^2 - 12xy + 9y^2}$

Solution

a) $\sqrt{64x^2} = \sqrt{(8x)^2} = 8x$ Write $64x^2$ as $(8x)^2$.

b) $\sqrt{81p^4} = \sqrt{(9p^2)^2} = 9p^2$ Write $81p^4$ as $(9p^2)^2$.

c) $\sqrt{49x^6} = \sqrt{(7x^3)^2} = 7x^3$ Write $49x^6$ as $(7x^3)^2$.

d) $\sqrt{4x^2 - 12xy + 9y^2} = \sqrt{(2x - 3y)^2}$ Write $4x^2 - 12xy + 9y^2$ as $(2x - 3y)^2$.
$= 2x - 3y$

Now Try Exercise 67

We only need to be concerned about using absolute value signs when discussing square (and other even) roots. We do not need to use absolute value signs when the index is odd.

EXERCISE SET 7.1

MathXL® MyMathLab

Warm-Up Exercises

Fill in the blanks with the appropriate word, phrase, or symbol(s) from the following list.

odd   index   principal   cube   rational   radical

square   radicand   even   irrational   negative

1. The symbol $\sqrt{}$ is called the _____ sign.

2. In the radical expression $\sqrt[3]{5}$ the 3 is the _____.

3. In the radical expression $\sqrt[3]{5}$ the 5 is the _____.

4. When we say "_____ root" we refer to a radical expression with an index of 2.

5. When we say "_____ root" we refer to a radical expression with an index of 3.

6. The _____ square root of a, written $\sqrt{a}$, is the positive number b such that $b^2 = a$.

7. The _____ square root of a, written $-\sqrt{a}$, is the opposite of the principal square root of a.

8. A _____ number can be written as either a terminating or a repeating decimal number.

9. The _____ root of a negative number is not a real number.

10. The _____ root of a negative number is a negative number.

Practice the Skills

Evaluate each radical expression if it is a real number. Use a calculator to approximate irrational numbers to the nearest hundredth. If the expression is not a real number, so state.

11. a) $\sqrt{9}$ **b)** $-\sqrt{9}$ **c)** $\sqrt{-9}$ **d)** $-\sqrt{-9}$

12. a) $\sqrt{16}$ **b)** $-\sqrt{16}$ **c)** $\sqrt{-16}$ **d)** $-\sqrt{-16}$

13. $\sqrt[3]{-64}$ **14.** $\sqrt[3]{125}$ **15.** $\sqrt[3]{-125}$ **16.** $-\sqrt[3]{-125}$

17. $\sqrt[5]{-1}$ **18.** $-\sqrt[5]{-1}$ **19.** $\sqrt[5]{1}$ **20.** $\sqrt[6]{64}$

21. $\sqrt[6]{-64}$ **22.** $\sqrt[4]{-81}$ **23.** $\sqrt[3]{-343}$ **24.** $\sqrt{121}$

25. $\sqrt{-36}$ **26.** $\sqrt{45.3}$ **27.** $\sqrt{-45.3}$ **28.** $\sqrt{53.9}$

29. $\sqrt{\dfrac{1}{25}}$ **30.** $\sqrt{-\dfrac{1}{25}}$ **31.** $\sqrt[3]{\dfrac{1}{8}}$ **32.** $\sqrt[3]{-\dfrac{1}{8}}$

33. $\sqrt{\dfrac{4}{49}}$ **34.** $\sqrt[3]{\dfrac{8}{27}}$ **35.** $\sqrt[3]{-\dfrac{8}{27}}$ **36.** $\sqrt[4]{-8.9}$

37. $-\sqrt[4]{18.2}$ **38.** $\sqrt[5]{93}$

Use absolute value to evaluate.

39. $\sqrt{7^2}$ **40.** $\sqrt{(-7)^2}$ **41.** $\sqrt{(-3)^2}$ **42.** $\sqrt{3^2}$

43. $\sqrt{119^2}$ **44.** $\sqrt{(-119)^2}$ **45.** $\sqrt{(235.23)^2}$ **46.** $\sqrt{(-201.5)^2}$

47. $\sqrt{(0.06)^2}$ **48.** $\sqrt{(-0.19)^2}$ **49.** $\sqrt{\left(\dfrac{12}{13}\right)^2}$ **50.** $\sqrt{\left(-\dfrac{101}{319}\right)^2}$

Write as an absolute value.

51. $\sqrt{(x-8)^2}$
52. $\sqrt{(a+10)^2}$
53. $\sqrt{(x-3)^2}$
54. $\sqrt{(7a-11b)^2}$

55. $\sqrt{(3x^2-1)^2}$
56. $\sqrt{(7y^2-3y)^2}$
57. $\sqrt{(6a^3-5b^4)^2}$
58. $\sqrt{(9y^4-2z^3)^2}$

Use absolute value to simplify. You may need to factor first.

59. $\sqrt{x^{10}}$
60. $\sqrt{y^{72}}$
61. $\sqrt{z^{32}}$
62. $\sqrt{x^{200}}$

63. $\sqrt{a^2-8a+16}$
64. $\sqrt{x^2-12x+36}$
65. $\sqrt{9a^2+12ab+4b^2}$
66. $\sqrt{4x^2+20xy+25y^2}$

Simplify. Assume that all variables represent positive values and that the radicand is nonnegative.

67. $\sqrt{25a^2}$
68. $\sqrt{100a^4}$
69. $\sqrt{16c^6}$
70. $\sqrt{121z^8}$

71. $\sqrt{x^2+4x+4}$
72. $\sqrt{9a^2-6a+1}$
73. $\sqrt{4x^2+4xy+y^2}$
74. $\sqrt{16b^2-40bc+25c^2}$

Find the indicated value of each function. Use your calculator to approximate irrational numbers. Round irrational numbers to the nearest thousandth.

75. $f(x)=\sqrt{5x-6}, f(2)$
76. $f(c)=\sqrt{7c+1}, f(5)$
77. $q(x)=\sqrt{76-3x}, q(4)$

78. $q(b)=\sqrt{9b+34}, q(-1)$
79. $t(a)=\sqrt{-15a-9}, t(-6)$
80. $f(a)=\sqrt{14a-36}, f(4)$

81. $g(x)=\sqrt{64-8x}, g(-3)$
82. $p(x)=\sqrt[3]{8x+9}, p(2)$
83. $h(x)=\sqrt[3]{9x^2+4}, h(4)$

84. $k(c)=\sqrt[4]{16c-5}, k(6)$
85. $f(x)=\sqrt[3]{-2x^2+x-6}, f(-3)$
86. $t(x)=\sqrt[4]{2x^3-3x^2+6x}, t(2)$

Problem Solving

87. Find $f(81)$ if $f(x)=x+\sqrt{x}+7$.

88. Find $g(25)$ if $g(x)=x^2+\sqrt{x}-13$.

89. Find $t(18)$ if $t(x)=\dfrac{x}{2}+\sqrt{2x}-4$.

90. Find $m(36)$ if $m(x)=\dfrac{x}{3}+\sqrt{4x}+10$.

91. Find $k(8)$ if $k(x)=x^2+\sqrt{\dfrac{x}{2}}-21$.

92. Find $r(45)$ if $r(x)=\dfrac{x}{9}+\sqrt{\dfrac{x}{5}}+13$.

93. Select a value for x for which $\sqrt{(2x+1)^2}\neq 2x+1$.

94. Select a value for x for which $\sqrt{(5x-3)^2}\neq 5x-3$.

By considering the domains of the functions in Exercises 95 through 98, match each function with its graph, labeled a) through d).

95. $f(x)=\sqrt{x}$
96. $f(x)=\sqrt{x^2}$
97. $f(x)=\sqrt{x-5}$
98. $f(x)=\sqrt{x+5}$

a)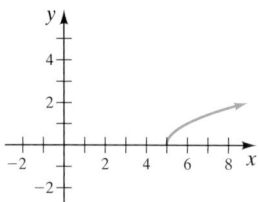
b)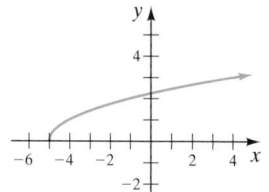
c)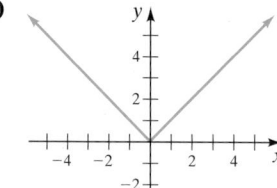
d)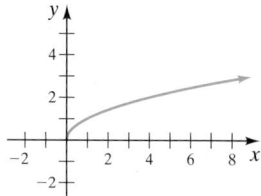

99. Graph $f(x)=\sqrt{x}+1$.
100. Graph $g(x)=-\sqrt{x}$.

101. Graph $g(x)=\sqrt{x}+1$.
102. Graph $f(x)=\sqrt{x-2}$.

103. Give a radical function whose domain is $\{x|x\geq 8\}$.

104. Give a radical function whose domain is $\{x|x\leq 5\}$.

105. Velocity of an Object The velocity, V, of an object, in feet per second, after it has fallen a distance, h, in feet, can be found by the formula $V=\sqrt{64.4h}$. A pile driver is a large mass that is used as a hammer to drive pilings into soft earth to support a building or other structure.

With what velocity will the hammer hit the piling if it falls from

a) 20 feet above the top of the piling?

b) 40 feet above the top of the piling?

106. Wave Action Scripps Institute of Oceanography in La Jolla, California, developed the formula for relating wind speed, u, in knots, with the height, H, in feet, of the waves the wind produces in certain areas of the ocean. This formula is

$$u=\sqrt{\dfrac{H}{0.026}}$$

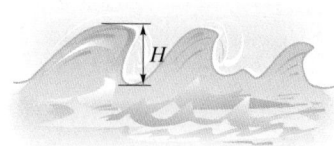

If waves produced by a storm have a height of 15 feet, what is the wind speed producing the waves?

Concept/Writing Exercises

107. Explain why $\sqrt{-81}$ is not a real number.

108. Will a radical expression with an odd index and a real number as the radicand always be a real number? Explain your answer.

109. Will a radical expression with an even index and a real number as the radicand always be a real number? Explain your answer.

110. **a)** To what is $\sqrt{a^2}$ equal?
　b) To what is $\sqrt{a^2}$ equal if we know $a \geq 0$?

111. **a)** Evaluate $\sqrt{a^2}$ for $a = 1.3$.
　b) Evaluate $\sqrt{a^2}$ for $a = -1.3$.

112. **a)** Evaluate $\sqrt[4]{16}$
　b) Evaluate $-\sqrt[4]{16}$
　c) Evaluate $\sqrt[4]{-16}$.

113. For what values of x will $\sqrt{(x-1)^2} = x - 1$? Explain how you determined your answer.

114. For what values of x will $\sqrt{(x+3)^2} = x + 3$? Explain how you determined your answer.

115. For what values of x will $\sqrt{(2x-6)^2} = 2x - 6$? Explain how you determined your answer.

116. For what values of x will $\sqrt{(3x-8)^2} = 3x - 8$? Explain how you determined your answer.

117. **a)** For what values of a is $\sqrt{a^2} = |a|$?
　b) For what values of a is $\sqrt{a^2} = a$?
　c) For what values of a is $\sqrt[3]{a^3} = a$?

118. Under what circumstances is the expression $\sqrt[n]{x}$ not a real number?

119. Explain why the expression $\sqrt[n]{x^n}$ is a real number for any real number x.

120. Under what circumstances is the expression $\sqrt[n]{x^m}$ not a real number?

121. Find the domain of $\dfrac{\sqrt{x+5}}{\sqrt[3]{x+5}}$. Explain how you determined your answer.

122. Find the domain of $\dfrac{\sqrt[3]{x-2}}{\sqrt[6]{x+1}}$. Explain how you determined your answer.

123. If $f(x) = -\sqrt{x}$, can $f(x)$ ever be
　a) greater than 0,
　b) equal to 0,
　c) less than 0?
　Explain your answers.

124. If $f(x) = \sqrt{x+5}$, can $f(x)$ ever be
　a) less than 0,
　b) equal to 0,
　c) greater than 0?
　Explain your answers.

Group Activity

In this activity, you will determine the conditions under which certain properties of radicals are true. We will discuss these properties later in this chapter. Discuss and answer these exercises as a group.

125. The property $\sqrt[n]{a} \cdot \sqrt[n]{b} = \sqrt[n]{ab}$, called the *multiplication property for radicals*, is true for certain real numbers a and b. By substituting values for a and b, determine under what conditions this property is true.

126. The property $\dfrac{\sqrt[n]{a}}{\sqrt[n]{b}} = \sqrt[n]{\dfrac{a}{b}}$, called the *division property for radicals*, is true for certain real numbers a and b. By substituting values for a and b, determine under what conditions this property is true.

Cumulative Review Exercises

Factor.

[5.4] **127.** $9ax - 3bx + 12ay - 4by$

[5.5] **128.** $3x^3 - 18x^2 + 24x$

　　129. $8x^4 + 10x^2 - 3$

[5.6] **130.** $x^3 - \dfrac{8}{27}y^3$

7.2 Rational Exponents

1 Change a radical expression to an exponential expression.

2 Simplify radical expressions.

3 Apply the rules of exponents to rational and negative exponents.

4 Factor expressions with rational exponents.

1 Change a Radical Expression to an Exponential Expression

So far, we have not discussed exponential expressions with rational exponents such as $5^{1/3}$, $x^{3/4}$, and $-27^{-4/3}$. In this section, we discuss the relationship between such expressions and radical expressions.

Consider $x = 5^{1/3}$. Now cube both sides of this equation and simplify using the rules of exponents.

$$x = 5^{1/3}$$
$$x^3 = (5^{1/3})^3 = 5^{(1/3)\cdot 3} = 5^1 = 5$$

Thus, $5^{1/3}$ is a number whose cube is 5. Recall that $\sqrt[3]{5}$ is also a number whose cube is 5. Thus, we can conclude that

$$5^{1/3} = \sqrt[3]{5}$$

The following rule shows that a radical expression can be rewritten as an expression with a rational exponent.

Exponential Form of $\sqrt[n]{a}$

$$\sqrt[n]{a} = a^{1/n}$$

When a is nonnegative, n can be any index.
When a is negative, n must be odd.

For the remainder of this chapter, unless you are instructed otherwise, assume that all variables in radicands represent nonnegative real numbers and that the radicand is nonnegative. This assumption will allow us to write many answers without absolute value signs.

EXAMPLE 1   Write each expression in exponential form (with rational exponents).

a) $\sqrt{7}$ **b)** $\sqrt[3]{15ab}$ **c)** $\sqrt[7]{-4x^2y^5}$ **d)** $\sqrt[8]{\dfrac{5x^7}{2z^{11}}}$

Solution

a) $\sqrt{7} = 7^{1/2}$ Recall that the index of a square root is 2.

b) $\sqrt[3]{15ab} = (15ab)^{1/3}$ **c)** $\sqrt[7]{-4x^2y^5} = (-4x^2y^5)^{1/7}$ **d)** $\sqrt[8]{\dfrac{5x^7}{2z^{11}}} = \left(\dfrac{5x^7}{2z^{11}}\right)^{1/8}$

Now Try Exercise 19

Exponential expressions can also be converted to radical expressions.

EXAMPLE 2   Write each expression in radical form (without rational exponents).

a) $9^{1/2}$ **b)** $(-8)^{1/3}$ **c)** $y^{1/4}$ **d)** $(10x^2y)^{1/7}$ **e)** $5rs^{1/2}$

Solution

a) $9^{1/2} = \sqrt{9} = 3$ **b)** $(-8)^{1/3} = \sqrt[3]{-8} = -2$ **c)** $y^{1/4} = \sqrt[4]{y}$

d) $(10x^2y)^{1/7} = \sqrt[7]{10x^2y}$ **e)** $5rs^{1/2} = 5r\sqrt{s}$

Now Try Exercise 33

2 Simplify Radical Expressions

We can expand the preceding rule so that radicals of the form $\sqrt[n]{a^m}$ can be written as exponential expressions. Consider $a^{2/3}$. We can write $a^{2/3}$ as $(a^{1/3})^2$ or $(a^2)^{1/3}$. This suggests $a^{2/3} = (\sqrt[3]{a})^2 = \sqrt[3]{a^2}$.

Exponential Form of $\sqrt[n]{a^m}$

For any nonnegative number a, and integers m and n,

$$\sqrt[n]{a^m} = \left(\sqrt[n]{a}\right)^m = a^{m/n}$$

Power — (pointing to m in $a^{m/n}$)
Index — (pointing to n in $a^{m/n}$)

This rule shows the relationship between radical expressions and exponential expressions with rational exponents.

When changing a radical expression to exponential expression

The power of the radicand becomes
the numerator of the rational exponent.

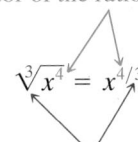

$$\sqrt[3]{x^4} = x^{4/3}$$

The index of the radical becomes
the denominator of the rational exponent.

Examples

$$\sqrt{y^3} = y^{3/2} \qquad \sqrt[3]{z^2} = z^{2/3} \qquad \sqrt[5]{2^8} = 2^{8/5}$$

$$\left(\sqrt{p}\right)^3 = p^{3/2} \qquad \left(\sqrt[4]{x}\right)^3 = x^{3/4} \qquad \left(\sqrt[4]{7}\right)^3 = 7^{3/4}$$

By this rule, for nonnegative values of the variable we can write

$$\sqrt{x^5} = \left(\sqrt{x}\right)^5 \qquad \left(\sqrt[4]{p}\right)^3 = \sqrt[4]{p^3}$$

EXAMPLE 3 Write each expression in exponential form (with rational exponents) and then simplify.

a) $\sqrt[4]{x^{12}}$ 　　　　　　　　**b)** $\left(\sqrt[3]{y}\right)^{15}$ 　　　　　　　　**c)** $\left(\sqrt[6]{x}\right)^{12}$

Solution

a) $\sqrt[4]{x^{12}} = x^{12/4} = x^3$ 　　**b)** $\left(\sqrt[3]{y}\right)^{15} = y^{15/3} = y^5$ 　　**c)** $\left(\sqrt[6]{x}\right)^{12} = x^{12/6} = x^2$

Now Try Exercise 45

When changing an exponential expression with a rational exponent to a radical expression

The numerator of the exponent
becomes the power of the radicand.

$$x^{2/3} = \sqrt[3]{x^2}$$

The denominator of the exponent
becomes the index of the radical.

Examples

$$x^{1/2} = \sqrt{x} \qquad\qquad\qquad 5^{1/3} = \sqrt[3]{5}$$

$$7^{2/3} = \sqrt[3]{7^2} \text{ or } \left(\sqrt[3]{7}\right)^2 \qquad\qquad y^{3/10} = \sqrt[10]{y^3} \text{ or } \left(\sqrt[10]{y}\right)^3$$

$$x^{9/5} = \sqrt[5]{x^9} \text{ or } \left(\sqrt[5]{x}\right)^9 \qquad\qquad z^{10/3} = \sqrt[3]{z^{10}} \text{ or } \left(\sqrt[3]{z}\right)^{10}$$

Notice that you may choose, for example, to write $6^{2/3}$ as either $\sqrt[3]{6^2}$ or $\left(\sqrt[3]{6}\right)^2$.

EXAMPLE 4 Write each expression in radical form (without rational exponents).

a) $x^{2/5}$ **b)** $(3ab)^{5/4}$

Solution

a) $x^{2/5} = \sqrt[5]{x^2}$ or $(\sqrt[5]{x})^2$ **b)** $(3ab)^{5/4} = \sqrt[4]{(3ab)^5}$ or $(\sqrt[4]{3ab})^5$

Now Try Exercise 35

EXAMPLE 5 Simplify.

a) $4^{3/2}$ **b)** $\sqrt[6]{(49)^3}$ **c)** $\sqrt[4]{(xy)^{20}}$ **d)** $(\sqrt[15]{z})^5$

Solution

a) Sometimes an expression with a rational exponent can be simplified more easily by writing the expression as a radical, as illustrated.

$$4^{3/2} = (\sqrt{4})^3 \quad \text{Write as a radical.}$$
$$= (2)^3$$
$$= 8$$

b) Sometimes a radical expression can be simplified more easily by writing the expression with rational exponents, as illustrated in parts **b)** through **d)**.

$$\sqrt[6]{(49)^3} = 49^{3/6} \quad \text{Write with a rational exponent.}$$
$$= 49^{1/2} \quad \text{Reduce exponent.}$$
$$= \sqrt{49} \quad \text{Write as a radical.}$$
$$= 7 \quad \text{Simplify.}$$

c) $\sqrt[4]{(xy)^{20}} = (xy)^{20/4} = (xy)^5$

d) $(\sqrt[15]{z})^5 = z^{5/15} = z^{1/3}$ or $\sqrt[3]{z}$

Now Try Exercise 51

Now let's consider $\sqrt[5]{x^5}$. When written in exponential form, this is $x^{5/5} = x^1 = x$. This leads to the following rule.

> **Exponential form of $\sqrt[n]{a^n}$**
>
> For any nonnegative real number a,
>
> $$\sqrt[n]{a^n} = (\sqrt[n]{a})^n = a^{n/n} = a$$

If n is an even index and a is a negative real number, $\sqrt[n]{a^n} = |a|$ and not a. For example, $\sqrt[6]{(-5)^6} = |-5| = 5$. *Since we are assuming, except where noted otherwise, that variables in radicands represent nonnegative real numbers,* we may write $\sqrt[6]{x^6} = x$ and not $|x|$. This assumption also lets us write $\sqrt{x^2} = x$ and $(\sqrt[4]{z})^4 = z$.

Examples

$$\sqrt{3^2} = 3 \qquad\qquad \sqrt[4]{y^4} = y$$
$$\sqrt[6]{(xy)^6} = xy \qquad\qquad (\sqrt[5]{z})^5 = z$$

3 Apply the Rules of Exponents to Rational and Negative Exponents

In Section 1.5, we introduced and discussed the rules of exponents where we only used exponents that were whole numbers. The rules still apply when the exponents are rational numbers. Let's review those rules now.

Rules of Exponents

For all real numbers a and b and all rational numbers m and n,

Product rule	$a^m \cdot a^n = a^{m+n}$
Quotient rule	$\dfrac{a^m}{a^n} = a^{m-n}, \quad a \neq 0$
Negative exponent rule	$a^{-m} = \dfrac{1}{a^m}, \quad a \neq 0$
Zero exponent rule	$a^0 = 1, \quad a \neq 0$
Raising a power to a power	$(a^m)^n = a^{m \cdot n}$
Raising a product to a power	$(ab)^m = a^m b^m$
Raising a quotient to a power	$\left(\dfrac{a}{b}\right)^m = \dfrac{a^m}{b^m}, \quad b \neq 0$

Using these rules, we will now work some problems in which the exponents are rational numbers.

EXAMPLE 6 Evaluate.   **a)** $8^{-2/3}$ **b)** $(-27)^{-5/3}$ **c)** $(-32)^{-6/5}$

Solution

a) Begin by using the negative exponent rule.

$$8^{-2/3} = \frac{1}{8^{2/3}} \qquad \text{Negative exponent rule}$$

$$= \frac{1}{(\sqrt[3]{8})^2} \qquad \text{Write the denominator as a radical.}$$

$$= \frac{1}{2^2} \qquad \text{Simplify the denominator.}$$

$$= \frac{1}{4}$$

b) $(-27)^{-5/3} = \dfrac{1}{(-27)^{5/3}} = \dfrac{1}{(\sqrt[3]{-27})^5} = \dfrac{1}{(-3)^5} = -\dfrac{1}{243}$

c) $(-32)^{-6/5} = \dfrac{1}{(-32)^{6/5}} = \dfrac{1}{(\sqrt[5]{-32})^6} = \dfrac{1}{(-2)^6} = \dfrac{1}{64}$

Now Try Exercise 81

Note that Example 6 **a)** could have been evaluated as follows:

$$8^{-2/3} = \frac{1}{8^{2/3}} = \frac{1}{\sqrt[3]{8^2}} = \frac{1}{\sqrt[3]{64}} = \frac{1}{4}$$

However, it is generally easier to evaluate the root before applying the power.

Consider the expression $(-16)^{3/4}$. This can be rewritten as $(\sqrt[4]{-16})^3$. Since $(\sqrt[4]{-16})^3$ is not a real number, the expression $(-16)^{3/4}$ is not a real number.

In Chapter 1, we indicated that

$$\left(\frac{a}{b}\right)^{-n} = \left(\frac{b}{a}\right)^n$$

We use this fact in the following example.

EXAMPLE 7 Evaluate.   **a)** $\left(\dfrac{9}{25}\right)^{-1/2}$ **b)** $\left(\dfrac{27}{8}\right)^{-1/3}$

Solution

a) $\left(\dfrac{9}{25}\right)^{-1/2} = \left(\dfrac{25}{9}\right)^{1/2} = \sqrt{\dfrac{25}{9}} = \dfrac{5}{3}$ **b)** $\left(\dfrac{27}{8}\right)^{-1/3} = \left(\dfrac{8}{27}\right)^{1/3} = \sqrt[3]{\dfrac{8}{27}} = \dfrac{2}{3}$

Now Try Exercise 83

EXAMPLE 8 Simplify each expression and write the answer without negative exponents.

a) $a^{1/2} \cdot a^{-2/3}$ **b)** $(6x^2y^{-4})^{-1/2}$ **c)** $3.2x^{1/3}(2.4x^{1/2} + x^{-1/4})$ **d)** $\left(\dfrac{9x^{-4}z^{2/5}}{z^{-3/5}}\right)^{1/8}$

Solution

a) $a^{1/2} \cdot a^{-2/3} = a^{(1/2)-(2/3)}$ .... Product rule

$= a^{-1/6}$ .... Find the LCD and subtract the exponents.

$= \dfrac{1}{a^{1/6}}$ .... Negative exponent rule

b) $(6x^2y^{-4})^{-1/2} = 6^{-1/2}x^{2(-1/2)}y^{-4(-1/2)}$ .... Raise the product to a power.

$= 6^{-1/2}x^{-1}y^2$ .... Multiply the exponents.

$= \dfrac{y^2}{6^{1/2}x}\left(\text{or } \dfrac{y^2}{x\sqrt{6}}\right)$ .... Negative exponent rule

c) Begin by using the distributive property.

$3.2x^{1/3}(2.4x^{1/2} + x^{-1/4}) = (3.2x^{1/3})(2.4x^{1/2}) + (3.2x^{1/3})(x^{-1/4})$ Distributive property

$= (3.2)(2.4)(x^{(1/3)+(1/2)}) + 3.2x^{(1/3)-(1/4)}$ Product rule

$= 7.68x^{5/6} + 3.2x^{1/12}$

d) $\left(\dfrac{9x^{-4}z^{2/5}}{z^{-3/5}}\right)^{1/8} = (9x^{-4}z^{(2/5)-(-3/5)})^{1/8}$ .... Quotient rule

$= (9x^{-4}z)^{1/8}$ .... Subtract the exponents.

$= 9^{1/8}x^{-4(1/8)}z^{1/8}$ .... Raise the product to a power.

$= 9^{1/8}x^{-4/8}z^{1/8}$ .... Multiply the exponents.

$= 9^{1/8}x^{-1/2}z^{1/8}$ .... Simplify exponent.

$= \dfrac{9^{1/8}z^{1/8}}{x^{1/2}}$ .... Negative exponent rule

Now Try Exercise 105

Understanding Algebra

When using the rules of exponents, there are many different ways to simplify exponential expressions. As long as you use the rules properly, you should arrive at the correct simplified expression.

EXAMPLE 9 Simplify.   **a)** $\sqrt[15]{(7y)^5}$ **b)** $\left(\sqrt[4]{a^2b^3c}\right)^{20}$ **c)** $\sqrt[4]{\sqrt[3]{x}}$

Solution

a) $\sqrt[15]{(7y)^5} = (7y)^{5/15}$ .... Write with a rational exponent.

$= (7y)^{1/3}$ .... Simplify the exponent.

$= \sqrt[3]{7y}$ .... Write as a radical.

b) $\left(\sqrt[4]{a^2b^3c}\right)^{20} = (a^2b^3c)^{20/4}$ .... Write with a rational exponent.

$= (a^2b^3c)^5$

$= a^{10}b^{15}c^5$ .... Raise the product to a power.

c) $\sqrt[4]{\sqrt[3]{x}} = \sqrt[4]{x^{1/3}}$ .... Write $\sqrt[3]{x}$ as $x^{1/3}$.

$= (x^{1/3})^{1/4}$ .... Write with a rational exponent.

$= x^{1/12}$ .... Raise the power to a power.

$= \sqrt[12]{x}$ .... Write as a radical.

Now Try Exercise 53

Simplify. Write the answer in exponential form without negative exponents. Assume that all variables represent positive real numbers.

91. $x^4 \cdot x^{1/2}$

92. $x^6 \cdot x^{1/2}$

93. $\dfrac{x^{1/2}}{x^{1/3}}$

94. $x^{-6/5}$

95. $\left(x^{1/2}\right)^{-2}$

96. $\left(a^{-1/3}\right)^{-1/2}$

97. $\left(9^{-1/3}\right)^0$

98. $\dfrac{x^4}{x^{-1/2}}$

99. $\dfrac{5y^{-1/3}}{60y^{-2}}$

100. $x^{-1/2}x^{-2/5}$

101. $4x^{5/3}3x^{-7/2}$

102. $\left(x^{-4/5}\right)^{1/3}$

103. $\left(\dfrac{3}{24x}\right)^{1/3}$

104. $\left(\dfrac{54}{2x^4}\right)^{1/3}$

105. $\left(\dfrac{22x^{3/7}}{2x^{1/2}}\right)^2$

106. $\left(\dfrac{x^{-1/3}}{x^{-2}}\right)^2$

107. $\left(\dfrac{a^4}{4a^{-2/5}}\right)^{-3}$

108. $\left(\dfrac{27z^{1/4}y^3}{3z^{1/4}}\right)^{1/2}$

109. $\left(\dfrac{x^{3/4}y^{-3}}{x^{1/2}y^2}\right)^4$

110. $\left(\dfrac{250a^{-3/4}b^5}{2a^{-2}b^2}\right)^{2/3}$

Multiply. Assume that all variables represent positive real numbers.

111. $4z^{-1/2}(2z^4 - z^{1/2})$

112. $-3a^{-4/9}(5a^{1/9} - a^2)$

113. $5x^{-1}(x^{-4} + 4x^{-1/2})$

114. $-9z^{3/2}(z^{3/2} - z^{-3/2})$

115. $-6x^{5/3}(-2x^{1/2} + 3x^{1/3})$

116. $\dfrac{1}{2}x^{-2}(10x^{4/3} - 38x^{-1/2})$

Use a calculator to evaluate each expression. Give the answer to the nearest hundredth.

117. $\sqrt{180}$

118. $\sqrt[3]{168}$

119. $\sqrt[5]{402.83}$

120. $\sqrt[4]{1096}$

121. $93^{2/3}$

122. $38.2^{3/2}$

123. $1000^{-1/2}$

124. $8060^{-3/2}$

Problem Solving

125. Under what conditions will $\sqrt[n]{a^n} = (\sqrt[n]{a})^n = a$?

126. By selecting values for a and b, show that $(a^2 + b^2)^{1/2}$ *is not equal to* $a + b$.

127. By selecting values for a and b, show that $(a^{1/2} + b^{1/2})^2$ *is not equal to* $a + b$.

128. By selecting values for a and b, show that $(a^3 + b^3)^{1/3}$ is not equal to $a + b$.

129. By selecting values for a and b, show that $(a^{1/3} + b^{1/3})^3$ is not equal to $a + b$.

130. Determine whether $\sqrt[3]{\sqrt{x}} = \sqrt{\sqrt[3]{x}}, x \geq 0$.

Factor. Write the answer without negative exponents. Assume that all variables represent positive real numbers.

131. $x^{3/2} + x^{1/2}$

132. $x^{1/4} - x^{5/4}$

133. $y^{1/3} - y^{7/3}$

134. $x^{-1/2} + x^{1/2}$

135. $y^{-2/5} + y^{8/5}$

136. $a^{6/5} + a^{-4/5}$

In Exercises 137 through 142, use a calculator where appropriate.

137. Growing Bacteria The function $B(t) = 2^{10} \cdot 2^t$ approximates the number of bacteria in a certain culture after t hours.

    a) The initial number of bacteria is determined when $t = 0$. What is the initial number of bacteria?

    b) How many bacteria are there after $\dfrac{1}{2}$ hour?

138. Carbon Dating Carbon dating is used by scientists to find the age of fossils, bones, and other items. The formula used in carbon dating is $P = P_0 2^{-t/5600}$, where P_0 represents the original amount of carbon 14 (C_{14}) present and P represents the amount of C_{14} present after t years. If 10 milligrams (mg) of C_{14} is present in an animal bone recently excavated, how many milligrams will be present in 5000 years?

139. Resting Metabolic Rate A person's resting metabolic rate (RMR) is the number of calories a person burns in one day of resting. A person's RMR can be estimated using the function $R(x) = 70x^{3/4}$ where $R(x)$ is measured in calories per day and x is the person's mass in kilo-

grams. If Ken Machol has a mass of 102 kg, estimate his RMR to the nearest calorie.

Source: www.bodybuilding.com

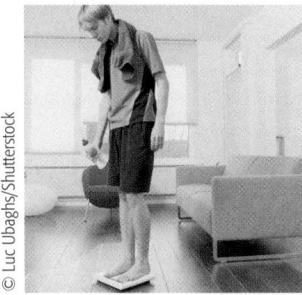

140. Resting Metabolic Rate If Martin Alexander has a mass of 82 kg, estimate his resting metabolic rate (see Exercise 139) to the nearest calorie.

141. Find the domain of $f(x) = (x - 7)^{1/2}(x + 3)^{-1/2}$.

142. Find the domain of $f(x) = (x + 4)^{1/2}(x - 3)^{-1/2}$.

Determine the index to be placed in the shaded area to make the statement true.

143. $\sqrt[4]{\sqrt[5]{\sqrt{\sqrt[3]{z}}}} = z^{1/120}$

144. $\sqrt[4]{\sqrt{\sqrt{x}}} = x^{1/24}$

Concept/Writing Exercises

145. a) Under what conditions is $\sqrt[n]{a}$ a real number?

  b) When $\sqrt[n]{a}$ is a real number, how can it be expressed with rational exponents?

146. a) Under what conditions is $\sqrt[n]{a^m}$ a real number?

  b) Under what conditions is $(\sqrt[n]{a})^m$ a real number?

  c) When $\sqrt[n]{a^m}$ is a real number, how can it be expressed with rational exponents?

147. a) Under what conditions is $\sqrt[n]{a^n}$ a real number?

  b) When n is even and $a \geq 0$, what is $\sqrt[n]{a^n}$ equal to?

  c) When n is odd, what is $\sqrt[n]{a^n}$ equal to?

  d) When n is even and a may be any real number, what is $\sqrt[n]{a^n}$ equal to?

148. a) Explain the difference between $-16^{1/2}$ and $(-16)^{1/2}$.

  b) Evaluate each expression in part **a)** if possible.

149. a) Is $(xy)^{1/2} = xy^{1/2}$? Explain.

  b) Is $(xy)^{-1/2} = \dfrac{x^{1/2}}{y^{-1/2}}$? Explain.

150. a) Is $\sqrt[6]{(3y)^3} = (3y)^{6/3}$? Explain.

  b) Is $\sqrt{(ab)^4} = (ab)^2$? Explain.

151. Evaluate $(3^{\sqrt{2}})^{\sqrt{2}}$. Explain how you determined your answer.

152. a) On your calculator, evaluate 3^π.

  b) Explain why your value from part **a)** does or does not make sense.

Cumulative Review Exercises

[3.2] **153.** Determine which of the following relations are also functions.

a) 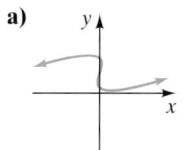   **b)** 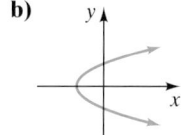   **c)**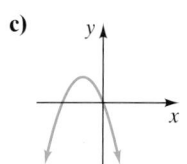

[6.3] **154.** Simplify $\dfrac{a^{-2} + ab^{-1}}{ab^{-2} - a^{-2}b^{-1}}$.

[6.4] **155.** Solve $\dfrac{3x - 2}{x + 4} = \dfrac{2x + 1}{3x - 2}$.

[6.5] **156. Flying a Plane** Amy Mayfield can fly her plane 500 miles against the wind in the same time it takes her to fly 560 miles with the wind. If the wind blows at 25 miles per hour, find the speed of the plane in still air.

7.3 Simplifying Radicals

1 Understand perfect powers.

2 Simplify radicals using the product rule for radicals.

3 Simplify radicals using the quotient rule for radicals.

1 Understand Perfect Powers

In this section, we will simplify radical expressions. We begin with a discussion of perfect powers.

Perfect Power, Perfect Square, Perfect Cube

- A **perfect power** is a number or expression that can be written as an expression raised to a power that is a whole number greater than 1.
- A **perfect square** is a number or expression that can be written as a square of an expression. A perfect square is a perfect *second* power.
- A **perfect cube** is a number or expression that can be written as a cube of an expression. A perfect cube is a perfect *third* power.

Examples of perfect squares are illustrated below.

Perfect squares   1,   4,   9,   16,   25,   36,... 49, 64, 81

Square of a number 1^2, 2^2, 3^2, 4^2, 5^2, 6^2,... 7, 8, 9

Variables with exponents may also be perfect squares, as illustrated below.

Perfect squares x^2, x^4, x^6, x^8, x^{10},...

Square of an expression $(x)^2$, $(x^2)^2$, $(x^3)^2$, $(x^4)^2$, $(x^5)^2$,...

Notice that the exponents on the variables in the perfect squares are all multiples of 2.

Examples of perfect cubes are illustrated below.

Perfect cubes   1,   8,   27,   64,   125,   216,...
                           ↓   ↓   ↓   ↓   ↓   ↓
Cube of a number 1^3, 2^3, 3^3, 4^3, 5^3, 6^3,...

Perfect cubes x^3, x^6, x^9, x^{12}, x^{15},...
                          ↓   ↓   ↓   ↓   ↓
Cube of an expression $(x)^3$, $(x^2)^3$, $(x^3)^3$, $(x^4)^3$, $(x^5)^3$,...

Notice that the exponents on the variables in the perfect cubes are all multiples of 3.

When simplifying radicals, we will be looking for perfect powers in the radicand. For example, if we are simplifying a square root, then we are interested in finding perfect squares. If we are simplifying a cube root, then we are interested in finding perfect cubes. If we are simplifying a fourth root, then we are interested in finding perfect fourth powers, and so on.

Understanding Algebra

A variable with an exponent is a

- Perfect square if the exponent is divisible by 2. For example, $x^2, x^4, x^6, \ldots$ are perfect squares.
- Perfect cube if the exponent is divisible by 3. For example, $x^3, x^6, x^9, \ldots$ are perfect cubes.
- Perfect fourth power if the exponent is divisible by 4. For example, $x^4, x^8, x^{12}, \ldots$ are perfect fourth powers.

This pattern continues for higher perfect powers.

Helpful Hint

A quick way to determine if a radicand x^m is a perfect power for an index is to determine if the exponent m is divisible by the index of the radical. For example, consider $\sqrt[5]{x^{20}}$. Since the exponent, 20, is divisible by the index, 5, x^{20} is a perfect fifth power. Now consider $\sqrt[6]{x^{20}}$. Since the exponent, 20, is not divisible by the index, 6, x^{20} is not a perfect sixth power. However, x^{18} and x^{24} are both perfect sixth powers since 6 divides both 18 and 24.

Notice that the square root of a perfect square simplifies to an expression without a radical sign, the cube root of a perfect cube simplifies to an expression without a radical sign, and so on.

Examples

$$\sqrt{36} = \sqrt{6^2} = 6^{2/2} = 6$$
$$\sqrt[3]{27} = \sqrt[3]{3^3} = 3^{3/3} = 3$$
$$\sqrt{x^6} = x^{6/2} = x^3$$
$$\sqrt[3]{z^{12}} = z^{12/3} = z^4$$
$$\sqrt[5]{n^{35}} = n^{35/5} = n^7$$

Now we are ready to discuss the product rule for radicals.

2 Simplify Radicals Using the Product Rule for Radicals

To introduce the **product rule for radicals**, observe that $\sqrt{4} \cdot \sqrt{9} = 2 \cdot 3 = 6$. Also, $\sqrt{4 \cdot 9} = \sqrt{36} = 6$. We see that $\sqrt{4} \cdot \sqrt{9} = \sqrt{4 \cdot 9}$. This is one example of the product rule for radicals.

Product Rule for Radicals

For nonnegative real numbers a and b,
$$\sqrt[n]{a} \cdot \sqrt[n]{b} = \sqrt[n]{ab}$$

Examples of the Product Rule for Radicals

$$\sqrt{20} = \begin{cases} \sqrt{1} \cdot \sqrt{20} \\ \sqrt{2} \cdot \sqrt{10} \\ \sqrt{4} \cdot \sqrt{5} \end{cases} \qquad \sqrt[3]{20} = \begin{cases} \sqrt[3]{1} \cdot \sqrt[3]{20} \\ \sqrt[3]{2} \cdot \sqrt[3]{10} \\ \sqrt[3]{4} \cdot \sqrt[3]{5} \end{cases}$$

$\sqrt{20}$ can be factored into any of these forms.　　$\sqrt[3]{20}$ can be factored into any of these forms.

$$\sqrt{x^7} = \begin{cases} \sqrt{x} \cdot \sqrt{x^6} \\ \sqrt{x^2} \cdot \sqrt{x^5} \\ \sqrt{x^3} \cdot \sqrt{x^4} \end{cases} \qquad \sqrt[3]{x^7} = \begin{cases} \sqrt[3]{x} \cdot \sqrt[3]{x^6} \\ \sqrt[3]{x^2} \cdot \sqrt[3]{x^5} \\ \sqrt[3]{x^3} \cdot \sqrt[3]{x^4} \end{cases}$$

$\sqrt{x^7}$ can be factored into any of these forms.　　$\sqrt[3]{x^7}$ can be factored into any of these forms.

Here is a general procedure that can be used to simplify radicals using the product rule.

To Simplify Radicals Using the Product Rule

1. If the radicand contains a coefficient other than 1, write it as a product of two numbers, one of which is the largest perfect power for the index.
2. Write each variable factor as a product of two factors, one of which is the largest perfect power of the variable for the index.
3. Use the product rule to write the radical expression as a product of radicals. Place all the perfect powers (numbers and variables) under the same radical.
4. Simplify the radical containing the perfect powers.

If we are simplifying a *square* root, we will write the radicand as the product of the largest *perfect square* and another expression. If we are simplifying a *cube* root, we will write the radicand as the product of the largest *perfect cube* and another expression, and so on.

EXAMPLE 1 Simplify. **a)** $\sqrt{32}$ **b)** $\sqrt{60}$ **c)** $\sqrt[3]{54}$ **d)** $\sqrt[4]{96}$

Solution   The radicands in this example contain no variables. We will follow step 1 of the procedure.

a) Since we are evaluating a square root, we look for the largest perfect square that divides 32. The largest perfect square that divides, or is a factor of, 32 is 16.

$$\sqrt{32} = \sqrt{16 \cdot 2} = \sqrt{16}\,\sqrt{2} = 4\sqrt{2}$$

b) The largest perfect square that is a factor of 60 is 4.

$$\sqrt{60} = \sqrt{4 \cdot 15} = \sqrt{4}\,\sqrt{15} = 2\sqrt{15}$$

c) The largest perfect cube that is a factor of 54 is 27.

$$\sqrt[3]{54} = \sqrt[3]{27 \cdot 2} = \sqrt[3]{27}\sqrt[3]{2} = 3\sqrt[3]{2}$$

d) The largest perfect fourth power that is a factor of 96 is 16.

$$\sqrt[4]{96} = \sqrt[4]{16 \cdot 6} = \sqrt[4]{16}\,\sqrt[4]{6} = 2\sqrt[4]{6}$$

Now Try Exercise 19

Understanding Algebra

To simplify $\sqrt{20}$ we write it as $\sqrt{4} \cdot \sqrt{5}$ because 4 is the largest perfect square that divides 20:

$$\sqrt{20} = \sqrt{4} \cdot \sqrt{5} = 2\sqrt{5}$$

Similarly, to simplify $\sqrt[3]{24}$ we write it as $\sqrt[3]{8} \cdot \sqrt[3]{3}$ because 8 is the largest perfect cube that divides 24:

$$\sqrt[3]{24} = \sqrt[3]{8} \cdot \sqrt[3]{3} = 2\sqrt[3]{3}$$

Helpful Hint

In Example 1 **a)**, if you first thought that 4 was the largest perfect square that divided 32, you could proceed as follows:

$$\sqrt{32} = \sqrt{4 \cdot 8} = \sqrt{4}\,\sqrt{8} = 2\sqrt{8}$$
$$= 2\sqrt{4 \cdot 2} = 2\sqrt{4}\,\sqrt{2} = 2 \cdot 2\sqrt{2} = 4\sqrt{2}$$

Note that the final result is the same, but you must perform more steps. The lists of perfect squares and perfect cubes on pages 442–443 can help you determine the largest perfect square or perfect cube that is a factor of a radicand.

In Example 1 **b)**, $\sqrt{15}$ can be factored as $\sqrt{5 \cdot 3}$; however, since neither 5 nor 3 is a perfect square, $\sqrt{15}$ cannot be simplified.

When the radicand is a perfect power for the index, the radical can be simplified by writing it in exponential form, as in Example 2.

EXAMPLE 2 Simplify. **a)** $\sqrt{x^4}$ **b)** $\sqrt[3]{x^{12}}$ **c)** $\sqrt[5]{z^{40}}$

Solution

a) $\sqrt{x^4} = x^{4/2} = x^2$ **b)** $\sqrt[3]{x^{12}} = x^{12/3} = x^4$ **c)** $\sqrt[5]{z^{40}} = z^{40/5} = z^8$

Now Try Exercise 33

EXAMPLE 3 Simplify.   **a)** $\sqrt{x^9}$ **b)** $\sqrt[5]{x^{23}}$ **c)** $\sqrt[4]{y^{33}}$

Solution   Because the radicands have coefficients of 1, we start with step 2 of the procedure.

a) The largest perfect square less than or equal to x^9 is x^8.

$$\sqrt{x^9} = \sqrt{x^8 \cdot x} = \sqrt{x^8} \cdot \sqrt{x} = x^{8/2}\sqrt{x} = x^4\sqrt{x}$$

b) The largest perfect fifth power less than or equal to x^{23} is x^{20}.

$$\sqrt[5]{x^{23}} = \sqrt[5]{x^{20} \cdot x^3} = \sqrt[5]{x^{20}}\,\sqrt[5]{x^3} = x^{20/5}\sqrt[5]{x^3} = x^4\sqrt[5]{x^3}$$

c) The largest perfect fourth power less than or equal to y^{33} is y^{32}.

$$\sqrt[4]{y^{33}} = \sqrt[4]{y^{32} \cdot y} = \sqrt[4]{y^{32}}\,\sqrt[4]{y} = y^{32/4}\sqrt[4]{y} = y^8\sqrt[4]{y}$$

Now Try Exercise 39

If you observe the answers to Example 3, you will see that the exponent on the variable in the radicand is always less than the index. *When a radical is simplified, the exponents on the variables in the radicand are less than the index.*

In Example 3 **b)**, we simplified $\sqrt[5]{x^{23}}$. If we divide 23, the exponent in the radicand, by 5, the index, we obtain

$$
\begin{array}{r}
4 \quad \longleftarrow \quad \text{Quotient} \\
5\overline{)23} \\
\underline{20} \quad\quad\quad\quad \\
3 \quad \longleftarrow \quad \text{Remainder}
\end{array}
$$

Notice that $\sqrt[5]{x^{23}}$ simplifies to $x^4\sqrt[5]{x^3}$ and

$$\text{Quotient} \longrightarrow x^4\sqrt[5]{x^3} \longleftarrow \text{Remainder}$$

When simplifying a radical, if you divide the exponent within the radical by the index, the quotient will be the exponent on the variable outside the radical sign and the remainder will be the exponent on the variable within the radical sign. Simplify Example 3 **c)** using this technique now.

EXAMPLE 4 Simplify.   **a)** $\sqrt{x^{12}y^{17}}$ **b)** $\sqrt[4]{x^6y^{23}}$

Solution

a) x^{12} is a perfect square. The largest perfect square that is a factor of y^{17} is y^{16}. Write y^{17} as $y^{16} \cdot y$.

$$
\begin{aligned}
\sqrt{x^{12}y^{17}} &= \sqrt{x^{12} \cdot y^{16} \cdot y} = \sqrt{x^{12}y^{16}}\,\sqrt{y} \\
&= \sqrt{x^{12}}\sqrt{y^{16}}\,\sqrt{y} \\
&= x^6 y^8 \sqrt{y}
\end{aligned}
$$

> ### Understanding Algebra
>
> Often the steps when simplifying a radical are done mentally. For example, in Example 4 **a)** we do not show $\sqrt{x^{12}} = x^{12/2} = x^6$ and $\sqrt{y^{16}} = y^{16/2} = y^8$.

b) We begin by finding the largest perfect fourth power factors of x^6 and y^{23}. For an index of 4, the largest perfect power that is a factor of x^6 is x^4. The largest perfect power that is a factor of y^{23} is y^{20}.

$$
\begin{aligned}
\sqrt[4]{x^6y^{23}} &= \sqrt[4]{x^4 \cdot x^2 \cdot y^{20} \cdot y^3} \\
&= \sqrt[4]{x^4y^{20} \cdot x^2y^3} \\
&= \sqrt[4]{x^4y^{20}}\,\sqrt[4]{x^2y^3} \\
&= xy^5\sqrt[4]{x^2y^3}
\end{aligned}
$$

Now Try Exercise 51

Often the steps where we change the radical expression to exponential form are done mentally, and those steps are not illustrated.

Helpful Hint

In Example 4 **b)**, we showed that $\sqrt[4]{x^6 y^{23}} = xy^5 \sqrt[4]{x^2 y^3}$. We can also simplify this radical by dividing the exponents on the variables in the radicand, 6 and 23, by the index, 4. Note the location of the quotients and the remainders.

6 ÷ 4 gives a quotient of 1 and a remainder of 2

$$\sqrt[4]{x^6 y^{23}} = x^1 y^5 \sqrt[4]{x^2 y^3}$$

23 ÷ 4 gives a quotient of 5 and a remainder of 3

EXAMPLE 5 Simplify.   **a)** $\sqrt{80x^5 y^{12} z^3}$ **b)** $\sqrt[3]{54x^{17} y^{25}}$

Solution

a) The largest perfect square that is a factor of 80 is 16. The largest perfect square that is a factor of x^5 is x^4. The expression y^{12} is a perfect square. The largest perfect square that is a factor of z^3 is z^2. Place all the perfect squares under the same radical, and then simplify.

$$\sqrt{80x^5 y^{12} z^3} = \sqrt{16 \cdot 5 \cdot x^4 \cdot x \cdot y^{12} \cdot z^2 \cdot z}$$
$$= \sqrt{16x^4 y^{12} z^2 \cdot 5xz}$$
$$= \sqrt{16x^4 y^{12} z^2} \cdot \sqrt{5xz}$$
$$= 4x^2 y^6 z \sqrt{5xz}$$

b) The largest perfect cube that is a factor of 54 is 27. The largest perfect cube that is a factor of x^{17} is x^{15}. The largest perfect cube that is a factor of y^{25} is y^{24}.

$$\sqrt[3]{54x^{17} y^{25}} = \sqrt[3]{27 \cdot 2 \cdot x^{15} \cdot x^2 \cdot y^{24} \cdot y}$$
$$= \sqrt[3]{27x^{15} y^{24} \cdot 2x^2 y}$$
$$= \sqrt[3]{27x^{15} y^{24}} \cdot \sqrt[3]{2x^2 y}$$
$$= 3x^5 y^8 \sqrt[3]{2x^2 y}$$

Now Try Exercise 57

3 Simplify Radicals Using the Quotient Rule for Radicals

In mathematics we sometimes need to simplify a quotient of two radicals. To do so we use the **quotient rule for radicals**.

Quotient Rule for Radicals

For nonnegative real numbers a and b,

$$\frac{\sqrt[n]{a}}{\sqrt[n]{b}} = \sqrt[n]{\frac{a}{b}}, \qquad b \neq 0$$

Examples of the Quotient Rule for Radicals

$$\frac{\sqrt{18}}{\sqrt{3}} = \sqrt{\frac{18}{3}} \qquad \sqrt{\frac{9}{25}} = \frac{\sqrt{9}}{\sqrt{25}}$$

$$\frac{\sqrt{x^3}}{\sqrt{x}} = \sqrt{\frac{x^3}{x}} \qquad \sqrt{\frac{x^4}{y^2}} = \frac{\sqrt{x^4}}{\sqrt{y^2}}$$

$$\frac{\sqrt[3]{y^5}}{\sqrt[3]{y^2}} = \sqrt[3]{\frac{y^5}{y^2}} \qquad \sqrt[3]{\frac{z^9}{27}} = \frac{\sqrt[3]{z^9}}{\sqrt[3]{27}}$$

Examples 6 and 7 illustrate how to use the quotient rule to simplify radical expressions.

EXAMPLE 6   Simplify.   **a)** $\dfrac{\sqrt{75}}{\sqrt{3}}$ **b)** $\dfrac{\sqrt[3]{24x}}{\sqrt[3]{3x}}$ **c)** $\dfrac{\sqrt[3]{x^4y^7}}{\sqrt[3]{xy^{-5}}}$

Solution   In each part we use the quotient rule to write the quotient of radicals as a single radical. Then we simplify.

a) $\dfrac{\sqrt{75}}{\sqrt{3}} = \sqrt{\dfrac{75}{3}} = \sqrt{25} = 5$

b) $\dfrac{\sqrt[3]{24x}}{\sqrt[3]{3x}} = \sqrt[3]{\dfrac{24x}{3x}} = \sqrt[3]{8} = 2$

c) $\dfrac{\sqrt[3]{x^4y^7}}{\sqrt[3]{xy^{-5}}} = \sqrt[3]{\dfrac{x^4y^7}{xy^{-5}}}$ Quotient rule for radicals

$\qquad\qquad = \sqrt[3]{x^3y^{12}}$ Simplify the radicand.

$\qquad\qquad = xy^4$

Now Try Exercise 93

In Section 7.1 when we introduced radicals, we indicated that $\sqrt{\dfrac{4}{9}} = \dfrac{2}{3}$ since $\dfrac{2}{3} \cdot \dfrac{2}{3} = \dfrac{4}{9}$.
The quotient rule may be helpful in evaluating square roots containing fractions as illustrated in Example 7 **a)**.

EXAMPLE 7   Simplify.   **a)** $\sqrt{\dfrac{121}{25}}$ **b)** $\sqrt[3]{\dfrac{8x^4y}{27xy^{10}}}$ **c)** $\sqrt[4]{\dfrac{18xy^5}{3x^9y}}$

Solution   In each part we first simplify the radicand, if possible. Then we use the quotient rule to write the given radical as a quotient of radicals.

a) $\sqrt{\dfrac{121}{25}} = \dfrac{\sqrt{121}}{\sqrt{25}} = \dfrac{11}{5}$

b) $\sqrt[3]{\dfrac{8x^4y}{27xy^{10}}} = \sqrt[3]{\dfrac{8x^3}{27y^9}} = \dfrac{\sqrt[3]{8x^3}}{\sqrt[3]{27y^9}} = \dfrac{2x}{3y^3}$

c) $\sqrt[4]{\dfrac{18xy^5}{3x^9y}} = \sqrt[4]{\dfrac{6y^4}{x^8}} = \dfrac{\sqrt[4]{6y^4}}{\sqrt[4]{x^8}} = \dfrac{\sqrt[4]{y^4}\,\sqrt[4]{6}}{x^2} = \dfrac{y\sqrt[4]{6}}{x^2}$

Now Try Exercise 97

Avoiding Common Errors

The following simplifications are correct because the numbers and variables divided out are not within square roots.

CORRECT

$\dfrac{\overset{2}{\cancel{6}}\,\sqrt{2}}{\underset{1}{\cancel{3}}} = 2\sqrt{2}$

CORRECT

$\dfrac{\cancel{x}\,\sqrt{2}}{\cancel{x}} = \sqrt{2}$

An expression within a square root cannot be divided by an expression not within the square root.

CORRECT

$\dfrac{\sqrt{2}}{2}$ Cannot be simplified further

$\dfrac{\sqrt{x^3}}{x} = \dfrac{\sqrt{x^2}\,\sqrt{x}}{x} = \dfrac{\cancel{x}\,\sqrt{x}}{\cancel{x}} = \sqrt{x}$

INCORRECT

$\dfrac{\sqrt{\cancel{2}^{\,1}}}{\underset{1}{\cancel{2}}} = \sqrt{1} = 1$

$\dfrac{\sqrt{x^{\cancel{3}\,2}}}{\cancel{x}} = \sqrt{x^2} = x$

EXERCISE SET 7.3

Math XL
MathXL®

MyMathLab
MyMathLab

Warm-Up Exercises

Fill in the blanks with the appropriate word, phrase, or symbol(s) from the following list.

| quotient | square | less | 3 | cube | greater | 2 | product | 4 |

1. A perfect _____ is a number or expression that can be written as a square of an expression.

2. A perfect _____ is a number or expression that can be written as a cube of an expression.

3. A variable with an exponent is a perfect cube if the exponent is divisible by _____.

4. A variable with an exponent is a perfect square if the exponent is divisible by _____.

5. A variable with an exponent is a perfect fourth power if the exponent is divisible by _____.

6. When a radical is simplified, the exponents on the variables in the radicand are _____ than the index.

7. The _____ rule for radicals states that for nonnegative real numbers, $\sqrt[n]{a} \cdot \sqrt[n]{b} = \sqrt[n]{ab}$.

8. The _____ rule for radicals states that for nonnegative real numbers, $\dfrac{\sqrt[n]{a}}{\sqrt[n]{b}} = \sqrt[n]{\dfrac{a}{b}}, b \neq 0$.

Practice the Skills

Simplify. Assume that all variables represent positive real numbers.

9. $\sqrt{49}$ **10.** $\sqrt{100}$ **11.** $\sqrt{24}$ **12.** $\sqrt{18}$

13. $\sqrt{32}$ **14.** $\sqrt{12}$ **15.** $\sqrt{50}$ **16.** $\sqrt{72}$

17. $\sqrt{75}$ **18.** $\sqrt{300}$ **19.** $\sqrt{40}$ **20.** $\sqrt{600}$

21. $\sqrt[3]{16}$ **22.** $\sqrt[3]{24}$ **23.** $\sqrt[3]{54}$ **24.** $\sqrt[3]{81}$

25. $\sqrt[3]{32}$ **26.** $\sqrt[3]{108}$ **27.** $\sqrt[3]{40}$ **28.** $\sqrt[4]{80}$

29. $\sqrt[4]{48}$ **30.** $\sqrt[4]{162}$ **31.** $-\sqrt[5]{64}$ **32.** $-\sqrt[5]{243}$

33. $\sqrt[3]{b^9}$ **34.** $6\sqrt{y^{12}}$ **35.** $\sqrt[3]{x^6}$ **36.** $\sqrt[5]{y^{20}}$

37. $\sqrt{x^3}$ **38.** $-\sqrt{x^5}$ **39.** $\sqrt{a^5}$ **40.** $\sqrt{b^7}$

41. $8\sqrt[3]{z^{32}}$ **42.** $\sqrt[3]{a^7}$ **43.** $\sqrt[4]{b^{23}}$ **44.** $\sqrt[5]{z^7}$

45. $\sqrt[6]{x^9}$ **46.** $\sqrt[7]{y^{15}}$ **47.** $3\sqrt[5]{y^{23}}$ **48.** $\sqrt{24x^3}$

49. $2\sqrt{50y^9}$ **50.** $\sqrt{75a^7b^{11}}$ **51.** $\sqrt[3]{x^3y^7}$ **52.** $\sqrt{x^5y^9}$

53. $\sqrt[5]{a^6b^{23}}$ **54.** $-\sqrt{20x^6y^7z^{12}}$ **55.** $\sqrt{24x^{15}y^{20}z^{27}}$ **56.** $\sqrt[3]{16x^3y^6}$

57. $\sqrt[3]{81a^6b^8}$ **58.** $\sqrt[3]{128a^{10}b^{11}c^{12}}$ **59.** $\sqrt[4]{32x^8y^9z^{19}}$ **60.** $\sqrt[4]{48x^{11}y^{21}}$

61. $\sqrt[4]{81a^8b^9}$ **62.** $-\sqrt[4]{32x^{18}y^{31}}$ **63.** $\sqrt[5]{32a^{10}b^{12}}$ **64.** $\sqrt[6]{64x^{12}y^{23}z^{50}}$

Simplify. Assume that all variables represent positive real numbers.

65. $\sqrt{\dfrac{18}{2}}$ **66.** $\sqrt{\dfrac{45}{5}}$ **67.** $\sqrt{\dfrac{81}{100}}$ **68.** $\sqrt{\dfrac{8}{50}}$

69. $\dfrac{\sqrt{27}}{\sqrt{3}}$ **70.** $\dfrac{\sqrt{72}}{\sqrt{2}}$ **71.** $\dfrac{\sqrt{3}}{\sqrt{48}}$ **72.** $\dfrac{\sqrt{15}}{\sqrt{60}}$

73. $\sqrt[3]{\dfrac{3}{24}}$ **74.** $\sqrt[3]{\dfrac{2}{54}}$ **75.** $\dfrac{\sqrt[3]{3}}{\sqrt[3]{81}}$ **76.** $\dfrac{\sqrt[3]{32}}{\sqrt[3]{4}}$

77. $\sqrt[4]{\dfrac{3}{48}}$ **78.** $\dfrac{\sqrt[4]{243}}{\sqrt[4]{3}}$ **79.** $\sqrt[5]{\dfrac{96}{3}}$ **80.** $\dfrac{\sqrt[5]{2}}{\sqrt[5]{64}}$

81. $\sqrt{\dfrac{x^2}{9}}$ **82.** $\sqrt{\dfrac{9y^4}{z^2}}$ **83.** $\sqrt{\dfrac{16x^4}{25y^{10}}}$ **84.** $\sqrt{\dfrac{49a^8b^{10}}{121c^{14}}}$

85. $\sqrt[3]{\dfrac{c^6}{64}}$ **86.** $\sqrt[3]{\dfrac{27x^6}{y^{12}}}$ **87.** $\sqrt[3]{\dfrac{a^8b^{12}}{b^{-8}}}$ **88.** $\sqrt[4]{\dfrac{16x^{16}y^{32}}{81x^{-4}}}$

89. $\dfrac{\sqrt{24}}{\sqrt{3}}$ **90.** $\dfrac{\sqrt{64x^5}}{\sqrt{2x^3}}$ **91.** $\dfrac{\sqrt{27x^6}}{\sqrt{3x^2}}$ **92.** $\dfrac{\sqrt{72x^3y^5}}{\sqrt{8x^3y^7}}$

93. $\dfrac{\sqrt{48x^6y^9}}{\sqrt{6x^2y^6}}$ **94.** $\dfrac{\sqrt{300a^{10}b^{11}}}{\sqrt{2ab^4}}$ **95.** $\sqrt[3]{\dfrac{5xy}{8x^{13}}}$ **96.** $\sqrt[3]{\dfrac{64a^5b^{12}}{27a^{14}b^5}}$

97. $\sqrt[3]{\dfrac{25x^2y^9}{5x^8y^2}}$ **98.** $\sqrt[3]{\dfrac{54xy^4z^{17}}{18x^{13}z^4}}$ **99.** $\sqrt[4]{\dfrac{10x^4y}{81x^{-8}}}$ **100.** $\sqrt[4]{\dfrac{3a^6b^5}{16a^{-6}b^{13}}}$

Concept/Writing Exercises

101. a) How do you obtain the numbers that are perfect squares?

   b) List the first six perfect squares.

102. a) How do you obtain the numbers that are perfect cubes?

   b) List the first six perfect cube numbers.

103. When we gave the product rule, we stated that for nonnegative real numbers a and b, $\sqrt[n]{a} \cdot \sqrt[n]{b} = \sqrt[n]{ab}$. Why is it necessary to specify that both a and b are nonnegative real numbers?

104. When we gave the quotient rule, we stated that for nonnegative real numbers a and b, $\dfrac{\sqrt[n]{a}}{\sqrt[n]{b}} = \sqrt[n]{\dfrac{a}{b}}, b \neq 0$. Why is it necessary to state that both a and b are nonnegative real numbers?

105. Prove $\sqrt{a \cdot b} = \sqrt{a}\,\sqrt{b}$ by converting $\sqrt{a \cdot b}$ to exponential form.

106. Will the product of two radicals always be a radical? Give an example to support your answer.

107. Will the quotient of two radicals always be a radical? Give an example to support your answer.

108. Prove $\sqrt[n]{\dfrac{a}{b}} = \dfrac{\sqrt[n]{a}}{\sqrt[n]{b}}$ by converting $\sqrt[n]{\dfrac{a}{b}}$ to exponential form.

109. a) Will $\dfrac{\sqrt[n]{x}}{\sqrt[n]{x}}$ always equal 1?

   b) If your answer to part **a)** was no, under what conditions does $\dfrac{\sqrt[n]{x}}{\sqrt[n]{x}}$ equal 1?

Cumulative Review Exercises

[2.2] **110.** Solve the formula $F = \dfrac{9}{5}C + 32$ for C.

[2.6] **111.** Solve for x: $\left|\dfrac{2x - 4}{5}\right| = 12$

[5.3] **112.** Divide $\dfrac{15x^{12} - 5x^9 + 20x^6}{5x^6}$.

[5.6] **113.** Factor $(x - 3)^3 + 8$.

7.4 Adding, Subtracting, and Multiplying Radicals

1 Add and subtract radicals.

2 Multiply radicals.

1 Add and Subtract Radicals

Like Radicals and Unlike Radicals

- **Like radicals** are radicals having the same radicand and index.

- **Unlike radicals** are radicals differing in either the radicand or the index.

Examples of Like Radicals	Examples of Unlike Radicals	
$\sqrt{5}, 3\sqrt{5}$	$\sqrt{5}, \sqrt[3]{5}$	Indices differ.
$6\sqrt{7}, -2\sqrt{7}$	$\sqrt{6}, \sqrt{7}$	Radicands differ.
$\sqrt{x}, 5\sqrt{x}$	$\sqrt{x}, \sqrt{2x}$	Radicands differ.
$\sqrt[3]{2x}, -4\sqrt[3]{2x}$	$\sqrt{x}, \sqrt[3]{x}$	Indices differ.
$\sqrt[4]{x^2y^5}, -\sqrt[4]{x^2y^5}$	$\sqrt[3]{xy}, \sqrt[3]{x^2y}$	Radicands differ.

> **Understanding Algebra**
>
> We add like radicals using the distributive property in the same way we add like terms. For example, we add like terms $3x$ and $4x$ as follows:
>
> $$3x + 4x = (3 + 4)x = 7x$$
>
> We add like radicals $3\sqrt{2}$ and $4\sqrt{2}$ as follows:
>
> $$3\sqrt{2} + 4\sqrt{2} =$$
> $$(3 + 4)\sqrt{2} = 7\sqrt{2}$$

The distributive property is used to add or subtract like radicals in the same way that like terms are added or subtracted.

Examples of Adding and Subtracting Like Radicals

$$3\sqrt{6} + 2\sqrt{6} = (3 + 2)\sqrt{6} = 5\sqrt{6}$$
$$5\sqrt{x} - 7\sqrt{x} = (5 - 7)\sqrt{x} = -2\sqrt{x}$$
$$\sqrt[3]{4x^2} + 5\sqrt[3]{4x^2} = (1 + 5)\sqrt[3]{4x^2} = 6\sqrt[3]{4x^2}$$
$$4\sqrt{5x} - y\sqrt{5x} = (4 - y)\sqrt{5x}$$

EXAMPLE 1 Simplify.

a) $6 + 4\sqrt{2} - \sqrt{2} + 7$ **b)** $2\sqrt[3]{x} + 8x + 4\sqrt[3]{x} - 3$

Solution

a) $6 + 4\sqrt{2} - \sqrt{2} + 7 = 6 + 7 + 4\sqrt{2} - \sqrt{2}$ Place like terms together.

$$= 13 + (4 - 1)\sqrt{2}$$

$$= 13 + 3\sqrt{2} \quad (\text{or } 3\sqrt{2} + 13)$$

b) $2\sqrt[3]{x} + 8x + 4\sqrt[3]{x} - 3 = 6\sqrt[3]{x} + 8x - 3$

Now Try Exercise 15

It is sometimes possible to convert unlike radicals into like radicals by simplifying one or more of the radicals, as was discussed in Section 7.3.

EXAMPLE 2 Simplify $\sqrt{3} + \sqrt{27}$.

Solution   Since $\sqrt{3}$ and $\sqrt{27}$ are unlike radicals, they cannot be added in their present form. We can simplify $\sqrt{27}$ to obtain like radicals.

$$\sqrt{3} + \sqrt{27} = \sqrt{3} + \sqrt{9}\sqrt{3}$$

$$= \sqrt{3} + 3\sqrt{3} = 4\sqrt{3}$$

Now Try Exercise 19

To Add or Subtract Radicals

1. Simplify each radical expression.

2. Combine like radicals (if there are any).

EXAMPLE 3 Simplify.

a) $5\sqrt{24} + \sqrt{54}$ **b)** $2\sqrt{45} - \sqrt{80} + \sqrt{20}$ **c)** $\sqrt[3]{27} + \sqrt[3]{81} - 7\sqrt[3]{3}$

Solution

a) $5\sqrt{24} + \sqrt{54} = 5 \cdot \sqrt{4} \cdot \sqrt{6} + \sqrt{9} \cdot \sqrt{6}$

$$= 5 \cdot 2\sqrt{6} + 3\sqrt{6}$$

$$= 10\sqrt{6} + 3\sqrt{6} = 13\sqrt{6}$$

b) $2\sqrt{45} - \sqrt{80} + \sqrt{20} = 2 \cdot \sqrt{9} \cdot \sqrt{5} - \sqrt{16} \cdot \sqrt{5} + \sqrt{4} \cdot \sqrt{5}$

$$= 2 \cdot 3\sqrt{5} - 4\sqrt{5} + 2\sqrt{5}$$

$$= 6\sqrt{5} - 4\sqrt{5} + 2\sqrt{5} = 4\sqrt{5}$$

c) $\sqrt[3]{27} + \sqrt[3]{81} - 7\sqrt[3]{3} = 3 + \sqrt[3]{27} \cdot \sqrt[3]{3} - 7\sqrt[3]{3}$

$$= 3 + 3\sqrt[3]{3} - 7\sqrt[3]{3} = 3 - 4\sqrt[3]{3}$$

Now Try Exercise 23

EXAMPLE 4 Simplify.   **a)** $\sqrt{x^2} - \sqrt{x^2 y} + x\sqrt{y}$ **b)** $\sqrt[3]{x^{13} y^2} - \sqrt[3]{x^4 y^8}$

Solution

a) $\sqrt{x^2} - \sqrt{x^2 y} + x\sqrt{y} = x - \sqrt{x^2} \cdot \sqrt{y} + x\sqrt{y}$

$$= x - x\sqrt{y} + x\sqrt{y}$$

$$= x$$

b) $\sqrt[3]{x^{13} y^2} - \sqrt[3]{x^4 y^8} = \sqrt[3]{x^{12}} \cdot \sqrt[3]{xy^2} - \sqrt[3]{x^3 y^6} \cdot \sqrt[3]{xy^2}$

$$= x^4 \sqrt[3]{xy^2} - xy^2 \sqrt[3]{xy^2}$$

Now factor out the common factor, $\sqrt[3]{xy^2}$.

$$= (x^4 - xy^2)\sqrt[3]{xy^2}$$

Now Try Exercise 35

> **Helpful Hint**
>
> The product rule and quotient rule for radicals presented in Section 7.3 are
>
> $$\sqrt[n]{a} \cdot \sqrt[n]{b} = \sqrt[n]{ab} \qquad \frac{\sqrt[n]{a}}{\sqrt[n]{b}} = \sqrt[n]{\frac{a}{b}}$$
>
> Students often incorrectly assume similar properties exist for addition and subtraction. They do not. To illustrate this, let n be a square root (index 2), $a = 9$, and $b = 16$.
>
> $$\sqrt[n]{a} + \sqrt[n]{b} \neq \sqrt[n]{a+b}$$
> $$\sqrt{9} + \sqrt{16} \neq \sqrt{9+16}$$
> $$3 + 4 \neq \sqrt{25}$$
> $$7 \neq 5$$

2 Multiply Radicals

To multiply radicals, we use the product rule given earlier. After multiplying, we can often simplify the new radical (see Examples 5 and 6).

EXAMPLE 5 Multiply and simplify.

a) $\sqrt{6x^3}\, \sqrt{8x^6}$ **b)** $\sqrt[3]{2x}\, \sqrt[3]{4x^2}$ **c)** $\sqrt[4]{4x^{11}y}\, \sqrt[4]{16x^6 y^{22}}$

Solution

a) $\sqrt{6x^3}\, \sqrt{8x^6} = \sqrt{6x^3 \cdot 8x^6}$ Product rule for radicals

$\qquad\qquad\quad = \sqrt{48x^9}$

$\qquad\qquad\quad = \sqrt{16x^8}\sqrt{3x}$ $16x^8$ is a perfect square.

$\qquad\qquad\quad = 4x^4\sqrt{3x}$

b) $\sqrt[3]{2x}\, \sqrt[3]{4x^2} = \sqrt[3]{2x \cdot 4x^2}$ Product rule for radicals

$\qquad\qquad\quad = \sqrt[3]{8x^3}$ $8x^3$ is a perfect cube.

$\qquad\qquad\quad = 2x$

c) $\sqrt[4]{4x^{11}y}\, \sqrt[4]{16x^6 y^{22}} = \sqrt[4]{4x^{11}y \cdot 16x^6 y^{22}}$ Product rule for radicals

$\qquad\qquad\qquad\qquad = \sqrt[4]{64x^{17}y^{23}}$

$\qquad\qquad\qquad\qquad = \sqrt[4]{16x^{16}y^{20}}\, \sqrt[4]{4xy^3}$ The largest perfect fourth root factors are $16, x^{16}$, and y^{20}.

$\qquad\qquad\qquad\qquad = 2x^4 y^5 \sqrt[4]{4xy^3}$

Now Try Exercise 47

As stated earlier, when a radical is simplified, the exponents on the variables in the radicand are less than the index.

EXAMPLE 6 Multiply and simplify $\sqrt{2x}\,(\sqrt{8x} - \sqrt{50})$.

Solution   Begin by using the distributive property.

$$\sqrt{2x}\,(\sqrt{8x} - \sqrt{50}) = (\sqrt{2x})(\sqrt{8x}) + (\sqrt{2x})(-\sqrt{50})$$
$$= \sqrt{16x^2} - \sqrt{100x}$$
$$= 4x - \sqrt{100}\,\sqrt{x}$$
$$= 4x - 10\sqrt{x}$$

Now Try Exercise 53

Understanding Algebra

Recall that FOIL is an acronym for **F**irst, **O**uter, **I**nner, **L**ast. The FOIL method is used when multiplying two binomials as follows:

F O I L
↓ ↓ ↓ ↓

$(a + b)(c + d) = ac + ad + bc + bd$

We also use the FOIL method when multiplying radical expressions such as $(\sqrt{a} + \sqrt{b})(\sqrt{c} + \sqrt{d})$.

Note in Example 6 that the same result could be obtained by first simplifying $\sqrt{8x}$ and $\sqrt{50}$ and then multiplying. You may wish to try this now.

We next will multiply radical expressions involving sums or differences of radicals such as $(\sqrt{a} + \sqrt{b})(\sqrt{c} + \sqrt{d})$. We will use the FOIL method just as we did when multiplying two binomials.

EXAMPLE 7 Multiply $(\sqrt{x} - \sqrt{y})(\sqrt{x} - y)$.

Solution   We will multiply using the FOIL method.

$$
\begin{array}{cccc}
\text{F} & \text{O} & \text{I} & \text{L} \\
\downarrow & \downarrow & \downarrow & \downarrow \\
(\sqrt{x})(\sqrt{x}) & + \;(\sqrt{x})(-y) & +\;(-\sqrt{y})(\sqrt{x}) & +\;(-\sqrt{y})(-y) \\
=\quad \sqrt{x^2} & -\quad y\sqrt{x} & -\quad \sqrt{xy} & +\quad y\sqrt{y}
\end{array}
$$

$$= x - y\sqrt{x} - \sqrt{xy} + y\sqrt{y}$$

Now Try Exercise 63

EXAMPLE 8 Simplify. **a)** $(2\sqrt{6} - \sqrt{3})^2$ **b)** $\left(\sqrt[3]{x} - \sqrt[3]{2y^2}\right)\left(\sqrt[3]{x^2} - \sqrt[3]{8y}\right)$

Solution

a) $(2\sqrt{6} - \sqrt{3})^2 = (2\sqrt{6} - \sqrt{3})(2\sqrt{6} - \sqrt{3})$

Now multiply using the FOIL method.

$$
\begin{array}{cccc}
\text{F} & \text{O} & \text{I} & \text{L}
\end{array}
$$
$$(2\sqrt{6})(2\sqrt{6}) + (2\sqrt{6})(-\sqrt{3}) + (-\sqrt{3})(2\sqrt{6}) + (-\sqrt{3})(-\sqrt{3})$$
$$= 4(6) - 2\sqrt{18} - 2\sqrt{18} + 3$$
$$= 24 - 2\sqrt{18} - 2\sqrt{18} + 3$$
$$= 27 - 4\sqrt{18}$$
$$= 27 - 4\sqrt{9}\sqrt{2}$$
$$= 27 - 12\sqrt{2}$$

b) Multiply using the FOIL method.

$$
\begin{array}{cccc}
 & \text{F} & \text{O} & \text{I} & \text{L}
\end{array}
$$
$$\left(\sqrt[3]{x} - \sqrt[3]{2y^2}\right)\left(\sqrt[3]{x^2} - \sqrt[3]{8y}\right) = (\sqrt[3]{x})(\sqrt[3]{x^2}) + (\sqrt[3]{x})(-\sqrt[3]{8y}) + \left(-\sqrt[3]{2y^2}\right)(\sqrt[3]{x^2}) + \left(-\sqrt[3]{2y^2}\right)(-\sqrt[3]{8y})$$
$$= \sqrt[3]{x^3} - \sqrt[3]{8xy} - \sqrt[3]{2x^2y^2} + \sqrt[3]{16y^3}$$
$$= \sqrt[3]{x^3} - \sqrt[3]{8}\,\sqrt[3]{xy} - \sqrt[3]{2x^2y^2} + \sqrt[3]{8y^3}\,\sqrt[3]{2}$$
$$= x - 2\sqrt[3]{xy} - \sqrt[3]{2x^2y^2} + 2y\sqrt[3]{2}$$

Now Try Exercise 99

EXAMPLE 9 Multiply $(3 + \sqrt{6})(3 - \sqrt{6})$.

Solution   We can multiply using the FOIL method.

$$
\begin{array}{cccc}
 & \text{F} & \text{O} & \text{I} & \text{L}
\end{array}
$$
$$(3 + \sqrt{6})(3 - \sqrt{6}) = 3(3) + 3(-\sqrt{6}) + (\sqrt{6})(3) + (\sqrt{6})(-\sqrt{6}).$$
$$= 9 \quad - \quad 3\sqrt{6} \quad + \quad 3\sqrt{6} \quad - \quad \sqrt{36}$$
$$= 9 - \sqrt{36}$$
$$= 9 - 6 = 3$$

Now Try Exercise 59

Note that in Example 9, we multiplied *the sum and difference of the same two radical expressions*. Recall from Section 5.6 that $(a + b)(a - b) = a^2 - b^2$. If we let $a = 3$ and $b = \sqrt{6}$, then we can multiply as follows.

$$(a + b)(a - b) = a^2 - b^2$$
$$(3 + \sqrt{6})(3 - \sqrt{6}) = 3^2 - (\sqrt{6})^2$$
$$= 9 - 6$$
$$= 3$$

When multiplying the sum and difference of the same two radical expressions, you may obtain the answer using the difference of the squares of the two radical expressions.

EXAMPLE 10 If $f(x) = \sqrt[3]{x^2}$ and $g(x) = \sqrt[3]{x^4} + \sqrt[3]{x^2}$, find **a)** $(f \cdot g)(x)$ and **b)** $(f \cdot g)(6)$.

Solution

a) From Section 3.6, we know that $(f \cdot g)(x) = f(x) \cdot g(x)$.

$$
\begin{aligned}
(f \cdot g)(x) &= f(x) \cdot g(x) \\
&= \sqrt[3]{x^2}\left(\sqrt[3]{x^4} + \sqrt[3]{x^2}\right) && \text{Substitute given values.} \\
&= \sqrt[3]{x^2}\,\sqrt[3]{x^4} + \sqrt[3]{x^2}\,\sqrt[3]{x^2} && \text{Distributive property} \\
&= \sqrt[3]{x^6} + \sqrt[3]{x^4} && \text{Product rule for radicals} \\
&= x^2 + x\sqrt[3]{x} && \text{Simplify radicals.}
\end{aligned}
$$

b) To compute $(f \cdot g)(6)$, substitute 6 for x in the answer obtained in part **a)**.

$$
\begin{aligned}
(f \cdot g)(x) &= x^2 + x\sqrt[3]{x} \\
(f \cdot g)(6) &= 6^2 + 6\sqrt[3]{6} && \text{Substitute 6 for } x. \\
&= 36 + 6\sqrt[3]{6}
\end{aligned}
$$

Now Try Exercise 77

As stated earlier in this chapter, unless we are instructed otherwise, we assume that variable expressions in radicands represent nonnegative real numbers. Example 11 demonstrates how we must use absolute value in the cases when the radicand can represent any real number.

EXAMPLE 11 Simplify $f(x)$ if

a) $f(x) = \sqrt{x^2 + 6x + 9}$ assuming $x \geq -3$.

b) $f(x) = \sqrt{x^2 + 6x + 9}$ assuming x can be any real number.

Solution

a) $f(x) = \sqrt{x^2 + 6x + 9}$

$\quad = \sqrt{(x + 3)^2}$ $\qquad x^2 + 6x + 9$ was factored as $(x + 3)^2$.

$\quad = x + 3$ $\qquad$ Since $x \geq -3$, $x + 3 \geq 0$ and no absolute value bars are needed.

b) $f(x) = \sqrt{x^2 + 6x + 9}$

$\quad = \sqrt{(x + 3)^2}$ $\qquad x^2 + 6x + 9$ was factored as $(x + 3)^2$.

$\quad = |x + 3|$ $\qquad$ Since x can be any real number, $x + 3$ may be negative and absolute value bars are required.

Now Try Exercise 105

EXERCISE SET 7.4

Math XL　MathXL®　　MyMathLab　MyMathLab

Warm-Up Exercises

Fill in the blanks with the appropriate word, phrase, or symbol(s) from the following list.

commutative　　　distributive　　　less　　　like　　　　FOIL　　　first　　　greater　　　unlike

1. Radicals having the same radicand and index are _____ radicals.

2. Radicals differing in either the radicand or the index are _____ radicals.

3. The _____ property is used to add or subtract like radicals in the same way that like terms are added or subtracted.

4. When a radical is simplified, the exponents on the variables in the radicand are _____ than the index.

5. When multiplying radical expressions involving sums or differences of radicals, such as $(\sqrt{a} + \sqrt{b})(\sqrt{c} + \sqrt{d})$, use the _____ method.

6. FOIL is an acronym for _____, outer, inner, last.

Practice the Skills

In this exercise set, assume that all variables represent positive real numbers.

Simplify.

7. $\sqrt{2} - \sqrt{2}$

8. $4\sqrt{3} - \sqrt{3}$

9. $6\sqrt{5} - 2\sqrt{5}$

10. $3\sqrt{2} + 7\sqrt{2} - 11$

11. $2\sqrt{3} - 2\sqrt{3} - 4\sqrt{3} + 5$

12. $6\sqrt[3]{7} - 8\sqrt[3]{7}$

13. $2\sqrt[4]{y} - 9\sqrt[4]{y}$

14. $3\sqrt[5]{a} + 7 + 5\sqrt[5]{a} - 2$

15. $3\sqrt{5} - \sqrt[3]{x} + 6\sqrt{5} + 3\sqrt[3]{x}$

16. $9 + 4\sqrt[4]{a} - 7\sqrt[4]{a} + 5$

17. $5\sqrt{x} - 8\sqrt{y} + 3\sqrt{x} + 2\sqrt{y} - \sqrt{x}$

18. $8\sqrt{a} + 4\sqrt[3]{b} + 7\sqrt{a} - 12\sqrt[3]{b}$

Simplify.

19. $\sqrt{3} + \sqrt{12}$

20. $\sqrt{8} + \sqrt{18}$

21. $-6\sqrt{75} + 5\sqrt{125}$

22. $3\sqrt{250} + 4\sqrt{160}$

23. $-4\sqrt{90} + 3\sqrt{40} + 2\sqrt{10}$

24. $3\sqrt{40x^2y} + 2x\sqrt{490y}$

25. $\sqrt{500xy^2} + y\sqrt{320x}$

26. $5\sqrt{8} + 2\sqrt{50} - 3\sqrt{72}$

27. $2\sqrt{5x} - 3\sqrt{20x} - 4\sqrt{45x}$

28. $3\sqrt{27c^2} - 2\sqrt{108c^2} - \sqrt{48c^2}$

29. $3\sqrt{50a^2} - 3\sqrt{72a^2} - 8a\sqrt{18}$

30. $4\sqrt[3]{5} - 5\sqrt[3]{40}$

31. $\sqrt[3]{108} + \sqrt[3]{32}$

32. $3\sqrt[3]{16} + \sqrt[3]{54}$

33. $\sqrt[3]{27} - 5\sqrt[3]{8}$

34. $3\sqrt{45x^3} + \sqrt{5x}$

35. $2\sqrt[3]{a^4b^2} + 4a\sqrt[3]{ab^2}$

36. $5y\sqrt[4]{48x^5} - x\sqrt[4]{3x^5y^4}$

37. $\sqrt{4r^7s^5} + 3r^2\sqrt{r^3s^5} - 2rs\sqrt{r^5s^3}$

38. $x\sqrt[3]{27x^5y^2} - x^2\sqrt[3]{x^2y^2} + 4\sqrt[3]{x^8y^2}$

39. $\sqrt[3]{128x^8y^{10}} - 2x^2y\sqrt[3]{16x^2y^7}$

40. $5\sqrt[3]{320x^5y^8} + 3x\sqrt[3]{135x^2y^8}$

Simplify.

41. $\sqrt{2}\sqrt{8}$

42. $\sqrt{3}\,\sqrt{27}$

43. $\sqrt[3]{4}\,\sqrt[3]{14}$

44. $\sqrt[3]{3}\,\sqrt[3]{54}$

45. $\sqrt{9m^3n^7}\,\sqrt{3mn^4}$

46. $\sqrt[3]{5ab^2}\,\sqrt[3]{25a^4b^{12}}$

47. $\sqrt[3]{9x^7y^{10}}\,\sqrt[3]{6x^4y^3}$

48. $\sqrt[4]{3x^9y^{12}}\,\sqrt[4]{54x^4y^7}$

49. $\sqrt[5]{x^{24}y^{30}z^9}\,\sqrt[5]{x^{13}y^8z^7}$

50. $\sqrt[4]{8x^4yz^3}\,\sqrt[4]{2x^2y^3z^7}$

51. $\left(\sqrt[3]{2x^3y^4}\right)^2$

52. $\sqrt{2}\,(\sqrt{6} + \sqrt{18})$

53. $\sqrt{5}\,(\sqrt{5} - \sqrt{3})$

54. $\sqrt{3}\,(\sqrt{12} + \sqrt{8})$

55. $\sqrt[3]{y}\,\left(2\sqrt[3]{y} - \sqrt[3]{y^8}\right)$

56. $\sqrt{3y}\,\left(\sqrt{27y^2} - \sqrt{y}\right)$

57. $2\sqrt[3]{x^4y^5}\left(\sqrt[3]{8x^{12}y^4} + \sqrt[3]{16xy^9}\right)$

58. $\sqrt[5]{16x^7y^6}\left(\sqrt[5]{2x^6y^9} - \sqrt[5]{10x^3y^7}\right)$

59. $(8 + \sqrt{5})(8 - \sqrt{5})$

60. $(9 - \sqrt{5})(9 + \sqrt{5})$

61. $(\sqrt{6} + x)(\sqrt{6} - x)$

62. $(\sqrt{x} + y)(\sqrt{x} - y)$

63. $(\sqrt{7} - \sqrt{z})(\sqrt{7} + \sqrt{z})$

64. $(3\sqrt{a} - 5\sqrt{b})(3\sqrt{a} + 5\sqrt{b})$

65. $(\sqrt{3} + 4)(\sqrt{3} + 5)$

66. $(1 + \sqrt{5})(8 + \sqrt{5})$

67. $(3 - \sqrt{2})(4 - \sqrt{8})$

68. $(5\sqrt{6} + 3)(4\sqrt{6} - 1)$

69. $(4\sqrt{3} + \sqrt{2})(\sqrt{3} - \sqrt{2})$

70. $(\sqrt{3} + 7)^2$

71. $(2\sqrt{5} - 3)^2$

72. $(\sqrt{y} + \sqrt{6z})(\sqrt{2z} - \sqrt{8y})$

73. $(2\sqrt{3x} - \sqrt{y})(3\sqrt{3x} + \sqrt{y})$

74. $(\sqrt[3]{9} + \sqrt[3]{2})(\sqrt[3]{3} + \sqrt[3]{4})$

75. $(\sqrt[3]{4} - \sqrt[3]{6})(\sqrt[3]{2} - \sqrt[3]{36})$

76. $(\sqrt[4]{4x} - \sqrt[4]{2y})(\sqrt[4]{4x} + \sqrt[4]{10})$

In Exercises 77–82, f(x) and g(x) are given. Find $(f \cdot g)(x)$.

77. $f(x) = \sqrt{x},\, g(x) = \sqrt{x} - \sqrt{3}$

78. $f(x) = \sqrt{6x},\, g(x) = \sqrt{6x} - \sqrt{10x}$

79. $f(x) = \sqrt[3]{x},\, g(x) = \sqrt[3]{x^5} + \sqrt[3]{x^4}$

80. $f(x) = \sqrt[3]{2x^2},\, g(x) = \sqrt[3]{4x} + \sqrt[3]{32x^2}$

81. $f(x) = \sqrt[4]{3x^2},\, g(x) = \sqrt[4]{9x^4} - \sqrt[4]{x^7}$

82. $f(x) = \sqrt[4]{2x^3},\, g(x) = \sqrt[4]{8x^5} - \sqrt[4]{5x^6}$

Simplify. These exercises are a combination of the types of exercises presented earlier in this exercise set.

83. $\sqrt{18}$

84. $\sqrt{300}$

85. $\sqrt{125} - \sqrt{20}$

86. $4\sqrt{7} + 2\sqrt{63} - 2\sqrt{28}$

87. $(3\sqrt{2} - 4)(\sqrt{2} + 5)$

88. $(\sqrt{5} + \sqrt{2})(\sqrt{2} + \sqrt{20})$

89. $\sqrt{6}(5 - \sqrt{2})$

90. $3\sqrt[3]{81} + 4\sqrt[3]{24}$

91. $\sqrt{150}\sqrt{3}$

92. $\sqrt[4]{2}\,\sqrt[4]{40}$

93. $\sqrt[3]{80x^{11}}$

94. $\sqrt[3]{x^9 y^{11} z}$

95. $\sqrt[6]{128ab^{17}c^9}$

96. $\sqrt[5]{14x^4 y^2}\,\sqrt[5]{3x^4 y^3}$

97. $2b\sqrt[4]{a^4 b} + ab\sqrt[4]{16b}$

98. $2\sqrt[3]{24a^3 y^4} + 4a\sqrt[3]{81y^4}$

99. $\left(\sqrt[3]{x^2} - \sqrt[3]{y}\right)\left(\sqrt[3]{x} - 2\sqrt[3]{y^2}\right)$

100. $\left(\sqrt[3]{a} + 5\right)\left(\sqrt[3]{a^2} - 6\right)$

101. $\sqrt[3]{3ab^2}\left(\sqrt[3]{4a^4 b^3} - \sqrt[3]{8a^5 b^4}\right)$

102. $\sqrt[4]{4st^2}\left(\sqrt[4]{2s^5 t^6} + \sqrt[4]{5s^9 t^2}\right)$

Simplify the following. In Exercises 105 and 106, assume the variable can be any real number. See Example 11.

103. $f(x) = \sqrt{2x - 5}\,\sqrt{2x - 5},\ x \ge \dfrac{5}{2}$

104. $g(a) = \sqrt{3a + 7}\,\sqrt{3a + 7},\ a \ge -\dfrac{7}{3}$

105. $h(r) = \sqrt{4r^2 - 32r + 64}$

106. $f(b) = \sqrt{20b^2 + 60b + 45}$

Problem Solving

Find the perimeter and area of the following figures. Write the perimeter and area in radical form with the radicals simplified. (Hint: See page 711 for the needed formulas.)

107.
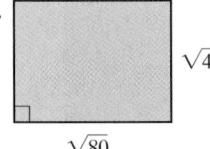

108.
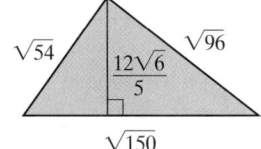

109.
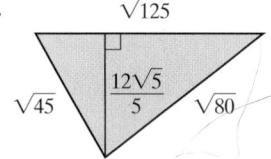

110.
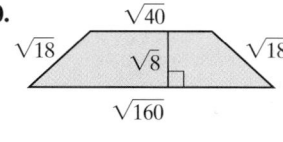

111. Skid Marks Law enforcement officials sometimes use the formula $s = \sqrt{30FB}$ to determine a car's speed, s, in miles per hour, from a car's skid marks. The F in the formula represents the "road factor," which is determined by the road's surface, and the B represents the braking distance, in feet. Officer Jenkins is investigating an accident. Find the car's speed if the skid marks are 80 feet long and **a)** the road was dry asphalt, whose road factor is 0.85, **b)** the road was wet gravel, whose road factor is 0.52.

112. Water through a Fire Hose The rate at which water flows through a particular fire hose, R, in gallons per minute, can be approximated by the formula $R = 28d^2\sqrt{P}$, where d is the diameter of the nozzle, in inches, and P is the nozzle pressure, in pounds per square inch. If a nozzle has a diameter of 2.5 inches and the nozzle pressure is 80 pounds per square inch, find the flow rate.

113. Height of Girls The function $f(t) = 3\sqrt{t} + 19$ can be used to approximate the median height, $f(t)$, in inches, for U.S. girls of age t, in months, where $1 \le t \le 60$. Estimate the median height of girls at age **a)** 36 months and **b)** 40 months.

114. Standard Deviation In statistics, the standard deviation of the population, σ, read "sigma," is a measure of the spread of a set of data about the mean of the data. One formula used to determine sigma is $\sigma = \sqrt{npq}$, where n represents the sample size, p represents the percent chance that something specific happens, and q represents the percent chance that the specific thing does not happen. In a sample of 600 people who purchase airline tickets, the percent that showed up for their flight, p, was 0.93, and the percent that did not show up for their flight, q, was 0.07. Use this information to find σ.

115. The graph of $f(x) = \sqrt{x}$ is shown.

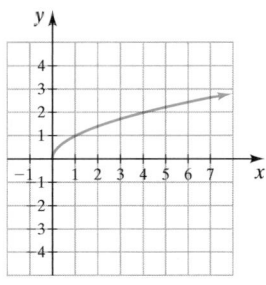

a) If $g(x) = 2$, sketch the graph of $(f + g)(x)$.

b) What effect does adding 2 have to the graph of $f(x)$?

117. You are given that $f(x) = \sqrt{x}$ and $g(x) = \sqrt{x} - 2$.

a) Sketch the graph of $(f - g)(x)$.

b) What is the domain of $(f - g)(x)$?

119. Graph the function $f(x) = \sqrt{x^2}$.

116. The graph of $f(x) = -\sqrt{x}$ is shown.

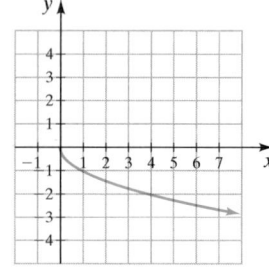

a) If $g(x) = 3$, sketch the graph of $(f + g)(x)$.

b) What effect does adding 3 have to the graph of $f(x)$?

118. You are given that $f(x) = \sqrt{x}$ and $g(x) = -\sqrt{x} - 3$.

a) Sketch the graph of $(f + g)(x)$.

b) What is the domain of $(f + g)(x)$?

120. Graph the function $f(x) = \sqrt{x^2} - 4$.

Concept/Writing Exercises

121. Does $\sqrt{a} + \sqrt{b} = \sqrt{a + b}$? Explain your answer and give an example supporting your answer.

122. Since $64 + 36 = 100$, does $\sqrt{64} + \sqrt{36} = \sqrt{100}$? Explain your answer.

123. Will the sum of two radicals always be a radical? Give an example to support your answer.

124. Will the difference of two radicals always be a radical? Give an example to support your answer.

Cumulative Review Exercises

[1.2] **125.** What is a rational number?

[1.3] **126.** What is a real number?

127. What is an irrational number?

128. What is the definition of $|a|$?

[2.2] **129.** Solve the formula $E = \dfrac{1}{2}mv^2$ for m.

[2.5] **130.** Solve the inequality $-4 < 2x - 3 \le 7$ and indicate the solution **a)** on the number line; **b)** in interval notation; **c)** in set builder notation.

Mid-Chapter Test: 7.1–7.4

To find out how well you understand the chapter material to this point, take this brief test. The answers, and the section where the material was initially discussed, are given in the back of the book. Review any questions that you answered incorrectly.

Find the indicated root.

1. $\sqrt{121}$

2. $\sqrt[3]{-\dfrac{27}{64}}$

Use absolute value to evaluate.

3. $\sqrt{(-16.3)^2}$

4. $\sqrt{(3a^2 - 4b^3)^2}$

5. Find $g(16)$ if $g(x) = \dfrac{x}{8} + \sqrt{4x} - 7$.

6. Write $\sqrt[5]{7a^4b^3}$ in exponential form.

7. Evaluate $-49^{1/2} + 81^{3/4}$.

Simplify each expression.

8. $\left(\sqrt[4]{a^2b^3c}\right)^{20}$

9. $7x^{-5/2} \cdot 2x^{3/2}$

10. Multiply $8x^{-2}(x^3 + 2x^{-1/2})$.

Simplify each radical.

11. $\sqrt{32x^4y^9}$

12. $\sqrt[6]{64a^{13}b^{23}c^{15}}$

13. $\dfrac{\sqrt[3]{3}}{\sqrt[3]{81}}$

14. $\dfrac{\sqrt{20x^5y^{12}}}{\sqrt{180x^{15}y^7}}$

Simplify. Assume that all variables represent positive real numbers.

15. $2\sqrt{x} - 3\sqrt{y} + 9\sqrt{x} + 15\sqrt{y}$

16. $2\sqrt{90x^2y} + 3x\sqrt{490y}$

17. $(x + \sqrt{5})(2x - 3\sqrt{5})$

18. $2\sqrt{3a}\left(\sqrt{27a^2} - 5\sqrt{4a}\right)$

19. $3b\sqrt[4]{a^5b} + 2ab\sqrt[4]{16ab}$

20. When simplifying the following square roots, in which parts will the answer contain an absolute value? Explain your answer and simplify parts **a)** and **b)**.

a) $\sqrt{(x - 3)^2}$

b) $\sqrt{64x^2}, \ x \ge 0$

7.5 Dividing Radicals

1 Rationalize denominators.

2 Rationalize a denominator using the conjugate.

3 Understand when a radical is simplified.

4 Use rationalizing the denominator in an addition problem.

5 Divide radical expressions with different indices.

1 Rationalize Denominators

When the denominator of a fraction contains radicals, it is customary to rewrite the fraction as an equivalent fraction in which the denominator does not contain radicals. The process we use to rewrite the fraction is known as **rationalizing the denominator.** In this section we will use the quotient rule for radicals, introduced in Section 7.3, to rationalize denominators.

> **To Rationalize a Denominator**
>
> Multiply both the numerator and the denominator of the fraction by a radical that will produce a radicand in the denominator that is a perfect power for the index.

When both the numerator and denominator are multiplied by the same radical expression, you are in effect multiplying the fraction by 1, which does not change its value. Also recall that we will assume that all variables in the radicands will represent nonnegative real numbers.

EXAMPLE 1 Simplify.   **a)** $\dfrac{1}{\sqrt{5}}$ **b)** $\dfrac{x}{4\sqrt{3}}$ **c)** $\dfrac{11}{\sqrt{2x}}$ **d)** $\dfrac{\sqrt[3]{16a^4}}{\sqrt[3]{b}}$

Solution   To simplify each expression, we must rationalize the denominator. We do so by multiplying both the numerator and denominator by a radical that will produce a denominator that is a perfect power for the given index.

a) $\dfrac{1}{\sqrt{5}} = \dfrac{1}{\sqrt{5}} \cdot \dfrac{\sqrt{5}}{\sqrt{5}} = \dfrac{\sqrt{5}}{\sqrt{25}} = \dfrac{\sqrt{5}}{5}$ **b)** $\dfrac{x}{4\sqrt{3}} = \dfrac{x}{4\sqrt{3}} \cdot \dfrac{\sqrt{3}}{\sqrt{3}} = \dfrac{x\sqrt{3}}{4\cdot 3} = \dfrac{x\sqrt{3}}{12}$

c) There are two factors in the radicand, 2 and x. Since 2^2 or 4 is a perfect square and x^2 is a perfect square, we multiply both numerator and denominator by $\sqrt{2x}$.

$$\frac{11}{\sqrt{2x}} = \frac{11}{\sqrt{2x}} \cdot \frac{\sqrt{2x}}{\sqrt{2x}}$$
$$= \frac{11\sqrt{2x}}{\sqrt{4x^2}}$$
$$= \frac{11\sqrt{2x}}{2x}$$

d) There are no common factors in the numerator and denominator. Before we rationalize the denominator, we will simplify the numerator.

$$\frac{\sqrt[3]{16a^4}}{\sqrt[3]{b}} = \frac{\sqrt[3]{8a^3}\sqrt[3]{2a}}{\sqrt[3]{b}} \quad \text{Product rule for radicals}$$
$$= \frac{2a\sqrt[3]{2a}}{\sqrt[3]{b}} \quad \text{Simplify the numerator.}$$

Now we rationalize the denominator. Since the denominator is a cube root, we need to produce a radicand that is a perfect cube. Since the denominator contains b and we want b^3, we need two more factors of b, or b^2. We therefore multiply both numerator and denominator by $\sqrt[3]{b^2}$.

$$= \frac{2a\sqrt[3]{2a}}{\sqrt[3]{b}} \cdot \frac{\sqrt[3]{b^2}}{\sqrt[3]{b^2}}$$
$$= \frac{2a\sqrt[3]{2ab^2}}{\sqrt[3]{b^3}}$$
$$= \frac{2a\sqrt[3]{2ab^2}}{b}$$

Understanding Algebra

Historically, denominators were rationalized so that obtaining a decimal approximation would be easier. Example 1 **a)** showed that $\dfrac{1}{\sqrt{5}} = \dfrac{\sqrt{5}}{5}$.

Approximating $\dfrac{1}{\sqrt{5}}$ using long division requires you to divide 1 by an approximation of $\sqrt{5}$ such as 2.236.

However, approximating $\dfrac{\sqrt{5}}{5}$ requires you to divide 2.236 by 5—a much easier task!

Now Try Exercise 15

EXAMPLE 2 Simplify.   **a)** $\sqrt{\dfrac{5}{7}}$ **b)** $\sqrt[3]{\dfrac{x}{2y^2}}$ **c)** $\sqrt[4]{\dfrac{32x^9y^6}{3z^2}}$

Solution   In each part, we will use the quotient rule to rewrite the radical as a quotient of two radicals.

a) $\sqrt{\dfrac{5}{7}} = \dfrac{\sqrt{5}}{\sqrt{7}} \cdot \dfrac{\sqrt{7}}{\sqrt{7}} = \dfrac{\sqrt{35}}{\sqrt{49}} = \dfrac{\sqrt{35}}{7}$

b) $\sqrt[3]{\dfrac{x}{2y^2}} = \dfrac{\sqrt[3]{x}}{\sqrt[3]{2y^2}}$

The denominator is $\sqrt[3]{2y^2}$. We will multiply both the numerator and denominator by the cube root of an expression that will produce the denominator $\sqrt[3]{2^3y^3}$. Since $2 \cdot 2^2 = 2^3$ and $y^2 \cdot y = y^3$, we multiply both numerator and denominator by $\sqrt[3]{2^2y}$.

$$\dfrac{\sqrt[3]{x}}{\sqrt[3]{2y^2}} = \dfrac{\sqrt[3]{x}}{\sqrt[3]{2y^2}} \cdot \dfrac{\sqrt[3]{2^2y}}{\sqrt[3]{2^2y}}$$

$$= \dfrac{\sqrt[3]{x}\,\sqrt[3]{4y}}{\sqrt[3]{2^3y^3}}$$

$$= \dfrac{\sqrt[3]{4xy}}{2y}$$

c) After using the quotient rule, we simplify the numerator.

$$\sqrt[4]{\dfrac{32x^9y^6}{3z^2}} = \dfrac{\sqrt[4]{32x^9y^6}}{\sqrt[4]{3z^2}} \qquad \text{Quotient rule for radicals}$$

$$= \dfrac{\sqrt[4]{16x^8y^4}\,\sqrt[4]{2xy^2}}{\sqrt[4]{3z^2}} \qquad \text{Product rule for radicals}$$

$$= \dfrac{2x^2y\,\sqrt[4]{2xy^2}}{\sqrt[4]{3z^2}} \qquad \text{Simplify the numerator.}$$

Now we rationalize the denominator. We need to multiply both the numerator and the denominator by a fourth root to produce a radicand that is a perfect fourth power. Since the denominator contains one factor of 3, we need 3 more factors of 3, or 3^3. Since there are two factors of z, we need 2 more factors of z, or z^2. Thus we will multiply both numerator and denominator by $\sqrt[4]{3^3z^2}$.

$$= \dfrac{2x^2y\,\sqrt[4]{2xy^2}}{\sqrt[4]{3z^2}} \cdot \dfrac{\sqrt[4]{3^3z^2}}{\sqrt[4]{3^3z^2}}$$

$$= \dfrac{2x^2y\,\sqrt[4]{2xy^2}\,\sqrt[4]{27z^2}}{\sqrt[4]{3z^2}\,\sqrt[4]{3^3z^2}}$$

$$= \dfrac{2x^2y\,\sqrt[4]{54xy^2z^2}}{\sqrt[4]{3^4z^4}} \qquad \text{Product rule for radicals}$$

$$= \dfrac{2x^2y\,\sqrt[4]{54xy^2z^2}}{3z}$$

Note: There are no perfect fourth power factors of 54, and each exponent in the radicand is less than the index.

Now Try Exercise 53

2 Rationalize a Denominator Using the Conjugate

The denominator of some fractions is a sum or difference involving radicals that cannot be simplified. One such fraction is $\dfrac{13}{4 + \sqrt{3}}$. To rationalize the denominator of this fraction, we multiply the numerator and the denominator by $4 - \sqrt{3}$. The expressions $4 + \sqrt{3}$ and $4 - \sqrt{3}$ are **conjugates** of each other. Other examples of conjugates are shown below.

Radical Expression	Conjugate
$9 + \sqrt{2}$	$9 - \sqrt{2}$
$8\sqrt{3} - \sqrt{5}$	$8\sqrt{3} + \sqrt{5}$
$\sqrt{x} + \sqrt{y}$	$\sqrt{x} - \sqrt{y}$
$-6a - \sqrt{b}$	$-6a + \sqrt{b}$

When a radical expression is multiplied by its conjugate using the FOIL method, the product simplifies to an expression that contains no square roots. Our next example shows the product of a radical expression and its conjugate.

EXAMPLE 3 Multiply $(6 + \sqrt{3})(6 - \sqrt{3})$.

Solution   Multiply using the FOIL method.

$$\overset{\text{F}\qquad\quad\text{O}\qquad\quad\text{I}\qquad\quad\text{L}}{(6 + \sqrt{3})(6 - \sqrt{3}) = 6(6) + 6(-\sqrt{3}) + 6(\sqrt{3}) + \sqrt{3}\,(-\sqrt{3})}$$

$$= 36 \;\boxed{-6\sqrt{3} + 6\sqrt{3}}\; - \sqrt{9}$$

$$= 36 - \sqrt{9}$$

$$= 36 - 3$$

$$= 33$$

Now Try Exercise 57

Understanding Algebra

In Chapter 5 we discussed the product

$$(a + b)(a - b) = a^2 - b^2$$

Notice how the sum of the outer and inner products, $-ab + ab$, is always 0. This same procedure can be used when multiplying a radical expression by its conjugate. For example,

$$(3 + \sqrt{2})(3 - \sqrt{2}) = 3^2 - (\sqrt{2})^2$$
$$= 9 - 2 = 7$$

In Example 3, we would get the same result using the formula for the product of the sum and difference of the same two terms. The product results in the difference of two squares, $(a + b)(a - b) = a^2 - b^2$. In Example 3, if we let $a = 6$ and $b = \sqrt{3}$, then using the formula we get the following.

$$(a + b)(a - b) = a^2 - b^2$$
$$(6 + \sqrt{3})(6 - \sqrt{3}) = 6^2 - (\sqrt{3})^2$$
$$= 36 - 3$$
$$= 33$$

To Rationalize the Denominator Using the Conjugate

Multiply the numerator and the denominator by the conjugate of the denominator. Simplify using the distributive property, FOIL, or the special product $(a + b)(a - b) = a^2 - b^2$.

Now let's work an example where we rationalize a denominator using the conjugate.

EXAMPLE 4 Simplify.   **a)** $\dfrac{13}{4 + \sqrt{3}}$ **b)** $\dfrac{6}{\sqrt{5} - \sqrt{2}}$ **c)** $\dfrac{a - \sqrt{b}}{a + \sqrt{b}}$

Solution   In each part, we rationalize the denominator by multiplying the numerator and the denominator by the conjugate of the denominator.

a)
$$\frac{13}{4 + \sqrt{3}} = \frac{13}{4 + \sqrt{3}} \cdot \frac{4 - \sqrt{3}}{4 - \sqrt{3}}$$

$$= \frac{13(4 - \sqrt{3})}{(4 + \sqrt{3})(4 - \sqrt{3})}$$

$$= \frac{13(4 - \sqrt{3})}{16 - 3}$$

$$= \frac{\overset{1}{\cancel{13}}(4 - \sqrt{3})}{\underset{1}{\cancel{13}}} \quad \text{or} \quad 4 - \sqrt{3}$$

b)
$$\frac{6}{\sqrt{5} - \sqrt{2}} = \frac{6}{\sqrt{5} - \sqrt{2}} \cdot \frac{\sqrt{5} + \sqrt{2}}{\sqrt{5} + \sqrt{2}}$$

$$= \frac{6(\sqrt{5} + \sqrt{2})}{5 - 2}$$

$$= \frac{\overset{2}{\cancel{6}}(\sqrt{5} + \sqrt{2})}{\underset{1}{\cancel{3}}}$$

$$= 2(\sqrt{5} + \sqrt{2}) \quad \text{or} \quad 2\sqrt{5} + 2\sqrt{2}$$

c)
$$\frac{a - \sqrt{b}}{a + \sqrt{b}} = \frac{a - \sqrt{b}}{a + \sqrt{b}} \cdot \frac{a - \sqrt{b}}{a - \sqrt{b}}$$

$$= \frac{a^2 - a\sqrt{b} - a\sqrt{b} + \sqrt{b^2}}{a^2 - b}$$

$$= \frac{a^2 - 2a\sqrt{b} + b}{a^2 - b}$$

Remember that you cannot divide out a^2 or b because they are terms, not factors.

Now Try Exercise 75

Now that we have illustrated how to rationalize denominators, let's discuss the criteria a radical must meet to be considered simplified.

3 Understand When a Radical Is Simplified

After you have simplified a radical expression, you should check it to make sure that it is simplified as far as possible.

A Radical Expression Is Simplified When the Following Are All True

1. No perfect powers are factors of the radicand and all exponents in the radicand are less than the index.

2. No radicand contains a fraction.

3. No denominator contains a radical.

EXAMPLE 5 Determine whether the following expressions are simplified. If not, explain why. Simplify the expressions if not simplified.

a) $\sqrt{48x^5}$ **b)** $\sqrt{\dfrac{1}{2}}$ **c)** $\dfrac{1}{\sqrt{6}}$

Solution

a) This expression is not simplified because 16 is a perfect square factor of 48 and because x^4 is a perfect square factor of x^5. Let's simplify the radical.

$$\sqrt{48x^5} = \sqrt{16x^4 \cdot 3x} = \sqrt{16x^4} \cdot \sqrt{3x} = 4x^2\sqrt{3x}$$

b) This expression is not simplified since the radicand contains the fraction $\dfrac{1}{2}$. We simplify by first using the quotient rule, and then we rationalize the denominator as follows.

$$\sqrt{\dfrac{1}{2}} = \dfrac{\sqrt{1}}{\sqrt{2}} \cdot \dfrac{\sqrt{2}}{\sqrt{2}} = \dfrac{\sqrt{2}}{2}$$

c) This expression is not simplified since the denominator, $\sqrt{6}$, contains a radical. We simplify by rationalizing the denominator as follows.

$$\dfrac{1}{\sqrt{6}} = \dfrac{1}{\sqrt{6}} \cdot \dfrac{\sqrt{6}}{\sqrt{6}} = \dfrac{\sqrt{6}}{6}$$

Now Try Exercise 7

> **Understanding Algebra**
>
> When written in simplified form, the following must be true about the radicand:
>
> - The coefficient of the radicand has no perfect power factors.
> - The exponents on the variables are all less than the index of the radical.

4 Use Rationalizing the Denominator in an Addition Problem

Now let's work an addition problem that requires rationalizing the denominator. This example makes use of the methods discussed in Sections 7.3 and 7.4 to add and subtract radicals.

EXAMPLE 6 Simplify $4\sqrt{2} - \dfrac{3}{\sqrt{8}} + \sqrt{32}$.

Solution Begin by rationalizing the denominator and by simplifying $\sqrt{32}$.

$$4\sqrt{2} - \dfrac{3}{\sqrt{8}} + \sqrt{32} = 4\sqrt{2} - \dfrac{3}{\sqrt{8}} \cdot \dfrac{\sqrt{2}}{\sqrt{2}} + \sqrt{16}\,\sqrt{2} \qquad \text{Rationalize denominator.}$$

$$= 4\sqrt{2} - \dfrac{3\sqrt{2}}{\sqrt{16}} + 4\sqrt{2} \qquad \text{Product rule}$$

$$= 4\sqrt{2} - \dfrac{3}{4}\sqrt{2} + 4\sqrt{2} \qquad \text{Write } \dfrac{3\sqrt{2}}{\sqrt{16}} \text{ as } \dfrac{3}{4}\sqrt{2}.$$

$$= \left(4 - \dfrac{3}{4} + 4\right)\sqrt{2} \qquad \text{Simplify.}$$

$$= \dfrac{29\sqrt{2}}{4}$$

Now Try Exercise 115

5 Divide Radical Expressions with Different Indices

Now we will divide radical expressions where the radicals have different indices. To divide such problems, write each radical in exponential form. Then use the rules of exponents with the rational exponents, as was discussed in Section 7.2, to simplify the expression. Example 7 illustrates this procedure.

EXAMPLE 7 Simplify.   **a)** $\dfrac{\sqrt[5]{(m+n)^7}}{\sqrt[3]{(m+n)^4}}$ **b)** $\dfrac{\sqrt[3]{a^5b^4}}{\sqrt{a^2b}}$

Solution   Begin by writing the numerator and denominator with rational exponents.

a) $\dfrac{\sqrt[5]{(m+n)^7}}{\sqrt[3]{(m+n)^4}} = \dfrac{(m+n)^{7/5}}{(m+n)^{4/3}}$ Write with rational exponents.

$= (m+n)^{(7/5)-(4/3)}$ Quotient rule for exponents

$= (m+n)^{1/15}$

$= \sqrt[15]{m+n}$ Write as a radical.

b) $\dfrac{\sqrt[3]{a^5b^4}}{\sqrt{a^2b}} = \dfrac{(a^5b^4)^{1/3}}{(a^2b)^{1/2}}$ Write with rational exponents.

$= \dfrac{a^{5/3}b^{4/3}}{ab^{1/2}}$ Raise the product to a power.

$= a^{(5/3)-1}b^{(4/3)-(1/2)}$ Quotient rule for exponents

$= a^{2/3}b^{5/6}$

$= a^{4/6}b^{5/6}$ Write the fractions with denominator 6.

$= (a^4b^5)^{1/6}$ Rewrite using the laws of exponents.

$= \sqrt[6]{a^4b^5}$ Write as a radical.

Now Try Exercise 133

EXERCISE SET 7.5

Math XL MathXL®   *MyMathLab* MyMathLab

Warm-Up Exercises

Fill in the blanks with the appropriate word, phrase, or symbol(s) from the following list.

conjugate   sum   power   product   rationalizing   radicand   denominator   numerator

1. The process we use to rewrite the fraction without any radicals in the denominator is known as _____ the denominator.

2. When using the FOIL method on the product $(a+b)(a-b)$ the _____ of the outer and inner products is always 0.

3. To rationalize the denominator of a fraction containing a radical expression, multiply both the numerator and the denominator of the fraction by a radical that will produce a radicand in the denominator that is a perfect _____ for the index.

4. To rationalize the denominator using the conjugate, multiply the numerator and the denominator by the _____ of the denominator.

5. The expression $\sqrt{\dfrac{1}{2}}$ is not considered simplified since the _____ contains a fraction.

6. The expression $\dfrac{\sqrt{2}}{\sqrt{5}}$ is not considered simplified since the _____ contains a radical.

Practice the Skills

Simplify. In this exercise set, assume all variables represent positive real numbers.

7. $\dfrac{1}{\sqrt{2}}$

8. $\dfrac{1}{\sqrt{11}}$

9. $\dfrac{4}{\sqrt{5}}$

10. $\dfrac{3}{\sqrt{7}}$

11. $\dfrac{6}{\sqrt{6}}$

12. $\dfrac{17}{\sqrt{17}}$

13. $\dfrac{1}{\sqrt{z}}$

14. $\dfrac{y}{\sqrt{y}}$

15. $\dfrac{p}{\sqrt{2}}$

16. $\dfrac{m}{\sqrt{13}}$

17. $\dfrac{\sqrt{y}}{\sqrt{7}}$

18. $\dfrac{\sqrt{19}}{\sqrt{q}}$

19. $\dfrac{6\sqrt{3}}{\sqrt{6}}$

20. $\dfrac{15x}{\sqrt{x}}$

21. $\dfrac{\sqrt{x}}{\sqrt{y}}$

22. $\dfrac{2\sqrt{3}}{\sqrt{a}}$

23. $\sqrt{\dfrac{5m}{8}}$

24. $\dfrac{9\sqrt{3}}{\sqrt{y^3}}$

25. $\dfrac{2n}{\sqrt{18n}}$

26. $\sqrt{\dfrac{120x}{4y^3}}$

27. $\sqrt{\dfrac{18x^4y^3}{2z^3}}$

28. $\sqrt{\dfrac{7pq^4}{2r}}$

29. $\sqrt{\dfrac{20y^4z^3}{3xy^{-4}}}$

30. $\sqrt{\dfrac{5xy^6}{3z}}$

31. $\sqrt{\dfrac{48x^6y^5}{3z^3}}$

32. $\sqrt{\dfrac{45y^{12}z^{10}}{2x}}$

Simplify.

33. $\dfrac{1}{\sqrt[3]{2}}$

34. $\dfrac{1}{\sqrt[3]{4}}$

35. $\dfrac{8}{\sqrt[3]{y}}$

36. $\dfrac{2}{\sqrt[3]{a^2}}$

37. $\dfrac{1}{\sqrt[4]{2}}$

38. $\dfrac{1}{\sqrt[4]{4}}$

39. $\dfrac{a}{\sqrt[4]{8}}$

40. $\dfrac{8}{\sqrt[4]{z}}$

41. $\dfrac{5}{\sqrt[4]{z^2}}$

42. $\dfrac{13}{\sqrt[4]{z^3}}$

43. $\dfrac{10}{\sqrt[5]{y^3}}$

44. $\dfrac{x}{\sqrt[5]{y^4}}$

45. $\dfrac{2}{\sqrt[7]{a^4}}$

46. $\sqrt[3]{\dfrac{4x}{y}}$

47. $\sqrt[3]{\dfrac{1}{2x}}$

48. $\sqrt[3]{\dfrac{7c}{9y^2}}$

49. $\dfrac{5m}{\sqrt[4]{2}}$

50. $\dfrac{3}{\sqrt[4]{a}}$

51. $\sqrt[4]{\dfrac{5}{3x^3}}$

52. $\sqrt[4]{\dfrac{2x^3}{4y^2}}$

53. $\sqrt[3]{\dfrac{3x^2}{2y^2}}$

54. $\sqrt[3]{\dfrac{15x^6y^7}{2z^2}}$

55. $\sqrt[3]{\dfrac{14xy^2}{2z^2}}$

56. $\sqrt[6]{\dfrac{r^4s^9}{2r^5}}$

Multiply.

57. $(4 + \sqrt{5})(4 - \sqrt{5})$

58. $(2 + \sqrt{7})(2 - \sqrt{7})$

59. $(8 + \sqrt{2})(8 - \sqrt{2})$

60. $(6 - \sqrt{7})(6 + \sqrt{7})$

61. $(2 - \sqrt{10})(2 + \sqrt{10})$

62. $(3 + \sqrt{17})(3 - \sqrt{17})$

63. $(\sqrt{a} - \sqrt{b})(\sqrt{a} + \sqrt{b})$

64. $(\sqrt{x} - \sqrt{y})(\sqrt{x} + \sqrt{y})$

65. $(2\sqrt{x} - 3\sqrt{y})(2\sqrt{x} + 3\sqrt{y})$

66. $(5\sqrt{c} - 4\sqrt{d})(5\sqrt{c} + 4\sqrt{d})$

Simplify by rationalizing the denominator.

67. $\dfrac{1}{\sqrt{2} + 1}$

68. $\dfrac{4}{\sqrt{3} - 1}$

69. $\dfrac{1}{2 + \sqrt{3}}$

70. $\dfrac{3}{5 - \sqrt{7}}$

71. $\dfrac{5}{\sqrt{2} - 7}$

72. $\dfrac{6}{\sqrt{2} + \sqrt{3}}$

73. $\dfrac{\sqrt{5}}{2\sqrt{5} - \sqrt{6}}$

74. $\dfrac{1}{\sqrt{17} - \sqrt{8}}$

75. $\dfrac{3}{6 + \sqrt{x}}$

76. $\dfrac{4\sqrt{5}}{\sqrt{a} - 3}$

77. $\dfrac{4\sqrt{x}}{\sqrt{x} - y}$

78. $\dfrac{\sqrt{8x}}{x + \sqrt{y}}$

79. $\dfrac{\sqrt{2} - 2\sqrt{3}}{\sqrt{2} + 4\sqrt{3}}$

80. $\dfrac{\sqrt{c} - \sqrt{2d}}{\sqrt{c} - \sqrt{d}}$

81. $\dfrac{\sqrt{a^3} + \sqrt{a^7}}{\sqrt{a}}$

82. $\dfrac{2\sqrt{xy} - \sqrt{xy}}{\sqrt{x} + \sqrt{y}}$

83. $\dfrac{4}{\sqrt{x + 2} - 3}$

84. $\dfrac{8}{\sqrt{y - 3} + 6}$

Simplify. These exercises are a combination of the types of exercises presented earlier in this exercise set.

85. $\sqrt{\dfrac{a^2}{25}}$

86. $\sqrt[3]{\dfrac{a^3}{8}}$

87. $\sqrt{\dfrac{2}{9}}$

88. $\sqrt{\dfrac{a}{b}}$

89. $(\sqrt{7} + \sqrt{6})(\sqrt{7} - \sqrt{6})$

90. $\sqrt[3]{\dfrac{1}{16}}$

91. $\sqrt{\dfrac{24x^3y^6}{5z}}$

92. $\dfrac{5}{4 - \sqrt{y}}$

93. $\sqrt{\dfrac{28xy^4}{2x^3y^4}}$

94. $\dfrac{8x}{\sqrt[3]{5y}}$

95. $\dfrac{1}{\sqrt{a} + 7}$

96. $\dfrac{\sqrt{x}}{\sqrt{x} + 6\sqrt{y}}$

97. $-\dfrac{7\sqrt{x}}{\sqrt{98}}$

98. $\sqrt{\dfrac{2xy^4}{50xy^2}}$

99. $\sqrt[4]{\dfrac{3y^2}{2x}}$

100. $\sqrt{\dfrac{49x^2y^5}{3z}}$

101. $\sqrt[3]{\dfrac{32y^{12}z^{10}}{2x}}$

102. $\dfrac{\sqrt{3} + 2}{\sqrt{2} + \sqrt{3}}$

103. $\dfrac{\sqrt{ar}}{\sqrt{a} - 2\sqrt{r}}$

104. $\sqrt[4]{\dfrac{2}{9x}}$

105. $\dfrac{\sqrt[3]{6x}}{\sqrt[3]{5xy}}$

106. $\dfrac{\sqrt[3]{16m^2n}}{\sqrt[3]{2mn^2}}$

107. $\sqrt[4]{\dfrac{2x^7y^{12}z^4}{3x^9}}$

108. $\dfrac{9}{\sqrt{y + 9} - \sqrt{y}}$

Simplify.

109. $\dfrac{1}{\sqrt{2}} + \dfrac{\sqrt{2}}{2}$

110. $\dfrac{1}{\sqrt{3}} + \dfrac{\sqrt{3}}{3}$

111. $\sqrt{5} - \dfrac{2}{\sqrt{5}}$

112. $\dfrac{\sqrt{6}}{2} - \dfrac{2}{\sqrt{6}}$

113. $4\sqrt{\dfrac{1}{6}} + \sqrt{24}$

114. $5\sqrt{3} - \dfrac{3}{\sqrt{3}} + 2\sqrt{18}$

115. $5\sqrt{2} - \dfrac{2}{\sqrt{8}} + \sqrt{50}$

116. $\dfrac{2}{3} + \dfrac{1}{\sqrt{3}} + \sqrt{75}$

117. $\sqrt{\dfrac{1}{2}} + 7\sqrt{2} + \sqrt{18}$

118. $\dfrac{1}{2}\sqrt{18} - \dfrac{3}{\sqrt{2}} - 9\sqrt{50}$

119. $\dfrac{2}{\sqrt{50}} - 3\sqrt{50} - \dfrac{1}{\sqrt{8}}$

120. $\dfrac{\sqrt{3}}{3} + \dfrac{5}{\sqrt{3}} + \sqrt{12}$

121. $\sqrt{\dfrac{3}{8}} + \sqrt{\dfrac{3}{2}}$

122. $2\sqrt{\dfrac{8}{3}} - 4\sqrt{\dfrac{100}{6}}$

123. $-2\sqrt{\dfrac{x}{y}} + 3\sqrt{\dfrac{y}{x}}$

124. $-5x\sqrt{\dfrac{y}{y^2}} + 9x\sqrt{\dfrac{1}{y}}$

125. $\dfrac{3}{\sqrt{a}} - \sqrt{\dfrac{9}{a}} + 2\sqrt{a}$

126. $6\sqrt{x} + \dfrac{1}{\sqrt{x}} + \sqrt{\dfrac{1}{x}}$

Simplify.

127. $\dfrac{\sqrt{(a+b)^4}}{\sqrt[3]{a+b}}$

128. $\dfrac{\sqrt[3]{c+2}}{\sqrt[4]{(c+2)^3}}$

129. $\dfrac{\sqrt[5]{(a+2b)^4}}{\sqrt[3]{(a+2b)^2}}$

130. $\dfrac{\sqrt[6]{(r+3)^5}}{\sqrt[3]{(r+3)^5}}$

131. $\dfrac{\sqrt[3]{r^2 s^4}}{\sqrt{rs}}$

132. $\dfrac{\sqrt{a^2 b^4}}{\sqrt[3]{ab^2}}$

133. $\dfrac{\sqrt[5]{x^4 y^6}}{\sqrt[3]{(xy)^2}}$

134. $\dfrac{\sqrt[6]{4m^8 n^4}}{\sqrt[4]{m^4 n^2}}$

Problem Solving

135. Illumination of a Light Under certain conditions the formula

$$d = \sqrt{\dfrac{72}{I}}$$

is used to show the relationship between the illumination on an object, I, in lumens per meter, and the distance, d, in meters, the object is from the light source. If the illumination on a person standing near a light source is 5.3 lumens per meter, how far is the person from the light source?

© Allen R. Angel

136. Strength of a Board When sufficient pressure is applied to a particular particle board, the particle board will break (or rupture). The thicker the particle board the greater will be the pressure that will need to be applied before the board breaks. The formula

$$T = \sqrt{\dfrac{0.05\,LB}{M}}$$

relates the thickness of a specific particle board, T, in inches, the board's length, L, in inches, the board's load that will cause the board to rupture, B, in pounds, and the modulus of rupture, M, in pounds per square inch. The modulus of rupture is a constant determined by sample tests on the specific type of particle board.

Find the thickness of a 36-inch-long particle board if the modulus of rupture is 2560 pounds per square inch and the board ruptures when 800 pounds are applied.

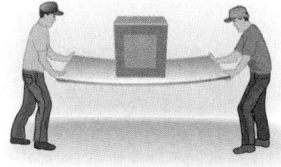

137. Volume of a Fish Tank A new restaurant wants to have a spherical fish tank in its lobby. The radius, r, in inches, of a spherical tank can be found by the formula

$$r = \sqrt[3]{\dfrac{3V}{4\pi}}$$

where V is the volume of the tank in cubic inches. Find the radius of a spherical tank whose volume is 7238.23 cubic inches.

138. Consecutive Numbers If we consider the set of consecutive natural numbers $1, 2, 3, 4, \ldots, n$ to be the population, the standard deviation, σ, which is a measure of the spread of the data from the mean, can be calculated by the formula

$$\sigma = \sqrt{\dfrac{n^2 - 1}{12}}$$

where n represents the number of natural numbers in the population. Find the standard deviation for the first 100 consecutive natural numbers.

139. U.S. Farms The number of farms in the United States is declining annually (however, the size of the remaining farms is increasing). A function that can be used to estimate the number of farms, $N(t)$, in millions, is

$$N(t) = \dfrac{6.21}{\sqrt[4]{t}}$$

where t is years since 1959 and $1 \le t \le 50$. Estimate the number of farms in the United States in **a)** 1960 and **b)** 2008.

© Allen R. Angel

140. Infant Mortality Rate The U.S. infant mortality rate has been declining steadily. The infant mortality rate, $N(t)$, defined as deaths per 1000 live births, can be estimated by the function

$$N(t) = \dfrac{28.46}{\sqrt[3]{t^2}}$$

where t is years since 1969 and $1 \le t \le 37$. Estimate the infant mortality rate in **a)** 1970 and **b)** 2006.

In higher math courses, it may be necessary to rationalize the numerators of radical expressions. Rationalize the numerators of the following expressions. (Your answers will contain radicals in the denominators.)

141. $\dfrac{5 - \sqrt{5}}{6}$

142. $\dfrac{\sqrt{7}}{3}$

143. $\dfrac{\sqrt{x + h} - \sqrt{x}}{h}$

144. $\dfrac{6\sqrt{x} - \sqrt{3}}{x}$

Concept/Writing Exercises

145. Which is greater, $\dfrac{2}{\sqrt{2}}$ or $\dfrac{3}{\sqrt{3}}$? Explain.

146. Which is greater, $\dfrac{\sqrt{3}}{2}$ or $\dfrac{2}{\sqrt{3}}$? Explain.

147. Which is greater, $\dfrac{1}{\sqrt{3} + 2}$ or $2 + \sqrt{3}$? (Do not use a calculator.) Explain how you determined your answer.

148. Which is greater, $\dfrac{1}{\sqrt{3}} + \sqrt{75}$ or $\dfrac{2}{\sqrt{12}} + \sqrt{48} + 2\sqrt{3}$? (Do not use a calculator.) Explain how you determined your answer.

149. Consider the functions $f(x) = x^{a/2}$ and $g(x) = x^{b/3}$.
 a) List three values for a that will result in $x^{a/2}$ being a perfect square.
 b) List three values for b that will result in $x^{b/3}$ being a perfect cube.
 c) If $x \geq 0$, find $(f \cdot g)(x)$.
 d) If $x \geq 0$, find $(f/g)(x)$.

Group Activity

Similar Figures *The following two exercises will reinforce many of the concepts presented in this chapter. Work each problem as a group. Make sure each member of the group understands each step in obtaining the solution. The figures in each exercise are similar. For each exercise, use a proportion to find the length of side x. Write the answer in radical form with a rationalized denominator.*

150.

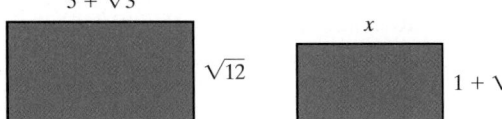

151.

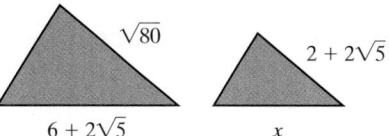

Cumulative Review Exercises

[2.2] **152.** Solve the equation $A = \dfrac{1}{2}h(b_1 + b_2)$ for b_2.

[2.4] **153. Moving Vehicles** Two cars leave from West Point at the same time traveling in opposite directions. One travels 10 miles per hour faster than the other. If the two cars are 270 miles apart after 3 hours, find the speed of each car.

[5.2] **154.** Multiply $(x - 2)(4x^2 + 9x - 2)$.

[6.4] **155.** Solve $\dfrac{x}{2} - \dfrac{4}{x} = -\dfrac{7}{2}$.

7.6 Solving Radical Equations

1 Solve equations containing one radical.

2 Solve equations containing two radicals.

3 Solve equations containing two radical terms and a nonradical term.

4 Solve applications using radical equations.

5 Solve for a variable in a radicand.

1 Solve Equations Containing One Radical

Radical Equation

A **radical equation** is an equation that contains a variable in a radicand.

Examples of Radical Equations

$$\sqrt{x} = 5, \qquad \sqrt[3]{y + 4} = 9, \qquad \sqrt{x - 2} = 7 + \sqrt{x + 8}$$

To Solve Radical Equations

1. Rewrite the equation so that one radical containing a variable is by itself (isolated) on one side of the equation.
2. Raise each side of the equation to a power equal to the index of the radical.
3. Combine like terms.
4. If the equation still contains a term with a variable in a radicand, repeat steps 1 through 3.
5. Solve the resulting equation for the variable.
6. Check all solutions in the original equation for extraneous solutions.

Whenever we raise both sides of an equation to an even power, we could be introducing an extraneous, or false, solution. Therefore, we must check the results in the original equation.

EXAMPLE 1   Solve the equation $\sqrt{x} = 5$.

Solution   The square root containing the variable is already by itself on one side of the equation. Square both sides of the equation.

$$\sqrt{x} = 5$$
$$(\sqrt{x})^2 = (5)^2$$
$$x = 25$$

Check $\sqrt{x} = 5$
$$\sqrt{25} \overset{?}{=} 5$$
$$5 = 5 \quad \text{True}$$

Now Try Exercise 11

EXAMPLE 2   Solve.

a) $\sqrt{x - 4} - 6 = 0$ **b)** $\sqrt[3]{x} + 10 = 8$ **c)** $\sqrt{x} + 3 = 0$

Solution   The first step in each part will be to isolate the term containing the radical.

a)
$$\sqrt{x - 4} - 6 = 0$$
$$\sqrt{x - 4} = 6 \qquad \text{Isolate the radical containing the variable.}$$
$$(\sqrt{x - 4})^2 = 6^2 \qquad \text{Square both sides.}$$
$$x - 4 = 36 \qquad \text{Solve for the variable.}$$
$$x = 40$$

A check will show that 40 is the solution.

b)
$$\sqrt[3]{x} + 10 = 8$$
$$\sqrt[3]{x} = -2 \qquad \text{Isolate the radical containing the variable.}$$
$$(\sqrt[3]{x})^3 = (-2)^3 \qquad \text{Cube both sides.}$$
$$x = -8$$

A check will show that -8 is the solution.

c)
$$\sqrt{x} + 3 = 0$$
$$\sqrt{x} = -3 \qquad \text{Isolate the radical containing the variable.}$$
$$(\sqrt{x})^2 = (-3)^2 \qquad \text{Square both sides.}$$
$$x = 9$$

Check $\sqrt{x} + 3 = 0$
$$\sqrt{9} + 3 \overset{?}{=} 0$$
$$3 + 3 \overset{?}{=} 0$$
$$6 = 0 \quad \text{False}$$

A check shows that 9 is not a solution. The answer to part **c)** is "no real solution." You may have realized there was no real solution to the problem when you obtained the equation $\sqrt{x} = -3$, because $\sqrt{x}$ cannot equal a negative real number.

Now Try Exercise 17

Understanding Algebra

Whenever we raise both sides of an equation to an *even* power, we could be introducing an extraneous or false solution. Therefore we must check our answers. Example 2 **c)** shows that when we square both sides of the equation $\sqrt{x} = -3$ we obtain the result $x = 9$. By checking this result in the original equation we are able to see that 9 is an extraneous solution.

EXAMPLE 3   Solve $\sqrt{2x - 3} = x - 3$.

Solution   Since the radical is already isolated, we square both sides of the equation, resulting in a quadratic equation.

$$(\sqrt{2x - 3})^2 = (x - 3)^2$$
$$2x - 3 = (x - 3)(x - 3)$$
$$2x - 3 = x^2 - 6x + 9$$
$$0 = x^2 - 8x + 12$$

Now we factor and use the zero-factor property to solve the quadratic equation.

$$x^2 - 8x + 12 = 0$$
$$(x - 6)(x - 2) = 0$$
$$x - 6 = 0 \quad \text{or} \quad x - 2 = 0$$
$$x = 6 \qquad\qquad x = 2$$

Check

$x = 6$	$x = 2$
$\sqrt{2x - 3} = x - 3$	$\sqrt{2x - 3} = x - 3$
$\sqrt{2(6) - 3} \stackrel{?}{=} 6 - 3$	$\sqrt{2(2) - 3} \stackrel{?}{=} 2 - 3$
$\sqrt{9} \stackrel{?}{=} 3$	$\sqrt{1} \stackrel{?}{=} -1$
$3 = 3$ True	$1 = -1$ False

Thus, 6 is a solution, but 2 is not a solution to the equation. The 2 is an extraneous solution because 2 satisfies the equation $(\sqrt{2x - 3})^2 = (x - 3)^2$, but not the original equation, $\sqrt{2x - 3} = x - 3$.

Now Try Exercise 43

Using Your Graphing Calculator ▪▪▪▫▫

We can use a graphing calculator to solve or check radical equations. In Example 3, we found the solution to $\sqrt{2x - 3} = x - 3$ to be 6. By graphing both

$$Y_1 = \sqrt{2x - 3} \qquad \text{and} \qquad Y_2 = x - 3$$

on a graphing calculator, we can see from the graph in **Figure 7.3** that the two functions appear to intersect at $x = 6$.

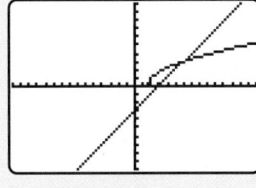

FIGURE 7.3

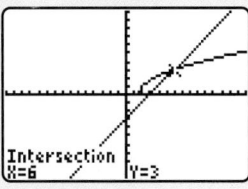

FIGURE 7.4

We can also use the *intersect* feature to determine the solutions to radical equations. **Figure 7.4** shows that the two graphs intersect when $x = 6$ and $y = 3$.

EXAMPLE 4   Solve $x - 2\sqrt{x} - 3 = 0$.

Solution   First, isolate the radical term by writing the radical term by itself on one side of the equation.

$$x - 2\sqrt{x} - 3 = 0$$
$$-2\sqrt{x} = -x + 3$$
$$2\sqrt{x} = x - 3$$

Now square both sides of the equation.

$$(2\sqrt{x})^2 = (x - 3)^2$$
$$4x = x^2 - 6x + 9$$
$$0 = x^2 - 10x + 9$$
$$0 = (x - 1)(x - 9)$$
$$x - 1 = 0 \quad \text{or} \quad x - 9 = 0$$
$$x = 1 \qquad\qquad x = 9$$

Check

$x = 1$	$x = 9$
$x - 2\sqrt{x} - 3 = 0$	$x - 2\sqrt{x} - 3 = 0$
$1 - 2\sqrt{1} - 3 \overset{?}{=} 0$	$9 - 2\sqrt{9} - 3 \overset{?}{=} 0$
$1 - 2(1) - 3 \overset{?}{=} 0$	$9 - 2(3) - 3 \overset{?}{=} 0$
$1 - 2 - 3 \overset{?}{=} 0$	$9 - 6 - 3 \overset{?}{=} 0$
$-4 = 0$ False	$3 - 3 \overset{?}{=} 0$
	$0 = 0$ True

The solution is 9. The value 1 is an extraneous solution.

Now Try Exercise 41

2 Solve Equations Containing Two Radicals

Now we will look at some equations that contain two radicals.

EXAMPLE 5 Solve $\sqrt{9x^2 + 6} = 3\sqrt{x^2 + x - 2}$.

Solution   Since the two radicals appear on different sides of the equation, we square both sides of the equation.

$$\left(\sqrt{9x^2 + 6}\right)^2 = \left(3\sqrt{x^2 + x - 2}\right)^2 \quad \text{Square both sides.}$$
$$9x^2 + 6 = 9(x^2 + x - 2)$$
$$9x^2 + 6 = 9x^2 + 9x - 18 \qquad \text{Distributive property.}$$
$$6 = 9x - 18 \qquad\qquad 9x^2 \text{ was subtracted from both sides.}$$
$$24 = 9x$$
$$\frac{8}{3} = x$$

A check will show that $\frac{8}{3}$ is the solution.

Now Try Exercise 27

Equations are sometimes given using exponents rather than radicals. Example 6 illustrates such an equation.

EXAMPLE 6 For $f(x) = 3(x - 2)^{1/3}$ and $g(x) = (17x - 14)^{1/3}$, find all values of x for which $f(x) = g(x)$.

Solution   We set the two functions equal to each other and solve for x.

$$f(x) = g(x)$$
$$3(x - 2)^{1/3} = (17x - 14)^{1/3}$$
$$[3(x - 2)^{1/3}]^3 = [(17x - 14)^{1/3}]^3 \quad \text{Cube both sides.}$$
$$3^3(x - 2) = 17x - 14$$
$$27(x - 2) = 17x - 14$$
$$27x - 54 = 17x - 14$$
$$10x - 54 = -14$$
$$10x = 40$$
$$x = 4$$

Understanding Algebra

The functions in Example 6 could also be written using radicals. Thus, we could also answer the question by solving the equation

$$3\sqrt[3]{x - 2} = \sqrt[3]{17x - 14}$$

A check will show that the solution is 4. If you substitute 4 into both $f(x)$ and $g(x)$, you will find they both simplify to $3\sqrt[3]{2}$. Check this now.

Now Try Exercise 69

3 Solve Equations Containing Two Radical Terms and a Nonradical Term

When a radical equation contains two radical terms and a third nonradical term, you will sometimes need to raise both sides of the equation to a given power twice to obtain the solution. This procedure is illustrated in Example 7.

EXAMPLE 7 Solve $\sqrt{5x - 1} - \sqrt{3x - 2} = 1$.

Solution   We must isolate one radical term on one side of the equation. We will begin by adding $\sqrt{3x - 2}$ to both sides of the equation to isolate $\sqrt{5x - 1}$. Then we will square both sides of the equation and combine like terms.

$\sqrt{5x - 1} = 1 + \sqrt{3x - 2}$	Isolate $\sqrt{5x - 1}$.
$(\sqrt{5x - 1})^2 = (1 + \sqrt{3x - 2})^2$	Square both sides.
$5x - 1 = (1 + \sqrt{3x - 2})(1 + \sqrt{3x - 2})$	Write as a product.
$5x - 1 = 1 + \sqrt{3x - 2} + \sqrt{3x - 2} + (\sqrt{3x - 2})^2$	Multiply.
$5x - 1 = 1 + 2\sqrt{3x - 2} + 3x - 2$	Combine like terms; simplify.
$5x - 1 = 3x - 1 + 2\sqrt{3x - 2}$	Combine like terms.
$2x = 2\sqrt{3x - 2}$	Isolate the radical term.
$x = \sqrt{3x - 2}$	Both sides were divided by 2.

We have isolated the remaining radical term. We now square both sides of the equation again and solve for x.

$$x = \sqrt{3x - 2}$$
$$x^2 = (\sqrt{3x - 2})^2 \quad \text{Square both sides.}$$
$$x^2 = 3x - 2$$
$$x^2 - 3x + 2 = 0$$
$$(x - 2)(x - 1) = 0$$
$$x - 2 = 0 \quad \text{or} \quad x - 1 = 0$$
$$x = 2 \qquad\qquad x = 1$$

A check will show that both 2 and 1 are solutions of the equation.

Now Try Exercise 61

EXAMPLE 8 For $f(x) = \sqrt{5x - 1} - \sqrt{3x - 2}$, find all values of x for which $f(x) = 1$.

Solution   Substitute 1 for $f(x)$. This gives

$$1 = \sqrt{5x - 1} - \sqrt{3x - 2}$$

Since this is the same equation we solved in Example 7, the answers are $x = 2$ and $x = 1$. Verify for yourself that $f(2) = 1$ and $f(1) = 1$.

Now Try Exercise 121

> **Avoiding Common Errors**
>
> In Chapter 5, we stated that $(a + b)^2 \neq a^2 + b^2$. Be careful when squaring a binomial like $1 + \sqrt{x}$. Look at the following computations carefully so that you do not make the mistake shown on the right.
>
> | CORRECT | INCORRECT |
> |---|---|
> | $(1 + \sqrt{x})^2 = (1 + \sqrt{x})(1 + \sqrt{x})$ | $\cancel{(1 + \sqrt{x})^2 = 1^2 + (\sqrt{x})^2}$ |
> | $\quad\quad\quad\quad$ F $\quad$ O $\quad$ I $\quad\quad$ L | $\cancel{= 1 + x}$ |
> | $\quad\quad = 1 + \sqrt{x} + \sqrt{x} + \sqrt{x}\,\sqrt{x}$ | |
> | $\quad\quad = 1 + 2\sqrt{x} + x$ | |

4 Solve Applications Using Radical Equations

Now we will look at a few of the many applications of radicals.

© Allen R. Angel

EXAMPLE 9 **The Green Monster** In Fenway Park, where the Boston Red Sox play baseball, the distance from home plate down the third base line to the bottom of the wall in left field is 310 feet. In left field at the end of the baseline there is a green wall perpendicular to the ground that is 37 feet tall. This green wall is commonly known as *the Green Monster* (see photo). Determine the distance from home plate to the top of the Green Monster along the third base line.

Solution Understand **Figure 7.5** illustrates the problem. We need to find the distance from home plate to the top of the wall in left field.

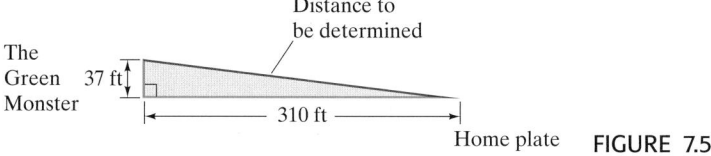

FIGURE 7.5

Translate To solve the problem we use the Pythagorean Theorem, which was discussed earlier, $\text{leg}^2 + \text{leg}^2 = \text{hyp}^2$, or $a^2 + b^2 = c^2$.

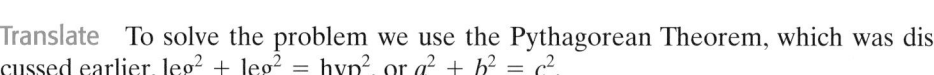

$$310^2 + 37^2 = c^2 \quad\quad \text{Substitute known values.}$$

Carry Out
$$96{,}100 + 1369 = c^2$$
$$97{,}469 = c^2$$
$$\sqrt{97{,}469} = \sqrt{c^2} \quad\quad \text{Take the square root of both sides.}$$
$$\sqrt{97{,}469} = c \quad\quad\quad \text{* See footnote.}$$
$$312.20 \approx c$$

Answer The distance from home plate to the top of the wall is about 312.20 feet.

Now Try Exercise 99

EXAMPLE 10 **Period of a Pendulum** The length of time it takes for a pendulum to make one complete swing back and forth is called the *period* of the pendulum. See **Figure 7.6.** The period of a pendulum, T, in seconds, can be found by the formula $T = 2\pi\sqrt{\dfrac{L}{32}}$, where L is the length of the pendulum in feet. Find the period of a pendulum if its length is 5 feet.

$^*c^2 = 97{,}469$ has two solutions: $c = \sqrt{97{,}469}$ and $c = -\sqrt{97{,}469}$. Since we are solving for a length, which must be a positive quantity, we use the positive root.

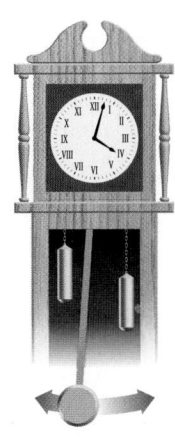

FIGURE 7.6

Solution Substitute 5 for L and 3.14 for π in the formula. If you have a calculator that has a $\boxed{\pi}$ key, use it to enter π.

$$T = 2\pi\sqrt{\frac{L}{32}}$$

$$\approx 2(3.14)\sqrt{\frac{5}{32}}$$

$$\approx 2(3.14)\sqrt{0.15625} \approx 2.48$$

Thus, the period is about 2.48 seconds. If you have a grandfather clock with a 5-foot pendulum, it will take about 2.48 seconds for it to swing back and forth.

Now Try Exercise 103

5 Solve for a Variable in a Radicand

Many formulas involve radicals that contain variables. It is sometimes necessary to solve for a variable that is part of a radicand. To do so, follow the same general procedure we have used thus far.

EXAMPLE 11 **Margin of Error** A formula in statistics for finding the margin of error is $E = t\dfrac{s}{\sqrt{n}}$.

a) Find E if $t = 2.064$, $s = 15$, and $n = 25$

b) Solve this equation for n.

Solution

a) $E = t\dfrac{s}{\sqrt{n}} = 2.064\left(\dfrac{15}{\sqrt{25}}\right) = 2.064\left(\dfrac{15}{5}\right) = 6.192$

b) First eliminate the fraction by multiplying both sides of the equation by $\sqrt{n}$.

$$E = t\frac{s}{\sqrt{n}}$$

$$E\sqrt{n} = \left(t\frac{s}{\sqrt{n}}\right)\sqrt{n} \qquad \text{Eliminate the fraction.}$$

$$E\sqrt{n} = ts$$

$$\sqrt{n} = \frac{ts}{E} \qquad \text{Isolate the radical by dividing both sides by } E.$$

$$(\sqrt{n})^2 = \left(\frac{ts}{E}\right)^2 \qquad \text{Square both sides.}$$

$$n = \left(\frac{ts}{E}\right)^2 \quad \text{or} \quad n = \frac{t^2 s^2}{E^2}$$

Now Try Exercise 75

EXERCISE SET 7.6

Math XL
MathXL®

MyMathLab
MyMathLab

Warm-Up Exercises

Fill in the blanks with the appropriate word, phrase, or symbol(s) from the following list.

original   radical   isolated   raise   square   index   cube   extraneous   root   radicand

1. A _____ equation is an equation that contains a variable in the radicand.

2. Whenever we raise both sides of an equation to an even power, we could be introducing an _____, or false, solution.

3. To determine whether the result of solving an equation is an extraneous solution, we must check the result in the _____ equation.

4. When solving a radical equation, we rewrite the equation so one radical containing a variable is _____.

5. To eliminate an isolated radical from an equation, raise each side of the equation to the _____ of the radical.

6. To eliminate the radical in the equation $\sqrt{2x-5}=7$, _____ both sides of the equation.

7. To eliminate the radical in the equation $\sqrt[3]{x+1}=2$, _____ both sides of the equation.

8. To eliminate the radical in the equation $\sqrt[4]{x-1}=3$, _____ both sides of the equation to the fourth power.

Practice the Skills

Solve and check your solution(s). If the equation has no real solution, so state.

9. $\sqrt{x}=4$

10. $\sqrt{x}=2$

11. $\sqrt{x}=-2$

12. $\sqrt[3]{x}=2$

13. $\sqrt[3]{x}=-4$

14. $\sqrt{a}+5=0$

15. $\sqrt{2x+3}=5$

16. $\sqrt[3]{7x-6}=4$

17. $\sqrt[3]{3x}+4=7$

18. $2\sqrt{4x+5}=14$

19. $\sqrt[3]{2x+29}=3$

20. $\sqrt[3]{6x+2}=-4$

21. $\sqrt[4]{x}=3$

22. $\sqrt[4]{x}=-3$

23. $\sqrt[4]{x+10}=3$

24. $\sqrt[4]{3x-2}=2$

25. $\sqrt{x-1}+3=8$

26. $\sqrt{x+3}+1=5$

27. $\sqrt{x+8}=\sqrt{x-8}$

28. $\sqrt{r+5}+7=10$

29. $2\sqrt[3]{x-1}=\sqrt[3]{x^2+2x}$

30. $\sqrt[3]{6t-1}=\sqrt[3]{2t+3}$

31. $\sqrt[4]{x+8}=\sqrt[4]{2x}$

32. $\sqrt[4]{3x-1}+4=0$

33. $\sqrt{5x+1}-6=0$

34. $\sqrt{x^2+12x+3}=-x$

35. $\sqrt{m^2+6m-4}=m$

36. $\sqrt{x^2+3x+12}=x$

37. $\sqrt{5c+1}-9=0$

38. $\sqrt{b^2-2}=b+4$

39. $\sqrt{z^2+5}=z+1$

40. $\sqrt{x+6x}=1$

41. $\sqrt{2y+5}+5-y=0$

42. $\sqrt{4x+1}=\dfrac{1}{2}x+2$

43. $\sqrt{5x+6}=2x-6$

44. $\sqrt{4b+5}+b=10$

45. $(2a+9)^{1/2}-a+3=0$

46. $(3x+4)^{1/2}-x=-2$

47. $(2x^2+4x+9)^{1/2}=(2x^2+9)^{1/2}$

48. $(2x+1)^{1/2}+7=x$

49. $(r+4)^{1/3}=(3r+10)^{1/3}$

50. $(7x+6)^{1/3}+4=0$

51. $(5x+7)^{1/4}=(9x+1)^{1/4}$

52. $(5b+3)^{1/4}=(2b+17)^{1/4}$

53. $\sqrt[4]{x+5}=-2$

54. $\sqrt{x^2+x-1}=-\sqrt{x+3}$

Solve. You will have to square both sides of the equation twice to eliminate all radicals.

55. $\sqrt{x}+1=\sqrt{x+1}$

56. $\sqrt{x}-1=\sqrt{x-1}$

57. $\sqrt{3a+1}=\sqrt{a-4}+3$

58. $\sqrt{x+1}=2-\sqrt{x}$

59. $\sqrt{x+3}=\sqrt{x}-3$

60. $\sqrt{y+1}=2+\sqrt{y-7}$

61. $\sqrt{x+7}=6-\sqrt{x-5}$

62. $\sqrt{b-3}=4-\sqrt{b+5}$

63. $\sqrt{4x-3}=2+\sqrt{2x-5}$

64. $\sqrt{r+10}+2+\sqrt{r-5}=0$

65. $\sqrt{y+1}=\sqrt{y+10}-3$

66. $3+\sqrt{x+1}=\sqrt{3x+12}$

For each pair of functions, find all real values of x where $f(x)=g(x)$.

67. $f(x)=\sqrt{2x+3}, g(x)=\sqrt{x+5}$

68. $f(x)=\sqrt{x^2-2x}, g(x)=\sqrt{x-2}$

69. $f(x)=\sqrt[3]{5x-19}, g(x)=\sqrt[3]{6x-23}$

70. $f(x)=(14x-8)^{1/2}, g(x)=2(3x+2)^{1/2}$

71. $f(x)=2(8x+24)^{1/3}, g(x)=4(2x-2)^{1/3}$

72. $f(x)=2\sqrt{x+2}, g(x)=8-\sqrt{x+14}$

Solve each formula for the indicated variable.

73. $p=\sqrt{2v}$, for v

74. $l=\sqrt{4r}$, for r

75. $v=\sqrt{2gh}$, for g

76. $v=\sqrt{\dfrac{2E}{m}}$, for E

77. $v=\sqrt{\dfrac{FR}{M}}$, for F

78. $\omega=\sqrt{\dfrac{a_0}{b_0}}$, for b_0

79. $x=\sqrt{\dfrac{m}{k}}V_0$, for m

80. $T=2\pi\sqrt{\dfrac{L}{32}}$, for L

81. $r=\sqrt{\dfrac{A}{\pi}}$, for A

82. $r=\sqrt[3]{\dfrac{3V}{4\pi}}$, for V

Problem Solving

Use the Pythagorean Theorem to find the length of the unknown side of each triangle. Write the answer as a radical in simplified form.

83.

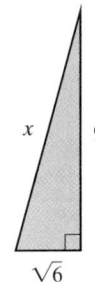

84.

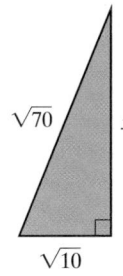

85.

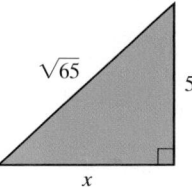

86.

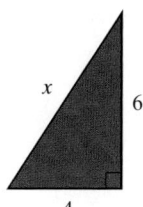

Solve. You will need to square both sides of the equation twice.

87. $\sqrt{x+3} + \sqrt{x} = \sqrt{x+8}$

88. $\sqrt{2x} - \sqrt{x-4} = \sqrt{12-x}$

89. $\sqrt{4y+6} + \sqrt{y+5} = \sqrt{y+1}$

90. $\sqrt{2b-2} + \sqrt{b-5} = \sqrt{4b}$

91. $\sqrt{c+1} + \sqrt{c-2} = \sqrt{3c}$

92. $\sqrt{2t-1} + \sqrt{t-4} = \sqrt{3t+1}$

93. $\sqrt{a+2} - \sqrt{a-3} = \sqrt{a-6}$

94. $\sqrt{r-1} - \sqrt{r+6} = \sqrt{r-9}$

Solve. You will need to square both sides of the equation twice.

95. $\sqrt{2 - \sqrt{x}} = \sqrt{x}$

96. $\sqrt{6 + \sqrt{x+4}} = \sqrt{2x-1}$

97. $\sqrt{2 + \sqrt{x+1}} = \sqrt{7-x}$

98. $\sqrt{1 + \sqrt{x-1}} = \sqrt{x-6}$

99. Baseball Diamond A regulation baseball diamond is a square with 90 feet between bases. How far is second base from home plate?

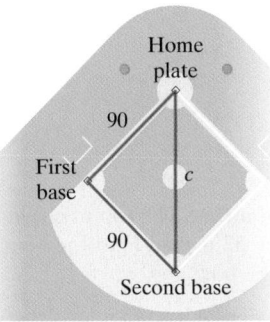

100. Little League Baseball Diamond A Little League baseball diamond is a square with 60 feet between bases (see Exercise 99). How far is second base from home plate?

101. Wire from Telephone Pole A telephone pole is at a right, or 90°, angle with the ground as shown in the figure. Find the length of the wire that connects to the pole 40 feet above the ground and is anchored to the ground 20 feet from the base of the pole.

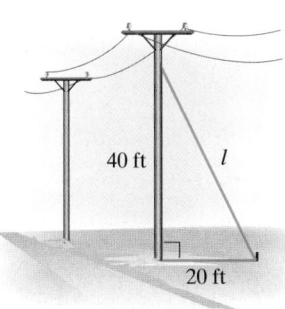

102. Radius of Basketball Hoop When you are given the area of a circle, its radius can be found by the formula $r = \sqrt{A/\pi}$.

© Allen R. Angel

a) Find the radius of a basketball hoop if the area enclosed by the hoop is 254.47 square inches.

b) If the diameter of a basketball is 9 inches, what is the minimum distance possible between the hoop and the ball when the center of the ball is in the center of the hoop?

103. Period of a Pendulum The formula for the period of a pendulum is

$$T = 2\pi\sqrt{\frac{l}{g}}$$

where T is the period in seconds, l is its length in feet, and g is the acceleration of gravity. On Earth, gravity is 32 ft/sec². The formula when used on Earth becomes

$$T = 2\pi\sqrt{\frac{l}{32}}$$

a) Find the period of a pendulum whose length is 8 feet.

b) If the length of a pendulum is doubled, what effect will this have on the period?

c) The gravity on the Moon is 1/6 that on Earth. If a pendulum has a period of 2 seconds on Earth, what will be the period of the same pendulum on the Moon?

104. Diagonal of a Suitcase A formula for the length of a diagonal from the upper corner of a box to the opposite lower corner is $d = \sqrt{L^2 + W^2 + H^2}$, where L, W, and H are the length, width, and height, respectively.

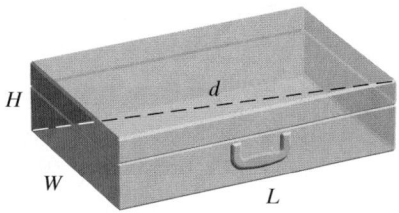

a) Find the length of the diagonal of a suitcase of length 22 inches, width 15 inches, and height 12 inches.

b) If the length, width, and height are all doubled, how will the diagonal change?

c) Solve the formula for W.

105. Blood Flowing in an Artery The formula

$$r = \sqrt[4]{\frac{8\mu l}{\pi R}}$$

is used in determining movement of blood through arteries. In the formula, R represents the resistance to blood flow, μ is the viscosity of blood, l is the length of the artery, and r is the radius of the artery. Solve this equation for R.

106. Falling Object The formula

$$t = \frac{\sqrt{19.6s}}{9.8}$$

can be used to tell the time, t, in seconds, that an object has been falling if it has fallen s meters. Suppose an object has been dropped from a helicopter and has fallen 100 meters. How long has it been in free fall?

107. Earth Days For any planet in our solar system, its "year" is the time it takes for the planet to revolve once around the Sun. The number of Earth days in a given planet's year, N, is approximated by the formula $N = 0.2(\sqrt{R})^3$, where R is the mean distance of the planet to the Sun in millions of kilometers. Find the number of Earth days in the year of the planet Earth, whose mean distance to the Sun is 149.4 million kilometers.

© Ragnarock/Shutterstock

108. Earth Days Find the number of Earth days in the year of the planet Mercury, whose mean distance to the Sun is 58 million kilometers. See Exercise 107.

109. Forces on a Car When two forces, F_1 and F_2, pull at right angles to each other as illustrated below, the resultant, or the effective force, R, can be found by the formula $R = \sqrt{F_1^2 + F_2^2}$. Two cars are trying to pull a third out of the mud, as illustrated. If car A is exerting a force of 60 pounds and car B is exerting a force of 80 pounds, find the resulting force on the car stuck in the mud.

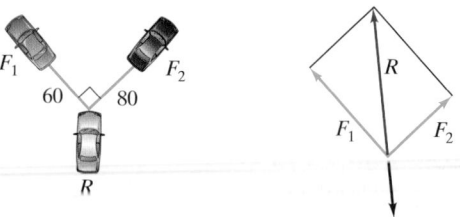

110. Escape Velocity The escape velocity, or the velocity needed for a spacecraft to escape a planet's gravitational field, is found by the formula $v_e = \sqrt{2gR}$, where g is the force of gravity of the planet and R is the radius of the planet. Find the escape velocity for Earth, in meters per second, where $g = 9.75$ meters per second squared and $R = 6{,}370{,}000$ meters.

111. Motion of a Wave A formula used in the study of shallow water wave motion is $c = \sqrt{gH}$, in which c is wave velocity, H is water depth, and g is the acceleration due to gravity. Find the wave velocity if the water's depth is 10 feet. (Use $g = 32$ ft/sec^2.)

112. Diagonal of a Box The top of a rectangular box measures 20 inches by 32 inches. Find the length of the diagonal for the top of the box.

113. Flower Garden A rectangular flower garden measures 25 meters by 32 meters. Find the length of the diagonal for the garden.

114. Speed of Sound When sound travels through air (or any gas), the velocity of the sound wave is dependent on the air (or gas) temperature. The velocity, v, in meters per second, at air temperature, t, in degrees Celsius, can be found by the formula

$$v = 331.3\sqrt{1 + \frac{t}{273}}$$

Find the speed of sound in air whose temperature is 20°C (equivalent to 68°F).

A formula that we have already mentioned and that we will be discussing in more detail shortly is the quadratic formula

$$x = \frac{-b \pm \sqrt{b^2 - 4ac}}{2a}$$

115. Find x when $a = 1, b = 0, c = -4$.

116. Find x when $a = 1, b = 1, c = -12$.

117. Find x when $a = -1, b = 4, c = 5$.

118. Find x when $a = 2, b = 5, c = -12$.

Given $f(x)$, find all values of x for which $f(x)$ is the indicated value.

119. $f(x) = \sqrt{x - 5}, f(x) = 5$

120. $f(x) = \sqrt[3]{2x + 3}, f(x) = 3$

121. $f(x) = \sqrt{3x^2 - 11} + 7, f(x) = 15$

122. $f(x) = 8 + \sqrt[3]{x^2 + 152}, f(x) = 14$

In Exercises 123 and 124, solve the equation.

123. $\sqrt{x^2 + 49} = (x^2 + 49)^{1/2}$

124. $\sqrt{x^2 - 16} = (x^2 - 16)^{1/2}$

Concept/Writing Exercises

125. Consider the equation $\sqrt{x + 3} = -\sqrt{2x - 1}$. Explain why this equation can have no real solution.

126. Consider the equation $-\sqrt{x^2} = \sqrt{(-x)^2}$. By studying the equation, can you determine its solution? Explain.

127. Consider the equation $\sqrt[3]{x^2} = -\sqrt[3]{x^2}$. By studying the equation, can you determine its solution? Explain.

128. Explain without solving the equation how you can tell that $\sqrt{x - 3} + 4 = 0$ has no solution.

129. Does the equation $\sqrt{x} = 5$ have one or two solutions? Explain.

130. Does the equation $x^2 = 9$ have one or two solutions? Explain

131. a) Consider the equation $\sqrt{4x - 12} = x - 3$. Setting each side of the equation equal to y yields the following system of equations.

$$y = \sqrt{4x - 12}$$

$$y = x \quad 3$$

The graphs of the equations in the system are illustrated in the figure.

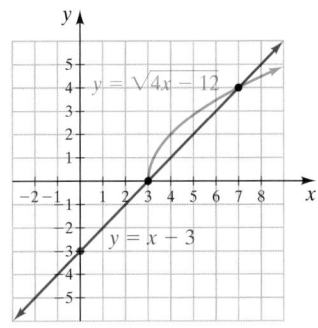

From the graphs, determine the values that appear to be solutions to the equation $\sqrt{4x - 12} = x - 3$. Explain how you determined your answer.

b) Substitute the values found in part **a)** into the original equation and determine whether they are the solutions to the equation.

c) Solve the equation $\sqrt{4x - 12} = x - 3$ algebraically and see if your solution agrees with the values obtained in part **a)**.

132. If the graph of a radical function, $f(x)$, does not intersect the x-axis, then the equation $f(x) = 0$ has no real solutions. Explain why.

133. Suppose we are given a rational function $g(x)$. If $g(4) = 0$, then the graph of $g(x)$ must intersect the x-axis at 4. Explain why.

134. The graph of the equation $y = \sqrt{x - 3} + 2$ is illustrated in the figure.

a) What is the domain of the function?

b) How many real solutions does the equation $\sqrt{x - 3} + 2 = 0$ have? List all the real solutions. Explain how you determined your answer.

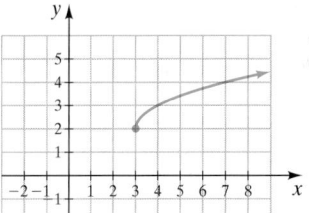

Challenge Problems

Solve.

135. $\sqrt{\sqrt{x + 25} - \sqrt{x}} = 5$

136. $\sqrt{\sqrt{x + 9} + \sqrt{x}} = 3$

Solve each equation for n.

137. $z = \dfrac{\bar{x} - \mu}{\dfrac{\sigma}{\sqrt{n}}}$

138. $z = \dfrac{p' - p}{\sqrt{\dfrac{pq}{n}}}$

139. Confidence Interval In statistics, a "confidence interval" is a range of values that is likely to contain the true value of the population. For a "95% confidence interval," the lower limit of the range, L_1, and the upper limit of the range, L_2, can be found by the formulas (above and to the right)

$$L_1 = p - 1.96\sqrt{\dfrac{p(1 - p)}{n}}$$

$$L_2 = p + 1.96\sqrt{\dfrac{p(1 - p)}{n}}$$

where p represents the percent obtained from a sample and n is the size of the sample. Francesco, a statistician, takes a sample of 36 families and finds that 60% of those surveyed use an answering machine in their home. He can be 95% certain that the true percent of families that use an answering machine in their home is between L_1 and L_2. Find the values of L_1 and L_2. Use $p = 0.60$ and $n = 36$ in the formulas.

140. Quadratic Mean The *quadratic mean* (or *root mean square, RMS*) is often used in physical applications. In power distribution systems, for example, voltages and currents are usually referred to in terms of their RMS values. The quadratic mean of a set of scores is obtained by squaring each score and adding the results (signified by Σx^2), then dividing the value obtained by the number of scores, n, and then taking the square root of this value. We may express the formula as

$$\text{quadratic mean} = \sqrt{\frac{\Sigma x^2}{n}}$$

Find the quadratic mean of the numbers 2, 4, and 10.

Group Activity

Discuss and answer Exercise 141 as a group.

141. Heron's Formula The area of a triangle is $A = \frac{1}{2}bh$. If the height is not known but we know the lengths of the three sides, we can use Heron's formula to find the area, A. Heron's formula is

$$A = \sqrt{S(S - a)(S - b)(S - c)}$$

where a, b, and c are the lengths of the three sides and

$$S = \frac{a + b + c}{2}$$

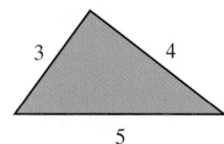

a) Have each group member use Heron's formula to find the area of a triangle whose sides are 3 inches, 4 inches, and 5 inches.

b) Compare your answers for part **a)**. If any member of the group did not get the correct answer, make sure they understand their error.

c) Have each member of the group do the following.
   1. Draw a triangle on the following grid. Place each vertex of the triangle at the intersection of two grid lines.

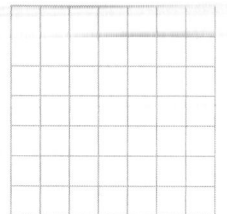

   2. Measure with a ruler the length of each side of your triangle.
   3. Use Heron's formula to find the area of your triangle.

d) Compare and discuss your work from part **c)**.

Cumulative Review Exercises

[2.2] **142.** Solve the formula $P_1P_2 - P_1P_3 = P_2P_3$ for P_2.

[6.1] **143.** Simplify $\dfrac{x(x - 5) + x(x - 2)}{2x - 7}$.

Perform each indicated operation.

[6.1] **144.** $\dfrac{4a^2 - 9b^2}{4a^2 + 12ab + 9b^2} \cdot \dfrac{6a^2b}{8a^2b^2 - 12ab^3}$

[   ] **145.** $(t^2 - 2t - 15) \div \dfrac{t^2 - 9}{t^2 - 3t}$

[6.2] **146.** $\dfrac{2}{x + 3} - \dfrac{1}{x - 3} + \dfrac{2x}{x^2 - 9}$

[6.4] **147.** Solve $2 + \dfrac{3x}{x - 1} = \dfrac{8}{x - 1}$.

7.7 Complex Numbers

1 Recognize a complex number.

2 Add and subtract complex numbers.

3 Multiply complex numbers.

4 Divide complex numbers.

5 Find powers of i.

1 Recognize a Complex Number

In Section 7.1 we mentioned that the square roots of negative numbers, such as $\sqrt{-4}$, are not real numbers. Numbers like $\sqrt{-4}$ are called **imaginary numbers**. Although they do not belong to the set of real numbers, the imaginary numbers do exist and are very useful in mathematics and science.

Every imaginary number has $\sqrt{-1}$ as a factor. The $\sqrt{-1}$, called the **imaginary unit**, is denoted by the letter i.

Imaginary Unit

$$i = \sqrt{-1}$$

Understanding
Algebra

The square root of any negative number is an imaginary number. For example, $\sqrt{-25}$ is an imaginary number that we write as follows:

$$\sqrt{-25} = \sqrt{-1 \cdot 25}$$
$$= \sqrt{-1} \cdot \sqrt{25}$$
$$= i \cdot 5 \text{ or } 5i$$

$\sqrt{-3}$ is an imaginary number that we write as follows:

$$\sqrt{-3} = \sqrt{-1 \cdot 3}$$
$$= \sqrt{-1} \cdot \sqrt{3}$$
$$= i\sqrt{3}$$

To write the square root of a negative number in terms of i, use the following property.

Square Root of a Negative Number

For any positive real number n,
$$\sqrt{-n} = \sqrt{-1 \cdot n} = \sqrt{-1}\,\sqrt{n} = i\sqrt{n}$$

Therefore, we can write
$$\sqrt{-4} = \sqrt{-1}\,\sqrt{4} = i2 \quad \text{or} \quad 2i$$
$$\sqrt{-9} = \sqrt{-1}\,\sqrt{9} = i3 \quad \text{or} \quad 3i$$
$$\sqrt{-7} = \sqrt{-1}\,\sqrt{7} = i\sqrt{7}$$

In this book we will generally write $i\sqrt{7}$ rather than $\sqrt{7}i$ to avoid confusion with $\sqrt{7i}$. Also, $3\sqrt{5}i$ is written as $3i\sqrt{5}$.

Examples

$$\sqrt{-81} = 9i \qquad \sqrt{-6} = i\sqrt{6}$$
$$\sqrt{-49} = 7i \qquad \sqrt{-10} = i\sqrt{10}$$

The real number system is a part of a larger number system, called the *complex number system*. Now we will discuss **complex numbers**.

Complex Number

Every number of the form

$$a + bi$$

where a and b are real numbers and i is the imaginary unit, is a **complex number**.

Every real number and every imaginary number are also complex numbers. A complex number has two parts: a real part, a, and an imaginary part, b.

Real part ⟶ ⟵ Imaginary part

$$a + b\,i$$

If $b = 0$, the complex number is a real number. If $a = 0$, the complex number is a **pure imaginary number**.

Examples of Complex Numbers

$3 + 2i$	$a = 3, b = 2$	
$5 - i\sqrt{6}$	$a = 5, b = -\sqrt{6}$	
4	$a = 4, b = 0$	(real number, $b = 0$)
$8i$	$a = 0, b = 8$	(imaginary number, $a = 0$)
$-i\sqrt{7}$	$a = 0, b = -\sqrt{7}$	(imaginary number, $a = 0$)

We stated that all real numbers and imaginary numbers are also complex numbers. The relationship between the various sets of numbers is illustrated in **Figure 7.7**.

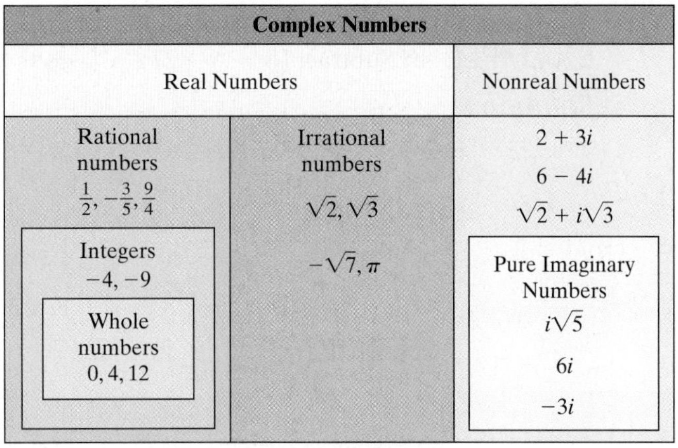

Complex Numbers		
Real Numbers		**Nonreal Numbers**
Rational numbers $\frac{1}{2}, -\frac{3}{5}, \frac{9}{4}$	Irrational numbers $\sqrt{2}, \sqrt{3}$	$2 + 3i$ $6 - 4i$ $\sqrt{2} + i\sqrt{3}$
Integers $-4, -9$	$-\sqrt{7}, \pi$	Pure Imaginary Numbers $i\sqrt{5}$ $6i$ $-3i$
Whole numbers $0, 4, 12$		

FIGURE 7.7

EXAMPLE 1 Write each complex number in the form $a + bi$.

a) $7 + \sqrt{-36}$ **b)** $4 - \sqrt{-12}$ **c)** 19   **d)** $\sqrt{-50}$ **e)** $6 + \sqrt{10}$

Solution

a) $7 + \sqrt{-36} = 7 + \sqrt{-1}\sqrt{36}$
$= 7 + i6 \quad \text{or} \quad 7 + 6i$

b) $4 - \sqrt{-12} = 4 - \sqrt{-1}\sqrt{12}$
$= 4 - \sqrt{-1}\sqrt{4}\sqrt{3}$
$= 4 - i(2)\sqrt{3} \quad \text{or} \quad 4 - 2i\sqrt{3}$

c) $19 = 19 + 0i$

d) $\sqrt{-50} = 0 + \sqrt{-50}$
$= 0 + \sqrt{-1}\sqrt{25}\sqrt{2}$
$= 0 + i(5)\sqrt{2} \quad \text{or} \quad 0 + 5i\sqrt{2}$

e) Both 6 and $\sqrt{10}$ are real numbers. Written as a complex number, the answer is $(6 + \sqrt{10}) + 0i$.

Now Try Exercise 23

Complex numbers can be added, subtracted, multiplied, and divided. To perform these operations, we use the definitions that $i = \sqrt{-1}$ and $i^2 = -1$.

> **Definition of i^2**
>
> If $i = \sqrt{-1}$, then
> $$i^2 = -1$$

2 Add and Subtract Complex Numbers

We now explain how to add or subtract complex numbers.

> **To Add or Subtract Complex Numbers**
>
> 1. Change all imaginary numbers to bi form.
> 2. Add (or subtract) the real parts of the complex numbers.
> 3. Add (or subtract) the imaginary parts of the complex numbers.
> 4. Write the answer in the form $a + bi$.

EXAMPLE 2 Add $(9 + 15i) + (-6 - 2i) + 18$.

Solution $(9 + 15i) + (-6 - 2i) + 18 = 9 + 15i - 6 - 2i + 18$
$= 9 - 6 + 18 + 15i - 2i$
$= 21 + 13i$

Now Try Exercise 27

EXAMPLE 3 Subtract $(8 - \sqrt{-27}) - (-3 + \sqrt{-48})$.

Solution
$(8 - \sqrt{-27}) - (-3 + \sqrt{-48}) = (8 - \sqrt{-1}\sqrt{27}) - (-3 + \sqrt{-1}\sqrt{48})$
$= (8 - \sqrt{-1}\sqrt{9}\sqrt{3}) - (-3 + \sqrt{-1}\sqrt{16}\sqrt{3})$
$= (8 - 3i\sqrt{3}) - (-3 + 4i\sqrt{3})$
$= 8 - 3i\sqrt{3} + 3 - 4i\sqrt{3}$
$= 8 + 3 - 3i\sqrt{3} - 4i\sqrt{3}$
$= 11 - 7i\sqrt{3}$

Now Try Exercise 35

3 Multiply Complex Numbers

Now let's discuss how to multiply complex numbers.

> **To Multiply Complex Numbers**
>
> 1. Change all imaginary numbers to bi form.
> 2. Multiply the complex numbers as you would multiply polynomials.
> 3. Substitute -1 for each i^2.
> 4. Combine the real parts and the imaginary parts. Write the answer in $a + bi$ form.

EXAMPLE 4 Multiply.

a) $5i(6 - 2i)$　　**b)** $\sqrt{-9}(\sqrt{-3} + 8)$　　**c)** $(2 - \sqrt{-18})(\sqrt{-2} + 5)$

Solution

a)
$$
\begin{aligned}
5i(6 - 2i) &= 5i(6) + 5i(-2i) && \text{Distributive property} \\
&= 30i - 10i^2 \\
&= 30i - 10(-1) && \text{Replace } i^2 \text{ with } -1. \\
&= 30i + 10 \quad \text{or} \quad 10 + 30i
\end{aligned}
$$

b)
$$
\begin{aligned}
\sqrt{-9}(\sqrt{-3} + 8) &= 3i(i\sqrt{3} + 8) && \text{Change imaginary numbers to } bi \text{ form.} \\
&= 3i(i\sqrt{3}) + 3i(8) && \text{Distributive property} \\
&= 3i^2\sqrt{3} + 24i \\
&= 3(-1)\sqrt{3} + 24i && \text{Replace } i^2 \text{ with } -1. \\
&= -3\sqrt{3} + 24i
\end{aligned}
$$

c)
$$
\begin{aligned}
(2 - \sqrt{-18})(\sqrt{-2} + 5) &= (2 - \sqrt{-1}\sqrt{18})(\sqrt{-1}\sqrt{2} + 5) \\
&= (2 - \sqrt{-1}\sqrt{9}\sqrt{2})(\sqrt{-1}\sqrt{2} + 5) \\
&= (2 - 3i\sqrt{2})(i\sqrt{2} + 5)
\end{aligned}
$$

Now use the FOIL method to multiply.
$$
\begin{aligned}
(2 - 3i\sqrt{2})(i\sqrt{2} + 5) &= (2)(i\sqrt{2}) + (2)(5) + (-3i\sqrt{2})(i\sqrt{2}) + (-3i\sqrt{2})(5) \\
&= 2i\sqrt{2} + 10 - 3i^2(2) - 15i\sqrt{2} \\
&= 2i\sqrt{2} + 10 - 3(-1)(2) - 15i\sqrt{2} \\
&= 2i\sqrt{2} + 10 + 6 - 15i\sqrt{2} \\
&= 16 - 13i\sqrt{2}
\end{aligned}
$$

Now Try Exercise 45

Avoiding Common Errors

What is $\sqrt{-4} \cdot \sqrt{-2}$?

CORRECT	INCORRECT
$\sqrt{-4} \cdot \sqrt{-2} = 2i \cdot i\sqrt{2}$	$\sqrt{-4} \cdot \sqrt{-2} = \sqrt{8}$
$\quad = 2i^2\sqrt{2}$	$= \sqrt{4} \cdot \sqrt{2}$
$\quad = 2(-1)\sqrt{2}$	$= 2\sqrt{2}$
$\quad = -2\sqrt{2}$	

Recall that $\sqrt{a} \cdot \sqrt{b} = \sqrt{ab}$ only for *nonnegative* real numbers a and b.

---

4 Divide Complex Numbers

The **conjugate of a complex number** $a + bi$ is $a - bi$. For example,

Complex Number	Conjugate
$3 + 7i$	$3 - 7i$
$1 - i\sqrt{3}$	$1 + i\sqrt{3}$
$2i$ (or $0 + 2i$)	$-2i$ (or $0 - 2i$)

When a complex number is multiplied by its conjugate using the FOIL method, the inner and outer products have a sum of 0, and the result is a real number. For example,

$$(5 + 3i)(5 - 3i) = 25 - 15i + 15i - 9i^2$$
$$= 25 - 9i^2$$
$$= 25 - 9(-1)$$
$$= 25 + 9 = 34$$

Now we explain how to divide complex numbers.

To Divide Complex Numbers

1. Change all imaginary numbers to bi form.
2. Multiply both the numerator and denominator by the conjugate of the denominator.
3. Write the answer in $a + bi$ form.

EXAMPLE 5 Divide $\dfrac{9 + i}{i}$.

Solution Begin by multiplying both numerator and denominator by $-i$, the conjugate of i.

$$\frac{9 + i}{i} \cdot \frac{-i}{-i} = \frac{(9 + i)(-i)}{-i^2}$$

$$= \frac{-9i - i^2}{-i^2} \qquad \text{Distributive property}$$

$$= \frac{-9i - (-1)}{-(-1)} \qquad \text{Replace } i^2 \text{ with } -1.$$

$$= \frac{-9i + 1}{1}$$

$$= 1 - 9i$$

Now Try Exercise 59

EXAMPLE 6 Divide $\dfrac{3 + 2i}{4 - i}$.

Solution Multiply both numerator and denominator by $4 + i$, the conjugate of $4 - i$.

$$\frac{3 + 2i}{4 - i} \cdot \frac{4 + i}{4 + i} = \frac{12 + 3i + 8i + 2i^2}{16 - i^2}$$

$$= \frac{12 + 11i + 2(-1)}{16 - (-1)}$$

$$= \frac{10 + 11i}{17} = \frac{10}{17} + \frac{11}{17}i$$

Now Try Exercise 65

EXAMPLE 7 **Impedance** A concept in the study of electronics is *impedance*. Impedance affects the current in a circuit. The impedance, Z, in a circuit is found by the formula $Z = \dfrac{V}{I}$, where V is voltage and I is current. Find Z when $V = 1.6 - 0.3i$ and $I = -0.2i$, where $i = \sqrt{-1}$.

Solution $Z = \dfrac{V}{I} = \dfrac{1.6 - 0.3i}{-0.2i}$. Now multiply both numerator and denominator by the conjugate of the denominator, $0.2i$.

$$Z = \frac{1.6 - 0.3i}{-0.2i} \cdot \frac{0.2i}{0.2i} = \frac{0.32i - 0.06i^2}{-0.04i^2}$$

$$= \frac{0.32i + 0.06}{0.04}$$

$$= \frac{0.32i}{0.04} + \frac{0.06}{0.04}$$

$$= 8i + 1.5 = 1.5 + 8i$$

Now Try Exercise 121

5 Find Powers of *i*

Using $i = \sqrt{-1}$ and $i^2 = -1$, we can find other **powers of *i***. For example,

$$i^3 = i^2 \cdot i = -1 \cdot i = -i \qquad i^6 = i^4 \cdot i^2 = 1(-1) = -1$$
$$i^4 = i^2 \cdot i^2 = (-1)(-1) = 1 \qquad i^7 = i^4 \cdot i^3 = 1(-i) = -i$$
$$i^5 = i^4 \cdot i^1 = 1 \cdot i = i \qquad i^8 = i^4 \cdot i^4 = (1)(1) = 1$$

Note that successive powers of *i* rotate through the four values $i, -1, -i,$ and 1 (see **Fig. 7.8**).

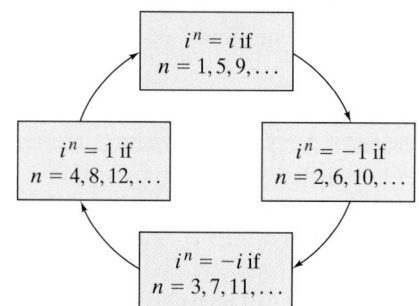

FIGURE 7.8

EXAMPLE 8 Evaluate. **a)** i^{35} **b)** i^{101}

Solution Write each expression as a product of factors such that the exponent of one factor is the largest multiple of 4 less than or equal to the given exponent. Then write this factor as i^4 raised to some power. Since i^4 has a value of 1, the expression i^4 raised to a power will also have a value of 1.

a) $i^{35} = i^{32} \cdot i^3 = \left(i^4\right)^8 \cdot i^3 = 1 \cdot i^3 = 1(-i) = -i$

b) $i^{101} = i^{100} \cdot i^1 = \left(i^4\right)^{25} \cdot i = 1 \cdot i = i$

Now Try Exercise 101

> **Helpful Hint**
>
> A quick way of evaluating i^n is to divide the exponent by 4 and observe the remainder.
>
> If the remainder is 0, the value is 1. If the remainder is 2, the value is -1.
>
> If the remainder is 1, the value is i. If the remainder is 3, the value is $-i$.
>
> For Example 8 **a)**
> $$\begin{array}{r} 8 \\ 4\overline{)35} \\ \underline{32} \\ 3 \end{array}$$ Answer is $-i$.
>
> For 8 **b)**
> $$\begin{array}{r} 25 \\ 4\overline{)101} \\ \underline{8} \\ 21 \\ \underline{20} \\ 1 \end{array}$$ Answer is i.
>
> $i^{35} = \left(i^4\right)^8 \cdot i^3 = (1)^8 \cdot i^3 = 1 \cdot i^3 = i^3 = -i$

EXAMPLE 9 Let $f(x) = x^2$. Find   **a)** $f(6i)$ **b)** $f(3 + 7i)$.

Solution

a) $f(x) = x^2$

$f(6i) = (6i)^2 = 36i^2 = 36(-1) = -36$

b) $f(x) = x^2$

$f(3 + 7i) = (3 + 7i)^2 = (3)^2 + 2(3)(7i) + (7i)^2$

$= 9 + 42i + 49i^2$

$= 9 + 42i + 49(-1)$

$= 9 + 42i - 49$

$= -40 + 42i$

Now Try Exercise 111

EXERCISE SET 7.7

Math XL   MyMathLab
MathXL®   MyMathLab

Warm-Up Exercises

Fill in the blanks with the appropriate word, phrase, or symbol(s) from the following list.

−1	pure	real	complex	number	−i
unit	part	1	conjugate	imaginary	rational

1. The square root of any negative number is an _____ number.

2. The imaginary _____ is denoted with the letter *i*.

3. Every number of the form $a + bi$ where a and b are real numbers and i is the imaginary unit, is a _____ number.

4. In a complex number, $a + bi$, a is known as the _____ part.

5. In a complex number, $a + bi$, b is known as the imaginary _____.

6. If $a = 0$, the complex number $a + bi$ is a _____ imaginary number.

7. If $b = 0$, the complex number $a + bi$ is a real _____.

8. For a complex number $a + bi$ the _____ is $a - bi$.

9. If i is the imaginary unit, then $i^2 = $ _____.

10. If i is the imaginary unit, then $i^3 = $ _____.

Practice the Skills

Write each expression as a complex number in the form $a + bi$.

11. 4

12. $2i$

13. $\sqrt{25}$

14. $\sqrt{-100}$

15. $21 - \sqrt{-36}$

16. $\sqrt{3} + \sqrt{-3}$

17. $\sqrt{-24}$

18. $\sqrt{49} - \sqrt{-49}$

19. $8 - \sqrt{-12}$

20. $\sqrt{-9} + \sqrt{-81}$

21. $3 + \sqrt{-98}$

22. $\sqrt{-9} + 7i$

23. $12 - \sqrt{-25}$

24. $10 + \sqrt{-32}$

25. $7i - \sqrt{-45}$

26. $\sqrt{144} + \sqrt{-96}$

Add or subtract.

27. $(3 + 7i) + (1 - 3i)$

28. $(-4 + 5i) + (2 - 4i)$

29. $(3 + 7i) - (1 - 3i)$

30. $(7 - \sqrt{-4}) - (-1 - \sqrt{-16})$

31. $(1 + \sqrt{-1}) + (-18 - \sqrt{-169})$

32. $(16 - i\sqrt{3}) + (17 - \sqrt{-3})$

33. $(\sqrt{3} + \sqrt{2}) + (3\sqrt{2} - \sqrt{-8})$

34. $(8 - \sqrt{2}) - (5 + \sqrt{-15})$

35. $(5 - \sqrt{-72}) + (6 + \sqrt{-8})$

36. $(29 + \sqrt{-75}) + (\sqrt{-147})$

37. $(\sqrt{4} - \sqrt{-45}) + (-\sqrt{25} + \sqrt{-5})$

38. $(\sqrt{20} - \sqrt{-12}) + (2\sqrt{5} + \sqrt{-75})$

Multiply.

39. $3(2 - 4i)$

40. $-2(-3 + i)$

41. $i(4 + 9i)$

42. $3i(6 - i)$

43. $\sqrt{-9}(6 + 11i)$

44. $\dfrac{1}{2}i\left(\dfrac{1}{3} - 18i\right)$

45. $\sqrt{-16}(\sqrt{3} - 7i)$

46. $-\sqrt{-24}(\sqrt{6} - \sqrt{-3})$

47. $\sqrt{-27}(\sqrt{3} - \sqrt{-3})$

48. $\sqrt{-32}(\sqrt{2} + \sqrt{-8})$

49. $(3 + 2i)(1 + i)$

50. $(6 - 2i)(3 + i)$

51. $(10 - 3i)(10 + 3i)$

52. $(-4 + 3i)(2 - 5i)$

53. $(7 + \sqrt{-2})(5 - \sqrt{-8})$

54. $(\sqrt{4} - 3i)(4 + \sqrt{-4})$

55. $\left(\dfrac{1}{2} - \dfrac{1}{3}i\right)\left(\dfrac{1}{4} + \dfrac{2}{3}i\right)$

56. $\left(\dfrac{3}{5} - \dfrac{1}{4}i\right)\left(\dfrac{2}{3} + \dfrac{2}{5}i\right)$

Divide.

57. $\dfrac{2}{3i}$

58. $\dfrac{-3}{4i}$

59. $\dfrac{2 + 3i}{2i}$

60. $\dfrac{7 - 3i}{2i}$

61. $\dfrac{6}{2 - i}$

62. $\dfrac{9}{5 + i}$

63. $\dfrac{3}{1 - 2i}$

64. $\dfrac{13}{-3 - 4i}$

65. $\dfrac{6 - 3i}{4 + 2i}$

66. $\dfrac{4 - 3i}{4 + 3i}$

67. $\dfrac{4}{6 - \sqrt{-4}}$

68. $\dfrac{2}{3 + \sqrt{-5}}$

69. $\dfrac{\sqrt{2}}{5 + \sqrt{-12}}$

70. $\dfrac{\sqrt{6}}{\sqrt{3} - \sqrt{-9}}$

71. $\dfrac{5 + \sqrt{-4}}{4 - \sqrt{-16}}$

72. $\dfrac{2 + \sqrt{-25}}{4 - \sqrt{-9}}$

73. $\dfrac{\sqrt{-75}}{\sqrt{-3}}$

74. $\dfrac{\sqrt{-30}}{\sqrt{-2}}$

75. $\dfrac{\sqrt{-32}}{\sqrt{-18}\,\sqrt{8}}$

76. $\dfrac{\sqrt{-40}\,\sqrt{-20}}{\sqrt{-4}}$

Perform each indicated operation. These exercises are a combination of the types of exercises presented earlier in this exercise set.

77. $(-5 + 7i) + (5 - 7i)$

78. $\left(\dfrac{4}{3} + \dfrac{1}{2}i\right) - \left(\dfrac{1}{3} - \dfrac{1}{2}i\right)$

79. $(\sqrt{50} - \sqrt{2}) - (\sqrt{-12} - \sqrt{-48})$

80. $(8 - \sqrt{-6}) - (2 - \sqrt{-24})$

81. $5.2(4 - 3.2i)$

82. $\sqrt{-6}(\sqrt{3} - \sqrt{-10})$

83. $(9 + 2i)(3 - 5i)$

84. $(\sqrt{3} + 2i)(\sqrt{6} - \sqrt{-8})$

85. $\dfrac{11 + 4i}{2i}$

86. $\dfrac{1}{4 + 3i}$

87. $\dfrac{6}{\sqrt{3} - \sqrt{-4}}$

88. $\dfrac{5 - 2i}{3 + 2i}$

89. $\left(11 - \dfrac{5}{9}i\right) - \left(4 - \dfrac{3}{5}i\right)$

90. $\dfrac{8}{7}\left(4 - \dfrac{2}{5}i\right)$

91. $\left(\dfrac{2}{3} - \dfrac{1}{5}i\right)\left(\dfrac{3}{5} - \dfrac{3}{4}i\right)$

92. $\sqrt{\dfrac{4}{9}}\left(\sqrt{\dfrac{25}{36}} - \sqrt{-\dfrac{4}{25}}\right)$

93. $\dfrac{\sqrt{-48}}{\sqrt{-12}}$

94. $\dfrac{2 - \sqrt{-9}}{4 + \sqrt{-49}}$

95. $(5.23 - 6.41i) - (9.56 + 4.5i)$

96. $(7 + \sqrt{-9})(4 - \sqrt{-4})$

For each imaginary number, indicate whether its value is i, -1, $-i$, or 1.

97. i^6

98. i^{63}

99. i^{160}

100. i^{231}

101. i^{93}

102. i^{103}

103. i^{811}

104. i^{1213}

Problem Solving

105. Consider the complex number $2 + 3i$.
   a) Find the additive inverse.
   b) Find the multiplicative inverse. Write the answer in simplified form.

106. Consider the complex number $4 - 5i$.
   a) Find the additive inverse.
   b) Find the multiplicative inverse. Write the answer in simplified form.

107. If $f(x) = x^2$, find $f(2i)$.

108. If $f(x) = x^2$, find $f(4i)$.

109. If $f(x) = x^4 - 2x$, find $f(2i)$.

110. If $f(x) = x^3 - 4x^2$, find $f(5i.)$

111. If $f(x) = x^2 + 2x$, find $f(3 + i)$.

112. If $f(x) = \dfrac{x^2}{x - 2}$, find $f(4 - i)$.

Evaluate each expression for the given value of x.

113. $x^2 - 2x + 5, x = 1 + 2i$

114. $x^2 - 2x + 5, x = 1 - 2i$

115. $x^2 + 2x + 7, x = -1 + i\sqrt{5}$

116. $x^2 + 2x + 9, x = -1 - i\sqrt{5}$

In Exercises 117–120, determine whether the given value of x is a solution to the equation.

117. $x^2 - 4x + 5 = 0, x = 2 - i$

118. $x^2 - 4x + 5 = 0, x = 2 + i$

119. $x^2 - 6x + 11 = 0, x = -3 + i\sqrt{3}$

120. $x^2 - 6x + 15 = 0, x = 3 - i\sqrt{3}$

121. Impedance Find the impedance, Z, using the formula $Z = \dfrac{V}{I}$ when $V = 1.8 + 0.5i$ and $I = 0.6i$. See Example 7.

122. Impedance Refer to Exercise 121. Find the impedance when $V = 2.4 - 0.6i$ and $I = -0.4i$.

123. Impedance Under certain conditions, the total impedance, Z_T, of a circuit is given by the formula

$$Z_T = \dfrac{Z_1 Z_2}{Z_1 + Z_2}$$

Find Z_T when $Z_1 = 2 - i$ and $Z_2 = 4 + i$.

124. Impedance Refer to Exercise 123. Find Z_T when $Z_1 = 3 - i$ and $Z_2 = 5 + i$.

125. Determine whether i^{-1} is equal to $i, -1, -i$, or 1. Show your work.

126. Determine whether i^{-5} is equal to $i, -1, -i$, or 1. Show your work.

*In Chapter 8, we will use the quadratic formula $x = \dfrac{-b \pm \sqrt{b^2 - 4ac}}{2a}$ to solve equations of the form $ax^2 + bx + c = 0$. **(a)** Use the quadratic formula to solve the following quadratic equations. **(b)** Check each of the two solutions by substituting the values found for x (one at a time) back into the original equation. In these exercises, the $\pm$ (read "plus or minus") results in two distinct complex answers.*

127. $x^2 - 2x + 6 = 0$

128. $x^2 - 4x + 6 = 0$

Given the complex numbers $a = 5 + 2i\sqrt{3}, b = 1 + i\sqrt{3}$, evaluate each expression.

129. $a + b$

130. $a - b$

131. ab

132. $\dfrac{a}{b}$

Concept/Writing Exercises

133. Are all of the following complex numbers? If any are not complex numbers, explain why.

a) 9   b) $-\dfrac{1}{2}$ c) $4 - \sqrt{-2}$

d) $7 - 3i$ e) $4.2i$ f) $11 + \sqrt{3}$

134. Is every real and every imaginary number a complex number?

135. Is every complex number a real number?

136. a) Is $i \cdot i$ a real number? Explain.

b) Is $i \cdot i \cdot i$ a real number? Explain.

In Exercises 137–140, answer true or false. Support your answer with an example.

137. The product of two pure imaginary numbers is always a real number.

138. The sum of two pure imaginary numbers is always an imaginary number.

139. The product of two complex numbers is always a real number.

140. The sum of two complex numbers is always a complex number.

141. What values of n will result in i^n being a real number? Explain.

142. What values of n will result in i^{2n} being a real number? Explain.

Cumulative Review Exercises

[4.3]   **143. Mixture** Berreda Coughlin, a grocer in Dallas, has two coffees, one selling for $5.50 per pound and the other for $6.30 per pound. How many pounds of each type of coffee should he mix to make 40 pounds of coffee to sell for $6.00 per pound?

[5.3]   **144.** Divide $\dfrac{8c^2 + 6c - 35}{4c + 9}$.

[6.2]   **145.** Add $\dfrac{b}{a - b} + \dfrac{a + b}{b}$.

[6.4]   **146.** Solve $\dfrac{x}{4} + \dfrac{1}{2} = \dfrac{x - 1}{2}$.

Chapter 7 Summary

IMPORTANT FACTS AND CONCEPTS	EXAMPLES

Section 7.1

A **radical expression** has the form $\sqrt[n]{x}$, where n is the index and x is the radicand.

In the radical expression $\sqrt[3]{x}$, 3 is the index and x is the radicand.

The **principal square root** of a positive number a, written $\sqrt{a}$, is the positive number b such that $b^2 = a$.

$$\sqrt{81} = 9, \text{ since } 9^2 = 81$$
$$\sqrt{0.36} = 0.6 \text{ since } (0.6)^2 = 0.36$$

The **square root function** is $f(x) = \sqrt{x}$. Its domain is $[0, \infty)$ and its range is $[0, \infty)$.

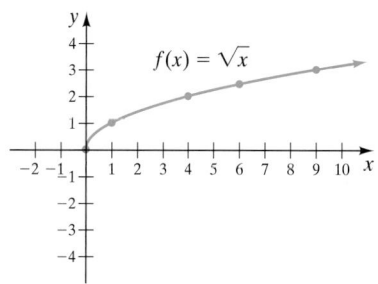

The **cube root** of a number a, written $\sqrt[3]{a}$, is the number b such that $b^3 = a$.

$$\sqrt[3]{27} = 3 \text{ since } 3^3 = 27$$
$$\sqrt[3]{-125} = -5 \text{ since } (-5)^3 = -125$$

The **cube root function** is $f(x) = \sqrt[3]{x}$. Its domain is $(-\infty, \infty)$ or $\mathbb{R}$ and its range is $(-\infty, \infty)$ or $\mathbb{R}$.

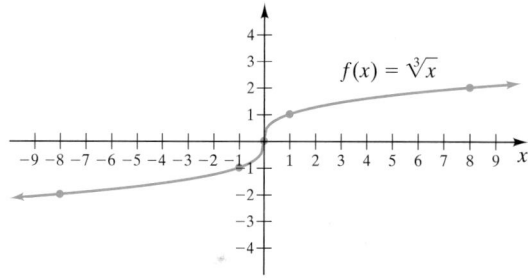

The nth root of a, $\sqrt[n]{a}$, where a is an *even index* and a is a nonnegative real number, is called an **even root** and is the nonnegative number b such that $b^n = a$.

$$\sqrt{4} = 2 \text{ since } 2^2 = 2 \cdot 2 = 4$$
$$\sqrt[4]{81} = 3 \text{ since } 3^4 = 3 \cdot 3 \cdot 3 \cdot 3 = 81$$

The nth root of a, $\sqrt[n]{a}$, where n is an *odd index* and a is any real number, is called an **odd root** and is the real number b such that $b^n = a$.

$$\sqrt[3]{27} = 3 \text{ since } 3^3 = 3 \cdot 3 \cdot 3 = 27$$
$$\sqrt[5]{-32} = -2 \text{ since } (-2)^5 = (-2)(-2)(-2)(-2)(-2) = -32$$

For any real number a, $\sqrt{a^2} = |a|$.

$$\sqrt{(-6)^2} = |-6| = 6$$
$$\sqrt{(y + 8)^2} = |y + 8|$$

Section 7.2

Rational Exponent

$$\sqrt[n]{a} = a^{1/n}$$

When a is nonnegative, n can be any index.

When a is negative, n must be odd.

$$\sqrt{17} = 17^{1/2}$$
$$\sqrt[4]{21x^3y^2} = (21x^3y^2)^{1/4}$$

For any nonnegative number a, and integers m and n,

$$\sqrt[n]{a^m} = \left(\sqrt[n]{a}\right)^m = a^{m/n} \leftarrow \text{Index}$$
$$\overset{\text{Power}}{}$$

$$\sqrt[4]{z^9} = \left(\sqrt[4]{z}\right)^9 = z^{9/4}$$

For any nonnegative real number a,

$$\sqrt[n]{a^n} = \left(\sqrt[n]{a}\right)^n = a^{n/n} = a$$

$$\sqrt[4]{y^4} = y, \qquad \sqrt[8]{14^8} = 14$$

IMPORTANT FACTS AND CONCEPTS	EXAMPLES

Section 7.2 (cont.)

Rules of Exponents

For all real numbers a and b and all rational numbers m and n,

Product rule $\qquad\qquad a^m \cdot a^n = a^{m+n}$

Quotient rule $\qquad\qquad \dfrac{a^m}{a^n} = a^{m-n}, \quad a \neq 0$

Negative exponent rule $\qquad a^{-m} = \dfrac{1}{a^m}, \quad a \neq 0$

Zero exponent rule $\qquad a^0 = 1, \quad a \neq 0$

Raising a power to a power $\quad (a^m)^n = a^{m \cdot n}$

Raising a product to a power $\quad (ab)^m = a^m b^m$

Raising a quotient to a power $\quad \left(\dfrac{a}{b}\right)^m = \dfrac{a^m}{b^m}, \quad b \neq 0$

$$\left(\dfrac{a}{b}\right)^{-n} = \left(\dfrac{b}{a}\right)^n = \dfrac{b^n}{a^n}, \quad a \neq 0, b \neq 0$$

$x^{1/3} \cdot x^{4/3} = x^{(1/3)+(4/3)} = x^{5/3}$

$\dfrac{x^{4/5}}{x^{1/2}} = x^{(4/5)-(1/2)} = x^{(8/10)-(5/10)} = x^{3/10}$

$x^{-1/7} = \dfrac{1}{x^{1/7}}$

$m^0 = 1$

$(c^{1/8})^{16} = c^{(1/8) \cdot 16} = c^2$

$(p^3 q^4)^{1/8} = p^{3/8} q^{1/2}$

$\left(\dfrac{81}{49}\right)^{-1/2} = \left(\dfrac{49}{81}\right)^{1/2} = \dfrac{49^{1/2}}{81^{1/2}} = \dfrac{7}{9}$

Section 7.3

A number or expression is a **perfect square** if it is the square of an expression.

A number or expression is a **perfect cube** if it is the cube of an expression.

Perfect squares:

	49	81	x^{12}	y^{50}
	↓	↓	↓	↓

Square of a number or expression: $\quad 7^2 \quad 9^2 \quad (x^6)^2 \quad (y^{25})^2$

Perfect cubes:

	27	-27	y^{18}	z^{30}
	↓	↓	↓	↓

Cube of a number or expression: $\quad 3^3 \quad (-3)^3 \quad (y^6)^3 \quad (z^{10})^3$

Product Rule for Radicals

For nonnegative real numbers a and b,

$$\sqrt[n]{a} \cdot \sqrt[n]{b} = \sqrt[n]{ab}$$

$\sqrt{2} \cdot \sqrt{8} = \sqrt{16} = 4, \quad \sqrt[3]{2x^3} = \sqrt[3]{x^3} \cdot \sqrt[3]{2} = x\sqrt[3]{2}$

To Simplify Radicals Using the Product Rule

1. If the radicand contains a coefficient other than 1, write it as a product of two numbers, one of which is the largest perfect power for the index.

2. Write each variable factor as a product of two factors, one of which is the largest perfect power of the variable for the index.

3. Use the product rule to write the radical expression as a product of radicals. Place all the perfect powers (numbers and variables) under the same radical.

4. Simplify the radical containing the perfect powers.

$\sqrt{24} = \sqrt{4 \cdot 6} = \sqrt{4}\,\sqrt{6} = 2\sqrt{6}$

$\sqrt[3]{16x^5 y^9} = \sqrt[3]{8x^3 y^9 \cdot 2x^2}$

$\qquad\qquad = \sqrt[3]{8x^3 y^9}\,\sqrt[3]{2x^2}$

$\qquad\qquad = 2xy^3 \sqrt[3]{2x^2}$

Quotient Rule for Radicals

For nonnegative real numbers a and b,

$$\dfrac{\sqrt[n]{a}}{\sqrt[n]{b}} = \sqrt[n]{\dfrac{a}{b}}, \quad b \neq 0$$

$\dfrac{\sqrt{32}}{\sqrt{2}} = \sqrt{\dfrac{32}{2}} = \sqrt{16} = 4, \quad \sqrt[3]{\dfrac{x^6}{y^{12}}} = \dfrac{\sqrt[3]{x^6}}{\sqrt[3]{y^{12}}} = \dfrac{x^2}{y^4}$

Section 7.4

Like radicals are radicals with the same radicand and index.
Unlike radicals are radicals with a different radicand or index.

Like Radicals	Unlike Radicals
$\sqrt{3}, \quad 12\sqrt{3}$	$\sqrt{3}, \quad 7\sqrt[4]{3}$
$2\sqrt[4]{xy^3}, \quad -3\sqrt[4]{xy^3}$	$\sqrt[5]{xy^3}, \quad x\sqrt[5]{y^3}$

To Add or Subtract Radicals

1. Simplify each radical expression.

2. Combine like radicals (if there are any).

$\sqrt{27} + \sqrt{48} - 2\sqrt{75} = \sqrt{9} \cdot \sqrt{3} + \sqrt{16} \cdot \sqrt{3} - 2 \cdot \sqrt{25} \cdot \sqrt{3}$

$\qquad\qquad\qquad\qquad = 3\sqrt{3} + 4\sqrt{3} - 10\sqrt{3}$

$\qquad\qquad\qquad\qquad = -3\sqrt{3}$

IMPORTANT FACTS AND CONCEPTS	EXAMPLES

Section 7.4 (cont.)

To Multiply Radicals

Use the product rule.

$$\sqrt[n]{a} \cdot \sqrt[n]{b} = \sqrt[n]{ab}$$

$\sqrt[4]{8c^2}\,\sqrt[4]{4c^3} = \sqrt[4]{32c^5} = \sqrt[4]{16c^4}\,\sqrt[4]{2c}$
$= 2c\,\sqrt[4]{2c}$

Section 7.5

To **rationalize a denominator** multiply both the numerator and the denominator of the fraction by a radical that will result in the radicand in the denominator becoming a perfect power.

$\dfrac{6}{\sqrt{3x}} \cdot \dfrac{\sqrt{3x}}{\sqrt{3x}} = \dfrac{6\sqrt{3x}}{\sqrt{9x^2}} = \dfrac{6\sqrt{3x}}{3x} = \dfrac{2\sqrt{3x}}{x}$

A Radical Expression Is Simplified When the Following Are All True

1. No perfect powers are factors of the radicand and all exponents in the radicand are less than the index.

2. No radicand contains a fraction.

3. No denominator contains a radical.

	Not Simplified	Simplified
1.	$\sqrt{x^3}$	$x\sqrt{x}$
2.	$\sqrt{\dfrac{1}{2}}$	$\dfrac{\sqrt{2}}{2}$
3.	$\dfrac{1}{\sqrt{2}}$	$\dfrac{\sqrt{2}}{2}$

Section 7.6

To Solve Radical Equations

1. Rewrite the equation so that one radical containing a variable is by itself (isolated) on one side of the equation.

2. Raise each side of the equation to a power equal to the index of the radical.

3. Combine like terms.

4. If the equation still contains a term with a variable in a radicand, repeat steps 1 through 3.

5. Solve the resulting equation for the variable.

6. Check all solutions in the original equation for extraneous solutions.

Solve $\sqrt{x} - 8 = 0$.

$$\sqrt{x} - 8 = 0$$
$$\sqrt{x} = 8$$
$$(\sqrt{x})^2 = 8^2$$
$$x = 64$$

A check shows that 64 is the solution.

Section 7.7

The **imaginary unit**, i is defined as $i = \sqrt{-1}$. (Also, $i^2 = -1$.)

$\sqrt{-25} = \sqrt{25}\,\sqrt{-1} = 5i$

Imaginary Number

For any positive number n,

$$\sqrt{-n} = i\sqrt{n}.$$

$\sqrt{-19} = i\sqrt{19}$

A **complex number** is a number of the form $a + bi$, where a and b are real numbers.

$3 + 2i$ and $26 - 15i$ are complex numbers.

To Add or Subtract Complex Numbers

1. Change all imaginary numbers to bi form.

2. Add (or subtract) the real parts of the complex numbers.

3. Add (or subtract) the imaginary parts of the complex numbers.

4. Write the answer in the form $a + bi$.

Add $(8 - 3i) + (12 + 5i)$.

$$(8 - 3i) + (12 + 5i)$$
$$= 8 + 12 - 3i + 5i$$
$$= 20 + 2i$$

To Multiply Complex Numbers

1. Change all imaginary numbers to bi form.

2. Multiply the complex numbers as you would multiply polynomials.

3. Substitute -1 for each i^2.

4. Combine the real parts and the imaginary parts. Write the answer in $a + bi$ form.

Multiply $(7 + 2i\sqrt{3})(5 - 4i\sqrt{3})$.

$$(7 + 2i\sqrt{3})(5 - 4i\sqrt{3})$$
$$= 35 - 28i\sqrt{3} + 10i\sqrt{3} - 8(i^2)(3)$$
$$= 35 - 28i\sqrt{3} + 10i\sqrt{3} + 24$$
$$= 59 - 18i\sqrt{3}$$

IMPORTANT FACTS AND CONCEPTS	EXAMPLES

Section 7.7 (cont.)

The **conjugate of a complex number** $a + bi$ is $a - bi$.	Complex Number	Conjugate
	$14 + 2i$	$14 - 2i$
	$-17 - 8i$	$-17 + 8i$

To Divide Complex Numbers

1. Change all imaginary numbers to bi form.
2. Rationalize the denominator by multiplying both the numerator and denominator by the conjugate of the denominator.
3. Write the answer in $a + bi$ form.

Divide $\dfrac{2 - i}{5 + 3i}$.

$$\frac{2 - i}{5 + 3i} \cdot \frac{5 - 3i}{5 - 3i} = \frac{10 - 6i - 5i + 3i^2}{25 - 9i^2} = \frac{7 - 11i}{34}$$

Powers of i

$$i^2 = -1, i^3 = -i, i^4 = 1, i^5 = i$$

$$i^{38} = i^{36} \cdot i^2 = (i^4)^9 \cdot i^2 = 1^9 \cdot (-1) = -1$$
$$i^{63} = i^{60} \cdot i^3 = (i^4)^{15} \cdot i^3 = 1^{15}(-i) = -i$$

Chapter 7 Review Exercises

[7.1] *Evaluate.*

1. $\sqrt{9}$ **2.** $\sqrt[3]{-27}$ **3.** $\sqrt[3]{-125}$ **4.** $\sqrt[4]{256}$

Use absolute value to evaluate.

5. $\sqrt{(-5)^2}$ **6.** $\sqrt{(38.2)^2}$

Write as an absolute value.

7. $\sqrt{x^2}$ **8.** $\sqrt{(x + 7)^2}$ **9.** $\sqrt{(x - y)^2}$ **10.** $\sqrt{(x^2 - 4x + 12)^2}$

11. Let $f(x) = \sqrt{10x + 9}$. Find $f(4)$.

12. Let $k(x) = 2x + \sqrt{\dfrac{x}{3}}$. Find $k(27)$.

13. Let $g(x) = \sqrt[3]{2x + 3}$. Find $g(4)$ and round the answer to the nearest tenth.

14. Area The area of a square is 144 square meters. Find the length of its side.

For the remainder of these review exercises, assume that all variables represent positive real numbers.

[7.2] *Write in exponential form.*

15. $\sqrt{x^3}$ **16.** $\sqrt[3]{x^5}$ **17.** $(\sqrt[4]{y})^{13}$ **18.** $\sqrt[7]{6^{-2}}$

Write in radical form.

19. $x^{1/2}$ **20.** $a^{2/3}$ **21.** $(8m^2n)^{7/4}$ **22.** $(x + y)^{-5/3}$

Simplify each radical expression by changing the expression to exponential form. Write the answer in radical form when appropriate.

23. $\sqrt[3]{4^6}$ **24.** $\sqrt{x^{12}}$ **25.** $(\sqrt[4]{9})^8$ **26.** $\sqrt[20]{a^5}$

Evaluate if possible. If the expression is not a real number, so state.

27. $-36^{1/2}$ **28.** $(-36)^{1/2}$ **29.** $\left(\dfrac{64}{27}\right)^{-1/3}$ **30.** $64^{-1/2} + 8^{-2/3}$

Simplify. Write the answer without negative exponents.

31. $x^{3/5} \cdot x^{-1/3}$ **32.** $\left(\dfrac{64}{y^9}\right)^{1/3}$ **33.** $\left(\dfrac{a^{-6/5}}{a^{2/5}}\right)^{2/3}$ **34.** $\left(\dfrac{20x^5y^{-3}}{4y^{1/2}}\right)^2$

Multiply.

35. $a^{1/2}(5a^{3/2} - 3a^2)$

36. $4x^{-2/3}\left(x^{-1/2} + \dfrac{11}{4}x^{2/3}\right)$

Factor each expression. Write the answer without negative exponents.

37. $x^{2/5} + x^{7/5}$

38. $a^{-1/2} + a^{3/2}$

For each function, find the indicated value of the function. Use your calculator to evaluate irrational numbers. Round irrational numbers to the nearest thousandth.

39. If $f(x) = \sqrt{6x - 11}$, find $f(6)$.

40. If $g(x) = \sqrt[3]{9x - 17}$, find $g(4)$.

Graph the following functions.

41. $f(x) = \sqrt{x}$

42. $f(x) = \sqrt{x} - 4$

[7.2–7.5] *Simplify.*

43. $\sqrt{18}$

44. $\sqrt[3]{16}$

45. $\sqrt{\dfrac{49}{9}}$

46. $\sqrt[3]{\dfrac{8}{125}}$

47. $-\sqrt{\dfrac{81}{49}}$

48. $\sqrt[3]{-\dfrac{27}{125}}$

49. $\sqrt{32}\,\sqrt{2}$

50. $\sqrt[3]{32}\,\sqrt[3]{2}$

51. $\sqrt{18x^2y^3z^4}$

52. $\sqrt{75x^3y^7}$

53. $\sqrt[3]{54a^7b^{10}}$

54. $\sqrt[3]{125x^8y^9z^{16}}$

55. $\left(\sqrt[6]{x^2y^3z^5}\right)^{42}$

56. $\left(\sqrt[5]{2ab^4c^6}\right)^{15}$

57. $\sqrt{5x}\,\sqrt{8x^5}$

58. $\sqrt[3]{2x^2y}\,\sqrt[3]{4x^9y^4}$

59. $\sqrt[3]{2x^4y^5}\,\sqrt[3]{16x^4y^4}$

60. $\sqrt[4]{4x^4y^7}\,\sqrt[4]{4x^5y^9}$

61. $\sqrt{3x}\left(\sqrt{12x}-\sqrt{20}\right)$

62. $\sqrt[3]{2x^2y}\left(\sqrt[3]{4x^4y^7}+\sqrt[3]{9x}\right)$

63. $\sqrt{\sqrt{a^3b^2}}$

64. $\sqrt{\sqrt[3]{x^5y^2}}$

65. $\left(\dfrac{4r^2p^{1/3}}{r^{1/2}p^{4/3}}\right)^3$

66. $\left(\dfrac{6y^{2/5}z^{1/3}}{x^{-1}y^{3/5}}\right)^{-1}$

67. $\sqrt{\dfrac{3}{5}}$

68. $\sqrt[3]{\dfrac{7}{9}}$

69. $\sqrt[4]{\dfrac{5}{4}}$

70. $\dfrac{x}{\sqrt{10}}$

71. $\dfrac{8}{\sqrt{x}}$

72. $\dfrac{m}{\sqrt[3]{25}}$

73. $\dfrac{10}{\sqrt[3]{y^2}}$

74. $\dfrac{9}{\sqrt[4]{z}}$

75. $\sqrt[3]{\dfrac{x^3}{27}}$

76. $\dfrac{\sqrt[3]{2x^{10}}}{\sqrt[3]{16x^7}}$

77. $\sqrt{\dfrac{32x^2y^5}{2x^8y}}$

78. $\sqrt[4]{\dfrac{48x^9y^{15}}{3xy^3}}$

79. $\sqrt{\dfrac{6x^4}{y}}$

80. $\sqrt{\dfrac{12a}{7b}}$

81. $\sqrt{\dfrac{18x^4y^5}{3z}}$

82. $\sqrt{\dfrac{125x^2y^5}{3z}}$

83. $\sqrt[3]{\dfrac{108x^3y^7}{2y^3}}$

84. $\sqrt[3]{\dfrac{3x}{5y}}$

85. $\sqrt[3]{\dfrac{9x^5y^3}{x^6}}$

86. $\sqrt[3]{\dfrac{y^6}{5x^2}}$

87. $\sqrt[4]{\dfrac{2a^2b^{11}}{a^5b}}$

88. $\sqrt[4]{\dfrac{3x^2y^6}{8x^3}}$

89. $(3-\sqrt{2})(3+\sqrt{2})$

90. $(\sqrt{3}+\sqrt{5})(\sqrt{3}-\sqrt{5})$

91. $(x-\sqrt{y})(x+\sqrt{y})$

92. $(\sqrt{3}+2)^2$

93. $(\sqrt{x}-\sqrt{3y})(\sqrt{x}+\sqrt{5y})$

94. $(\sqrt[3]{2x}-\sqrt[3]{3y})(\sqrt[3]{3x}-\sqrt[3]{2y})$

95. $\dfrac{6}{2+\sqrt{5}}$

96. $\dfrac{x}{4+\sqrt{x}}$

97. $\dfrac{a}{4-\sqrt{b}}$

98. $\dfrac{x}{\sqrt{y}-7}$

99. $\dfrac{\sqrt{x}}{\sqrt{x}+\sqrt{y}}$

100. $\dfrac{\sqrt{x}-3\sqrt{y}}{\sqrt{x}-\sqrt{y}}$

101. $\dfrac{2}{\sqrt{a-1}-2}$

102. $\dfrac{5}{\sqrt{y+2}-3}$

103. $\sqrt[3]{x}+10\sqrt[3]{x}-2\sqrt[3]{x}$

104. $\sqrt{3}+\sqrt{27}-\sqrt{192}$

105. $\sqrt[3]{16}-5\sqrt[3]{54}+3\sqrt[3]{64}$

106. $\sqrt{2}-\dfrac{3}{\sqrt{32}}+\sqrt{50}$

107. $9\sqrt{x^5y^6}-\sqrt{16x^7y^8}$

108. $8\sqrt[3]{x^7y^8}-\sqrt[3]{x^4y^2}+3\sqrt[3]{x^{10}y^2}$

In Exercises 109 and 110, $f(x)$ and $g(x)$ are given. Find $(f \cdot g)(x)$.

109. $f(x)=\sqrt{3x},\ g(x)=\sqrt{6x}-\sqrt{15}$

110. $f(x)=\sqrt[3]{2x^2},\ g(x)=\sqrt[3]{4x^4}+\sqrt[3]{16x^5}$

Simplify. In Exercise 112, assume that the variable can be any real number.

111. $f(x)=\sqrt{2x+7}\,\sqrt{2x+7},\ x\geq-\dfrac{7}{2}$

112. $g(a)=\sqrt{20a^2+100a+125}$

Simplify.

113. $\dfrac{\sqrt[3]{(x+5)^5}}{\sqrt{(x+5)^3}}$

114. $\dfrac{\sqrt[3]{a^3b^2}}{\sqrt[4]{a^4b}}$

Perimeter and Area *For each figure, find* **a)** *the perimeter and* **b)** *the area. Write the perimeter and area in radical form with the radicals simplified.*

115.

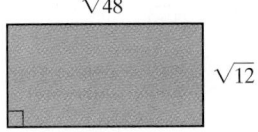

$\sqrt{48}$

$\sqrt{12}$

Rectangle

116.

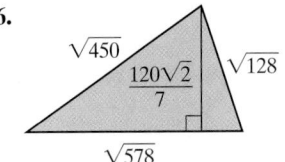

$\sqrt{450}$ $\sqrt{128}$

$\dfrac{120\sqrt{2}}{7}$

$\sqrt{578}$

117. The graph of $f(x) = \sqrt{x} + 2$ is given.

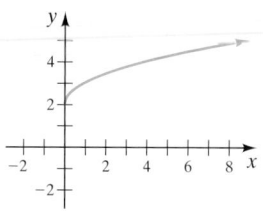

a) For $g(x) = -3$, sketch the graph of $(f + g)(x)$.

b) What is the domain of $(f + g)(x)$?

118. The graph of $f(x) = -\sqrt{x}$ is given.

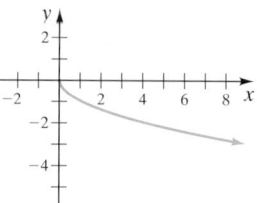

a) For $g(x) = \sqrt{x} + 2$, sketch the graph of $(f + g)(x)$.

b) What is the domain of $(f + g)(x)$?

[7.6]   *Solve each equation and check your solutions.*

119. $\sqrt{x} = 4$

120. $\sqrt{x} = -4$

121. $\sqrt[3]{x} = 4$

122. $\sqrt[3]{x} = -5$

123. $7 + \sqrt{x} = 10$

124. $7 + \sqrt[3]{x} = 12$

125. $\sqrt{3x + 4} = \sqrt{5x + 14}$

126. $\sqrt{x^2 + 2x - 8} = x$

127. $\sqrt[3]{x - 9} = \sqrt[3]{5x + 3}$

128. $(x^2 + 7)^{1/2} = x + 1$

129. $\sqrt{x + 3} = \sqrt{3x + 9}$

130. $\sqrt{6x - 5} - \sqrt{2x + 6} - 1 = 0$

For each pair of functions, find all values of x for which $f(x) = g(x)$.

131. $f(x) = \sqrt{3x + 4}$, $g(x) = 2\sqrt{2x - 4}$

132. $f(x) = (4x + 5)^{1/3}$, $g(x) = (6x - 7)^{1/3}$

Solve the following for the indicated variable.

133. $V = \sqrt{\dfrac{2L}{w}}$, for L

134. $r = \sqrt{\dfrac{A}{\pi}}$, for A

Find the length of the unknown side of each right triangle. Write the answer as a radical in simplified form.

135.

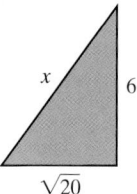

136.

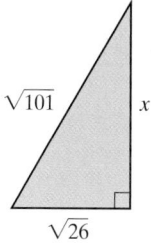

Solve.

137. Telephone Pole How long a wire does a phone company need to reach the top of a 5-meter telephone pole from a point on the ground 2 meters from the base of the pole?

138. Velocity Use the formula $v = \sqrt{2gh}$ to find the velocity of an object after it has fallen 20 feet ($g = 32$ ft/s^2).

139. Pendulum Use the formula

$$T = 2\pi\sqrt{\dfrac{L}{32}}$$

to find the period of a pendulum, T, if its length, L, is 64 feet.

140. Kinetic Energy Kinetic energy is the energy of an object in motion. The formula

$$V = \sqrt{\dfrac{2K}{m}}$$

can be used to determine the velocity, V, in meters per second, when a mass, m, in kilograms, has a kinetic energy, K, in joules. A 0.145-kg baseball is thrown. If the kinetic energy of the moving ball is 45 joules, at what speed is the ball moving?

141. Speed of Light Albert Einstein found that if an object at rest, with mass m_0, is made to travel close to the speed of light, its mass increases to m, where

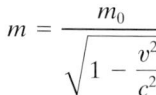

In the formula, v is the velocity of the moving object and c is the speed of light.* In an accelerator used for cancer therapy, particles travel at speeds of $0.98c$, that is, at 98% the speed of light. At a speed of $0.98c$, determine a particle's mass, m, in terms of its resting mass, m_0. Use $v = 0.98c$ in the above formula.

© National Archives and Records Administration

Albert Einstein

*The speed of light is 3.00×10^8 meters per second. However, you do not need this information to solve this problem.

[7.7] *Write each expression as a complex number in the form a + bi.*

142. 1 **143.** -8 **144.** $7 - \sqrt{-256}$ **145.** $9 + \sqrt{-16}$

Perform each indicated operation.

146. $(3 + 2i) + (10 - i)$ **147.** $(4 + i) - (7 - 2i)$ **148.** $(\sqrt{3} + \sqrt{-5}) + (11\sqrt{3} - \sqrt{-7})$

149. $\sqrt{-6}(\sqrt{6} + \sqrt{-6})$ **150.** $(4 + 3i)(2 - 3i)$ **151.** $(6 + \sqrt{-3})(4 - \sqrt{-15})$

152. $\dfrac{8}{3i}$ **153.** $\dfrac{2 + \sqrt{3}}{2i}$ **154.** $\dfrac{4}{3 + 2i}$

155. $\dfrac{\sqrt{3}}{5 - \sqrt{-6}}$

Evaluate each expression for the given value of x.

156. $x^2 - 2x + 9, x = 1 + 2i\sqrt{2}$ **157.** $x^2 - 2x + 12, x = 1 - 2i$

For each imaginary number, indicate whether its value is i, −1, −i, or 1.

158. i^{33} **159.** i^{59} **160.** i^{404} **161.** i^{802}

Chapter 7 Practice Test

CHAPTER
Test Prep
VIDEOS

*Chapter Test Prep Videos provide fully worked-out solutions to any of the exercises you want to review. Chapter Test Prep Videos are available via **MyMathLab**, or on **You Tube** (search "Angel Intermediate Algebra" and click on "Channels")*

1. Write $\sqrt{(5x - 3)^2}$ as an absolute value.

2. Simplify $\left(\dfrac{x^{2/5} \cdot x^{-1}}{x^{3/5}}\right)^2$.

3. Factor $x^{-2/3} + x^{4/3}$.

4. Graph $g(x) = \sqrt{x} + 1$.

In Exercises 5–14, simplify. Assume that all variables represent positive real numbers.

5. $\sqrt{54x^7y^{10}}$

6. $\sqrt[3]{25x^5y^2} \, \sqrt[3]{10x^6y^8}$

7. $\sqrt{\dfrac{7x^6y^3}{8z}}$

8. $\dfrac{9}{\sqrt[3]{x}}$

9. $\dfrac{\sqrt{3}}{3 + \sqrt{27}}$

10. $2\sqrt{24} - 6\sqrt{6} + 3\sqrt{54}$

11. $\sqrt[3]{8x^3y^5} + 4\sqrt[3]{x^6y^8}$

12. $(\sqrt{3} - 2)(6 - \sqrt{8})$

13. $\sqrt[4]{\sqrt{x^5y^3}}$

14. $\dfrac{\sqrt[4]{(7x + 2)^5}}{\sqrt[3]{(7x + 2)^2}}$

In Exercises 15–17, solve the equation.

15. $\sqrt{2x + 19} = 3$

16. $\sqrt{x^2 - x - 12} = x + 3$

17. $\sqrt{a - 8} = \sqrt{a} - 2$

18. For $f(x) = (9x + 37)^{1/3}$ and $g(x) = 2(2x + 2)^{1/3}$, find all values of x such that $f(x) = g(x)$.

19. Solve the formula $w = \dfrac{\sqrt{2gh}}{4}$ for g.

20. Falling Object The velocity, V, in feet per second, after an object has fallen a distance, h, in feet, can be found by the formula $V = \sqrt{64.4h}$. Find the velocity of a pen after it has fallen 200 feet.

21. Ladder A ladder is placed against a house. If the base of the ladder is 5 feet from the house and the ladder rests on the house 12 feet above the ground, find the length of the ladder.

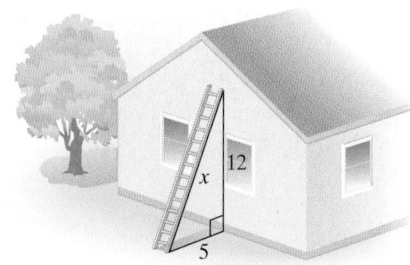

22. Springs A formula used in the study of springs is

$$T = 2\pi\sqrt{\dfrac{m}{k}}$$

where T is the period of a spring (the time for the spring to stretch and return to its resting point), m is the mass on the spring, in kilograms, and k is the spring's constant, in newtons/meter. A mass of 1400 kilograms rests on a spring. Find the period of the spring if the spring's constant is 65,000 newtons/meter.

23. Multiply $(6 - \sqrt{-4})(2 + \sqrt{-16})$.

24. Divide $\dfrac{5 - i}{7 + 2i}$.

25. Evaluate $x^2 + 6x + 12$ for $x = -3 + i$.

Cumulative Review Test

Take the following test and check your answers with those given in the back of the book. Review any questions that you answered incorrectly. The section where the material was covered is indicated after the answer.

1. Solve $\frac{1}{5}(x - 3) = \frac{3}{4}(x + 3) - x$.

2. Solve $3(x - 4) = 6x - (4 - 5x)$.

3. Sweater When the price of a sweater is decreased by 60%, it costs $16. Find the original price of the sweater.

4. Find the solution set of $|3 - 2x| < 5$.

5. Graph $y = \frac{3}{2}x - 3$.

6. Determine whether the graphs of the given equations are parallel lines, perpendicular lines, or neither.

$$y = 3x - 8$$

$$6y = 18x + 12$$

7. Given $f(x) = x^2 - 3x + 4$ and $g(x) = 2x - 9$, find $(g - f)(x)$.

8. Find the equation of the line through $(1, -4)$ that is perpendicular to the graph of $3x - 2y = 6$.

9. Solve the system of equations.

$$x + 2y = 12$$

$$4x = 8$$

$$3x - 4y + 5z = 20$$

10. Evaluate the determinant.

$$\begin{vmatrix} 3 & -6 & -1 \\ 2 & 1 & -2 \\ 1 & 3 & 1 \end{vmatrix}$$

11. Volume The volume of the box that follows is $6r^3 + 5r^2 + r$. Find w in terms of r.

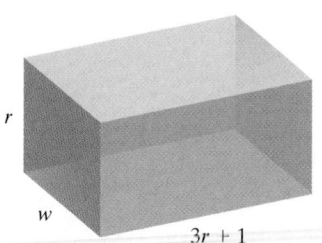

12. Multiply $(5xy - 3)(5xy + 3)$.

13. Solve $\sqrt{2x^2 + 7} + 3 = 8$.

14. Factor $4x^3 - 9x^2 + 5x$.

15. Factor $(x + 1)^3 - 27$.

16. Solve $8x^2 - 3 = -10x$.

17. Multiply $\frac{4x + 4y}{x^2y} \cdot \frac{y^3}{12x}$.

18. Add $\frac{x - 4}{x - 5} - \frac{3}{x + 5} - \frac{10}{x^2 - 25}$.

19. Solve $\frac{4}{x} - \frac{1}{6} = \frac{1}{x}$.

20. Falling Object The distance, d, an object drops in free fall is directly proportional to the square of the time, t. If an object falls 16 feet in 1 second, how far will the object fall in 5 seconds?   400 ft

8

Quadratic Functions

Goals of This Chapter

We introduced quadratic functions in Section 5.1. Now we expand on the concepts. We introduce completing the square and the quadratic formula. After studying these sections, you will know three techniques for solving quadratic equations: factoring (when possible), completing the square, and the quadratic formula. We will also discuss techniques for graphing quadratic functions and study nonlinear inequalities in one variable.

There are many real-life situations that can be represented or approximated with quadratic equations. As you read through this chapter, you will see several interesting applications of quadratic equations and quadratic functions. For instance, in Exercises 99 and 100 on page 514, we will use quadratic equations and the quadratic formula to discuss gravity on the moon.

© NASA/Johnson Space Center

493

8.1 Solving Quadratic Equations by Completing the Square

1 Use the square root property to solve equations.

2 Understand perfect square trinomials.

3 Solve quadratic equations by completing the square.

Understanding Algebra

Recall that a quadratic equation is an equation of the form $ax^2 + bx + c = 0$, where a, b, and c are real numbers and $a \neq 0$.

Understanding Algebra

A convenient way to write both the positive and the negative square roots is to use the symbol $\pm$, which is read "plus or minus." The solutions to the equation $x^2 = 25$ can be written as $\pm\sqrt{25} = \pm 5$. Thus the solutions are 5 and -5.

In Chapter 5 we solved quadratic equations by factoring. However, not every quadratic equation can be solved by factoring. In this section we introduce two additional procedures used to solve quadratic equations: the square root property and completing the square.

1 Use the Square Root Property to Solve Quadratic Equations

In Chapter 7 we learned that every positive number has two square roots—a positive square root and a negative square root. For example, 25 has two square roots;

Positive Square Root of 25	Negative Square Root of 25
$\sqrt{25} = 5$	$-\sqrt{25} = -5$

We can refer to both square roots of 25 using the symbol $\pm$, which is read "plus or minus." Thus, ± 5 refers to both 5 and -5.

Next, consider the quadratic equation $x^2 = 25$. Since $5^2 = 25$ and $(-5)^2 = 25$, both $x = 5$ and $x = -5$ are solutions to the equation. We can summarize the solution process as follows

$$x^2 = 25$$
$$x = \pm\sqrt{25} \qquad \pm \text{ means take the positive and the negative square roots.}$$
$$x = \pm 5 \qquad \text{The solutions are 5 and } -5.$$

In general, when solving equations of the form $x^2 = a$, we can use the *square root property*.

Square Root Property

If $x^2 = a$, where a is a real number, then $x = \pm\sqrt{a}$.

EXAMPLE 1   Solve the following equations.

a) $x^2 = 49$ **b)** $x^2 - 9 = 0$ **c)** $x^2 + 10 = 85$

Solution

a) Use the square root property to solve the equation.

$$x^2 = 49 \qquad \text{Use the square root property.}$$
$$x = \pm\sqrt{49} \qquad \text{The positive and the negative square roots of 49.}$$
$$x = \pm 7 \qquad \text{The solutions are 7 and } -7.$$

Check the solutions in the original equation.

$x = 7$	$x = -7$
$x^2 = 49$	$x^2 = 49$
$(7)^2 = 49$	$(-7)^2 = 49$
$49 = 49$ True	$49 = 49$ True

In both cases we have true statements. Therefore, both 7 and -7 are solutions to the equation.

b)

$$x^2 - 9 = 0 \qquad \text{Add 9 to both sides.}$$
$$x^2 = 9 \qquad \text{Use the square root property.}$$
$$x = \pm\sqrt{9}$$
$$= \pm 3 \qquad \text{The solutions are 3 and } -3.$$

Check the solutions in the original equation.

$x = 3$	$x = -3$
$x^2 - 9 = 0$	$x^2 - 9 = 0$
$3^2 - 9 \stackrel{?}{=} 0$	$(-3)^2 - 9 \stackrel{?}{=} 0$
$0 = 0$ True	$0 = 0$ True

In both cases the check is true, which means that both 3 and -3 are solutions to the equation.

c)

$$x^2 + 10 = 85 \qquad \text{Subtract 10 from both sides.}$$
$$x^2 = 75 \qquad \text{Use the square root property.}$$
$$x = \pm\sqrt{75} \qquad \text{Simplify } \sqrt{75}.$$
$$= \pm\sqrt{25}\sqrt{3}$$
$$= \pm 5\sqrt{3}$$

The solutions are $5\sqrt{3}$ and $-5\sqrt{3}$.

Now Try Exercise 13

Not all quadratic equations have real number solutions, as is illustrated in Example 2.

EXAMPLE 2 Solve the equation $x^2 + 7 = 0$.

Solution

$$x^2 + 7 = 0 \qquad \text{Subtract 7 from both sides.}$$
$$x^2 = -7 \qquad \text{Use the square root property.}$$
$$x = \pm\sqrt{-7} \qquad \text{Simplify the imaginary solutions.}$$
$$= \pm i\sqrt{7}$$

The solutions are $i\sqrt{7}$ and $-i\sqrt{7}$, both of which are imaginary numbers.

Now Try Exercise 15

> **Understanding Algebra**
>
> Recall from Chapter 7 that the square root of a negative number is an *imaginary number*. In general, for any positive real number, n,
>
> $$\sqrt{-n} = \sqrt{-1}\sqrt{n} = i\sqrt{n}$$
>
> where $i = \sqrt{-1}$.

EXAMPLE 3 Solve **a)** $(a - 5)^2 = 32$ **b)** $(z + 3)^2 + 28 = 0$.

Solution

a) Begin by using the square root property.

$$(a - 5)^2 = 32 \qquad \text{Use the square root property.}$$
$$a - 5 = \pm\sqrt{32} \qquad \text{Add 5 to both sides.}$$
$$a = 5 \pm \sqrt{32} \qquad \text{Simplify } \sqrt{32}.$$
$$= 5 \pm \sqrt{16}\sqrt{2}$$
$$= 5 \pm 4\sqrt{2}$$

The solutions are $5 + 4\sqrt{2}$ and $5 - 4\sqrt{2}$.

b) Begin by subtracting 28 from both sides of the equation to isolate the term containing the variable.

$$(z + 3)^2 + 28 = 0 \qquad \text{Subtract 28 from both sides.}$$
$$(z + 3)^2 = -28 \qquad \text{Use the square root property.}$$
$$z + 3 = \pm\sqrt{-28} \qquad \text{Subtract 3 from both sides.}$$
$$z = -3 \pm \sqrt{-28} \qquad \text{Simplify } \sqrt{-28}.$$
$$= -3 \pm \sqrt{28}\sqrt{-1}$$
$$= -3 \pm i\sqrt{4}\sqrt{7}$$
$$= -3 \pm 2i\sqrt{7}$$

> **Understanding Algebra**
>
> Recall that *complex numbers* are numbers of the form
>
> $$a + bi$$
>
> where a and b are real numbers and
>
> $$i = \sqrt{-1}.$$

The solutions are $-3 + 2i\sqrt{7}$ and $-3 - 2i\sqrt{7}$. Note that the solutions to the equation $(z + 3)^2 + 28 = 0$ are complex numbers.

Now Try Exercise 23

2 Understand Perfect Square Trinomials

Recall that a **perfect square trinomial** is a trinomial that can be expressed as the square of a binomial. Some examples follow.

Perfect Square Trinomials		Factors		Square of a Binomial
$x^2 + 8x + 16$	$=$	$(x + 4)(x + 4)$	$=$	$(x + 4)^2$
$x^2 - 8x + 16$	$=$	$(x - 4)(x - 4)$	$=$	$(x - 4)^2$
$x^2 + 10x + 25$	$=$	$(x + 5)(x + 5)$	$=$	$(x + 5)^2$
$x^2 - 10x + 25$	$=$	$(x - 5)(x - 5)$	$=$	$(x - 5)^2$

In a perfect square trinomial with a leading coefficient of 1, the constant term is the square of one-half the coefficient of the first-degree term.

Let's examine some perfect square trinomials for which the leading coefficient is 1.

$$x^2 + 8x + 16 = (x + 4)^2$$
$$\left[\tfrac{1}{2}(8)\right]^2 = (4)^2$$

$$x^2 - 10x + 25 = (x - 5)^2$$
$$\left[\tfrac{1}{2}(-10)\right]^2 = (-5)^2$$

When a perfect square trinomial with a leading coefficient of 1 is written as the square of a binomial, the constant in the binomial is one-half the coefficient of the first-degree term in the trinomial. For example,

$$x^2 + 8x + 16 = (x + 4)^2$$
$$\tfrac{1}{2}(8)$$

$$x^2 - 10x + 25 = (x - 5)^2$$
$$\tfrac{1}{2}(-10)$$

3 Solve Quadratic Equations by Completing the Square

To solve a quadratic equation by **completing the square**, we add a constant to both sides of the equation so the resulting trinomial is a perfect square trinomial. Then we use the square root property to solve the resulting equation.

> **To Solve a Quadratic Equation by Completing the Square**
>
> 1. Use the multiplication (or division) property of equality if necessary to make the leading coefficient 1.
> 2. Rewrite the equation with the constant by itself on the right side of the equation.
> 3. Take one-half the numerical coefficient of the first-degree term, square it, and add this quantity to both sides of the equation.
> 4. Factor the trinomial as the square of a binomial.
> 5. Use the square root property to take the square root of both sides of the equation.
> 6. Solve for the variable.
> 7. Check your solutions in the *original* equation.

EXAMPLE 4 Solve the equation $x^2 + 6x + 5 = 0$ by completing the square.

Solution Since the leading coefficient is 1, step 1 is not necessary.

Step 2 Subtract 5 from both sides of the equation.

$$x^2 + 6x + 5 = 0$$
$$x^2 + 6x = -5$$

Step 3 Determine the square of one-half the numerical coefficient of the first-degree term, 6.

$$\frac{1}{2}(6) = 3, \qquad 3^2 = 9$$

Add this value to both sides of the equation.

$$x^2 + 6x + 9 = -5 + 9$$
$$x^2 + 6x + 9 = 4$$

Step 4 By following this procedure, we produce a perfect square trinomial on the left side of the equation. The expression $x^2 + 6x + 9$ is a perfect square trinomial that can be factored as $(x + 3)^2$.

$$\overbrace{(x + 3)}^{\frac{1}{2} \text{ the numerical coefficient of the first-degree term is } \frac{1}{2}(6) = +3}{}^2 = 4$$

Step 5 Use the square root property.

$$x + 3 = \pm\sqrt{4}$$
$$x + 3 = \pm 2$$

Step 6 Finally, solve for x by subtracting 3 from both sides of the equation.

$$x + 3 - 3 = -3 \pm 2$$
$$x = -3 \pm 2$$
$$x = -3 + 2 \quad \text{or} \quad x = -3 - 2$$
$$x = -1 \qquad\qquad x = -5$$

Step 7 Check both solutions in the original equation.

$x = -1$	$x = -5$
$x^2 + 6x + 5 = 0$	$x^2 + 6x + 5 = 0$
$(-1)^2 + 6(-1) + 5 \overset{?}{=} 0$	$(-5)^2 + 6(-5) + 5 \overset{?}{=} 0$
$1 - 6 + 5 \overset{?}{=} 0$	$25 - 30 + 5 \overset{?}{=} 0$
$0 = 0$ True	$0 = 0$ True

Since each number checks, both -1 and -5 are solutions to the original equation.

Now Try Exercise 49

EXAMPLE 5 Solve the equation $-x^2 = -3x - 18$ by completing the square.

Solution Begin by multiplying both sides of the equation by -1 to make the coefficient of the squared term equal to 1.

$$-x^2 = -3x - 18$$
$$-1(-x^2) = -1(-3x - 18) \qquad \text{Multiply both sides by } -1.$$
$$x^2 = 3x + 18 \qquad\qquad \text{Subtract } 3x \text{ from both sides.}$$
$$x^2 - 3x = 18$$

Take half the numerical coefficient of the first-degree term, square it, and add this product to both sides of the equation.

$$\frac{1}{2}(-3) = -\frac{3}{2} \qquad \left(-\frac{3}{2}\right)^2 = \frac{9}{4}$$

$$x^2 - 3x + \frac{9}{4} = 18 + \frac{9}{4} \qquad \text{Complete the square.}$$

$$\left(x - \frac{3}{2}\right)^2 = 18 + \frac{9}{4} \qquad \text{Factor the trinomial as the square of a binomial.}$$

$$\left(x - \frac{3}{2}\right)^2 = \frac{72}{4} + \frac{9}{4}$$

$$\left(x - \frac{3}{2}\right)^2 = \frac{81}{4}$$

$$x - \frac{3}{2} = \pm\sqrt{\frac{81}{4}} \qquad \text{Square root property}$$

$$x - \frac{3}{2} = \pm\frac{9}{2} \qquad \text{Simplify.}$$

$$x = \frac{3}{2} \pm \frac{9}{2} \qquad \text{Added } \frac{3}{2} \text{ to both sides.}$$

$$x = \frac{3}{2} + \frac{9}{2} \qquad \text{or} \qquad x = \frac{3}{2} - \frac{9}{2}$$

$$x = \frac{12}{2} = 6 \qquad\qquad x = -\frac{6}{2} = -3$$

The solutions are 6 and −3.

Now Try Exercise 53

Understanding Algebra

When a quadratic equation has a leading coefficient that is not 1, you can multiply each term on both sides of the equation by the reciprocal of the leading coefficient. This will result in an equivalent equation with a leading coefficient of 1. For example, the equation

$$-\frac{1}{4}x^2 + 2x - 8 = 0$$

can be multiplied by −4 to obtain the equivalent equation

$$x^2 - 8x + 32 = 0$$

which has a leading coefficient of 1.

In the following examples we will not show some of the intermediate steps.

EXAMPLE 6 Solve $x^2 - 8x + 34 = 0$ by completing the square.

Solution

$$x^2 - 8x + 34 = 0 \qquad \text{Subtract 34 from both sides.}$$

$$x^2 - 8x = -34$$

$$x^2 - 8x + 16 = -34 + 16 \qquad \text{Complete the square.}$$

$$(x - 4)^2 = -18 \qquad \text{Factor the trinomial as the square of a binomial.}$$

$$x - 4 = \pm\sqrt{-18} \qquad \text{Square root property}$$

$$x - 4 = \pm 3i\sqrt{2} \qquad \text{Simplify.}$$

$$x = 4 \pm 3i\sqrt{2} \qquad \text{Solve for } x.$$

The solutions are $4 + 3i\sqrt{2}$ and $4 - 3i\sqrt{2}$.

Now Try Exercise 61

EXAMPLE 7 Solve the equation $-4x^2 + 8x + 32 = 0$ by completing the square.

Solution

$$-4x^2 + 8x + 32 = 0$$

$$-\frac{1}{4}(-4x^2 + 8x + 32) = -\frac{1}{4}(0) \qquad \begin{array}{l}\text{Multiply by } -1/4 \text{ to obtain}\\ \text{a leading coefficient of 1.}\end{array}$$

$$x^2 - 2x - 8 = 0$$

$$x^2 - 2x = 8 \qquad \text{Added 8 to both sides.}$$

$$x^2 - 2x + 1 = 8 + 1 \qquad \text{Complete the square.}$$

$$(x - 1)^2 = 9 \qquad \begin{array}{l}\text{Factor the trinomial as the}\\ \text{square of a binomial.}\end{array}$$

$$x - 1 = \pm 3 \qquad \text{Square root property}$$

$$x = 1 \pm 3 \qquad \text{Solve for } x.$$

$$x = 1 + 3 \qquad \text{or} \qquad x = 1 - 3$$

$$x = 4 \qquad\qquad x = -2$$

Now Try Exercise 75

◆*Using Your Graphing Calculator* ▨▨▨▨

We can verify the *real* solutions to quadratic equations using a graphing calculator. Recall that the x-intercepts of a function occur when $y = 0$. Thus, to verify the solutions to the equation in Example 7, $-4x^2 + 8x + 32 = 0$, we will observe the x-intercepts of the function $y = -4x^2 + 8x + 32$.

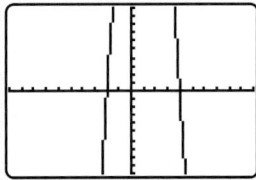

The figure above verifies that the solutions to $-4x^2 + 8x + 32 = 0$ are -2 and 4. Remember that the x-intercepts only represent *real* solutions to an equation. Thus, we cannot use a graphing calculator to verify the solutions to the equation in Example 6 since the solutions are complex numbers.

Generally, quadratic equations that cannot be easily solved by factoring will be solved by the *quadratic formula*, which we present in the next section.

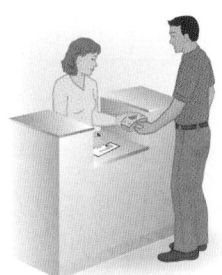

EXAMPLE 8 **Compound Interest** The compound interest formula $A = p\left(1 + \dfrac{r}{n}\right)^{nt}$ can be used to find the amount, A, when an initial principal, p, is invested at an annual interest rate, r, compounded n times a year for t years.

a) Josh Adams initially invested \$1000 in a savings account where interest is compounded annually (once a year). If after 2 years the amount, or balance, in the account is \$1102.50, find the annual interest rate, r.

b) Trisha McDowell initially invested \$1000 in a savings account where interest is compounded quarterly (four times per year). If after 3 years the amount in the account is \$1195.62, find the annual interest rate, r.

Solution

a) Understand   We are given the following information:

$$p = \$1000, \qquad A = \$1102.50, \qquad n = 1, \qquad t = 2$$

We are asked to find the annual interest rate, r. To do so, we substitute the appropriate values into the formula and solve for r.

Translate
$$A = p\left(1 + \frac{r}{n}\right)^{nt}$$

$$1102.50 = 1000\left(1 + \frac{r}{1}\right)^{1(2)}$$

Carry Out
$$1102.50 = 1000(1 + r)^2 \qquad \text{Divide both sides by 1000.}$$

$$1.10250 = (1 + r)^2 \qquad\qquad \text{Square root property}$$

$$\pm\sqrt{1.10250} = 1 + r \qquad\qquad \begin{array}{l}\text{We will only use the positive} \\ \text{square root since } r \text{ must be positive.}\end{array}$$

$$1.05 = 1 + r \qquad\qquad \text{Subtract 1 from both sides.}$$

$$0.05 = r$$

Answer   The annual interest rate is 0.05 or 5%.

Understanding Algebra

Recall from Chapter 7 that

$$\sqrt[n]{a} = a^{1/n}.$$

Therefore, from Example 8,

$$\sqrt[12]{1.19562} = 1.19562^{1/12}$$

b) Understand We are given

$$p = 1000, \qquad A = \$1195.62, \qquad n = 4, \qquad t = 3$$

To find r, we substitute the appropriate values into the formula and solve for r.

Translate

$$A = p\left(1 + \frac{r}{n}\right)^{nt}$$

$$1195.62 = 1000\left(1 + \frac{r}{4}\right)^{4(3)}$$

Carry Out

$$1.19562 = \left(1 + \frac{r}{4}\right)^{12} \qquad \text{Divided both sides by 1000.}$$

$$\sqrt[12]{1.19562} = 1 + \frac{r}{4} \qquad \begin{array}{l}\text{Take the 12th root of both}\\ \text{sides (or raise both sides to}\\ \text{the } 1/12 \text{ power).}\end{array}$$

$$1.015 \approx 1 + \frac{r}{4} \qquad \begin{array}{l}\text{Approximated } 1.19562^{1/12}\\ \text{on a calculator.}\end{array}$$

$$0.015 \approx \frac{r}{4} \qquad \text{Subtracted 1 from both sides.}$$

$$0.06 \approx r \qquad \text{Multiplied both sides by 4.}$$

Answer The annual interest rate is approximately 0.06 or 6%.

Now Try Exercise 103

> **Helpful Hint**
>
> **Study Tip**
>
> In this chapter, you will be working with roots and radicals. This material was discussed in Chapter 7. If you do not remember how to evaluate or simplify radicals, review Chapter 7 now.

EXERCISE SET 8.1

Math XL   MyMathLab
MathXL®   MyMathLab

Warm-Up Exercises

Fill in the blanks with the appropriate word, phrase, or symbol(s) from the following list.

reciprocal	imaginary	quadratic	square root property	perfect square trinomial	$\dfrac{2}{3}$	$\dfrac{3}{2}$
25	10	100	one-half	complex	binomial	

1. A _____ equation is an equation of the form $ax^2 + bx + c = 0$, where a, b, and c are real numbers and $a \neq 0$.

2. The square root of a negative number is an _____ number.

3. A number in the form $a + bi$ is a _____ number.

4. A perfect square trinomial is a trinomial that can be expressed as the square of a _____.

5. To solve a quadratic equation by completing the square, we add a constant to both sides of the equation so the resulting trinomial is a _____.

6. Once one side of a quadratic equation is rewritten as a perfect square trinomial, the equation can be solved by using the _____.

7. When a quadratic equation has a leading coefficient that is not 1, you can multiply each term on both sides of the equation by the _____ of the leading coefficient.

8. In a perfect square trinomial with a leading coefficient of 1, the constant term is the square of _____ the coefficient of the first-degree term.

9. To solve the quadratic equation $x^2 - 10x = 24$ by completing the square, you would add _____ to both sides of the equation.

10. To obtain an equivalent equation with a leading coefficient of 1, multiply both sides of the equation $\frac{2}{3}x^2 + 3x - 5 = 0$ by _____.

Practice the Skills

For Exercises 11–36, use the square root property to solve each equation.

11. $x^2 = 81$

12. $x^2 = 100$

13. $x^2 - 25 = 0$

14. $x^2 - 49 = 0$

15. $x^2 + 49 = 0$

16. $x^2 - 24 = 0$

17. $x^2 + 24 = 0$

18. $y^2 - 10 = 51$

19. $y^2 + 10 = -51$

20. $(x - 3)^2 = 49$

21. $(p - 4)^2 = 16$

22. $(x + 3)^2 = 49$

23. $(x + 3)^2 + 25 = 0$

24. $(a - 3)^2 = 45$

25. $(a - 2)^2 + 45 = 0$

26. $(a + 2)^2 + 45 = 0$

27. $\left(b + \dfrac{1}{3}\right)^2 = \dfrac{4}{9}$

28. $\left(b - \dfrac{1}{3}\right)^2 = \dfrac{4}{9}$

29. $\left(b - \dfrac{2}{3}\right)^2 + \dfrac{4}{9} = 0$

30. $(x - 0.2)^2 = 0.64$

31. $(x + 0.8)^2 = 0.81$

32. $\left(x + \dfrac{1}{2}\right)^2 = \dfrac{16}{9}$

33. $(2a - 5)^2 = 18$

34. $(4y + 1)^2 = 12$

35. $\left(2y + \dfrac{1}{2}\right)^2 = \dfrac{4}{25}$

36. $\left(3x - \dfrac{1}{4}\right)^2 = \dfrac{9}{25}$

Solve each equation by completing the square.

37. $x^2 + 6x + 5 = 0$

38. $x^2 - 2x - 3 = 0$

39. $x^2 + 8x + 15 = 0$

40. $x^2 - 8x + 15 = 0$

41. $x^2 + 6x + 8 = 0$

42. $x^2 - 6x + 8 = 0$

43. $x^2 - 7x + 6 = 0$

44. $x^2 + 9x + 18 = 0$

45. $2x^2 + x - 1 = 0$

46. $3c^2 - 4c - 4 = 0$

47. $2z^2 - 7z - 4 = 0$

48. $4a^2 + 9a = 9$

49. $x^2 - 13x + 40 = 0$

50. $x^2 + x - 12 = 0$

51. $-x^2 + 6x + 7 = 0$

52. $-a^2 - 5a + 14 = 0$

53. $-z^2 + 9z - 20 = 0$

54. $-z^2 - 4z + 12 = 0$

55. $b^2 = 3b + 28$

56. $-x^2 = 6x - 27$

57. $x^2 + 10x = 11$

58. $-x^2 + 40 = -3x$

59. $x^2 - 4x - 10 = 0$

60. $x^2 - 6x + 2 = 0$

61. $r^2 + 8r + 5 = 0$

62. $a^2 + 4a - 8 = 0$

63. $c^2 - c - 3 = 0$

64. $p^2 - 5p = 4$

65. $x^2 + 3x + 6 = 0$

66. $z^2 - 5z + 7 = 0$

67. $9x^2 - 9x = 0$

68. $4y^2 + 12y = 0$

69. $-\dfrac{3}{4}b^2 - \dfrac{1}{2}b = 0$

70. $\dfrac{1}{3}a^2 - \dfrac{5}{3}a = 0$

71. $36z^2 - 6z = 0$

72. $x^2 = \dfrac{9}{2}x$

73. $-\dfrac{1}{2}p^2 - p + \dfrac{3}{2} = 0$

74. $2x^2 + 6x = 20$

75. $2x^2 = 8x + 64$

76. $3x^2 + 33x + 72 = 0$

77. $2x^2 + 18x + 4 = 0$

78. $\dfrac{2}{3}x^2 + \dfrac{4}{3}x + 1 = 0$

79. $\dfrac{3}{4}w^2 + \dfrac{1}{2}w - \dfrac{1}{4} = 0$

80. $\dfrac{3}{4}c^2 - 2c + 1 = 0$

81. $2x^2 - x = -5$

82. $\dfrac{5}{2}x^2 + \dfrac{3}{2}x - \dfrac{5}{4} = 0$

83. $-3x^2 + 6x = 6$

84. $x^2 + 2x = -5$

Problem Solving

Area *In Exercises 85–88, the area, A, of each rectangle is given.* **a)** *Write an equation for the area.* **b)** *Solve the equation for x.*

85.

$A = 21$, $x - 2$, $x + 2$

86.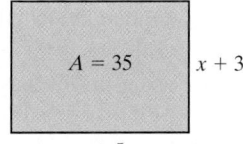

$A = 35$, $x + 3$, $x + 5$

87.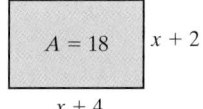

$A = 18$, $x + 2$, $x + 4$

88.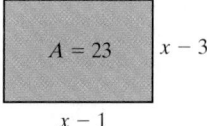

$A = 23$, $x - 3$, $x - 1$

89. Stopping Distance on Snow The formula for approximating the stopping distance, d, in feet, for a specific car on snow is $d = \dfrac{1}{6}x^2$, where x is the speed of the car, in miles per hour, before the brakes are applied. If the car's stopping distance was 24 feet, what was the car's speed before the brakes were applied?

90. Stopping Distance on Dry Pavement The formula for approximating the stopping distance, d, in feet, for a specific car on dry pavement is $d = \frac{1}{10}x^2$, where x is the speed of the car, in miles per hour, before the brakes are applied. If the car's stopping distance was 90 feet, what was the car's speed before the brakes were applied?

91. Integers The product of two consecutive positive odd integers is 35. Find the two odd integers.

92. Integers The larger of two integers is 2 more than twice the smaller. Find the two numbers if their product is 12.

93. Rectangular Garden Donna Simm has marked off an area in her yard where she will plant tomato plants. Find the dimensions of the rectangular area if the length is 2 feet more than twice the width and the area is 60 square feet.

94. Driveway Manuel Cortez is planning to blacktop his driveway. Find the dimensions of the rectangular driveway if its area is 381.25 square feet and its length is 18 feet greater than its width.

95. Patio Bill Justice is designing a square patio whose diagonal is 6 feet longer than the length of a side. Find the dimensions of the patio.

96. Wading Pool The Lakeside Hotel is planning to build a shallow wading pool for children. If the pool is to be square and the diagonal of the square is 7 feet longer than a side, find the dimensions of the pool.

97. Inscribed Triangle When a triangle is inscribed in a semicircle where a diameter of the circle is a side of the triangle, the triangle formed is always a right triangle. If an isosceles triangle (two equal sides) is inscribed in a semicircle of radius 10 inches, find the length of the other two sides of the triangle.

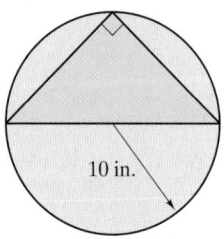

10 in.

98. Inscribed Triangle Refer to Exercise 97. Suppose a triangle is inscribed in a semicircle whose diameter is 12 meters. If one side of the inscribed triangle is 6 meters, find the third side.

99. Area of Circle The area of a circle is 24π square feet. Use the formula $A = \pi r^2$ to find the radius of the circle.

100. Area of Circle The area of a circle is 16.4π square meters. Find the radius of the circle.

Use the formula $A = p\left(1 + \dfrac{r}{n}\right)^{nt}$ *to answer Exercises 101–104.*

101. Savings Account Frank Dipalo initially invested $500 in a savings account where interest is compounded annually. If after 2 years the amount in the account is $540.80, find the annual interest rate.

102. Savings Account Margret Chang initially invested $1500 in a savings account where interest is compounded annually. If after 3 years the amount in the account is $1687.30, find the annual interest rate.

103. Savings Account Steve Rodi initially invested $1200 in a savings account where interest is compounded semiannually. If after 3 years the amount in the account is $1432.86, find the annual interest rate.

104. Savings Account Angela Reyes initially invested $1500 in a savings account where interest is compounded semiannually. If after 4 years the amount in the account is $2052.85, find the annual interest rate.

105. Surface Area and Volume The surface area, S, and volume, V, of a right circular cylinder of radius, r, and height, h, are given by the formulas

$$S = 2\pi r^2 + 2\pi rh, \qquad V = \pi r^2 h$$

a) Find the surface area of the cylinder if its height is 10 inches and its volume is 160 cubic inches.

b) Find the radius if the height is 10 inches and the volume is 160 cubic inches.

c) Find the radius if the height is 10 inches and the surface area is 160 square inches.

Group Activity

Discuss and answer Exercise 106 as a group.

106. On the following grid, the points (x_1, y_1), (x_2, y_2), and (x_1, y_2) are plotted.

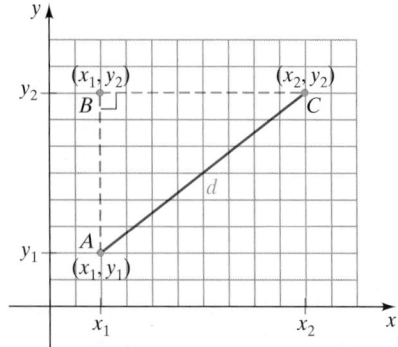

a) Explain why (x_1, y_2) is placed where it is and not somewhere else on the graph.

b) Express the length of the orange dashed line in terms of y_2 and y_1. Explain how you determined your answer.

c) Express the length of the green dashed line in terms of x_2 and x_1.

d) Using the Pythagorean Theorem and the right triangle ABC, derive a formula for the distance, d, between points (x_1, y_1) and (x_2, y_2).* Explain how you determined the formula.

e) Use the formula you determined in part **d)** to find the distance of the line segment between the points $(1, 4)$ and $(3, 7)$.

*The distance formula will be discussed in a later chapter.

Cumulative Review Exercises

[2.1]   **107.** Solve $-4(2z - 6) = -3(z - 4) + z$.

[2.4]   **108. Investment** Thea Prettyman invested $10,000 for 1 year, part at 7% and part at $6\frac{1}{4}$%. If she earned a total interest of $656.50, how much was invested at each rate?

[2.6]   **109.** Solve $|x + 3| = |2x - 7|$.

[3.4]   **110.** Find the slope of the line through $(-2, 5)$ and $(0, 5)$.

[5.2]   **111.** Multiply $(x - 2)(4x^2 + 9x - 3)$.

8.2 Solving Quadratic Equations by the Quadratic Formula

1. Derive the quadratic formula.

2. Use the quadratic formula to solve equations.

3. Determine a quadratic equation given its solutions.

4. Use the discriminant to determine the number of real solutions to a quadratic equation.

5. Study applications that use quadratic equations.

Understanding Algebra

Although any quadratic equation can also be solved by completing the square, we generally prefer to use the quadratic formula because it is easier and more efficient to use. However, completing the square is a useful skill that can be used in other areas of algebra including the study of conic sections in Chapter 10.

1 Derive the Quadratic Formula

The quadratic formula can be used to solve any quadratic equation. *It is the most useful and most versatile method of solving quadratic equations.*

The standard form of a quadratic equation is $ax^2 + bx + c = 0$, where a is the coefficient of the squared term, b is the coefficient of the first-degree term, and c is the constant.

Quadratic Equation in Standard Form	Values of Coefficients
$x^2 - 3x + 4 = 0$	$a = 1, \quad b = -3, \quad c = 4$
$1.3x^2 - 7.9 = 0$	$a = 1.3, \quad b = 0, \quad c = -7.9$
$-\frac{5}{6}x^2 + \frac{3}{8}x = 0$	$a = -\frac{5}{6}, \quad b = \frac{3}{8}, \quad c = 0$

We can derive the quadratic formula by starting with a quadratic equation in standard form and completing the square, as discussed in the preceding section.

$$ax^2 + bx + c = 0$$

$$\frac{ax^2}{a} + \frac{b}{a}x + \frac{c}{a} = 0 \qquad \text{Divide both sides by } a.$$

$$x^2 + \frac{b}{a}x = -\frac{c}{a} \qquad \text{Subtract } c/a \text{ from both sides.}$$

$$x^2 + \frac{b}{a}x + \frac{b^2}{4a^2} = -\frac{c}{a} + \frac{b^2}{4a^2} \qquad \text{Take } 1/2 \text{ of } b/a \text{ (than is, } b/2a\text{), and square it to get } b^2/4a^2. \text{ Then add this expression to both sides.}$$

$$\left(x + \frac{b}{2a}\right)^2 = \frac{b^2}{4a^2} - \frac{c}{a} \qquad \text{Rewrite the left side of the equation as the square of a binomial.}$$

$$\left(x + \frac{b}{2a}\right)^2 = \frac{b^2 - 4ac}{4a^2} \qquad \text{Write the right side with a common denominator.}$$

$$x + \frac{b}{2a} = \pm\sqrt{\frac{b^2 - 4ac}{4a^2}} \qquad \text{Square root property}$$

$$x + \frac{b}{2a} = \pm\frac{\sqrt{b^2 - 4ac}}{2a} \qquad \text{Quotient rule for radicals}$$

$$x = -\frac{b}{2a} \pm \frac{\sqrt{b^2 - 4ac}}{2a} \qquad \text{Subtract } b/2a \text{ from both sides.}$$

$$x = \frac{-b \pm \sqrt{b^2 - 4ac}}{2a} \qquad \text{Write with a common denominator to get the quadratic formula.}$$

2 Use the Quadratic Formula to Solve Equations

Now that we have derived the quadratic formula, we will use it to solve quadratic equations.

To Solve a Quadratic Equation by the Quadratic Formula

1. Write the quadratic equation in standard form, $ax^2 + bx + c = 0$, and determine the numerical values for a, b, and c.

2. Substitute the values for a, b, and c into the quadratic formula and then evaluate the formula to obtain the solution.

The Quadratic Formula

$$x = \frac{-b \pm \sqrt{b^2 - 4ac}}{2a}$$

EXAMPLE 1 Solve $x^2 + 2x - 8 = 0$ by using the quadratic formula.

Solution   In this equation, $a = 1, b = 2$, and $c = -8$.

$$x = \frac{-b \pm \sqrt{b^2 - 4ac}}{2a}$$

$$x = \frac{-2 \pm \sqrt{2^2 - 4(1)(-8)}}{2(1)}$$

$$= \frac{-2 \pm \sqrt{4 + 32}}{2}$$

$$= \frac{-2 \pm \sqrt{36}}{2}$$

$$= \frac{-2 \pm 6}{2}$$

$$x = \frac{-2 + 6}{2} \quad \text{or} \quad x = \frac{-2 - 6}{2}$$

$$x = \frac{4}{2} = 2 \qquad\qquad x = \frac{-8}{2} = -4$$

A check will show that both 2 and −4 are solutions to the equation. Note that the solutions to the equation $x^2 + 2x - 8 = 0$ are two real numbers.

Now Try Exercise 23

Helpful Hint

The solution to Example 1 could also be obtained by factoring, as follows:

$$x^2 + 2x - 8 = 0$$

$$(x + 4)(x - 2) = 0$$

$$x + 4 = 0 \quad \text{or} \quad x - 2 = 0$$

$$x = -4 \qquad\qquad x = 2$$

When you are given a quadratic equation to solve and the method to solve it has not been specified, you may try solving by factoring first (as we discussed in Section 5.8). If the equation cannot be easily factored, use the quadratic formula.

EXAMPLE 2 Solve $-9x^2 = -6x + 1$ by the quadratic formula.

Solution   Begin by adding $9x^2$ to both sides of the equation to obtain

$$0 = 9x^2 - 6x + 1$$

$$\text{or} \quad 9x^2 - 6x + 1 = 0$$

Understanding Algebra

When using the quadratic formula, the calculations are easier if the leading coefficient, a, is a positive integer. Thus if solving the equation

$$-x^2 + 3x = 2,$$

we can add x^2 to both sides and subtract $3x$ from both sides to get the equivalent equation

$$0 = x^2 - 3x + 2$$

or $x^2 - 3x + 2 = 0$

$$a = 9, \qquad b = -6, \qquad c = 1$$

$$
\begin{aligned}
x &= \frac{-b \pm \sqrt{b^2 - 4ac}}{2a} \\[2mm]
&= \frac{-(-6) \pm \sqrt{(-6)^2 - 4(9)(1)}}{2(9)} \\[2mm]
&= \frac{6 \pm \sqrt{36 - 36}}{18} = \frac{6 \pm \sqrt{0}}{18} = \frac{6}{18} = \frac{1}{3}
\end{aligned}
$$

Note that the solution to the equation $-9x^2 = -6x + 1$ is $\frac{1}{3}$, a single value. Some quadratic equations have just one value as the solution.

Now Try Exercise 39

Avoiding Common Errors

The entire numerator of the quadratic formula must be divided by $2a$.

CORRECT	INCORRECT
$x = \dfrac{-b \pm \sqrt{b^2 - 4ac}}{2a}$	$x = -b \pm \dfrac{\sqrt{b^2 - 4ac}}{2a}$
	$x = \dfrac{-b}{2a} \pm \sqrt{b^2 - 4ac}$

EXAMPLE 3 Solve $p^2 + \frac{1}{3}p + \frac{5}{6} = 0$ by using the quadratic formula.

Solution   Begin by clearing the fractions from the equation by multiplying both sides of the equation by the least common denominator, 6.

$$6\left(p^2 + \frac{1}{3}p + \frac{5}{6}\right) = 6(0)$$

$$6p^2 + 2p + 5 = 0$$

Now we can use the quadratic formula with $a = 6, b = 2,$ and $c = 5$.

$$
\begin{aligned}
p &= \frac{-b \pm \sqrt{b^2 - 4ac}}{2a} \\[2mm]
&= \frac{-2 \pm \sqrt{2^2 - 4(6)(5)}}{2(6)} \\[2mm]
&= \frac{-2 \pm \sqrt{-116}}{12} \\[2mm]
&= \frac{-2 \pm \sqrt{-4}\sqrt{29}}{12} \\[2mm]
&= \frac{-2 \pm 2i\sqrt{29}}{12} \\[2mm]
&= \frac{\overset{1}{2}(-1 \pm i\sqrt{29})}{\underset{6}{\cancel{12}}} \\[2mm]
&= \frac{-1 \pm i\sqrt{29}}{6}
\end{aligned}
$$

The solutions are $\dfrac{-1 + i\sqrt{29}}{6}$ and $\dfrac{-1 - i\sqrt{29}}{6}$. Note that neither solution is a real number. Both solutions are complex numbers.

Now Try Exercise 53

Avoiding Common Errors

Some students use the quadratic formula correctly until the last step, where they make an error. Below are illustrated both the correct and incorrect procedures for simplifying an answer.

When *both* terms in the numerator *and* the denominator have a common factor, that common factor may be divided out, as follows:

CORRECT

$$\frac{2 + 4\sqrt{3}}{2} = \frac{\overset{1}{\cancel{2}}(1 + 2\sqrt{3})}{\underset{1}{\cancel{2}}} = 1 + 2\sqrt{3}$$

$$\frac{6 + 3\sqrt{3}}{6} = \frac{\overset{1}{\cancel{3}}(2 + \sqrt{3})}{\underset{2}{\cancel{6}}} = \frac{2 + \sqrt{3}}{2}$$

Below are some common errors. Study them carefully so you will not make them. Can you explain why each of the following procedures is incorrect?

INCORRECT

$$\frac{\cancel{2} + 3}{2} = \frac{\overset{1}{\cancel{2}} + 3}{\underset{1}{\cancel{2}}} \qquad\qquad \frac{3 + 2\sqrt{5}}{2} = \frac{3 + \overset{1}{\cancel{2}}\sqrt{5}}{\underset{1}{\cancel{2}}}$$

$$\frac{3 + \cancel{\sqrt{6}}}{2} = \frac{3 + \sqrt{\cancel{6}^3}}{\underset{1}{\cancel{2}}} \qquad\qquad \frac{4 + 3\sqrt{5}}{2} = \frac{\overset{2}{\cancel{4}} + 3\sqrt{5}}{\underset{1}{\cancel{2}}}$$

Note that $\dfrac{2 + 3}{2}$ simplifies to $\dfrac{5}{2}$. However, $\dfrac{3 + 2\sqrt{5}}{2}$, $\dfrac{3 + \sqrt{6}}{2}$, and $\dfrac{4 + 3\sqrt{5}}{2}$ cannot be simplified any further.

EXAMPLE 4 Given $f(x) = 2x^2 + 4x$, find all real values of x for which $f(x) = 5$.

Solution   We wish to determine all real values of x for which $2x^2 + 4x = 5$

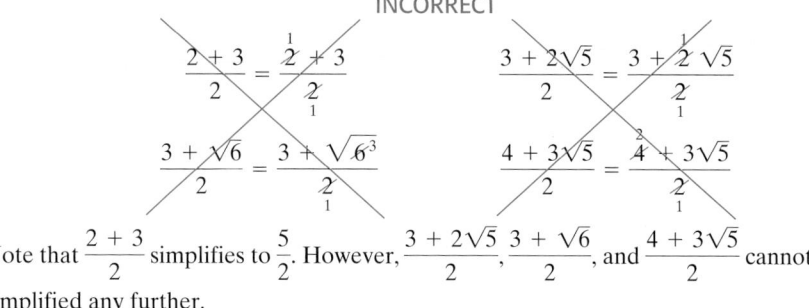

$$\begin{array}{ll}
2x^2 + 4x = 5 & \text{Subtract 5 from both sides.}\\
2x^2 + 4x - 5 = 0 & \text{Use quadratic formula with } a = 2, b = 4, c = -5.
\end{array}$$

$$x = \frac{-b \pm \sqrt{b^2 - 4ac}}{2a}$$

$$= \frac{-4 \pm \sqrt{4^2 - 4(2)(-5)}}{2(2)} = \frac{-4 \pm \sqrt{56}}{4} = \frac{-4 \pm 2\sqrt{14}}{4}$$

Next, factor out 2 from both terms in the numerator, and then divide out the common factor.

$$x = \frac{\overset{1}{\cancel{2}}(-2 \pm \sqrt{14})}{\underset{2}{\cancel{4}}} = \frac{-2 \pm \sqrt{14}}{2}\;^*$$

Thus, the solutions are $\dfrac{-2 + \sqrt{14}}{2}$ and $\dfrac{-2 - \sqrt{14}}{2}$.

Note that the expression in Example 4, $2x^2 + 4x - 5$, is not factorable. Therefore, Example 4 could not be solved by factoring.

Now Try Exercise 69

Understanding Algebra

If the numerical coefficients of a quadratic equation have a common factor, divide each term by the common factor. For example, for the equation $3x^2 + 12x + 3 = 0$, first divide each term by the common factor, 3, and simplify:

$$\frac{3x^2}{3} + \frac{12x}{3} + \frac{3}{3} = \frac{0}{3}$$
$$x^2 + 4x + 1 = 0$$

This new equation is equivalent to the original equation and is easier to solve.

---

*Solutions will be given in this form in the Answer Section.

3 Determine a Quadratic Equation Given Its Solutions

If we are given the solutions of an equation, we can find the equation by working backward. This procedure is illustrated in Example 5.

EXAMPLE 5 Determine an equation that has the following solutions:

a) -5 and 1 **b)** $3 + 2i$ and $3 - 2i$

Solution

a) If the solutions are -5 and 1 we write

$$x = -5 \quad \text{or} \quad x = 1$$
$$x + 5 = 0 \qquad x - 1 = 0 \quad \text{Set equations equal to 0.}$$
$$(x + 5)(x - 1) = 0 \quad \text{Zero-factor property}$$
$$x^2 - x + 5x - 5 = 0 \quad \text{Multiply factors.}$$
$$x^2 + 4x - 5 = 0 \quad \text{Combine like terms.}$$

Thus, the equation is $x^2 + 4x - 5 = 0$. Many other equations have solutions -5 and 1. In fact, any equation of the form $k(x^2 + 4x - 5) = 0$, where k is a non-zero constant, has those solutions.

b)
$$x = 3 + 2i \quad \text{or} \quad x = 3 - 2i$$
$$x - (3 + 2i) = 0 \qquad x - (3 - 2i) = 0 \quad \text{Set equations equal to 0.}$$
$$[x - (3 + 2i)][x - (3 - 2i)] = 0 \quad \text{Zero-factor property}$$
$$x \cdot x - x(3 - 2i) - x(3 + 2i) + (3 + 2i)(3 - 2i) = 0 \quad \text{Multiply.}$$
$$x^2 - 3x + 2xi - 3x - 2xi + (9 - 4i^2) = 0 \quad \text{Distributive property; multiply}$$
$$x^2 - 6x + 9 - 4i^2 = 0 \quad \text{Combine like terms.}$$
$$x^2 - 6x + 9 - 4(-1) = 0 \quad \text{Substitute: } i^2 = -1.$$
$$x^2 - 6x + 13 = 0 \quad \text{Simplify.}$$

The equation $x^2 - 6x + 13 = 0$ has the complex solutions $3 + 2i$ and $3 - 2i$.

Now Try Exercise 75

In Example 5 **a)**, the equation $x^2 + 4x - 5 = 0$ had the real number solutions -5 and 1. The solutions correspond to the x-intercepts $(-5, 0)$ and $(1, 0)$ of the graph of the function $f(x) = x^2 + 4x - 5$, as shown in **Figure 8.1**.

Understanding Algebra

Recall from Chapter 3 that an *x-intercept* of a graph is a point $(x, 0)$ where a graph crosses the x-axis. To find the x-intercept of a graph, we set $y = 0$ or set $f(x) = 0$ and solve for x. If x is a real number then the point $(x, 0)$ is an x-intercept.

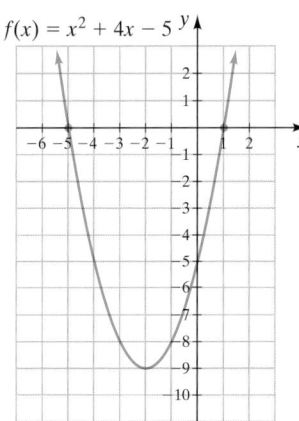

FIGURE 8.1

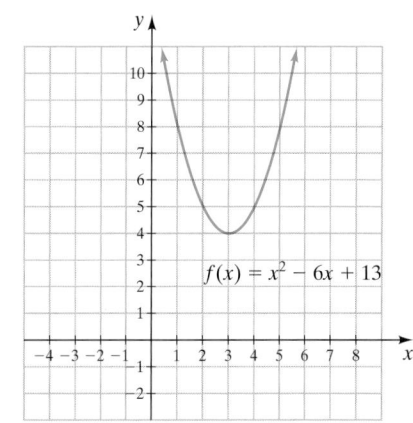

FIGURE 8.2

In Example 5 **b)**, the equation $x^2 - 6x + 13 = 0$ had only complex solutions and no real number solutions. Thus the graph of the function $f(x) = x^2 - 6x + 13$ has no x-intercepts, as shown in **Figure 8.2**.

4 Use the Discriminant to Determine the Number of Real Solutions to a Quadratic Equation

Discriminant

The **discriminant** of a quadratic equation is the expression under the radical sign in the quadratic formula.

$$b^2 - 4ac$$

Discriminant

The discriminant provides information to determine the number and kinds of solutions of a quadratic equation.

Solutions of a Quadratic Equation

For a quadratic equation of the form $ax^2 + bx + c = 0, a \neq 0$:

If $b^2 - 4ac > 0$, the quadratic equation has two distinct real number solutions.
If $b^2 - 4ac = 0$, the quadratic equation has a single real number solution.
If $b^2 - 4ac < 0$, the quadratic equation has no real number solution.

EXAMPLE 6

a) Find the discriminant of the equation $x^2 - 8x + 16 = 0$.

b) How many real number solutions does the given equation have?

c) Use the quadratic formula to find the solution(s).

Solution

a) $a = 1, \qquad b = -8, \qquad c = 16$

$$b^2 - 4ac = (-8)^2 - 4(1)(16)$$
$$= 64 - 64 = 0$$

b) Since the discriminant equals 0, there is a single real number solution.

c)
$$x = \frac{-b \pm \sqrt{b^2 - 4ac}}{2a}$$
$$= \frac{-(-8) \pm \sqrt{0}}{2(1)} = \frac{8 \pm 0}{2} = \frac{8}{2} = 4$$

The only solution is 4.

Now Try Exercise 9

EXAMPLE 7

Without actually finding the solutions, determine whether the following equations have two distinct real number solutions, a single real number solution, or no real number solution.

a) $2x^2 - 4x + 6 = 0$ **b)** $x^2 - 5x - 3 = 0$ **c)** $4x^2 - 12x = -9$

Solution We use the discriminant of the quadratic formula to answer these questions.

a) $b^2 - 4ac = (-4)^2 - 4(2)(6) = 16 - 48 = -32$

Since the discriminant is negative, this equation has no real number solution.

b) $b^2 - 4ac = (-5)^2 - 4(1)(-3) = 25 + 12 = 37$

Since the discriminant is positive, this equation has two distinct real number solutions.

c) First, rewrite $4x^2 - 12x = -9$ as $4x^2 - 12x + 9 = 0$.

$$b^2 - 4ac = (-12)^2 - 4(4)(9) = 144 - 144 = 0$$

Since the discriminant is 0, this equation has a single real number solution.

Now Try Exercise 15

The discriminant can be used to find the number of real solutions to an equation of the form $ax^2 + bx + c = 0$. Since the x-intercepts of a quadratic function, $f(x) = ax^2 + bx + c$, occur where $f(x) = 0$, the discriminant can also be used to find the number of x-intercepts of a quadratic function. **Figure 8.3** shows the relationship between the discriminant and the number of x-intercepts for a function of the form $f(x) = ax^2 + bx + c$.

Graphs of $f(x) = ax^2 + bx + c$

If $b^2 - 4ac > 0$, $f(x)$ has two distinct x-intercepts.

If $b^2 - 4ac = 0$, $f(x)$ has a single x-intercept.

If $b^2 - 4ac < 0$, $f(x)$ has no x-intercepts.

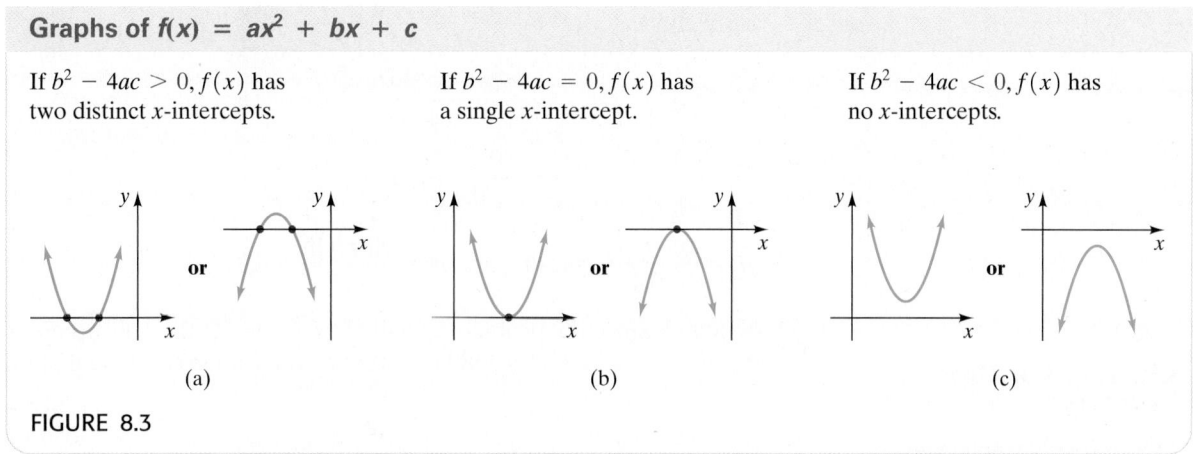

(a)   (b)   (c)

FIGURE 8.3

---

5 Study Applications That Use Quadratic Equations

We will now look at some applications of quadratic equations.

EXAMPLE 8 **Cell Phones** Mary Olson owns a business that sells cell phones. The revenue, $R(n)$, from selling the cell phones is determined by multiplying the number of cell phones by the selling price per phone. Suppose the revenue from selling n cell phones, $n \leq 50$, is

$$R(n) = n(50 - 0.2n)$$

where $(50 - 0.2n)$ is the price per cell phone, in dollars.

a) Find the revenue when 30 cell phones are sold.

b) How many cell phones must be sold to have a revenue of $480?

Solution

a) To find the revenue when 30 cell phones are sold, we evaluate the revenue function for $n = 30$.

$$R(n) = n(50 - 0.2n)$$
$$R(30) = 30[50 - 0.2(30)]$$
$$= 30(50 - 6)$$
$$= 30(44)$$
$$= 1320$$

The revenue from selling 30 cell phones is $1320.

b) Understand   We want to find the number of cell phones that need to be sold to have $480 in revenue. Thus, we need to let $R(n) = 480$ and solve for n.

$$R(n) = n(50 - 0.2n)$$
$$480 = n(50 - 0.2n)$$
$$480 = 50n - 0.2n^2$$
$$0.2n^2 - 50n + 480 = 0$$

Now we can use the quadratic formula to solve the equation.

Translate $a = 0.2,$ $b = -50,$ $c = 480$

$$n = \frac{-b \pm \sqrt{b^2 - 4ac}}{2a}$$

$$= \frac{-(-50) \pm \sqrt{(-50)^2 - 4(0.2)(480)}}{2(0.2)}$$

Carry Out $= \frac{50 \pm \sqrt{2500 - 384}}{0.4}$

$$= \frac{50 \pm \sqrt{2116}}{0.4}$$

$$= \frac{50 \pm 46}{0.4}$$

$$n = \frac{50 + 46}{0.4} = 240 \quad \text{or} \quad n = \frac{50 - 46}{0.4} = 10$$

Answer   Since the problem specified that $n \le 50$, the only acceptable solution is $n = 10$. Thus, to obtain $480 in revenue, Mary must sell 10 cell phones.

Now Try Exercise 87

Understanding Algebra

In the projectile motion equation, the variables v_0 and h_0 have *subscripts* of 0. Subscripts of 0 usually refer to the *initial* value for a variable. Thus, v_0 refers to the *initial* velocity with which the object is projected upward and h_0 refers to the *initial* height from which the object is projected.

An important equation in physics relates the height of an object to the time after the object is projected upward.

Projectile Motion Equation

The height, h, of an object t seconds after it is projected upward can be found by solving the equation

$$h = \frac{1}{2}gt^2 + v_0 t + h_0, \text{ where}$$

- g is the acceleration due to gravity,
- v_0 is the initial velocity of the object, and
- h_0 is the initial height of the object.

Before we use the equation, a few notes about g, the acceleration due to gravity.

- When we measure the height of the object in feet, the acceleration due to gravity on earth is -32 ft/sec^2 or $g = -32$.
- When we measure the height of an object in meters, the acceleration due to gravity on earth is -9.8 m/sec^2 or $g = -9.8$.
- The value of g will be different on the moon or on a planet other than Earth, but we can still use the projectile motion formula.

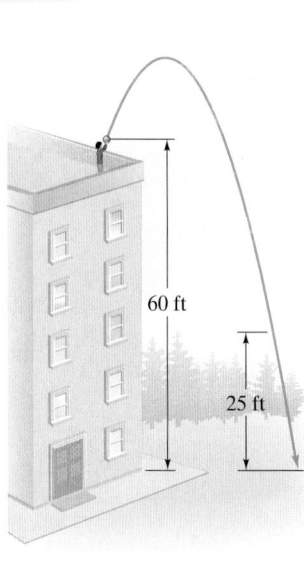

FIGURE 8.4

EXAMPLE 9   **Throwing a Ball** Betty Heller is on top of a building and throws a baseball upward from an initial height of 60 feet with an initial velocity of 30 feet per second. Use the projectile motion equation to answer the following questions. Round your answers to the nearest tenth of a second.

a) How long after Betty throws the ball will the ball be 25 feet above the ground?

b) How long after Betty throws the ball will the ball hit the ground?

c) What is the height of the ball after 2 seconds?

Solution

a) Understand   The problem is illustrated in **Figure 8.4**. We are asked to find the time, t, it takes for the object to reach a height, h, of 25 feet. Thus we have the following values to substitute into the projectile motion equation: $h = 25$, $g = -32$, $v_0 = 30$, and $h_0 = 60$.

Translate $$h = \frac{1}{2}gt^2 + v_0t + h_0,$$

$$25 = \frac{1}{2}(-32)t^2 + (30)t + 60$$

Carry Out $25 = -16t^2 + 30t + 60$ Add $16t^2$, subtract $30t$, and subtract 60 from both sides to get a quadratic equation in standard form with a positive leading coefficient.

$16t^2 - 30t - 35 = 0$

$$t = \frac{-b \pm \sqrt{b^2 - 4ac}}{2a}$$ Solve using the quadratic formula with $a = 16, b = -30$, and $c = -35$.

$$= \frac{-(-30) \pm \sqrt{(-30)^2 - 4(16)(-35)}}{2(16)}$$

$$= \frac{30 \pm \sqrt{3140}}{32}$$

$$t = \frac{30 + \sqrt{3140}}{32} \quad \text{or} \quad t = \frac{30 - \sqrt{3140}}{32}$$

$$\approx 2.7 \qquad\qquad\qquad \approx -0.8$$

Answer   Since time cannot be negative, the only reasonable solution is 2.7 seconds. Thus, about 2.7 seconds after the ball is thrown upward, it will be 25 feet above the ground.

b) Understand   When the ball strikes the ground, its distance above the ground is 0. We substitute $h = 0$ into the formula and solve for t.

Translate $$h = \frac{1}{2}gt^2 + v_0t + h_0$$

$$0 = \frac{1}{2}(-32)t^2 + 30t + 60$$

Carry Out $0 = -16t^2 + 30t + 60$

$16t^2 - 30t - 60 = 0$ Multiply both sides of the equation by -1 to get a positive leading coefficient.

$$t = \frac{-b \pm \sqrt{b^2 - 4ac}}{2a}$$ Solve using the quadratic formula with $a = 16, b = -30$, and $c = -60$

$$= \frac{-(-30) \pm \sqrt{(-30)^2 - 4(16)(-60)}}{2(16)}$$

$$= \frac{30 \pm \sqrt{4740}}{32}$$

$$t = \frac{30 + \sqrt{4740}}{32} \quad \text{or} \quad t = \frac{30 - \sqrt{4740}}{32}$$

$$\approx 3.1 \qquad\qquad\qquad \approx -1.2$$

Answer   Since time cannot be negative, the only reasonable solution is 3.1 seconds. Thus, the ball will strike the ground after about 3.1 seconds.

c) Understand   We are asked to find the height, h, after 2 seconds. We will substitute $t = 2$ into the projectile motion equation and solve for h.

$$h = \frac{1}{2}gt^2 + v_0t + h_0$$

Translate and Carry Out $$h = \frac{1}{2}(-32)(2)^2 + 30(2) + 60$$

$$= -64 + 60 + 60 = 56$$

Answer   Thus, the ball will be at a height of 56 feet after 2 seconds.

Now Try Exercise 97

EXERCISE SET 8.2

MathXL MathXL®   MyMathLab MyMathLab

Warm-Up Exercises

Fill in the blanks with the appropriate word, phrase, or symbol(s) from the following list.

no	one	two	completing the square	discriminant	positive integer
projectile motion	divide	multiply	factoring		linear

1. Although any quadratic equation can be solved by _____ we generally prefer to use the quadratic formula because it is easier and more efficient to use.

2. When using the quadratic formula, the calculations are generally easier if the leading coefficient, a, is a _____.

3. If the numerical coefficients of all of the terms in a quadratic equation have a common factor, _____ each term by the common factor.

4. The _____ of a quadratic equation is the expression under the radical sign in the quadratic formula: $b^2 - 4ac$.

5. If the discriminant of a quadratic equation is positive, the equation has _____ distinct real number solution(s).

6. If the discriminant of a quadratic equation is equal to zero, the equation has _____ real number solution(s).

7. If the discriminant of a quadratic equation is negative, the equation has _____ real number solution(s).

8. The height, h, of an object t seconds after it is projected upward can be found using the _____ equation.

Practice the Skills

Use the discriminant to determine whether each equation has two distinct real solutions, a single real solution, or no real solution.

9. $x^2 + 6x + 2 = 0$

10. $2x^2 + x + 3 = 0$

11. $4z^2 + 6z + 5 = 0$

12. $-a^2 + 3a - 6 = 0$

13. $5p^2 + 3p - 7 = 0$

14. $2x^2 = 16x - 32$

15. $-5x^2 + 5x - 8 = 0$

16. $4.1x^2 - 3.1x - 2.8 = 0$

17. $x^2 + 10.2x + 26.01 = 0$

18. $\frac{1}{2}x^2 + \frac{2}{3}x + 10 = 0$

19. $b^2 = -3b - \frac{9}{4}$

20. $\frac{x^2}{3} = \frac{2x}{7}$

Solve each equation by using the quadratic formula.

21. $x^2 + 7x + 10 = 0$

22. $x^2 + 9x + 18 = 0$

23. $a^2 - 6a + 8 = 0$

24. $a^2 + 6a + 8 = 0$

25. $x^2 = -6x + 7$

26. $-a^2 - 9a + 10 = 0$

27. $-b^2 = 4b - 20$

28. $a^2 - 16 = 0$

29. $b^2 - 64 = 0$

30. $2x^2 = 4x + 1$

31. $3w^2 - 4w + 5 = 0$

32. $x^2 - 6x = 0$

33. $c^2 - 5c = 0$

34. $-t^2 - t - 1 = 0$

35. $4s^2 - 8s + 6 = 0$

36. $-3r^2 = 9r + 6$

37. $a^2 + 2a + 1 = 0$

38. $y^2 + 16y + 64 = 0$

39. $16x^2 - 8x + 1 = 0$

40. $100m^2 + 20m + 1 = 0$

41. $x^2 - 2x - 1 = 0$

42. $2 - 3r^2 = -4r$

43. $-n^2 = 3n + 6$

44. $-9d - 3d^2 = 5$

45. $2x^2 + 5x - 3 = 0$

46. $(r - 3)(3r + 4) = -10$

47. $(2a + 3)(3a - 1) = 2$

48. $6x^2 = 21x + 27$

49. $\frac{1}{2}t^2 + t - 12 = 0$

50. $\frac{2}{3}x^2 = 8x - 18$

51. $9r^2 + 3r - 2 = 0$

52. $2x^2 - 4x - 2 = 0$

53. $\frac{1}{2}x^2 + 2x + \frac{2}{3} = 0$

54. $x^2 - \frac{11}{3}x = \frac{10}{3}$

55. $a^2 - \frac{a}{5} - \frac{1}{3} = 0$

56. $b^2 = -\frac{b}{2} + \frac{2}{3}$

57. $c = \frac{c - 6}{4 - c}$

58. $3y = \frac{5y + 6}{2y + 3}$

59. $2x^2 - 4x + 5 = 0$

60. $3a^2 - 4a = -5$

61. $y^2 + \frac{y}{2} = -\frac{3}{2}$

62. $2b^2 - \frac{7}{3}b + \frac{4}{3} = 0$

63. $0.1x^2 + 0.6x - 1.2 = 0$

64. $2.3x^2 - 5.6x - 0.4 = 0$

For each function, determine all real values of the variable for which the function has the value indicated.

65. $f(x) = 2x^2 - 3x + 7, f(x) = 7$

66. $g(x) = x^2 + 3x + 8, g(x) = 8$

67. $k(x) = x^2 - x - 15, k(x) = 15$

68. $p(r) = r^2 + 17r + 81, p(r) = 9$

69. $h(t) = 2t^2 - 7t + 6, h(t) = 2$

70. $t(x) = x^2 + 5x - 4, t(x) = 3$

71. $g(a) = 2a^2 - 3a + 16, g(a) = 14$

72. $h(x) = 6x^2 + 3x + 1, h(x) = -7$

Determine a quadratic equation that has the given solutions.

73. $1, 6$

74. $-3, 4$

75. $1, -9$

76. $-2, -6$

77. $-\dfrac{3}{5}, \dfrac{2}{3}$

78. $-\dfrac{1}{3}, \dfrac{3}{4}$

79. $\sqrt{2}, -\sqrt{2}$

80. $\sqrt{5}, -\sqrt{5}$

81. $3i, -3i$

82. $8i, -8i$

83. $3 + \sqrt{2}, 3 - \sqrt{2}$

84. $5 - \sqrt{3}, 5 + \sqrt{3}$

85. $2 + 3i, 2 - 3i$

86. $5 - 4i, 5 + 4i$

Problem Solving

In Exercises 87–90, **a)** *set up a revenue function, R(n), that can be used to solve the problem, and then,* **b)** *solve the problem. See Example 8.*

87. Selling Lamps A business sells n lamps, $n \le 65$, at a price of $(10 - 0.02n)$ dollars per lamp. How many lamps must be sold to have a revenue of $450?

88. Selling Batteries A business sells n batteries, $n \le 26$, at a price of $(25 - 0.1n)$ dollars per battery. How many batteries must be sold to have a revenue of $460?

89. Selling Chairs A business sells n chairs, $n \le 50$, at a price of $(50 - 0.4n)$ dollars per chair. How many chairs must be sold to have a revenue of $660?

90. Selling Watches A business sells n watches, $n \le 75$, at a price of $(30 - 0.15n)$ dollars per watch. How many watches must be sold to have a revenue of $1260?

© Tatiana Popova/Shutterstock

In Exercises 91–108, use a calculator as needed to give the solution in decimal form. Round irrational numbers to the nearest hundredth.

91. Numbers Twice the square of a positive number increased by three times the number is 27. Find the number.

92. Numbers Three times the square of a positive number decreased by twice the number is 21. Find the number.

93. Rectangular Garden The length of a rectangular garden is 1 foot less than 3 times its width. Find the length and width if the area of the garden is 24 square feet.

94. Rectangular Region Lora Wallman wishes to fence in a rectangular region along a riverbank by constructing fencing as il-

lustrated in the diagram. If she has only 400 feet of fencing and wishes to enclose an area of 15,000 square feet, find the dimensions of the rectangular region.

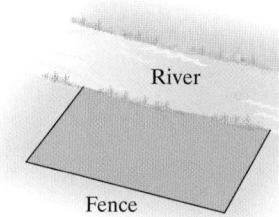

River

Fence

95. Photo John Williams, a professional photographer, has a 6-inch-by-8-inch photo. He wishes to reduce the photo by the same amount on each side so that the resulting photo will have half the area of the original photo. By how much will he have to reduce the length of each side?

96. Rectangular Garden Bart Simmons has a 12-meter-by-9-meter flower garden. He wants to build a gravel path of uniform width along the inside of the garden on each side so that the resulting garden will have half the area of the original garden. What will be the width of the gravel path?

In Exercises 97–100, use the projectile motion equation $h = \dfrac{1}{2}gt^2 + v_0t + h_0$. *If necessary, round your answers to the nearest tenth of a second. See Example 9.*

97. Hitting a Tennis Ball A tennis ball is hit upward from an initial height of 4 feet with an initial velocity of 40 feet per second.

a) How long after the ball is hit will it be 20 feet above the ground?

b) How long after the ball is hit will it hit the ground?

c) What is the height of the ball after 1 second?

98. Throwing a Horseshoe A horseshoe is thrown upward from an initial height of 3 feet with an initial velocity of 25 feet per second.

a) How long after the horseshoe is thrown will it be 10 feet above the ground?

b) How long after the horseshoe is thrown will it hit the ground?

c) What is the height of the horseshoe after 1 second?

99. Golfing on the Moon On February 6, 1971, astronaut Alan Shepard hit a golf ball while standing on the moon. The acceleration due to gravity, g, on the moon is about -5.3 ft/sec². Assume the ball was hit upward from an initial height of 0 feet and with an initial velocity of 26.5 ft/sec.

a) How long after the golf ball was hit would it reach a height of 23.85 feet?

(*Hint:* Substitute the appropriate values into the projectile motion equation and then divide each term by the leading coefficient.)

b) How long after the golf ball was hit would it return to the surface of the moon?

c) What would be the height of the golf ball after 5 seconds?

100. Baseball on the Moon Suppose Neil Armstrong had hit a baseball when he first set foot on the moon on July 20, 1969. Assume the ball was hit upward from an initial height of 2.65 feet and with an initial velocity of 132.5 ft/sec. See Exercise 99.

a) How long after the baseball was hit would it reach a height of 132.5 feet?

b) How long after the baseball was hit would it hit the ground?

c) What would be the height of the baseball after 25 seconds?

Neil Armstrong on the moon

Concept/Writing Exercises

101. Consider the two equations $-6x^2 + \dfrac{1}{2}x - 5 = 0$ and $6x^2 - \dfrac{1}{2}x + 5 = 0$. Must the solutions to these two equations be the same? Explain your answer.

102. Consider $12x^2 - 15x - 6 = 0$ and $3(4x^2 - 5x - 2) = 0$.

a) Will the solution to the two equations be the same? Explain.

b) Solve $12x^2 - 15x - 6 = 0$.

c) Solve $3(4x^2 - 5x - 2) = 0$.

103. a) Explain how to find the discriminant.

b) What is the discriminant for the equation $3x^2 - 6x + 10 = 0$?

c) Write a paragraph or two explaining the relationship between the value of the discriminant and the number of real solutions to a quadratic equation. In your paragraph, explain *why* the value of the discriminant determines the number of real solutions.

104. Write a paragraph or two explaining the relationship between the value of the discriminant and the number of x-intercepts of $f(x) = ax^2 + bx + c$. In your paragraph, explain when the function will have no, one, and two x-intercepts.

105. Give your own example of a quadratic equation that can be solved by the quadratic formula but not by factoring over the set of integers.

106. Are there any quadratic equations that **a)** can be solved by the quadratic formula that cannot be solved by completing the square? **b)** can be solved by completing the square that cannot be solved by factoring over the set of integers?

107. When solving a quadratic equation by the quadratic formula, if the discriminant is a perfect square, must the equation be factorable over the set of integers?

108. When solving a quadratic equation by the quadratic formula, if the discriminant is a natural number, must the equation be factorable over the set of integers?

Challenge Problems

109. Heating a Metal Cube A metal cube expands when heated. If each edge increases 0.20 millimeter after being heated and the total volume increases by 6 cubic millimeters, find the original length of a side of the cube.

110. Six Solutions The equation $x^n = 1$ has n solutions (including the complex solutions). Find the six solutions to $x^6 = 1$. (*Hint:* Rewrite the equation as $x^6 - 1 = 0$, then factor using the formula for the difference of two squares.)

111. Throwing a Rock Travis Hawley is on the fourth floor of an eight-story building and Courtney Prenzlow is on the roof. Travis is 60 feet above the ground while Courtney is 120 feet above the ground.

a) If Travis drops a rock out of a window, determine the time it takes for the rock to strike the ground.

b) If Courtney drops a rock off the roof, determine the time it takes for the rock to strike the ground.

c) If Travis throws a rock upward with an initial velocity of 100 feet per second at the same time that Courtney throws a rock upward at 60 feet per second, whose rock will strike the ground first? Explain.

d) Will the rocks ever be at the same distance above the ground? If so, at what time?

Cumulative Review Exercises

[1.6]   **112.** Evaluate $\dfrac{5.55 \times 10^3}{1.11 \times 10^1}$.

[3.2]   **113.** If $f(x) = x^2 + 2x - 8$, find $f(3)$.

[4.1]   **114.** Solve the system of equations.

$$3x + 4y = 2$$
$$2x = -5y - 1$$

[6.3]   **115.** Simplify $2x^{-1} - (3y)^{-1}$.

[7.6]   **116.** Solve $\sqrt{x^2 - 6x - 4} = x$.

8.3 Quadratic Equations: Applications and Problem Solving

1 Solve additional applications.

2 Solve for a variable in a formula.

1   Solve Additional Applications

In this section, we will explore several more applications of quadratic equations. We start by investigating the profit of a new company.

EXAMPLE 1   Company Profit   Laserox, a start-up company, projects that its annual profit, $p(t)$, in thousands of dollars, over the first 6 years of operation can be approximated by the function $p(t) = 1.2t^2 + 4t - 8$, where t is the number of years completed.

a) Estimate the profit (or loss) of the company after the first year.

b) Estimate the profit (or loss) of the company after 6 years.

c) Estimate the time needed for the company to break even.

Solution

a) To estimate the profit after 1 year, we evaluate the function at 1.

$$p(t) = 1.2t^2 + 4t - 8$$
$$p(1) = 1.2(1)^2 + 4(1) - 8 = -2.8$$

Thus, at the end of the first year the company projects a loss of $2.8 thousand or a loss of $2800.

b) $p(6) = 1.2(6)^2 + 4(6) - 8 = 59.2$

Thus, at the end of the sixth year the company's projected profit is $59.2 thousand, or a profit of $59,200.

c) Understand   The company will break even when the profit is 0. Thus, to find the break-even point (no profit or loss) we solve the equation

$$1.2t^2 + 4t - 8 = 0$$

We can use the quadratic formula to solve this equation.

Translate $a = 1.2, \qquad b = 4, \qquad c = -8$

$$t = \frac{-b \pm \sqrt{b^2 - 4ac}}{2a}$$

$$= \frac{-4 \pm \sqrt{4^2 - 4(1.2)(-8)}}{2(1.2)}$$

Carry Out

$$= \frac{-4 \pm \sqrt{16 + 38.4}}{2.4}$$

$$= \frac{-4 \pm \sqrt{54.4}}{2.4}$$

$$\approx \frac{-4 \pm 7.376}{2.4}$$

$$t \approx \frac{-4 + 7.376}{2.4} \approx 1.4 \qquad \text{or} \qquad t \approx \frac{-4 - 7.376}{2.4} \approx -4.74$$

Answer   Since time cannot be negative, the break-even time is about 1.4 years.

NowTry Exercise 29

EXAMPLE 2 Life Expectancy The function $N(t) = 0.0054t^2 - 1.46t + 95.11$ can be used to estimate the average number of years of life expectancy remaining for a person of age t years where $30 \le t \le 100$.

a) Estimate the remaining life expectancy of a person of age 40.

b) If a person has a remaining life expectancy of 14.3 years, estimate the age of the person.

Solution

a) Understand   To determine the remaining life expectancy for a 40-year-old, we substitute 40 for t in the function and evaluate.

Translate $\qquad\qquad N(t) = 0.0054t^2 - 1.46t + 95.11$

$$N(40) = 0.0054(40)^2 - 1.46(40) + 95.11$$

Carry Out $\qquad\qquad\qquad = 0.0054(1600) - 58.4 + 95.11$

$$= 8.64 - 58.4 + 95.11$$

$$= 45.35$$

Answer and Check   The answer appears reasonable. Thus, on the average, a 40-year-old can expect to live another 45.35 years to an age of 85.35 years.

b) Understand   Here we are given the remaining life expectancy, $N(t)$, and asked to find the age of the person, t. To solve this problem, we substitute 14.3 for $N(t)$ and solve for t. We will use the quadratic formula.

Translate $\qquad\qquad N(t) = 0.0054t^2 - 1.46t + 95.11$

$$14.3 = 0.0054t^2 - 1.46t + 95.11$$

Carry Out $\qquad\qquad 0 = 0.0054t^2 - 1.46t + 80.81$

$$a = 0.0054, \quad b = -1.46, \quad c = 80.81$$

$$t = \frac{-b \pm \sqrt{b^2 - 4ac}}{2a}$$

$$= \frac{-(-1.46) \pm \sqrt{(-1.46)^2 - 4(0.0054)(80.81)}}{2(0.0054)}$$

$$= \frac{1.46 \pm \sqrt{2.1316 - 1.745496}}{0.0108}$$

$$= \frac{1.46 \pm \sqrt{0.386104}}{0.0108}$$

$$\approx \frac{1.46 \pm 0.6214}{0.0108}$$

$$t \approx \frac{1.46 + 0.6214}{0.0108} \qquad \text{or} \qquad t \approx \frac{1.46 - 0.6214}{0.0108}$$

$$\approx 192.72 \qquad\qquad\qquad\qquad \approx 77.65$$

Answer   Since 192.72 is not a reasonable age, we can exclude that as a possibility. Thus, the average person who has a life expectancy of 14.3 years is about 77.65 years old.

Now Try Exercise 33

Motion Problems   We first discussed motion problems in Section 2.4.

EXAMPLE 3 Motorboat Ride Charles Curtis travels in his motorboat downstream for 12 miles with the current. He then turns around and heads back to the starting point going upstream against the current. The total time of his trip is 5 hours and the river current is 2 miles per hour. If during the entire trip he did not touch the throttle to change the speed, find the speed the boat would have traveled in still water.

Solution   Understand   We are asked to find the rate of the boat in still water. Let r = the rate of the boat in still water. We know that the total time of the trip is 5 hours. Thus, the time downriver plus the time upriver must total 5 hours. Since distance = rate · time, we can find the time by dividing the distance by the rate.

Direction	Distance	Rate	Time
Downriver (with current)	12	$r + 2$	$\dfrac{12}{r + 2}$
Upriver (against current)	12	$r - 2$	$\dfrac{12}{r - 2}$

Translate   time downriver + time upriver = total time

$$\frac{12}{r + 2} + \frac{12}{r - 2} = 5$$

Carry Out $(r + 2)(r - 2)\left(\dfrac{12}{r + 2} + \dfrac{12}{r - 2}\right) = (r + 2)(r - 2)\,(5)$ Multiply by the LCD.

$(r + 2)(r - 2)\left(\dfrac{12}{r + 2}\right) + (r + 2)(r - 2)\left(\dfrac{12}{r - 2}\right) = (r + 2)(r - 2)(5)$ Distributive property

$$12(r - 2) + 12(r + 2) = 5(r^2 - 4)$$

$$12r - 24 + 12r + 24 = 5r^2 - 20$$ Distributive property

$$24r = 5r^2 - 20$$ Simplify.

$$\text{or}\quad 5r^2 - 24r - 20 = 0$$

Using the quadratic formula with $a = 5, b = -24$, and $c = -20$, we obtain

$$r = \frac{24 \pm \sqrt{976}}{10}$$

$$r \approx 5.5 \quad \text{or} \quad r \approx -0.7$$

Answer   Since the rate cannot be negative, the rate or speed of the boat in still water is about 5.5 miles per hour.

Now Try Exercise 43

Work Problems Let's do an example involving a work problem, first discussed in Section 6.5. You may wish to review that section before studying the next example.

EXAMPLE 4 **Pumping Water** The Durals need to pump water from their flooded basement. They have a pump and they borrow another from their neighbors, the Sullivans. With both pumps working together, the job can be done in 6 hours. The Sullivans' pump working alone would do the job in 2 hours less than the Durals' pump working alone. How long would it take for each pump to do the job working alone?

Solution   Understand   Recall from Section 6.5 that the rate of work multiplied by the time worked gives the part of the task completed. Let t = number of hours for the Durals' (slower) pump to complete the job by itself, then $t - 2$ = number of hours for the Sullivan's pump to complete the job by itself.

Pump	Rate of Work	Time Worked	Part of Task Completed
Durals' pump	$\dfrac{1}{t}$	6	$\dfrac{6}{t}$
Sullivans' pump	$\dfrac{1}{t - 2}$	6	$\dfrac{6}{t - 2}$

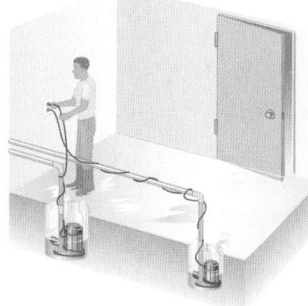

Translate

$$\left(\begin{array}{c} \text{part of task} \\ \text{by Durals' pump} \end{array} \right) + \left(\begin{array}{c} \text{part of task} \\ \text{by Sullivans' pump} \end{array} \right) = 1$$

$$\frac{6}{t} + \frac{6}{t-2} = 1$$

Carry Out

$$t(t-2)\left(\frac{6}{t} + \frac{6}{t-2} \right) = t(t-2)(1) \qquad \text{Multiply both sides by the LCD, } t(t-2).$$

$$t(t-2)\left(\frac{6}{t} \right) + t(t-2)\left(\frac{6}{t-2} \right) = t^2 - 2t \qquad \text{Distributive property}$$

$$6(t-2) + 6t = t^2 - 2t$$

$$6t - 12 + 6t = t^2 - 2t$$

$$t^2 - 14t + 12 = 0$$

Using the quadratic formula, we obtain

$$t = \frac{14 \pm \sqrt{148}}{2}$$

$$t \approx 13.1 \quad \text{or} \quad t \approx 0.9$$

Answer   Both $t = 13.1$ and $t = 0.9$ satisfy the original equation. However, if $t = 0.9$ then the Durals' pump could complete the job in 0.9 hours and the Sullivans' pump could complete the job in $0.9 - 2$ or -1.1 hours, which is not possible. Therefore, the only acceptable solution is $t = 13.1$. Thus, the Durals' pump takes about 13.1 hours by itself and the Sullivans' pump takes about $13.1 - 2$ or 11.1 hours by itself to complete the job.

Now Try Exercise 45

2 Solve for a Variable in a Formula

When solving for a variable in a formula, it may be necessary to use the square root property to isolate the variable. However, often the variable can only represent a non-negative number. Therefore, *when using the square root property to solve for a variable in most formulas, we will use only the positive square root.* This will be demonstrated in our next example.

EXAMPLE 5

a) The formula for the area of a circle is $A = \pi r^2$. Solve this equation for the radius, r.

b) *Newton's law of universal gravity* is

$$F = G\frac{m_1 m_2}{r^2}$$

Solve the equation for r, which measures distance.

Solution

a)

$$A = \pi r^2$$

$$\frac{A}{\pi} = r^2 \qquad \text{Isolate } r^2 \text{ by dividing both sides by } \pi.$$

$$\sqrt{\frac{A}{\pi}} = r \qquad \text{Square root property; } r \text{ must be positive.}$$

b)

$$F = G\frac{m_1 m_2}{r^2}$$

$$r^2 F = Gm_1 m_2 \qquad \text{Multiply both sides of formula by } r^2.$$

$$r^2 = \frac{Gm_1 m_2}{F} \qquad \text{Isolate } r^2 \text{ by dividing both sides by } F.$$

$$r = \sqrt{\frac{Gm_1 m_2}{F}} \qquad \text{Square root property; } r \text{ must be positive.}$$

Now Try Exercise 23

In both parts of Example 5, since r must be greater than 0, when we used the square root property, we listed only the positive square root.

EXAMPLE 6 **Diagonal of a Suitcase** The diagonal of a box can be calculated by the formula

$$d = \sqrt{L^2 + W^2 + H^2}$$

where L is the length, W is the width, and H is the height of the box. See **Figure 8.5**.

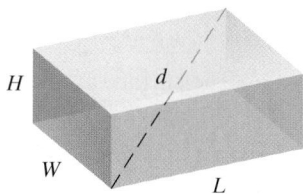

FIGURE 8.5

a) Find the diagonal of a suitcase of length 30 inches, width 15 inches, and height 10 inches.

b) Solve the equation for the width, W.

Solution

a) Understand   To find the diagonal, we need to substitute the appropriate values into the formula and solve for the diagonal, d.

Translate $d = \sqrt{L^2 + W^2 + H^2}$

                             $d = \sqrt{(30)^2 + (15)^2 + (10)^2}$

Carry Out $= \sqrt{900 + 225 + 100}$

                             $= \sqrt{1225}$

                             $= 35$

Answer   Thus, the diagonal of the suitcase is 35 inches.

b) Our first step in solving for W is to square both sides of the formula.

$$d = \sqrt{L^2 + W^2 + H^2}$$

$$d^2 = \left(\sqrt{L^2 + W^2 + H^2}\right)^2 \qquad \text{Square both sides.}$$

$$d^2 = L^2 + W^2 + H^2 \qquad \text{Use } (\sqrt{a})^2 = a, a \geq 0.$$

$$d^2 - L^2 - H^2 = W^2 \qquad \text{Isolate } W^2.$$

$$\sqrt{d^2 - L^2 - H^2} = W \qquad \text{Square root property}$$

Now Try Exercise 15

EXAMPLE 7 **Traffic Cones** The surface area of a right circular cone is

$$s = \pi r \sqrt{r^2 + h^2}$$

© Stephen Aaron Rees/Shutterstock

a) An orange traffic cone used on roads is 18 inches high with a radius of 12 inches. Find the surface area of the cone.

b) Solve the formula for h.

Solution

a) Understand and Translate   To find the surface area, we substitute the appropriate values into the formula.

$$s = \pi r \sqrt{r^2 + h^2}$$
$$= \pi(12)\sqrt{(12)^2 + (18)^2}$$

Carry Out
$$= 12\pi\sqrt{144 + 324}$$
$$= 12\pi\sqrt{468}$$
$$\approx 815.56$$

Answer   The surface area is about 815.56 square inches.

b) To solve for h we need to isolate h on one side of the equation. There are various ways to solve the equation for h.

$$s = \pi r \sqrt{r^2 + h^2}$$

$$\frac{s}{\pi r} = \sqrt{r^2 + h^2} \qquad \text{Divide both sides by } \pi r.$$

$$\left(\frac{s}{\pi r}\right)^2 = \left(\sqrt{r^2 + h^2}\right)^2 \qquad \text{Square both sides.}$$

$$\frac{s^2}{\pi^2 r^2} = r^2 + h^2 \qquad \text{Use } (\sqrt{a})^2 = a, a \geq 0.$$

$$\frac{s^2}{\pi^2 r^2} - r^2 = h^2 \qquad \text{Subtract } r^2 \text{ from both sides.}$$

$$\sqrt{\frac{s^2}{\pi^2 r^2} - r^2} = h \qquad \text{Square root property}$$

Other acceptable answers are $h = \sqrt{\dfrac{s^2 - \pi^2 r^4}{\pi^2 r^2}}$ and $h = \dfrac{\sqrt{s^2 - \pi^2 r^4}}{\pi r}$.

Now Try Exercise 27

EXERCISE SET 8.3

Math XL
MathXL®

MyMathLab
MyMathLab

Warm-Up Exercises

Fill in the blanks with the appropriate word, phrase, or symbol(s) from the following list.

square root	profit	square	$\dfrac{\text{rate}}{\text{distance}}$	$\dfrac{\text{distance}}{\text{rate}}$	revenue	one whole task	part of the task

1. A company will break even when it has a _____ of 0.

2. If we solve the distance formula for *time*, we get time = _____.

3. To solve work problems we use the fact that the part of the task done by the first person (or machine) plus the part of the task done by the second person (or machine) is equal to _____ completed.

4. When we need to eliminate a square root from an equation, we can isolate the square root and then _____ both sides of the equation.

Practice the Skills

Solve for the indicated variable. Assume the variable you are solving for must be greater than 0.

5. $A = s^2$, for s (area of a square)

6. $A = (s + 1)^2$, for s (area of a square)

7. $E = i^2 r$, for i (current in electronics)

8. $A = 4\pi r^2$, for r (surface area of a sphere)

9. $d = 16t^2$, for t (distance of a falling object)

10. $d = \dfrac{1}{9}x^2$, for x (stopping distance on pavement)

11. $E = mc^2$, for c (Einstein's famous energy formula)

12. $V = \pi r^2 h$, for r (volume of a right circular cylinder)

13. $V = \frac{1}{3}\pi r^2 h$, for r (volume of a right circular cone)

14. $d = \sqrt{L^2 + W^2}$, for L (diagonal of a rectangle)

15. $d = \sqrt{L^2 + W^2}$, for W (diagonal of a rectangle)

16. $a^2 + b^2 = c^2$, for a (Pythagorean Theorem)

17. $a^2 + b^2 = c^2$, for b (Pythagorean Theorem)

18. $d = \sqrt{L^2 + W^2 + H^2}$, for L (diagonal of a box)

19. $d = \sqrt{L^2 + W^2 + H^2}$, for H (diagonal of a box)

20. $A = P(1 + r)^2$, for r (compound interest formula)

21. $h = -16t^2 + s_0$, for t (height of an object)

22. $h = -4.9t^2 + s_0$, for t (height of an object)

23. $E = \frac{1}{2}mv^2$, for v (kinetic energy)

24. $f_x^2 + f_y^2 = f^2$, for f_x (forces acting on an object)

25. $a = \frac{v_2^2 - v_1^2}{2d}$, for v_1 (acceleration of a vehicle)

26. $A = 4\pi(R^2 - r^2)$, for R (surface area of two spheres)

27. $v' = \sqrt{c^2 - v^2}$, for c (relativity; v' is read "v prime")

28. $L = L_0\sqrt{1 - \dfrac{v^2}{c^2}}$, for v (art, a painting's contraction)

Problem Solving

29. Coffee Shop Profit Benny's Beans, a new coffee shop, projects that its annual profit, $p(t)$, in thousands of dollars, over the first 6 years in business can be approximated by the function $p(t) = 0.2t^2 + 5.4t - 8$, where t is measured in years.

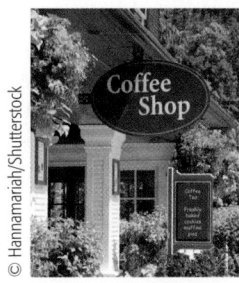

© Hannamariah/Shutterstock

a) Estimate the profit (or loss) of the coffee shop after the first year.

b) Estimate the profit (or loss) of the coffee shop after the fifth year.

c) Estimate the time needed for the coffee shop to break even.

30. Organic Food Store Profit Hatty's Health Foods, an organic food store, projects that its annual profit, $p(t)$, in thousands of dollars, over the first 6 years in business can be approximated by the function $p(t) = 0.3t^2 + 4.8t - 9$, where t is measured in years.

a) Estimate the profit (or loss) of the food store after the first year.

b) Estimate the profit (or loss) of the food store after the fifth year.

c) Estimate the time needed for the food store to break even.

31. Souvenir Shop Profit Bull Snort, a University of South Florida souvenir shop, projects that its annual profit, $p(t)$, in thousands of dollars, over the first 7 years in business can be approximated by the function $p(t) = 0.1t^2 + 3.9t - 6$, where t is measured in years.

a) Estimate the profit (or loss) of the souvenir shop after the first year.

b) Estimate the profit (or loss) of the souvenir shop after the sixth year.

c) Estimate the time needed for the souvenir shop to break even.

32. Watch Repair Shop Profit Repair Time, a watch and clock repair shop, projects that its annual profit, $p(t)$, in thousands of dollars, over the first 7 years in business can be approximated by the function $p(t) = 0.4t^2 + 4.4t - 8$, where t is measured in years.

a) Estimate the profit (or loss) of the repair shop after the first year.

b) Estimate the profit (or loss) of the repair shop after the sixth year.

c) Estimate the time needed for the repair shop break even.

33. Temperature The temperature, T, in degrees Fahrenheit, in a car's radiator during the first 4 minutes of driving is a function of time, t. The temperature can be found by the formula $T = 6.2t^2 + 12t + 32, 0 \le t \le 4$.

a) What is the car radiator's temperature at the instant the car is turned on?

b) What is the car radiator's temperature after the car has been driven for 2 minutes?

c) How long after the car has begun operating will the car radiator's temperature reach 120°F?

34. Grade Point Average At a college, records show that the average person's grade point average, G, is a function of the number of hours he or she studies and does homework per week, h. The grade point average can be estimated by the equation $G = 0.01h^2 + 0.2h + 1.2, 0 \le h \le 8$.

a) What is the GPA of the average student who studies for 0 hours a week?

b) What is the GPA of the average student who studies 3 hours per week?

c) To obtain a 3.2 GPA, how many hours per week would the average student need to study?

35. Age of Drivers and Accidents The number of fatal automobile crashes in the United States per 100 million miles driven, $A(t)$, for drivers age t years old, can be estimated by the function $A(t) = 0.013t^2 - 1.19t + 28.24$.

Source: Insurance Institute for Highway Safety

a) Estimate the number of fatal crashes per 100 million miles driven for drivers age $t = 20$ years old.

b) Estimate the number of fatal crashes per 100 million miles driven for drivers age $t = 75$ years old.

c) Estimate the ages at which about 10 fatal crashes per 100 million miles driven occur.

36. Drug-Free Schools The following graph summarizes data on the percent of students at various ages who say their school is not drug free.

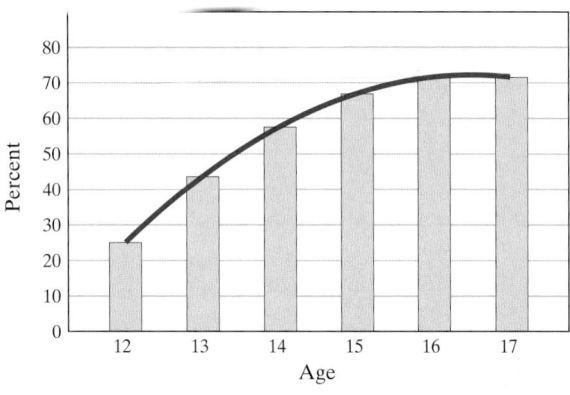

Students Who Say Their School Is Not Drug Free

Source: National Center on Addiction and Substance Abuse

The function $f(a) = -2.32a^2 + 76.58a - 559.87$ can be used to estimate the percent of students who say their school is not drug free. In the function, a represents the student's age, where $12 \le a \le 17$. Use the function to answer the following questions.

a) Estimate the percent of 14-year-olds who say their school is not drug free.

b) At what age do 70% of the students say their school is not drug free?

37. Social Security Assets The projected assets for Social Security can be estimated by the function $f(t) = -20.57t^2 + 758.9t - 3140$, where $f(t)$ is the total amount of assets in billions of dollars and t represents the number of years since 2000.

Source: Social Security Administration

a) Estimate the projected assets for Social Security in 2015.

b) Estimate the projected assets for Social Security in 2030.

c) During what years will the projected assets for Social Security be about $2,000 billion?

38. Profit A video store's weekly profit, P, in thousands of dollars, is a function of the rental price of the tapes, t. The profit equation is $P = 0.2t^2 + 1.5t - 1.2$, $0 \le t \le 5$.

a) What is the store's weekly profit or loss if they charge $3 per tape?

b) What is the weekly profit if they charge $5 per tape?

c) At what tape rental price will their weekly profit be $1.4 thousand?

39. Playground The area of a children's rectangular playground is 600 square meters. The length is 10 meters longer than the width. Find the length and width of the playground.

© Jupiter Unlimited

40. Travel Hana Juarez drove for 80 miles in heavy traffic. She then reached the highway where she drove 260 miles at an average speed that was 25 miles per hour greater than the average speed she drove in heavy traffic. If the total trip took 6 hours, determine her average speed in heavy traffic and on the highway.

41. Drilling a Well Paul and Rima Jones hired the Ruth Cardiff Drilling Company to drill a well. The company had to drill 64 feet to hit water. The company informed the Joneses that they had just ordered new drilling equipment that drills at an average of 1 foot per hour faster, and that with their new equipment, they would have hit water in 3.2 hours less time. Find the rate at which their present equipment drills.

42. Car Carrier Frank Sims, a truck driver, was transporting a heavy load of new cars on a car carrier from Detroit, Michigan, to Indianapolis, Indiana. On his return trip to Detroit, he averaged 10 miles per hour faster than on his trip to Indianapolis. If the total distance traveled each way was 300 miles and the total time he spent driving was 11 hours, find his average speed going and returning.

43. Runner Latoya Williams, a long-distance runner, starts jogging at her house. She jogs 6 miles and then turns around and jogs back to her house. The first part of her jog is mostly uphill, so her speed averages 2 miles per hour less than her returning speed. If the total time she spends jogging is $1\frac{3}{4}$ hours, find her speed going and her speed returning.

44. Red Rock Canyon Kathy Nickell traveled from the Red Rock Canyon Conservation Area, just outside Las Vegas, to Phoenix, Arizona. The total distance she traveled was 300 miles. After she got to Phoenix, she figured out that had she averaged 10 miles per hour faster, she would have arrived 1 hour earlier. Find the average speed that Kathy drove.

Red Rock Canyon

45. **Build an Engine** Two mechanics, Bonita Rich and Pamela Pearson, take 6 hours to rebuild an engine when they work together. If each worked alone, Bonita, the more experienced mechanic, could complete the job 1 hour faster than Pamela. How long would it take each of them to rebuild the engine working alone?

46. **Riding a Bike** Ricky Bullock enjoys riding his bike from Washington, D.C., to Bethesda, Maryland, and back, a total of 30 miles on the Capital Crescent path. The trip to Bethesda is uphill most of the distance. The bike's average speed going to Bethesda is 5 miles per hour slower than the average speed returning to D.C. If the round trip takes 4.5 hours, find the average speed in each direction.

47. **Flying a Plane** Dole Rohm flew his single-engine Cessna airplane 80 miles with the wind from Jackson Hole, Wyoming, to above Blackfoot, Idaho. He then turned around and flew back to Jackson Hole against the wind. If the wind was a constant 30 miles per hour, and the total time going and returning was 1.3 hours, find the speed of the plane in still air.

48. **Ships** After a small oil spill, two cleanup ships are sent to siphon off the oil floating in Baffin Bay. The newer ship can clean up the entire spill by itself in 3 hours less time than the older ship takes by itself. Working together the two ships can clean up the oil spill in 8 hours. How long will it take the newer ship by itself to clean up the spill?

49. **Janitorial Service** The O'Connors own a small janitorial service. John requires $\frac{1}{2}$ hour more time to clean the Moose Club by himself than Chris does working by herself. If together they can clean the club in 6 hours, find the time required by each to clean the club.

50. **Electric Heater** A small electric heater requires 6 minutes longer to raise the temperature in an unheated garage to a comfortable level than does a larger electric heater. Together the two heaters can raise the garage temperature to a comfortable level in 42 minutes. How long would it take each heater by itself to raise the temperature in the garage to a comfortable level?

51. **Travel** Shywanda Moore drove from San Antonio, Texas, to Austin, Texas, a distance of 75 miles. She then stopped for 2 hours to see a friend in Austin before continuing her journey from Austin to Dallas, Texas, a distance of 195 miles. If she drove 10 miles per hour faster from San Antonio to Austin and the total time of the trip was 6 hours, find her average speed from San Antonio to Austin.

River Walk, San Antonio, Texas

52. **Travel** Lewis and his friend George are traveling from Nashville to Baltimore. Lewis travels by car while George travels by train. The train and car leave Nashville at the same time from the same point. During the trip Lewis and George speak by cellular phone, and Lewis informs George that he has just stopped for the evening after traveling for 500 miles. One and two-thirds hours later, George calls Lewis and informs him that the train had just reached Baltimore, a distance of 800 miles from Nashville. Assuming the train averaged 20 miles per hour faster than the car, find the average speed of the car and the train.

53. **Widescreen TVs** A widescreen television (see figure) has an aspect ratio of 16 : 9. This means that the ratio of the length to the height of the screen is 16 to 9. The figure drawn on the photo illustrates how the length and the height of the screen of a 40-inch widescreen television can be found. Determine the length and height of a 40-inch widescreen television.

54. **Standard TVs** Many picture tube televisions have a screen aspect ratio of 4:3. Determine the length and height of a television screen that has an aspect ratio of 4:3 and whose diagonal is 36 inches. See Exercise 53.

Concept/Writing Exercises

55. Write your own motion problem and solve it.

56. Write your own work problem and solve it.

57. In general, when solving for a variable in a formula, whether you use the square root property or the quadratic formula, you use only the positive square root. Explain why.

58. Suppose $P = ☺^2 + ☐^2$ is a real formula. Solving for ☺ gives $☺ = \sqrt{P - ☐^2}$. If ☺ is to be a real number, what relationship must exist between P and ☐?

Challenge Problems

59. Area The area of a rectangle is 18 square meters. When the length is increased by 2 meters and the width by 3 meters, the area becomes 48 square meters. Find the dimensions of the smaller rectangle.

60. Area The area of a rectangle is 35 square inches. When the length is decreased by 1 inch and the width is increased by 1 inch, the area of the new rectangle is 36 square inches. Find the dimensions of the original rectangle.

Cumulative Review Exercises

[1.4]   **61.** Evaluate $-[4(5 - 3)^3] + 2^4$.

[2.2]   **62.** Solve $IR + Ir = E$ for R.

[6.2]   **63.** Add $\dfrac{r}{r - 4} - \dfrac{r}{r + 4} + \dfrac{32}{r^2 - 16}$.

[7.2]   **64.** Simplify $\left(\dfrac{x^{3/4}y^{-2}}{x^{1/2}y^2}\right)^8$.

[7.6]   **65.** Solve $\sqrt{x^2 + 3x + 12} = x$.

Mid-Chapter Test: 8.1–8.3

To find out how well you understand the chapter material to this point, take this brief test. The answers, and the section where the material was initially discussed, are given in the back of the book. Review any questions that you answered incorrectly.

Use the square root property to solve each equation.

1. $x^2 - 12 = 86$

2. $(a - 3)^2 + 20 = 0$

3. $(2m + 7)^2 = 36$

Solve each equation by completing the square.

4. $y^2 + 4y - 12 = 0$

5. $3a^2 - 12a - 30 = 0$

6. $4c^2 + c = -9$

7. Patio The patio of a house is a square where the diagonal is 6 meters longer than a side. Find the length of one side of the patio.

8. a) Give the formula for the discriminant of a quadratic equation.

   b) Explain how to determine if a quadratic equation has two distinct real solutions, a single real solution, or no real solution.

9. Use the discriminant to determine if the equation $2b^2 - 6b - 11 = 0$ has two distinct real solutions, a single real solution, or no real solution.

Solve each equation by the quadratic formula.

10. $6n^2 + n = 15$

11. $p^2 = -4p + 8$

12. $3d^2 - 2d + 5 = 0$

In Exercises 13 and 14, determine an equation that has the given solutions.

13. $7, -2$

14. $2 + \sqrt{5}, 2 - \sqrt{5}$

15. Lamps A business sells n lamps, $n \leq 20$, at a price of $(60 - 0.5n)$ dollars per lamp. How many lamps must be sold to have revenue of $550?

In Exercises 16–18, solve for the indicated variable. Assume all variables are positive.

16. $y = x^2 - r^2$, for r

17. $A = \dfrac{1}{3}kx^2$, for x

18. $D = \sqrt{x^2 + y^2}$, for y

19. Area The length of a rectangle is two feet more than twice the width. Find the dimensions if its area is 60 square feet.

20. Clocks The profit from a company selling n clocks is $p(n) = 2n^2 + n - 35$, where $p(n)$ is hundreds of dollars. How many clocks must be sold to have a profit of $2000?

8.4 Writing Equations in Quadratic Form

1. Solve equations that are quadratic in form.

2. Solve equations with rational exponents.

1 Solve Equations That Are Quadratic in Form

In this section, we will solve equations that are not quadratic, but can be rewritten in the form of a quadratic equation.

Equations Quadratic in Form

An equation that can be rewritten in the form $au^2 + bu + c = 0$ for $a \neq 0$, where u is an algebraic expression, is called **quadratic in form**.

When given an equation quadratic in form, we will make a substitution to rewrite the equation in the form $au^2 + bu + c = 0$. For example, consider the equation $2(x - 3)^2 + 5(x - 3) - 7 = 0$.

We will let $u = x - 3$. Then $u^2 = (x - 3)^2$ and the equation can then be rewritten as follows

$$2\underline{(x - 3)}^2 + 5\underline{(x - 3)} - 7 = 0$$

$$2 \quad u^2 \quad + 5 \quad u \quad - 7 = 0$$

The equation $2u^2 + 5u - 7 = 0$ can be solved by factoring, completing the square, or by using the quadratic formula. Other examples are shown in the table below.

Equation Quadratic in Form	Substitution	Equation with Substitution
$y^4 - y^2 - 6 = 0$	$u = y^2$	$u^2 - u - 6 = 0$
$2(x + 5)^2 - 5(x + 5) - 12 = 0$	$u = x + 5$	$2u^2 - 5u - 12 = 0$
$x^{2/3} + 4x^{1/3} - 3 = 0$	$u = x^{1/3}$	$u^2 + 4u - 3 = 0$

To solve equations quadratic in form, we use the following procedure.

To Solve Equations Quadratic in Form

1. Make a substitution that will result in an equation of the form $au^2 + bu + c = 0$, $a \neq 0$, where u is a function of the original variable.

2. Solve the equation $au^2 + bu + c = 0$ for u.

3. Replace u with the function of the original variable from step 1 and solve the resulting equation for the original variable.

4. Check for extraneous solutions by substituting the apparent solutions into the original equation.

EXAMPLE 1

a) Solve $x^4 - 5x^2 + 4 = 0$.

b) Find the x-intercepts of the graph of the function $f(x) = x^4 - 5x^2 + 4$.

Solution

a) To obtain an equation quadratic in form, we will let $u = x^2$. Then $u^2 = (x^2)^2 = x^4$.

$$x^4 - 5x^2 + 4 = 0$$

$$\downarrow$$

$$(x^2)^2 - 5x^2 + 4 = 0 \quad x^4 \text{ was replaced with } (x^2)^2.$$

$$\downarrow \qquad \downarrow$$

$$u^2 - 5u + 4 = 0 \quad u \text{ was substituted for } x^2.$$

We now have a quadratic equation we can solve by factoring

$$u^2 - 5u + 4 = 0$$
$$(u - 4)(u - 1) = 0$$
$$u - 4 = 0 \quad \text{or} \quad u - 1 = 0$$
$$u = 4 \qquad\qquad u = 1$$

Next, we replace u with x^2 and solve for x.

$$u = 4 \qquad\qquad u = 1$$
$$\downarrow \qquad\qquad\quad \downarrow$$
$$x^2 = 4 \qquad\qquad x^2 = 1 \qquad u \text{ was replaced with } x^2.$$
$$x = \pm\sqrt{4} \qquad\quad x = \pm\sqrt{1} \quad \text{Square root property}$$
$$x = \pm 2 \qquad\qquad x = \pm 1$$

Check the four possible solutions in the original equation.

$x = 2$	$x = -2$	$x = 1$	$x = -1$
$x^4 - 5x^2 + 4 = 0$	$x^4 - 5x^2 + 4 = 0$	$x^4 - 5x^2 + 4 = 0$	$x^4 - 5x^2 + 4 = 0$
$2^4 - 5(2)^2 + 4 \overset{?}{=} 0$	$(-2)^4 - 5(-2)^2 + 4 \overset{?}{=} 0$	$1^4 - 5(1)^2 + 4 \overset{?}{=} 0$	$(-1)^4 - 5(-1)^2 + 4 \overset{?}{=} 0$
$16 - 20 + 4 \overset{?}{=} 0$	$16 - 20 + 4 \overset{?}{=} 0$	$1 - 5 + 4 \overset{?}{=} 0$	$1 - 5 + 4 \overset{?}{=} 0$
$0 = 0$	$0 = 0$	$0 = 0$	$0 = 0$
True	True	True	True

Understanding Algebra

Recall that the *x*-intercepts of a function are the points where the graph crosses the *x*-axis. The *x*-intercepts always have a *y*-coordinate of 0. To determine the *x*-intercepts of a function, set *y* or $f(x) = 0$ and solve for *x*.

Thus, the solutions are 2, −2, 1, and −1.

b) The x-intercepts occur where $f(x) = 0$. Therefore, the graph will cross the x-axis at the solutions to the equation $x^4 - 5x^2 + 4 = 0$.

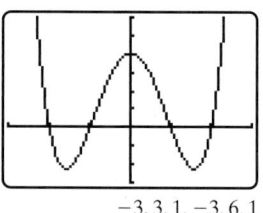

FIGURE 8.6

$-3, 3, 1, -3, 6, 1$

From part **a)**, we know the solutions are 2, −2, 1, and −1. Thus, the x-intercepts are $(2, 0), (-2, 0), (1, 0),$ and $(-1, 0)$. **Figure 8.6** is the graph of $f(x) = x^4 - 5x^2 + 4 = 0$ as illustrated on a graphing calculator. Notice that the graph crosses the x-axis at $x = 2, x = -2, x = 1,$ and $x = -1$.

Now Try Exercise 7

EXAMPLE 2 Solve $p^4 + 2p^2 = 8$.

Solution   We will let $u = p^2$, then $u^2 = (p^2)^2 = p^4$.

$$p^4 + 2p^2 - 8 = 0 \qquad \text{Equation was set equal to 0.}$$

$$(p^2)^2 + 2p^2 - 8 = 0 \qquad p^4 \text{ was written as } (p^2)^2.$$

$$\downarrow \qquad\quad \downarrow$$

$$u^2 + 2u - 8 = 0 \qquad u \text{ was substituted for } p^2.$$

$$(u + 4)(u - 2) = 0$$

$$u + 4 = 0 \quad \text{or} \quad u - 2 = 0$$

$$u = -4 \qquad\qquad u = 2$$

Next, we substitute back p^2 for u and solve for p.

$$p^2 = -4 \qquad\qquad p^2 = 2 \qquad u \text{ was replaced with } p^2.$$

$$p = \pm\sqrt{-4} \qquad\qquad p = \pm\sqrt{2} \qquad \text{Square root property}$$

$$p = \pm 2i$$

Check the four possible solutions in the *original* equation.

$p = 2i$	$p = -2i$	$p = \sqrt{2}$	$p = -\sqrt{2}$
$p^4 + 2p^2 = 8$	$p^4 + 2p^2 = 8$	$p^4 + 2p^2 = 8$	$p^4 + 2p^2 = 8$
$(2i)^4 + 2(2i)^2 \stackrel{?}{=} 8$	$(-2i)^4 + 2(-2i)^2 \stackrel{?}{=} 8$	$(\sqrt{2})^4 + 2(\sqrt{2})^2 \stackrel{?}{=} 8$	$(-\sqrt{2})^4 + 2(-\sqrt{2})^2 \stackrel{?}{=} 8$
$2^4 i^4 + 2(2^2)(i^2) \stackrel{?}{=} 8$	$(-2)^4 i^4 + 2(-2)^2 i^2 \stackrel{?}{=} 8$	$4 + 2(2) \stackrel{?}{=} 8$	$4 + 2(2) \stackrel{?}{=} 8$
$16(1) + 8(-1) \stackrel{?}{=} 8$	$16(1) + 8(-1) \stackrel{?}{=} 8$	$8 = 8$	$8 = 8$
$16 - 8 = 8$	$16 - 8 = 8$	True	True
True	True		

Thus, the solutions are $2i$, $-2i$, $\sqrt{2}$, and $-\sqrt{2}$.

Now Try Exercise 17

> **Helpful Hint**
>
> Students sometimes solve the equation for u but then forget to complete the problem by solving for the original variable. Remember that if the original equation is in x you must obtain values for x.

EXAMPLE 3 Solve $4(2w + 1)^2 - 16(2w + 1) + 15 = 0$.

Solution   We will let $u = 2w + 1$, then $u^2 = (2w + 1)^2$ and the equation becomes

$$4(2w + 1)^2 - 16(2w + 1) + 15 = 0$$

$$4u^2 - 16u + 15 = 0 \quad u \text{ was substituted for } 2w + 1.$$

Now we can factor and solve.

$$(2u - 3)(2u - 5) = 0$$

$$2u - 3 = 0 \qquad \text{or} \qquad 2u - 5 = 0$$

$$2u = 3 \qquad\qquad\qquad 2u = 5$$

$$u = \frac{3}{2} \qquad\qquad\qquad u = \frac{5}{2}$$

Next, we substitute back $2w + 1$ for u and solve for w.

$$2w + 1 = \frac{3}{2} \qquad\qquad 2w + 1 = \frac{5}{2} \quad 2w + 1 \text{ was substituted for } u.$$

$$2w = \frac{1}{2} \qquad\qquad\qquad 2w = \frac{3}{2}$$

$$w = \frac{1}{4} \qquad\qquad\qquad w = \frac{3}{4}$$

A check will show that both $\frac{1}{4}$ and $\frac{3}{4}$ are solutions to the original equation.

Now Try Exercise 29

EXAMPLE 4 Find the x-intercepts of the graph of the function $f(x) = 2x^{-2} + x^{-1} - 1$.

Solution   The x-intercepts occur where $f(x) = 0$. Therefore, to find the x-intercepts we must solve the equation

$$2x^{-2} + x^{-1} - 1 = 0$$

We will let $u = x^{-1}$, then $u^2 = (x^{-1})^2 = x^{-2}$.

$$2(x^{-1})^2 + x^{-1} - 1 = 0$$
$$2u^2 + u - 1 = 0 \qquad u \text{ was substituted for } x^{-1}.$$
$$(2u - 1)(u + 1) = 0$$
$$2u - 1 = 0 \qquad \text{or} \qquad u + 1 = 0$$
$$u = \frac{1}{2} \qquad\qquad u = -1$$

Now we substitute x^{-1} for u.

$$x^{-1} = \frac{1}{2} \qquad \text{or} \qquad x^{-1} = -1$$
$$\frac{1}{x} = \frac{1}{2} \qquad\qquad \frac{1}{x} = -1$$
$$x = 2 \qquad\qquad x = -1$$

A check will show that both 2 and -1 are solutions to the original equation. Thus, the x-intercepts are $(2, 0)$ and $(-1, 0)$.

Now Try Exercise 61

2 Solve Equations with Rational Exponents

In our next two examples, while solving the equations that are quadratic in form, we will raise both sides of the equation to a power to eliminate rational exponents (or radicals). Recall from Chapter 7 that whenever we raise both sides of an equation to a power, it is possible to introduce extraneous solutions. *Therefore, whenever you raise both sides of an equation with rational exponents to a power, you must check all apparent solutions in the original equation.*

EXAMPLE 5 Solve $x^{2/5} + x^{1/5} - 6 = 0$.

Solution We will let $u = x^{1/5}$, then $u^2 = (x^{1/5})^2 = x^{2/5}$. The equation becomes

$$(x^{1/5})^2 + x^{1/5} - 6 = 0$$
$$u^2 + u - 6 = 0 \qquad u \text{ was substituted for } x^{1/5}.$$
$$(u + 3)(u - 2) = 0$$
$$u + 3 = 0 \qquad \text{or} \qquad u - 2 = 0$$
$$u = -3 \qquad\qquad u = 2$$

Now substitute $x^{1/5}$ for u and raise both sides of the equation to the fifth power to remove the rational exponents.

$$x^{1/5} = -3 \qquad \text{or} \qquad x^{1/5} = 2$$
$$(x^{1/5})^5 = (-3)^5 \qquad (x^{1/5})^5 = 2^5$$
$$x = -243 \qquad\qquad x = 32$$

The two *possible* solutions are -243 and 32. Remember that whenever you raise both sides of an equation to a power, as you did here, you need to check for extraneous solutions.

Check

$x = -243$	$x = 32$
$x^{2/5} + x^{1/5} - 6 = 0$	$x^{2/5} + x^{1/5} - 6 = 0$
$(-243)^{2/5} + (-243)^{1/5} - 6 \stackrel{?}{=} 0$	$(32)^{2/5} + (32)^{1/5} - 6 \stackrel{?}{=} 0$
$(\sqrt[5]{-243})^2 + \sqrt[5]{-243} - 6 \stackrel{?}{=} 0$	$(\sqrt[5]{32})^2 + \sqrt[5]{32} - 6 \stackrel{?}{=} 0$
$(-3)^2 - 3 - 6 \stackrel{?}{=} 0$	$2^2 + 2 - 6 \stackrel{?}{=} 0$
$9 - 3 - 6 \stackrel{?}{=} 0$	$4 + 2 - 6 \stackrel{?}{=} 0$
$0 = 0$ True	$0 = 0$ True

Since both values check, the solutions are -243 and 32.

Now Try Exercise 63

EXAMPLE 6 Solve $2p - \sqrt{p} - 10 = 0$.

Solution   Since $\sqrt{p} = p^{1/2}$ we can express the equation as

$$2p - p^{1/2} - 10 = 0$$

We will let $u = p^{1/2}$, then $u^2 = (p^{1/2})^2 = p$ and the equation becomes

$$2p - p^{1/2} - 10 = 0$$
$$2(p^{1/2})^2 - p^{1/2} - 10 = 0$$

If we let $u = p^{1/2}$, this equation is quadratic in form.

$$2u^2 - u - 10 = 0$$
$$(2u - 5)(u + 2) = 0$$

$$2u - 5 = 0 \qquad \text{or} \qquad u + 2 = 0$$
$$2u = 5 \qquad\qquad\qquad u = -2$$
$$u = \frac{5}{2}$$

Next, we substitute $p^{1/2}$ for u.

$$p^{1/2} = \frac{5}{2} \qquad\qquad p^{1/2} = -2$$

Now we square both sides of the equation.

$$(p^{1/2})^2 = \left(\frac{5}{2}\right)^2 \qquad (p^{1/2})^2 = (-2)^2$$

$$p = \frac{25}{4} \qquad\qquad p = 4$$

We must now check both apparent solutions in the original equation.

Check
$$p = \frac{25}{4} \qquad\qquad\qquad\qquad p = 4$$

$$2p - \sqrt{p} - 10 = 0 \qquad\qquad 2p - \sqrt{p} - 10 = 0$$

$$2\left(\frac{25}{4}\right) - \sqrt{\frac{25}{4}} - 10 \overset{?}{=} 0 \qquad 2(4) - \sqrt{4} - 10 \overset{?}{=} 0$$

$$\frac{25}{2} - \frac{5}{2} - 10 \overset{?}{=} 0 \qquad\qquad 8 - 2 - 10 \overset{?}{=} 0$$

$$0 = 0 \quad \text{True} \qquad\qquad\qquad -4 = 0 \quad \text{False}$$

Since 4 does not check, it is an extraneous solution. The only solution is $\frac{25}{4}$.

Now Try Exercise 25

Understanding Algebra

When solving an equation involving variables with rational exponents or radicals, we will often raise both sides of the equation to a power in order to eliminate the rational exponent or radical. Whenever we do this, it is possible to introduce an extraneous or false solution. Therefore, we must be careful to check our answers in the original equation.

EXERCISE SET 8.4

MathXL®   MyMathLab

Warm-Up Exercises

Fill in the blanks with the appropriate word, phrase, or symbol(s) from the following list.

$u = h - 2$ extraneous solution v quadratic in form $u = \sqrt{v}$ $(h - 2)^2$ x-intercepts $u = c^2$

1. An equation that can be written in the form $au^2 + bu + c = 0$ for $a \neq 0$, where u is an algebraic expression, is called _____.

2. The _____ of a function are the points where the graph crosses the x-axis.

3. Whenever we solve an equation by raising both sides of the equation to a power we may introduce an _____.

4. To solve the equation $c^4 + c^2 - 2 = 0$, the best choice for u to obtain an equation quadratic in form is _____.

5. To solve the equation $(h - 2)^2 + (h - 2) - 42 = 0$, the best choice for u to obtain an equation quadratic in form is _____.

6. To solve the equation $v - 3\sqrt{v} - 28 = 0$, the best choice for u to obtain an equation quadratic in form is _____.

Practice the Skills

Solve each equation.

7. $x^4 - 10x^2 + 9 = 0$

8. $x^4 - 5x^2 + 4 = 0$

9. $x^4 + 13x^2 + 36 = 0$

10. $x^4 + 50x^2 + 49 = 0$

11. $x^4 - 13x^2 + 36 = 0$

12. $x^4 + 13x^2 + 36 = 0$

13. $a^4 - 7a^2 + 12 = 0$

14. $b^4 + 7b^2 + 12 = 0$

15. $4x^4 - 17x^2 + 4 = 0$

16. $9d^4 - 13d^2 + 4 = 0$

17. $r^4 - 8r^2 = -15$

18. $p^4 - 8p^2 = -12$

19. $z^4 - 7z^2 = 18$

20. $a^4 + a^2 = 42$

21. $-c^4 = 4c^2 - 5$

22. $9b^4 = 57b^2 - 18$

23. $\sqrt{x} = 2x - 6$

24. $x - 2\sqrt{x} = 8$

25. $x - \sqrt{x} = 6$

26. $x - 4 = -3\sqrt{x}$

27. $9x + 3\sqrt{x} = 2$

28. $8x + 2\sqrt{x} = 1$

29. $(x + 3)^2 + 2(x + 3) = 24$

30. $(x + 1)^2 + 4(x + 1) + 3 = 0$

31. $6(a - 2)^2 = -19(a - 2) - 10$

32. $10(z + 2)^2 = 3(z + 2) + 1$

33. $(x^2 - 3)^2 - (x^2 - 3) - 6 = 0$

34. $(a^2 - 1)^2 - 5(a^2 - 1) - 14 = 0$

35. $2(b + 3)^2 + 5(b + 3) - 3 = 0$

36. $(z^2 - 6)^2 + 2(z^2 - 6) - 24 = 0$

37. $18(x^2 - 5)^2 + 27(x^2 - 5) + 10 = 0$

38. $28(x^2 - 8)^2 - 23(x^2 - 8) - 15 = 0$

39. $a^{-2} + 4a^{-1} + 4 = 0$

40. $x^{-2} + 10x^{-1} + 25 = 0$

41. $12b^{-2} - 7b^{-1} + 1 = 0$

42. $5x^{-2} + 4x^{-1} - 1 = 0$

43. $2b^{-2} = 7b^{-1} - 3$

44. $10z^{-2} - 3z^{-1} - 1 = 0$

45. $x^{-2} + 9x^{-1} = 10$

46. $6a^{-2} = a^{-1} + 12$

47. $x^{-2} = 4x^{-1} + 12$

48. $x^{2/3} - 5x^{1/3} + 6 = 0$

49. $x^{2/3} - 4x^{1/3} = -3$

50. $x^{2/3} = 3x^{1/3} + 4$

51. $b^{2/3} - 9b^{1/3} + 18 = 0$

52. $c^{2/3} - 4 = 0$

53. $-2a - 5a^{1/2} + 3 = 0$

54. $r^{2/3} - 7r^{1/3} + 10 = 0$

55. $c^{2/5} + 3c^{1/5} + 2 = 0$

56. $x^{2/5} - 5x^{1/5} + 6 = 0$

Find all x-intercepts of each function.

57. $f(x) = x - 5\sqrt{x} + 6$

58. $g(x) = x - 15\sqrt{x} + 56$

59. $h(x) = x + 14\sqrt{x} + 45$

60. $k(x) = x + 7\sqrt{x} + 12$

61. $p(x) = 4x^{-2} - 19x^{-1} - 5$

62. $g(x) = 4x^{-2} + 12x^{-1} + 9$

63. $f(x) = x^{2/3} - x^{1/3} - 6$

64. $f(x) = x^{1/2} + 6x^{1/4} - 7$

65. $g(x) = (x^2 - 3x)^2 + 2(x^2 - 3x) - 24$

66. $g(x) = (x^2 - 6x)^2 - 5(x^2 - 6x) - 24$

67. $f(x) = x^4 - 29x^2 + 100$

68. $h(x) = x^4 - 4x^2 + 3$

Problem Solving

69. Solve the equation $\dfrac{3}{x^2} - \dfrac{3}{x} = 60$ by

   a) multiplying both sides of the equation by the LCD.

   b) writing the equation with negative exponents.

70. Solve the equation $1 = \dfrac{2}{x} - \dfrac{2}{x^2}$ by

   a) multiplying both sides of the equation by the LCD.

   b) writing the equation with negative exponents.

Find all real solutions to each equation.

71. $15(r + 2) + 22 = -\dfrac{8}{r + 2}$

72. $2(p + 3) + 5 = \dfrac{3}{p + 3}$

73. $4 - (x - 1)^{-1} = 3(x - 1)^{-2}$

74. $3(x - 4)^{-2} = 16(x - 4)^{-1} + 12$

75. $x^6 - 9x^3 + 8 = 0$

76. $x^6 - 28x^3 + 27 = 0$

77. $(x^2 + 2x - 2)^2 - 7(x^2 + 2x - 2) + 6 = 0$

78. $(x^2 + 3x - 2)^2 - 10(x^2 + 3x - 2) + 16 = 0$

Find all solutions to each equation.

79. $2n^4 - 6n^2 - 3 = 0$

80. $3x^4 + 8x^2 - 1 = 0$

Concept/Writing Exercises

81. Give a general procedure for solving an equation of the form $ax^4 + bx^2 + c = 0$.

82. Give a general procedure for solving an equation of the form $ax^{2n} + bx^n + c = 0$.

83. Give a general procedure for solving an equation of the form $ax^{-2} + bx^{-1} + c = 0$.

84. Give a general procedure for solving an equation of the form $a(x - r)^2 + b(x - r) - c = 0$.

85. Determine an equation of the form $ax^4 + bx^2 + c = 0$ that has solutions ± 2 and ± 1. Explain how you obtained your answer.

86. Determine an equation of the form $ax^4 + bx^2 + c = 0$ that has solutions ± 3 and $\pm 2i$. Explain how you obtained your answer.

87. Determine an equation of the form $ax^4 + bx^2 + c = 0$ that has solutions $\pm\sqrt{2}$ and $\pm\sqrt{5}$. Explain how you obtained your answer.

88. Determine an equation of the form $ax^4 + bx^2 + c = 0$ that has solutions $\pm 2i$ and $\pm 5i$. Explain how you obtained your answer.

89. Is it possible for an equation of the form $ax^4 + bx^2 + c = 0$ to have exactly one imaginary solution? Explain.

90. Is it possible for an equation of the form $ax^4 + bx^2 + c = 0$ to have exactly one real solution? Explain.

Cumulative Review Exercises

[1.3] **91.** Evaluate $\dfrac{4}{5} - \left(\dfrac{3}{4} - \dfrac{2}{3}\right)$.

[2.1] **92.** Solve $3(x + 2) - 2(3x + 3) = -3$

[3.2] **93.** State the domain and range for $y = (x - 3)^2$.

[7.3] **94.** Simplify $\sqrt[3]{16x^3y^6}$.

[7.4] **95.** Add $\sqrt{75} + \sqrt{48}$.

8.5 Graphing Quadratic Functions

1 Determine when a parabola opens upward or downward.

2 Find the axis of symmetry, vertex, and x-intercepts of a parabola.

3 Graph quadratic functions using the axis of symmetry, vertex, and intercepts.

4 Solve maximum and minimum problems.

5 Understand translations of parabolas.

6 Write functions in the form $f(x) = a(x - h)^2 + k$.

In Chapters 3 and 5 we briefly discussed graphs of quadratic functions. In this section we will study how to graph quadratic functions using the axis of symmetry, vertex, and intercepts. We will also use translations to graph quadratic functions.

1 Determine When a Parabola Opens Upward or Downward

We begin with the definition of a quadratic function.

> **Quadratic Function**
>
> A **quadratic function** is a function that can be written in the form
>
> $$f(x) = ax^2 + bx + c \qquad \text{for real numbers } a, b, \text{ and } c, \text{ with } a \neq 0.$$

The graph of quadratic function is a **parabola**. A parabola has a shape that resembles, but is not the same as, the letter U. For a quadratic function, the sign of the leading coefficient, a, determines whether the parabola opens upward (see **Fig. 8.7a**) or downward (see **Fig. 8.7b**).

$a > 0$,
Parabola
opens upward

$f(x) = ax^2 + bx + c$

$a < 0$,
Parabola
opens downward

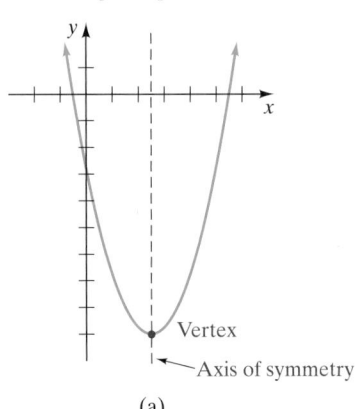

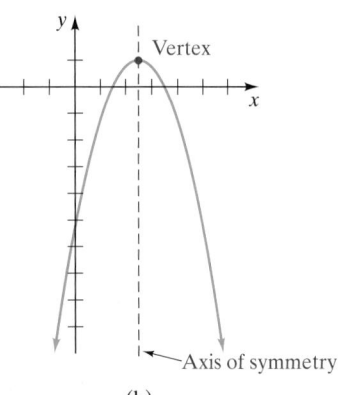

(a)

(b)

- When $a > 0$, the parabola opens upward.
- The **vertex** is the lowest point on the curve.
- The **minimum value of the function** is the y-coordinate of the vertex.

- When $a < 0$, the parabola opens downward.
- The **vertex** is the highest point on the curve.
- The **maximum value of the function** is the y-coordinate of the vertex.

FIGURE 8.7

2 Find the Axis of Symmetry, Vertex, and x-Intercepts of a Parabola

Graphs of quadratic functions will have **symmetry** about a vertical line, called the **axis of symmetry**, which goes through the vertex. This means that if we fold the paper along this imaginary line, the right and left sides of the parabola will coincide (see both graphs in **Fig. 8.7** on p. 531).

Now we will derive the formula for the axis of symmetry, and find the coordinates of the vertex of a parabola, by beginning with a quadratic function of the form $f(x) = ax^2 + bx + c$ and completing the square on the first two terms.

$$f(x) = ax^2 + bx + c$$

$$= a\left(x^2 + \frac{b}{a}x\right) + c \quad \text{Factor out } a.$$

One half the coefficient of x is $\frac{b}{2a}$. Its square is $\frac{b^2}{4a^2}$. Add and subtract this term inside the parentheses. The sum of these two terms is zero.

$$f(x) = a\left(x^2 + \frac{b}{a}x + \frac{b^2}{4a^2} - \frac{b^2}{4a^2}\right) + c$$

Now rewrite the function as follows.

$$f(x) = a\left(x^2 + \frac{b}{a}x + \frac{b^2}{4a^2}\right) - a\left(\frac{b^2}{4a^2}\right) + c$$

$$= a\left(x + \frac{b}{2a}\right)^2 - \frac{b^2}{4a} + c \quad \begin{array}{l}\text{Replace the trinomial with the} \\ \text{square of a binomial.}\end{array}$$

$$= a\left(x + \frac{b}{2a}\right)^2 - \frac{b^2}{4a} + \frac{4ac}{4a} \quad \begin{array}{l}\text{Write fractions with a common} \\ \text{denominator.}\end{array}$$

$$= a\left(x + \frac{b}{2a}\right)^2 + \frac{4ac - b^2}{4a} \quad \begin{array}{l}\text{Combine the last two terms;} \\ \text{write with the variable } a \text{ first.}\end{array}$$

$$= a\left(x - \left(-\frac{b}{2a}\right)\right)^2 + \frac{4ac - b^2}{4a}$$

Now consider the following.

- The expression $\left(x - \left(-\frac{b}{2a}\right)\right)^2$ will always be greater than or equal to 0.

- If $a > 0$, the parabola opens upward and the function has a minimum value. This minimum value will occur when $x = -\frac{b}{2a}$.

- If $a < 0$, the parabola opens downward and the function has a maximum value. This maximum value will occur when $x = -\frac{b}{2a}$.

- Therefore, the x-coordinate of the vertex can be found using the formula $x = -\frac{b}{2a}$. To determine the y-coordinate of the vertex, evaluate $f\left(-\frac{b}{2a}\right)$. We find that the y-coordinate of the vertex is $y = \frac{4ac - b^2}{4a}$.

Vertex of a Parabola

The parabola represented by the function $f(x) = ax^2 + bx + c$ will have vertex

$$\left(-\frac{b}{2a}, \frac{4ac - b^2}{4a} \right)$$

Since we often find the y-coordinate of the vertex by substituting the x-coordinate of the vertex into $f(x)$, the vertex may also be designated as

$$\left(-\frac{b}{2a}, f\left(-\frac{b}{2a} \right) \right)$$

Since the axis of symmetry is the vertical line through the vertex, its equation is found using the same formula used to find the x-coordinate of the vertex.

Axis of Symmetry of a Parabola

For a quadratic function of the form $f(x) = ax^2 + bx + c$, the equation of the **axis of symmetry** of the parabola is

$$x = -\frac{b}{2a}$$

Recall that to find the x-intercepts of a function, we set y or $f(x) = 0$ and solve for x.

x-Intercepts of a Parabola

To find the x-intercepts (if there are any) of a quadratic function, solve the equation $ax^2 + bx + c = 0$ for x.

This equation may be solved by factoring, by using the quadratic formula, or by completing the square.

As we mentioned in Section 8.2, the discriminant, $b^2 - 4ac$, may be used to determine the *number of x-intercepts*. The following table summarizes information about the discriminant.

Discriminant, $b^2 - 4ac$	Number of x-Intercepts	Possible Graphs of $f(x) = ax^2 + bx + c$
> 0	Two	
$= 0$	One	
< 0	None	

3 Graph Quadratic Functions Using the Axis of Symmetry, Vertex, and Intercepts

Now we will draw graphs of quadratic functions.

EXAMPLE 1 Consider the quadratic function $y = -x^2 + 8x - 12$.

a) Determine whether the parabola opens upward or downward.

b) Find the y-intercept.

c) Find the vertex.

d) Find the x-intercepts, if any.

e) Draw the graph.

Solution

a) Since a is -1, which is less than 0, the parabola opens downward.

b) To find the y-intercept, set $x = 0$ and solve for y.

$$y = -(0)^2 + 8(0) - 12 = -12$$

The y-intercept is $(0, -12)$.

c) First, find the x-coordinate, then find the y-coordinate of the vertex. From the function, $a = -1, b = 8$, and $c = -12$.

$$x = -\frac{b}{2a} = -\frac{8}{2(-1)} = 4$$

$$y = \frac{4ac - b^2}{4a} = \frac{4(-1)(-12) - 8^2}{4(-1)} = \frac{48 - 64}{-4} = 4$$

The vertex is at $(4, 4)$. The y-coordinate of the vertex could also be found by substituting 4 for x in the function and finding the corresponding value of y, which is 4.

d) To find the x-intercepts, set $y = 0$.

$$0 = -x^2 + 8x - 12$$

$$\text{or} \quad x^2 - 8x + 12 = 0$$

$$(x - 6)(x - 2) = 0$$

$$x - 6 = 0 \quad \text{or} \quad x - 2 = 0$$

$$x = 6 \quad \text{or} \quad x = 2$$

Thus, the x-intercepts are $(2, 0)$ and $(6, 0)$. These values could also be found by the quadratic formula (or by completing the square).

e) Use all this information to draw the graph (**Fig. 8.8**).

Now Try Exercise 15

Now Try Exercise 15

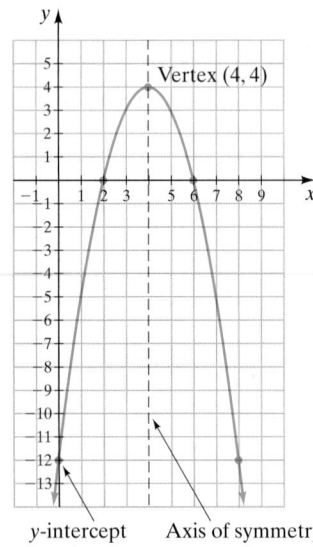

FIGURE 8.8

Understanding Algebra

Recall from Chapter 3, y is a function of x, and we write $y = f(x)$. Therefore, when graphing the function in Example 1, we can write $f(x) = -x^2 + 8x - 12$, or equivalently $y = -x^2 + 8x - 12$.

Notice that in Example 1, the equation is $y = -x^2 + 8x - 12$ and the y-intercept is $(0, -12)$. In general, for any equation of the form $y = ax^2 + bx + c$, the y-intercept will be $(0, c)$.

If you obtain irrational values when finding x-intercepts by the quadratic formula, use your calculator to estimate these values, and then plot these decimal values. For example, if you obtain $x = \frac{2 \pm \sqrt{10}}{2}$, you would evaluate $\frac{2 + \sqrt{10}}{2}$ and $\frac{2 - \sqrt{10}}{2}$ on your calculator and obtain 2.58 and -0.58, respectively, to the nearest hundredth. The x-intercepts would therefore be $(2.58, 0)$ and $(-0.58, 0)$.

EXAMPLE 2 Consider the quadratic function $f(x) = 2x^2 + 6x + 5$.

a) Determine whether the parabola opens upward or downward.

b) Find the y-intercept.   **c)** Find the vertex.

d) Find the x-intercepts, if any.   **e)** Draw the graph.

Solution

a) Since a is 2, which is greater than 0, the parabola opens upward.

b) Since $f(x)$ is the same as y, to find the y-intercept, set $x = 0$ and solve for $f(x)$, or y.

$$f(0) = 2(0)^2 + 6(0) + 5 = 5$$

The y-intercept is $(0, 5)$.

c) Here $a = 2$, $b = 6$, and $c = 5$.

$$x = -\frac{b}{2a} = -\frac{6}{2(2)} = -\frac{6}{4} = -\frac{3}{2}$$

$$y = \frac{4ac - b^2}{4a} = \frac{4(2)(5) - 6^2}{4(2)} = \frac{40 - 36}{8} = \frac{4}{8} = \frac{1}{2}$$

The vertex is $\left(-\frac{3}{2}, \frac{1}{2}\right)$. The y-coordinate of the vertex can also be found by evaluating $f\left(-\frac{3}{2}\right)$.

d) To find the x-intercepts, set $f(x) = 0$.

$$0 = 2x^2 + 6x + 5$$

This trinomial cannot be factored. To determine whether this equation has any real solutions, evaluate the discriminant.

$$b^2 - 4ac = 6^2 - 4(2)(5) = 36 - 40 = -4$$

Since the discriminant is less than 0, this equation has no real solutions and the graph has no x-intercepts.

e) The graph is given in **Figure 8.9**.

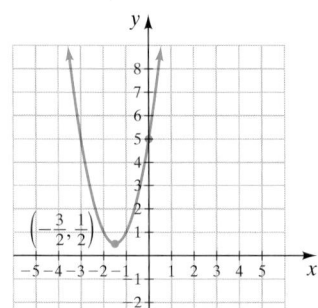

FIGURE 8.9

Now Try Exercise 39

4 Solve Maximum and Minimum Problems

A parabola that opens upward has a **minimum value** at its vertex, as illustrated in **Figure 8.10a**. A parabola that opens downward has a **maximum value** at its vertex, as shown in **Figure 8.10b**. If you are given a function of the form $f(x) = ax^2 + bx + c$, the maximum or minimum value will occur at $-\frac{b}{2a}$, and the value will be $\frac{4ac - b^2}{4a}$.

There are many real-life problems that require finding maximum and minimum values.

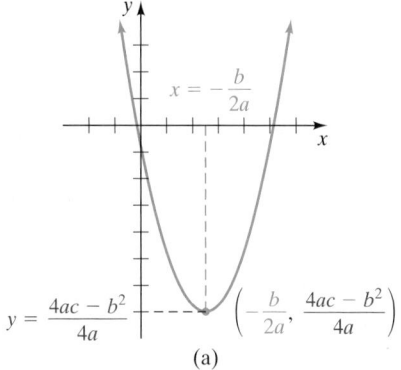

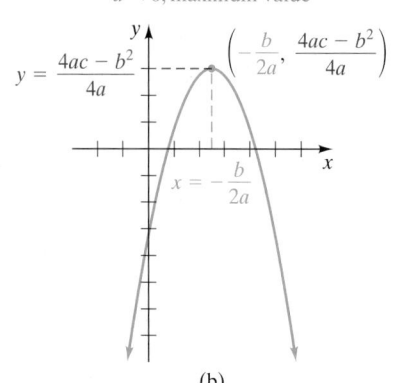

FIGURE 8.10

(a)   (b)

EXAMPLE 3 **Baseball** Mark DeRosa hits a baseball at a height of 3 feet above the ground. The height of the ball above the ground, $f(t)$, in feet, at time, t, in seconds, can be estimated by the function

$$f(t) = -16t^2 + 52t + 3$$

a) Find the maximum height attained by the baseball.

b) Find the time it takes for the baseball to reach its maximum height.

c) Find the time it takes for the baseball to strike the ground.

Solution

a) Understand   The baseball will follow the path of a parabola that opens downward ($a < 0$). The baseball will rise to a maximum height, then begin its fall back to the ground due to gravity. To find the maximum height, we use the formula $y = \dfrac{4ac - b^2}{4a}$.

Translate

$$a = -16, \qquad b = 52, \qquad c = 3$$

$$y = \frac{4ac - b^2}{4a}$$

Carry Out

$$= \frac{4(-16)(3) - (52)^2}{4(-16)}$$

$$= \frac{-192 - 2704}{-64}$$

$$= \frac{-2896}{-64}$$

$$= 45.25$$

Answer   The maximum height attained by the baseball is 45.25 feet.

b) The baseball reaches its maximum height at

$$t = -\frac{b}{2a} = -\frac{52}{2(-16)} = -\frac{52}{-32} = \frac{13}{8} \quad \text{or} \quad 1\frac{5}{8} \quad \text{or} \quad 1.625 \text{ seconds}$$

c) Understand and Translate   When the baseball strikes the ground, its height, y, above the ground is 0. Thus, to determine when the baseball strikes the ground, we solve the equation

$$-16t^2 + 52t + 3 = 0$$

We will use the quadratic formula to solve the equation.

$$t = \frac{-b \pm \sqrt{b^2 - 4ac}}{2a}$$

$$= \frac{-52 \pm \sqrt{(52)^2 - 4(-16)(3)}}{2(-16)}$$

Carry Out

$$= \frac{-52 \pm \sqrt{2704 + 192}}{-32}$$

$$= \frac{-52 \pm \sqrt{2896}}{-32}$$

$$\approx \frac{-52 \pm 53.81}{-32}$$

$$t \approx \frac{-52 + 53.81}{-32} \quad \text{or} \quad t \approx \frac{-52 - 53.81}{-32}$$

$$\approx -0.06 \text{ second} \qquad\qquad \approx 3.31 \text{ seconds}$$

Answer   The only acceptable value is 3.31 seconds. The baseball strikes the ground in about 3.31 seconds. Notice in part **b)** that the time it takes the baseball to reach its maximum height, 1.625 seconds, is not quite half the total time the baseball was in flight, 3.31 seconds. The reason for this is that the baseball was hit from a height of 3 feet and not at ground level.

Now Try Exercise 93

EXAMPLE 4 Area of a Rectangle Consider the rectangle below where the length is $x + 3$ and the width is $10 - x$.

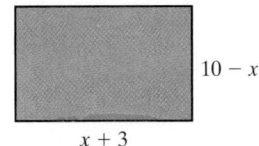

$10 - x$

$x + 3$

a) Find an equation for the area, $A(x)$.

b) Find the value for x that gives the largest (maximum) area.

c) Find the maximum area.

Solution

a) Area is length times width. The area function is

$$A(x) = (x + 3)(10 - x)$$
$$= -x^2 + 7x + 30$$

b) Understand and Translate   The graph of the function is a parabola that opens downward. Thus, the maximum value occurs at the vertex. Therefore, the maximum area occurs at $x = -\dfrac{b}{2a}$. Here $a = -1$ and $b = 7$.

Carry Out
$$x = -\frac{b}{2a} = -\frac{7}{2(-1)} = \frac{7}{2} = 3.5$$

Answer   The maximum area occurs when x is 3.5 units.

c) To find the maximum area, substitute 3.5 for each x in the equation determined in part **a)**.

$$A(x) = -x^2 + 7x + 30$$

$$A(3.5) = -(3.5)^2 + 7(3.5) + 30$$

$$= -12.25 + 24.5 + 30$$

$$= 42.25$$

Observe that for this rectangle, the length is $x + 3 = 3.5 + 3 = 6.5$ units and the width is $10 - x = 10 - 3.5 = 6.5$ units. The rectangle is actually a square, and its area is $(6.5)(6.5) = 42.25$ square units. Therefore, the maximum area is 42.25 square units.

Now Try Exercise 75

In Example 4 **c)**, the maximum area could have been determined by using the formula $y = \dfrac{4ac - b^2}{4a}$. Determine the maximum area now using this formula. You should obtain the same answer, 42.25 square units.

FIGURE 8.11

EXAMPLE 5 **Rectangular Corral** Greg Fierro is building a corral for newborn calves in the shape of a rectangle (see **Fig. 8.11**). If he plans to use 160 meters of fencing, find the dimensions of the corral that will give the greatest area.

Solution Understand We are given the perimeter of the corral, 160 meters. The formula for the perimeter of a rectangle is $P = 2l + 2w$. For this problem, $160 = 2l + 2w$. We are asked to maximize the area, A, where

$$A = lw$$

We need to express the area in terms of one variable, not two. To express the area in terms of l, we solve the perimeter formula, $160 = 2l + 2w$, for w, then make a substitution.

Translate
$$160 = 2l + 2w$$
$$160 - 2l = 2w$$
$$80 - l = w$$

Carry Out Now we substitute $80 - l$ for w into $A = lw$. This gives

$$A = lw$$
$$A = l(80 - l)$$
$$A = -l^2 + 80l$$

In this quadratic equation, $a = -1$, $b = 80$, and $c = 0$. The maximum area will occur at

$$l = -\frac{b}{2a} = -\frac{80}{2(-1)} = 40$$

Answer The length that will give the largest area is 40 meters. The width, $w = 80 - l$, will also be 40 meters. Thus, a square with dimensions 40 by 40 meters will give the largest area.

The largest area can be found by substituting $l = 40$ into the formula $A = l(80 - l)$ or by using $A = \frac{4ac - b^2}{4a}$. In either case, we obtain an area of 1600 square meters.

Now Try Exercise 91

5 Understand Translations of Parabolas

Our next method of graphing parabolas begins with the graphs of quadratic functions of the form $f(x) = ax^2$. Consider the functions $f(x) = x^2$, $g(x) = 2x^2$, and $h(x) = \frac{1}{2}x^2$ shown in **Figure 8.12**. Notice that the *value* of a determines the width of the parabola.

Next, consider the functions $f(x) = -x^2$, $g(x) = -2x^2$, and $h(x) = -\frac{1}{2}x^2$ shown in **Figure 8.13**. Although the parabolas open downward, the value of a still determines the width of the parabola.

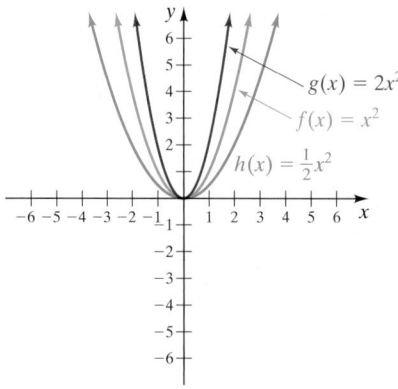

FIGURE 8.12

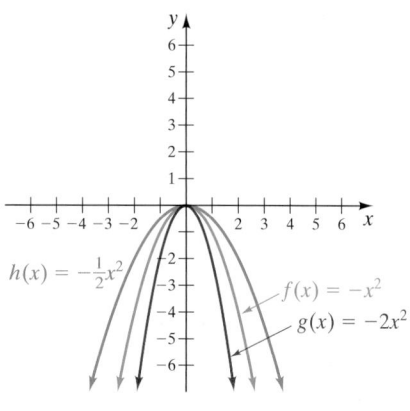

FIGURE 8.13

From Figure 8.12 and Figure 8.13 we can see that, *in general, as |a| gets larger, the parabola gets narrower. As |a| gets smaller, the parabola gets wider.*

We now will **translate**, or shift, the position of graphs of the form $f(x) = ax^2$ to obtain the graphs of other quadratic functions. For example, consider the three functions $f(x) = x^2$, $g(x) = (x - 2)^2$, and $h(x) = (x + 2)^2$ shown in **Figure 8.14**. Notice that the three graphs are identical in *shape* but are in different positions.

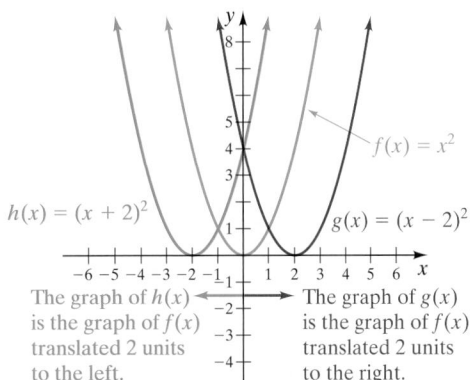

FIGURE 8.14

Note also that the graph of $g(x)$ is translated, or shifted, 2 units to the right of the graph of $f(x)$. The graph of $h(x)$ is translated 2 units to the left of the graph of $f(x)$.

In general, the graph of $g(x) = a(x - h)^2$ will have the same shape as the graph of $f(x) = ax^2$. If h is positive, then the graph of $g(x)$ will be shifted h units to the right of the graph of $f(x)$. If h is negative, then the graph of $g(x)$ will be shifted |h| units to the left of the graph of $f(x)$.

Next, consider the graphs of $f(x) = x^2$, $g(x) = x^2 + 3$, and $h(x) = x^2 - 3$ shown in **Figure 8.15**. Notice again that the three graphs are identical in *shape* but are in different positions.

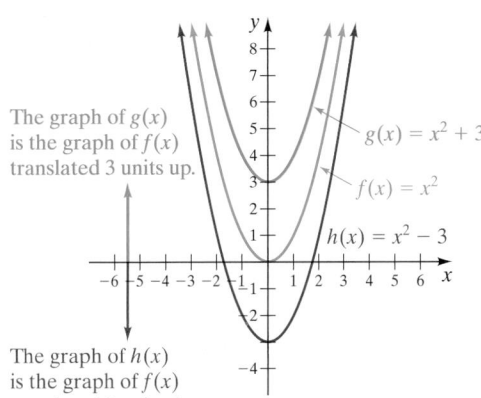

FIGURE 8.15

Note also that the graph of $g(x)$ is translated up from the graph of $f(x)$. The graph of $h(x)$ is translated 3 units down from the graph of $f(x)$.

In general, the graph of $g(x) = ax^2 + k$ will have the same shape as the graph of $f(x) = ax^2$. If k is positive, then the graph of $g(x)$ will be shifted k units up from the graph of $f(x)$. If k is negative, then the graph of $g(x)$ will be shifted |k| units down from the graph of $f(x)$.

Now consider the graphs of $f(x) = x^2$ and $g(x) = (x - 2)^2 + 3$, shown in **Figure 8.16**. Notice that the graph of $g(x)$ has the same shape as that of $f(x)$. The graph of $g(x)$ is the graph of $f(x)$ translated 2 units to the right and 3 units up. This graph and the discussion preceding it lead to the following important facts.

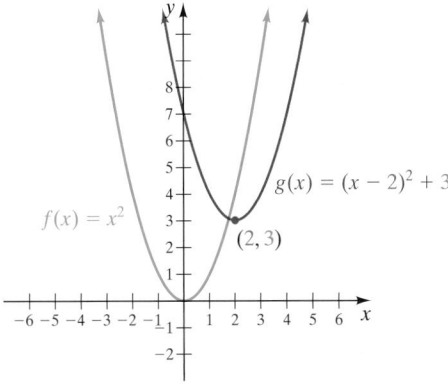

FIGURE 8.16

Parabola Shifts

For any function $f(x) = ax^2$, the graph of $g(x) = a(x - h)^2 + k$ will have the same shape as the graph of $f(x)$. The graph of $g(x)$ will be the graph of $f(x)$ shifted as follows:

- If h is a positive real number, the graph will be shifted h units to the right.
- If h is a negative real number, the graph will be shifted $|h|$ units to the left.
- If k is a positive real number, the graph will be shifted k units up.
- If k is a negative real number, the graph will be shifted $|k|$ units down.

Examine the graph of $g(x) = (x - 2)^2 + 3$ in **Figure 8.16** on page 539. Notice that its axis of symmetry is $x = 2$ and its vertex is $(2, 3)$.

Axis of Symmetry and Vertex of a Parabola

The graph of any function of the form

$$f(x) = a(x - h)^2 + k$$

will be a parabola with axis of symmetry $x = h$ and vertex at (h, k).

Example	Axis of Symmetry	Vertex	Parabola Opens
$f(x) = 2(x - 5)^2 + 7$	$x = 5$	$(5, 7)$	upward, $a > 0$
$f(x) = -\dfrac{1}{2}(x - 6)^2 - 3$	$x = 6$	$(6, -3)$	downward, $a < 0$

Now consider $f(x) = 2(x + 5)^2 + 3$. We can rewrite this as $f(x) = 2[x - (-5)]^2 + 3$. Therefore, h has a value of -5 and k has a value of 3. The graph of this function has axis of symmetry $x = -5$ and vertex at $(-5, 3)$.

Example	Axis of Symmetry	Vertex	Parabola Opens
$f(x) = 3(x + 4)^2 - 2$	$x = -4$	$(-4, -2)$	upward, $a > 0$
$f(x) = -\dfrac{1}{2}\left(x + \dfrac{1}{3}\right)^2 + \dfrac{1}{4}$	$x = -\dfrac{1}{3}$	$\left(-\dfrac{1}{3}, \dfrac{1}{4}\right)$	downward, $a < 0$

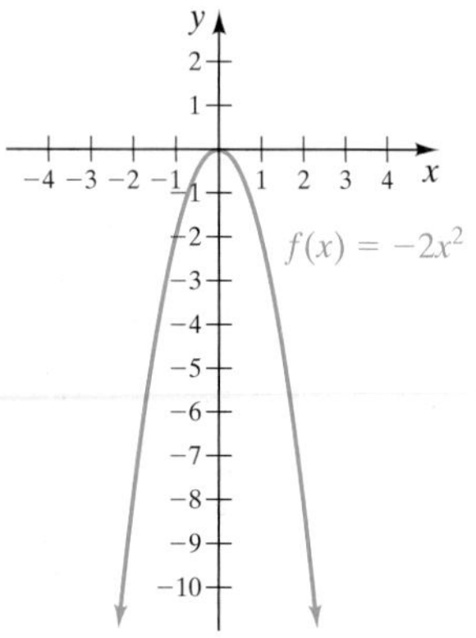

Now we are ready to graph parabolas using translations.

EXAMPLE 6 The graph of $f(x) = -2x^2$ is illustrated in **Figure 8.17**. Using this graph as a guide, graph $g(x) = -2(x + 3)^2 - 4$.

Solution The function $g(x)$ may be written $g(x) = -2[x - (-3)]^2 - 4$. Therefore, in the function, h has a value of -3 and k has a value of -4. The graph of $g(x)$ will therefore be the graph of $f(x)$ translated 3 units to the left (because $h = -3$) and 4 units down (because $k = -4$). The graphs of $f(x)$ and $g(x)$ are illustrated in **Figure 8.18**.

FIGURE 8.17

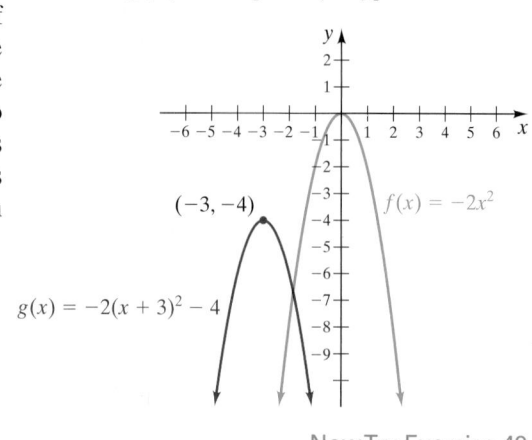

$g(x) = -2(x + 3)^2 - 4$

FIGURE 8.18

Now Try Exercise 49

In objective 2, we started with a function of the form $f(x) = ax^2 + bx + c$ and completed the square to obtain

$$f(x) = a\left[x - \left(-\frac{b}{2a}\right)\right]^2 + \frac{4ac - b^2}{4a}$$

We stated that the vertex of the parabola of this function is $\left(-\frac{b}{2a}, \frac{4ac - b^2}{4a}\right)$.

Suppose we substitute h for $-\frac{b}{2a}$ and k for $\frac{4ac - b^2}{4a}$ in the function. We then get

$$f(x) = a(x - h)^2 + k$$

which we know is a parabola with vertex at (h, k). Therefore, both functions $f(x) = ax^2 + bx + c$ and $f(x) = a(x - h)^2 + k$ yield the same vertex and axis of symmetry for any given function.

6 Write Functions in the Form $f(x) = a(x - h)^2 + k$

If we wish to graph parabolas using translations, we need to change the form of a function from $f(x) = ax^2 + bx + c$ to $f(x) = a(x - h)^2 + k$. To do this, we *complete the square* as was discussed in Section 8.1.

EXAMPLE 7 Given $f(x) = x^2 - 6x + 10$,

a) Write $f(x)$ in the form $f(x) = a(x - h)^2 + k$.
b) Graph $f(x)$.

Solution

a) We use the x^2 and $-6x$ terms to obtain a perfect square trinomial.

$$f(x) = (x^2 - 6x) + 10$$

Now we take half the coefficient of the x-term and square it.

$$\left[\frac{1}{2}(-6)\right]^2 = \boxed{9}$$

We then add this value, 9, within the parentheses. Since we are adding 9 within parentheses, we add -9 outside parentheses. Adding 9 and -9 to an expression is the same as adding 0, which does not change the value of the expression.

$$f(x) = (x^2 - 6x \boxed{+ 9}) \boxed{- 9} + 10$$

By doing this we have created a perfect square trinomial within the parentheses, plus a constant outside the parentheses. We express the perfect square trinomial as the square of a binomial.

$$f(x) = (x - 3)^2 + 1$$

The function is now in the form we are seeking.

b) Since $a = 1$, which is greater than 0, the parabola opens upward. The axis of symmetry of the parabola is $x = 3$ and the vertex is at $(3, 1)$. The y-intercept can be obtained by substituting $x = 0$ and finding $f(x)$. When $x = 0$, $f(x) = (-3)^2 + 1 = 10$. Thus, the y-intercept is at 10. By plotting the vertex, y-intercept, and a few other points, we obtain the graph in **Figure 8.19**. The figure also shows the graph of $y = x^2$ for comparison.

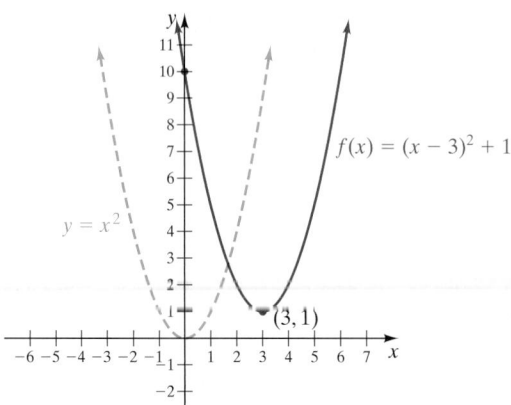

FIGURE 8.19

Now Try Exercise 59

EXAMPLE 8　Given $f(x) = -2x^2 - 10x - 13$,

a) Write $f(x)$ in the form $f(x) = a(x - h)^2 + k$.

b) Graph $f(x)$.

Solution

a) When the leading coefficient is not 1, we factor out the leading coefficient from the terms containing the variable.

$$f(x) = -2(x^2 + 5x) - 13$$

Now we complete the square.

Half of coefficient of
first-degree-term squared

$$\left[\frac{1}{2}(5)\right]^2 = \frac{25}{4}$$

If we add $\frac{25}{4}$ within the parentheses, we are actually adding $-2\left(\frac{25}{4}\right)$ or $-\frac{25}{2}$, since each term in parentheses is multiplied by -2. Therefore, to compensate, we must add $\frac{25}{2}$ outside the parentheses.

$$f(x) = -2\left(x^2 + 5x + \frac{25}{4}\right) + \frac{25}{2} - 13$$

$$= -2\left(x + \frac{5}{2}\right)^2 - \frac{1}{2}$$

b) Since $a = -2$, the parabola opens downward. The axis of symmetry is $x = -\frac{5}{2}$ and the vertex is $\left(-\frac{5}{2}, -\frac{1}{2}\right)$. The y-intercept is at $f(0) = -13$. We plot a few points and draw the graph in **Figure 8.20** on page 453. In the figure, we also show the graph of $y = -2x^2$ for comparison.

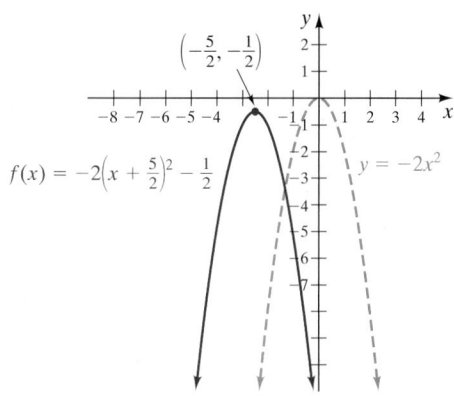

FIGURE 8.20

Notice that $f(x) = -2\left(x + \dfrac{5}{2}\right)^2 - \dfrac{1}{2}$ has no x-intercepts. Therefore, there are no real values of x for which $f(x) = 0$.

Now Try Exercise 63

A second way to change the equation from $f(x) = ax^2 + bx + c$ to $f(x) = a(x - h)^2 + k$ form is to let $h = -\dfrac{b}{2a}$ and $k = \dfrac{4ac - b^2}{4a}$. Find the values for h and k and then substitute the values obtained into $f(x) = a(x - h)^2 + k$. For example, for the function $f(x) = -2x^2 - 10x - 13$ in Example 8, $a = -2, b = -10$, and $c = -13$. Then

$$h = -\frac{b}{2a} = -\frac{-10}{2(-2)} = -\frac{5}{2}$$

$$k = \frac{4ac - b^2}{4a} = \frac{4(-2)(-13) - (-10)^2}{4(-2)} = -\frac{1}{2}$$

Therefore,

$$f(x) = a(x - h)^2 + k$$

$$= -2\left[x - \left(-\frac{5}{2}\right)\right]^2 - \frac{1}{2}$$

$$= -2\left(x + \frac{5}{2}\right)^2 - \frac{1}{2}$$

This answer checks with that obtained in Example 8.

EXERCISE SET 8.5

Math XL
MathXL®

MyMathLab
MyMathLab

Warm-Up Exercises

Fill in the blanks with the appropriate word, phrase, or symbol(s) from the following list.

down	axis of symmetry	y-intercept	$-\dfrac{b}{2a}$	opens downward
vertex	narrower	right	left	x-intercepts
$\dfrac{4ac - b^2}{4a}$	parabola	up	line	

1. The graph of a quadratic function is a _____.

2. When $a > 0$, the graph of $f(x) = ax^2 + bx + c$ is a parabola that opens upward and the _____ is the lowest point on the curve.

3. When $a < 0$, the graph of $f(x) = ax^2 + bx + c$ is a parabola that _____ and the vertex is the highest point on the curve.

4. Graphs of quadratic functions will have symmetry about a vertical line, called the _____.

5. The equation $x =$ _____ gives the equation, of the axis of symmetry and the x-coordinate of the vertex.

6. To find the _____ of a quadratic function (if there are any), set y or $f(x)$ equal to 0 and solve for x.

7. To find the _____ of a quadratic function, set $x = 0$ and solve for y or $f(x)$.

8. In general, when graphing the parabola that corresponds to a quadratic function of the form $f(x) = ax^2$, as $|a|$ gets larger, the parabola gets _____.

9. If h is positive, then the graph of $g(x) = a(x - h)^2$ will have the same shape as the graph of $f(x) = ax^2$ but will be shifted h units to the _____ of the graph of $f(x)$.

10. If h is negative, then the graph of $g(x) = a(x - h)^2$ will have the same shape as the graph of $f(x) = ax^2$ but will be shifted $|h|$ units to the _____ of the graph of $f(x)$.

11. If k is positive, then the graph of $g(x) = ax^2 + k$ will have the same shape as the graph of $f(x) = ax^2$ but will be shifted k units _____ from the graph of $f(x)$.

12. If k is negative, then the graph of $g(x) = ax^2 + k$ will have the same shape as the graph of $f(x) = ax^2$ but will be shifted $|k|$ units _____ from the graph of $f(x)$.

Practice the Skills

a) *Determine whether the parabola opens upward or downward.* b) *Find the y-intercept.* c) *Find the vertex.* d) *Find the x-intercepts (if any).* e) *Draw the graph.*

13. $f(x) = x^2 + 6x + 8$

14. $f(x) = x^2 - 6x + 5$

15. $f(x) = x^2 + 8x + 15$

16. $g(x) = x^2 + 2x - 3$

17. $f(x) = x^2 - 4x + 3$

18. $h(x) = x^2 - 2x - 8$

19. $f(x) = -x^2 - 2x + 8$

20. $p(x) = -x^2 + 8x - 15$

21. $g(x) = -x^2 + 4x + 5$

22. $n(x) = -x^2 - 2x + 24$

23. $t(x) = -x^2 + 4x - 5$

24. $g(x) = x^2 + 6x + 13$

25. $f(x) = x^2 - 4x + 4$

26. $r(x) = -x^2 + 10x - 25$

27. $r(x) = x^2 + 2$

28. $f(x) = x^2 + 4x$

29. $l(x) = -x^2 + 5$

30. $g(x) = -x^2 + 6x$

31. $f(x) = -2x^2 + 4x - 8$

32. $g(x) = -2x^2 - 6x + 4$

33. $m(x) = 3x^2 + 4x + 3$

34. $p(x) = -2x^2 + 5x + 4$

35. $y = 3x^2 + 4x - 6$

36. $y = x^2 - 6x + 4$

37. $y = 2x^2 - x - 6$

38. $g(x) = -4x^2 + 6x - 9$

39. $f(x) = -x^2 + 3x - 5$

40. $h(x) = -2x^2 + 4x - 5$

*Using the graphs in **Figures 8.12** through **8.16** as a guide, graph each function and label the vertex.*

41. $f(x) = (x - 3)^2$ **42.** $f(x) = (x - 4)^2$ **43.** $f(x) = (x + 1)^2$ **44.** $f(x) = (x + 2)^2$

45. $f(x) = x^2 + 3$ **46.** $f(x) = x^2 + 5$ **47.** $f(x) = x^2 - 1$ **48.** $f(x) = x^2 - 4$

49. $f(x) = (x - 2)^2 + 3$ **50.** $f(x) = (x - 3)^2 - 4$ **51.** $f(x) = (x + 4)^2 + 4$ **52.** $h(x) = (x + 4)^2 - 1$

53. $g(x) = -(x + 3)^2 - 2$ **54.** $g(x) = (x - 1)^2 + 4$ **55.** $y = -2(x - 2)^2 + 2$ **56.** $y = -2(x - 3)^2 + 1$

57. $h(x) = -2(x + 1)^2 - 3$ **58.** $f(x) = -(x - 5)^2 + 2$

*In Exercises 59–68, **a)** express each function in the form $f(x) = a(x - h)^2 + k$ and **b)** draw the graph of each function and label the vertex.*

59. $f(x) = x^2 - 6x + 8$ **60.** $g(x) = x^2 + 6x + 2$

61. $g(x) = x^2 - x - 3$ **62.** $f(x) = x^2 - x + 1$

63. $f(x) = -x^2 - 4x - 6$ **64.** $h(x) = -x^2 + 6x + 1$

65. $g(x) = x^2 - 4x - 1$ **66.** $p(x) = x^2 - 2x - 6$

67. $f(x) = 2x^2 + 5x - 3$ **68.** $k(x) = 2x^2 + 7x - 4$

Problem Solving

*Match the functions in Exercises 69–72 with the appropriate graphs labeled **a)** through **d)**.*

a) 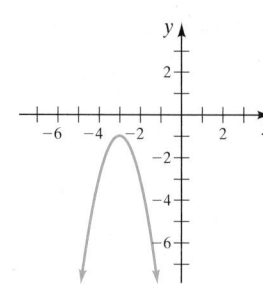   **b)** 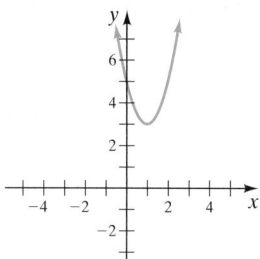   **c)** 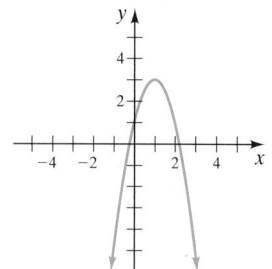   **d)**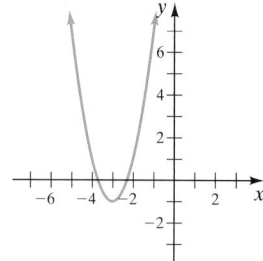

69. $f(x) = 2(x - 1)^2 + 3$ **70.** $f(x) = -2(x + 3)^2 - 1$ **71.** $f(x) = 2(x + 3)^2 - 1$ **72.** $f(x) = -2(x - 1)^2 + 3$

Area *For each rectangle, **a)** find the value for x that gives the maximum area and **b)** find the maximum area.*

73.
$26 - x$
$x + 5$

74.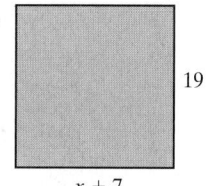
$19 - x$
$x + 7$

75.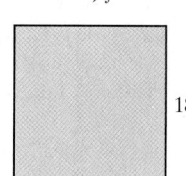
$18 - x$
$x + 4$

76.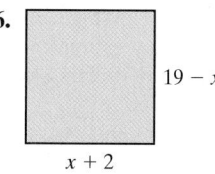
$19 - x$
$x + 2$

77. Selling Batteries The revenue function for selling n batteries is $R(n) = n(8 - 0.02n) = -0.02n^2 + 8n$. Find **a)** the number of batteries that must be sold to obtain the maximum revenue and **b)** the maximum revenue.

© Allen R. Angel

78. Selling Watches The revenue function for selling n watches is $R(n) = n(25 - 0.1n) = -0.1n^2 + 25n$. Find **a)** the number

of watches that must be sold to obtain the maximum revenue and **b)** the maximum revenue.

79. Enrollment The enrollment in a high school in the Naplewood School District can be approximated by the function

$$N(t) = -0.043t^2 + 1.82t + 46.0$$

where t is the number of years since 1989 and $1 \le t \le 22$. In what year will the maximum enrollment be obtained?

© Michael Chamberlin/Shutterstock

80. Drug-Free Schools The percent of students in schools in the United States who say their school is not drug free can be approximated by the function

$$f(a) = -2.32a^2 + 76.58a - 559.87$$

where a is the student's age and $12 < a < 20$. What age has the highest percent of students who say their school is not drug free?

81. What is the distance between the vertices of the graphs of $f(x) = (x - 2)^2 + \dfrac{5}{2}$ and $g(x) = (x - 2)^2 - \dfrac{3}{2}$?

82. What is the distance between the vertices of the graphs of $f(x) = 2(x - 4)^2 - 3$ and $g(x) = -3(x - 4)^2 + 2$?

83. What is the distance between the vertices of the graphs of $f(x) = 2(x + 4)^2 - 3$ and $g(x) = -(x + 1)^2 - 3$?

84. What is the distance between the vertices of the graphs of $f(x) = -\dfrac{1}{3}(x - 3)^2 - 2$ and $g(x) = 2(x + 5)^2 - 2$?

85. Write the function whose graph has the shape of the graph of $f(x) = 2x^2$ and has a vertex at $(3, -2)$.

86. Write the function whose graph has the shape of the graph of $f(x) = -\dfrac{1}{2}x^2$ and has a vertex at $\left(\dfrac{2}{3}, -5\right)$.

87. Write the function whose graph has the shape of the graph of $f(x) = -4x^2$ and has a vertex at $\left(-\dfrac{3}{5}, -\sqrt{2}\right)$.

88. Write the function whose graph has the shape of the graph of $f(x) = \dfrac{3}{5}x^2$ and has a vertex at $(-\sqrt{3}, \sqrt{5})$.

89. Selling Tickets The Johnson High School Theater Club is trying to set the price of tickets for a play. If the price is too low, they will not make enough money to cover expenses, and if the price is too high, not enough people will pay the price of a ticket. They estimate that their total income per play, I, in hundreds of dollars, can be approximated by the formula

$$I = -x^2 + 24x - 44, 0 \le x \le 24$$

where x is the cost of a ticket.

a) Draw a graph of income versus the cost of a ticket.

b) Determine the minimum cost of a ticket for the theater club to break even.

c) Determine the maximum cost of a ticket that the theater club can charge and break even.

d) How much should they charge to receive the maximum income?

e) Find the maximum income.

90. Throwing an Object An object is projected upward with an initial velocity of 192 feet per second. The object's distance above the ground, d, after t seconds may be found by the formula $d = -16t^2 + 192t$.

a) Find the object's distance above the ground after 3 seconds.

b) Draw a graph of distance versus time.

c) What is the maximum height the object will reach?

d) At what time will it reach its maximum height?

e) At what time will the object strike the ground?

91. Profit The Fulton Bird House Company earns a weekly profit according to the function $f(x) = -0.4x^2 + 80x - 200$, where x is the number of bird feeders built and sold.

a) Find the number of bird feeders that the company must sell in a week to obtain the maximum profit.

b) Find the maximum profit.

© Chris Alcock/Shutterstock

92. Profit The A. B. Bronson Company earns a weekly profit according to the function $f(x) = -1.2x^2 + 180x - 280$, where x is the number of rocking chairs built and sold.

a) Find the number of rocking chairs that the company must sell in a week to obtain the maximum profit.

b) Find the maximum profit.

93. Firing a Cannon If a certain cannon is fired from a height of 9.8 meters above the ground, at a certain angle, the height of the cannonball above the ground, h, in meters, at time, t, in seconds, is found by the function

$$h(t) = -4.9t^2 + 24.5t + 9.8$$

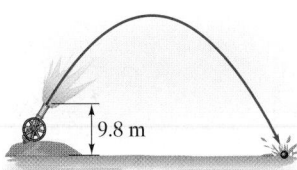

9.8 m

a) Find the maximum height attained by the cannonball.

b) Find the time it takes for the cannonball to reach its maximum height.

c) Find the time it takes for the cannonball to strike the ground.

94. Throwing a Ball Ramon Loomis throws a ball into the air with an initial velocity of 32 feet per second. The height of the ball at any time t is given by the formula $h = 96t - 16t^2$. At what time does the ball reach its maximum height? What is the maximum height?

95. Room in a House Jake Kishner is designing plans for his house. What is the maximum possible area of a room if its perimeter is to be 80 feet?

96. Greatest Area What are the dimensions of a rectangular garden that will have the greatest area if its perimeter is to be 70 feet?

© Imageman/Shutterstock

97. Minimum Product What is the minimum product of two numbers that differ by 8? What are the numbers?

98. Minimum Product What is the minimum product of two numbers that differ by 10? What are the numbers?

99. Maximum Product What is the maximum product of two numbers that add to 60? What are the numbers?

100. Maximum Product What is the maximum product of two numbers that add to 5? What are the numbers?

The profit of a company, in dollars, is the difference between the company's revenue and cost. Exercises 101 and 102 give cost, $C(x)$, and revenue, $R(x)$, functions for a particular company. The x represents the number of items produced and sold to distributors. Determine **a)** *the maximum profit of the company and* **b)** *the number of items that must be produced and sold to obtain the maximum profit.*

101. $C(x) = 2000 + 40x$
$R(x) = 800x - x^2$

102. $C(x) = 5000 + 12x$
$R(x) = 2000x - x^2$

Concept/Writing Exercises

103. Consider the graph of $f(x) = ax^2$. What is the general shape of $f(x)$ if **a)** $a > 0$, **b)** $a < 0$?

104. Consider the graph of $f(x) = ax^2$. Explain how the shape of $f(x)$ changes as $|a|$ increases and as $|a|$ decreases.

105. Does the function $f(x) = 3x^2 - 4x + 2$ have a maximum or a minimum value? Explain.

106. Does the function $g(x) = -\frac{1}{2}x^2 + 2x - 7$ have a maximum or a minimum value? Explain.

107. Consider $f(x) = x^2 - 8x + 12$ and $g(x) = -x^2 + 8x - 12$.

a) Without graphing, can you explain how the graphs of the two functions compare?

b) Will the graphs have the same x-intercepts? Explain.

c) Will the graphs have the same vertex? Explain.

d) Graph both functions on the same axes.

108. By observing the leading coefficient in a quadratic function and by determining the coordinates of the vertex of its graph, explain how you can determine the number of x-intercepts the parabola has.

Challenge Problems

109. Baseball In Example 3 of this section, we used the function $f(t) = -16t^2 + 52t + 3$ to find that the maximum height, y, attained by a baseball hit by Mark DeRosa was 45.25 feet. The ball reached this height at 1.625 seconds after the baseball was hit.

Review Example 3 now.

a) Write $f(t)$ in the form $f(t) = a(t - h)^2 + k$ by completing the square.

b) Using the function you obtained in part **a)**, determine the maximum height attained by the baseball, and the time after it was hit that the baseball attained its maximum value.

c) Are the answers you obtained in part **b)** the same answers obtained in Example 3? If not, explain why not.

Group Activity

Discuss and answer Exercise 110 as a group.

110. a) Group member 1: Write two quadratic functions $f(x)$ and $g(x)$ so that the functions will not intersect.

b) Group member 2: Write two quadratic functions $f(x)$ and $g(x)$ so that neither function will have x-intercepts, and the vertices of the functions are on opposite sides of the x-axis.

c) Group member 3: Write two quadratic functions $f(x)$ and $g(x)$ so that both functions have the same vertex but one function opens upward and the other opens downward.

d) As a group, review each answer in parts **a)–c)** and decide whether each answer is correct. Correct any answer that is incorrect.

Cumulative Review Exercises

[2.2]   **111.** Find the area shaded blue in the figure.

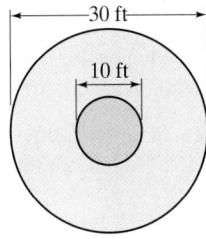

[3.7]   **112.** Graph $y \le \dfrac{2}{3}x + 3$.

[4.2]   **113.** Solve the system of equations.

$$x - y = -5$$
$$2x + 2y - z = 0$$
$$x + y + z = 3$$

[4.5]   **114.** Evaluate the determinant.

$$\begin{vmatrix} \dfrac{1}{2} & 3 \\ 2 & -4 \end{vmatrix}$$

[6.1]   **115.** Divide $(x - 3) \div \dfrac{x^2 + 3x - 18}{x}$.

8.6 Quadratic and Other Inequalities in One Variable

① Solve quadratic inequalities.

② Solve other polynomial inequalities.

③ Solve rational inequalities.

In Section 2.5, we discussed linear inequalities in one variable. Now we discuss quadratic inequalities in one variable.

Quadratic Inequality

A **quadratic inequality*** is an inequality that can be written in one of the following forms

$$ax^2 + bx + c < 0 \qquad ax^2 + bx + c > 0$$
$$ax^2 + bx + c \le 0 \qquad ax^2 + bx + c \ge 0$$

where a, b, and c are real numbers, with $a \ne 0$.

Examples of Quadratic Inequalities

$$x^2 + x - 12 > 0, \qquad 2x^2 - 9x - 5 \le 0$$

Solution to a Quadratic Inequality

The **solution to a quadratic inequality** is the set of all values that make the inequality a true statement.

For example, substitute 5 for x in $x^2 + x - 12 > 0$.

$$x^2 + x - 12 > 0$$
$$5^2 + 5 - 12 \overset{?}{>} 0$$
$$18 > 0 \qquad \text{True}$$

The inequality is true when x is 5, so 5 is a solution to (or satisfies) the inequality. However, 5 is not the only solution. There are other values that are solutions to the inequality.

---

*An inequality of the form $ax^2 + bx + c \ne 0$ is also a quadratic inequality that we will not be discussing in this book.

1   Solve Quadratic Inequalities

We will begin our study of quadratic inequalities by studying the graphs of quadratic functions. For example, consider once again the inequality $x^2 + x - 12 > 0$. **Figure 8.21a** shows the graph of the function $f(x) = x^2 + x - 12$. Notice that when $x < -4$ or when $x > 3$, $f(x) > 0$ (shown in red in **Figure 8.21b**). Thus, when $x < -4$ or when $x > 3$, $x^2 + x - 12 > 0$.

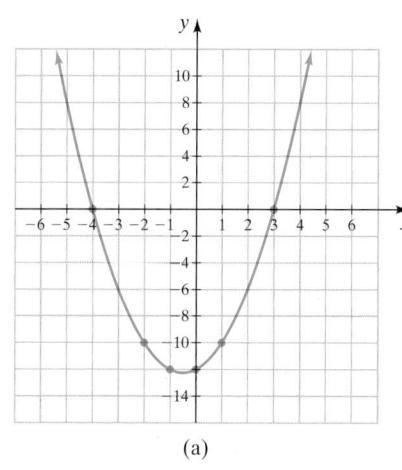

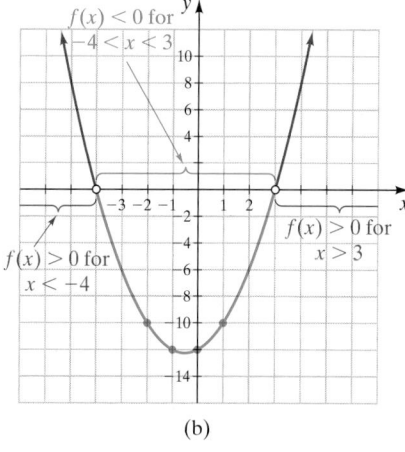

FIGURE 8.21

(a)   (b)

Notice also that when $-4 < x < 3$, $f(x) < 0$ (shown in green in **Figure 8.21b**). Thus, when $-4 < x < 3$, $x^2 + x - 12 < 0$.

Although the method we just described to solve quadratic inequalities will work for any quadratic inequality, for many inequalities it may be inconvenient or take too much time to graph the function. In our next example, we will outline a more efficient procedure to find the solution to a quadratic inequality.

EXAMPLE 1   Solve the inequality $x^2 + x - 12 > 0$. Give the solution **a)** on a number line, **b)** in interval notation, and **c)** in set builder notation.

Solution   First, set the inequality equal to 0 and solve the equation.

$$x^2 + x - 12 = 0$$

$$(x + 4)(x - 3) = 0$$

$$x + 4 = 0 \quad \text{or} \quad x - 3 = 0$$

$$x = -4 \qquad\qquad x = 3$$

The values $x = -4$ and $x = 3$ are called **boundary values** since they will be on the boundaries of the intervals that make up the solution to the inequality. The boundary values are used to break up a number line into intervals. Whenever the original inequality symbol is $<$ or $>$, we will indicate the boundary values on the number line using open circles, $\circ$. This indicates the boundary values are not part of the solution. Whenever the original inequality symbol is $\leq$ or $\geq$, we will indicate the boundary values on the number line using closed circles, $\bullet$. This indicates the boundary values are part of the solution.

$$
\begin{array}{ccc}
A & B & C \\
(-\infty, -4) & (-4, 3) & (3, \infty)
\end{array}
$$

$\xleftarrow{\qquad\quad} \underset{-4}{\circ} \qquad \underset{3}{\circ} \xrightarrow{\qquad\quad}$

boundary values indicated
with open circles

FIGURE 8.22

In **Figure 8.22**, we have labeled the intervals A, B, and C. Next, we select one test value in *each* interval. Then we substitute each of those numbers, one at a time, into the *inequality* $x^2 + x - 12 > 0$. If the test value results in a true statement, all

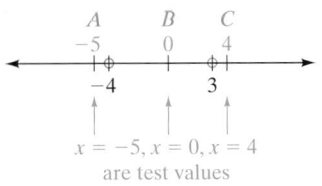

$x = -5, x = 0, x = 4$
are test values

FIGURE 8.23

values in that interval will also satisfy the inequality. If the test value results in a false statement, no numbers in that interval will satisfy the inequality.

In this example, we will use the test values of -5 in interval A, 0 in interval B, and 4 in interval C (see **Fig. 8.23**).

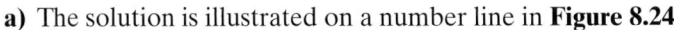

Interval A	Interval B	Interval C
$(-\infty, -4)$	$(-4, 3)$	$(3, \infty)$
Test value, -5	Test value, 0	Test value, 4
Is $x^2 + x - 12 > 0$?	Is $x^2 + x - 12 > 0$?	Is $x^2 + x - 12 > 0$?
$(-5)^2 - 5 - 12 \overset{?}{>} 0$	$0^2 + 0 - 12 \overset{?}{>} 0$	$4^2 + 4 - 12 \overset{?}{>} 0$
$8 > 0$	$-12 > 0$	$8 > 0$
True	False	True

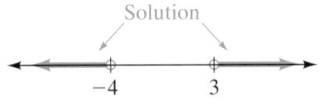

FIGURE 8.24

Since the test values in both intervals A and C satisfy the inequality, the solution is all real numbers in intervals A or C. Since the inequality symbol is $>$, the values -4 and 3 are not included in the solution because they make the inequality equal to 0.

The answers to parts **a)**, **b)**, and **c)** follow.

a) The solution is illustrated on a number line in **Figure 8.24**.

b) The solution in interval notation is $(-\infty, -4) \cup (3, \infty)$.

c) The solution in set builder notation is $\{x \mid x < -4 \text{ or } x > 3\}$.

Note that the solution, in any form, is consistent with the red portion of the graph in **Figure 8.21b**.

Now Try Exercise 15

To Solve Quadratic and Other Inequalities

1. Write the inequality as an equation and solve the equation. The solutions are the boundary values.

2. Construct a number line and mark each boundary value from step 1 as follows:
   - If the inequality symbol is $<$ or $>$, use an open circle $\circ$.
   - If the inequality symbol is $\leq$ or $\geq$, use a closed circle $\bullet$.

3. If solving a rational inequality, determine the values that make any denominator 0. These values are also boundary values. Indicate these boundary values on your number line with an open circle $\circ$.

4. Select a test value in each interval and determine whether it satisfies the inequality.

5. The solution is the set of points that satisfy the inequality.

6. Write the solution in the form instructed.

EXAMPLE 2 Solve the inequality $x^2 - 4x \geq -4$. Give the solution **a)** on a number line, **b)** in interval notation, and **c)** in set builder notation.

Solution   Write the inequality as an equation, then solve the equation.

$$x^2 - 4x = -4$$
$$x^2 - 4x + 4 = 0$$
$$(x - 2)(x - 2) = 0$$
$$x - 2 = 0 \quad \text{or} \quad x - 2 = 0$$
$$x = 2 \qquad\qquad x = 2$$

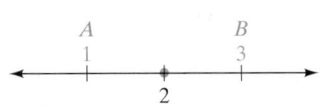

FIGURE 8.25

Since both factors are the same there is only one boundary value, 2, indicated in **Figure 8.25** with a closed circle. Both test values, 1 and 3, result in true statements.

Interval A	Interval B
$(-\infty, 2]$	$[2, \infty)$
Test value, 1	Test value, 3

$$x^2 - 4x \geq -4 \qquad\qquad x^2 - 4x \geq -4$$
$$1^2 - 4(1) \overset{?}{\geq} -4 \qquad\qquad 3^2 - 4(3) \overset{?}{\geq} -4$$
$$1 - 4 \overset{?}{\geq} -4 \qquad\qquad 9 - 12 \overset{?}{\geq} -4$$
$$-3 \geq -4 \quad \text{True} \qquad\qquad -3 \geq -4 \quad \text{True}$$

The solution set includes both intervals and the boundary value, 2. The solution set is the set of real numbers, $\mathbb{R}$. The answers to parts **a)**, **b)**, and **c)** follow.

a) $\xleftarrow{\qquad\underset{2}{+}\qquad}$ **b)** $(-\infty, \infty)$ **c)** $\{x | -\infty < x < \infty\}$

Now Try Exercise 11

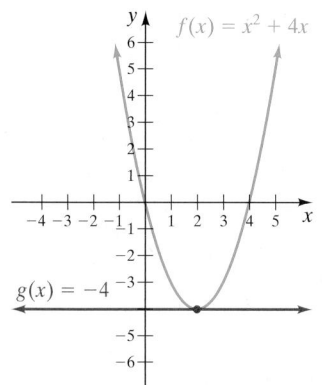

FIGURE 8.26

We can check the solution to Example 2 using graphing. Let $f(x) = x^2 - 4x$ and $g(x) = -4$. For $x^2 - 4x \geq -4$ to be true, we want $f(x) \geq g(x)$. The graphs of $f(x)$ and $g(x)$ are given in **Figure 8.26**.

Observe that $f(x) = g(x)$ at $x = 2$ and $f(x) > g(x)$ for all other values of x. Thus, $f(x) \geq g(x)$ for all values of x, and the solution set is the set of real numbers.

In Example 2, if we rewrite the inequality $x^2 - 4x \geq -4$ as $x^2 - 4x + 4 \geq 0$ and then as $(x - 2)^2 \geq 0$ we can see that the solution must be the set of real numbers, since $(x - 2)^2$ must be greater than or equal to 0 for any real number x.

EXAMPLE 3 Solve the inequality $x^2 - 2x - 4 \leq 0$. Express the solution in interval notation.

Solution   First we need to solve the equation $x^2 - 2x - 4 = 0$. Since this equation is not factorable, we use the quadratic formula to solve. Here $a = 1$, $b = -2$, and $c = -4$.

$$x = \frac{-b \pm \sqrt{b^2 - 4ac}}{2a}$$

$$= \frac{-(-2) \pm \sqrt{(-2)^2 - 4(1)(-4)}}{2(1)} = \frac{2 \pm \sqrt{20}}{2} = \frac{2 \pm 2\sqrt{5}}{2} = 1 \pm \sqrt{5}$$

The boundary values are $1 - \sqrt{5}$ and $1 + \sqrt{5}$. The value of $1 - \sqrt{5}$ is about -1.24 and the value of $1 + \sqrt{5}$ is about 3.24. We indicate the boundary values on the number line with closed circles and we will select test values of -2, 0, and 4 (see **Fig. 8.27**).

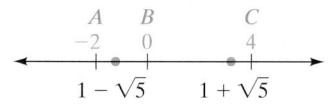

FIGURE 8.27

Interval A	Interval B	Interval C
$(-\infty, 1 - \sqrt{5}]$	$[1 - \sqrt{5}, 1 + \sqrt{5}]$	$[1 + \sqrt{5}, \infty)$
Test value, -2	Test value, 0	Test value, 4

$$x^2 - 2x - 4 \leq 0 \qquad\quad x^2 - 2x - 4 \leq 0 \qquad\quad x^2 - 2x - 4 \leq 0$$
$$(-2)^2 - 2(-2) - 4 \overset{?}{\leq} 0 \qquad 0^2 - 2(0) - 4 \overset{?}{\leq} 0 \qquad 4^2 - 2(4) - 4 \overset{?}{\leq} 0$$
$$4 + 4 - 4 \overset{?}{\leq} 0 \qquad\quad 0 - 0 - 4 \overset{?}{\leq} 0 \qquad\quad 16 - 8 - 4 \overset{?}{\leq} 0$$
$$4 \leq 0 \qquad\qquad\qquad -4 \leq 0 \qquad\qquad\qquad 4 \leq 0$$
$$\text{False} \qquad\qquad\qquad \text{True} \qquad\qquad\qquad \text{False}$$

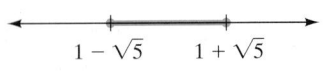

FIGURE 8.28

The boundary values are part of the solution because the inequality symbol is $\leq$ and the boundary values make the inequality equal to 0. Thus, the solution in interval notation is $[1 - \sqrt{5}, 1 + \sqrt{5}]$. The solution is illustrated on the number line in **Figure 8.28**.

Now Try Exercise 19

> ### Helpful Hint
>
> If $ax^2 + bx + c = 0$, with $a > 0$, has two distinct real solutions, then:
>
> | Inequality of Form | Solution Is | Solution on Number Line |
> |---|---|---|
> | $ax^2 + bx + c \geq 0$ | End intervals | ← —— ——→ |
> | $ax^2 + bx + c \leq 0$ | Center interval | ← ——— → |
>
> Example 1 is an inequality of the form $ax^2 + bx + c > 0$, and Example 3 is an inequality of the form $ax^2 + bx + c \leq 0$. Example 2 does not have two distinct real solutions, so this Helpful Hint does not apply.

2 Solve Other Polynomial Inequalities

The same procedure we used to solve quadratic inequalities can be used to solve other **polynomial inequalities**, as illustrated in the following examples.

EXAMPLE 4 Solve the polynomial inequality $(3x - 2)(x + 3)(x + 5) < 0$. Illustrate the solution on a number line and write the solution in both interval notation and set builder notation.

Solution We use the zero-factor property to solve the equation $(3x - 2)(x + 3)(x + 5) = 0$.

$$3x - 2 = 0 \quad \text{or} \quad x + 3 = 0 \quad \text{or} \quad x + 5 = 0$$

$$x = \frac{2}{3} \qquad\qquad x = -3 \qquad\qquad x = -5$$

The solutions -5, -3, and $\frac{2}{3}$ are indicated with open circles and break the number line into four intervals (see **Fig. 8.29**). The test values we will use are $-6, -4, 0$, and 1. We show the results in the following table.

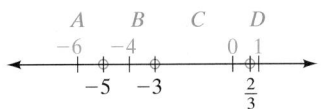

FIGURE 8.29

Interval	Test value	$(3x - 2)(x + 3)(x + 5)$	< 0
A: $(-\infty, -5)$	-6	-60	True
B: $(-5, -3)$	-4	14	False
C: $\left(-3, \frac{2}{3}\right)$	0	-30	True
D: $\left(\frac{2}{3}, \infty\right)$	1	24	False

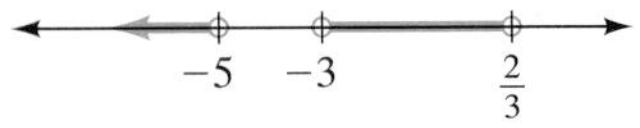

Since the original inequality symbol is $<$, the boundary values are not part of the solution. The solution, intervals A and C, is illustrated on the number line in **Figure 8.30**. The solution in interval notation is $(-\infty, -5) \cup \left(-3, \frac{2}{3}\right)$. The solution in set builder notation is $\left\{ x \,\middle|\, x < -5 \text{ or } -3 < x < \frac{2}{3} \right\}$.

Now Try Exercise 27

FIGURE 8.30

EXAMPLE 5 Given $f(x) = 3x^3 - 3x^2 - 6x$, find all values of x for which $f(x) \geq 0$. Illustrate the solution on a number line and give the solution in interval notation.

Solution We need to solve the inequality

$$3x^3 - 3x^2 - 6x \geq 0$$

We start by solving the equation $3x^3 - 3x^2 - 6x = 0$.

$$3x(x^2 - x - 2) = 0$$
$$3x(x - 2)(x + 1) = 0$$

$$3x = 0 \quad \text{or} \quad x - 2 = 0 \quad \text{or} \quad x + 1 = 0$$
$$x = 0 \qquad\qquad x = 2 \qquad\qquad x = -1$$

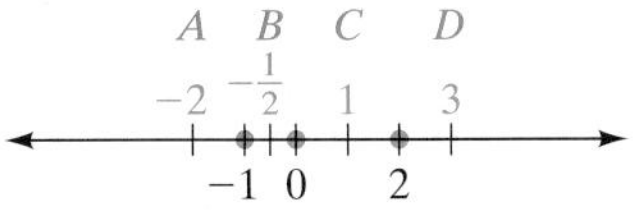

The solutions $-1, 0$, and 2 are indicated with closed circles and break the number line into four intervals (see **Fig. 8.31**). The test values that we will use are -2, $-\dfrac{1}{2}$, 1, and 3.

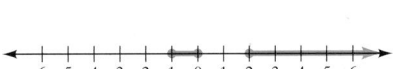

Interval	Test Value	$3x^3 - 3x^2 - 6x$	≥ 0
$A: (-\infty, -1]$	-2	-24	False
$B: [-1, 0]$	$-\dfrac{1}{2}$	$\dfrac{15}{8}$	True
$C: [0, 2]$	1	-6	False
$D: [2, \infty)$	3	36	True

Since the original inequality is $\geq$, the boundary values are part of the solution. The solution, intervals B and D, is illustrated on the number line in **Figure 8.32a**. The solution in interval notation is $[-1, 0] \cup [2, \infty)$. **Figure 8.32b** shows the graph of $f(x) = 3x^3 - 3x^2 - 6x$. Notice $f(x) \geq 0$ for $-1 \leq x \leq 0$ and for $x \geq 2$, which agrees with our solution.

FIGURE 8.31

Understanding Algebra

Consider the inequality

$$-3x^3 + 3x^2 + 6x \leq 0.$$

It is generally easier to solve a polynomial inequality with a positive leading coefficient. We can change the leading coefficient to a positive number by multiplying both sides of the inequality by -1. When doing this, we must change the direction of the inequality symbol.

$$-3x^3 + 3x^2 + 6x \leq 0$$
$$-1(-3x^3 + 3x^2 + 6x) \geq -1(0)$$
$$3x^3 - 3x^2 - 6x \geq 0$$

This inequality was solved in Example 5.

FIGURE 8.32A

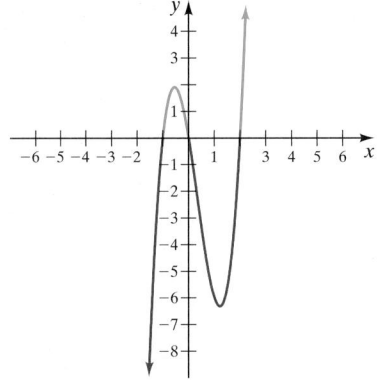

FIGURE 8.32B

Now Try Exercise 41

3 Solve Rational Inequalities

In Examples 6 and 7, we solve **rational inequalities**, which are inequalities that contain at least one rational expression.

EXAMPLE 6 Solve the inequality $\dfrac{x-1}{x+3} \geq 2$ and graph the solution on a number line.

Solution   Change the $\geq$ to $=$ and solve the resulting equation.

$$\frac{x-1}{x+3} = 2$$

$$\cancel{x+3} \cdot \frac{x-1}{\cancel{x+3}} = 2(x+3) \qquad \text{Multiply both sides by } x+3.$$

$$x - 1 = 2x + 6$$
$$-1 = x + 6$$
$$-7 = x$$

The solution -7 is a boundary value and is indicated with a closed circle on the number line (see **Fig. 8.33**).

When solving rational inequalities, we also need to determine the value or values that make any denominator 0. We set the denominator equal to 0 and solve.

$$x + 3 = 0$$
$$x = -3$$

Since -3 cannot be a solution, we indicate it on the number line with an open circle (see **Fig. 8.33**).

We use the solution to the equation, -7, and the value that makes the denominator 0, -3, to determine the intervals, shown in **Figure 8.33**. We will use -8, -5, and 0 as our test values.

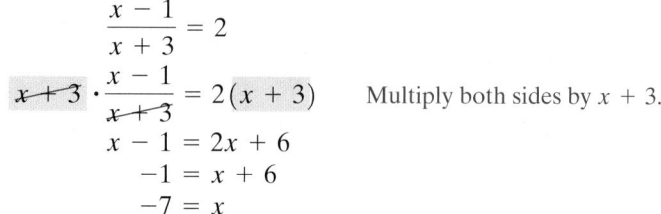

Interval A	Interval B	Interval C
$(-\infty, -7]$	$[-7, -3)$	$(-3, \infty)$
Test value, -8	Test value, -5	Test value, 0
$\dfrac{x-1}{x+3} \geq 2$	$\dfrac{x-1}{x+3} \geq 2$	$\dfrac{x-1}{x+3} \geq 2$
$\dfrac{-8-1}{-8+3} \overset{?}{\geq} 2$	$\dfrac{-5-1}{-5+3} \overset{?}{\geq} 2$	$\dfrac{0-1}{0+3} \overset{?}{\geq} 2$
$\dfrac{9}{5} \geq 2$ False	$3 \geq 2$ True	$-\dfrac{1}{3} \geq 2$ False

Now check the boundary values -7 and -3. Since -7 results in the inequality $-2 \geq -2$, which is true, -7 is a solution. Since division by 0 is not permitted, -3 is not a solution. The solution is illustrated on the number line in **Figure 8.34**, and the solution is $[-7, -3)$.

Now Try Exercise 81

In Example 6, we solved $\dfrac{x-1}{x+3} \geq 2$. Suppose we graphed $f(x) = \dfrac{x-1}{x+3}$.

Figure 8.35 shows the graph of $f(x) = \dfrac{x-1}{x+3}$ and the graph of $y = 2$. Notice $f(x) \geq 2$ when $-7 \leq x < -3$.

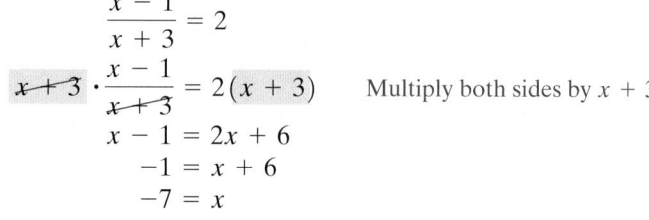

FIGURE 8.33

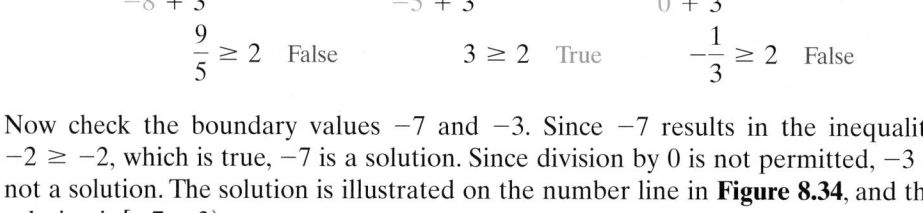

FIGURE 8.34

Understanding Algebra

When solving a rational inequality, the values at which any denominator is equal to 0 are boundary values. However, since a value that makes the denominator 0 can never be a solution, we will *always* indicate these values with an open circle. This shows they are not included in the solution set.

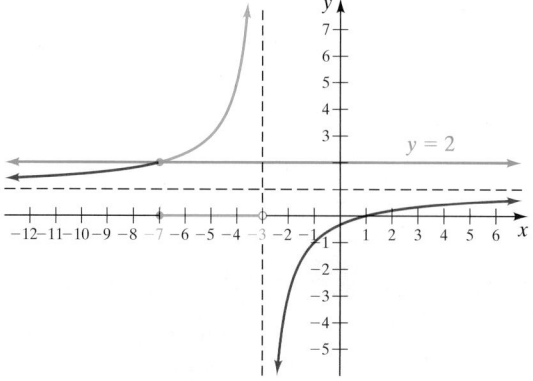

FIGURE 8.35

EXAMPLE 7 Solve the inequality $\dfrac{(x-3)(x+4)}{x+1} \ge 0$. Graph the solution on a number line and give the solution in interval notation.

Solution   The solutions to the equation $\dfrac{(x-3)(x+4)}{x+1} = 0$ are 3 and −4. We indicate the solutions on the number line with closed circles (see **Fig. 8.36**) since the inequality symbol is $\ge$. The inequality is not defined at −1. We indicate this value on the number line with an open circle (see **Fig. 8.36**) since −1 cannot be a solution to the inequality.

We therefore use the values −4, −1, and 3 to determine the intervals on the number line (see **Fig. 8.36**). Checking test values at −5, −2, 0, and 4, we find that the values in intervals B and D, $-4 \le x < -1$ and $x \ge 3$, satisfy the inequality. Check the test values yourself to verify this. The values 3 and −4 make the inequality equal to 0 and are part of the solution. The inequality is not defined at −1, so −1 is not part of the solution. The solution is $[-4, -1) \cup [3, \infty)$. The solution is illustrated on the number line in **Figure 8.37**.

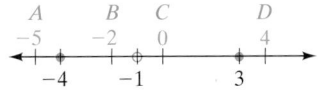

FIGURE 8.36

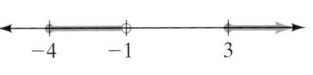

FIGURE 8.37

Now Try Exercise 71

EXERCISE SET 8.6

Math XL
MathXL®

MyMathLab
MyMathLab

Warm-Up Exercises

Fill in the blanks with the appropriate word, phrase, or symbol(s) from the following list.

not included   boundary values   closed   included   solution   open   vertices   divisor

1. The _____ to an inequality is the set of all values that make the inequality a true statement.

2. Consider the inequality $x^2 - 3x - 4 \le 0$. The solutions to the equation $x^2 - 3x - 4 = 0$ are called _____

3. In solving the inequality $(x - 5)(x + 3) \ge 0$ the boundary values 5 and −3 are _____ in the solution set.

4. In solving the inequality $(x - 2)(x + 4) < 0$ the boundary values 2 and −4 are _____ in the solution set.

5. In solving the inequality $\dfrac{1}{x - 3} \le 0$ the boundary value of 3 is indicated on the number line with a(n) _____ circle.

6. In solving the inequality $\dfrac{x + 1}{x - 3} \le 0$ the boundary value of −1 is indicated on the number line with a(n) _____ circle.

Practice the Skills

Solve each quadratic inequality and graph the solution on a number line.

7. $x^2 - 2x - 3 \ge 0$

8. $x^2 - 2x - 3 < 0$

9. $x^2 + 7x + 6 > 0$

10. $x^2 + 8x + 7 < 0$

11. $n^2 - 6n + 9 \ge 0$

12. $x^2 - 8x \ge 0$

13. $x^2 - 16 < 0$

14. $r^2 - 5r < 0$

15. $2x^2 + 5x - 3 \ge 0$

16. $3n^2 - 7n \le 6$

17. $5x^2 + 6x \le 8$

18. $3x^2 + 5x - 3 \le 0$

19. $2x^2 - 12x + 9 \le 0$

20. $5x^2 \le -20x - 4$

Solve each polynomial inequality and give the solution in interval notation.

21. $(x - 1)(x + 2)(x - 3) \le 0$

22. $(x - 2)(x + 2)(x + 5) \le 0$

23. $(a - 3)(a + 2)(a + 4) < 0$

24. $(r - 1)(r + 2)(r + 7) < 0$

25. $(2c + 5)(3c - 6)(c + 6) > 0$

26. $(a - 4)(a - 2)(a + 8) > 0$

27. $(3x + 5)(x - 3)(x + 1) > 0$

28. $(3c - 1)(c + 4)(3c + 6) \leq 0$

29. $(x + 2)(x + 2)(3x - 8) \geq 0$

30. $(x + 3)^2(4x - 7) \leq 0$

31. $x^3 - 6x^2 + 9x < 0$

32. $x^3 + 3x^2 - 40x > 0$

For each function provided, find all values of x for which f(x) satisfies the indicated conditions. Graph the solution on a number line.

33. $f(x) = x^2 - 2x, f(x) \geq 0$

34. $f(x) = x^2 - 7x, f(x) > 0$

35. $f(x) = x^2 + 4x, f(x) > 0$

36. $f(x) = x^2 + 8x, f(x) \leq 0$

37. $f(x) = x^2 - 14x + 48, f(x) < 0$

38. $f(x) = x^2 - 2x - 15, f(x) < 0$

39. $f(x) = 2x^2 + 9x - 1, f(x) \leq 5$

40. $f(x) = x^2 + 5x - 3, f(x) \leq 4$

41. $f(x) = 2x^3 + 9x^2 - 35x, f(x) \geq 0$

42. $f(x) = x^3 - 9x, f(x) \leq 0$

Solve each rational inequality and give the solution in set builder notation.

43. $\dfrac{x + 1}{x - 3} \leq 0$

44. $\dfrac{x + 2}{x - 4} \geq 0$

45. $\dfrac{x - 1}{x + 5} < 0$

46. $\dfrac{x - 1}{x + 5} \leq 0$

47. $\dfrac{x + 3}{x - 2} \geq 0$

48. $\dfrac{x - 4}{x + 6} > 0$

49. $\dfrac{a - 9}{a + 5} < 0$

50. $\dfrac{b + 7}{b + 1} \leq 0$

51. $\dfrac{c - 10}{c - 4} > 0$

52. $\dfrac{2d - 6}{d - 1} < 0$

53. $\dfrac{3y + 6}{y + 4} \leq 0$

54. $\dfrac{4z - 8}{z - 9} \geq 0$

55. $\dfrac{5a + 10}{3a - 1} \geq 0$

56. $\dfrac{x + 4}{x - 4} \leq 0$

57. $\dfrac{3x + 4}{2x - 1} < 0$

58. $\dfrac{k + 3}{k} \geq 0$

59. $\dfrac{3x + 8}{x - 2} \leq 0$

60. $\dfrac{4x - 2}{2x - 8} > 0$

Solve each rational inequality and give the solution in interval notation.

61. $\dfrac{(x - 2)(x - 4)}{x + 1} < 0$

62. $\dfrac{(x + 1)(x - 6)}{x + 3} \leq 0$

63. $\dfrac{(x - 2)(x + 3)}{x - 5} > 0$

64. $\dfrac{(x - 2)(x + 3)}{x - 5} \geq 0$

65. $\dfrac{(a - 1)(a - 7)}{a + 2} \geq 0$

66. $\dfrac{(b - 2)(b + 4)}{b} < 0$

67. $\dfrac{c}{(c - 3)(c + 8)} \leq 0$

68. $\dfrac{z - 5}{(z + 6)(z - 9)} \geq 0$

69. $\dfrac{x - 6}{(x + 4)(x - 1)} \leq 0$

70. $\dfrac{x + 9}{(x - 2)(x + 4)} > 0$

71. $\dfrac{(x - 3)(2x + 5)}{x - 4} \geq 0$

72. $\dfrac{r(r - 8)}{2r + 6} < 0$

Solve each rational inequality and graph the solution on a number line.

73. $\dfrac{2}{x - 4} \geq 1$

74. $\dfrac{2}{x - 4} > 1$

75. $\dfrac{3}{x - 1} > -1$

76. $\dfrac{3}{x + 1} \geq -1$

77. $\dfrac{5}{x + 2} \leq 1$

78. $\dfrac{5}{x + 2} < 1$

79. $\dfrac{2p - 5}{p - 4} \le 1$

80. $\dfrac{2}{2a - 1} > 2$

81. $\dfrac{4}{x + 2} \ge 2$

82. $\dfrac{x + 6}{x + 2} > 1$

83. $\dfrac{w}{3w - 2} > -2$

84. $\dfrac{x - 1}{2x + 6} \le -3$

Concept/Writing Exercises

85. The graph of $f(x) = x^2 - 7x + 10$ is given. Find the solution to **a)** $f(x) > 0$ and **b)** $f(x) < 0$.

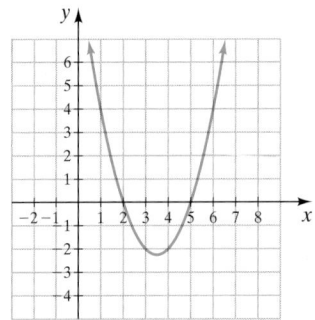

86. The graph of $f(x) = -x^2 - 4x + 5$ is given. Find the solution to **a)** $f(x) \ge 0$ and **b)** $f(x) \le 0$.

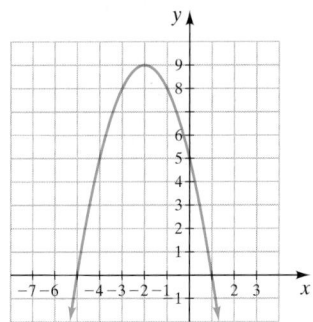

87. The graph of $y = \dfrac{x^2 - 4x + 4}{x - 4}$ is illustrated. Determine the solution to the following inequalities.

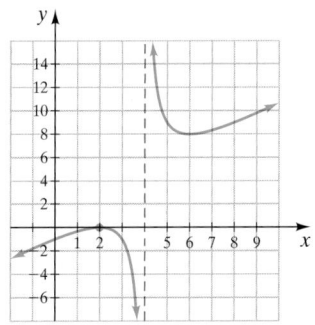

a) $\dfrac{x^2 - 4x + 4}{x - 4} > 0$

b) $\dfrac{x^2 - 4x + 4}{x - 4} < 0$

Explain how you determined your answer.

88. The graph of $y = \dfrac{x^2 + x - 6}{x - 4}$ is illustrated. Determine the solution to the following inequalities.

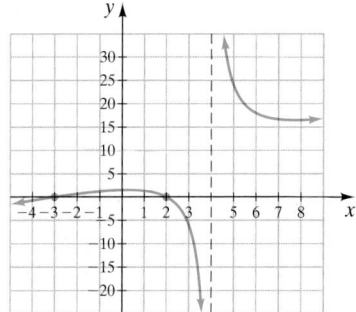

a) $\dfrac{x^2 + x - 6}{x - 4} \ge 0$

b) $\dfrac{x^2 + x - 6}{x - 4} < 0$

Explain how you determined your answer.

89. Write a quadratic inequality whose solution is

90. Write a quadratic inequality whose solution is

91. Write a rational inequality whose solution is

92. Write a rational inequality whose solution is

93. What is the solution to the inequality $(x + 3)^2(x - 1)^2 \ge 0$? Explain your answer.

94. What is the solution to the inequality $x^2(x - 3)^2 (x + 4)^2 < 0$? Explain your answer.

95. What is the solution to the inequality $\dfrac{x^2}{(x + 2)^2} \ge 0$? Explain your answer.

96. What is the solution to the inequality $\dfrac{x^2}{(x - 3)^2} > 0$? Explain your answer.

97. If $f(x) = ax^2 + bx + c$ where $a > 0$ and the discriminant is negative, what is the solution to $f(x) < 0$? Explain.

98. If $f(x) = ax^2 + bx + c$ where $a < 0$ and the discriminant is negative, what is the solution to $f(x) > 2$? Explain.

Challenge Problems

Solve each inequality and graph the solution on the number line.

99. $(x + 1)(x - 3)(x + 5)(x + 8) \geq 0$

100. $\dfrac{(x - 4)(x + 2)}{x(x + 9)} \geq 0$

Write a quadratic inequality with the following solutions. Many answers are possible. Explain how you determined your answers.

101. $(-\infty, 0) \cup (3, \infty)$ **102.** $\{2\}$ **103.** $\varnothing$ **104.** $\mathbb{R}$

In Exercises 105 and 106, solve each inequality and give the solution in interval notation. Use techniques from Section 8.5 to help you find the solution.

105. $x^4 - 10x^2 + 9 > 0$ **106.** $x^4 - 26x^2 + 25 \leq 0$

In Exercises 107 and 108, solve each inequality using factoring by grouping. Give the solution in interval notation.

107. $x^3 + x^2 - 4x - 4 \geq 0$ **108.** $2x^3 + x^2 - 32x - 16 < 0$

Group Activity

Discuss and answer Exercises 109 and 110 as a group.

109. Consider the number line below, where $a, b,$ and c are distinct real numbers.

   a) In which intervals will the real numbers satisfy the inequality $(x - a)(x - b)(x - c) > 0$? Explain.

   b) In which intervals will the real numbers satisfy the inequality $(x - a)(x - b)(x - c) < 0$? Explain.

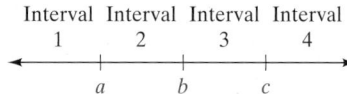

110. Consider the number line below where $a, b, c,$ and d are distinct real numbers.

   Interval   Interval   Interval   Interval   Interval
      1   2   3   4   5

            a b c d

   a) In which intervals do the real numbers satisfy the inequality $(x - a)(x - b)(x - c)(x - d) > 0$? Explain.

   b) In which interval do the real numbers satisfy the inequality $(x - a)(x - b)(x - c)(x - d) < 0$? Explain.

Cumulative Review Exercises

[2.4] **111. Antifreeze** How many quarts of a 100% antifreeze solution should Paul Simmons add to 10 quarts of a 20% antifreeze solution to make a 50% antifreeze solution?

[3.2] **112.** If $h(x) = \dfrac{x^2 + 4x}{x + 9}$, find $h(-3)$.

[5.1] **113.** Add $(6r + 5s - t) + (-3r - 2s - 8t)$.

[6.3] **114.** Simplify $\dfrac{1 + \dfrac{x}{x + 1}}{\dfrac{2x + 1}{x - 3}}$.

[7.7] **115.** Multiply $(3 - 4i)(6 + 5i)$.

Chapter 8 Summary

IMPORTANT FACTS AND CONCEPTS	EXAMPLES

Section 8.1

Square Root Property

If $x^2 = a$, where a is a real number, then $x = \pm\sqrt{a}$.

Solve $x^2 - 36 = 0$.

$$x^2 - 36 = 0$$
$$x^2 = 36$$
$$x = \pm\sqrt{36} = \pm 6$$

The solutions are -6 and 6.

A **perfect square trinomial** is a trinomial that can be expressed as the square of a binomial.

$$x^2 - 10x + 25 = (x - 5)^2$$

To Solve a Quadratic Equation by Completing the Square

1. Use the multiplication (or division) property of equality if necessary to make the leading coefficient 1.
2. Rewrite the equation with the constant by itself on the right side of the equation.
3. Take one-half the numerical coefficient of the first-degree term, square it, and add this quantity to both sides of the equation.
4. Factor the trinomial as the square of a binomial.
5. Use the square root property to take the square root of both sides of the equation.
6. Solve for the variable.

Solve $x^2 + 4x - 12 = 0$ by completing the square.

$$x^2 + 4x - 12 = 0$$
$$x^2 + 4x = 12$$
$$x^2 + 4x + 4 = 12 + 4$$
$$(x + 2)^2 = 16$$
$$x + 2 = \pm\sqrt{16}$$
$$x + 2 = \pm 4$$
$$x = -2 \pm 4$$
$$x = -2 - 4 = -6 \quad \text{or} \quad x = -2 + 4 = 2$$

The solutions are -6 and 2.

Section 8.2

The **standard form of a quadratic equation** is $ax^2 + bx + c = 0$, $a \neq 0$.

$$x^2 - 5x + 17 = 0$$

To Solve a Quadratic Equation by the Quadratic Formula

1. Write the quadratic equation in standard form, $ax^2 + bx + c = 0$, and determine the numerical values for a, b, and c.
2. Substitute the values for a, b, and c into the quadratic formula and then evaluate the formula to obtain the solution.

The Quadratic Formula
$$x = \frac{-b \pm \sqrt{b^2 - 4ac}}{2a}$$

Solve $x^2 - 2x - 15 = 0$ by the quadratic formula.

$$a = 1, \quad b = -2, \quad c = -15$$
$$x = \frac{-b \pm \sqrt{b^2 - 4ac}}{2a}$$
$$= \frac{-(-2) \pm \sqrt{(-2)^2 - 4(1)(-15)}}{2(1)}$$
$$= \frac{2 \pm \sqrt{64}}{2}$$
$$= \frac{2 \pm 8}{2}$$

$$x = \frac{2 + 8}{2} = \frac{10}{2} = 5 \quad \text{or} \quad x = \frac{2 - 8}{2} = \frac{-6}{2} = -3$$

The solutions are 5 and -3.

Solutions of a Quadratic Equation

For a quadratic equation of the form $ax^2 + bx + c = 0, a \neq 0$, the **discriminant** is $b^2 - 4ac$.

If $b^2 - 4ac > 0$, the quadratic equation has two distinct real number solutions.

If $b^2 - 4ac = 0$, the quadratic equation has a single real number solution.

If $b^2 - 4ac < 0$, the quadratic equation has no real number solution.

Determine the number of solutions of $3x^2 - x + 7 = 0$.

$$a = 3, b = -1, c = 7$$
$$b^2 - 4ac = (-1)^2 - 4(3)(7)$$
$$= 1 - 84$$
$$= -83$$

Since the discriminant is negative, the equation has no real number solution.

IMPORTANT FACTS AND CONCEPTS	EXAMPLES

Section 8.4

An equation that can be written in the form $au^2 + bu + c = 0$ for $a \neq 0$, where u is an algebraic expression, is called **quadratic in form**.

To Solve Equations Quadratic in Form

1. Make a substitution that will result in an equation of the form $au^2 + bu + c = 0, a \neq 0$, where u is a function of the original variable.
2. Solve the equation $au^2 + bu + c = 0$ for u.
3. Replace u with the function of the original variable from step 1 and solve the resulting equation for the original variable.
4. Check for extraneous solutions by substituting the apparent solutions into the original equation.

Solve $x^4 - 17x^2 + 16 = 0$.

$$\text{Let } u = x^2.$$

Then,
$$u^2 - 17u + 16 = 0$$
$$(u - 16)(u - 1) = 0$$
$$u - 16 = 0 \quad \text{or} \quad u - 1 = 0$$
$$u = 16 \qquad\qquad u = 1$$
$$x^2 = 16 \qquad\qquad x^2 = 1$$
$$x = \pm 4 \qquad\qquad x = \pm 1$$

A check will show the solutions are $4, -4, 1,$ and -1.

Section 8.5

Quadratic Functions

Functions of the form $f(x) = ax^2 + bx + c$ are called **quadratic functions** and have graphs that are parabolas.

a) The parabola opens upward when $a > 0$ and downward when $a < 0$.

b) The axis of symmetry is the line $x = -\dfrac{b}{2a}$.

c) The vertex is the point $\left(-\dfrac{b}{2a}, \dfrac{4ac - b^2}{4a}\right)$ or $\left[-\dfrac{b}{2a}, f\left(-\dfrac{b}{2a}\right)\right]$.

d) The y-intercept is the point $(0, c)$.

e) To obtain the x-intercept(s) (if there are any), solve the equation $ax^2 + bx + c = 0$ for x.

The graph of the quadratic function $f(x) = x^2 - 2x - 3$ is a parabola.

a) It opens upward since $a > 0$.

b) The axis of symmetry is $x = -\dfrac{-2}{2(1)} = 1$.

c) The vertex is $(1, -4)$.

d) The y-intercept is $(0, -3)$.

e)
$$x^2 - 2x - 3 = 0$$
$$(x - 3)(x + 1) = 0$$
$$x - 3 = 0 \quad \text{or} \quad x + 1 = 0$$
$$x = 3 \qquad\qquad x = -1$$

The x-intercepts are $(3, 0)$ and $(-1, 0)$.

The graph of $f(x) = x^2 - 2x - 3$ is shown below.

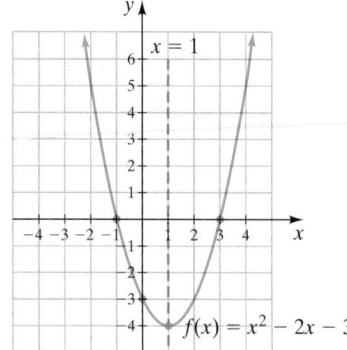

Section 8.6

A **quadratic inequality** is an inequality that can be written in one of the following forms

$$ax^2 + bx + c < 0 \quad ax^2 + bx + c > 0$$
$$ax^2 + bx + c \leq 0 \quad ax^2 + bx + c \geq 0$$

where $a, b,$ and c are real numbers, with $a \neq 0$.

The **solution to a quadratic inequality** is the set of all values that make the inequality a true statement.

$$x^2 - 5x + 7 > 0$$

IMPORTANT FACTS AND CONCEPTS	EXAMPLES

Section 8.6 (cont.)

To Solve Quadratic, Polynomial, and Rational Inequalities

1. Write the inequality as an equation and solve the equation. The solutions are the boundary values.

2. Construct a number line and mark each boundary value from step 1 as follows:

   • If the inequality symbol is $<$ or $>$, use an open circle, $\circ$.
   • If the inequality symbol is $\leq$ or $\geq$, use a closed circle, $\bullet$.

3. If solving a rational inequality, determine the values that make any denominator 0. These values are also boundary values. Indicate these boundary values on your number line with an open circle, $\circ$.

4. Select a test value in each interval and determine whether it satisfies the inequality.

5. The solution is the set of points that satisfy the inequality.

6. Write the solution in the form instructed.

Solve $(2x - 1)(x - 3)(x + 1) < 0$.

$$(2x - 1)(x - 3)(x + 1) < 0$$

$$(2x - 1)(x - 3)(x + 1) = 0$$

$2x - 1 = 0$ or $x - 3 = 0$ or $x + 1 = 0$

$x = \dfrac{1}{2}$ $x = 3$ $x = -1$

The intervals and test values selected are shown below.

Interval	Test value	$(2x - 1)(x - 3)(x + 1) < 0$	
$(-\infty, -1)$	-2	-25	True
$\left(-1, \dfrac{1}{2}\right)$	0	3	False
$\left(\dfrac{1}{2}, 3\right)$	1	-4	True
$(3, \infty)$	5	108	False

Solution on a number line:

Solution in interval notation: $(-\infty, -1) \cup \left(\dfrac{1}{2}, 3\right)$

Solution in set builder notation:

$$\left\{ x \,\middle|\, x < -1 \ \text{ or } \ \frac{1}{2} < x < 3 \right\}$$

Chapter 8 Review Exercises

[8.1]   *Use the square root property to solve each equation.*

1. $(x - 5)^2 = 16$ **2.** $(2x + 1)^2 = 60$ **3.** $\left(x - \dfrac{1}{3}\right)^2 = \dfrac{4}{9}$ **4.** $\left(2x - \dfrac{1}{2}\right)^2 = 4$

Solve each equation by completing the square.

5. $x^2 - 8x + 12 = 0$ **6.** $x^2 + 4x - 32 = 0$ **7.** $a^2 + 2a - 9 = 0$

8. $z^2 + 6z = 12$ **9.** $x^2 - 2x + 10 = 0$ **10.** $2r^2 - 8r = -64$

Area   *In Exercises 11 and 12, the area, A, of each rectangle is given.* **a)** *Write an equation for the area.* **b)** *Solve the equation for x.*

11.

$A = 32$ | $x + 1$

$x + 5$

12.

$A = 63$ | $x + 2$

$x + 4$

13. Consecutive Integers The product of two consecutive positive integers is 56. Find the two integers.

14. Living Room Nedal Williams just moved into a new house where the living room is a square whose diagonal is 7 feet longer than the length of a side. Find the dimensions of the living room.

[8.2]   Determine whether each equation has two distinct real solutions, a single real solution, or no real solution.

15. $3x^2 - 7x + 1 = 0$

16. $3x^2 + 2x = -6$

17. $r^2 + 16r = -64$

18. $5x^2 - x + 2 = 0$

19. $a^2 - 14a = -49$

20. $\frac{1}{2}x^2 - 3x = 8$

Solve each equation by using the quadratic formula.

21. $6x^2 - 5x - 50 = 0$

22. $x^2 - 11x = -18$

23. $r^2 = 3r + 40$

24. $7x^2 = 9x$

25. $6a^2 + a - 15 = 0$

26. $4x^2 + 11x = 3$

27. $x^2 + 8x + 5 = 0$

28. $b^2 + 4b = 8$

29. $2x^2 + 4x - 3 = 0$

30. $3y^2 - 6y = 8$

31. $x^2 - x + 13 = 0$

32. $x^2 - 2x + 11 = 0$

33. $2x^2 - \frac{5}{3}x = \frac{25}{3}$

34. $4x^2 + 5x - \frac{3}{2} = 0$

For the given function, determine all real values of the variable for which the function has the value indicated.

35. $f(x) = x^2 - 4x - 35, f(x) = 25$

36. $g(x) = 6x^2 + 5x, g(x) = 6$

37. $h(r) = 5r^2 - 7r - 10, h(r) = -8$

38. $f(x) = -2x^2 + 6x + 7, f(x) = -2$

Determine an equation that has the given solutions.

39. $3, -1$

40. $\frac{2}{3}, -2$

41. $-\sqrt{11}, \sqrt{11}$

42. $3 - 2i, 3 + 2i$

[8.1–8.3]

43. Rectangular Garden Sophia Yang is designing a rectangular flower garden. If the area is to be 96 square feet and the length is to be 4 feet greater than the width, find the dimensions of the garden.

44. Triangle and Circle Find the length of side x in the figure.

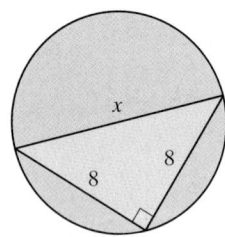

45. Savings Account Samuel Rivera invested $1000 in a savings account where the interest is compounded annually. If after 2 years the amount in the account is $1081.60, find the annual interest rate.

46. Numbers The larger of two positive numbers is 4 greater than the smaller. Find the two numbers if their product is 77.

47. Rectangle The length of a rectangle is 4 inches less than twice its width. Find the dimensions if its area is 96 square inches.

48. Wheat Crop The value, V, in dollars per acre of a wheat crop d days after planting is given by the formula $V = 12d - 0.05d^2, 20 < d < 80$. Find the value of an acre of wheat 60 days after it has been planted.

49. Redwood Growth The growth, $f(x)$, in inches per year of a redwood tree less than 30 years old is given by the function $f(x) = -0.02x^2 + x + 1$, where x represents the number of inches of rainfall per year.

Source: www.humboldtredwoods.org

   a) Find the growth per year of a redwood tree in a year in which there is 12 inches of rainfall.

   b) If the growth of a redwood tree in one year is 10 inches, find the number of inches of rainfall for that year (round your answer to 2 decimal places).

50. Falling Object The distance, d, in feet, that an object is from the ground t seconds after being dropped from an airplane is given by the formula $d = -16t^2 + 784$.

   a) Find the distance the object is from the ground 2 seconds after it has been dropped.

   b) When will the object hit the ground?

51. Oil Leak A tractor has an oil leak. The amount of oil, $L(t)$, in milliliters per hour that leaks out is a function of the tractor's operating temperature, t, in degrees Celsius. The function is

$$L(t) = 0.0004t^2 + 0.16t + 20, 100°C \le t \le 160°C$$

   a) How many milliliters of oil will leak out in 1 hour if the operating temperature of the tractor is 100°C?

   b) If oil is leaking out at 53 milliliters per hour, what is the operating temperature of the tractor?

52. Molding Machines Two molding machines can complete an order in 12 hours. The larger machine can complete the order by itself in 1 hour less time than the smaller machine can by itself. How long will it take each machine to complete the order working by itself?

53. Travel Time Steve Forrester drove 25 miles at a constant speed, and then he increased his speed by 15 miles per hour for the next 65 miles. If the time required to travel 90 miles was 1.5 hours, find the speed he drove during the first 25 miles.

54. Canoe Trip Joan Banker canoed downstream going with the current for 3 miles, then turned around and canoed upstream against the current to her starting point. If the total time she spent canoeing was 4 hours and the current was 0.4 mile per hour, what is the speed she canoes in still water?

In Exercises 57–60, solve each equation for the variable indicated.

57. $a^2 + b^2 = c^2$, for a (Pythagorean Theorem)

59. $v_x^2 + v_y^2 = v^2$, for v_y (vectors)

55. Area The area of a rectangle is 80 square units. If the length is x units and the width is $x - 2$ units, find the length and the width. Round your answer to the nearest tenth of a unit.

$$A = 80 \qquad x - 2$$

$$x$$

56. Selling Tables A business sells n tables, $n \le 40$, at a price of $(60 - 0.3n)$ dollars per table. How many tables must be sold to have a revenue of \$1080?

58. $h = -4.9t^2 + c$, for t (height of an object)

60. $a = \dfrac{v_2^2 - v_1^2}{2d}$, for v_2

[8.4]   *Solve each equation.*

61. $x^4 - 10x^2 + 9 = 0$

63. $a^4 = 5a^2 + 24$

65. $3r + 11\sqrt{r} - 4 = 0$

67. $6(x - 2)^{-2} = -13(x - 2)^{-1} + 8$

62. $x^4 - 21x^2 + 80 = 0$

64. $3y^{-2} + 16y^{-1} = 12$

66. $2p^{2/3} - 7p^{1/3} + 6 = 0$

68. $10(r + 1) = \dfrac{12}{r + 1} - 7$

Find all x-intercepts of the given function.

69. $f(x) = x^4 - 82x^2 + 81$

71. $f(x) = x - 6\sqrt{x} + 12.$

70. $f(x) = 30x + 13\sqrt{x} - 10$

72. $f(x) = (x^2 - 6x)^2 - 5(x^2 - 6x) - 24$

[8.5]   **a)** *Determine whether the parabola opens upward or downward.* **b)** *Find the y-intercept.* **c)** *Find the vertex.* **d)** *Find the x-intercepts, if they exist.* **e)** *Draw the graph.*

73. $f(x) = x^2 + 5x$

75. $g(x) = -x^2 - 2$

74. $f(x) = x^2 - 2x - 8$

76. $g(x) = -2x^2 - x + 15$

77. Selling Tickets The Hamilton Outdoor Theater estimates that its total income, I, in hundreds of dollars, for its production of a play, can be approximated by the formula $I = -x^2 + 22x - 45, 2 \le x \le 20$, where x is the cost of a ticket.

a) How much should the theater charge to maximize its income?

b) What is the maximum income?

78. Tossing a Ball Josh Vincent tosses a ball upward from the top of a 75-foot building. The height, $s(t)$, of the ball at any time t can be determined by the function $s(t) = -16t^2 + 80t + 75$.

a) At what time will the ball attain its maximum height?

b) What is the maximum height?

Graph each function.

79. $f(x) = (x - 3)^2$ **80.** $f(x) = -(x + 2)^2 - 3$ **81.** $g(x) = -2(x + 4)^2 - 1$ **82.** $h(x) = \dfrac{1}{2}(x - 1)^2 + 3$

[8.6]   *Graph the solution to each inequality on a number line.*

83. $x^2 + 4x + 3 \geq 0$ **84.** $x^2 + 3x - 10 \leq 0$

85. $x^2 \leq 11x - 20$ **86.** $3x^2 + 8x > 16$

87. $4x^2 - 9 \leq 0$ **88.** $6x^2 - 30 > 0$

Solve each inequality and give the solution in set builder notation.

89. $\dfrac{x + 1}{x - 5} > 0$ **90.** $\dfrac{x - 3}{x + 2} \leq 0$ **91.** $\dfrac{2x - 4}{x + 3} \geq 0$

92. $\dfrac{3x + 5}{x - 6} < 0$ **93.** $(x + 4)(x + 1)(x - 2) > 0$ **94.** $x(x - 3)(x - 6) \leq 0$

Solve each inequality and give the solution in interval notation.

95. $(3x + 4)(x - 1)(x - 3) \geq 0$ **96.** $2x(x + 2)(x + 4) < 0$

97. $\dfrac{x(x - 4)}{x + 2} > 0$ **98.** $\dfrac{(x - 2)(x - 8)}{x + 3} < 0$

99. $\dfrac{x - 3}{(x + 2)(x - 7)} \geq 0$ **100.** $\dfrac{x(x - 6)}{x + 3} \leq 0$

Solve each inequality and graph the solution on a number line.

101. $\dfrac{5}{x + 4} \geq -1$ **102.** $\dfrac{2x}{x - 2} \leq 1$ **103.** $\dfrac{2x + 3}{3x - 5} < 4$

Chapter 8 Practice Test

CHAPTER
Test Prep
VIDEOS

Chapter Test Prep Videos provide fully worked-out solutions to any of the exercises you want to review. Chapter Test Prep Videos are available via MyMathLab, or on YouTube (search "Angel Intermediate Algebra" and click on "Channels")

Solve by completing the square.

1. $x^2 + 2x - 15 = 0$ **2.** $a^2 + 7 = 6a$

Solve by the quadratic formula.

3. $x^2 - 6x - 16 = 0$ **4.** $x^2 - 4x = -11$

Solve by the method of your choice.

5. $3r^2 + r = 2$ **6.** $p^2 + 4 = -7p$

7. Write an equation that has x-intercepts $4, -\dfrac{2}{5}$.

8. Solve the formula $K = \dfrac{1}{2}mv^2$ for v.

9. Cost The cost, c, of a house in Du Quoin, Illinois, is a function of the number of square feet, s, of the house. The cost of the house can be approximated by

$$c(s) = -0.01s^2 + 78s + 22{,}000, \quad 1300 \leq s \leq 3900$$

a) Estimate the cost of a 1600-square-foot house.

b) How large a house can Sharon Hamsa purchase if she wishes to spend $160,000 on a house?

10. Trip in a Park David Price drove his 4-wheel-drive Jeep from Anchorage, Alaska, to the Chena River State Recre-

ation Park, a distance of 520 miles. Had he averaged 15 miles per hour faster, the trip would have taken 2.4 hours less. Find the average speed that David drove.

© Michael Klenetsky /Shutterstock

Chena River State Recreation Park

Solve.

11. $2x^4 + 15x^2 - 50 = 0$

12. $3r^{2/3} + 11r^{1/3} - 42 = 0$

13. Find all x-intercepts of $f(x) = 16x - 24\sqrt{x} + 9$.

Graph each function.

14. $f(x) = (x - 3)^2 + 2$

15. $h(x) = -\dfrac{1}{2}(x - 2)^2 - 2$

16. Determine whether $6x^2 = 2x + 3$ has two distinct real solutions, a single real solution, or no real solution. Explain your answer.

17. Consider the quadratic equation $y = x^2 + 2x - 8$.

    a) Determine whether the parabola opens upward or downward.

    b) Find the y-intercept.

    c) Find the vertex.

    d) Find the x-intercepts (if they exist).

    e) Draw the graph.

18. Write a quadratic equation whose x-intercepts are $(-7, 0)$ and $\left(\dfrac{1}{2}, 0\right)$.

Solve each inequality and graph the solution on a number line.

19. $x^2 - x \geq 42$

20. $\dfrac{(x + 5)(x - 4)}{x + 1} \geq 0$

*Solve the following inequality. Write the answer in **a)** interval notation and **b)** set builder notation.*

21. $\dfrac{x + 3}{x + 2} \leq -1$

22. **Carpet** The length of a rectangular Persian carpet is 3 feet greater than twice its width. Find the length and width of the carpet if its area is 65 square feet.

23. **Throwing a Ball** Jose Ramirez throws a ball upward from the top of a building. The distance, d, of the ball from the ground at any time, t, is $d = -16t^2 + 80t + 96$. How long will it take for the ball to strike the ground?

24. **Profit** The Leigh Ann Sims Company earns a weekly profit according to the function $f(x) = -1.4x^2 + 56x - 70$, where x is the number of wood carvings made and sold each week.

    a) Find the number of carvings the company must sell in a week to maximize its profit.

    b) What is its maximum weekly profit?

25. **Selling Brooms** A business sells n brooms, $n \leq 32$, at a price of $(10 - 0.1n)$ dollars per broom. How many brooms must be sold to have a revenue of $210?

Cumulative Review Test

Take the following test and check your answers with those given in the back of the book. Review any questions that you answered incorrectly. The section where the material was covered is indicated after the answer.

1. Evaluate $-4 \div (-2) + 18 - \sqrt{49}$.

2. Evaluate $2x^2 + 3x + 4$ when $x = 2$.

3. Express 2,540,000 in scientific notation.

4. Find the solution set for the equation $|4 - 2x| = 5$.

5. Simplify $6x - \{3 - [2(x - 2) - 5x]\}$.

6. Solve the equation $-\dfrac{1}{2}(4x - 6) = \dfrac{1}{3}(3 - 6x) + 2$.

7. Solve the inequality $-4 < \dfrac{x + 4}{2} < 6$. Write the solution in interval notation.

8. Find the slope and y-intercept of the graph of $9x + 7y = 15$.

9. **Small Orchard** The number of baskets of apples, N, that are produced by x trees in a small orchard is given by the function $N(x) = -0.2x^2 + 40x$. How many baskets of apples are produced by 50 trees?

10. Write the equation in slope-intercept form of a line passing through the points $(6, 5)$ and $(4, 3)$.

11. **a)** Determine whether the following graph represents a function. Explain your answer.

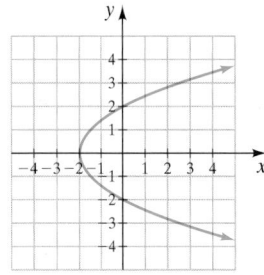

    b) Determine the domain and range of the function or relation.

12. Graph each equation.

    a) $x = -4$ **b)** $y = 2$

13. Evaluate the following determinant.

$$\begin{vmatrix} 4 & 0 & -2 \\ 3 & 5 & 1 \\ 1 & -1 & 7 \end{vmatrix}$$

14. Solve the system of equations.

$$4x - 3y = 10$$

$$2x + y = 5$$

15. Factor $(x + 3)^2 + 10(x + 3) + 24$.

16. a) Write an expression for the shaded area in the figure and

    b) write the expression in factored form.

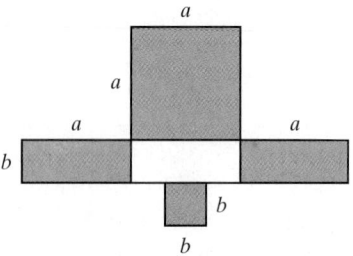

17. Add $\dfrac{x + 2}{x^2 - x - 6} + \dfrac{x - 3}{x^2 - 8x + 15}$.

18. Solve the equation

$$\frac{1}{a - 2} = \frac{4a - 1}{a^2 + 5a - 14} + \frac{2}{a + 7}.$$

19. Wattage Rating The wattage rating of an appliance, w, varies jointly as the square of the current, I, and the resistance, R. If the wattage is 12 when the current is 2 amperes and the resistance is 100 ohms, find the wattage when the current is 0.8 ampere and the resistance is 600 ohms.

20. Simplify $\dfrac{3 - 4i}{2 + 5i}$.

9 Exponential and Logarithmic Functions

Goals of This Chapter

Exponential and logarithmic functions have a wide variety of uses, some of which you will see as you read through this chapter. You often read in newspaper and magazine articles that health care spending, use of the Internet, and the world population, to list just a few, are growing exponentially. By the time you finish this chapter you should have a clear understanding of just what this means.

We also introduce two special functions, the natural exponential function and the natural logarithmic function. Many natural phenomena, such as carbon dating, radioactive decay, and the growth of savings invested in an account compounding interest continuously, can be described by natural exponential functions.

The populations of many species can be modeled using either exponential or logarithmic functions. Specific examples include modeling the human populations of cities, countries, and of the entire world. Other examples involve the population of bears in a forest, manatees in an estuary, and bald eagles in a national park. In Exercise 74 on page 622, we use an exponential function to model the population of trout in a lake.

© Katrina Brown/Shutterstock

9.1 Composite and Inverse Functions

1 Find composite functions.

2 Understand one-to-one functions.

3 Find inverse functions.

4 Find the composition of a function and its inverse.

We begin with composite functions, one-to-one functions, and inverse functions. These concepts will be used in our discussion of logarithmic and exponential functions.

1 Find Composite Functions

Often one variable is a function of another variable which in turn is a function of a third variable. We describe the relationship among such functions as a **composition of functions**. For example, suppose that 1 U.S. dollar can be converted into 1.20 Canadian dollars, and 1 Canadian dollar can be converted into 11.40 Mexican pesos. Using this information, we can convert 20 U.S. dollars into Mexican pesos. We have the following functions.

$$g(x) = 1.20x \qquad (x \text{ U.S. dollars to Canadian dollars})$$

$$f(x) = 11.40x \qquad (x \text{ Canadian dollars to Mexican pesos})$$

If we let $x = 20$, for $20 U.S., then it can be converted into $24 Canadian using function g:

$$g(x) = 1.20x$$

$$g(20) = 1.20(20) = \$24 \text{ Canadian}$$

The $24 Canadian can, in turn, be converted into 273.60 Mexican pesos using function f:

$$f(x) = 11.40x$$

$$f(24) = 11.40(24) = 273.60 \text{ Mexican pesos}$$

This conversion can be found directly without performing this string of calculations. One U.S. dollar can be converted into Mexican pesos by substituting the $1.20x$ found in function $g(x)$ for the x in $f(x)$. This gives a new function, h, which converts U.S. dollars directly into Mexican pesos.

$$g(x) = 1.20x \qquad f(x) = 11.40x$$

$$h(x) = f[g(x)]$$

$$= 11.40(\,1.20x\,) \qquad \text{Substitute } g(x) \text{ for } x \text{ in } f(x).$$

$$= 13.68x$$

Thus, each U.S. dollar, x, is worth 13.68 Mexican pesos. If we substitute $20 for x, we get 273.60 pesos, which is what we expected.

$$h(x) = 13.68x$$

$$h(20) = 13.68(20) = 273.60$$

Function h is called a **composition of f with g** or the **composite function of f with g**. We denote the composite function $(f \circ g)(x)$ and is read "f of g of x" or "f composed with g of x." **Figure 9.1** shows how the composite function h relates to functions f and g.

Understanding Algebra

We observe composite functions in our daily lives. For example, your weekly budget is a function of the price of gasoline. The price of gasoline is a function of the price of oil. Thus, in the big picture, we see that your weekly budget is a function of the price of oil. This is an example of a composite function.

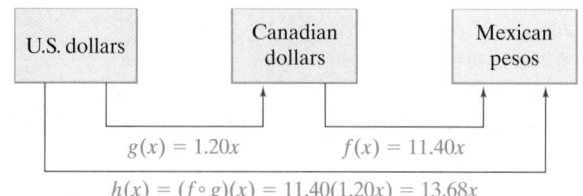

FIGURE 9.1

We now define the **composite function**.

Composite Function

The **composite function** $(f \circ g)(x)$ is defined as

$$(f \circ g)(x) = f[g(x)]$$

When we are given $f(x)$ and $g(x)$, to find $(f \circ g)(x)$ we substitute $g(x)$ for x in $f(x)$ to get $f[g(x)]$.

EXAMPLE 1 Given $f(x) = x^2 - 2x + 3$ and $g(x) = x - 5$, find

a) $f(4)$ **b)** $f(a)$ **c)** $(f \circ g)(x)$ **d)** $(f \circ g)(3)$

Solution

a) To find $f(4)$, we substitute 4 for each x in $f(x)$.

$$f(x) = x^2 - 2x + 3$$
$$f(4) = 4^2 - 2 \cdot 4 + 3 = 16 - 8 + 3 = 11$$

b) To find $f(a)$, we substitute a for each x in $f(x)$.

$$f(x) = x^2 - 2x + 3$$
$$f(a) = a^2 - 2a + 3$$

c) $(f \circ g)(x) = f[g(x)]$. To find $(f \circ g)(x)$, we substitute $g(x)$, which is $x - 5$, for each x in $f(x)$.

$$f(\,x\,) = x^2 - 2x + 3$$
$$f[g(x)] = [g(x)]^2 - 2[g(x)] + 3$$

Since $g(x) = x - 5$, we substitute as follows

$$
\begin{aligned}
f[g(x)] &= (\,x - 5\,)^2 - 2(\,x - 5\,) + 3 \\
&= (x - 5)(x - 5) - 2x + 10 + 3 \\
&= x^2 - 10x + 25 - 2x + 13 \\
&= x^2 - 12x + 38
\end{aligned}
$$

Therefore, the composite function of f with g is $x^2 - 12x + 38$.

$$(f \circ g)(x) = f[g(x)] = x^2 - 12x + 38$$

d) To find $(f \circ g)(3)$, we substitute 3 for x in $(f \circ g)(x)$.

$$(f \circ g)(x) = x^2 - 12x + 38$$
$$(f \circ g)(3) = 3^2 - 12(3) + 38 = 11$$

Now Try Exercise 9

Understanding Algebra

Several examples and exercises in this section show us that composition of functions is not a commutative operation. That is, in general,

$$(f \circ g)(x) \neq (g \circ f)(x)$$

To determine $(g \circ f)(x)$ or $g[f(x)]$, substitute $f(x)$ for each x in $g(x)$. Using $f(x)$ and $g(x)$ as given in Example 1, we find $(g \circ f)(x)$ as follows.

$$g(x) = x - 5, \qquad f(x) = x^2 - 2x + 3$$
$$g[f(x)] = f(x) - 5$$
$$g[f(x)] = (x^2 - 2x + 3) - 5$$
$$= x^2 - 2x + 3 - 5$$
$$= x^2 - 2x - 2$$

Therefore, the composite function of g with f is $x^2 - 2x - 2$.

$$(g \circ f)(x) = g[f(x)] = x^2 - 2x - 2$$

By comparing the illustrations above, we see that in this example, $f[g(x)] \neq g[f(x)]$.

EXAMPLE 2 Given $f(x) = x^2 + 4$ and $g(x) = \sqrt{x - 1}$, find

a) $(f \circ g)(x)$ **b)** $(g \circ f)(x)$

Solution

a) To find $(f \circ g)(x)$, we substitute $g(x)$, which is $\sqrt{x - 1}$ for each x in $f(x)$. You should realize that $\sqrt{x - 1}$ is a real number only when $x \geq 1$.

$$f(x) = \boxed{x}^2 + 4$$

$$(f \circ g)(x) = f[g(x)] = (\boxed{\sqrt{x - 1}})^2 + 4 = x - 1 + 4 = x + 3, x \geq 1$$

Since values of $x < 1$ are not in the domain of $g(x)$, values of $x < 1$ are not in the domain of $(f \circ g)(x)$.

b) To find $(g \circ f)(x)$, we substitute $f(x)$, which is $x^2 + 4$, for each x in $g(x)$.

$$g(x) = \sqrt{\boxed{x} - 1}$$

$$(g \circ f)(x) = g[f(x^2 + 4)] = \sqrt{(\boxed{x^2 + 4}) - 1} = \sqrt{x^2 + 3}$$

Now Try Exercise 19

EXAMPLE 3 Given $f(x) = x - 3$ and $g(x) = x + 7$, find

a) $(f \circ g)(x)$ **b)** $(f \circ g)(2)$ **c)** $(g \circ f)(x)$ **d)** $(g \circ f)(2)$

Solution

a)
$$f(x) = \boxed{x} - 3$$
$$(f \circ g)(x) = f[g(x)] = (\boxed{x + 7}) - 3 = x + 4$$

b) We find $(f \circ g)(2)$ by substituting 2 for each x in $(f \circ g)(x)$.

$$(f \circ g)(x) = x + 4$$
$$(f \circ g)(2) = 2 + 4 = 6$$

c)
$$g(x) = \boxed{x} + 7$$
$$(g \circ f)(x) = g[f(x)] = (\boxed{x - 3}) + 7 = x + 4$$

d) Since $(g \circ f)(x) = x + 4$, $(g \circ f)(2) = 2 + 4 = 6$.

Now Try Exercise 11

In general, $(f \circ g)(x) \neq (g \circ f)(x)$ as we saw at the end of Example 1. In Example 3, $(f \circ g)(x) = (g \circ f)(x)$, but this is only due to the specific functions used.

Helpful Hint

Do not confuse finding the product of two functions with finding a composite function.

Product of functions f and g: $(fg)(x) = (f \cdot g)(x) = f(x) \cdot g(x)$

Composite function of f with g: $(f \circ g)(x) = f[g(x)]$

When multiplying functions f and g, we can use a dot between the f and g. When finding the composite function of f with g, we use a small *open* circle.

2 Understand One-to-One Fuctions

Before discussing one-to-one functions, take a moment to review the definitions of relation, domain, range, and function shown in the Understanding Algebra box.

One-to-One Function

A function is a **one-to-one function** if each element in the range corresponds to exactly one element in the domain.

Consider the following two sets of ordered pairs.

$$A = \{(1, 2), (3, 5), (4, 6), (-2, 1)\}$$

$$B = \{(1, 2), (3, 5), (4, 6), (-2, 5)\}$$

- Both sets A and B represent functions since each x-coordinate corresponds to exactly one y-coordinate.
- Set A is also a one-to-one function since each y-coordinate also corresponds to exactly one x-coordinate.
- Set B is not a one-to-one function since the y-coordinate 5 corresponds to two x-coordinates, 3 and -2.

Figure 9.2 shows the graph of function A and **Figure 9.3** shows the graph of function B.

Function *A*

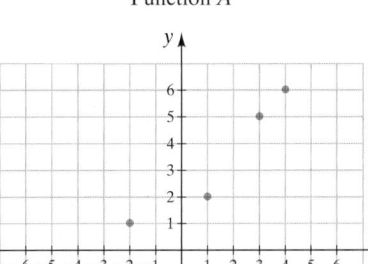

Function *B*

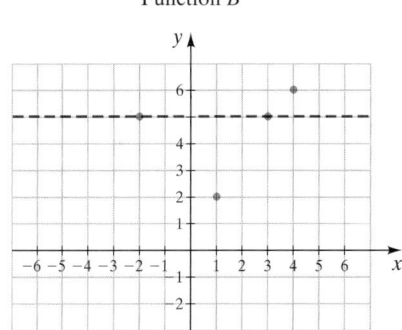

- Each *x*-coordinate corresponds to exactly one *y*-coordinate.

  Therefore, set A is a function.

- Each *y*-coordinate also corresponds to exactly one *x*-coordinate. *Therefore, set A is a one-to-one function.*

FIGURE 9.2

- Each *x*-coordinate corresponds to exactly one *y*-coordinate.

  Therefore, set B is a function.

- The *y*-coordinate 5 corresponds to two *x*-coordinates, 3 and -2. *Therefore, set B is not a one-to-one function.*

FIGURE 9.3

For a graph to represent a function, it must pass the *vertical* line test (see Understanding Algebra box). For a graph to represent a one-to-one function it must also pass the *horizontal* line test.

Horizontal Line Test

If a horizontal line can be drawn so that it intersects the graph of a function at more than one point, the graph is not a one-to-one function.

Observe the graph of the function $f(x) = x^2$ shown in **Figure 9.4** First, it is a function since it passes the *vertical* line test. However, it is not a one-to-one function since it fails the *horizontal* line test. **Figure 9.5** shows that $f(x) = x^2$ has a y-coordinate that corresponds to two different x-coordinates.

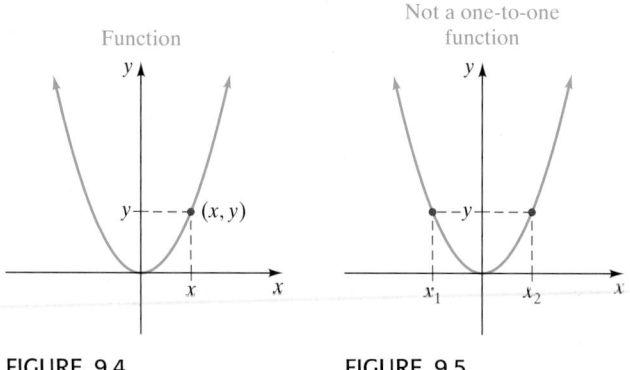

FIGURE 9.4   FIGURE 9.5

Often we can restrict the domain of a function that is not a one-to-one function so that the resulting function is a one-to-one function. For example, both $f(x) = x^2, x \geq 0$ **(Fig. 9.6a)** and $f(x) = x^2, x \leq 0$ **(Figure 9.6b)** are examples of one-to-one functions. Note that both graphs pass the horizontal line test.

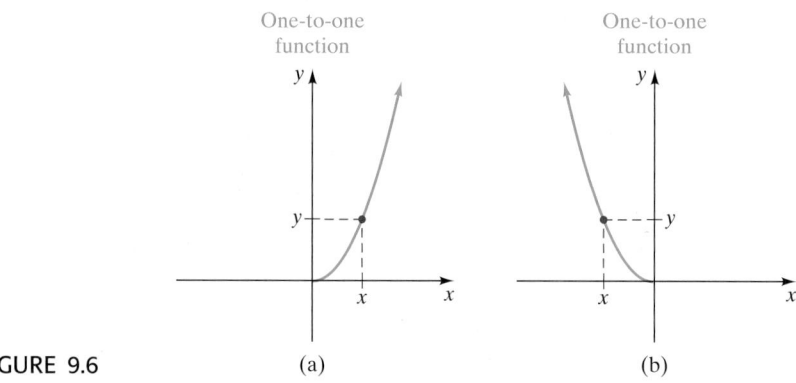

FIGURE 9.6   (a)   (b)

Figure 9.7 shows several graphs and whether or not they represent one-to-one functions. Notice that the graph in part (f) does not represent a function since it fails the vertical line test. Therefore this graph cannot represent a one-to-one function even though the graph passes the horizontal line test.

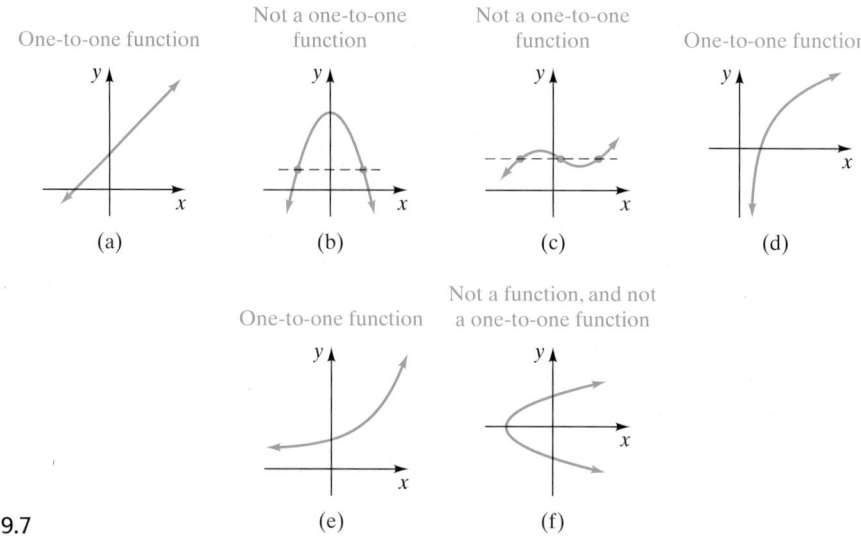

FIGURE 9.7   (e)   (f)

3 Find Inverse Functions

One-to-one functions consist of ordered pairs in which each x-coordinate corresponds to exactly one y-coordinate *and* each y-coordinate corresponds to exactly one x-coordinate. Such a relationship allows us to create a new function called the *inverse* function. *Only one-to-one functions have inverses.*

> **Inverse Function**
>
> If $f(x)$ is a one-to-one function with ordered pairs of the form (x, y), the **inverse function**, $f^{-1}(x)$, is a one-to-one function with ordered pairs of the form (y, x).

We begin with an example.

Function, $f(x)$: $\{(1, 4), (2, 0), (3, 7), (-2, 1), (-1, -5)\}$

Inverse function, $f^{-1}(x)$: $\{(4, 1), (0, 2), (7, 3), (1, -2), (-5, -1)\}$

If we graph the points in the function and the points in the inverse function **(Fig. 9.8)**, we see that the points are symmetric with respect to the line $y = x$.

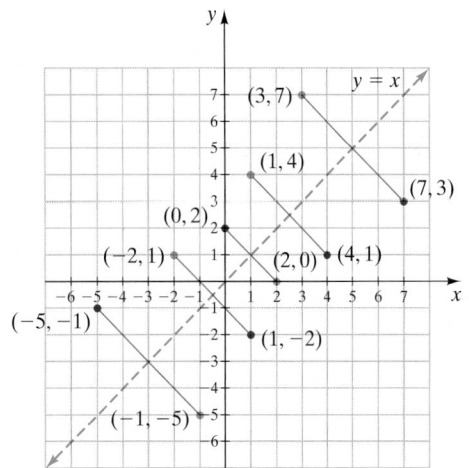

● Ordered pair in function, $f(x)$

● Ordered pair in inverse function, $f^{-1}(x)$

FIGURE 9.8

Some important information about inverse functions:

- Only one-to-one functions have inverse functions.
- The domain of $f(x)$ is the range of $f^{-1}(x)$.
- The range of $f(x)$ is the domain of $f^{-1}(x)$.
- When the graphs of $f(x)$ and $f^{-1}(x)$ are graphed on the same axes, the graphs of $f(x)$ and $f^{-1}(x)$ are symmetric about the line $y = x$.

When a one-to-one function is given as an equation, its inverse function can be found by the following procedure.

> **To Find the Inverse Function of a One-to-One Function**
>
> 1. Replace $f(x)$ with y.
> 2. Interchange the two variables x and y.
> 3. Solve the equation for y.
> 4. Replace y with $f^{-1}(x)$ (this gives the inverse function using inverse function notation).

EXAMPLE 4

a) Find the inverse function of $f(x) = 4x + 2$.

b) On the same axes, graph both $f(x)$ and $f^{-1}(x)$.

Solution

a) This function is one-to-one, therefore we will follow the four-step procedure.

$$f(x) = 4x + 2 \qquad \text{Original function}$$

Step 1 $\qquad\qquad y = 4x + 2 \qquad$ Replace $f(x)$ with y.

Step 2 $\qquad\qquad x = 4y + 2 \qquad$ Interchange x and y.

Step 3 $\qquad\qquad x - 2 = 4y \qquad$ Solve for y.

$$\frac{x-2}{4} = y$$

or $\qquad y = \dfrac{x-2}{4}$

Step 4 $\qquad\qquad f^{-1}(x) = \dfrac{x-2}{4} \qquad$ Replace y with $f^{-1}(x)$.

b) Below we show tables of values for $f(x)$ and $f^{-1}(x)$. The graphs of $f(x)$ and $f^{-1}(x)$ are shown in **Figure 9.9**.

x	$y = f(x)$
0	2
1	6

x	$y = f^{-1}(x)$
2	0
6	1

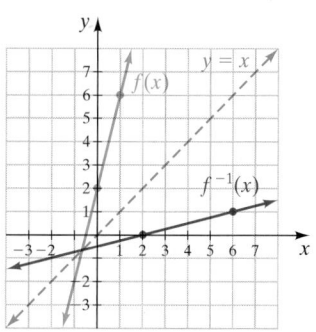

FIGURE 9.9

Note the symmetry of $f(x)$ and $f^{-1}(x)$ about the line $y = x$. Also note that both the domain and range of both $f(x)$ and $f^{-1}(x)$ are the set of real numbers, $\mathbb{R}$.

Now Try Exercise 67

EXAMPLE 5

a) Find the inverse function of $f(x) = x^3 + 2$.

b) On the same axes, graph both $f(x)$ and $f^{-1}(x)$.

Solution

a) This function is one-to-one; therefore we will follow the four-step procedure to find its inverse.

$$f(x) = x^3 + 2 \qquad \text{Original function}$$

Step 1 $\qquad\qquad y = x^3 + 2 \qquad$ Replace $f(x)$ with y.

Step 2 $\qquad\qquad x = y^3 + 2 \qquad$ Interchange x and y.

Step 3 $\qquad\qquad x - 2 = y^3 \qquad$ Solve for y.

$$\sqrt[3]{x-2} = \sqrt[3]{y^3} \qquad \text{Take the cube root of both sides.}$$

$$\sqrt[3]{x-2} = y$$

or $\qquad y = \sqrt[3]{x-2}$

Step 4 $\qquad\qquad f^{-1}(x) = \sqrt[3]{x-2} \qquad$ Replace y with $f^{-1}(x)$.

b) Below we show tables of values for $f(x)$ and $f^{-1}(x)$.

x	$y = f(x)$
-2	-6
-1	1
0	2
1	3
2	10

x	$y = f^{-1}(x)$
-6	-2
1	-1
2	0
3	1
10	2

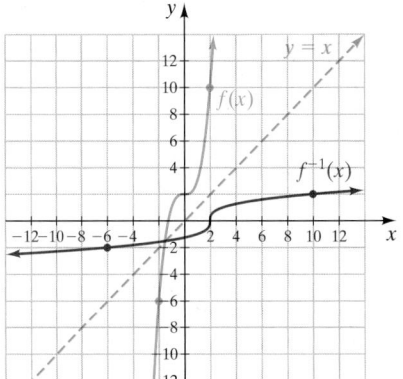

FIGURE 9.10

The graphs of $f(x)$ and $f^{-1}(x)$ are shown in **Figure 9.10**. Notice for each point (a, b) on the graph of $f(x)$, the point (b, a) appears on the graph of $f^{-1}(x)$. For example, the points $(2, 10)$ and $(-2, -6)$, indicated in blue, appear on the graph of $f(x)$, and the points $(10, 2)$ and $(-6, -2)$, indicated in red, appear on the graph of $f^{-1}(x)$.

Now Try Exercise 61

4 Find the Composition of a Function and Its Inverse

To help reinforce the relationship between a function and its inverse, we will evaluate the composition of $f^{-1}(x)$ and $f(x)$ from Example 5 for the values of $x = -2$, $x = 0$, and $x = 2$.

$$(f^{-1} \circ f)(-2) = f^{-1}[f(-2)] = f^{-1}(-6) = -2$$
$$(f^{-1} \circ f)(0) = f^{-1}[f(0)] = f^{-1}(2) = 0$$
$$(f^{-1} \circ f)(2) = f^{-1}[f(2)] = f^{-1}(10) = 2$$

Notice that the output of the composition of a function and its inverse is always equal to the input. Similarly, we will evaluate the composition of $f(x)$ and $f^{-1}(x)$ from Example 5 for the values of $x = -6$, $x = 2$, and $x = 10$.

$$(f \circ f^{-1})(-6) = f[f^{-1}(-6)] = f(-2) = -6$$
$$(f \circ f^{-1})(2) = f[f^{-1}(2)] = f(0) = 2$$
$$(f \circ f^{-1})(10) = f[f^{-1}(10)] = f(2) = 10$$

Notice again that the output is always equal to the input. This relationship is summarized as follows.

The Composition of a Function and Its Inverse

For any one-to-one function $f(x)$ and its inverse $f^{-1}(x)$,

$$(f \circ f^{-1})(x) = x \quad \text{and} \quad (f^{-1} \circ f)(x) = x.$$

EXAMPLE 6 In Example 4, we determined that for $f(x) = 4x + 2$, $f^{-1}(x) = \dfrac{x-2}{4}$. Show that

a) $(f \circ f^{-1})(x) = x$ **b)** $(f^{-1} \circ f)(x) = x$

Solution

a) To determine $(f \circ f^{-1})(x)$, substitute $f^{-1}(x)$ for each x in $f(x)$.

$$f(x) = 4\,x\, + 2$$

$$(f \circ f^{-1})(x) = 4\left(\frac{x-2}{4}\right) + 2$$

$$= x - 2 + 2 = x$$

b) To determine $(f^{-1} \circ f)(x)$, substitute $f(x)$ for each x in $f^{-1}(x)$.

$$f^{-1}(x) = \frac{x-2}{4}$$

$$(f^{-1} \circ f)(x) = \frac{4x + 2 - 2}{4}$$

$$= \frac{4x}{4} = x$$

Thus, $(f \circ f^{-1})(x) = (f^{-1} \circ f)(x) = x$.

Now Try Exercise 77

EXAMPLE 7 In Example 5, we determined that $f(x) = x^3 + 2$ and $f^{-1}(x) = \sqrt[3]{x - 2}$ are inverse functions. Show that

a) $(f \circ f^{-1})(x) = x$ **b)** $(f^{-1} \circ f)(x) = x$

Solution

a) To determine $(f \circ f^{-1})(x)$, substitute $f^{-1}(x)$ for each x in $f(x)$.

$$f(x) = x^{\,3} + 2$$

$$(f \circ f^{-1})(x) = (\sqrt[3]{x - 2})^3 + 2$$

$$= x - 2 + 2 = x$$

b) To determine $(f^{-1} \circ f)(x)$, substitute $f(x)$ for each x in $f^{-1}(x)$.

$$f^{-1}(x) = \sqrt[3]{x - 2}$$

$$(f^{-1} \circ f)(x) = \sqrt[3]{(x^3 + 2) - 2}$$

$$= \sqrt[3]{x^3} = x$$

Thus, $(f \circ f^{-1})(x) = (f^{-1} \circ f)(x) = x$.

Now Try Exercise 79

Because a function and its inverse "undo" each other, the composite of a function with its inverse results in the given value from the domain. For example, for any function $f(x)$ and its inverse $f^{-1}(x)$, $(f^{-1} \circ f)(3) = 3$, and $(f \circ f^{-1})\left(-\dfrac{1}{2}\right) = -\dfrac{1}{2}$.

EXERCISE SET 9.1   *Math* XL   *MyMathLab*
MathXL®   MyMathLab

Warm-Up Exercises

Fill in the blanks with the appropriate word, phrase, or symbol(s) from the following list.

inverse	domain	horizontal	x	$f(x)$
vertical	range	one-to-one	composition	y

1. When one variable is a function of another variable which in turn is a function of a third variable, we describe the relationship among such functions as a _____ of functions.

2. A function is a _____ function if each element in the range corresponds to exactly one element in the domain.

3. The _____ line test states that, if a _____ line can be drawn so that it intersects a graph at more than one point, then the graph is not the graph of a function.

4. The _____ line test states that, if a _____ line can be drawn so that it intersects the graph of a function at more than one point, the function is not a one-to-one function.

5. If a one-to-one function has ordered pairs to the form (x, y), the _____ function is a one-to-one function with ordered pairs of the form (y, x).

6. For a one-to-one function $f(x)$ and its inverse function $f^{-1}(x)$, the domain of $f(x)$ is the _____ of $f^{-1}(x)$.

7. For a one-to-one function $f(x)$ and its inverse function $f^{-1}(x)$, the range of $f(x)$ is the _____ of $f^{-1}(x)$.

8. For any one-to-one function $f(x)$ and its inverse $f^{-1}(x)$, $(f \circ f^{-1})(x) = $ _____ and $(f^{-1} \circ f)(x) = $ _____.

Practice the Skills

For each pair of functions, find a) $(f \circ g)(x)$, b) $(f \circ g)(4)$, c) $(g \circ f)(x)$, and d) $(g \circ f)(4)$.

9. $f(x) = x + 4, g(x) = 2x - 3$

10. $f(x) = 3x - 2, g(x) = x + 1$

11. $f(x) = x + 3, g(x) = x^2 + x - 4$

12. $f(x) = x + 2, g(x) = x^2 + 4x - 2$

13. $f(x) = \dfrac{1}{x}, g(x) = 2x + 3$

14. $f(x) = \dfrac{2}{x}, g(x) = x^2 + 1$

15. $f(x) = 3x + 1, g(x) = \dfrac{3}{x}$

16. $f(x) = x^2 - 5, g(x) = \dfrac{4}{x}$

17. $f(x) = x^2 + 1, g(x) = x^2 + 5$

18. $f(x) = x^2 - 4, g(x) = x^2 + 3$

19. $f(x) = x - 4, g(x) = \sqrt{x + 5}, x \geq -5$

20. $f(x) = \sqrt{x + 6}, x \geq -6, g(x) = x + 7$

In Exercises 21–42, determine whether each function is a one-to-one function.

21.
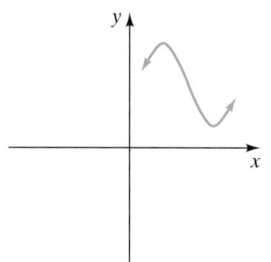

22.
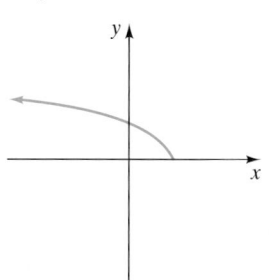

23.
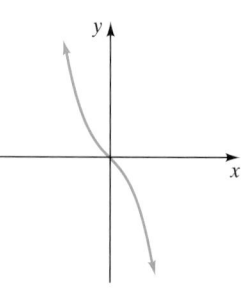

24.
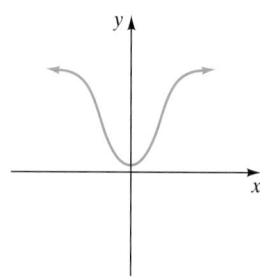

25. $\{(1, 2), (2, 3), (3, 4), (4, 5)\}$

26. $\{(1, 1), (2, 2), (3, 3), (4, 4)\}$

27. $\{(-4, 2), (5, 3), (0, 2), (4, 8)\}$

28. $\{(0, 5), (1, 4), (-3, 5), (4, 2)\}$

29. $y = 2x + 5$

30. $y = 3x - 8$

31. $y = x^2 - 1$

32. $y = -x^2 + 3$

33. $y = x^2 - 2x + 5$

34. $y = x^2 - 2x + 6, x \geq 1$

35. $y = x^2 - 9, x \geq 0$

36. $y = x^2 - 9, x \leq 0$

37. $y = \sqrt{x}$

38. $y = -\sqrt{x}$

39. $y = |x|$

40. $y = -|x|$

41. $y = \sqrt[3]{x}$

42. $y = x^3$

In Exercises 43–48, for the given function, find the domain and range of both $f(x)$ and $f^{-1}(x)$.

43. $\{(4, 0), (8, 9), (2, 7), (-1, 6), (-2, 4)\}$

44. $\left\{(-2, -3), (-4, 0), (5, 3), (6, 2), \left(2, \dfrac{1}{2}\right)\right\}$

45.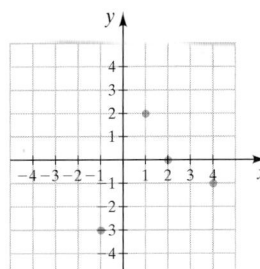

46.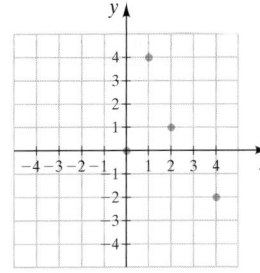

47.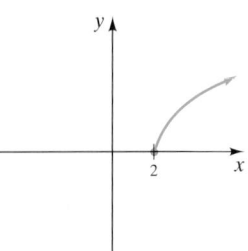

48.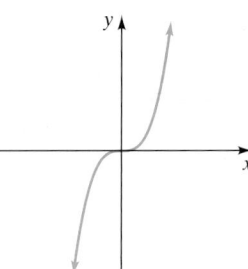

For each function, **a)** *determine whether it is one-to-one;* **b)** *if it is one-to-one, find its inverse function.*

49. $f(x) = x + 3$

50. $f(x) = x - 4$

51. $h(x) = 4x$

52. $k(x) = 2x - 7$

53. $p(x) = 3x^2$

54. $r(x) = |x|$

55. $t(x) = x^2 + 3$

56. $m(x) = x^2 + x + 8$

57. $g(x) = \dfrac{1}{x}$

58. $h(x) = \dfrac{5}{x}$

59. $f(x) = x^2 + 10$

60. $g(x) = x^3 + 9$

61. $g(x) = x^3 - 6$

62. $f(x) = \sqrt{x}, x \geq 0$

63. $g(x) = \sqrt{x + 2}, x \geq -2$

64. $f(x) = x^2 - 3, x \geq 0$

65. $h(x) = x^2 - 4, x \geq 0$

66. $h(x) = |x|$

For each one-to-one function, **a)** *find* $f^{-1}(x)$ *and* **b)** *graph* $f(x)$ *and* $f^{-1}(x)$ *on the same axes.*

67. $f(x) = 2x + 8$

68. $f(x) = -3x + 6$

69. $f(x) = \sqrt{x}, x \geq 0$

70. $f(x) = -\sqrt{x}, x \geq 0$

71. $f(x) = \sqrt{x - 1}, x \geq 1$

72. $f(x) = \sqrt{x + 4}, x \geq -4$

73. $f(x) = \sqrt[3]{x}$

74. $f(x) = \sqrt[3]{x + 3}$

75. $f(x) = \dfrac{1}{x}, x > 0$

76. $f(x) = \dfrac{1}{x}$

For each pair of inverse functions, show that $(f \circ f^{-1})(x) = x$ *and* $(f^{-1} \circ f)(x) = x.$

77. $f(x) = x + 5, f^{-1}(x) = x - 5$

78. $f(x) = 3x, f^{-1}(x) = \dfrac{x}{3}$

79. $f(x) = \dfrac{1}{2}x + 3, f^{-1}(x) = 2x - 6$

80. $f(x) = -\dfrac{1}{3}x + 2, f^{-1}(x) = -3x + 6$

81. $f(x) = \sqrt[3]{x - 2}, f^{-1}(x) = x^3 + 2$

82. $f(x) = \sqrt[3]{x + 9}, f^{-1}(x) = x^3 - 9$

83. $f(x) = \dfrac{3}{x}, f^{-1}(x) = \dfrac{3}{x}$

84. $f(x) = \sqrt{x + 5}, f^{-1}(x) = x^2 - 5, x \geq 0$

Problem Solving

85. The function $f(x) = 3x$ converts yards, x, into feet. Find the inverse function that converts feet into yards. In the inverse function, what do x and $f^{-1}(x)$ represent?

86. The function $f(x) = 12x$ converts feet, x, into inches. Find the inverse function that converts inches into feet. In the inverse function, what do x and $f^{-1}(x)$ represent?

87. The function $f(x) = \dfrac{5}{9}(x - 32)$ converts degrees Fahrenheit, x, to degrees Celsius. Find the inverse function that changes degrees Celsius into degrees Fahrenheit.

88. The function $f(x) = \dfrac{22}{15}x$ converts miles per hour, x, into feet per second. Find the inverse function that converts feet per second into miles per hour.

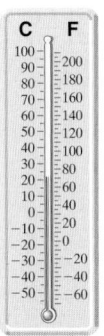

See Exercise 87.

In Exercises 89–92, the functions f(x) and g(x) are given. Determine the composition $(f \circ g)(x)$. For the composition function, what does x represent and what does $(f \circ g)(x)$ represent?

89. $f(x) = 16x$ converts pounds, x, to ounces. $g(x) = 28.35x$ converts ounces, x to grams.

90. $f(x) = 2000x$ converts tons, x, to pounds. $g(x) = 16x$ converts pounds, x, to ounces.

91. $f(x) = 3x$ converts yards, x, to feet. $g(x) = 0.305x$ converts feet, x, to meters.

92. $f(x) = 1760x$ converts miles, x, to yards. $g(x) = 0.915x$ converts yards, x, to meters.

Concept/Writing Exercises

93. Is $(f \circ g)(x) = (g \circ f)(x)$ for all values of x? Explain and give an example to support your answer.

94. Consider the functions $f(x) = \sqrt{x + 5}, x \geq -5,$ and $g(x) = x^2 - 5, x \geq 0.$

   a) Show that $(f \circ g)(x) = (g \circ f)(x)$ for $x \geq 0.$

   b) Explain why we need to stipulate that $x \geq 0$ for part **a)** to be true.

95. Consider the functions $f(x) = x^3 + 2$ and $g(x) = \sqrt[3]{x - 2}.$

   a) Show that $(f \circ g)(x) = (g \circ f)(x).$

   b) What are the domains of $f(x)$, $g(x)$, $(f \circ g)(x)$, and $(g \circ f)(x)$? Explain.

96. For the function $f(x) = x^3, f(2) = 2^3 = 8.$ Explain why $f^{-1}(8) = 2.$

97. For the function $f(x) = x^4, x > 0, f(2) = 16.$ Explain why $f^{-1}(16) = 2.$

98. **a)** Does the function $f(x) = |x|$ have an inverse? Explain.

   b) If the domain is limited to $x \geq 0$, does the function have an inverse? Explain.

   c) Find the inverse function of $f(x) = |x|, x \geq 0.$

Challenge Problems

99. **Area** When a pebble is thrown into a pond, the circle formed by the pebble hitting the water expands with time. The area of the expanding circle may be found by the formula $A = \pi r^2.$ The radius, r, of the circle, in feet, is a function of time, t, in seconds. Suppose that the function is $r(t) = 2t.$

© Daniel Rybkin/Shutterstock

   a) Find the radius of the circle at 3 seconds.

   b) Find the area of the circle at 3 seconds.

   c) Express the area as a function of time by finding $A \circ r.$

   d) Using the function found in part **c)**, find the area of the circle at 3 seconds.

   e) Do your answers in parts **b)** and **d)** agree? If not, explain.

100. **Surface Area** The surface area, S, of a spherical balloon of radius r, in inches, is found by $S(r) = 4\pi r^2.$ If the balloon is being blown up at a constant rate by a machine, then the radius of the balloon is a function of time. Suppose that this function is $r(t) = 1.2t$, where t is in seconds.

   a) Find the radius of the balloon at 2 seconds.

   b) Find the surface area at 2 seconds.

   c) Express the surface area as a function of time by finding $S \circ r.$

   d) Using the function found in part **c)**, find the surface area after 2 seconds.

   e) Do your answers in parts **b)** and **d)** agree? If not, explain why not.

Group Activity

Discuss and answer Exercise 101 as a group.

101. Consider the function $f(x) = 2^x.$ This is an example of an *exponential function*, which we will discuss in the next section.

   a) Graph this function by substituting values for x and finding the corresponding values of $f(x).$

   b) Do you think this function has an inverse? Explain your answer.

   c) Using the graph in part **a)**, draw the inverse function, $f^{-1}(x)$ on the same axes.

   d) Explain how you obtained the graph of $f^{-1}(x).$

Cumulative Review Exercises

[1.3]   **102.** Divide $\left| \dfrac{-9}{4} \right| \div \left| \dfrac{-4}{9} \right|.$

[3.5]   **103.** Determine the equation of a line in standard form that passes through $\left(\dfrac{1}{2}, 3 \right)$ and is parallel to the graph of $2x + 3y - 9 = 0.$

[6.3]   **104.** Simplify $\dfrac{\dfrac{3}{x^2} - \dfrac{2}{x}}{\dfrac{x}{6}}.$

[6.4]   **105.** Solve the formula $\dfrac{1}{f} = \dfrac{1}{p} + \dfrac{1}{q}$ for p.

[8.1]   **106.** Solve $x^2 + 2x - 10 = 0$ by completing the square.

9.2 Exponential Functions

1 Graph exponential functions.

2 Solve applications of exponential functions.

1 Graph Exponential Functions

There are many applications of exponential functions. Some examples include the growth of populations, the doubling of bacteria in a biology experiment, the value of money in a bank account with compound interest, the declining amount of carbon 14 remaining in a fossil, and many others. The graphs shown in **Figure 9.11** and **Figure 9.12** show two examples of exponential functions.

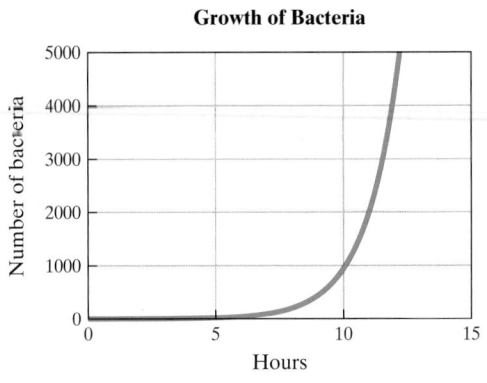

FIGURE 9.11   **FIGURE 9.12**

As seen in the definition below, an exponential function will always have a variable exponent.

Exponential Function

For any real number $a > 0$ and $a \neq 1$,

$$f(x) = a^x \quad \text{or} \quad y = a^x$$

is an **exponential function**.

An exponential function is a function of the form $f(x) = a^x$, or $y = a^x$, where a is a positive real number not equal to 1. Notice the variable is in the exponent.

Examples of Exponential Functions

$$f(x) = 2^x, \qquad y = 5^x, \qquad g(x) = \left(\frac{1}{2}\right)^x$$

Exponential functions can be graphed by selecting values for x, finding the corresponding values of y [or $f(x)$], and plotting the points.

Before we graph exponential functions, let's discuss some characteristics of the graphs of exponential functions.

Graphs of Exponential Functions

For all exponential functions of the form $y = a^x$ or $f(x) = a^x$, where $a > 0$ and $a \neq 1$,

1. The domain of the function is $(-\infty, \infty)$.

2. The range of the function is $(0, \infty)$.

3. The graph of the function passes through the points $\left(-1, \frac{1}{a}\right)$, $(0, 1)$, and $(1, a)$.

EXAMPLE 1 Graph the exponential function $y = 2^x$. State the domain and range of the function.

Solution The function is of the form $y = a^x$, where $a = 2$. First, construct a table of values. In the table, the three points listed in item 3 of the box on page 580 are shown in red.

x	-4	-3	-2	-1	0	1	2	3	4
y	$\frac{1}{16}$	$\frac{1}{8}$	$\frac{1}{4}$	$\frac{1}{2}$	1	2	4	8	16

Now plot these points and connect them with a smooth curve (**Fig. 9.13**). The three ordered pairs in red in the table are marked in red on the graph.

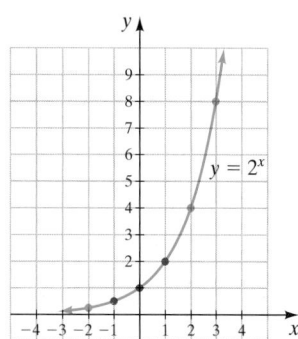

FIGURE 9.13

The domain of this function is the set of real numbers, $\mathbb{R}$. The range is $\{y \mid y > 0\}$.

Now Try Exercise 7

EXAMPLE 2 Graph $y = \left(\frac{1}{2}\right)^x$. State the domain and range of the function.

Solution This function is of the form $y = a^x$, where $a = \frac{1}{2}$. Construct a table of values and plot the curve (**Fig. 9.14**).

x	-4	-3	-2	-1	0	1	2	3	4
y	16	8	4	2	1	$\frac{1}{2}$	$\frac{1}{4}$	$\frac{1}{8}$	$\frac{1}{16}$

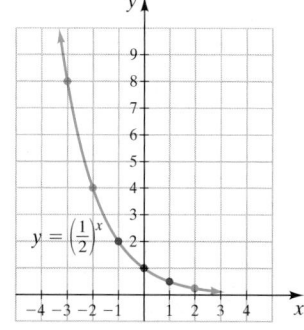

FIGURE 9.14

The domain is the set of real numbers, $\mathbb{R}$. The range is $\{y \mid y > 0\}$.

Now Try Exercise 13

Note that the graphs in **Figures 9.13** and **9.14** represent one-to-one functions since each graph passes the horizontal line test.

Helpful Hint

When graphing exponential functions of the form $y = a^x$ where $x > 0$, if

- $a > 1$, the graph will rise from left to right. See the graph of $y = 2^x$ in **Figure 9.13**.

- $0 < a < 1$, the graph will fall from left to right. See the graph of $y = \left(\frac{1}{2}\right)^x$ in **Figure 9.14**.

Whenever we encounter an exponential function with a negative exponent such as $y - 2^{-x}$, we can recall our rules of exponents to see that

$$y = 2^{-x}$$
$$= \frac{1}{2^x}$$
$$= \left(\frac{1}{2}\right)^x$$

Thus, the graph of $y = 2^{-x}$ is the graph of $y = \left(\frac{1}{2}\right)^x$ shown in **Figure 9.14** on page 581.

Similarly, when we encounter a function such as $y = \left(\frac{1}{2}\right)^{-x}$, we can use the rules of exponents to see that

$$y = \left(\frac{1}{2}\right)^{-x}$$
$$= \left(\frac{2}{1}\right)^x$$
$$= 2^x$$

Thus, the graph of $y = \left(\frac{1}{2}\right)^{-x}$ is the graph of $y = 2^x$ shown in **Figure 9.13** on page 581.

2 Solve Applications of Exponential Functions

Exponential functions are often used to describe the growth and decay of certain quantities. We illustrate exponential functions in the next five examples.

EXAMPLE 3 **Pennies Add Up** Jennifer Hewlett told her young son that if he did his chores, she would give him 2 cents the first week and double the amount each week for the next 10 weeks. The number of cents her son would receive in any given week, w, can be determined by the function $n(w) = 2^w$. Determine the number of cents Jennifer would give her son in week 8.

Solution By evaluating 2^8, we determine that in week 8 Jennifer would give her son 256 cents, or $2.56.

Now Try Exercise 29

EXAMPLE 4 **Value of a Jeep** Ronald Yates just bought a new Jeep Compass for $22,000. Assume the value of the Jeep depreciates at a rate of 20% per year. Therefore, the value of the Jeep is 80% of the previous year's value. One year from now, its value will be $22,000(0.80). Two years from now, its value will be $22,000(0.80)(0.80) = $22,000(0.80)^2$ and so on. Therefore, the formula for the value of the Jeep is

$$v(t) = 22,000(0.80)^t$$

where t is time in years. Find the value of the Jeep **a)** 1 year from now and **b)** 5 years from now.

Solution

a) To find the value 1 year from now, substitute 1 for t.

$$v(t) = 22,000(0.80)^t$$
$$v(1) = 22,000(0.80)^1 \quad \text{Substitute 1 for } t.$$
$$= 17,600$$

One year from now, the value of the Jeep will be $17,600.

b) To find the value 5 years from now, substitute 5 for t.

$$v(t) = 22{,}000(0.80)^t$$

$$v(5) = 22{,}000(0.80)^5 \qquad \text{Substitute 5 for } t.$$

$$= 22{,}000(0.32768)$$

$$= 7208.96$$

Five years from now, the value of the Jeep will be $7208.96.

Now Try Exercise 45

Previously, we have used the compound interest formula to determine the amount of money we accumulate in a savings or other investment account.

Compound Interest Formula

The accumulated amount, A, in a compound interest account can be found using the formula

$$A = p\left(1 + \frac{r}{n}\right)^{nt}$$

where p is the principal or the initial investment amount, r is the interest rate as a decimal, n is the number of compounding periods per year, and t is the time in years.

EXAMPLE 5   **Compound Interest** Nancy Johnson invests $10,000 in a certificate of deposit (CD) with 5% interest compounded quarterly for 6 years. Determine the value of the CD after 6 years.

Solution   Understand   We are given that the principal, p, is $10,000. We are also given that the interest rate, r, is 5%. Because the interest is compounded quarterly, the number of compounding periods, n, is 4. The money is invested for 6 years. Therefore, t is 6.

Translate   Now we substitute these values into the formula

$$A = p\left(1 + \frac{r}{n}\right)^{nt}$$

$$= 10{,}000\left(1 + \frac{0.05}{4}\right)^{4(6)}$$

Carry Out

$$= 10{,}000(1 + 0.0125)^{24}$$

$$= 10{,}000(1.0125)^{24}$$

$$\approx 10{,}000(1.347351) \qquad \text{From a calculator}$$

$$\approx 13{,}473.51$$

Answer   The original $10,000 has grown to about $13,473.51 after 6 years.

Now Try Exercise 33

> **Understanding Algebra**
>
> When using the compound interest formula you need to be careful to write the interest rate as a decimal. For example, an interest rate of 5% means that $r = 0.05$. An interest rate of 2.75% means that $r = 0.0275$, and so on.
>
> When writing n, the number of compounding periods per year, here are the most commonly used values:
>
> semiannually: $n = 2$
>
> quarterly: $n = 4$
>
> monthly: $n = 12$

EXAMPLE 6   **Carbon 14 Dating** Carbon 14 dating is used by scientists to find the age of fossils and artifacts. The formula used in carbon dating is

$$A = A_0 \cdot 2^{-t/5600}$$

where A_0 represents the amount of carbon 14 present when the fossil was formed and A represents the amount of carbon 14 present after t years. If 500 grams of carbon 14 were present when an organism died, how many grams will be found in the fossil 2000 years later?

Solution   Understand   When the fossil died, 500 grams of carbon 14 were present. Therefore, $A_0 - 500$. To find out how many grams of carbon 14 will be present 2000 years later, we substitute 2000 for t in the formula.

Translate

$$A = A_0 \cdot 2^{-t/5600}$$
$$= 500(2)^{-2000/5600}$$

Carry Out

$$\approx 500(0.7807092) \qquad \text{From a calculator}$$
$$\approx 390.35 \text{ grams}$$

Answer   After 2000 years, about 390.35 of the original 500 grams of carbon 14 are still present.

Now Try Exercise 39

EXAMPLE 7   **Exports to China. Figure 9.15** shows the annual amount of exports from the United States to China for the years 2000 through 2007 in billions of dollars.

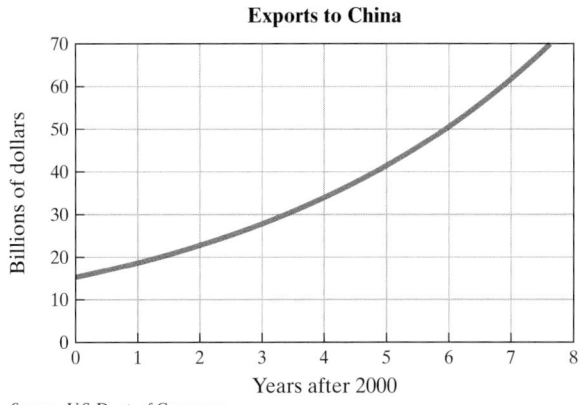

Exports to China

Source: U.S. Dept. of Commerce

FIGURE 9.15

An exponential function that closely approximates this curve is $f(t) = 15.37(1.22)^t$. In this function, $f(t)$ is the total value of exports from the United States to China and t is the number of years since 2000. Assume that this trend continues. Use this function to estimate the value of the exports to China in **a)** 2010 and **b)** 2015. Round your answers to nearest billion dollars.

Solution

a) Understand   In this function, t is years since 2000. Thus, the year 2010 is represented by $t = 10$. To estimate the value of the exports to China in 2010, we need to evaluate this function for $t = 10$.

Translate and Carry Out $f(t) = 15.37(1.22)^t$

$$f(10) = 15.37(1.22)^{10} \approx 112.27$$

Answer   Therefore, if this trend continues, the exports to China in 2010 will be about $112 billion.

b) The year 2015 is represented by $t = 15$ and we need to evaluate $f(15)$.

$$f(t) = 15.37(1.22)^t$$

$$f(15) = 15.37(1.22)^{15} \approx 303.44$$

Answer   Therefore, if this trend continues, the exports to China in 2015 will be about $303 billion.

Now Try Exercise 47

EXERCISE SET 9.2

Math XL   MyMathLab
MathXL®   MyMathLab

Warm-Up Exercises

Fill in the blanks with the appropriate word, phrase, or symbol(s) from the following list.

periods   base   fall   exponent   principal   rise   rate   time

1. In an exponential function, the variable is in the _____ position.

2. In a quadratic function, the variable is in the _____ position.

3. When graphing exponential functions of the form $y = a^x$, if $a > 1$, the graph will _____ from left to right.

4. When graphing exponential functions of the form $y = a^x$, if $0 < a < 1$, the graph will _____ from left to right.

5. In the compound interest formula $A = p\left(1 + \dfrac{r}{n}\right)^{nt}$, p is the _____ or the initial investment amount.

6. In the compound interest formula $A = p\left(1 + \dfrac{r}{n}\right)^{nt}$, n is the number of compounding _____ per year.

Practice the Skills

Graph each exponential function.

7. $y = 2^x$

8. $y = 3^x$

9. $y = \left(\dfrac{1}{2}\right)^x$

10. $y = \left(\dfrac{1}{3}\right)^x$

11. $y = 4^x$

12. $y = 5^x$

13. $y = \left(\dfrac{1}{4}\right)^x$

14. $y = \left(\dfrac{1}{5}\right)^x$

15. $y = 3^{-x}$

16. $y = 4^{-x}$

17. $y = \left(\dfrac{1}{3}\right)^{-x}$

18. $y = \left(\dfrac{1}{4}\right)^{-x}$

19. $y = 2^{x-1}$

20. $y = 2^{x+1}$

21. $y = \left(\dfrac{1}{3}\right)^{x+1}$

22. $y = \left(\dfrac{1}{3}\right)^{x-1}$

23. $y = 2^x + 1$

24. $y = 2^x - 1$

25. $y = 3^x - 1$

26. $y = 3^x + 2$

Problem Solving

27. U.S. Population The following graph shows the growth in the U.S. population of people age 85 and older for the years from 1960 to 2000 and projected to 2050. The exponential function that closely approximates this graph is

$$f(t) = 0.592(1.042)^t$$

In this function, $f(t)$ is the population, in millions, of people age 85 and older and t is the number of years since 1960. Assuming this trend continues, use this function to estimate the number of U.S. people age 85 and older in **a)** 2060. **b)** 2100.

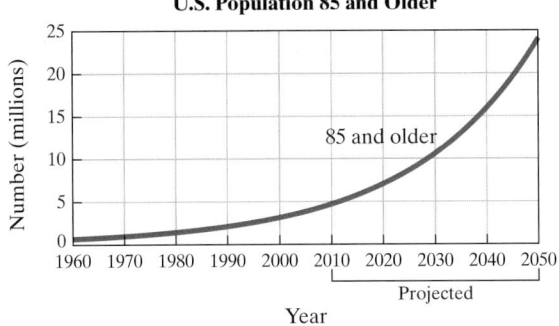

U.S. Population 85 and Older

Source: U.S. Census Bureau

28. World Population Since about 1650 the world population has been growing exponentially. The exponential function that closely approximates the world population from 1650 and projected to 2015 is

$$f(t) = \frac{1}{2}(2.718)^{0.0072t}$$

In this function, $f(t)$ is the world population, in billions of people, and t is the number of years since 1650. If this trend continues, estimate the world population in **a)** 2010. **b)** 2015.

29. Doubling If $2 is doubled each day for 9 days, determine the amount on day 9.

30. Doubling If $2 is doubled each day for 12 days, determine the amount on day 12.

31. Bacteria in a Petri Dish Five bacteria are placed in a petri dish. The population will triple every day. The formula for the number of bacteria in the dish on day t is

$$N(t) = 5(3)^t$$

where t is the number of days after the five bacteria are placed in the dish. How many bacteria are in the dish 2 days after the five bacteria are placed in the dish?

32. Bacteria in a Petri Dish Refer to Exercise 31. How many bacteria are in the dish 6 days after the five bacteria are placed in the dish?

33. Compound Interest If Don Gecewicz invests $5000 at 6% interest compounded quarterly, find the amount after 4 years (see Example 5).

34. Compound Interest If Don Treadwell invests $8000 at 4% interest compounded quarterly, find the amount after 5 years.

35. Certificate of Deposit Joni Burnette receives a bonus of $5000 for meeting her annual sales quota. She invests the bonus in a certificate of deposit (CD) that pays 4.2% interest compounded monthly. Determine the value of the CD after 5 years.

36. Money Market Account Martha Goshaw invests $2500 in a money market account that pays 3.6% interest compounded quarterly. Determine the accumulated amount after 2 years.

37. Savings Account Byron Dyce deposits $3000 in a savings account that pays 2.4% interest compounded quarterly. Determine the accumulated amount after 2 years.

38. Retirement Account To invest for his retirement, John Salak invests $10,000 in an account paying 6% interest compounded semiannually. Determine the accumulated amount after 25 years.

39. Carbon 14 Dating If 12 grams of carbon 14 are originally present in a certain animal bone, how much will remain at the end of 1000 years? Use $A = A_0 \cdot 2^{-t/5600}$ (see Example 6).

40. Carbon 14 Dating If 60 grams of carbon 14 are originally present in the fossil Tim Jonas found at an archeological site, how much will remain after 10,000 years?

41. Radioactive Substance The amount of a radioactive substance present, in grams, at time t, in years, is given by the formula $y = 80(2)^{-0.4t}$. Find the number of grams present in **a)** 10 years. **b)** 100 years.

42. Radioactive Substance The amount of a radioactive substance present, in grams, at time t, in years, is given by the formula $y = 20(3)^{-0.6t}$. Find the number of grams present in 4 years.

43. Population The expected future population of Ackworth, which now has 2000 residents, can be approximated by the formula $y = 2000(1.2)^{0.1t}$, where t is the number of years in the future. Find the expected population of the town in **a)** 10 years. **b)** 50 years.

44. Population The expected future population of Antwerp, which currently has 6800 residents, can be approximated by the formula $y = 6800(1.4)^{-0.2t}$, where t is the number of years in the future. Find the expected population of the town 30 years in the future.

45. Value of an SUV The cost of a new SUV is $24,000. If it depreciates at a rate of 18% per year, the value of the SUV in t years can be approximated by the formula

$$V(t) = 24{,}000(0.82)^t$$

Find the value of the SUV in 4 years.

46. Value of an ATV The cost of a new all-terrain vehicle is $6200. If it depreciates at a rate of 15% per year, the value of the ATV in t years can be approximated by the formula

$$V(t) = 6200(0.85)^t$$

Find the value of the ATV in 10 years.

See Exercise 46.

47. Atmospheric Pressure Atmospheric pressure varies with altitude. The greater the altitude, the lower the pressure, as shown in the following graph.

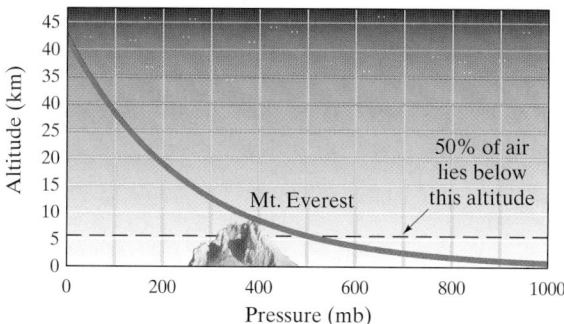

The equation $A = 41.97(0.996)^x$ can be used to estimate the altitude, A, in kilometers, for a given pressure, x, in millibars (mb). If the atmospheric pressure on top of Mt. Everest is about 389 mb, estimate the altitude of the top of Mt. Everest.

48. Centenarians Based on projections of the U.S. Census Bureau, the number of centenarians (people age 100 or older) will grow exponentially beyond 1995 (see the graph below). The function

$$f(t) = 71.24(1.045)^t$$

can be used to approximate the number of centenarians, in thousands, in the United States where t is time in years since 1995. Use this function to estimate the number of centenarians in **a)** 2060. **b)** 2070.

Number of Centenarians in the United States

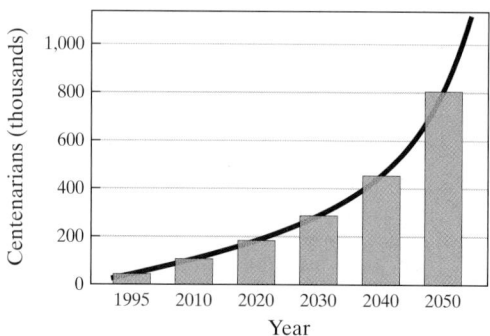

Source: U.S. Bureau of the Census

49. Bike Shop Sales Spokes for Folks, a bike shop, has annual sales for years 2006–2010 (in thousands of dollars) as shown in the graph below.

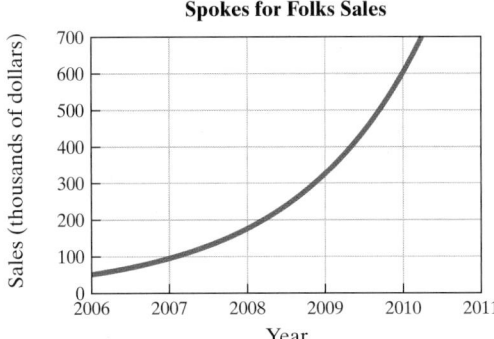

Spokes for Folks Sales

Source: U.S. Dept. of Commerce

The annual sales can be estimated by the function $S(t) = 51.4(1.85)^t$, where $S(t)$ is annual sales in thousands of dollars and t is the number of years after 2006. Assuming this trend continues, determine the annual sales for the following years. Round your answers to the nearest thousand dollars.

a) 2015

b) 2020

50. Sign Sales Signs 2 Go, a sign-making print shop, has annual sales for years 2006–2010 (in thousands of dollars) as shown in the graph below.

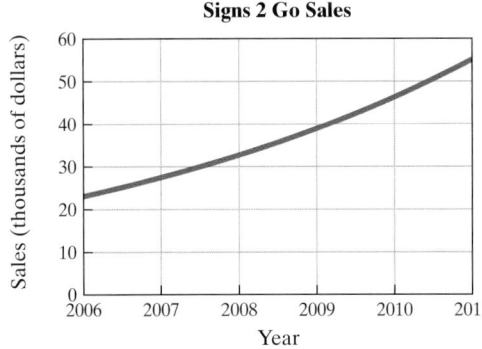

Signs 2 Go Sales

The annual sales can be estimated by the function $S(t) = 23.1(1.19)^t$, where $S(t)$ is annual sales in thousands of dollars and t is the number of years after 2006. Assuming this trend

continues, determine the annual sales for the following years. Round your answers to the nearest thousand dollars.

a) 2015

b) 2020

51. Simple and Compound Interest The following graph indicates linear growth of $100 invested at 7% simple interest and exponential growth at 7% interest compounded annually. In the formulas, A represents the amount in dollars and t represents the time in years.

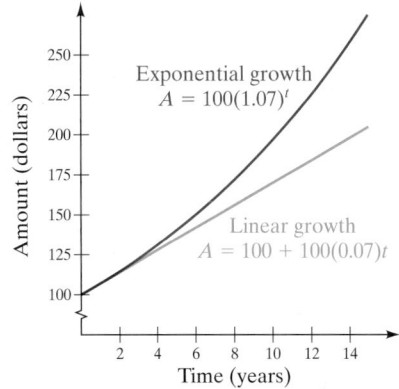

a) Use the graph to estimate the doubling time for $100 invested at 7% simple interest.

b) Estimate the doubling time for $100 invested at 7% interest compounded annually.

c) Estimate the difference in amounts after 10 years for $100 invested by each method.

d) Most banks compound interest daily instead of annually. What effect does this have on the total amount? Explain.

52. In Exercise 51, we graphed the amount for various years when $100 is invested at 7% simple interest and at 7% interest compounded annually.

a) Use the compound interest formula to determine the amount if $100 is compounded daily at 7% for 10 years (assume 365 days per year).

b) Estimate the difference in the amount in 10 years for the $100 invested at 7% simple interest versus the 7% interest compounded daily.

Concept/Writing Exercises

53. Consider the exponential function $y = \left(\dfrac{1}{2}\right)^x$.

a) As x increases, what happens to y?

b) Can y ever be 0? Explain.

c) Can y ever be negative? Explain.

54. Consider the exponential function $y = 2^x$.

a) As x increases, what happens to y?

b) Can y ever be 0? Explain.

c) Can y ever be negative? Explain.

55. Consider the equations $y = 2^x$ and $y = 3^x$.

a) Will both graphs have the same or different y-intercepts? Determine their y-intercepts.

b) How will the graphs of the two functions compare?

56. Consider the equations $y = \left(\dfrac{1}{2}\right)^x$ and $y = \left(\dfrac{1}{3}\right)^x$.

a) Will both graphs have the same or different y-intercepts? Determine their y-intercepts.

b) How will the graphs of the two functions compare?

57. We stated earlier that, for exponential functions $f(x) = a^x$, the value of a cannot equal 1.

a) What does the graph of $f(x) = a^x$ look like when $a = 1$?

b) Is $f(x) = a^x$ a function when $a = 1$?

c) Does $f(x) = a^x$ have an inverse function when $a = 1$? Explain your answer.

58. How will the graphs of $y = a^x$ and $y = a^x + k, k > 0$, compare?

59. How will the graphs of $y = a^x$ and $y = a^x - k, k > 0$, compare?

60. For $a > 1$, how will the graphs of $y = a^x$ and $y = a^{x+1}$ compare?

61. For $a > 1$, how will the graphs of $y = a^x$ and $y = a^{x+2}$ compare?

62. a) Is $y = x^\pi$ an exponential function? Explain.

b) Is $y = \pi^x$ an exponential function? Explain.

Challenge Problem

63. Suppose Bob Jenkins gives Carol Dantuma $1 on day 1, $2 on day 2, $4 on day 3, $8 on day 4, and continues this doubling process for 30 days.

a) Determine how much Bob will give Carol on day 15.

b) Determine how much Bob will give Carol on day 20.

c) Express the amount, using exponential form, that Bob gives Carol on day n.

d) How much, in dollars, will Bob give Carol on day 30? Write the amount in exponential form. Then use a calculator to evaluate.

e) Express the total amount Bob gives Carol over the 30 days as a sum of exponential terms. (Do not find the actual value.)

Group Activity

64. Functions that are exponential or are approximately exponential are commonly seen.

a) Have each member of the group individually determine a function not given in this section that may approximate an exponential function. You may use newspapers, books, or other sources.

b) As a group, discuss one another's functions. Determine whether each function presented is an exponential function.

c) As a group, write a paper that discusses each of the exponential functions and state why you believe each function is exponential.

Cumulative Review Exercises

[5.1]   **65.** Consider the polynomial

$$2.3x^4y - 6.2x^6y^2 + 9.2x^5y^2$$

a) Write the polynomial in descending order of the variable x.

b) What is the degree of the polynomial?

c) What is the leading coefficient?

[5.2]   **66.** If $f(x) = x + 5$ and $g(x) = x^2 - 2x + 4$, find $(f \cdot g)(x)$.

[7.1]   **67.** Write $\sqrt{a^2 - 8a + 16}$ as an absolute value.

[7.3]   **68.** Simplify $\sqrt[4]{\dfrac{32x^5y^9}{2y^3z}}$.

9.3 Logarithmic Functions

1 Define a Logarithm

2 Convert from exponential form to logarithmic form.

3 Graph logarithmic functions.

4 Compare the graphs of exponential and logarithmic functions.

5 Solve applications of logarithmic functions.

1 **Define a Logarithm**

Consider the exponential function $y = 2^x$. In **Figure 9.13** on page 581, we see that the graph of this function passes the horizontal line test and therefore this function is a one-to-one function and has an inverse. To find the inverse of $y = 2^x$ we interchange x and y to get the equation $x = 2^y$. To solve this equation for y, we introduce a new definition.

> **Logarithm**
>
> For $x > 0$ and $a > 0, a \neq 1$
>
> $$y = \log_a x \text{ means } x = a^y$$
>
> The expression $\log_a x$ is read "the logarithm of x to the base a" or simply "log, base a, of x."

Using this definition, we can solve the equation $x = 2^y$ for y by rewriting the exponential equation as an equation involving a logarithm.

$$\underbrace{x = 2^y}_{\substack{\text{Exponential} \\ \text{form}}} \text{ is an equivalent equation to } \underbrace{y = \log_2 x}_{\substack{\text{Logarithmic} \\ \text{form}}}$$

Thus, $y = 2^x$ and $y = \log_2 x$ are inverse functions. In general, $y = a^x$ and $y = \log_a x$ are inverse functions.

2 Convert from Exponential Form to Logarithmic Form

To convert from an exponential equation to a logarithmic equation consider the following diagram:

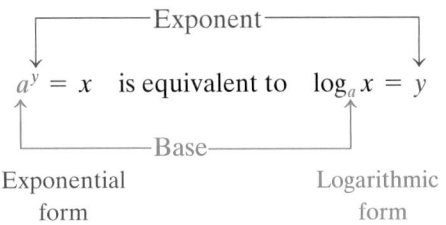

$$\underbrace{a^y = x}_{\substack{\text{Exponential} \\ \text{form}}} \text{ is equivalent to } \underbrace{\log_a x = y}_{\substack{\text{Logarithmic} \\ \text{form}}}$$

For example:

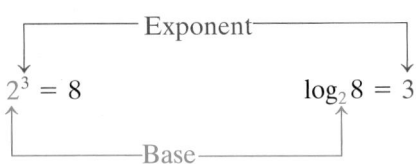

$$2^3 = 8 \qquad\qquad \log_2 8 = 3$$

Understanding Algebra

A useful sentence when working with logarithms is:

A logarithm is an exponent.

The expression $\log_a x$ represents the exponent to which the base a must be raised to obtain x.

For example, $\log_2 8$ represents the exponent to which 2 must be raised to obtain 8. Therefore,

$\log_2 8 = 3$ since $2^3 = 8$

Thus the exponential equation $2^3 = 8$ is equivalent to the logarithmic equation $\log_2 8 = 3$ or $3 = \log_2 8$.

Following are some examples of how an exponential expression can be converted to a logarithmic expression.

Exponential Form	Logarithmic Form
$10^0 = 1$	$\log_{10} 1 = 0$
$4^2 = 16$	$\log_4 16 = 2$
$\left(\dfrac{1}{2}\right)^5 = \dfrac{1}{32}$	$\log_{1/2} \dfrac{1}{32} = 5$
$5^{-2} = \dfrac{1}{25}$	$\log_5 \dfrac{1}{25} = -2$

Now let's do a few examples involving conversion from exponential form into logarithmic form, and vice versa.

EXAMPLE 1 Write each equation in logarithmic form.

a) $3^4 = 81$ **b)** $\left(\dfrac{1}{5}\right)^3 = \dfrac{1}{125}$ **c)** $2^{-5} = \dfrac{1}{32}$

Solution

a) $\log_3 81 = 4$ **b)** $\log_{1/5} \dfrac{1}{125} = 3$ **c)** $\log_2 \dfrac{1}{32} = -5$

Now Try Exercise 19

EXAMPLE 2 Write each equation in exponential form.

a) $\log_7 49 = 2$ **b)** $\log_4 64 = 3$ **c)** $\log_{1/3} \dfrac{1}{81} = 4$

Solution

a) $7^2 = 49$ **b)** $4^3 = 64$ **c)** $\left(\dfrac{1}{3}\right)^4 = \dfrac{1}{81}$

Now Try Exercise 35

EXAMPLE 3 Write each equation in exponential form; then find the unknown value.

a) $y = \log_5 25$ **b)** $2 = \log_a 16$ **c)** $3 = \log_{1/2} x$

Solution

a) $5^y = 25$. Since $5^2 = 25$, $y = 2$.

b) $a^2 = 16$. Since $4^2 = 16$, $a = 4$. Note that a must be greater than 0, so -4 is not a possible answer for a.

c) $\left(\dfrac{1}{2}\right)^3 = x$. Since $\left(\dfrac{1}{2}\right)^3 = \dfrac{1}{8}$, $x = \dfrac{1}{8}$.

Now Try Exercise 53

EXAMPLE 4 Evaluate the following

a) $\log_5 25$ **b)** $\log_5 625$ **c)** $\log_5 5$

d) $\log_5 1$ **e)** $\log_5\left(\dfrac{1}{5}\right)$ **f)** $\log_5 \sqrt{5}$

Solution

a) $\log_5 25 = 2$ since $5^2 = 25$.

b) $\log_5 625 = 4$ since $5^4 = 625$.

c) $\log_5 5 = 1$ since $5^1 = 5$.

d) $\log_5 1 = 0$ since $5^0 = 1$.

e) $\log_5\left(\dfrac{1}{5}\right) = -1$ since $5^{-1} = \dfrac{1}{5}$.

f) $\log_5 \sqrt{5} = \dfrac{1}{2}$ since $5^{1/2} = \sqrt{5}$.

Now Try Exercise 75

Helpful Hint

Since a logarithm is an exponent, it is very important to know the rules of exponents when evaluating logarithmic expressions. It would be helpful to review the rules of exponents from Section 1.5 on page 45 as well as the relationship between radicals and rational exponents from Section 7.2 on page 435.

3 Graph Logarithmic Functions

We are now ready to introduce logarithmic functions.

Logarithmic Function

For any real number $a > 0$, $a \neq 1$, and $x > 0$,

$$f(x) = \log_a x \text{ or } y = \log_a x$$

is a **logarithmic function**.

Examples of Logarithmic Functions
$$f(x) = \log_5 x \qquad y = \log_{1/2} x \qquad g(x) = \log_5 x$$

Logarithmic functions can be graphed by converting the logarithmic equation to an exponential equation and then plotting points.

Before we graph logarithmic functions, let's discuss some characteristics of the graphs of logarithmic functions.

Graphs of Logarithmic Functions

For all logarithmic functions of the form $y = \log_a x$ or $f(x) = \log_a x$, where $a > 0$, $a \neq 1$, and $x > 0$

   1. The domain of the function is $(0, \infty)$.

   2. The range of the function is $(-\infty, \infty)$.

   3. The graph passes through the points $\left(\dfrac{1}{a}, -1\right)$, $(1, 0)$, and $(a, 1)$.

EXAMPLE 5 Graph $y = \log_2 x$. State the domain and range of the function.

Solution   This is an equation of the form $y = \log_a x$, where $a = 2$. $y = \log_2 x$ means $x = 2^y$. Using $x = 2^y$, construct a table of values. The table will be easier to develop by selecting values for y and finding the corresponding values for x. In the table, the three points listed in item 3 in the box are shown in blue.

x	$\dfrac{1}{16}$	$\dfrac{1}{8}$	$\dfrac{1}{4}$	$\dfrac{1}{2}$	1	2	4	8	16
y	-4	-3	-2	-1	0	1	2	3	4

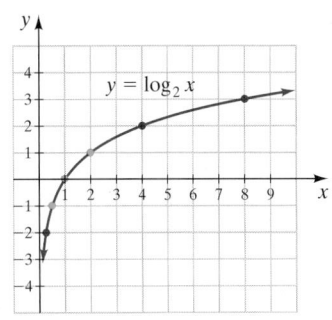

FIGURE 9.16

Now draw the graph (**Fig. 9.16**). The three ordered pairs in blue in the table are marked in blue on the graph. The domain, the set of x-values, is $\{x \mid x > 0\}$. The range, the set of y-values, is the set of all real numbers, $\mathbb{R}$.

Now Try Exercise 91

EXAMPLE 6 Graph $y = \log_{1/2} x$. State the domain and range of the function.

Solution   This is an equation of the form $y = \log_a x$, where $a = \dfrac{1}{2}$. $y = \log_{1/2} x$ means $x = \left(\dfrac{1}{2}\right)^y$. Construct a table of values by selecting values for y and finding the corresponding values of x.

x	16	8	4	2	1	$\dfrac{1}{2}$	$\dfrac{1}{4}$	$\dfrac{1}{8}$	$\dfrac{1}{16}$
y	-4	-3	-2	-1	0	1	2	3	4

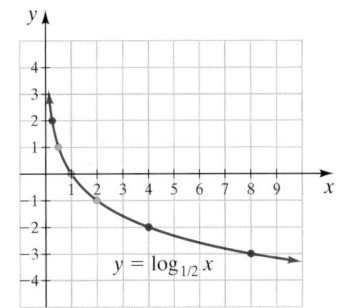

FIGURE 9.17

The graph is illustrated in **Figure 9.17**. The domain is $\{x \mid x > 0\}$. The range is the set of real numbers, $\mathbb{R}$.

Now Try Exercise 93

If we study the domains in Examples 5 and 6, we see that the domains of both $y = \log_2 x$ and $y = \log_{1/2} x$ are $\{x \mid x > 0\}$. In fact, *for any logarithmic function* $y = \log_a x$, *the domain is* $\{x \mid x > 0\}$. Also note that the graphs in Examples 5 and 6 are both graphs of one-to-one functions.

Understanding Algebra

When graphing functions of the form $y = \log_a x$ or $f(x) = \log_a x$, we can predict the shape of the graph by observing three points:

$\left(\dfrac{1}{a}, -1\right)$, $(1, 0)$, and $(a, 1)$.

- When $a > 1$, the graph becomes almost vertical to left of $\left(\dfrac{1}{a}, -1\right)$ and somewhat horizontal to the right of $(a, 1)$; see Example 5.
- When $0 < a < 1$, the graph becomes almost vertical to the left of $(a, 1)$ and somewhat horizontal to the right of $\left(\dfrac{1}{a}, -1\right)$; see Example 6.

4 Compare the Graphs of Exponential and Logarithmic Functions

Recall that $y = a^x$ and $y = \log_a x$ are *inverse functions*. We may therefore write: if $f(x) = a^x$ then $f^{-1}(x) = \log_a x$. In the box below we highlight some of the characteristics of the graphs of the general exponential function $y = a^x$ and the general logarithmic function $y = \log_a x$.

Graph Characteristics

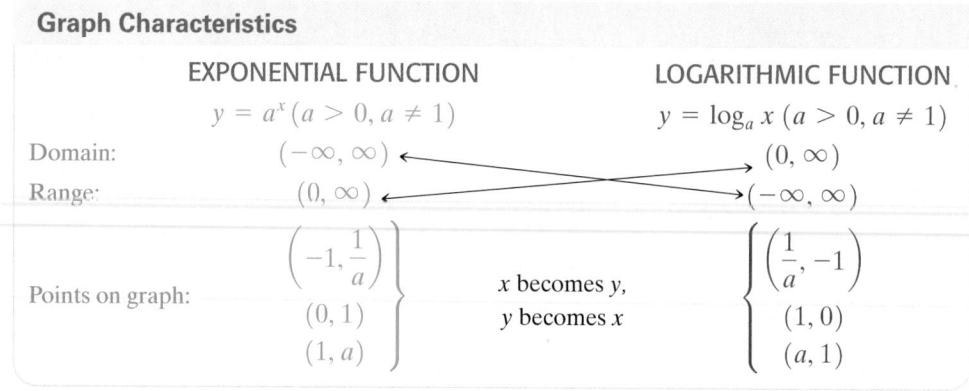

	EXPONENTIAL FUNCTION $y = a^x \, (a > 0, a \neq 1)$		LOGARITHMIC FUNCTION $y = \log_a x \, (a > 0, a \neq 1)$
Domain:	$(-\infty, \infty)$		$(0, \infty)$
Range:	$(0, \infty)$		$(-\infty, \infty)$
Points on graph:	$\left(-1, \dfrac{1}{a}\right)$ $(0, 1)$ $(1, a)$	x becomes y, y becomes x	$\left(\dfrac{1}{a}, -1\right)$ $(1, 0)$ $(a, 1)$

In **Figure 9.18**, we show the graphs of $y = a^x$ and $y = \log_a x$ for $a > 1$. Notice the graphs are symmetric about the line $y = x$.

Understanding Algebra

Since $y = a^x$ and $y = \log_a x$ are inverse functions, we note the following:

- The range of the exponential function is the domain of the logarithmic function and vice versa.

- If (a, b) is a point on the graph of the exponential function, then (b, a) is a point on the graph of the logarithmic function.

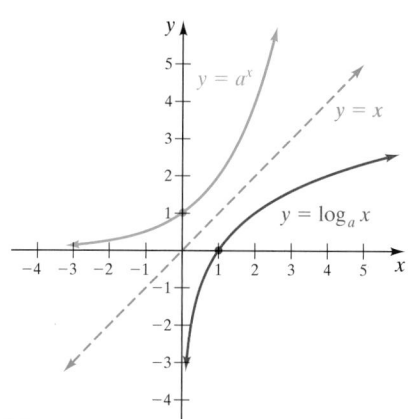

FIGURE 9.18

The graphs of $y = 2^x$ and $y = \log_2 x$ are illustrated in **Figure 9.19**. The graphs of $y = \left(\dfrac{1}{2}\right)^x$ and $y = \log_{1/2} x$ are illustrated in **Figure 9.20**. In each figure, the graphs are inverses of each other and are symmetric with respect to the line $y = x$.

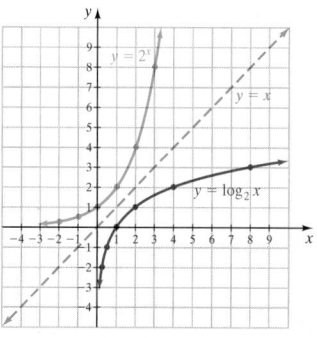

FIGURE 9.19

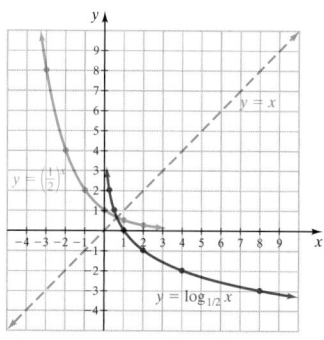

FIGURE 9.20

5 Solve Applications of Logarithmic Functions

We will see many applications of logarithms later, but let's look at one important application now.

EXAMPLE 7 **Earthquakes** Logarithms are used to measure the magnitude of earthquakes. The Richter scale for measuring earthquakes was developed by Charles R. Richter. The magnitude, R, of an earthquake on the Richter scale is given by the formula

$$R = \log_{10} I$$

where I represents the number of times greater (or more intense) the earthquake is than the smallest measurable activity that can be measured on a seismograph.

a) If an earthquake measures 4 on the Richter scale, how many times more intense is it than the smallest measurable activity?

b) How many times more intense is an earthquake that measures 5 on the Richter scale than an earthquake that measures 4?

Solution

a) Understand   The Richter number, R, is 4. To find how many times more intense the earthquake is than the smallest measurable activity, I, we substitute $R = 4$ into the formula and solve for I.

Translate $\qquad\qquad\qquad\qquad\qquad R = \log_{10} I$
$\qquad\qquad\qquad\qquad\qquad\qquad\qquad 4 = \log_{10} I$

Carry Out $\qquad\qquad\qquad\qquad 10^4 = I \qquad$ Change to exponential form.
$\qquad\qquad\qquad\qquad\qquad 10{,}000 = I$

Answer   Therefore, an earthquake that measures 4 on the Richter scale is 10,000 times more intense than the smallest measurable activity.

b) $\qquad\qquad\qquad\qquad\qquad\qquad 5 = \log_{10} I$
$\qquad\qquad\qquad\qquad\qquad 10^5 = I \qquad$ Change to exponential form.
$\qquad\qquad\qquad\qquad 100{,}000 = I$

Since $(10{,}000)(10) = 100{,}000$, an earthquake that measures 5 on the Richter scale is 10 times more intense than an earthquake that measures 4 on the Richter scale .

Now Try Exercise 99

© robert paul van beets/Shutterstock

EXERCISE SET 9.3

MathXL
MathXL®

MyMathLab
MyMathLab

Warm-Up Exercises

Fill in the blanks with the appropriate word, phrase, or symbol(s) from the following list.

logarithmic   exponent   symmetric   base   inverse   domain   range   composite

1. The expression $\log_a x$ represents the _____ to which the base a must be raised to obtain x.

2. In general, $y = a^x$ and $y = \log_a x$ are _____ functions.

3. In both of the equations $a^y = x$ and $\log_a x = y$, a is called the _____.

4. $f(x) = \log_5 x$ is an example of a _____ function.

5. For any logarithmic function, the _____ is $\{x \mid x > 0\}$.

6. The graphs of $y = a^x$ and $y = \log_a x$ are _____ with respect to the line $y = x$.

Practice the Skills

Write each equation in logarithmic form.

7. $5^2 = 25$ **8.** $4^2 = 16$ **9.** $3^2 = 9$

10. $2^6 = 64$ **11.** $16^{1/2} = 4$ **12.** $49^{1/2} = 7$

13. $8^{1/3} = 2$ **14.** $16^{1/4} = 2$ **15.** $\left(\dfrac{1}{2}\right)^5 = \dfrac{1}{32}$

16. $\left(\dfrac{1}{3}\right)^4 = \dfrac{1}{81}$ **17.** $2^{-3} = \dfrac{1}{8}$ **18.** $6^{-3} = \dfrac{1}{216}$

19. $4^{-3} = \dfrac{1}{64}$ **20.** $81^{1/2} = 9$ **21.** $64^{1/3} = 4$

22. $5^{-4} = \dfrac{1}{625}$ **23.** $8^{-1/3} = \dfrac{1}{2}$ **24.** $16^{-1/2} = \dfrac{1}{4}$

25. $81^{-1/4} = \dfrac{1}{3}$ **26.** $32^{-1/5} = \dfrac{1}{2}$ **27.** $10^{0.8451} = 7$

28. $10^{1.0792} = 12$ **29.** $e^2 = 7.3891$ **30.** $e^{-1/2} = 0.6065$

31. $a^n = b$ **32.** $c^b = w$

Write each equation in exponential form.

33. $\log_3 9 = 2$ **34.** $\log_4 64 = 3$ **35.** $\log_{1/3} \dfrac{1}{27} = 3$

36. $\log_{1/2} \dfrac{1}{64} = 6$ **37.** $\log_5 \dfrac{1}{25} = -2$ **38.** $\log_5 \dfrac{1}{625} = -4$

39. $\log_{49} 7 = \dfrac{1}{2}$ **40.** $\log_{64} 4 = \dfrac{1}{3}$ **41.** $\log_9 \dfrac{1}{81} = -2$

42. $\log_{10} \dfrac{1}{100} = -2$ **43.** $\log_{10} \dfrac{1}{1000} = -3$ **44.** $\log_{10} 1000 = 3$

45. $\log_6 216 = 3$ **46.** $\log_4 1024 = 5$ **47.** $\log_{10} 0.62 = -0.2076$

48. $\log_{10} 8 = 0.9031$ **49.** $\log_e 6.52 = 1.8749$ **50.** $\log_e 30 = 3.4012$

51. $\log_w s = -p$ **52.** $\log_r c = -a$

Write each equation in exponential form; then find the unknown value.

53. $\log_4 64 = y$ **54.** $\log_5 25 = y$ **55.** $\log_a 125 = 3$ **56.** $\log_a 81 = 4$

57. $\log_3 x = 3$ **58.** $\log_2 x = 5$ **59.** $\log_2 \dfrac{1}{16} = y$ **60.** $\log_8 \dfrac{1}{64} = y$

61. $\log_{1/2} x = 6$ **62.** $\log_{1/3} x = 4$ **63.** $\log_a \dfrac{1}{27} = -3$ **64.** $\log_9 \dfrac{1}{81} = y$

65. $\log_{25} 5 = y$ **66.** $\log_{36} 6 = y$

Evaluate the following.

67. $\log_{10} 1$ **68.** $\log_{10} 10$ **69.** $\log_{10} 100$ **70.** $\log_{10} 1000$

71. $\log_{10} \dfrac{1}{100}$ **72.** $\log_{10} \dfrac{1}{1000}$ **73.** $\log_{10} 10,000$ **74.** $\log_{10} 100,000$

75. $\log_4 256$ **76.** $\log_{13} 169$ **77.** $\log_3 \dfrac{1}{81}$ **78.** $\log_5 \dfrac{1}{125}$

79. $\log_8 \dfrac{1}{64}$ **80.** $\log_{14} \dfrac{1}{14}$ **81.** $\log_7 \sqrt{7}$ **82.** $\log_7 \sqrt[3]{7}$

83. $\log_9 9$ **84.** $\log_{12} 12$ **85.** $\log_{100} 10$ **86.** $\log_{1000} 10$

Graph the logarithmic function.

87. $y = \log_2 x$ **88.** $y = \log_3 x$ **89.** $y = \log_{1/2} x$ **90.** $y = \log_{1/3} x$

91. $y = \log_5 x$ **92.** $y = \log_4 x$ **93.** $y = \log_{1/5} x$ **94.** $y = \log_{1/4} x$

Graph each pair of functions on the same axes.

95. $y = 2^x$, $y = \log_{1/2} x$ **96.** $y = \left(\dfrac{1}{2}\right)^x$, $y = \log_2 x$ **97.** $y = 2^x$, $y = \log_2 x$ **98.** $y = \left(\dfrac{1}{2}\right)^x$, $y = \log_{1/2} x$

Problem Solving

99. Earthquake If the magnitude of an earthquake is 7 on the Richter scale, how many times more intense is the earthquake than the smallest measurable activity? Use $R = \log_{10} I$ (see Example 7).

100. Earthquake If the magnitude of an earthquake is 5 on the Richter scale, how many times more intense is the earthquake than the smallest measurable activity? Use $R = \log_{10} I$.

101. Earthquake How many times more intense is an earthquake that measures 6 on the Richter scale than an earthquake that measures 2?

102. Earthquake How many times more intense is an earthquake that measures 4 on the Richter scale than an earthquake that measures 1?

103. Graph $y = \log_2 (x - 1)$.

104. Graph $y = \log_3 (x - 2)$.

Concept/Writing Exercises

105. Consider the logarithmic function $y = \log_a x$.

  a) What are the restrictions on a?

  b) What is the domain of the function?

  c) What is the range of the function?

106. For the logarithmic function $y = \log_a(x - 3)$, what must be true about x? Explain.

107. If some points on the graph of the exponential function $f(x) = a^x$ are $\left(-3, \frac{1}{27}\right), \left(-2, \frac{1}{9}\right), \left(-1, \frac{1}{3}\right), (0, 1), (1, 3),$ $(2, 9),$ and $(3, 27),$ list some points on the graph of the logarithmic function $g(x) = \log_a x$. Explain how you determined your answer.

108. What is the x-intercept of the graph of an equation of the form $y = \log_a x$?

109. If $f(x) = 5^x$, what is $f^{-1}(x)$?

110. If $f(x) = \log_6 x$, what is $f^{-1}(x)$?

111. Between which two integers must $\log_3 62$ lie? Explain.

112. Between which two integers must $\log_{10} 0.672$ lie? Explain.

113. Between which two integers must $\log_{10} 425$ lie? Explain.

114. Between which two integers must $\log_5 0.3256$ lie? Explain.

115. For $x > 1$, which will grow faster as x increases, 2^x or $\log_{10} x$? Explain.

116. For $x > 1$, which will grow faster as x increases, x or $\log_{10} x$? Explain.

Cumulative Review Exercises

[5.4–5.7]   *Factor.*

117. $2x^3 - 6x^2 - 36x$

118. $x^4 - 16$

119. $40x^2 + 52x - 12$

120. $6r^2s^2 + rs - 1$

9.4 Properties of Logarithms

1 Use the product rule for logarithms.

2 Use the quotient rule for logarithms.

3 Use the power rule for logarithms.

4 Use additional properties of logarithms.

In this section we will study several properties of logarithms. We begin with an important definition.

> **Argument**
>
> In the logarithmic expression $\log_a x$, x is called the **argument** of the logarithm.

Logarithmic Expression	Argument
$\log_{10} 3$	3
$\log_2 (x - 5)$	$x - 5$
$\log_7 (x^2 - 4x + 2)$	$x^2 - 4x + 2$

Since we can only take the logarithm of positive numbers, when an argument contains a variable, we will assume that the argument represents a positive number.

1 Use the Product Rule for Logarithms

Product Rule for Logarithms

For positive real numbers x, y, and $a, a \neq 1$,

$$\log_a xy = \log_a x + \log_a y \qquad \text{Property 1}$$

The logarithm of a product equals the sum of the logarithms of the factors.

Examples of Property 1

$$\log_3 (6 \cdot 7) = \log_3 6 + \log_3 7$$

$$\log_4 3z = \log_4 3 + \log_4 z$$

$$\log_8 x^2 = \log_8 (x \cdot x) = \log_8 x + \log_8 x \text{ or } 2 \log_8 x$$

Property 1, the product rule, can be expanded to three or more factors, for example, $\log_a xyz = \log_a x + \log_a y + \log_a z$.

2 Use the Quotient Rule for Logarithms

Understanding Algebra

Remember that *a logarithm is an exponent*. Thus, the rules of exponents we studied earlier are related to the logarithm rules we present here.

- The exponent rule

$$a^m \cdot a^n = a^{m+n}$$

is related to the logarithm rule

$$\log_a xy = \log_a x + \log_a y$$

- The exponent rule

$$\frac{a^m}{a^n} = a^{m-n}$$

is related to the logarithm rule

$$\log_a \frac{x}{y} = \log_a x - \log_a y$$

- The exponent rule

$$(a^m)^n = a^{m \cdot n}$$

is related to the logarithm rule

$$\log_a x^n = n \log_a x$$

Quotient Rule for Logarithms

For positive real numbers x, y, and, $a, a \neq 1$,

$$\log_a \frac{x}{y} = \log_a x - \log_a y \qquad \text{Property 2}$$

The logarithm of a quotient equals the logarithm of the numerator minus the logarithm of the denominator.

Examples of Property 2

$$\log_3 \frac{19}{4} = \log_3 19 - \log_3 4$$

$$\log_6 \frac{x}{3} = \log_6 x - \log_6 3$$

$$\log_5 \frac{z}{z+2} = \log_5 z - \log_5 (z + 2)$$

3 Use the Power Rule for Logarithms

Power Rule for Logarithms

If x and a are positive real numbers, $a \neq 1$, and n is any real number, then

$$\log_a x^n = n \log_a x \qquad \text{Property 3}$$

The logarithm of a number raised to an exponent equals the exponent times the logarithm of the number.

Examples of Property 3

$$\log_2 4^3 = 3 \log_2 4$$

$$\log_3 x^2 = 2 \log_3 x$$

$$\log_5 \sqrt{12} = \log_5 (12)^{1/2} = \frac{1}{2} \log_5 12$$

$$\log_8 \sqrt[5]{z + 3} = \log_8 (z + 3)^{1/5} = \frac{1}{5} \log_8 (z + 3)$$

EXAMPLE 1 Use properties 1 through 3 to expand.

a) $\log_8 \dfrac{29}{43}$ **b)** $\log_4 (64 \cdot 180)$ **c)** $\log_{10} (22)^{1/5}$

Solution

a) $\log_8 \dfrac{29}{43} = \log_8 29 - \log_8 43$ Quotient rule

b) $\log_4 (64 \cdot 180) = \log_4 64 + \log_4 180$ Product rule

c) $\log_{10} (22)^{1/5} = \dfrac{1}{5} \log_{10} 22$ Power rule

Now Try Exercise 11

Often we will have to use two or more of these properties in the same problem.

EXAMPLE 2 Expand.

a) $\log_{10} 4(x + 2)^3$

b) $\log_5 \dfrac{(4 - a)^2}{3}$

c) $\log_5 \left(\dfrac{4 - a}{3}\right)^2$

d) $\log_5 \dfrac{[x(x + 4)]^3}{8}$

Solution

a) $\log_{10} 4(x + 2)^3 = \log_{10} 4 + \log_{10} (x + 2)^3$ Product rule

$\qquad\qquad\qquad\; = \log_{10} 4 + 3 \log_{10} (x + 2)$ Power rule

b) $\log_5 \dfrac{(4 - a)^2}{3} = \log_5 (4 - a)^2 - \log_5 3$ Quotient rule

$\qquad\qquad\qquad = 2 \log_5 (4 - a) - \log_5 3$ Power rule

c) $\log_5 \left(\dfrac{4 - a}{3}\right)^2 = 2 \log_5 \left(\dfrac{4 - a}{3}\right)$ Power rule

$\qquad\qquad\qquad = 2[\log_5 (4 - a) - \log_5 3]$ Quotient rule

$\qquad\qquad\qquad = 2 \log_5 (4 - a) - 2 \log_5 3$ Distributive property

d) $\log_5 \dfrac{[x(x + 4)]^3}{8} = \log_5 [x(x + 4)]^3 - \log_5 8$ Quotient rule

$\qquad\qquad\qquad = 3 \log_5 x(x + 4) - \log_5 8$ Power rule

$\qquad\qquad\qquad = 3[\log_5 x + \log_5 (x + 4)] - \log_5 8$ Product rule

$\qquad\qquad\qquad = 3 \log_5 x + 3 \log_5 (x + 4) - \log_5 8$ Distributive property

Now Try Exercise 21

Helpful Hint

In Example 2 **b)**, when we expanded $\log_5 \frac{(4-a)^2}{3}$, we first used the quotient rule. In Example 2 **c)**, when we expanded $\log_5 \left(\frac{4-a}{3}\right)^2$, we first used the power rule. Do you see the difference in the two problems? In $\log_5 \frac{(4-a)^2}{3}$, just the numerator of the argument is squared; therefore, we use the quotient rule first. In $\log_5 \left(\frac{4-a}{3}\right)^2$, the entire argument is squared, so we use the power rule first.

EXAMPLE 3   Write each of the following as the logarithm of a single expression.

a) $3 \log_8 (z + 2) - \log_8 z$

b) $\log_7 (x + 1) + 2 \log_7 (x + 4) - 3 \log_7 (x - 5)$

Solution

a) $3 \log_8 (z + 2) - \log_8 z = \log_8 (z + 2)^3 - \log_8 z$ Power rule

$$= \log_8 \frac{(z + 2)^3}{z}$$ Quotient rule

b) $\log_7 (x + 1) + 2 \log_7 (x + 4) - 3 \log_7 (x - 5)$

$= \log_7 (x + 1) + \log_7 (x + 4)^2 - \log_7 (x - 5)^3$ Power rule

$= \log_7 (x + 1)(x + 4)^2 - \log_7 (x - 5)^3$ Product rule

$= \log_7 \dfrac{(x + 1)(x + 4)^2}{(x - 5)^3}$ Quotient rule

Now Try Exercise 39

Understanding Algebra

Our last two logarithm properties are related to the properties of inverse functions. Earlier in this chapter we found that if two functions $f(x)$ and $f^{-1}(x)$ are inverses of each other, then

$$(f^{-1} \circ f)(x) = x$$

and

$$(f \circ f^{-1})(x) = x$$

We also discussed that the functions

$f(x) = a^x$ and $f^{-1}(x) = \log_a x$

are inverse functions. Therefore, we have

$(f^{-1} \circ f)(x) = f^{-1}[f(x)]$

$= f^{-1}(a^x)$

$= \log_a a^x = x$

and

$(f \circ f^{-1})(x) = f[f^{-1}(x)]$

$= f(\log_a x)$

$= a^{\log_a x} = x$

Avoiding Common Errors

THE CORRECT RULES ARE

$$\log_a xy = \log_a x + \log_a y$$

$$\log_a \frac{x}{y} = \log_a x - \log_a y$$

Note that

$\log_a (x + y) \neq \log_a x + \log_a y$ $\log_a xy \neq (\log_a x)(\log_a y)$

$\log_a (x - y) \neq \log_a x - \log_a y$ $\log_a \dfrac{x}{y} \neq \dfrac{\log_a x}{\log_a y}$

4   Use Additional Properties of Logarithms

The last properties we discuss in this section will be used to solve equations in Section 9.6.

Additional Properties of Logarithms

If $a > 0$, and $a \neq 1$, then

$$\log_a a^x = x$$ Property 4

and $a^{\log_a x} = x \, (x > 0)$ Property 5

Examples of Property 4   Examples of Property 5

$\log_6 6^5 = 5$ $3^{\log_3 7} = 7$

$\log_9 9^x = x$ $5^{\log_5 x} = x \, (x > 0)$

EXAMPLE 4 Evaluate. **a)** $\log_5 25$ **b)** $\sqrt{16}^{\log_4 9}$

Solution

a) $\log_5 25$ may be written as $\log_5 5^2$. By property 4,

$$\log_5 25 = \log_5 5^2 = 2$$

b) $\sqrt{16}^{\log_4 9}$ may be written $4^{\log_4 9}$. By property 5,

$$\sqrt{16}^{\log_4 9} = 4^{\log_4 9} = 9$$

Now Try Exercise 55

EXERCISE SET 9.4

Math XL
MathXL®

MyMathLab
MyMathLab

Warm-Up Exercises

Fill in the blanks with the appropriate word, phrase, or symbol(s) from the following list.

times	plus	minus	positive	inverses	argument	sum	composites

1. We can only take the logarithm of _____ numbers.

2. In the logarithmic expression $\log_a x$, the x is called the _____ of the logarithm.

3. The logarithm of a product equals the _____ of the logarithms of the factors.

4. The logarithm of a quotient equals the logarithm of the numerator _____ the logarithm of the denominator.

5. The logarithm of a number raised to an exponent equals the exponent _____ the logarithm of the number.

6. If two functions $f(x)$ and $f^{-1}(x)$ are _____ of each other, then $(f^{-1} \circ f)(x) = x$ and $(f \circ f^{-1})(x) = x$.

Practice the Skills

Use properties 1–3 to expand.

7. $\log_2 (3 \cdot 5)$

8. $\log_3 (2 \cdot 11)$

9. $\log_8 7(x + 3)$

10. $\log_9 x(x + 2)$

11. $\log_2 \dfrac{27}{11}$

12. $\log_5 (41 \cdot 9)$

13. $\log_{10} \dfrac{\sqrt{x}}{x - 9}$

14. $\log_5 3^8$

15. $\log_6 x^7$

16. $\log_9 12(4)^6$

17. $\log_4 (r + 7)^5$

18. $\log_8 b^3(b - 2)$

19. $\log_4 \sqrt{\dfrac{a^3}{a + 2}}$

20. $\log_9 (x - 6)^3 x^2$

21. $\log_3 \dfrac{d^6}{(a - 8)^4}$

22. $\log_7 x^2(x - 13)$

23. $\log_8 \dfrac{y(y + 4)}{y^3}$

24. $\log_{10} \left(\dfrac{z}{6}\right)^2$

25. $\log_{10} \dfrac{9m}{8n}$

26. $\log_5 \dfrac{\sqrt{a}\sqrt[3]{b}}{\sqrt[4]{c}}$

Write as a logarithm of a single expression.

27. $\log_2 3 + \log_2 7$

28. $\log_7 4 + \log_7 3$

29. $\log_2 9 - \log_2 5$

30. $\log_7 17 - \log_7 3$

31. $6 \log_4 2$

32. $\dfrac{1}{3} \log_8 7$

33. $\log_{10} x + \log_{10} (x + 3)$

34. $\log_5 (a + 1) - \log_5 (a + 10)$

35. $2 \log_9 z - \log_9 (z - 2)$

36. $3 \log_8 y + 2 \log_8 (y - 9)$

37. $4(\log_5 p - \log_5 3)$

38. $\dfrac{1}{2} [\log_6 (r - 1) - \log_6 r]$

39. $\log_{\prime} n + \log_{\prime}(n + 4) - \log_{\prime}(n - 3)$

40. $2\log_5 t + 5\log_5(t - 6) + \log_5(3t + 7)$

41. $\dfrac{1}{2}[\log_5(x - 8) - \log_5 x]$

42. $6\log_7(a + 3) + 2\log_7(a - 1) - \dfrac{1}{2}\log_7 a$

43. $2\log_9 4 + \dfrac{1}{3}\log_9(r - 6) - \dfrac{1}{2}\log_9 r$

44. $5\log_6(x + 3) - [2\log_6(7x + 1) + 3\log_6 x]$

45. $4\log_6 3 - [2\log_6(x + 3) + 4\log_6 x]$

46. $2\log_7(m - 4) + 3\log_7(m + 3) - [5\log_7 2 + 3\log_7(m - 2)]$

Find the value by writing each argument using the numbers 2 and/or 5 and using the values $\log_a 2 = 0.3010$ *and* $\log_a 5 = 0.6990$.

47. $\log_a 10$

48. $\log_a 2.5$

49. $\log_a 0.4$

50. $\log_a \dfrac{1}{8}$

51. $\log_a 25$

52. $\log_a \sqrt[3]{5}$

Evaluate (see Example 4).

53. $7^{\log_7 2}$

54. $\log_4 4$

55. $(2^3)^{\log_8 7}$

56. $\log_8 64$

57. $\log_3 27$

58. $2\log_9 \sqrt{9}$

59. $5(\sqrt[3]{27})^{\log_3 5}$

60. $\dfrac{1}{2}\log_6 \sqrt[3]{6}$

Problem Solving

61. Express $\log_a(x^2 - 4) - \log_a(x + 2)$ as a single logarithm and simplify.

62. Express $\log_a(x - 3) - \log_a(x^2 + 5x - 24)$ as a single logarithm and simplify.

Use properties 1–3 to expand.

63. $\log_2 \dfrac{\sqrt[4]{xy}\,\sqrt[3]{a}}{\sqrt[5]{a - b}}$

64. $\log_3\left[\dfrac{(a^2 + b^2)(c^2)}{(a - b)(b + c)(c + d)}\right]^2$

If $\log_{10} x = 0.4320$, *find the following.*

65. $\log_{10} x^2$

66. $\log_{10} \sqrt[3]{x}$

67. $\log_{10} \sqrt[4]{x}$

68. $\log_{10} x^{11}$

If $\log_{10} x = 0.5000$ *and* $\log_{10} y = 0.2000$, *find the following.*

69. $\log_{10} xy$

70. $\log_{10}\left(\dfrac{x}{y}\right)$

71. Using the information given in the instructions for Exercises 69 and 70, is it possible to find $\log_{10}(x + y)$? Explain.

72. Are the graphs of $y = \log_b x^2$ and $y = 2\log_b x$ the same? Explain your answer by discussing the domain of each equation.

Concept/Writing Exercises

73. Is $\log_a(xyz) = \log_a x + \log_a y + \log_a z$? Explain.

74. Is $\log_b(x + y + z) = \log_b x + \log_b y + \log_b z$? Explain.

75. Is $\log_a(x^2 + 8x + 16) = 2\log_a(x + 4)$? Explain.

76. Is $\log_a(4x^2 - 20x + 25) = 2\log_a(2x - 5)$? Explain.

Challenge Problems

77. For $x > 0$ and $y > 0$, is $\log_a \dfrac{x}{y} = \log_a xy^{-1}$

$= \log_a x + \log_a y^{-1} = \log_a x + \log_a \dfrac{1}{y}$?

78. Read Exercise 77. By the quotient rule, $\log_a \dfrac{x}{y} = \log_a x - \log_a y$. Can we therefore conclude that $\log_a x - \log_a y = \log_a x + \log_a \dfrac{1}{y}$?

79. Use the product rule to show that

$$\log_a \dfrac{x}{y} = \log_a x + \log_a \dfrac{1}{y}$$

80. a) Explain why

$$\log_a \dfrac{3}{xy} \neq \log_a 3 - \log_a x + \log_a y$$

b) Expand $\log_a \dfrac{3}{xy}$ correctly.

Group Activity

Discuss and answer Exercise 81 as a group.

81. Consider $\log_a \dfrac{\sqrt{x^4 y}}{\sqrt{xy^3}}$, where $x > 0$ and $y > 0$.

a) Group member 1: Expand the expression using the quotient rule.

b) Group member 2: Expand the expression using the product rule.

c) Group member 3: First simplify $\dfrac{\sqrt{x^4 y}}{\sqrt{xy^3}}$, then expand the resulting logarithm.

d) Check each other's work and make sure all answers are correct. Can this expression be simplified by all three methods?

Cumulative Review Exercises

[2.5] **82.** Solve the inequality $\dfrac{x-4}{2} - \dfrac{2x-5}{5} > 3$ and indicate the solution in

    a) set builder notation.

    b) interval notation.

[5.7] **83. a)** Write an expression for the shaded area of the figure.

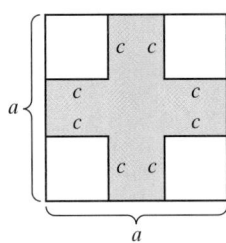

    b) Write the expression in part **a)** in factored form.

[6.4] **84.** Solve $\dfrac{15}{x} + \dfrac{9x-7}{x+2} = 9$ for x.

[7.7] **85.** Multiply $(3i+4)(2i-5)$.

[8.4] **86.** Solve $a - 6\sqrt{a} = 7$ for a.

Mid-Chapter Test: 9.1–9.4

To find out how well you understand the chapter material to this point, take this brief test. The answers, and the section where the material was initially discussed, are given in the back of the book. Review any questions you answered incorrectly.

1. a) Explain how to find $(f \circ g)(x)$.

    b) If $f(x) = 3x + 3$ and $g(x) = 2x + 5$, find $(f \circ g)(x)$.

2. Let $f(x) = x^2 + 5$ and $g(x) = \dfrac{6}{x}$; find

    a) $(f \circ g)(x)$

    b) $(f \circ g)(3)$

    c) $(g \circ f)(x)$

    d) $(g \circ f)(3)$

3. a) Explain what it means when a function is a one-to-one function.

    b) Is the function represented by the following graph a one-to-one function? Explain.

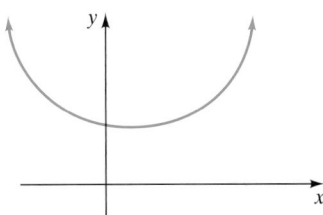

In Exercises 4–6, for each function, **a)** *determine whether it is a one-to-one function;* **b)** *if it is a one-to-one function, find its inverse function.*

4. $\{(-3, 2), (2, 3), (5, 1), (6, 8)\}$

5. $p(x) = \dfrac{1}{3}x - 5$

6. $k(x) = \sqrt{x - 4}, \quad x \ge 4$

7. Let $m(x) = -2x + 4$. Find $m^{-1}(x)$ and then graph $m(x)$ and $m^{-1}(x)$ on the same axes.

Graph each exponential function.

8. $y = 2^x$

9. $y = 3^{-x}$

10. Graph the logarithmic function $y = \log_2 x$.

11. Bacteria The number of bacteria in a petri-dish is $N(t) = 5(2)^t$, where t is the number of hours after the 5 original bacteria are placed in the dish. How many bacteria are in the dish

    a) 1 hour later?

    b) 6 hours later?

12. Write $27^{2/3} = 9$ in logarithmic form.

13. Write $\log_2 \dfrac{1}{64} = -6$ in exponential form.

14. Evaluate $\log_5 125$.

15. Solve the equation $\log_{1/4} \dfrac{1}{16} = x$ for x.

16. Solve the equation $\log_x 64 = 3$ for x.

Use properties 1–3 to write as a sum or difference of logarithms.

17. $\log_9 x^2(x - 5)$

18. $\log_5 \dfrac{7m}{\sqrt{n}}$

Write as a single logarithm.

19. $3 \log_2 x + \log_2 (x + 7) - 4 \log_2 (x + 1)$

20. $\dfrac{1}{2}[\log_7 (x + 2) - \log_7 x]$

9.5 Common Logarithms

1 Find common logarithms of powers of 10.

2 Approximate common logarithms.

3 Approximate powers of 10.

The properties of logarithms discussed in Section 9.4 apply to any logarithms with a real number base a with $a > 0$ and $a \neq 1$. Since our number system is based on the number 10, we often use logarithms with a base of 10, which are called *common logarithms*.

Common Logarithm

A **common logarithm** is a logarithm with a base of 10. When the base of a logarithm is not indicated, we assume the base is 10. Thus

$$\log x = \log_{10} x \qquad (x > 0)$$

The five properties we studied in Section 9.4 can be rewritten as properties of common logarithms.

Common Logarithm Properties

1. $\log xy = \log x + \log y$

2. $\log \dfrac{x}{y} = \log x - \log y$

3. $\log x^n = n \log x$

4. $\log 10^x = x$

5. $10^{\log x} = x$

Understanding Algebra

Remember that *a logarithm is an exponent.* A common logarithm is the exponent to which you would raise 10 in order to get the argument. For example, log 100 is the exponent to which you would raise 10 in order to get 100. Thus, $\log 100 = 2$.

1 Find Common Logarithms of Powers of 10

We can use the fourth property of common logarithms to evaluate common logarithms of numbers that are powers of 10. We begin with some examples of common logarithms of non-negative powers of 10.

Common Logarithms of Non-Negative Powers of 10

$\log 1 = \log 10^0 = 0$ $\qquad\qquad$ $\log 10 = \log 10^1 = 1$

$\log 100 = \log 10^2 = 2$ $\qquad\qquad$ $\log 1000 = \log 10^3 = 3$

$\log 10{,}000 = \log 10^4 = 4$ $\qquad\quad$ $\log 100{,}000 = \log 10^5 = 5$

Understanding Algebra

To find common logarithms of 10 raised to a negative power, we need to remember the following rule of exponents:

$$a^{-m} = \frac{1}{a^m}.$$

We can also evaluate common logarithms of negative powers of 10.

Common Logarithms of Negative Powers of 10

$$\log 0.1 = \log \frac{1}{10} = \log 10^{-1} = -1$$

$$\log 0.01 = \log \frac{1}{100} = \log 10^{-2} = -2$$

$$\log 0.001 = \log \frac{1}{1000} = \log 10^{-3} = -3$$

$$\log 0.0001 = \log \frac{1}{10{,}000} = \log 10^{-4} = -4$$

$$\log 0.00001 = \log \frac{1}{100{,}000} = \log 10^{-5} = -5$$

2 Approximate Common Logarithms

We will use a scientific or graphing calculator to approximate most common logarithms. Before we do this, we introduce a method to estimate the value of a common logarithm between two integers. For example, suppose we wish to estimate the value of log 5. Since 5 is between 1 and 10, we can conclude that log 5 is between log 1 and log 10 and we see the following:

$$1 < 5 < 10$$

$$\log 1 < \log 5 < \log 10$$

$$0 < \log 5 < 1$$

Thus we can conclude that log 5 is a number between 0 and 1.

EXAMPLE 1 Without using a calculator, estimate between which two integers each common logarithm will be.

a) log 82 **b)** log 5091

c) log 0.7 **d)** log 0.03

Solution

a) Since 82 is between 10 and 100, we have the following

$$10 < 82 < 100$$

$$\log 10 < \log 82 < \log 100$$

$$1 < \log 82 < 2$$

Thus, log 82 is a number between 1 and 2.

b) Since 5091 is between 1000 and 10,000, we have the following

$$1000 < 5091 < 10{,}000$$

$$\log 1000 < \log 5091 < \log 10{,}000$$

$$3 < \log 5091 < 4$$

Thus, log 5091 is a number between 3 and 4.

c) Since 0.7 is between 0.1 and 1, we have the following

$$0.1 < 0.7 < 1$$

$$\log 0.1 < \log 0.7 < \log 1$$

$$-1 < \log 0.7 < 0$$

Thus, log 0.7 is a number between −1 and 0.

d) Since 0.03 is between 0.01 and 0.1, we have the following

$$0.01 < 0.03 < 0.1$$

$$\log 0.01 < \log 0.03 < \log 0.1$$

$$-2 < \log 0.03 < -1$$

Thus, log 0.03 is a number between −2 and −1.

Now Try Exercise 15

We now will approximate common logarithms using a scientific or graphing calculator using the key $\boxed{\text{LOG}}$ as shown on page 604.

Using Your Calculator ▪▪▪▪▪

Approximating Common Logarithms

Scientific Calculator
To approximate common logarithms on many scientific calculators, enter the argument, then press the logarithm key.

EXAMPLE	KEYS TO PRESS	ANSWER DISPLAYED
Approximate log 400.	400 [LOG]	2.60206

Graphing Calculator
On graphing calculators and on many scientific calculators, you first press the [LOG] key and then you enter the number. For example, on the TI-84 Plus, you would do the following:

EXAMPLE	KEYS TO PRESS	ANSWER DISPLAYED
Approximate log 400.	[LOG] (400 [)] [ENTER]	2.602059991

↑
Generated by calculator

EXAMPLE 2 Use a calculator to approximate the following common logarithms. Round your answers to four decimal places. Compare your answers to those from Example 1 on page 603.

a) $\log 82$ **b)** $\log 5091$

c) $\log 0.7$ **d)** $\log 0.03$

Solution   Using a calculator as shown above, we obtain the following.

a) $\log 82 \approx 1.9138$. Note in Example 1 **a)** we correctly estimated that $\log 82$ is a number between 1 and 2.

b) $\log 5091 \approx 3.7068$. Note in Example 1 **b)** we correctly estimated that $\log 5091$ is a number between 3 and 4.

c) $\log 0.7 \approx -0.1549$. Note in Example 1 **c)** we correctly estimated that $\log 0.7$ is a number between -1 and 0.

d) $\log 0.03 \approx -1.5229$. Note in Example 1 **d)** we correctly estimated that $\log 0.03$ is a number between -2 and -1.

Now Try Exercise 27

Recall from our definition of logarithm from Section 9.3 that $y = \log_a x$ means that $x = a^y$. We restate this definition for common logarithms.

Definition of a Common Logarithm

For all positive numbers x

$$y = \log x \quad \text{means} \quad x = 10^y$$

The **common logarithm** of a positive number x is the exponent to which the base 10 must be raised to obtain the number x.

EXAMPLE 3 Find the exponent to which 10 must be raised to obtain each of the following numbers. Round your answer to four decimal places.

a) 75   **b)** 3594   **c)** 0.00324

Solution   We are asked to find the *exponent* of 10. Using the definition of a common logarithm, we can see that we are asked to find the common logarithm of each of these numbers.

a) $\log 75 \approx 1.8751$ Note: $10^{1.8751} \approx 75$

b) $\log 3594 \approx 3.5556$ Note: $10^{3.5556} \approx 3594$

c) $\log 0.00324 \approx -2.4895$ Note: $10^{-2.4895} \approx 0.00324$

Now Try Exercise 37

3 Approximate Powers of 10

While solving equations that involve common logarithms, we will often need to evaluate a power of 10. For example, if $\log x = 3$ then, using the definition of a common logarithm, we see that $x = 10^3$ or 1000. In many equations the power of 10 will not be an integer. We will use a scientific or graphing calculator to approximate such powers of 10.

Using Your Calculator

Approximating Powers of 10

To approximate powers of 10 on your calculator, use the 10^x function which is located directly above the $\boxed{\text{LOG}}$ key. To access this function, press the $\boxed{\text{2ND}}$, $\boxed{\text{INV}}$, or $\boxed{\text{Shift}}$ key prior to pressing the $\boxed{\text{LOG}}$ key.

EXAMPLE 4 Approximate the following powers of 10. Round your answers to four decimal places.

a) $10^{1.394}$ **b)** $10^{2.827}$ **c)** $10^{-0.356}$

Solution   Use a calculator to approximate each power of 10.

a) $10^{1.394} \approx 24.7742$

b) $10^{2.827} \approx 671.4289$

c) $10^{-0.356} \approx 0.4406$

Now Try Exercise 49

When we evaluate a power of 10, the number obtained may be referred to as an **antilogarithm** or an **antilog**. For example, in Example 4 **a)**, we found that $10^{1.394} \approx 24.7742$. Thus, we may write antilog $1.394 \approx 24.7742$. Note that $\log 24.7742 \approx 1.394$.

EXAMPLE 5 Solve for x in each of the following equations. Round your answers to four decimal places.

a) $\log x = 0.132$ **b)** $\log x = -1.203$

Solution

a) Using the definition of a common logarithm, we know that

$$\log x = 0.132 \quad \text{means} \quad x = 10^{0.132} \approx 1.3552$$

b) $$\log x = -1.203 \quad \text{means} \quad x = 10^{-1.203} \approx 0.0627$$

Now Try Exercise 65

In Example 5 **a)**, we solved the equation $\log x = 0.132$ to get the solution $x = 10^{0.132} \approx 1.3552$. We may also have written $x =$ antilog $0.132 \approx 1.3552$. Note that $\log 1.3552 \approx 0.132$.

EXAMPLE 6 **Earthquake** The magnitude of an earthquake on the Richter scale is given by the formula $R = \log I$, where I is the number of times more intense the quake is than the smallest measurable activity. How many times more intense is an earthquake measuring 6.2 on the Richter scale than the smallest measurable activity?

Solution We want to find the value for I. We are given that $R = 6.2$. Substitute 6.2 for R in the formula $R = \log I$, and then solve for I.

$$R = \log I$$
$$6.2 = \log I \quad \text{Substitute 6.2 for } R.$$

To find I, we will rewrite the logarithmic equation as an exponential equation using the definition of a common logarithm.

$$6.2 = \log I \quad \text{means} \quad I = 10^{6.2}$$
$$\text{and we get} \quad I \approx 1{,}584{,}893.$$

Thus, this earthquake is about 1,584,893 times more intense than the smallest measurable activity.

Now Try Exercise 89

EXERCISE SET 9.5

Math XL
MathXL®

MyMathLab
MyMathLab

Warm-Up Exercises

Fill in the blanks with the appropriate word, phrase, or symbol(s) from the following list.

10^y	common	4	y^{10}	exponent	2	10^x

1. A _____ logarithm is a logarithm with a base of 10.

2. A common logarithm is the _____ to which you would raise 10 to get the argument.

3. Using the definition of a common logarithm, $y = \log x$ means $x =$ _____.

4. Since 2010 is between 1000 and 10,000, log 2010 is between 3 and _____.

Practice the Skills

Evaluate the common logarithm of each power of 10 without the use of a calculator.

5. log 1
6. log 100
7. log 0.1
8. log 1000
9. log 0.01
10. log 10
11. log 0.001
12. 0.0001

Without using a calculator, estimate between which two integers each logarithm will be. See Example 1.

13. 86
14. 352
15. 19,200
16. 1001
17. 0.0613
18. 941,000
19. 101
20. 0.000835
21. 3.75
22. 0.375
23. 0.0173
24. 0.00872

Use a calculator to approximate the following common logarithms. Round your answers to four decimal places. Compare your answers to those from Exercises 13–24, See Example 2.

25. 86
26. 352
27. 19,200
28. 1001
29. 0.0613
30. 941,000
31. 101
32. 0.000835
33. 3.75
34. 0.375
35. 0.0173
36. 0.00872

Find the exponent to which 10 must be raised to obtain each of the following numbers. Round your answers to four decimal places. See Example 3.

37. 3560
38. 817,000
39. 0.0727
40. 0.00612
41. 243
42. 8.16
43. 0.00592
44. 73,700,000
45. 0.0098
46. 0.0037
47. 15.491
48. 10.892

Find the following powers of 10. Round your answers to four decimal places. See Example 4.

49. $10^{0.2137}$
50. $10^{1.3845}$
51. $10^{4.6283}$
52. $10^{3.5527}$
53. $10^{-1.7086}$
54. $10^{-2.7431}$
55. $10^{0.001}$
56. $10^{-0.001}$
57. $10^{2.7625}$
58. $10^{-0.1543}$
59. $10^{-2.014}$
60. $10^{5.5922}$

Solve for x in each of the following equations. If necessary, round your answers to four decimal places. See Example 5.

61. $\log x = 2.0000$
62. $\log x = 1.4612$
63. $\log x = 3.3817$
64. $\log x = 1.9330$
65. $\log x = 4.1409$
66. $\log x = -2.103$
67. $\log x = -1.06$
68. $\log x = -3.1469$
69. $\log x = -0.6218$
70. $\log x = 1.5177$
71. $\log x = -0.1256$
72. $\log x = -1.3206$

Use the common logarithm properties 4 and 5 on page 602 to evaluate the following.

73. $\log 10^7$

74. $\log 10^{3.4}$

75. $10^{\log 7}$

76. $10^{\log 3.4}$

77. $4 \log 10^{5.2}$

78. $8 \log 10^{1.2}$

79. $5(10^{\log 8.3})$

80. $2.3(10^{\log 5.2})$

Problem Solving

Solve Exercises 81–84 using $R = \log I$ (see Example 6). Round your answers to four decimal places.

81. Find I if $R = 3.4$

82. Find I if $R = 4.9$

83. Find I if $R = 5.7$

84. Find I if $R = 0.1$

85. Astronomy In astronomy, a formula used to find the diameter, in kilometers, of minor planets (also called asteroids) is $\log d = 3.7 - 0.2g$, where g is a quantity called the absolute magnitude of the minor planet. Find the diameter of a minor planet if its absolute magnitude is **a)** 11 and **b)** 20. **c)** Find the absolute magnitude of the minor planet whose diameter is 5.8 kilometers.

© serjoe/Shutterstock

86. Standardized Test The average score on a standardized test is a function of the number of hours studied for the test. The average score, $f(x)$, in points, can be approximated by $f(x) = \log 0.3x + 1.8$, where x is the number of hours studied for the test. The maximum possible score on the test is 4.0. Find the score received by the average person who studied for **a)** 15 hours. **b)** 55 hours.

87. Learning Retention Sammy Barcia just finished a course in physics. The percent of the course he will remember t months later can be approximated by the function

$$R(t) = 94 - 46.8 \log (t + 1)$$

for $0 \le t \le 48$. Find the percent of the course Sammy will remember **a)** 2 months later. **b)** 48 months later.

88. Learning Retention Karen Frye just finished a course in psychology. The percent of the course she will remember t months later can be approximated by the function

$$R(t) = 85 - 41.9 \log (t + 1)$$

for $0 \le t \le 48$. Find the percent of the course she will remember **a)** 10 months later. **b)** 25 months later.

89. Earthquake How many times more intense is an earthquake having a Richter scale number of 3.8 than the smallest measurable activity? See Example 6.

90. Earthquake The strongest earthquake ever recorded took place in Chile on May 22, 1960. It measured 9.5 on the Richter scale. How many times more intense was this earthquake than the smallest measurable activity?

91. Energy of an Earthquake A formula sometimes used to estimate the seismic energy released by an earthquake is $\log E = 11.8 + 1.5m_s$, where E is the seismic energy and m_s is the surface wave magnitude.

a) Find the energy released in an earthquake whose surface wave magnitude is 6.

b) If the energy released during an earthquake is 1.2×10^{15}, what is the magnitude of the surface wave?

92. Sound Pressure The sound pressure level, s_p, is given by the formula $s_p = 20 \log \dfrac{p_r}{0.0002}$, where p_r is the sound pressure in dynes/cm^2.

a) Find the sound pressure level if the sound pressure is 0.0036 dynes/cm^2

b) If the sound pressure level is 10.0, find the sound pressure.

93. Earthquake The Richter scale, used to measure the strength of earthquakes, relates the magnitude, M, of the earthquake to the release of energy, E, in ergs, by the formula

$$M = \frac{\log E - 11.8}{1.5}$$

An earthquake releases 1.259×10^{21} ergs of energy. What is the magnitude of such an earthquake on the Richter scale?

94. pH of a Solution The pH is a measure of the acidity or alkalinity of a solution. The pH of water, for example, is 7. In general, acids have pH numbers less than 7 and alkaline solutions have pH numbers greater than 7. The pH of a solution is defined as pH $= -\log[H_3O^+]$, where H_3O^+ represents the hydronium ion concentration of the solution. Find the pH of a solution whose hydronium ion concentration is 2.8×10^{-3}.

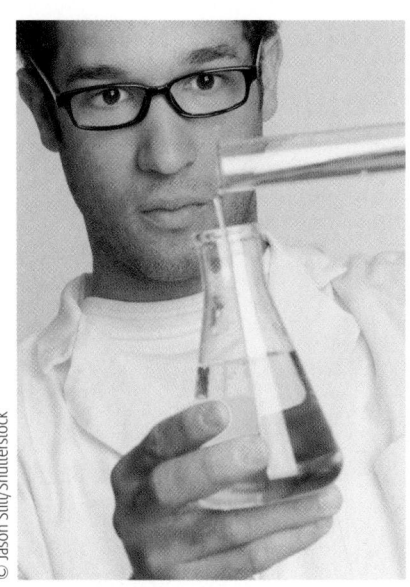

© Jason Stitt/Shutterstock

Concept/Writing Exercises

95. On your calculator, you find $\log 462$ and obtain the value 1.6646. Can this value be correct? Explain.

96. On your calculator, you find $\log 6250$ and obtain the value 2.7589. Can this value be correct? Explain.

97. On your calculator, you find $\log 0.163$ and obtain the value -2.7878. Can this value be correct? Explain.

98. On your calculator, you find $\log(-1.23)$ and obtain the value 0.08991. Can this value be correct? Explain.

99. Is $\log \dfrac{y}{4x} = \log y - \log 4 + \log x$? Explain.

100. Is $\log \dfrac{5x^2}{3} = 2(\log 5 + \log x) - \log 3$? Explain.

Challenge Problems

101. Solve the formula $R = \log I$ for I.

102. Solve the formula $\log E = 11.8 + 1.5m$ for E.

103. Solve the formula $R = 26 - 41.9 \log(t + 1)$ for t.

104. Solve the formula $f = 76 - \log x$ for x.

Group Activity

105. In Section 9.7, we introduce the *change of base formula*, $\log_a x = \dfrac{\log_b x}{\log_b a}$, where a and b are bases and x is a positive number.

    a) Group member 1: Use the change of base formula to evaluate $\log_3 45$. (*Hint:* Let $b = 10$.)

    b) Group member 2: Repeat part **a)** for $\log_5 30$.

    c) Group member 3: Repeat part **a)** for $\log_6 40$.

    d) As a group, use the fact that $\log_a x = \dfrac{\log_b x}{\log_b a}$, where $b = 10$, to graph the equation $y = \log_2 x$ for $x > 0$. Use a graphing calculator if available.

Cumulative Review Exercises

[4.3] **106. Cars** Two cars start at the same point in Alexandria, Virginia, and travel in opposite directions. One car travels 5 miles per hour faster than the other car. After 4 hours, the two cars are 420 miles apart. Find the speed of each car.

[4.5] **107.** Solve the system of equations.

$$3r = -4s - 6$$
$$3s = -5r + 1$$

[5.8] **108.** Solve $3x^3 + 3x^2 - 36x = 0$ for x.

[7.1] **109.** Write $\sqrt{(3x^2 - y)^2}$ as an absolute value.

[8.6] **110.** Solve $(x - 5)(x + 4)(x - 2) \le 0$ and give the solution in interval notation.

9.6 Exponential and Logarithmic Equations

1 Solve exponential and logarithmic equations.

2 Solve applications.

1 Solve Exponential and Logarithmic Equations

In this section, we will further study *exponential equations* and *logarithmic equations*. Below we list some of the properties we will use when solving such equations.

Properties for Solving Exponential and Logarithmic Equations

    a. If $x = y$, then $a^x = a^y$.

    b. If $a^x = a^y$, then $x = y$.

    c. If $x = y$, then $\log_b x = \log_b y$ $(x > 0, y > 0)$.

    d. If $\log_b x = \log_b y$, then $x = y$ $(x > 0, y > 0)$.   Properties 6a–6d

---

EXAMPLE 1   Solve the equation $8^x = \dfrac{1}{2}$.

Solution   To solve this equation, we will write both sides of the equation with the same base, 2, and then use property 6b.

$$8^x = \frac{1}{2}$$

$$(2^3)^x = \frac{1}{2} \qquad \text{Write 8 as } 2^3.$$

$$2^{3x} = 2^{-1} \qquad \text{Write } \frac{1}{2} \text{ as } 2^{-1}.$$

Using property 6b, we can write

$$3x = -1$$

$$x = -\frac{1}{3}$$

Now Try Exercise 7

When both sides of the exponential equation cannot be written as a power of the same base, we often begin by taking the logarithm of both sides of the equation, as in Example 2. In the following examples, we will round logarithms to four decimal places.

EXAMPLE 2   Solve the equation $5^n = 28$.

Solution   Take the logarithm of both sides of the equation and solve for n.

$$\log 5^n = \log 28$$

$$n \log 5 = \log 28 \qquad \qquad \text{Power rule}$$

$$n = \frac{\log 28}{\log 5} \qquad \qquad \text{Divide both sides by } \log 5.$$

$$\approx \frac{1.4472}{0.6990} \approx 2.0704$$

Now Try Exercise 23

Some logarithmic equations can be solved by expressing the equation in exponential form. *It is necessary to check logarithmic equations for extraneous solutions.* When checking a solution, if you obtain the logarithm of a nonpositive number, the solution is extraneous.

EXAMPLE 3   Solve the equation $\log_2 (x + 3)^3 = 4$.

Solution   Write the equation in exponential form.

$$(x + 3)^3 = 2^4 \qquad \text{Write in exponential form.}$$

$$(x + 3)^3 = 16$$

$$x + 3 = \sqrt[3]{16} \qquad \text{Take the cube root of both sides.}$$

$$x = -3 + \sqrt[3]{16} \qquad \text{Solve for } x.$$

Check

$$\log_2 (x + 3)^3 = 4$$

$$\log_2 [(-3 + \sqrt[3]{16}) + 3]^3 \overset{?}{=} 4$$

$$\log_2 (\sqrt[3]{16})^3 \overset{?}{=} 4$$

$$\log_2 16 \overset{?}{=} 4 \qquad (\sqrt[3]{16})^3 = 16$$

$$2^4 \overset{?}{=} 16 \qquad \text{Write in exponential form.}$$

$$16 = 16 \qquad \text{True}$$

Now Try Exercise 43

Other logarithmic equations can be solved using the properties of logarithms given in earlier sections.

EXAMPLE 4 Solve the equation $\log(3x + 2) + \log 9 = \log(x + 5)$.

Solution

$$\log(3x + 2) + \log 9 = \log(x + 5)$$
$$\log[(3x + 2)(9)] = \log(x + 5) \quad \text{Product rule}$$
$$(3x + 2)(9) = (x + 5) \quad \text{Property 6d}$$
$$27x + 18 = x + 5$$
$$26x + 18 = 5$$
$$26x = -13$$
$$x = -\frac{1}{2}$$

Check for yourself that the solution is $-\frac{1}{2}$.

Now Try Exercise 51

EXAMPLE 5 Solve the equation $\log x + \log(x + 1) = \log 12$.

Solution

$$\log x + \log(x + 1) = \log 12$$
$$\log x(x + 1) = \log 12 \quad \text{Product rule}$$
$$x(x + 1) = 12 \quad \text{Property 6d}$$
$$x^2 + x = 12$$
$$x^2 + x - 12 = 0$$
$$(x + 4)(x - 3) = 0$$
$$x + 4 = 0 \quad \text{or} \quad x - 3 = 0$$
$$x = -4 \qquad\qquad x = 3$$

Understanding Algebra

Recall from Section 9.3 that the domain of logarithmic functions is $(0, \infty)$. In other words, *we cannot take the logarithm of zero or a negative number.* Thus, we must check our results from solving logarithmic equations to make sure the argument of the logarithm is positive.

Check

$x = -4$

$$\log x + \log(x + 1) = \log 12$$
$$\log(-4) + \log(-4 + 1) \overset{?}{=} \log 12$$
$$\log(-4) + \log(-3) \overset{?}{=} \log 12$$

Stop.   ↑   ↑
Logarithms of negative numbers are not real numbers.

$x = 3$

$$\log x + \log(x + 1) = \log 12$$
$$\log 3 + \log(3 + 1) \overset{?}{=} \log 12$$
$$\log 3 + \log 4 \overset{?}{=} \log 12$$
$$\log[(3)(4)] \overset{?}{=} \log 12$$
$$\log 12 = \log 12 \quad \text{True}$$

Thus, -4 is an extraneous solution. The only solution is 3.

Now Try Exercise 65

Using Your Graphing Calculator

Previously, we have shown how to solve equations in one variable using a graphing calculator. To solve the equation $\log x + \log(x + 1) = \log 12$ (see Example 5) using a TI-84 plus, we can graph

$Y_1 = \log(x) + \log(x + 1)$
$Y_2 = \log(12)$

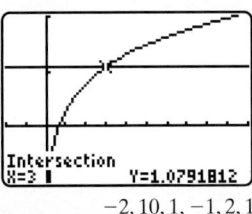

Intersection
X=3   Y=1.0791812

$-2, 10, 1, -1, 2, 1$

FIGURE 9.21

The point of intersection of these graphs can be found using the *intersect* function from the CALC menu.
Figure 9.21 shows that the x-coordinate of the point of intersection, 3, is the solution to the equation.

EXAMPLE 6 Solve the equation $\log(5x - 3) - \log(2x) = 1$.

Solution

$$\log(5x - 3) - \log(2x) = 1$$

$$\log\left(\frac{5x - 3}{2x}\right) = 1 \qquad \text{Quotient rule}$$

$$\left(\frac{5x - 3}{2x}\right) = 10^1 \qquad \text{Write in exponential form.}$$

$$\frac{5x - 3}{2x} = 10$$

$$5x - 3 = 20x \qquad \text{Multiply both sides by } 2x.$$

$$-3 = 15x \qquad \text{Subtract } 5x \text{ from both sides.}$$

$$x = -\frac{3}{15} = -\frac{1}{5} = -0.2$$

Check

$$\log(5x - 3) - \log(2x) = 1$$

$$\log[5(-0.2) - 3] - \log[2(-0.2)] = 1$$

$$\log(-4) - \log(-0.4) = 1$$

Since we get the logarithms of negative numbers, -0.2 is an extraneous solution. Thus, the equation has no solution. Its solution set is the empty set, $\varnothing$.

Now Try Exercise 57

2 Solve Applications

Now we will look at an application that involves an exponential equation.

EXAMPLE 7 **Bacteria** If there are initially 1000 bacteria in a culture, and the number of bacteria doubles each hour, the number of bacteria after t hours can be found by the formula

$$N = 1000(2)^t$$

How long will it take for the culture to grow to 30,000 bacteria?

Solution

$$N = 1000(2)^t$$

$$30,000 = 1000(2)^t \qquad \text{Substitute 30,000 for } N.$$

$$30 = (2)^t \qquad \text{Divide both sides by 1000.}$$

We want to find the value for t. To accomplish this we will use logarithms. Begin by taking the logarithm of both sides of the equation.

$$\log 30 = \log(2)^t$$

$$\log 30 = t \log 2 \qquad \text{Power rule}$$

$$\frac{\log 30}{\log 2} = t \qquad \text{Divide both sides by log 2.}$$

$$\frac{1.4771}{0.3010} \approx t$$

$$4.91 \approx t$$

It will take about 4.91 hours for the culture to grow to 30,000 bacteria.

Now Try Exercise 69

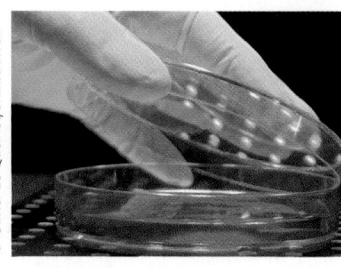
© Olivier Le Queinec/Shutterstock

EXERCISE SET 9.6

Math XL MyMathLab
MathXL®   MyMathLab

Warm-Up Exercises

Fill in the blanks with the appropriate word, phrase, or symbol(s) from the following list.

| extraneous | exponential | base | logarithm | quotient | sum | product | difference |

1. To solve the equation $9^x = \dfrac{1}{3}$, write both sides with the same _____, 3, and then use property 6b.

2. To solve the equation $3^x = 41$, take the common _____ of both sides.

3. It is necessary to check logarithmic equations for _____ solutions.

4. To solve the equation $\log_3 (x + 1) = 2$, first write the equation in _____ form.

5. To solve the equation $\log_3 x + \log_3 (x - 2) = 1$, first rewrite the left side of the equation as a single logarithm using the _____ rule for logarithms.

6. To solve the equation $\log_3 x - \log_3 (x - 2) = 1$, first rewrite the left side of the equation as a single logarithm using the _____ rule for logarithms.

Practice the Skills

Solve each exponential equation without using a calculator.

7. $5^x = 125$

8. $2^x = 128$

9. $3^x = 81$

10. $4^x = 256$

11. $64^x = 8$

12. $81^x = 3$

13. $7^{-x} = \dfrac{1}{49}$

14. $6^{-x} = \dfrac{1}{216}$

15. $27^x = \dfrac{1}{3}$

16. $25^x = \dfrac{1}{5}$

17. $2^{x+2} = 64$

18. $3^{x-6} = 81$

19. $2^{3x-2} = 128$

20. $64^x = 4^{4x+1}$

21. $27^x = 3^{2x+3}$

22. $\left(\dfrac{1}{2}\right)^x = 16$

Solve each exponential equation. Use a calculator and round your answers to the nearest hundredth.

23. $7^x = 50$

24. $1.05^x = 23$

25. $4^{x-1} = 35$

26. $2.3^{x-1} = 26.2$

27. $1.63^{x+1} = 25$

28. $4^x = 9^{x-2}$

29. $3^{x+4} = 6^x$

30. $5^x = 2^{x+5}$

Solve each logarithmic equation. Use a calculator where appropriate. If the answer is irrational, round the answer to the nearest hundredth.

31. $\log_{36} x = \dfrac{1}{2}$

32. $\log_{81} x = \dfrac{1}{2}$

33. $\log_{125} x = \dfrac{1}{3}$

34. $\log_{81} x = \dfrac{1}{4}$

35. $\log_2 x = -4$

36. $\log_7 x = -2$

37. $\log x = 2$

38. $\log x = 4$

39. $\log_2 (5 - 3x) = 3$

40. $\log_4 (3x + 7) = 3$

41. $\log_5 (x + 1)^2 = 2$

42. $\log_3 (a - 2)^2 = 2$

43. $\log_2 (r + 4)^2 = 4$

44. $\log_2 (p - 3)^2 = 6$

45. $\log (x + 8) = 2$

46. $\log (3x - 8) = 1$

47. $\log_2 x + \log_2 5 = 2$

48. $\log_3 2x + \log_3 x = 4$

49. $\log (r + 2) = \log (3r - 1)$

50. $\log 2a = \log (1 - a)$

51. $\log (2x + 1) + \log 4 = \log (7x + 8)$

52. $\log (x - 5) + \log 3 = \log (2x)$

53. $\log n + \log (3n - 5) = \log 2$

54. $\log (x + 4) - \log x = \log (x + 1)$

55. $\log 6 + \log y = 0.72$

56. $\log (x + 4) - \log x = 1.22$

57. $2 \log x - \log 9 = 2$

58. $\log 6000 - \log (x + 2) = 3.15$

59. $\log x + \log (x - 3) = 1$

60. $2 \log_2 x = 4$

61. $\log x = \dfrac{1}{3} \log 64$

62. $\log_7 x = \dfrac{3}{2} \log_7 9$

63. $\log_8 x = 4 \log_8 2 - \log_8 8$

64. $\log_4 x + \log_4 (6x - 7) = \log_4 5$

65. $\log_5 (x + 3) + \log_5 (x - 2) = \log_5 6$

66. $\log_7 (x + 6) - \log_7 (x - 3) = \log_7 4$

67. $\log_2 (x + 3) - \log_2 (x - 6) = \log_2 4$

68. $\log (x - 7) - \log (x + 3) = \log 6$

Problem Solving

Solve each problem. Round your answers to the nearest hundredth.

69. **Bacteria** If the initial number of bacteria in the culture in Example 7 is 4500, when will the number of bacteria in the culture reach 50,000? Use $N = 4500(2)^t$.

70. **Bacteria** If after 4 hours the culture in Example 7 contains 2224 bacteria, how many bacteria were present initially?

71. **Radioactive Decay** The amount, A, of 200 grams of a certain radioactive material remaining after t years can be found by the equation $A = 200(0.75)^t$. When will 80 grams remain?

72. **Radioactive Decay** The amount, A, of 70 grams of a certain radioactive material remaining after t years can be found by the equation $A = 70(0.62)^t$. When will 10 grams remain?

73. **Savings Account** Paul Trapper invests $2000 in a savings account earning interest at a rate of 5% compounded annually. How long will it take for the $2000 to grow to $4600? Use the compound interest formula, $A = p\left(1 + \dfrac{r}{n}\right)^{nt}$, which was discussed on page 583.

74. **Savings Account** If Tekar Werner invests $600 in a savings account earning interest at a rate of 6% compounded semi-annually, how long will it take for the $600 to grow to $1800?

75. **Money Market Account** If Jacci White invests $2500 in a money market account earning interest at a rate of 4% compounded quarterly, how long will it take for the $2500 to grow to $4000?

76. **Credit Union Account** Charlotte Newsome invests $1000 in a share savings account at her credit union. If her money earns interest at a rate of 3% compounded monthly, how long will it take for the $1000 to grow to $1500?

77. **Depreciation** A machine purchased for business use can be depreciated to reduce income tax. The value of the machine at the end of its useful life is called its *scrap value*. When the machine depreciates by a constant percentage annually, its scrap value, S, is $S = c(1 - r)^n$, where c is the original cost, r is the annual rate of depreciation as a decimal, and n is the useful life in years. Find the scrap value of a machine that costs $50,000, has a useful life of 12 years, and has an annual depreciation rate of 15%.

78. **Depreciation** If the machine in Exercise 77 costs $100,000, has a useful life of 15 years, and has an annual depreciation rate of 8%, find its scrap value.

79. **Power Gain of an Amplifier** The power gain, P, of an amplifier is defined as

$$P = 10 \log\left(\frac{P_{out}}{P_{in}}\right)$$

where P_{out} is the output power in watts and P_{in} is the input power in watts. If an amplifier has an output power of 12.6 watts and an input power of 0.146 watts, find the power gain.

80. **Earthquake** Measured on the Richter scale, the magnitude, R, of an earthquake of intensity I is defined by $R = \log I$, where I is the number of times more intense the earthquake is than the minimum level for comparison.

a) How many times more intense was the 1906 San Francisco earthquake, which measured 8.25 on the Richter scale, than the minimum level for comparison?

b) How many times more intense is an earthquake that measures 8.3 on the Richter scale than one that measures 4.7?

81. **Magnitude of Sound** The decibel scale is used to measure the magnitude of sound. The magnitude d, in decibels, of a sound is defined to be $d = 10 \log I$, where I is the number of times greater (or more intense) the sound is than the minimum intensity of audible sound.

a) An airplane engine (nearby) measures 120 decibels. How many times greater than the minimum level of audible sound is the airplane engine?

b) The intensity of the noise in a busy city street is 50 decibels. How many times greater is the intensity of the sound of the airplane engine than the sound of the city street?

82. In the following procedure, we begin with a true statement and end with a false statement. Can you find the error?

$2 < 3$	True
$2 \log (0.1) < 3 \log (0.1)$	Multiply both sides by $\log (0.1)$.
$\log (0.1)^2 < \log (0.1)^3$	Property 3
$(0.1)^2 < (0.1)^3$	Property 6d
$0.01 < 0.001$	False

83. Solve $8^x = 16^{x-2}$.

84. Solve $27^x = 81^{x-3}$.

85. Use equations that are quadratic in form to solve the equation $2^{2x} - 6(2^x) + 8 = 0$.

86. Use equations that are quadratic in form to solve the equation $2^{2x} - 18(2^x) + 32 = 0$.

Change the exponential or logarithmic equation to the form $ax + by = c$, and then solve the system of equations.

87. $2^x = 8^y$
$x + y = 4$

88. $3^{2x} = 9^{y+1}$
$x - 2y = -3$

89. $\log (x + y) = 2$
$x - y = 8$

90. $\log (x + y) = 3$
$2x - y = 5$

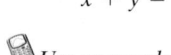 *Use your calculator to estimate the solutions to the nearest tenth. If a real solution does not exist, so state.*

91. $\log (x + 3) + \log x = \log 16$

92. $\log (3x + 5) = 2.3x - 6.4$

93. $5.6 \log (5x - 12) - 2.3 \log (x - 5.4)$

94. $5.6 \log (x + 12.2) - 1.6 \log (x - 4) = 20.3 \log (2x - 6)$

Concept/Writing Exercises

95. How can you tell quickly that $\log (x + 4) = \log (-2)$ has no real solution?

96. In properties 6c and 6d, we specify that both x and y must be positive. Explain why.

Cumulative Review Exercises

[2.2] **97.** Consider the following two figures. Which has a greater volume, and by how much?

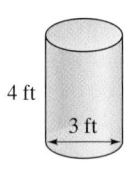 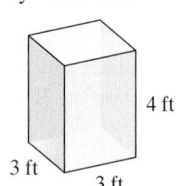

[3.6] **98.** Let $f(x) = x^2 - x$ and $g(x) = x - 1$. Find $(g - f)(3)$.

[4.6] **99.** Determine the solution set to the system of inequalities.
$$3x - 4y \le 6$$
$$y > -x + 4$$

[7.5] **100.** Simplify $\dfrac{2\sqrt{xy} - \sqrt{xy}}{\sqrt{x} + \sqrt{y}}$.

[8.3] **101.** Solve $E = mc^2$ for c.

[8.5] **102.** Determine the function for the parabola that has the shape of $f(x) = 2x^2$ and has its vertex at $(3, -5)$.

9.7 Natural Exponential and Natural Logarithmic Functions

1 Identify the natural exponential function.

2 Identify the natural logarithmic function.

3 Approximate natural logarithms and powers of e on a calculator.

4 Use the change of base formula.

5 Solve natural logarithmic and natural exponential equations.

6 Solve applications.

In this section we discuss the *natural exponential function*, and its inverse, *the natural logarithmic function*. These functions are frequently used to describe naturally occurring events, thus the adjective *natural*. Both of these functions rely on an irrational number designated by the letter e.

> **The Natural Base, e**
>
> The **natural base, e**, is an irrational number that serves as the base for the *natural exponential function* and the *natural logarithmic function*.
>
> $$e \approx 2.7183$$

1 Identify the Natural Exponential Function

In Section 9.2, we discussed exponential functions. Recall that exponential functions are of the form $f(x) = a^x, a > 0$ and $a \ne 1$. We now introduce the *natural exponential function*.

Understanding Algebra

There are two very important types of logarithms whose notation differs from other logarithms.

- *Common* logarithms have a base of 10 and are written using the notation log x. Thus,

$$\log_{10} x = \log x$$

- *Natural* logarithms have a base of e, the natural base, and are written using the notation ln x. Thus,

$$\log_e x = \ln x$$

The Natural Exponential Function

The **natural exponential function** is

$$f(x) = e^x$$

where e is the natural base.

2 Identify the Natural Logarithmic Function

We begin by defining natural logarithms.

Natural Logarithms

Natural logarithms are logarithms with a base of e, the natural base. We indicate natural logarithms with the notation ln.

$$\log_e x = \ln x \qquad (x > 0)$$

ln x is read "the natural logarithm of x."

In Section 9.3, we discussed logarithmic functions. Recall that logarithmic functions are of the form $f(x) = \log_a x, a > 0, a \neq 1, x > 0$. We now introduce the *natural logarithmic function*.

The Natural Logarithmic Function

The **natural logarithmic function** is

$$f(x) = \ln x \quad (x > 0)$$

where $\ln x = \log_e x$ and e is the natural base.

When we change a natural logarithm to exponential form, the base of the exponential expression is the natural base, e.

Natural Logarithm in Exponential Form

For $x > 0$, if $y = \ln x$, then $e^y = x$.

EXAMPLE 1 Find the value of the expression by changing the natural logarithm to exponential form.

a) ln 1   **b)** ln e

Solution

a) Let $y = \ln 1$; then $e^y = 1$. Since any nonzero value to the 0th power equals 1, y must equal 0. Thus, ln 1 = 0.

b) Let $y = \ln e$; then $e^y = e$. For e^y to equal e, y must equal 1. Thus, ln e = 1.

Now Try Exercise 1

Recall from Section 9.3 that the functions $y = a^x$ and $y = \log_a x$ are inverse functions. Thus, $y = e^x$ and $y = \ln x$ are inverse functions. The graphs are shown in **Figure 9.22**. Notice that the graphs are symmetric about the line $y = x$. Notice also that the graph of $y = e^x$ is similar to the graph of $y = a^x$ where $a > 1$ and the graph of $y = \ln x$ is similar to the graph of $y = \log_a x$ where $a > 1$.

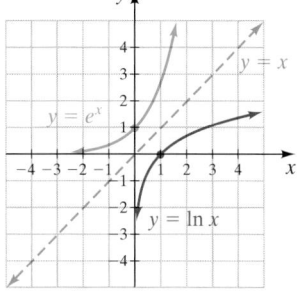

FIGURE 9.22

3 Approximate Natural Logarithms and Powers of e on a Calculator

In a manner similar to how we approximated common logarithms and powers of 10 in Section 9.5, we can use a scientific or graphing calculator to approximate natural logarithms and powers of e.

 Using Your Calculator ■■■■

Approximating Natural Logarithms

Scientific Calculator
To approximate natural logarithms on many scientific calculators, enter the argument, then press the $\boxed{\text{LN}}$ key.

EXAMPLE	KEYS TO PRESS	ANSWER DISPLAYED
Approximate ln 31	31 $\boxed{\text{LN}}$	3.433987204

Graphing Calculator
To approximate natural logarithms on a graphing calculator and on many scientific calculators, first press $\boxed{\text{LN}}$ and then enter the number. For example, on the TI-84 Plus, you would do the following:

EXAMPLE	KEYS TO PRESS	ANSWER DISPLAYED
Approximate ln 31	$\boxed{\text{LN}}$ (31) $\boxed{\text{ENTER}}$	3.433987204

↑
Generated by calculator

Remember that *a logarithm is an exponent*. Thus, the natural logarithm of a number is the exponent to which the natural base e must be raised to obtain that number. For example:

$$\ln 242 \approx 5.488937726 \qquad \text{therefore} \qquad e^{5.488937726} \approx 242$$
$$\ln 0.85 \approx -0.1625189295 \qquad \text{therefore} \qquad e^{-0.1625189295} \approx 0.85$$

Using Your Calculator ■■■■

Approximating Powers of e

To evaluate powers of the natural base e on your calculator, use the e^x function, which is located directly above the $\boxed{\text{LN}}$ key. To access this function, press the $\boxed{\text{2ND}}$, $\boxed{\text{INV}}$, or $\boxed{\text{SHIFT}}$ key prior to pressing the $\boxed{\text{LN}}$ key.

EXAMPLE 2 Approximate the following powers of the natural base e. Round your answers to four decimal places

a) $e^{1.394}$ **b)** $e^{2.827}$ **c)** $e^{-0.356}$

Solution   Use a calculator to approximate each power of e.

a) $e^{1.394} \approx 4.0309$ **b)** $e^{2.827} \approx 16.8947$ **c)** $e^{-0.356} \approx 0.7005$

Now Try Exercise 15

EXAMPLE 3 Solve for x in each of the following equations. Round your answers to four decimal places.

a) $\ln x = 0.132$ **b)** $\ln x = -1.203$

Solution

a) Using the definition of a logarithm, we know that
$$\ln x = 0.132 \quad \text{means} \quad x = e^{0.132} \approx 1.1411$$

b) $\ln x = -1.203 \quad \text{means} \quad x = e^{-1.203} \approx 0.3002$

Now Try Exercise 21

4 Use the Change of Base Formula

If you are given a logarithm in a base other than 10 or e, you will not be able to evaluate it on your calculator directly. When this occurs, you can use the **change of base formula**.

Change of Base Formula

For any logarithm bases a and b, and positive number x,

$$\log_a x = \frac{\log_b x}{\log_b a}$$

In the change of base formula, 10 is often used in place of base b because we can approximate common logarithms on a calculator. Replacing base b with 10, we get

$$\log_a x = \frac{\log_{10} x}{\log_{10} a} \quad \text{or} \quad \log_a x = \frac{\log x}{\log a}$$

EXAMPLE 4   Use the change of base formula to approximate $\log_3 24$.

Solution   If we substitute 3 for a and 24 for x in $\log_a x = \dfrac{\log x}{\log a}$, we obtain

$$\log_3 24 = \frac{\log 24}{\log 3} \approx 2.8928$$

Note that $3^{2.8928} \approx 24$.

Now Try Exercise 27

5 Solve Natural Logarithmic and Natural Exponential Equations

The properties of logarithms discussed in Section 9.4 also hold true for natural logarithms.

Properties for Natural Logarithms

$\ln xy = \ln x + \ln y$	$(x > 0 \text{ and } y > 0)$	Product rule
$\ln \dfrac{x}{y} = \ln x - \ln y$	$(x > 0 \text{ and } y > 0)$	Quotient rule
$\ln x^n = n \ln x$	$(x > 0)$	Power rule

Properties 4 and 5 from page 598 can also be written using natural logarithms. Thus, $\ln e^x = x$ and $e^{\ln x} = x$. We will refer to these properties as properties 7 and 8, respectively.

Additional Properties for Natural Logarithms and Natural Exponential Expressions

$\ln e^x = x$		Property 7
$e^{\ln x} = x,$	$x > 0$	Property 8

Using property 7, $\ln e^x = x$, we can state, for example, that $\ln e^{kt} = kt$, and $\ln e^{-2.06t} = -2.06t$. Using property 8, $e^{\ln x} = x$, we can state, for example, that $e^{\ln (t+2)} = t + 2$ and $e^{\ln kt} = kt$.

EXAMPLE 5   Solve the equation $\ln y - \ln (x + 9) = t$ for y.

Solution

$$\ln y - \ln (x + 9) = t$$

$$\ln \frac{y}{x + 9} = t \qquad \text{Quotient rule}$$

$$\frac{y}{x + 9} = e^t \qquad \text{Write in exponential form.}$$

$$y = e^t(x + 9) \qquad \text{Solve for } y.$$

Now Try Exercise 63

EXAMPLE 6   Solve the equation $225 = 450e^{-0.4t}$ for t.

Solution   Begin by dividing both sides of the equation by 450 to isolate $e^{-0.4t}$.

$$\frac{225}{450} = \frac{\cancel{450}e^{-0.4t}}{\cancel{450}}$$

$$0.5 = e^{-0.4t}$$

Now take the natural logarithm of both sides of the equation to eliminate the exponential expression on the right side of the equation.

$$\ln 0.5 = \ln e^{-0.4t}$$

$$\ln 0.5 = -0.4t \qquad \text{Property 7}$$

$$-0.6931472 = -0.4t$$

$$\frac{-0.6931472}{-0.4} = t$$

$$1.732868 = t$$

Now Try Exercise 49

EXAMPLE 7   Solve the equation $P = P_0 e^{kt}$ for t.

Solution   We can follow the same procedure as used in Example 6.

$$P = P_0 e^{kt}$$

$$\frac{P}{P_0} = \frac{\cancel{P_0} e^{kt}}{\cancel{P_0}} \qquad \text{Divide both sides by } P_0.$$

$$\frac{P}{P_0} = e^{kt}$$

$$\ln \frac{P}{P_0} = \ln e^{kt} \qquad \text{Take natural log of both sides.}$$

$$\ln P - \ln P_0 = \ln e^{kt} \qquad \text{Quotient rule}$$

$$\ln P - \ln P_0 = kt \qquad \text{Property 7}$$

$$\frac{\ln P - \ln P_0}{k} = t \qquad \text{Solve for } t.$$

Now Try Exercise 59

6 Solve Applications

We now introduce some applications involving the natural base and natural logarithms. We begin with a formula used when a quantity increases or decreases at an *exponential rate*.

Exponential Growth or Decay Formula

When a quantity P increases (grows) or decreases (decays) at an exponential rate, the value of P after time t can be found using the formula

$$P = P_0 e^{kt},$$

where P_0 is the initial or starting value of the quantity P, and k is the constant growth rate or decay rate.

When $k > 0$, P increases as t increases.

When $k < 0$, P decreases and gets closer to 0 as t increases.

EXAMPLE 8 **Interest Compounded Continuously** When interest is compounded continuously, the balance, P, in the account at any time, t, can be calculated by the exponential growth formula $P = P_0 e^{kt}$, where P_0 is the principal initially invested and k is the interest rate.

a) Suppose the interest rate is 6% compounded continuously and $1000 is initially invested. Determine the balance in the account after 3 years.

b) How long will it take the account to double in value?

Solution

a) Understand and Translate   We are told that the principal initially invested, P_0, is $1000. We are also given that the time, t, is 3 years and that the interest rate, k, is 6% or 0.06. We substitute these values into the given formula and solve for P.

$$P = P_0 e^{kt}$$
$$P = 1000\,e^{(0.06)(3)}$$

Carry Out
$$= 1000 e^{0.18} = 1000(1.1972174) \qquad \text{From a calculator}$$
$$\approx 1197.22$$

Answer   After 3 years, the balance in the account is $\approx$ $1197.22.

b) Understand and Translate   For the value of the account to double, the balance in the account would have to reach $2000. Therefore, we substitute 2000 for P and solve for t.

$$P = P_0 e^{kt}$$
$$2000 = 1000\,e^{0.06t}$$
$$2 = e^{0.06t} \qquad \text{Divide both sides by 1000.}$$

Carry Out
$$\ln 2 = \ln e^{0.06t} \qquad \text{Take natural log of both sides.}$$
$$\ln 2 = 0.06t \qquad \text{Property 7}$$
$$\frac{\ln 2}{0.06} = t$$
$$\frac{0.6931472}{0.06} = t$$
$$11.552453 \approx t$$

Answer   Thus, with an interest rate of 6% compounded continuously, the account will double in about 11.6 years.

Now Try Exercise 69

EXAMPLE 9 **Radioactive Decay** Strontium 90 is a radioactive isotope that decays exponentially at 2.8% per year. Suppose there are initially 1000 grams of strontium 90 in a substance.

a) Find the number of grams of strontium 90 left after 50 years.

b) Find the half-life of strontium 90.

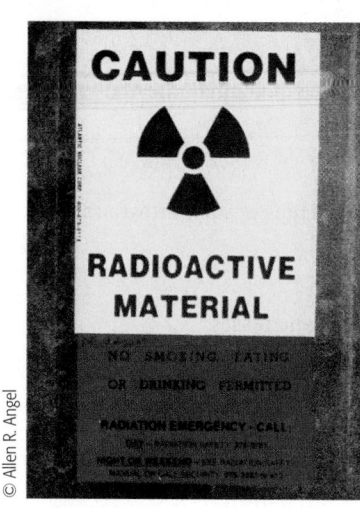

© Allen R. Angel

Solution

a) Understand   Since the strontium 90 is decaying over time, the value of k in the formula $P = P_0 e^{kt}$ is negative. Since the rate of decay is 2.8% per year, we use $k = -0.028$. Therefore, the formula we use is $P = P_0 e^{-0.028t}$.

Translate $P = P_0 e^{-0.028t}$

                            $= 1000 e^{-0.028(50)}$

Carry Out $= 1000 e^{-1.4} = 1000(0.246597) = 246.597$

Answer   Thus, after 50 years, 246.597 grams of strontium 90 remain.

b) To find the half-life, we need to determine when 500 grams of strontium 90 are left.

$$P = P_0 e^{-0.028t}$$
$$500 = 1000 e^{-0.028t}$$
$$0.5 = e^{-0.028t} \qquad \text{Divide both sides by 1000.}$$
$$\ln 0.5 = \ln e^{-0.028t} \qquad \text{Take natural log of both sides.}$$
$$-0.6931472 = -0.028t \qquad \text{Property 7}$$
$$\frac{-0.6931472}{-0.028} = t$$
$$24.755257 \approx t$$

Thus, the half-life of strontium 90 is about 24.8 years.

Now Try Exercise 71

EXAMPLE 10 Selling Toys   The formula for estimating the amount of money, A, spent on advertising a certain toy is $A = 350 + 650 \ln n$ where n is the expected number of toys to be sold.

a) If the company wishes to sell 2200 toys, how much money should the company expect to spend on advertising?

b) How many toys can be expected to be sold if $6000 is spent on advertising?

Solution

a) $A = 350 + 650 \ln n$

      $= 350 + 650 \ln 2200 \qquad$ Substitute 2200 for n.

      $= 350 + 650(7.6962126)$

      $= 5352.54$

Thus, $5352.54 should be expected to be spent on advertising.

b) Understand and Translate   We are asked to find the number of toys expected to be sold, n, if $6000 is spent on advertising. We substitute the given values into the equation and solve for n.

                    $A = 350 + 650 \ln n$

Carry Out $6000 = 350 + 650 \ln n \qquad$ Substitute 6000 for A.

                    $5650 = 650 \ln n \qquad$ Subtract 350 from both sides.

                    $\dfrac{5650}{650} = \ln n \qquad$ Divide both sides by 650.

             $8.69231 \approx \ln n$

             $e^{8.69231} \approx n \qquad$ Change to exponential form.

                $5957 \approx n \qquad$ Obtain answer from a calculator.

Answer   Thus, about 5957 toys can be expected to be sold if $6000 is spent on advertising.

Now Try Exercise 75

Using Your Graphing Calculator ▦▦▦▦

To approximate the solutions to the equation $4e^{0.3x} - 5 = x + 3$ using a TI-84 Plus, we can graph

$$Y_1 = 4e^\wedge(0.3x) - 5$$
$$Y_2 = x + 3$$

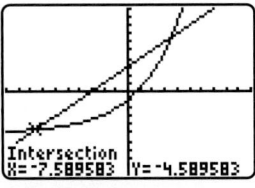

FIGURE 9.23

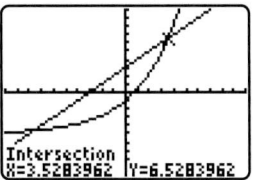

FIGURE 9.24

The points of intersection of these graphs can be found using the *intersect* function from the CALC menu. **Figure 9.23** shows that the x-coordinate of one point of intersection, $x \approx -7.5896$, is one solution. **Figure 9.24** shows that the x-coordinate of the other point of intersection, $x \approx 3.5284$, is the other solution.

EXERCISE SET 9.7

MathXL
MathXL®

MyMathLab
MyMathLab

Warm-Up Exercises

Fill in the blanks with the appropriate word, phrase, or symbol(s) from the following list.

write	natural	inverse	quarterly	exponential	composite
logarithmic	continuously	base	change	common	growth

1. The natural _____, e, is an irrational number that serves as the base for the natural exponential function and the natural logarithmic function.

2. The natural exponential function and the natural logarithmic function are _____ functions.

3. The natural _____ function is $f(x) = e^x$, where e is the natural base.

4. The natural _____ function is $f(x) = \ln x$, $x > 0$, where e is the natural base.

5. The formula $\log_a x = \dfrac{\log_b x}{\log_b a}$ is the _____ of base formula.

6. In the change of base formula, 10 is often used in place of b because we can approximate _____ logarithms on a calculator.

7. To solve the equation $5.7 = e^x$, take the _____ logarithm of both sides of the equation.

8. To solve the equation $\ln x = 1.239$, _____ the equation in exponential form.

9. The exponential _____ or decay formula states that when a quantity P increases or decreases at an exponential rate, the value of P after time t can be found using the formula $P = P_0 e^{kt}$.

10. When interest is compounded _____, the balance in the account can be calculated using the exponential growth formula.

Practice the Skills

Approximate the following values. Round your answers to four decimal places.

11. $\ln 27$ **12.** $\ln 810$ **13.** $\ln 0.415$ **14.** $\ln 0.000176$

Approximate the following values. Round your answers to four decimal places.

15. $e^{1.2}$ **16.** $e^{4.8}$ **17.** $e^{-0.56}$ **18.** $e^{-2.6}$

Approximate the value of x. Round your answers to four decimal places.

19. $\ln x = 1.6$ **20.** $\ln x = 5.2$ **21.** $\ln x = -2.85$ **22.** $\ln x = -0.674$

Approximate the value of x. Round your answers to four decimal places.

23. $e^x = 98$ **24.** $e^x = 2010$ **25.** $e^x = 0.0574$ **26.** $e^x = 0.000231$

Use the change of base formula to approximate the value of the following logarithms. Round your answers to four decimal places.

27. $\log_2 21$ **28.** $\log_2 89$ **29.** $\log_4 11$ **30.** $\log_4 316$

31. $\log_5 82$

32. $\log_5 1893$

33. $\log_6 185$

34. $\log_6 806$

35. $\log_5 0.463$

36. $\log_3 0.0365$

Solve the following logarithmic equations.

37. $\ln x + \ln (x - 1) = \ln 12$

38. $\ln (x + 4) + \ln (x - 2) = \ln 16$

39. $\ln x + \ln (x + 4) = \ln 5$

40. $\ln (x + 3) + \ln (x - 3) = \ln 40$

41. $\ln x = 5 \ln 2 - \ln 8$

42. $\ln x = \dfrac{3}{2} \ln 16$

43. $\ln (x^2 - 4) - \ln (x + 2) = \ln 4$

44. $\ln (x + 12) - \ln (x - 4) = \ln 5$

Each of the following equations is in the form $P = P_0 e^{kt}$. Solve each equation for the remaining variable. Remember, e is a constant. Round your answers to four decimal places.

45. $P = 120e^{2.3(1.6)}$

46. $900 = P_0 e^{(0.4)(3)}$

47. $50 = P_0 e^{-0.5(3)}$

48. $18 = 9e^{2t}$

49. $60 = 20e^{1.4t}$

50. $29 = 58e^{-0.5t}$

51. $86 = 43e^{k(3)}$

52. $15 = 75e^{k(4)}$

53. $20 = 40e^{k(2.4)}$

54. $100 = A_0 e^{-0.02(3)}$

55. $A = 6000e^{-0.08(3)}$

56. $51 = 68e^{-0.04t}$

Solve for the indicated variable.

57. $V = V_0 e^{kt}$, for V_0

58. $P = P_0 e^{kt}$, for P_0

59. $P = 150e^{7t}$, for t

60. $361 = P_0 e^{kt}$, for t

61. $A = A_0 e^{kt}$, for k

62. $167 = R_0 e^{kt}$, for k

63. $\ln y - \ln x = 2.3$, for y

64. $\ln y + 9 \ln x = \ln 2$, for y

65. $\ln y - \ln (x + 6) = 5$ for y

66. $\ln (x + 2) - \ln (y - 1) = \ln 5$, for y

Problem Solving

Use a calculator to solve.

67. Interest Compounded Continuously If $5000 is invested at 6% compounded continuously,

  a) determine the balance in the account after 2 years.

  b) How long would it take the value of the account to double? (See Example 8.)

68. Interest Compounded Continuously If $3000 is invested at 3% compounded continuously,

  a) determine the balance in the account after 30 years.

  b) How long would it take the value of the account to double?

69. Credit Union Account Lucy Alfonso invested $2500 in her credit union savings account earning 4% interest compounded continuously.

  a) Determine the balance in Lucy's account after 3 years.

  b) How long will it take for the value of Lucy's account to double?

70. Retirement Account Kristie Paulson invested $10,000 in a retirement savings account earning 5% interest compounded continuously.

  a) Determine the balance in Kristie's account after 25 years.

  b) How long will it take for the value of Kristie's account to triple?

71. Radioactive Decay Refer to Example 9. Determine the amount of strontium 90 remaining after 20 years if there were originally 70 grams.

72. Strontium 90 Refer to Example 9. Determine the amount of strontium 90 remaining after 40 years if there were originally 200 grams.

73. Soft Drinks For a certain soft drink, the percent of a target market, $f(t)$, that buys the soft drink is a function of the number of days, t, that the soft drink is advertised. The function that describes this relationship is $f(t) = 1 - e^{-0.04t}$.

  a) What percent of the target market buys the soft drink after 50 days of advertising?

  b) How many days of advertising are needed if 75% of the target market is to buy the soft drink?

© Allen R. Angel

74. Trout in a Lake In 2010, Thomas Pearse Lake had 300 trout. The growth in the number of trout is estimated by the function $g(t) = 300e^{0.07t}$, where t is the number of years after 2010. How many trout will be in the lake in **a)** 2015? **b)** 2020?

© Allen R. Angel

75. Walking Speed It was found in a psychological study that the average walking speed, $f(P)$, of a person living in a city is a function of the population of the city. For a city of population P, the average walking speed in feet per second is given by $f(P) = 0.37 \ln P + 0.05$. The population of Nashville, Tennessee, is 972,000.

a) What is the average walking speed of a person living in Nashville?

b) What is the average walking speed of a person living in New York City, population 8,567,000?

c) If the average walking speed of the people in a certain city is 5.0 feet per second, what is the population of the city?

76. Advertising For a certain type of tie, the number of ties sold, $N(a)$, is a function of the dollar amount spent on advertising, a (in thousands of dollars). The function that describes this relationship is $N(a) = 800 + 300 \ln a$.

a) How many ties were sold after $1500 (or $1.5 thousand) was spent on advertising?

b) How much money must be spent on advertising to sell 1000 ties?

77. Assume that the value of the island of Manhattan has grown at an exponential rate of 8% per year since 1626 when Peter Minuet of the Dutch West India Company purchased it for $24. Then the value of Manhattan can be determined by the equation $V = 24e^{0.08t}$, where t is the number of years since 1626. Determine the value of the island of Manhattan in 2010, that is when $t = 384$ years.

Times Square in Manhattan

78. Prescribing a Drug The percent of doctors who accept and prescribe a new drug is given by the function $P(t) = 1 - e^{-0.22t}$, where t is the time in months since the drug was placed on the market. What percent of doctors accept a new drug 2 months after it is placed on the market?

79. World Population The word population in February 2009 was about 6.76 billion people. Assume that the world population continues to grow exponentially at the growth rate of about 1.2% per year. Thus, the expected world population, in billions of people, in t years, is given by the function $P(t) = 6.76e^{0.012t}$, where t is the number of years since 2009.

a) Use this function to estimate the world population in 2015.

b) At the current rate, how long will it take for the world population to double?

80. United States Population The United States population in February 2009 was about 305.8 million people. Assume that the United States population continues to grow exponentially at the growth rate of about 1.0% per year. Thus, the expected United States population, in millions of people, in t years, is given by the function $P(t) = 305.8e^{0.010t}$, where t is the number of years since 2009.

a) Use this function to estimate the United States population in 2015.

b) At the current rate, how long will it take for the United States population to double?

81. China Population The population of China in February 2009 was about 1.34 billion people. Assume that the population of China continues to grow exponentially at the growth rate of about 0.6% per year. Thus, the expected population of China, in billions of people, in t years, is given by the function $P(t) = 1.34e^{0.006t}$, where t is the number of years since 2009.

a) Use this function to estimate the population of China in 2015.

b) At the current rate, how long will it take for the China population to double?

82. Mexico Population The population of Mexico in February 2009 was about 111.2 million people. Assume that the population of Mexico continues to grow exponentially at the growth rate of about 1.1% per year. Thus, the expected population of Mexico, in millions of people, in t years, is given by the function $P(t) = 111.2e^{0.011t}$, where t is the number of years since 2009.

a) Use this function to estimate the population of Mexico in 2015.

b) At the current rate, how long will it take for the population of Mexico to double?

Mexico City

83. Boys' Heights The shaded area of the following graph shows the normal range (from the 5th to the 95th percentile) of heights for boys from birth up to 36 months. The median, or 50th percentile, of heights is indicated by the green line. The function $y = 15.29 + 5.93 \ln x$ can be used to estimate the median height of boys from 3 months to 36 months. Use this function to estimate the median height for boys at age **a)** 18 months. **b)** 30 months.

Boys

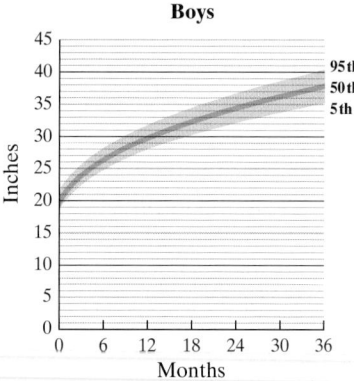

Source: Newsweek

84. Girls' Weights The shaded area of the following graph shows the normal range (from the 5th to the 95th percentile) of weights for girls from birth up to 36 months. The median, or 50th percentile, of weights is indicated by the orange line. The function $y = 3.17 + 7.32 \ln x$ can be used to estimate the median weight of girls from 3 months to 36 months. Use this function to estimate the median weight for girls at **a)** 18 months. **b)** 30 months.

Girls

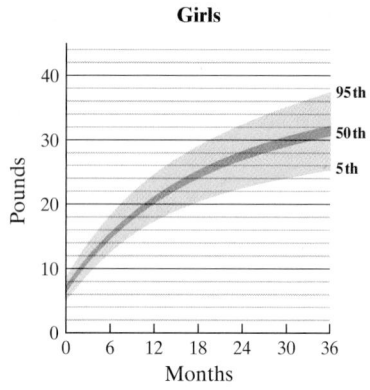

Source: Newsweek

85. Radioactive Decay Uranium 234 (U-234) decays exponentially at a rate of 0.0003% per year. Thus, the formula $A = A_0 e^{-0.000003t}$ can be used to determine the amount of U-234 remaining from an initial amount A_0, after t years.

a) If 1000 grams of U-234 are present in 2010, how many grams will remain in 2110, 100 years later?

b) Find the half-life of U-234.

86. Radioactive Decay Plutonium, which is commonly used in nuclear reactors, decays exponentially at a rate of 0.003% per year. The formula $A = A_0 e^{kt}$ can be used to find the amount of plutonium remaining from an initial amount, A_0, after t years. In the formula, the k is replaced with -0.00003.

a) If 1000 grams of plutonium are present in 2009, how many grams of plutonium will remain in the year 2109, 100 years later?

b) Find the half-life of plutonium.

87. Carbon Dating Carbon dating is used to estimate the age of ancient plants and objects. The radioactive element, carbon 14, is most often used for this purpose. Carbon 14 decays exponentially at a rate of 0.01205% per year. The amount of carbon 14 remaining in an object after t years can be found by the function $f(t) = v_0 e^{-0.0001205t}$, where v_0 is the initial amount present.

a) If an ancient animal bone originally had 20 grams of carbon 14, and when found it had 9 grams of carbon 14, how old is the bone?

b) How old is an item that has 50% of its original carbon 14 remaining?

88. Compound Interest At what rate, compounded continuously, must a sum of money be invested if it is to double in 13 years?

89. Compound Interest How much money must be deposited today to become $20,000 in 18 years if invested at 6% compounded continuously?

90. Radioisotope The power supply of a satellite is a radioisotope. The power P, in watts, remaining in the power supply is a function of the time the satellite is in space.

a) If there are 50 grams of the isotope originally, the power remaining after t days is $P = 50e^{-0.002t}$. Find the power remaining after 50 days.

b) When will the power remaining drop to 10 watts?

91. Radioactive Decay During the nuclear accident at Chernobyl in Ukraine in 1986, two of the radioactive materials that escaped into the atmosphere were cesium 137, with a decay rate of 2.3% and strontium 90, with a decay rate of 2.8%.

a) Which material will decompose more quickly? Explain.

b) What percentage of the cesium will remain in 2036, 50 years after the accident?

92. Radiometric Dating In the study of radiometric dating (using radioactive isotopes to determine the age of items), the formula

$$t = \frac{t_h}{0.693} \ln\left(\frac{N_0}{N}\right)$$

is often used. In the formula, t is the age of the item, t_h is the half-life of the radioactive isotope used, N_0 is the original number of radioactive atoms present, and N is the number remaining at time t. Suppose a rock originally contained 5×10^{12} atoms of uranium 238. Uranium 238 has a half-life of 4.5×10^9 years. If at present there are 4×10^{12} atoms, how old is the rock?

Challenge Exercises

In Exercises 93–96, when you solve for the given variable, write the answer without using the natural logarithm.

93. Velocity The distance traveled by a train initially moving at velocity v_0 after the engine is shut off can be calculated by the formula $x = \frac{1}{k} \ln (k v_0 t + 1)$. Solve this equation for v_0.

94. Intensity of Light The intensity of light as it passes through a certain medium is found by the formula $x = k(\ln I_0 - \ln I)$. Solve this equation for I_0.

95. Electric Circuit An equation relating the current and time in an electric circuit is $\ln i - \ln I = \frac{-t}{RC}$. Solve this equation for i.

96. Molecule A formula used in studying the action of a protein molecule is $\ln M = \ln Q - \ln (1 - Q)$. Solve this equation for Q.

Cumulative Review Exercises

[3.3] **97.** Let $h(x) = \dfrac{x^2 + 4x}{x + 6}$. Find **a)** $h(-4)$. **b)** $h\left(\dfrac{2}{5}\right)$.

[4.3] **98. Tickets** The admission at an ice hockey game is $15 for adults and $11 for children. A total of 550 tickets were sold. Determine how many children's tickets and how many adults' tickets were sold if a total of $7290 was collected.

[5.2] **99.** Multiply $(3xy^2 + y)(4x - 3xy)$.

[5.6] **100.** Find two values of b that will make $4x^2 + bx + 25$ a perfect square trinomial.

[7.4] **101.** Multiply $\sqrt[3]{x}\left(\sqrt[3]{x^2} + \sqrt[3]{x^5}\right)$.

Chapter 9 Summary

IMPORTANT FACTS AND CONCEPTS	EXAMPLES

Section 9.1

The **composite function** $(f \circ g)(x)$ is defined as

$$(f \circ g)(x) = f[g(x)]$$

Given $f(x) = x^2 + 3x - 1$ and $g(x) = x - 4$, then
$$(f \circ g)(x) = f[g(x)] = (x - 4)^2 + 3(x - 4) - 1$$
$$= x^2 - 8x + 16 + 3x - 12 - 1$$
$$= x^2 - 5x + 3$$
$$(g \circ f)(x) = g[f(x)] = (x^2 + 3x - 1) - 4$$
$$= x^2 + 3x - 5$$

A function is a **one-to-one function** if each element in the range corresponds with exactly one element in the domain.

The set $\{(1, 3), (-2, 5), (6, 2), (4, -1)\}$ is a one-to-one function since each element in the range corresponds with exactly one element in the domain.

For a function to be a one-to-one function, its graph must pass the **vertical line test** (the test to ensure it is a function) and the **horizontal line test** (to test the one-to-one criteria).

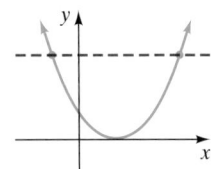

Function, but not a one-to-one function

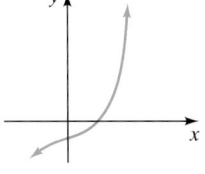

One-to-one function

If $f(x)$ is a one-to-one function with ordered pairs of the form (x, y), its **inverse function**, $f^{-1}(x)$, is a one-to-one function with ordered pairs of the form (y, x). Only one-to-one functions have inverse functions.

To Find the Inverse Function of a One-to-One Function

1. Replace $f(x)$ with y.

2. Interchange the two variables x and y.

3. Solve the equation for y.

4. Replace y with $f^{-1}(x)$ (this gives the inverse function using inverse function notation).

Find the inverse function for $f(x) = 2x + 5$. Graph $f(x)$ and $f^{-1}(x)$ on the same set of axes.

Solution:
$$f(x) = 2x + 5$$
$$y = 2x + 5$$
$$x = 2y + 5$$
$$x - 5 = 2y$$
$$\frac{1}{2}x - \frac{5}{2} = y$$
$$\text{or} \quad f^{-1}(x) = \frac{1}{2}x - \frac{5}{2}$$

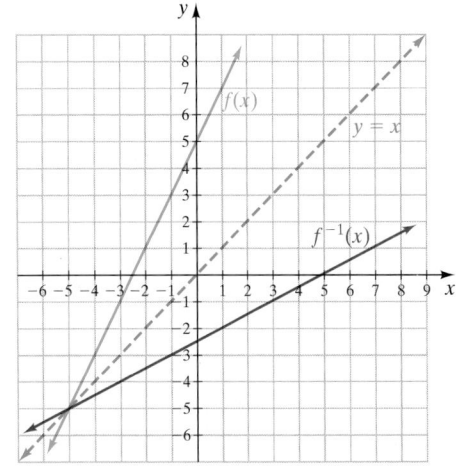

IMPORTANT FACTS AND CONCEPTS	EXAMPLES

Section 9.1 (cont.)

If two functions $f(x)$ and $f^{-1}(x)$ are inverses of each other, $(f \circ f^{-1})(x) = x$ and $(f^{-1} \circ f)(x) = x$.

For the previous example with $f(x) = 2x + 5$ and $f^{-1}(x) = \dfrac{1}{2}x - \dfrac{5}{2}$, then

$$(f \circ f^{-1})(x) = f[f^{-1}(x)] = 2\left(\frac{1}{2}x - \frac{5}{2}\right) + 5$$

$$= x - 5 + 5 = x$$

and

$$(f^{-1} \circ f)(x) = f^{-1}[f(x)] = \frac{1}{2}(2x + 5) - \frac{5}{2}$$

$$= x + \frac{5}{2} - \frac{5}{2} = x$$

Section 9.2

For any real number $a > 0$ and $a \neq 1$,

$$f(x) = a^x \quad \text{or} \quad y = a^x$$

is an **exponential function**.

For all exponential functions of the form $y = a^x$ or $f(x) = a^x$, where $a > 0$ and $a \neq 1$,

1. The domain of the function is $(-\infty, \infty)$.

2. The range of the function is $(0, \infty)$.

3. The graph of the function passes through the points $\left(-1, \dfrac{1}{a}\right)$, $(0, 1)$, and $(1, a)$.

Graph $y = 3^x$.

x	y
-2	$1/9$
-1	$1/3$
0	1
1	3
2	9

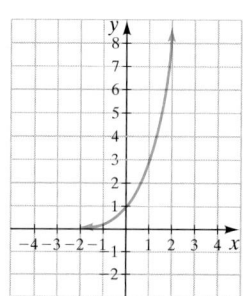

Section 9.3

Logarithms

For $x > 0$ and $a > 0$, $a \neq 1$

$$y = \log_a x \quad \text{means} \quad x = a^y$$

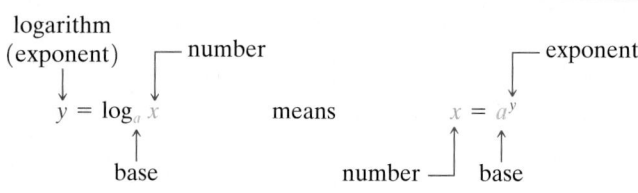

Exponential Form	Logarithmic Form
$9^2 = 81$	$\log_9 81 = 2$
$\left(\dfrac{1}{4}\right)^3 = \dfrac{1}{64}$	$\log_{1/4} \dfrac{1}{64} = 3$

Logarithmic Functions

For all logarithmic functions of the form $y = \log_a x$ or $f(x) = \log_a x$, where $a > 0$, $a \neq 1$, and $x > 0$,

1. The domain of the function is $(0, \infty)$.

2. The range of the function is $(-\infty, \infty)$.

3. The graph of the function passes through the points $\left(\dfrac{1}{a}, -1\right)$, $(1, 0)$, and $(a, 1)$.

Graph $y = \log_4 x$.

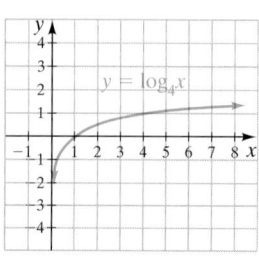

IMPORTANT FACTS AND CONCEPTS	EXAMPLES

Section 9.3 (cont.)

Characteristics of Exponential and Logarithmic Functions

$y = a^x$ and $y = \log_a x$ are inverse functions.

	Exponential Function	Logarithmic Function
	$y = a^x \ (a > 0, a \neq 1)$	$y = \log_a x \ (a > 0, a \neq 1)$
Domain:	$(-\infty, \infty)$	$(0, \infty)$
Range:	$(0, \infty)$	$(-\infty, \infty)$

Points on graph:
$$\left.\begin{array}{l} \left(-1, \dfrac{1}{a}\right) \\ (0, 1) \\ (1, a) \end{array}\right\}$$
x becomes y
y becomes x
$$\left\{\begin{array}{l} \left(\dfrac{1}{a}, -1\right) \\ (1, 0) \\ (a, 1) \end{array}\right.$$

Graph $y = 3^x$ and $y = \log_3 x$ on the same set of axes.

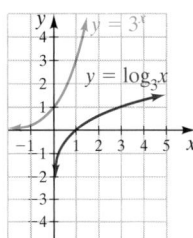

Section 9.4

Product Rule for Logarithms

For positive real numbers x, y, and $a, a \neq 1$,

$$\log_a xy = \log_a x + \log_a y \qquad \text{Property 1}$$

$$\log_5 (9 \cdot 13) = \log_5 9 + \log_5 13$$
$$\log_7 mn = \log_7 m + \log_7 n$$

Quotient Rule for Logarithms

For positive real numbers x, y, and $a, a \neq 1$,

$$\log_a \frac{x}{y} = \log_a x - \log_a y \qquad \text{Property 2}$$

$$\log_3 \frac{15}{4} = \log_3 15 - \log_3 4$$

$$\log_8 \frac{z + 1}{z + 3} = \log_8 (z + 1) - \log_8 (z + 3)$$

Power Rule for Logarithms

If x and a are positive real numbers, $a \neq 1$, and n is any real number, then

$$\log_a x^n = n \log_a x \qquad \text{Property 3}$$

$$\log_9 23^5 = 5 \log_9 23$$

$$\log_6 \sqrt[3]{x + 4} = \log_6 (x + 4)^{1/3} = \frac{1}{3} \log_6 (x + 4)$$

Additional Properties of Logarithms

If $a > 0$, and $a \neq 1$, then

$$\log_a a^x = x \qquad \text{Property 4}$$

$$\text{and} \qquad a^{\log_a x} = x \ (x > 0) \qquad \text{Property 5}$$

$$\log_4 16 = \log_4 4^2 = 2$$

$$7^{\log_7 3} = 3$$

Section 9.5

Common Logarithm

Base 10 logarithms are called common logarithms.

$$\log x \text{ means } \log_{10} x$$

The **common logarithm** of a positive real number is the *exponent* to which the base 10 is raised to obtain the number.

$$\text{If } \log N = L, \quad \text{then} \quad 10^L = N.$$

To find a common logarithm, use a scientific or graphing calculator. Here we round the answer to four decimal places.

$\log 17$ means $\log_{10} 17$

$\log (b + c)$ means $\log_{10} (b + c)$

If $\log 14 \approx 1.1461$, then $10^{1.1461} \approx 14$.

If $\log 0.6 \approx -0.2218$, then $10^{-0.2218} \approx 0.6$.

$\log 183 \approx 2.2625$

$\log 0.42 \approx -0.3768$

Powers of Ten

$$\text{If } \log x = y \quad \text{then} \quad x = 10^y$$

To find powers of 10, use a scientific or graphing calculator.

If $\log 1890.1662 \approx 3.2765$, then $10^{3.2765} \approx 1890.1662$.

If $\log 0.0143 \approx -1.8447$, then $10^{-1.8447} \approx 0.0143$.

IMPORTANT FACTS AND CONCEPTS	EXAMPLES

Section 9.6

Properties for Solving Exponential and Logarithmic Equations

a) If $x = y$, then $a^x = a^y$.

b) If $a^x = a^y$, then $x = y$.

c) If $x = y$, then $\log_b x = \log_b y$ $(x > 0, y > 0)$.

d) If $\log_b x = \log_b y$, then $x = y$ $(x > 0, y > 0)$.

Properties 6a–6d

a) If $x = 5$, then $3^x = 3^5$.

b) If $3^x = 3^5$, then $x = 5$.

c) If $x = 2$, then $\log x = \log 2$.

d) If $\log x = \log 2$, then $x = 2$.

Section 9.7

The **natural exponential function** is
$$f(x) = e^x$$
where $e \approx 2.7183$.

Natural logarithms are logarithms to the base e. Natural logarithms are indicated by the notation ln.

$$\log_e x = \ln x$$

For $x > 0$, if $y = \ln x$, then $e^y = x$.

The **natural logarithmic function** is

$$g(x) = \ln x$$

where the base $e \approx 2.7183$.

To approximate natural exponential and natural logarithmic values, use a scientific or graphing calculator.

The natural exponential function, $f(x) = e^x$, and the natural logarithmic function, $f(x) = \ln x$, are inverses of each other.

Graph $f(x) = e^x$ and $g(x) = \ln x$ on the same set of axes.

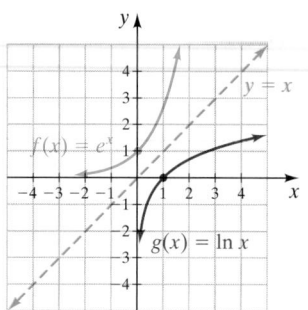

$\ln 5.83 \approx 1.7630$

If $\ln x = -2.09$, then $x = e^{-2.09} = 0.1237$.

Change of Base Formula

For any logarithm bases a and b, and positive number x,
$$\log_a x = \frac{\log_b x}{\log_b a}$$

$$\log_5 98 = \frac{\log 98}{\log 5} \approx 2.8488$$

Properties for Natural Logarithms

$\ln xy = \ln x + \ln y$ $(x > 0 \text{ and } y > 0)$ Product rule

$\ln \dfrac{x}{y} = \ln x - \ln y$ $(x > 0 \text{ and } y > 0)$ Quotient rule

$\ln x^n = n \ln x$ $(x > 0)$ Power rule

$\ln 7 \cdot 30 = \ln 7 + \ln 30$

$\ln \dfrac{x+1}{x+8} = \ln(x+1) - \ln(x+8)$

$\ln m^5 = 5 \ln m$

Additional Properties for Natural Logarithms and Natural Exponential Expressions

$\ln e^x = x$ — Property 7

$e^{\ln x} = x, \quad x > 0$ — Property 8

$\ln e^{19} = 19$

$e^{\ln 2} = 2$

Chapter 9 Review Exercises

[9.1] *Given $f(x) = x^2 - 3x + 4$ and $g(x) = 2x - 5$, find the following.*

1. $(f \circ g)(x)$ **2.** $(f \circ g)(3)$ **3.** $(g \circ f)(x)$ **4.** $(g \circ f)(-3)$

Given $f(x) = 6x + 7$ and $g(x) = \sqrt{x - 3}$, $x \geq 3$, find the following.

5. $(f \circ g)(x)$ **6.** $(g \circ f)(x)$

Determine whether each function is a one-to-one function.

7.

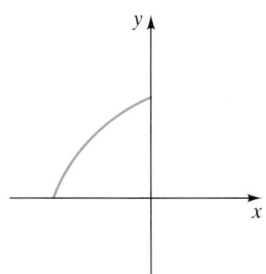

8.

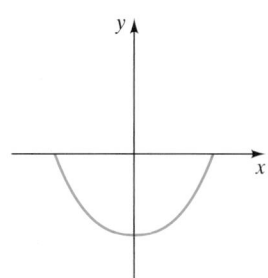

9. $\{(6, 2), (4, 0), (-5, 7), (3, 8)\}$

10. $\left\{(0, -2), (6, 1), (3, -2), \left(\frac{1}{2}, 4\right)\right\}$

11. $y = \sqrt{x + 8}, x \geq -8$

12. $y = x^2 - 9$

In Exercises 13 and 14, for each function, find the domain and range of both $f(x)$ and $f^{-1}(x)$.

13. $\{(5, 3), (6, 2), (-4, -3), (-1, 8)\}$

14.

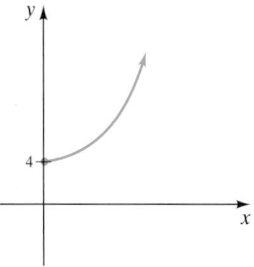

In Exercises 15 and 16, find $f^{-1}(x)$ and graph $f(x)$ and $f^{-1}(x)$ on the same axes.

15. $y = f(x) = 4x - 2$

16. $y = f(x) = \sqrt[3]{x - 1}$

17. Yards to Feet The function $f(x) = 36x$ converts yards, x, into inches. Find the inverse function that converts inches into yards. In the inverse function, what do x and $f^{-1}(x)$ represent?

18. Gallons to Quarts The function $f(x) = 4x$ converts gallons, x, into quarts. Find the inverse function that converts quarts into gallons. In the inverse function, what do x and $f^{-1}(x)$ represent?

[9.2] *Graph the following functions.*

19. $y = 2^x$

20. $y = \left(\frac{1}{2}\right)^x$

21. Compound Interest Jim Marino invested $1500 in a savings account that earns 4% interest compounded quarterly. Use the compound interest formula $A = p\left(1 + \frac{r}{n}\right)^{nt}$ to determine the amount in Jim's account after 5 years.

© Jupiter Unlimited

[9.3] *Write each equation in logarithmic form.*

22. $8^2 = 64$

23. $81^{1/4} = 3$

24. $5^{-3} = \frac{1}{125}$

Write each equation in exponential form.

25. $\log_2 32 = 5$

26. $\log_{1/4} \frac{1}{16} = 2$

27. $\log_6 \frac{1}{36} = -2$

Write each equation in exponential form and find the missing value.

28. $3 = \log_4 x$

29. $4 = \log_a 81$

30. $-3 = \log_{1/5} x$

Graph the following functions.

31. $y = \log_3 x$

32. $y = \log_{1/2} x$

[9.4] *Use the properties of logarithms to expand each expression.*

33. $\log_5 17^8$

34. $\log_3 \sqrt{x - 9}$

35. $\log \dfrac{6(a + 1)}{19}$

36. $\log \dfrac{x^4}{7(2x + 3)^5}$

Write the following as the logarithm of a single expression.

37. $5 \log x - 3 \log (x + 1)$

38. $4(\log 2 + \log x) - \log y$

39. $\dfrac{1}{3}[\ln x - \ln (x + 2)] - \ln 2$

40. $3 \ln x + \dfrac{1}{2} \ln (x + 1) - 6 \ln (x + 4)$

Evaluate.

41. $8^{\log_8 10}$

42. $\log_4 4^5$

43. $11^{\log_9 81}$

44. $9^{\log_8 \sqrt{8}}$

[9.5, 9.7] *Use a calculator to approximate each logarithm. Round your answers to four decimal places.*

45. $\log 819$

46. $\ln 0.0281$

Use a calculator to approximate each power of 10. Round your answers to four decimal places.

47. $10^{3.159}$

48. $10^{-3.157}$

Use a calculator to find x. Round your answers to four decimal places.

49. $\log x = 4.063$

50. $\log x = -1.2262$

Evaluate.

51. $\log 10^5$

52. $10^{\log 9}$

53. $7 \log 10^{3.2}$

54. $2(10^{\log 4.7})$

[9.6] *Solve without using a calculator.*

55. $625 = 5^x$

56. $49^x = \dfrac{1}{7}$

57. $2^{3x-1} = 32$

58. $27^x = 3^{2x+5}$

Solve using a calculator. Round your answers to four decimal places.

59. $7^x = 152$

60. $3.1^x = 856$

61. $12.5^{x+1} = 381$

62. $3^{x+2} = 8^x$

Solve the logarithmic equation.

63. $\log_7 (2x - 3) = 2$

64. $\log x + \log (4x - 19) = \log 5$

65. $\log_3 x + \log_3 (2x + 1) = 1$

66. $\ln (x + 1) - \ln (x - 2) = \ln 4$

[9.7] *Solve each exponential equation for the remaining variable. Round your answers to four decimal places.*

67. $50 = 25e^{0.6t}$

68. $100 = A_0 e^{-0.42(3)}$

Solve for the indicated variable.

69. $A = A_0 e^{kt}$, for t

70. $200 = 800e^{kt}$, for k

71. $\ln y - \ln x = 6$, for y

72. $\ln (y + 1) - \ln (x + 8) = \ln 3$, for y

Use the change of base formula to evaluate. Write the answers rounded to four decimal places.

73. $\log_2 196$

74. $\log_3 47$

[9.2–9.7]

75. Compound Interest Find the amount of money accumulated if Justine Elwood puts \$12,000 in a savings account yielding 6% interest compounded annually for 8 years. Use $A = p\left(1 + \dfrac{r}{n}\right)^{nt}$.

76. Interest Compounded Continuously If \$6000 is placed in a savings account paying 4% interest compounded continuously, find the time needed for the account to double in value.

77. Bacteria The bacteria *Escherichia coli* are commonly found in the bladders of humans. Suppose that 2000 bacteria are present at time 0. Then the number of bacteria present t minutes later may be found by the function $N(t) = 2000(2)^{0.05t}$.

a) When will 50,000 bacteria be present?

b) Suppose that a human bladder infection is classified as a condition with 120,000 bacteria. When would a person develop a bladder infection if he or she started with 2000 bacteria?

78. Atmospheric Pressure The atmospheric pressure, P, in pounds per square inch at an elevation of x feet above sea level can be found by the formula $P = 14.7e^{-0.00004x}$. Find the atmospheric pressure at the top of Half Dome in Yosemite National Park, an elevation of 8842 feet.

Yosemite National Park

79. Retention A class of history students is given a final exam at the end of the course. As part of a research project, the students are also given equivalent forms of the exam each month for n months. The average grade of the class after n months may be found by the function $A(n) = 72 - 18 \log(n + 1), n \geq 0$.

a) What was the class average when the students took the original exam ($n = 0$)?

b) What was the class average for the exam given 3 months later?

c) After how many months was the class average 58.0?

Chapter 9 Practice Test

CHAPTER
Test Prep
VIDEOS

Chapter Test Prep Videos provide fully worked-out solutions to any of the exercises you want to review. Chapter Test Prep Videos are available via **MyMathLab** *, or on* **You Tube** *(search "Angel Intermediate Algebra" and click on "Channels")*

1. a) Determine whether the following function is a one-to-one function.
$$\{(4, 2), (-3, 8), (-1, 3), (6, -7)\}$$

b) List the set of ordered pairs in the inverse function.

2. Given $f(x) = x^2 - 3$ and $g(x) = x + 2$, find **a)** $(f \circ g)(x)$. **b)** $(f \circ g)(6)$.

3. Given $f(x) = x^2 + 8$ and $g(x) = \sqrt{x - 5}, x \geq 5$, find **a)** $(g \circ f)(x)$. **b)** $(g \circ f)(7)$.

In Exercises 4 and 5, **a)** *find $f^{-1}(x)$ and* **b)** *graph $f(x)$ and $f^{-1}(x)$ on the same axes.*

4. $y = f(x) = -3x - 5$

5. $y = f(x) = \sqrt{x - 1}, x \geq 1$

6. What is the domain of $y = \log_5 x$?

7. Evaluate $\log_4 \dfrac{1}{256}$.

8. Graph $y = 3^x$.

9. Graph $y = \log_2 x$.

10. Write $2^{-5} = \dfrac{1}{32}$ in logarithmic form.

11. Write $\log_5 125 = 3$ in exponential form.

Write Exercises 12 and 13 in exponential form and find the missing value.

12. $4 = \log_2 (x + 3)$

13. $y = \log_{64} 16$

14. Expand $\log_2 \dfrac{x^3(x - 4)}{x + 2}$.

15. Write as the logarithm of a single expression.
$$7 \log_6 (x - 4) + 2 \log_6 (x + 3) - \frac{1}{2} \log_6 x.$$

16. Evaluate $10 \log_9 \sqrt{9}$.

17. a) Find $\log 4620$ rounded to 4 decimal places.

b) Find $\ln 0.0692$ rounded to 4 decimal places.

18. Solve $3^x = 19$ for x.

19. Solve $\log 4x = \log (x + 3) + \log 2$ for x.

20. Solve $\log (x + 5) - \log (x - 2) = \log 6$ for x.

21. Find N, rounded to 4 decimal places, if $\ln N = 2.79$.

22. Evaluate $\log_6 40$, rounded to 4 decimal places, using the change of base formula.

23. Solve $100 = 250e^{-0.03t}$ for t, rounded to 4 decimal places.

24. Savings Account What amount of money accumulates if Kim Lee puts $3500 in a savings account yielding 4% interest compounded quarterly for 10 years?

25. Carbon 14 The amount of carbon 14 remaining after t years is found by the formula $v = v_0 e^{-0.0001205t}$, where v_0 is the original amount of carbon 14. If a fossil originally contained 60 grams of carbon 14, and now contains 40 grams of carbon 14, how old is the fossil?

Cumulative Review Test

Take the following test and check your answers with those given in the back of the book. Review any questions that you answered incorrectly. The section where the material is covered is indicated after the answer.

1. Simplify $\dfrac{(2xy^2z^{-3})^2}{(3x^{-1}yz^2)^{-1}}$.

2. Evaluate $5^2 - (2 - 3^2)^2 + 4^3$.

3. **Dinner** Thomas Furgeson took his wife out to dinner. The cost of the meal before tax was $92. If the total price, including tax, was $98.90, find the tax rate.

4. Solve the inequality $-3 \le 2x - 7 < 8$ and write the answer as a solution set and in interval notation.

5. Solve $2x - 3y = 8$ for y.

6. Let $h(x) = \dfrac{x^2 + 4x}{x + 6}$. Find $h(-4)$.

7. Find the slope of the line in the figure below. Then write the equation of the given line.

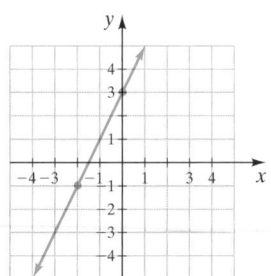

8. Graph $4x = 3y - 3$.

9. Graph $y \le \dfrac{1}{3}x + 6$.

10. Solve the system of equations
$$\frac{1}{2}x + \frac{1}{3}y = 13$$
$$\frac{1}{5}x + \frac{1}{8}y = 5$$

11. Divide $\dfrac{x^3 + 3x^2 + 5x + 9}{x + 1}$.

12. Factor $x^2 - 2xy + y^2 - 64$.

13. Solve $(2x + 1)^2 - 9 = 0$.

14. Solve $\dfrac{2x + 3}{x + 1} = \dfrac{3}{2}$.

15. Solve $a_n = a_1 + nd - d$, for d.

16. L varies inversely as the square of P. Find L when $P = 4$ and $k = 100$.

17. Simplify $4\sqrt{45x^3} + \sqrt{5x}$.

18. Solve $\sqrt{2a + 9} - a + 3 = 0$.

19. Solve $(x^2 - 5)^2 + 3(x^2 - 5) - 10 = 0$.

20. Let $g(x) = x^2 - 4x - 5$.

    a) Express $g(x)$ in the form $g(x) = a(x - h)^2 + k$.

    b) Draw the graph and label the vertex.

10 Conic Sections

Goals of This Chapter

The focus of this chapter is on graphing conic sections. These include the circle, ellipse, parabola, and hyperbola. We have already discussed parabolas. In this chapter, we will learn more about them. We will also solve nonlinear systems of equations both algebraically and graphically.

The shape of an ellipse gives it an unusual feature. Anything bounced off the wall of an elliptical shape from one focal point will ricochet to the other focal point. This feature has been used in architecture and medicine. One example is the National Statuary Hall in the Capitol Building, which has an elliptically shaped domed ceiling. If you whisper at one focal point, your whisper can be heard at the other focal point. Similarly, a ball hit from one focus of an elliptical billiard table will rebound to the other focal point. In Exercise 54 on page 649, you will determine the location of the foci of an elliptical billiard table.

© Santi Visalli Inc./Getty Images Inc.–Hulton Archive Photos

10.1 The Parabola and the Circle

1 Identify and describe the conic sections.

2 Review parabolas.

3 Graph parabolas of the form $x = a(y - k)^2 + h$.

4 Learn the distance and midpoint formulas.

5 Graph circles with centers at the origin.

6 Graph circles with centers at (h, k).

1 Identify and Describe the Conic Sections

A *parabola* is one type of conic section. Other conic sections are circles, ellipses, and hyperbolas. Each of these shapes is called a conic section because each can be made by slicing a cone and observing the shape of the slice. The methods used to slice the cone to obtain each conic section are illustrated in **Figure 10.1**.

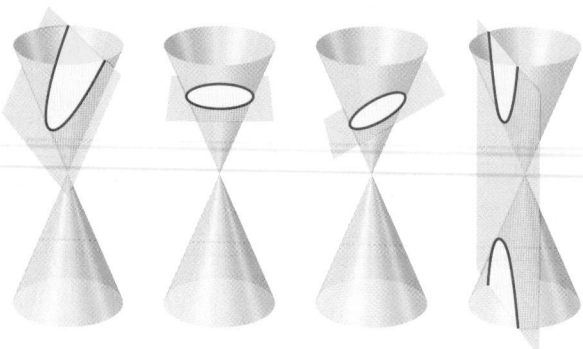

FIGURE 10.1   Parabola   Circle   Ellipse   Hyperbola

2 Review Parabolas

Parabolas were first discussed in Section 8.5. They will be discussed further in this chapter. Example 1 will refresh your memory on how to graph parabolas in the forms $y = ax^2 + bx + c$ and $y = a(x - h)^2 + k$.

EXAMPLE 1   Consider $y = 2x^2 + 4x - 6$. ────────

a) Write the equation in $y = a(x - h)^2 + k$ form.

b) Determine whether the parabola opens upward or downward.

c) Determine the vertex of the parabola.

d) Determine the y-intercept of the parabola.

e) Determine the x-intercepts of the parabola.

f) Graph the parabola.

Solution

a) First, factor 2 from the two terms containing the variable to make the coefficient of the squared term 1. (Do not factor 2 from the constant, -6.)

$$y = 2x^2 + 4x - 6$$
$$= 2(x^2 + 2x) - 6 \qquad \text{2 was factored from the first two terms.}$$
$$= 2\,(x^2 + 2x\,+\,1) - 2 - 6 \quad \text{Complete the square.}$$
$$= 2(x + 1)^2 - 8 \qquad \text{Simplify.}$$

b) The parabola opens upward because $a = 2$, which is greater than 0.

c) The vertex of the graph of an equation in the form $y = a(x - h)^2 + k$ is (h, k). Therefore, the vertex of the graph of $y = 2(x + 1)^2 - 8$ is $(-1, -8)$. The vertex of a parabola can also be found using

$$\left(-\frac{b}{2a}, \frac{4ac - b^2}{4a}\right) \quad \text{or} \quad \left(-\frac{b}{2a}, f\left(-\frac{b}{2a}\right)\right)$$

Show that both of these procedures give $(-1, -8)$ as the vertex of the parabola now.

d) To determine the *y*-intercept, let $x = 0$ and solve for *y*.

$$y = 2(x + 1)^2 - 8$$
$$= 2(0 + 1)^2 - 8$$
$$= -6$$

The *y*-intercept is $(0, -6)$.

e) To determine the *x*-intercepts, let $y = 0$ and solve for *x*.

$$y = 2(x + 1)^2 - 8$$

$0 = 2(x + 1)^2 - 8$ 0 was substituted for *y*.

$8 = 2(x + 1)^2$ 8 was added to both sides.

$4 = (x + 1)^2$ Both sides were divided by 2.

$\pm 2 = x + 1$ Square root property

$-1 \pm 2 = x$ 1 was subtracted from both sides.

$x = -1 - 2$ or $x = -1 + 2$

$x = -3$ $x = 1$

The *x*-intercepts are $(-3, 0)$ and $(1, 0)$. The *x*-intercepts could also be found by substituting 0 for *y* in $y = 2x^2 + 4x - 6$ and solving for *x* using factoring or the quadratic formula. Do this now and show that you get the same *x*-intercepts.

f) We use the vertex and the *x*- and *y*-intercepts to draw the graph, which is shown in **Figure 10.2**.

Now Try Exercise 9

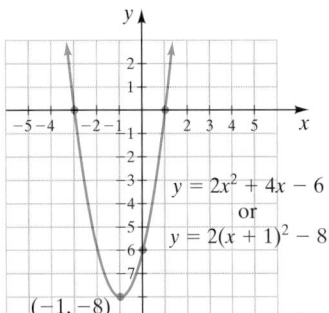

FIGURE 10.2

$y = 2x^2 + 4x - 6$
or
$y = 2(x + 1)^2 - 8$

$(-1, -8)$

3 Graph Parabolas of the Form $x = a(y - k)^2 + h$

Parabolas can also open to the right or left. The graph of an equation of the form $x = a(y - k)^2 + h$ will be a parabola whose vertex is at the point (h, k). If *a* is a positive number, the parabola will open to the right, and if *a* is a negative number, the parabola will open to the left.

The four different forms of a parabola are shown in **Figure 10.3**.

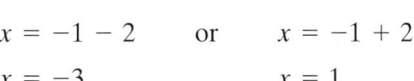

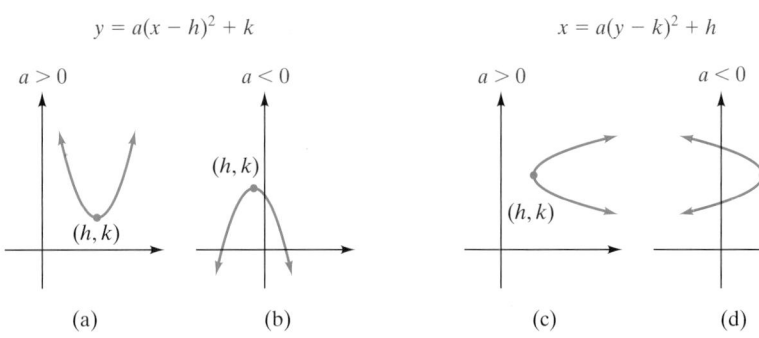

FIGURE 10.3   (a)   (b)   (c)   (d)

Understanding Algebra

With a parabola of the form $x = a(y - k)^2 + h$

- If $a > 0$, the parabola opens to the right.
- If $a < 0$, the parabola opens to the left.

Parabola with Vertex at (h, k)

1. $y = a(x - h)^2 + k, a > 0$ (opens upward)   Figure 10.3a
2. $y = a(x - h)^2 + k, a < 0$ (opens downward)   Figure 10.3b
3. $x = a(y - k)^2 + h, a > 0$ (opens to the right)   Figure 10.3c
4. $x = a(y - k)^2 + h, a < 0$ (opens to the left)   Figure 10.3d

---

Note that equations of the form $y = a(x - h)^2 + k$ are functions since their graphs pass the vertical line test. However, equations of the form $x = a(y - k)^2 + h$ are not functions since their graphs do not pass the vertical line test.

EXAMPLE 2 Sketch the graph of $x = -2(y + 4)^2 - 1$.

Solution The graph opens to the left since the equation is of the form $x = a(y - k)^2 + h$ and $a = -2$, which is less than 0. The equation can be expressed as $x = -2[y - (-4)]^2 - 1$. Thus, $h = -1$ and $k = -4$. The vertex of the graph is $(-1, -4)$. See **Figure 10.4**.

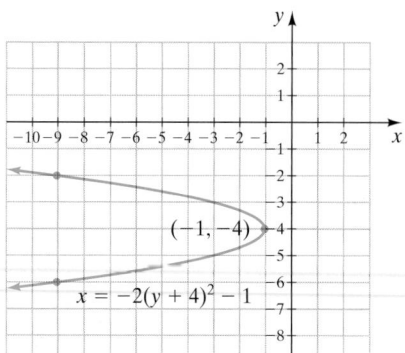

FIGURE 10.4

If we set $y = 0$, we see that the x-intercept is at $-2(0 + 4)^2 - 1 = -2(16) - 1$ or -33. By substituting values for y you can find the corresponding values of x. When $y = -2$, $x = -9$, and when $y = -6$, $x = -9$. These points are marked on the graph. Notice that this graph has no y-intercept.

Now Try Exercise 21

EXAMPLE 3

a) Write the equation $x = 2y^2 + 12y + 13$ in the form $x = a(y - k)^2 + h$.

b) Graph $x = 2y^2 + 12y + 13$.

Solution

a) First factor 2 from the first two terms. Then, complete the square on the expression within the parentheses.

$$x = 2y^2 + 12y + 13$$
$$= \boxed{2}\ (y^2 + 6y) + 13 \qquad \text{2 was factored from the first two terms.}$$
$$= \boxed{2}\ (y^2 + 6y + 9) + (\boxed{2})(-9) + 13 \quad \text{Complete the square.}$$
$$= 2(y^2 + 6y + 9) - 18 + 13 \qquad \text{Simplify.}$$
$$= 2(y + 3)^2 - 5$$

b) Since $a > 0$, the parabola opens to the right. Note that when $y = 0$, $x = 2(0)^2 + 12(0) + 13 = 13$. Thus, the x-intercept is $(13, 0)$.

   The vertex of the parabola is $(-5, -3)$. When $y = -6$, we find that $x = 13$. Thus, another point on the graph is $(13, -6)$.

   Using the quadratic formula, we can determine that the y-intercepts are about $(0, -4.6)$ and $(0, -1.4)$. The graph is shown in **Figure 10.5**.

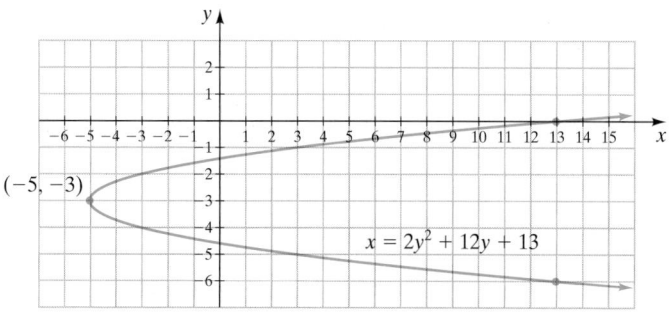

FIGURE 10.5

Now Try Exercise 35

4 Learn the Distance and Midpoint Formulas

Now we will derive a formula to find the *distance* between two points on a line. Consider **Figure 10.6**.

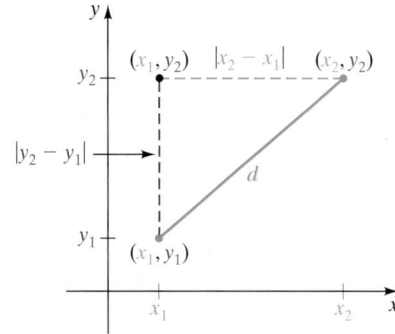

FIGURE 10.6

The horizontal distance between the two points (x_1, y_2) and (x_2, y_2), indicated by the red dashed line, is $|x_2 - x_1|$.

The vertical distance between the points (x_1, y_1) and (x_1, y_2), indicated by the green dashed line, is $|y_2 - y_1|$.

Using the Pythagorean Theorem where d is the distance between the two points (x_1, y_1) and (x_2, y_2), we get

$$d^2 = |x_2 - x_1|^2 + |y_2 - y_1|^2$$

Since any nonzero number squared is positive, we do not need absolute value signs. We can therefore write

$$d^2 = (x_2 - x_1)^2 + (y_2 - y_1)^2$$

Using the square root property, with the principal square root, we get the distance between the points (x_1, y_1) and (x_2, y_2), which is $d = \sqrt{(x_2 - x_1)^2 + (y_2 - y_1)^2}$.

Distance Formula

The distance, d, between any two points (x_1, y_1) and (x_2, y_2) can be found by the distance formula:

$$d = \sqrt{(x_2 - x_1)^2 + (y_2 - y_1)^2}$$

Helpful Hint

The distance between any two points will always be a positive number. When finding the distance, it makes no difference which point we designate as point 1, (x_1, y_1), or point 2, (x_2, y_2). Note that the square of any real number will always be greater than or equal to 0. For example, $(5 - 2)^2 = (2 - 5)^2 = 9$.

EXAMPLE 4 Determine the distance between the points $(4, 5)$ and $(-2, 3)$.

Solution We plot the points (**Fig. 10.7**). Label $(4, 5)$ point 1 and $(-2, 3)$ point 2. Thus, (x_2, y_2) represents $(-2, 3)$ and (x_1, y_1) represents $(4, 5)$. Now use the distance formula to find the distance, d.

$$
\begin{aligned}
d &= \sqrt{(x_2 - x_1)^2 + (y_2 - y_1)^2} \\
&= \sqrt{(-2 - 4)^2 + (3 - 5)^2} \\
&= \sqrt{(-6)^2 + (-2)^2} \\
&= \sqrt{36 + 4} \\
&= \sqrt{40} \quad \text{or} \quad \approx 6.32
\end{aligned}
$$

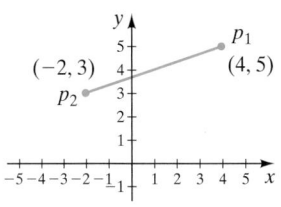

FIGURE 10.7

Thus, the distance between the points $(4, 5)$ and $(-2, 3)$ is $\sqrt{40}$ or about 6.32 units.

Now Try Exercise 47

Avoiding Common Errors

Students will sometimes begin finding the distance correctly using the distance formula but will forget to take the square root of the sum $(x_2 - x_1)^2 + (y_2 - y_1)^2$ to obtain the correct answer. When taking the square root, remember that $\sqrt{a^2 + b^2} \neq a + b$.

It is often necessary to find the **midpoint** of a line segment between two given endpoints.

Midpoint Formula

Given any two points (x_1, y_1) and (x_2, y_2), the point halfway between the given points can be found by the midpoint formula:

$$\text{midpoint} = \left(\frac{x_1 + x_2}{2}, \frac{y_1 + y_2}{2} \right)$$

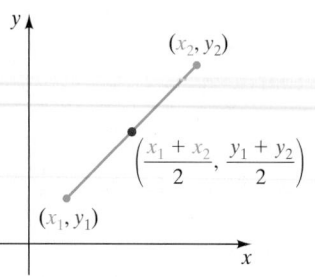

To find the midpoint, we take the average (the mean) of the x-coordinates and of the y-coordinates.

EXAMPLE 5 A line segment through the center of a circle intersects the circle at the points $(-3, 6)$ and $(4, 1)$. Find the center of the circle.

Solution   To find the center of the circle, we find the midpoint of the line segment between $(-3, 6)$ and $(4, 1)$. It makes no difference which points we label (x_1, y_1) and (x_2, y_2). We will let $(-3, 6)$ be (x_1, y_1) and $(4, 1)$ be (x_2, y_2). See **Figure 10.8**.

$$\text{midpoint} = \left(\frac{x_1 + x_2}{2}, \frac{y_1 + y_2}{2} \right)$$

$$= \left(\frac{-3 + 4}{2}, \frac{6 + 1}{2} \right) = \left(\frac{1}{2}, \frac{7}{2} \right)$$

The point $\left(\frac{1}{2}, \frac{7}{2} \right)$ is halfway between the points $(-3, 6)$ and $(4, 1)$. It is also the center of the circle.

Now Try Exercise 59

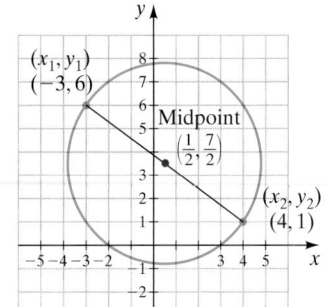

FIGURE 10.8

5 Graph Circles with Centers at the Origin

Circle

A **circle** is the set of points in a plane that are the same distance, called the **radius**, from a fixed point, called the **center**.

The *standard form* of the equation of a circle whose center is at the origin may be derived using the distance formula. Let (x, y) be a point on a circle of radius r with center at $(0, 0)$. See **Figure 10.9**. Using the distance formula, we have

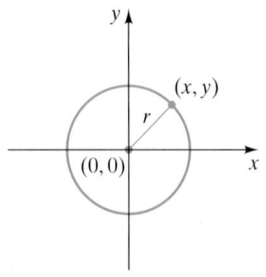

FIGURE 10.9

$$d = \sqrt{(x_2 - x_1)^2 + (y_2 - y_1)^2} \qquad \text{Distance formula}$$

$$\text{or} \quad r = \sqrt{(x - 0)^2 + (y - 0)^2} \qquad \begin{array}{l} \text{Substitute } r \text{ for } d, (x, y) \text{ for} \\ (x_2, y_2), \text{ and } (0, 0) \text{ for } (x_1, y_1). \end{array}$$

$$r = \sqrt{x^2 + y^2} \qquad \text{Simplify the radicand.}$$

$$r^2 = x^2 + y^2 \qquad \text{Square both sides.}$$

Circle with Its Center at the Origin and Radius *r*

$$x^2 + y^2 = r^2$$

For example, $x^2 + y^2 = 16$ is a circle with its center at the origin and radius 4, and $x^2 + y^2 = 10$ is a circle with its center at the origin and radius $\sqrt{10}$. Note that $4^2 = 16$ and $(\sqrt{10})^2 = 10$. Both circles are illustrated below.

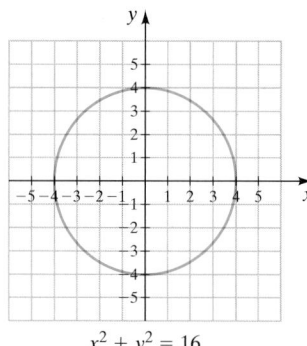
$x^2 + y^2 = 16$

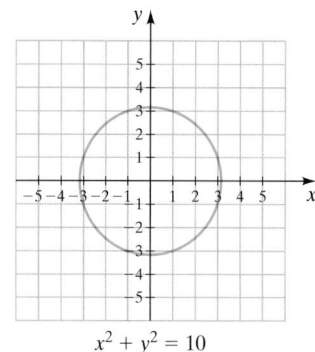
$x^2 + y^2 = 10$

EXAMPLE 6 Graph the following equations.

a) $x^2 + y^2 = 64$ **b)** $y = \sqrt{64 - x^2}$ **c)** $y = -\sqrt{64 - x^2}$

Solution

a) If we rewrite the equation as

$$x^2 + y^2 = 8^2$$

we see that this is the equation of a circle with center at the origin and radius 8. The graph is illustrated is **Figure 10.10**.

b) If we solve the equation $x^2 + y^2 = 64$ for y, we obtain

$$y^2 = 64 - x^2$$
$$y = \pm\sqrt{64 - x^2}$$

In the equation $y = \pm\sqrt{64 - x^2}$, the equation $y = +\sqrt{64 - x^2}$ or, simply, $y = \sqrt{64 - x^2}$, represents the top half of the circle, while the equation $y = -\sqrt{64 - x^2}$ represents the bottom half of the circle. Thus, the graph of $y = \sqrt{64 - x^2}$, where y is the principal square root, lies above and on the x-axis. The graph is the semicircle shown in **Figure 10.11**.

c) The graph of $y = -\sqrt{64 - x^2}$ is also a semicircle. However, this graph lies below and on the x-axis. The graph is shown in **Figure 10.12**.

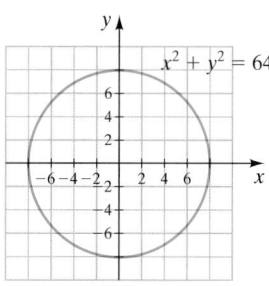

FIGURE 10.10

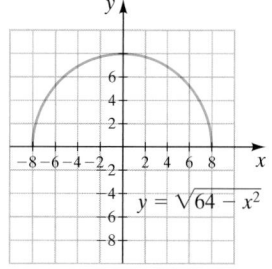

FIGURE 10.11

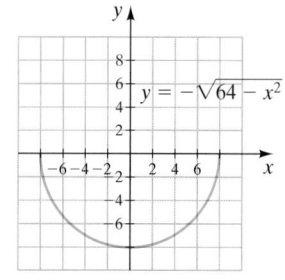

FIGURE 10.12

Now Try Exercise 91

Consider the equations $y = \sqrt{64 - x^2}$ and $y = -\sqrt{64 - x^2}$ in Example **6 b)** and **6 c)**. If you square both sides of the equations and rearrange the terms, you will obtain $x^2 + y^2 = 64$.

When using your calculator, you insert the function you wish to graph to the right of $y =$. Circles are not functions since they do not pass the vertical line test. To graph the equation $x^2 + y^2 = 64$, which is a circle of radius 8, we solve the equation for y to obtain $y = \pm\sqrt{64 - x^2}$. We then graph the two functions $Y_1 = \sqrt{64 - x^2}$ and $Y_2 = -\sqrt{64 - x^2}$ on the same axes to obtain the circle. These graphs are illustrated in **Figure 10.13**.

Because of the distortion due to unequal scales on the axes, the graph does not appear to be a circle. When you use the Z Square feature of the calculator found in the ZOOM menu, the figure appears as a circle (see **Fig. 10.14**).

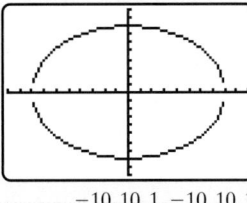

$-10, 10, 1, -10, 10, 1$

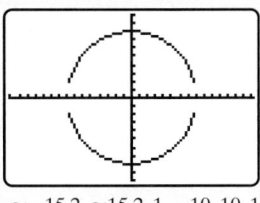

$\approx -15.2, \approx 15.2, 1, -10, 10, 1$

FIGURE 10.13   **FIGURE 10.14**

6 Graph Circles with Centers at (*h*, *k*)

The standard form of a circle with center at (h, k) and radius r can be derived using the distance formula. Let (h, k) be the center of the circle and let (x, y) be any point on the circle (see **Fig. 10.15**). If the radius, r, represents the distance between a point, (x, y), on the circle and its center, (h, k), then by the distance formula

$$r = \sqrt{(x - h)^2 + (y - k)^2}$$

We now square both sides of the equation to obtain the standard form of a circle with center at (h, k) and radius r.

$$r^2 = (x - h)^2 + (y - k)^2$$

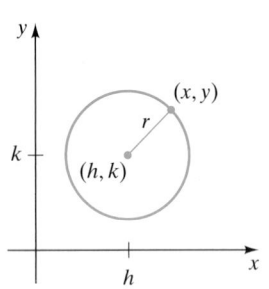

FIGURE 10.15

> **Circle with Its Center at (*h*, *k*) and Radius *r***
>
> $$(x - h)^2 + (y - k)^2 = r^2$$

EXAMPLE 7 Determine the equation of the circle shown in **Figure 10.16**.

Solution   The center is $(-3, 2)$ and the radius is 3.

$$(x - h)^2 + (y - k)^2 = r^2$$
$$[x - (-3)]^2 + (y - 2)^2 = 3^2$$
$$(x + 3)^2 + (y - 2)^2 = 9$$

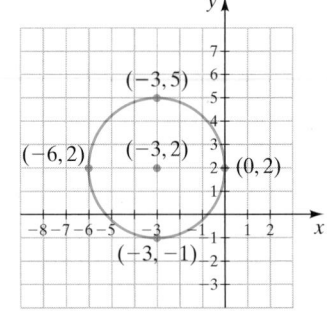

FIGURE 10.16

Now Try Exercise 77

EXAMPLE 8

a) Show that the graph of the equation $x^2 + y^2 + 6x - 2y - 6 = 0$ is a circle.

b) Determine the center and radius of the circle and then draw the circle.

c) Find the area of the circle.

Solution

a) We will write this equation in standard form by completing the square. First we rewrite the equation, placing all the terms containing x's together and the terms containing y's together.

$$x^2 + 6x + y^2 - 2y - 6 = 0$$

Understanding Algebra

The area of a circle is

$$A = \pi r^2.$$

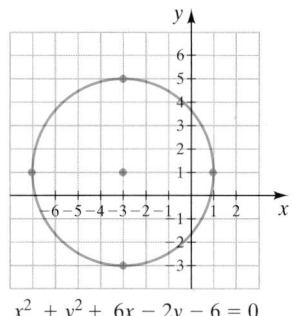

$$x^2 + y^2 + 6x - 2y - 6 = 0$$

FIGURE 10.17

Then we rewrite the equation with the constant on the right side of the equation.

$$x^2 + 6x + y^2 - 2y = 6 \qquad \text{Rewrite equation.}$$

Now we complete the square twice, once for each variable.

$$x^2 + 6x \boxed{+ 9} + y^2 - 2y = 6 \boxed{+ 9} \qquad \text{Complete the square on the } x\text{-terms.}$$

$$x^2 + 6x + 9 + y^2 - 2y \boxed{+ 1} = 6 + 9 \boxed{+ 1} \qquad \text{Complete the square on the } y\text{-terms.}$$

or

$$\underbrace{x^2 + 6x + 9}_{(x + 3)^2} + \underbrace{y^2 - 2y + 1}_{(y - 1)^2} = 16$$

$$(x + 3)^2 + (y - 1)^2 = 16$$

$$(x + 3)^2 + (y - 1)^2 = 4^2$$

b) The center of the circle is at $(-3, 1)$ and the radius is 4. The circle is sketched in **Figure 10.17**.

c) The area is

$$A = \pi r^2 = \pi(4)^2 = 16\pi \approx 50.3 \text{ square units}$$

Now Try Exercise 101

EXERCISE SET 10.1

 **MathXL®**   **MyMathLab**
MathXL®   MyMathLab

Warm-Up Exercises

Fill in the blanks with the appropriate word, phrase, or symbol(s) from the following list.

distance $(0, 0)$ parabola   circle (k, h) (h, k) (x, y) midpoint

1. The formula, midpoint $= \left(\dfrac{x_1 + x_2}{2}, \dfrac{y_1 + y_2}{2} \right)$, is called the _____ formula.

2. The formula, $d = \sqrt{(x_2 - x_1)^2 + (y_2 - y_1)^2}$, is called the _____ formula.

3. The equation $y = a(x - h)^2 + k$ is the equation of a _____.

4. The equation $x^2 + y^2 = r^2$ is the equation of a circle with center at _____ and radius r.

5. The equation $(x - h)^2 + (y - k)^2 = r$ is the equation of a circle with center at _____ and radius r.

6. The four conic sections are the parabola, _____, ellipse, and hyperbola.

Practice the Skills

Graph each equation.

7. $y = (x - 2)^2 + 3$

8. $y = (x - 2)^2 - 3$

9. $y = (x + 3)^2 + 2$

10. $y = (x + 3)^2 - 4$

11. $y = (x - 2)^2 - 1$

12. $y = (x + 2)^2 + 1$

13. $y = -(x - 1)^2 + 1$

14. $y = -(x + 4)^2 - 5$

15. $y = -(x + 3)^2 + 4$

16. $y = 2(x + 1)^2 - 3$

17. $y = -3(x - 5)^2 + 3$

18. $x = (y - 1)^2 + 1$

19. $x = (y - 4)^2 - 3$

20. $x = -(y - 2)^2 + 1$

21. $x = -(y - 5)^2 + 4$

22. $x = -2(y + 4)^2 - 3$

23. $x = -5(y + 3)^2 - 6$

24. $x = 3(y + 1)^2 + 5$

25. $y = -2\left(x + \dfrac{1}{2} \right)^2 + 6$

26. $y = -\left(x - \dfrac{5}{2} \right)^2 + \dfrac{1}{2}$

In Exercises 27–40, **a)** *Write the equation in the form $y = a(x - h)^2 + k$ or $x = a(y - k)^2 + h$.* **b)** *Graph the equation.*

27. $y = x^2 + 2x$

28. $y = x^2 - 2x$

29. $y = x^2 + 6x$

30. $y = x^2 - 4x$

31. $x = y^2 + 4y$

32. $x = y^2 - 6y$

33. $y = x^2 + 7x + 10$

34. $y = x^2 + 2x - 7$

35. $x = -y^2 + 6y - 9$

36. $x = -y^2 - 2y + 5$

37. $y = -x^2 + 4x - 4$

38. $y = 2x^2 - 4x - 4$

39. $x = -y^2 + 3y - 4$

40. $x = 3y^2 - 12y - 36$

Determine the distance between each pair of points. Use a calculator where appropriate and round your answers to the nearest hundredth.

41. $(5, -1)$ and $(5, -6)$

42. $(-7, 2)$ and $(-3, 2)$

43. $(-1, 6)$ and $(8, 6)$

44. $(1, 11)$ and $(4, 15)$

45. $(-1, -3)$ and $(4, 9)$

46. $(-4, -7)$ and $(2, 1)$

47. $(-4, -5)$ and $(5, -2)$ **48.** $(6, 7)$ and $(11, 0)$ **49.** $(3, -1)$ and $\left(\frac{1}{2}, 4\right)$

50. $\left(-\frac{1}{4}, 2\right)$ and $\left(-\frac{3}{2}, 6\right)$ **51.** $(-1.6, 3.5)$ and $(-4.3, -1.7)$ **52.** $(5.2, -3.6)$ and $(-1.6, 2.3)$

53. $(\sqrt{7}, \sqrt{3})$ and $(0, 0)$ **54.** $(-\sqrt{2}, -\sqrt{5})$ and $(0, 0)$

Determine the midpoint of the line segment between each pair of points.

55. $(1, 3)$ and $(5, 7)$ **56.** $(0, 8)$ and $(4, -6)$ **57.** $(-7, 3)$ and $(7, -3)$

58. $(4, 9)$ and $(1, -5)$ **59.** $(-1, 4)$ and $(4, 6)$ **60.** $(-8, -9)$ and $(-6, -3)$

61. $\left(3, \frac{1}{2}\right)$ and $(2, -4)$ **62.** $\left(\frac{5}{2}, 1\right)$ and $\left(2, \frac{9}{2}\right)$ **63.** $(\sqrt{3}, 2)$ and $(\sqrt{2}, 5)$

64. $(-\sqrt{7}, 8)$ and $(\sqrt{5}, \sqrt{3})$

Write the equation of each circle with the given center and radius.

65. Center $(0, 0)$, radius 5   **66.** Center $(0, 0)$, radius 9

67. Center $(2, 0)$, radius 8   **68.** Center $(-3, 0)$, radius 7

69. Center $(0, 5)$, radius 2   **70.** Center $(0, -6)$, radius 10

71. Center $(3, 4)$, radius 4   **72.** Center $(-5, 2)$, radius 3

73. Center $(7, -6)$, radius 12   **74.** Center $(-6, -1)$, radius 15

75. Center $(1, 2)$, radius $\sqrt{5}$ **76.** Center $(-7, -2)$, radius $\sqrt{13}$

Write the equation of each circle. Assume the radius is a whole number.

77.
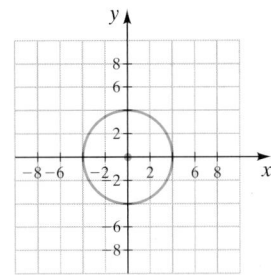

78.
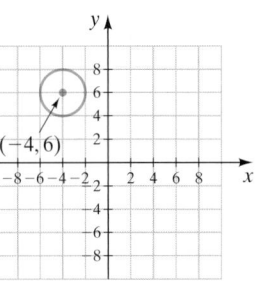

79.
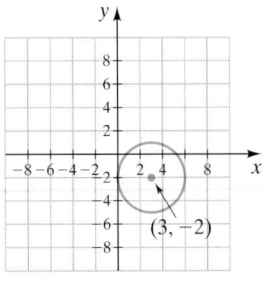

80.
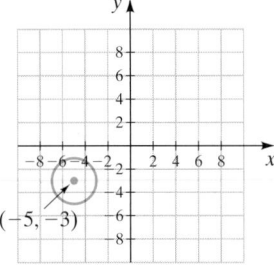

Graph each equation.

81. $x^2 + y^2 = 16$ **82.** $x^2 + y^2 = 5$ **83.** $x^2 + y^2 = 10$ **84.** $(x - 1)^2 + y^2 = 7$

85. $(x + 4)^2 + y^2 = 25$ **86.** $x^2 + (y + 1)^2 = 9$ **87.** $x^2 + (y - 3)^2 = 4$ **88.** $(x - 2)^2 + (y + 3)^2 = 16$

89. $(x + 8)^2 + (y + 2)^2 = 9$ **90.** $(x + 3)^2 + (y - 4)^2 = 36$ **91.** $y = \sqrt{25 - x^2}$ **92.** $y = \sqrt{16 - x^2}$

93. $y = -\sqrt{4 - x^2}$ **94.** $y = -\sqrt{49 - x^2}$

In Exercises 95–102, **a)** *use the method of completing the square to write each equation in standard form.* **b)** *Draw the graph.*

95. $x^2 + y^2 + 8x + 15 = 0$ **96.** $x^2 + y^2 + 4y = 0$

97. $x^2 + y^2 + 6x - 4y + 4 = 0$ **98.** $x^2 + y^2 + 2x - 4y - 4 = 0$

99. $x^2 + y^2 + 6x - 2y + 6 = 0$ **100.** $x^2 + y^2 + 4x - 6y - 3 = 0$

101. $x^2 + y^2 - 8x + 2y + 13 = 0$ **102.** $x^2 + y^2 - x + 3y - \frac{3}{2} = 0$

Problem Solving

103. Find the area of the circle in Exercise 85.   **104.** Find the area of the circle in Exercise 87.

In Exercises 105–108, find the x- and y-intercepts, if they exist, of the graph of each equation.

105. $x = y^2 - 6y - 7$

106. $x = -y^2 + 8y - 12$

107. $x = 2(y - 3)^2 + 6$

108. $x = -(y + 2)^2 - 8$

109. If you know the midpoint of a line segment, is it possible to determine the length of the line segment? Explain.

110. If you know one endpoint of a line segment and the length of the line segment, can you determine the other endpoint? Explain.

111. Find the length of the line segment whose midpoint is $(4, -6)$ with one endpoint at $(7, -2)$.

112. Find the length of the line segment whose midpoint is $(-2, 4)$ with one endpoint at $(3, 6)$.

113. Find the equation of a circle with center at $(-6, 2)$ that is tangent to the x-axis (that is, the circle touches the x-axis at only one point).

114. Find the equation of a circle with center at $(-3, 5)$ that is tangent to the y-axis.

*In Exercises 115 and 116, find **a)** the radius of the circle whose diameter is along the line shown, **b)** the center of the circle, and **c)** the equation of the circle.*

115.

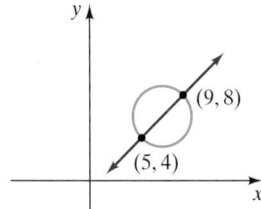

116.

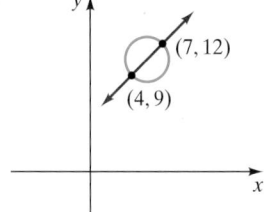

117. Points of Intersection What is the maximum number and the minimum number of points of intersection possible for the graphs of $y = a(x - h_1)^2 + k_1$ and $x = a(y - k_2)^2 + h_2$? Explain.

118. Triangle Consider the figure below.

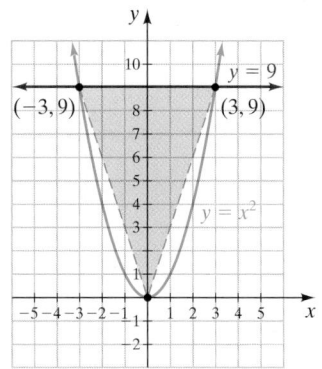

a) Find the area of the triangle outlined in green.

b) When a triangle is inscribed within a parabola, as in the figure, the area within the parabola from the base of the triangle is $\frac{4}{3}$ the area of the triangle. Find the area within the parabola from $x = -3$ to $x = 3$.

119. Ferris Wheel The Ferris wheel at Navy Pier in Chicago is 150 feet tall. The radius of the wheel itself is 68.2 feet.

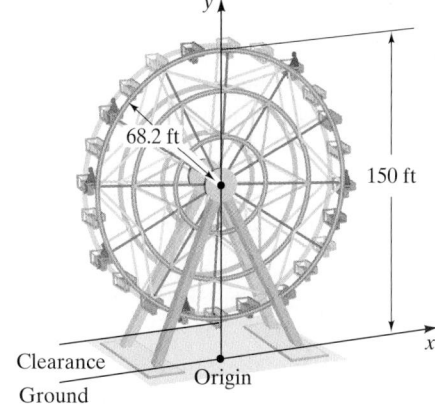

a) What is the clearance below the wheel?

b) How high is the center of the wheel from the ground?

c) Find the equation of the wheel. Assume the origin is on the ground directly below the center of the wheel.

120. Shaded Area Find the shaded area of the square in the figure. The equation of the circle is $x^2 + y^2 = 9$.

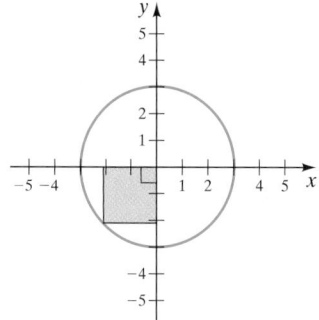

121. Shaded Area Consider the figure below. Write an equation for

a) the blue circle,

b) the red circle, and

c) the green circle.

d) Find the shaded area.

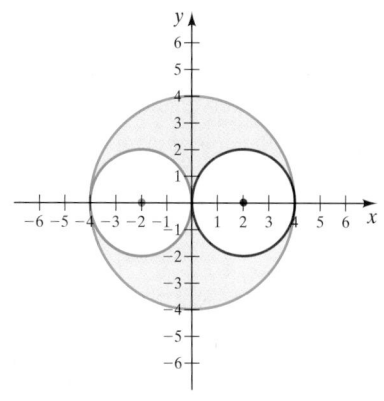

122. Points of Intersection Consider the equations $x^2 + y^2 = 16$ and $(x - 2)^2 + (y - 2)^2 = 16$. By considering the center and radius of each circle, determine the number of points of intersection of the two circles.

123. Concentric Circles Find the area between the two concentric circles whose equations are $(x - 2)^2 + (y + 4)^2 = 16$ and $(x - 2)^2 + (y + 4)^2 = 64$. *Concentric circles* are circles that have the same center.

124. Tunnel A highway department is planning to construct a semicircular one-way tunnel through a mountain. The tunnel is to be large enough so that a truck 8 feet wide and 10 feet tall will pass through the center of the tunnel with 1 foot to spare directly above the corner of the truck when it

is driving down the center of the tunnel (as shown in the figure below). Determine the minimum radius of the tunnel.

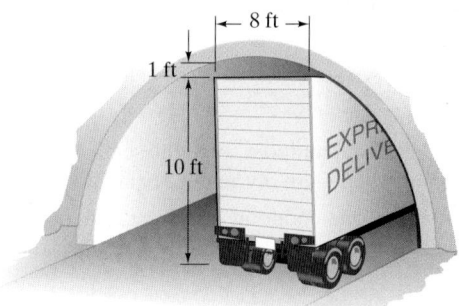

Concept/Writing Exercises

125. Will all parabolas of the form $y = a(x - h)^2 + k, a > 0$ be functions? Explain. What will be the domain and range of $y = a(x - h)^2 + k, a > 0$?

126. Will all parabolas of the form $x = a(y - k)^2 + h, a > 0$ be functions? Explain. What will be the domain and range of $x = a(y - k)^2 + h, a > 0$?

127. How will the graphs of $y = 2(x - 3)^2 + 4$ and $y = -2(x - 3)^2 + 4$ compare?

128. What is the definition of a circle?

129. Is $x^2 - y^2 = 9$ an equation for a circle? Explain.

130. Is $-x^2 + y^2 = 25$ an equation for a circle? Explain.

131. Is $2x^2 + 3y^2 = 6$ an equation for a circle? Explain.

132. Is $x = y^2 - 6y + 3$ an equation for a parabola? Explain.

133. Is $x^2 = y^2 - 6y + 3$ an equation for a parabola? Explain.

134. Is $x = y + 2$ an equation for a parabola? Explain.

Group Activity

Discuss and answer Exercise 135 as a group.

135. Equation of a Parabola The equation of a parabola can be found if three points on the parabola are known. To do so, start with $y = ax^2 + bx + c$. Then substitute the x- and y-coordinates of the first point into the equation. This will result in an equation in a, b, and c. Repeat the procedure for the other two points. This process yields a system of three equations in three variables. Next solve the system for a, b, and c. To find the equation of the parabola, substitute the values found for a, b, and c into the equation $y = ax^2 + bx + c$.

Three points on a parabola are $(0, 12)$, $(3, -3)$, and $(-2, 32)$.

a) Individually, find a system of equations in three variables that can be used to find the equation of the parabola. Then compare your answers. If each member

of the group does not have the same system, determine why.

b) Individually, solve the system and determine the values of a, b, and c. Then compare your answers.

c) Individually, write the equation of the parabola passing through $(0, 12)$, $(3, -3)$, and $(-2, 32)$. Then compare your answers.

d) Individually, write the equation in

$$y = a(x - h)^2 + k$$

form. Then compare your answers.

e) Individually, graph the equation in part **d)**. Then compare your answers.

Cumulative Review Exercises

[1.5] **136.** Simplify $\dfrac{6x^{-3}y^4}{18x^{-2}y^3}$.

[2.5] **137.** Solve the inequality $-4 < 3x - 4 < 17$. Write the solution in interval notation.

[4.5] **138.** Evaluate the determinant.

$$\begin{vmatrix} 4 & 0 & 3 \\ 5 & 2 & -1 \\ 3 & 6 & 4 \end{vmatrix}$$

[5.2] **139. a)** Write expressions to represent each of the four areas shown in the figure.

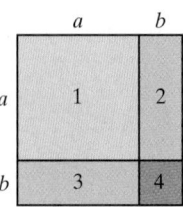

b) Express the total area shown as the square of a binomial.

[10.1] **140.** Graph $y = (x - 4)^2 + 1$.

10.2 The Ellipse

1 Graph ellipses.

2 Graph ellipses with centers at (h, k).

1 Graph Ellipses

Ellipse

An **ellipse** is a set of points in a plane, the sum of whose distances from two fixed points is a constant. The two fixed points are called the **foci** (each is a **focus**) of the ellipse.

In **Figure 10.18**, F_1 and F_2 represent the two foci of an ellipse.

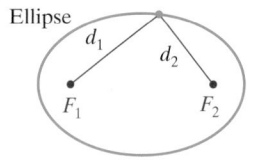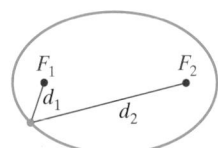

Ellipse

FIGURE 10.18

We can construct an ellipse using a length of string and two thumbtacks. Place the two thumbtacks fairly close together (**Fig. 10.19**). Then tie the ends of the string to the thumbtacks. With a pencil or pen pull the string taut, and, while keeping the string taut, draw the ellipse by moving the pencil around the thumbtacks.

In **Figure 10.20**, the line segment from $-a$ to a on the x-axis is the *longer* or **major axis** and the line segment from $-b$ to b is the *shorter* or **minor axis** of the ellipse. The major axis of an ellipse may also be on the y-axis. **Figure 10.20** also shows the *center* of the ellipse and the two vertices (the red dots). The vertices are the endpoints of the major axis.

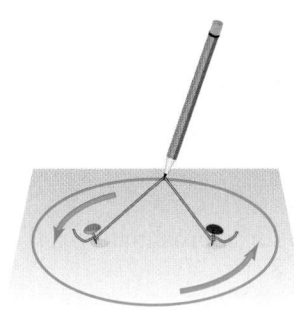

FIGURE 10.19

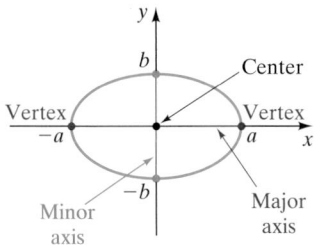

FIGURE 10.20

Ellipse with Its Center at the Origin

The standard form of an ellipse with its center at the origin is

$$\frac{x^2}{a^2} + \frac{y^2}{b^2} = 1$$

where $(a, 0)$ and $(-a, 0)$ are the x-intercepts and $(0, b)$ and $(0, -b)$ are the y-intercepts.

Observe that:

- The x-intercepts are obtained from the constant in the denominator of the x-term.
- The y-intercepts are obtained from the constant in the denominator of the y-term.
- If $a^2 > b^2$, the major axis is along the x-axis.
- If $b^2 > a^2$, the major axis is along the y-axis.

In Example 1, the major axis of the ellipse is along the x-axis.

EXAMPLE 1   Graph $\dfrac{x^2}{9} + \dfrac{y^2}{4} = 1$.

Solution   We can rewrite the equation as

$$\frac{x^2}{3^2} + \frac{y^2}{2^2} = 1$$

Thus, $a = 3$ and the x-intercepts are $(3, 0)$ and $(-3, 0)$. Since $b = 2$, the y-intercepts are $(0, 2)$ and $(0, -2)$. The ellipse is illustrated in **Figure 10.21**.

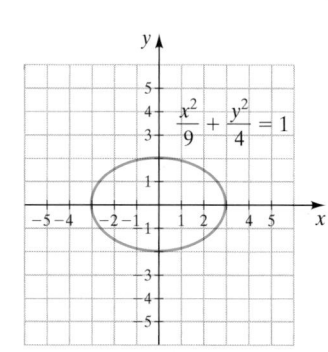

FIGURE 10.21

Now Try Exercise 15

An equation may be written so that it may not be obvious that its graph is an ellipse, as illustrated in Example 2.

EXAMPLE 2 Graph $20x^2 + 9y^2 = 180$.

Solution

$$20x^2 + 9y^2 = 180$$

$$\frac{20x^2 + 9y^2}{180} = \frac{180}{180} \qquad \text{Both sides were divided by 180.}$$

$$\frac{20x^2}{180} + \frac{9y^2}{180} = 1 \qquad \text{Simplify}$$

$$\frac{x^2}{9} + \frac{y^2}{20} = 1 \qquad \text{Equation of ellipse}$$

The equation can now be recognized as an ellipse in standard form.

$$\frac{x^2}{a^2} + \frac{y^2}{b^2} = 1$$

Since $a^2 = 9$, $a = 3$. We know that $b^2 = 20$; thus $b = \sqrt{20}$ (or approximately 4.47).

$$\frac{x^2}{3^2} + \frac{y^2}{(\sqrt{20})^2} = 1$$

The x-intercepts are $(3, 0)$ and $(-3, 0)$. The y-intercepts are $(0, -\sqrt{20})$ and $(0, \sqrt{20})$. The graph is illustrated in **Figure 10.22**. Note that the major axis lies along the y-axis.

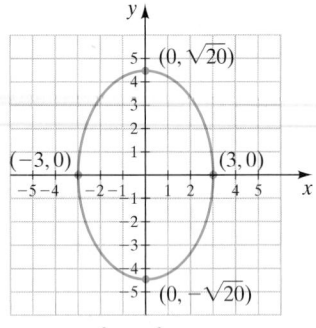

$20x^2 + 9y^2 = 180$

FIGURE 10.22

Now Try Exercise 19

In Example 1, since $a^2 = 9$ and $b^2 = 4$ and $a^2 > b^2$, the major axis is along the x-axis. In Example 2, since $a^2 = 9$ and $b^2 = 20$ and $b^2 > a^2$, the major axis is along the y-axis. In the specific case where $a^2 = b^2$, the figure is a circle. *Thus, the circle is a special case of an ellipse.*

EXAMPLE 3 Write the equation of the ellipse shown in **Figure 10.23**.

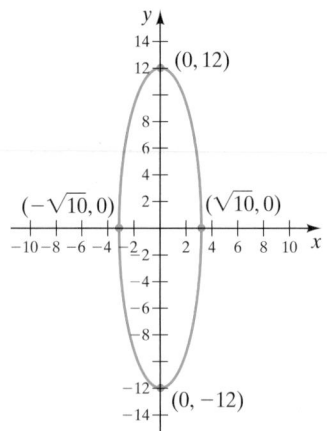

FIGURE 10.23

Solution   The x-intercepts are $(-\sqrt{10}, 0)$ and $(\sqrt{10}, 0)$; thus, $a = \sqrt{10}$ and $a^2 = 10$. The y-intercepts are $(0, -12)$ and $(0, 12)$; thus, $b = 12$ and $b^2 = 144$.

$$\frac{x^2}{a^2} + \frac{y^2}{b^2} = 1$$

$$\frac{x^2}{10} + \frac{y^2}{144} = 1$$

Now Try Exercise 45

The formula for the **area of an ellipse** is $A = \pi ab$. In Example 1, where $a = 3$ and $b = 2$, the area is $A = \pi(3)(2) = 6\pi \approx 18.8$ square units.

In Example 2, where $a = 3$ and $b = \sqrt{20}$, the area is $A = \pi(3)(\sqrt{20}) = \pi(3)(2\sqrt{5}) = 6\pi\sqrt{5} \approx 42.1$ square units.

FIGURE 10.24

2 Graph Ellipses with Centers at (h, k)

Horizontal and vertical translations, similar to those used in Chapter 8, may be used to obtain the equation of an ellipse with center at (h, k).

Ellipse with Its Center at (h, k)

$$\frac{(x - h)^2}{a^2} + \frac{(y - k)^2}{b^2} = 1$$

In the formula, the h shifts the graph left or right from the origin and k shifts the graph up or down from the origin, as shown in **Figure 10.24**.

EXAMPLE 4   Graph $\dfrac{(x - 2)^2}{25} + \dfrac{(y + 3)^2}{16} = 1$.

Solution   This is the graph of $\dfrac{x^2}{25} + \dfrac{y^2}{16} = 1$ or $\dfrac{x^2}{5^2} + \dfrac{y^2}{4^2} = 1$ translated so that its center is at $(2, -3)$. Note that $a = 5$ and $b = 4$. The graph is shown in **Figure 10.25**.

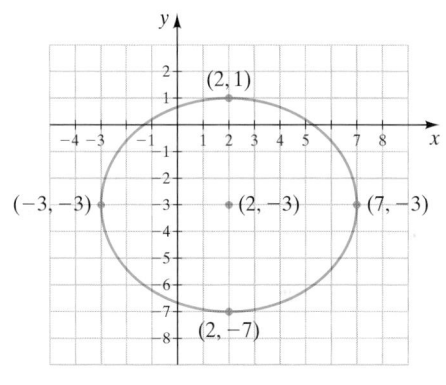

FIGURE 10.25

Now Try Exercise 33

An understanding of ellipses is useful in many areas. Astronomers know that planets revolve in elliptical orbits around the Sun. Communications satellites move in elliptical orbits around Earth (see **Fig. 10.26**).

Ellipses are used in medicine to smash kidney stones. When a signal emerges from one focus of an ellipse, the signal is reflected to the other focus. In kidney stone machines, the person is situated so that the stone to be smashed is at one focus of an elliptically shaped chamber called a lithotripter (see **Fig. 10.27** and Exercises 55 and 56).

In certain buildings with ellipsoidal ceilings, a person standing at one focus can whisper something and a person standing at the other focus can clearly hear what the person whispered. There are many other uses for ellipses, including lamps that are made to concentrate light at a specific point.

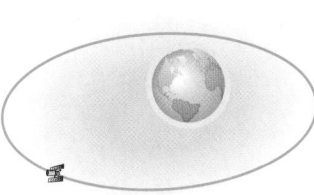

FIGURE 10.26

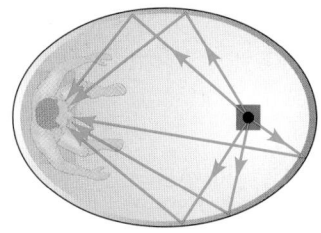

FIGURE 10.27

✎ Using Your Graphing Calculator ▩▩▩

Ellipses are not functions. To graph ellipses on a graphing calculator, we solve the equation for y. This will give the two equations that we use to graph the ellipse.

In Example 1, we graphed $\dfrac{x^2}{9} + \dfrac{y^2}{4} = 1$. Solving this equation for y, we get

$$\frac{x^2}{9} + \frac{y^2}{4} = 1$$

$$36 \cdot \frac{x^2}{9} + 36 \cdot \frac{y^2}{4} = 1 \cdot 36 \qquad \text{Multiply by the LCD.}$$

$$4x^2 + 9y^2 = 36$$

$$9y^2 = 36 - 4x^2$$

$$y^2 = \frac{36 - 4x^2}{9}$$

$$y^2 = \frac{4(9 - x^2)}{9} \qquad \text{Factor 4 from the numerator.}$$

$$y = \pm\frac{2}{3}\sqrt{9 - x^2} \qquad \text{Square root property}$$

To graph the ellipse, we let $Y_1 = \dfrac{2}{3}\sqrt{9 - x^2}$ and $Y_2 = -\dfrac{2}{3}\sqrt{9 - x^2}$ and graph both equations. The graphs of Y_1 and Y_2 are illustrated in **Figure 10.28**.

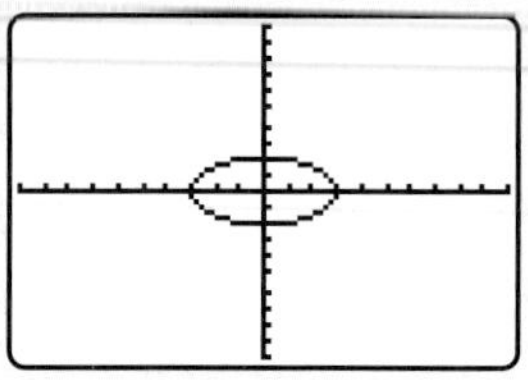

FIGURE 10.28

▨▨▨

EXERCISE SET 10.2   Math XL   MyMathLab
MathXL®   MyMathLab

Warm-Up Exercises

Fill in the blanks with the appropriate word, phrase, or symbol(s) from the following list.

major	y-intercepts	foci	divide	subtract	$(0,0)$
multiply	minor	x-intercepts	(k, h)	ellipse	(h, k)

1. $\dfrac{x^2}{a^2} + \dfrac{y^2}{b^2} = 1$ is an equation of an ellipse with center at _____.

2. $\dfrac{(x - h)^2}{a^2} + \dfrac{(y - k)^2}{b^2} = 1$ is an equation of an ellipse with center at _____.

3. A set of points the sum of whose distances from two fixed points is a constant is an _____.

4. An ellipse is the set of points in a plane whose distance from two fixed points, called the _____, is a constant.

5. For the ellipse $\dfrac{x^2}{81} + \dfrac{y^2}{16} = 1$, the points $(-9, 0)$ and $(9, 0)$ are the _____.

6. For the ellipse $\dfrac{x^2}{81} + \dfrac{y^2}{16} = 1$, the points $(0, -4)$ and $(0, 4)$ are the _____.

7. For the ellipse $\dfrac{x^2}{49} + \dfrac{y^2}{25} = 1$, the line from -7 to 7 on the x-axis is the _____ axis.

8. For the ellipse $\dfrac{x^2}{49} + \dfrac{y^2}{25} = 1$, the line from -5 to 5 on the y-axis is the _____ axis.

9. The first step to graph the ellipse $25x^2 + 50y^2 = 100$ is to _____ both sides by 100.

10. The first step to graph the ellipse $-\dfrac{x^2}{4} - \dfrac{y^2}{9} = -1$ is to _____ both sides by -1.

Practice the Skills

Graph each equation.

11.  $\dfrac{x^2}{4} + \dfrac{y^2}{1} = 1$

12. $\dfrac{x^2}{1} + \dfrac{y^2}{4} = 1$

13. $\dfrac{x^2}{4} + \dfrac{y^2}{9} = 1$

14. $\dfrac{x^2}{9} + \dfrac{y^2}{4} = 1$

15. $\dfrac{x^2}{25} + \dfrac{y^2}{9} = 1$

16. $\dfrac{x^2}{100} + \dfrac{y^2}{16} = 1$

17. $\dfrac{x^2}{16} + \dfrac{y^2}{25} = 1$

18. $\dfrac{x^2}{36} + \dfrac{y^2}{64} = 1$

19. $x^2 + 16y^2 = 16$

20. $x^2 + 25y^2 = 25$

21. $49x^2 + y^2 = 49$

22. $9x^2 + 25y^2 = 225$

23. $4x^2 + 9y^2 = 36$

24. $9x^2 + 16y^2 = 144$

25. $25x^2 + 100y^2 = 400$

26. $100x^2 + 25y^2 = 400$

27. $x^2 + 2y^2 = 8$

28. $x^2 + 36y^2 = 36$

29. $\dfrac{x^2}{16} + \dfrac{(y - 2)^2}{9} = 1$

30. $\dfrac{(x - 1)^2}{16} + \dfrac{y^2}{1} = 1$

31. $\dfrac{(x+3)^2}{9} + \dfrac{y^2}{25} = 1$ **32.** $\dfrac{(x-3)^2}{25} + \dfrac{(y+2)^2}{49} = 1$ **33.** $\dfrac{(x+1)^2}{9} + \dfrac{(y-2)^2}{4} = 1$ **34.** $\dfrac{(x-3)^2}{16} + \dfrac{(y-4)^2}{25} = 1$

35. $(x+3)^2 + 9(y+1)^2 = 81$ **36.** $18(x-1)^2 + 2(y+3)^2 = 72$ **37.** $(x-5)^2 + 4(y+4)^2 = 4$

38. $4(x-2)^2 + 9(y+2)^2 = 36$ **39.** $12(x+4)^2 + 3(y-1)^2 = 48$ **40.** $25(x-2)^2 + 9(y-1)^2 = 225$

Problem Solving

41. Find the area of the ellipse in Exercise 11.
42. Find the area of the ellipse in Exercise 15.
43. Find the area of the ellipse in Exercise 17.
44. Find the area of the ellipse in Exercise 29.

In Exercises 45–48, find the equation of the ellipse that has the four points as endpoints of the major and minor axes.

45. $(5,0), (-5,0), (0,4), (0,-4)$

46. $(6,0), (-6,0), (0,5), (0,-5)$

47. $(2,0), (-2,0), (0,3), (0,-3)$

48. $(1,0), (-1,0), (0,9), (0,-9)$

In Exercises 49 and 50, write the equation in standard form. Determine the center of each ellipse.

49. $x^2 + 4y^2 + 6x + 16y - 11 = 0$

50. $x^2 + 4y^2 - 4x - 8y - 92 = 0$

51. Art Gallery An art gallery has an elliptical hall. The maximum distance from one focus to the wall is 90.2 feet and the minimum distance is 20.7 feet. Find the distance between the foci.

52. Communications Satellite A space shuttle transported a communications satellite to space. The satellite travels in an elliptical orbit around Earth. The maximum distance of the satellite from Earth is 23,200 miles and the minimum distance is 22,800 miles. Earth is at one focus of the ellipse. Find the distance from Earth to the other focus.

53. Tunnel through a Mountain The tunnel in the photo is the top half of an ellipse. The width of the tunnel is 20 feet and the height is 24 feet.

© Allen R. Angel

a) If you pictured a completed ellipse with the center of the ellipse being at the center of the road, determine the equation of the ellipse.

b) Find the area of the ellipse found in part **a)**.

c) Find the area of the opening of the tunnel.

54. Billiard Table An elliptical billiard table is 8 feet long by 5 feet wide. Determine the location of the foci. On such a table, if a ball is put at each focus and one ball is hit with enough force, it would hit the ball at the other focus no matter where it banks on the table.

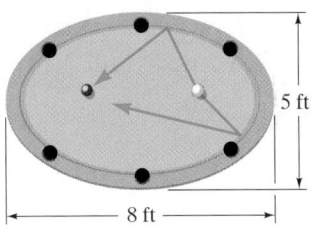

55. Lithotripter Machine Suppose the lithotripter machine described on page 647 is 6 feet long and 4 feet wide. Describe the location of the foci.

56. Lithotripter On page 647 we gave a brief introduction to the lithotripter, which uses ultrasound waves to shatter kidney stones. Do research and write a detailed report describing the procedure used to shatter kidney stones. Make sure that you explain how the waves are directed onto the stone.

57. Whispering Gallery The National Statuary Hall in the Capitol Building in Washington, D.C., is a "whispering gallery." Do research and explain why one person standing at a certain point can whisper something and someone standing a considerable distance away can hear it.

58. Check your answer to Exercise 11 on your graphing calculator.

59. Check your answer to Exercise 17 on your graphing calculator.

Concept/Writing Exercises

60. Discuss the graphs of $\dfrac{x^2}{a^2} + \dfrac{y^2}{b^2} = 1$ when $a > b$, $a < b$, and $a = b$.

61. Explain why the circle is a special case of the ellipse.

62. Is $\dfrac{x^2}{36} - \dfrac{y^2}{49} = 1$ an equation for an ellipse? Explain.

63. Is $-\dfrac{x^2}{49} + \dfrac{y^2}{81} = 1$ an equation for an ellipse? Explain.

64. How many points are on the graph of $16x^2 + 25y^2 = 0$? Explain.

65. Consider the graph of the equation $\dfrac{x^2}{a^2} + \dfrac{y^2}{b^2} = 1$. Explain what will happen to the shape of the graph as the value of b gets closer to the value of a. What is the shape of the graph when $a = b$?

66. How many points of intersection will the graphs of the equations $x^2 + y^2 = 49$ and $\dfrac{x^2}{16} + \dfrac{y^2}{25} = 1$ have? Explain.

67. How many points of intersection will the graphs of the equations $y = 2(x-2)^2 - 3$ and $\dfrac{(x-2)^2}{4} + \dfrac{(y+3)^2}{9} = 1$ have? Explain.

Challenge Problems

Determine the equation of the ellipse that has the following four points as endpoints of the major and minor axes.

68. $(-7, 3), (5, 3), (-1, 5), (-1, 1)$

69. $(-3, 2), (11, 2), (4, 5), (4, -1)$

Group Activity

Work Exercise 70 individually. Then compare your answers.

70. Tunnel The photo shows an elliptical tunnel (with the bottom part of the ellipse not shown) near Rockefeller Center in New York City. The maximum width of the tunnel is 18 feet and the maximum height *from the ground to the top* is 10.5 feet.

© Allen R. Angel

a) If the *completed ellipse* would have a maximum height of 15 feet, how high from the ground is the center of the elliptical tunnel?

b) Consider the following graph, which could be used to represent the tunnel.

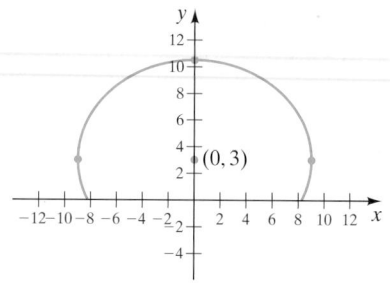

If the ellipse were continued, what would be the other *y*-intercept of the graph?

c) Write the equation of the ellipse, if completed, in part **b)**.

Cumulative Review Exercises

[2.2] **71.** Solve the formula $S = \dfrac{n}{2}(f + l)$ for *l*.

[5.4] **72.** Divide $\dfrac{2x^2 + 2x - 7}{2x - 3}$.

[7.6] **73.** Solve $\sqrt{3b - 2} = 10 - b$.

[8.6] **74.** Solve $\dfrac{3x + 5}{x - 4} \le 0$, and give the solution in interval notation.

[9.7] **75.** Find $\log_8 321$.

Mid-Chapter Test: 10.1–10.2

To find out how well you understand the chapter material to this point, take this brief test. The answers, and the section where the material was initially discussed, are given in the back of the book. Review any questions you answered incorrectly.

Graph each equation.

1. $y = (x - 2)^2 - 1$

2. $y = -(x + 1)^2 + 3$

3. $x = -(y - 4)^2 + 1$

4. $x = 2(y + 3)^2 - 2$

5. $y = x^2 + 6x + 10$

Find the distance between each pair of points. Where appropriate, round your answers to the nearest hundredth.

6. $(-7, 4)$ and $(-2, -8)$

7. $(5, -3)$ and $(2, 9)$

Find the midpoint of the line segment between each pair of points.

8. $(9, -1)$ and $(-11, 6)$

9. $\left(-\dfrac{5}{2}, 7\right)$ and $\left(8, \dfrac{1}{2}\right)$

10. Write the equation of the circle with center at $(-3, 2)$ and a radius of 5 units.

Graph each equation.

11. $x^2 + (y - 1)^2 = 16$

12. $y = \sqrt{36 - x^2}$

13. $x^2 + y^2 - 2x + 4y - 4 = 0$

14. What is the definition of a circle?

Graph each equation.

15. $\dfrac{x^2}{4} + \dfrac{y^2}{9} = 1$

16. $\dfrac{x^2}{81} + \dfrac{y^2}{25} = 1$

17. $\dfrac{(x - 1)^2}{49} + \dfrac{(y + 2)^2}{4} = 1$

18. $36(x + 3)^2 + (y - 4)^2 = 36$.

19. Find the area of the ellipse in Exercise 15.

20. Find the equation of the ellipse that has the four points $(8, 0), (-8, 0), (0, 5),$ and $(0, -5)$ as the endpoints of the major and minor axes.

10.3 The Hyperbola

1 Graph hyperbolas.

2 Review conic sections.

1 Graph Hyperbolas

> **Hyperbola**
>
> A **hyperbola** is the set of points in a plane, the difference of whose distances from two fixed points, called the **foci**, is a constant.

A hyperbola is illustrated in **Figure 10.29a**. In the figure, for every point on the hyperbola, the difference $M - N$ is the same constant. A hyperbola may look like a pair of parabolas. However, the shapes are actually quite different. A hyperbola has two **vertices**. The vertices are the points where the graph crosses the x-axis (**Figure 10.29b**) or the y-axis (**Figure 10.29c**). The point halfway between the two vertices is the **center** of the hyperbola. The line through the vertices is called the **transverse axis**. In **Figure 10.29b** the transverse axis lies along the x-axis, and in **Figure 10.29c**, the transverse axis lies along the y-axis.

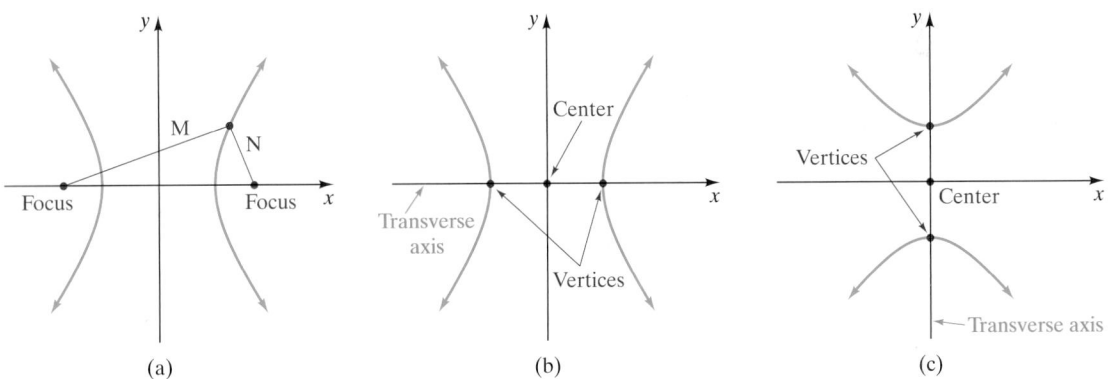

FIGURE 10.29   (a)   (b)   (c)

Figure 10.30 shows the graphs of the standard forms of the equations for both hyperbolas. In **Figure 10.30a**, both vertices are a units from the origin. In **Figure 10.30b**, both vertices are b units from the origin. Note that in the standard form of the equation, the denominator of the x^2 term is always a^2 and the denominator of the y^2 term is always b^2.

Hyperbola
with transverse axis
along the x-axis

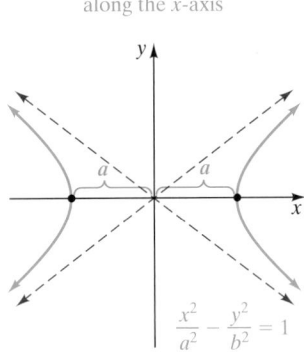

$$\frac{x^2}{a^2} - \frac{y^2}{b^2} = 1$$

- Hyperbola centered at origin.
- Vertices $(-a, 0)$ and $(a, 0)$.
- Transverse axis is along the x-axis.

Hyperbola
with transverse axis
along the y-axis

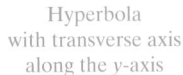

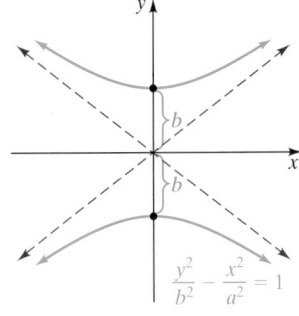

$$\frac{y^2}{b^2} - \frac{x^2}{a^2} = 1$$

- Hyperbola centered at origin.
- Vertices $(0, -b)$ and $(0, b)$.
- Transverse axis is along the y-axis.

FIGURE 10.30

When written in standard form, the intercepts will be on the axis indicated by the variable with the positive coefficient. The intercepts will be at the positive and the negative square root of the denominator of the positive term.

Examples	Intercepts on	Intercepts
$\dfrac{x^2}{49} - \dfrac{y^2}{16} = 1$	x-axis	$(-7, 0)$ and $(7, 0)$
$\dfrac{y^2}{16} - \dfrac{x^2}{49} = 1$	y-axis	$(0, -4)$ and $(0, 4)$

Understanding Algebra

Asymptotes are *not* part of the graph, but are used to show values the graph approaches, but does not touch.

The dashed lines in **Figure 10.30** are called **asymptotes**. The asymptotes are not a part of the hyperbola but are used as an aid in graphing hyperbolas. The asymptotes are two straight lines that go through the center of the hyperbola (see **Fig. 10.30**). As the values of x and y get larger, the graph of the hyperbola approaches the asymptotes. The equations of the asymptotes of a hyperbola whose center is the origin are

$$y = \frac{b}{a}x \quad \text{and} \quad y = -\frac{b}{a}x$$

The asymptotes can be drawn quickly by plotting the four points (a, b), $(-a, b)$, $(a, -b)$, and $(-a, -b)$, and then connecting these points with dashed lines to form a rectangle. Next, draw dashed lines through the opposite corners of the rectangle to obtain the asymptotes.

Hyperbola with Its Center at the Origin

TRANSVERSE AXIS ALONG x-AXIS (OPENS TO THE RIGHT AND LEFT)	TRANSVERSE AXIS ALONG y-AXIS (OPENS UPWARD AND DOWNWARD)
$\dfrac{x^2}{a^2} - \dfrac{y^2}{b^2} = 1$	$\dfrac{y^2}{b^2} - \dfrac{x^2}{a^2} = 1$

ASYMPTOTES

$$y = \frac{b}{a}x \quad \text{and} \quad y = -\frac{b}{a}x$$

EXAMPLE 1

a) Determine the equations of the asymptotes of the hyperbola with equation

$$\frac{x^2}{9} - \frac{y^2}{16} = 1$$

b) Draw the hyperbola using the asymptotes.

Solution

a) The value of a^2 is 9; the positive square root of 9 is 3. The value of b^2 is 16; the positive square root of 16 is 4. The asymptotes are

$$y = \frac{b}{a}x \quad \text{and} \quad y = -\frac{b}{a}x$$

or

$$y = \frac{4}{3}x \quad \text{and} \quad y = -\frac{4}{3}x$$

b) To graph the hyperbola, we first graph the asymptotes. First plot the points $(3, 4)$, $(-3, 4)$, $(3, -4)$, and $(-3, -4)$ and draw the rectangle as illustrated in **Figure 10.31**. The asymptotes are the dashed lines through the opposite corners of the rectangle.

Since the x-term in the original equation is positive, the graph intersects the x-axis. Since the denominator of the positive term is 9, the vertices are at $(3, 0)$ and $(-3, 0)$. Now draw the hyperbola by letting the hyperbola approach its asymptotes (**Fig. 10.32**). The asymptotes are drawn using dashed lines since they are not part of the hyperbola. They are used merely to help draw the graph.

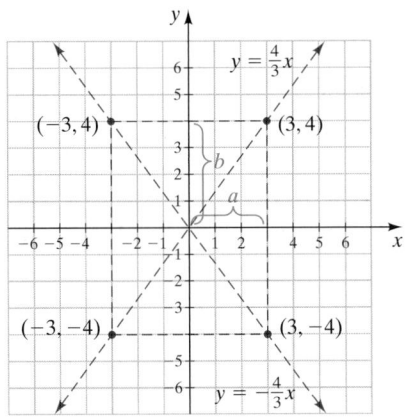

FIGURE 10.31

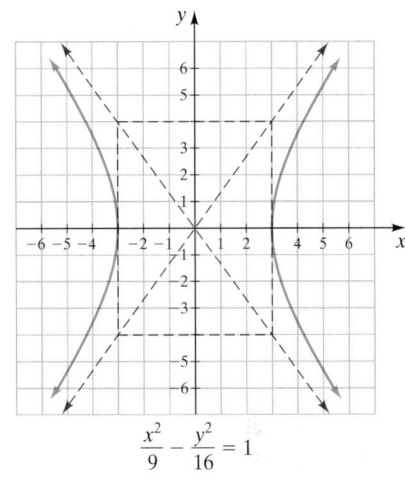

$$\frac{x^2}{9} - \frac{y^2}{16} = 1$$

FIGURE 10.32

Now Try Exercise 21

EXAMPLE 2

a) Show that the equation $-25x^2 + 4y^2 = 100$ is a hyperbola by expressing the equation in standard form.

b) Determine the equations of the asymptotes of the graph.

c) Draw the graph.

Solution

a) We divide both sides of the equation by 100 to obtain 1 on the right side of the equation.

$$\frac{-25x^2 + 4y^2}{100} = \frac{100}{100}$$

$$\frac{-25x^2}{100} + \frac{4y^2}{100} = 1$$

$$\frac{-x^2}{4} + \frac{y^2}{25} = 1$$

Rewriting the equation in standard form (positive term first), we get

$$\frac{y^2}{25} - \frac{x^2}{4} = 1$$

b) Since $a = 2$ and $b = 5$, the equations of the asymptotes are

$$y = \frac{5}{2}x \quad \text{and} \quad y = -\frac{5}{2}x$$

c) The graph intersects the y-axis at $(0, 5)$ and $(0, -5)$. **Figure 10.33a** illustrates the asymptotes, and **Figure 10.33b** illustrates the hyperbola.

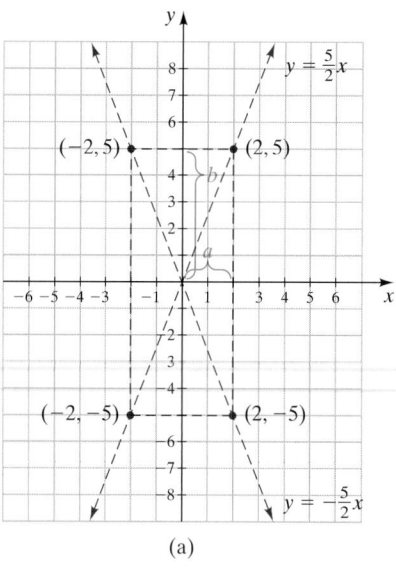

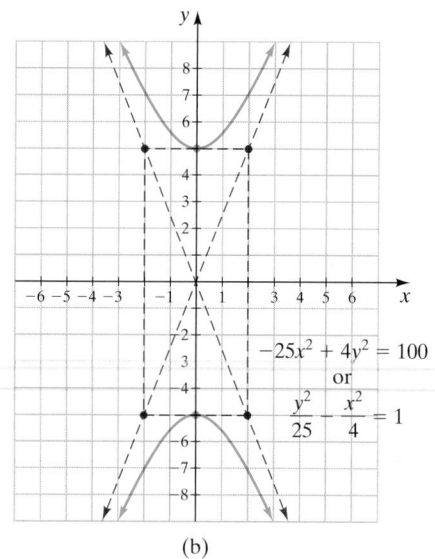

FIGURE 10.33   (a)   (b)

Now Try Exercise 29

We have discussed hyperbolas with their centers at the origin. Hyperbolas do not have to be centered at the origin. In this book, we will not discuss such hyperbolas.

Using Your Graphing Calculator ▒▒▒▒

To graph hyperbolas on the graphing calculator, solve the equation for y and graph each part. Consider Example 1,

$$\frac{x^2}{9} - \frac{y^2}{16} = 1.$$

Show that if you solve this equation for y you get $y = \pm\frac{4}{3}\sqrt{x^2 - 9}$. Let $Y_1 = \frac{4}{3}\sqrt{x^2 - 9}$ and $Y_2 = -\frac{4}{3}\sqrt{x^2 - 9}$. **Figures 10.34a, 10.34b, 10.34c,** and **10.34d,** shown below, give the graphs of Y_1 and Y_2 for different window settings. The window settings used are indicated above each graph.

			Set window as shown below figure (called "friendly window settings")
Standard setting	ZOOM: option 5 ZSquare setting	ZOOM: option 4, ZDecimal setting	

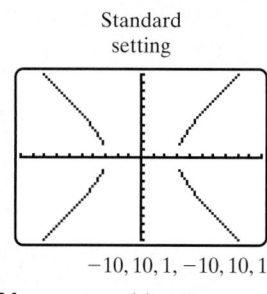

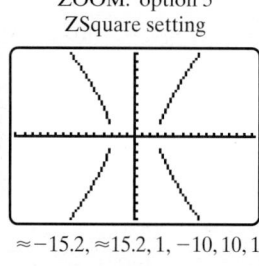

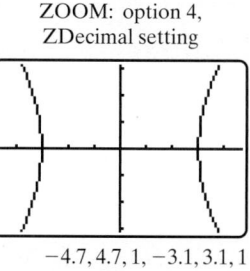

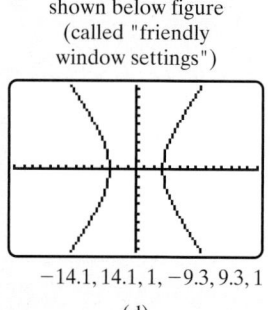

$-10, 10, 1, -10, 10, 1$	$\approx-15.2, \approx15.2, 1, -10, 10, 1$	$-4.7, 4.7, 1, -3.1, 3.1, 1$	$-14.1, 14.1, 1, -9.3, 9.3, 1$

FIGURE 10.34   (a)   (b)   (c)   (d)

In part (d), the "friendly window setting," the ratio of the length of the x-axis (28.2 units) to the length of the y-axis (18.6 units) is about 1.516. This is the same ratio as the length to the width of the display window of the calculator on the TI-84 Plus.

2 Review Conic Sections

The following chart summarizes conic sections.

Parabola

$y = a(x - h)^2 + k$

or $y = ax^2 + bx + c$

$a > 0$

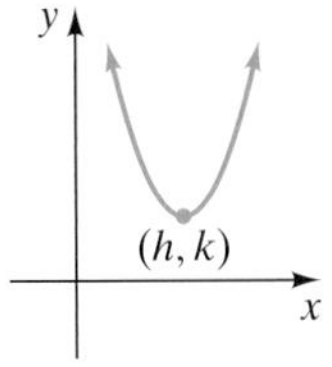

$a < 0$

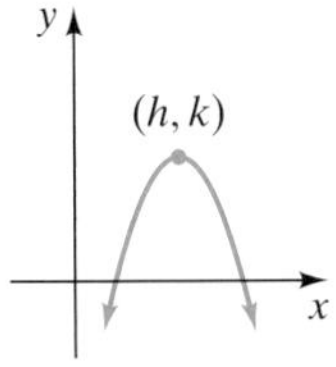

$x = a(y - k)^2 + h$

or $x = ay^2 + by + c$

$a > 0$

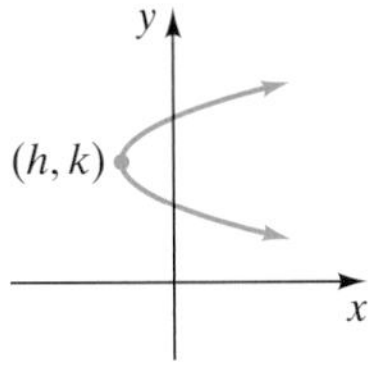

$a < 0$

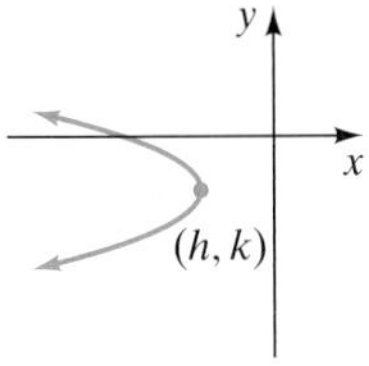

Circle

$x^2 + y^2 = r^2$

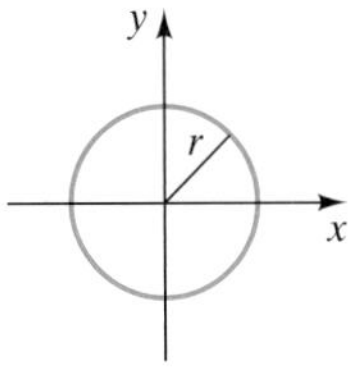

$(x - h)^2 + (y - k)^2 = r^2$

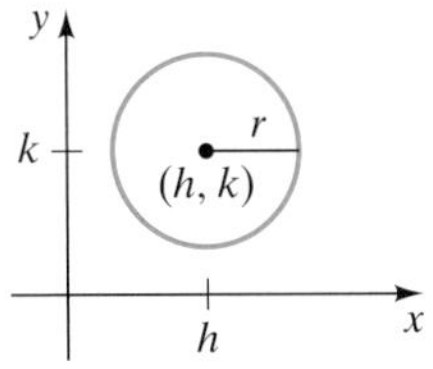

Ellipse

$$\frac{x^2}{a^2} + \frac{y^2}{b^2} = 1$$

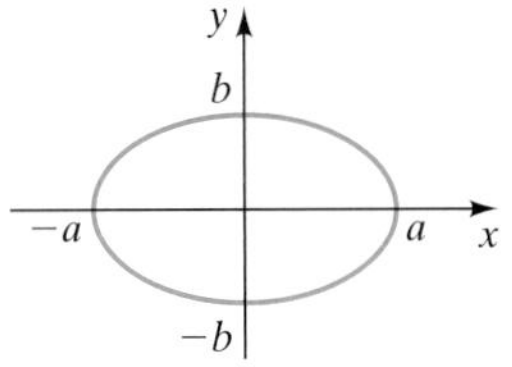

$$\frac{(x - h)^2}{a^2} + \frac{(y - k)^2}{b^2} = 1$$

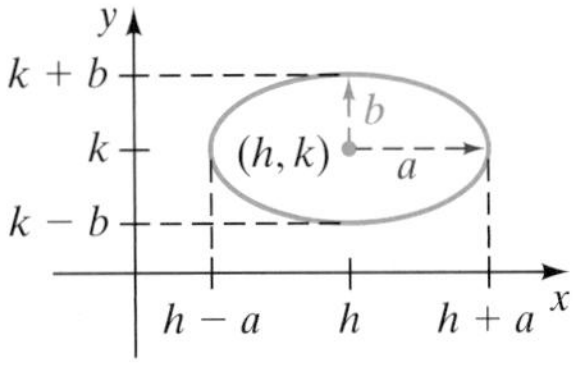

Hyperbola

$$\frac{x^2}{a^2} - \frac{y^2}{b^2} = 1$$

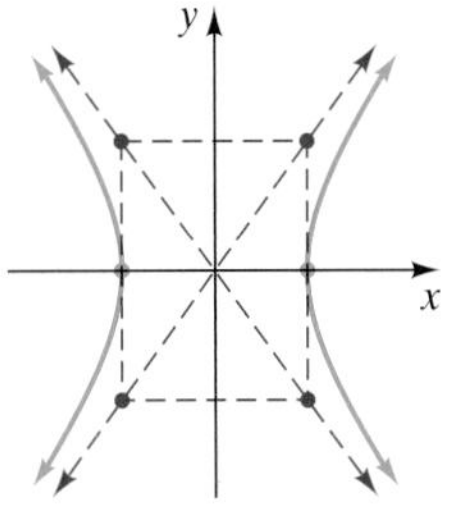

$$\frac{y^2}{b^2} - \frac{x^2}{a^2} = 1$$

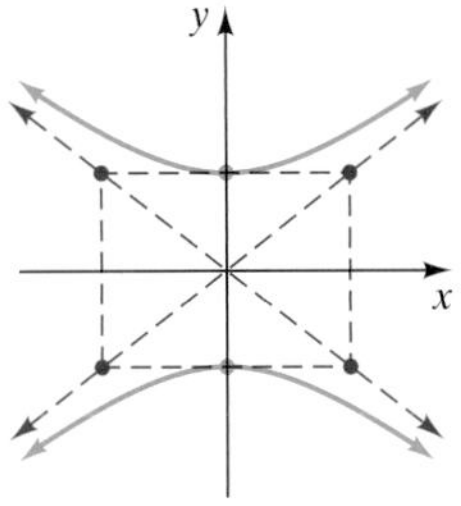

Asymptotes

$y = \frac{b}{a}x$ and $y = -\frac{b}{a}x$

EXAMPLE 3 Indicate whether each equation represents a parabola, a circle, an ellipse, or a hyperbola.

a) $6x^2 = -6y^2 + 48$ **b)** $x - y^2 = 9y + 3$ **c)** $2x^2 = 8y^2 + 72$

Solution

a) This equation has an x-squared term and a y-squared term. Let's place all the squared terms on the left side of the equation.

$$6x^2 = -6y^2 + 48$$

$$6x^2 + 6y^2 = 48 \qquad \text{Add } 6y^2 \text{ to both sides.}$$

Since the coefficients of both squared terms are the same number, we divide both sides of the equation by this number. Divide both sides by 6.

$$\frac{6x^2 + 6y^2}{6} = \frac{48}{6}$$

$$x^2 + y^2 = 8$$

This equation is of the form $x^2 + y^2 = r^2$ where $r^2 = 8$.

The equation $6x^2 = -6y^2 + 48$ represents a circle.

b) This equation has a y-squared term, but no x-squared term. Let's solve the equation for x.

$$x - y^2 = 9y + 3$$

$$x = y^2 + 9y + 3 \qquad \text{Add } y^2 \text{ to both sides.}$$

This equation is of the form $x = ay^2 + by + c$ where $a = 1, b = 9,$ and $c = 3$. The equation $x - y^2 = 9y + 3$ represents a parabola that opens to the right.

c) This equation has an x-squared term and a y-squared term. Let's place all the squared terms on the left side of the equation.

$$2x^2 = 8y^2 + 72$$

$$2x^2 - 8y^2 = 72 \qquad \text{Subtract } 8y^2 \text{ from both sides.}$$

Since the coefficients of both squared terms are different numbers, we want to divide the equation by the constant on the right side. Divide both sides by 72.

$$\frac{2x^2 - 8y^2}{72} = \frac{72}{72}$$

$$\frac{2x^2}{72} - \frac{8y^2}{72} = 1$$

$$\frac{x^2}{36} - \frac{y^2}{9} = 1$$

This equation is of the form $\dfrac{x^2}{a^2} - \dfrac{y^2}{b^2} = 1$ where $a^2 = 36$ (or $a = 6$) and $b^2 = 9$ (or $b = 3$).

The equation $2x^2 = 8y^2 + 72$ represents a hyperbola.

Now Try Exercise 53

EXERCISE SET 10.3   Math XP   MyMathLab
MathXL®   MyMathLab

Warm-Up Exercises

Fill in the blanks with the appropriate word, phrase, or symbol(s) from the following list.

divide	asymptotes	x	y	foci	transverse
horizontal	center	vertical	vertices	hyperbola	multiply

1. A set of points, the difference of whose distances from two fixed points is a constant, is a _____.

2. The fixed points for a hyperbola are called the _____.

3. For the hyperbola $\dfrac{x^2}{36} - \dfrac{y^2}{100} = 1$, the _____ is at $(0, 0)$.

4. For the hyperbola $\dfrac{x^2}{9} - \dfrac{y^2}{36} = 1$, the _____ are at the points $(-3, 0)$ and $(3, 0)$.

5. The line through the vertices of a hyperbola is the _____ axis.

6. For the hyperbola $\dfrac{x^2}{81} - \dfrac{y^2}{4} = 1$, the vertices are on the _____-axis.

7. For the hyperbola $\dfrac{y^2}{49} - \dfrac{x^2}{121} = 1$, the vertices are on the _____-axis.

8. For the hyperbola $\dfrac{x^2}{4} - \dfrac{y^2}{25} = 1$, the lines $y = \dfrac{5}{2}x$ and $y = -\dfrac{5}{2}x$ are the _____.

9. The first step in graphing the hyperbola $25x^2 - 9y^2 = 225$ is to _____ both sides by 225.

10. The first step in graphing the hyperbola $-\dfrac{x^2}{2} + \dfrac{y^2}{50} = -\dfrac{1}{2}$ is to _____ both sides by -2.

Practice the Skills

a) *Determine the equations of the asymptotes for each equation.* **b)** *Graph the equation.*

11. $\dfrac{x^2}{9} - \dfrac{y^2}{4} = 1$

12. $\dfrac{y^2}{4} - \dfrac{x^2}{9} = 1$

13. $\dfrac{x^2}{4} - \dfrac{y^2}{1} = 1$

14. $\dfrac{y^2}{1} - \dfrac{x^2}{4} = 1$

15. $\dfrac{x^2}{4} - \dfrac{y^2}{16} = 1$

16. $\dfrac{x^2}{4} - \dfrac{y^2}{4} = 1$

17. $\dfrac{x^2}{25} - \dfrac{y^2}{16} = 1$

18. $\dfrac{y^2}{16} - \dfrac{x^2}{25} = 1$

19. $\dfrac{y^2}{25} - \dfrac{x^2}{36} = 1$

20. $\dfrac{x^2}{36} - \dfrac{y^2}{25} = 1$

21. $\dfrac{y^2}{4} - \dfrac{x^2}{36} = 1$

22. $\dfrac{x^2}{16} - \dfrac{y^2}{9} = 1$

23. $\dfrac{y^2}{25} - \dfrac{x^2}{4} = 1$

24. $\dfrac{x^2}{4} - \dfrac{y^2}{25} = 1$

25. $\dfrac{x^2}{81} - \dfrac{y^2}{16} = 1$

26. $\dfrac{y^2}{16} - \dfrac{x^2}{81} = 1$

In Exercises 27–36, **a)** *write each equation in standard form and determine the equations of the asymptotes.* **b)** *Draw the graph.*

27. $25y^2 - x^2 = 25$

28. $x^2 - 25y^2 = 25$

29. $4y^2 - 16x^2 = 64$

30. $16x^2 - 4y^2 = 64$

31. $9y^2 - x^2 = 9$

32. $x^2 - 9y^2 = 9$

33. $25x^2 - 9y^2 = 225$

34. $4y^2 - 25x^2 = 100$

35. $4y^2 - 36x^2 = 144$

36. $25x^2 - 16y^2 = 400$

In Exercises 37–60, indicate whether the equation represents a parabola, a circle, an ellipse, or a hyperbola. See Example 3.

37. $10x^2 + 10y^2 = 40$

38. $30x^2 - 6y^2 = 180$

39. $x^2 + 16y^2 = 64$

40. $x = 5y^2 + 15y + 1$

41. $4x^2 - 4y^2 = 29$

42. $1.2x^2 + 1.2y^2 = 24$

43. $2y = 12x^2 - 8x + 16$

44. $3y^2 - 9x^2 = 54$

45. $5x^2 + 12y^2 = 60$

46. $9.2x^2 + 9.2y^2 = 46$

47. $3x = -2y^2 + 9y - 30$

48. $12x^2 - 3y^2 = 48$

49. $6x^2 + 6y^2 = 36$

50. $11x^2 = -11y^2 + 77$

51. $14y^2 = 7x^2 + 35$

52. $9x^2 = -18y^2 + 36$

53. $x + y = 2y^2 + 6$

54. $4x^2 = -4y^2 + 400$

55. $13x^2 = 7y^2 + 91$

56. $-8x^2 = -9y^2 - 72$

57. $y - x + 12 = x^2$

58. $17x^2 = -2y^2 + 34$

59. $-3x^2 - 3y^2 = -27$

60. $x - y^2 = 49$

Problem Solving

61. Determine an equation of the hyperbola whose vertices are $(0, 2)$ and $(0, -2)$ and whose asymptotes are $y = \dfrac{1}{2}x$ and $y = -\dfrac{1}{2}x$.

62. Determine an equation of the hyperbola whose vertices are $(0, 6)$ and $(0, -6)$ and whose asymptotes are $y = \dfrac{3}{2}x$ and $y = -\dfrac{3}{2}x$.

63. Determine an equation of the hyperbola whose vertices are $(-3, 0)$ and $(3, 0)$ and whose asymptotes are $y = 2x$ and $y = -2x$.

64. Determine an equation of the hyperbola whose vertices are $(7, 0)$ and $(-7, 0)$ and whose asymptotes are $y = \dfrac{4}{7}x$ and $y = -\dfrac{4}{7}x$.

65. Determine an equation of a hyperbola whose transverse axis is along the x-axis and whose equations of the asymptotes are $y = \dfrac{5}{3}x$ and $y = -\dfrac{5}{3}x$. Is this the only possible answer? Explain.

66. Determine an equation of a hyperbola whose transverse axis is along the y-axis and whose equations of the asymptotes are $y = \dfrac{2}{3}x$ and $y = -\dfrac{2}{3}x$. Is this the only possible answer? Explain.

67. Considering the graph of $\dfrac{x^2}{25} - \dfrac{y^2}{4} = 1$, determine the domain and range of the relation.

68. Considering the graph of $\dfrac{y^2}{36} - \dfrac{x^2}{9} = 1$, determine the domain and range of the relation.

69. Check your answer to Exercise 15 on your graphing calculator.

70. Check your answer to Exercise 21 on your graphing calculator.

71. Are any hyperbolas of the form $\dfrac{x^2}{a^2} - \dfrac{y^2}{b^2} = 1$ functions? Explain.

72. Are any hyperbolas of the form $\dfrac{y^2}{b^2} - \dfrac{x^2}{a^2} = 1$ functions? Explain.

73. If the equation $\dfrac{x^2}{a^2} - \dfrac{y^2}{b^2} = 1$, where $a > b$, is graphed, and then the values of a and b are interchanged, and the new equation is graphed, how will the two graphs compare? Explain your answer.

74. If the equation $\dfrac{x^2}{a^2} - \dfrac{y^2}{b^2} = 1$, where $a > b$, is graphed, and then the signs of each term on the left side of the equation are changed, and the new equation is graphed, how will the two graphs compare? Explain your answer.

Concept/Writing Exercises

75. Discuss the graph of $\dfrac{x^2}{a^2} - \dfrac{y^2}{b^2} = 1$ for nonzero real numbers a and b. Include the transverse axis, vertices, and asymptotes.

76. Discuss the graph of $\dfrac{y^2}{b^2} - \dfrac{x^2}{a^2} = 1$ for nonzero real numbers a and b. Include the transverse axis, vertices, and asymptotes.

77. Is $\dfrac{x^2}{81} + \dfrac{y^2}{64} = 1$ an equation for a hyperbola? Explain.

78. Is $-\dfrac{x^2}{81} - \dfrac{y^2}{64} = 1$ an equation for a hyperbola? Explain.

79. Is $4x^2 - 25y^2 = 100$ an equation for a hyperbola? Explain.

80. Is $36x^2 - 9y^2 = -324$ an equation for a hyperbola? Explain.

Cumulative Review Exercises

[3.4] **81.** Write the equation, in slope-intercept form, of the line that passes through the points $(-6, 4)$ and $(-2, 2)$.

[3.6] **82.** Let $f(x) = 3x^2 - x + 5$ and $g(x) = 6 - 4x^2$. Find $(f + g)(x)$.

[4.4] **83.** Solve the system of equations.

$$-4x + 9y = 7$$
$$5x + 6y = -3$$

[6.2] **84.** Add $\dfrac{3x}{2x - 3} + \dfrac{2x + 4}{2x^2 + x - 6}$.

[8.3] **85.** Solve the formula $E = \dfrac{1}{2}mv^2$ for v.

[9.6] **86.** Solve the equation $\log(x + 4) = \log 5 - \log x$.

10.4 Nonlinear Systems of Equations and Their Applications

1 Solve nonlinear systems using substitution.

2 Solve nonlinear systems using addition.

3 Solve applications.

1 Solve Nonlinear Systems Using Substitution

In Chapter 4, we discussed systems of linear equations. Here we discuss nonlinear systems of equations.

> **Nonlinear System of Equations**
>
> A **nonlinear system of equations** is a system of equations in which at least one equation is not linear (that is, one whose graph is not a straight line).

The solution to a system of equations is the point or points that satisfy all equations in the system. Consider the system of equations

$$x^2 + y^2 = 25$$
$$3x + 4y = 0$$

Both equations are graphed on the same axes in **Figure 10.35**. Note that the graphs appear to intersect at the points $(-4, 3)$ and $(4, -3)$. The check shows that these points satisfy both equations in the system and are therefore solutions to the system.

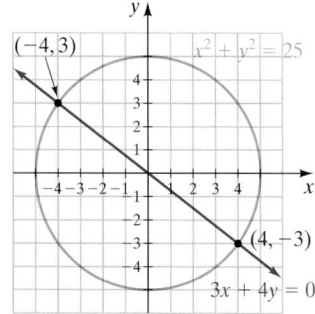

FIGURE 10.35

Check $(-4, 3)$

$$x^2 + y^2 = 25$$
$$(-4)^2 + 3^2 \stackrel{?}{=} 25$$
$$16 + 9 \stackrel{?}{=} 25$$
$$25 = 25 \quad \text{True}$$

$$3x + 4y = 0$$
$$3(-4) + 4(3) \stackrel{?}{=} 0$$
$$-12 + 12 \stackrel{?}{=} 0$$
$$0 = 0 \quad \text{True}$$

Check $(4, -3)$

$$4^2 + (-3)^2 = 25$$
$$16 + 9 \stackrel{?}{=} 25$$
$$25 = 25 \quad \text{True}$$

$$3(4) + 4(-3) = 0$$
$$12 - 12 \stackrel{?}{=} 0$$
$$0 = 0 \quad \text{True}$$

The graphical procedure for solving a system of equations may be inaccurate since we have to estimate the point or points of intersection. An exact answer may be obtained algebraically.

To solve a system of equations algebraically, we often solve one or more of the equations for one of the variables and then use substitution.

EXAMPLE 1   Solve the previous system of equations algebraically using the substitution method.

$$x^2 + y^2 = 25$$
$$3x + 4y = 0$$

Solution   We first solve the linear equation $3x + 4y = 0$ for either x or y. We will solve for y.

$$3x + 4y = 0$$
$$4y = -3x$$
$$y = -\frac{3x}{4}$$

Now we substitute $-\dfrac{3x}{4}$ for y in the equation $x^2 + y^2 = 25$ and solve for the remaining variable, x.

$$x^2 + y^2 = 25$$
$$x^2 + \left(-\frac{3x}{4}\right)^2 = 25 \qquad \text{Substitute } -\frac{3x}{4} \text{ for } y.$$
$$x^2 + \frac{9x^2}{16} = 25$$
$$16\left(x^2 + \frac{9x^2}{16}\right) = 16(25) \qquad \text{Multiply both sides by 16.}$$
$$16x^2 + 9x^2 = 400$$
$$25x^2 = 400$$
$$x^2 = \frac{400}{25} = 16 \qquad \text{Both sides were divided by 25.}$$
$$x = \pm\sqrt{16} = \pm 4 \quad \text{Square root property}$$

Next, we find the corresponding value of y for each value of x by substituting each value of x (one at a time) into the equation solved for y.

$$
\begin{array}{ll}
x = 4 & x = -4 \\[4pt]
y = -\dfrac{3x}{4} & y = -\dfrac{3x}{4} \\[10pt]
 = -\dfrac{3(4)}{4} & = -\dfrac{3(-4)}{4} \\[10pt]
 = -3 & = 3
\end{array}
$$

The solutions are $(4, -3)$ and $(-4, 3)$. This checks with the solution obtained graphically in **Figure 10.35**, on page 658.

Now Try Exercise 9

Our objective in using substitution is to obtain a single equation containing only one variable.

Helpful Hint

Study Tip

In this section, we will be using the substitution method and addition method to solve nonlinear systems of equations. If you do not remember how to use both methods to solve linear systems of equations, now is a good time to review Chapter 4.

In Examples 1 and 2, we solve systems using the substitution method, while in Examples 3 and 4, we solve systems using the addition method.

You may choose to solve a system by the substitution method if addition of the two equations will not lead to an equation that can be easily solved, as is the case with the systems in Examples 1 and 2.

EXAMPLE 2 Solve the system of equations using the substitution method.

$$y = x^2 - 3$$
$$x^2 + y^2 = 9$$

Solution   Since both equations contain x^2, we will solve one of the equations for x^2. We will choose to solve $y = x^2 - 3$ for x^2.

$$y = x^2 - 3$$
$$y + 3 = x^2$$

Now substitute $y + 3$ for x^2 in the equation $x^2 + y^2 = 9$.

$$x^2 + y^2 = 9$$
$$y + 3 + y^2 = 9 \qquad \text{Substitute } y + 3 \text{ for } x^2.$$
$$y^2 + y + 3 = 9 \qquad \text{Rewrite.}$$
$$y^2 + y - 6 = 0$$
$$(y + 3)(y - 2) = 0 \qquad \text{Factor.}$$
$$y + 3 = 0 \qquad \text{or} \qquad y - 2 = 0$$
$$y = -3 \qquad\qquad y = 2$$

Now find the corresponding values of x by substituting the values found for y.

$y = -3$	$y = 2$
$y = x^2 - 3$	$y = x^2 - 3$
$-3 = x^2 - 3$	$2 = x^2 - 3$
$0 = x^2$	$5 = x^2$
$0 = x$	$\pm\sqrt{5} = x$

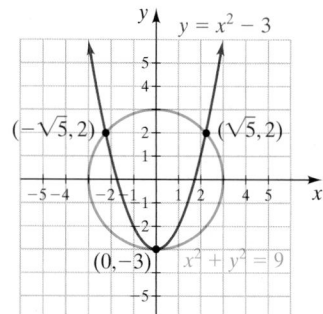

FIGURE 10.36

This system has three solutions: $(0, -3)$, $(\sqrt{5}, 2)$, and $(-\sqrt{5}, 2)$.

Note that the graph of the equation $y = x^2 - 3$ is a parabola and the graph of the equation $x^2 + y^2 = 9$ is a circle. The graphs of both equations are illustrated in **Figure 10.36**.

Now Try Exercise 19

Helpful Hint

Students will sometimes solve for one variable and assume that they have the solution. Remember that the solution, if one exists, to a system of equations in two variables consists of one or more ordered pairs.

2 Solve Nonlinear Systems Using Addition

We can often solve systems of equations more easily using the addition method. As with the substitution method, our objective is to obtain a single equation containing only one variable.

EXAMPLE 3 Solve the system of equations using the addition method.

$$x^2 + y^2 = 9$$
$$2x^2 - y^2 = -6$$

Solution   If we add the two equations, we will obtain one equation containing only one variable.

$$
\begin{aligned}
x^2 + y^2 &= 9 \\
2x^2 - y^2 &= -6 \\
\hline
3x^2 &= 3 \\
x^2 &= 1 \\
x &= \pm 1
\end{aligned}
$$

Now solve for the variable y by substituting $x = \pm 1$ into *either* of the original equations.

$x = 1$	$x = -1$
$x^2 + y^2 = 9$	$x^2 + y^2 = 9$
$1^2 + y^2 = 9$	$(-1)^2 + y^2 = 9$
$1 + y^2 = 9$	$1 + y^2 = 9$
$y^2 = 8$	$y^2 = 8$
$y = \pm\sqrt{8}$	$y = \pm\sqrt{8}$
$= \pm 2\sqrt{2}$	$= \pm 2\sqrt{2}$

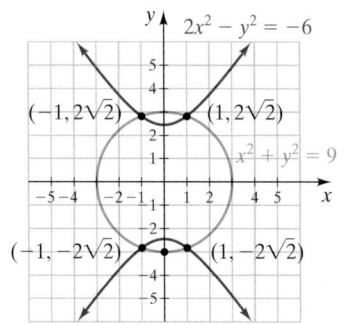

FIGURE 10.37

There are four solutions to this system of equations:

$$(1, 2\sqrt{2}), (1, -2\sqrt{2}), (-1, 2\sqrt{2}), \text{ and } (-1, -2\sqrt{2})$$

The graphs of the equations in the system are given in **Figure 10.37**. Notice the four points of intersection of the two graphs.

Now Try Exercise 25

It is possible that a system of equations has no real solution (therefore, the graphs do not intersect).

EXAMPLE 4 Solve the system of equations using the addition method.

$$x^2 + 4y^2 = 16 \quad (\text{eq. 1})$$
$$x^2 + y^2 = 1 \quad (\text{eq. 2})$$

Solution   Multiply (eq. 2) by -1 and add the resulting equation to (eq. 1).

$$
\begin{aligned}
x^2 + 4y^2 &= 16 \\
-x^2 - y^2 &= -1 \qquad (\text{eq. 2}) \text{ multiplied by } -1 \\
\hline
3y^2 &= 15 \\
y^2 &= 5 \\
y &= \pm\sqrt{5}
\end{aligned}
$$

Now solve for x.

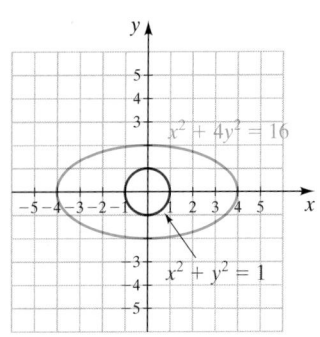

FIGURE 10.38

$$y = \sqrt{5} \qquad\qquad y = -\sqrt{5}$$
$$x^2 + y^2 = 1 \qquad\qquad x^2 + y^2 = 1$$
$$x^2 + (\sqrt{5})^2 = 1 \qquad\qquad x^2 + (-\sqrt{5})^2 = 1$$
$$x^2 + 5 = 1 \qquad\qquad x^2 + 5 = 1$$
$$x^2 = -4 \qquad\qquad x^2 = -4$$
$$x = \pm\sqrt{-4} \qquad\qquad x = \pm\sqrt{-4}$$
$$x = \pm 2i \qquad\qquad x = \pm 2i$$

Since x is an imaginary number for both values of y, this system of equations has no real solution. In solving nonlinear systems of equations, we are interested in finding all real number solutions.

The graphs of the equations are shown in **Figure 10.38**. Notice that the two graphs do not intersect; therefore, there is no real solution. This agrees with the answer we obtained algebraically.

Now Try Exercise 37

3 Solve Applications

EXAMPLE 5 **Flower Garden** Fred and Judy Vespucci want to build a rectangular flower garden behind their house. Fred went to a local nursery and bought enough topsoil to cover 150 square meters of land. Judy went to the local hardware store and purchased 50 meters of fence for the perimeter of the garden. How should they build the garden to use all the topsoil he bought and all the fence she purchased?

Solution   Understand and Translate   We begin by drawing a sketch (see **Fig 10.39**).

FIGURE 10.39

Let x = length of garden.

Let y = width of garden.

Since $A = xy$ and Fred bought topsoil to cover 150 square meters, we have

$$xy = 150$$

Since $P = 2x + 2y$ and Judy purchased 50 meters of fence for the perimeter of the garden, we have

$$2x + 2y = 50$$

The system of equations is

$$xy = 150$$
$$2x + 2y = 50$$

Carry Out   We will solve the system using substitution. The equation $2x + 2y = 50$ is a linear equation. We will solve this equation for y.

$$2x + 2y = 50$$
$$2y = 50 - 2x$$
$$y = \frac{50 - 2x}{2} = \frac{50}{2} - \frac{2x}{2} = 25 - x$$

Now substitute $25 - x$ for y in the equation $xy = 150$.

$$xy = 150$$
$$x(25 - x) = 150$$
$$25x - x^2 = 150$$
$$0 = x^2 - 25x + 150$$
$$0 = (x - 10)(x - 15)$$
$$x - 10 = 0 \quad \text{or} \quad x - 15 = 0$$
$$x = 10 \qquad\qquad x = 15$$

Answer   If $x = 10$, then $y = 25 - 10 = 15$. And, if $x = 15$, then $y = 25 - 15 = 10$. Thus, in either case, the dimensions of the flower garden are 10 meters by 15 meters.

Now Try Exercise 41

EXAMPLE 6 **Bicycles** Hike 'n' Bike Company produces and sells bicycles. Its weekly cost equation is $C = 50x + 400$, $0 \leq x \leq 160$, and its weekly revenue equation is $R = 100x - 0.3x^2$, $0 \leq x \leq 160$, where x is the number of bicycles produced and sold each week. Find the number of bicycles that must be produced and sold for Hike 'n' Bike to break even.

Solution   Understand and Translate   A company breaks even when its cost equals its revenue. When its cost is greater than its revenue, the company has a loss. When its revenue exceeds its cost, the company makes a profit.

The system of equations is

$$C = 50x + 400$$
$$R = 100x - 0.3x^2$$

For Hike 'n' Bike to break even, its cost must equal its revenue. Thus, we write

$$C = R$$
$$50x + 400 = 100x - 0.3x^2$$

Carry Out   Writing this quadratic equation in standard form, we obtain

$$0.3x^2 - 50x + 400 = 0, \quad 0 \leq x \leq 160$$

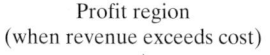
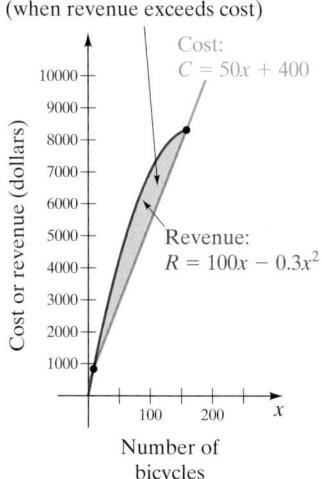

Profit region
(when revenue exceeds cost)

Cost:
$C = 50x + 400$

Revenue:
$R = 100x - 0.3x^2$

Cost or revenue (dollars)

10000
9000
8000
7000
6000
5000
4000
3000
2000
1000

100   200 x

Number of bicycles

FIGURE 10.40

We will solve this equation using the quadratic formula.

$$a = 0.3, \quad b = -50, \quad c = 400$$

$$x = \frac{-b \pm \sqrt{b^2 - 4ac}}{2a}$$

$$= \frac{-(-50) \pm \sqrt{(-50)^2 - 4(0.3)(400)}}{2(0.3)}$$

$$= \frac{50 \pm \sqrt{2020}}{0.6}$$

$$x = \frac{50 + \sqrt{2020}}{0.6} \approx 158.2 \quad \text{or} \quad x = \frac{50 - \sqrt{2020}}{0.6} \approx 8.4$$

Answer   The cost will equal the revenue and the company will break even when approximately 8 bicycles or 158 bicycles are sold. The company will make a profit when between 9 and 158 bicycles are sold. When fewer than 9 or more than 158 bicycles are sold, the company will have a loss (see **Fig. 10.40**).

Now Try Exercise 53

EXERCISE SET 10.4   *Math XL* MathXL®   **MyMathLab** MyMathLab

Warm-Up Exercises

Fill in the blanks with the appropriate word, phrase, or symbol(s) from the following list.

intersection   substitution   1   addition   2   3   nonlinear   no

1. A system of equations in which at least one equation is not linear is a _____ system of equations.

2. The easiest way to solve the nonlinear system of equations

$$x^2 + y^2 = 100$$
$$6x + 8y = 0$$

algebraically, is to use the _____ method.

3. The easiest way to solve the nonlinear system of equations

$$x^2 + y^2 = 25$$
$$4x^2 - y^2 = 100$$

algebraically, is to use the _____ method.

4. If the graphs of the equations in the nonlinear system do not intersect, the system has _____ real solutions.

5. If the graphs of the equations in the nonlinear system of equations have three points of intersection, the system has _____ real solutions.

6. Graphically, the solution to a system of equations occurs at the point or points of _____ of the graphs of the equations.

Practice the Skills

Find all real solutions to each system of equations using the substitution method.

7. $x^2 + y^2 = 18$
$x + y = 0$

8. $x^2 + y^2 = 18$
$x - y = 0$

9. $x^2 + y^2 = 9$
$y = 3x + 9$

10. $x^2 + y^2 = 4$
$x - 2y = 4$

11. $y = x^2 - 5$
$3x + 2y = 10$

12. $x^2 = y^2 + 4$
$x + y = 4$

13. $x^2 + y = 6$
$y = x^2 + 4$

14. $y - x = 2$
$x^2 - y^2 = 4$

15. $x^2 + 2y^2 = 25$
$x^2 - 3y^2 = 25$

16. $x + y^2 = 4$
$x^2 + y^2 = 6$

17. $x^2 + y^2 = 4$
$y = x^2 - 6$

18. $x^2 - 6y^2 = 36$
$x^2 + 2y^2 = 9$

19. $x^2 + y^2 = 9$
$y = x^2 - 3$

20. $x^2 + y^2 = 16$
$y = x^2 - 4$

21. $2x^2 - y^2 = -8$
$x - y = 6$

22. $x^2 + y^2 = 1$
$y - x = 3$

Find all real solutions to each system of equations using the addition method.

23. $x^2 - y^2 = 4$
$2x^2 + y^2 = 8$

24. $x^2 + y^2 = 36$
$x^2 - y^2 = 36$

25. $x^2 + y^2 = 25$
$x^2 - 2y^2 = -2$

26. $x^2 + y^2 = 25$
$x^2 - 2y^2 = 7$

27. $3x^2 - y^2 = 4$
$x^2 + 4y^2 = 10$

28. $3x^2 + 2y^2 = 30$
$x^2 + y^2 = 13$

29. $4x^2 + 9y^2 = 36$
$2x^2 - 9y^2 = 18$

30. $x^2 + 4y^2 = 16$
$-9x^2 + y^2 = 4$

31. $2x^2 - y^2 = 7$
$x^2 + 2y^2 = 6$

32. $5x^2 - 2y^2 = -13$
$3x^2 + 4y^2 = 39$

33. $x^2 + y^2 = 25$
$2x^2 - 3y^2 = -30$

34. $x^2 - 2y^2 = 7$
$x^2 + y^2 = 34$

35. $x^2 + y^2 = 9$
$16x^2 - 4y^2 = 64$

36. $3x^2 + 2y^2 = 6$
$4x^2 + 5y^2 = 15$

37. $x^2 + y^2 = 4$
$16x^2 + 9y^2 = 144$

38. $x^2 + y^2 = 1$
$9x^2 - 4y^2 = 36$

39. $x^2 + 4y^2 = 4$
$10y^2 - 9x^2 = 90$

40. $-4x^2 + y^2 = -15$
$8x^2 + 3y^2 = -5$

Problem Solving

41. Dance Floor Kris Hundley wants to build a dance floor at her gym. The dance floor is to have a perimeter of 90 meters and an area of 500 square meters. Find the dimensions of the dance floor.

42. Rectangular Region Ellen Dupree fences in a rectangular area along a riverbank as illustrated. If 20 feet of fencing encloses an area of 48 square feet, find the dimensions of the enclosed area.

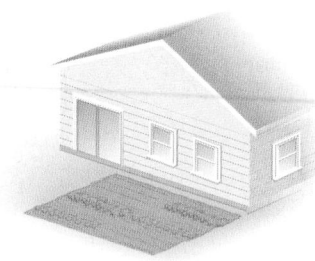

44. Rectangular Region A rectangular area is to be fenced along a river as illustrated in Exercise 42. If 20 feet of fencing encloses an area of 50 square feet, find the dimensions of the enclosed area.

45. Currency A country's currency includes a bill that has an area of 112 square centimeters with a diagonal of $\sqrt{260}$ centimeters. Find the length and width of the bill.

43. Vegetable Garden James Cannon is planning to build a rectangular flower garden in his yard. The garden is to have a perimeter of 54 feet and an area of 170 square feet. Find the dimensions of the vegetable garden.

46. Ice Rink A rectangular ice rink has an area of 3000 square feet. If the diagonal across the rink is 85 feet, find the dimensions of the rink.

Rockefeller Plaza, New York City

47. Piece of Wood Frank Samuelson, a carpenter, has a rectangular piece of plywood. When he measures the diagonal it measures 34 inches. When he cuts the wood along the diagonal, the perimeter of each triangle formed is 80 inches. Find the dimensions of the original piece of wood.

48. Sailboat A sail on a sailboat is shaped like a right triangle with a perimeter of 36 meters and a hypotenuse of 15 meters. Find the length of the legs of the triangle.

49. Baseball and Football Paul Martin throws a football upward from the ground. Its height above the ground at any time, t, is given by the formula $d = -16t^2 + 64t$. At the same time that the football is thrown, Shannon Ryan throws a baseball upward from the top of an 80-foot-tall building. Its height above the ground at any time, t, is given by the formula $d = -16t^2 + 16t + 80$. Find the time at which the two balls will be the same height above the ground.

50. Tennis Ball and Snowball Robert Snell throws a tennis ball downward from a helicopter flying at a height of 950 feet. The height of the ball above the ground at any time, t, is found by the formula $d = -16t^2 - 10t + 950$. At the instant the ball is thrown from the helicopter, Ramon Sanchez throws a snowball upward from the top of an 750-foot-tall building. The height above the ground of the snowball at any time, t, is found by the formula $d = -16t^2 + 80t + 750$. At what time will the ball and snowball pass each other?

51. Simple Interest Simple interest is calculated using the simple interest formula, interest = principal · rate · time or $i = prt$. If Seana Hayden invests a certain principal at a specific interest rate for 1 year, the interest she obtains is $7.50. If she increases the principal by $25 and the interest rate is decreased by 1%, the interest remains the same. Find the principal and the interest rate.

52. Simple Interest If Claire Brooke invests a certain principal at a specific interest rate for 1 year, the interest she obtains is $72. If she increases the principal by $120 and the interest rate is decreased by 2%, the interest remains the same. Find the principal and the interest rate. Use $i = prt$.

For the given cost and revenue equations, find the break-even point(s).

53. $C = 10x + 300$, $R = 30x - 0.1x^2$

54. $C = 0.6x^2 + 9$, $R = 12x - 0.2x^2$

55. $C = 12.6x + 150$, $R = 42.8x - 0.3x^2$

56. $C = 80x + 900$, $R = 120x - 0.2x^2$

Solve the following systems using your graphing calculator. Round your answers to the nearest hundredth.

57. $3x - 5y = 12$
$x^2 + y^2 = 10$

58. $y = 2x^2 - x + 2$
$4x^2 + y^2 = 36$

Concept/Writing Exercises

59. Make up your own nonlinear system of equations whose solution is the empty set. Explain how you know the system has no solution.

60. If a system of equations consists of an ellipse and a hyperbola, what is the maximum number of points of intersection? Make a sketch to illustrate this.

61. Can a nonlinear system of equations have exactly one real solution? If so, give an example. Explain.

62. Can a nonlinear system of equations have exactly two real solutions? If so, give an example. Explain.

63. Can a nonlinear system of equations have exactly three real solutions? If so, give an example. Explain.

64. Can a nonlinear system of equations have no real solutions? If so, give an example. Explain.

Challenge Problems

65. Intersecting Roads The intersection of three roads forms a right triangle, as shown in the figure.

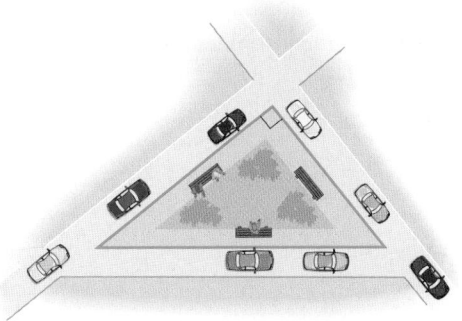

If the hypotenuse is 26 yards and the area is 120 square yards, find the length of the two legs of the triangle.

66. In the figure shown, R represents the radius of the larger orange circle and r represents the radius of the smaller orange circles. If $R = 2r$ and if the shaded area is 122.5π, find r and R.

Cumulative Review Exercises

[1.4] **67.** List the order of operations we follow when evaluating an expression.

[5.6] **68.** Factor $(x + 1)^3 + 1$.

[6.6] **69.** x varies inversely as the square of P. If $x = 10$ when P is 6, find x when $P = 20$.

[7.5] **70.** Simplify $\dfrac{5}{\sqrt{x + 2} - 3}$.

[9.7] **71.** Solve $A = A_0 e^{kt}$ for k.

Chapter 10 Summary

IMPORTANT FACTS AND CONCEPTS	EXAMPLES

Section 10.1

The four **conic sections** are the parabola, circle, ellipse, and hyperbola, which are obtained by slicing a cone.

Parabola   Circle   Ellipse   Hyperbola

The four different forms for equations of parabolas are summarized below.

Parabola with Vertex at (h, k)

1. $y = a(x - h)^2 + k, a > 0$ (opens upward)
2. $y = a(x - h)^2 + k, a < 0$ (opens downward)
3. $x = a(y - k)^2 + h, a > 0$ (opens to the right)
4. $x = a(y - k)^2 + h, a < 0$ (opens to the left)

$y = -(x - 2)^2 + 3$ $x = 2(y + 1)^2 - 4$

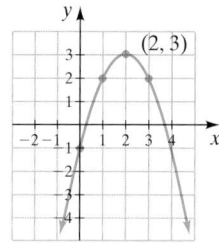 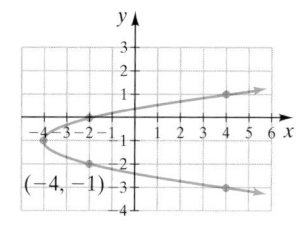

Distance Formula

The distance, d, between any two points (x_1, y_1) and (x_2, y_2) can be found by the distance formula:
$$d = \sqrt{(x_2 - x_1)^2 + (y_2 - y_1)^2}$$

The distance between $(-1, 3)$ and $(4, 15)$ is
$$d = \sqrt{[4 - (-1)]^2 + (15 - 3)^2} = \sqrt{5^2 + 12^2} = \sqrt{169} = 13$$

Midpoint Formula

Given any two points (x_1, y_1) and (x_2, y_2), the point halfway between the given points can be found by the midpoint formula:
$$\text{midpoint} = \left(\frac{x_1 + x_2}{2}, \frac{y_1 + y_2}{2}\right)$$

The midpoint of the line segment joining $(7, 6)$ and $(-11, 10)$ is
$$\text{midpoint} = \left(\frac{7 + (-11)}{2}, \frac{6 + 10}{2}\right) = \left(\frac{-4}{2}, \frac{16}{2}\right) = (-2, 8)$$

A **circle** is a set of points in a plane that are the same distance, called the **radius**, from a fixed point, called the **center**.

Circle with Its Center at the Origin and Radius r
$$x^2 + y^2 = r^2$$

Sketch the graph of $x^2 + y^2 = 9$.

The graph is a circle with its center at $(0, 0)$ and radius $r = 3$.

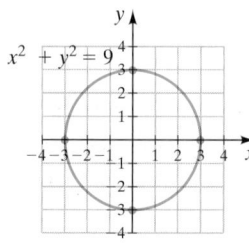

IMPORTANT FACTS AND CONCEPTS	EXAMPLES

Section 10.1 (cont.)

Circle with Its center at (h, k) and Radius r

$$(x - h)^2 + (y - k)^2 = r^2$$

Sketch the graph of $(x - 3)^2 + (y + 5)^2 = 25$.
The graph is a circle with its center at $(3, -5)$ and radius $r = 5$.

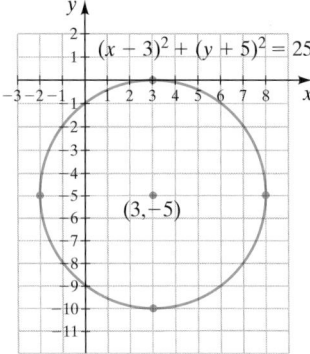

Section 10.2

An **ellipse** is a set of points in a plane, the sum of whose distances from two fixed points (called the **foci**) is a constant.

Ellipse with Its Center at the Origin

$$\frac{x^2}{a^2} + \frac{y^2}{b^2} = 1$$

where $(a, 0)$ and $(-a, 0)$ are the x-intercepts and $(0, b)$ and $(0, -b)$ are the y-intercepts.

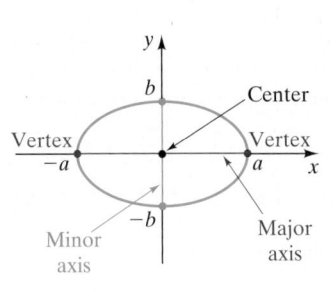
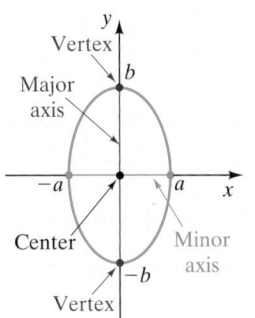

Sketch the graph of $\dfrac{x^2}{25} + \dfrac{y^2}{16} = 1$.

The graph is an ellipse. Since $a = 5$, the x-intercepts are $(-5, 0)$ and $(5, 0)$. Since $b = 4$, the y-intercepts are $(0, -4)$ and $(0, 4)$.

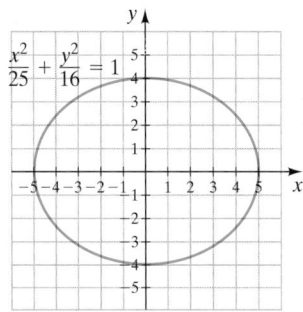

Ellipse with Its Center at (h, k)

$$\frac{(x - h)^2}{a^2} + \frac{(y - k)^2}{b^2} = 1$$

Sketch the graph of $\dfrac{(x - 2)^2}{9} + \dfrac{(y + 1)^2}{16} = 1$.

The graph is an ellipse with its center at $(2, -1)$, where $a = 3$ and $b = 4$.

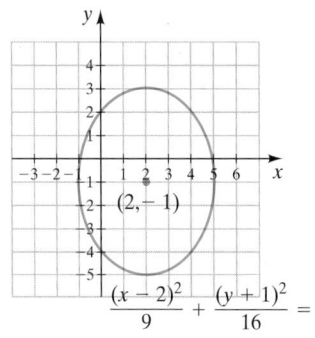

IMPORTANT FACTS AND CONCEPTS	EXAMPLES

Section 10.2 (cont.)

The area, A, of an ellipse is $A = \pi ab$.

The area of the second ellipse from page 667 is

$$A = \pi ab = \pi \cdot 3 \cdot 4 = 12\pi \approx 37.70 \text{ square units.}$$

Section 10.3

A **hyperbola** is a set of points in a plane, the difference of whose distances from two fixed points (called **foci**) is a constant.

Hyperbola with Its Center at the Origin

Hyperbola
with transverse axis
along the x-axis

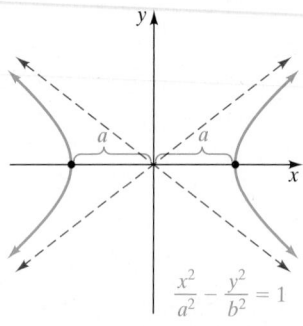

$$\frac{x^2}{a^2} - \frac{y^2}{b^2} = 1$$

Asymptotes

$$y = \frac{b}{a}x \quad \text{and} \quad y = -\frac{b}{a}x$$

Hyperbola
with transverse axis
along the y-axis

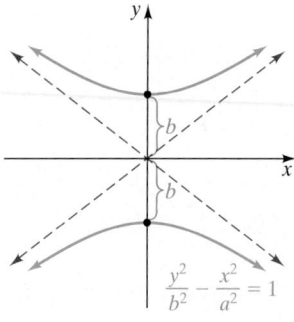

$$\frac{y^2}{b^2} - \frac{x^2}{a^2} = 1$$

Asymptotes

$$y = \frac{b}{a}x \quad \text{and} \quad y = -\frac{b}{a}x$$

Determine the equations of the asymptotes and sketch a graph of $\dfrac{x^2}{4} - \dfrac{y^2}{9} = 1$.

The graph is a hyperbola with $a = 2$ and $b = 3$.

The equations for the asymptotes are $y = \dfrac{3}{2}x$ and $y = -\dfrac{3}{2}x$.

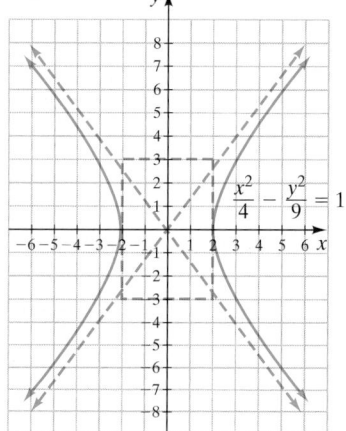

Determine the equations of the asymptotes and sketch a graph of $\dfrac{y^2}{25} - \dfrac{x^2}{16} = 1$.

The graph is a hyperbola with $a = 4$ and $b = 5$. The equations of the asymptotes are $y = \dfrac{5}{4}x$ and $y = -\dfrac{5}{4}x$.

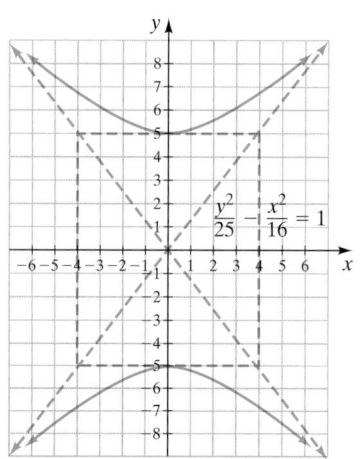

IMPORTANT FACTS AND CONCEPTS	EXAMPLES

Section 10.4

A **nonlinear system of equations** is a system of equations where at least one equation is not linear. The solution to a nonlinear system of equations is the point or points that satisfy all equations in the system.

Solve the system of equations.

$$x^2 + y^2 = 14$$
$$5x^2 - y^2 = -2$$

We will solve this system using the addition method.

$$x^2 + y^2 = 14$$
$$\underline{5x^2 - y^2 = -2}$$
$$6x^2 = 12$$
$$x^2 = 2$$
$$x = \pm\sqrt{2}$$

To obtain the value(s) for y, use the equation $x^2 + y^2 = 14$.

$x = \sqrt{2}$	$x = -\sqrt{2}$
$x^2 + y^2 = 14$	$x^2 + y^2 = 14$
$(\sqrt{2})^2 + y^2 = 14$	$(-\sqrt{2})^2 + y^2 = 14$
$2 + y^2 = 14$	$2 + y^2 = 14$
$y^2 = 12$	$y^2 = 12$
$y = \pm\sqrt{12}$	$y = \pm\sqrt{12}$
$= \pm 2\sqrt{3}$	$= \pm 2\sqrt{3}$

The system has four solutions:

$$(\sqrt{2}, 2\sqrt{3}), (\sqrt{2}, -2\sqrt{3}), (-\sqrt{2}, 2\sqrt{3}), (-\sqrt{2}, -2\sqrt{3})$$

Chapter 10 Review Exercises

[10.1]   *Find the length and the midpoint of the line segment between each pair of points.*

1. $(0,0), (5,-12)$ **2.** $(-4,1), (-1,5)$ **3.** $(-9,-5), (-1,10)$ **4.** $(-4,3), (-2,5)$

Graph each equation.

5. $y = (x-2)^2 + 1$ **6.** $y = (x+3)^2 - 2$ **7.** $x = (y-1)^2 + 4$ **8.** $x = -2(y-4)^2 + 4$

In Exercises 9–12, **a)** *write each equation in the form* $y = a(x-h)^2 + k$ *or* $x = a(y-k)^2 + h$. **b)** *Graph the equation.*

9. $y = x^2 - 8x + 22$ **10.** $x = -y^2 - 5y - 4$ **11.** $x = y^2 + 5y + 4$ **12.** $y = 2x^2 - 8x - 24$

In Exercises 13–18, **a)** *write the equation of each circle in standard form.* **b)** *Draw the graph.*

13. Center $(0,0)$, radius 4   **14.** Center $(-3,4)$, radius 1   **15.** $x^2 + y^2 - 4y = 0$
16. $x^2 + y^2 - 2x + 6y + 1 = 0$ **17.** $x^2 - 8x + y^2 - 10y + 40 = 0$ **18.** $x^2 + y^2 - 4x + 10y + 17 = 0$

Graph each equation.

19. $y = \sqrt{9 - x^2}$ **20.** $y = -\sqrt{36 - x^2}$

Determine the equation of each circle.

21.

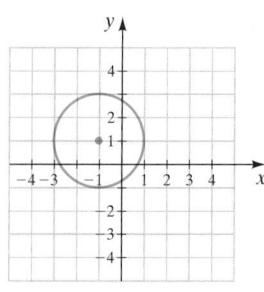

22.

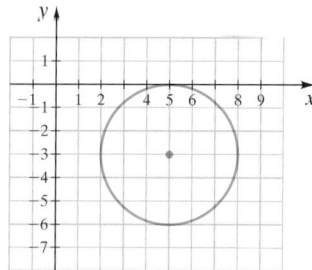

[10.2]　*Graph each equation.*

23. $\dfrac{x^2}{4} + \dfrac{y^2}{9} = 1$

24. $\dfrac{x^2}{81} + \dfrac{y^2}{49} = 1$

25. $9x^2 + 16y^2 = 144$

26. $25x^2 + 4y^2 = 100$

27. $\dfrac{(x-3)^2}{16} + \dfrac{(y+2)^2}{4} = 1$

28. $\dfrac{(x-4)^2}{9} + \dfrac{(y+3)^2}{25} = 1$

29. $16(x-2)^2 + 4(y+3)^2 = 16$

30. For the ellipse in Exercise 23, find the area.

[10.3]　*In Exercises 31–34,* **a)** *determine the equations of the asymptotes for each equation.* **b)** *Draw the graph.*

31. $\dfrac{x^2}{9} - \dfrac{y^2}{25} = 1$

32. $\dfrac{y^2}{25} - \dfrac{x^2}{9} = 1$

33. $\dfrac{y^2}{9} - \dfrac{x^2}{16} = 1$

34. $\dfrac{y^2}{25} - \dfrac{x^2}{16} = 1$

In Exercises 35–38, **a)** *write each equation in standard form.* **b)** *Determine the equations of the asymptotes.* **c)** *Draw the graph.*

35. $x^2 - 9y^2 = 9$

36. $64y^2 - 25x^2 = 1600$

37. $9y^2 - 25x^2 = 225$

38. $49y^2 - 9x^2 = 441$

[10.1–10.3]　*Identify the graph of each equation as a circle, ellipse, parabola, or hyperbola.*

39. $\dfrac{x^2}{49} - \dfrac{y^2}{16} = 1$

40. $4x^2 + 8y^2 = 32$

41. $6x^2 + 6y^2 = 96$

42. $4x^2 - 36y^2 = 36$

43. $\dfrac{x^2}{18} + \dfrac{y^2}{9} = 1$

44. $y = (x-3)^2 + 4$

45. $12x^2 + 9y^2 = 108$

46. $x = -y^2 + 8y - 9$

[10.4]　*Find all real solutions to each system of equations using the substitution method.*

47. $2x^2 + y^2 = 16$
$x^2 - y^2 = -4$

48. $x^2 - y^2 = 4$
$x + y = 4$

49. $x^2 + y^2 = 9$
$x + 2y = 3$

50. $x^2 + 2y^2 = 5$
$x^2 - 4y^2 = 36$

Find all real solutions to each system of equations using the addition method.

51. $x^2 + y^2 = 36$
$x^2 - y^2 = 36$

52. $x^2 + y^2 = 16$
$2x^2 - 5y^2 = 25$

53. $x^2 + y^2 = 81$
$25x^2 + 4y^2 = 100$

54. $3x^2 + 4y^2 = 35$
$2x^2 + 5y^2 = 42$

55. Pool Table Jerry and Denise have a pool table in their house. It has an area of 45 square feet and a perimeter of 28 feet. Find the dimensions of the pool table.

56. Bottles of Glue The Dip and Dap Company has a cost equation of $C = 20.3x + 120$ and a revenue equation of $R = 50.2x - 0.2x^2$, where x is the number of bottles of glue sold. Find the number of bottles of glue the company must sell to break even.

57. Savings Account If Kien Kempter invests a certain principal at a specific interest rate for 1 year, the interest is $120. If he increases the principal by $2000 and the interest rate is decreased by 1%, the interest remains the same. Find the principal and interest rate. Use $i = prt$.

Chapter 10 Practice Test

Chapter Test Prep Videos provide fully worked-out solutions to any of the exercises you want to review. Chapter Test Prep Videos are available via **MyMathLab** *, or on* **You Tube** *(search "Angel Intermediate Algebra" and click on "Channels")*

1. Why are parabolas, circles, ellipses, and hyperbolas called conic sections?

2. Determine the length of the line segment whose endpoints are $(-1, 8)$ and $(6, 7)$.

3. Determine the midpoint of the line segment whose endpoints are $(-9, 4)$ and $(7, -1)$.

4. Determine the vertex of the graph of $y = -2(x + 3)^2 + 1$, and then graph the equation.

5. Graph $x = y^2 - 2y + 4$.

6. Write the equation $x = -y^2 - 4y - 5$ in the form $x = a(y - k)^2 + h$, and then draw the graph.

7. Write the equation of a circle with center at $(2, 4)$ and radius 3 and then draw the graph of the circle.

8. Find the area of the circle whose equation is $(x + 2)^2 + (y - 8)^2 = 9$.

9. Write the equation of the circle shown.

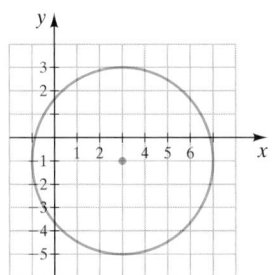

10. Graph $y = -\sqrt{16 - x^2}$.

11. Write the equation $x^2 + y^2 + 2x - 6y + 1 = 0$ in standard form, and then draw the graph.

12. Graph $4x^2 + 25y^2 = 100$.

13. Is the following graph the graph of $\dfrac{(x + 2)^2}{4} + \dfrac{(y + 1)^2}{16} = 1$? Explain your answer.

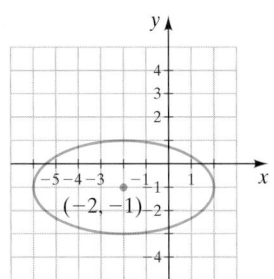

14. Graph $4(x - 4)^2 + 36(y + 2)^2 = 36$.

15. Find the center of the ellipse given by the equation $3(x - 8)^2 + 6(y + 7)^2 = 18$.

16. Explain how to determine whether the transverse axis of a hyperbola lies on the x- or y-axis.

17. What are the equations of the asymptotes of the graph of $\dfrac{x^2}{16} - \dfrac{y^2}{49} = 1$?

18. Graph $\dfrac{y^2}{25} - \dfrac{x^2}{1} = 1$.

19. Graph $\dfrac{x^2}{4} - \dfrac{y^2}{9} = 1$.

In Exercises 20 and 21, determine whether the graph of the equation is a parabola, circle, ellipse, or hyperbola.

20. $4x^2 - 15y^2 = 30$

21. $25x^2 + 4y^2 = 100$

Solve each system of equations.

22. $x^2 + y^2 = 7$
$2x^2 - 3y^2 = -1$

23. $x + y = 8$
$x^2 + y^2 = 4$

24. **Vegetable Garden** Tom Wilson has a rectangular vegetable garden on his farm that has an area of 1500 square meters. Find the dimensions of the garden if the perimeter is 160 meters.

25. **Truck Bed** Gina Chang owns a truck. The rectangular bed of the truck has an area of 60 square feet, and the diagonal across the bed measures 13 feet. Find the dimensions of the bed of the truck.

Cumulative Review Test

Take the following test and check your answers with those given in the back of the book. Review any questions that you answered incorrectly. The section where the material was covered is indicated after the answer.

1. Simplify $(9x^2y^5)(-3xy^4)$.

2. Solve $4x - 2(3x - 7) = 2x - 5$.

3. Find the solution set: $2(x - 5) + 2x = 4x - 7$.

4. Find the solution set: $|3x + 1| > 4$

5. Graph $y = -2x + 2$.

6. If $f(x) = x^2 + 3x + 9$, find $f(10)$.

7. Solve the system of equations.
$$\frac{1}{2}x - \frac{1}{3}y = 2$$
$$\frac{1}{4}x + \frac{2}{3}y = 6$$

8. Factor $x^4 - x^2 - 42$.

9. A large triangular sign has a height that is 6 feet less than its base. If the area of the sign is 56 square feet, find the length of the base and the height of the sign.

10. Multiply $\dfrac{3x^2 - x - 4}{4x^2 + 7x + 3} \cdot \dfrac{2x^2 - 5x - 12}{6x^2 + x - 12}$.

11. Subtract $\dfrac{x}{x + 3} - \dfrac{x + 5}{2x^2 - 2x - 24}$.

12. Solve $\dfrac{3}{x + 3} + \dfrac{5}{x + 4} = \dfrac{12x + 19}{x^2 + 7x + 12}$.

13. Simplify $\left(\dfrac{18x^{1/2}y^3}{2x^{3/2}}\right)^{1/2}$.

14. Simplify $\dfrac{6\sqrt{x}}{\sqrt{x} - y}$.

15. Solve $3\sqrt[3]{2x + 2} = \sqrt[3]{80x - 24}$.

16. Solve $3x^2 - 4x + 5 = 0$ by the quadratic formula.

17. Solve $\log(3x - 4) + \log 4 = \log(x + 6)$.

18. Solve $35 = 70e^{-0.3t}$.

19. Graph $9x^2 + 4y^2 = 36$.

20. Graph $\dfrac{y^2}{25} - \dfrac{x^2}{16} = 1$.

Sequences, Series, and the Binomial Theorem

Goals of This Chapter

Sequences and series are discussed in this chapter. A sequence is a list of numbers in a specific order and a series is the sum of the numbers in a sequence. In this book, we discuss two types of sequences and series: arithmetic and geometric. Sequences and series can be used to solve many real-life problems as illustrated in this chapter.

In this chapter, we introduce the summation symbol, Σ, which is often used in statistics and other mathematics courses. We also discuss the binomial theorem for expanding an expression of the form $(a + b)^n$.

If a ball rebounds 9 feet when dropped from 10 feet, it has rebounded 90% of its original height. Theoretically, every rebound will have a rebound and the ball will never stop bouncing. In Exercise 78 on page 687, we determine the height of a ping-pong ball after it falls off a table.

© Jonathan Larsen/Shutterstock

11.1 Sequences and Series

1 Find the Terms of a Sequence

Many times we see patterns in numbers. For example, suppose you are given a job offer with a starting salary of $30,000. You are given two options for your annual salary increases. One option is an annual salary increase of $2000 per year. The salary you would receive under this option is shown below.

Year	1	2	3	4	$\cdots$
	$\downarrow$	$\downarrow$	$\downarrow$	$\downarrow$	$\downarrow$
Salary	$30,000	$32,000	$34,000	$36,000	$\cdots$

Each year the salary is $2000 greater than the previous year. The three dots on the right of the lists of numbers indicate that the list continues in the same manner.

The second option is a 5% salary increase each year. The salary you would receive under this option is shown below.

Year	1	2	3	4	$\cdots$
	$\downarrow$	$\downarrow$	$\downarrow$	$\downarrow$	$\downarrow$
Salary	$30,000	$31,500	$33,075	$34,728.75	$\cdots$

With this option, the salary in a given year after year 1 is 5% greater than the previous year's salary.

The two lists of numbers that illustrate the salaries are examples of sequences. A **sequence** of numbers is a list of numbers arranged in a specific order. Consider the list of numbers given below, which is a sequence.

$$5, 10, 15, 20, 25, 30, \ldots$$

Each number is called a **term** of the sequence. The first term is 5. We indicate this by writing $a_1 = 5$. Since the second term is 10, $a_2 = 10$, and so on.

$$
\begin{array}{cccccc}
5 & 10 & 15 & 20 & 25 & 30\ldots \\
\downarrow & \downarrow & \downarrow & \downarrow & \downarrow & \downarrow \\
a_1 & a_2 & a_3 & a_4 & a_5 & a_6\ldots
\end{array}
$$

The three dots, called an **ellipsis**, indicate that the sequence continues indefinitely and is an **infinite sequence**.

> **Infinite Sequence**
>
> An **infinite sequence** is a function whose domain is the set of natural numbers.

Understanding Algebra

An *infinite* sequence continues on forever and never ends.

Consider the infinite sequence $5, 10, 15, 20, 25, 30, 35, \ldots$

Domain: $\{1, \quad 2, \quad 3, \quad 4, \quad 5, \quad 6, \quad 7, \quad \ldots, \quad n, \quad \ldots\}$ Natural numbers

$\quad\quad\quad\;\; \downarrow \;\; \downarrow \;\; \downarrow \;\; \downarrow \;\; \downarrow \;\; \downarrow \;\; \downarrow \quad\quad\quad\; \downarrow$

Range: $\{5, \quad 10, \quad 15, \quad 20, \quad 25, \quad 30, \quad 35, \quad \ldots, \quad 5n, \quad \ldots\}$ Function value

Note that the terms of the sequence $5, 10, 15, 20, \ldots$ are found by multiplying each natural number by 5. For any natural number, n, the corresponding term in the sequence is $5 \cdot n$ or $5n$. The **general term of the sequence**, a_n, which defines the sequence, is $a_n = 5n$.

$$a_n = f(n) = 5n$$

To find the twelfth term of the sequence, substitute 12 for n in the general term of the sequence: $a_{12} = 5 \cdot 12 = 60$. Thus, the twelfth term of the sequence is 60. Note that the terms in the sequence are the function values, or the numbers in the range of the function. The general form of a sequence is

$$a_1, a_2, a_3, a_4, \ldots, a_n, \ldots$$

n	$a_n = 2^n$
1	$a_1 = 2^1 = 2$
2	$a_2 = 2^2 = 4$
3	$a_3 = 2^3 = 8$
4	$a_4 = 2^4 = 16$
5	$a_5 = 2^5 = 32$
6	$a_6 = 2^6 = 64$
⋮	⋮

For the infinite sequence $2, 4, 8, 16, 32, \ldots, 2^n, \ldots$ we can write

$$a_n = f(n) = 2^n$$

The table to the left shows the first six terms of this sequence.

Finite Sequence

A **finite sequence** is a function whose domain includes only the first n natural numbers. A finite sequence has only a finite number of terms.

Examples of Finite Sequences

$5, 10, 15, 20$ domain is $\{1, 2, 3, 4\}$
$2, 4, 8, 16, 32$ domain is $\{1, 2, 3, 4, 5\}$

EXAMPLE 1 Write the finite sequence defined by $a_n = 2n + 3$, for $n = 1, 2, 3, 4$.

Solution

$$a_n = 2n + 3$$
$$a_1 = 2(1) + 3 = 5$$
$$a_2 = 2(2) + 3 = 7$$
$$a_3 = 2(3) + 3 = 9$$
$$a_4 = 2(4) + 3 = 11$$

Thus, the sequence is $5, 7, 9, 11$.

Now Try Exercise 13

> ## Understanding Algebra
>
> A sequence is *increasing* when each term is greater than the preceding term.

Since each term of the sequence in Example 1 is greater than the preceding term, it is called an **increasing sequence**.

EXAMPLE 2 Given $a_n = \dfrac{2n + 3}{n^2}$,

a) find the first term in the sequence.   b) find the third term in the sequence.
c) find the fifth term in the sequence.   d) find the tenth term in the sequence.

Solution

a) When $n = 1$, $a_1 = \dfrac{2(1) + 3}{1^2} = \dfrac{5}{1} = 5$.

b) When $n = 3$, $a_3 = \dfrac{2(3) + 3}{3^2} = \dfrac{9}{9} = 1$.

c) When $n = 5$, $a_5 = \dfrac{2(5) + 3}{5^2} = \dfrac{13}{25} = 0.52$.

d) When $n = 10$, $a_{10} = \dfrac{2(10) + 3}{10^2} = \dfrac{23}{100} = 0.23$.

Now Try Exercise 29

> ## Understanding Algebra
>
> A sequence is *decreasing* when each term is less than the preceding term.

Note in Example 2 that since there is no restriction on n, a_n is the general term of an infinite sequence.

In Example 2, the first four terms of the sequence are $5, \dfrac{7}{4} = 1.75, 1, \dfrac{11}{16} = 0.6875$.

Since each term of the sequence generated by $a_n = \dfrac{2n + 3}{n^2}$ will be less than the preceding term, the sequence is called a **decreasing sequence**.

EXAMPLE 3 Find the first four terms of the sequence whose general term is $a_n = (-1)^n(n)$.

Solution

$$a_n = (-1)^n(n)$$
$$a_1 = (-1)^1(1) = -1$$
$$a_2 = (-1)^2(2) = 2$$
$$a_3 = (-1)^3(3) = -3$$
$$a_4 = (-1)^4(4) = 4$$

Now Try Exercise 21

Understanding Algebra

To *alternate* means to go back and forth with + and − signs. In an *alternating sequence*, the terms alternate in sign.

If we write the sequence, we get $-1, 2, -3, 4, \ldots, (-1)^n(n), \ldots$. Notice that each term alternates in sign. We call this an **alternating sequence**.

2 Write a Series

Series

A **series** is the expressed sum of the terms of a sequence.

A series may be finite or infinite, depending on whether the sequence it is based on is finite or infinite.

Understanding Algebra

A *series* is obtained by adding the terms from its corresponding *sequence*.

Examples

Finite Sequence

$$a_1, a_2, a_3, a_4, a_5$$

Finite Series

$$a_1 + a_2 + a_3 + a_4 + a_5$$

Infinite Sequence

$$a_1, a_2, a_3, a_4, a_5, \ldots, a_n, \ldots$$

Infinite Series

$$a_1 + a_2 + a_3 + a_4 + a_5 + \cdots + a_n + \cdots$$

EXAMPLE 4 Write the first eight terms of the sequence given by each general term, then write the series that represents the sum of that sequence.

a) $a_n = \left(\dfrac{1}{2}\right)^n$ **b)** $a_n = (-2)^n$

Solution

a) We begin with $n = 1$; thus, the first eight terms of the sequence whose general term is $a_n = \left(\dfrac{1}{2}\right)^n$ are

$$\left(\frac{1}{2}\right)^1, \left(\frac{1}{2}\right)^2, \left(\frac{1}{2}\right)^3, \left(\frac{1}{2}\right)^4, \left(\frac{1}{2}\right)^5, \left(\frac{1}{2}\right)^6, \left(\frac{1}{2}\right)^7, \left(\frac{1}{2}\right)^8$$

or

$$\frac{1}{2}, \frac{1}{4}, \frac{1}{8}, \frac{1}{16}, \frac{1}{32}, \frac{1}{64}, \frac{1}{128}, \frac{1}{256}$$

The series that represents the sum of the sequence is

$$\frac{1}{2} + \frac{1}{4} + \frac{1}{8} + \frac{1}{16} + \frac{1}{32} + \frac{1}{64} + \frac{1}{128} + \frac{1}{256} = \frac{255}{256}$$

b) We again begin with $n = 1$; thus, the first eight terms of the sequence whose general term is $a_n = (-2)^n$ are

$$(-2)^1, (-2)^2, (-2)^3, (-2)^4, (-2)^5, (-2)^6, (-2)^7, (-2)^8$$

or

$$-2, 4, -8, 16, -32, 64, -128, 256$$

The series that represents the sum of this sequence is

$$-2 + 4 + (-8) + 16 + (-32) + 64 + (-128) + 256 = 170$$

Now Try Exercise 45

3 Find Partial Sums

Partial Sum

For an infinite sequence with the terms $a_1, a_2, a_3, \ldots, a_n, \ldots$, a **partial sum** is the sum of a finite number of consecutive terms of the sequence, beginning with the first term.

$$
\begin{aligned}
s_1 &= a_1 & &\text{First partial sum} \\
s_2 &= a_1 + a_2 & &\text{Second partial sum} \\
s_3 &= a_1 + a_2 + a_3 & &\text{Third partial sum} \\
&\;\;\vdots \\
s_n &= a_1 + a_2 + a_3 + \cdots + a_n & &n\text{th partial sum}
\end{aligned}
$$

The sum of all the terms of the infinite sequence is called an **infinite series** and is given by the following:

$$s = a_1 + a_2 + a_3 + \cdots + a_n + \cdots$$

EXAMPLE 5 Given the infinite sequence defined by $a_n = \dfrac{3 + n^2}{n}$, find the indicated partial sums.

a) s_1 **b)** s_4

Solution

a) $s_1 = a_1 = \dfrac{3 + 1^2}{1} = \dfrac{3 + 1}{1} = 4$

b) $s_4 = a_1 + a_2 + a_3 + a_4$

$$= \frac{3 + 1^2}{1} + \frac{3 + 2^2}{2} + \frac{3 + 3^2}{3} + \frac{3 + 4^2}{4}$$

$$= 4 + \frac{7}{2} + \frac{12}{3} + \frac{19}{4}$$

$$= \frac{48}{12} + \frac{42}{12} + \frac{48}{12} + \frac{57}{12}$$

$$= \frac{195}{12} \quad \text{or} \quad 16\frac{1}{4}$$

Now Try Exercise 35

4 Use Summation Notation, Σ

When the general term of a sequence is known, the Greek letter **sigma**, Σ, can be used to write a series. The sum of the first n terms of the sequence whose nth term is a_n is represented by

$$\sum_{i=1}^{n} a_i = a_1 + a_2 + a_3 + \cdots + a_n$$

where i is called the **index of summation** or simply the **index**, n is the **upper limit of summation**, and 1 is the **lower limit of summation**. In this illustration, we used i for the index; however, any letter can be used for the index.

Consider the sequence $7, 9, 11, 13, \ldots, 2n + 5, \ldots$. The sum of the first five terms can be represented using **summation notation**.

$$\sum_{i=1}^{5} (2i + 5)$$

This notation is read "the sum as i goes from 1 to 5 of $2i + 5$."

To evaluate the series represented by $\displaystyle\sum_{i=1}^{5} (2i + 5)$,

1. First substitute 1 for i in $2i + 5$ to obtain the value 7.
2. Then, substitute 2 for i in $2i + 5$ to obtain the value 9.
3. Continue to follow this procedure for $i = 3, i = 4$, and $i = 5$.
4. Finally, add these values to evaluate the series. The results are summarized as follows:

$$
\begin{array}{ccccc}
i = 1 & i = 2 & i = 3 & i = 4 & i = 5 \\
\downarrow & \downarrow & \downarrow & \downarrow & \downarrow
\end{array}
$$

$$\sum_{i=1}^{5} (2i + 5) = (2 \cdot 1 + 5) + (2 \cdot 2 + 5) + (2 \cdot 3 + 5) + (2 \cdot 4 + 5) + (2 \cdot 5 + 5)$$

$$= 7 + 9 + 11 + 13 + 15$$

$$= 55$$

EXAMPLE 6 Write out the series $\displaystyle\sum_{i=1}^{6} (i^2 + 1)$ and evaluate it.

Solution

$$\sum_{i=1}^{6} (i^2 + 1) = (1^2 + 1) + (2^2 + 1) + (3^2 + 1) + (4^2 + 1) + (5^2 + 1) + (6^2 + 1)$$

$$= 2 + 5 + 10 + 17 + 26 + 37$$

$$= 97$$

Now Try Exercise 57

EXAMPLE 7 Consider the general term of a sequence $a_n = 2n^2 - 9$. Represent the third partial sum, s_3, in summation notation.

Solution The third partial sum will be the sum of the first three terms, $a_1 + a_2 + a_3$.

We can represent the third partial sum as $\displaystyle\sum_{i=1}^{3} (2i^2 - 9)$.

Now Try Exercise 65

EXAMPLE 8 For the following set of values $x_1 = 3, x_2 = 4, x_3 = 5, x_4 = 6$, and $x_5 = 7$, does $\displaystyle\sum_{i=1}^{5} (x_i)^2 = \left(\sum_{i=1}^{5} x_i \right)^2$?

Solution

$$\sum_{i=1}^{5} (x_i)^2 = (x_1)^2 + (x_2)^2 + (x_3)^2 + (x_4)^2 + (x_5)^2$$

$$= 3^2 + 4^2 + 5^2 + 6^2 + 7^2$$

$$= 9 + 16 + 25 + 36 + 49 = 135$$

$$\left(\sum_{i=1}^{5} x_i \right)^2 = (x_1 + x_2 + x_3 + x_4 + x_5)^2$$

$$= (3 + 4 + 5 + 6 + 7)^2 = (25)^2 = 625$$

Since $135 \neq 625$, $\displaystyle\sum_{i=1}^{5} (x_i)^2 \neq \left(\sum_{i=1}^{5} x_i \right)^2$.

Now Try Exercise 71

When a summation symbol is written without any upper and lower limits, it means that all the given data are to be summed.

EXAMPLE 9 A formula used to find the arithmetic mean, $\bar{x}$ (read x bar), of a set of data is $\bar{x} = \dfrac{\Sigma x}{n}$, where n is the number of pieces of data.

Joan Sally's five test scores are 70, 95, 83, 74, and 92. Find the arithmetic mean of her scores.

Solution $\bar{x} = \dfrac{\Sigma x}{n} = \dfrac{70 + 95 + 83 + 74 + 92}{5} = \dfrac{414}{5} = 82.8$

Now Try Exercise 75

EXERCISE SET 11.1

MathXL® MathXL   MyMathLab MyMathLab

Warm-Up Exercises

Fill in the blanks with the appropriate word, phrase, or symbol(s) from the following list.

finite	series	greater	upper	sequence	less
constant	infinite	alternating	index	lower	

1. A list of numbers arranged in a specific order is a _____.

2. An _____ sequence is a function whose domain is the set of natural numbers.

3. A _____ sequence is a function whose domain includes only the first n natural numbers.

4. In an increasing sequence, each term is _____ than the preceding term.

5. In a decreasing sequence, each term is _____ than the preceding term.

6. A sequence in which each term has the opposite sign of the preceding term is called an _____ sequence.

7. The sum of the terms in a sequence is a _____.

8. Consider the summation $\displaystyle\sum_{k=1}^{50}(k^2 + 3)$. 1 is the _____ limit.

9. Consider the summation $\displaystyle\sum_{k=1}^{50}(k^2 + 3)$. 50 is the _____ limit.

10. Consider the summation $\displaystyle\sum_{k=1}^{50}(k^2 + 3)$. k is the _____.

Practice the Skills

Write the first five terms of the sequence whose nth term is shown.

11. $a_n = 6n$

12. $a_n = -7n$

13. $a_n = 4n - 1$

14. $a_n = 2n + 5$

15. $a_n = \dfrac{9}{n}$

16. $a_n = \dfrac{8}{n^2}$

17. $a_n = \dfrac{n + 2}{n + 1}$

18. $a_n = \dfrac{n - 3}{n + 6}$

19. $a_n = (-1)^n$

20. $a_n = (-1)^{2n}$

21. $a_n = (-2)^{n+1}$

22. $a_n = 3^{n-1}$

Find the indicated term of the sequence whose nth term is shown.

23. $a_n = 2n + 7$, twelfth term

24. $a_n = 3n + 2$, sixth term

25. $a_n = \dfrac{n}{4} + 8$, sixteenth term

26. $a_n = \dfrac{n}{2} - 13$, fourteenth term

27. $a_n = (-1)^n$, eighth term

28. $a_n = (-2)^n$, seventh term

29. $a_n = n(n + 2)$, eleventh term

30. $a_n = (n - 1)(n + 4)$, fifth term

31. $a_n = \dfrac{n^2}{2n + 7}$, ninth term

32. $a_n = \dfrac{n(n + 6)}{n^2}$, tenth term

Find the first and third partial sums, s_1 and s_3, for each sequence.

33. $a_n = 3n - 1$

34. $a_n = 2n + 5$

35. $a_n = 2^n + 1$

36. $a_n = 3^n - 8$

37. $a_n = \dfrac{n - 1}{n + 2}$

38. $a_n = \dfrac{n}{n + 4}$

39. $a_n = (-1)^n$

40. $a_n = (-3)^n$

41. $a_n = \dfrac{n^2}{2}$

42. $a_n = \dfrac{n^2}{n + 4}$

Write the next three terms of each sequence.

43. $2, 4, 8, 16, 32, \ldots$

44. $7, 12, 17, 22, 27, \ldots$

45. $7, 9, 11, 13, 15, \ldots$

46. $\dfrac{1}{2}, \dfrac{1}{3}, \dfrac{1}{4}, \dfrac{1}{5}, \ldots$

47. $1, \dfrac{1}{2}, \dfrac{1}{3}, \dfrac{1}{4}, \dfrac{1}{5}, \ldots$

48. $\dfrac{2}{3}, \dfrac{3}{4}, \dfrac{4}{5}, \dfrac{5}{6}, \dfrac{6}{7}, \ldots$

49. $-1, 1, -1, 1, -1, \ldots$

50. $-10, -20, -30, -40, \ldots$

51. $1, \dfrac{1}{3}, \dfrac{1}{9}, \dfrac{1}{27}, \ldots$

52. $\dfrac{1}{4}, \dfrac{2}{4}, \dfrac{3}{4}, \dfrac{4}{4}, \ldots$

53. $1, -\dfrac{1}{2}, \dfrac{1}{4}, -\dfrac{1}{8}, \ldots$

54. $\dfrac{1}{3}, -\dfrac{1}{6}, \dfrac{1}{12}, -\dfrac{1}{24}, \ldots$

55. $42, 38, 34, 30, \ldots$

56. $7, -1, -9, -17, \ldots$

Write out each series, then evaluate it.

57. $\displaystyle\sum_{i=1}^{5} (3i - 1)$

58. $\displaystyle\sum_{i=1}^{4} (4i + 9)$

59. $\displaystyle\sum_{i=1}^{6} (i^2 + 1)$

60. $\displaystyle\sum_{i=1}^{5} (2i^2 - 7)$

61. $\displaystyle\sum_{i=1}^{4} \dfrac{i^2}{3}$

62. $\displaystyle\sum_{i=1}^{3} \dfrac{i^2}{4}$

63. $\displaystyle\sum_{i=4}^{9} \dfrac{i^2 + i}{i + 1}$

64. $\displaystyle\sum_{i=2}^{5} \dfrac{i^3}{i + 1}$

For the given general term a_n, write an expression using Σ to represent the indicated partial sum.

65. $a_n = n + 10$, fifth partial sum

66. $a_n = n^2 + 5$, fourth partial sum

67. $a_n = \dfrac{n^2}{4}$, third partial sum

68. $a_n = \dfrac{n^2 + 7}{n + 9}$, third partial sum

For the set of values $x_1 = 2$, $x_2 = 3$, $x_3 = 5$, $x_4 = -1$, and $x_5 = 4$, find each of the following.

69. $\displaystyle\sum_{i=1}^{5} x_i$

70. $\displaystyle\sum_{i=1}^{5} (x_i + 5)$

71. $\left(\displaystyle\sum_{i=1}^{5} x_i \right)^2$

72. $\displaystyle\sum_{i=1}^{5} 4x_i$

73. $\displaystyle\sum_{i=1}^{5} (x_i)^2$

74. $\displaystyle\sum_{i=1}^{4} (x_i^2 + 3)$

Find the arithmetic mean, $\overline{x}$, of the following sets of data.

75. $15, 20, 25, 30, 35$

76. $16, 22, 96, 18, 48$

77. $72, 83, 4, 60, 18, 20$

78. $12, 13, 9, 19, 23, 36, 70$

Problem Solving

In Exercises 79 and 80, consider the following rectangles. For the nth rectangle, the length is 2n and the width is n.

79. Perimeter

a) Find the perimeters for the first four rectangles, and then list the perimeters in a sequence.

b) Find the general term for the perimeter of the *n*th rectangle in the sequence. Use p_n for perimeter.

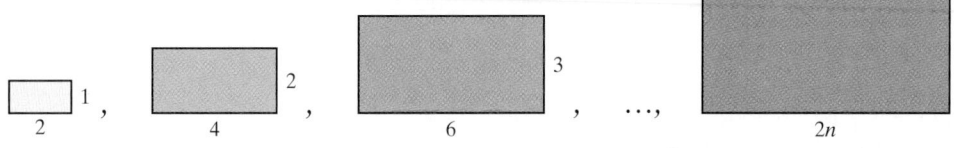

80. Area

a) Find the areas for the four rectangles, and then list the areas in a sequence.

b) Find the general term for the area of the *n*th rectangle in the sequence. Use a_n for area.

81. Write

a) $\displaystyle\sum_{i=1}^{n} x_i$ as a sum of terms.

b) $\displaystyle\sum_{j=1}^{n} x_j$ as a sum of terms.

c) For a given set of values of x, from x_1 to x_n, will $\displaystyle\sum_{i=1}^{n} x_i = \displaystyle\sum_{j=1}^{n} x_j$? Explain.

82. Solve $\overline{x} = \dfrac{\Sigma x}{n}$ for Σx.

83. Solve $\overline{x} = \dfrac{\Sigma x}{n}$ for n.

84. Is $\displaystyle\sum_{i=1}^{n} 4x_i = 4\displaystyle\sum_{i=1}^{n} x_i$? Illustrate your answer with an example.

85. Is $\displaystyle\sum_{i=1}^{n} \dfrac{x_i}{3} = \dfrac{1}{3}\displaystyle\sum_{i=1}^{n} x_i$? Illustrate your answer with an example.

86. Let $x_1 = 3$, $x_2 = 5$, $x_3 = 2$, and $y_1 = 4$, $y_2 = 1$, $y_3 = 6$. Find the following. Note that $\Sigma x = x_1 + x_2 + x_3$, $\Sigma y = y_1 + y_2 + y_3$, and $\Sigma xy = x_1 y_1 + x_2 y_2 + x_3 y_3$.

a) Σx

b) Σy

c) $\Sigma x \cdot \Sigma y$

d) Σxy

e) Is $\Sigma x \cdot \Sigma y = \Sigma xy$?

Concept/Writing Exercises

87. What is the nth partial sum of a series?

88. Write the following notation in words: $\sum\limits_{i=1}^{5} (i + 4)$.

89. Let $a_n = 2n - 1$. Is this an increasing sequence or a decreasing sequence? Explain.

90. Let $a_n = -3n + 7$. Is this an increasing sequence or a decreasing sequence? Explain.

91. Let $a_n = 1 + (-2)^n$. Is this an alternating sequence? Explain.

92. Let $a_n = (-1)^{2n}$. Is this an alternating sequence? Explain.

93. Create your own sequence that is an increasing sequence and list the first five terms.

94. Create your own sequence that is a decreasing sequence and list the first five terms.

95. Create your own sequence that is an alternating sequence and list the first five terms.

Cumulative Review Exercises

[2.6]   **96.** Solve $\left| \dfrac{1}{2}x + \dfrac{3}{5} \right| = \left| \dfrac{1}{2}x - 1 \right|$.

[5.6]   **97.** Factor $8y^3 - 64x^6$.

[7.6]   **98.** Solve $\sqrt{x + 5} - 1 = \sqrt{x - 2}$.

[8.3]   **99.** Solve $V = \pi r^2 h$ for r.

11.2 Arithmetic Sequences and Series

1 Find the common difference in an arithmetic sequence.

2 Find the nth term of an arithmetic sequence.

3 Find the nth partial sum of an arithmetic sequence.

1 Find the Common Difference in an Arithmetic Sequence

In the previous section, we started our discussion by assuming you got a job with a starting salary of $30,000. One option for salary increases was an increase of $2000 each year. This would result in the sequence

$$\$30{,}000, \$32{,}000, \$34{,}000, \$36{,}000, \dots$$

This is an example of an arithmetic sequence.

> **Arithmetic Sequence**
>
> An **arithmetic sequence** is a sequence in which each term after the first differs from the preceding term by a constant amount. The constant amount by which each pair of successive terms differs is called the **common difference**, d.

Understanding Algebra

In an *arithmetic sequence*, you can get the next term by adding the *common difference* to the previous term.

The common difference can be found by subtracting any term from the term that directly follows it.

Arithmetic Sequence	Common Difference
$1, 3, 5, 7, 9, \dots$	$d = 3 - 1 = 2$
$5, 1, -3, -7, -11, -15, \dots$	$d = 1 - 5 = -4$
$\dfrac{7}{2}, \dfrac{2}{2}, -\dfrac{3}{2}, -\dfrac{8}{2}, -\dfrac{13}{2}, -\dfrac{18}{2}, \dots$	$d = \dfrac{2}{2} - \dfrac{7}{2} = -\dfrac{5}{2}$

Notice that the common difference can be a positive number or a negative number. If the sequence is increasing, then d is a positive number. If the sequence is decreasing, then d is a negative number.

EXAMPLE 1 Write the first five terms of the arithmetic sequence with

a) first term 6 and common difference 4.

b) first term 3 and common difference -2.

c) first term 1 and common difference $\dfrac{1}{3}$.

Solution

a) Start with 6 and keep adding 4. The sequence is $6, 10, 14, 18, 22$.

b) $3, 1, -1, -3, -5$

c) $1, \dfrac{4}{3}, \dfrac{5}{3}, 2, \dfrac{7}{3}$

Now Try Exercise 13

2 Find the *n*th Term of an Arithmetic Sequence

In general, an arithmetic sequence with first term, a_1, and common difference, d, has the following terms:

$$a_1 = a_1, \quad a_2 = a_1 + d, \quad a_3 = a_1 + 2d, \quad a_4 = a_1 + 3d, \quad \text{and so on}$$

If we continue this process, we can see that the *n*th term, a_n, can be found by the following formula:

nth Term of an Arithmetic Sequence

$$a_n = a_1 + (n - 1)d$$

EXAMPLE 2

a) Write an expression for the general (or *n*th) term, a_n, of the arithmetic sequence whose first term is -3 and whose common difference is 2.

b) Find the twelfth term of the sequence.

Solution

a) The *n*th term of the sequence is $a_n = a_1 + (n - 1)d$.

$$
\begin{aligned}
a_n &= a_1 + (n - 1)d && \text{nth term of sequence} \\
&= -3 + (n - 1)2 && \text{Substitute } a_1 = -3 \text{ and } d = 2. \\
&= -3 + 2(n - 1) && \text{Commutative property} \\
&= -3 + 2n - 2 && \text{Distributive property} \\
&= 2n - 5 && \text{Simplify.}
\end{aligned}
$$

Thus, $a_n = 2n - 5$.

b) $a_n = 2n - 5$

$$a_{12} = 2(12) - 5 = 24 - 5 = 19$$

The twelfth term in the sequence is 19.

Now Try Exercise 11

EXAMPLE 3 Find the number of terms in the arithmetic sequence $5, 9, 13, 17, \ldots, 41$.

Solution The first term, a_1, is 5; the *n*th term is 41, and the common difference, d, is 4. Substitute the appropriate values into the formula for the *n*th term and solve for *n*.

$$
\begin{aligned}
a_n &= a_1 + (n - 1)d \\
41 &= 5 + (n - 1)4 \\
41 &= 5 + 4n - 4 \\
41 &= 4n + 1 \\
40 &= 4n \\
10 &= n
\end{aligned}
$$

The sequence has 10 terms.

Now Try Exercise 51

Understanding Algebra

In an arithmetic sequence, to get the value for the common difference, *d*, *subtract* a term from the next term in the sequence.

3 Find the *n*th Partial Sum of an Arithmetic Sequence

An **arithmetic series** is the sum of the terms of an arithmetic sequence. A finite arithmetic series can be written

$$s_n = a_1 + (a_1 + d) + (a_1 + 2d) + (a_1 + 3d) + \cdots + (a_n - 2d) + (a_n - d) + a_n$$

If we consider the last term as a_n, the term before the last term will be $a_n - d$, the second before the last term will be $a_n - 2d$, and so on.

A formula for the *n*th partial sum, s_n, can be obtained by adding the reverse of s_n to itself.

$$
\begin{array}{l}
s_n = \quad a_1 \quad + (a_1 + d) + (a_1 + 2d) + \cdots + (a_n - 2d) + (a_n - d) + \quad a_n \\
s_n = \quad a_n \quad + (a_n - d) + (a_n - 2d) + \cdots + (a_1 + 2d) + (a_1 + d) + \quad a_1 \\
\hline
2s_n = (a_1 + a_n) + (a_1 + a_n) + (a_1 + a_n) + \cdots + (a_1 + a_n) + (a_1 + a_n) + (a_1 + a_n)
\end{array}
$$

Since the right side of the equation contains *n* terms of $(a_1 + a_n)$, we can write

$$2s_n = n(a_1 + a_n)$$

Now divide both sides of the equation by 2 to obtain the following formula.

> **_n_th Partial Sum of an Arithmetic Sequence**
>
> $$s_n = \frac{n(a_1 + a_n)}{2}$$

EXAMPLE 4 Find the sum of the first 25 natural numbers.

Solution   The arithmetic sequence is $1, 2, 3, 4, 5, 6, \ldots, 25$. The first term, a_1, is 1; the last term, a_n, is 25. There are 25 terms; thus, $n = 25$. Using the formula for the *n*th partial sum, we have

$$
\begin{aligned}
s_n &= \frac{n(a_1 + a_n)}{2} \\
&= \frac{25(1 + 25)}{2} \\
&= \frac{25(26)}{2} \\
&= 25(13) = 325
\end{aligned}
$$

The sum of the first 25 natural numbers is 325. Thus, $s_{25} = 325$.

Now Try Exercise 57

EXAMPLE 5   The first term of an arithmetic sequence is 4, and the last term is 31. If $s_n = 175$, find the number of terms in the sequence and the common difference.

Solution   We substitute the appropriate values, $a_1 = 4$, $a_n = 31$, and $s_n = 175$, into the formula for the *n*th partial sum and solve for *n*.

$$
\begin{aligned}
s_n &= \frac{n(a_1 + a_n)}{2} \\
175 &= \frac{n(4 + 31)}{2} \\
175 &= \frac{35n}{2} \\
350 &= 35n \\
10 &= n
\end{aligned}
$$

There are 10 terms in the sequence. We can now find the common difference by using the formula for the nth term of an arithmetic sequence.

$$a_n = a_1 + (n - 1)d$$
$$31 = 4 + (10 - 1)d$$
$$31 = 4 + 9d$$
$$27 = 9d$$
$$3 = d$$

The common difference is 3. The sequence is 4, 7, 10, 13, 16, 19, 22, 25, 28, 31.

Now Try Exercise 31

EXAMPLE 6 Salary Mary Tufts is given a starting salary of $35,000 and is promised a $1200 raise after each of the next 8 years. Find her salary during her eighth year of work.

Solution   Understand   Her salaries during the first few years would be

$$\$35,000, \$36,200, \$37,400, \$38,600, \ldots$$

Since we are adding a constant amount each year, this is an arithmetic sequence. The general term of an arithmetic sequence is $a_n = a_1 + (n - 1)d$.

Translate   In this example, $a_1 = 35{,}000$ and $d = 1200$. Thus, for $n = 8$, Mary's salary would be

$$a_n = \ a_1 \ + (n - 1) \ d$$
$$a_8 = 35{,}000 + (8 - 1)1200$$

Carry Out
$$= 35{,}000 + 7(1200)$$
$$= 35{,}000 + 8400$$
$$= 43{,}400$$

Answer   During her eighth year of work, Mary's salary would be $43,400. If we listed all the salaries for the 8-year period, they would be $35,000, $36,200, $37,400, $38,600, $39,800, $41,000, $42,200, $43,400.

Now Try Exercise 83

EXAMPLE 7 Pendulum Each swing of a pendulum (left to right or right to left) is 3 inches shorter than the preceding swing. The first swing is 8 feet.

a) Find the length of the tenth swing.

b) Determine the distance traveled by the pendulum during the first 10 swings.

Solution

a) Understand   Since each swing is decreasing by a constant amount, this problem can be represented as an arithmetic series. Since the first swing is given in feet and the decrease in swing in inches, we will change 3 inches to 0.25 feet ($3 \div 12 = 0.25$). The tenth swing can be considered a_{10}. The difference, d, is negative since the distance is decreasing with each swing.

Translate
$$a_n = a_1 + (n - 1)d$$
$$a_{10} = 8 + (10 - 1)(-0.25)$$

Carry Out
$$= 8 + 9(-0.25)$$
$$= 8 - 2.25$$
$$= 5.75 \text{ feet}$$

Answer   The tenth swing is 5.75 feet.

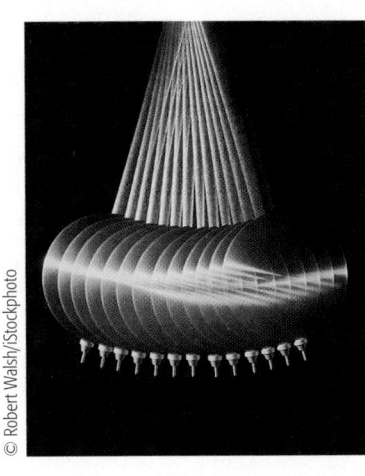

b) Understand and Translate The distance traveled during the first 10 swings can be found using the formula for the nth partial sum. The first swing, a_1, is 8 feet and the twelfth swing, a_{10}, is 5.75 feet.

$$s_n = \frac{n(a_1 + a_n)}{2}$$

$$s_{10} = \frac{10(a_1 + a_{10})}{2}$$

Carry Out $= \dfrac{10(8 + 5.75)}{2} = \dfrac{10(13.75)}{2} = 5(13.75) = 68.75$ feet

Answer The pendulum travels 68.75 feet during its first 10 swings.

Now Try Exercise 75

EXERCISE SET 11.2

Math XL
MathXL®

MyMathLab
MyMathLab

Warm-Up Exercises

Fill in the blanks with the appropriate word, phrase, or symbol(s) from the following list.

sum	series	constant	0	1
n	adding	subtracting	d	a_n

positive	difference	negative	arithmetic
term	adding	-1	2

1. A sequence in which each term after the first term differs from the preceding term by a constant amount is an _____ sequence.

2. The constant amount by which each pair of successive terms differs in an arithmetic sequence is called the common _____.

3. The common difference in an arithmetic sequence is identified by the letter _____.

4. The common difference can be found by _____ any term from the next term in the sequence.

5. The formula $a_n = a_1 + (n - 1)d$ is used to find the nth _____ of an arithmetic sequence.

6. The sum of the terms of an arithmetic sequence is an arithmetic _____.

7. The formula $s_n = \dfrac{n(a_1 + a_n)}{2}$ is used to find the nth partial _____ of an arithmetic sequence.

8. If an arithmetic sequence is increasing, the value for d is a _____ number.

9. If the terms of an arithmetic sequence are decreasing, the value for d is a _____ number.

10. If all the terms of an arithmetic sequence are the same, the value for d is _____.

Practice the Skills

Write the first five terms of the arithmetic sequence with the given first term and common difference. Write the expression for the general (or nth) term, a_n, of the arithmetic sequence.

11. $a_1 = 4, d = 3$

12. $a_1 = -8, d = 4$

13. $a_1 = 7, d = -2$

14. $a_1 = 3, d = -5$

15. $a_1 = \dfrac{1}{2}, d = \dfrac{3}{2}$

16. $a_1 = \dfrac{7}{4}, d = -\dfrac{3}{4}$

17. $a_1 = 100, d = -5$

18. $a_1 = 50, d = -9$

Find the indicated quantity of the arithmetic sequence.

19. $a_1 = 5, d = 3$; find a_4

20. $a_1 = -6, d = 5$; find a_5

21. $a_1 = -9, d = 4$; find a_{10}

22. $a_1 = -1, d = -2$; find a_{12}

23. $a_1 = -8, d = \dfrac{5}{3}$; find a_{13}

24. $a_1 = 5, a_8 = 10$; find d

25. $a_1 = -\dfrac{11}{2}, a_9 = \dfrac{21}{2}$, find d

26. $a_1 = \dfrac{1}{2}, a_7 = \dfrac{19}{2}$; find d

27. $a_1 = 4, a_n = 28, d = 3$; find n

28. $a_1 = 1, a_n = -17, d = -3$; find n

29. $a_1 = 82, a_n = 42, d = -8$; find n

30. $a_1 = -\dfrac{4}{3}, a_n = -\dfrac{14}{3}, d = -\dfrac{2}{3}$; find n

Find the sum, s_n, and common difference, d, of each sequence.

31. $a_1 = 1, a_{10} = 19, n = 10$

32. $a_1 = -8, a_7 = 10, n = 7$

33. $a_1 = \dfrac{3}{5}, a_8 = 2, n = 8$

34. $a_1 = 12, a_9 = -28, n = 9$

35. $a_1 = -5, a_6 = 13.5, n = 6$

36. $a_1 = -\dfrac{3}{5}, a_5 = \dfrac{13}{5}, n = 5$

37. $a_1 = 7, a_{11} = 67, n = 11$

38. $a_1 = 14.25, a_{31} = 18.75, n = 31$

Write the first four terms of each sequence; then find a_{10} and s_{10}.

39. $a_1 = 4, d = 3$

40. $a_1 = 1, d = 7$

41. $a_1 = -6, d = 2$

42. $a_1 = -7, d = -4$

43. $a_1 = -8, d = -5$

44. $a_1 = 5, d = 4$

45. $a_1 = \dfrac{7}{2}, d = \dfrac{5}{2}$

46. $a_1 = \dfrac{9}{5}, d = \dfrac{3}{5}$

47. $a_1 = 100, d = -7$

48. $a_1 = 35, d = 6$

Find the number of terms in each sequence and find s_n.

49. $1, 4, 7, 10, \ldots, 43$

50. $-10, -8, -6, -4, \ldots, 40$

51. $-9, -5, -1, 3, \ldots, 31$

52. $6, 13, 20, 27, \ldots, 55$

53. $\dfrac{1}{2}, \dfrac{2}{2}, \dfrac{3}{2}, \dfrac{4}{2}, \dfrac{5}{2}, \ldots, \dfrac{17}{2}$

54. $-\dfrac{5}{6}, -\dfrac{7}{6}, -\dfrac{9}{6}, -\dfrac{11}{6}, \ldots, -\dfrac{21}{6}$

55. $7, 10, 13, 16, \ldots, 91$

56. $-11, -15, -19, \ldots, -51$

Problem Solving

57. Find the sum of the first 50 natural numbers.

58. Find the sum of the first 50 even numbers.

59. Find the sum of the first 50 odd numbers.

60. Find the sum of the first 40 multiples of 5.

61. Find the sum of the first 30 multiples of 3.

62. Find the sum of the numbers between 50 and 150, inclusive.

63. Determine how many numbers between 7 and 1610 are divisible by 6.

64. Determine how many numbers between 14 and 1470 are divisible by 8.

Pyramids occur everywhere. At athletic events, cheerleaders may form a pyramid where the people above stand on the shoulders of the people below. The illustration below shows a pyramid with 1 cheerleader on the top row, 2 cheerleaders in the middle row, and 3 cheerleaders in the bottom row. Notice that $a_1 = 1$, $a_2 = 2$, and $a_3 = 3$. Also, observe that $d = 1$, $n = 3$, and $s_3 = 6$.

At a bowling alley, the pins at the end of the bowling lane form a pyramid. The first row has 1 pin, the second row has 2 pins, the third row has 3 pins, and the fourth row has 4 pins. Thus, $a_1 = 1$, $d = 1$, $n = 4$, and $s_4 = 10$.

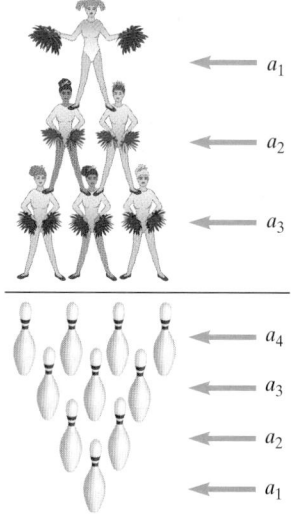

Use the idea of a pyramid to solve Exercises 65–70.

65. **Auditorium** An auditorium has 20 seats in the first row. Each successive row has two more seats than the previous row. How many seats are in the twelfth row? How many seats are in the first 12 rows?

66. **Auditorium** An auditorium has 22 seats in the first row. Each successive row has four more seats than the previous row. How many seats are in the ninth row? How many seats are in the first nine rows?

67. **Logs** Wolfgang Schmidt stacks logs so that there are 26 logs in the bottom layer, and each layer contains one log less than the layer below it. How many logs are in the pile?

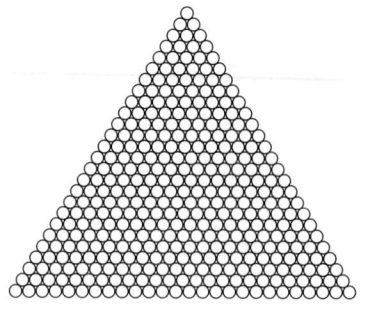

68. **Logs** Suppose Wolfgang, in Exercise 67, stopped stacking the logs after completing the layer containing eight logs. How many logs are in the pile?

69. **Glasses in a Stack** At their fiftieth wedding anniversary, Mr. and Mrs. Carlson are about to pour champagne into the top glass in the photo shown on the next page. The top row has 1 glass, the second row has 3 glasses, the third row has 5 glasses, and so on. Each row has 2 more glasses than the row above it. This pyramid has 14 rows.

a) How many glasses are in the fourteenth row (bottom row)?

b) How many glasses are there total?

See Exercise 69 on page 686.

70. Candies in a Stack Individually wrapped candies are stacked in rows such that the top row has 1 candy, the second row has 3 candies, the third row has 5 candies, and so on. Each row has 2 more candies than the row above it. There are 7 rows of candies.

a) How many candies are in the seventh row (bottom row)?

b) How many candies are there total?

71. Sum of Numbers Karl Friedrich Gauss (1777–1855), a famous mathematician, as a child found the sum of the first 100 natural numbers quickly in his head $(1 + 2 + 3 + \cdots + 100)$. Explain how he might have done this and find the sum of the first 100 natural numbers as you think Gauss might have. (*Hint:* $1 + 100 = 101, 2 + 99 = 101$, etc.)

Karl Friedrich Gauss

72. Sum of Numbers Use the same process from Exercise 71 to find the sum of the numbers from 101 to 150.

73. Sum of Numbers Find a formula for the sum of the first n consecutive odd numbers starting with 1.

$$1 + 3 + 5 + \cdots + (2n - 1)$$

74. Sum of Even Numbers Find a formula for the sum of the first n consecutive even numbers starting with 2.

$$2 + 4 + 6 + 8 + \cdots + 2n$$

75. Swinging on a Vine A long vine is attached to the branch of a tree. Sally Wynn swings from the vine, and each swing (left

to right or right to left) is $\frac{1}{2}$ foot less than the previous swing. If her first swing is 22 feet, find

a) the length of the seventh swing, and

b) the distance traveled during her seven swings.

76. Pendulum Each swing of a pendulum is 2 inches shorter than the previous swing (left to right or right to left). The first swing is 6 feet. Find

a) the length of the eighth swing, and

b) the distance traveled by the pendulum during the eight swings.

77. Bouncing Ball Frank Holyton drops a ball from a second-story window. Each time the ball bounces, the height reached is 6 inches less than on the previous bounce. If the first bounce reaches a height of 6 feet, find the height attained on the ninth bounce.

78. Ping-Pong Ball A ping-pong ball falls from a table and bounces to a height of 3 feet. If each successive bounce is 3 inches less than the previous bounce, find the height attained on the tenth bounce.

79. Packages On Monday, March 17, Brian Wallin started a new job at a packing company. On that day, he was able to prepare 105 packages for shipment. His boss expects Brian to be more productive with experience. Each day for six days, Brian is expected to prepare 10 more packages than the previous day's total.

a) How many packages is Brian expected to prepare on March 22?

b) How many packages is Brian expected to prepare during his first six days of employment?

80. Salary Marion Nickelson is making an annual salary of $37,500 at the Thompson Frozen Food Factory. Her boss has promised her an increase of $1500 in her salary each year over the next 10 years.

a) What will be Marion's salary 10 years from now?

b) What will be her total salary for these 11 years?

81. Money If Craig Campanella saves $1 on day 1, $2 on day 2, $3 on day 3, and so on, how much money, in total, will he have saved on day 31?

82. Money If Dan Currier saves 50¢ on day 1, $1.00 on day 2, $1.50 on day 3, and so on, how much, in total, will he have saved by the end of 1 year (365 days)?

83. Money Carrie Dereshi recently retired and met with her financial planner. She arranged to receive $42,000 the first year. Because of inflation, each year she will get $400 more than she received the previous year.

a) What income will she receive in her tenth year of retirement?

b) How much money will she have received in total during her first 10 years of retirement?

84. Salary Susan Forman is given a starting salary of $25,000 and is told she will receive a $1000 raise at the end of each year.

a) Find her salary during year 12.

b) How much will she receive in total during her first 12 years?

85. Angles The sum of the interior angles of a triangle, a quadrilateral, a pentagon, and a hexagon are $180°, 360°, 540°$, and $720°$, respectively. Use the pattern here to find the formula for the sum of the interior angles of a polygon with n sides.

86. Another formula that may be used to find the nth partial sum of an arithmetic series is

$$s_n = \frac{n}{2}[2a_1 + (n-1)d]$$

Derive this formula using the two formulas presented in this section.

Concept/Writing Exercises

87. Can an arithmetic sequence consist of only negative numbers? Explain.

88. Can an arithmetic sequence consist of only odd numbers? Explain.

89. Can an arithmetic sequence consist of only even numbers? Explain.

90. Can an alternating sequence be an arithmetic sequence? Explain.

Group Activity

In calculus, a topic of importance is limits. Consider $a_n = \frac{1}{n}$. *The first five terms of this sequence are* $\frac{1}{1}, \frac{1}{2}, \frac{1}{3}, \frac{1}{4}, \frac{1}{5}$. *Since the value of* $\frac{1}{n}$

gets closer and closer to 0 as n gets larger and larger, we say that the limit of $\frac{1}{n}$ *as n approaches infinity is 0. We write this as* $\lim\limits_{n \to +\infty} \frac{1}{n} = 0$

or $\lim\limits_{n \to +\infty} a_n = 0$. *Notice that* $\frac{1}{n}$ *can never equal 0, but its value approaches 0 as n gets larger and larger.*

a) *Group member 1: Find* $\lim\limits_{n \to +\infty} a_n$ *for Exercises 91 and 92.*

b) *Group member 2: Find* $\lim\limits_{n \to +\infty} a_n$ *for Exercises 93 and 94.*

c) *Group member 3: Find* $\lim\limits_{n \to +\infty} a_n$ *for Exercises 95 and 96.*

d) *Exchange work and check each other's answers.*

91. $a_n = \dfrac{1}{n-2}$

92. $a_n = \dfrac{n}{n+1}$

93. $a_n = \dfrac{1}{n^2+2}$

94. $a_n = \dfrac{2n+1}{n}$

95. $a_n = \dfrac{4n-3}{3n+1}$

96. $a_n = \dfrac{n^2}{n+1}$

Cumulative Review Exercises

[2.2] **97.** Solve $A = P + Prt$ for r.

[4.1] **98.** Solve the system of equations.

$$y = 2x + 1$$
$$3x - 2y = 1$$

[5.4] **99.** Factor $12n^2 - 6n - 30n + 15$.

[10.1] **100.** Graph $(x+4)^2 + y^2 = 25$.

11.3 Geometric Sequences and Series

1 Find the common ratio in a geometric sequence.

2 Find the nth term of a geometric sequence.

3 Find the nth partial sum of a geometric sequence.

4 Identify infinite geometric series.

5 Find the sum of an infinite geometric series.

6 Applications of geometric series.

1 Find the Common Ratio in a Geometric Sequence

In Section 11.1, we assumed you got a job with a starting salary of $30,000. We also mentioned that an option for salary increases was a 5% salary increase each year. This would result in the following sequence.

$$\$30{,}000, \ \$31{,}500, \ \$33{,}075, \ \$34{,}728.75, \ldots$$

This is an example of a geometric sequence.

> **Geometric Sequence**
>
> A **geometric sequence** is a sequence in which each term after the first term is a common multiple of the preceding term. The common multiple is called the **common ratio**.

Understanding Algebra

In a *geometric sequence*, you get the next term by multiplying the previous term by the *common ratio*.

The common ratio, r, in any geometric sequence can be found by dividing any term, except the first, by the preceding term. The common ratio of the previous geometric sequence is $\dfrac{31{,}500}{30{,}000} = 1.05$ (or 105%).

Consider the geometric sequence

$$1, 3, 9, 27, 81, \ldots, 3^{n-1}, \ldots$$

The common ratio is 3 since $3 \div 1 = 3, 9 \div 3 = 3, 27 \div 9 = 3$, and so on.

Geometric Sequence	Common Ratio
$4, 8, 16, 32, 64, \ldots, 4(2)^{n-1}, \ldots$	2
$3, 12, 48, 192, 768, \ldots, 3(4)^{n-1}, \ldots$	4
$7, \dfrac{7}{2}, \dfrac{7}{4}, \dfrac{7}{8}, \dfrac{7}{16}, \ldots, 7\left(\dfrac{1}{2}\right)^{n-1}, \ldots$	$\dfrac{1}{2}$
$5, -\dfrac{5}{3}, \dfrac{5}{9}, -\dfrac{5}{27}, \dfrac{5}{81}, \ldots, 5\left(-\dfrac{1}{3}\right)^{n-1}, \ldots$	$-\dfrac{1}{3}$

EXAMPLE 1   Determine the first five terms of the geometric sequence if $a_1 = 6$ and $r = \dfrac{1}{2}$.

Solution $a_1 = 6,\ a_2 = 6 \cdot \dfrac{1}{2} = 3,\ a_3 = 3 \cdot \dfrac{1}{2} = \dfrac{3}{2},\ a_4 = \dfrac{3}{2} \cdot \dfrac{1}{2} = \dfrac{3}{4},\ a_5 = \dfrac{3}{4} \cdot \dfrac{1}{2} = \dfrac{3}{8}$

Thus, the first five terms of the geometric sequence are

$$6, 3, \dfrac{3}{2}, \dfrac{3}{4}, \dfrac{3}{8}$$

Now Try Exercise 15

2 Find the *n*th Term of a Geometric Sequence

Understanding Algebra

In a geometric sequence, to get the value for the common ratio, r, divide a term by the previous term.

In general, a geometric sequence with first term, a_1, and common ratio, r, has the following terms:

$$a_1, \qquad a_1 r, \qquad a_1 r^2, \qquad a_1 r^3, \qquad a_1 r^4, \ldots, \ a_1 r^{n-1}, \ldots$$

↑	↑	↑	↑	↑	↑
1st	2nd	3rd	4th	5th	nth
term, a_1	term, a_2	term, a_3	term, a_4	term, a_5	term, a_n

Thus, we can see that the nth term of a geometric sequence is given by the following formula:

nth Term of a Geometric Sequence

$$a_n = a_1 r^{n-1}$$

EXAMPLE 2

a) Write an expression for the general (or nth) term, a_n, of the geometric sequence with $a_1 = 3$ and $r = -2$.

b) Find the twelfth term of this sequence.

Solution

a) The nth term of the sequence is $a_n = a_1 r^{n-1}$. Substituting $a_1 = 3$ and $r = -2$, we obtain

$$a_n = a_1 r^{n-1} = 3(-2)^{n-1}$$

Thus, $a_n = 3(-2)^{n-1}$.

b)
$$a_n = 3(-2)^{n-1}$$
$$a_{12} = 3(-2)^{12-1} = 3(-2)^{11} = 3(-2048) = -6144$$

The twelfth term of the sequence is -6144. The first 12 terms of the sequence are $3, -6, 12, -24, 48, -96, 192, -384, 768, -1536, 3072, -6144$.

Now Try Exercise 35

Helpful Hint

Study Tip

In this chapter, you will be working with exponents and using rules for exponents. The rules for exponents were discussed in Section 1.5 and again in Chapter 6. If you do not remember the rules for exponents, now is a good time to review Section 1.5.

EXAMPLE 3   Find r and a_1 for the geometric sequence with $a_2 = 12$ and $a_5 = 324$.

Solution   The sequence can be represented with blanks for the missing terms.

$$\underset{\underset{a_2}{\uparrow}}{\_\_}, 12, \_\_, \_\_, \underset{\underset{a_5}{\uparrow}}{324}$$

If we assume that a_2 is the first term of a sequence with the same common ratio, we obtain

$$\underset{\underset{\substack{\text{1st} \\ \text{term}}}{\uparrow}}{12}, \_\_, \_\_, \underset{\underset{\substack{\text{4th} \\ \text{term}}}{\uparrow}}{324}$$

Now we use the formula for the nth term of a geometric sequence to find r. We let the first term, a_1, be 12 and the number of terms, n, be 4.

$$a_n = a_1 r^{n-1}$$
$$324 = 12 r^{4-1}$$
$$324 = 12 r^3$$
$$\frac{324}{12} = r^3$$
$$27 = r^3$$
$$3 = r$$

Thus, the common ratio is 3.

The first term of the original sequence must be $12 \div 3$ or 4. Thus, $a_1 = 4$. The first term, a_1, could also be found using the formula with $a_n = 324$, $r = 3$, and $n = 5$. Find a_1 by the formula now.

Now Try Exercise 83

3   Find the nth Partial Sum of a Geometric Sequence

Geometric Series

A **geometric series** is the sum of the terms of a geometric sequence.

The sum of the first n terms, s_n, of a geometric sequence can be expressed as

$$s_n = a_1 + a_1 r + a_1 r^2 + a_1 r^3 + \cdots + a_1 r^{n-2} + a_1 r^{n-1} \quad \text{(eq. 1)}$$

If we multiply both sides of the equation by r, we obtain

$$r s_n = a_1 r + a_1 r^2 + a_1 r^3 + \cdots + a_1 r^{n-1} + a_1 r^n \quad \text{(eq. 2)}$$

Now we subtract the corresponding sides of $(eq.\,2)$ from $(eq.\,1)$. The red-colored terms drop out, leaving

$$s_n - rs_n = a_1 - a_1 r^n$$

Now we solve the equation for s_n.

$$s_n(1 - r) = a_1(1 - r^n) \qquad \text{Factor.}$$

$$s_n = \frac{a_1(1 - r^n)}{1 - r} \qquad \text{Divide both sides by } 1 - r.$$

Thus, we have the following formula for the nth partial sum of a geometric sequence.

---

nth Partial Sum of a Geometric Sequence

$$s_n = \frac{a_1(1 - r^n)}{1 - r}, \qquad r \neq 1$$

EXAMPLE 4 Find the seventh partial sum of a geometric sequence whose first term is 16 and whose common ratio is $-\dfrac{1}{2}$.

Solution   Substitute the appropriate value for $a_1, r,$ and n.

$$s_n = \frac{a_1(1 - r^n)}{1 - r}$$

$$s_7 = \frac{16\left[1 - \left(-\dfrac{1}{2}\right)^7\right]}{1 - \left(-\dfrac{1}{2}\right)} = \frac{16\left(1 + \dfrac{1}{128}\right)}{\dfrac{3}{2}} = \frac{16\left(\dfrac{129}{128}\right)}{\dfrac{3}{2}} = \frac{\dfrac{129}{8}}{\dfrac{3}{2}} = \frac{129}{8} \cdot \frac{2}{3} = \frac{43}{4}$$

Thus, $s_7 = \dfrac{43}{4}$.

Now Try Exercise 41

EXAMPLE 5 Given $s_n = 93$, $a_1 = 3$, and $r = 2$, find n.

Solution

$$s_n = \frac{a_1(1 - r^n)}{1 - r}$$

$$93 = \frac{3(1 - 2^n)}{1 - 2} \qquad \text{Substitute values for } s_n, a_1, \text{ and } r.$$

$$93 = \frac{3(1 - 2^n)}{-1}$$

$$-93 = 3(1 - 2^n) \qquad \text{Both sides were multiplied by } -1.$$

$$-31 = 1 - 2^n \qquad \text{Both sides were divided by 3.}$$

$$-32 = -2^n \qquad \text{1 was subtracted from both sides.}$$

$$32 = 2^n \qquad \text{Both sides were divided by } -1.$$

$$2^5 = 2^n \qquad \text{Write 32 as } 2^5.$$

Therefore, $n = 5$.

Now Try Exercise 65

When working with a geometric series, r can be a positive number as we saw in Example 5 or a negative number as we saw in Example 4.

4 Identify Infinite Geometric Series

All the geometric sequences that we have examined thus far have been finite since they have had a last term. The following sequence is an example of an infinite geometric sequence.

$$1, \frac{1}{2}, \frac{1}{4}, \frac{1}{8}, \frac{1}{16}, \ldots, \left(\frac{1}{2}\right)^{n-1}, \ldots$$

Note that the three dots at the end of the sequence indicate that the sequence continues indefinitely.

> **Infinite Geometric Series**
>
> An **infinite geometric series** is the sum of the terms in an infinite geometric sequence.

For example,

$$1 + \frac{1}{2} + \frac{1}{4} + \frac{1}{8} + \frac{1}{16} + \cdots + \left(\frac{1}{2}\right)^{n-1} + \cdots$$

is an infinite geometric series. Let's find some partial sums.

Partial Sum	Series	Sum
Second	$1 + \frac{1}{2}$	1.5
Third	$1 + \frac{1}{2} + \frac{1}{4}$	1.75
Fourth	$1 + \frac{1}{2} + \frac{1}{4} + \frac{1}{8}$	1.875
Fifth	$1 + \frac{1}{2} + \frac{1}{4} + \frac{1}{8} + \frac{1}{16}$	1.9375
Sixth	$1 + \frac{1}{2} + \frac{1}{4} + \frac{1}{8} + \frac{1}{16} + \frac{1}{32}$	1.96875

With each successive partial sum, the amount being added is less than with the previous partial sum. Also, the sum seems to be getting closer and closer to 2. In Example 6, we will show that the sum of this infinite geometric series is indeed 2.

5 Find the Sum of an Infinite Geometric Series

Consider the formula for the sum of the first n terms of an infinite geometric series:

$$s_n = \frac{a_1(1 - r^n)}{1 - r}, \qquad r \neq 1$$

What happens to r^n if $|r| < 1$ and n gets larger and larger? Suppose that $r = \frac{1}{2}$; then

$$\left(\frac{1}{2}\right)^1 = 0.5, \quad \left(\frac{1}{2}\right)^2 = 0.25, \quad \left(\frac{1}{2}\right)^3 = 0.125, \quad \left(\frac{1}{2}\right)^{20} \approx 0.000001$$

We can see that when $|r| < 1$, the value of r^n gets exceedingly close to 0 as n gets larger and larger. Thus, when considering the sum of an infinite geometric series, symbolized s_∞, the expression r^n approaches 0 when $|r| < 1$. Therefore, replacing r^n with 0 in the formula $s_n = \frac{a_1(1 - r^n)}{1 - r}$ leads to the following formula.

> **Sum of an Infinite Geometric Series**
>
> $$s_\infty = \frac{a_1}{1 - r} \quad \text{where} \quad |r| < 1$$

EXAMPLE 6 Find the sum of the infinite geometric series

$$1 + \frac{1}{2} + \frac{1}{4} + \frac{1}{8} + \cdots + \left(\frac{1}{2}\right)^{n-1} + \cdots$$

Solution $a_1 = 1$ and $r = \frac{1}{2}$. Note that $\left|\frac{1}{2}\right| < 1$.

$$S_\infty = \frac{a_1}{1 - r}$$

$$= \frac{1}{1 - \frac{1}{2}}$$

$$= \frac{1}{\frac{1}{2}}$$

$$= 2$$

Thus, $1 + \frac{1}{2} + \frac{1}{4} + \frac{1}{8} + \frac{1}{16} + \cdots + \left(\frac{1}{2}\right)^{n-1} + \cdots = 2$.

Now Try Exercise 69

EXAMPLE 7 Find the sum of the infinite geometric series

$$3 - \frac{6}{5} + \frac{12}{25} - \frac{24}{125} + \frac{48}{625} + \cdots$$

Solution The terms of the corresponding sequence are $3, -\frac{6}{5}, \frac{12}{25}, -\frac{24}{125}, \ldots$. Note that $a_1 = 3$. To find the common ratio, r, we can divide the second term, $-\frac{6}{5}$, by the first term, 3.

$$r = -\frac{6}{5} \div 3$$

$$= -\frac{6}{5} \cdot \frac{1}{3}$$

$$= -\frac{2}{5}.$$

Since $\left|-\frac{2}{5}\right| < 1$,

$$S_\infty = \frac{a_1}{1 - r}$$

$$= \frac{3}{1 - \left(-\frac{2}{5}\right)}$$

$$= \frac{3}{1 + \frac{2}{5}}$$

$$= \frac{3}{\frac{7}{5}}$$

$$= \frac{15}{7}$$

Now Try Exercise 71

EXAMPLE 8 Write $0.343434\ldots$ as a ratio of integers.

Solution We can write this decimal as

$$0.34 + 0.0034 + 0.000034 + \cdots + (0.34)(0.01)^{n-1} + \cdots$$

This is an infinite geometric series with $r = 0.01$. Since $|r| < 1$,

$$s_\infty = \frac{a_1}{1 - r} = \frac{0.34}{1 - 0.01} = \frac{0.34}{0.99} = \frac{34}{99}$$

If you divide 34 by 99, you obtain $0.343434\ldots$.

Now Try Exercise 81

What is the sum of a geometric series when $|r| > 1$? Consider the geometric sequence in which $a_1 = 1$ and $r = 2$.

$$1, 2, 4, 8, 16, 32, \ldots, 2^{n-1}, \ldots$$

The sum of its terms is

$$1 + 2 + 4 + 8 + 16 + 32 + \cdots + 2^{n-1} + \cdots$$

What is the sum of this series? As n gets larger and larger, the sum gets larger and larger. We therefore say that the sum "does not exist." For $|r| > 1$, the sum of an infinite geometric series does not exist.

6 Applications of Geometric Series

EXAMPLE 9 **Savings Account** Gene Simmons invests $1000 at 5% interest compounded annually in a savings account. Determine the amount in his account and the amount of interest earned at the end of 10 years.

Solution   Understand   Suppose we let P represent any principal invested. At the beginning of the second year, the amount grows to $P + 0.05P$ or $1.05P$. This amount will be the principal invested for year 2. At the beginning of the third year the second year's principal will grow by 5% to $(1.05P)(1.05)$, or $(1.05)^2P$. The amount in Gene's account at the beginning of successive years is

Year 1	Year 2	Year 3	Year 4
P	$1.05P$	$(1.05)^2P$	$(1.05)^3P$

Savings account
5% interest annually

and so on. This is a geometric series with $r = 1.05$. The amount in his account at the end of 10 years will be the same as the amount in his account at the beginning of year 11. We will therefore use the formula

$$a_n = a_1 r^{n-1}, \quad \text{with} \quad r = 1.05 \quad \text{and} \quad n = 11$$

Translate   We have a geometric sequence with $a_1 = 1000$, $r = 1.05$, and $n = 11$. Substituting these values into the formula, we obtain the following.

Carry Out

$$a_n = a_1 r^{n-1}$$
$$a_{11} = 1000(1.05)^{11-1}$$
$$= 1000(1.05)^{10}$$
$$\approx 1000(1.62889)$$
$$\approx 1628.89$$

Answer   After 10 years, the amount in the account is about $1628.89. The amount of interest is $1628.89 - \$1000 = \$628.89$.

Now Try Exercise 95

EXAMPLE 10 Money Suppose someone offered you $1000 a day for each day of a 30-day month. Or, you could elect to take a penny on day 1, 2¢ on day 2, 4¢ on day 3, 8¢ on day 4, and so on. The amount would continue to double each day for 30 days.

a) Without doing any calculations, take a guess at which of the two offerings would provide the greater total return for 30 days.

b) Calculate the total amount you would receive by selecting $1000 a day for 30 days.

c) Calculate the amount you would receive on day 30 by selecting 1¢ on day 1 and doubling the amount each day for 30 days.

d) Calculate the total amount you would receive for 30 days by selecting 1¢ on day 1 and doubling the amount each day for 30 days.

Solution

a) Each of you will have your own answer to part **a)**.

b) If you received $1000 a day for 30 days, you would receive 30($1000) = $30,000.

c) Understand   Since the amount is doubled each day, this represents a geometric sequence with $r = 2$. The chart that follows shows the amount you would receive in each of the first 7 days. We also show the amounts written with base 2, the common ratio.

Day	1	2	3	4	5	6	7
Amount (cents)	1	2	4	8	16	32	64
Amount (cents)	2^0	2^1	2^2	2^3	2^4	2^5	2^6

Notice that for any given day, the exponent on 2 is 1 less than the given day. For example, on day 7, the amount is 2^6. In general, the amount on day n is 2^{n-1}.

Translate   To find the amount received on day 30, we evaluate $a_n = a_1 r^{n-1}$ for $n = 30$.

$$a_n = a_1 r^{n-1}$$
$$a_{30} = 1(2)^{30-1}$$

Carry Out
$$a_{30} = 1(2)^{29}$$
$$= 1(536,870,912)$$
$$= 536,870,912$$

Answer   On day 30, the amount that you would receive is 536,870,912 cents or $5,368,709.12

d) Understand and Translate   To find the total amount received over the 30 days, we find the thirtieth partial sum.

$$s_n = \frac{a_1(1 - r^n)}{1 - r}$$
$$s_{30} = \frac{1(1 - 2^{30})}{1 - 2}$$

Carry Out
$$= \frac{1(1 - 1,073,741,824)}{-1}$$
$$= 1,073,741,823$$

Answer   Therefore, over 30 days the total amount you would receive by this method would be 1,073,741,823 cents or $10,737,418.23. The amount received by this method greatly surpasses the $30,000 received by selecting $1000 a day for 30 days.

Now Try Exercise 87

EXAMPLE 11 Pendulum On each swing (left to right or right to left), a certain pendulum travels 90% as far as on its previous swing. For example, if the swing to the right is 10 feet, the swing back to the left is $0.9 \times 10 = 9$ feet (see **Fig. 11.1**). If the first swing is 10 feet, determine the total distance traveled by the pendulum by the time it comes to rest.

Solution   Understand   This problem may be considered an infinite geometric series with $a_1 = 10$ and $r = 0.9$. We can therefore use the formula $s_\infty = \dfrac{a_1}{1 - r}$ to find the total distance traveled by the pendulum.

Translate and Carry Out

$$s_\infty = \frac{a_1}{1 - r} = \frac{10}{1 - 0.9} = \frac{10}{0.1} = 100 \text{ feet}$$

Answer   By the time the pendulum comes to rest, it has traveled 100 feet.

Now Try Exercise 99

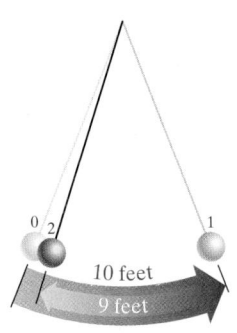

10 feet
9 feet

FIGURE 11.1

EXERCISE SET 11.3   MathXL   MyMathLab

Warm-Up Exercises

Fill in the blanks with the appropriate word, phrase, or symbol(s) from the following list.

sum	difference	d	r	infinite	term	ratio	less
0	1	dividing	multiplying	geometric	series	greater	

1. A sequence in which each term after the first term is a common multiple of the preceding term is a _____ sequence.

2. The common multiple of a geometric sequence is called the common _____.

3. The common ratio of a geometric sequence is identified by the letter _____.

4. The common ratio can be found by _____ any term, except the first term, by the preceding term.

5. The formula $a_n = a_1 r^{n-1}$ is used to find the nth _____ of a geometric sequence.

6. The sum of the terms of a geometric sequence is a geometric _____.

7. The formula $s_n = \dfrac{a_1(1 - r^n)}{1 - r}$ is used to find the nth partial _____ of a geometric sequence.

8. If the terms of a geometric sequence are increasing, the value for r is _____ than 1.

9. If the terms of a geometric sequence are the same, the value for r is _____.

10. The formula $s_\infty = \dfrac{a_1}{1 - r}, |r| < 1$ is used to find the sum of an _____ geometric series.

Practice the Skills

Determine the first five terms of each geometric sequence.

11. $a_1 = 2, r = 3$

12. $a_1 = 5, r = 2$

13. $a_1 = 6, r = -\dfrac{1}{2}$

14. $a_1 = 6, r = \dfrac{1}{2}$

15. $a_1 = 72, r = \dfrac{1}{3}$

16. $a_1 = 50, r = \dfrac{1}{5}$

17. $a_1 = 90, r = -\dfrac{1}{3}$

18. $a_1 = 32, r = -\dfrac{1}{4}$

19. $a_1 = -1, r = 3$

20. $a_1 = -1, r = -3$

21. $a_1 = 7, r = -2$

22. $a_1 = -13, r = -1$

23. $a_1 = \dfrac{1}{3}, r = \dfrac{1}{2}$

24. $a_1 = \dfrac{1}{2}, r = -\dfrac{1}{3}$

25. $a_1 = 3, r = \dfrac{3}{2}$

26. $a_1 = 60, r = -\dfrac{2}{5}$

Find the indicated term of each geometric sequence.

27. $a_1 = 4, r = 2$; find a_6

28. $a_1 = 4, r = -2$; find a_6

29. $a_1 = -24, r = \dfrac{1}{2}$; find a_9

30. $a_1 = 27, r = \dfrac{1}{3}$; find a_7

31. $a_1 = \dfrac{1}{8}, r = 2$; find a_{10}

32. $a_1 = 3, r = 3$; find a_5

33. $a_1 = -3, r = -2$; find a_{12}

34. $a_1 = -10, r = -2$; find a_4

35. $a_1 = 2, r = \dfrac{1}{2}$; find a_8

36. $a_1 = 5, r = \dfrac{2}{3}$; find a_9

37. $a_1 = 50, r = \dfrac{1}{3}$; find a_7

38. $a_1 = -7, r = -\dfrac{3}{4}$; find a_7

Find the indicated sum.

39. $a_1 = 5, r = 2$; find s_5

40. $a_1 = 7, r = -3$; find s_4

41. $a_1 = 2, r = 5$; find s_6

42. $a_1 = 9, r = \dfrac{1}{2}$; find s_6

43. $a_1 = 80, r = 2$; find s_7

44. $a_1 = 2, r = -2$; find s_{12}

45. $a_1 = -15, r = -\dfrac{1}{2}$; find s_9

46. $a_1 = \dfrac{3}{4}, r = 3$; find s_7

47. $a_1 = -9, r = \dfrac{2}{5}$; find s_5

48. $a_1 = 35, r = \dfrac{1}{5}$; find s_{12}

For each geometric sequence, find the common ratio, r, and then write an expression for the general (or nth) term, a_n.

49. $7, \dfrac{7}{2}, \dfrac{7}{4}, \dfrac{7}{8}, \ldots$

50. $3, -\dfrac{3}{2}, \dfrac{3}{4}, -\dfrac{3}{8}, \ldots$

51. $9, 18, 36, 72, \ldots$

52. $2, 6, 18, 54, \ldots$

53. $2, -6, 18, -54, \ldots$

54. $-1, -4, -16, -64, -256, \ldots$

55. $\dfrac{3}{4}, \dfrac{1}{2}, \dfrac{1}{3}, \dfrac{2}{9}$

56. $\dfrac{2}{3}, \dfrac{10}{3}, \dfrac{50}{3}, \dfrac{250}{3}, \ldots$

Find the sum of the terms in each geometric sequence.

57. $1, \dfrac{1}{2}, \dfrac{1}{4}, \dfrac{1}{8}, \dfrac{1}{16}, \ldots$

58. $1, -\dfrac{1}{2}, \dfrac{1}{4}, -\dfrac{1}{8}, \dfrac{1}{16}, \ldots$

59. $1, \dfrac{1}{5}, \dfrac{1}{25}, \dfrac{1}{125}, \dfrac{1}{625}, \ldots$

60. $1, -\dfrac{1}{5}, \dfrac{1}{25}, -\dfrac{1}{125}, \dfrac{1}{625}, \ldots$

61. $12, 6, 3, \dfrac{3}{2}, \dfrac{3}{4}, \ldots$

62. $\dfrac{1}{4}, \dfrac{1}{16}, \dfrac{1}{64}, \dfrac{1}{128}, \ldots$

63. $5, 2, \dfrac{4}{5}, \dfrac{8}{25}, \ldots$

64. $-\dfrac{4}{3}, -\dfrac{4}{9}, -\dfrac{4}{27}, -\dfrac{4}{81}, \ldots$

Given s_n, a_1, and r, find n in each geometric series.

65. $s_n = 93, a_1 = 3$, and $r = 2$

66. $s_n = 80, a_1 = 2$, and $r = 3$

67. $s_n = \dfrac{189}{32}, a_1 = 3$, and $r = \dfrac{1}{2}$

68. $s_n = \dfrac{104}{9}, a_1 = 8$, and $r = \dfrac{1}{3}$

Find the sum of each infinite geometric series.

69. $2 + 1 + \dfrac{1}{2} + \dfrac{1}{4} + \dfrac{1}{8} + \cdots$

70. $16 + 8 + 4 + 2 + \cdots$

71. $4 + \dfrac{8}{3} + \dfrac{16}{9} + \dfrac{32}{27} + \cdots$

72. $6 - 2 + \dfrac{2}{3} - \dfrac{4}{9} + \cdots$

73. $-60 + 20 - \dfrac{20}{3} + \dfrac{20}{9} - \cdots$

74. $2 + \dfrac{4}{3} + \dfrac{8}{9} + \dfrac{16}{27} + \cdots$

75. $-12 - \dfrac{12}{5} - \dfrac{12}{25} - \dfrac{12}{125} - \cdots$

76. $7 - 1 + \dfrac{1}{7} - \dfrac{1}{49} + \cdots$

Write each repeating decimal as a ratio of integers.

77. $0.242424\ldots$

78. $0.454545\ldots$

79. $0.7777\ldots$

80. $0.375375\ldots$

81. $0.515151\ldots$

82. $0.742742\ldots$

Problem Solving

83. In a geometric sequence, $a_2 = 15$ and $a_5 = 405$; find r and a_1.

84. In a geometric sequence, $a_2 = 27$ and $a_5 = 1$; find r and a_1.

85. In a geometric sequence, $a_3 = 28$ and $a_5 = 112$, find r and a_1.

86. In a geometric sequence, $a_2 = 12$ and $a_5 = -324$; find r and a_1.

87. **Loaf of Bread** A loaf of bread currently costs $1.40. Determine the cost of a loaf of bread after 8 years (the start of the 9th year) if inflation were to grow at a constant rate of 3% per year. *Hint*: After year 1 (the start of year 2), the cost of a loaf of bread is $1.40(1.03). After year 2 (the start of year 3), the cost would be $1.40(1.03)^2$, and so on.

88. **Bicycle** A specific bicycle currently costs $300. Determine the cost of the bicycle after 12 years if inflation were to grow at a constant rate of 4% per year.

89. **Mass** A substance loses half its mass each day. If there are initially 600 grams of the substance, find
  a) the number of days after which only 37.5 grams of the substance remain.
  b) the amount of the substance remaining after 9 days.

90. **Bacteria** The number of a certain type of bacteria doubles every hour. If there are initially 500 bacteria, after how many hours will the number of bacteria reach 32,000?

91. Population On July 1, 2008, the population of the United States was about 303.5 million people. If the population grows at a rate of 1.1% per year, find

a) the population after 10 years.

b) the number of years for the population to double.

92. Farm Equipment A piece of farm equipment that costs $105,000 decreases in value by 15% per year. Find the value of the equipment after 4 years.

© Jose Ignacio Soto/Shutterstock

93. Filtered Light The amount of light filtering through a lake diminishes by one-half for each meter of depth.

a) Write a sequence indicating the amount of light remaining at depths of 1, 2, 3, 4, and 5 meters.

b) What is the general term for this sequence?

c) What is the remaining light at a depth of 7 meters?

94. Pendulum On each swing (left to right or right to left), a pendulum travels 80% as far as on its previous swing. If the first swing is 10 feet, determine the total distance traveled by the pendulum by the time it comes to rest.

95. Investment You invest $10,000 in a savings account paying 6% interest annually. Find the amount in your account at the end of 8 years.

96. Injected Dye A tracer dye is injected into Mark Damion for medical reasons. After each hour, two-thirds of the previous hour's dye remains. How much dye remains in Mark's system after 10 hours?

97. Bungee Jumping Shawna Kelly goes bungee jumping off a bridge above water. On the initial jump, the bungee cord stretches to 220 feet. Assume the first bounce reaches a height of 60% of the original jump and that each additional bounce reaches a height of 60% of the previous bounce.

a) What will be the height of the fourth bounce?

b) Theoretically, Shawna would never stop bouncing, but realistically, she will. Use the infinite geometric series to estimate the total distance Shawna travels in a *downward* direction.

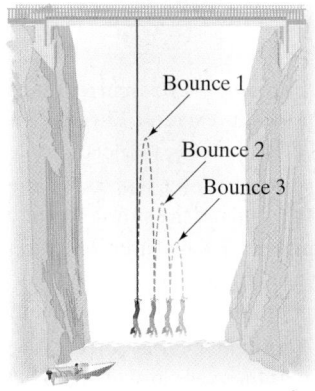

Bounce 1

Bounce 2

Bounce 3

98. Bungee Jumping Repeat Exercise 97 **b)**, but this time find the total distance traveled in an *upward* direction.

99. Ping-Pong Ball A ping-pong ball falls off a table 30 inches high. Assume that the first bounce reaches a height of 70% of the distance the ball fell and each additional bounce reaches a height of 70% of the previous bounce.

a) How high will the ball bounce on the third bounce?

b) Theoretically, the ball would never stop bouncing, but realistically, it will. Estimate the total distance the ball travels in the *downward* direction.

100. Ping-Pong Ball Repeat Exercise 99 **b)**, but this time find the total distance traveled in the *upward* direction.

101. Stack of Chips Suppose that you form stacks of blue chips such that there is one blue chip in the first stack and in each successive stack you double the number of chips. Thus, you have stacks of 1, 2, 4, 8, and so on, of blue chips. You also form stacks of red chips, starting with one red chip and then tripling the number in each successive stack. Thus the stacks will contain 1, 3, 9, 27 and so on, red chips. How many more would the sixth stack of red chips have than the sixth stack of blue chips?

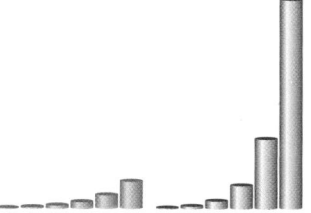

102. Stack of Money If you start with $1 and double your money each day, how many days will it take to surpass $1,000,000?

103. Depreciation One method of depreciating an item on an income tax return is the declining balance method. With this method, a given fraction of the remaining value of the item is depreciated each year. Suppose that an item has a 5-year life and is depreciated using the declining balance method. Then, at the end of its first year, it loses $\frac{1}{5}$ of its value and $\frac{4}{5}$ of its value remains. At the end of the second year it loses $\frac{1}{5}$ of the remaining $\frac{4}{5}$ of its value, and so on. A car has a 5-year life expectancy and costs $15,000.

a) Write a sequence showing the value of the car remaining for each of the first 3 years.

b) What is the general term of this sequence?

c) Find the value of the car at the end of 5 years.

104. Scrap Value On page 613, Exercise Set 9.6, Exercise 77, a formula for scrap value was given. The scrap value, S, is found by $S = c(1 - r)^n$ where c is the original cost, r is the annual depreciation rate and n is the number of years the object is depreciated.

a) If you have not already done so, do Exercise 103 above to find the value of the car remaining at the end of 5 years.

b) Use the formula given to find the scrap value of the car at the end of 5 years and compare this answer with the answer found in part **a)**.

105. Bouncing Ball A ball is dropped from a height of 10 feet. The ball bounces to a height of 9 feet. On each successive bounce, the ball rises to 90% of its previous height. Find the *total vertical distance* traveled by the ball when it comes to rest.

106. Wave Action A particle follows the path indicated by the wave shown. Find the *total vertical distance* traveled by the particle.

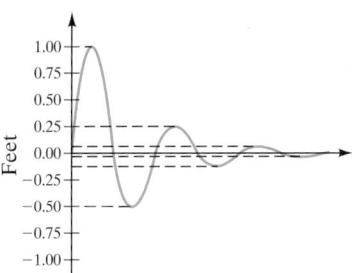

107. The formula for the nth term of a geometric sequence is $a_n = a_1 r^{n-1}$. If $a_1 = 1$, $a_n = r^{n-1}$.

a) How do you think the graphs of $y_1 = 2^{n-1}$ and $y_2 = 3^{n-1}$ will compare?

b) Graph both y_1 and y_2 and determine whether your answer to part a) was correct.

108. Use your graphing calculator to approximate the value of n to the nearest hundredth, where $100 = 3 \cdot 2^{n-1}$.

Concept/Writing Exercises

109. In a geometric series, if $|r| < 1$, what does r^n approach as n gets larger and larger?

110. Does the sum of an infinite geometric series when $|r| > 1$ exist?

111. In a geometric sequence, can the value of r be a negative number?

112. In a geometric sequence, can the value of r be a positive number?

113. In a geometric series, if $a_1 = 6$ and $r = \dfrac{1}{4}$ does s_∞ exist? If so, what is its value? Explain.

114. In a geometric series, if $a_1 = 6$ and $r = -2$, does s_∞ exist? If so, what is its value? Explain.

Challenge Problem

115. Find the sum of the sequence $1, 2, 4, 8, \ldots, 1{,}048{,}576$ and the number of terms in the sequence.

Cumulative Review Exercises

[3.6] **116.** Let $f(x) = x^2 - 4$ and $g(x) = x - 3$. Find $(f \cdot g)(4)$.

[5.2] **117.** Multiply $(2x - 3y)(3x^2 + 4xy - 2y^2)$.

[6.4] **118.** Solve $S = \dfrac{2a}{1 - r}$ for r.

[9.1] **119.** Let $g(x) = x^3 + 9$. Find $g^{-1}(x)$.

[9.6] **120.** Solve $\log x + \log (x - 1) = \log 20$.

[10.4] **121. Sail on a Sailboat** A sail on a sailboat is shaped like a right triangle with a perimeter of 36 meters and a hypotenuse of 15 meters. Find the length of each leg of the triangle.

Mid-Chapter Test: 11.1–11.3

To find out how well you understand the chapter material to this point, take this brief test. The answers, and the section where the material was initially discussed, are given in the back of the book. Review any questions you answered incorrectly.

1. Write the first five terms of the sequence whose nth term is $a_n = -3n + 7$.

2. If $a_n = n(n + 6)$, find the seventh term.

3. Find the first and third partial sums, s_1 and s_3, for the sequence whose nth term is $a_n = 2^n - 1$.

4. Write the next three terms of the sequence $5, 1, -3, -7, -11, \ldots$.

5. Evaluate the series $\displaystyle\sum_{i=1}^{5} (4i - 3)$.

6. If the general term of a sequence is $a_n = \dfrac{1}{4}n + 9$, write an expression using Σ to represent the fifth partial sum.

7. Write the first four terms of the arithmetic sequence with $a_1 = -6$ and $d = 5$. Find an expression for the general term a_n.

8. Find d for the arithmetic sequence with $a_1 = \dfrac{11}{2}$ and $a_7 = -\dfrac{1}{2}$.

9. Find n for the arithmetic sequence with $a_1 = 22$, $a_n = -3$, and $d = -5$.

10. Find the common difference, d, and the sum, s_6, for the arithmetic sequence with $a_1 = -8$ and $a_6 = 7$.

11. Find s_{10} for the arithmetic sequence with $a_1 = \dfrac{5}{2}$ and $d = \dfrac{1}{2}$.

12. Find the number of terms in the arithmetic sequence $-7, 0, 7, 14, \ldots, 70$.

13. Logs are stacked in a pile with 16 logs on the bottom row, 15 on the next row, 14 on the next row, and so on to the top, with the top row having one log. Each row has one log less than the row below it. How many logs are on the pile?

14. Write the first five terms of the geometric series with $a_1 = 100$ and $r = -\dfrac{1}{2}$.

15. Find a_7 for the geometric sequence with $a_1 = 27$ and $r = \dfrac{1}{3}$.

(continued next page)

16. Find s_6 for the geometric sequence with $a_1 = 5$ and $r = 2$.

17. For the geometric sequence $8, -\dfrac{16}{3}, \dfrac{32}{9}, -\dfrac{64}{27}, \ldots$, find r

18. Find the sum of the infinite series $12, 4, \dfrac{4}{3}, \dfrac{4}{9}, \ldots$.

19. Write the repeating decimal $0.878787\ldots$ as a ratio of two integers.

20. a) What is a sequence?

    b) What is an arithmetic sequence?

    c) What is geometric sequence?

    d) What is a series?

11.4 The Binomial Theorem

1 Evaluate factorials.

2 Use Pascal's triangle.

3 Use the binomial theorem.

1 Evaluate Factorials

To understand the binomial theorem, you must have an understanding of what **factorials** are.

> ### n Factorial
>
> The notation $n!$ is read "n **factorial**."
>
> $$n! = n(n-1)(n-2)(n-3)\cdots(1)$$
>
> for any positive integer n.

Examples
$$6! = 6 \cdot 5 \cdot 4 \cdot 3 \cdot 2 \cdot 1 = 720$$
$$7! = 7 \cdot 6 \cdot 5 \cdot 4 \cdot 3 \cdot 2 \cdot 1 = 5040$$

Note that by definition **0! is 1**.

Using Your Calculator

Scientific Calculator

Factorials can be found on calculators that contain an $\boxed{n!}$ or $\boxed{x!}$ key. Often, the factorial key is a second function key.

In the following examples, the answers appear after $\boxed{n!}$.

Evaluate 6! $6\;\boxed{2^{nd}}\;\boxed{n!}$ 720

Evaluate 9! $9\;\boxed{2^{nd}}\;\boxed{n!}$ 362880

Graphing Calculator

On most graphing calculators, factorials are found under the $\boxed{\text{MATH}}$, Probability function menu (PRB).

On the TI-84 Plus calculator, press $\boxed{\text{MATH}}$, and then scroll to the right using the right arrow key, $\boxed{\blacktriangleright}$, three times, until you get to PRB. The $n!$ (or !) is the fourth item on the menu.

To find 5! or 6!, the keystrokes are as follows.

KEYSTROKES   ANSWER

$5\;\boxed{\text{MATH}}\;\boxed{\blacktriangleright}\;\boxed{\blacktriangleright}\;\boxed{\blacktriangleright}\;\boxed{4}\;\boxed{\text{ENTER}}$ 120

$6\;\boxed{\text{MATH}}\;\boxed{\blacktriangleright}\;\boxed{\blacktriangleright}\;\boxed{\blacktriangleright}\;\boxed{4}\;\boxed{\text{ENTER}}$ 720

Understanding Algebra

A *two-term* polynomial such as $a + b$ or $2x + 3y$ is called a *binomial*.

2 Use Pascal's Triangle

Using polynomial multiplication, we can obtain the following powers of the binomial $a + b$: These multiplications are called expansions of the binomial.

$$(a + b)^0 = 1$$
$$(a + b)^1 = a + b$$
$$(a + b)^2 = a^2 + 2ab + b^2$$
$$(a + b)^3 = a^3 + 3a^2b + 3ab^2 + b^3$$
$$(a + b)^4 = a^4 + 4a^3b + 6a^2b^2 + 4ab^3 + b^4$$
$$(a + b)^5 = a^5 + 5a^4b + 10a^3b^2 + 10a^2b^3 + 5ab^4 + b^5$$
$$(a + b)^6 = a^6 + 6a^5b + 15a^4b^2 + 20a^3b^3 + 15a^2b^4 + 6ab^5 + b^6$$

Note that when expanding a binomial of the form $(a + b)^n$,

1. There are $n + 1$ terms in the expansion.
2. The first term is a^n and the last term is b^n.
3. Reading from left to right, the exponents on a decrease by 1 from term to term, while the exponents on b increase by 1 from term to term.
4. The sum of the exponents on the variables in each term is n.
5. The coefficients of the terms equidistant from the ends are the same.

If we examine just the variables in $(a + b)^5$, we have a^5, a^4b, a^3b^2, a^2b^3, ab^4, and b^5.

The numerical coefficients of each term in the expansion of $(a + b)^n$ can be found by using **Pascal's triangle**. For example, if $n = 5$, we can determine the numerical coefficients of $(a + b)^5$ as follows.

Blaise Pascal

The French mathematician Blaise Pascal (1623–1662) is responsible for developing the array of numbers in Pascal's Triangle.

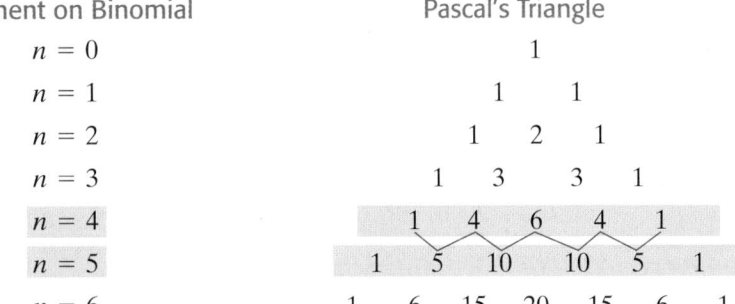

Exponent on Binomial	Pascal's Triangle
$n = 0$	1
$n = 1$	1 1
$n = 2$	1 2 1
$n = 3$	1 3 3 1
$n = 4$	1 4 6 4 1
$n = 5$	1 5 10 10 5 1
$n = 6$	1 6 15 20 15 6 1

Examine row 5 ($n = 4$) and row 6 ($n = 5$).

1 + 4 + 6 + 4 + 1
1   5   10   10   5   1
—— Sum of two numbers

Notice that the first and last numbers in each row are 1, and the inner numbers are obtained by adding the two numbers in the row above (to the right and left). The numerical coefficients of $(a + b)^5$ are 1, 5, 10, 10, 5, and 1. Thus, we can write the expansion of $(a + b)^5$ by using the information in 1–5 above for the variables and their exponents, and by using Pascal's triangle for the coefficients.

$$(a + b)^5 = a^5 + 5a^4b + 10a^3b^2 + 10a^2b^3 + 5ab^4 + b^5$$

This method of expanding a binomial is not practical when n is large.

3 Use the Binomial Theorem

We will shortly introduce a more practical method, called the binomial theorem, to expand expressions of the form $(a + b)^n$. However, before we introduce this formula, we need to explain how to find *binomial coefficients* of the form $\binom{n}{r}$.

Binomial Coefficients

For n and r nonnegative integers, $n \geq r$,

$$\binom{n}{r} = \frac{n!}{r! \cdot (n - r)!}$$

The binomial coefficient $\binom{n}{r}$ is read "the number of *combinations* of n items taken r at a time." Combinations are used in many areas of mathematics, including the study of probability.

EXAMPLE 1 Evaluate $\binom{6}{2}$.

Solution Using the definition, if we substitute 6 for n and 2 for r, we obtain

$$\binom{6}{2} = \frac{6!}{2! \cdot (6-2)!} = \frac{6!}{2! \cdot 4!} = \frac{6 \cdot 5 \cdot \cancel{4 \cdot 3 \cdot 2 \cdot 1}}{(2 \cdot 1) \cdot (\cancel{4 \cdot 3 \cdot 2 \cdot 1})} = 15$$

Thus, $\binom{6}{2}$ equals 15.

Now Try Exercise 9

EXAMPLE 2 Evaluate.

a) $\binom{7}{4}$ **b)** $\binom{8}{8}$ **c)** $\binom{5}{0}$

Solution

a) $\binom{7}{4} = \dfrac{7!}{4! \cdot (7-4)!} = \dfrac{7!}{4! \cdot 3!} = \dfrac{7 \cdot 6 \cdot 5 \cdot \cancel{4 \cdot 3 \cdot 2 \cdot 1}}{(\cancel{4 \cdot 3 \cdot 2 \cdot 1})(3 \cdot 2 \cdot 1)} = 35$

b) $\binom{8}{8} = \dfrac{8!}{8! (8-8)!} = \dfrac{\cancel{8!}}{\cancel{8!} \cdot 0!} = \dfrac{1}{1} = 1$ Remember that $0! = 1$

c) $\binom{5}{0} = \dfrac{5!}{0! \cdot (5-0)!} = \dfrac{\cancel{5!}}{0! \cdot \cancel{5!}} = \dfrac{1}{1} = 1$

Now Try Exercise 17

By studying Examples 2 **b)** and **c)**, you can reason that, for any positive integer n,

$$\binom{n}{n} = 1 \quad \text{and} \quad \binom{n}{0} = 1$$

🖩 Using Your Graphing Calculator ▪▪▪

On most graphing calculators the notation $_nC_r$ is used instead of $\binom{n}{r}$. Thus, $\binom{7}{4}$ would be represented as $_7C_4$.

On the TI-84 Plus calculator, the notation $_nC_r$ can be found under the PRB menu. This time it is item 3, $_nC_r$. To find $_7C_4$ or $_8C_2$, use the following keystrokes:

	KEYSTROKES	ANSWER
$_7C_4$	[7] [MATH] [▶] [▶] [▶] [3] [4] [ENTER]	35
$_8C_2$	[8] [MATH] [▶] [▶] [▶] [3] [2] [ENTER]	28

If you are using a different graphing calculator, consult the manual to learn how to evaluate combinations.

▪▪▪

Now we introduce the binomial theorem.

Binomial Theorem

For any positive integer n,

$$(a+b)^n = \binom{n}{0}a^n b^0 + \binom{n}{1}a^{n-1}b^1 + \binom{n}{2}a^{n-2}b^2 + \binom{n}{3}a^{n-3}b^3 + \cdots + \binom{n}{n}a^0 b^n$$

Notice in the binomial theorem that the sum of the exponents on the variables in each term is n. In the combination, the top number is always n and the bottom number is always the same as the exponent on the second variable in the term.

For example, consider the term $\binom{n}{3}a^{n-3}b^3$.

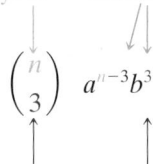

The top number is always n.   The sum of the exponents always equals n.

$$\binom{n}{3} \quad a^{n-3}b^3$$

The bottom number always matches the power of the second variable.

Now we will expand $(a + b)^5$ using the binomial theorem and see if we get the same expression as we did when we used polynomial multiplication and Pascal's triangle to obtain the expansion.

$$(a + b)^5 = \binom{5}{0}a^5b^0 + \binom{5}{1}a^{5-1}b^1 + \binom{5}{2}a^{5-2}b^2 + \binom{5}{3}a^{5-3}b^3 + \binom{5}{4}a^{5-4}b^4 + \binom{5}{5}a^{5-5}b^5$$

$$= \binom{5}{0}a^5b^0 + \binom{5}{1}a^4b^1 + \binom{5}{2}a^3b^2 + \binom{5}{3}a^2b^3 + \binom{5}{4}a^1b^4 + \binom{5}{5}a^0b^5$$

$$= \frac{5!}{0!\cdot 5!}a^5 + \frac{5!}{1!\cdot 4!}a^4b + \frac{5!}{2!\cdot 3!}a^3b^2 + \frac{5!}{3!\cdot 2!}a^2b^3 + \frac{5!}{4!\cdot 1!}ab^4 + \frac{5!}{5!\cdot 0!}b^5$$

$$= a^5 + 5a^4b + 10a^3b^2 + 10a^2b^3 + 5ab^4 + b^5$$

This is the same expression as we obtained earlier.

EXAMPLE 3   Use the binomial theorem to expand $(2x + 3)^6$.

Solution   If we use $2x$ for a and 3 for b, we obtain

$$(2x+3)^6 = \binom{6}{0}(2x)^6(3)^0 + \binom{6}{1}(2x)^5(3)^1 + \binom{6}{2}(2x)^4(3)^2 + \binom{6}{3}(2x)^3(3)^3 + \binom{6}{4}(2x)^2(3)^4 + \binom{6}{5}(2x)^1(3)^5 + \binom{6}{6}(2x)^0(3)^6$$

$$= 1(2x)^6 + 6(2x)^5(3) + 15(2x)^4(9) + 20(2x)^3(27) + 15(2x)^2(81) + 6(2x)(243) + 1(729)$$

$$= 64x^6 + 576x^5 + 2160x^4 + 4320x^3 + 4860x^2 + 2916x + 729$$

Now Try Exercise 19

EXAMPLE 4   Use the binomial theorem to expand $(5x - 2y)^4$.

Solution   Write $(5x - 2y)^4$ as $[5x + (-2y)]^4$. Use $5x$ in place of a and $-2y$ in place of b in the binomial theorem.

$$[5x + (-2y)]^4 = \binom{4}{0}(5x)^4(-2y)^0 + \binom{4}{1}(5x)^3(-2y)^1 + \binom{4}{2}(5x)^2(-2y)^2 + \binom{4}{3}(5x)^1(-2y)^3 + \binom{4}{4}(5x)^0(-2y)^4$$

$$= 1(5x)^4 + 4(5x)^3(-2y) + 6(5x)^2(-2y)^2 + 4(5x)(-2y)^3 + 1(-2y)^4$$

$$= 625x^4 - 1000x^3y + 600x^2y^2 - 160xy^3 + 16y^4$$

Now Try Exercise 25

EXERCISE SET 11.4   Math XL   MyMathLab

MathXL®   MyMathLab

Warm Up Exercises

Fill in the blanks with the appropriate word, phrase, or symbol(s) from the following list.

| binomial | 0 | 1 | 2 | 4 | product |
| Pascal's | 10 | 20 | factorial | sum | quotient |

1. The notation $n!$ is read as "n _____."

2. The value of $n!(n \neq 0)$ is the _____ of the whole numbers from 1 to n.

3. The value of $0!$ is _____.

4. The value of $2!$ is _____.

5. The number of terms in the expansion of $(a + b)^{19}$ is _____.

6. The formula $\binom{n}{r} = \dfrac{n!}{r! \cdot (n - r)!}$ is used to find the _____ coefficients in the expansion of $(a + b)^n$.

7. The value of $\binom{5}{3}$ is _____.

8. The numerical coefficients of the terms in the expansion of $(a + b)^n$ can be found using _____ triangle.

Practice the Skills

Evaluate each combination.

9. $\binom{5}{2}$

10. $\binom{6}{3}$

11. $\binom{5}{5}$

12. $\binom{9}{6}$

13. $\binom{7}{0}$

14. $\binom{12}{10}$

15. $\binom{8}{4}$

16. $\binom{11}{8}$

17. $\binom{8}{2}$

18. $\binom{10}{4}$

Use the binomial theorem to expand each expression.

19. $(x + 4)^3$

20. $(x - 4)^3$

21. $(2x - 3)^3$

22. $(2x + 3)^3$

23. $(a - b)^4$

24. $(2r + s^2)^4$

25. $(3a - b)^5$

26. $(x + 2y)^5$

27. $\left(2x + \dfrac{1}{2}\right)^4$

28. $\left(\dfrac{2}{3}x + \dfrac{3}{2}\right)^4$

29. $\left(\dfrac{x}{2} - 3\right)^4$

30. $(3x^2 + y)^5$

Write the first four terms of each expansion.

31. $(x + 10)^{10}$

32. $(2x + 3)^8$

33. $(3x - y)^7$

34. $(3p - 2q)^{11}$

35. $(x^2 - 3y)^8$

36. $\left(2x + \dfrac{y}{7}\right)^9$

Problem Solving

37. What are the first, second, next-to-last, and last terms of the expansion $(x + 3)^8$?

38. What are the first, second, next-to-last, and last terms of the expansion of $(x + 2)^{10}$?

39. What are the first, second, next-to-last, and last terms of the expansion $(2x + 5)^6$?

40. What are the first, second, next-to-last, and last terms of the expansion of $(3x - 4)^5$?

Concept/Writing Exercises

41. Is $n!$ equal to $n \cdot (n - 1)!$? Explain and give an example to support your answer.

42. Is $(n + 1)!$ equal to $(n + 1) \cdot n!$? Explain and give an example to support your answer.

43. Is $(n - 3)!$ equal to $(n - 3)(n - 4)(n - 5)!$ for $n \geq 5$? Explain and give an example to support your answer.

44. Is $(n + 2)!$ equal to $(n + 2)(n + 1)(n)(n - 1)!$ for $n \geq 1$? Explain and give an example to support your answer.

45. Under what conditions will $\binom{n}{m}$, where n and m are nonnegative integers, have a value of 1?

46. Can $\binom{n}{m}$ ever have a value of 0? Explain.

47. Write the binomial theorem using summation notation.

48. Prove that $\binom{n}{r} = \binom{n}{n - r}$ for any whole numbers n and r, and $r \leq n$.

Cumulative Review Exercises

[3.4] **49.** Find the y-intercept for the line $2x + y = 10$

[4.1] **50.** Solve the system of equations.

$$\frac{1}{5}x + \frac{1}{2}y = 4$$

$$\frac{2}{3}x - y = \frac{8}{3}$$

[5.8] **51.** Solve $x(x - 11) = -18$.

[7.4] **52.** Simplify $\sqrt{20xy^4}\ \sqrt{6x^5y^7}$.

[9.1] **53.** Find $f^{-1}(x)$ if $f(x) = 3x + 8$.

Chapter 11 Summary

IMPORTANT FACTS AND CONCEPTS	EXAMPLES

Section 11.1

A **sequence** of numbers is a list of numbers arranged in a specific order. Each number is called a **term** of the sequence.	$2, 6, 10, 14, 18, 22, \ldots$ is a sequence $7, 14, 21, 28, 35, 42, \ldots$ is a sequence
An **infinite sequence** is a function whose domain is the set of natural numbers.	Domain: $\{1,\quad 2,\quad 3,\quad 4,\quad \ldots,\quad n,\quad \ldots\}$ $\qquad\quad\ \downarrow\quad \downarrow\quad \downarrow\quad \downarrow\qquad\qquad \downarrow$ Range: $\{7,\quad 14,\quad 21,\quad 28,\quad \ldots\quad 7n,\quad \ldots\}$ The infinite sequence is $7, 14, 21, 28, \ldots$.
A **finite sequence** is a function whose domain includes only the first n natural numbers.	Domain: $\{1,\quad 2,\quad 3,\quad 4\}$ $\qquad\quad\ \downarrow\quad \downarrow\quad \downarrow\quad \downarrow$ Range: $\{4,\quad 8,\quad 12,\quad 16\}$ The finite sequence is $4, 8, 12, 16$.
The **general term of a sequence**, a_n, can determine the sequence.	Let $a_n = n^2 - 3$. Write the first three terms of this sequence $$a_1 = 1^2 - 3 = -2$$ $$a_2 = 2^2 - 3 = 1$$ $$a_3 = 3^2 - 3 = 6$$ The first three terms of the sequence are $-2, 1, 6$.
An **increasing sequence** is a sequence where each term is greater than the preceding term.	$-2, 5, 7, 11$ is an increasing sequence.
A **decreasing sequence** is a sequence where each term is less than the preceding term.	$50, 48, 46, 44$ is a decreasing sequence.
A **series** is the sum of the terms of a sequence. A series may be finite or infinite.	If the sequence is $1, 3, 5, 7, 9$, then the series is $1 + 3 + 5 + 7 + 9 = 25$. If the sequence is $\dfrac{1}{3}, \dfrac{1}{9}, \dfrac{1}{27}, \cdots, \left(\dfrac{1}{3}\right)^n, \ldots$ then the series is $\dfrac{1}{3} + \dfrac{1}{9} + \dfrac{1}{27} + \cdots + \left(\dfrac{1}{3}\right)^n + \cdots$
A **partial sum**, s_n, of an infinite sequence $a_1, a_2, a_3, \ldots, a_n, \ldots$ is the sum of the first n terms. That is, $$s_1 = a_1$$ $$s_2 = a_1 + a_2$$ $$s_3 = a_1 + a_2 + a_3$$ $$\vdots$$ $$s_n = a_1 + a_2 + a_3 + \cdots + a_n$$	Let $a_n = \dfrac{5 + n}{n^2}$. Compute s_1 and s_3. $$s_1 = a_1 = \frac{5 + 1}{1^2} = \frac{6}{1} = 6$$ $$s_3 = a_1 + a_2 + a_3$$ $$= \frac{5 + 1}{1^2} + \frac{5 + 2}{2^2} + \frac{5 + 3}{3^2}$$ $$= \frac{6}{1} + \frac{7}{4} + \frac{8}{9} = 8\frac{23}{36}$$

IMPORTANT FACTS AND CONCEPTS	EXAMPLES

Section 11.1 (cont.)

A series can be written using **summation notation**:

$$\sum_{i=1}^{n} a_1 = a_1 + a_2 + a_3 + \cdots + a_n.$$

i is the **index of summation**, n is the **upper limit of summation**, and 1 is the **lower limit of summation**.

$$\sum_{i=1}^{4}(3i - 7) = (3 \cdot 1 - 7) + (3 \cdot 2 - 7) + (3 \cdot 3 - 7) + (3 \cdot 4 - 7)$$

$$= -4 - 1 + 2 + 5 = 2$$

If $a_n = 6n^2 + 11$, the third partial sum, s_3, in summation notation, is written as $\sum_{i=1}^{3}(6i^2 + 11)$.

Section 11.2

An **arithmetic sequence** is a sequence in which each term after the first differs from the preceding term by a **common difference, d**.

Arithmetic Sequence	Common Difference, d
$3, 8, 13, 18, 23, \ldots$	$d = 8 - 3 = 5$
$20, 14, 8, 2, -4, \ldots$	$d = 14 - 20 = -6$

The **nth term, a_n**, of an arithmetic sequence is

$$a_n = a_1 + (n - 1)d$$

The nth term of the arithmetic sequence with $a_1 = 7$ and $d = -5$ is

$$a_n = 7 + (n - 1)(-5)$$
$$= 7 - 5n + 5$$
$$= -5n + 12$$

For this sequence, the 20th term is

$$a_{20} = -5(20) + 12 = -100 + 12$$
$$= -88$$

An **arithmetic series** is the sum of the terms of an arithmetic sequence. The sum of the first n terms, s_n, of an arithmetic sequence, also known as the **nth partial sum**, is

$$s_n = a_1 + a_2 + a_3 + \cdots + a_n$$

For an arithmetic series, this sum is determined by the formula

$$s_n = \frac{n(a_1 + a_n)}{2}$$

Find the sum of the first 30 natural numbers. That is, find the sum of

$$1 + 2 + 3 + \cdots + 30$$

Since $a_1 = 1, a_{30} = 30$, and $n = 30$, the sum is

$$s_{30} = \frac{30(1 + 30)}{2} = \frac{30(31)}{2} = 465$$

Section 11.3

A **geometric sequence** is a sequence in which each term after the first term is a common multiple of the preceding term. The common multiple is called the **common ration, r**.

Geometric Sequence	Common Ratio, r
$2, 6, 18, 54, 162, \ldots$	$r = \dfrac{6}{2} = 3$
$8, -2, \dfrac{1}{2}, -\dfrac{1}{8}, \dfrac{1}{32}, \ldots$	$r = \dfrac{-2}{8} = -\dfrac{1}{4}$

The **nth term, a_n**, of a geometric sequence is

$$a_n = a_1 r^{n-1}$$

For the geometric sequence with $a_1 = 5, r = \dfrac{1}{2}$, and $n = 6$, a_6 is found as follows.

$$a_6 = 5\left(\frac{1}{2}\right)^{6-1} = 5\left(\frac{1}{2}\right)^5 = \frac{5}{32}$$

A **geometric series** is the sum of the terms of a geometric sequence. The sum of the first n terms, s_n, of a geometric sequence, also known as the **nth partial sum**, is

$$s_n = a_1 + a_2 + a_3 + \cdots + a_n$$

For a geometric series, this sum is determined by the formula

$$s_n = \frac{a_1(1 - r^n)}{1 - r}, r \neq 1$$

To find the sum of the six terms of a geometric sequence with $a_1 = 12$ and $r = \dfrac{1}{3}$, use the formula with $n = 6$ to obtain

$$s_6 = \frac{12\left[1 - \left(\frac{1}{3}\right)^6\right]}{1 - \frac{1}{3}} = \frac{12\left[1 - \frac{1}{729}\right]}{\frac{2}{3}} = \frac{12\left(\frac{728}{729}\right)}{\frac{1}{3}}$$

$$= 12\left(\frac{728}{729}\right)\left(\frac{3}{1}\right) = \frac{2912}{81} \text{ or } 35\frac{77}{81}$$

IMPORTANT FACTS AND CONCEPTS	EXAMPLES

Section 11.3 (cont.)

The sum of an infinite geometric series is

$$s_\infty = \frac{a_1}{1 - r} \text{ where } |r| < 1$$

To find the sum of the infinite series $4 - 2 + 1 - \frac{1}{2} + \frac{1}{4} + \cdots$,

use the formula with $a_1 = 4$ and $r = -\frac{1}{2}$ to obtain

$$s_\infty = \frac{4}{1 - \left(-\frac{1}{2}\right)} = \frac{4}{\frac{3}{2}} = 4 \cdot \frac{2}{3} = \frac{8}{3} \text{ or } 2\frac{2}{3}$$

Section 11.4

n Factorial

$$n! = n(n - 1)(n - 2)(n - 3)\cdots(1)$$

for any positive integer n.

Note that 0! Is defined to be 1.

$5! = 5 \cdot 4 \cdot 3 \cdot 2 \cdot 1 = 120$

$8! = 8 \cdot 7 \cdot 6 \cdot 5 \cdot 4 \cdot 3 \cdot 2 \cdot 1 = 40{,}320$

Binomial Coefficients

For n and r non-negative integers, $n \geq r$,

$$\binom{n}{r} = \frac{n!}{r! \cdot (n - r)!}$$

$$\binom{n}{n} = 1 \text{ and } \binom{n}{0} = 1$$

$$\binom{7}{3} = \frac{7!}{3! \cdot (7 - 3)!} = \frac{7!}{3! \cdot 4!} = \frac{7 \cdot 6 \cdot 5 \cdot 4 \cdot 3 \cdot 2 \cdot 1}{3 \cdot 2 \cdot 1 \cdot 4 \cdot 3 \cdot 2 \cdot 1} = 35$$

$$\binom{10}{10} = 1, \binom{10}{0} = 1$$

Binomial Theorem

For any positive integer n,

$$(a + b)^n = \binom{n}{0}a^n b^0 + \binom{n}{1}a^{n-1}b^1 + \binom{n}{2}a^{n-2}b^2 +$$
$$\binom{n}{3}a^{n-3}b^3 + \cdots + \binom{n}{n}a^0 b^n$$

$$(x + 2y)^4 = \binom{4}{0}x^4 + \binom{4}{1}x^3(2y) + \binom{4}{2}x^2(2y)^2$$
$$+ \binom{4}{3}x(2y)^3 + \binom{4}{4}(2y)^4$$
$$= 1 \cdot x^4 + 4 \cdot x^3(2y) + 6 \cdot x^2(4y^2) + 4 \cdot x(8y^3) + 1 \cdot 16y^4$$
$$= x^4 + 8x^3 y + 24x^2 y^2 + 32xy^3 + 16y^4$$

Chapter 11 Review Exercises

[11.1]   *Write the first five terms of each sequence.*

1. $a_n = n + 5$ **2.** $a_n = n^2 + n - 3$ **3.** $a_n = \frac{12}{n}$ **4.** $a_n = \frac{n^2}{n + 4}$

Find the indicated term of each sequence.

5. $a_n = 3n - 10$, seventh term   **6.** $a_n = (-1)^n + 5$, tenth term

7. $a_n = \frac{n + 17}{n^2}$, ninth term   **8.** $a_n = (n)(n - 3)$, eleventh term

For each sequence, find the first and third partial sums, s_1 and s_3.

9. $a_n = 2n + 5$ **10.** $a_n = n^2 + 11$

11. $a_n = \frac{n + 3}{n + 2}$ **12.** $a_n = (-1)^n(n + 8)$

Write the next three terms of each sequence. Then write an expression for the general term, a_n.

13. $2, 4, 8, 16, \ldots$ **14.** $-27, 9, -3, 1, \ldots$

15. $\frac{1}{9}, \frac{2}{9}, \frac{4}{9}, \frac{8}{9}, \ldots$ **16.** $13, 9, 5, 1, \ldots$

Write out each series. Then find the sum of the series.

17. $\sum_{i=1}^{3} i^2 + 10$ **18.** $\sum_{i=1}^{4} i(i + 5)$

19. $\sum_{i=1}^{5} \frac{i^2}{6}$ **20.** $\sum_{i=1}^{4} \frac{i}{i + 1}$

For the set of values $x_1 = 3$, $x_2 = 9$, $x_3 = 7$, $x_4 = 10$, evaluate the indicated sum.

21. $\displaystyle\sum_{i=1}^{4} x_i$

22. $\displaystyle\sum_{i=1}^{4} (x_i)^2$

23. $\displaystyle\sum_{i=2}^{3} (x_i^2 + 1)$

24. $\displaystyle\left(\sum_{i=1}^{4} x_i\right)^2$

In Exercises 25 and 26, consider the following rectangles. For the nth rectangle, the length is $n + 3$ and the width is n.

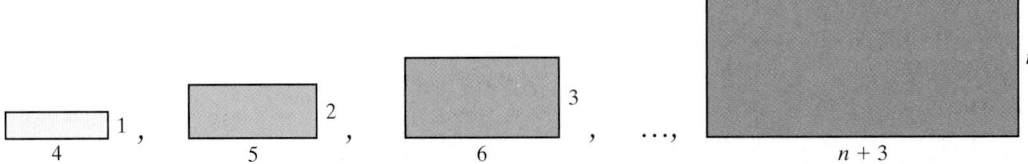

25. Perimeter

   a) Find the perimeters for the first four rectangles, and then list the perimeters in a sequence.

   b) Find the general term for the perimeter of the nth rectangle in the sequence. Use p_n for perimeter.

26. Area

   a) Find the areas for the first four rectangles, and then list the areas in a sequence.

   b) Find the general term for the area of the nth rectangle in the sequence. Use a_n for area.

[11.2]   *Write the first five terms of the arithmetic sequence with the indicated first term and common difference.*

27. $a_1 = 5$, $d = 3$

28. $a_1 = 5$, $d = -\dfrac{1}{3}$

29. $a_1 = \dfrac{1}{2}$, $d = -2$

30. $a_1 = -20$, $d = \dfrac{1}{5}$

For each arithmetic sequence, find the indicated value.

31. $a_1 = 6$, $d = 3$; find a_9

32. $a_1 = 26$, $a_8 = -9$; find d

33. $a_1 = -3$, $a_{11} = 2$; find d

34. $a_1 = 22$, $a_n = -3$, $d = -5$; find n

Find s_n and d for each arithmetic sequence.

35. $a_1 = 7$, $a_8 = 21$, $n = 8$

36. $a_1 = -2$, $a_7 = -38$, $n = 7$

37. $a_1 = \dfrac{3}{5}$, $a_6 = \dfrac{13}{5}$, $n = 6$

38. $a_1 = -\dfrac{10}{3}$, $a_9 = -6$, $n = 9$

Write the first four terms of each arithmetic sequence. Then find a_{10} and s_{10}.

39. $a_1 = -7$, $d = 4$

40. $a_1 = 8$, $d = -3$

41. $a_1 = \dfrac{5}{6}$, $d = \dfrac{2}{3}$

42. $a_1 = -60$, $d = 5$

Find the number of terms in each arithmetic sequence. Then find s_n.

43. $4, 9, 14, \ldots, 64$

44. $-7, -4, -1, \ldots, 14$

45. $\dfrac{6}{10}, \dfrac{9}{10}, \dfrac{12}{10}, \ldots, \dfrac{36}{10}$

46. $-9, -3, 3, 9, \ldots, 45$

[11.3]   *Determine the first five terms of each geometric sequence.*

47. $a_1 = 6$, $r = 2$

48. $a_1 = -56$, $r = \dfrac{1}{2}$

49. $a_1 = 20$, $r = -\dfrac{2}{3}$

50. $a_1 = -20$, $r = \dfrac{1}{5}$

Find the indicated term of each geometric sequence.

51. $a_1 = 12$, $r = \dfrac{1}{3}$; find a_5

52. $a_1 = 15$, $r = 2$; find a_6

53. $a_1 = -8$, $r = -3$; find a_4

54. $a_1 = \dfrac{1}{12}$, $r = \dfrac{2}{3}$; find a_5

Find each sum.

55. $a_1 = 7$, $r = 2$; find s_6

56. $a_1 = -84$, $r = -\dfrac{1}{4}$; find s_5

57. $a_1 = 9$, $r = \dfrac{3}{2}$; find s_4

58. $a_1 = 8$, $r = \dfrac{1}{2}$; find s_7

For each geometric sequence, find the common ratio, r, and then write an expression for the general term, a_n.

59. $6, 12, 24, \ldots$

60. $-6, -30, -150, \ldots$

61. $10, \dfrac{10}{3}, \dfrac{10}{9}, \ldots$

62. $\dfrac{9}{5}, \dfrac{18}{15}, \dfrac{36}{45}, \ldots$

Find the sum of the terms in each infinite geometric sequence.

63. $5, \dfrac{5}{2}, \dfrac{5}{4}, \dfrac{5}{8}, \ldots$

64. $\dfrac{7}{2}, 1, \dfrac{2}{7}, \dfrac{4}{49}, \ldots$

65. $-8, \dfrac{8}{3}, -\dfrac{8}{9}, \dfrac{8}{27}, \ldots$

66. $-6, -4, -\dfrac{8}{3}, -\dfrac{16}{9}, \ldots$

Find the sum of each infinite series.

67. $16 + 8 + 4 + 2 + 1 + \cdots$

68. $11 + \dfrac{11}{3} + \dfrac{11}{9} + \dfrac{11}{27} + \cdots$

69. $5 - 1 + \dfrac{1}{5} - \dfrac{1}{25} + \cdots$

70. $-4, -\dfrac{8}{3}, -\dfrac{16}{9}, -\dfrac{32}{27}, \ldots$

Write the repeating decimal as a ratio of integers.

71. $0.363636\ldots$

72. $0.783783783\ldots$

[11.4]   *Use the binomial theorem to expand the expression.*

73. $(3x + y)^4$

74. $(2x - 3y^2)^3$

Write the first four terms of the expansion.

75. $(x - 2y)^9$

76. $(2a^2 + 3b)^8$

[11.2]

77. Sum of Integers Find the sum of the integers between 101 and 200, inclusive.

78. Barrels of Oil Barrels of oil are stacked with 21 barrels in the bottom row, 20 barrels in the second row, 19 barrels in the third row, and so on, to the top row, which has only 1 barrel. How many barrels are there?

79. Salary Ahmed Mocanda just started a new job with an annual salary of $36,000. He has been told that his salary will increase by $1000 per year for the next 10 years.

a) Write a sequence showing his salary for the first 4 years.

b) Write a general term of this sequence.

c) What will his salary be 6 years from now?

d) How much money will he make in the first 11 years?

[11.3]

80. Money You begin with $50, double that to get $100, double that again to get $200, and so on. How much will you have after you perform this process 10 times?

81. Salary Gertude Dibble started a new job on January 1, 2010, with a monthly salary of $1600. Her boss has agreed to give her a 4% raise each month for the remainder of the year.

a) What is Gertude's salary in July?

b) What is Gertude's salary in December?

c) How much money does Gertude make in 2010?

82. Inflation If the inflation rate was a constant 8% per year (each year the cost of living is 8% greater than the previous year), how much would a product that costs $200 now cost after 12 years?

83. Pendulum On each swing (left to right or right to left), a pendulum travels 92% as far as on its previous swing. If the first swing is 12 feet, find the distance traveled by the pendulum by the time it comes to rest.

Chapter 11 Practice Test

Chapter Test Prep Videos provide fully worked-out solutions to any of the exercises you want to review. Chapter Test Prep Videos are available via **MyMathLab**, *or on* **YouTube** *(search "Angel Intermediate Algebra" and click on "Channels")*

1. What is a series?

2. a) What is an arithmetic sequence?

  b) What is a geometric sequence?

3. Write the first five terms of the sequence if $a_n = \dfrac{n - 2}{3n}$.

4. Find the first and third partial sums if $a_n = \dfrac{2n + 1}{n^2}$.

5. Write out the following series and find the sum of the series.

$$\sum_{i=1}^{5} (2i^2 + 3)$$

6. For $x_1 = 4$, $x_2 = 2$, $x_3 = 8$, and $x_4 = 10$ find $\displaystyle\sum_{i=1}^{4} (x_i)^2$.

7. Write the general term for the following arithmetic sequence.

$$\frac{1}{3}, \frac{2}{3}, \frac{3}{3}, \frac{4}{3}, \ldots$$

8. Write the general term for the following geometric sequence.

$$5, 10, 20, 40, \ldots$$

In Exercises 9 and 10, write the first four terms of each sequence.

9. $a_1 = 15, d = -6$

10. $a_1 = \dfrac{5}{12}, r = \dfrac{2}{3}$

11. Find a_{11} when $a_1 = 40$ and $d = -8$.

12. Find s_8 for the arithmetic sequence with $a_1 = 7$ and $a_8 = -12$.

13. Find the number of terms in the arithmetic sequence $-4, -16, -28, \ldots, -136$.

14. Find a_6 when $a_1 = 8$ and $r = \dfrac{2}{3}$.

15. Find s_7 when $a_1 = \dfrac{3}{5}$ and $r = -5$.

16. Find the common ratio and write an expression for the general term of the sequence $15, 5, \dfrac{5}{3}, \dfrac{5}{9}, \ldots$.

17. Find the sum of the following infinite geometric series.

$$4 + \frac{8}{3} + \frac{16}{9} + \frac{32}{27} + \cdots$$

18. Write $0.3939\ldots$ as a ratio of integers.

19. Evaluate $\dbinom{8}{3}$.

20. Use the binomial theorem to expand $(x + 2y)^4$.

21. Arithmetic Mean Paul Misselwitz's five test scores are 76, 93, 83, 87, and 71. Use $\bar{x} = \dfrac{\Sigma x}{n}$ to find the arithmetic mean of Paul's scores.

22. A Pile of Logs Logs are piled with 13 logs in the bottom row, 12 logs in the second row, 11 logs in the third row, and so on, to the top. How many logs are there?

23. Saving for Retirement To save for retirement, Jamie Monroe plans to save $1000 the first year, $2000 the second year, $3000 the third year, and to increase the amount saved by $1000 in each successive year. How much will she have saved by the end of her twentieth year of savings?

24. Earnings Yolanda Rivera makes $700 per week working at an insurance office. Her boss has guaranteed her an increase of 4% per week for the next 7 weeks. How much is she making in the sixth week?

25. Culture of Bacteria The number of bacteria in a culture is tripling every hour. If there are initially 500 bacteria in the culture, how many bacteria will be in the culture by the end of the sixth hour?

Cumulative Review Test

Take the following test and check your answers with those given in the back of the book. Review any questions that you answered incorrectly. The section where the material was covered is indicated after the answer.

1. Solve $A = \dfrac{1}{2}bh$, for b.

2. Find an equation of the line through $(4, -2)$ and $(1, 9)$. Write the equation in slope-intercept form.

3. Solve the system of equations.

$$x + y + z = 1$$
$$2x + 2y + 2z = 2$$
$$3x + 3y + 3z = 3$$

4. Multiply $(5x^3 + 4x^2 - 6x + 2)(x + 5)$.

5. Factor $x^3 + 2x - 6x^2 - 12$.

6. Factor $(a + b)^2 + 8(a + b) + 16$.

7. Subtract $5 - \dfrac{x - 1}{x^2 + 3x - 10}$.

8. y varies directly as the square of z. If y is 80 when z is 20, find y when z is 50.

9. If $f(x) = 2\sqrt[3]{x - 3}$ and $g(x) = \sqrt[3]{5x - 15}$, find all values of x for which $f(x) = g(x)$.

10. Solve $\sqrt{6x - 5} - \sqrt{2x + 6} - 1 = 0$.

11. Solve by completing the square.

$$x^2 + 2x + 15 = 0$$

12. Solve by the quadratic formula.

$$x^2 - \frac{x}{5} - \frac{1}{3} = 0$$

13. Numbers Twice the square of a positive number decreased by nine times the number is 5. Find the number.

14. Graph $y = x^2 - 4x$ and label the vertex.

15. Solve $\log_a \dfrac{1}{64} = 6$ for a.

16. Graph $y = 2^x - 1$.

17. Find an equation of a circle with center at $(-6, 2)$ and radius 7.

18. Graph $(x + 3)^2 + (y + 1)^2 = 16$.

19. Graph $9x^2 + 16y^2 = 144$.

20. Find the sum of the infinite geometric series.

$$6 + 4 + \frac{8}{3} + \frac{16}{9} + \frac{32}{27} + \cdots$$

Appendix

Geometric Formulas

Areas and Perimeters

Figure	Sketch	Area	Perimeter
Square		$A = s^2$	$P = 4s$
Rectangle		$A = lw$	$P = 2l + 2w$
Parallelogram		$A = lh$	$P = 2l + 2w$
Trapezoid		$A = \frac{1}{2}h(b_1 + b_2)$	$P = s_1 + s_2 + b_1 + b_2$
Triangle		$A = \frac{1}{2}bh$	$P = s_1 + s_2 + b$

Area and Circumference of a Circle

Circle		$A = \pi r^2$	$C = 2\pi r$

Volumes and Surface Areas of Three-Dimensional Figures

Figure	Sketch	Volume	Surface Area
Rectangular solid		$V = lwh$	$s = 2lh + 2wh + 2wl$
Right circular cylinder		$V = \pi r^2 h$	$s = 2\pi rh + 2\pi r^2$
Sphere		$V = \dfrac{4}{3}\pi r^3$	$s = 4\pi r^2$
Right circular cone		$V = \dfrac{1}{3}\pi r^2 h$	$s = \pi r \sqrt{r^2 + h^2}$
Square or rectangular pyramid		$V = \dfrac{1}{3}lwh$	

Answers

Chapter 1

Exercise Set 1.1   **1–11.** Answers will vary.   **13.** Do all the homework and preview the new material to be covered in class.   **15.** See the steps on page 4 of your text.   **17.** The more you put into the course, the more you will get out of it.   **19.** Answers will vary.

Exercise Set 1.2   **1.** Variable   **3.** Algebraic expression   **5.** Elements   **7.** Subset   **9.** Intersection   **11.** Irrational   **13.** $>$
15. $=$ **17.** $>$ **19.** $<$ **21.** $>$ **23.** $<$ **25.** $>$ **27.** $>$ **29.** $A = \{0\}$ **31.** $C = \{18, 20\}$ **33.** $E = \{0, 1, 2\}$

35. $H = \{0, 7, 14, 21, \dots\}$ **37.** $J = \{1, 2, 3, 4, \dots\}$ or $J = N$ **39. a)** 4 **b)** $4, 0$ **c)** $-2, 4, 0$ **d)** $-2, 4, \dfrac{1}{2}, \dfrac{5}{9}, 0, -1.23, \dfrac{78}{79}$
e) $\sqrt{2}, \sqrt{8}$ **f)** $-2, 4, \dfrac{1}{2}, \dfrac{5}{9}, 0, \sqrt{2}, \sqrt{8}, -1.23, \dfrac{78}{79}$ **41.** $A \cup B = \{1, 2, 3, 4, 5, 6\}$; $A \cap B = \{\ \}$ **43.** $A \cup B = \{-4, -3, -2,$
$-1, 0, 1, 3\}$; $A \cap B = \{-3, -1\}$ **45.** $A \cup B = \{2, 4, 6, 8, 10\}$; $A \cap B = \{\ \}$ **47.** $A \cup B = \{0, 5, 10, 15, 20, 25, 30\}$; $A \cap B = \{\ \}$
49. $A \cup B = \{-1, -0, 1, e, i, \pi\}$; $A \cap B = \{-1, 0, 1\}$ **51.** The set of natural numbers   **53.** The set of whole number multiples of 3
55. The set of odd integers   **57. a)** Set A is the set of all x such that x is a natural number less than 7   **b)** $A = \{1, 2, 3, 4, 5, 6\}$
59. ←●——→   **61.** ←——●→   **63.** ←●—⊕→   **65.** ←●●●●···→   **67.** ←●●●●→   **69.** $\{x | x \ge 1\}$
     0   2   -6   3   1 2 3 4   0 1 2 3

71. $\{x | x < 5 \text{ and } x \in I\}$ or $\{x | x \le 4 \text{ and } x \in I\}$ **73.** $\{x | -3 < x \le 5\}$ **75.** $\{x | -2.5 \le x < 4.2\}$ **77.** $\{x | -3 \le x \le 1 \text{ and } x \in I\}$

79. Yes   **81.** No   **83.** Yes   **85.** No   **87.** One example is $\left\{\dfrac{3}{2}, \dfrac{4}{3}, \dfrac{5}{4}, \dfrac{6}{5}, \dfrac{7}{6}\right\}$ **89.** One example is $A = \{2, 4, 5, 8, 9\}$, $B = \{4, 5, 6, 9\}$

91. a) {Johnson, Earnhardt Jr, Kahn, Kenseth, Vickers, Edwards, Busch, Regan, Hamlin}   **b)** Union   **c)** {Earnhardt Jr, Kenseth, Edwards}   **d)** Intersection   **93. a)** {Albert, Carmen, Frank, Linda, Barbara, Jason, David, Earl, Kate, Ingrid}   **b)** Union **c)** {Frank, Linda}   **d)** Intersection   **95. a)** {China, India, United States, Indonesia, Brazil, Nigeria}   **b)** {China, India, United States, Russia, Japan, Indonesia, Nigeria}   **c)** {China, India, United States}   **d)** {China, India, United States, Indonesia}   **e)** {China, India, United States}   **97. a)** $A = \{\text{Alex, James}\}$, $B = \{\text{Alex, James, George, Connor}\}$, $C = \{\text{Alex, Stephen}\}$, $D = \{\text{Alex, George, Connor}\}$ **b)** {Alex}   **c)** Only Alex   **99. a)** $\{1, 3, 4, 5, 6, 7\}$ **b)** $\{2, 3, 4, 6, 8, 9\}$ **c)** $\{1, 2, 3, 4, 5, 6, 7, 8, 9\}$ **d)** $\{3, 4, 6\}$ **101. a)** $\{x | x > 1\}$ includes fractions and decimal numbers which the other set does not contain.   **b)** $\{2, 3, 4, 5, \dots\}$ **c)** No, since it is not possible to list all real numbers greater than 1 in roster form.   **103.**
105. $<, \le, >, \ge, \ne$
107. $\{-1, 0, 1, 2, 3\}$
109. Answers will vary.
111. False   **113.** True
115. True   **117.** False   **119.** True

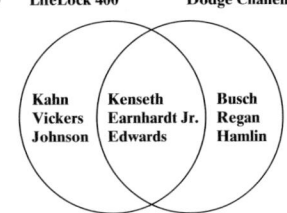

LifeLock 400   Dodge Challenger 500

Kahn   Kenseth   Busch
Vickers   Earnhardt Jr.   Regan
Johnson   Edwards   Hamlin

Exercise Set 1.3   **1.** Positive   **3.** Additive inverse   **5.** Add   **7.** Absolute value   **9.** Distributive   **11.** 5   **13.** 7   **15.** $\dfrac{7}{8}$ **17.** 0
19. -7 **21.** $-\dfrac{5}{9}$ **23.** $=$ **25.** $>$ **27.** $>$ **29.** $>$ **31.** $>$ **33.** $<$ **35.** $-|5|, -2, -1, |-3|, 4$ **37.** $-32, -|4|, 4, |-7|, 15$
39. $-|-6.5|, -6.1, |-6.3|, |6.4|, 6.8$ **41.** $-2, \dfrac{1}{3}, \left|-\dfrac{1}{2}\right|, \left|\dfrac{3}{5}\right|, \left|-\dfrac{3}{4}\right|$ **43.** 3   **45.** -22 **47.** -4 **49.** $-\dfrac{2}{35}$ **51.** -0.99 **53.** 7.92
55. -16.2 **57.** 2   **59.** -2 **61.** $\dfrac{17}{20}$ **63.** -40 **65.** $\dfrac{5}{4}$ **67.** 12   **69.** 235.9192   **71.** 11   **73.** 1   **75.** $-\dfrac{3}{64}$ **77.** $\dfrac{7}{3}$ **79.** -4 **81.** 18
83. 7   **85.** -20.6 **87.** 11   **89.** -6 **91.** $\dfrac{81}{16}$ **93.** -1 **95.** $-\dfrac{17}{45}$ **97.** 77   **99.** -39 **101.** 0   **103.** Commutative property of addition   **105.** Multiplicative property of zero   **107.** Associative property of addition   **109.** Identity property of multiplication **111.** Associative property of multiplication   **113.** Distributive property   **115.** Identity property of addition   **117.** Inverse property of addition   **119.** Double negative property   **121.** $-6, \dfrac{1}{6}$ **123.** $\dfrac{22}{9}, -\dfrac{9}{22}$ **125.** 49°F   **127.** 148.2 feet below the starting point, or -148.2 feet   **129.** 10.1°F   **131.** Gain of $1207   **133.** Answers will vary.   **135.** $24,000   **137.** All real numbers, $\mathbb{R}$ **139.** $6, -6$
141. $\{\ \}$ **143.** Answers will vary.   **145.** Answers will vary.   **147.** $-\dfrac{a}{b}$ or $\dfrac{-a}{b}$ **149. a)** $a + b = b + a$ **b)** Answers will vary.
151. $2 + (3 \cdot 4) \ne (2 + 3) \cdot (2 + 4), 14 \ne 30$ **153.** 84   **155.** 1   **156.** True   **157.** $\{1, 2, 3, 4, \dots\}$ **158. a)** $3, 4, -2, 0$
b) $3, 4, -2, \dfrac{5}{6}, 0$ **c)** $\sqrt{11}$ **d)** $3, 4, -2, \dfrac{5}{6}, \sqrt{11}, 0$ **159. a)** $\{1, 4, 7, 9, 12, 19\}$ **b)** $\{4, 7\}$ **160.** ←⊕————●→
              -4   5

Exercise Set 1.4   **1.** Factors   **3.** Exponent   **5.** 64   **7.** Radicand   **9.** Index   **11.** Real   **13.** 9   **15.** -9 **17.** 25   **19.** $-\dfrac{81}{625}$

21. 7   **23.** -6 **25.** -3 **27.** 0.1   **29.** 0.015   **31.** 1.897   **33.** 76,183.335   **35.** 2.962   **37.** 3.250   **39.** -0.723 **41. a)** 9   **b)** -9

43. a) 100   **b)** -100 **45. a)** 1   **b)** -1 **47. a)** $\dfrac{1}{9}$ **b)** $-\dfrac{1}{9}$ **49. a)** 27   **b)** -27 **51. a)** -125 **b)** 125   **53. a)** -8 **b)** 8

55. a) $\dfrac{8}{125}$ **b)** $-\dfrac{8}{125}$ **57.** -7 **59.** -19 **61.** -22.221 **63.** $-\dfrac{5}{16}$ **65.** 43   **67.** 25   **69.** 0   **71.** $\dfrac{1}{2}$ **73.** -10 **75.** 10   **77.** 64

79. 12   **81.** $\dfrac{27}{5}$ **83.** Undefined   **85.** -4 **87.** 0   **89.** $-\dfrac{10}{3}$ **91.** $\dfrac{242}{5}$ **93.** $\dfrac{1}{4}$ **95.** 34   **97.** -41 **99.** -9 **101.** -100 **103.** 33

105. -6 **107.** $\dfrac{3}{2}$ **109.** $\dfrac{7y-14}{2}$, 14   **111.** $6(3x+6)-9$, 81   **113.** $\left(\dfrac{x+3}{2y}\right)^2-3$, 1   **115. a)** 24.6 miles   **b)** 57.4 miles

117. a) 102 feet   **b)** 54 feet   **119. a)** \$837.97   **b)** \$972.30   **121. a)** 9.51 billion trips   **b)** 22.51 billion trips   **123. a)** \$297.83 billion **b)** \$405.83 billion   **125. a)** 7.62%   **b)** 21.78%   **127. a)** \$1.262 billion   **b)** \$26.38 billion   **129.** n factors of a **131.** The positive number whose square equals the radicand   **133.** A negative number raised to an odd power is a negative number. **135.** Parentheses; exponents and radicals; multiplication or division from left to right; addition or subtraction from left to right. **137. a)** Answers will vary.   **b)** 24   **139. a)** $A\cap B=\{b,c,f\}$ **b)** $A\cup B=\{a,b,c,d,f,g,h\}$ **140.** All real numbers, $\mathbb{R}$ **141.** $a\ge0$ **142.** $8,-8$ **143.** $-|6|,-4,-|-2|,0,|-5|$ **144.** Associative property of addition

Mid-Chapter Test*   **1.** Answers will vary. [1.1]   **2.** $A\cup B=\{-3,-2,-1,0,1,2,3,5\}$, $A\cap B=\{-1,1\}$ [1.2]   **3.** The set of whole number multiples of 5. [1.2]   **4.** ◄━━━●━━━► [1.2]   **5.** $>$ [1.2]   **6.** $|x|-5\le x<2$ [1.2]   **7.** No [1.2]   **8.** $-15,|-6|,7,$
                       3

$|-17|$ [1.2]   **9.** 9.2 [1.3]   **10.** $\dfrac{7}{30}$ [1.3]   **11.** 256 [1.3]   **12.** $-\dfrac{4}{13}$ [1.3]   **13.** -3 [1.3]   **14.** Distributive property [1.3]   **15.** 0.9 [1.4]

16. a) -121 **b)** 121 [1.4]   **17. a)** 1) Grouping symbols, 2) exponents and radicals, 3) multiplications or divisions leftto right, 4) additions or subtractions left to right   **b)** -14 [1.4]   **18.** 23 [1.4]   **19.** 4 [1.4]   **20.** $\dfrac{5}{2}$ [1.4]

Exercise Set 1.5   **1.** Product   **3.** Negative exponent   **5.** Undefined   **7.** raising a product   **9.** Reciprocal   **11.** $\dfrac{1}{9}$

13. 32   **15.** 9   **17.** $\dfrac{1}{81}$ **19.** 125   **21.** 1   **23.** 64   **25.** 64   **27.** $\dfrac{16}{49}$ **29. a)** $\dfrac{1}{9}$ **b)** $\dfrac{1}{9}$ **c)** $-\dfrac{1}{9}$ **d)** $-\dfrac{1}{9}$ **31. a)** 2   **b)** -2 **c)** -2

d) 2   **33. a)** 5   **b)** -5 **c)** 1   **d)** -1 **35. a)** $3xy$ **b)** 1   **c)** $3x$ **d)** 3   **37.** $\dfrac{7}{y^3}$ **39.** $9x^4$ **41.** $3ab^3$ **43.** $\dfrac{17}{2m^2n^3}$ **45.** $\dfrac{5z^4}{x^2y^3}$

47. $\dfrac{1}{9xy}$ **49.** $\dfrac{1}{4}$ **51.** x^2 **53.** 64   **55.** $\dfrac{1}{49}$ **57.** $\dfrac{1}{m^{11}}$ **59.** $5w^5$ **61.** $\dfrac{12}{a^8}$ **63.** $3p$ **65.** $-10r^7$ **67.** $8x^7y^2$ **69.** $\dfrac{3x^2}{y^6}$ **71.** $-\dfrac{3x^3z^2}{y^5}$

73. a) 4   **b)** 8   **c)** 1   **d)** 0   **75. a)** $-\dfrac{1}{12}$ **b)** $\dfrac{7}{12}$ **c)** $\dfrac{11}{10}$ **d)** $\dfrac{23}{120}$ **77.** 81   **79.** $\dfrac{1}{81}$ **81.** b^6 **83.** $-c^3$ **85.** $\dfrac{25}{x^6}$ **87.** $\dfrac{3}{16}$ **89.** $\dfrac{21}{16}$

91. $\dfrac{9}{16b^2}$ **93.** $\dfrac{16x^4}{y^4}$ **95.** $\dfrac{q^{12}}{125p^6}$ **97.** $-\dfrac{g^{12}}{27h^9}$ **99.** $\dfrac{25j^2}{16k^4}$ **101.** $8r^6s^{15}$ **103.** $\dfrac{y^6}{64x^3}$ **105.** $125x^9y^3$ **107.** $\dfrac{z^3}{8x^3y^3}$ **109.** $\dfrac{x^{20}}{y^{10}}$ **111.** $\dfrac{x^4y^8}{4z^{12}}$

113. $\dfrac{64b^{12}}{a^6c^3}$ **115.** $\dfrac{27}{8x^{21}y^9}$ **117.** x^{7a+3} **119.** w^{5a-7} **121.** x^{w+7} **123.** x^{5p+2} **125.** x^{2m+2} **127.** $\dfrac{5m^{2b}}{n^{2a}}$ **129. a)** $x<0$ or $x>1$

b) $0<x<1$ **c)** $x=0$ or $x=1$ **d)** Not true for $0\le x\le1$ **131. a)** The product of an even number of negative factors is positive.   **b)** The product of an odd number of negative factors is negative.   **133. a)** Yes   **b)** Yes, because $x^{-2}=\dfrac{1}{x^2}$ and

$(-x)^{-2}=\dfrac{1}{(-x)^2}=\dfrac{1}{x^2}$ **135.** -3, because $\left(\dfrac{y^{-2}}{y^{-3}}\right)^2=y^2$ **137.** $-1,3$, because $\left(\dfrac{x^{-1}}{x^4}\right)^{-1}=x^5$, and $\left(\dfrac{y^5}{y^3}\right)^{-1}=\dfrac{1}{y^2}$ **139.** $x^{9/8}$

141. $\dfrac{1}{x^{9/2}y^{19/6}}$ **144. a)** $A\cup B=\{1,2,3,4,5,6,9\}$ **b)** $A\cap B=\{\,\}$ **145.** ◄━●━━━○►   **146.** -4 **147.** -5
                                                     -3 2

Exercise Set 1.6   **1.** Scientific notation   **3.** 3   **5.** 3.7×10^3 **7.** 4.3×10^{-2} **9.** 7.6×10^5 **11.** 1.86×10^{-6} **13.** 5.78×10^6
15. 1.06×10^{-4} **17.** 31,000   **19.** 0.0000213   **21.** 0.917   **23.** 3,000,000   **25.** 203,000   **27.** 1,000,000   **29.** 240,000,000   **31.** 0.021
33. 0.000027   **35.** 11,480   **37.** 0.0003   **39.** 0.0000006734   **41.** 1.5×10^{-5} **43.** 5.0×10^3 **45.** 3.0×10^{-8} **47.** 1.645×10^{12}
49. 9.6×10^5 **51.** 3.0×10^0 **53.** 9.369×10^{14} **55.** 1.056×10^3 **57.** 5.337×10^2 **59.** 3.115×10^{-25} **61.** 3.802×10^{-27}
63. 3.333×10^{60} **65.** 8.5×10^8 **67.** 2.7×10^6 **69.** 6.2×10^{10} **71.** 9.5×10^{12} **73.** 1.0×10^{-4} **75.** 1.58×10^{-5} **77.** 1.0×10^{-9}
79. a) Subtract 1 from the exponent.   **b)** Subtract 2 from the exponent.   **c)** Subtract 6 from the exponent.   **d)** 6.58×10^{-10}
81. a) 1.0×10^4 or 10,000   **b)** 4.725×10^5 or 472,500   **c)** The error in part **b)** because the answer is off by more.   **83.** 30,000 hours
85. a) $\approx6.406\times10^9$ people   **b)** $\approx4.539\%$ **87. a)** 1.1750×10^{13}, 3.022×10^8 **b)** $\approx\$38,881.54$ **89.** 135 people/square kilometer

---

*Numbers in blue brackets after the answers indicate the sections where the material was discussed.

91. a) 2.1×10^8 pounds   **b)** 3.99×10^9 pounds   **93. a)** 1020 million   **b)** $\approx 20.02\%$ **c)** ≈ 357 people/square mile
d) ≈ 83.1 people/square mile   **95. a)** $\$5.31 \times 10^{10}$ **b)** $\$9.588 \times 10^{11}$ **c)** $\$2.4426 \times 10^{12}$
97. a) 6.03×10^7 square kilometers   **b)** 4.4×10^6 square kilometers

Chapter 1 Review Exercises
1. $\{4, 5, 6, 7, 8, 9\}$ **2.** $[0, 3, 6, 9, \ldots]$ **3.** Yes   **4.** Yes   **5.** No   **6.** Yes   **7.** $4, 6$ **8.** $4, 6, 0$
9. $-2, 4, 6, 0$ **10.** $-2, 4, 6, \dfrac{1}{2}, 0, \dfrac{15}{27}, -\dfrac{1}{5}, 1.47$ **11.** $\sqrt{7}, \sqrt{3}$ **12.** $-2, 4, 6, \dfrac{1}{2}, \sqrt{7}, \sqrt{3}, 0, \dfrac{15}{27}, -\dfrac{1}{5}, 1.47$ **13.** False   **14.** True   **15.** True
16. True   **17.** $A \cup B = \{1, 2, 3, 4, 5, 6, 8, 10\}; A \cap B = \{2, 4, 6\}$ **18.** $A \cup B = \{2, 3, 4, 5, 6, 7, 8, 9\}; A \cap B = \{\ \}$
19. $A \cup B = \{1, 2, 3, 4, \ldots\}; A \cap B = \{\ \}$ **20.** $A \cup B = \{3, 4, 5, 6, 9, 10, 11, 12\}; A \cap B = \{9, 10\}$ **21.** ←⊕——→ 5
22. ←●——→ -2 **23.** ←⊕——●→ -1.3 2.4 **24.** ←+●●●●+→ $0\ 1\ 2\ 3\ 4$ **25.** $<$ **26.** $<$ **27.** $<$ **28.** $=$ **29.** $<$ **30.** $>$ **31.** $>$ **32.** $>$
33. $-\pi, -3, 3, \pi$ **34.** $0, \dfrac{3}{5}, 2.7, |-3|$ **35.** $-2, 3, |-5|, |-10|$ **36.** $-7, -3, |-3|, |-7|$ **37.** $-4, -|-3|, 5, 6$ **38.** $-2, 0, |1.6|, |-2.3|$
39. Distributive property   **40.** Commutative property of multiplication   **41.** Associative property of addition
42. Identity property of addition   **43.** Associative property of multiplication   **44.** Double negative property
45. Multiplicative property of zero   **46.** Inverse property of addition   **47.** Inverse property of multiplication   **48.** Identity property
of multiplication   **49.** 11   **50.** 9   **51.** 12   **52.** -3 **53.** 2   **54.** 21   **55.** 9   **56.** -59 **57.** 15   **58.** 42   **59.** 4   **60.** 64
61. Undefined   **62.** $\dfrac{8}{3}$ **63.** 22   **64.** -67 **65. a)** $\$816.37$ million   **b)** $\$7,223.73$ million   **66. a)** 944.53 ton-miles   **b)** 2135.65 ton-
miles   **67.** 32   **68.** x^5 **69.** a^8 **70.** y^7 **71.** b^9 **72.** $\dfrac{1}{c^3}$ **73.** $\dfrac{1}{125}$ **74.** 8   **75.** $81m^6$ **76.** $\dfrac{7}{5}$ **77.** $\dfrac{27}{8}$ **78.** $\dfrac{y^2}{x}$ **79.** $-15x^3y^4$
80. $\dfrac{14}{v^3w^3}$ **81.** $\dfrac{3y^7}{x^5}$ **82.** $\dfrac{3}{xy^9}$ **83.** $\dfrac{g^5}{h^5j^{14}}$ **84.** $\dfrac{3m}{n^4}$ **85.** $64a^3b^3$ **86.** $\dfrac{x^{10}}{9y^2}$ **87.** $\dfrac{p^{14}}{q^{12}}$ **88.** $-\dfrac{8a^3}{b^9c^6}$ **89.** $\dfrac{z^4}{25x^2y^6}$ **90.** $\dfrac{m^9}{27}$ **91.** $\dfrac{n^6}{4m^4}$
92. $\dfrac{625x^4y^4}{z^{20}}$ **93.** $\dfrac{9x^{10}}{4y^{14}z^{12}}$ **94.** $-\dfrac{x^6z^2}{10y^2}$ **95.** 7.42×10^{-5} **96.** 4.6×10^5 **97.** 1.83×10^5 **98.** 2.0×10^{-6} **99.** 30,000   **100.** 0.03
101. 300,000,000   **102.** 2000   **103. a)** 2.8×10^7 **b)** 7.45×10^8 **c)** 2.548×10^9 **104. a)** 14,000,000,000   **b)** 14 billion kilometers
c) 5.0×10^8 or 500,000,000 kilometers   **d)** 8.4×10^9 or 8,400,000,000 miles

Chapter 1 Practice Test
1. $A = \{6, 7, 8, 9, \ldots\}$ [1.2]   **2.** False [1.2]   **3.** True [1.2]   **4.** $-\dfrac{3}{5}, 2, -4, 0, \dfrac{19}{12}, 2.57, -1.92$ [1.2]
5. $-\dfrac{3}{5}, 2, -4, 0, \dfrac{19}{12}, 2.57, \sqrt{8}, \sqrt{2}, -1.92$ [1.2]   **6.** $A \cup B = \{5, 7, 8, 9, 10, 11, 14\}; A \cap B = \{8, 10\}$ [1.2]
7. $A \cup B = \{1, 3, 5, 7, \ldots\}; A \cap B = \{3, 5, 7, 9, 11\}$ [1.2]   **8.** ←●——⊕→ -2.3 5.2 [1.2]   **9.** ←●●●●→ -2 -1 0 1 [1.2]   **10.** $-|4|, -2, |3|, 9$ [1.3]
11. Associative property of addition [1.3]   **12.** Commutative property of addition [1.3]   **13.** 2 [1.4]   **14.** 33 [1.4]
15. Undefined [1.4]   **16.** $-\dfrac{37}{22}$ [1.4]   **17.** 17 [1.4]   **18. a)** 304 feet   **b)** 400 feet [1.4]   **19.** $\dfrac{1}{9}$ [1.5]   **20.** $\dfrac{16}{m^6n^4}$ [1.5]   **21.** $\dfrac{4c^2}{5ab^5}$ [1.5]
22. $-\dfrac{y^{21}}{27x^{12}}$ [1.5]   **23.** 3.89×10^8 [1.6]   **24.** 260,000,000 [1.6]   **25. a)** 9.2×10^9 **b)** 0–14: 1.794×10^9, 15–64: 5.8052×10^9,
65 and older: 1.6008×10^9 [1.6]

Chapter 2

Exercise Set 2.1   **1.** Terms   **3.** Isolate   **5.** Identity   **7.** Contradiction   **9.** ∅   **11.** Symmetric property   **13.** Transitive prop-
erty   **15.** Reflexive property   **17.** Addition property of equality   **19.** Multiplication property of equality   **21.** Multiplication prop-
erty of equality   **23.** Addition property   **25.** One   **27.** Three   **29.** Two   **31.** Zero   **33.** One   **35.** Seven   **37.** Twelve
39. Cannot be simplified   **41.** $-2x^2 + 2x - 3$ **43.** $8.7c^2 + 3.6c$ **45.** Cannot be simplified   **47.** $-pq + p + q$ **49.** $8d + 2$
51. $\dfrac{8}{3}x + \dfrac{13}{2}$ **53.** $-17x - 4$ **55.** $11x - 6y$ **57.** $-9b + 93$ **59.** $4r^2 - 2rs + 3r + 4s$ **61.** 3   **63.** $\dfrac{15}{2}$ **65.** 2   **67.** 16   **69.** 5
71. $\dfrac{3}{5}$ **73.** 1   **75.** 0   **77.** 3   **79.** -1 **81.** 5   **83.** 5   **85.** -1 **87.** $-\dfrac{1}{2}$ **89.** 6   **91.** 2   **93.** 68   **95.** -35 **97.** -4 **99.** 24   **101.** 10
103. -4 **105.** $\dfrac{15}{16}$ **107.** 5   **109.** 1.00   **111.** 1.18   **113.** 0.43   **115.** 1701.39   **117.** -1.85 **119.** ∅; contradiction
121. $\left\{-\dfrac{4}{3}\right\}$; conditional   **123.** $\mathbb{R}$; identity   **125.** $\mathbb{R}$; identity   **127.** ∅; contradiction   **129. a)** ≈ 85 people per square mile
b) ≈ 2026 **131. a)** $\$3$ trillion   **b)** 2014   **133.** $\Delta = \dfrac{\odot + \square}{*}$ **135.** $\odot = \dfrac{\otimes - \triangle}{\square}$ **137.** Answers will vary. One possible answer:
$x = \dfrac{5}{2}, 2x - 4 = 1, 4x = 10$ **139.** Answers will vary. One possible answer: $2x - 4 = 5x - 3(1 + x)$ **141.** Answers will vary. One
possible answer: $3p + 3 = \dfrac{3}{2}p + p + 6$ **143.** -22, substitute -2 for a and solve for n.   **145. a)** Answers will vary.
b) $|a| = \begin{cases} a \text{ if } a \geq 0 \\ -a \text{ if } a < 0 \end{cases}$ **146. a)** -9 **b)** 9   **147.** -5 **148.** $\dfrac{4}{49}$

Exercise Set 2.2　**1.** Mathematical model　**3.** Translate　**5.** Check　**7.** 300　**9.** 300　**11.** 201.06　**13.** 70　**15.** 176　**17.** $\dfrac{7}{4}$
19. 66.67　**21.** 4　**23.** 119.10　**25.** $y = -3x + 5$　**27.** $y = -\dfrac{3}{2}x + 3$　**29.** $y = 3x - 8$　**31.** $y = \dfrac{3}{4}x - 5$　**33.** $y = x + 2$

35. $y = -\dfrac{4}{3}x + 11$　**37.** $I = \dfrac{E}{R}$　**39.** $d = \dfrac{C}{\pi}$　**41.** $l = \dfrac{P - 2w}{2}$　**43.** $h = \dfrac{V}{lw}$　**45.** $r = \dfrac{A - P}{Pt}$　**47.** $l = \dfrac{3V}{wh}$　**49.** $m = \dfrac{y - b}{x}$

51. $m = \dfrac{y - y_1}{x - x_1}$　**53.** $\mu = x - z\sigma$　**55.** $T_2 = \dfrac{T_1 P_2}{P_1}$　**57.** $h = \dfrac{2A}{b_1 + b_2}$　**59.** $n = \dfrac{2S}{f + l}$　**61.** $F = \dfrac{9}{5}C + 32$　**63.** $m_1 = \dfrac{Fd^2}{km_2}$

65. a) $e = 0.68d$　**b)** $d = \dfrac{e}{0.68}$ or $d \approx 1.47e$　**67.** \$308　**69.** 6.5 years　**71. a)** 3.14 square inches　**b)** 78.54 square inches

73. a) 75 cubic feet　**b)** 2.78 cubic yards　**c)** \$105　**75.** The cylinder, difference is 0.22 cubic inch　**77.** \$11,264.93　**79.** \$4958.41
81. $\approx 4.12\%$　**83. a)** $\approx 7.08\%$　**b)** $\approx 6.39\%$　**85. a)** 4 pounds per week　**b)** 2500 calories　**87. a)** $S = 100 - a$　**b)** 40%

89. a) $s = \dfrac{rt^2}{u}$　**b)** $u = \dfrac{rt^2}{s}$　**90.** -40　**91.** 1　**92.** -125　**93.** $\dfrac{4}{3}$

Exercise Set 2.3　**1.** $x + 3$　**3.** $7 - x$　**5.** Less than　**7.** $19.95y$　**9.** $11n - 7.5$　**11.** $x, 12 - x$　**13.** $w, w + 29$　**15.** $p, 165 - p$
17. $z, z + 1.3$　**19.** $e, e + 0.22e$　**21.** $A = 72°, B = 18°$　**23.** $A = 36°, B = 144°$　**25.** $40°, 60°, 80°$　**27.** \$32　**29.** 25 rides　**31.** 225
miles　**33.** 13 times　**35.** 10 times　**37.** \$1600　**39.** Northwest: \$2.145 million, Southeast: \$2.455 million　**41.** 4 grams　**43.** \$6.55
per hour　**45.** grasses: 12, weeds: 19, trees: 26　**47.** \$16.25　**49. a)** ≈ 63.49 months or 5.29 years　**b)** First National　**51. a)** ≈ 28
months or 2.33 years　**b)** Yes　**53.** Phelps: 8, Coughlin: 6, Lochte: 4, Grevers: 3　**55.** animals: 250,000, plants: 350,000, nonbeetle
insects: 540,000, beetles: 360,000　**57.** 9 inches, 12 inches, 15 inches　**59.** 10 feet, 24 feet, 26 feet　**61.** 13 meters by 13 meters

63. 3 feet by 6 feet　**65.** \$60　**67.** 3　**69.** \$16　**71. a)** $\dfrac{88 + 92 + 97 + 96 + x}{5} = 90$　**b)** Answer will vary.　**c)** 77

73. a), b) Answers will vary.　**75.** 220 miles　**78.** $\dfrac{32}{5}$　**79.** -2.7　**80.** $\dfrac{5}{32}$　**81.** -10　**82.** $\dfrac{y^{18}}{8x^{12}}$

Mid-Chapter Test　**1.** 12 [2.1]　**2.** $5x^2 - 2x - 11$ [2.1]　**3.** $6.4a - 9.6$ [2.1]　**4.** -6 [2.1]　**5.** 14 [2.1]　**6.** $-\dfrac{11}{3}$ [2.1]　**7.** 5 [2.1]

8. $\mathbb{R}$, identity [2.1]　**9.** $\varnothing$, contradiction [2.1]　**10.** 80 [2.2]　**11.** $\dfrac{100}{3}$ [2.2]　**12.** $x = \dfrac{y - 13}{7}$ [2.2]　**13.** $x_3 = nA - 2x_1 - x_2$ [2.2]
14. \$942.80 [2.2]　**15.** $A = 62°, B = 28°$ [2.3]　**16.** 10 days [2.3]　**17.** 20 ft, 40 ft, 40 ft [2.3]　**18.** 4.5% [2.3]　**19.** 40 months
[2.3]　**20.** Multiply both sides by the same number, 12; $-\dfrac{10}{3}$ [2.3]

Exercise Set 2.4　**1.** 11.4 miles　**3.** 4 hours　**5.** 6 hours　**7. a)** 6 miles per hour　**b)** 12 miles per hour　**9. a)** 0.15 hour or 9
minutes　**b)** 3.6 miles　**11.** 13.8 hours　**13.** 3.5 hours　**15.** \$12,570 at 3%, \$17,430 at 4.1%　**17.** 54 pounds　**19.** 4 basset hounds
and 8 black cats　**21.** 30 ounces　**23.** 2.8 teaspoons 30%, 1.2 teaspoons 80%　**25.** 35%　**27.** 4 pounds leaves, 8 pounds slices

29. ≈ 25.77 hours　**31.** 500 minutes or $8\dfrac{1}{3}$ hours　**33.** 6 quarts　**35. a)** ≈ 3.71 hours　**b)** ≈ 2971.43 miles　**37.** 8 small, 4 large

paintings　**39.** 9.6 ounces of 80% solution, 118.4 ounces of water　**41. a)** 248 tickets　**b)** 3002 tickets　**43.** 3 miles

45. ≈ 11.4 ounces　**47. a), b), c)** Answers will vary.　**49.** ≈ 149 miles　**51.** 6 quarts　**52.** 6.0×10^{-4}　**53.** -5.7　**54.** $\dfrac{21}{4}$
55. $y = \dfrac{x - 42}{30}$　**56.** 300 miles

Exercise Set 2.5　**1.** Compound　**3.** Closed　**5.** Union　**7. a)** 　**b)** $(-3, \infty)$　**c)** $\{x \mid x > -3\}$

9. a) 　**b)** $(-\infty, \pi]$　**c)** $\{w \mid w \le \pi\}$　**11. a)** 　**b)** $\left(-3, \dfrac{4}{5}\right]$　**c)** $\left\{q \mid -3 < q \le \dfrac{4}{5}\right\}$

13. a) 　**b)** $(-7, -4]$　**c)** $\{x \mid -7 < x \le -4\}$　**15.** 　**17.** 　**19.**

21. 　**23.** 　**25.** 　**27.** $\left(-\infty, \dfrac{3}{2}\right)$　**29.** $[2, \infty)$　**31.** $\left(-\infty, \dfrac{3}{2}\right]$　**33.** $(-\infty, \infty)$

35. $[-5, 1)$　**37.** $[-4, 5]$　**39.** $\left[4, \dfrac{11}{2}\right)$　**41.** $\left(-\dfrac{13}{3}, -4\right]$　**43.** $\{x \mid 3 \le x < 7\}$　**45.** $\{x \mid 0 < x \le 3\}$　**47.** $\left\{u \mid 4 \le u \le \dfrac{19}{3}\right\}$
49. $\{c \mid -3 < c \le 1\}$　**51.** $\varnothing$　**53.** $\{x \mid -5 < x < 2\}$　**55.** $(-\infty, 2) \cup [7, \infty)$　**57.** $[0, 2]$　**59.** $(-\infty, 0) \cup (6, \infty)$　**61.** $[0, \infty)$
63. a) $l + g \le 130$　**b)** $l + 2w + 2d \le 130$　**c)** 24.5 inches　**65.** 11 boxes　**67.** 300 text messages　**69.** 1881 books
71. 12 ounces　**73.** For sales over \$5,000 per week　**75.** 24　**77.** $76 \le x \le 100$　**79. a)** \$12,242.75　**b)** \$78,135.89　**81. a)** $[0, 3)$
b) $(3, 10]$　**83. a)** $[0, 5)$　**b)** $(5, 13]$　**85. a)** $[0, 8]$　**b)** None　**87.** $6.97 < x < 8.77$　**89.** Answers will vary.
91. a) $[17.5, 23.5]$　**b)** $[23.5, 31]$　**c)** $[27.2, 36.5]$　**93.** $84 \le x \le 100$　**95. a)** Answers will vary.　**b)** $(-3, \infty)$

97. a) $A \cup B = \{1, 2, 3, 4, 5, 6, 8, 9\}$　**b)** $A \cap B = \{1, 8\}$　**98. a)** 4　**b)** 0, 4　**c)** $-3, 4, \dfrac{5}{2}, 0, -\dfrac{13}{29}$　**d)** $-3, 4, \dfrac{5}{2}, \sqrt{7}, 0, -\dfrac{13}{29}$

99. Associative property of addition　**100.** Commutative property of addition　**101.** $V = \dfrac{R - L + Dr}{r}$

Exercise Set 2.6   **1.** $|x| \le 4$ **3.** $|x| = 4$ **5.** $|x| > 4$ **7.** $|x| \ge 5$ **9.** $|x| < 5$ **11.** $|x| > -6$ **13.** $\{-7, 7\}$ **15.** $\left\{-\frac{1}{2}, \frac{1}{2}\right\}$

17. $\varnothing$ **19.** $\{-13, 3\}$ **21.** $\{-7\}$ **23.** $\left\{\frac{3}{2}, \frac{11}{6}\right\}$ **25.** $\{-17, 23\}$ **27.** $\{3\}$ **29.** $\varnothing$ **31.** $\{w | -11 < w < 11\}$ **33.** $\{q | -13 \le q \le 3\}$

35. $\{b | 1 < b < 5\}$ **37.** $\{x | -9 \le x \le 6\}$ **39.** $\left\{x \Big| \frac{1}{3} < x < \frac{13}{3}\right\}$ **41.** $\varnothing$ **43.** $\{j | -22 < j < 6\}$ **45.** $\{x | -1 \le x \le 7\}$

47. $\{y | y < -8 \text{ or } y > 8\}$ **49.** $\{x | x < -9 \text{ or } x > 1\}$ **51.** $\left\{b \Big| b < \frac{2}{3} \text{ or } b > 4\right\}$ **53.** $\{h | h < 1 \text{ or } h > 4\}$ **55.** $\{x | x < 2 \text{ or } x > 6\}$

57. $\{x | x \le -18 \text{ or } x \ge 2\}$ **59.** $\mathbb{R}$ **61.** $\{x | x < 2 \text{ or } x > 2\}$ **63.** $\{-1, 15\}$ **65.** $\{-3, 1\}$ **67.** $\left\{-23, \frac{13}{7}\right\}$ **69.** $\{10\}$ **71.** $\{-9, 9\}$

73. $\{q | q < -8 \text{ or } q > -4\}$ **75.** $\{w | -1 \le w \le 8\}$ **77.** $\left\{-\frac{8}{5}, 2\right\}$ **79.** $\left\{x \Big| x < -\frac{5}{2} \text{ or } x > -\frac{5}{2}\right\}$ **81.** $\left\{x \Big| -\frac{13}{3} \le x \le \frac{5}{3}\right\}$ **83.** $\varnothing$

85. $\{w | -16 < w < 8\}$ **87.** $\mathbb{R}$ **89.** $\left\{2, \frac{22}{3}\right\}$ **91.** $\left\{-\frac{3}{2}, \frac{9}{7}\right\}$ **93. a)** $[0.085, 0.093]$ **b)** 0.085 inch   **c)** 0.093 inch

95. a) $[132, 188]$ **b)** 132 to 188 feet below sea level, inclusive   **97. a)** Two   **b)** Infinite number   **c)** Infinite number

99. a) None   **b)** One   **c)** Two   **101.** $x = -\frac{b}{a}$; $|ax + b|$ is never less than 0, so set $|ax + b| = 0$ and solve for x.

103. a) Set $ax + b = -c$ or $ax + b = c$ and solve each equation for x.   **b)** $x = \frac{-c - b}{a}$ or $x = \frac{c - b}{a}$

105. a) Write $ax + b < -c$ or $ax + b > c$ and solve each inequality for x.   **b)** $x < \frac{-c - b}{a}$ or $x > \frac{c - b}{a}$

107. $\mathbb{R}$; Since $4 - x = -(x - 4)$ **109.** $\{x | x \ge 0\}$; by definition of absolute value   **111.** $\{2\}$; set $x + 1 = 2x - 1$ or $x + 1 = -(2x - 1)$

113. $\{x | x \le 4\}$; by definition $|x - 4| = -(x - 4)$ if $x \le 4$ **115.** $\{4\}$ **117.** $\varnothing$ **119.** $\frac{29}{72}$ **120.** 25   **121.** ≈ 1.33 miles   **122.** $\{x | x \le 3\}$

Chapter 2 Review Exercises   **1.** Eight   **2.** One   **3.** Seven   **4.** $5z + 13$ **5.** $7x^2 + 2xy - 13$ **6.** Cannot be simplified

7. $4x - 3y + 10$ **8.** -2 **9.** 20   **10.** $-\frac{13}{3}$ **11.** -10 **12.** $-\frac{9}{2}$ **13.** No solution   **14.** $\mathbb{R}$ **15.** 5   **16.** $\frac{1}{4}$ **17.** 69   **18.** -4

19. $t = \frac{D}{r}$ **20.** $w = \frac{P - 2l}{2}$ **21.** $h = \frac{A}{\pi r^2}$ **22.** $h = \frac{2A}{b}$ **23.** $m = \frac{y - b}{x}$ **24.** $y = \frac{2x - 5}{3}$ **25.** $R_2 = R_T - R_1 - R_3$

26. $a = \frac{2S - b}{3}$ **27.** $l = \frac{K - 2d}{2}$ **28.** \$700   **29.** 7 years   **30.** \$6800   **31.** 150 miles   **32.** \$260   **33.** \$2570 at 3.5%, \$2430 at 4%

34. 187.5 gallons of 20%, 62.5 gallons of 60%   **35.** $6\frac{1}{2}$ hours   **36. a)** 3000 miles per hour   **b)** 16,500 miles   **37.** 15 pounds of \$6.00 coffee; 25 pounds of \$6.80 coffee   **38.** \$36   **39. a)** 1 hour   **b)** 14.4 miles   **40.** $40°, 65°, 75°$ **41.** 300 gallons per hour, 450 gallons per hour   **42.** $40°, 50°$ **43.** 7.5 ounces   **44.** \$4500 at 10%, \$7500 at 6%   **45.** More than 5   **46.** 40 miles per hour, 50 miles per hour   **47.**   **48.**   **49.**   **50.**   **51.**

52.   **53.**   **54.**   **55.** 6 boxes   **56.** 8 hours   **57.** ≈ 15.67 weeks   **58.** $\{x | 81 \le x \le 100\}$

59. $(3, 8)$ **60.** $(-3, 5]$ **61.** $\left(\frac{7}{2}, 8\right)$ **62.** $\left(\frac{8}{3}, 6\right)$ **63.** $(-3, 1]$ **64.** $(2, 14)$ **65.** $\{x | 2 < x \le 3\}$ **66.** $\mathbb{R}$ **67.** $\{x | x \le -4\}$

68. $\{g | g < -6 \text{ or } g \ge 11\}$ **69.** $\{-4, 4\}$ **70.** $\{x | -8 < x < 8\}$ **71.** $\{x | x \le -9 \text{ or } x \ge 9\}$ **72.** $\{-18, 8\}$ **73.** $\{x | x \le -3 \text{ or } x \ge 7\}$

74. $\left\{-\frac{1}{2}, \frac{9}{2}\right\}$ **75.** $\{q | 1 < q < 8\}$ **76.** $\{-1, 4\}$ **77.** $\{x | -14 < x < 22\}$ **78.** $\left\{-5, -\frac{4}{5}\right\}$ **79.** $\mathbb{R}$ **80.** $(-\infty, -1] \cup [4, \infty)$

81. $(4, 8]$ **82.** $\left(-\frac{17}{2}, \frac{27}{2}\right]$ **83.** $[-2, 6)$ **84.** $(-\infty, \infty)$ **85.** $\left(\frac{2}{3}, 10\right]$

Chapter 2 Practice Test   **1.** Seven [2.1]   **2.** $16p - 3q - 4pq$ [2.1]   **3.** $10q + 42$ [2.1]   **4.** -26 [2.1]   **5.** $\frac{4}{3}$ [2.1]   **6.** $-\frac{35}{11}$ [2.1]

7. $\varnothing$ [2.1]   **8.** $\mathbb{R}$ [2.1]   **9.** $\frac{13}{3}$ [2.2]   **10.** $b = \frac{a - 2c}{5}$ [2.2]   **11.** $b_2 = \frac{2A - hb_1}{h}$ [2.2]   **12.** \$625 [2.3–2.4]   **13.** 80 visits [2.3–2.4]

14. 4.2 hours [2.3–2.4]   **15.** 6.25 liters [2.3–2.4]   **16.** \$7000 at 8%, \$5000 at 7% [2.3–2.4]   **17.** [2.5]

18. [2.5]   **19.** $\left(\frac{9}{2}, 7\right]$ [2.5]   **20.** $[13, 16)$ [2.6]   **21.** $\{-7, 2\}$ [2.6]   **22.** $\left\{-\frac{14}{3}, \frac{26}{5}\right\}$ [2.6]   **23.** $\{-3\}$ [2.6]

24. $\{x | x < -1 \text{ or } x > 4\}$ [2.6]   **25.** $\left\{x \Big| \frac{1}{2} \le x \le \frac{5}{2}\right\}$ [2.6]

Cumulative Review Test   1. a) $\{1, 2, 3, 5, 7, 9, 11, 13, 15\}$ **b)** $\{3, 5, 7, 11, 13\}$ [1.2]   **2. a)** Commutative property of addition
b) Associative property of multiplication   **c)** Distributive property [1.3]   **3.** -63 [1.4]   **4.** -6 [1.4]   **5.** 7 [1.4]   **6.** $\dfrac{1}{25x^8 y^6}$ [1.5]

7. $\dfrac{16m^{10}}{n^{12}}$ [1.5]   **8.** ≈ 545.8 times [1.6]   **9.** 5 [2.1]   **10.** 1.15 [2.1]   **11.** $\dfrac{3}{4}$ [2.1]   **12.** A conditional linear equation is true for only one
value; a linear equation that is an identity is always true; a linear equation that is a contradiction is never true. [2.1]   **13.** 3 [2.2]

14. $x = \dfrac{y - y_1 + mx_1}{m}$ [2.2]   **15. a)** 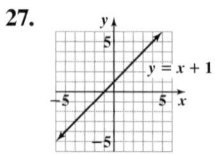   **b)** $\left\{ x \mid -2 < x < \dfrac{8}{5} \right\}$ **c)** $\left(-2, \dfrac{8}{5} \right)$ [2.5]   **16.** $\left\{ -\dfrac{7}{3}, 3 \right\}$ [2.6]

17. $\{x \mid x \le -10 \text{ or } x \ge 14\}$ [2.6]   **18.** \$35 [2.3]   **19.** 40 miles per hour, 60 miles per hour [2.4]   **20.** Cashews: 15 pounds, peanuts:
25 pounds [2.4]

Chapter 3

Exercise Set 3.1   **1.** Ordered pair   **3.** Collinear

5. $A(3, 2), B(-6, 0), C(2, -4), D(-2, -4), E(0, 3), F(-8, 1), G\left(\dfrac{3}{2}, -1 \right) H(2, 3)$ **7**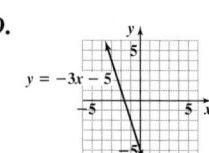

9. I   **11.** IV   **13.** II   **15.** III   **17.** Yes   **19.** No   **21.** Yes   **23.** Yes   **25.** No

27. 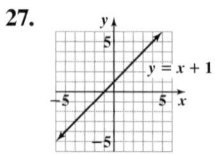   **29.**   **31.** 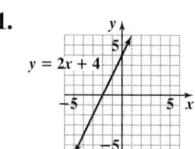   **33.** 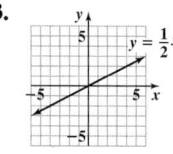   **35.**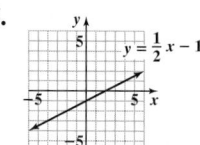

37. 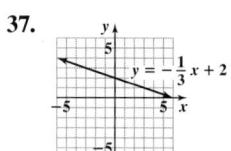   **39.** 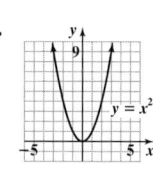   **41.** 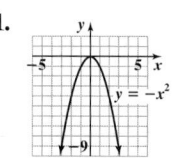   **43.** 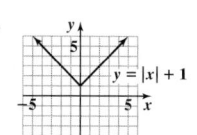   **45.**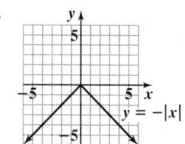

47. 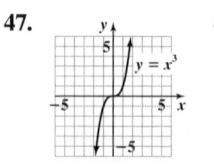   **49.** 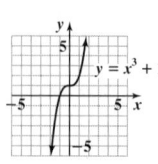   **51.** 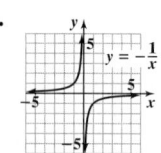   **53.** 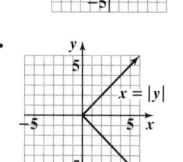   **55.**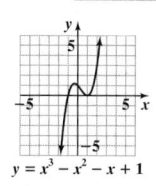

57. 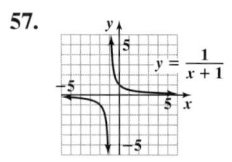   **59.** 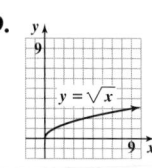   **61.** 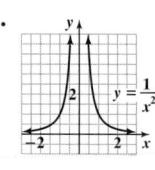   **63.** Yes, the coordinates satisfy the equation

65. a) 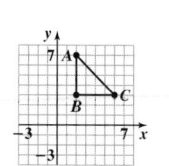   **b)** 8 square units

67. a) About 800,000 passengers   **b)** About 667,000 passengers   **c)** 1890–1894, 1900–1904, 1905–1909, 1910–1914   **d)** No

69. a   **71.** c   **73.** b   **75.** d   **77.** b   **79.** d   **81.** b   **83.** d

85. a) 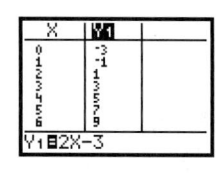   **b)** 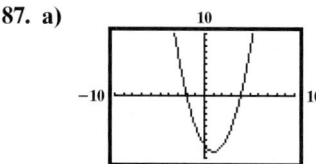   **87. a)** 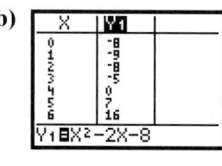   **b)**

89. a) 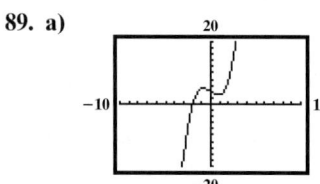   **b)** 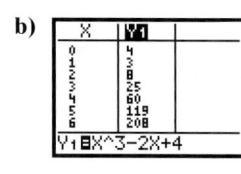   **91.** 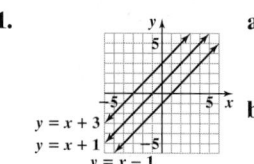   **a)** Each graph crosses the y-axis at the point corresponding to the constant term in the graph's equation.   **b)** Yes

93. The rate of change is 2.   **95.** The rate of change is 3.   **97.** $(4, -3), (5, 1)$, other answers possible.

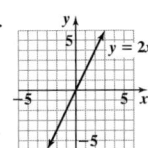   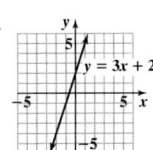

99. 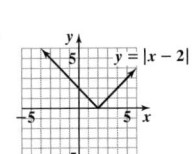   **103.** $\dfrac{3}{2}$ **104.** ≈ 71 miles   **105.** $\{x | -2 < x \le 2\}$ **106.** $\left\{ x \mid x < -3 \text{ or } x > \dfrac{5}{3} \right\}$

Exercise Set 3.2   1. Domain   **3.** Ordered pairs   **5.** Vertical line   **7.** Range   **9.** Function   **11.** Independent
13. a) Function   **b)** Domain: {Idaho, Texas, Georgia} Range: {15, 16, 18}   **15. a)** Function   **b)** Domain: {3, 5, 11},
Range: {6, 10, 22}   **17. a)** Function   **b)** Domain: {Cameron, Tyrone, Vishnu}, Range: {3, 6}   **19. a)** Not a function
b) Domain: {1990, 2001, 2002}; Range: {20, 34, 37}   **21. a)** Function   **b)** Domain: {1, 2, 3, 4, 5}, Range: {1, 2, 3, 4, 5}
23. a) Function   **b)** Domain: {1, 2, 3, 4, 5, 7},   Range: {-1, 0, 2, 4, 9}   **25. a)** Not a function   **b)** Domain: {1, 2, 3},
Range: {1, 2, 4, 5, 6}   **27. a)** Not a function   **b)** Domain: {0, 1, 2}, Range: {-7, -1, 2, 3}   **29. a)** Function   **b)** Domain: $\mathbb{R}$, Range: $\mathbb{R}$
c) 2   **31. a)** Not a function   **b)** Domain: $\{x | 0 \le x \le 2\}$, Range: $\{y | -3 \le y \le 3\}$ **c)** ≈ 1.5 **33. a)** Function   **b)** Domain: $\mathbb{R}$,
Range: $\{y | y \ge 0\}$ **c)** $-3, -1$ **35. a)** Function   **b)** Domain: {-1, 0, 1, 2, 3}, Range: {-1, 0, 1, 2, 3}   **c)** 2   **37. a)** Not a function
b) Domain: $\{x | x \ge 2\}$, Range: $\mathbb{R}$ **c)** 3   **39. a)** Function   **b)** Domain: $\{x | -2 \le x \le 2\}$, Range: $\{y | -1 \le y \le 2\}$ **c)** $-2, 2$
41. a) 5 **b)** -11 **43. a)** -6 **b)** -4 **45. a)** 2 **b)** 2   **47. a)** 7 **b)** 0   **49. a)** 0 **b)** 3   **51. a)** 1   **b)** Undefined
53. a) 16 square feet   **b)** 26 square feet   **55. a)** $A(r) = \pi r^2$ **b)** ≈ 452.4 square yards   **57. a)** $C(F) = \dfrac{5}{9}(F - 32)$
b) $-35°C$ **59. a)** $18.23°C$ **b)** $27.68°C$ **61. a)** $78.32°$ **b)** $73.04°$ **63. a)** 91 oranges   **b)** 204 oranges
65. a) 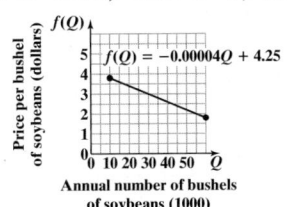   **b)** \$2.65 per bushel

67. Answers will vary. One possible interpretation: The person warms up slowly, possibly by walking, for 5 minutes. Then the person begins jogging slowly over a period of 5 minutes. For the next 15 minutes the person jogs. For the next 5 minutes the person walks slowly and his heart rate decreases to his normal heart rate. The rate stays the same for the next 5 minutes.
69. Answers will vary. One possible interpretation: The man walks on level ground, about 30 feet above sea level, for 5 minutes. For the next 5 minutes he walks uphill to 45 feet above sea level. For 5 minutes he walks on level ground, then walks quickly downhill for 3 minutes to an elevation of 20 feet above sea level. For 7 minutes he walks on level ground. Then he walks quickly uphill for 5 minutes.
71. Answers will vary. One possible interpretation: Driver is in stop-and-go traffic, then gets on highway for about 15 minutes, then drives on a country road for a few minutes, then stops car for a couple of minutes, then stop-and-go traffic.
73. a) Yes   **b)** Year   **c)** $\approx \$22,000$ **d)** $\approx \$6,000$ **e)** $\approx 100\%$
75. a) 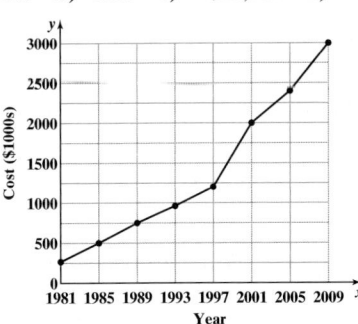 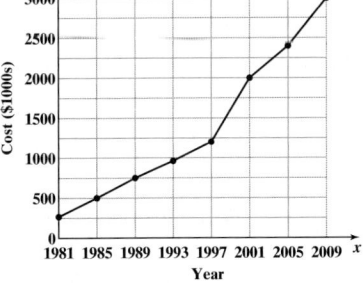 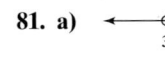   **b)** No. It is not a straight line.   **c)** \$2,300,000

79. $\dfrac{1}{2}$ **80.** $p_2 = \dfrac{E - a_1 p_1 - a_3 p_3}{a_2}$ **81. a)** 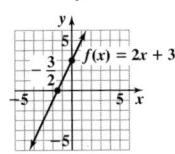   **b)** $(3, \infty)$ **c)** $\{x | x > 3\}$ **82.** $-2, 10$

Exercise Set 3.3   1. Real numbers   **3.** x-intercept.   **5.** Horizontal   **7.** Fail   **9.** Linear.   **11.** $4x + y = 3$
13. $3x - 4y = -14$
15. 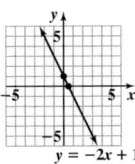   **17.** 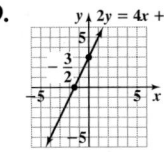   **19.** 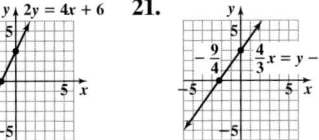   **21.** 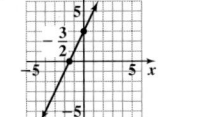   **23.**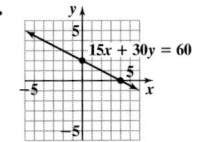

25. 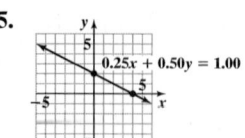   **27.** 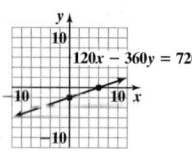   **29.** 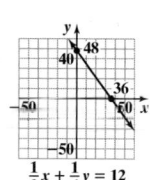   **31.** 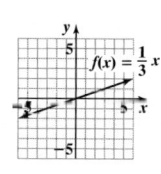   **33.**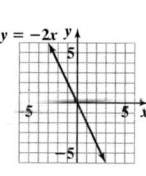

35. 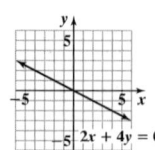   **37.** 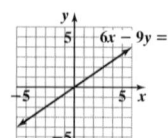   **39.** 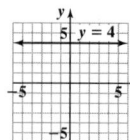   **41.** 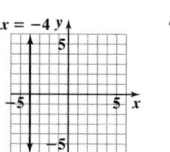   **43.** 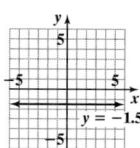   **45.**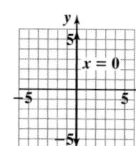

47. 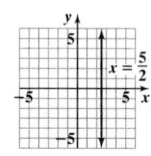   **49.** 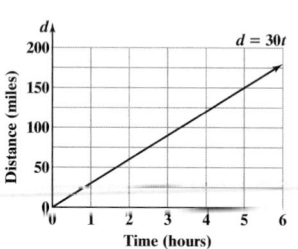   **51. a)** 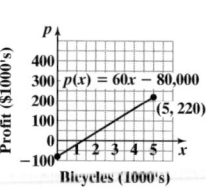   **b)** 1300 bicycles   **c)** 3800 bicycles

53. **a)** $s(x) = 500 + 0.15x$ **b)** 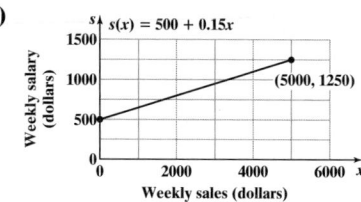   **c)** $950   **d)** $4000

55. a) There is only one y-value for each x-value.
b) Independent: length, Dependent: weight
c) Yes
d) 11.5 kilograms
e) 65 centimeters
f) 12.0–15.5 kilograms
g) Increases; yes, as babies get older their weights vary more.

57. When the graph goes through the origin, because at the origin both x and y are zero.
59. Answers may vary. One example: $f(x) = 4$.   **61.** Both intercepts will be at 0.   **63.** $(-3.2, 0), (0, 6.4)$

65. $(-2, 0), (0, -2.5)$ **69.** 96   **70.** $-\dfrac{18}{13}$ **71. a)** Answers will vary.   **b)** $x = a + b$ or $x = a - b$

72. a) Answers will vary.   **b)** $a - b < x < a + b$ **73. a)** Answers will vary.   **b)** $x < a - b$ or $x > a + b$ **74.** $\{-2, 2\}$

Exercise Set 3.4   **1.** Slope   **3.** Standard form   **5.** Rise, Run   **7.** Negative   **9.** Vertical   **11.** Rate of change   **13.** -2

15. $-\dfrac{1}{2}$ **17.** -1 **19.** Undefined   **21.** 0   **23.** $-\dfrac{2}{3}$ **25.** $j = 3$ **27.** $k = -2$ **29.** $x = 6$ **31.** $r = 0$ **33.** $m = -3, y = -3x$

35. $m = -\dfrac{1}{3}, y = -\dfrac{1}{3}x + 2$ **37.** m is undefined, $x = -2$ **39.** $m = 0, y = 3$ **41.** $m = -\dfrac{3}{2}, y = -\dfrac{3}{2}x + 15$

43. $y = -x + 2, -1, (0, 2)$ 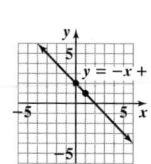   **45.** $y = -\dfrac{1}{3}x + 2, -\dfrac{1}{3}, (0, 2)$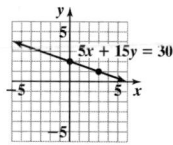

47. $y = \dfrac{5}{2}x + 2, \dfrac{5}{2}, (0, 2)$ 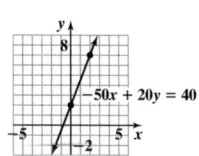   **49.** 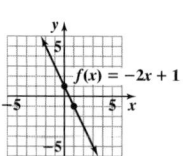   **51.** 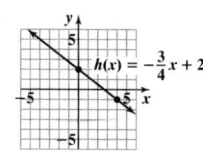   **53. a)** 2   **b)** 4   **c)** 1   **d)** 3

55. If the slopes are the same and the y-intercepts are different, the lines are parallel.
57. $(0, -5)$

59. a) $y = 3x$ **b)** $y = 3x - 2$ **61. a)** 1   **b)** $(0, 4)$ **c)** $y = x + 4$ **63.** $y = \dfrac{3}{2}x - 7$ **65.** 0.2   **67. a)** 11.3   **b)** Positive

c) 7.075   **69. a–b)** 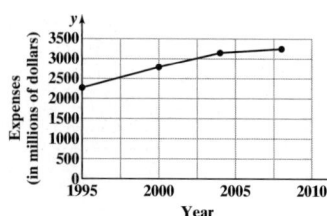   **c)** $123.8, 64.25, 31.75$ **d)** 1995–2000, because its line segment has the greatest slope   **71. a)** $h(x) = -x + 200$
b) 186 beats per minute   **73. a)** $S(t) = 210t + 37,550$ **b)** $38,180
c) $39,650   **d)** 12 years   **75. a)** $S(t) = 1853.6t + 37,855$
b) $43,415.80   **c)** $56,391   **d)** 8 years

77. The y-intercept is wrong.   **79.** The slope is wrong.   **81.** Height: 14.2 inches, width: 6.4 inches   **84.** 19   **85.** 5   **86.** 2.4   **87.** First: 75 miles per hour, second: 60 miles per hour   **88. a)** $x < -3$ or $x > 2$ **b)** $-3 < x < 2$

Mid-Chapter Test   **1.** III [3.1]   **2.**   [3.1]   **3.**   [3.1]   **4.**   [3.1]

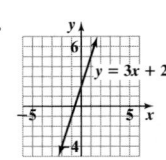 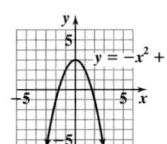 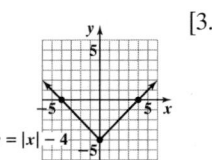

5. 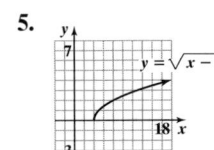 [3.1]   **6. a)** A relation is any set of ordered pairs.   **b)** A function is a correspondence between a first set of elements, the domain, and a second set of elements, the range, such that each element of the domain corresponds to exactly one element in the range.   **c)** No   **d)** Yes [3.2]   **7.** Function; Domain: $\{1, 2, 7, -5\}$, Range: $\{5, -3, -1, 6\}$ [3.2]   **8.** Not a function; Domain: $\{x \mid -2 \le x \le 2\}$, Range: $\{y \mid -4 \le y \le 4\}$ [3.2]   **9.** Function; Domain: $\{x \mid -5 \le x \le 3\}$, Range: $\{y \mid -1 \le y \le 3\}$ [3.2]

10. -21 [3.2]   **11.** 105 feet [3.2]   **12.** $7x - y = -6$ [3.3]   **13.**   [3.3]   **14.**   [3.3]

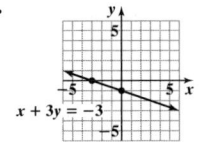 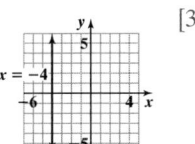

15. 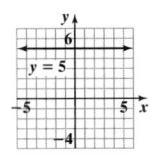 [3.3]   **16. a)** 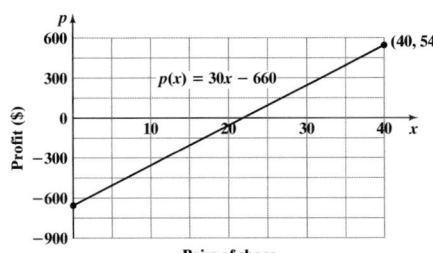   **b)** 22 pairs of shoes   **c)** 34 pairs of shoes [3.3]

17. $-\dfrac{5}{8}$ [3.4]

18. $y = -2x + 2$ [3.4]

19. $y = \dfrac{3}{2}x + 9; \dfrac{3}{2}; (0, 9)$ [3.4]

20. a) 5   **b)** $(0, 1)$ **c)** $y = 5x + 1$ [3.4]

Exercise Set 3.5   **1.** Parallel   **3.** Point-slope form   **5.** $y = 3x - 5$ **7.** $y = -\dfrac{1}{2}x + 1$ **9.** $y = \dfrac{1}{2}x - \dfrac{9}{2}$ **11.** $y = -\dfrac{3}{2}x$

13. $y = \dfrac{1}{2}x - 5$ **15.** Parallel   **17.** Neither   **19.** Perpendicular   **21.** Perpendicular   **23.** Parallel   **25.** Neither

27. Perpendicular   **29.** Parallel   **31.** Neither   **33.** $y = 2x + 1$ **35.** $2x - 5y = 19$ **37.** $y = -\dfrac{5}{3}x + 5$ **39.** $f(x) = 4x - 2$

41. $y = -\dfrac{2}{3}x + 6$ **43. a)** $C(s) = 45.7s + 95.8$ **b)** 324.3 calories   **45. a)** $d(p) = -0.20p + 90$ **b)** 38 iPods   **c)** \$225

47. a) $s(p) = 95p - 60$ **b)** 206 kites   **c)** \$3.00   **49. a)** $i(t) = 12.5t$ **b)** \$1500   **c)** 176 tickets   **51. a)** $r(w) = 0.01w + 10$ **b)** \$46.13   **c)** 5000 pounds   **53. a)** $y(a) = -0.865a + 79.25$ **b)** 47.2 years   **c)** 62.7 years old   **55. a)** $w(a) \approx 0.189a + 10.6$

b) 14.758 kilograms   **58.** $\left(-\infty, \dfrac{2}{5}\right)$ **59.** Reverse the direction of the inequality symbol.   **60. a)** Any set of ordered pairs **b)** A correspondence where each member of the domain corresponds to a unique member in the range   **c)** Answers will vary. **61.** Domain: $\{3, 4, 5, 6\}$, Range: $\{-4, -1, 2, 7\}$

Exercise Set 3.6   **1.** $x^2 - 2x + 5$ **3.** $x^2 + 2x - 5$ **5.** $2x^3 - 5x^2$ **7.** all real numbers   **9. a)** $4x + 3$ **b)** $4a + 3$ **c)** 11 **11. a)** $x^3 + x - 4$ **b)** $a^3 + a - 4$ **c)** 6   **13. a)** $4x^3 - x + 4$ **b)** $4a^3 - a + 4$ **c)** 34   **15.** -7 **17.** 29   **19.** -60

21. Undefined   **23.** 13   **25.** $-\dfrac{3}{4}$ **27.** $2x^2 - 6$ **29.** -4 **31.** 18   **33.** 0   **35.** $-\dfrac{3}{7}$ **37.** $-\dfrac{1}{45}$ **39.** $-2x^2 + 2x - 6$ **41.** 3

43. -4 **45.** 1   **47.** Undefined   **49.** 0   **51.** 0   **53.** -3 **55.** -2 **57. a)** 2008   **b)** \$800   **c)** \$7900   **d)** \$900 **59. a)** Summer A, 36   **b)** Summer B, 10   **c)** 11 male students   **d)** 19 female students   **61. a)** ≈ 20 **b)** ≈ 8 **c)** ≈ 12 **d)** ≈ 23

63. a) 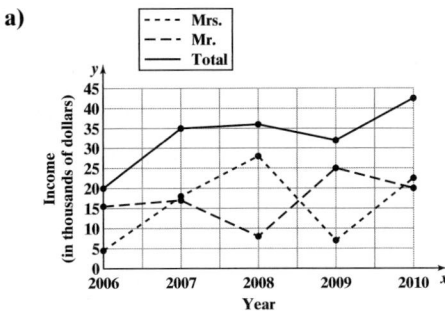   **b)** 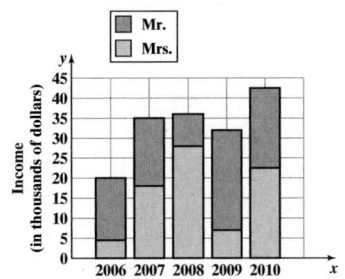   **a)**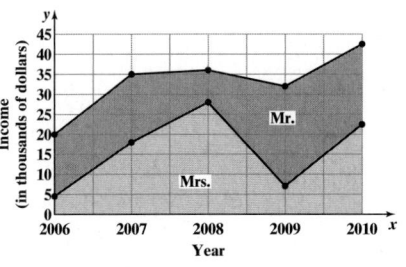

65. a)

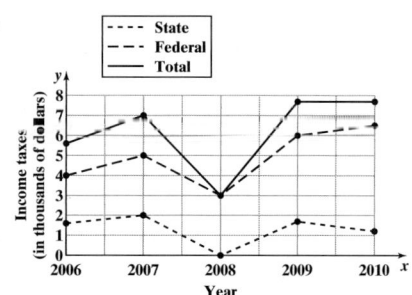

b)

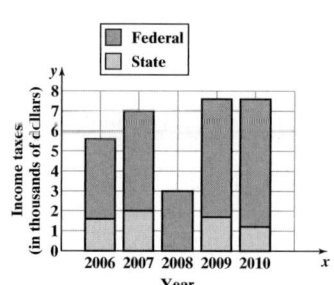

c)

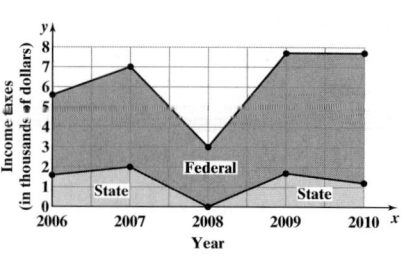

67. $g(x) \neq 0$ since division by zero is undefined.   **69.** No, subtraction is not commutative. One example is $5 - 3 = 2$ but $3 - 5 = -2$

71. $f(a)$ and $g(a)$ must either be opposites or both be equal to 0.   **73.** $f(a) = g(a)$ **75.** $f(a)$ and $g(a)$ must have opposite signs.

78. $-\dfrac{1}{64}$ **79.** 2.96×10^6 **80.** $h = \dfrac{2A}{b}$ **81.** \$450   **82.**

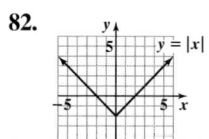

83.

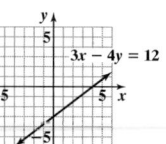

Exercise Set 3.7   **1.** Boundary line   **3.** Solid   **5.**

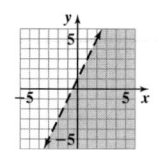

7.

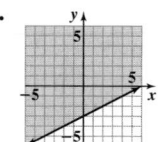

9.

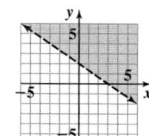

11.

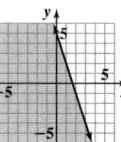

13.

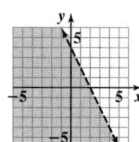

15.

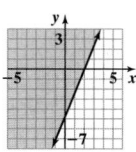

17.

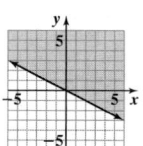

19.

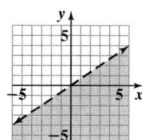

21.

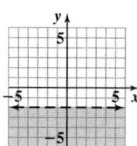

23.

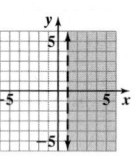

25.

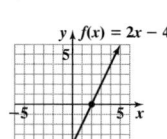

27. a) $8x + 15y \leq 175$ **b)** Yes   **c)** No

29. a)

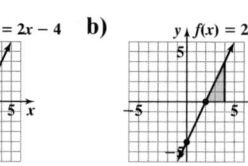

b)

31. Points on the line are solutions to the corresponding equation, and are not solutions if the symbol used is $<$ or $>$.

33. $(0, 0)$ cannot be used as a checkpoint if the line passes through the origin.   **35.**

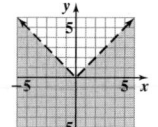

37.

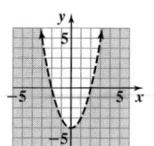

38. 9

39. 81.176   **40.** \$15.72   **41.** -4

42. $x + 2y = 2$ (other answers are possible)   **43.** -2

Chapter 3 Review Exercises   **1.**

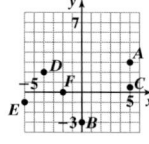

2.

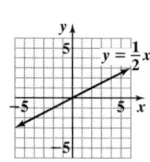

3.

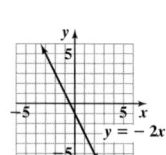

4.

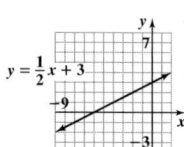

5.

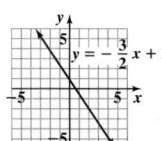

6.

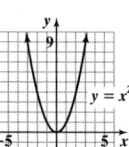

7.

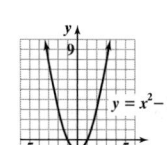

8.

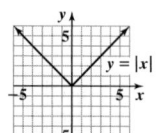

9.

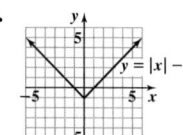

10.

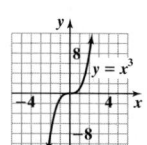

11.

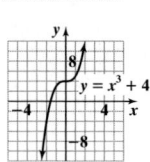

12. A function is a correspondence where each member of the domain corresponds to exactly one member of the range.   **13.** No, every relation is not a function. $\{(4, 2), (4, -2)\}$ is a relation but not a function. Yes, every function is a relation because it is a set of ordered pairs.   **14.** Yes, each member of the domain corresponds to exactly one member of the range.   **15.** No, the domain element 1 corresponds to more than one member of the range (1 and -1).   **16. a)** Yes, the relation is a function.   **b)** Domain: $\mathbb{R}$, Range: $\mathbb{R}$

17. a) Yes, the relation is a function.   **b)** Domain: $\mathbb{R}$, Range: $\{y | y \le 0\}$ **18. a)** No, the relation is not a function.
b) Domain: $\{x | -3 \le x \le 3\}$,   Range: $\{y | -3 \le y \le 3\}$ **19. a)** No, the relation is not a function.   **b)** Domain: $\{x | -2 \le x \le 2\}$,
Range: $\{y | -1 \le y \le 1\}$ **20. a)** -1 **b)** $3h - 7$ **21. a)** 1   **b)** $2a^3 - 3a^2 + 6$ **22.** Answers will vary. Here is one possible interpretation: The car speeds up to 50 mph. Stays at this speed for about 11 minutes. Speeds up to about 68 mph. Stays at that speed for about 5 minutes, then stops quickly. Stopped for about 5 minutes. Then in stop-and-go traffic for about 5 minutes.
23. a) 1020 baskets   **b)** 1500 baskets   **24. a)** 180 feet   **b)** 52 feet
25.   **26.**   **27.**   **28.**   **29.**

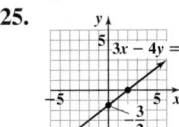 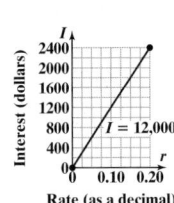

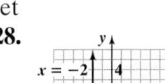

b) 50,000 bagels
c) 270,000 bagels

30.   **31.** $m = \dfrac{1}{2}, (0, 6)$ **32.** $m = -2, (0, 3)$ **33.** $m = -\dfrac{3}{5}, \left(0, \dfrac{13}{5}\right)$ **34.** $m = -\dfrac{3}{4}, \left(0, \dfrac{5}{2}\right)$

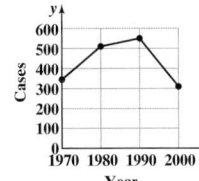

35. m is undefined, no y-intercept   **36.** $m = 0, (0, 8)$ **37.** 2   **38.** $-\dfrac{1}{3}$ **39.** $m = 0, y = 3$
40. m is undefined, $x = 2$ **41.** $m = -\dfrac{1}{2}, y = -\dfrac{1}{2}x + 2$ **42. a)** -2 **b)** $(0, 1)$ **c)** $y = -2x + 1$
43. $(0, 0)$

44. a)   **b)** 1970–1980: 16.4, 1980–1990: 4.2, 1990–2000: -23.5 **c)** 1970–1980   **45.** $n(t) \approx 0.7t + 35.6$

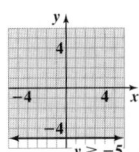

46. Parallel   **47.** Perpendicular   **48.** Neither   **49.** $y = \dfrac{1}{2}x + 2$ **50.** $y = -x - 2$
51. $y = -\dfrac{2}{3}x + 6$ **52.** $y = \dfrac{5}{2}x + 3$ **53.** $y = -\dfrac{5}{3}x - 4$ **54.** $y = -\dfrac{1}{2}x + 7$ **55.** Neither
56. Parallel   **57.** Perpendicular   **58.** Neither   **59. a)** $r(a) = 0.61a - 10.59$ **b)** $13.81
60. a) $C(r) = 1.8r + 435$ **b)** 507 calories   **c)** ≈ 91.7 yards per minute   **61.** $x^2 - x - 1$ **62.** 11   **63.** $-x^2 + 5x - 9$
64. -15 **65.** -56 **66.** 4   **67.** $-\dfrac{2}{3}$ **68.** -2 **69.**   **a)** ≈ 4.6 billion   **b)** ≈ 2.1 billion   **c)** ≈ 0.8 billion   **d)** $\approx 33\%$
70. a) $\approx \$47,000$ **b)** $\approx \$28,000$ **c)** $\approx \$3000$

71.   **72.**   **73.**   **74.**

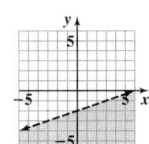

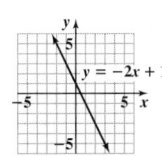 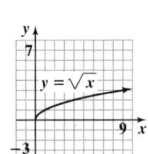 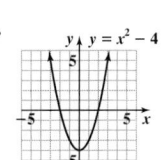 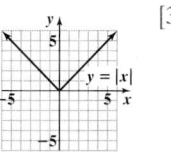

Chapter 3 Practice Test   **1.**   [3.1]   **2.**   [3.1]   **3.**   [3.1]   **4.**   [3.1]

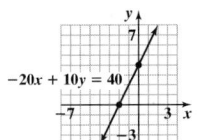 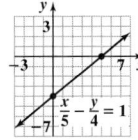 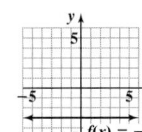 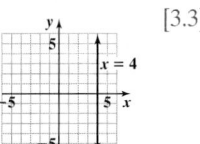

5. A function is a correspondence where each element in the domain corresponds with exactly one element in the range. [3.2]
6. Yes, because each element in the domain corresponds to exactly one element in the range. [3.2]   **7.** Yes; Domain: $\mathbb{R}$,
Range: $\{y | y \le 4\}$ [3.2]   **8.** No; Domain: $\{x | -3 \le x \le 3\}$, Range: $\{y | -2 \le y \le 2\}$ [3.2]   **9.** 29 [3.2]

10.   [3.3]   **11.**   [3.3]   **12.**   [3.3]   **13.**   [3.3]

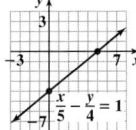 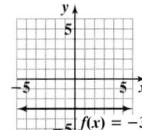 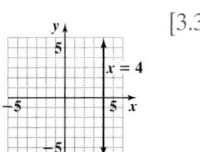

14. a)   **b)** 4900 books   **c)** 14,700 books [3.3]   **15.** $m = \dfrac{4}{3}, (0, -5)$ [3.4]   **16.** $y = 3x - 7$ [3.4]
17. $y = -2x + 7$ [3.4]   **18.** $p(t) = 2.9044t + 274.634$ [3.4]
19. Parallel, the slope of both lines is the same, $\dfrac{2}{3}$. [3.5]   **20. a)** $r(t) = -3t + 266$
b) 248 per 100,000   **c)** 206 per 100,000 [3.5]   **21.** 12 [3.6]   **22.** $-\dfrac{3}{7}$ [3.6]

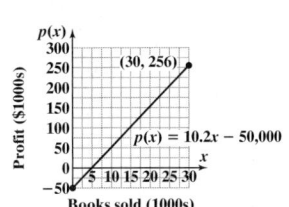

23. $2a^2 - a$ [3.6]   **24. a)** ≈ 44 million tons   **b)** ≈ 18 million tons   **c)** ≈ 26 million tons [3.6]   **25.**   [3.7]

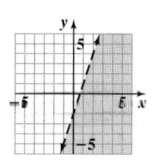

Cumulative Review Test   **1. a)** $\{3, 5, 7\}$ **b)** $\{1, 2, 3, 5, 7, 9, 11, 14\}$ [1.2]   **2. a)** None

b) $-6, -4, \dfrac{1}{3}, 0, \sqrt{3}, 4.67, \dfrac{37}{2}, -\sqrt{5}$ [1.2]   **3.** 100 [1.4]   **4.** $25x^4y^6$ [1.5]   **5.** $\dfrac{x^9}{8y^{15}}$ [1.5]

6. a) 3.052×10^{12} cubic feet   **b)** 7.412×10^{12} cubic feet   **c)** 2.398×10^{13} cubic feet [1.6]   **7.** 0 [2.1]   **8.** $-\dfrac{138}{5}$ [2.1]

9. $9x - 7$ [2.1]   **10.** $b_1 = \dfrac{2A}{h} - b_2$ [2.2]   **11.** 12 gallons [2.4]   **12.** $x > -\dfrac{10}{3}$ [2.5]   **13.** $2 < x < 6$ [2.5]   **14.** $\{-15, 1\}$ [2.6]

15. $\{x | -1 \le x \le 2\}$ [2.6]   **16.**   [3.1]   **17. a)** Not a function   **b)** Domain: $\{x | x \le 2\}$, Range: $\mathbb{R}$ [3.2]

$y = -\dfrac{3}{2}x - 4$

18. $-\dfrac{4}{9}$ [3.4]   **19.** Neither [3.5]   **20.** $x^2 + 7x - 11$ [3.6]

Chapter 4

Exercise Set 4.1   **1.** One variable   **3.** Dependent   **5.** Consistent   **7.** Parallel lines   **9.** Ordered pair   **11.** None   **13. b)**
15. b)   **17.** Consistent; one solution   **19.** Dependent; infinite number of solutions   **21.** Inconsistent; no solution   **23.** Inconsistent; no solution   **25.** 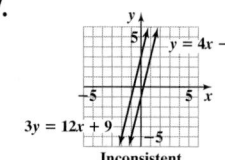   **27.** 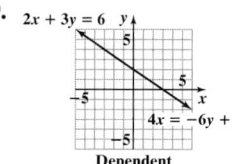   **29.** $2x + 3y = 6$ Dependent   **31.**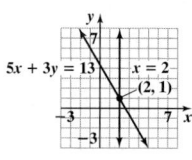

33.   **35.** 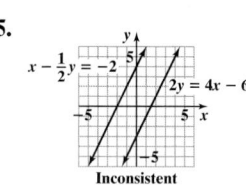   **37.** $(1, 0)$ **39.** $(-3, -3)$ **41.** $(2, 1)$ **43.** $(0.5, 0.7)$ **45.** Infinite number of solutions   **47.** No solution   **49.** $\left(-\dfrac{19}{5}, -3\right)$ **51.** $(8, 6)$ **53.** $(5, 2)$ **55.** $\left(-1, -\dfrac{5}{3}\right)$ **57.** Infinite number of solutions   **59.** No solution   **61.** $(1, 1)$ **63.** Infinite number of solutions.

65. $\left(\dfrac{14}{5}, -\dfrac{12}{5}\right)$ **67.** $\left(\dfrac{37}{7}, \dfrac{19}{7}\right)$ **69.** $(3, 2)$ **71.** $(4, 0)$ **73.** $(4, 3)$ **75.** $\left(\dfrac{192}{25}, \dfrac{144}{25}\right)$ **77.** 2025, \$53,000

79. a) One example is: $x + y = 7, x - y = -3$.   **b)** Choose coefficients for x and y, then use the given coordinates to find the constants.   **81.** $A = 2$ and $B = 5$ **83.** $m = 4, b = -2$ **85.** Compare the slopes and y-intercepts of the equations. If the slopes are different, the system is consistent. If the slopes and y-intercepts are the same, the system is dependent. If the slopes are the same and the y-intercepts are different, the system is inconsistent.   **87.** You will get a true statement, like $0 = 0$.   **89.** Multiply the first equation by 2 and notice that the new equation is identical to the second equation.   **91. a), b),** and **c)** Answers will vary.
93. a) Infinite number, because a system of equations can have no solution, one solution, or an infinite number of solutions
b) $m = -4, y = -4x - 13, (0, -13)$ **c)** Yes   **95.** One example is: $x + y = 1, 2x + 2y = 2$, create one equation and then multiply it by a constant to get the second equation   **97.** The system is dependent or one graph is not in the viewing window.

99. $(8, -1)$ **101.** $(-1, 2)$ **103.** $\left(\dfrac{1}{a}, 5\right)$ **106.** Rational numbers can be expressed as quotients of two integers, denominator not 0.
Irrational numbers cannot.   **107. a)** Yes, the set of real numbers includes the set of rational numbers.   **b)** Yes, the set of real numbers includes the set of irrational numbers.

108. $-\dfrac{17}{4}$ **109.** $\mathbb{R}$ **110.** 520.20   **111.** No, the points $(-3, 4)$ and $(-3, -1)$ have the same first coordinate but different second coordinates.   **112.** Undefined

Exercise Set 4.2

1. Ordered triple.   **3.** Inconsistent   **5.** $(2, 0, 1)$ **7.** $\left(-7, -\dfrac{35}{4}, -3\right)$ **9.** $(0, 3, 6)$ **11.** $(1, 1, 1)$ **13.** $(-3, 15, -7)$

15. $(3, 1, -2)$ **17.** $(2, -1, 3)$ **19.** $\left(\dfrac{2}{3}, -\dfrac{1}{3}, 1\right)$ **21.** $(0, -1, 0)$ **23.** $\left(-\dfrac{11}{17}, \dfrac{7}{34}, -\dfrac{49}{17}\right)$ **25.** $(0, 0, 0)$ **27.** $(4, 6, 8)$

29. $\left(\frac{2}{3}, \frac{23}{15}, \frac{37}{15}\right)$ **31.** $(1, 1, 2)$ **33.** Inconsistent **35.** Dependent **37.** Inconsistent **39.** $A = 9, B = 6, C = 2; 9x + 6y + 2z = 1$
41. Answers will vary. One example is $x + y + z = 10, x + 2y + z = 11, x + y + 2z = 16$
43. a) $a = 1, b = 2, c = -4$ **b)** $y = x^2 + 2x - 4$, Substitute 1 for a, 2 for b, and -4 for c into $y = ax^2 + bx + c$
45. No point is common to all three planes. Therefore, the system is inconsistent. **47.** One point is common to all three planes; therefore, the system is consistent. **49. a)** Yes, the 3 planes can be parallel **b)** Yes, the 3 planes can intersect at one point
c) No, the 3 planes cannot intersect at exactly two points **51.** $(1, 2, 3, 4)$ **53. a)** $\frac{1}{4}$ hour or 15 minutes **b)** 1.25 miles
54. $\left\{x \mid x < -\frac{3}{2} \text{ or } x > \frac{27}{2}\right\}$ **55.** $\left\{x \mid -\frac{8}{3} < x < \frac{16}{3}\right\}$ **56.** $\varnothing$

Exercise Set 4.3
1. Ireland: 70,273 square kilometers, Georgia: 69,700 square kilometers **3.** Hamburger: 21 grams, fries: 67 grams **5.** Hot dog: \$2, soda: \$1 **7.** 512 MB: 109 photos, 4 GB: 887 photos **9.** $25°, 65°$ **11.** $52°, 128°$ **13.** 12.2 miles per hour, 3.4 miles per hour **15.** \$500, 4% **17.** 1.2 ounces of 5%, 1.8 ounces of 30% **19.** 10 gallons concentrate, 190 gallons water
21. $17\frac{1}{3}$ pounds birdseed, $22\frac{2}{3}$ pounds sunflower seeds **23.** Adult: \$29, child: \$18 **25.** \$6000 at 5%, \$4000 at 6%
27. 160 gallons whole, 100 gallons skim milk **29.** 7 pounds Season's Choice, 13 pounds Garden Mix **31.** 50 miles per hour, 55 miles per hour **33.** Cabrina: 8 hours, Dabney: 3.4 hours **35.** 80 grams A, 60 grams B **37.** 200 grams first alloy, 100 grams second alloy
39. 2012 **41.** Tom: 60 miles per hour, Melissa: 75 miles per hour **43.** Personal: 3, bills and statements: 4, advertisements: 17
45. Alabama: 54, Tennessee: 47, Texas: 46 **47.** Bandits: 32 wins, Force: 31 wins, Glory: 30 wins **49.** *Their Greatest Hits 1971–1975*: 29 million albums, *Thriller*: 27 million albums, *The Wall*: 23.5 million albums **51.** Florida: 15, California: 11, Louisiana: 9
53. $30°, 45°, 105°$ **55.** \$1500 at 3%, \$3000 at 5%, \$5500 at 6% **57.** 4 liters of 10% solution, 2 liters of 12% solution, 2 liters of 20% solution **59.** 10 children's chairs, 12 standard chairs, 8 executive chairs **61.** $I_A = \frac{27}{38}; I_B = -\frac{15}{38}; I_C = -\frac{6}{19}$ **64.** $-\frac{35}{8}$ **65.** 4
66. Use the vertical line test. **67.** $y = x - 10$

Mid-Chapter Test
1. a) $y = 7x - 13$, $y = -\frac{2}{3}x + 3$, **b)** Consistent **c)** One solution [4.1] **2.** $(1, 2)$ [4.1]
3. $(-1, -3)$ [4.1] **4.** $(-4, 1)$ [4.1] **5.** $\left(\frac{1}{2}, -2\right)$ [4.1] **6.** $(-3, 4)$ [4.1] **7.** $\left(\frac{1}{3}, \frac{1}{2}\right)$ [4.1] **8.** $(6, 12)$ [4.1] **9.** Inconsistent, no solution [4.1] **10.** Dependent, infinite number of solutions [4.1] **11.** $(1, 2, -1)$ [4.2] **12.** $(2, 0, 3)$ [4.2] **13.** Solution must have values for y and z in addition to a value for x. The solution is $(1, -1, 4)$ or $x = 1, y = -1, z = 4$. [4.2] **14.** 10 pounds cashews, 5 pounds of pecans [4.3] **15.** 5, 7, 20 [4.3]

Exercise Set 4.4
1. Row transformations **3.** Dimensions **5.** Inconsistent **7.** Square matrix **9.** $\begin{bmatrix} 1 & -2 & | & -5 \\ 3 & -7 & | & -4 \end{bmatrix}$

11. $\begin{bmatrix} 1 & 1 & 3 & | & -8 \\ 3 & 2 & 1 & | & -5 \\ 4 & 7 & 2 & | & -1 \end{bmatrix}$ **13.** $\begin{bmatrix} 1 & 3 & | & 12 \\ 0 & 23 & | & 42 \end{bmatrix}$ **15.** $\begin{bmatrix} 1 & 0 & 8 & | & \frac{1}{4} \\ 0 & 2 & -38 & | & -\frac{13}{4} \\ 6 & -3 & 1 & | & 0 \end{bmatrix}$ **17.** $(3, 0)$ **19.** $(-5, 1)$ **21.** $(0, 1)$ **23.** Dependent system

25. $\left(-\frac{1}{3}, 3\right)$ **27.** Inconsistent system **29.** $\left(\frac{2}{3}, \frac{1}{4}\right)$ **31.** $\left(\frac{4}{5}, -\frac{7}{8}\right)$ **33.** $(1, 2, 0)$ **35.** $(3, 1, 2)$ **37.** $\left(1, -1, \frac{1}{2}\right)$

39. Dependent system **41.** $\left(\frac{1}{2}, 2, 4\right)$ **43.** Inconsistent system **45.** $\left(5, \frac{1}{3}, -\frac{1}{2}\right)$ **47.** $\angle x = 30°, \angle y = 65°, \angle z = 85°$
49. Washington: \$1100 million (or \$1.1 billion), Dallas: \$923 million, Houston: \$905 million **51.** Dependent
53. No, this is the same as switching the order of the equations. **55. a)** $\{1, 2, 3, 4, 5, 6, 9, 10\}$ **b)** $\{4, 6\}$ **56. a)** ◄─○────●─► $\begin{smallmatrix} -1 & & 4 \end{smallmatrix}$

b) $\{x \mid -1 < x \le 4\}$ **c)** $(-1, 4]$ **57.** A graph is an illustration of the set of points whose coordinates satisfy an equation.
58. -71

Exercise Set 4.5
1. Determinant **3.** D_y **5.** Dependent **7.** 11 **9.** -8 **11.** -12 **13.** 44 **15.** $(2, 1)$ **17.** $(6, -4)$

19. $\left(\frac{1}{2}, -1\right)$ **21.** $(-7, -2)$ **23.** Infinite number of solutions **25.** $(2, -3)$ **27.** No solution **29.** $(2, 5)$ **31.** $(1, 2, 1)$

33. $\left(\frac{1}{2}, -\frac{1}{2}, 2\right)$ **35.** $\left(\frac{1}{2}, -\frac{1}{8}, 2\right)$ **37.** $(-1, 0, 2)$ **39.** Infinite number of solutions **41.** $(1, -1, 2)$ **43.** No solution
45. $(3, 4, 1)$ **47.** $(-1, 5, -2)$ **49.** 5 **51.** 6 **53.** adult: \$10, student: \$5 **55.** It will have the opposite sign. This can be seen by comparing $a_1b_2 - a_2b_1$ to $a_2b_1 - a_1b_2$ **57.** 0 **59.** 0 **61.** Yes, it will have the opposite sign. **63.** No, same value as original value
65. Yes, value is double original value **67. a)** $x = \dfrac{c_1b_2 - c_2b_1}{a_1b_2 - a_2b_1}$ **b)** $y = \dfrac{a_1c_2 - a_2c_1}{a_1b_2 - a_2b_1}$ **68.** $\left(-\infty, \frac{14}{11}\right)$

69. 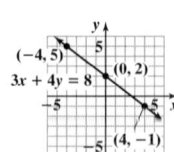 **70.** 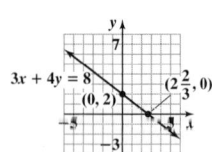 **71.**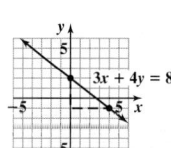

Exercise Set 4.6   **1.** Satisfy   **3.** Constraints   **5.**

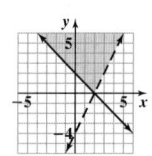   **7.** 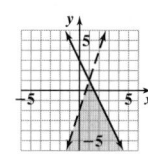   **9.**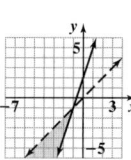

11. 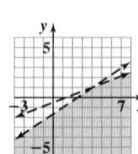   **13.** 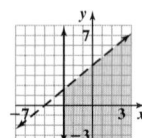   **15.** 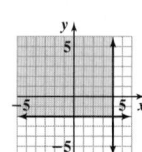   **17.** 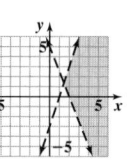   **19.**   **21.**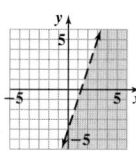

23. 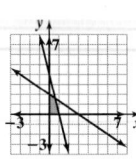   **25.** 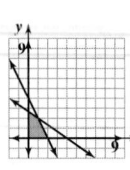   **27.** 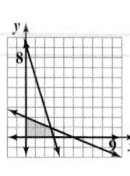   **29.** 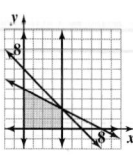   **31.** 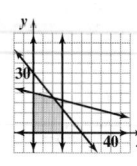   **33.**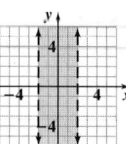

35. 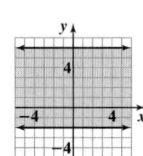   **37.** 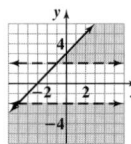   **39.** 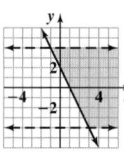   **41.** 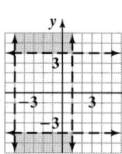   **43.**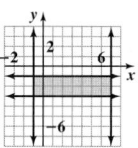

45. No; the point is not contained in the solution to the $<$ inequality.   **47.** No; the point is not a solution to either inequality.
49. There is no solution. Opposite sides of the same line are being shaded and only one inequality includes the line.   **51.** There are
an infinite number of solutions. Both inequalities include the line $5x - 2y = 3$.   **53.** There are an infinite number of solutions. The
lines are not parallel or identical.

55. 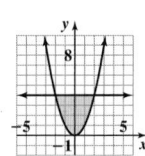   **57.** 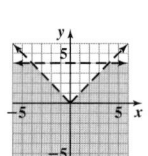   **59.** $f_2 = \dfrac{f_3 d_3 - f_1 d_1}{d_2}$ **60.** Domain: $\{-1, 0, 4, 5\}$, Range: $\{-5, -2, 2, 3\}$
61. Domain: $\mathbb{R}$, Range: $\mathbb{R}$
62. Domain: $\mathbb{R}$, Range: $\{y | y \geq -1\}$

Chapter 4 Review Exercises   **1.** Inconsistent; no solution   **2.** Consistent; one solution   **3.** Consistent; one solution

4. Consistent; one solution   **5.** 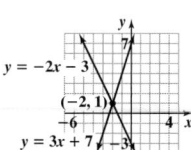   **6.** 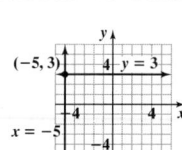   **7.** 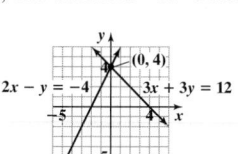   **8.**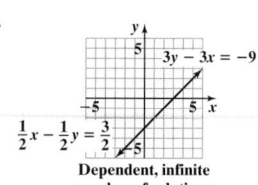

Dependent, infinite
number of solutions

9. $(1, -1)$ **10.** $(-1, -1)$ **11.** $(2, 5)$ **12.** $(5, 2)$ **13.** $(2, 1)$ **14.** $(-8, 11)$ **15.** $(-1, 3)$ **16.** $(3, -2)$ **17.** $\left(\dfrac{32}{13}, \dfrac{8}{13}\right)$
18. $\left(-1, \dfrac{13}{3}\right)$ **19.** $(1, 2)$ **20.** $\left(\dfrac{7}{5}, \dfrac{13}{5}\right)$ **21.** $(6, -2)$ **22.** $\left(-\dfrac{78}{7}, -\dfrac{48}{7}\right)$ **23.** Infinite number of solutions   **24.** No solution
25. $(2, 0, 1)$ **26.** $(-1, 3, -2)$ **27.** $(-5, 1, 2)$ **28.** $(3, -2, -2)$ **29.** $\left(\dfrac{8}{3}, \dfrac{2}{3}, 3\right)$ **30.** $(0, 2, -3)$ **31.** No solution
32. Infinite number of solutions   **33.** Luan: 38, Jennifer: 28   **34.** Airplane: 520 mph, wind: 40 mph   **35.** Combine 2 liters of the 20%
acid solution with 4 liters of the 50% acid solution.   **36.** 410 adult tickets and 240 children tickets were sold.   **37.** His ages were 41
years and 77 years.   **38.** $20,000 invested at 7%, $15,000 invested at 5%, and $5000 was invested at 3%.   **39.** $(3, -2)$ **40.** $(3, 1)$
41. Infinite number of solutions   **42.** $(2, 1, -2)$ **43.** No solution   **44.** Infinite number of solutions   **45.** $(1, 5)$ **46.** $(-3, 2)$
47. $(-1, 2)$ **48.** $(-2, 3, 4)$ **49.** $(1, 1, 2)$ **50.** No solution   **51.** 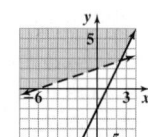   **52.** 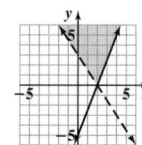   **53.**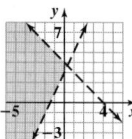

54. No solution 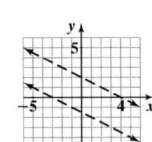   **55.** 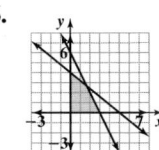   **56.**   **57.**   **58.**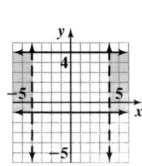

Chapter 4 Practice Test   **1.** Answers will vary [4.1]   **2.** Consistent; one solution [4.1]   **3.** Dependent; infinite number of solutions [4.1]   **4.** Inconsistent; no solution [4.1]   **5.** 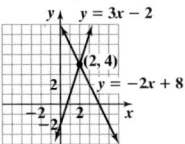 [4.1]   **6.** 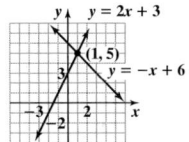 [4.1]   **7.** $(1, 1)$ [4.1]   **8.** $(-3, 2)$ [4.1]

9. $\left(-\dfrac{1}{2}, 4\right)$ [4.1]   **10.** Infinite number of solutions [4.1]   **11.** $\left(\dfrac{44}{19}, \dfrac{48}{19}\right)$ [4.1]   **12.** $(1, -1, 2)$ [4.2]   **13.** $\begin{bmatrix} -2 & 3 & 7 & 5 \\ 3 & -2 & 1 & -2 \\ 1 & -6 & 9 & -13 \end{bmatrix}$ [4.4]

14. $\begin{bmatrix} 6 & -2 & 4 & 4 \\ 0 & 5 & -3 & 12 \\ 2 & -1 & 4 & -3 \end{bmatrix}$ [4.4]   **15.** $(4, -1)$ [4.4]   **16.** $(3, -1, 2)$ [4.4]   **17.** -1 [4.5]   **18.** 165 [4.5]   **19.** $(-3, 2)$ [4.5]

20. $(3, 1, -1)$ [4.5]   **21.** 8 pounds sunflower; 12 pounds bird mix [4.3]   **22.** $6\dfrac{2}{3}$ liters 6% solution; $3\dfrac{1}{3}$ liters 15% solution [4.3]

23. 4, 9, and 16 [4.3]   **24.** 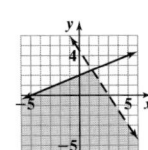 [4.6]   **25.** 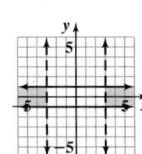 [4.6]

Cumulative Review Test   **1.** 3 [1.4]   **2. a)** $9, 1$ **b)** $\dfrac{1}{2}, -4, 9, 0, -4.63, 1$ **c)** $\dfrac{1}{2}, -4, 9, 0, \sqrt{3}, -4.63, 1$ [1.2]

3. $-|-8|, -1, \dfrac{5}{8}, \dfrac{3}{4}, |-4|, |-12|$ [1.3]   **4.** 7 [2.1]   **5.** $\dfrac{17}{4}$ [2.1]   **6.** $6, -3$ [2.6]   **7.** $x = 2M - a$ [2.2]   **8.** $\left\{x \mid \dfrac{2}{3} < x \le \dfrac{34}{3}\right\}$ [2.5]

9. $\dfrac{y^{10}}{9x^4}$ [1.5]   **10.** 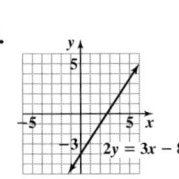 [3.3]   **11.** $y = \dfrac{2}{3}x + \dfrac{5}{3}$ [3.5]   **12.** 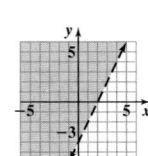 [3.7]   **13. a)** Function   **b)** Function   **c)** Not a function [3.2]

14. a) $-\dfrac{1}{7}$ **b)** $\dfrac{h + 3}{h^2 - 9}$ **c)** Undefined [3.2]

15. $(1, 3)$ [4.1]   **16.** $(7, -1)$ [4.1]   **17.** $(2, 1, 3)$ [4.2]   **18.** $10°, 80°, 90°$ [2.3]   **19.** 1 hour [2.4]   **20.** 600 at \$20, 400 at \$16 [4.3]

Chapter 5

Exercise Set 5.1
1. Terms   **3.** 5   **5.** Leading   **7.** Binomial   **9.** Monomial   **11.** Monomial   **13.** Not a polynomial; -1 exponent

15. Not a polynomial; $\dfrac{1}{2}$ exponent   **17.** $-x^2 + 2x - 5, 2$ **19.** $10x^2 + 3xy + 9y^2, 2$ **21.** In descending order, 4   **23. a)** 6   **b)** 3

25. a) 6   **b)** 9   **27. a)** 17   **b)** $-\dfrac{1}{3}$ **29.** -5 **31.** -7 **33.** -2.0312 **35.** $x^2 + 9x - 6$ **37.** $x^2 - 13x + 2$ **39.** $2y^2 + 9y - 11$

41. $-\dfrac{2}{3}a^2 - \dfrac{29}{36}a + 5$ **43.** $-3.5x^2 - 2.1x - 19.6$ **45.** $-\dfrac{4}{3}x^3 - \dfrac{1}{4}x^2y + 9xy^2$ **47.** $5a - 10b + 13c$ **49.** $8a^2b - 10ab + 11b^2$

51. $7r^2 - 4rt - 3t^2$ **53.** $10x^2 - 8x - 9$ **55.** $-3w^2 + 6w$ **57.** $3x + 19$ **59.** $-3x^2 + 2x - 12$ **61.** $-5.4a^2 - 5.7a - 26.4$

63. $-\dfrac{11}{2}x^2y + xy^2 + \dfrac{2}{45}$ **65.** $5x^{2r} - 10x^r + 3$ **67.** $-x^{2s} - 4x^s + 19$ **69.** $7b^{4n} - 3b^{3n} - 4b^{2n} + 1$ **71.** $4x^2 + 8x + 4$

73. $3x^2 + 4x + 19$ **75.** $2x^2 + 12x + 9$ **77.** 144 square meters   **79.** $A \approx 113.10$ square inches   **81.** 674 feet   **83.** 105 committees
85. a) \$674   **b)** \$1010   **87. a)** $P(x) = 2x^2 + 360x - 8050$ **b)** \$47,950   **89. c)** The y-intercept is $(0, -4)$ and the leading coefficient is positive   **91. c)** The y-intercept is $(0, -6)$ and the leading coefficient is negative   **93. a)** 3.5%   **b)** Yes   **c)** 12.5%   **95.** \$88,210
97. a) 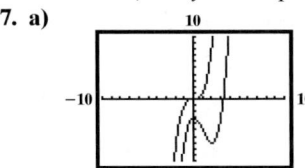   **b)** Increase   **c)** Answers will vary.   **d)** Decrease   **e)** Answers will vary.
99. 0   **101.** The leading coefficient is the coefficient of the leading term.
103. a) It is the same as that of the highest degree term.   **b)** 7
105. a) A polynomial is linear if its degree is 0 or 1.   **b)** Answer will vary. One example is $x + 4$.
107. a) A polynomial is cubic if it has degree 3 and is in one variable.   **b)** Answer will vary. One example is $x^3 + x - 4$.   **109.** Answers will vary. One example is $x^5 + x + 1$.

111. No, for example $(x^2 + x + 1) + (x^3 - 2x^2 + x) = x^3 - x^2 + 2x + 1$ **113.** No, for example
$(x^2 + 3x - 5) + (-x^2 - 4x + 2) = -x - 3$ **115. b)** The y-intercept is $(0, -5)$ and the leading coefficient is negative   **119.** 3
120. $\dfrac{15}{16}$ **121.** 6 hours   **122.** $-\dfrac{2}{11}$ **123.** $(-4, 0, -1)$

Exercise Set 5.2

1. Second   **3.** Factors   **5.** First   **7.** Square   **9.** $42x^2y^5$ **11.** $\dfrac{1}{9}x^7y^8z^2$ **13.** $6x^6y^3 - 15x^3y^4 - 18x^2y$ **15.** $2xyz + \dfrac{8}{3}y^2z - 8y^3z$

17. $0.6x^2 - 1.5x + 3.3y$ **19.** $2.85a^{11}b^5 - 1.38a^9b^7 + 0.36a^6b^9$ **21.** $12x^2 - 38x + 30$ **23.** $-2x^3 + 8x^2 - 3x + 12$

25. $x^2 + \dfrac{23}{6}xy - \dfrac{2}{3}y^2$ **27.** $0.09a^2 - 0.25b^2$ **29.** $x^3 - x^2 - 11x - 4$ **31.** $2a^3 - 7a^2b + 5ab^2 - 6b^3$ **33.** $x^4 + 2x^2 + 10x + 7$

35. $5x^4 + 29x^3 + 14x^2 - 28x + 10$ **37.** $3m^4 - 11m^3 - 5m^2 - 2m - 20$ **39.** $8x^3 - 12x^2 + 6x - 1$
41. $10r^4 - 2r^3s - r^2s^2 + rs^3 - 2s^4$ **43.** $x^2 + 4x + 4$ **45.** $4x^2 - 36x + 81$ **47.** $16x^2 - 24xy + 9y^2$ **49.** $25m^4 - 4n^2$
51. $y^2 + 8y - 4xy + 16 - 16x + 4x^2$ **53.** $25x^2 + 20xy + 10x + 4y^2 + 4y + 1$ **55.** $a^2 - b^2 - 8b - 16$

57. $2x^3y + 2x^2y^2 + 24xy^3$ **59.** $2x^3y^2 + \dfrac{3}{2}x^2y^3 - \dfrac{7}{2}xy^6$ **61.** $\dfrac{3}{5}x^2y^5z^7 + 3x^2y^4z^2 - \dfrac{1}{15}x^2y^3z^9$ **63.** $21a^2 + 10a - 24$ **65.** $64x^2 - \dfrac{1}{25}$

67. $x^3 - \dfrac{3}{2}x^2y + \dfrac{3}{4}xy^2 - \dfrac{1}{8}y^3$ **69.** $2x^3 + 10x^2 + 9x - 9$ **71.** $6p^3 - p^2q - 16pq^2 + 6q^3$ **73.** $9x^2 + 12x + 4 - y^2$

75. $a^4 - 2a^2b^2 + b^4$ **77.** $2x^3 - 4x^2 - 64x + 192$ **79. a)** $x^2 + x - 30$ **b)** -10 **81.   a)** $10x^3 + 36x^2 - 2x - 12$ **b)** 1196
83. a) $-x^4 + 3x^3 - 2x^2 + 6x$ **b)** -72 **85.** $x^2 + 5x$ **87.** $x^2 + y^2$ **89. a)** and   **b)** $x^2 + 7x + 12$ **91.** $36 - x^2$
93. a) $11x + 12$ **b)** 117 square inches, 50 square inches.   **95.** $(x + 7)(x - 7)$, product of the sum and difference of the same two
terms.   **97.** $(x + 6)(x + 6)$, square of a binomial formula   **99.** $a(x - n)(x - n)(x - n)$ **101. a)** Answers will vary.
b) $a^2 + 2ab + b^2$ **c)** $a^2 + 2ab + b^2$ **d)** Same   **103. a)** $A = P(1 + r)^t$ **b)** \$1123.60   **105. a)** 110 ways   **b)** $P(n) = n^2 - n$
c) 110 ways   **d)** Yes   **107.** $a^2 + 2ab + b^2 - 3a - 3b + 5$ **109.** $15x^{3t-1} + 18x^{4t}$ **111.** $12x^{3m} - 18x^m - 10x^{2m} + 15$ **113.** $y^{a^2-b^2}$
115. $x^4 - 12x^3y + 54x^2y^2 - 108xy^3 + 81y^4$ **117. a)** Answers will vary.   **b)**

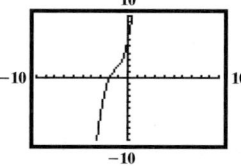

It is correct.

119. a)–d) Answer will vary.   **121. a)** Answer will vary.
b) $x^3 - 2x^2 - 21x + 12$ **123. a)** Answer will vary.   **b)** Answer will vary.
One possible answer is $(x + 4)(x - 4)$.   **c)** Answer will vary.   **d)** Answer
will vary. One possible answer is $x^2 - 16$.   **125.** Yes, for example
$(x + 2)(x - 1) = x^2 + x - 2$ **127.** $y^2 - 2y - 2xy + 2x + x^2 + 1$

129. $\dfrac{43}{60}$ **130.** $8r^6s^{15}$ **131.** $\left(-\dfrac{7}{3}, \dfrac{4}{3}\right]$ **132.** $\dfrac{15}{4}$

Exercise Set 5.3

1. Dividend   **3.** Quotient rule   **5.** Term   **7.** Synthetic   **9.** $P(7)$ **11.** x^2 **13.** a^4 **15.** z^8 **17.** $4r^6s^2$ **19.** $5x^8y^{11}$ **21.** $2x + 9$

23. $2x + 1$ **25.** $\dfrac{5}{3}y^2 + 2y - 4$ **27.** $x^3 - \dfrac{3}{2}x^2 + 3x - 2$ **29.** $4x^2 - 5xy - \dfrac{5}{2y}$ **31.** $\dfrac{9x}{2y} - 6x^2 + \dfrac{25y}{2x}$ **33.** $\dfrac{z}{2} + z^2 - \dfrac{3}{2}x^2y^4z^7$

35. $x + 2$ **37.** $2x + 4$ **39.** $3x + 2$ **41.** $x + 5 - \dfrac{2}{x + 1}$ **43.** $2b + 5 + \dfrac{2}{b - 2}$ **45.** $4x + 9 + \dfrac{2}{2x - 3}$ **47.** $2x + 6$

49. $x^2 + 2x + 3 + \dfrac{6}{x + 1}$ **51.** $2y^2 + 3y - 1 - \dfrac{6}{2y + 3}$ **53.** $2a^2 + a - 2 - \dfrac{2}{2a - 1}$ **55.** $3x^3 + 6x + 2$ **57.** $x + 4$

59. $2c^2 - 6c + 3$ **61.** $x + 6$ **63.** $x + 3$ **65.** $x - 7$ **67.** $x + 8 + \dfrac{10}{x - 3}$ **69.** $3x + 5 + \dfrac{10}{x - 4}$ **71.** $4x^2 + x + 3 + \dfrac{3}{x - 1}$

73. $3c^2 - 2c + 2 + \dfrac{10}{c + 3}$ **75.** $y^3 + y^2 + y + 1$ **77.** $x^3 - 4x^2 + 16x - 64 + \dfrac{272}{x + 4}$ **79.** $x^4 - \dfrac{9}{x + 1}$

81. $b^4 + 3b^3 - 3b^2 + 3b - 3 - \dfrac{11}{b + 1}$ **83.** $3x^2 + 3x - 3$ **85.** $2x^3 + 2x - 2 + \dfrac{6}{x - \dfrac{1}{2}}$ **87.** 12   **89.** 0, factor   **91.** $-\dfrac{19}{4}$ or -4.75

93. $3x + 2$ **95.** 3 times greater, find the areas by multiplying the polynomials, then compare   **97.** $x^2 - 2x - 8$ **99.** $x^2 + 9x + 26$

101. $2x^2 + 3xy - y^2$ **103.** $x + \dfrac{5}{2} + \dfrac{11}{2(2x - 3)}$ **105.** $w = r + 1$ **107.** $x^3 - 6x^2 + 13x - 7$; multiply $(x - 3)(x^2 - 3x + 4)$ and

then add 5.   **109.** $2x + 1 - \dfrac{3}{2x} - \dfrac{1}{2x^2}$ **111.** Not a factor; compute $P(1)$. $P(1) = 101$, which is not 0, so $x - 1$ is not a factor

113. Factor; compute $P(-1)$. $P(-1) = 0$, so $x + 1$ is a factor.   **115. a)** Answer will vary.   **b)** $\dfrac{5}{3}x^3 - 2x^2 - \dfrac{4}{3}x - 4 + \dfrac{1}{3x}$

117. Yes; answers will vary.   **119.** Place them in descending order of the variable.   **121. a)** Answers will vary.   **b)** $x + 8 + \dfrac{36}{x - 5}$

123. No; because the remainder is not equal to 0.   **125.** No, the dividend is a binomial   **127.** If the remainder is 0, $x - a$ is a factor.

129. 2.0×10^{10} **130.** $30°, 60°, 90°$ **131.** $\left\{-1, \dfrac{11}{5}\right\}$ **132.** -864 **133.** $3r + 3s - 8t$

Exercise Set 5.4

1. Multiplying   **3.** Grouping   **5.** Common   **7.** $4(x - 2)^3$ **9.** $7(n + 2)$ **11.** $2(x^2 - 2x + 8)$ **13.** $4(3y^2 - 4y + 7)$

15. $x^2(9x^2 - 3x + 11)$ **17.** $-3a^2(8a^5 - 3a^4 + 1)$ **19.** $3xy(x + 2xy + 1)$ **21.** $8a^2c(10a^3b^4 - 2a^2b^2c + 3)$
23. $3pq^2r(3p^3q^3 - pr + 4q^3r^2)$ **25.** $-2(11p^2q^2 + 8pq^3 - 13r)$ **27.** $-4(2x - 1)$ **29.** $-(x^2 + 4x - 23)$ **31.** $-3(r^2 + 2r - 3)$
33. $-2rs^3(3r^3 - 2rs - s^2)$ **35.** $-a^2b(a^2bc - 5ac^2 - 3)$ **37.** $(a + 9)(x + 1)$ **39.** $(x - 4)(9x - 8)$ **41.** $-(x - 2)(2x - 9)$
43. $-2(a + 2)(a + 2)$ or $-2(a + 2)^2$ **45.** $(x + 4)(x - 5)$ **47.** $2(2y - 1)(2y - 5)$ **49.** $(a + b)(m + n)$ **51.** $(x - 3)(x^2 + 4)$
53. $(5m - 6n)(2m - 5n)$ **55.** $5(a + 3)(a^2 - 2)$ **57.** $c^2(c - 1)(c^2 + 1)$ **59.** $(2x + 1)(6x - 5)$ **61.** $(x + 4)(3x - 2)$
63. $(3x + 2)(9x - 5)$ **65. a)** 96 feet   **b)** $h(t) = -16t(t - 5)$ **c)** 96 feet   **67. a)** ≈ 2856.64 square feet   **b)** $A = r(\pi r + 2l)$
c) ≈ 2856.64 square feet   **69. a)** \$525   **b)** $A(t) = 75(13 - t)$ **c)** \$525   **71. a)** $(1 - 0.06)(x + 0.06x) = 0.94(1.06x)$ **b)** $0.9964x$;
slightly lower than the price of the 2009 model (99.64% of the original cost)   **73.**   **a)** $(x + 0.15x) - 0.20(x + 0.15x) = 0.80(x + 0.15x)$
b) $0.80(1.15x) = 0.92x$; 92% of the regular price   **75.** $(3x + 2)^4(15ax + 10a + 4)$ **77.** $2(x - 3)(2x^4 - 12x^3 + 15x^2 + 9x + 2)$
79. $(x^2 + 2x - 4)(a + b)$ **81.** $x^{4m}(x^{2m} - 5)$ **83.** $x^{2m}(3x^{2m} - 2x^m + 1)$ **85.** $(a^r + c^r)(b^r - d^r)$ **87. a)** Yes   **b)** 0; subtracting
the same quantity from itself   **c)** Answers will vary.   **89. a)** They should be the same graph; they represent the same function
89. b)

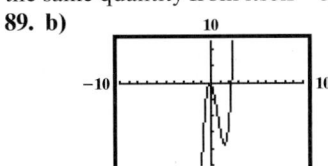

   c) Answers will vary.   **d)** Factoring is not correct.
91. Determine if all the terms contain a greatest common factor and, if so, factor it out.
93. a) Answers will vary.   **b)** $2x^2y$ **c)** $2x^2y(3y^4 - x + 6x^7y^2)$
95. $3(x - 4)^3$ **97. a)** Answers will vary.   **b)** $(3x^2 - y^3)(2x + y^2)$
99. $-\dfrac{15}{72} = -\dfrac{5}{24}$

100. 2   **101.**

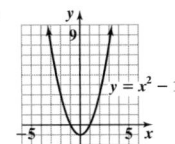

   102. 0.4 hour   **103.** $-14a^3 - 22a^2 + 19a - 3$

Mid-Chapter Test

1. $5x^4 - 1.5x^3 + 2x - 7, 4$ [5.1]   **2.** $\dfrac{3}{2}$ or $1\dfrac{1}{2}$ [5.1]   **3.** $-n^2 - 7n - 4$ [5.1]   **4.** $-16x^2y + 14xy$ [5.1]   **5.** $9x^2 - 4x + 13$ [5.1]

6. $6x^6y^4 + 10x^7 - 14x^8y$ [5.1]   **7.** $21x^2 - 4xy - 12y^2$ [5.2]   **8.** $6x^4 - x^3 + 14x^2 + 32x + 9$ [5.2]   **9.** $64p^2 - \dfrac{1}{25}$ [5.2]

10. $12m^3 - m^2n - 30mn^2 + 18n^3$ [5.2]   **11.** $x^2 - 14x + 49 = (x - 7)^2$ [5.2]   **12.** $2x^2y + 3 - \dfrac{11}{2xy^2}$ [5.3]

13. $3x + 5 + \dfrac{2}{4x + 1}$ [5.3]   **14.** $y^2 + y + 5 + \dfrac{5}{2y - 3}$ [5.3]   **15.** $x - 9$ [5.3]   **16.** $3a^3 + 4a^2 - 6a - 1$ [5.3]

17. $8b^2c(4bc^2 + 2 + 3b^3c^3)$ [5.4]   **18.** $(2x + 9)(7b - 3c)$ [5.4]   **19.** $b^2(b + 2c)(2b - c)$ [5.4]   **20.** $(3x - 2)^5(5a - 12x + 8)$ [5.4]

Exercise Set 5.5

1. Prime   **3.** Trial and error   **5.** Using substitution   **7.** -3 **9.** Both are $+$ **11.** One is $+$, one is $-$ **13.** $(x + 3)(x + 4)$
15. $(b - 1)(b + 9)$ **17.** $(z + 2)^2$ **19.** $(r + 12)^2$ **21.** $(x + 32)(x - 2)$ **23.** $(x + 2)(x - 15)$ **25.** $-(a - 15)(a - 3)$
27. Prime   **29.** $-2(m + 2)(m + 5)$ **31.** $4(r + 4)(r - 1)$ **33.** $x(x + 6)(x - 3)$ **35.** $(a - 1)(5a - 3)$ **37.** $3(x - 2)(x + 1)$
39. $(3c + 7)(2c - 9)$ **41.** $(2b + 1)(4b - 3)$ **43.** $(3c - 2)(2c + 5)$ **45.** $4(2p - 3q)(2p + q)$ **47.** Prime
49. $2(3a + 4b)(3a - b)$ **51.** $(4x - 3y)(2x + 9y)$ **53.** $10(5b - 2)(2b - 1)$ **55.** $ab^5(a - 4)(a + 3)$ **57.** $3b^2c(b - 3c)^2$
59. $4m^6n^3(m + 2n)(2m - 3n)$ **61.** $(6x - 5)(5x + 4)$ **63.** $8x^2y^5(x + 4)(x - 1)$ **65.** $(x^2 + 3)(x^2 - 2)$ **67.** $(b^2 + 5)(b^2 + 4)$
69. $(2a^2 + 5)(3a^2 - 5)$ **71.** $(2x + 5)(2x + 3)$ **73.** $(3a + 1)(2a + 5)$ **75.** $(xy + 7)(xy + 2)$ **77.** $(2xy - 11)(xy + 1)$
79. $(2 - y)(y - 1)(2y - 5)$ **81.** $(p - 4)(2p + 3)(p + 2)$ **83.** $(a^3 - 10)(a^3 + 3)$ **85.** $(x + 5)(x + 2)(x + 1)$
87. $a^3b^2(5a - 3b)(a - b)$ **89.** $(x + 6)(x + 1)$ **91.** $(x + 6)(x + 3)$ **93.** $2x^2 - 5xy - 12y^2$, multiply $(2x + 3y)(x - 4y)$
95. Divide, $x + 7$ **97. a)** Answers will vary.   **b)** $(6x - 5)(5x + 8), (7x - 1)(7x - 13)$ **99.** $\pm 3, \pm 9$
101. 6 or -6, b is the sum of the factors of 5   **103. a)** 4   **b)** $(x - 3)(x - 5)$ **105. a)** -8 **b)** Not factorable
107. Answers will vary. One example is $x^2 + 2x + 1$.   **109.** $(2a^n + 3)(2a^n - 5)$ **111.** $(x + y)^2(x - 4y)(x - 3y)$
113. $(x^n - 2)(x^n + 5)$ **115. a)** Answers will vary.   **b)** Correct   **117.** Factor out the greatest common factor if it is present.
119. a) Answers will vary.   **b)** $(2x + 3)(3x - 4)$ **121.** No; $(x + 3)(x + 1)$; $(2x + 2)$ has the GCF 2

123. No, $3x(x + 4)(x - 2)$; $(3x - 6)$ has the GCF 3   **125.** $C = \dfrac{5}{9}(F - 32)$ **126.**

127. 4   **128.** $x^2 + 2xy + y^2 + 12x + 12y + 36$
129. $(2x^2 - 5)(x + 2)$

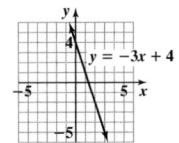

Exercise Set 5.6

1. Two   **3.** Perfect square   **5.** $(x + 7)^2$ **7.** Sum   **9.** $(a + b)(a^2 - ab + b^2)$ **11.** $(x + 9)(x - 9)$ **13.** $(a + 10)(a - 10)$
15. $(1 + 7b)(1 - 7b)$ **17.** $(5 + 4y^2)(5 - 4y^2)$ **19.** $\left(\dfrac{1}{10} + y\right)\left(\dfrac{1}{10} - y\right)$ **21.** $(xy + 11c)(xy - 11c)$
23. $(0.2x + 0.3)(0.2x - 0.3)$ **25.** $x(12 - x)$ **27.** $(a + 3b + 2)(a - 3b - 2)$ **29.** $(x + 5)^2$ **31.** $(7 - t)^2$ **33.** $(6pq + 1)^2$
35. $(0.9x - 0.2)^2$ **37.** $(y^2 + 2)^2$ **39.** $(a + b + 3)^2$ **41.** $(y + 1)^2$ **43.** $(x + 3 + y)(x + 3 - y)$ **45.** $(x + 7)(3 - x)$
47. $(3a - 2b + 3)(3a - 2b - 3)$ **49.** $(y^2 - 3)^2$ **51.** $(a + 5)(a^2 - 5a + 25)$ **53.** $(4 - a)(16 + 4a + a^2)$
55. $(p - 3a)(p^2 + 3ap + 9a^2)$ **57.** $(3y - 2x)(9y^2 + 6xy + 4x^2)$ **59.** $2(2a - 3b)(4a^2 + 6ab + 9b^2)$

61. $(x^2 + y^3)(x^4 - x^2y^3 + y^6)$ **63.** $(x + 2)(x^2 + x + 1)$ **65.** $(a - b - 3)(a^2 - 2ab + b^2 + 3a - 3b + 9)$
67. $-9(b^2 + 3b + 3)$ **69.** $(a^2 + 2b^2)(a^2 - 2b^2)$ **71.** $(7 + 8xy)(7 - 8xy)$ **73.** $(x + y + 4)(x + y - 4)$
75. $(x - 4)(x^2 + 4x + 16)$ **77.** $(3xy + 4)^2$ **79.** $(a^2 + b^2)^2$ **81.** $(x - 1 + y)(x - 1 - y)$
83. $(x + y + 1)(x^2 + 2xy + y^2 - x - y + 1)$ **85.** $3m(-m + 2n)$ **87.** $(3x - 2)(9x^2 + 6x + 4)$
89. $(6a - b)(36a^2 + 6ab + b^2)$ **91.** $(4x + 3a)(16x^2 - 12ax + 9a^2)$ **93. a)** $a^2 - b^2$ **b)** $(a + b)(a - b)$ **95. a)** $6a^3 - 6ab^2$
b) $6a(a + b)(a - b)$ **97.   a)** $\frac{4}{3}\pi R^3 - \frac{4}{3}\pi r^3$ **b)** $\frac{4}{3}\pi(R - r)(R^2 + Rr + r^2)$ **99.** $12, -12$; write $4x^2 + bx + 9$ as
$(2x)^2 + bx + (3)^2$; $bx = 2(2x)(3)$ or $bx = -2(2x)(3)$ **101.** 4; write $25x^2 + 20x + c$ as $(5x)^2 + 20x + (a)^2$ then $20x = 2(5x)(a)$.
Since $a = 2, c = 4$ **103.   a)** Find an expression whose square is $25x^2 - 30x + 9$ **b)** $s(x) = 5x - 3$ **c)** 7
105. $(x^2 + 4x + 8)(x^2 - 4x + 8)$ **107.** $h(2a + h)$ **109. a)** 16   **b)** $x^2 + 8x + 16$ **c)** $(x + 4)^2$ **111.** $(8x^{2a} + 3y^{3a})(8x^{2a} - 3y^{3a})$
113. $(a^n - 8)^2$ **115.** $(x^n - 2)(x^{2n} + 2x^n + 4)$ **117.** Correct   **119. a)** Answers will vary.   **b)** $(x + 4)(x - 4)$
121. $a^3 + b^3 = (a + b)(a^2 - ab + b^2)$ **123.** Answers will vary.   **125.** No. $(x + 7)(x - 7) = x^2 - 49$
127. No. $x^2 - 81 = (x + 9)(x - 9)$ **129. a)** $(x^3 + 1)(x^3 - 1)$ **b)** $(x^2 - 1)(x^4 + x^2 + 1)$ **131.** $4x + 7y - 2$ **132.** -17
133. $20°, 30°, 40°$ **134.** $15y^{10}(3y^2 + 4)$ **135.** $(4x - 3y)(3x + y)$

Exercise Set 5.7

1. -5 **3.** $3(x + 5)(x - 5)$ **5.** $(5s - 3)(2s + 5)$ **7.** $2x^2y^2(3x + 5y + 7)$ **9.** $0.8(x + 0.3)(x - 0.3)$ **11.** $6x(x^2 + 3)(x^2 - 3)$
13. $3x^4(x - 1)(x + 4)$ **15.** $5x^2y^2(x + 4)(x + 3)$ **17.** $x^2(x + y)(x - y)$ **19.** $x^4y^2(x - 1)(x^2 + x + 1)$
21. $x(x^2 + 4)(x + 2)(x - 2)$ **23.** $4(x^2 + 2y)(x^4 - 2x^2y + 4y^2)$ **25.** $5(a + b + 2)(a + b - 2)$ **27.** $6(x + 3y)^2$ **29.** $x(x + 4)$
31. $3(2x - y)(x + 4y)$ **33.** $(y + 7)^2$ **35.** $(b^2 + 1)^2$ **37.** $\left(x + \frac{1}{4}\right)\left(x^2 - \frac{1}{4}x + \frac{1}{16}\right)$ **39.** $2y(3y + 1)(y + 2)$
41. $ab(a + 9b)(a - 9b)$ **43.** $(7 + x + y)(7 - x - y)$ **45.** $2(3x - 2)(4x - 3)$ **47.** $3(3x - 1)(2x + 5)$
49. $(x^2 + 4)(x + 2)(x - 2)$ **51.** $(b - 2x)(5c - 7y)$ **53.** $(3x^2 - 4)(x^2 + 1)$ **55.** $(z + x - 6)(z - x + 6)$ **57.** $(2y + 5)(y + 8)$
59. $(a + 6b + 4c)(a + 6b - 4c)$ **61.** $5x^2y(x + 3)(2x - 1)$ **63.** $(x + y)^2(x - y)^2$ **65. e)**   **67. d)**   **69. f)**   **71. c)**
73. $2(x + 3)(x + 2)$ **75.** $(x + 6)(x + 2)$ **77.** $(y + 3)(y - 3)$ **79.** $(5x - 3)(25x^2 + 15x + 9)$
81. a) $a(a + b) - b(a + b) = a^2 - b^2$ **b)** $(a + b)(a - b)$ **83. a)** $a^2 + 2ab + b^2$ **b)** $(a + b)^2$
85. a) $a(a - b) + a(a - b) + b(a - b) + b(a - b)$ or $2a(a - b) + 2b(a - b)$ **b)** $2(a + b)(a - b)$ **87. a)** Answers will vary.
b) Answers will vary.   **89. a)** $x^{-5}(x^2 - 2x - 3)$ **b)** $x^{-5}(x - 3)(x + 1)$ **91. a)** $x^{-3/2}(5x^2 + 2x - 3)$ **b)** $x^{-3/2}(5x - 3)(x + 1)$
92. 1   **93.** $\{z | z < -6 \text{ or } z > 0\}$ **94.** ≈ 17.3 pounds at $5.20, ≈ 12.7 pounds at $6.30   **95.** $5x^3 - x^2 + 16x + 16$
96. $(x + 3)(2x^2 - 5)$

Exercise Set 5.8

1. Equation   **3.** $ax^2 + bx + c = 0$ **5.** x-intercept   **7.** $0, -3$ **9.** $0, 1$ **11.** $-1, 7$ **13.** $0, 4, 9$ **15.** $\frac{2}{3}, \frac{1}{7}$ **17.** $0, 3$ **19.** $0, -5$
21. $0, 6$ **23.** $0, 9$ **25.** $-1, -5$ **27.** $-4, 3$ **29.** -4 **31.** $-\frac{1}{2}, 5$ **33.** $\frac{3}{2}, -2$ **35.** $-4, 6$ **37.** $0, 6, -3$ **39.** $0, -4, 3$ **41.** $0, -\frac{1}{2}, \frac{4}{3}$
43. $5, -5$ **45.** $-\frac{3}{4}, \frac{3}{4}$ **47.** $0, -3, 3$ **49.** $-11, 9$ **51.** $-3, -11$ **53.** $-1, -4$ **55.** $\frac{5}{2}, \frac{4}{3}$ **57.** $0, -3, -5$ **59.** $-2, -\frac{1}{3}$ **61.** $\frac{3}{5}, \frac{5}{2}$
63. $6, -5$ **65.** $(4, 0), (6, 0)$ **67.** $(-8, 0)$ **69.** $(0, 0), \left(\frac{4}{3}, 0\right), \left(\frac{5}{2}, 0\right)$ **71.** $x = 1$ **73.** $x = 5$ **75.** $x = 9$ **77. d)**   **79. b)**
81. $y = x^2 - 6x + 5$ **83.** $y = x^2 - 2x - 8$ **85.** $y = 6x^2 - 7x - 10$ **87.** Width = 2 feet, length = 5 feet
89. Base = 10 feet, height = 16 feet   **91.** 2 feet   **93.** 3 feet   **95.** 2 seconds   **97.** Tim: 5 miles, Bob: 12 miles   **99.** 13 feet
101. 50 bicycles   **103.** 13 inches by 13 inches   **105. a)** $V = a^3 - ab^2$ **b)** $V = a(a + b)(a - b)$ **c)** 3 inches
107. a) $f(x) = x^2 + 7x + 10$ **b)** $x^2 + 7x + 10 = 0$ **c)** An infinite number; any function of the form
$f(x) = a(x^2 + 7x + 10), a \neq 0$ **d)** An infinite number; any equation of the form $a(x^2 + 7x + 10) = 0, a \neq 0$
109.   a) Answers will vary. Examples are:

b) None (no x-intercepts), one (one x-intercept), or two (two x-intercepts)
111. ≈ 73.721949 mph
113. A second-degree equation in one variable
115. a) Answers will vary.   **b)** $\frac{7}{3}, -\frac{3}{2}$

117. Constant term does not contain a variable   **119. a)** Factor out -1 **b)** $7, -5$ **121.** $a^2 + b^2 = c^2$ **123.** $\left(\frac{3}{2}, 0\right), (6, 0)$

125. Yes   **127.** No   **129.** $\pm 2, \pm 3$ **134.** $\frac{x^4}{16y^6}$ **135.**   **136.** $(2, -1)$ **137.** -84 **138.** $(x + 3)(x - 2)$

Chapter 5 Review Exercises

1. a) Binomial   **b)** $3x^2 + 9$ **c)** 2   **2. a)** Trinomial   **b)** $4x^3 + 5x - 7$ **c)** 3   **3.** Not a polynomial   **4. a)** Polynomial
b) $2x^4 - 10x^2y + 6xy^3 - 3$ **c)** 4   **5.** $x^2 - 3x + 14$ **6.** $5x^2 + 11x - 4$ **7.** $3a - 8b + 7$ **8.** $6x^3 - 9x + 13$
9. $3x^2y + 3xy - 9y^2$ **10.** $-3ab + b^2 - 2a$ **11.** $5x^2 + 7x + 3$ **12.** $-10a^2b + ab$ **13.** 21   **14.** -76 **15.** $3x^2 + 27$
16. $2x^2 + 24x + 23$ **17. a)** $780.46 billion   **b)** Yes   **18. a)** $773.13 billion   **b)** Yes   **19.** $6x^3 - 14x^2 + 10x$

20. $-3x^4y^2 - 3x^2y^6 + 12xy^7$ **21.** $6x^2 + 17x - 45$ **22.** $50a^2 - 5a - 3$ **23.** $x^2 + 16xy + 64y^2$ **24.** $a^2 - 22ab + 121b^2$
25. $10x^2y + 8xy^2 - 5x - 4y$ **26.** $6p^2q^2 + 11pqr - 7r^2$ **27.** $4a^2 + 36ab + 81b^2$ **28.** $16x^2 - 24xy + 9y^2$ **29.** $49x^2 - 25y^2$
30. $4a^2 - 25b^4$ **31.** $16x^2y^2 - 36$ **32.** $81a^4 - 4b^4$ **33.** $x^2 + 6xy + 9y^2 + 4x + 12y + 4$ **34.** $4p^2 - 4pq + q^2 - 20p + 10q + 25$
35. $6x^3 - x^2 - 24x + 18$ **36.** $4x^4 + 12x^3 + 6x^2 + 16x - 6$ **37.** $x^2 + 8x + 10$ **38.** $x^2 + xy + 4y + xz$ **39. a)** $x^2 - 2x - 3$
b) 0 **40. a)** $2x^3 - 4x^2 - 6x + 12$ **b)** 12 **41. a)** $x^3 - x^2 - 5x + 6$ **b)** 9 **42. a)** $x^4 - 4$ **b)** 77 **43.** $\frac{1}{5}x^6y^2$ **44.** $\frac{1}{4}t^5$

45. $9p - 5q - 3$ **46.** $\frac{7}{4}a^2 - 4a + 8$ **47.** $\frac{x^2}{4y} + x + \frac{3y}{2}$ **48.** $4x - 3$ **49.** $x^3 - 2x^2 + 3x + 7$ **50.** $2a^3 + a^2 - 3a - 4$

51. $x + 4 - \dfrac{10}{x - 3}$ **52.** $2x^2 + 3x - 4 + \dfrac{3}{2x + 3}$ **53.** $3x^2 + 7x + 21 + \dfrac{73}{x - 3}$ **54.** $2y^4 - 2y^3 - 8y^2 + 8y - 7 + \dfrac{5}{y + 1}$

55. $x^4 + 2x^3 + 4x^2 + 8x + 16 + \dfrac{14}{x - 2}$ **56.** $2x^2 + 2x + 6$ **57.** 10 **58.** -236 **59.** $-\dfrac{53}{9}$ or $-5.\overline{8}$ **60.** 0, factor

61. $4(x^2 + 2x + 8)$ **62.** $3x^4(5x + 2 - 4xy^3)$ **63.** $2a^2b^3(5a - 7b^3)$ **64.** $6xy^2z^2(4y^2z + 2xy - 5x^2z)$ **65.** $(5x - y)(x + 6y)$
66. $(3a + 2b)(4a + 5b)$ **67.** $(2x - 5)(x + 9)$ **68.** $(3x - 7)(16x - 21)$ **69.** $(5x + 2)(13x - 7)$ **70.** $(7x + 9)(2x - 1)$
71. $(17x + 3)(9x - 7)$ **72.** $(4x + 5)(5x - 2)$ **73.** $(x + 6)(x + 3)$ **74.** $(x - 2)(x + 5)$ **75.** $(x - 7)(x + 4)$
76. $(x - 8)(x - 2)$ **77.** $-(x - 15)(x + 3)$ **78.** $-(x - 12)(x - 1)$ **79.** $x(2x + 1)(x + 6)$ **80.** $x^2(4x - 5)(2x + 5)$
81. $a^3(4a - 5)(a - 1)$ **82.** $y^3(12y + 1)(y + 5)$ **83.** $(x - 18y)(x + 3y)$ **84.** $(2p - 5q)(3p - 2q)$ **85.** $(x^2 + 3)(x^2 + 7)$
86. $(x^2 - 7)(x^2 + 9)$ **87.** $(x + 9)(x + 7)$ **88.** $x(x - 9)$ **89.** $(x + 10)(x + 1)$ **90.** $(x + 10)(x + 2)$ **91.** $(x + 6)(x - 6)$
92. $(x + 11)(x - 11)$ **93.** $(x^2 + 9)(x + 3)(x - 3)$ **94.** $(x^2 + 4)(x + 2)(x - 2)$ **95.** $(2a + 1)^2$ **96.** $(4y - 3)^2$
97. $(x - 2)(x + 6)$ **98.** $(3y + 5)(3y - 7)$ **99.** $(p^2 + 9)^2$ **100.** $(m^2 - 10)^2$ **101.** $(x + 4 + y)(x + 4 - y)$
102. $(a + 3b + 6c)(a + 3b - 6c)$ **103.** $(4x + y)^2$ **104.** $(6b - 5c)^2$ **105.** $(x - 3)(x^2 + 3x + 9)$
106. $(y + 4z)(y^2 - 4yz + 16z^2)$ **107.** $(5x - 1)(25x^2 + 5x + 1)$ **108.** $(2a + 3b)(4a^2 - 6ab + 9b^2)$
109. $(y - 4z)(y^2 + 4yz + 16z^2)$ **110.** $(x - 5)(x^2 - x + 7)$ **111.** $(x - 1)(x^2 + 4x + 7)$ **112.** $(a + 5)(a^2 + 7a + 13)$
113. $(x + 3)(x - 3)$ **114.** $(a + 2b)(a - 2b)$ **115.** $(2x - y)(4x^2 + 2xy + y^2)$ **116.** $4a(a + c)(a - c)$ **117.** $y^4(x + 3)(x - 5)$
118. $5x(x - 4)(x - 2)$ **119.** $3xy^4(x - 2)(x + 6)$ **120.** $3y(y^2 + 5)(y^2 - 5)$ **121.** $4y(x + 2)(x^2 - 2x + 4)$
122. $5x^2y(x + 2)^2$ **123.** $3x(2x + 1)(x - 4)$ **124.** $(x + 5 + z)(x + 5 - z)$ **125.** $5(x + 2y)(x^2 - 2xy + 4y^2)$
126. $(x + 4)(x - 1)(x + 6)$ **127.** $(4x + 1)(4x + 5)$ **128.** $(2x^2 - 1)(2x^2 + 3)$ **129.** $(x + 1)^2(x - 2)$ **130.** $(3a - b)(3x + 7y)$
131. $(2pq - 3)(3pq + 2)$ **132.** $(3x^2 - 2)^2$ **133.** $(4y + x + 2)(4y - x - 2)$ **134.** $2(3a + 5)(4a + 3)$

135. $3x^2y^5(x + 3)(2x - 3)$ **136.** $\left(x - \dfrac{2}{3}y^2\right)\left(x^2 + \dfrac{2}{3}xy^2 + \dfrac{4}{9}y^4\right)$ **137.** $(x + 9)(x + 2)$ **138.** $(y + 10)(y + 5)$

139. $(a + 2b)(a - 2b)$ **140.** $2b(a + b)$ **141.** $(2a + b)(a + 3b)$ **142.** $(a + b)^2$ **143.** $2, -\dfrac{1}{4}$ **144.** $-\dfrac{5}{2}, -\dfrac{10}{3}$ **145.** $0, 2$

146. $0, -\dfrac{4}{3}$ **147.** $-4, -3$ **148.** $5, -6$ **149.** $7, 1$ **150.** $0, 2, 4$ **151.** $4, -4$ **152.** $2, -3$ **153.** $\dfrac{4}{3}, -\dfrac{1}{4}$ **154.** $\dfrac{3}{4}, -\dfrac{2}{5}$

155. $(-3, 0), (6, 0)$ **156.** $\left(\dfrac{6}{5}, 0\right), \left(\dfrac{5}{4}, 0\right)$ **157.** $y = x^2 - 2x - 24$ **158.** $y = 12x^2 + 32x + 5$

159. Width = 9 feet, length = 12 feet   **160.** Height = 4 feet, base = 13 feet   **161.** 3 inches, 7 inches   **162.** 9 seconds   **163.** 9

Chapter 5 Practice Test

1. a) Trinomial   **b)** $-6x^4 - 4x^2 + 3x$ **c)** 4 **d)** -6 [5.1]   **2.** $4x^2y - 14y^2 + 4x + 6y$ [5.1]   **3.** $-8x^8y^3 + 24x^6y^4 - 12x^4y^2$ [5.2]

4. $10a^2 - 13ab - 3b^2$ [5.2]   **5.** $4x^3 + 8x^2y - 9xy^2 - 6y^3$ [5.2]   **6.** $4x^4 - 5y + \dfrac{7}{x^2}$ [5.3]   **7.** $x - 5 + \dfrac{24}{2x + 3}$ [5.3]

8. $3x^3 + 3x^2 + 15x + 15 + \dfrac{76}{x - 5}$ [5.3]   **9.** -85 [5.3]   **10.** $2xy(6x^2 + 5xy^3 - 7y^2)$ [5.4]   **11.** $x(x - 3)(x + 1)$ [5.5]

12. $(a + 2b)(2a + 3b)$ [5.4]   **13.** $(2b^2 + 9)(b^2 - 2)$ [5.5]   **14.** $4x(x - 5)$ [5.5]   **15.** $(x + 7)(x + 3)$ [5.5]
16. $q^6(3p - 2)(9p^2 + 6p + 4)$ [5.6]   **17. a)** $3x^2 - 19x + 20$ **b)** -6 [5.2]   **18.** $4(x + y)(x - y)$ [5.5]   **19.** $(x + 11)(x + 4)$ [5.5]
20. $\dfrac{3}{7}, -4$ [5.8]   **21.** $0, -5, 2$ [5.8]   **22.** $\left(\dfrac{1}{4}, 0\right), \left(-\dfrac{3}{2}, 0\right)$ [5.8]   **23.** $y = x^2 - 9x + 14$ [5.8]
24. Height = 4 meters, base = 11 meters [5.8]   **25.** 7 seconds [5.8]

Cumulative Review Test

1. $A \cup B = \{2, 3, 4, 5, 6, 8\}$ [1.2]   **2.** 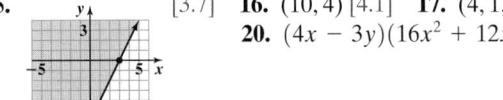 [1.2]   **3.** $-\dfrac{3}{32}$ [1.3]   **4.** -34 [1.4]   **5.** $8r^6s^{15}$ [1.5]   **6.** 5 [2.1]   **7.** $e = \dfrac{k - 2d}{2}$ [2.2]

8. 13 meters by 13 meters [2.3]   **9.** 620 pages [2.3]   **10.** $34 \le x < 84$ [2.5]   **11.** No [3.1]   **12.** $6x - 3y = 2$ [3.3]   **13.** $-\dfrac{2}{9}$ [3.4]

14. -180 [3.6]   **15.** [3.7]   **16.** $(10, 4)$ [4.1]   **17.** $(4, 1, 2)$ [4.2]   **18.** 18 [4.5]   **19.** $2x^2 + 12x + 63 + \dfrac{393}{x - 6}$ [5.3]
20. $(4x - 3y)(16x^2 + 12xy + 9y^2)$ [5.6]

Chapter 6

Exercise Set 6.1

1. Rational expression **3.** Domain **5.** Multiply **7.** $\dfrac{x+3}{x+8}$ **9.** $\dfrac{x^3}{y^2}$ **11.** 0 **13.** $5, \dfrac{5}{2}$ **15.** None **17.** $9, -9$ **19.** $\{x \mid x \neq 5\}$

21. $\{x \mid x \neq -3 \text{ and } x \neq 2\}$ **23.** $\{a \mid a \neq -3 \text{ and } a \neq -1\}$ **25.** $\{x \mid x \text{ is a real number}\}$ **27.** $\{a \mid a \neq -6 \text{ and } a \neq 6\}$ **29.** $x + 1$

31. $\dfrac{x - 4y}{3}$ **33.** x **35.** -1 **37.** $-(p + 4)$ **39.** $\dfrac{a - 5}{a + 3}$ **41.** $4x^2 + 10xy + 25y^2$ **43.** $\dfrac{2x - 5}{2}$ **45.** $\dfrac{a + 7}{a + 5}$ **47.** $\dfrac{x - 4}{x^2 - 3x + 9}$

49. $\dfrac{x}{15y}$ **51.** $12x^3y^2$ **53.** 1 **55.** $\dfrac{x + 5}{4}$ **57.** $\dfrac{r^3}{r - 8}$ **59.** $\dfrac{7}{x - 4}$ **61.** $\dfrac{(a + 1)^2}{9(a + b)^2}$ **63.** $\dfrac{x - 4}{4x + 1}$ **65.** $\dfrac{(x + 2)(x - 2)}{(x^2 + 2x + 4)(x^2 + 4)}$

67. $\dfrac{x - y}{x + y}$ **69.** $\dfrac{x^3}{x + 2}$ **71.** $\dfrac{(a - b)(a + b)}{a^2 + ab + b^2}$ **73.** 1 **75.** $\dfrac{p - q}{p + q}$ **77.** $\dfrac{r + 5s}{2r + 5s}$ **79.** $x + 5$; numerator must be $x + 5$

81. $y^2 - 4y - 5$, factors must be $(y - 5)(y + 1)$ **83.** $x^2 + x - 2$; factors must be $(x - 1)(x + 2)$

85. $2x^2 + x - 6$; factors must be $(x + 2)(2x - 3)$. **87.** $\dfrac{3a + b}{2}$ **89.** $2(a + b)$ **91.** $\dfrac{(x + 2)(3x + 1)}{(2x - 3)(x + 1)}$ **93.** $\dfrac{x - 1}{x + 3}$ **95.** $\dfrac{1}{x^4(x - p)^n}$

97. x^y **99. a)** $\{x \mid x \neq 2\}$ **b)** 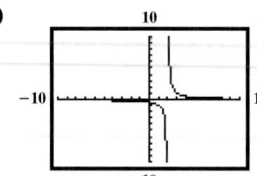 **101. a)** $\{x \mid x \neq 2\}$ **b)** 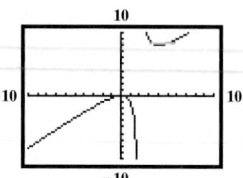

103. a) $\{x \mid x \neq 0\}$ **b)** $-0.1, -1, -2, -10, -100, 100, 10, 2, 1, 0.1$ **c)** **d)** No; numerator can never be 0

105. One possible answer is $\dfrac{1}{(x - 2)(x + 3)}$; denominator 0 at $x = 2$ and $x = -3$ **107.** Numerator is never 0.

109. a) 4, makes numerator 0 **b)** 6 and -6, makes denominator 0

111. One possible answer is $f(x) = \dfrac{x - 2}{(x - 3)(x + 1)}$; numerator 0 at $x = 2$, denominator 0 at $x = 3$ and $x = -1$ **113.** $y = x + 2$

114. $\left(-\infty, \dfrac{3}{2}\right)$ **115.** $-28, 32$ **116.** 0.1 **117.** $(2, -1)$ **118.** $(3x + y + 2)(3x + y - 2)$

Exercise Set 6.2

1. Least common denominator **3.** Highest **5.** $\dfrac{4x + 1}{x + 7}$ **7.** $\dfrac{7x - 2}{x - 5}$ **9.** $\dfrac{x + 7}{x + 3}$ **11.** $\dfrac{7x - 11}{x - 8}$ **13.** $\dfrac{x - 3}{x - 1}$ **15.** $x - 5$

17. $\dfrac{x + 5}{x + 3}$ **19.** $12a^2$ **21.** $40x^4y^6$ **23.** $6a^4b^5$ **25.** $(x + 3)(x + 9)$ **27.** $z - 6$ **29.** $x^4(x - 2)^3$ **31.** $(a - 8)(a + 3)(a + 8)$

33. $(x - 3)(2x - 1)(2x + 3)$ **35.** $\dfrac{13}{2x}$ **37.** $\dfrac{5x - 3}{12x^2}$ **39.** $\dfrac{15y^2 + 8x^2}{40x^4y^3}$ **41.** $\dfrac{2b^2 - a^2}{b(a - b)}$ **43.** $\dfrac{2a}{a - b}$ **45.** $\dfrac{5x^2 + 3x - 12}{(x - 4)(x + 1)}$

47. $\dfrac{6a + 7}{(a + 2)^2}$ **49.** $\dfrac{2x^2 + 4x + 4}{(x - 1)(x + 4)(x - 2)}$ **51.** $\dfrac{2x + 3}{(x - 8)(x - 1)}$ **53.** $\dfrac{4x^2 + 11x - 39}{(x + 5)(x - 2)}$ **55.** $\dfrac{3a - 1}{4a + 1}$ **57.** $\dfrac{2x^2 - 4xy + 4y^2}{(x - 2y)^2(x + 2y)}$

59. $\dfrac{16}{r - 4}$ **61.** 0 **63.** $\dfrac{15x^2 - 70x + 30}{(3x - 2)(x - 4)}$ **65.** $\dfrac{18r^2 + 11r - 25}{(4r - 5)(2r + 3)}$ **67.** $\dfrac{x^2 - 18x - 30}{(5x + 6)(x - 2)}$ **69.** $\dfrac{12m^2 + 7mn}{(2m + 3n)(3m + 2n)(2m + n)}$

71. 0 **73.** $\dfrac{1}{2x + 3y}$ **75. a)** $\{x \mid x \neq 3\}$ **b)** $\{x \mid x \neq -4\}$ **c)** $\dfrac{2x^2 + 3x + 8}{(x - 3)(x + 4)}$ **d)** $\{x \mid x \neq 3 \text{ and } x \neq -4\}$ **77.** $\dfrac{2x^2 + 8x - 3}{(x + 1)(x + 2)}$

79. $\dfrac{3x^2 + 19x + 7}{(x + 2)(x + 3)}$ **81.** $\dfrac{x^2 + 5x + 4}{(x + 2)(x - 2)(x + 3)}$ **83.** $\dfrac{2x}{x^4 + x^3 - 10x^2 - 4x + 24}$ **85.** $\dfrac{ad}{bd} + \dfrac{bc}{bd} = \dfrac{ad + bc}{bd}$

87. a) 4 **b)** $\dfrac{a^2 - b^2}{a^2}$ **89.** $7x^2 - 6x + 6$; $5x^2 - (7x^2) = -2x^2, -(-6x) = 6x, -6 - (6) = -12$ **91.** $\dfrac{3x + 10}{x - 2}$ **93.** $\dfrac{1}{(a - 5)(a + 3)}$

95. $-x^2 + 4x + 5$ **97. a)** $\dfrac{ax + bn - bx}{n}$ **b)** 79.2 **99.** $\dfrac{a - b + 1}{(a - b)^2}$ **101. a)** The entire numerator was not substracted.

b) $\dfrac{x^2 - 4x - x^2 - x + 2}{(x + 3)(x - 2)}$ **103.** No **105.** Yes, if you multiply either fraction by $\dfrac{-1}{-1}$ you get the other fraction.

107. a) $\dfrac{x + 1}{x}$ **b)** $\dfrac{x^2 + x + 1}{x^2}$ **c)** $\dfrac{x^4 + x^3 + x^2 + x + 1}{x^4}$ **d)** $\dfrac{x^n + x^{n-1} + x^{n-2} + \cdots + 1}{x^n}$ **109.** $\dfrac{-h}{(a + 1)(a + h + 1)}$

110. a) 6 minutes **b)** 960 boxes **111.** $\{x \mid -2 < x < 8\}$ **112.** $-\dfrac{2}{3}$ **113.** -11 **114.** $3x - 7 + \dfrac{27}{2x + 3}$ **115.** $\dfrac{1}{3}, 7$

Exercise Set 6.3

1. Complex fraction **3.** Denominator **5.** $\dfrac{75a}{b^5}$ **7.** $\dfrac{12x^3}{y^6}$ **9.** $\dfrac{3z^4}{4xy^4}$ **11.** $\dfrac{x(y - 1)}{8 + x}$ **13.** $\dfrac{xy + 5}{y + x}$ **15.** $\dfrac{5}{3a^2}$ **17.** $-\dfrac{a}{b}$ **19.** $\dfrac{x - y}{y}$

21. -1 **23.** $\dfrac{3(x+2)}{x^5}$ **25.** $\dfrac{-a+1}{(a+1)(2a+1)}$ **27.** $\dfrac{x-1}{x+1}$ **29.** $\dfrac{a^2+1}{2a}$ **31.** $\dfrac{x}{5(x-3)}$ **33.** $\dfrac{x^2+5x-6}{x(x-2)}$ **35.** $\dfrac{a(a+3)}{(a-2)(a+1)}$

37. $\dfrac{ab}{b+a}$ **39.** $\dfrac{2}{y-x}$ **41.** $\dfrac{b(1+a)}{a(1-b)}$ **43.** $\dfrac{b^2-a^3b}{a^3+ab}$ **45.** $\dfrac{9a^2+b}{b(b+1)}$ **47.** $\dfrac{(a+b)^2}{ab}$ **49.** $\dfrac{15y-x}{3xy}$ **51.** $\dfrac{2y-8xy+5y^2}{3y^2-4x}$

53. $\dfrac{x^2+9x+14}{x+1}$ **55.** $\dfrac{x^2+3x-4}{x+1}$ **57. a)** $\dfrac{2}{9}$ **b)** $\dfrac{1}{5}$ **59.** $R_T=\dfrac{R_1R_2R_3}{R_2R_3+R_1R_3+R_1R_2}$ **61.** a **63.** $\dfrac{-1}{a(a+h)}$

65. $\dfrac{-1}{(a+1)(a+h+1)}$ **67.** $\dfrac{-2a-h}{a^2(a+h)^2}$ **69.** $\dfrac{5}{12}$ **71.** $\dfrac{4a^2+1}{4a(2a^2+1)}$ **72.** $\dfrac{13}{48}$ **73.** $\left(-23,-\dfrac{34}{5}\right)$ **74.** $\left\{3,\dfrac{5}{3}\right\}$ **75.** Neither

Exercise Set 6.4

1. Least common denominator **3.** Extraneous solution **5.** Cross-multiplication **7.** Factoring **9.** 5 **11.** $\dfrac{11}{2}$ **13.** 3 **15.** -5

17. All real numbers **19.** $\dfrac{1}{4}$ **21.** $\dfrac{11}{3}$ **23.** No solution **25.** $\dfrac{6}{5}$ **27.** ≈-1.63 **29.** 8 **31.** $-\dfrac{4}{3}$, 1 **33.** 3.76 **35.** $-1,-6$ **37.** -5

39. $-\dfrac{5}{2}$ **41.** 5 **43.** No solution **45.** -5 **47.** $\dfrac{17}{4}$ **49.** 12, 2 **51.** 12, 4 **53.** 3, -1 **55.** $\dfrac{25}{2}$ **57.** $\dfrac{3}{2}$ **59.** $P_1=\dfrac{P_2V_2}{V_1}$

61. $V_2=\dfrac{V_1P_1}{P_2}$ **63.** $y=y_1+m(x-x_1)$ **65.** $x=zs+\bar{x}$ **67.** $w=\dfrac{fl-df}{d}$ **69.** $q=\dfrac{pf}{p-f}$ **71.** $a=\dfrac{v_2-v_1}{t_2-t_1}$ **73.** $d=\dfrac{a_n-a_1}{n-1}$

75. $G=\dfrac{Fd^2}{m_1m_2}$ **77.** $T_1=\dfrac{T_2P_1V_1}{P_2V_2}$ **79.** $V_0=\dfrac{S-S_0-gt^2}{t}$ **81. a)** $\dfrac{3x+7}{(x+3)(x+1)}$ **b)** $-\dfrac{7}{3}$ **83. a)** $\dfrac{4}{b+5}$ **b)** No solution

85. a) 12.5% **b)** $T_f=T_a(1-f)$ **c)** 8.64% **87. a)** \$6250 **b)** $R=\dfrac{AC}{0.80I}$ **89. a)** 20 feet per minute squared

b) $t_1=t_2+\dfrac{v_1-v_2}{a}$ **91. a)** $\approx22.5\%$ **b)** $D=PR$ **c)** $R=\dfrac{D}{P}$ **93.** 150 ohms **95.** ≈0.101 meter **97. a)** $\approx9.71\%$ **b)** Tax-Free Money Market Portfolio since 9.71% > 7.68%. **99.** $c\neq0$ **101.** One answer is $\dfrac{1}{x-4}+\dfrac{1}{x+2}=0$; 4 or -2 make a fraction undefined

103. One answer is $\dfrac{1}{x}+\dfrac{1}{x}=\dfrac{2}{x}$. **105.** $-1<x\le3$ **106.** $m=-\dfrac{1}{3}$; y-intercept, $\left(0,\dfrac{14}{3}\right)$ **107.** $3x^2y-7xy-4y^2-9x$ **108.** 2 feet

Mid-Chapter Test

1. $\{x\,|\,x\neq0, x\neq-5, \text{and } x\neq5\}$ [6.1] **2.** $\dfrac{x+5}{2x-3}$ [6.1] **3.** $\dfrac{55b}{a^2-ab+b^2}$ [6.1] **4.** $\dfrac{x-3}{x+1}$ [6.1]

5. $\dfrac{(2a+1)(2a+3)}{(2a-1)(a-9)}$ or $\dfrac{4a^2+8a+3}{2a^2-19a+9}$ [6.1] **6.** $\dfrac{4a+3b}{6}$ [6.1] **7.** $(x+5)(x-6)(x+2)$ [6.2] **8.** 5 [6.2] **9.** $\dfrac{20y^2+ax}{6x^2y^3}$ [6.2]

10. $\dfrac{-2x-7}{(x-4)(x+4)(2x-3)}$ [6.2] **11.** $\dfrac{9b+a}{3-c}$ [6.3] **12.** $\dfrac{5x-8}{6x^2-x}$ [6.3] **13.** y^2 [6.3]

14. An extraneous solution is a number obtained when solving an equation that is not a solution to the original equation. Whenever a variable appears in the denominator, you must check the apparent solution(s). [6.4] **15.** 5 [6.4] **16.** No solution [6.4]

17. $4, -3$ [6.4] **18.** $a=\dfrac{bc}{b+c}$ [6.4] **19.** $r=\dfrac{x-4}{x}$ [6.4] **20.** 14 and 5 [6.4]

Exercise Set 6.5

1. One whole task **3.** Part of the task completed **5.** 1.5 hours **7.** 2 hours **9.** 18.75 minutes **11.** 4 hours **13.** ≈2.48 days

15. 2.4 hours **17.** ≈3.08 hours **19.** 100 hours **21.** 7.8 months **23.** 75 minutes **25.** ≈15.27 minutes **27.** ≈1.62 hours

29. 12 hours **31.** 3 **33.** 2, 4 **35.** 4, 6 **37.** 20 **39.** $\dfrac{2}{3}$, 1 **41.** ≈0.064 mile per hour **43.** ≈1.53 feet per second **45.** 7.5 miles

47. 36 miles per hour **49.** ≈30.59 yards **51.** Local: ≈10.93 miles per hour, express: ≈16.13 miles per hour.
53. Car: 60 miles per hour, train: 30 miles per hour **55.** 60 miles per hour **57.** 120 kilometers per hour

59. 2 hours at 6 miles per hour, $\dfrac{1}{2}$ hour at 10 miles per hour **61.** 18 feet per minute **63.** 108,000 miles **65.** Answers will vary.

67. a) 10 minutes **b)** 15 miles **c)** 165 miles per hour **68.** $\dfrac{1}{72x^5y^{12}}$ **69.** 9.26×10^9 **70.** \$2500 **71.**
72. $a(2a^2-5)(a-1)$

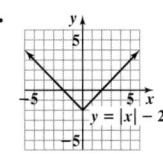

Exercise Set 6.6

1. Combined **3.** Direct **5.** Constant **7.** Direct **9.** Inverse **11.** Direct **13.** Direct **15.** Direct **17.** Inverse **19.** Direct

21. Inverse **23.** Inverse **25. a)** $x=ky$ **b)** 72 **27. a)** $y=kR$ **b)** 306 **29. a)** $R=\dfrac{k}{W}$ **b)** $\dfrac{1}{20}$ **31. a)** $A=\dfrac{kB}{C}$ **b)** 9

33. a) $x=ky$ **b)** 20 **35. a)** $y=kR^2$ **b)** 20 **37. a)** $S=\dfrac{k}{G}$ **b)** 0.96 **39. a)** $x=\dfrac{k}{P^2}$ **b)** 25 **41. a)** $F=\dfrac{kM_1M_2}{d}$ **b)** 40

43. Doubled **45.** Halved **47.** Doubled **49.** Unchanged **51.** Doubled **53.** $y = \dfrac{k}{x}; k = 5$ **55.** $8814 **57.** 3096 milligrams

59. 1.05 inches **61.** 6400 cubic centimeters **63.** 3.12 hours **65.** 45 foot-candles **67.** 117.6 feet **69.** 126 cubic meters

71. 4600 DVDs **73.** ≈ 133.25 pounds **75.** $\approx 121,528$ calls **77.** $\dfrac{1}{49}$ of the light of the flash **79. a)** $P = 14.7 + kx$ **b)** 0.43

c) ≈ 337.9 feet **80.** $h = \dfrac{3V}{4\pi r^2}$ **81.** 132 **82.** $-14x^3 - 22x^2 + 47x - 15$ **83.** $(x + 3)(x - 2)$

Chapter 6 Review Exercises

1. 4 **2.** -1 **3.** None **4.** $\{x | x \ne 1\}$ **5.** $\{x | x \ne 0\}$ **6.** $\{x | x \ne 2 \text{ and } x \ne -6\}$ **7.** $\dfrac{2}{5}$ **8.** $x - 6$ **9.** -1 **10.** $\dfrac{x - 1}{x - 2}$ **11.** $\dfrac{x - 3}{x + 1}$

12. $\dfrac{a^2 + 2ab + 4b^2}{a + 2b}$ **13.** $\dfrac{9x^2 - 3xy + y^2}{3x - y}$ **14.** $\dfrac{2x - 3}{x^2 - 2x + 4}$ **15.** $x(x - 4)$ **16.** $(x + 2y)(x - 2y)$ **17.** $(x + 7)(x - 5)(x + 2)$

18. $(x + 2)^2(x - 2)(x + 3)$ **19.** $\dfrac{4xyz}{3}$ **20.** $-\dfrac{x}{6}$ **21.** $9x^3z^5$ **22.** $\dfrac{11x + 6}{3x^2}$ **23.** $\dfrac{(x - y)y^2}{4x^3}$ **24.** $3x + 2$ **25.** $\dfrac{30x + 3y^2}{5x^2y}$ **26.** 1

27. $\dfrac{2x + 1}{3x + 1}$ **28.** $\dfrac{6a + 7}{a + 1}$ **29.** $\dfrac{6b - 8}{b - 1}$ **30.** $\dfrac{a^2 - b^2}{a^2}$ **31.** $\dfrac{1}{3(a + 3)}$ **32.** $\dfrac{a^2 + c^2}{ac}$ **33.** $\dfrac{x + 1}{2x - 1}$ **34.** 1 **35.** $4x(x - 5y)$

36. $\dfrac{2a^2 + 9a + 4}{4a(a + 2)}$ **37.** $\dfrac{x^2 + 5}{(x + 5)(x - 5)}$ **38.** $-\dfrac{2(x + 1)}{x^2 - 4}$ **39.** $\dfrac{x + 5}{x + 6}$ **40.** $\dfrac{-x + 5}{(x + 2)(x - 2)(x - 3)}$ **41.** $\dfrac{16(x - 2y)}{3(x + 2y)}$ **42.** $\dfrac{3}{a^3}$

43. $\dfrac{22x + 5}{(x - 5)(x - 10)(x + 5)}$ **44.** $\dfrac{2(x - 4)}{(x - 3)(x - 5)}$ **45.** $-\dfrac{1}{x - 3}$ **46.** $\dfrac{a + 3}{a + 5}$ **47.** $\dfrac{x + 6}{x - 4}$ **48.** $\dfrac{a + 2b^2}{3}$ **49.** $\dfrac{x^2 + 6x - 24}{(x - 1)(x + 9)}$

50. $\dfrac{x - 4}{x - 6}$ **51. a)** $\{x | x \ne -2\}$ **b)** $\{x | x \ne -4\}$ **c)** $\dfrac{2x^2 + 7x + 4}{(x + 2)(x + 4)}$ **d)** $\{x | x \ne -2 \text{ and } x \ne -4\}$ **52. a)** $\{x | x \ne 3 \text{ and } x \ne -3\}$

b) $\{x | x \ne 3\}$ **c)** $\dfrac{x^2 + 8x + 12}{(x + 3)(x - 3)}$ **d)** $\{x | x \ne 3 \text{ and } x \ne -3\}$ **53.** $\dfrac{3ac^2}{b^3}$ **54.** $\dfrac{4x + 2y}{x^2 + xy^3}$ **55.** $\dfrac{3y - 1}{7y^2 + 1}$ **56.** $\dfrac{5a + 1}{2}$ **57.** $\dfrac{3x + 1}{-x + 1}$

58. $\dfrac{3x^2 - 29x + 68}{4x^2 - 6x - 54}$ **59.** $\dfrac{x^2 + 3x + 2}{x + 5}$ **60.** $\dfrac{x^2 + 6x + 8}{x + 3}$ **61.** 16 **62.** -2 **63.** 52 **64.** 2.4 **65.** 5 **66.** -9 **67.** -18 **68.** -28

69. -6 **70.** -10 **71.** $b = \dfrac{ac}{a - c}$ **72.** $\bar{x} = x - sz$ **73.** 60 ohms **74.** 2 centimeters **75.** 10, 2 **76.** 21, 3 **77.** ≈ 17.14 minutes

78. 14 hours **79.** 3 **80.** $\dfrac{5}{6}$ **81.** 5 miles per hour **82.** car: 50 miles per hour, plane: 150 miles per hour **83.** 20 **84.** $\dfrac{25}{2}$

85. ≈ 426.7 **86.** $8.40 **87.** 1600 feet **88.** 200.96 square units **89.** 2.38 minutes

Chapter 6 Practice Test

1. -7 and 4 [6.1] **2.** $\left\{ x \mid x \ne -4 \text{ and } x \ne \dfrac{1}{2} \right\}$ [6.1] **3.** $5x^5y + 8 + 11xy^2$ [6.1] **4.** $\dfrac{x - 6y}{x + y}$ [6.1] **5.** $\dfrac{1}{x^4y^2}$ [6.1] **6.** $\dfrac{1}{x + 2}$ [6.1]

7. $\dfrac{7}{a(a + b)}$ [6.1] **8.** $x^2 + y^2$ [6.1] **9.** $\dfrac{5x^2 + 2x + 2}{x^2(x + 1)}$ [6.2] **10.** $\dfrac{-3x - 1}{(x - 3)(x + 3)(x + 1)}$ [6.2]

11. $\dfrac{m(6m + n)}{(6m + 5n)(2m - n)(2m + 3n)}$ [6.2] **12.** $\dfrac{x(x + 10)}{(2x - 1)^2(x + 3)}$ [6.2] **13.** $x + 3$ [6.1] **14. a)** $\dfrac{3x^2 + 2x - 9}{(x + 5)(2x + 3)}$

b) $\left\{ x \mid x \ne -5 \text{ and } x \ne -\dfrac{3}{2} \right\}$ [6.2] **15.** $\dfrac{x + 5}{x + 2}$ [6.1] **16.** $\dfrac{y + 2x}{y - 3x}$ [6.3] **17.** $\dfrac{b(a - b)}{a}$ [6.3] **18.** $\dfrac{7x - 6}{4x^2 - x}$ [6.3] **19.** 20 [6.4]

20. 12 [6.4] **21.** $C = \dfrac{2b + Ad}{A}$ [6.4] **22.** 0.75 watt [6.6] **23.** 6 [6.6] **24.** ≈ 4.44 hours [6.5] **25.** $6\dfrac{2}{3}$ miles [6.5]

Cumulative Review Test

1. [1.2] **2.** $-27\dfrac{3}{4}$ [1.4] **3.** -3 [2.1] **4. a)** 28% **b)** $\approx 44,000$ [1.3] **5.** 62 [1.4] **6.** $\dfrac{x^3}{8y^3}$ [1.5] **7.** $m = \dfrac{rF}{v^2}$ [2.2]

8. 6% [2.2] **9.** 3 hr [2.4] **10.** $\left\{ -\dfrac{32}{3}, \dfrac{22}{3} \right\}$ [2.6] **11.** [3.1] **12.** 5 [3.2] **13.** $-\dfrac{1}{7}$ [3.4] **14.** $2x + 3y = 4$ [3.5]

15. $\left(\dfrac{1}{2}, 3 \right)$ [4.1] **16.** $9x^4 - 25y^2$ [5.2] **17.** $3(x - 5)^2$ [5.6] **18.** [3.1] **19.** $\dfrac{3x - 4}{(x - 1)(x - 2)}$ [6.2] **20.** 4 [6.4]

Chapter 7

Exercise Set 7.1 **1.** Radical **3.** Radicand **5.** Cube **7.** Negative **9.** Even **11. a)** 3 **b)** -3 **c)** Not a real number
d) Not a real number **13.** -4 **15.** -5 **17.** -1 **19.** 1 **21.** Not a real number **23.** -7 **25.** Not a real number
27. Not a real number **29.** $\frac{1}{5}$ **31.** $\frac{1}{2}$ **33.** $\frac{2}{7}$ **35.** $-\frac{2}{3}$ **37.** ≈ -2.07 **39.** 7 **41.** 3 **43.** 119 **45.** 235.23 **47.** 0.06 **49.** $\frac{12}{13}$
51. $|x-8|$ **53.** $|x-3|$ **55.** $|3x^2-1|$ **57.** $|6a^3-5b^4|$ **59.** $|x^5|$ **61.** $|z^{16}|$ **63.** $|a-4|$ **65.** $|3a+2b|$ **67.** $5a$ **69.** $4c^3$
71. $x+2$ **73.** $2x+y$ **75.** 2 **77.** 8 **79.** 9 **81.** ≈ 9.381 **83.** ≈ 5.290 **85.** -3 **87.** 97 **89.** 11 **91.** 45

93. Select a value less than $-\frac{1}{2}$. **95.** d **97.** a **99.**

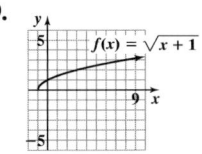

101.

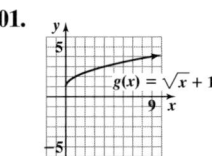

103. One answer is $f(x)=\sqrt{x-8}$ **105. a)** $\sqrt{1288}\approx 35.89$ feet per second **b)** $\sqrt{2576}\approx 50.75$ feet per second
107. There is no real number which when squared gives -81. **109.** No; if the radicand is negative, the answer is not a real number.
111. a) 1.3 **b)** 1.3 **113.** $x\geq 1$ **115.** $x\geq 3$ **117. a)** All real numbers **b)** $a\geq 0$ **c)** All real numbers **119.** If n is even, you are
finding an even root of a positive number. If n is odd, the expression is real. **121.** $x>-5$ **123. a)** No **b)** Yes, when $x=0$
c) Yes **127.** $(3a-b)(3x+4y)$ **128.** $3x(x-4)(x-2)$ **129.** $(2x+1)(2x-1)(2x^2+3)$

130. $\left(x-\frac{2}{3}y\right)\left(x^2+\frac{2}{3}xy+\frac{4}{9}y^2\right)$

Exercise Set 7.2 **1.** Numerator **3.** Index **5.** Add **7.** $x^{5/2}$ **9.** $9^{5/2}$ **11.** $z^{5/3}$ **13.** $7^{10/3}$ **15.** $9^{7/4}$ **17.** $y^{14/3}$ **19.** $(a^3b)^{1/4}$

21. $(x^9z^5)^{1/4}$ **23.** $(3a+8b)^{1/6}$ **25.** $\left(\frac{2x^6}{11y^7}\right)^{1/5}$ **27.** $\sqrt{a}$ **29.** $\sqrt[3]{a^2}$ **31.** $\sqrt[3]{18^5}$ **33.** $\sqrt{24x^3}$ **35.** $\left(\sqrt[5]{11b^2c}\right)^3$ **37.** $\sqrt[5]{6a+5b}$

39. $\frac{1}{\sqrt[3]{b^3-d}}$ **41.** a **43.** x^3 **45.** $\sqrt[3]{y}$ **47.** $\sqrt{y}$ **49.** 19.3 **51.** x^5y^{10} **53.** $\sqrt{xyz}$ **55.** $\sqrt[4]{x}$ **57.** $\sqrt[8]{y}$ **59.** $\sqrt[9]{x^2y}$ **61.** $\sqrt[10]{a^9}$

63. 3 **65.** 4 **67.** 16 **69.** Not a real number **71.** $\frac{5}{3}$ **73.** $\frac{1}{2}$ **75.** -9 **77.** -4 **79.** $\frac{1}{4}$ **81.** $\frac{1}{64}$ **83.** $\frac{3}{4}$ **85.** Not a real number

87. 24 **89.** $\frac{11}{28}$ **91.** $x^{9/2}$ **93.** $x^{1/6}$ **95.** $\frac{1}{x}$ **97.** 1 **99.** $\frac{y^{5/3}}{12}$ **101.** $\frac{12}{x^{11/6}}$ **103.** $\frac{1}{2x^{1/3}}$ **105.** $\frac{121}{x^{1/7}}$ **107.** $\frac{64}{a^{66/5}}$ **109.** $\frac{x}{y^{20}}$

111. $8z^{7/2}-4$ **113.** $\frac{5}{x^5}+\frac{20}{x^{3/2}}$ **115.** $12x^{13/6}-18x^2$ **117.** ≈ 13.42 **119.** ≈ 3.32 **121.** ≈ 20.53 **123.** ≈ 0.03 **125.** n is odd, or n is

even and $a\geq 0$. **127.** $(4^{1/2}+9^{1/2})^2\neq 4+9;\ 25\neq 13$ **129.** $(1^{1/3}+1^{1/3})^3\neq 1+1;\ 8\neq 2$ **131.** $x^{1/2}(x+1)$

133. $y^{1/3}(1-y)(1+y)$ **135.** $\frac{1+y^2}{y^{2/5}}$ **137. a)** $2^{10}=1024$ bacteria **b)** $2^{10}\sqrt{2}\approx 1448$ bacteria **139.** 2247 calories **141.** $\{x|x\geq 7\}$

143. $2;\ z^{\frac{1}{4}\cdot\frac{1}{5}\cdot\frac{1}{a}\cdot\frac{1}{3}}=z^{\frac{1}{60a}};\ z^{\frac{1}{60a}}=z^{\frac{1}{120}},\ 60a=120;\ a=2$ **145. a)** When n is even and $a\geq 0$, or n is odd **b)** $a^{1/n}$ **147. a)** Always real

b) a **c)** a **d)** $|a|$ **149. a)** No; $(xy)^{1/2}=x^{1/2}y^{1/2}$ **b)** No; $(xy)^{-1/2}=x^{-1/2}y^{-1/2}=\frac{1}{x^{1/2}y^{1/2}}$ **151.** 9 **153. c)** is a function

154. $\frac{b^2+a^3b}{a^3-b}$ **155.** $0,3$ **156.** ≈ 441.67 miles per hour.

Exercise Set 7.3 **1.** Square **3.** 3 **5.** 4 **7.** Product **9.** 7 **11.** $2\sqrt{6}$ **13.** $4\sqrt{2}$ **15.** $5\sqrt{2}$ **17.** $5\sqrt{3}$ **19.** $2\sqrt{10}$ **21.** $2\sqrt[3]{2}$
23. $3\sqrt[3]{2}$ **25.** $2\sqrt[4]{4}$ **27.** $2\sqrt[5]{5}$ **29.** $2\sqrt[4]{3}$ **31.** $-2\sqrt[5]{2}$ **33.** b^3 **35.** x^2 **37.** $x\sqrt{x}$ **39.** $a^2\sqrt{a}$ **41.** $8z^{10}\sqrt[3]{z^2}$ **43.** $b^5\sqrt[4]{b^3}$
45. $x\sqrt[6]{x^3}$ or $x\sqrt{x}$ **47.** $3y^4\sqrt[5]{y^3}$ **49.** $10y^4\sqrt{2y}$ **51.** $xy^2\sqrt[3]{y}$ **53.** $ab^4\sqrt[5]{ab^3}$ **55.** $2x^7y^{10}z^{13}\sqrt{6xz}$ **57.** $3a^2b^2\sqrt[3]{3b^2}$
59. $2x^2y^2z^4\sqrt[4]{2yz^3}$ **61.** $3a^2b^2\sqrt[4]{b}$ **63.** $2a^2b^2\sqrt[5]{b^2}$ **65.** 3 **67.** $\frac{9}{10}$ **69.** 3 **71.** $\frac{1}{4}$ **73.** $\frac{1}{2}$ **75.** $\frac{1}{3}$ **77.** $\frac{1}{2}$ **79.** 2 **81.** $\frac{x}{3}$ **83.** $\frac{4x^2}{5y^5}$
85. $\frac{c^2}{4}$ **87.** $a^2b^6\sqrt[3]{a^2b^2}$ **89.** $2\sqrt{2}$ **91.** $3x^2$ **93.** $2x^2y\sqrt{2y}$ **95.** $\frac{\sqrt[3]{5y}}{2x^4}$ **97.** $\frac{y^2\sqrt[3]{5y}}{x^2}$ **99.** $\frac{x^3\sqrt[4]{10y}}{3}$
101. a) Square the natural numbers. **b)** $1,4,9,16,25,36$ **103.** If n is even and a or b are negative, the numbers are not real
numbers. **105.** $(a\cdot b)^{1/2}=a^{1/2}b^{1/2}=\sqrt{a}\sqrt{b}$ **107.** No; One example is $\sqrt{18}/\sqrt{2}=3$. **109. a)** No
b) When $\sqrt[n]{x}$ is a real number and not equal to 0. **110.** $C=\frac{5}{9}(F-32)$ **111.** $\{-28,32\}$ **112.** $3x^6-x^3+4$
113. $(x-1)(x^2-8x+19)$

Exercise Set 7.4 **1.** Like **3.** Distributive **5.** FOIL **7.** 0 **9.** $4\sqrt{5}$ **11.** $-4\sqrt{3}+5$ **13.** $-7\sqrt[4]{y}$ **15.** $2\sqrt[3]{x}+9\sqrt{5}$
17. $7\sqrt{x}-6\sqrt{y}$ **19.** $3\sqrt{3}$ **21.** $-30\sqrt{3}+25\sqrt{5}$ **23.** $-4\sqrt{10}$ **25.** $18y\sqrt{5x}$ **27.** $-16\sqrt{5x}$ **29.** $-27a\sqrt{2}$ **31.** $5\sqrt[3]{4}$ **33.** -7
35. $6a\sqrt[3]{ab^2}$ **37.** $3r^3s^2\sqrt{rs}$ **39.** 0 **41.** 4 **43.** $2\sqrt[3]{7}$ **45.** $3m^2n^5\sqrt[3]{3n}$ **47.** $3x^3y^4\sqrt[3]{2x^2y}$ **49.** $x^7y^7z^3\sqrt[5]{x^2y^3z}$ **51.** $x^2y^2\sqrt[3]{4y^2}$

53. $5 - \sqrt{15}$ **55.** $2\sqrt[3]{y^2} - y^3$ **57.** $4x^5y^3\sqrt[3]{x} + 4xy^4\sqrt[3]{2x^2y^2}$ **59.** 59 **61.** $6 - x^2$ **63.** $7 - z$ **65.** $23 + 9\sqrt{3}$ **67.** $16 - 10\sqrt{2}$
69. $10 - 3\sqrt{6}$ **71.** $29 - 12\sqrt{5}$ **73.** $18x - \sqrt{3xy} - y$ **75.** $8 - 2\sqrt[3]{18} - \sqrt[3]{12}$ **77.** $x - \sqrt{3x}$ **79.** $x^2 + x\sqrt[3]{x^2}$
81. $x\sqrt[4]{27x^2} - x^2\sqrt[4]{3x}$ **83.** $3\sqrt{2}$ **85.** $3\sqrt{5}$ **87.** $-14 + 11\sqrt{2}$ **89.** $5\sqrt{6}$ $2\sqrt{3}$ **91.** $15\sqrt{2}$ **93.** $2x^3\sqrt[3]{10x^2}$ **95.** $2b^2c\sqrt[6]{2ab^5c^3}$
97. $4ab\sqrt[4]{b}$ **99.** $x - 2\sqrt[3]{x^2y^2} - \sqrt[3]{xy} + 2y$ **101.** $ab\sqrt[3]{12a^2b^2} - 2a^2b^2\sqrt[3]{3}$ **103.** $2x - 5$ **105.** $2|r - 4|$ **107.** $P = 14\sqrt{5}, A = 60$
109. $P = 12\sqrt{5}, A = 30$ **111. a)** ≈ 45.17 miles per hour **b)** ≈ 35.33 miles per hour **113. a)** 37 inches **b)** ≈ 37.97 inches

115. a) 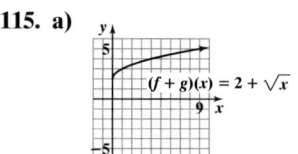   **117. a)** 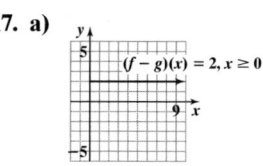   **119.**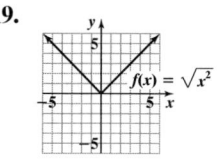

b) Raises the graph 2 units   **b)** $\{x|x \geq 0\}$

121. No: one example is $\sqrt{9} + \sqrt{16} \neq \sqrt{9 + 16}, 3 + 4 \neq 5, 7 \neq 5$. **123.** No, $-\sqrt{2} + \sqrt{2} = 0$ **125.** A quotient of two integers, denominator not 0. **126.** A number that can be represented on a real number line. **127.** A real number that cannot be expressed as a quotient of two integers. **128.** $|a| = \begin{cases} a, & a \geq 0 \\ -a, & a < 0 \end{cases}$ **129.** $m = \dfrac{2E}{v^2}$ **130. a)** (number line with open circle at $-\frac{1}{2}$ and closed at 5) **b)** $\left(-\frac{1}{2}, 5\right]$ **c)** $\left\{x \middle| -\frac{1}{2} < x \leq 5\right\}$

Mid-Chapter Test
1. 11 [7.1] **2.** $-\dfrac{3}{4}$ [7.1] **3.** 16.3 [7.1] **4.** $|3a^2 - 4b^3|$ [7.1] **5.** 3 [7.1] **6.** $(7a^4b^3)^{1/5}$ [7.2] **7.** 20 [7.2]
8. $a^{10}b^{15}c^5$ [7.3] **9.** $\dfrac{14}{x}$ [7.3] **10.** $8x + \dfrac{16}{x^{5/2}}$ [7.3] **11.** $4x^2y^4\sqrt{2y}$ [7.3] **12.** $2a^2b^3c^2\sqrt[6]{ab^5c^3}$ [7.3] **13.** $\dfrac{1}{3}$ [7.3] **14.** $\dfrac{y^2\sqrt{y}}{3x^5}$ [7.3]
15. $11\sqrt{x} + 12\sqrt{y}$ [7.4] **16.** $27x\sqrt{10y}$ [7.4] **17.** $2x^2 - x\sqrt{5} - 15$ [7.4] **18.** $18a\sqrt{a} - 20a\sqrt{3}$ [7.4] **19.** $7ab\sqrt[4]{ab}$ [7.4]
20. Part **a)** will have an absolute value. **a)** $|x - 3|$, **b)** $8x$ [7.1]

Exercise Set 7.5
1. Rationalizing **3.** Power **5.** Radicand **7.** $\dfrac{\sqrt{2}}{2}$ **9.** $\dfrac{4\sqrt{5}}{5}$ **11.** $\sqrt{6}$ **13.** $\dfrac{\sqrt{z}}{z}$ **15.** $\dfrac{p\sqrt{2}}{2}$ **17.** $\dfrac{\sqrt{7y}}{7}$
19. $3\sqrt{2}$ **21.** $\dfrac{\sqrt{xy}}{y}$ **23.** $\dfrac{\sqrt{10m}}{4}$ **25.** $\dfrac{\sqrt{2n}}{3}$ **27.** $\dfrac{3x^2y\sqrt{yz}}{z^2}$ **29.** $\dfrac{2y^4z\sqrt{15xz}}{3x}$ **31.** $\dfrac{4x^3y^2\sqrt{yz}}{z^2}$ **33.** $\dfrac{\sqrt[3]{4}}{2}$ **35.** $\dfrac{8\sqrt[3]{y^2}}{y}$ **37.** $\dfrac{\sqrt[4]{8}}{2}$
39. $\dfrac{a\sqrt[4]{2}}{2}$ **41.** $\dfrac{5\sqrt[4]{z^2}}{z}$ **43.** $\dfrac{10\sqrt[5]{y^2}}{y}$ **45.** $\dfrac{2\sqrt[7]{a^3}}{a}$ **47.** $\dfrac{\sqrt[3]{4x^2}}{2x}$ **49.** $\dfrac{5m\sqrt[4]{8}}{2}$ **51.** $\dfrac{\sqrt[4]{135x}}{3x}$ **53.** $\dfrac{\sqrt[3]{12x^2y}}{2y}$ **55.** $\dfrac{\sqrt[3]{7xy^2z}}{z}$ **57.** 11
59. 62 **61.** -6 **63.** $a - b$ **65.** $4x - 9y$ **67.** $\sqrt{2} - 1$ **69.** $2 - \sqrt{3}$ **71.** $\dfrac{-5\sqrt{2} - 35}{47}$ **73.** $\dfrac{10 + \sqrt{30}}{14}$ **75.** $\dfrac{18 - 3\sqrt{x}}{36 - x}$
77. $\dfrac{4x + 4y\sqrt{x}}{x - y^2}$ **79.** $\dfrac{-13 + 3\sqrt{6}}{23}$ **81.** $a + a^3$ **83.** $\dfrac{4\sqrt{x + 2} + 12}{x - 7}$ **85.** $\dfrac{a}{5}$ **87.** $\dfrac{\sqrt{2}}{3}$ **89.** 1 **91.** $\dfrac{2xy^3\sqrt{30xz}}{5z}$ **93.** $\dfrac{\sqrt{14}}{x}$
95. $\dfrac{\sqrt{a} - 7}{a - 49}$ **97.** $-\dfrac{\sqrt{2x}}{2}$ **99.** $\dfrac{\sqrt[4]{24x^3y^2}}{2x}$ **101.** $\dfrac{2y^4z^3\sqrt[3]{2x^2z}}{x}$ **103.** $\dfrac{a\sqrt{r} + 2r\sqrt{a}}{a - 4r}$ **105.** $\dfrac{\sqrt[3]{150y^2}}{5y}$ **107.** $\dfrac{y^3z\sqrt[4]{54x^2}}{3x}$ **109.** $\sqrt{2}$
111. $\dfrac{3\sqrt{5}}{5}$ **113.** $\dfrac{8\sqrt{6}}{3}$ **115.** $\dfrac{19\sqrt{2}}{2}$ **117.** $\dfrac{21\sqrt{2}}{2}$ **119.** $-\dfrac{301\sqrt{2}}{20}$ **121.** $\dfrac{3\sqrt{6}}{4}$ **123.** $\left(-\dfrac{2}{y} + \dfrac{3}{x}\right)\sqrt{xy}$ **125.** $2\sqrt{a}$ **127.** $\sqrt[3]{(a + b)^5}$
129. $\sqrt[15]{(a + 2b)^2}$ **131.** $\sqrt[6]{rs^5}$ **133.** $\sqrt[15]{x^2y^8}$ **135.** ≈ 3.69 meters **137.** ≈ 12 inches **139. a)** 6.21 million **b)** ≈ 2.35 million
141. $\dfrac{10}{15 + 3\sqrt{5}}$ **143.** $\dfrac{1}{\sqrt{x + h} + \sqrt{x}}$ **145.** $\dfrac{3}{\sqrt{3}}; \dfrac{2}{\sqrt{2}} = \sqrt{2}, \dfrac{3}{\sqrt{3}} = \sqrt{3}$ **147.** $2 + \sqrt{3}$; rationalize the denominator
and compare **149. a)** $4, 8, 12$ **b)** $9, 18, 27$ **c)** $x^{(3a+2b)/6}$ **d)** $x^{(3a-2b)/6}$ **152.** $b_2 = \dfrac{2A}{h} - b_1$ **153.** 40 miles per hour, 50 miles
per hour **154.** $4x^3 + x^2 - 20x + 4$ **155.** $-8, 1$

Exercise Set 7.6
1. Radical **3.** Original **5.** Index **7.** Cube **9.** 16 **11.** No real solution **13.** -64 **15.** 11 **17.** 9
19. -1 **21.** 81 **23.** 71 **25.** 26 **27.** No real solution **29.** $2, 4$ **31.** 8 **33.** 7 **35.** $\dfrac{2}{3}$ **37.** 16 **39.** 2 **41.** 10 **43.** 6 **45.** 8
47. 0 **49.** -3 **51.** $\dfrac{3}{2}$ **53.** No real solution **55.** 0 **57.** $5, 8$ **59.** No real solution **61.** 9 **63.** $3, 7$ **65.** -1 **67.** 2 **69.** 4
71. 5 **73.** $v = \dfrac{p^2}{2}$ **75.** $g = \dfrac{v^2}{2h}$ **77.** $F = \dfrac{Mv^2}{R}$ **79.** $m = \dfrac{x^2k}{V_0^2}$ **81.** $A = \pi r^2$ **83.** $\sqrt{87}$ **85.** $2\sqrt{10}$ **87.** 1 **89.** No solution
91. 3 **93.** 7 **95.** 1 **97.** 3 **99.** $\sqrt{16,200} \approx 127.28$ feet **101.** $\sqrt{2000} \approx 44.7$ feet **103. a)** ≈ 3.14 seconds
b) $\sqrt{2} \cdot T$; compare $\sqrt{\dfrac{l}{32}}$ with $\sqrt{\dfrac{l}{16}}$ **c)** $\sqrt{24} \approx 4.90$ seconds **105.** $R = \dfrac{8\mu l}{\pi r^4}$ **107.** $0.2(\sqrt{149.4})^3 \approx 365.2$ days

109. $\sqrt{10,000} = 100$ pounds **111.** $\sqrt{320} \approx 17.89$ feet per second **113.** $\sqrt{1649} \approx 40.61$ meters **115.** $2, -2$ **117.** $5, -1$
119. 30 **121.** $5, -5$ **123.** All real numbers **125.** Answers will vary. **127.** 0 **129.** 1; Answers will vary.
131. a) $3, 7$; points of intersection **b)** Yes **c)** $3, 7$; yes **133.** At $x = 4$, $g(x)$ or $y = 0$. Therefore the graph must have an x-intercept at 4.

135. No solution **137.** $n = \dfrac{z^2\sigma^2}{(\bar{x} - \mu)^2}$ **139.** $L_1 \approx 0.44, L_2 \approx 0.76$ **142.** $P_2 = \dfrac{P_1 P_3}{P_1 - P_3}$ **143.** x **144.** $\dfrac{3a}{2b(2a + 3b)}$

145. $t(t - 5)$ **146.** $\dfrac{3}{x + 3}$ **147.** 2

Exercise Set 7.7
1. Imaginary **3.** Complex **5.** Part **7.** Number **9.** -1 **11.** $4 + 0i$ **13.** $5 + 0i$ **15.** $21 - 6i$
17. $0 + 2i\sqrt{6}$ **19.** $8 - 2i\sqrt{3}$ **21.** $3 + 7i\sqrt{2}$ **23.** $12 - 5i$ **25.** $0 + (7 - 3\sqrt{5})i$ **27.** $4 + 4i$ **29.** $2 + 10i$ **31.** $-17 - 12i$
33. $(4\sqrt{2} + \sqrt{3}) - 2i\sqrt{2}$ **35.** $11 - 4i\sqrt{2}$ **37.** $-3 - 2i\sqrt{5}$ **39.** $6 - 12i$ **41.** $-9 + 4i$ **43.** $-33 + 18i$ **45.** $28 + 4i\sqrt{3}$
47. $9 + 9i$ **49.** $1 + 5i$ **51.** 109 **53.** $39 - 9i\sqrt{2}$ **55.** $\dfrac{25}{72} + \dfrac{1}{4}i$ **57.** $-\dfrac{2}{3}i$ **59.** $\dfrac{3}{2} - i$ **61.** $\dfrac{12}{5} + \dfrac{6}{5}i$ **63.** $\dfrac{3}{5} + \dfrac{6}{5}i$ **65.** $\dfrac{9}{10} - \dfrac{6}{5}i$
67. $\dfrac{3}{5} + \dfrac{1}{5}i$ **69.** $\dfrac{5\sqrt{2}}{37} - \dfrac{2\sqrt{6}}{37}i$ **71.** $\dfrac{3}{8} + \dfrac{7}{8}i$ **73.** 5 **75.** $\dfrac{\sqrt{2}}{3}$ **77.** 0 **79.** $4\sqrt{2} + 2i\sqrt{3}$ **81.** $20.8 - 16.64i$ **83.** $37 - 39i$
85. $2 - \dfrac{11}{2}i$ **87.** $\dfrac{6\sqrt{3}}{7} + \dfrac{12}{7}i$ **89.** $7 + \dfrac{2}{45}i$ **91.** $\dfrac{1}{4} - \dfrac{31}{50}i$ **93.** 2 **95.** $-4.33 - 10.91i$ **97.** -1 **99.** 1 **101.** i **103.** $-i$
105. a) $-2 - 3i$ **b)** $\dfrac{2}{13} - \dfrac{3}{13}i$ **107.** -4 **109.** $16 - 4i$ **111.** $14 + 8i$ **113.** 0 **115.** 1 **117.** Yes **119.** No **121.** $\approx 0.83 - 3i$
123. $\approx 1.5 - 0.33i$ **125.** $-i$ **127.** $1 + i\sqrt{5}, 1 - i\sqrt{5}$ **129.** $6 + 3i\sqrt{3}$ **131.** $-1 + 7i\sqrt{3}$ **133.** Yes **135.** No
137. True; $(2i)(2i) = -4$ **139.** False; $(1 + i)(1 + 2i) = -1 + 3i$ **141.** Even values; i^n where n is even will either be -1 or 1
143. 15 pounds at \$5.50, 25 pounds at \$6.30 **144.** $2c - 3 - \dfrac{8}{4c + 9}$ **145.** $\dfrac{a^2}{b(a - b)}$ **146.** 4

Chapter 7 Review Exercises
1. 3 **2.** -3 **3.** -5 **4.** 4 **5.** 5 **6.** 38.2 **7.** $|x|$ **8.** $|x + 7|$ **9.** $|x - y|$
10. $|x^2 - 4x + 12|$ **11.** 7 **12.** 57 **13.** ≈ 2.2 **14.** 12 meters **15.** $x^{3/2}$ **16.** $x^{5/3}$ **17.** $y^{13/4}$ **18.** $6^{-2/7}$ **19.** $\sqrt{x}$
20. $\sqrt[3]{a^2}$ **21.** $\left(\sqrt[4]{8m^2n}\right)^7$ **22.** $\dfrac{1}{\left(\sqrt[3]{x + y}\right)^5}$ **23.** 16 **24.** x^6 **25.** 81 **26.** $\sqrt[4]{a}$ **27.** -6 **28.** Not a real number **29.** $\dfrac{3}{4}$ **30.** $\dfrac{3}{8}$
31. $x^{4/15}$ **32.** $\dfrac{4}{y^3}$ **33.** $\dfrac{1}{a^{16/15}}$ **34.** $\dfrac{25x^{10}}{y^7}$ **35.** $5a^2 - 3a^{5/2}$ **36.** $\dfrac{4}{x^{7/6}} + 11$ **37.** $x^{2/5}(1 + x)$ **38.** $\dfrac{1 + a^2}{a^{1/2}}$ **39.** 5 **40.** ≈ 2.668
41. **42.** **43.** $3\sqrt{2}$ **44.** $2\sqrt[3]{2}$ **45.** $\dfrac{7}{3}$ **46.** $\dfrac{2}{5}$ **47.** $-\dfrac{9}{7}$ **48.** $-\dfrac{3}{5}$ **49.** 8 **50.** 4 **51.** $3xyz^2\sqrt{2y}$

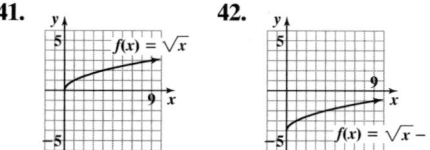

52. $5xy^3\sqrt{3xy}$ **53.** $3a^2b^3\sqrt[3]{2ab}$ **54.** $5x^2y^3z^5\sqrt[4]{x^2z}$ **55.** $x^{14}y^{21}z^{35}$ **56.** $8a^3b^{12}c^{18}$ **57.** $2x^3\sqrt{10}$ **58.** $2x^3y\sqrt[3]{x^2y^2}$ **59.** $2x^2y^3\sqrt[4]{4x^2}$
60. $2x^2y^4\sqrt[4]{x}$ **61.** $6x - 2\sqrt{15x}$ **62.** $2x^2y^2\sqrt[3]{y^2} + x\sqrt[3]{18y}$ **63.** $\sqrt[4]{a^3b^2}$ **64.** $\sqrt[6]{x^5y^2}$ **65.** $\dfrac{64r^{9/2}}{p^3}$ **66.** $\dfrac{y^{1/5}}{6xz^{1/3}}$ **67.** $\dfrac{\sqrt{15}}{5}$ **68.** $\dfrac{\sqrt[3]{21}}{3}$
69. $\dfrac{\sqrt[4]{20}}{2}$ **70.** $\dfrac{x\sqrt{10}}{10}$ **71.** $\dfrac{8\sqrt{x}}{x}$ **72.** $\dfrac{m\sqrt[3]{5}}{5}$ **73.** $\dfrac{10\sqrt[3]{y}}{y}$ **74.** $\dfrac{9\sqrt[4]{z^3}}{z}$ **75.** $\dfrac{x}{3}$ **76.** $\dfrac{x}{2}$ **77.** $\dfrac{4y^2}{x^3}$ **78.** $2x^2y^3$ **79.** $\dfrac{x^2\sqrt{6y}}{y}$
80. $\dfrac{2\sqrt{21ab}}{7b}$ **81.** $\dfrac{x^2y^2\sqrt{6yz}}{z}$ **82.** $\dfrac{5xy^2\sqrt{15yz}}{3z}$ **83.** $3xy\sqrt[3]{2y}$ **84.** $\dfrac{\sqrt[3]{75xy^2}}{5y}$ **85.** $\dfrac{y\sqrt[3]{9x^2}}{x}$ **86.** $\dfrac{y^2\sqrt[3]{25x}}{5x}$ **87.** $\dfrac{b^2\sqrt[4]{2ab^2}}{a}$
88. $\dfrac{y\sqrt[4]{6x^3y^2}}{2x}$ **89.** 7 **90.** -2 **91.** $x^2 - y$ **92.** $7 + 4\sqrt{3}$ **93.** $x + \sqrt{5xy} - \sqrt{3xy} - y\sqrt{15}$
94. $\sqrt[3]{6x^2} - \sqrt[3]{4xy} - \sqrt[3]{9xy} + \sqrt[3]{6y^2}$ **95.** $-12 + 6\sqrt{5}$ **96.** $\dfrac{4x - x\sqrt{x}}{16 - x}$ **97.** $\dfrac{4a + a\sqrt{b}}{16 - b}$ **98.** $\dfrac{x\sqrt{y} + 7x}{y - 49}$ **99.** $\dfrac{x - \sqrt{xy}}{x - y}$
100. $\dfrac{x - 2\sqrt{xy} - 3y}{x - y}$ **101.** $\dfrac{2\sqrt{a - 1} + 4}{a - 5}$ **102.** $\dfrac{5\sqrt{y + 2} + 15}{y - 7}$ **103.** $9\sqrt[3]{x}$ **104.** $-4\sqrt{3}$ **105.** $12 - 13\sqrt[3]{2}$ **106.** $\dfrac{45\sqrt{2}}{8}$
107. $(9x^2y^3 - 4x^3y^4)\sqrt{x}$ **108.** $(8x^2y^2 - x + 3x^3)\sqrt[3]{xy^2}$ **109.** $3x\sqrt{2} - 3\sqrt{5x}$ **110.** $2x^2 + 2x^2\sqrt[3]{4x}$ **111.** $2x + 7$
112. $\sqrt{5}|2a + 5|$ **113.** $\sqrt[6]{x + 5}$ **114.** $\sqrt[12]{b^5}$ **115. a)** $12\sqrt{3}$ **b)** 24 **116. a)** $40\sqrt{2}$ **b)** $\dfrac{2040}{7} \approx 291.4$
117. a) **b)** $x \geq 0$ **118. a)** **b)** $x \geq 0$ **119.** 16 **120.** No solution **121.** 64 **122.** -125

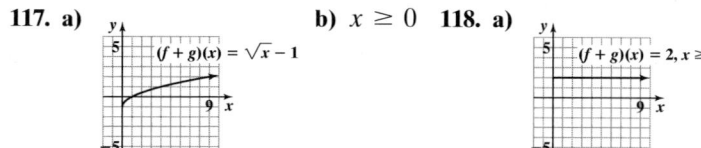

123. 9 **124.** 125 **125.** No solution **126.** 4 **127.** −3 **128.** 3 **129.** 0, 9 **130.** 5 **131.** 4 **132.** 6 **133.** $L = \dfrac{V^2 w}{2}$

134. $A = \pi r^2$ **135.** $2\sqrt{14}$ **136.** $5\sqrt{3}$ **137.** $\sqrt{29} \approx 5.39$ meters **138.** $\sqrt{1280} \approx 35.78$ feet per second

139. $2\pi\sqrt{2} \approx 2.83\pi \approx 8.89$ seconds **140.** $\sqrt{\dfrac{90}{0.145}} \approx 24.91$ meters per second **141.** $m \approx 5m_0$. Thus, it is ≈ 5 times its original mass.

142. $1 + 0i$ **143.** $-8 + 0i$ **144.** $7 - 16i$ **145.** $9 + 4i$ **146.** $13 + i$ **147.** $-3 + 3i$ **148.** $12\sqrt{3} + (\sqrt{5} - \sqrt{7})i$ **149.** $-6 + 6i$

150. $17 - 6i$ **151.** $(24 + 3\sqrt{5}) + (4\sqrt{3} - 6\sqrt{15})i$ **152.** $-\dfrac{8i}{3}$ **153.** $-1 - \dfrac{\sqrt{3}}{2}i$ **154.** $\dfrac{12}{13} - \dfrac{8}{13}i$ **155.** $\dfrac{5\sqrt{3}}{31} + \dfrac{3\sqrt{2}}{31}i$ **156.** 0

157. 7 **158.** i **159.** $-i$ **160.** 1 **161.** -1

Chapter 7 Practice Test

1. $|5x - 3|$ [7.1] **2.** $\dfrac{1}{x^{12/5}}$ [7.2] **3.** $\dfrac{1 + x^2}{x^{2/3}}$ [7.2] **4.** [7.1] **5.** $3x^3 y^5 \sqrt{6x}$ [7.3]

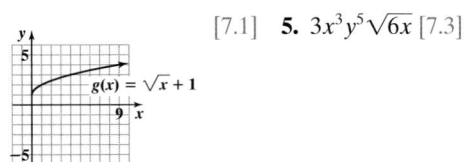

$g(x) = \sqrt{x} + 1$

6. $5x^3 y^3 \sqrt[3]{2x^2 y}$ [7.4] **7.** $\dfrac{x^3 y \sqrt{14yz}}{4z}$ [7.5] **8.** $\dfrac{9\sqrt[3]{r^2}}{x}$ [7.5] **9.** $\dfrac{3 - \sqrt{3}}{6}$ [7.5] **10.** $7\sqrt{6}$ [7.3] **11.** $(2xy + 4x^2 y^2)\sqrt[3]{y^2}$ [7.4]

12. $6\sqrt{3} - 2\sqrt{6} - 12 + 4\sqrt{2}$ [7.4] **13.** $\sqrt[8]{x^5 y^3}$ [7.2] **14.** $\sqrt[12]{(7x + 2)^7}$ [7.5] **15.** -5 [7.6] **16.** -3 [7.6] **17.** 9 [7.6] **18.** 3 [7.6]

19. $g = \dfrac{8w^2}{h}$ [7.6] **20.** $\sqrt{12,880} \approx 113.49$ feet per second [7.6] **21.** 13 feet [7.6] **22.** $2\pi\sqrt{\dfrac{1400}{65,000}} \approx 0.92$ second [7.6]

23. $20 + 20i$ [7.7] **24.** $\dfrac{33}{53} - \dfrac{17}{53}i$ [7.7] **25.** 2 [7.7]

Cumulative Review Test

1. $\dfrac{57}{9}$ [2.1] **2.** -1 [2.1] **3.** 40 [2.3] **4.** $\{x | -1 < x < 4\}$ [2.6] **5.** [3.4]

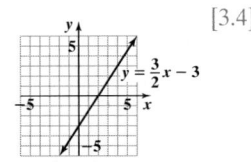

$y = \dfrac{3}{2}x - 3$

6. Parallel [3.5] **7.** $-x^2 + 5x - 13$ [3.6] **8.** $y = -\dfrac{2}{3}x - \dfrac{10}{3}$ [3.5] **9.** $\left(2, 5, \dfrac{34}{5}\right)$ [4.2] **10.** 40 [4.5] **11.** $w = 2r + 1$ [5.3]

12. $25x^2 y^2 - 9$ [5.2] **13.** $3, -3$ [7.6] **14.** $x(4x - 5)(x - 1)$ [5.5] **15.** $(x - 2)(x^2 + 5x + 13)$ [5.6] **16.** $\dfrac{1}{4}, -\dfrac{3}{2}$ [5.8]

17. $\dfrac{(x + y)y^2}{3x^3}$ [6.1] **18.** $\dfrac{x + 3}{x + 5}$ [6.2] **19.** 18 [6.4] **20.** 400 feet [6.6]

Chapter 8

Exercise Set 8.1 **1.** Quadratic **3.** Complex **5.** Perfect square trinomial **7.** Reciprocal **9.** 25 **11.** ± 9 **13.** ± 5 **15.** $\pm 7i$

17. $\pm 2i\sqrt{6}$ **19.** $\pm i\sqrt{61}$ **21.** $8, 0$ **23.** $-3 \pm 5i$ **25.** $2 \pm 3i\sqrt{5}$ **27.** $-1, \dfrac{1}{3}$ **29.** $\dfrac{2 \pm 2i}{3}$ **31.** $0.1, -1.7$ **33.** $\dfrac{5 \pm 3\sqrt{2}}{2}$

35. $-\dfrac{1}{20}, -\dfrac{9}{20}$ **37.** $-1, -5$ **39.** $-3, -5$ **41.** $-2, -4$ **43.** $1, 6$ **45.** $-1, \dfrac{1}{2}$ **47.** $-\dfrac{1}{2}, 4$ **49.** $5, 8$ **51.** $-1, 7$ **53.** $4, 5$ **55.** $7, -4$

57. $1, -11$ **59.** $2 \pm \sqrt{14}$ **61.** $-4 \pm \sqrt{11}$ **63.** $\dfrac{1 \pm \sqrt{13}}{2}$ **65.** $\dfrac{-3 \pm i\sqrt{15}}{2}$ **67.** $0, 1$ **69.** $0, -\dfrac{2}{3}$ **71.** $0, \dfrac{1}{6}$ **73.** $1, -3$ **75.** $8, -4$

77. $\dfrac{-9 \pm \sqrt{73}}{2}$ **79.** $\dfrac{1}{3}, -1$ **81.** $\dfrac{1 \pm i\sqrt{39}}{4}$ **83.** $1 \pm i$ **85. a)** $21 = (x + 2)(x - 2)$ **b)** 5 **87. a)** $18 = (x + 4)(x + 2)$

b) $-3 + \sqrt{19}$ **89.** 12 mph **91.** $5, 7$ **93.** 5 feet by 12 feet **95.** $\dfrac{12 + \sqrt{288}}{2} \approx 14.49$ feet by 14.49 feet **97.** $\sqrt{200} \approx 14.14$ inches

99. $\sqrt{24} \approx 4.90$ feet **101.** 4% **103.** $\approx 6\%$ **105. a)** $S = 32 + 80\sqrt{\pi} \approx 173.80$ square inches **b)** $r = \dfrac{4\sqrt{\pi}}{\pi} \approx 2.26$ inches

c) $r = -5 + \sqrt{\dfrac{80 + 25\pi}{\pi}} \approx 2.1$ inches **107.** 2 **108.** \$4200 at 7%, \$5800 at $6\dfrac{1}{4}$% **109.** $\left\{10, \dfrac{4}{3}\right\}$ **110.** 0 **111.** $4x^3 + x^2 - 21x + 6$

Exercise Set 8.2 **1.** Completing the square **3.** Divide **5.** Two **7.** No **9.** Two real solutions **11.** No real solution
13. Two real solutions **15.** No real solution **17.** One real solution **19.** One real solution **21.** $-2, -5$ **23.** $2, 4$ **25.** $1, -7$

27. $-2 \pm 2\sqrt{6}$ **29.** ± 8 **31.** $\dfrac{2 \pm i\sqrt{11}}{3}$ **33.** $0, 5$ **35.** $\dfrac{2 \pm i\sqrt{2}}{2}$ **37.** -1 **39.** $\dfrac{1}{4}$ **41.** $1 \pm \sqrt{2}$ **43.** $\dfrac{-3 \pm i\sqrt{15}}{2}$ **45.** $-3, \dfrac{1}{2}$

47. $\dfrac{1}{2}, -\dfrac{5}{3}$ **49.** $4, -6$ **51.** $\dfrac{1}{3}, -\dfrac{2}{3}$ **53.** $\dfrac{-6 \pm 2\sqrt{6}}{3}$ **55.** $\dfrac{3 \pm \sqrt{309}}{30}$ **57.** $\dfrac{3 \pm \sqrt{33}}{2}$ **59.** $\dfrac{2 \pm i\sqrt{6}}{2}$ **61.** $\dfrac{-1 \pm i\sqrt{23}}{4}$

63. $\dfrac{-0.6 \pm \sqrt{0.84}}{0.2}$ or $-3 \pm \sqrt{21}$ **65.** $0, \dfrac{3}{2}$ **67.** $-5, 6$ **69.** $\dfrac{7 \pm \sqrt{17}}{4}$ **71.** No real number **73.** $x^2 - 7x + 6 = 0$

75. $x^2 + 8x - 9 = 0$ **77.** $15x^2 - x - 6 = 0$ **79.** $x^2 - 2 = 0$ **81.** $x^2 + 9 = 0$ **83.** $x^2 - 6x + 7 = 0$ **85.** $x^2 - 4x + 13 = 0$

87. a) $n(10 - 0.02n) = 450$ **b)** 50 **89. a)** $n(50 - 0.4n) = 660$ **b)** 15 **91.** 3 **93.** $w = 3$ feet, $l = 8$ feet **95.** 2 inches

97. a) 0.5 second and 2 seconds **b)** 2.6 seconds **c)** 28 feet **99. a)** 1 second and 9 seconds **b)** 10 seconds **c)** 66.25 feet

101. Yes; if you multiply both sides of one equation by -1 you get the other equation. **103. a)** $b^2 - 4ac$ **b)** -84

c) Answers will vary. **105.** Answers will vary. **107.** Yes **109.** $(-0.12 + \sqrt{14.3952})/1.2 \approx 3.0618$ millimeters

111. a) ≈ 1.94 seconds **b)** ≈ 2.74 seconds **c)** Courtney's **d)** Yes, at 1.5 seconds **112.** 5.0×10^2 or 500 **113.** 7 **114.** $(2, -1)$

115. $\dfrac{6y - x}{3xy}$ **116.** No real solution

Exercise Set 8.3
1. Profit **3.** One whole task **5.** $s = \sqrt{A}$ **7.** $i = \sqrt{\dfrac{E}{r}}$ **9.** $t = \dfrac{\sqrt{d}}{4}$ **11.** $c = \sqrt{\dfrac{E}{m}}$ **13.** $r = \sqrt{\dfrac{3V}{\pi h}}$

15. $W = \sqrt{d^2 - L^2}$ **17.** $b = \sqrt{c^2 - a^2}$ **19.** $H = \sqrt{d^2 - L^2 - W^2}$ **21.** $t = \sqrt{\dfrac{h - s_0}{-16}}$ or $t = \dfrac{\sqrt{s_0 - h}}{4}$ **23.** $v = \sqrt{\dfrac{2E}{m}}$

25. $v_1 = \sqrt{v_2^2 - 2ad}$ **27.** $c = \sqrt{(v')^2 + v^2}$ **29. a)** loss of \$2400 **b)** profit of \$24,000 **c)** about 1.4 years **31. a)** loss of \$2000

b) profit of \$21,000 **c)** about 1.5 years **33. a)** $32°$F **b)** $80.8°$F **c)** ≈ 2.92 minutes **35. a)** about 9.64 crashes

b) about 12.115 crashes **c)** about 19 years old and 72 years old **37. a)** \$3,615.25 billion **b)** \$1,114 billion

c) 2009 and 2028 **39.** $l = 30$ meters, $w = 20$ meters **41.** 4 feet per hour **43.** Going: 6 mph, returning: 8 mph

45. Bonita: ≈ 11.52 hours, Pamela: ≈ 12.52 hours **47.** 130 mph **49.** Chris ≈ 11.76 hours, John: ≈ 12.26 hours **51.** 75 mph

53. $l \approx 34.86$ inches, $h \approx 19.61$ inches **55.** Answers will vary. **57.** Answers will vary. **59.** 6 meters by 3 meters or 2 meters by 9 meters

61. -16 **62.** $R = \dfrac{E - Ir}{I}$ **63.** $\dfrac{8}{r - 4}$ **64.** $\dfrac{x^2}{y^{32}}$ **65.** No solution

Mid-Chapter Test
1. $\pm 7\sqrt{2}$ [8.1] **2.** $3 \pm 2i\sqrt{5}$ [8.1] **3.** $-\dfrac{1}{2}, -\dfrac{13}{2}$ [8.1] **4.** $-6, 2$ [8.1] **5.** $2 \pm \sqrt{14}$ [8.1] **6.** $\dfrac{-1 \pm i\sqrt{143}}{8}$ [8.1]
7. $(6 + 6\sqrt{2})$ meters [8.1] **8. a)** $b^2 - 4ac$ **b)** Two distinct real solutions: $b^2 - 4ac > 0$; single real solution: $b^2 - 4ac = 0$;

no real solution: $b^2 - 4ac < 0$ [8.2] **9.** Two distinct real solutions [8.2] **10.** $-\dfrac{5}{3}, \dfrac{3}{2}$ [8.2] **11.** $-2 \pm 2\sqrt{3}$ [8.2] **12.** $\dfrac{1 \pm i\sqrt{14}}{3}$ [8.2]

13. $x^2 - 5x - 14 = 0$ [8.2] **14.** $x^2 - 4x - 1 = 0$ [8.2] **15.** 10 lamps [8.2] **16.** $r = \sqrt{x^2 - y}$ [8.3] **17.** $x = \sqrt{\dfrac{3A}{k}}$ [8.3]
18. $y = \sqrt{D^2 - x^2}$ [8.3] **19.** 5 feet by 12 feet [8.3] **20.** 5 clocks [8.3]

Exercise Set 8.4
1. Quadratic in form **3.** Extraneous solution **5.** $u = h - 2$ **7.** $\pm 1, \pm 3$ **9.** $\pm 2i, \pm 3i$ **11.** $\pm 2, \pm 3$

13. $\pm 2, \pm\sqrt{3}$ **15.** $\pm\dfrac{1}{2}, \pm 2$ **17.** $\pm\sqrt{3}, \pm\sqrt{5}$ **19.** $\pm 3, \pm i\sqrt{2}$ **21.** $\pm 1, \pm i\sqrt{5}$ **23.** 4 **25.** 9 **27.** $\dfrac{1}{9}$ **29.** $1, -9$ **31.** $\dfrac{4}{3}, -\dfrac{1}{2}$

33. $\pm\sqrt{6}, \pm 1$ **35.** $-6, -\dfrac{5}{2}$ **37.** $\pm\dfrac{5\sqrt{6}}{6}, \pm\dfrac{\sqrt{39}}{3}$ **39.** $-\dfrac{1}{2}$ **41.** $3, 4$ **43.** $2, \dfrac{1}{3}$ **45.** $1, -\dfrac{1}{10}$ **47.** $-\dfrac{1}{2}, \dfrac{1}{6}$ **49.** $1, 27$ **51.** $27, 216$ **53.** $\dfrac{1}{4}$

55. $-32, -1$ **57.** $(4, 0), (9, 0)$ **59.** None **61.** $(-4, 0), \left(\dfrac{1}{5}, 0\right)$ **63.** $(-8, 0), (27, 0)$ **65.** $(-1, 0), (4, 0)$ **67.** $(\pm 2, 0), (\pm 5, 0)$

69. a) and b) $\dfrac{1}{5}, -\dfrac{1}{4}$ **71.** $-\dfrac{14}{5}, -\dfrac{8}{3}$ **73.** $2, \dfrac{1}{4}$ **75.** $2, 1$ **77.** $-3, 1, 2, -4$ **79.** $\pm\sqrt{\dfrac{3 \pm \sqrt{15}}{2}}$ **81.** Let $u = x^2$ **83.** Let $u = x^{-1}$

85. $x^4 - 5x^2 + 4 = 0$; start with $(x - 2)(x + 2)(x - 1)(x + 1) = 0$

87. $x^4 - 7x^2 + 10 = 0$; start with $(x + \sqrt{2})(x - \sqrt{2})(x + \sqrt{5})(x - \sqrt{5}) = 0$ **89.** No; imaginary solutions always occur in pairs.

91. $\dfrac{43}{60}$ **92.** 1 **93.** D: $\mathbb{R}$, R: $\{y | y \geq 0\}$ **94.** $2xy^2\sqrt[3]{2}$ **95.** $9\sqrt{3}$

Exercise Set 8.5
1. Parabola **3.** Opens downward **5.** $-\dfrac{b}{2a}$ **7.** y-intercept **9.** Right **11.** Up

13. a) Upward **b)** $(0, 8)$
c) $(-3, -1)$ **d)** $(-4, 0), (-2, 0)$
e)

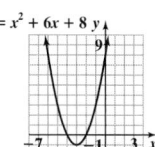

$f(x) = x^2 + 6x + 8$

15. a) Upward **b)** $(0, 15)$
c) $(-4, -1)$ **d)** $(-5, 0), (-3, 0)$
e)

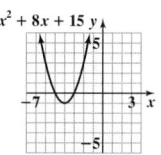

$f(x) = x^2 + 8x + 15$

17. a) Upward **b)** $(0, 3)$
c) $(2, -1)$ **d)** $(1, 0), (3, 0)$
e)

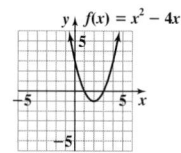

$f(x) = x^2 - 4x + 3$

19. a) Downward **b)** $(0, 8)$
c) $(-1, 9)$ **d)** $(-4, 0), (2, 0)$
e)

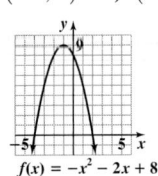
$f(x) = -x^2 - 2x + 8$

21. a) Downward **b)** $(0, 5)$ **c)** $(2, 9)$
d) $(-1, 0), (5, 0)$ **e)**

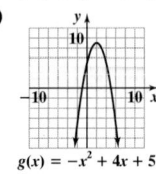
$g(x) = -x^2 + 4x + 5$

23. a) Downward **b)** $(0, -5)$ **c)** $(2, -1)$
d) No x-intercepts **e)**

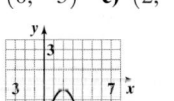
$t(x) = -x^2 + 4x - 5$

25. a) Upward **b)** $(0, 4)$ **c)** $(2, 0)$
d) $(2, 0)$ **e)**

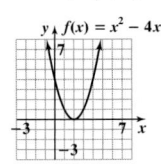
$f(x) = x^2 - 4x + 4$

27. a) Upward **b)** $(0, 2)$ **c)** $(0, 2)$
d) No x-intercepts **e)**

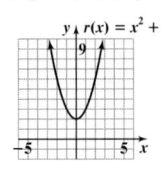
$r(x) = x^2 + 2$

29. a) Downward **b)** $(0, 5)$
c) $(0, 5)$ **d)** $(-\sqrt{5}, 0), (\sqrt{5}, 0)$
e)

$l(x) = -x^2 + 5$

31. a) Downward **b)** $(0, -8)$
c) $(1, -6)$ **d)** No x-intercepts
e)

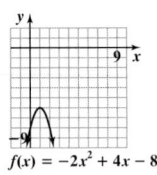
$f(x) = -2x^2 + 4x - 8$

33. a) Upward **b)** $(0, 3)$
c) $\left(-\dfrac{2}{3}, \dfrac{5}{3}\right)$ **d)** No x-intercepts
e)

$m(x) = 3x^2 + 4x + 3$

35. a) Upward **b)** $(0, -6)$
c) $\left(-\dfrac{2}{3}, -\dfrac{22}{3}\right)$ **d)** $\left(\dfrac{2 + \sqrt{22}}{3}, 0\right), \left(\dfrac{2 - \sqrt{22}}{3}, 0\right)$
e)

$y = 3x^2 + 4x - 6$

37. a) Upward **b)** $(0, -6)$
c) $\left(\dfrac{1}{4}, -\dfrac{49}{8}\right)$ **d)** $\left(-\dfrac{3}{2}, 0\right), (2, 0)$
e)

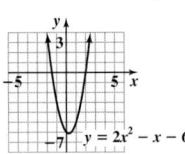
$y = 2x^2 - x - 6$

39. a) Downward **b)** $(0, -5)$
c) $\left(\dfrac{3}{2}, -\dfrac{11}{4}\right)$ **d)** No x-intercepts
e)

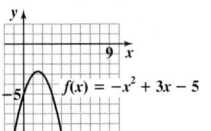
$f(x) = -x^2 + 3x - 5$

41.

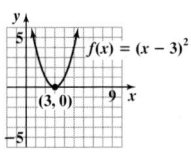
$f(x) = (x - 3)^2$
$(3, 0)$

43.

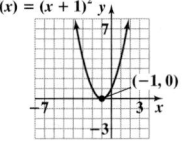
$f(x) = (x + 1)^2$
$(-1, 0)$

45.

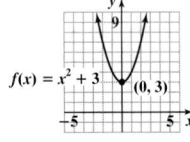
$f(x) = x^2 + 3$
$(0, 3)$

47.

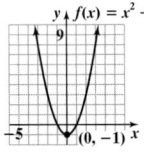
$f(x) = x^2 - 1$
$(0, -1)$

49.

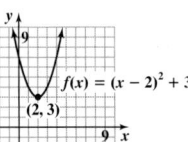
$f(x) = (x - 2)^2 + 3$
$(2, 3)$

51.

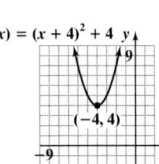
$f(x) = (x + 4)^2 + 4$
$(-4, 4)$

53.

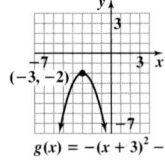
$(-3, -2)$
$g(x) = -(x + 3)^2 - 2$

55.

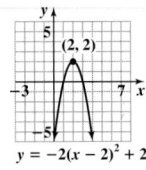
$(2, 2)$
$y = -2(x - 2)^2 + 2$

57.

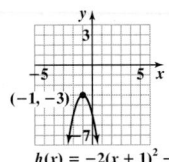
$(-1, -3)$
$h(x) = -2(x + 1)^2 - 3$

59. a) $f(x) = (x - 3)^2 - 1$
b)

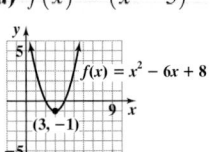
$f(x) = x^2 - 6x + 8$
$(3, -1)$

61. a) $g(x) = \left(x - \dfrac{1}{2}\right)^2 - \dfrac{13}{4}$
b)

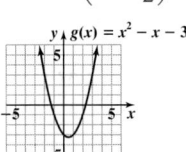
$g(x) = x^2 - x - 3$

63. a) $f(x) = -(x + 2)^2 - 2$
b)

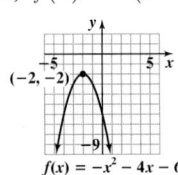
$(-2, -2)$
$f(x) = -x^2 - 4x - 6$

65. a) $g(x) = (x - 2)^2 - 5$
b)

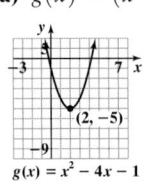
$(2, -5)$
$g(x) = x^2 - 4x - 1$

67. a) $f(x) = 2\left(x + \dfrac{5}{4}\right)^2 - \dfrac{49}{8}$
b)

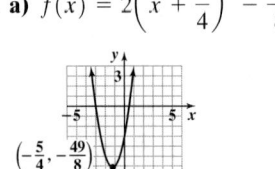
$\left(-\dfrac{5}{4}, -\dfrac{49}{8}\right)$
$f(x) = 2x^2 + 5x - 3$

69. b)   **71.** d)   **73. a)** $x = 10.5$ **b)** $A = 240.25$ **75. a)** $x = 7$ **b)** $A = 121$ **77. a)** $n = 200$ **b)** $R = \$800$

79. 2010   **81.** 4 units   **83.** 3 units   **85.** $f(x) = 2(x - 3)^2 - 2$ **87.** $f(x) = -4\left(x + \dfrac{3}{5}\right)^2 - \sqrt{2}$

89. a)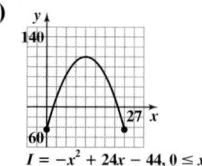
$I = -x^2 + 24x - 44, 0 \le x \le 24$

b) $2 **c)** $22 **d)** $12 **e)** $10,000 **91. a)** 100 **b)** $3800 **93. a)** 40.425 meters **b)** 2.5 seconds **c)** ≈ 5.37 seconds **95.** 400 square feet **97.** $-16, 4$ and -4 **99.** 900, 30 and 30 **101. a)** $142,400 **b)** 380 **103. a)** 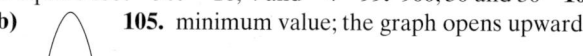 **b)**   **105.** minimum value; the graph opens upward

107. a) The graphs will have the same x-intercepts but $f(x) = x^2 - 8x + 12$ will open upward and $g(x) = -x^2 + 8x - 12$ will open downward. **b)** Yes, both at $(6, 0)$ and $(2, 0)$ **c)** No; vertex of $f(x)$ at $(4, -4)$, vertex of $g(x)$ at $(4, 4)$
d)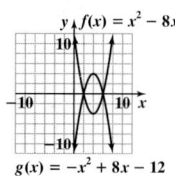
$g(x) = -x^2 + 8x - 12$

109. a) $f(t) = -16(t - 1.625)^2 + 45.25$ **b)** 45.25 feet, 1.625 seconds **c)** Same
111. 200π square feet **112.** 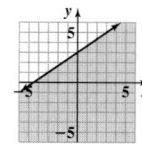 **113.** $(-2, 3, 2)$ **114.** -8 **115.** $\dfrac{x}{x + 6}$

Exercise Set 8.6

1. Solution **3.** Included **5.** Open **7.**  **9.**   **11.**
13.   **15.**   **17.**   **19.**   **21.** $(-\infty, -2] \cup [1, 3]$ **23.** $(-\infty, -4) \cup (-2, 3)$

25. $\left(-6, -\dfrac{5}{2}\right) \cup (2, \infty)$ **27.** $\left(-\dfrac{5}{3}, -1\right) \cup (3, \infty)$ **29.** $[-2, -2] \cup \left[\dfrac{8}{3}, \infty\right)$ **31.** $(-\infty, 0)$ **33.**

35.   **37.**   **39.**   **41.**   **43.** $\{x | -1 \le x < 3\}$ **45.** $\{x | -5 < x < 1\}$

47. $\{x | x \le -3 \text{ or } x > 2\}$ **49.** $\{a | -5 < a < 9\}$ **51.** $\{c | c < 4 \text{ or } c > 10\}$ **53.** $\{y | -4 < y \le -2\}$ **55.** $\left\{a \middle| a \le -2 \text{ or } a > \dfrac{1}{3}\right\}$

57. $\left\{x \middle| -\dfrac{4}{3} < x < \dfrac{1}{2}\right\}$ **59.** $\left\{x \middle| -\dfrac{8}{3} \le x < 2\right\}$ **61.** $(-\infty, -1) \cup (2, 4)$ **63.** $(-3, 2) \cup (5, \infty)$ **65.** $(-2, 1] \cup [7, \infty)$

67. $(-\infty, -8) \cup [0, 3)$ **69.** $(-\infty, -4) \cup (1, 6]$ **71.** $\left[-\dfrac{5}{2}, 3\right] \cup (4, \infty)$ **73.**   **75.**

77.   **79.**   **81.**   **83.**   **85. a)** $x < 2$ or $x > 5$ **b)** $2 < x < 5$

87. a) $(4, \infty)$; $y > 0$ in this interval **b)** $(-\infty, 2) \cup (2, 4)$; $y < 0$ in this interval **89.** $x^2 + 2x - 8 > 0$ **91.** $\dfrac{x + 3}{x - 4} \ge 0$
93. All real numbers; for any value of x, the expression is ≥ 0.
95. All real numbers except -2; for any value of x except -2, the expression is ≥ 0.
97. No solution; the graph opens upward and has no x-intercepts, so it is always above the x-axis. **99.**

101. $x^2 - 3x > 0$; multiply factors containing boundary values. **103.** $x^2 < 0$; x^2 is always ≥ 0. **105.** $(-\infty, -3) \cup (-1, 1) \cup (3, \infty)$

107. $[-2, -1] \cup [2, \infty)$ **111.** 6 quarts **112.** $-\dfrac{1}{2}$ **113.** $3r + 3s - 9t$ **114.** $\dfrac{x - 3}{x + 1}$ **115.** $38 - 9i$

Chapter 8 Review Exercises

1. $1, 9$ **2.** $\dfrac{-1 \pm 2\sqrt{15}}{2}$ **3.** $-\dfrac{1}{3}, 1$ **4.** $\dfrac{5}{4}, -\dfrac{3}{4}$ **5.** $2, 6$ **6.** $4, -8$ **7.** $-1 \pm \sqrt{10}$

8. $-3 \pm \sqrt{21}$ **9.** $1 \pm 3i$ **10.** $2 \pm 2i\sqrt{7}$ **11. a)** $32 = (x + 1)(x + 5)$ **b)** 3 **12. a)** $63 = (x + 2)(x + 4)$ **b)** 5
13. $7, 8$ **14.** ≈ 16.90 feet by ≈ 16.90 feet **15.** Two real solutions **16.** No real solution **17.** One real solution **18.** No real solution
19. One real solution **20.** Two real solutions **21.** $-\dfrac{5}{2}, \dfrac{10}{3}$ **22.** $2, 9$ **23.** $8, -5$ **24.** $0, \dfrac{9}{7}$ **25.** $\dfrac{3}{2}, -\dfrac{5}{3}$ **26.** $\dfrac{1}{4}, -3$

27. $-4 \pm \sqrt{11}$ **28.** $-2 \pm 2\sqrt{3}$ **29.** $\dfrac{-2 \pm \sqrt{10}}{2}$ **30.** $\dfrac{3 \pm \sqrt{33}}{3}$ **31.** $\dfrac{1 \pm i\sqrt{51}}{2}$ **32.** $1 \pm i\sqrt{10}$ **33.** $\dfrac{5}{2}, -\dfrac{5}{3}$ **34.** $\dfrac{1}{4}, -\dfrac{3}{2}$

35. $10, -6$ **36.** $\dfrac{2}{3}, -\dfrac{3}{2}$ **37.** $\dfrac{7 \pm \sqrt{89}}{10}$ **38.** $\dfrac{3 \pm 3\sqrt{3}}{2}$ **39.** $x^2 - 2x - 3 = 0$ **40.** $3x^2 + 4x - 4 = 0$ **41.** $x^2 - 11 = 0$

42. $x^2 - 6x + 13 = 0$ **43.** 8 feet by 12 feet **44.** $\sqrt{128} \approx 11.31$ **45.** 4% **46.** $7, 11$ **47.** 8 inches by 12 inches **48.** $540
49. a) 10.12 inches **b)** 11.77 inches or 38.23 inches **50. a)** 720 feet **b)** 7 seconds **51. a)** 40 milliliters **b)** $150°C$
52. larger: ≈ 23.51 hours, smaller: ≈ 24.51 hours **53.** 50 miles per hour **54.** 1.6 miles per hour **55.** $l = 10$ units, $w = 8$ units
56. 20 tables **57.** $a = \sqrt{c^2 - b^2}$ **58.** $t = \sqrt{\dfrac{c - h}{4.9}}$ **59.** $v_y = \sqrt{v^2 - v_x^2}$ **60.** $v_2 = \sqrt{v_1^2 + 2ad}$ **61.** $\pm 1, \pm 3$ **62.** $\pm 4, \pm \sqrt{5}$

63. $\pm 2\sqrt{2}, \pm i\sqrt{3}$ **64.** $\dfrac{3}{2}, -\dfrac{1}{6}$ **65.** $\dfrac{1}{9}$ **66.** $\dfrac{27}{8}, 8$ **67.** $4, \dfrac{13}{8}$ **68.** $-\dfrac{1}{5}, -\dfrac{5}{2}$ **69.** $(\pm 1, 0), (\pm 9, 0)$ **70.** $\left(\dfrac{4}{25}, 0\right)$ **71.** None

72. $(3 \pm \sqrt{17}, 0), (3 \pm \sqrt{6}, 0)$

73. a) Upward **b)** $(0,0)$
c) $\left(-\dfrac{5}{2}, -\dfrac{25}{4}\right)$ **d)** $(0,0), (-5,0)$
e)

74. a) Upward **b)** $(0, -8)$
c) $(1, -9)$ **d)** $(-2, 0), (4, 0)$
e)

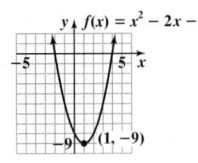

75. a) Downward **b)** $(0, -2)$
c) $(0, -2)$ **d)** No x-intercepts
e)

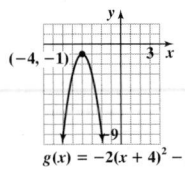

76. a) Downward **b)** $(0, 15)$
c) $\left(-\dfrac{1}{4}, \dfrac{121}{8}\right)$ **d)** $(-3, 0), \left(\dfrac{5}{2}, 0\right)$
e)

77. a) \$11 **b)** \$7600
78. a) 2.5 seconds **b)** 175 feet
79.

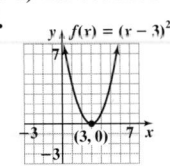

80.

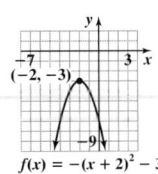

81.

82.

83. $-3 \quad -1$

84. $-5 \quad 2$

85. $\dfrac{11-\sqrt{41}}{2} \quad \dfrac{11+\sqrt{41}}{2}$

86. $-4 \quad \dfrac{4}{3}$

87. $-\dfrac{3}{2} \quad \dfrac{3}{2}$

88. $-\sqrt{5} \quad \sqrt{5}$

89. $\{x \mid x < -1 \text{ or } x > 5\}$ **90.** $\{x \mid -2 < x \le 3\}$ **91.** $\{x \mid x < -3 \text{ or } x \ge 2\}$ **92.** $\left\{x \mid -\dfrac{5}{3} < x < 6\right\}$

93. $\{x \mid -4 < x < -1 \text{ or } x > 2\}$ **94.** $\{x \mid x \le 0 \text{ or } 3 \le x \le 6\}$ **95.** $\left[-\dfrac{4}{3}, 1\right] \cup [3, \infty)$ **96.** $(-\infty, -4) \cup (-2, 0)$

97. $(-2, 0) \cup (4, \infty)$ **98.** $(-\infty, -3) \cup (2, 8)$ **99.** $(-2, 3] \cup (7, \infty)$ **100.** $(-\infty, -3) \cup [0, 6]$

101. $-9 \quad -4$ **102.** $-2 \quad 2$ **103.** $\dfrac{5}{3} \quad \dfrac{23}{10}$

Chapter 8 Practice Test   **1.** $3, -5$ [8.1]   **2.** $3 \pm \sqrt{2}$ [8.1]   **3.** $8, -2$ [8.2]   **4.** $2 \pm i\sqrt{7}$ [8.2]   **5.** $\dfrac{2}{3}, -1$ [8.1–8.2]

6. $\dfrac{-7 \pm \sqrt{33}}{2}$ [8.1–8.2]   **7.** $5x^2 - 18x - 8 = 0$ [8.2]   **8.** $v = \sqrt{\dfrac{2K}{m}}$ [8.3]   **9. a)** \$121,200 **b)** ≈ 2712.57 square feet [8.1–8.3]

10. 50 mph [8.1–8.3]   **11.** $\pm \dfrac{\sqrt{10}}{2}, \pm i\sqrt{10}$ [8.4]   **12.** $\dfrac{343}{27}, -216$ [8.4]   **13.** $\left(\dfrac{9}{16}, 0\right)$ [8.4]   **14.** [8.5]

15. [8.5]   **16.** Two real solutions [8.5]   **17. a)** Upward **b)** $(0, -8)$
c) $(-1, -9)$ **d)** $(-4, 0), (2, 0)$ **e)** [8.5]

18. $2x^2 + 13x - 7 = 0$ [8.5]   **19.** $-6 \quad 7$ [8.6]   **20.** $-5 \quad -1 \quad 4$ [8.6]   **21. a)** $\left[-\dfrac{5}{2}, -2\right)$

b) $\left\{x \mid -\dfrac{5}{2} \le x < -2\right\}$ [8.6]   **22.** $w = 5$ feet, $l = 13$ feet [8.5]   **23.** 6 seconds [8.5]   **24. a)** 20 **b)** \$490 [8.5]   **25.** 30 [8.5]

Cumulative Review Test **1.** 13 [1.4] **2.** 18 [1.4] **3.** 2.54×10^6 [1.6] **4.** $\left\{-\frac{1}{2}, \frac{9}{2}\right\}$ [2.6] **5.** $3x - 7$ [2.1]

6. All real numbers, $\mathbb{R}$ [2.1] **7.** $(-12, 8)$ [2.5] **8.** $m = -\frac{9}{7}, \left(0, \frac{15}{7}\right)$ [3.4] **9.** 1500 [3.2] **10.** $y = x - 1$ [3.5]

11. a) No, the graph does not pass the vertical line test **b)** Domain: $\{x | x \geq -2\}$, Range: $\mathbb{R}$ [3.2]

12. a) **b)** [3.3] **13.** 160 [4.5] **14.** $\left(\frac{5}{2}, 0\right)$ [4.1] **15.** $(x + 7)(x + 9)$ [5.5]

16. a) $a^2 + 2ab + b^2$ **b)** $(a + b)^2$ [5.7] **17.** $\dfrac{2(x - 4)}{(x - 3)(x - 5)}$ [6.2]

18. $\dfrac{12}{5}$ [6.4] **19.** 11.52 watts [6.6] **20.** $-\dfrac{14}{29} - \dfrac{23}{29}i$ [7.7]

Chapter 9

Exercise Set 9.1 **1.** Composition **3.** Vertical **5.** Inverse **7.** Domain **9. a)** $2x + 1$ **b)** 9 **c)** $2x + 5$ **d)** 13

11. a) $x^2 + x - 1$ **b)** 19 **c)** $x^2 + 7x + 8$ **d)** 52 **13. a)** $\dfrac{1}{2x + 3}$ **b)** $\dfrac{1}{11}$ **c)** $\dfrac{2}{x} + 3$ **d)** $3\dfrac{1}{2}$ **15. a)** $\dfrac{9}{x} + 1$ **b)** $3\dfrac{1}{4}$ **c)** $\dfrac{3}{3x + 1}$ **d)** $\dfrac{3}{13}$

17. a) $x^4 + 10x^2 + 26$ **b)** 442 **c)** $x^4 + 2x^2 + 6$ **d)** 294 **19. a)** $\sqrt{x + 5} - 4$ **b)** -1 **c)** $\sqrt{x + 1}$ **d)** $\sqrt{5}$ **21.** No **23.** Yes

25. Yes **27.** No **29.** Yes **31.** No **33.** No **35.** Yes **37.** Yes **39.** No **41.** Yes **43.** $f(x)$: Domain: $\{-2, -1, 2, 4, 8\}$,

Range: $\{0, 4, 6, 7, 9\}$; $f^{-1}(x)$: Domain: $\{0, 4, 6, 7, 9\}$, Range: $\{-2, -1, 2, 4, 8\}$ **45.** $f(x)$: Domain: $\{-1, 1, 2, 4\}$,

Range: $\{-3, -1, 0, 2\}$; $f^{-1}(x)$: Domain: $\{-3, -1, 0, 2\}$, Range: $\{-1, 1, 2, 4\}$ **47.** $f(x)$: Domain: $\{x | x \geq 2\}$, Range: $\{y | y \geq 0\}$; $f^{-1}(x)$;

Domain: $\{x | x \geq 0\}$; Range: $\{y | y \geq 2\}$ **49. a)** Yes **b)** $f^{-1}(x) = x - 3$ **51. a)** Yes **b)** $h^{-1}(x) = \dfrac{x}{4}$ **53. a)** No **55. a)** No

57. a) Yes **b)** $g^{-1}(x) = \dfrac{1}{x}$ **59. a)** No **61. a)** Yes **b)** $g^{-1}(x) = \sqrt[3]{x + 6}$ **63. a)** Yes **b)** $g^{-1}(x) = x^2 - 2, x \geq 0$ **65. a)** Yes

b) $h^{-1}(x) = \sqrt{x + 4}, x \geq -4$

67. a) $f^{-1}(x) = \dfrac{x - 8}{2}$ **69. a)** $f^{-1}(x) = x^2, x \geq 0$ **71. a)** $f^{-1}(x) = x^2 + 1, x \geq 0$ **73. a)** $f^{-1}(x) = x^3$

b) **b)** **b)** **b)**

75. a) $f^{-1}(x) = \dfrac{1}{x}, x > 0$ **77.** $(f \circ f^{-1})(x) = x, (f^{-1} \circ f)(x) = x$ **79.** $(f \circ f^{-1})(x) = x, (f^{-1} \circ f)(x) = x$

b) **81.** $(f \circ f^{-1})(x) = x, (f^{-1} \circ f)(x) = x$ **83.** $(f \circ f^{-1})(x) = x, (f^{-1} \circ f)(x) = x$

85. $f^{-1}(x) = \dfrac{x}{3}$; x is feet and $f^{-1}(x)$ is yards **87.** $f^{-1}(x) = \dfrac{9}{5}x + 32$.

89. $(f \circ g)(x) = 453.6x$, x is pounds, $(f \circ g)(x)$ is grams **91.** $(f \circ g)(x) = 0.915x$, x is

yards, $(f \circ g)(x)$ is meters **93.** No, composition of functions is not commutative. Let $f(x) = x^2$

and $g(x) = x + 1$. Then $(f \circ g)(x) = x^2 + 2x + 1$, while $(g \circ f)(x) = x^2 + 1$. **95. a)** $(f \circ g)(x) = x; (g \circ f)(x) = x$

b) The domain is $\mathbb{R}$ for all of them. **97.** The range of $f^{-1}(x)$ is the domain of $f(x)$. **99. a)** 6 feet **b)** $36\pi \approx 113.10$ square feet

c) $A(t) = 4\pi t^2$ **d)** $36\pi \approx 113.10$ square feet **e)** Answers should agree **102.** $\dfrac{81}{16}$ **103.** $2x + 3y = 10$ **104.** $\dfrac{18 - 12x}{x^3}$

105. $p = \dfrac{fq}{q - f}$ **106.** $-1 \pm \sqrt{11}$

Exercise Set 9.2 **1.** Exponent **3.** Rise **5.** Principal

7. **9.** **11.** **13.** **15.** **17.**

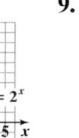
$y = 2^x$

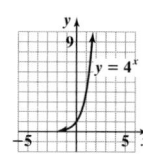
$y = \left(\frac{1}{2}\right)^x$

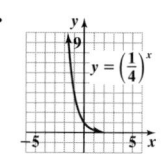
$y = 4^x$

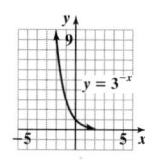
$y = \left(\frac{1}{4}\right)^x$

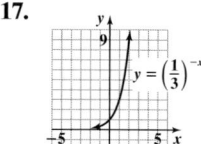
$y = 3^{-x}$

$y = \left(\frac{1}{3}\right)^{-x}$

19. 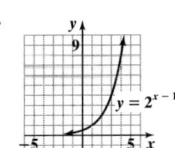 **21.** 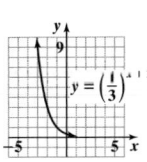 **23.** 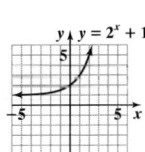 **25.** 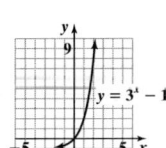 **27. a)** $\approx$36.232 million **b)** $\approx$187.846 million

29. \$512 **31.** 45 **33.** $\approx$\$6344.93 **35.** \$6166.13 **37.** \$3147.06 **39.** $\approx$10.6 grams **41. a)** 5 grams **b)** $\approx 7.28 \times 10^{-11}$ grams

43. a) 2400 **b)** $\approx$4977 **45.** $\approx$\$10,850.92 **47.** $\approx$8.83 kilometers **49. a)** \$13,047 thousand **b)** \$282,727 thousand **51. a)** 14 years

b) 10 years **c)** \$25 **d)** Increases it **53. a)** As x increases, y decreases. **b)** No, $\left(\frac{1}{2}\right)^x$ can never be 0. **c)** No, $\left(\frac{1}{2}\right)^x$ can never be

negative. **55. a)** Same; $(0, 1)$. **b)** $y = 3^x$ will be steeper than $y = 2^x$ for $x > 0$. **57. a)** It is a horizontal line through $y = 1$. **b)** Yes

c) No, not a one-to-one function **59.** $y = a^x - k$ is $y = a^x$ lowered k units. **61.** The graph of $y = a^{x+2}$ is the graph of $y = a^x$

shifted 2 units to the left. **63. a)** \$16,384 **b)** \$524,288 **c)** 2^{n-1} **d)** $\$2^{29} = \$536,870,912$ **e)** $2^0 + 2^1 + 2^2 + \cdots + 2^{29}$

65. a) $-6.2x^6y^2 + 9.2x^5y^2 + 2.3x^4y$ **b)** 8 **c)** -6.2 **66.** $x^3 + 3x^2 - 6x + 20$ **67.** $|a - 4|$ **68.** $\dfrac{2xy\sqrt[4]{xy^2z^3}}{z}$

Exercise Set 9.3

1. Exponent **3.** Base **5.** Domain **7.** $\log_5 25 = 2$ **9.** $\log_3 9 = 2$ **11.** $\log_{16} 4 = \dfrac{1}{2}$ **13.** $\log_8 2 = \dfrac{1}{3}$

15. $\log_{1/2} \dfrac{1}{32} = 5$ **17.** $\log_2 \dfrac{1}{8} = -3$ **19.** $\log_4 \dfrac{1}{64} = -3$ **21.** $\log_{64} 4 = \dfrac{1}{3}$ **23.** $\log_8 \dfrac{1}{2} = -\dfrac{1}{3}$ **25.** $\log_{81} \dfrac{1}{3} = -\dfrac{1}{4}$

27. $\log_{10} 7 = 0.8451$ **29.** $\log_e 7.3891 = 2$ **31.** $\log_a b = n$ **33.** $3^2 = 9$ **35.** $\left(\dfrac{1}{3}\right)^3 = \dfrac{1}{27}$ **37.** $5^{-2} = \dfrac{1}{25}$ **39.** $49^{1/2} = 7$

41. $9^{-2} = \dfrac{1}{81}$ **43.** $10^{-3} = \dfrac{1}{1000}$ **45.** $6^3 = 216$ **47.** $10^{-0.2076} = 0.62$ **49.** $e^{1.8749} = 6.52$ **51.** $w^{-p} = s$ **53.** 3 **55.** 5 **57.** 27

59. -4 **61.** $\dfrac{1}{64}$ **63.** 3 **65.** $\dfrac{1}{2}$ **67.** 0 **69.** 2 **71.** -2 **73.** 4 **75.** 4 **77.** -4 **79.** -2 **81.** $\dfrac{1}{2}$ **83.** 1 **85.** $\dfrac{1}{2}$

87. 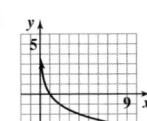 **89.** 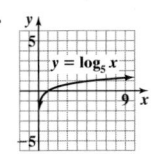 **91.** 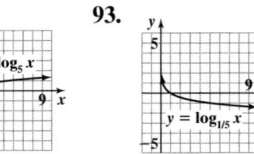 **93.** 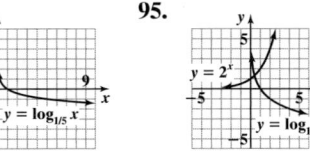 **95.** 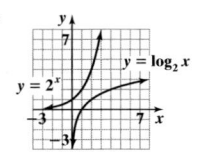 **97.**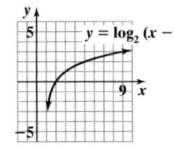

99. 10,000,000 **101.** 10,000 **103.** **105. a)** $a > 0$ and $a \neq 1$ **b)** $\{x | x > 0\}$ **c)** $\mathbb{R}$

107. $\left(\dfrac{1}{27}, -3\right) \left(\dfrac{1}{9}, -2\right), \left(\dfrac{1}{3}, -1\right)$ $(1, 0), (3, 1), (9, 2)$, and $(27, 3)$; the functions $f(x) = a^x$ and $g(x) = \log_a x$ are inverses.

109. $f^{-1}(x) = \log_5 x$ **111.** 3 and 4, since 62 lies between $3^3 = 27$ and $3^4 = 81$. **113.** 2 and 3, since 425 lies between $10^2 = 100$ and

$10^3 = 1000$. **115.** 2^x; Note that for $x = 10$, $2^x = 1024$ while $\log_{10} x = 1$. **117.** $2x(x + 3)(x - 6)$ **118.** $(x - 2)(x + 2)(x^2 + 4)$

119. $4(2x + 3)(5x - 1)$ **120.** $(3rs - 1)(2rs + 1)$

Exercise Set 9.4

1. Positive **3.** Sum **5.** Times **7.** $\log_2 3 + \log_2 5$ **9.** $\log_8 7 + \log_8 (x + 3)$ **11.** $\log_2 27 - \log_2 11$

13. $\dfrac{1}{2}\log_{10} x - \log_{10} (x - 9)$ **15.** $7 \log_6 x$ **17.** $5 \log_4 (r + 7)$ **19.** $\dfrac{3}{2}\log_4 a - \dfrac{1}{2}\log_4 (a + 2)$ **21.** $6 \log_3 d - 4 \log_3 (a - 8)$

23. $\log_8 (y + 4) - 2 \log_8 y$ **25.** $\log_{10} 9 + \log_{10} m - \log_{10} 8 - \log_{10} n$ **27.** $\log_2 21$ **29.** $\log_2 \dfrac{9}{5}$ **31.** $\log_4 64$ **33.** $\log_{10} x(x + 3)$

35. $\log_9 \dfrac{z^2}{z - 2}$ **37.** $\log_5 \left(\dfrac{p}{3}\right)^4$ **39.** $\log_2 \dfrac{n(n + 4)}{n - 3}$ **41.** $\log_5 \sqrt{\dfrac{x - 8}{x}}$ **43.** $\log_9 \dfrac{16\sqrt[3]{r} - 6}{\sqrt{r}}$ **45.** $\log_6 \dfrac{81}{(x + 3)^2 x^4}$ **47.** 1 **49.** -0.3980

51. 1.3980 **53.** 2 **55.** 7 **57.** 3 **59.** 25 **61.** $\log_a (x - 2)$ **63.** $\dfrac{1}{4}\log_2 x + \dfrac{1}{4}\log_2 y + \dfrac{1}{3}\log_2 a - \dfrac{1}{5}\log_2 (a - b)$ **65.** 0.8640

67. 0.1080 **69.** 0.7000 **71.** No, there is no relationship between $\log_{10} (x + y)$ and $\log_{10} xy$ or $\log_{10} \left(\dfrac{x}{y}\right)$ **73.** Yes, it is an expansion

of property 1. **75.** Yes, $\log_a (x^2 + 8x + 16) = \log_a (x + 4)^2 = 2 \log_a (x + 4)$ **77.** Yes **79.** $\log_a \dfrac{x}{y} = \log_a \left(x \cdot \dfrac{1}{y}\right) = \log_a x + \log_a \dfrac{1}{y}$

82. a) $\{x | x > 40\}$ **b)** $(40, \infty)$ **83. a)** $a^2 - 4c^2$ **b)** $(a + 2c)(a - 2c)$ **84.** 3 **85.** $-26 - 7i$ **86.** 49

Mid-Chapter Test   1. a) In $f(x)$, replace x by $g(x)$. **b)** $6x + 18$ [9.1]   **2. a)** $\left(\dfrac{6}{x}\right)^2 + 5$ or $\dfrac{36}{x^2} + 5$ **b)** 9 **c)** $\dfrac{6}{x^2 + 5}$ **d)** $\dfrac{3}{7}$ [9.1]

3. a) Answers will vary. **b)** No [9.1]   **4. a)** Yes **b)** $\{(2, -3), (3, 2), (1, 5), (8, 6)\}$ [9.1]   **5. a)** Yes **b)** $p^{-1}(x) = 3x + 15$ [9.1]

6. a) Yes **b)** $k^{-1}(x) = x^2 + 4\ x \geq 0$ [9.1]   **7.** $m^{-1}(x) = -\dfrac{1}{2}x + 2$ [9.1] **8.**   [9.2]

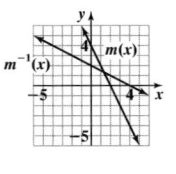

9.   [9.2]   **10.**   [9.3]

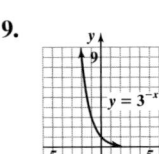   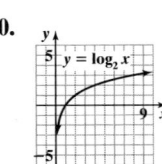

11. a) 10 **b)** 320 [9.3]   **12.** $\log_{27} 9 = \dfrac{2}{3}$ [9.3]   **13.** $2^{-6} = \dfrac{1}{64}$ [9.3]   **14.** 3 [9.3]   **15.** 2 [9.3]   **16.** 4 [9.3]   **17.** $2 \log_9 x + \log_9 (x - 5)$ [9.4]

18. $\log_5 7 + \log_5 m - \dfrac{1}{2}\log_5 n$ [9.4]   **19.** $\log_2 \dfrac{x^3(x + 7)}{(x + 1)^4}$ [9.4]   **20.** $\log_7 \sqrt{\dfrac{x + 2}{x}}$ [9.4]

Exercise Set 9.5   1. Common   **3.** 10^y **5.** 0   **7.** -1 **9.** -2 **11.** -3 **13.** 1 and 2   **15.** 4 and 5   **17.** -2 and -1
19. 2 and 3   **21.** 0 and 1   **23.** -2 and -1 **25.** 1.9345   **27.** 4.2833   **29.** -1.2125 **31.** 2.0004   **33.** 0.5740   **35.** -1.7620
37. 3.5514   **39.** -1.1385 **41.** 2.3856   **43.** -2.2277 **45.** -2.0088 **47.** 1.1901   **49.** 1.6357   **51.** 42,491.2982   **53.** 0.0196
55. 1.0023   **57.** 578.7620   **59.** 0.0097   **61.** 100   **63.** 2408.2413   **65.** 13,832.4784   **67.** 0.0871   **69.** 0.2389   **71.** 0.7489   **73.** 7
75. 7   **77.** 20.8   **79.** 41.5   **81.** 2511.8864   **83.** 501,187.2336   **85. a)** ≈ 31.62 kilometers **b)** ≈ 0.50 kilometer **c)** ≈ 14.68
87. a) $\approx 72\%$ **b)** $\approx 15\%$ **89.** ≈ 6310 times more intense   **91. a)** $\approx 6.31 \times 10^{20}$ **b)** ≈ 2.19 **93.** ≈ 6.2 **95.** No; $10^2 = 100$ and
since $462 > 100$, $\log 462$ must be greater than 2.   **97.** No; $10^0 = 1$ and $10^{-1} = 0.1$ and since $1 > 0.163 > 0.1$, $\log 0.163$ must be
between 0 and -1.   **99.** No; $\log \dfrac{y}{4x} = \log y - \log 4 - \log x$ **101.** $I = 10^R$ **103.** $t = 10^{\left(\frac{26 - R}{41.9}\right)} - 1$

106. 50 miles per hour, 55 miles per hour   **107.** $(2, -3)$ **108.** $0, -4, 3$ **109.** $|3x^2 - y|$ **110.** $(-\infty, -4] \cup [2, 5]$

Exercise Set 9.6   1. Base   **3.** Extraneous   **5.** Product   **7.** 3   **9.** 4   **11.** $\dfrac{1}{2}$ **13.** 2   **15.** $-\dfrac{1}{3}$ **17.** 4   **19.** 3   **21.** 3   **23.** 2.01

25. 3.56   **27.** 5.59   **29.** 6.34   **31.** 6   **33.** 5   **35.** $\dfrac{1}{16}$ **37.** 100   **39.** -1 **41.** $-6, 4$ **43.** $0, -8$ **45.** 92   **47.** $\dfrac{4}{5}$ **49.** $\dfrac{3}{2}$ **51.** 4   **53.** 2
55. 0.87   **57.** 30   **59.** 5   **61.** 4   **63.** 2   **65.** 3   **67.** 9   **69.** ≈ 3.47 hours   **71.** ≈ 3.19 years   **73.** ≈ 17.07 years   **75.** ≈ 11.81 years
77. $\approx \$7112.09$ **79.** ≈ 19.36 **81. a)** 1,000,000,000,000 times greater **b)** 10,000,000 times greater   **83.** 8   **85.** $x = 1$ and $x = 2$
87. $(3, 1)$ **89.** $(54, 46)$ **91.** 2.8   **93.** No solution   **95.** $\log (-2)$ is not a real number.   **97.** The box is greater by ≈ 7.73 cubic feet
98. -4 **99.**   **100.** $\dfrac{x\sqrt{y} - y\sqrt{x}}{x - y}$ **101.** $c = \sqrt{\dfrac{E}{m}}$ **102.** $f(x) = 2(x - 3)^2 - 5$

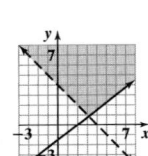

Exercise Set 9.7   1. Base   **3.** Exponential   **5.** Change   **7.** Natural   **9.** Growth   **11.** 3.2958   **13.** -0.8795 **15.** 3.3201
17. 0.5712   **19.** 4.95   **21.** 0.0578   **23.** 4.5850   **25.** -2.8577 **27.** 4.3923   **29.** 1.7297   **31.** 2.7380   **33.** 2.9135   **35.** -0.4784
37. 4   **39.** 1   **41.** 4   **43.** 6   **45.** $P = 4757.5673$ **47.** $P_0 = 224.0845$ **49.** $t = 0.7847$ **51.** $k = 0.2310$ **53.** $k = -0.2888$
55. $A = 4719.7672$ **57.** $V_0 = \dfrac{V}{e^{kt}}$ **59.** $t = \dfrac{\ln P - \ln 150}{7}$ **61.** $k = \dfrac{\ln A - \ln A_0}{t}$ **63.** $y = xe^{2.3}$ **65.** $y = (x + 6)e^5$
67. a) $\approx \$5637.48$ **b)** ≈ 11.55 years   **69. a)** $\approx \$2818.74$ **b)** ≈ 17.33 years   **71.** ≈ 39.98 grams   **73. a)** $\approx 86.47\%$ **b)** ≈ 34.66 days
75. a) ≈ 5.15 feet per second **b)** ≈ 5.96 feet per second **c)** $\approx 646,000$ **77.** $\approx \$526,911,558,800,000$ **79. a)** ≈ 7.26 billion people
b) ≈ 57.76 years   **81. a)** ≈ 1.39 billion people **b)** ≈ 115.52 years   **83. a)** ≈ 32.43 inches **b)** ≈ 35.46 inches   **85. a)** ≈ 999.7 grams
b) 231,049 years   **87. a)** ≈ 6626.62 years **b)** ≈ 5752.26 years   **89.** $\approx \$6791.91$ **91. a)** Strontium 90, since it has a higher decay rate
b) $\approx 31.66\%$ of original amount   **93.** $v_0 = \dfrac{e^{xk} - 1}{kt}$ **95.** $i = Ie^{-t/RC}$ **97. a)** 0 **b)** $\dfrac{11}{40}$ or 0.275   **98.** 240 children, 310 adults
99. $-9x^2y^3 + 12x^2y^2 - 3xy^2 + 4xy$ **100.** $-20, 20$ **101.** $x + x^2$

Chapter 9 Review Exercises   1. $4x^2 - 26x + 44$ **2.** 2   **3.** $2x^2 - 6x + 3$ **4.** 39   **5.** $6\sqrt{x - 3} + 7, x \geq 3$
6. $\sqrt{6x + 4}, x \geq -\dfrac{2}{3}$ **7.** One-to-one   **8.** Not one-to-one   **9.** One-to-one   **10.** Not one-to-one   **11.** One-to-one
12. Not one-to-one   **13.** $f(x)$: Domain: $\{-4, -1, 5, 6\}$, Range: $\{-3, 2, 3, 8\}$; $f^{-1}(x)$: Domain: $\{-3, 2, 3, 8\}$, Range: $\{-4, -1, 5, 6\}$

14. $f(x)$: Domain: $\{x|x \geq 0\}$, Range: $\{y|y \geq 4\}$; $f^{-1}(x)$: Domain: $\{x|x \geq 4\}$, Range: $\{y|y \geq 0\}$

15. $f^{-1}(x) = \dfrac{x+2}{4}$; 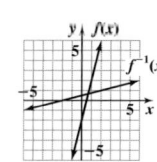   **16.** $f^{-1}(x) = x^3 + 1$; 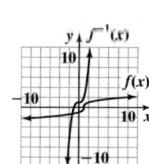   **17.** $f^{-1}(x) = \dfrac{x}{36}$, x is inches, $f^{-1}(x)$ is yards.

18. $f^{-1}(x) = \dfrac{x}{4}$, x is quarts, $f^{-1}(x)$ is gallons   **19.** 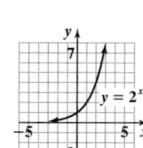   **20.** 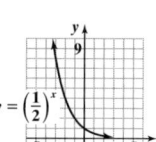   **21.** $\approx\$1830.29$ **22.** $\log_8 64 = 2$

23. $\log_{81} 3 = \dfrac{1}{4}$ **24.** $\log_5 \dfrac{1}{125} = -3$ **25.** $2^5 = 32$ **26.** $\left(\dfrac{1}{4}\right)^2 = \dfrac{1}{16}$ **27.** $6^{-2} = \dfrac{1}{36}$ **28.** $4^3 = x; 64$ **29.** $a^4 = 81; 3$

30. $\left(\dfrac{1}{5}\right)^{-3} = x; 125$ **31.** 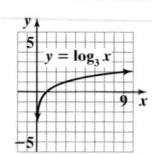   **32.** 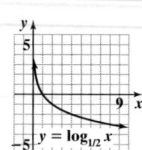   **33.** $8 \log_5 17$ **34.** $\dfrac{1}{2}\log_3(x-9)$ **35.** $\log 6 + \log(a+1) - \log 19$

36. $4\log x - \log 7 - 5\log(2x+3)$ **37.** $\log \dfrac{x^5}{(x+1)^3}$ **38.** $\log \dfrac{(2x)^4}{y}$ **39.** $\ln \dfrac{\sqrt[3]{\dfrac{x}{x+2}}}{2}$ **40.** $\ln \dfrac{x^3\sqrt{x+1}}{(x+4)^6}$ **41.** 10   **42.** 5   **43.** 121

44. 3   **45.** 2.9133   **46.** -3.5720 **47.** 1442.1154   **48.** 0.0007   **49.** 11,561.1224   **50.** 0.0594   **51.** 5   **52.** 9   **53.** 22.4   **54.** 9.4

55. 4   **56.** $-\dfrac{1}{2}$ **57.** 2   **58.** 5   **59.** 2.5818   **60.** 5.9681   **61.** 1.3529   **62.** 2.2402   **63.** 26   **64.** 5   **65.** 1   **66.** 3   **67.** $t \approx 1.1552$

68. $A_0 \approx 352.5421$ **69.** $t = \dfrac{\ln A - \ln A_0}{k}$ **70.** $k = \dfrac{\ln 0.25}{t}$ **71.** $y = xe^6$ **72.** $y = 3x + 23$ **73.** 7.6147   **74.** 3.5046

75. $\approx\$19,126.18$ **76.** ≈17.3 years   **77. a)** ≈92.88 minutes **b)** ≈118.14 minutes   **78.** ≈10.32 pounds per square inch   **79. a)** 72 **b)** ≈61.2 **c)** ≈5 months

Chapter 9 Practice Test

1. a) Yes **b)** $\{(2,4),(8,-3),(3,-1),(-7,6)\}$ [9.1]   **2. a)** $x^2 + 4x + 1$ **b)** 61 [9.1]

3. a) $\sqrt{x^2+3}$, **b)** $2\sqrt{13}$ [9.1]   **4. a)** $f^{-1}(x) = -\dfrac{1}{3}(x+5)$ [9.1]   **5. a)** $f^{-1}(x) = x^2 + 1$, $x \geq 0$

b) 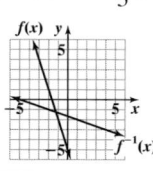   **b)** 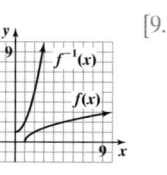 [9.1]

6. $\{x|x > 0\}$ [9.3]   **7.** -4 [9.4]   **8.** 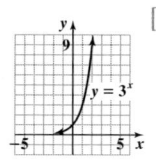 [9.2]   **9.** 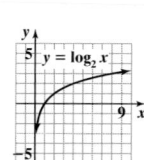 [9.3]   **10.** $\log_2 \dfrac{1}{32} = -5$ [9.4]   **11.** $5^3 = 125$ [9.3]

12. $2^4 = x + 3, 13$ [9.3]   **13.** $64^y = 16, \dfrac{2}{3}$ [9.3]   **14.** $3\log_2 x + \log_2(x-4) - \log_2(x+2)$ [9.4]   **15.** $\log_6 \dfrac{(x-4)^7(x+3)^2}{\sqrt{x}}$ [9.4]

16. 5 [9.4]   **17. a)** 3.6646 **b)** -2.6708 [9.5]   **18.** ≈2.68 [9.6]   **19.** 3 [9.6]   **20.** $\dfrac{17}{5}$ [9.6]   **21.** 16.2810 [9.7]   **22.** 2.0588 [9.7]

23. 30.5430 [9.7]   **24.** $\approx\$5211.02$ [9.7]   **25.** ≈3364.86 years old [9.7]

Cumulative Review Test

1. $\dfrac{12xy^5}{z^4}$ [1.5]   **2.** 40 [1.4]   **3.** 7.5% [2.3]   **4.** $\left\{x\,\middle|\,2 \leq x < \dfrac{15}{2}\right\}, \left[2, \dfrac{15}{2}\right)$ [2.5]   **5.** $y = \dfrac{2x-8}{3}$ [2.2]

6. 0 [3.2]   **7.** $m = 2, y = 2x + 3$ [3.4]   **8.** 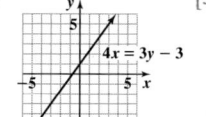 [3.4]   **9.** 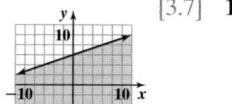 [3.7]   **10.** $(10, 24)$ [4.1]

11. $x^2 + 2x + 3 + \dfrac{6}{x+1}$ [5.3] **12.** $(x - y + 8)(x - y - 8)$ [5.6] **13.** $1, -2$ [8.1] **14.** -3 [6.4] **15.** $d = \dfrac{a_n - a_1}{n-1}$ [6.4]

16. 6.25 [6.6] **17.** $(12x + 1)\sqrt{5x}$ [7.4] **18.** 8 [7.6] **19.** $0, \pm\sqrt{7}$ [8.4] **20. a)** $g(x) = (x - 2)^2 - 9$ **b)** [8.5]

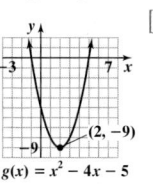

$g(x) = x^2 - 4x - 5$

Chapter 10

Exercise Set 10.1 **1.** Midpoint **3.** Parabola **5.** (h, k)

7.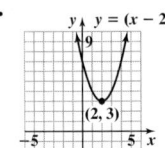
$y = (x - 2)^2 + 3$
$(2, 3)$

9.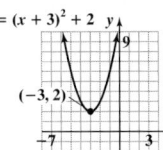
$y = (x + 3)^2 + 2$
$(-3, 2)$

11.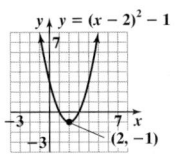
$y = (x - 2)^2 - 1$
$(2, -1)$

13.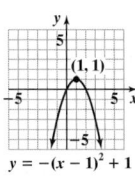
$(1, 1)$
$y = -(x - 1)^2 + 1$

15.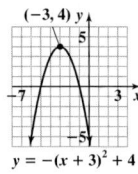
$(-3, 4)$
$y = -(x + 3)^2 + 4$

17.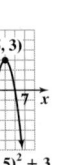
$(5, 3)$
$y = -3(x - 5)^2 + 3$

19.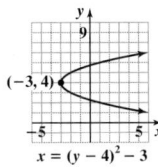
$(-3, 4)$
$x = (y - 4)^2 - 3$

21.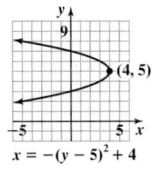
$(4, 5)$
$x = -(y - 5)^2 + 4$

23.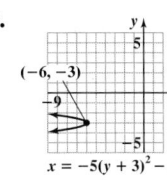
$(-6, -3)$
$x = -5(y + 3)^2 - 6$

25. $\left(-\frac{1}{2}, 6\right)$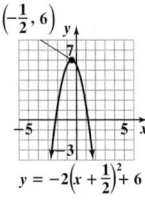
$y = -2\left(x + \frac{1}{2}\right) + 6$

27. a) $y = (x + 1)^2 - 1$
b)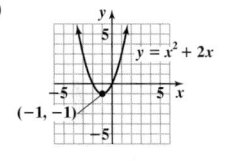
$y = x^2 + 2x$
$(-1, -1)$

29. a) $y = (x + 3)^2 - 9$
b)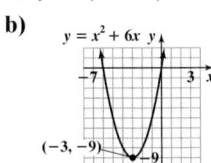
$y = x^2 + 6x$
$(-3, -9)$

31. a) $x = (y + 2)^2 - 4$
b)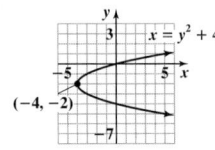
$x = y^2 + 4y$
$(-4, -2)$

33. a) $y = \left(x + \frac{7}{2}\right)^2 - \frac{9}{4}$
b)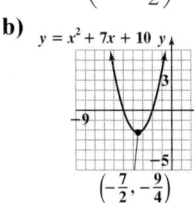
$y = x^2 + 7x + 10$
$\left(-\frac{7}{2}, -\frac{9}{4}\right)$

35. a) $x = -(y - 3)^2$
b)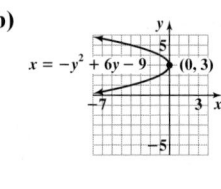
$x = -y^2 + 6y - 9$
$(0, 3)$

37. a) $y = -(x - 2)^2$
b)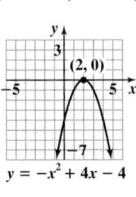
$(2, 0)$
$y = -x^2 + 4x - 4$

39. a) $x = -\left(y - \frac{3}{2}\right)^2 - \frac{7}{4}$
b)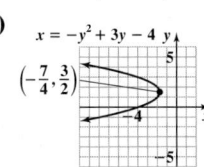
$x = -y^2 + 3y - 4$
$\left(-\frac{7}{4}, \frac{3}{2}\right)$

41. 5 **43.** 9 **45.** 13 **47.** $\sqrt{90} \approx 9.49$

49. $\sqrt{\dfrac{125}{4}} \approx 5.59$ **51.** $\sqrt{34.33} \approx 5.86$ **53.** $\sqrt{10} \approx 3.16$

55. $(3, 5)$ **57.** $(0, 0)$ **59.** $\left(\dfrac{3}{2}, 5\right)$ **61.** $\left(\dfrac{5}{2}, -\dfrac{7}{4}\right)$

63. $\left(\dfrac{\sqrt{3} + \sqrt{2}}{2}, \dfrac{7}{2}\right)$ **65.** $x^2 + y^2 = 25$ **67.** $(x - 2)^2 + y^2 = 64$

69. $x^2 + (y - 5)^2 = 4$ **71.** $(x - 3)^2 + (y - 4)^2 = 16$ **73.** $(x - 7)^2 + (y + 6)^2 = 144$ **75.** $(x - 1)^2 + (y - 2)^2 = 5$

77. $x^2 + y^2 = 16$ **79.** $(x - 3)^2 + (y + 2)^2 = 9$

81.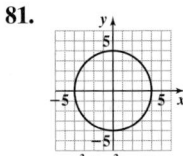
$x^2 + y^2 = 16$

83.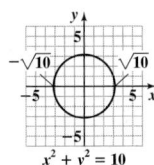
$-\sqrt{10} \quad \sqrt{10}$
$x^2 + y^2 = 10$

85.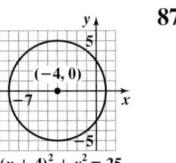
$(-4, 0)$
$(x + 4)^2 + y^2 = 25$

87.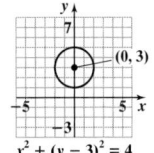
$(0, 3)$
$x^2 + (y - 3)^2 = 4$

89.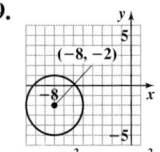
$(-8, -2)$
$(x + 8)^2 + (y + 2)^2 = 9$

91.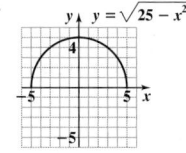
$y = \sqrt{25 - x^2}$

93.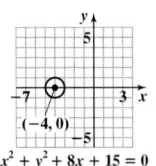
$y = -\sqrt{4 - x^2}$

95. a) $(x + 4)^2 + y^2 = 1^2$
b)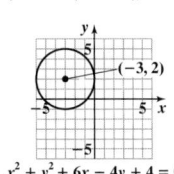
$(-4, 0)$
$x^2 + y^2 + 8x + 15 = 0$

97. a) $(x + 3)^2 + (y - 2)^2 = 3^2$
b)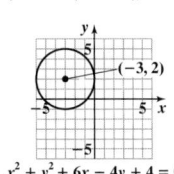
$(-3, 2)$
$x^2 + y^2 + 6x - 4y + 4 = 0$

99. a) $(x + 3)^2 + (y - 1)^2 = 2^2$
b)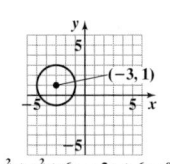
$(-3, 1)$
$x^2 + y^2 + 6x - 2y + 6 = 0$

101. a) $(x - 4)^2 + (y + 1)^2 = 2^2$ **b)**

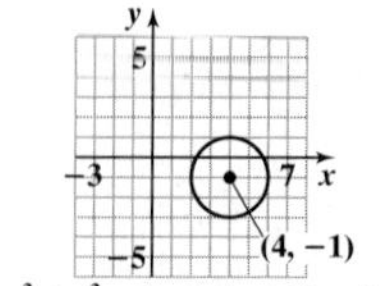
(4, −1)
$x^2 + y^2 - 8x + 2y + 13 = 0$

103. $25\pi \approx 78.5$ square units **105.** x-intercept: $(-7, 0)$, y-intercepts: $(0, -1)$, $(0, 7)$ **107.** x-intercept: $(24, 0)$, no y-intercepts **109.** No, different line segments can have the same midpoint **111.** 10 **113.** $(x + 6)^2 + (y - 2)^2 = 4$ **115. a)** $2\sqrt{2}$ **b)** $(7, 6)$ **c)** $(x - 7)^2 + (y - 6)^2 = 8$ **117.** 4, 0; a parabola opening up or down and a parabola opening right or left can be drawn to have a maximum of 4 intersections, or a minimum of 0 intersections.

119. a) 13.6 feet **b)** 81.8 feet **c)** $x^2 + (y - 81.8)^2 = 4651.24$ **121. a)** $x^2 + y^2 = 16$ **b)** $(x - 2)^2 + y^2 = 4$ **c)** $(x + 2)^2 + y^2 = 4$ **d)** 8π square units **123.** 48π square units **125.** Yes; D: $\mathbb{R}$, R: $\{y | y \geq k\}$

127. Same vertex; the first graph opens up and the second opens down. **129.** No **131.** No **133.** No **136.** $\dfrac{y}{3x}$ **137.** $(0, 7)$

138. 128 **139. a)** 1. a^2, 2. ab, 3. ab, 4. b^2 **b)** $(a + b)^2$ **140.**

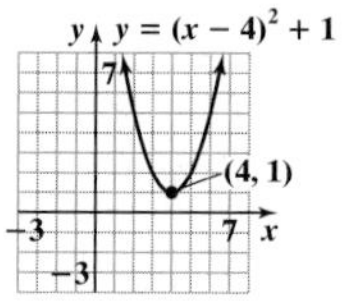
$y = (x - 4)^2 + 1$
(4, 1)

Exercise Set 10.2

1. $(0, 0)$ **3.** Ellipse **5.** x-intercepts **7.** Major **9.** Divide

11.

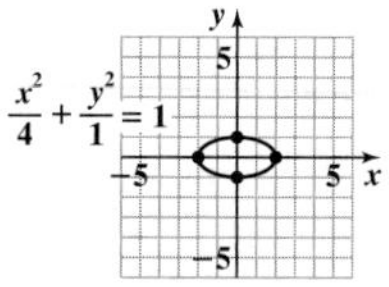
$\frac{x^2}{4} + \frac{y^2}{1} = 1$

13.

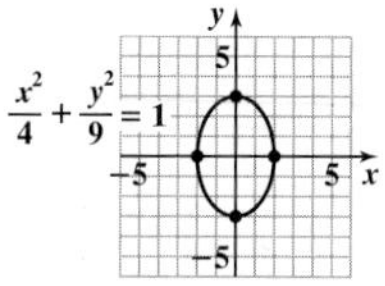
$\frac{x^2}{4} + \frac{y^2}{9} = 1$

15.

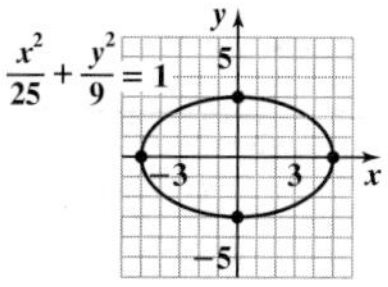
$\frac{x^2}{25} + \frac{y^2}{9} = 1$

17.

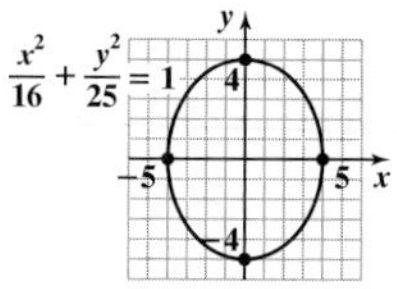
$\frac{x^2}{16} + \frac{y^2}{25} = 1$

19.

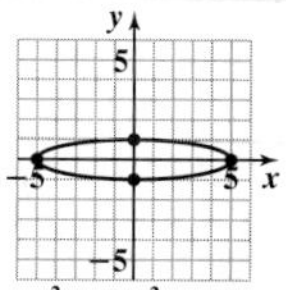
$x^2 + 25y^2 = 25$

21.

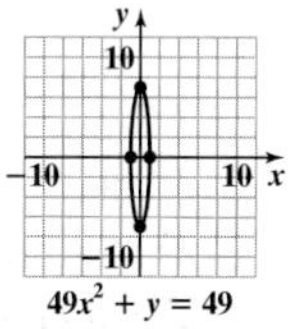
$49x^2 + y = 49$

23.

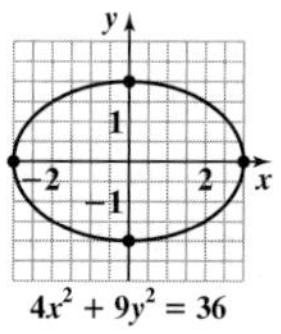
$4x^2 + 9y^2 = 36$

25.

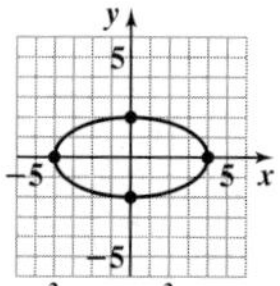
$25x^2 + 100y^2 = 400$

27.

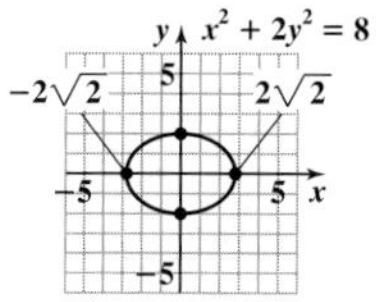
$x^2 + 2y^2 = 8$
$-2\sqrt{2}$
$2\sqrt{2}$

29.

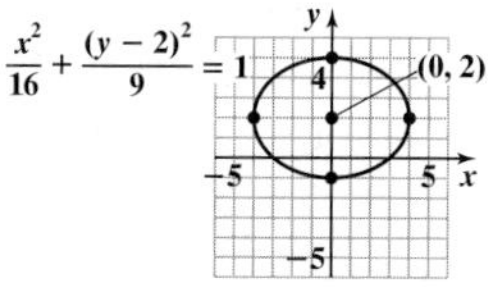
$\frac{x^2}{16} + \frac{(y - 2)^2}{9} = 1$
(0, 2)

31.

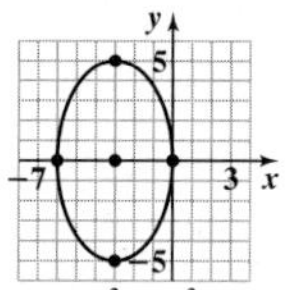
$\frac{(x + 3)^2}{9} + \frac{y^2}{25} = 1$

33.

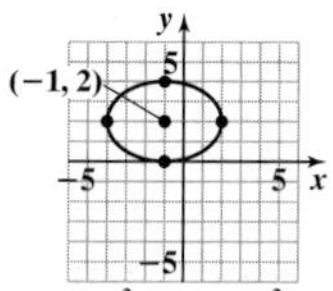
(−1, 2)
$\frac{(x + 1)^2}{9} + \frac{(y - 2)^2}{4} = 1$

35.

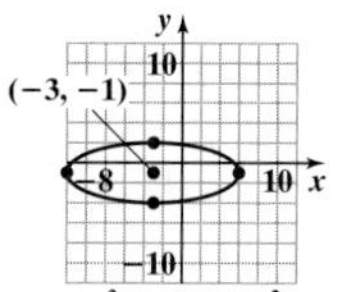
(−3, −1)
$(x + 3)^2 + 9(y + 1)^2 = 81$

37.

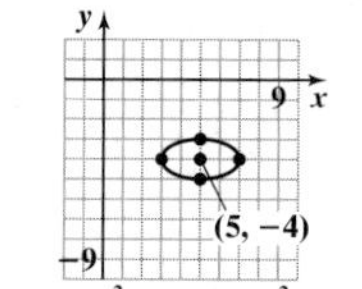
(5, −4)
$(x - 5)^2 + 4(y + 4)^2 = 4$

39.

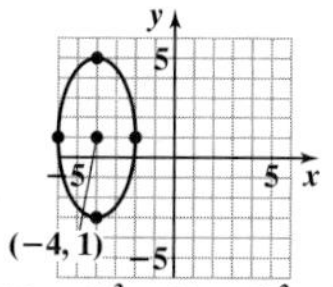
(−4, 1)
$12(x + 4)^2 + 3(y - 1)^2 = 48$

41. $2\pi \approx 6.3$ square units **43.** $20\pi \approx 62.8$ square units **45.** $\dfrac{x^2}{25} + \dfrac{y^2}{16} = 1$ **47.** $\dfrac{x^2}{4} + \dfrac{y^2}{9} = 1$

49. $\dfrac{(x + 3)^2}{36} + \dfrac{(y + 2)^2}{9} = 1$; $(-3, -2)$ **51.** 69.5 feet **53. a)** $\dfrac{x^2}{100} + \dfrac{y^2}{576} = 1$ **b)** $240\pi \approx 753.98$ square feet **c)** ≈ 376.99 square feet **55.** $\sqrt{5} \approx 2.24$ feet, in both directions, from the center of the ellipse, along the major axis. **57.** Answers will vary. **59.** Answers will vary. **61.** If $a = b$, the formula for a circle is obtained. **63.** No

65. It becomes more circular. When $a = b$, the graph is a circle. **67.** 2 **69.** $\dfrac{(x - 4)^2}{49} + \dfrac{(y - 2)^2}{9} = 1$ **71.** $l = \dfrac{2S - nf}{n}$ **72.** $x + \dfrac{5}{2} + \dfrac{1}{2(2x - 3)}$ **73.** 6 **74.** $\left[-\dfrac{5}{3}, 4\right)$ **75.** ≈ 2.7755

Mid-Chapter Test

1.

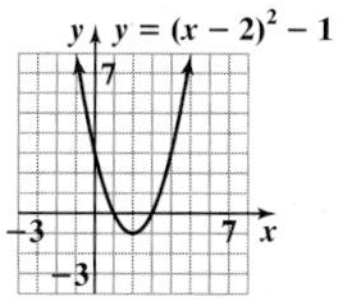
$y = (x - 2)^2 - 1$

[10.1] **2.**

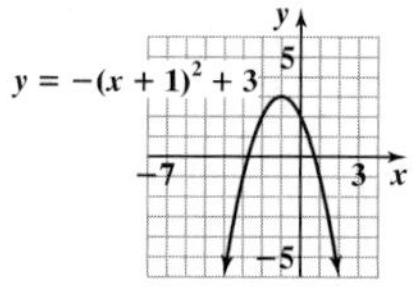
$y = -(x + 1)^2 + 3$

[10.1] **3.**

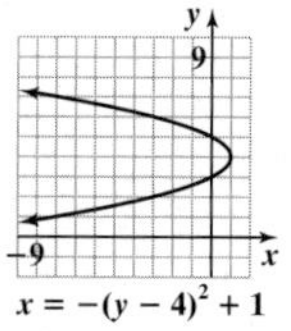
$x = -(y - 4)^2 + 1$

[10.1] **4.**

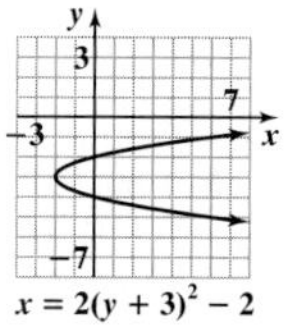
$x = 2(y + 3)^2 - 2$

[10.1]

5.

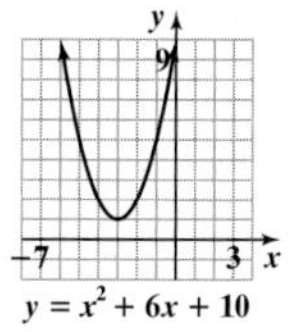
$y = x^2 + 6x + 10$

[10.1] **6.** 13 [10.1] **7.** $\sqrt{153} \approx 12.37$ [10.1] **8.** $\left(-1, \dfrac{5}{2}\right)$ [10.1] **9.** $\left(\dfrac{11}{4}, \dfrac{15}{4}\right)$ [10.1]

10. $(x + 3)^2 + (y - 2)^2 = 25$ [10.1]   **11.** [10.1]   **12.** [10.1]

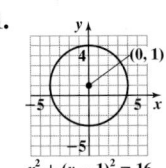

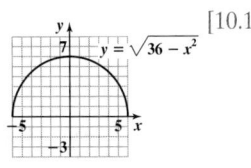

13. [10.1]   **14.** A circle is a set of points in a plane that are the same distance from a fixed point called its center. [10.1]   **15.** [10.2]   **16.** [10.2]

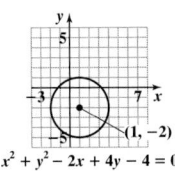

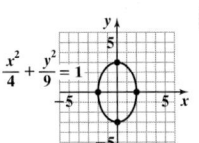

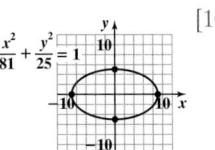

17. [10.2]   **18.** [10.2]   **19.** $6\pi \approx 18.85$ square units [10.2]   **20.** $\dfrac{x^2}{64} + \dfrac{y^2}{25} = 1$ [10.2]

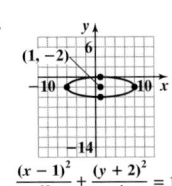

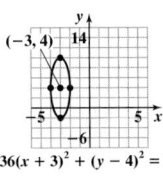

Exercise Set 10.3

1. Hyperbola   **3.** Center   **5.** Transverse   **7.** y **9.** Divide

11. a) $y = \pm \dfrac{2}{3}x$ **13.** a) $y = \pm \dfrac{1}{2}x$ **15.** a) $y = \pm 2x$ **17.** a) $y = \pm \dfrac{4}{5}x$ **19.** a) $y = \pm \dfrac{5}{6}x$ **21.** a) $y = \pm \dfrac{1}{3}x$

b) 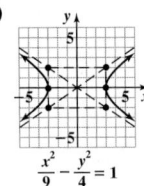   b) 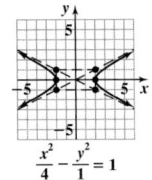   b) 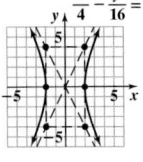   b) 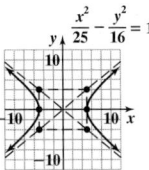   b) 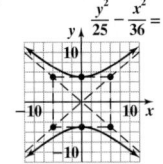   b)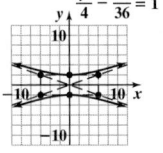

23. a) $y = \pm \dfrac{5}{2}x$ **25.** a) $y = \pm \dfrac{4}{9}x$ **27.** a) $\dfrac{y^2}{1} - \dfrac{x^2}{25} = 1$, $y = \pm \dfrac{1}{5}x$ **29.** a) $\dfrac{y^2}{16} - \dfrac{x^2}{4} = 1$, $y = \pm 2x$

b) 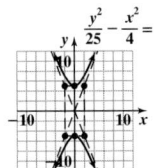   b) 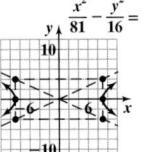   b) 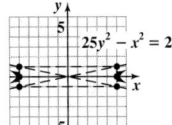   b)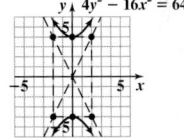

31. a) $\dfrac{y^2}{1} - \dfrac{x^2}{9} = 1$, $y = \pm \dfrac{1}{3}x$ **33.** a) $\dfrac{x^2}{9} - \dfrac{y^2}{25} = 1$, $y = \pm \dfrac{5}{3}x$ **35.** a) $\dfrac{y^2}{36} - \dfrac{x^2}{4} = 1$, $y = \pm 3x$

b) 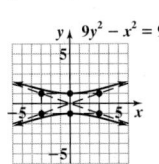   b) 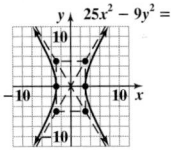   b)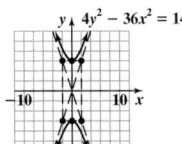

37. Circle   **39.** Ellipse   **41.** Hyperbola   **43.** Parabola   **45.** Ellipse   **47.** Parabola   **49.** Circle   **51.** Hyperbola   **53.** Parabola

55. Hyperbola   **57.** Parabola   **59.** Circle   **61.** $\dfrac{y^2}{4} - \dfrac{x^2}{16} = 1$ **63.** $\dfrac{x^2}{9} - \dfrac{y^2}{36} = 1$ **65.** $\dfrac{x^2}{9} - \dfrac{y^2}{25} = 1$, no, $\dfrac{x^2}{18} - \dfrac{y^2}{50} = 1$ and others will

work. The ratio of $\dfrac{b}{a}$ must be $\dfrac{5}{3}$.   **67.** D: $(-\infty, -5] \cup [5, \infty)$; R: $\mathbb{R}$ **69.** Answers will vary.   **71.** No, graphs of hyperbolas of this form do not pass the vertical line test.   **73.** The transverse axes of both graphs are along the x-axis. The vertices of the second graph will be closer to the origin, and the second graph will open wider.   **75.** It is a hyperbola with vertices at $(a, 0)$ and $(-a, 0)$. The transverse axis is along the x-axis. The asymptotes are $y = \pm \dfrac{b}{a}x$.   **77.** No   **79.** Yes   **81.** $y = -\dfrac{1}{2}x + 1$ **82.** $-x^2 - x + 11$ **83.** $\left(-1, \dfrac{1}{3}\right)$

84. $\dfrac{3x + 2}{2x - 3}$ **85.** $v = \sqrt{\dfrac{2E}{m}}$ **86.** 1

Exercise Set 10.4   **1.** Nonlinear   **3.** Addition   **5.** 3   **7.** $(3, -3), (-3, 3)$ **9.** $(-3, 0), \left(-\dfrac{12}{5}, \dfrac{9}{5}\right)$ **11.** $(-4, 11), \left(\dfrac{5}{2}, \dfrac{5}{4}\right)$

13. $(-1, 5), (1, 5)$ **15.** $(5, 0), (\ 5, 0)$ **17.** No real solution   **19.** $(0, -3), (\sqrt{5}, 2)(-\sqrt{5}, 2)$ **21.** $(2, -4), (-14, -20)$

23. $(2, 0), (-2, 0)$ **25.** $(4, 3), (4, -3), (-4, 3), (-4, -3)$ **27.** $(\sqrt{2}, \sqrt{2}), (\sqrt{2}, -\sqrt{2}), (-\sqrt{2}, \sqrt{2}), (-\sqrt{2}, -\sqrt{2})$
29. $(3, 0), (-3, 0)$ **31.** $(2, 1), (2, -1), (-2, 1), (-2, -1)$ **33.** $(3, 4), (3, -4), (-3, 4), (-3, -4)$

35. $(\sqrt{5}, 2), (\sqrt{5}, -2), (-\sqrt{5}, 2), (-\sqrt{5}, -2)$ **37.** No real solution   **39.** No real solution   **41.** 20 meters by 25 meters

43. 10 feet by 17 feet   **45.** length: 14 centimeters, width: 8 centimeters   **47.** 16 inches by 30 inches   **49.** ≈ 1.67 seconds

51. $r = 6\%, p = \$125$ **53.** ≈ 16 and ≈ 184 **55.** ≈ 5 and ≈ 95 **57.** $(-1, -3), (3.12, -0.53)$ **59.** Answers will vary

61. Yes, for example ⚲   **63.** Yes, for example ⚲   **65.** 10 yards, 24 yards

67. Parentheses, exponents, multiplication or division, addition or subtraction   **68.** $(x + 2)(x^2 + x + 1)$ **69.** 0.9

70. $\dfrac{5\sqrt{x + 2} + 15}{x - 7}$ **71.** $k = \dfrac{\ln A - \ln A_0}{t}$

Chapter 10 Review Exercises   **1.** $13; \left(\dfrac{5}{2}, -6\right)$ **2.** $5; \left(-\dfrac{5}{2}, 3\right)$ **3.** $17; \left(-5, \dfrac{5}{2}\right)$ **4.** $\sqrt{8} \approx 2.83; (-3, 4)$

5. 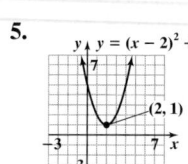   **6.** 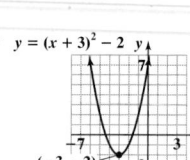   **7.** 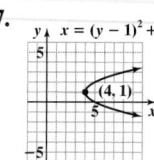   **8.** 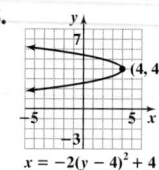   **9. a)** $y = (x - 4)^2 + 6$

b)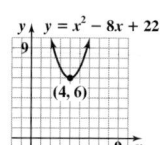
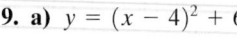

10. a) $x = -\left(y + \dfrac{5}{2}\right)^2 + \dfrac{9}{4}$ **11. a)** $x = \left(y + \dfrac{5}{2}\right)^2 - \dfrac{9}{4}$ **12. a)** $y = 2(x - 2)^2 - 32$ **13. a)** $x^2 + y^2 = 4^2$

b) 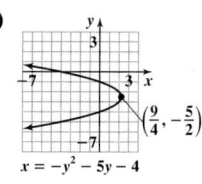   **b)** 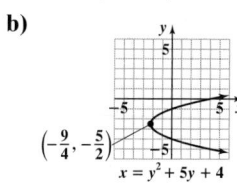   **b)** 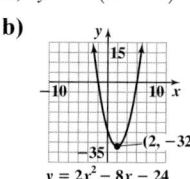   **b)**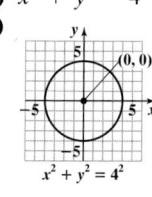

14. a) $(x + 3)^2 + (y - 4)^2 = 1^2$ **15. a)** $x^2 + (y - 2)^2 = 2^2$ **16. a)** $(x - 1)^2 + (y + 3)^2 = 3^2$

b) 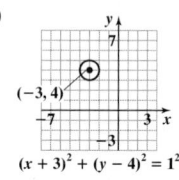   **b)** 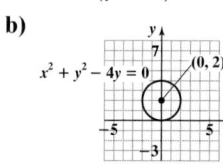   **b)**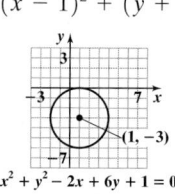

17. a) $(x - 4)^2 + (y - 5)^2 = 1^2$ **18. a)** $(x - 2)^2 + (y + 5)^2 = (\sqrt{12})^2$ **19.**

b) 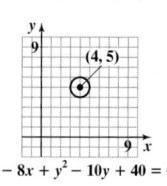   **b)**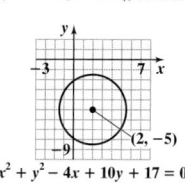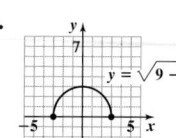

20. 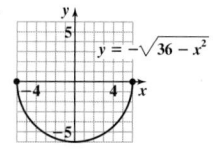   **21.** $(x + 1)^2 + (y - 1)^2 = 4$ **22.** $(x - 5)^2 + (y + 3)^2 = 9$ **23.**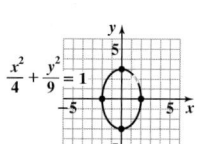

24. 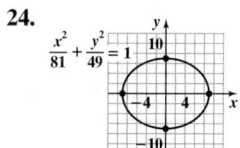   **25.** 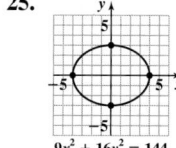   **26.** 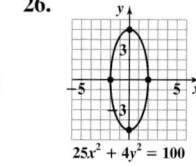   **27.** 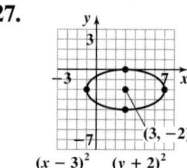   **28.**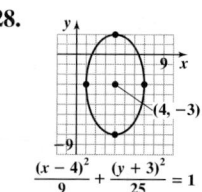

29.

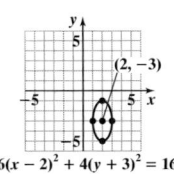

$16(x - 2)^2 + 4(y + 3)^2 = 16$

30. $6\pi \approx 18.85$ square units

31. a) $y = \pm\dfrac{5}{3}x$

b)

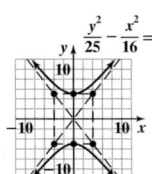

$\dfrac{x^2}{9} - \dfrac{y^2}{25} = 1$

32. a) $y = \pm\dfrac{5}{3}x$

b)

$\dfrac{y^2}{25} - \dfrac{x^2}{9} = 1$

33. a) $y = \pm\dfrac{3}{4}x$

b)

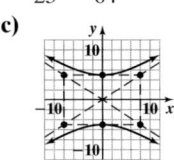

$\dfrac{y^2}{9} - \dfrac{x^2}{16} = 1$

34. a) $y = \pm\dfrac{5}{4}x$

b)

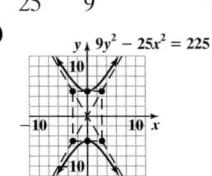

$\dfrac{y^2}{25} - \dfrac{x^2}{16} = 1$

35. a) $\dfrac{x^2}{9} - \dfrac{y^2}{1} = 1$ **b)** $y = \pm\dfrac{1}{3}x$

c)

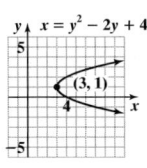

$x^2 - 9y^2 = 9$

36. a) $\dfrac{y^2}{25} - \dfrac{x^2}{64} = 1$ **b)** $y = \pm\dfrac{5}{8}x$

c)

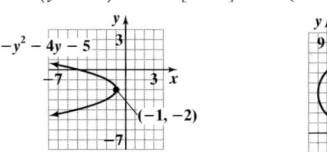

$64y^2 - 25x^2 = 1600$

37. a) $\dfrac{y^2}{25} - \dfrac{x^2}{9} = 1$ **b)** $y = \pm\dfrac{5}{3}x$

c)

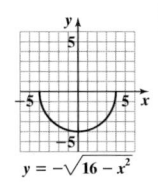

$9y^2 - 25x^2 = 225$

38. a) $\dfrac{y^2}{9} - \dfrac{x^2}{49} = 1$ **b)** $y = \pm\dfrac{3}{7}x$

c)

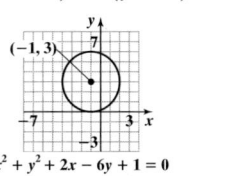

$49y^2 - 9x^2 = 441$

39. Hyperbola
40. Ellipse
41. Circle
42. Hyperbola
43. Ellipse

44. Parabola **45.** Ellipse **46.** Parabola
47. $(2, 2\sqrt{2}), (2, -2\sqrt{2}), (-2, 2\sqrt{2}),$
$(-2, -2\sqrt{2})$
48. $\left(\dfrac{5}{2}, \dfrac{3}{2}\right)$ **49.** $(3, 0), \left(-\dfrac{9}{5}, \dfrac{12}{5}\right)$
50. No real solution **51.** $(6, 0), (-6, 0)$

52. $(\sqrt{15}, 1), (-\sqrt{15}, 1), (\sqrt{15}, -1), (-\sqrt{15}, -1)$ **53.** No real solution **54.** $(1, 2\sqrt{2}), (-1, 2\sqrt{2}), (1, -2\sqrt{2}), (-1, -2\sqrt{2})$
55. 5 feet by 9 feet **56.** ≈ 4 and ≈ 145 **57.** $r = 3\%, p = \$4000$

Chapter 10 Practice Test
1. They are formed by cutting a cone or a pair of cones. [10.1] **2.** $\sqrt{50} \approx 7.07$ [10.1]
3. $\left(-1, \dfrac{3}{2}\right)$ [10.1] **4.** $(-3, 1)$, [10.1] **5.**

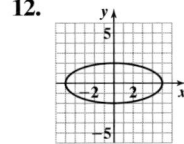

$x = y^2 - 2y + 4$ [10.1] **6.** $x = -(y + 2)^2 - 1$ [10.1] **7.** $(x - 2)^2 + (y - 4)^2 = 9$ [10.1]

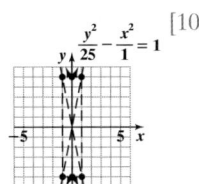

$y = -2(x + 3)^2 + 1$

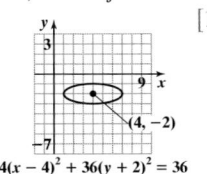

$x = -y^2 - 4y - 5$

8. $9\pi \approx 28.27$ square units [10.1] **9.** $(x - 3)^2 + (y + 1)^2 = 16$ [10.1] **10.**

$y = -\sqrt{16 - x^2}$

[10.1] **11.** $(x + 1)^2 + (y - 3)^2 = 9$ [10.1]

$(-1, 3)$

$x^2 + y^2 + 2x - 6y + 1 = 0$

12.

$4x^2 + 25y^2 = 100$

[10.2] **13.** No, the major axis should be along the y-axis. [10.2]
14.

$(4, -2)$

$4(x - 4)^2 + 36(y + 2)^2 = 36$

[10.2]

15. $(8, -7)$ [10.2]
16. The transverse axis lies along the axis corresponding to the positive term of the equation in standard form. [10.3]
17. $y = \pm\dfrac{7}{4}x$ [10.3]

18.

$\dfrac{y^2}{25} - \dfrac{x^2}{1} = 1$

[10.3] **19.**

$\dfrac{x^2}{4} - \dfrac{y^2}{9} = 1$

[10.3] **20.** Hyperbola, divide both sides of the equation by 30. [10.3]
21. Ellipse, divide both sides of the equation by 100. [10.3]
22. $(2, \sqrt{3}), (2, -\sqrt{3}), (-2, \sqrt{3}), (-2, -\sqrt{3})$ [10.4]
23. No real solution [10.4]
24. 30 meters by 50 meters [10.4]
25. 5 feet by 12 feet [10.4]

Cumulative Review Test   **1.** $-27x^3y^9$ [1.5]   **2.** $\dfrac{19}{4}$ [2.1]   **3.** $\varnothing$ [2.1]   **4.** $\left\{x \,\middle|\, x < -\dfrac{5}{3} \text{ or } x > 1\right\}$ [2.6]

5.
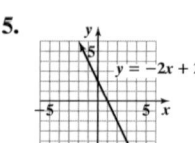
[3.1]   **6.** 139 [3.2]   **7.** $(8, 6)$ [4.1]   **8.** $(x^2 + 6)(x^2 - 7)$ [5.5]   **9.** base: 14 feet, height: 8 feet [5.8]

10. $\dfrac{x - 4}{4x + 3}$ [6.1]   **11.** $\dfrac{2x^2 - 9x - 5}{2(x + 3)(x - 4)}$ [6.2]   **12.** 2 [6.4]   **13.** $\dfrac{3y^{3/2}}{x^{1/2}}$ [7.2]   **14.** $\dfrac{6x + 6y\sqrt{x}}{x - y^2}$ [7.5]

15. 3 [7.6]   **16.** $\dfrac{2 \pm i\sqrt{11}}{3}$ [8.2]   **17.** 2 [9.6]   **18.** ≈ 2.31 [9.7]   **19.**
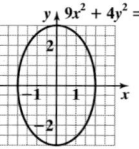
$9x^2 + 4y^2 = 36$ [10.2]

20.
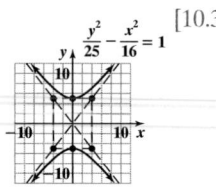
[10.3]

Chapter 11

Exercise Set 11.1   **1.** Sequence   **3.** Finite   **5.** Less   **7.** Series   **9.** Upper   **11.** 6, 12, 18, 24, 30   **13.** 3, 7, 11, 15, 19

15. $9, \dfrac{9}{2}, \dfrac{9}{3}, \dfrac{9}{4}, \dfrac{9}{5}$ **17.** $\dfrac{3}{2}, \dfrac{4}{3}, \dfrac{5}{4}, \dfrac{6}{5}, \dfrac{7}{6}$ **19.** $-1, 1, -1, 1, -1$ **21.** $4, -8, 16, -32, 64$ **23.** 31   **25.** 12   **27.** 1   **29.** 143   **31.** $\dfrac{81}{25}$

33. 2, 15   **35.** 3, 17   **37.** $0, \dfrac{13}{20}$ **39.** $-1, -1$ **41.** $\dfrac{1}{2}, 7$ **43.** 64, 128, 256   **45.** 17, 19, 21   **47.** $\dfrac{1}{6}, \dfrac{1}{7}, \dfrac{1}{8}$ **49.** $1, -1, 1$

51. $\dfrac{1}{81}, \dfrac{1}{243}, \dfrac{1}{729}$ **53.** $\dfrac{1}{16}, -\dfrac{1}{32}, \dfrac{1}{64}$ **55.** 26, 22, 18   **57.** $2 + 5 + 8 + 11 + 14 = 40$ **59.** $2 + 5 + 10 + 17 + 26 + 37 = 97$

61. $\dfrac{1}{3} + \dfrac{4}{3} + \dfrac{9}{3} + \dfrac{16}{3} = \dfrac{30}{3} = 10$ **63.** $4 + 5 + 6 + 7 + 8 + 9 = 39$ **65.** $\displaystyle\sum_{i=1}^{5}(i + 10)$ **67.** $\displaystyle\sum_{i=1}^{3}\dfrac{i^2}{4}$ **69.** 13   **71.** 169   **73.** 55

75. 25   **77.** ≈ 42.83 **79. a)** 6, 12, 18, 24   **b)** $p_n = 6n$ **81. a)** $x_1 + x_2 + x_3 + \cdots + x_n$ **b)** $x_1 + x_2 + x_3 + \cdots + x_n$ **c)** Yes

83. $n = \dfrac{\sum x}{\bar{x}}$ **85.** Yes   **87.** The nth partial sum of a series is the sum of the first n consecutive terms of a series.

89. Increasing sequence   **91.** Yes   **93.** Answers will vary.   **95.** Answers will vary.   **96.** $\dfrac{2}{5}$ **97.** $8(y - 2x^2)(y^2 + 2x^2y + 4x^4)$

98. 11   **99.** $r = \sqrt{\dfrac{V}{\pi h}}$

Exercise Set 11.2   **1.** Arithmetic   **3.** d **5.** Term   **7.** Sum   **9.** Negative   **11.** $4, 7, 10, 13, 16; a_n = 3n + 1$

13. $7, 5, 3, 1, -1; a_n = -2n + 9$ **15.** $\dfrac{1}{2}, 2, \dfrac{7}{2}, 5, \dfrac{13}{2}; a_n = \dfrac{3}{2}n - 1$ **17.** $100, 95, 90, 85, 80; a_n = -5n + 105$ **19.** 14   **21.** 27   **23.** 12

25. 2   **27.** 9   **29.** 6   **31.** $s_{10} = 100, d = 2$ **33.** $s_8 = \dfrac{52}{5}, d = \dfrac{1}{5}$ **35.** $s_6 = 25.5, d = 3.7$ **37.** $s_{11} = 407, d = 6$

39. $4, 7, 10, 13; a_{10} = 31, s_{10} = 175$ **41.** $-6, -4, -2; 0; a_{10} = 12, s_{10} = 30$ **43.** $-8, -13, -18, -23; a_{10} = -53, s_{10} = -305$

45. $\dfrac{7}{2}, 6, \dfrac{17}{2}, 11; a_{10} = 26, s_{10} = 147.5$ **47.** $100, 93, 86, 79; a_{10} = 37, s_{10} = 685$ **49.** $n = 15, s_{15} = 330$ **51.** $n = 11, s_{11} = 121$

53. $n = 17, s_{17} = \dfrac{153}{2}$ **55.** $n = 29, s_{29} = 1421$ **57.** 1275   **59.** 2500   **61.** 1395   **63.** 267   **65.** 42; 372   **67.** 351   **69. a)** 27   **b)** 196

71. $101 \cdot 50 = 5050$ **73.** $s_n = n^2$ **75. a)** 19 feet   **b)** 143.5 feet   **77.** 2 feet   **79. a)** 155   **b)** 780   **81.** \$496   **83. a)** \$45,600

b) \$438,000   **85.** $a_n = 180°(n - 2)$ **87.** Yes   **89.** Yes   **97.** $r = \dfrac{A - P}{Pt}$ **98.** $(-3, -5)$ **99.** $3(2n - 5)(2n - 1)$

100.
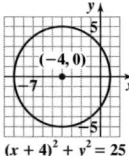
$(x + 4)^2 + y^2 = 25$

Exercise Set 11.3 **1.** Geometric **3.** r **5.** Term **7.** Sum **9.** 1 **11.** $2, 6, 18, 54, 162$ **13.** $6, -3, \dfrac{3}{2}, -\dfrac{3}{4}, \dfrac{3}{8}$ **15.** $72, 24, 8, \dfrac{8}{3}, \dfrac{8}{9}$

17. $90, -30, 10, -\dfrac{10}{3}, \dfrac{10}{9}$ **19.** $-1, -3, -9, -27, -81$ **21.** $7, -14, 28, -56, 112$ **23.** $\dfrac{1}{3}, \dfrac{1}{6}, \dfrac{1}{12}, \dfrac{1}{24}, \dfrac{1}{48}$ **25.** $3, \dfrac{9}{2}, \dfrac{27}{4}, \dfrac{81}{8}, \dfrac{243}{16}$

27. 128 **29.** $-\dfrac{3}{32}$ **31.** 64 **33.** 6144 **35.** $\dfrac{1}{64}$ **37.** $\dfrac{50}{729}$ **39.** 155 **41.** 7812 **43.** 10,160 **45.** $-\dfrac{2565}{256}$ **47.** $-\dfrac{9279}{625}$

49. $r = \dfrac{1}{2}, a_n = 7\left(\dfrac{1}{2}\right)^{n-1}$ **51.** $r = 2, a_n = 9(2)^{n-1}$ **53.** $r = -3, a_n = 2(-3)^{n-1}$ **55.** $r = \dfrac{2}{3}, a_n = \dfrac{3}{4}\left(\dfrac{2}{3}\right)^{n-1}$ **57.** 2 **59.** $\dfrac{5}{4}$ **61.** 24

63. $\dfrac{25}{3}$ **65.** 5 **67.** 6 **69.** 4 **71.** 12 **73.** -45 **75.** -15 **77.** $\dfrac{8}{33}$ **79.** $\dfrac{7}{9}$ **81.** $\dfrac{17}{33}$ **83.** $r = 3, a_1 = 5$ **85.** $r = 2$ or $r = -2, a_1 = 7$

87. $\approx\$1.77$ **89. a)** 4 days **b)** ≈ 1.172 grams **91. a)** ≈ 338.59 million **b)** ≈ 63.4 years **93. a)** $\dfrac{1}{2}, \dfrac{1}{4}, \dfrac{1}{8}, \dfrac{1}{16}, \dfrac{1}{32}$

b) $a_n = \dfrac{1}{2}\left(\dfrac{1}{2}\right)^{n-1} = \left(\dfrac{1}{2}\right)^n$ **c)** $\dfrac{1}{128} \approx 0.78\%$ **95.** $\approx\$15,938.48$ **97. a)** 28.512 feet **b)** 550 feet **99. a)** 10.29 inches **b)** 100 inches

101. 211 **103. a)** $\$12,000, \$9600, \$7680$ **b)** $a_n = 12,000\left(\dfrac{4}{5}\right)^{n-1}$ **c)** $\approx \$4915.20$ **105.** 190 feet **107. a)** y_2 goes up more steeply.

b)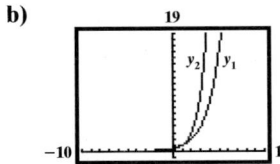

109. 0 **111.** Yes **113.** Yes, $s_\infty = 8$ **115.** $n = 21, s_n = 2,097,151$ **116.** 12

117. $6x^3 - x^2y - 16xy^2 + 6y^3$

118. $r = \dfrac{S - 2a}{S}$ **119.** $g^{-1}(x) = \sqrt[3]{x-9}$

120. 5 **121.** 9 meters, 12 meters

Mid-Chapter Test **1.** $4, 1, -2, -5, -8$ [11.1] **2.** 91 [11.1] **3.** $1, 11$ [11.1] **4.** $-15, -19, -23$ [11.1] **5.** 45 [11.1]

6. $\displaystyle\sum_{i=1}^{5}\left(\dfrac{1}{4}i + 9\right)$ [11.1] **7.** $-6, -1, 4, 9; a_n = -11 + 5n$ [11.2] **8.** -1 [11.2] **9.** 6 [11.2] **10.** $3, -3$ [11.2] **11.** $47\dfrac{1}{2}$ [11.2]

12. 12 [11.2] **13.** 136 [11.2] **14.** $100, -50, 25, -\dfrac{25}{2}, \dfrac{25}{4}$ [11.3] **15.** $\dfrac{1}{27}$ [11.3] **16.** 315 [11.3] **17.** $-\dfrac{2}{3}$ [11.3] **18.** 18 [11.3]

19. $\dfrac{29}{33}$ [11.3] **20. a)** A sequence is a list of numbers arranged in a specific order. **b)** An arithmetic sequence is a sequence where each term differs by a constant amount. **c)** A geometric sequence is a sequence where the terms differ by a common multiple. **d)** A series is the sum of the terms of a sequence. [11.1–11.3]

Exercise Set 11.4 **1.** Factorial **3.** 1 **5.** 20 **7.** 10 **9.** 10 **11.** 1 **13.** 1 **15.** 70 **17.** 28 **19.** $x^3 + 12x^2 + 48x + 64$
21. $8x^3 - 36x^2 + 54x - 27$ **23.** $a^4 - 4a^3b + 6a^2b^2 - 4ab^3 + b^4$ **25.** $243a^5 - 405a^4b + 270a^3b^2 - 90a^2b^3 + 15ab^4 - b^5$

27. $16x^4 + 16x^3 + 6x^2 + x + \dfrac{1}{16}$ **29.** $\dfrac{1}{16}x^4 - \dfrac{3}{2}x^3 + \dfrac{27}{2}x^2 - 54x + 81$ **31.** $x^{10} + 100x^9 + 4500x^8 + 120,000x^7$

33. $2187x^7 - 5103x^6y + 5103x^5y^2 - 2835x^4y^3$ **35.** $x^{16} - 24x^{14}y + 252x^{12}y^2 - 1512x^{10}y^3$ **37.** $x^8, 24x^7, 17,496x, 6561$
39. $64x^6, 960x^5, 37,500x, 15,625$ **41.** Yes, $4! = 4 \cdot 3!$ **43.** Yes, $(7 - 3)! = (7 - 3)(7 - 4)(7 - 5)! = 4 \cdot 3 \cdot 2!$ **45.** $m = n$ or $m = 0$
47. $(a + b)^n = \displaystyle\sum_{i=0}^{n}\binom{n}{i}a^{n-i}b^i$ **49.** $(0, 10)$ **50.** $(10, 4)$ **51.** $2, 9$ **52.** $2x^3y^5\sqrt{30y}$ **53.** $f^{-1}(x) = \dfrac{x - 8}{3}$

Chapter 11 Review Exercises **1.** $6, 7, 8, 9, 10$ **2.** $-1, 3, 9, 17, 27$ **3.** $12, 5, 4, 3, \dfrac{12}{5}$ **4.** $\dfrac{1}{5}, \dfrac{2}{3}, \dfrac{9}{7}, 2, \dfrac{25}{9}$ **5.** 11 **6.** 6 **7.** $\dfrac{26}{81}$

8. 88 **9.** $s_1 = 7, s_3 = 27$ **10.** $s_1 = 12, s_3 = 47$ **11.** $s_1 = \dfrac{4}{3}, s_3 = \dfrac{227}{60}$ **12.** $s_1 = -9, s_3 = -10$ **13.** $32, 64, 128; a_n = 2^n$

14. $-\dfrac{1}{3}, \dfrac{1}{9}, -\dfrac{1}{27}; a_n = (-1)^n(3^{4-n})$ **15.** $\dfrac{16}{9}, \dfrac{32}{9}, \dfrac{64}{9}; a_n = \dfrac{2^n - 1}{9}$ **16.** $-3, -7, -11; a_n = 17 - 4n$ **17.** $11 + 14 + 19 = 44$

18. $6 + 14 + 24 + 36 = 80$ **19.** $\dfrac{1}{6} + \dfrac{4}{6} + \dfrac{9}{6} + \dfrac{16}{6} + \dfrac{25}{6} = \dfrac{55}{6}$ **20.** $\dfrac{1}{2} + \dfrac{2}{3} + \dfrac{3}{4} + \dfrac{4}{5} = \dfrac{163}{60}$ **21.** 29 **22.** 239 **23.** 132 **24.** 841

25. a) $10, 14, 18, 22$ **b)** $p_n = 4n + 6$ **26. a)** $4, 10, 18, 28$ **b)** $a_n = n(n + 3) = n^2 + 3n$ **27.** $5, 8, 11, 14, 17$ **28.** $5, \dfrac{14}{3}, \dfrac{13}{3}, 4, \dfrac{11}{3}$

29. $\dfrac{1}{2}, -\dfrac{3}{2}, -\dfrac{7}{2}, -\dfrac{11}{2}, -\dfrac{15}{2}$ **30.** $-20, -\dfrac{99}{5}, -\dfrac{98}{5}, -\dfrac{97}{5}, -\dfrac{96}{5}$ **31.** 30 **32.** -5 **33.** $\dfrac{1}{2}$ **34.** 6 **35.** $s_8 = 112, d = 2$

36. $s_7 = -140, d = -6$ **37.** $s_7 = \dfrac{48}{5}, d = \dfrac{2}{5}$ **38.** $s_9 = -42, d = -\dfrac{1}{3}$ **39.** $-7, -3, 1, 5; a_{10} = 29, s_{10} = 110$

40. $8, 5, 2, -1; a_{10} = -19, s_{10} = -55$ **41.** $\dfrac{5}{6}, \dfrac{3}{2}, \dfrac{13}{6}, \dfrac{17}{6}; a_{10} = \dfrac{41}{6}, s_{10} = \dfrac{115}{3}$ **42.** $-60, -55, -50, -45; a_{10} = -15, s_{10} = -375$

43. $n = 13, s_{13} = 442$ **44.** $n = 8, s_8 = 28$ **45.** $n = 11, s_{11} = \dfrac{231}{10}$ **46.** $n = 10, s_{10} = 180$ **47.** $6, 12, 24, 48, 96$

48. $-28, -14, -7, -\dfrac{7}{2}, -\dfrac{7}{14}$ **49.** $20, -\dfrac{40}{3}, \dfrac{80}{9}, -\dfrac{160}{27}, \dfrac{320}{81}$ **50.** $-20, -4, -\dfrac{4}{5}, -\dfrac{4}{25}, -\dfrac{4}{125}$ **51.** $\dfrac{4}{27}$ **52.** 480   **53.** 216

54. $\dfrac{4}{243}$ **55.** 441   **56.** $-\dfrac{4305}{64}$ **57.** $\dfrac{585}{8}$ **58.** $\dfrac{127}{8}$ **59.** $r = 2; a_n = 6(2)^{n-1}$ **60.** $r = 5, a_n = -6(5)^{n-1}$ **61.** $r = \dfrac{1}{3}, a_n = 10\left(\dfrac{1}{3}\right)^{n-1}$

62. $r = \dfrac{2}{3}, a_n = \dfrac{9}{5}\left(\dfrac{2}{3}\right)^{n-1}$ **63.** 10   **64.** $\dfrac{49}{10}$ **65.** -6 **66.** -18 **67.** 32   **68.** $\dfrac{33}{2}$ **69.** $\dfrac{25}{6}$ **70.** -12 **71.** $\dfrac{4}{11}$ **72.** $\dfrac{29}{37}$

73. $81x^4 + 108x^3y + 54x^2y^2 + 12xy^3 + y^4$ **74.** $8x^3 - 36x^2y^2 + 54xy^4 - 27y^6$ **75.** $x^9 - 18x^8y + 144x^7y^2 - 672x^6y^3$

76. $256a^{16} + 3072a^{14}b + 16{,}128a^{12}b^2 + 48{,}384a^{10}b^3$ **77.** 15,050   **78.** 231   **79. a)** \$36,000, \$37,000, \$38,000, \$39,000

b) $a_n = \$35{,}000 + 1000n$ **c)** \$41,000   **d)** \$451,000   **80.** \$51,200   **81. a)** $\approx\$2024.51$ **b)** ≈ 2463.13 **c)** $\approx\$24{,}378.78$

82. $\approx\$503.63$ **83.** 150 feet

Chapter 11 Practice Test

1. A series is the sum of the terms of a sequence. [11.1]   **2. a)** An arithmetic sequence is one whose terms differ by a constant amount.   **b)** A geometric sequence is one whose terms differ by a common multiple [11.2–11.3]

3. $-\dfrac{1}{3}, 0, \dfrac{1}{9}, \dfrac{1}{6}, \dfrac{1}{5}$ [11.1]   **4.** $s_1 = 3, s_3 = \dfrac{181}{36}$ [11.1]   **5.** $5 + 11 + 21 + 35 + 53 = 125$ [11.1]   **6.** 184 [11.1]

7. $a_n = \dfrac{1}{3} + \dfrac{1}{3}(n - 1) = \dfrac{1}{3}n$ [11.1]   **8.** $a_n = 5(2)^{n-1}$ [11.3]   **9.** $15, 9, 3, -3$ [11.2]   **10.** $\dfrac{5}{12}, \dfrac{5}{18}, \dfrac{5}{27}, \dfrac{10}{81}$ [11.3]   **11.** -40 [11.2]

12. -20 [11.2]   **13.** 12 [11.2]   **14.** $\dfrac{256}{243}$ [11.3]   **15.** $\dfrac{39{,}063}{5}$ [11.3]   **16.** $r = \dfrac{1}{3}, a_n = 15\left(\dfrac{1}{3}\right)^{n-1}$ [11.3]   **17.** 12 [11.3]   **18.** $\dfrac{13}{33}$ [11.3]

19. 56 [11.4]   **20.** $x^4 + 8x^3y + 24x^2y^2 + 32xy^3 + 16y^4$ [11.4]   **21.** 82 [11.2]   **22.** 91 [11.2]   **23.** \$210,000 [11.2]   **24.** $\approx\$851.66$ [11.3]

25. 364,500 [11.3]

Cumulative Review Test

1. $b = \dfrac{2A}{h}$ [2.2]   **2.** $y = -\dfrac{11}{3}x + \dfrac{38}{3}$ [3.5]   **3.** Infinite number of solutions [4.2]

4. $5x^4 + 29x^3 + 14x^2 - 28x + 10$ [5.2]   **5.** $(x^2 + 2)(x - 6)$ [5.4]   **6.** $(a + b + 4)^2$ [5.5]   **7.** $\dfrac{5x^2 + 14x - 49}{(x + 5)(x - 2)}$ [6.2]   **8.** 500 [6.6]

9. 3 [7.6]   **10.** 5 [7.6]   **11.** $-1 \pm i\sqrt{14}$ [8.1]   **12.** $\dfrac{3 \pm \sqrt{309}}{30}$ [8.2]   **13.** 5 [5.8]   **14.** 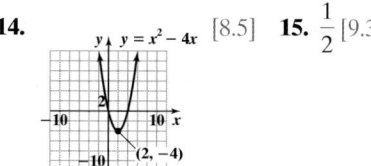 [8.5]   **15.** $\dfrac{1}{2}$ [9.3]

16. 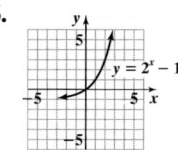 [9.2]   **17.** $(x + 6)^2 + (y - 2)^2 = 49$ [10.1]

18. 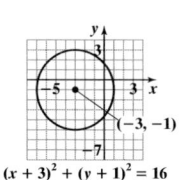 [10.1]   **19.** 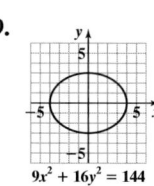 [10.2]   **20.** 18 [11.3]

Applications Index

Subject Index

Chapter 1   Basic Concepts

Commutative properties: $a + b = b + a, ab = ba$

Associative properties: $(a + b) + c = a + (b + c), (ab)c = a(bc)$

Distributive property: $a(b + c) = ab + ac$

Identity properties: $a + 0 = 0 + a = a, a \cdot 1 = 1 \cdot a = a$

Inverse properties: $a + (-a) = -a + a = 0, a \cdot \frac{1}{a} = \frac{1}{a} \cdot a = 1$

Multiplication property of 0: $a \cdot 0 = 0 \cdot a = 0$

Double negative property: $-(-a) = a$

$>$ means is greater than, $\geq$ means is greater than or equal to
$<$ means is less than, $\leq$ means is less than or equal to

$$|a| = \begin{cases} a, & a \geq 0 \\ -a, & a < 0 \end{cases} \qquad a - b = a + (-b)$$

$$\sqrt[n]{a} = b \text{ if } \underbrace{b \cdot b \cdot b \cdots \cdot b}_{n \text{ factors of } b} = a \qquad b^n = \underbrace{b \cdot b \cdot b \cdots \cdot b}_{n \text{ factors of } b}$$

Order of Operations: Parentheses, exponent and radicals, multiplication and division, addition and subtraction.

Rules of Exponents

$a^m \cdot a^n = a^{m+n}$ $(a^m)^n = a^{m \cdot n}$

$a^m/a^n = a^{m-n}, a \neq 0$ $(ab)^m = a^m b^m$

$a^{-m} = \dfrac{1}{a^m}, a \neq 0$ $\left(\dfrac{a}{b}\right)^m = \dfrac{a^m}{b^m}, b \neq 0$

$a^0 = 1, a \neq 0$ $\left(\dfrac{a}{b}\right)^{-m} = \left(\dfrac{b}{a}\right)^m, a \neq 0, b \neq 0$

Chapter 2   Equations and Inequalities

Addition property of equality: If $a = b$, then $a + c = b + c$.

Multiplication property of equality: If $a = b$, then $a \cdot c = b \cdot c$.

Problem-Solving Procedure for Solving Application Problems

1. **Understand the problem.** Identify the quantity or quantities you are being asked to find.

2. **Translate the problem into mathematical language** (express the problem as an equation).

   a) Choose a variable to represent one quantity, and **write down exactly what it represents**. Represent any other quantity to be found in terms of this variable.

   b) Using the information from step a), write an equation that represents the word problem.

3. **Carry out the mathematical calculations** (solve the equation).
4. **Check the answer** (using the original wording of the problem).
5. **Answer the question asked**.

Distance formula: $d = rt$

Inequalities

If $a > b$, then $a + c > b + c$.
If $a > b$, then $a - c > b - c$.
If $a > b$ and $c > 0$, then $a \cdot c > b \cdot c$.
If $a > b$ and $c > 0$, then $a/c > b/c$.
If $a > b$ and $c < 0$, then $a \cdot c < b \cdot c$.
If $a > b$ and $c < 0$, then $a/c < b/c$.

Absolute Value

If $|x| = a$, then $x = a$ or $x = -a$.   If $|x| > a$, then $x < -a$ or $x > a$.
If $|x| < a$, then $-a < x < a$.   If $|x| = |y|$, then $x = y$ or $x = -y$.

Chapter 3   Graphs and Functions

A **relation** is any set of ordered pairs.

A **function** is a correspondence between a first set of elements, the domain, and a second set of elements, the range, such that each element of the domain corresponds to exactly one element in the range.

Functions

Sum: $(f + g)(x) = f(x) + g(x)$
Difference: $(f - g)(x) = f(x) - g(x)$
Product: $(f \cdot g)(x) = f(x) \cdot g(x)$

Quotient: $\left(\dfrac{f}{g}\right)(x) = \dfrac{f(x)}{g(x)}, g(x) \neq 0$

A **graph of an equation** is an illustration of the set of points that satisfy an equation.

To find the y-intercept of a graph, set $x = 0$ and solve the equation for y.
To find the x-intercept of a graph, set $y = 0$ and solve the equation for x.

Slope of a line: $m = \dfrac{\Delta y}{\Delta x} = \dfrac{y_2 - y_1}{x_2 - x_1}$

Standard form of a linear equation: $ax + by = c$
Slope–intercept form of a linear equation: $y = mx + b$
Point–slope form of a linear equation: $y - y_1 = m(x - x_1)$

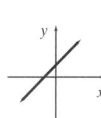

Positive slope
(rises to right)

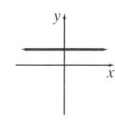

Slope is 0.
(horizontal line)

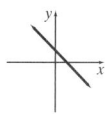

Negative slope
(falls to right)

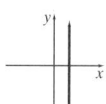

Slope is undefined.
(vertical line)

Chapter 4   Systems of Equations and Inequalities

Exactly 1 solution
(Nonparallel lines)

Line 1   Line 2

Consistent system

No solution
(Parallel lines)

Inconsistent system

Infinite number of solutions
(Same line)

Dependent system

$$\begin{vmatrix} a_1 & b_1 \\ a_2 & b_2 \end{vmatrix} = a_1 b_2 - a_2 b_1$$

Cramer's Rule:
Given a system of equations of the form

$$\begin{matrix} a_1 x + b_1 y = c_1 \\ a_2 x + b_2 y = c_2 \end{matrix} \quad \text{then } x = \frac{\begin{vmatrix} c_1 & b_1 \\ c_2 & b_2 \end{vmatrix}}{\begin{vmatrix} a_1 & b_1 \\ a_2 & b_2 \end{vmatrix}} \text{ and } y = \frac{\begin{vmatrix} a_1 & c_1 \\ a_2 & c_2 \end{vmatrix}}{\begin{vmatrix} a_1 & b_1 \\ a_2 & b_2 \end{vmatrix}}$$

A system of linear equations may be solved: (a) graphically, (b) by the substitution method, (c) by the addition or elimination method, (d) by matrices, or (e) by determinants.

Chapter 5   Polynomials and Polynomial Functions

FOIL method to multiply two binomials:

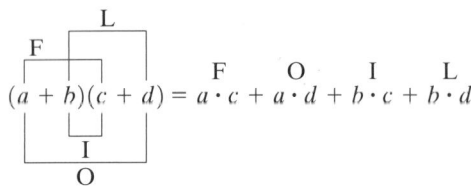

$$(a + b)(c + d) = a \cdot c + a \cdot d + b \cdot c + b \cdot d$$

Pythagorean Theorem:

$$\text{leg}^2 + \text{leg}^2 = \text{hyp}^2 \quad \text{or} \quad a^2 + b^2 = c^2$$

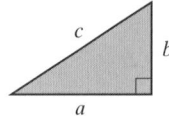

Square of a binomial:

$$(a + b)^2 = a^2 + 2ab + b^2 \qquad (a - b)^2 = a^2 - 2ab + b^2$$

Difference of two squares: $(a + b)(a - b) = a^2 - b^2$

Perfect square trinomials:

$$a^2 + 2ab + b^2 = (a + b)^2, \quad a^2 - 2ab + b^2 = (a - b)^2$$

Sum of two cubes: $a^3 + b^3 = (a + b)(a^2 - ab + b^2)$

Difference of two cubes: $a^3 - b^3 = (a - b)(a^2 + ab + b^2)$

Standard form of a quadratic equation: $ax^2 + bx + c = 0, a \neq 0$

Zero-factor property: If $a \cdot b = 0$, then either $a = 0$ or $b = 0$, or both $a = 0$ and $b = 0$.

Chapter 6   Rational Expressions and Equations

To Multiply Rational Expressions:

1. Factor all numerators and denominators.
2. Divide out any common factors.
3. Multiply numerators and multiply denominators.
4. Simplify the answer when possible.

To Divide Rational Expressions:

Invert the divisor and then multiply the resulting rational expressions.

To Add or Subtract Rational Expressions:

1. Write each fraction with a common denominator.
2. Add or subtract the numerators while keeping the common denominator.
3. When possible factor the numerator and simplify the fraction.

Similar Figures: Corresponding angles are equal and corresponding sides are in proportion.

Proportion: If $\dfrac{a}{b} = \dfrac{c}{d}$, then $ad = bc$.

Variation: direct, $y = kx$; inverse, $y = \dfrac{k}{x}$; joint, $y = kxz$

Chapter 7   Roots, Radicals, and Complex Numbers

If n is even and $a \geq 0$: $\sqrt[n]{a} = b$ if $b^n = a$

If n is odd: $\sqrt[n]{a} = b$ if $b^n = a$

Rules of radicals

$$\sqrt{a^2} = |a|$$

$$\sqrt{a^2} = a, \ a \geq 0$$

$$\sqrt[n]{a^n} = a, \ a \geq 0$$

$$\sqrt[n]{a} = a^{1/n}, \ a \geq 0$$

$$\sqrt[n]{a^m} = \left(\sqrt[n]{a}\right)^m = a^{m/n}, a \geq 0$$

$$\sqrt[n]{a} \, \sqrt[n]{b} = \sqrt[n]{ab}, \ a \geq 0, b \geq 0$$

$$\frac{\sqrt[n]{a}}{\sqrt[n]{b}} = \sqrt[n]{\frac{a}{b}}, a \geq 0, b > 0$$

A radical is simplified when the following are all true:

1. No perfect powers are factors of any radicand.
2. No radicand contains a fraction.
3. No denominator contains a radical.

Complex numbers: numbers of the form $a + bi$

Powers of i: $i = \sqrt{-1}, i^2 = -1, i^3 = -i, i^4 = 1$

Chapter 8   Quadratic Functions

Square Root Property:

If $x^2 = a$, where a is a real number, then $x = \pm\sqrt{a}$.

A quadratic equation may be solved by factoring, completing the square, or the quadratic formula.

Quadratic formula: $x = \dfrac{-b \pm \sqrt{b^2 - 4ac}}{2a}$

Discriminant: $b^2 - 4ac$

If $b^2 - 4ac > 0$, then equation has two distinct real number solutions.

If $b^2 - 4ac = 0$, then equation has a single real number solution.

If $b^2 - 4ac < 0$, then equation has no real number solution.

Parabolas

For $f(x) = ax^2 + bx + c$, the vertex of the parabola is

$$\left(-\frac{b}{2a}, \frac{4ac - b^2}{4a}\right) \text{ or } \left(-\frac{b}{2a}, f\left(-\frac{b}{2a}\right)\right).$$

For $f(x) = a(x - h)^2 + k$, the vertex of the parabola is (h, k).

If $f(x) = ax^2 + bx + c, a > 0$, the function will have a minimum value of $\dfrac{4ac - b^2}{4a}$ at $x = -\dfrac{b}{2a}$.

If $f(x) = ax^2 + bx + c, a < 0$, the function will have a maximum value of $\dfrac{4ac - b^2}{4a}$ at $x = -\dfrac{b}{2a}$.